AF633471

HUMAN FACTORS AND ERGONOMICS DESIGN HANDBOOK

Human Factors and Ergonomics Design Handbook

Barry Tillman

David J. Fitts

Wesley E. Woodson

Rhonda Rose-Sundholm

Peggy Tillman

Third Edition

McGraw Hill Education

New York Chicago San Francisco Athens London Madrid
Mexico City Milan New Delhi Singapore Sydney Toronto

Library of Congress Control Number: 2015957146

Human Factors and Ergonomics Design Handbook, Third Edition

Copyright © 2016, 1992, 1981 by McGraw-Hill Education. All rights reserved. Printed in the United States of America. Except as permitted under the United States Copyright Act of 1976, no part of this publication may be reproduced or distributed in any form or by any means, or stored in a data base or retrieval system, without the prior written permission of the publisher.

1 2 3 4 5 6 7 8 9 0 DOW/DOW 1 2 1 0 9 8 7 6

ISBN 978-0-07-170287-4
MHID 0-07-170287-3

Sponsoring Editor: Robert Argentieri
Editing Supervisor: Stephen M. Smith
Production Supervisor: Lynn M. Messina
Acquisitions Coordinator: Lauren Rogers
Project Manager: Tanya Punj, Cenveo® Publisher Services
Copy Editor: Manoj Bachheti, Cenveo Publisher Services
Proofreader: Barnali Ojha, Cenveo Publisher Services
Indexer: Ariel O. Tuplano
Art Director, Cover: Jeff Weeks
Composition: Cenveo Publisher Services

McGraw-Hill Education books are available at special quantity discounts to use as premiums and sales promotions or for use in corporate training programs. To contact a representative, please visit the Contact Us page at www.mhprofessional.com.

This book is printed on acid-free paper.

Information contained in this work has been obtained by McGraw-Hill Education from sources believed to be reliable. However, neither McGraw-Hill Education nor its authors guarantee the accuracy or completeness of any information published herein, and neither McGraw-Hill Education nor its authors shall be responsible for any errors, omissions, or damages arising out of use of this information. This work is published with the understanding that McGraw-Hill Education and its authors are supplying information but are not attempting to render engineering or other professional services. If such services are required, the assistance of an appropriate professional should be sought.

About the Authors

As a human factors engineer, **Barry Tillman** has worked directly with designers, engineering drawings, and software developers for nearly 50 years to achieve safer and more-cost-effective human-system interfaces. He has been involved with a broad range of systems including coal mining equipment, military vehicles, ships, manufacturing equipment, and space habitats. Mr. Tillman has developed human factors analytical tools and written design standards for the military, NASA, and the shipping industry. He has also developed and taught numerous human factors engineering and ergonomics courses.

David J. Fitts worked for the National Aeronautics and Space Administration (NASA) for 27 years before retiring in 2015. In his career with NASA he was deeply engaged in the three major human spaceflight programs of his era: the Space Shuttle Program, the International Space Station (ISS) Program, and the next-generation Constellation Program for deep space exploration. During ISS and Constellation development phases, Mr. Fitts served the agency as chief of the Habitability and Human Factors Branch, which focuses on ensuring that human capabilities and limitations are strongly integrated into human spaceflight programs' systems engineering and management processes.

Wesley E. Woodson (deceased) was president of Man Factors Associates, a human factors consulting firm in El Cajon, California. Experienced in both military and domestic systems and product design in a broad range of applications, he was the author of the first edition of *Human Factors Design Handbook: Human Engineering Guide for Equipment Designers* (University of California Press, 1964).

Rhonda Rose-Sundholm has over 26 years of communications experience in the aerospace industry. She has worked directly with scientists, engineers, and physicians to organize their information and communicate it to diverse audiences. She has a background in business administration, configuration management, and information systems, including website design and management.

Peggy Tillman was president of a human factors/ergonomics company for over 20 years. She has written, illustrated, and designed numerous books, presentations, and reports. Ms. Tillman has a degree in behavioral psychology and a teaching credential. She is listed in Marquis *Who's Who in the West* and *Who's Who in Science and Engineering.*

CONTENTS

Introduction

The purpose of this handbook is to provide a general reference for addressing key human factors questions and human-product interface design suggestions in a form that engineers and designers can use with a minimum of searching or study. The handbook is also intended to serve as an introduction to human factors data and practices for both students and practicing system designers and developers.

The definition of human factors engineering accepted by the Human Factors and Ergonomics Society of the United States and the International Ergonomics Association is

> Ergonomics (or human factors) is the scientific discipline concerned with the understanding of interactions among humans and other elements of a system, and the profession that applies theory, principles, data, and other methods to design in order to optimize human well-being and overall system performance. [The term "ergonomics" is often used interchangeably with "human factors."]

This definition is broad and the various human factors specialists might give different definitions. For example, if you talk with a human factors researcher, human factors engineering generally is defined as the investigation of the input-output aspects of human-machine systems, with the primary objective of developing quantifiable measures of human performance under discretely specified conditions of the system environment. The researcher is interested in developing basic data that can be applied to many situations. More applications-oriented human factors engineering specialists typically address themselves to interpreting basic human factors research data as they apply to specific human-product interface situations. Unfortunately, in many cases this results in a specialist who is familiar with only a limited variety of system categories, such as aircraft flight control, communications systems display and control, computer interface, or vehicular safety.

In spite of these seeming restrictions resulting from specialization, the intent of human factors is to focus on and resolve human-system interface problems and solutions. Human factors engineering looks at a design from the standpoint of user total output effectiveness. Inherent in this philosophy are the following objectives:

1. To make the user's contribution to product output as efficient as possible so that the basic product output is not compromised by human failures.

2. To make the combined user-product involvement as safe as possible so that neither human nor product failures will compromise the user's health or damage the hardware. Inherent in this objective also is the avoidance of injury to others and of damage to adjacent hardware.

3. To minimize the stress that a product imposes on the user as he or she uses, operates, services, or maintains it. This includes such stresses as an undue energy demand, frustration in trying to deal with the product at any point in the human-product interaction, and worry about whether one is using the product properly.

4. To maximize the acceptability of the product, not only in terms of its attractiveness (to encourage its purchase or maintain its satisfactoriness throughout ownership) but also in terms of giving users the feeling that the product allows them to use it efficiently and keep it in good working order with a minimum of effort.

Another common tie across the breadth of human factors specialists is the logical process to achieve the aforementioned objectives:

1. Define system objectives

2. Allocate the accomplishment of these objectives to humans or automation

3. Analyze human tasks and help design a system that optimizes task performance

HOW TO USE THIS HANDBOOK

The *Human Factors and Ergonomics Design Handbook* is organized for immediate reference. It has five parts. The following is a brief overview of each part's contents.

Part 1, *Systems*, gives human factors data needed for initial planning of major systems. This part contains planning models and checklists that address important human factors concerns. By using human factors information early in design decisions, you can avoid costly redesign or an ineffective, uncomfortable, and unsafe system.

Part 2, *Subsystems Design*, contains human factors data for specific subsystem layout and preliminary design. For example, this part has recommendations for room layout, guidelines for control panel arrangement, and design standards for the work environment.

Part 3, *Component and Product Design*, provides detailed figures, charts, and tables of human factors standards required to design specific hardware and software items.

Design specifics cover items including computers, hand tools, furniture, road signs, and extensive controls and displays lists.

Part 4, *Human Factors Data*, contains information on human characteristics and human capabilities. It covers human size, information processing capabilities, reaction times, strength, vision, and hearing.

Part 5, *Human Factors Engineering Methods*, gives information on how to conduct human factors engineering programs from the systems analysis stage to final design. This part discusses analytical techniques and gives detailed examples.

At the start of each part is a series of Points to Consider. These points address the overall concepts in the part and provide seeds for discussion and further research.

Acknowledgments

One could not hope to acknowledge individually all the people who directly or indirectly have contributed to the fund of knowledge we normally attribute to human factors engineering. The information in this handbook, for instance, is the combined product of many professional disciplines, including engineering, design, medicine, physiology, anthropology, biology, personal hygiene, and psychology. The information is representative of many years of research and design practice. Therefore, as the authors, we can pay tribute only to the dedicated concern and perseverance of the multitude of professional and technical specialists who, individually and collectively, have generated the bits and pieces of information that make up the content of this handbook. We would also like to pay tribute to the various special organizations that examine and generate guidance information related to the human-use aspects of design.

In addition, our gratitude and recognition are given to the original author of this handbook, Wesley Woodson. He was one of the founders of the practice of human factors engineering and user interface design. Not only did he have an incredibly broad knowledge of human factors, but he also knew how best to condense and communicate this knowledge to designers and engineers. We constantly strived to preserve his standard for information clarity and utility.

Human Factors and Ergonomics Design Handbook

PART 1

SYSTEMS

POINTS TO CONSIDER

In order to understand systems and the human role in them, it is helpful to think of them in ways that relate to your experiences. The following exercises may help.

1. Think of a morning, from the time you wake up to the time you are ready to start the day. Identify the functions that occur. Put the functions in terms of what is done, not how it is done. For example, think of the function waking up. Don't say the alarm clock buzzes, simply say "wake up"—that's the core function. Continue naming the functions defecate and urinate, clean teeth, clean face, nourish body, clean body, groom hair, etc. Notice it doesn't say "take a shower," it says "clean body." That can be done in any number of ways; in space they have to do it with a washcloth. When naming a function, it helps to think of the goal of the function. Now think of how you accomplish the function, and how your function is allocated. Now imagine other possible ways of accomplishing the function. What other pieces of equipment, or software, or humans might be used to accomplish function? In this stage of a system development, designers and human factors professionals can begin to decide what roles the human plays in accomplishing the system functions. For example, would it be better to rely on a human to wake you, or better to simply use a reliable electronic alarm clock?

2. Systems can be big or small. A fast food restaurant is a system and the national railroads are a system. No matter what their size, however, all systems have people, procedures, equipment/software, a working environment, and a goal. Can you think of a system and identify the elements of that system?

3. When designing a system, particularly a large system, specialists may be called in to design the different elements of the system. For example, an automobile may require electrical, mechanical, and hydraulic engineers to design the engine. An acoustical engineer may design the stereo system. A human factors engineer will represent the driver and passengers. Each specialist has his or her own concerns. The primary concern of the electrical engineer may be that the wires are routed so that they don't get damaged and the equipment does not overheat. Likewise, the human factors engineer will be primarily concerned with the comfort, safety, and performance of the driver. All of the system developers must coordinate to create an optimum system. Human factors engineers must remember that, although the human is their primary concern, it is not always the primary concern of other design specialists. Think of a system and all of the specialists who are responsible for the different elements of the system: mechanical and software elements, the environment, the mission or goal, the procedures, and the people. Think of what the goal of each of these specialists might be.

SELECTED READING LIST

Bass, L. *Products Liability,* Shepard's McGraw-Hill, Colorado Springs, 1986.

DeGreene, K. B. *Systems Psychology,* McGraw-Hill Book Company, New York, 1970.

Gagne, R. M. (ed.). *Psychological Principles in System Development,* Holt, Rinehart and Winston, Inc., New York, 1962.

Hammer, W. *Handbook of System and Product Safety,* Prentice-Hall, Inc., Englewood Cliffs, NJ, 1972.

Harrigan, J. E., and J. R. Harrigan. *Human Factors Program for Architects, Interior Designers and Clients,* Blake Printing and Publishing, 1976.

Meister, D. *Behavioral Foundations of System Development,* John Wiley & Sons, Inc., New York, 1976.

Meister, D., and G. F. Rabideau. *Human Factors Evaluation in System Development,* John Wiley & Sons, Inc., New York, 1965.

Chapter 1

Systems Introduction

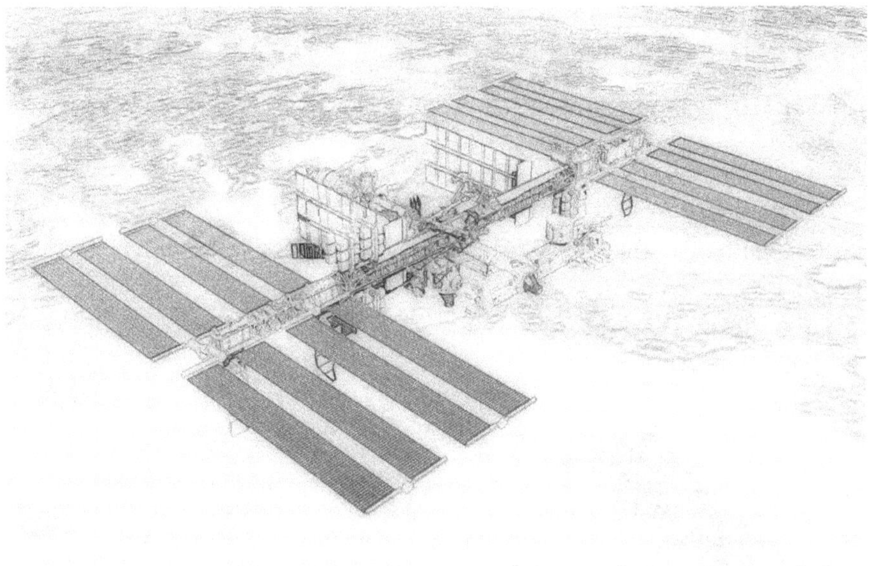

GENERAL

This chapter provides general guidelines appropriate to all systems. It deals with such areas as defining the scope of the human factors problems and efforts, defining system mission requirements, determining where to use humans within the system, and organizing to accomplish human factors research and application tasks.

Following this general introductory material is specific information organized into eight gross systems categories:

1. Architectural systems
2. Transportation systems
3. Military systems
4. Space systems
5. Industrial systems
6. Agricultural systems
7. Communication systems
8. Consumer product systems

SYSTEMS CONCEPT

A "system" as used in this chapter (as opposed to "subsystem" or "component" as used in later chapters) refers to larger, more complex, mission-oriented groupings of elements into an integrated, functional whole. As shown in Fig. 1.1, the

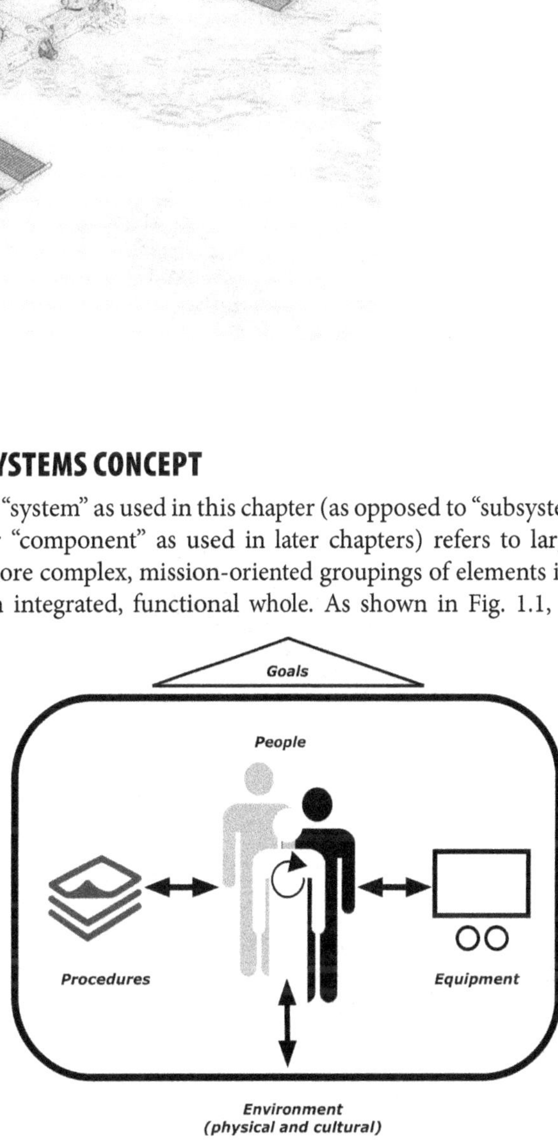

FIGURE 1.1 Elements of a system.

elements of a system are the people who operate and maintain it, the hardware and software, the operating environment, procedures, and the mission goal.

HUMAN FACTORS ROLE IN SYSTEMS DEVELOPMENT

Human factors focuses on the links between the systems elements: people to people, people to hardware and software, people to the environment, and people to the procedures (including learning). Human factors should be considered at the concept formation stage and at all the succeeding development stages. However, at the systems concept level, the principal considerations should include the following:

1. Deciding what roles humans are to be assigned in system operation, i.e., administrative management, operation, maintenance, and general use
2. Deciding where, when, and how humans will interact with subsystems and components, directly or indirectly
3. Deciding what has to be done to provide humans satisfactory living and working environments to ensure not only their safety but also their efficient performance and comfort
4. Determining what human constraints impact the system design and eventual system performance and deciding how to ensure that humans will not become the weak link in the system

The principal objectives should be to design a system that:

1. Is "adapted" to the human, as opposed to creating a system in which the human has to do all the adapting
2. Provides the human in the system the ability to perform in the best manner of which he or she is capable
3. Does not subject the human to extreme physical or mental stress or to possible injury or death as a result of either an equipment malfunction or unpremeditated operator error
4. Provides personal satisfaction for the user in terms of both successful operation and pride of ownership

The human factors analyst must identify and analyze three system elements that are as follows:

1. System functions: Define what the system must do. Define the mission goals and constraints.
2. System operating environment: Define the environment the system must operate in: indoors, outside, day or night, at sea, in space, etc.
3. Personnel and equipment: Define the potential human roles and major equipment items in the system.

Once the human factors analyst knows the mission, the work environment, and the humans and machines available, he or she can participate in allocating the functions to human, equipment, or software. A more detailed human factors analysis process description is given in Part 3 of this book. Figures 1.2

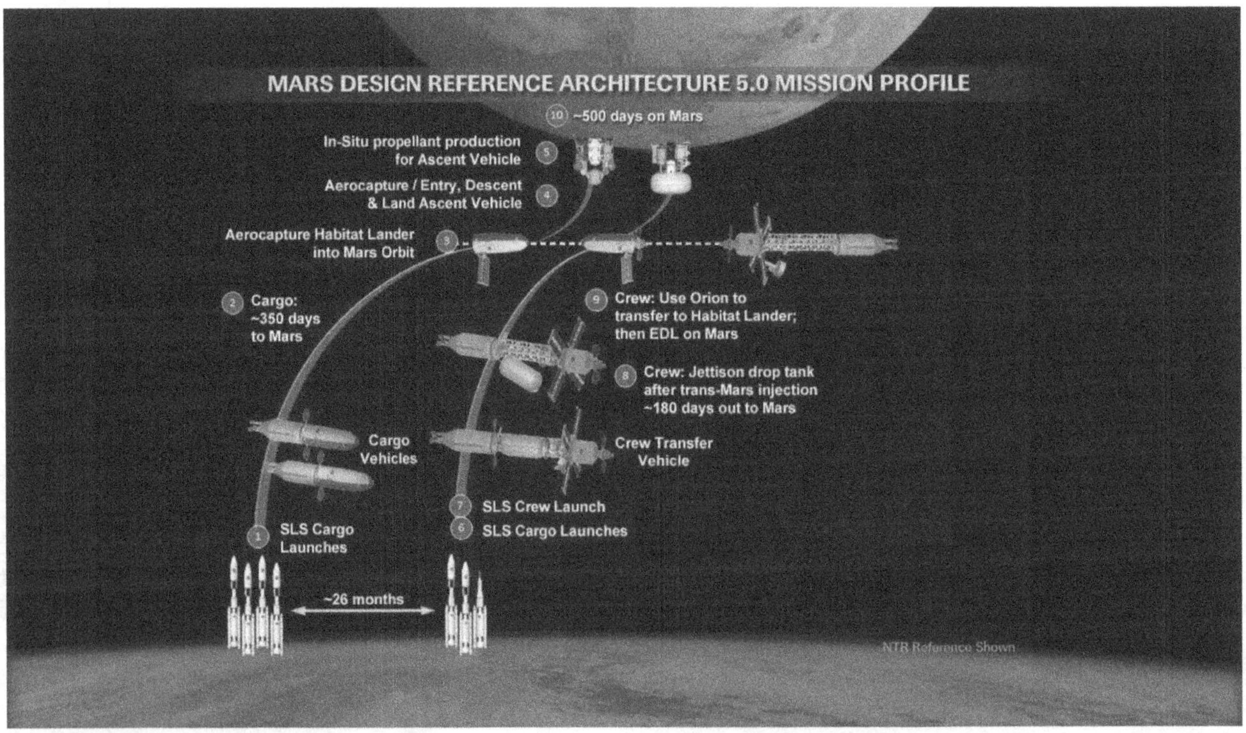

Figure 1.2 Mars mission profile.

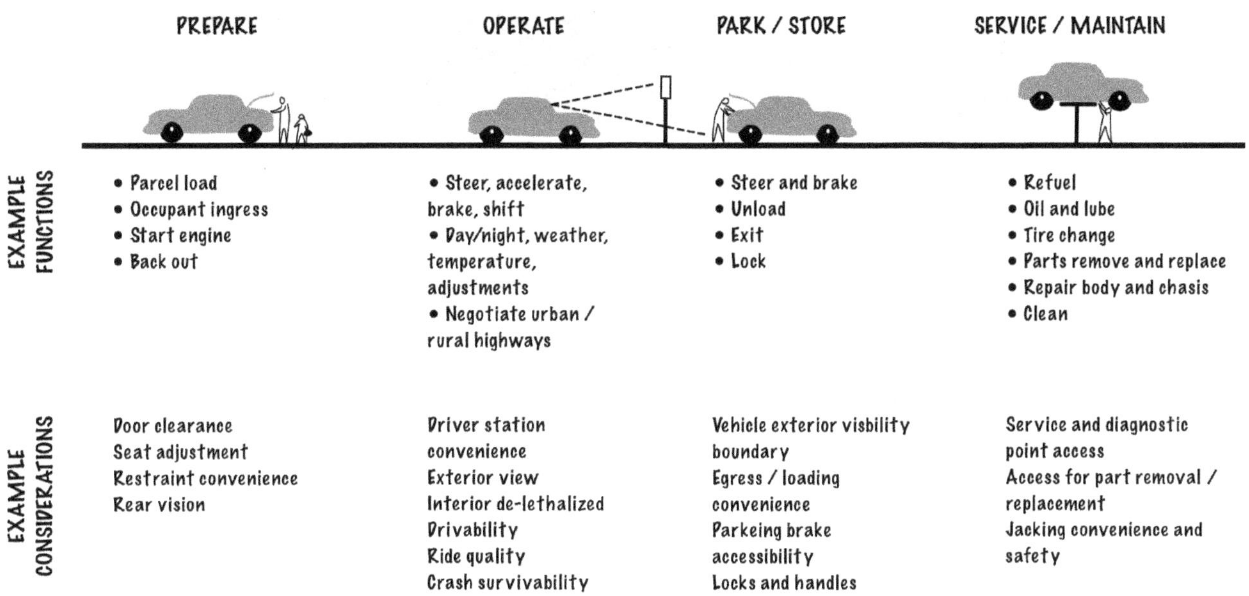

Figure 1.3 Personal vehicle system mission profile.

and 1.3 show functional definition examples. The missions are defined at the top level. The figures show the mission in graphic format. The graphic format helps envision mission constraints, potential hazards, and potential human roles.

FUNCTIONAL DEFINITION

All major systems concepts are developed around some stated mission. The proposed mission should be analyzed in terms of clarifying its purpose and objectives. These will be the foundation for all succeeding decisions regarding both the projected hardware and the facility and personnel requirements.

Once the general mission purpose and objectives are firmed up and everyone agrees on them, reasonably detailed operating requirements should be defined in order to clarify the demands that will be made on the eventual elements of the system. In the initial stages, these requirements may have to be of a fairly qualitative nature. But in the final stages, requirements should be refined to the point where quantitative values are established for each requirement. These requirements define the functions that have to be performed by physical elements, such as hardware, facilities, or software, and/or by operators, technicians, maintainers, or managers.

The general mission should be further studied, and the initial operating requirements should be reviewed with respect to the conditions under which operations must take place; the rules and regulations that the system must adhere to; and the physical, political, social, and economic factors to which the system must be responsive. These constraints, along with the basic operating requirements, are used to define the functions the system must include and for analyzing alternative means to accomplish those functions.

Functions (or processes) should be defined to fulfill each of the system requirements. They should be defined in such a way that there is minimal bias regarding how a function will be accomplished. That is, functions should not be described as either "machine" or "human"; rather, there should be independent process descriptions that allow one to consider alternative methods for accomplishing a given process. These descriptions will serve as the basis for examining alternative techniques for performing functions using objective evaluative criteria rather than subjective opinions (which are subject to argument).

THE ELEMENTS OF A SYSTEM

The human factors analyst must consider many factors in defining the missions. Table 1.1 defines some of these factors.

SYSTEM OPERATING ENVIRONMENT DEFINITION

The environment the system operates in is crucial to defining the human role and the equipment used by the operators. For example, vacuum and temperature extremes outside space capsules limit human performance and make robots a better choice for some tasks. Table 1.2 gives some important environmental considerations.

PERSONNEL AND EQUIPMENT DEFINITION

In the systems analysis phase, the human factors analyst gathers data on potential human roles in the system: What can people, machines, and software do? How many people are likely to operate, maintain, and use the system? What will their roles be? What major equipment items and software will they work with? In considering the potential human role in a system, consider all people involved directly or indirectly. Table 1.3 provides some examples.

TABLE 1.1 Mission Requirements and Goals Checklist
Answers to the below and other pertinent questions should be quantified whenever possible.

Topic	Questions
Scope	When and how will the mission begin and end?
Location	Where will the mission take place?
Interactions	How will the mission interact with other system missions?
Time constraints	What are the time constraints for the mission?
Environmental	What are the environmental conditions under which the mission must operate?
Sociopolitical	What are the sociopolitical conditions or terms under which the mission must operate?
Costs	What are the cost constraints within which the system mission must be conceived and operate?
Personnel	What are the personnel availability constraints within which the system or mission must operate?
Ecological	What are the ecological constraints within which the system must operate?
Signals	What signals must be sensed, detected, and processed and what information must be transferred and stored to meet mission objectives?
Hazards	What natural and man-made hazards may occur during the mission (e.g., extraterrestrial hazards, undersea hazards, earthquakes, floods, acceleration, explosions, fire, and shock)?
Movement	What physical movements of people, equipment, or materials will occur during the mission?
Human roles	What probable human roles are directly or indirectly associated with the mission (e.g., management, operation, maintenance, and support)?
Dependencies	What current systems or subsystems, if any, are to be used as part of the proposed mission or system?

TABLE 1.2 Operating Environment Considerations

	Comfort Range	Discomfort Range	Injury or Death Range
Light	50 foot candles for control room lighting	18 foot candles minimum for machinery	Glare can cause effective blinding when brighter than the point of interest
Sound	49 dBA max for sleep	58 dBA max for work areas	140 dBA max impulse noise; 85 dBA max for extended time
Heat	65° to 81° F for living comfort	Performance impairment at 100° F for more than 1 hour	Body core temperature above 104° F is fatal
Cold	65° to 81° F for living comfort	Work can be done between 10° F and 20° F using special gloves and protective gear	Body core temperatures below 78° F can be fatal
Thermal Radiation	Up to 1.0 kW/sq meter—solar radiation	0.8 to 1.6 kW/sq	More than 6.3 kW/sq meter for 30 sec
Ventilation	18 cubic ft. per minute per person (2/3 outdoor air)	At least 7 cubic ft. per minute per person	Approximately 0.3 cubic feet per min per person for breathing
Vibration	Discomfort if greater than 1.0 foot/minute	Performance drop if greater than 1.4 feet/minute	Body damage if greater than 2.8 feet/minute

TABLE 1.3 Potential Human Roles

Managers and supervisors	Delivery personnel
Administrators and assistants	Weapons operators
Equipment operators	Communications personnel
Technical and maintenance personnel	Visitors and VIPs
Housekeepers and laborers	First-aid and hygiene personnel
Customers and clients	Counselors and training personnel
Production personnel	Security personnel
Service personnel	Disaster control personnel

Because of environmental or performance constraints, some system functions are best done by machines and computers. Others should be done by people. The human factors analyst will participate in this functional allocation decision process. During the system analysis phase the human factors analyst should define the capabilities and limitations of people working in the system. Factors to consider about the people in the system are listed in Table 1.4.

TABLE 1.4 Potential User Characteristics

Age
Sex
Cultural background
Training and experience
Size
Mobility
Dexterity
Coordination
Reaction time
Sensory response
Cognitive response
Motor response
Health and handicaps
Physiological tolerance
Motivation
Fatigue limits
Strength
Metabolic requirements
Adaptive limits
Equilibrium
Intelligence level

FUNCTION ALLOCATION

Machines, software, and people differ in capabilities and limitations. These differences must be taken into account when allocating functions. Table 1.5 defines human and machine capabilities and limitations.

As machines become more sophisticated, their capabilities increase. It is important for the human factors analyst to keep up with advances in machine and software capabilities.

TABLE 1.5 Human and Machine Capabilities
Some Comparisons of the Limitations of Machine and Human Functions, in Relative Terms*

Conditions for Function	Limitations	
	Human	Machine
Sensing Display	Limited to certain ranges of energy change affecting human senses Sensitivity: very good	Range extends far beyond human senses (X rays, infrared, etc.) Sensitivity: excellent
Filtering	Easy to reprogram	Difficult to reprogram
Identifying Display	Can be varied over relatively wide range of physical dimensions Channel capacity: small	Can be varied only in very narrow range of physical dimensions Channel capacity: large
Filtering	Easy to reprogram	Difficult to reprogram
Memory	Limits to complexity of models probably fairly high, but not precisely known Limits to length of sequential routines fairly high, but time-consuming to train	Potential limits of complexity very high Potential limits of length of routines very high
Interpreting Display	Same as Identifying	Same as Identifying
Filtering	Easy to reprogram Highly flexible, i.e., adaptable May be reprogrammed by self-instruction following input changes contingent on previous response (dynamic decision making)	Difficult to reprogram Relatively inflexible
Shunting	Can be readily reprogrammed to lower levels of functioning	Difficult to reprogram
Memory	Limitation to rule storage not known Speed of reinstatement of rule sequences relatively low (as in computing) Use of novel rules possible (inventing)	limits of rule storage ("logic") quite high Speed of using rule sequences high (e.g., computing) limited in use of novel rules

*Gagne, R. M., *Psychological Principles in System Development*, Holt, Rinehart & Winston, New York, 1962.

CHAPTER 2

ARCHITECTURAL SYSTEMS

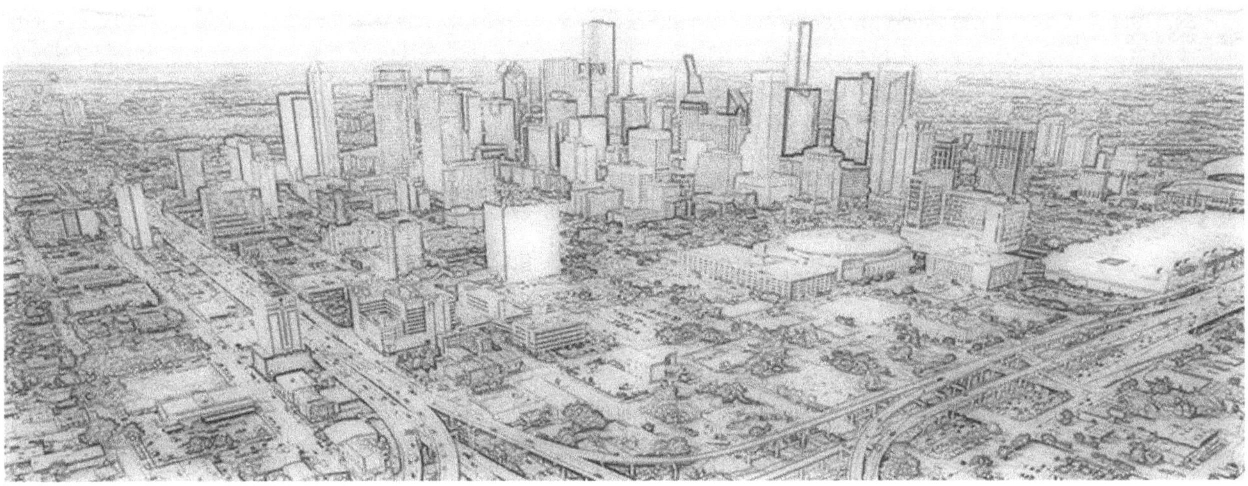

GENERAL INTRODUCTION

User-Oriented Planning

Step 1: Define and examine the needs of the total user population; i.e., do not concentrate only on the primary resident, for example, but look at the needs of his or her visitors or clients and the people who will serve the primary resident in the proposed facility.

Step 2: Examine and define the various tasks that each user has to perform. Determine what these tasks imply in terms of space, environmental control, supporting furnishings, and utilities.

Step 3: Explore the interactive as well as the isolative needs of the various users and their furnishings and equipment. Examine alternative arrangements to determine the most convenient organization of people, furnishings, spaces, buildings, etc.

Step 4: Create an enclosure for the most effective alternative defined in Step 3 and add appropriate partitioning to provide desired environmental control, privacy, and security.

Step 5: Select an appropriate site that will accommodate the building or buildings defined in Step 4 and locate, position, and arrange the building or buildings with respect to appropriate site and building access.

Now you are ready to examine the concept in terms of aesthetic features, including architectural style, special material effects, and landscaping. These features are generally cosmetic, and although they are important in terms of making the system pleasant to look at, they can for the most part wait until the five steps outlined above have been completed.

Meeting User Needs

A clear definition of the needs, characteristics, and limitations of the expected user population should be sought before any architectural concept is frozen. Otherwise, the system may not be completely compatible with user expectations or abilities to use the system effectively. Because people are different, it is a mistake to assume that a system can be designed for the so-called average person. Understanding these differences and accommodating the proposed concept to them are vital to the eventual operation of the system.

Table 2.1 lists the important characteristics that should be addressed in defining the user population.

Sociocultural Variables. It is important to recognize that people with different cultural backgrounds view architectural features differently. Some of the more important differences to evaluate are as follows:

1. Attitudes toward Privacy: Some cultural groups demand complete privacy, others appreciate privacy, and still others rely merely on subtle cues that signal a desire for privacy.

2. Family Structure: In certain cultural groups, several generations live together within a single dwelling; in others, several families live in separate but joined dwellings; and in still others, the family members function as individuals partially separated within the single dwelling.

3. The Role of Sexes: In some cultural groups, sexes are isolated; in others, the distinction is related entirely to functional factors (e.g., the mother–child relationship); and in still others, completely nonsegregated attitudes prevail.

4. Recreational Patterns: Some families are oriented toward more formal and sedentary recreational pursuits; others are more physically oriented, usually

TABLE 2.1 Architectural Human Factors Planning Considerations

User Population Characteristics	
Characteristic	**Architectural Implication**
Cultural factors	Considerable variation exists among people with respect to their cultural background, including social mores, religious attitudes, intellectual development, skill development, attitudes toward others, and where and how they live in terms of spatial features and modern technological amenities. Language differences create an important barrier to communications in many system operational settings.
Body size	People of different nationalities, as well as individuals of the same nationality, vary considerable in terms of size. There are also differences in size between children and adults, between men and women, and between members of special user populations. Differences in size have an impact on architectural space, including clearances and reach distances.
Mobility	The agility of individuals varies considerably (e.g., between the young and the old and between the handicapped and nonhandicapped persons), and mobility may be restricted by the garments people wear. The impact of restricted mobility on human–architectural interfaces may be critical to the operational utility of a system concept.
Strength	Very young and very old people have considerably less strength than those in the middle range, women generally have less strength than men, and handicapped persons may have virtually no strength. Architectural features that require lifting, pushing, pulling, or twisting must be tailored to the weakest member of the expected user population.
Sensory factors	Principal sensory factors associated with architectural systems relate to vision, hearing, and touch. Although only persons with so-called normal capacities may make up the expected user populations of special systems (because of operator selection restrictions), most general system concepts require consideration of the more limited capacities of elderly and handicapped individuals, especially the visually and aurally handicapped.
Motor skills	A limited number of people have superior motor skill capabilities as a result of either innate capability or training. Others are limited both innately and by lack of training. Still others are even more limited by physical handicaps.
Cognitive skills	Variation in cognitive skills occurs because of age differences, differences in education and/or technological opportunity, and innate mental handicaps. Understanding the operational aspects of the proposed architectural concept is critical to its effective use.
Human–Architectural Interfaces	
Interface	**Consideration**
Site location	Distance to related facilities; residential, commercial, and industrial areas; airport and bus terminals; etc. Accessibility (or nonaccessibility) to highway, roadway, street, rail, and waterway systems: vehicular or pedestrian, etc. Environmental factors: noise, air pollution, possible natural disasters, traffic problems, and aesthetic factors
Site amenities	Parking, public transit access, walk-in customer exposure, landscape illumination, security, and emergency access
Building, external	Number, size, and location; entrances (number, location, and type); identification and illumination; special amenities (utilities, sidewalks, and stairs); external maintenance requirements; etc.
Building, internal	Compartmentalization (number, size, and arrangement); doors and windows (type, number, size, and location); stairs, ramps, elevators, and escalators; heating, air conditioning, ventilation, and illumination; built-in features (cabinets, plumbing fixtures, and electrical and pneumatic systems); floor covering; safety equipment; communications systems; acoustics; and internal maintenance requirements
Safety	Fire protection, emergency escape, first aid, and disaster control

toward outdoor activities; and still others are travel oriented and treat facilities merely as a temporary base of operations.

5. Shopping Habits: Some families shop on the basis of day-to-day replenishment, and others shop infrequently but store for the long run.

6. Job Patterns: To some, the job is a means to an end (in which case, ease of getting back home is extremely important); others are job oriented and would just as soon the separation between home and job remain clear and distinct; and still others prefer to work at home and thus desire a single living and working setting.

7. Technological Experience: Although technology continues to advance even to the most primitive areas, care must be exercised not to assume that all cultures either want or will appreciate many of the amenities offered by the more technologically advanced societies, especially if these destroy certain living patterns held sacred for ethnic or religious reasons.

Personal Space. Individuals perceive their relationships with others in terms of the distance between themselves and the people they can see. At least four distinct territorial categories have been defined by various researchers:

1. Public: Those areas where the individual has freedom of access, but not necessarily of action.

2. Home: Those areas where the regular participants have regular freedom of behavior and a sense of control over the area.

3. Interactional: Those areas where social gatherings may occur. An invisible boundary and a territorial claim are implicit, though not officially promulgated by the people present.

4. Body: The area immediately surrounding the individual's body. This area is most private and inviolate to the individual.

Although absolute distance criteria are probably not pertinent, the following approximations are useful for architectural space consideration:

1. Intimate—0 to 18 in. (46 cm): People desire involvement with each other, both physically and emotionally.

2. Casual-personal—30 to 48 in. (76 to 122 cm): People will accept minimal contact (a handshake), but wish to retain freedom from domination.

3. Social-consultative—7 to 12 ft (2.1 to 3.7 m): People desire to maintain a certain formality and the freedom to break away when desired.

4. Public—30 to 1500 ft (9 to 450 m): People want to be able to remain unnoticed and to flee or be independent from the crowd.

Design for the Handicapped. Practically all aspects of architectural facility concepts, including the overall community planning, should be examined from the standpoint of use by elderly and handicapped persons.

According to the year 2000 U.S. Census, nearly 20 percent of the noninstitutionalized population over 5 years old are disabled in one way or another. (See Table 2.2.) The disabilities are shown in Fig. 2.1.

1. Considerations for the Blind: Blind or partially sighted individuals (many of whom are elderly) get about by depending on sound signals and tactile cues. They require the following special features.

   a. Well-defined, rectilinear walkways, street corners, and curbs that the blind person can touch with a cane

   b. Pathway obstructions that go all the way to the floor or ground so that the blind person's cane does not pass beneath the object and thus allow the person to run into the object

   c. Nothing at head height, such as signs, guy wires that support telephone poles, and trees with low branches

   d. Special braille signs for key public locations for identifying a building name and number, a street corner, or a bus stop

   e. Sound signals so that the blind person will know when a DON'T WALK signal is on or whether an elevator is going up or down

   f. Guardrails and/or special tactile identification of pathways to keep the blind person from veering into the street or dangerous areas.

2. Considerations for the Deaf: Whenever an audio signaling device (warning) is used for the general population because of the chance that people may not be looking in the direction of a hazard, an accompanying visual and/or tactile (vibration) signal should be devised for deaf persons to also draw their attention to the hazard.

3. Considerations for the Orthopedically Handicapped: Architectural mobility for the orthopedically handicapped can be increased by the following:

   a. Adequate clearance; level ground and floor surfaces, especially at thresholds of doorways, curbings, and ramps for change of elevation; and reachable heights for such items as drinking fountains, telephones, and built-in worktable tops and shelves

   b. Limited force application requirements for opening doors

   c. Door handles and cabinet handles that can be pushed rather than grasped or squeezed and turned

   d. Stairs that are not steep and railings that can be grasped and held firmly in the arthritic hand

TABLE 2.2 Selected Types of Disability for the Civilian Noninstitutionalized Population 5 Years and Over by Age: 2000

Census 2000 PHC-T-32. Disability status of the civilian noninstitutionalized population by sex and selected characteristics for the United States and Puerto Rico: 2000

	5 Years and Over		5 to 15 Years		16 to 64 Years		65 Years and Over	
Sex and Disability Status	Number	Percentage	Number	Percentage	Number	Percentage	Number	Percentage
Both Sexes	257,167,525	100.0	45,133,665	100.0	178,687,235	100.0	33,346,625	100.0
With a sensory, physical, mental, or self-care disability	32,178,220	12.5	2,614,920	5.8	17,300,050	9.7	12,263,250	36.8
With a self-care disability	6,752,735	2.6	419,020	0.9	3,149,875	1.8	3,183,840	9.5
No self-care disability	25,425,485	9.9	2,195,900	4.9	14,150,175	7.9	9,079,410	27.2
No disability of specified type	224,989,310	87.5	42,518,750	94.2	161,387185	90.3	21,083,375	63.2
With a sensory, physical, or mental disability	31,681,270	12.3	2,538,610	5.6	17,006,860	9.5	12,135,800	36.4
All three disabilities	2,027,750	0.8	74,615	0.2	743,160	0.4	1,209,975	3.6
With two disabilities: sensory and physical	2,486,615	1.0	18,905	0.0	809,655	0.5	1,658,050	5.0
With two disabilities: sensory and mental	688,150	0.3	108,350	0.2	328,260	0.2	251,540	0.8
With two disabilities: physical and mental	3,981,105	1.5	161,755	0.4	2,407,615	1.3	1,411,735	4.2
With sensory disability only	4,102,765	1.6	241,020	0.5	2,242,830	1.3	1,618,915	4.9
With physical disability only	12,656,040	4.9	200,185	0.4	7,189,935	4.0	5,265,925	15.8
With mental disability only	5,738,850	2.2	1,733,780	3.8	3,285,405	1.8	719,665	2.2
Men	124,636,825	100.0	23,125,325	100.0	87,570,585	100.0	13,940,920	100.0
With a sensory, physical, mental, or self-care disability	15,321,025	12.3	1,666,230	7.2	8,710,290	9.9	4,944,505	35.5
With a self-care disability	2,752,920	2.2	244,825	1.1	1,463,185	1.7	1,044,910	7.5
No self-care disability	12,568,110	10.1	1,421,405	6.1	7,247,105	8.3	3,899,595	28.0
No disability of specified type	109,315,800	87.7	21,459,095	92.8	78,860,290	90.1	8,996,410	64.5
With a sensory, physical, or mental disability	15,092,275	12.1	1,625,940	7.0	8,565,310	9.8	4,901,025	35.2
All three disabilities	904,040	0.7	42,190	0.2	399,310	0.5	462,540	3.3
With two disabilities: sensory and physical	1,154,085	0.9	9,495	0.0	450,845	0.5	693,745	5.0
With two disabilities: sensory and mental	388,610	0.3	66,840	0.3	190,660	0.2	131,115	0.9
With two disabilities: physical and mental	1,688,800	1.4	95,300	0.4	1,097,045	1.3	496,450	3.6
With sensory disability only	2,361,310	1.9	124,185	0.5	1,347,305	1.5	889,815	6.4
With physical disability only	5,374,800	4.3	104,870	0.5	3,332,530	3.8	1,937,400	13.9
With mental disability only	3,220,635	2.6	1,183,065	5.1	1,747,615	2.0	289,955	2.1
Women	132,530,700	100.0	22,008,345	100.0	91,116,650	100.0	19,405,710	100.0
With a sensory, physical, mental, or self-care disability	16,857,190	12.7	948,690	4.3	8,589,760	9.4	7,318,745	37.7
With a self-care disability	3,999,815	3.0	174,195	0.8	1,686,690	1.9	2,138,930	11.0
No self-care disability	12,857,375	9.7	774,495	3.5	6,903,070	7.6	5,179,815	26.7
No disability of specified type	115,673,510	87.3	21,059,655	95.7	82,526,890	90.6	12,086,965	62.3

(*Continued*)

TABLE 2.2 Selected Types of Disability for the Civilian Noninstitutionalized Population 5 Years and Over by Age: 2000 (*Continued*)

Census 2000 PHC-T-32. Disability status of the civilian noninstitutionalized population by sex and selected characteristics for the United States and Puerto Rico: 2000

	5 YEARS AND OVER		5 TO 15 YEARS		16 TO 64 YEARS		65 YEARS AND OVER	
SEX AND DISABILITY STATUS	NUMBER	PERCENTAGE	NUMBER	PERCENTAGE	NUMBER	PERCENTAGE	NUMBER	PERCENTAGE
With a sensory, physical, or mental disability	16,588,995	12.5	912,670	4.1	8,441,545	9.3	7,234,780	37.3
All three disabilities	1,123,710	0.8	32,430	0.1	343,850	0.4	747,435	3.9
With two disabilities: sensory and physical	1,332,530	1.0	9,415	0.0	358,810	0.4	964,300	5.0
With two disabilities: sensory and mental	299,535	0.2	41,510	0.2	137,600	0.2	120,425	0.6
With two disabilities: physical and mental	2,292,305	1.7	66,455	0.3	1,310,570	1.4	915,280	4.7
With sensory disability only	1,741,460	1.3	116,835	0.5	895,525	1.0	729,100	3.8
With physical disability only	7,281,240	5.5	95,315	0.4	3,857,405	4.2	3,328,525	17.2
With mental disability only	2,518,215	1.9	550,715	2.5	1,537,790	1.7	429,710	2.2

Data based on a sample. For information on confidentiality, sampling errors, nonsampling error, and definitions, see www.census.gov/prod/cen2000/doc/sf3.pdf.

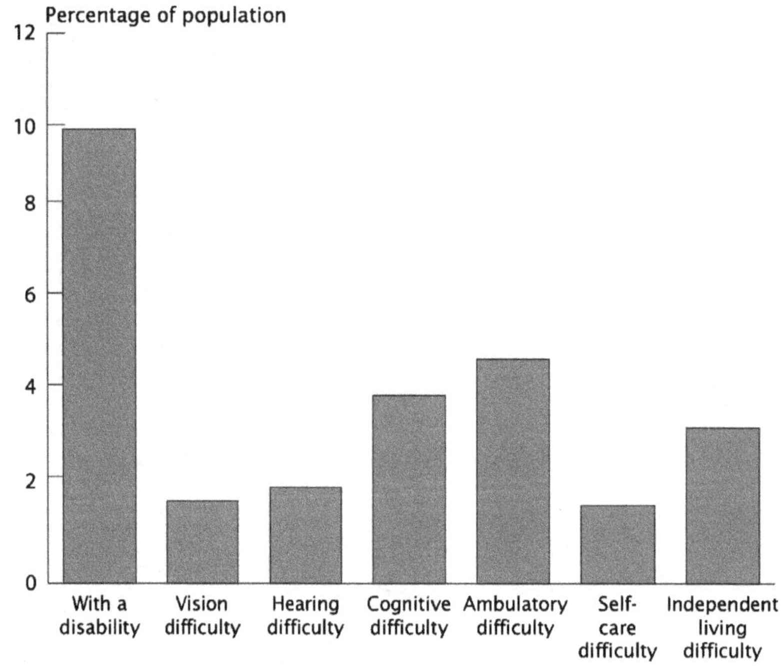

Note: Data are for the civilian noninstitutionalized population aged 16 to 64 years. This excludes people in correctional facilities, nursing homes, other institutions, and the armed forces.

Source: U.S. Census Bureau, American Community Survey, 2009.

FIGURE 2.1 Disability status and type: 2009.

4. Considerations for People with Different Handicaps: Extreme care should be exercised in developing mobility aids or concepts for people with one specific handicap, since the same facility may have to be used by people with other handicaps (e.g., a smooth, ramped intersection corner designed to aid the wheelchair user may remove the very tactile cues that tell the blind person where the street begins). In addition, care should be exercised in terms of how some aids to the handicapped may affect the use of the facility or device by nonhandicapped persons. In many cases, however, the aid may help both the handicapped and the nonhandicapped person. For example, larger, clearer street signs are needed in most cities today both for the normally sighted motorist and for the partially sighted person.

Functional Considerations

Specific human–architectural interfaces and the level of critically of each interface may vary from one system to another.

Considering Ease of Maintenance during the Planning Stage of Design. Too often the question of ease of maintenance does not come up when key architectural configuration decisions are being made. All facilities have to be maintained, and by human beings. Some common human factors problems associated with maintenance-related design include the following:

1. One cannot get to the spot that requires inspection, adjustment, cleaning, removal, replacement, or refurbishment.
2. There is insufficient space to do the job once a person has reached or located the maintenance problem.
3. There is insufficient illumination to see what needs to be seen.
4. There is a lack of appropriate service connections to enable use of the necessary tools at the work site.
5. The device to be repaired or replaced is buried into the structure, requiring major destruction and eventual repair.
6. Main service shutoffs (e.g., water, electricity, or gas) are variously distributed, hidden, and/or inaccessible, requiring an inordinate amount of time to find them.
7. The composite land-site–structural relationship precludes the normal and safe use of common maintenance aids such as ladders or scaffolds.

Architectural Safety. An objective during conceptual planning should be to create an environment in which the user can be as safe as possible. Although this is challenging, many of the accidents that frequently occur in homes, offices, schools, factories, and elsewhere are due as much to the facility design as they are to user errors. The following typical safety considerations are applicable to all architectural systems:

1. Use nonflammable, nontoxic materials.
2. Eliminate sharp edges, corners, etc., that could cause injury.
3. Create properly designed stairs, ramps, and walkways.
4. Do not use large ceiling-to-floor glass windows or doors without appropriate barriers to prevent people from walking through them when they are closed.
5. Ground all electrical controls, cover outlets, and otherwise prevent people from receiving electric shocks.
6. Provide adequate illumination so that people can see where they are going and avoid tripping over a walkway obstruction or step.
7. Use nonskid materials on floors, walkways, and stairs, especially if there is a possibility of their becoming wet.
8. Provide appropriate handrails around balconies and alongside stairs and use railing designs that children cannot fall through or get their heads caught in.
9. Cover moving parts of machines to prevent people from getting their hands or clothing caught.
10. Avoid locating heaters where they can be touched inadvertently or where pilot lights could ignite the structure or adjacent materials or cause an explosion as a result of gas fumes from a nearby vehicle.
11. Provide adequate emergency escape routes that can be used in the event that normal passageways and exits are impassable.
12. Consider the problems of window washing and of house or building repair in terms of typical unsafe practices associated with ladders and scaffolds.
13. Provide appropriate fencing around special facilities from which children should be barred (e.g., swimming pools, high-voltage wires, and heaters).
14. Provide fire sprinkler and alarm systems

COMMUNITY PLANNING

Introduction

In planning a new community or modifications to an older one, the needs of the community residents and their visitors should be uppermost in the mind of the planner. The human needs addressed in this section pertain to more down-to-earth factors, such as finding one's way around, identifying buildings and streets, and not having to travel a long distance to the market or having to go to several parts of town to complete a simple business transaction involving local government offices.

Unfortunately, the needs of each individual vary, and so do the needs of different individuals as they play out the special roles of housewife, businessman or businesswoman, factory worker, police officer, plumber, and clergyman, for example. (See Table 2.3.) In order to sort out these varied needs and thus be better prepared to understand how they can be accommodated best for the benefit of all concerned, planners should develop written or verbal descriptions of the behavioral circuits for all the expected community citizenry. These are, in a sense, scenarios of activity as each individual carries out his or her own personal mission. The scenarios need not be long or detailed, but they should be complete enough to give some idea of the importance of events, probable frequency, time implications, and so on.

With such information at hand, the planner is better prepared to consider alternative ways in which common elements of the community can be put together to serve all users in an approximately balanced manner. The following deals with a number of planning and design principles and guidelines that should be useful to the community planner and architect.

Community Facilities User Identification

Arrange community facilities on the basis of the primary facility user or tenant. Also consider other facility users such as repair or service individuals. Avoid the tendency to arrange community facilities purely on the basis of general facility relationships. Convenience should be thought of from the following points of view:

TABLE 2.3 Potential Users and Their Special Needs

User	Special Need
Primary tenants (residential)	Convenience to, but isolation from, heavy traffic arteries
Primary tenants (hotel and motel)	Convenience to business, public transportation, and special entertainment facilities
Primary tenants (others, e.g., employers and employees)	Convenience to key traffic arteries, related business, and public transit
Facility visitors (clients and customers)	Convenience to key traffic arteries and public transit
Facility service personnel (mail and package delivery)	Convenience to related service facilities
Facility repair personnel (structural, utility, etc.)	Convenience to related service facilities
Emergency service personnel (police, fire, and ambulance)	Convenience to both key traffic arteries and local streets and to related service facilities

1. Distance
2. Directness of access
3. Mode of access (by foot, private vehicle, public transit, or special vehicle, e.g., heavy fire or construction equipment or helicopter)

Community Planning Model

The key community activity elements shown in Fig. 2.2 must be arranged to provide balanced interaction, isolation, and accessibility. If a priority is given, it should be to the primary community resident.

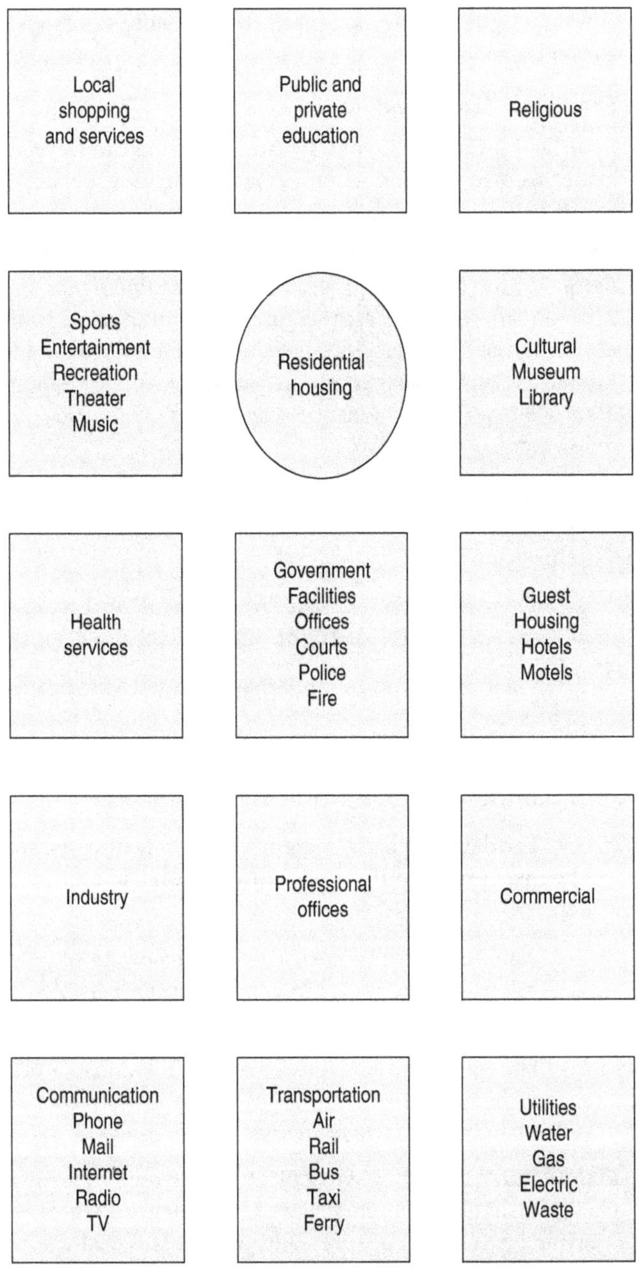

FIGURE 2.2 Community planning model.

Location Considerations for Other Key Community Features and Functions

Airports

1. There should be sufficient flat topography so that aircraft are not required to take off or land across obstructing hills.

2. Noise isolation should be provided. Although adequate access roadways and reasonable distances are important, the airport should not be located close to residential or other special community functions, e.g., hospitals, schools, recreational facilities, or churches. Early zoning policies are necessary to keep these latter functions and/or residential development from surrounding the planned airport site.

Government Offices

1. Centralize government offices with respect to principal user functions, i.e., business, legal services, law enforcement services, etc.

2. Local citizens should be able to reach government offices by means of public transit services.

Emergency and Disaster Services

1. Decentralize fire stations and ambulance and hospital services; i.e., equalize travel requirements for all principal community areas.

2. Provide access to main thoroughfares; i.e., minimize the time required to pass congested business or residential streets and potential traffic barriers.

Shopping Centers

1. Locate regional shopping centers so that they serve large residential areas.

2. Provide local shopping-center sites to serve specific areas such as residential neighborhoods, center-city areas, and separate commercial areas.

Churches

1. Set aside specific sites for a range of religious denominations for each defined community area.

2. Set aside specific sites within each planned downtown area where center-city residential facilities are planned.

3. Avoid choosing a site that is adjacent to businesses or other facilities that are known to be objectionable to parishioners, e.g., bars, restaurants, supper clubs, theaters, or the local jail.

Railroad Terminals. Although the site for a passenger terminal should be generally centralized with respect to access from all parts of the planned community, rail lines should be as far from residential, school, church, hospital, and recreational functions as practicable because of the potential noise and air pollution generated by trains.

Heavy Industry. Like railroads, heavy industry often produces undesirable noise and air pollution. In addition, certain hazards may be associated with specific industrial facilities and operations, e.g., fires, explosions, and chemical contamination. Set aside sites for expected heavy industry that will minimize these problems for the entire community, but especially for residential, school, church, hospital, and recreational areas.

Planning Checklist for Minimizing Architectural Barriers to the Mobility of Handicapped Citizens

1. Provide wheelchair-use ramps at all intersections, footbridges, parking-lot-to-sidewalk interfaces, public buildings, etc.

2. Provide wheelchair-use clearance for all public-use doorways, elevator doors, restroom entries, and interior stall entrances and hallways.

3. Provide smooth thresholds and crossing surfaces at all public-use doorways, railway crossings, etc.

4. Provide optimized identification signs for the partially sighted at all intersections used by these handicapped individuals (i.e., position signs at eye level so that users can get close to the signs and provide optimized lettering in terms of font, stroke width, letter height, and figure-ground contrast).

5. Provide tactile identification devices at all intersections used by the totally blind and/or audio signaling devices that both identify the intersection and give information about when to cross safely.

6. Establish a general policy regarding placing obstructions within normal pathways (i.e., mailboxes, trash containers, telephone booths, utility poles, and guy wires within the sidewalk perimeter) and/or provide suitable guard devices around these obstructions so that the blind can identify and avoid them.

7. Position any controls used by the handicapped so that they are accessible both to the handicapped person in a wheelchair and to the nonhandicapped or handicapped user who may be standing.

8. Establish a policy regarding the positioning of public-use devices such as drinking fountains and telephones so that they are accessible to both the handicapped person in a wheelchair and the nonhandicapped or handicapped person who normally stands to use these devices.

9. Make public sidewalks wide enough so that handicapped persons are not forced to walk or maneuver their wheelchairs too close to the street.

10. Provide surface guide markers that can be easily identified by the blind or partially blind by touching these markers with a cane.

11. Consider the effect that inclement weather conditions may have on particular barrier-removal systems, e.g., snow and ice that could cover surface markers.

12. Do not use or allow to be used any gratings in sidewalk or street surfaces into which the wheel of a wheelchair could drop or become caught.

Activity Centers

The community layout should be responsive to the needs of individuals in terms of the activities they pursue, primarily in going from facility to facility. Individuals should be able to identify where they are ad where they want to go first. Then they should be able to get there by the most direct route possible with a minimum of interference. The following suggestions are made for various facility activities, starting with the key user category, the local citizen.

Educational Facilities. Public educational facilities include both the precollege institution and the college or university. Although planners must also consider other private institutions, they may have less control over the location of these facilities. The key issue for locating public schools is access between home and school for students; access for faculty members and administrators is an important but less critical consideration. The following is a general checklist for determining the location of public schools:

1. Grade schools should be located in terms of minimized walking distance between the school and the related residential neighborhood. It is important to arrange school–residence relationships to minimize the necessity for small children to cross heavily traveled highways and streets. Grade schools should be considered primarily a single neighborhood facility.

2. Junior high schools should be conveniently located with respect to related residential neighborhoods and should be arranged to minimize the necessity to cross heavily traveled highways and streets, although this need not be given the same high priority as in the case of grade schools. Since more real estate is usually required for these facilities, consider a more central location with respect to several neighborhoods.

3. High schools should be centrally located with respect to several neighborhoods, but also should be arranged to minimize unusually heavy traffic-crossing requirements. In most communities, many high school students drive their own automobiles to and from school. Thus, special consideration should be given not only to the vehicular routing between residential and school areas but also to the effect this may have on other types of traffic, such as general "going-to-work traffic," commercial traffic, and transient traffic.

4. Colleges and universities should be planned and located more or less as independent student communities, and both resident and off-campus students must be taken into account. Depending on the nature of the institution, consideration should be given to the plan for a total college–residential arrangement that will provide satisfactory local convenience with a minimum of interaction with the rest of the community.

5. Common factors

   a. Public transit access

   b. Off-street parking

   c. Security services

   d. Special community usage (local community sharing of the auditorium and sports facilities)

   e. Exclusions (isolation from certain types of businesses such as liquor stores, bars, and adult entertainment centers)

Religious Facilities. Although local governmental bodies do not always exercise control over the location of churches and other religious facilities, there will be fewer problems in the future if there is a logical plan for including suggested locations for these facilities and, preferably, for holding these land sites for potential church use. The following is a general checklist for determining the location of religious facilities:

1. Set aside future church property in, or adjacent to, residential neighborhoods, rather than commercial, industrial, or recreational neighborhoods.

2. Set aside sufficient property for off-street parking.

3. Provide reasonable separation between church property and public schools (to minimize potential vandalism) and commercial property of an antireligious character.

4. Provide reasonable separation between church property and noise-producing environments (heavily traveled roads, airports, railroad rights-of-way, etc.).

5. Locate religious facilities near planned public transit routes.

Health Facilities. Advanced planning for the location of hospitals, convalescent homes, and similar health-oriented facilities is important for several reasons:

1. Because emergencies may happen in the home, at work, or on the highway, there should be rapid access to health facilities from all parts of the community. Although centralization is often suggested, this may create considerable conflict with downtown traffic.

2. Health facilities should be isolated from typical noise-generating aspects of the community, e.g., traffic,

schoolchildren, factory equipment, aircraft, and railroads. Although modern techniques make it possible to lower noise levels within a facility, an improper location adds immeasurably to the cost.

3. Large hospitals, especially, generate tremendous parking needs; therefore, sufficiently large land sites must be identified early in the planning stage.

4. Health facilities should be located near planned public transit routes.

5. Health facilities should not be located in disaster-prone areas (near earthquake faults, below watersheds or dams, etc.) because of the difficulty of moving patients.

The following is a general checklist for determining the location of various health facilities:

1. Major medical centers should be located somewhat centrally but peripherally to primary residential areas, with immediate access from main thoroughfares and highways.

2. Convalescent homes should be located in suburban areas, preferably where adequate open space can be retained around the convalescent home. Convenient, but not necessarily major, street access should be considered so that the facility will be isolated from traffic and noise, which tend to build as the community is further developed.

3. Retirement homes should be located similarly to convalescent homes, except that convenient public transit availability is perhaps more critical. Land-site considerations should include the possible need for outdoor recreational facilities such as a golf course, swimming pool, or picnic area.

Recreational Facilities. The increasing demand for recreational facilities of all types, places a considerable burden on the community planner, principally because of the amount of land that such facilities require. All recreational facilities should be planned with adequate parking, and special consideration should also be given to the peak traffic problems associated with special holidays and sporting events. The following is a checklist for determining the location of various types of recreational facilities:

1. Public Parks

   a. Neighborhood Parks: These are a combination of playground and picnic ground within walking distance of a defined residential neighborhood. They should not require crossing heavily traveled streets.

   b. General Community Parks: These consist of one (or more than one) larger park areas that might include an open-air Theater, equestrian paths, zoological facilities, etc., in addition to playgrounds and picnic grounds. They require easy access by vehicle and local bus and also adequate parking, especially for holiday occasions.

   c. Downtown Parks: One or more small downtown parks can provide noontime relief for business and professional workers. They should be accessible with minimum requirement to cross heavily traveled downtown streets.

2. Amusement Parks

   a. Public and/or Commercial Educational Parks (Including Botanical Gardens, Wild-Animal Parks, and Sea-Life Parks): These should be located outside the main periphery of the community, but with good access by private and public conveyance.

   b. Carnival and Concession Parks: These should be located outside the main periphery of the community, but with good access by private and public conveyance.

3. Sports Parks

   a. Amateur Sports Parks (Including Golf, Tennis, Baseball, etc.): These should be located on the periphery of the community, with repetition to provide convenience to all residential areas of the community. Good access is required for vehicular, bike, and public transit traffic.

   b. Professional Sports Parks (Including Golf, Tennis, Baseball, Football, Basketball, Hockey, etc.): These should be located on the periphery of the community or just outside.

Entertainment Facilities

1. Local Motion Picture Theaters: These should be located strategically within or near typical residential neighborhoods that are large enough to support such facilities. Public parking is an important factor in land-site selection. Such facilities should also be convenient for pedestrian and public transit traffic.

2. Downtown Theaters (Including Motion Picture Theaters, Music Halls, Opera Houses, etc.): These should be located downtown and should be accessible to both residents and visitors. Adequate off-street parking, space for dispensing and picking up private and public transit patrons, and access for delivery of entertainers' equipment are important factors to consider.

3. Dinner Theaters: These should be located outside the periphery of the community, but with good access and parking for private conveyance.

Convention Facilities. Publicly supported convention facilities should be located downtown, wherever feasible, in order to provide good access between these facilities and local hotels. However, an equally important factor to consider is the accessibility of such facilities to key outside transportation (e.g., the airport), major expressways or freeways, etc. Similar consideration should be given to convenience between convention facilities and entertainment facilities.

Commercial Facilities

1. Neighborhood Shopping and Services: Proper zoning for commercial shopping and services within various residential neighborhoods is needed, not only to provide convenience for residents, but also to preclude the proliferation of sometimes unsightly commercial buildings in an otherwise purely residential atmosphere. As a general rule, smaller shopping areas should be limited to those residential areas in which people typically walk to stores. On the other hand, larger shopping complexes should be located centrally, but peripherally to several residential neighborhoods, assuming that the general pattern would be to shop by private vehicle—in which case parking becomes a major consideration.

2. Commercial Services: Services required by commercial establishments should be located for maximum convenience to these establishments in terms of delivery, but there should not be direct individual interface. This suggests a more or less central location away from residential neighborhoods. However, there is a considerable advantage to confining different types of commercial services to specific areas within the community so that individuals, when necessary, can find a particular kind of service in one location, rather than spread out in different parts of the city. Most commercial services require easy delivery access, both incoming and outgoing, which means that there must be good access to major external transportation terminals and to key arterial streets and highways both for delivering within the city and for shipping intercity.

3. Industrial Facilities: Industrial facilities, whether light or heavy, should be located away from both residential and light commercial areas because of the unique nature of their traffic and possible environmental pollution. Heavy industry should be located the farthest out, but it should be convenient to major highways for trucking access, to railroads, to waterways, and to air terminals. Consideration should be given to locating heavy industrial facilities where any expelled fumes, smoke, or dust will not be blown across the city by prevailing winds. However, industry must also be located so that workers do not spend considerable time getting to and from their homes.

Community Administrative Facilities. Whereas, in the past, the "central courthouse" concept was extremely popular for the small town, modern community planners are warned not to consider the location of their jobs in the completed facility the central issue in community planning. The convenience of the local citizen should still be the key issue, and the convenience of the community administrative facility should be attuned to the needs of the typical citizen who wants to make contact with public officials. Public administrative facilities should be located so that citizens can find them easily; consideration should also be given to how easily public servants, such as police and fire fighters, can service the public. The following are guidelines to locating public offices:

a. Facilities with a Direct Public Interface: Facilities to which the individual citizen must go (for building permits, licenses, etc.) should be located not in crowded downtown areas but in peripheral areas where public and private access and parking are available.

b. Facilities from Which Special Services Emanate to Various Parts of the Community (Police and Fire Fighters): These facilities should be distributed variously to increase the efficiency of response, i.e., to control crime, put out fires, etc.

c. Facilities without a Direct Public Interface (Mayors and City Council's Offices and Technical Departments): These facilities should be located together primarily for effective administrative interaction and secondarily for convenience of public interaction. The actual site is perhaps of less importance in terms of general accessibility to various parts of the community than in terms of the almost inevitable expansion that will come with community growth. Other special administrative and service facilities should be located on the periphery of the community, according to the shape and size of the community area. These include such services as road and street maintenance and maintenance of public transit system vehicles, for which storage requirements continually expand as the community expands. Detention facilities such as jails or prisons should be isolated from the rest of the community wherever feasible. Although such facilities are often considered a part of typical administrative facilities, the average citizen fears the possible escape of a prisoner and reacts negatively to such facilities being near his or her residence, schools, or church.

d. Public Disaster Command Post: Although city planners often think of their regular administrative facility as the primary command post for control of all activities within the community, consideration should be given to the location and development of a separate emergency command post to which key disaster control personnel could go to oversee emergency procedures. The site for such a facility should be chosen on the basis of the invulnerability of communications and personnel to interference by man-made or natural causes, rather than on the basis of convenience, other than rapid access by key personnel.

Linking Activity Centers

The layout of the street and roadway system is important to both local users and visitors, since it determines how easily they can locate themselves and decide where they want to go. The important considerations are as follows.

Use a Rectilinear North–South–East–West Plan

1. Most people relate where they are going to a north–south reference. They use this reference as they follow a projected path on a road map. They expect to turn left or right as they progress north or south or east or west; therefore, they are more positive when the roadways are at right angles to each other.

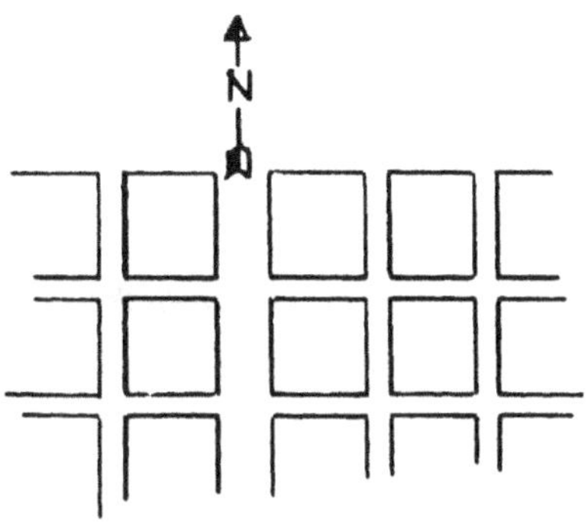

2. People become confused as they traverse curved or angled streets. Not only do they lose track of what is north and south or east and west, but all intersecting cues become distorted—nothing is "equal or square anymore."

3. Multiple-entry intersections (five or more streets) make it difficult for people to follow their course across the intersection.

4. Traffic circles or "town squares" interrupt people's perception of where a street continues on the other side, making it difficult to find the continuation.

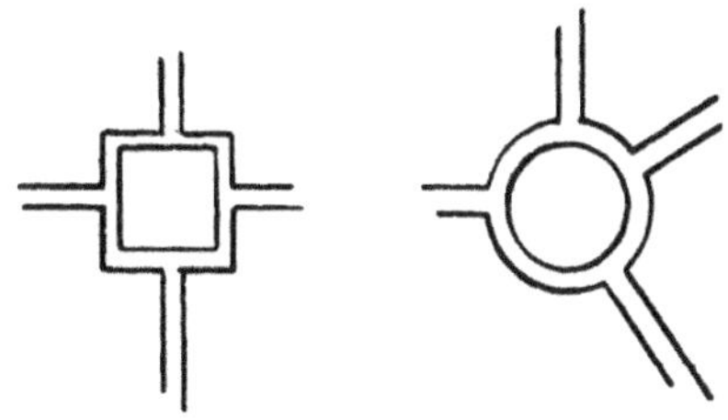

5. Street jogs confuse people who are trying to follow through across an intersection street. Similarly, a person using a cross street is confused by a cross street that appears sooner on one side than on the other.

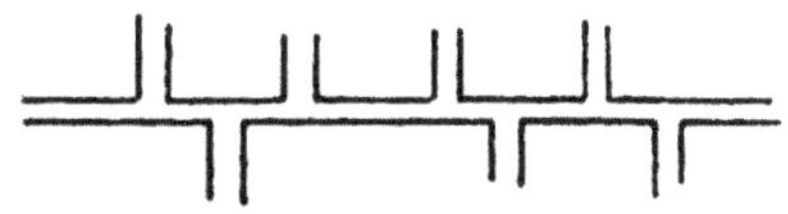

Develop and Maintain Constant Roadway Width

1. People soon recognize the importance and nature of a given street or roadway by its size (and sometimes by its surface characteristics). They soon associate wider streets with through traffic, and narrower streets with cross traffic and typically interrupted traffic (i.e., frequent stops). A plan should be developed in which street widths are standardized for various traffic and operating categories; otherwise, people become confused with respect to street identification and often fail to note stop signs or signals.

2. Changing width of a continuing roadway is not only confusing but also extremely hazardous. If street paving cannot be continued beyond a certain point because of the cost involved, the roadway width should continue, even though only the center portion of the street is paved for the time being.

3. Multilaned freeways, boulevards, etc., should continue the same number of lanes (although expansion is less serious than contraction). Unless such changes are properly signed way in advance, motorists become trapped when their particular lane runs out.

4. Bike lanes should be "added," rather than taken away from the width of the regular street or highway. They should also be isolated by a berm or raised surface to keep motorists off the bike route and cyclists out of the street.

Sidewalk Configuration. Regardless of local standards, several user-related considerations are important regarding the design of community sidewalks. Some of these are as follows:

1. Sidewalk Setback: For at least two important reasons, all sidewalks should be set back from the street. First, the sidewalk should be kept level as it crosses typical driveways, which must be sloped to the street. Second, pedestrians should be kept away from the curb. Children are especially

vulnerable because, as they play, they may inadvertently step off the sidewalk and into the street. (See Fig. 2.3.)

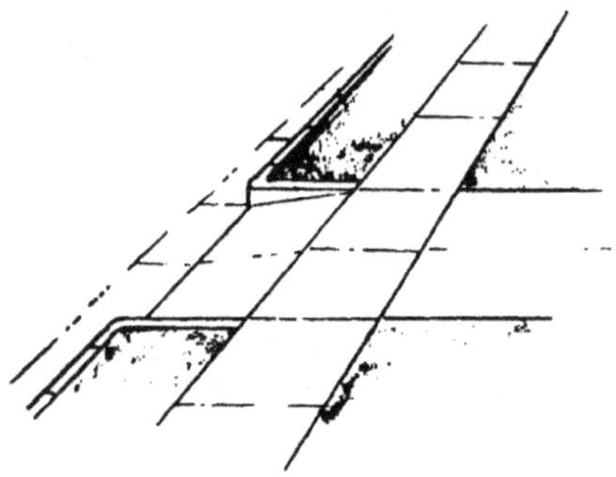

FIGURE 2.3 Sidewalk setback.

2. Ramps for Wheelchairs: Most states require special wheelchair ramps so that handicapped persons can negotiate between sidewalk and street with a minimum chance of tipping over their wheelchairs. The threshold of the ramp should be serrated so that blind persons can feel the ramp threshold and thus not inadvertently step into the street in front of a passing vehicle. (See Fig. 2.4.)

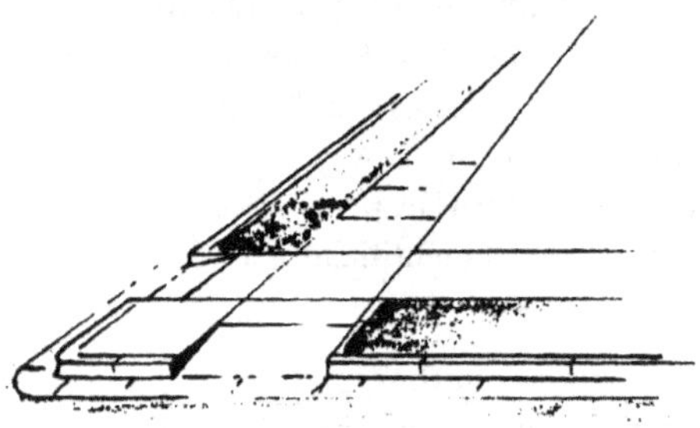

FIGURE 2.4 Ramps for wheelchairs.

3. Downtown Sidewalks. Wide sidewalks are preferred in order to "carry" the larger volume of foot traffic. The width is especially important at corners where people collect to wait for traffic signals to change. A nonsidewalk setback should be provided to keep people from being accidentally pushed into the street. (See Fig. 2.5.)

FIGURE 2.5 Downtown sidewalks.

Street and Roadway Identification and Address System. Although the planner has considerable latitude in devising a name and numbering system for streets and addresses, certain systems are more easily understood by people who are trying to find their way around the community. One of the best systems is to use proper names for streets running in one direction (preferably north and south) and numbers for streets running in the other direction (i.e., "First," "Second," etc., for streets running east and west). (See Fig. 2.6.) Some factors to consider

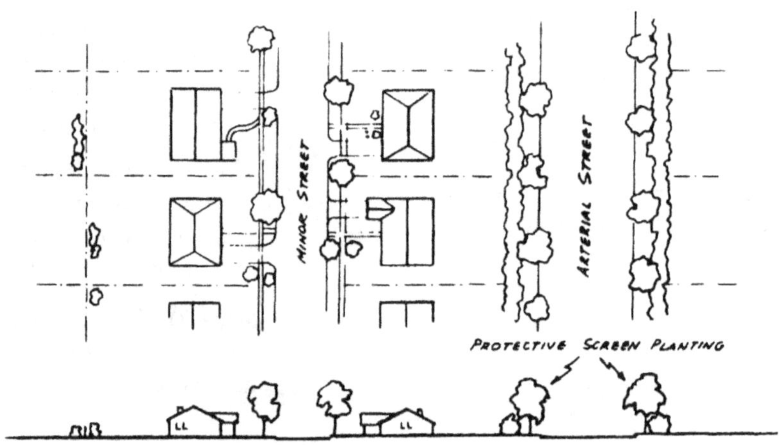

FIGURE 2.6 Planning traffic for a subdivision.

in creating a viable and nonconfusing system are as follows:

1. Avoid using a community name for a freeway; observers become confused if they do not intend to take the freeway to the community of the same name. That is, if they want to go to another area, they may believe that the freeway will not take them in the correct direction. For example, if the community at one end of the freeway is called "Santa Ana," do not call the freeway "the Santa Ana Freeway." Although this may not confuse local residents, it may cause people who want to go in the opposite direction not to take that particular freeway, even though they should. People also may not differentiate between roadway and community designations when they are the same.

2. Keep street names short and use names that are easy to pronounce. Motorists are easily confused when they have to remember long, hard-to-pronounce street names as they try to follow a path through a new community.

3. Do not use a geographically related name for a street that is not coincident with the name; e.g., "North Ventura" should not be used if the street does not run generally in a northerly direction.

4. Avoid using street names that sound the same but are spelled differently, e.g., "Sleighter" and "Slater." A motorist may have obtained directions over the phone and not realize that there are two different streets with the same-sounding name.

5. Do not use the same proper name for two different streets, one of which is called "Street," and the other "Avenue." Many people give their address using only the proper name and fail to let the other person know that there is more than one street with the same proper name.

6. Do not select an encoding system that uses letters or proper names arranged alphabetically if it can be anticipated that the system will run out as a result of future expansion. In other words, there are only 26 letters in the alphabet.

7. Try to maintain consistency in address numbering; i.e., numbers should progress in the same direction from a common reference point, and odd numbers should always appear on the same side of the street, with even numbers on the other.

8. Avoid changing the number sequence in the middle of a block, even though there may be an intervening street breaking into the block on just one side. Starting from a reference point, the first block should be numbered from 1000, the second block from 2000, and so on. This helps observers keep track of where they are without having to look at every number in the block.

9. Develop a consistent policy with regard to how streets and roadways will be "signed." That is, the size and shape and the location of identification signs and address numbers should be consistent throughout the community; people should not have to hunt for the sign or address number. Although there are many standards for signing, these are not always appropriate in terms of size and other legibility factors. The alphanumeric font and size should be compatible with the needs of observers relative to how far away they must be able to read the sign.

Environmental Control. Habitability within the proposed community will be influenced by how well the overall plan provides control over both natural and man-made pollution. The most serious possibilities include traffic noise; odors, fumes, or smoke from factories; and pollution of adjacent waterways. These factors should be considered as early as possible because their possible impact can be mitigated much more easily before the community is laid out. Although corrective devices can be added to control air and water pollution, it is difficult to control noise and odors except by properly routing traffic away from residential areas or by placing odor-producing factories in positions where the odor is blown away from, rather than toward, the community.

Effective noise control, as well as a general improvement in the quality of life, can be obtained by planned green belts, parks, and earthen barriers between heavily traveled roadways and nearby dwellings or facilities within which noise is undesirable. Figure 2.7 illustrates how a major freeway can be lowered with land barriers and foliage on each side to block and/or muffle the traffic noise. Obviously, the proper foliage treatment can also provide desirable park-like surroundings that people can use for outdoor recreation, picnicking, or resting. To have sufficient land space, such planning has to be considered as early as possible. This is even more critical with respect to locating airports.

Highway Classification. In the most basic classification system for design work, highways and streets are grouped into (1) interstate, primary (excluding interstate), secondary and tertiary road classes in rural areas, and (2) expressway, arterial, collector, and local road classes in urban areas. These classifications usually carry with them a set of suggested minimum design standards that are in keeping with the importance of the system and are governed by the specific transportation services the system is to perform. The principal considerations for designating roads into systems are the travel desires of the public, land-access requirements based on existing and future land use, and continuity of the system. Four basic purposes of urban street systems have been suggested:

1. Expressway system (including freeways and parkways)—Providing for expeditious movement or large volumes of through traffic between areas and across the city, and not intended to provide land-access service

2. Major arterial system—Providing for the through traffic movement between areas and across the city, and direct access to adjacent property; subject to necessary control of entrances, exists, and curb use

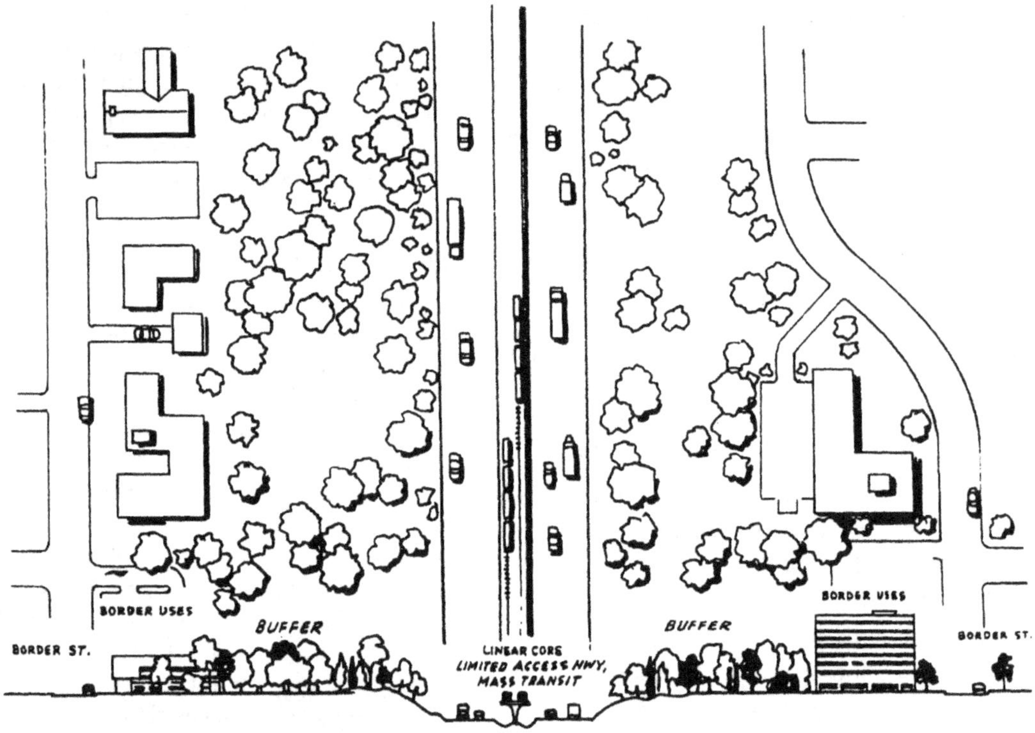

FIGURE 2.7 Buffering a major freeway.

3. Collect or street system—Providing for traffic movement between major arterials and local streets, and direct access to adjacent property

4. Local street system—Providing for direct access to adjacent land, and for local traffic movements

RESIDENTIAL

Residential Systems

Whether planning a single-family or a multifamily dwelling, a separate house, or an apartment building, planners should approach residential systems from the standpoint of optimizing dynamic activity relationships. The human is, of course, the focal point of all such activities, and the objective should be to satisfy human needs for all expected activities, such as the following:

1. Sleep and rest
2. Personal hygiene and dressing and undressing
3. Preparing and eating food
4. Performing household chores (e.g., laundry, sewing, and cleaning)
5. Performing exterior maintenance and gardening
6. Studying and record keeping
7. Entertaining guests
8. Child care
9. Performing hobby tasks
10. Performing shopping chores

In addition to these activities of residents, attention should also be given to the following supportive activities of nonresidents:

1. Deliveries (newspapers, furnishings and appliances, packages, etc.)
2. Special services (plumbing, electrical work, building repair, etc.)
3. Parking (for both residents and guests)

The principal objectives in creating a residential system plan should be as follows:

1. Convenience (i.e., arrangement of both external and internal activity areas so that unnecessary steps, traffic interferences, and communication difficulties between household members are minimized)
2. Privacy (i.e., appropriate visual and sound barriers to provide individuals with the necessary isolation for specific tasks such as resting, sleeping, and studying; separation of activities for adults and children; and specific areas for personal activities such as bathing)
3. Personal interaction (i.e., for group and family activities)

4. Security (i.e., protection from vandalism and from natural or man-induced disasters such as high winds, inclement weather, floods, earthquakes, and fire)

Finally, considerable attention should be given to the storage requirements associated with all activities, including the following:

1. Clothing
2. Food
3. Books and records
4. Housekeeping equipment and materials
5. Exterior maintenance and gardening equipment and materials
6. Linens
7. Eating and cooking utensils
8. Hobby, entertainment, and play equipment
9. Family automobiles and recreational vehicles and equipment
10. Waste and garbage
11. Other special storage requirements (for drugs and first-aid supplies, weapons, suitcases, empty boxes, etc.)

Whether the residence plan calls for a fully partitioned interior or a more "open space" configuration, all the above activities and requirements should be addressed, starting with the individual and his or her activities and working "outward." That is, planning should begin with a thorough analysis of the individual and the tasks that he or she must perform; then the planner must determine what materials and equipment or furnishings the individual needs to perform each task. This analysis establishes the space requirements for each activity center. Activity centers should then be analyzed to determine the most effective arrangement of activity spaces. This determines the size and shape of the total dwelling. A final analysis should then be made of external activities; this analysis determines the size and shape of the land site upon which the dwelling is to be placed. This is referred to as "designing from the human-out" and is quite different from typical planning approaches. That is, most architectural plans are developed from the opposite point of view. Land sites are selected first, and then a building size and form are chosen that will fit the land site; this is followed by an attempt to divide the building into workable activity modules. Whether the human and his or her activities are properly accommodated is often a matter of sheer coincidence or luck.

Obviously, some compromise between the two approaches is usually required since available land, costs, and other factors make the ideal "human-out" approach somewhat hypothetical. However, unless this approach is considered, designers seldom know what they are compromising or how much they may have compromised the human-activity objective until it is too late to do much about it. Experience indicates that a better compromise is made when the design is approached from the user-activity reference point than when it is approached from the other end of the reference chain.

Residential Functional Planning Model

The key family, guest, and service activity elements must be arranged to provide balanced interaction, isolation, and accessibility. If priority is given, it should be given to residents. Among the more important layout considerations are the following:

1. Ease of ingress and egress
2. Ease of traffic flow
3. Privacy
4. Security
5. Emergency disaster control
6. Environmental control

Functional Space Layout Model for a Single-Family Dwelling

Certain adjacency implications are illustrated in Fig. 2.8 (e.g., the family sleeping quarters are associated with a bathroom, and the kitchen is associated with both indoor and outdoor dining), certain separations are illustrated (e.g., the guest areas and entrance are separated from the family areas and entrances), and differentiation is made between guest and family parking. Other implications to draw from the model are that, although some things should be located adjacent to each other and although some should purposely be separated, a number of typical functions have no strong relationships that require special attention during the layout process; i.e., many alternatives are available to match the client's preferences. (See Fig. 2.9.)

Multifamily Residential Planning

The key family, guest, and service activity elements of the single-family residence are similar to those of multifamily residential facilities. (See Fig. 2.10.) In addition, however, the following special requirements should be addressed:

Common-Use Interior Activity Areas

1. Reception area and lobby, central mail box, call box, and paper delivery area
2. Central laundry facility
3. Central recreational and hobby facilities
4. Resident parking
5. Central trash and garbage collection
6. Stairs and elevator

In addition, consideration must be given to noise attenuation between neighbors, building management, and maintenance (for larger complexes); building security; fire protection; and signing and marking.

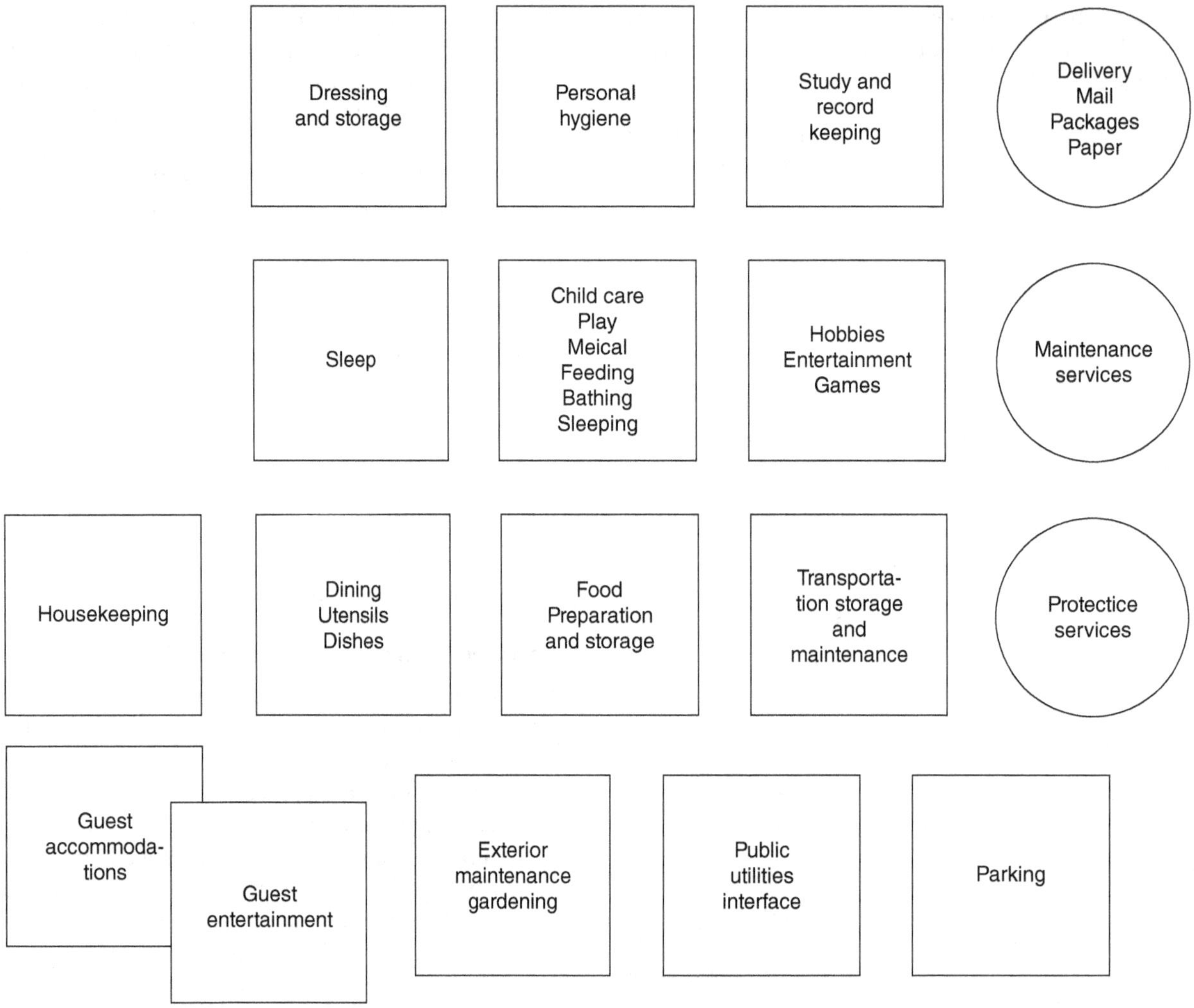

FIGURE 2.8 Residential functional model.

Common-Use External Activity Areas

1. Guest and delivery parking
2. Guest and delivery ingress and egress
3. Children's play area
4. External maintenance (electric and water outlets and tool and equipment storage)

Also, consideration must be given to external lighting, signing and marking, vehicular approach, and accessibility for emergency activities (police, fire, ambulance, etc.). In the case of firefighting, consider the availability of water supply and the ease of ladder use (for multistory units).

Common Planning Errors

1. Lack of adequate guest parking
2. Lack of adequate noise control
3. Lack of dual emergency escape routes
4. Lack of convenient delivery features (for delivering heavy, large furnishings)
5. Lack of sufficient "dead storage" space per tenant
6. Maintenance inconvenience (e.g., window cleaning)
7. Barriers for handicapped users
8. Hidden identification and address signs
9. Camouflage of main entrance
10. Lack of adequate, general trash-collection space (including the maneuvering space for garbage trucks)
11. Lack of play space for tenant children, which increases their chances of being exposed to street or internal area traffic, swimming pool hazards, etc.

Unique Layout Considerations

Although it is recognized that considerable cost savings accrue by arranging multitenant facilities in a somewhat regimented fashion, this often leads to interior arrangement of each tenant's residence in a less-than-efficient configuration. The guidelines for a single home should be applied wherever practical. In addition, the following unique layout considerations relating to the total complex should be addressed (see Table 2.4.)

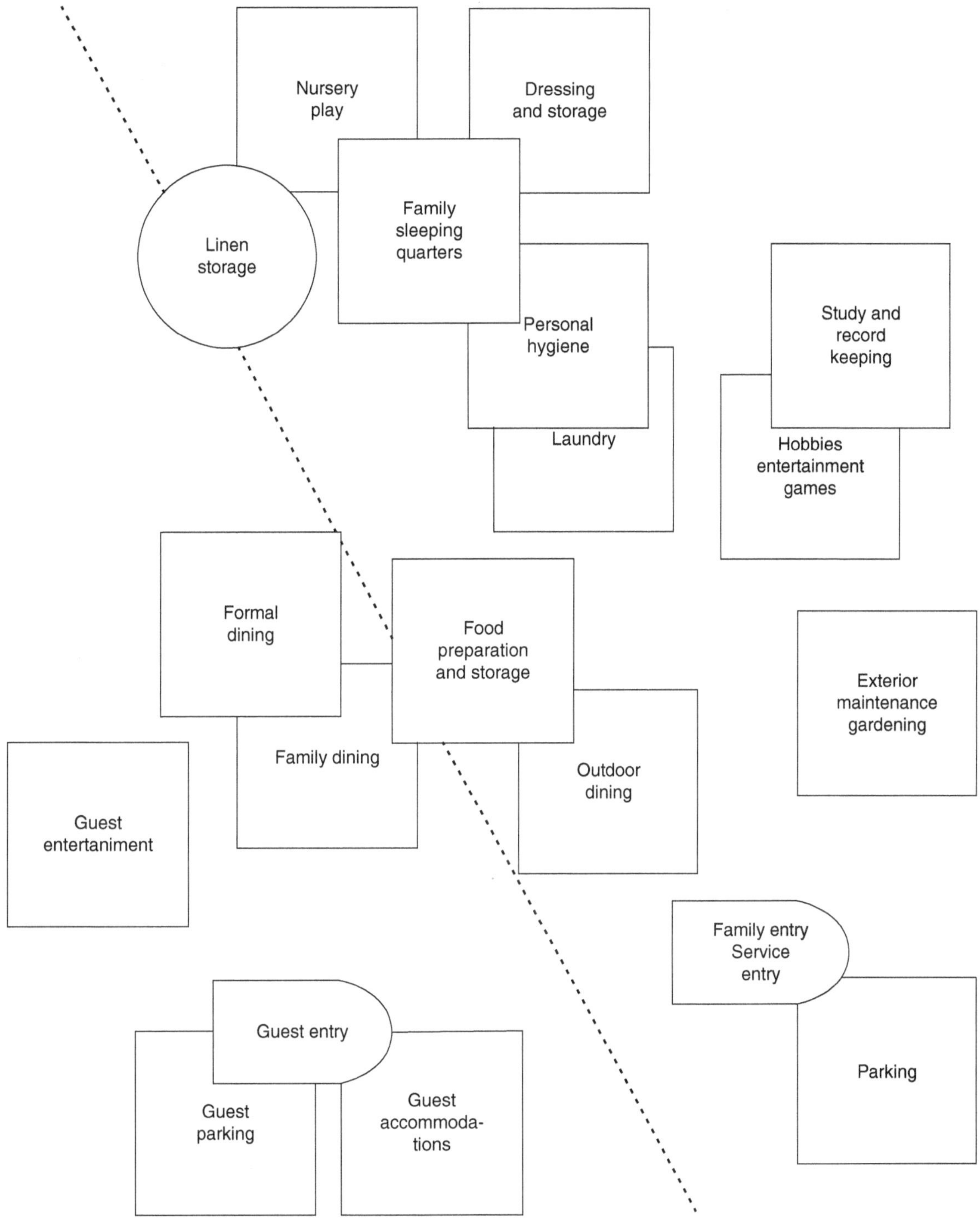

FIGURE 2.9 Arrangement of residential functional elements.

Visitor Parking and Main Entrance. Both the visitor parking area and the main building entrance should be conspicuous from typical street approaches, which implies an obvious adjacency between these two elements. Guests should not be required to walk great distances from their parked vehicle to the entrance. The parking and walkway both should have adequate night illumination.

Main Entrance and Lobby, Resident Manager's Quarters, and Lounge or Recreation Area. All these facilities are important to infrequent guests as well as to guests whom a tenant may want to entertain outside his or her apartment or condominium; thus, these facilities need to be fairly close to one another. Such facilities should be located generally outside a main security perimeter to prevent unwanted visitors from penetrating the private tenant area.

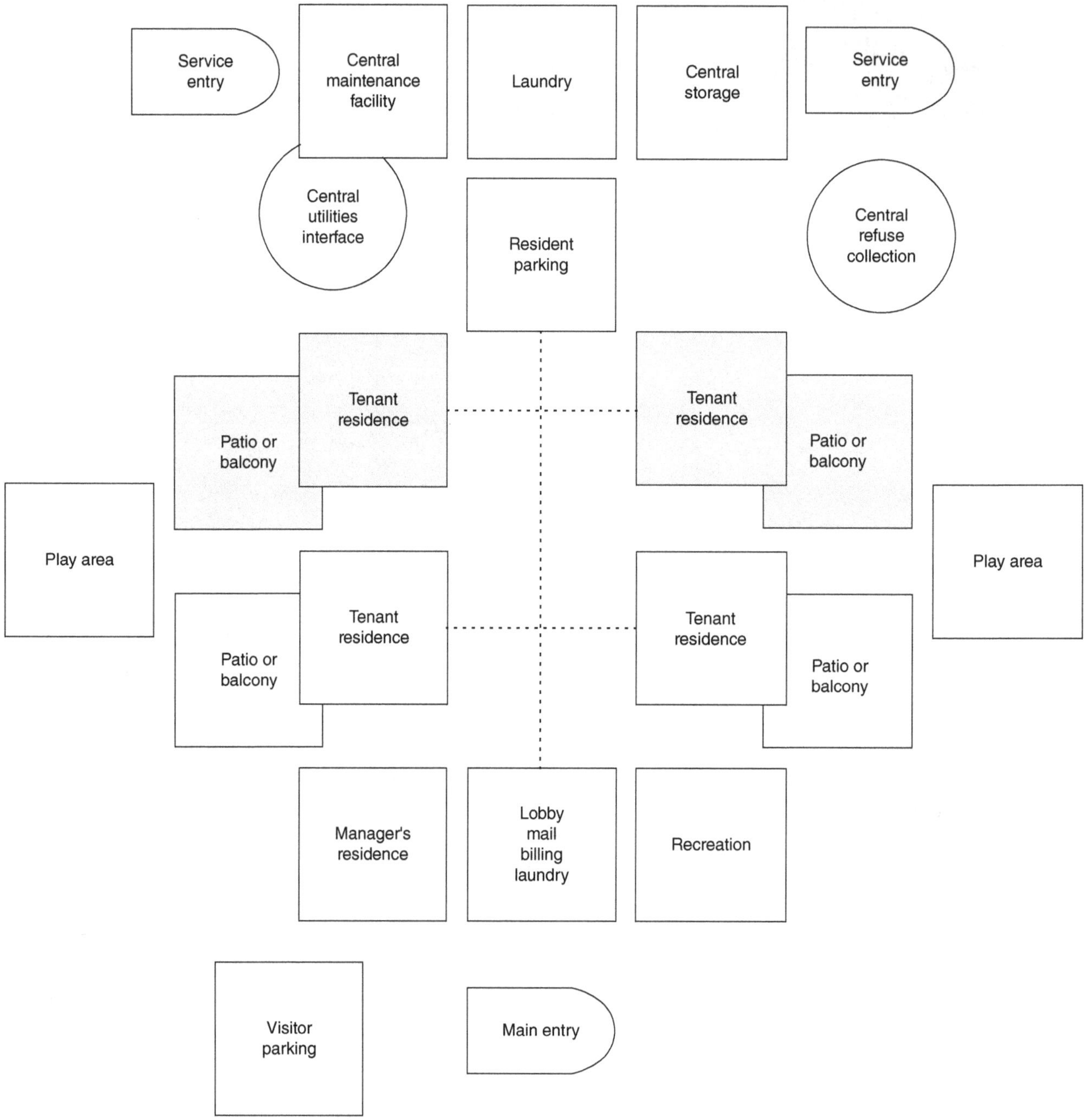

Figure 2.10 Multifamily residential planning model.

Indoor Resident Parking. When indoor parking is provided, it should have an entrance that is not generally exposed to the public, and if practical it should have an automatic gate controlled from the resident's car.

Central Maintenance, Laundry, and Storage Facilities. These facilities should be readily accessible only to building maintenance personnel and residents via special stairs and/or elevators, and they should be convenient to service entries that are not available to guests. Similarly, a central utilities facility should be directly accessible from one of the service entries.

Play Areas. If the multifamily complex is designed for use by families with children, special enclosed play areas should be provided with convenient interior access from tenant residences. Other special facilities such as a pool, sauna, hobby shops, etc., should be centrally located as a general rule, and accessible only to residents.

Stairs and Elevators. More than one stair and elevator system is recommended for complexes of more than two stories (although each community generally sets forth standards for such installations). The location of the dual

TABLE 2.4 Site Information Requirements Checklist for Apartment Complex Planning and Development[a]

	Information Required	Details
1. Marketability	Demand for multifamily housing	At what rents; distribution (0-BR, 1-BR, 2-BR, 3-BR); size of rooms
	Existing population and potential growth	
	Type of existing tenants living in apartments	With children and how many; without children; elderly; single occupancy; income brackets; age brackets
	Industries in the area and their future plans	
2. Pertinent information of the surrounding area	Existing street layout and how it may affect the parcel in question	
	Proposed street changes	Widening; elimination of streets; map changes
	Location of main arteries (parkways, freeways, and highways	
	Mobility from site (in all directions)	
	Zoning and proposed changes	
	Kind of buildings	Single family; multifamily, commercial, and industrial
	Appearance and general character	Design of the exteriors; condition of buildings, grounds, and streets
	Off-street parking	Is the surrounding area provided with adequate off-street parking? Are the existing streets wide enough for street parking and easy access for cars and service vehicles? Proximity of parks, public playgrounds, other recreation areas, and waterways, if any, to site. (See 7, Community facilities.)
	Hazards	Noise; proximity of airports, railroads, and trucking highways; smoke and fumes; high-tension wires; ravines
	General trend	Stability of area; building expansion; deterioration
3. Transportation available	Other than automobile	Rapid transit; bus; railroad; taxis and other vehicles, such as helicopters, hydrofoils, ferryboats, airplanes
	Time of travel to center city and to job location	
	Automobile travel	To center city; to job location
	Cost of daily travel	Daily fares; gas and oil, parking charges, if by car
	Schedule of transport services	
4. Zoning of site	Density coverage, height, yard requirements, and parking	
	Proposed changes, if any	
5. Planning boards	Rules and regulations that control land development other than zoning	
6. Deed restrictions		
7. Community facilities (distance from site and methods of getting there)	Schools	Public, parochial, or preschool; grammar, junior high, high school
	Shopping	Necessities; all others
	Religious buildings	Denominations
	Recreation	Theaters; playgrounds, beaches, swimming pools, bowling alleys, others
	Hospitals, medical centers, and clinics	
	Cultural	Libraries, art galleries, museums, and others

(*Continued*)

TABLE 2.4 Site Information Requirements Checklist for Apartment Complex Planning and Development[a] (*Continued*)

	Information Required	Details
8. Community services	Garbage and refuse collection	
	Police and fire protection	
	Snow removal	
	Street cleaning	
	Street maintenance	
	Street lighting	
9. Size and shape	If irregular, can plot be utilized efficiently?	
	If small, can economical project be built?	
10. Topography	Rugged terrain, gently sloping or flat	
	Rock exposure or filled-in land	
	Type of surface soil	
	Surface drainage and ground water	
	Natural features	Trees, streams, lakes, adjoining parks, and rock outcroppings
11. Subsurface conditions (information usually received from borings)	Composition of soil	
	Evidence of rock or filled-in land	
	Soil bearing capacity (necessity of piling)	
	Underground streams	
	Water level	
	Percolation of soil	
12. Utilities	Storm and sanitary	Combined or separate; nearby body of water or drainage ditch; depth of sewers; adequacy of sizes and pitch for additional loads anticipated; public or private
	Water supply	Pressure; reservoir well or other; rates; who pays for installation (from what point to what point)?
	Gas	High or low pressure; natural or manufactured; rates, who pays for installation (from what point to what point)?
	Electricity	Overhead or underground; current available; rates, who pays for installation (from what point to what point)?
	Telephone service	
13. Features	Views	
	Trees, streams lakes, and parks	
14. Cost of site	Potential yield (number of families)	Land cost per family; land cost per room
	Rent limitations for area	Rent per room per month; relation of total rent to cost of site
	Cost of abnormal site conditions	Excessive fill and grading; piling; rock excavation; possible retaining walls; cost of bringing utilities to the site

[a]Data from Samuel Paul, *Apartments: Their Design and Development*, Reinhold Book Corporation, New York, 1967.

facilities should be selected to create the best balance between travel distances from each tenant residence, the principal concern being to provide adequate emergency escape in the event of fire. Avoid adjacent stairs and elevators, which can cause traffic jams.

Exterior Privacy. Most multistoried residential facilities provide some sort of balcony for each tenant. The need for privacy as well as freedom from other annoyances (people throwing debris onto a lower balcony, smoke drifting into a balcony from a neighboring tenant's barbecue, etc.) should be considered in initial planning. Considerable privacy can be obtained by means of proper overhangs, directional orientation, and/or special barriers designed into the original structure. Avoid configurations in which balconies on opposite sides of a courtyard face directly toward each other.

Internal Privacy. In addition to the obvious need to insulate between residences for acoustic reasons, consider methods for providing reasonable visual privacy for individuals who may stand momentarily outside a friend's apartment. One method is to provide an alcove or setback at the entrance to each apartment so that visitors do not have to stand directly in the hallway.

Special Safety Considerations. The following are unique objectives for the typical high-density residential complex:

1. Each apartment should have more than one exit. The common practice of considering a balcony an emergency exit is not an acceptable compromise. Even though an agile tenant may be able to use an emergency rope device, the elderly, the handicapped, and especially small children cannot do so safely.

2. Do not arrange kitchens so that they are immediately next to primary hallways (prime escape routes). Because the majority of internally generated fires may occur in the kitchen, such fires typically can block the primary exit.

3. Provide emergency devices. Fire codes usually require the location of fire extinguishers at strategic points both internally and externally. Of equal importance is provision of emergency lighting and emergency warning communications.

4. Do not use flammable materials. Although it is difficult to control the materials used by tenants, the materials used in common areas of the multifamily complex should be a nonflammable variety.

5. Provide for building maintenance and security monitoring. Systems for monitoring general maintenance features within the building and general security should be considered for the large complex. A careful study is, of course, necessary to devise a system that does not encroach upon the privacy of tenants.

Land-Site and Facilities Layout Objectives

Figure 2.11 shows an idealized (although perhaps impractical) layout of an apartment building and grounds and illustrates a number of important user-oriented planning objectives:

1. Identifiable main and guest entrance and convenient parking.

2. Semi-isolated tenant parking and multiple entrances—also for service parking and entrance.

3. General separation of adult- and child-oriented outdoor recreation areas (although children would also associate with parents at times in area D).

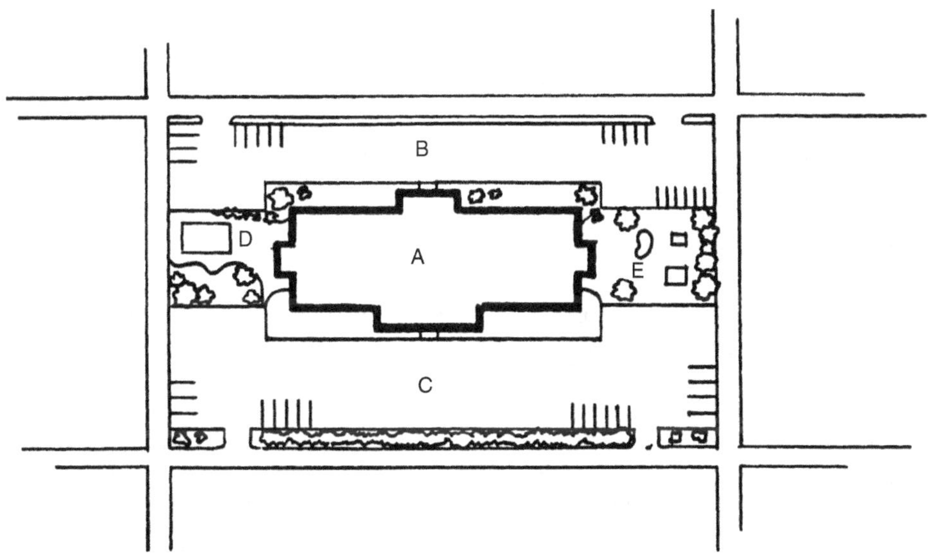

FIGURE 2.11 Idealized site layout (A: apartment building; B: tenant and service parking; C: guest parking; D: mixed recreational area; and E: children's play area).

4. Multiple street and site approaches to reduce traffic conflicts.

5. Parking and building relationships readily apparent even before the guest motorist enters the site.

6. Regularity of layout to simplify visitors' immediate recognition of where they should go to enter the site. (Note that it is easier to sign and mark sites that are rectilinear.)

7. Building accessible to emergency vehicles and equipment (firefighters, police, etc.) from more than one entrance and side.

8. Block plan easily approached from typical public transportation.

Multibuilding Layout. Figure 2.12 is an actual plan of an apartment complex and site layout; it illustrates some important points:

1. It may be desirable to separate buildings in order to accommodate more than one lifestyle, e.g., families with no children and families with children.

2. A separate garage facility is provided.

3. There is lower apparent density because the buildings are separate.

4. Many of the objectives identified for the idealized site layout are still maintained, i.e., separate visitor and guest entrances and parking, separation of play areas for small children and recreation areas used either by adults and older children or by small children under adult supervision, multiple access for emergency vehicles and equipment, and convenient service parking and entry.

Multibuilding Estate Layout. Very large apartment or condominium complexes often are laid out to create an almost estate or park-like atmosphere. Although some of the advantages of the more regular or rectilinear patterns noted in the previous examples are lost, there is still an opportunity to retain reasonable visitor orientation. Some of the special features that emphasize user-related objectives are as follows:

1. The cluster concept creates certain neighborhood identification for residents and also minimizes the general high-density, "row-on-row" effect that might develop when so many units are aligned in regular patterns.

2. The cluster concept provides a simpler means for separating childless family groups from those with children.

3. Parking areas can be much smaller and, because they are separated, do not give the impression of large masses of paving and a sea of vehicles.

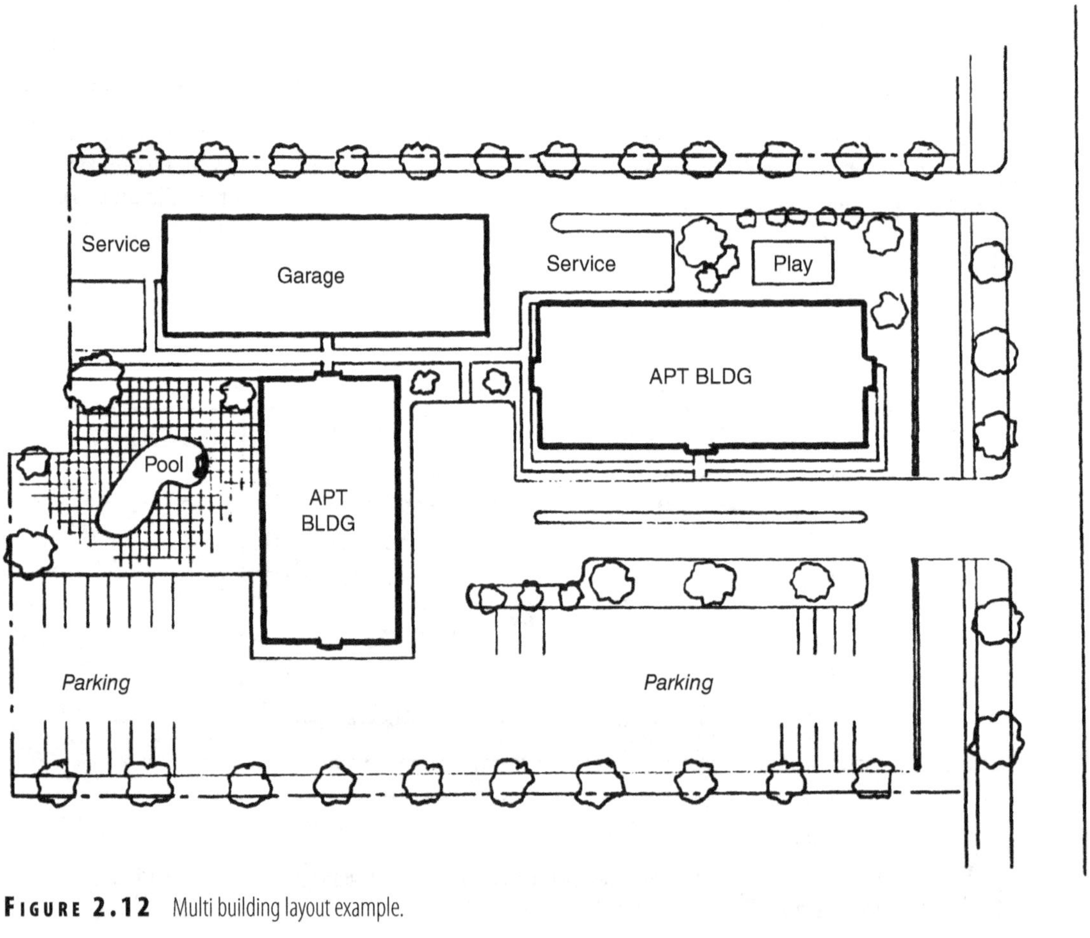

FIGURE 2.12 Multi building layout example.

TABLE 2.5 Apartment Layout Objectives	
Objective	**Considerations**
Minimum distance between apartment entrance and the kitchen	Tenants often have their arms full of heavy groceries, or they may have heavy trash or garbage containers to dispose of.
Minimum traffic through the living room	Not only is the wear and tear on the living-room carpet undesirable, but also most people prefer not to have other family members annoy guests by walking through the living room.
Isolation of bathroom noises	Noises emanating from the bathroom are often embarrassing when guests are present in the living-room area.
Natural light and view of the exterior	All rooms (including the kitchen) should have an exterior view and the opportunity for introducing natural light into spaces.
Patios and balconies	Each apartment owner generally enjoys his or her own private outdoor facility, which is visually inaccessible to the immediate neighbors. These should normally be accessible from a living or dining area or both.
Bedrooms isolated from noise	Generally, people prefer bedrooms to be isolated from areas where the noise of entertaining could interrupt the sleep of another member of the family.
Two exits	For safety reasons, at least two exits should be available in the event of an emergency. Since most internally initiated fires start in the kitchen, at least one exit should bypass this potential barrier.

4. Although internal roadways may be slightly curved, a reasonably clear pattern should be maintained so that visitors can easily find their way to the area they are looking for. Signing and marking should not be difficult.

Individual Apartment Layout Objectives. Because of the many physical constraints associated with multifamily facility planning, it is obviously difficult to create so-called optimized layouts for each of the apartments within the facility. However, it is helpful to have a set of objectives in order to avoid overlooking certain desirable features. Table 2.5 lists some of the more important objectives to be considered in analyzing various layout alternatives.

Although few, if any, ideal examples of apartment layouts could be presented, Figs. 2.13–2.15 show reasonably good compromises illustrating desirable and undesirable layout features.

In Example 1, the following points should be noted (see Fig. 2.13):

1. Multiple access and escape routes via two staircases and two elevators are a desirable feature. By distributing these facilities, the architect has maximized convenience and also provided the extra safety feature of minimizing traffic jams.

2. The architect has provided a reasonably good solution to the problems of access between apartment entrance and kitchen, minimal traffic through the living room, and convenience between kitchen and dining areas.

3. Balconies have been inset rather than extended, providing maximum individual privacy for each tenant.

Typical floor plan

Figure 2.13 Individual apartment layout—Example 1.

4. In some apartments, guests have access to a bathroom only by passing through family bedrooms. Although not completely objectionable, this is not as desirable as a less isolated location.

In Example 2 (See Fig. 2.14.), the architect has provided many at the same desirable features found in Example 1, e.g., multiple access and escape routes via stairs and elevators and almost equal distance between these and individual apartments (considerably shorter distances). One significantly new improvement is provided in this plan; i.e., there are more spaces with exterior viewing capability, especially in the case of kitchens in the corner apartments. Note also that bathrooms are located for convenience to all other spaces, as opposed to some of the inconvenience observed in Example 1.

In Example 3 (See Fig. 2.15.), the architect has provided unusually good isolation of guest and family areas, excellent convenience between entry and kitchen and between kitchen and dining area, and very desirable bathroom centralization. In addition, balcony privacy is assured, although perhaps not as well as in Example 1; i.e., extending the balconies may make visual isolation less satisfactory. In addition, this plan provides exterior views and natural light for every space except bathrooms and kitchens in two of the apartments.

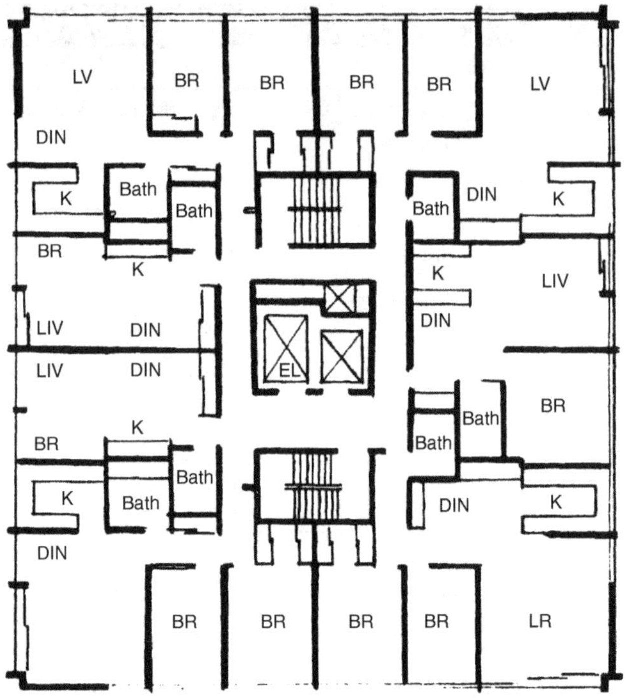

FIGURE 2.14 Individual apartment layout—Example 2.

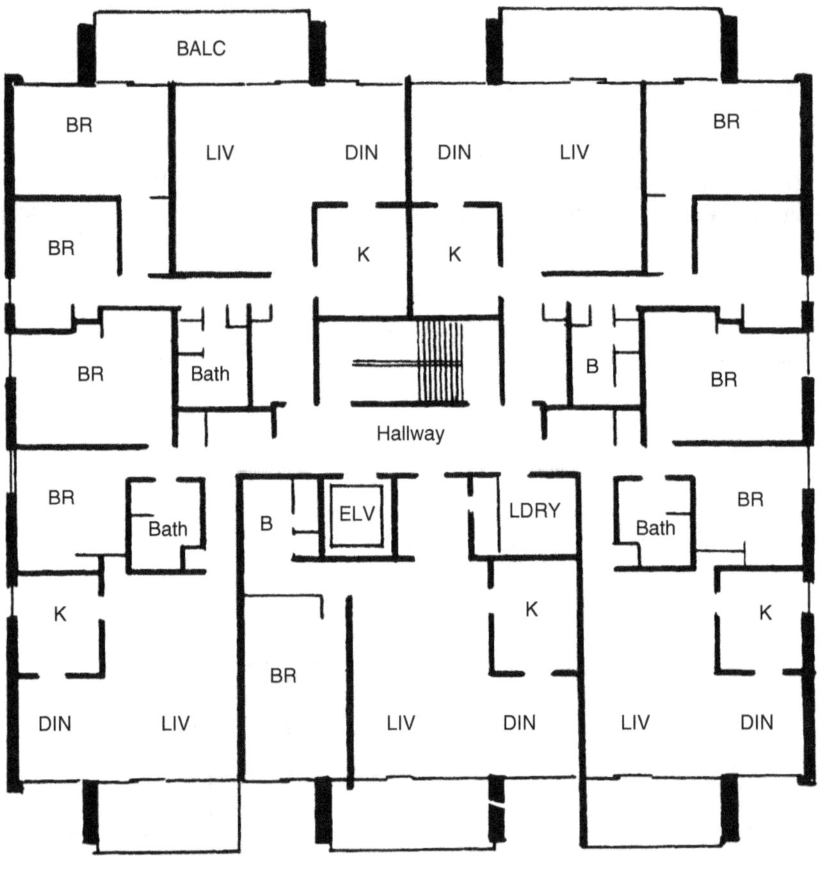

FIGURE 2.15 Individual apartment layout—Example 3.

The questionable feature of this plan is the general concentration of stairway traffic; i.e., although there would be little problem for individual tenants on one floor (because there would be so few people involved), if the building contained a considerable number of floors, there would be the potential for traffic jams at the bottom-floor level.

It is hoped that these three examples will provide useful ideas to readers in evaluating their own plans. It is possible, for example, to utilize the good features of the several layouts and, by an appropriate rearrangement of general building shape, to overcome some of the less desirable features that have been noted.

Special Amenities Considerations

Multiunit facilities require a number of unique amenities not necessarily required for the single-family dwelling. The following checklist identifies these requirements so that the planner can make sure they have been included in planning and design specifications. Consult local ordinances for the latest community building codes and standards.

1. Exterior lighting
    a. Parking area, both interior and exterior
    b. Walkways
    c. Entrances
    d. Recreational areas
    e. Building security (automatic timer system desirable)
2. Central mailbox or mailboxes and tenant directory
3. Visitor–tenant communication system
4. Common-area fire-extinguishing system and separate hand extinguishers
5. Manager–tenant communication system
6. Special security and key system
7. General and/or individual dead storage facility
8. Building identification signs
    a. Individual buildings
    b. Individual apartments within each building
    c. Emergency exit signs
    d. Parking signs
    e. Other safety signs, e.g., for play and pool areas
9. Fire hydrants

PUBLIC ACCOMMODATIONS

Public Accommodation Systems

The principal functional considerations for public accommodations systems relate to the overnight guest and include the following:

1. Sleeping
2. Personal hygiene
3. Dressing and temporary apparel and baggage storage
4. Communications (telephone, etc.)

In addition, depending on the nature of the proposed system, the following functions should be considered:

1. Dining
2. Vehicle parking
3. Recreational facilities (pool, sauna, tennis courts, etc.)
4. Special group meeting facilities (business, conventions, industry exhibits, etc.)

In support of the above guest-oriented facilities, many other functions must be considered, including the following (see Fig. 2.16):

1. Administrative (guest register, record keeping, etc.)
2. Maid service and laundry
3. Facility maintenance and security
4. Guest assistance (baggage handling, room service, transportation, etc.)
5. Other (cleaning, drugs and notions, gifts and flowers, liquor and other shops such as a barber shop and beauty salon, travel agency, tour agency, etc.)

The following are among the more important layout considerations (see Fig. 2.17):

1. Initial guest convenience to main entrance, checkin area, and guest quarters
2. Secondary guest convenience to dining, recreation, and parking
3. Guest service efficiency
4. Hotel administrative efficiency
5. Acoustic isolation of guest quarters

In conceptualizing the public accommodations facility, whether all the functions are contained and regulated as a whole or are subleased and possibly detached but adjacent facilities, each user's needs and conveniences should be analyzed before launching into some preconceived facility layout. In general, the needs of guests have priority since it is their patronage that creates a successful operation. It is the overnight guests that should be considered first, starting with their arrival, proceeding through the series of typical activities they and their party will participate in, and ending finally with their departure.

Obviously, however, unless other participants or users associated with serving guests are also efficient (i.e., unless all their functional needs are satisfied reasonably well),

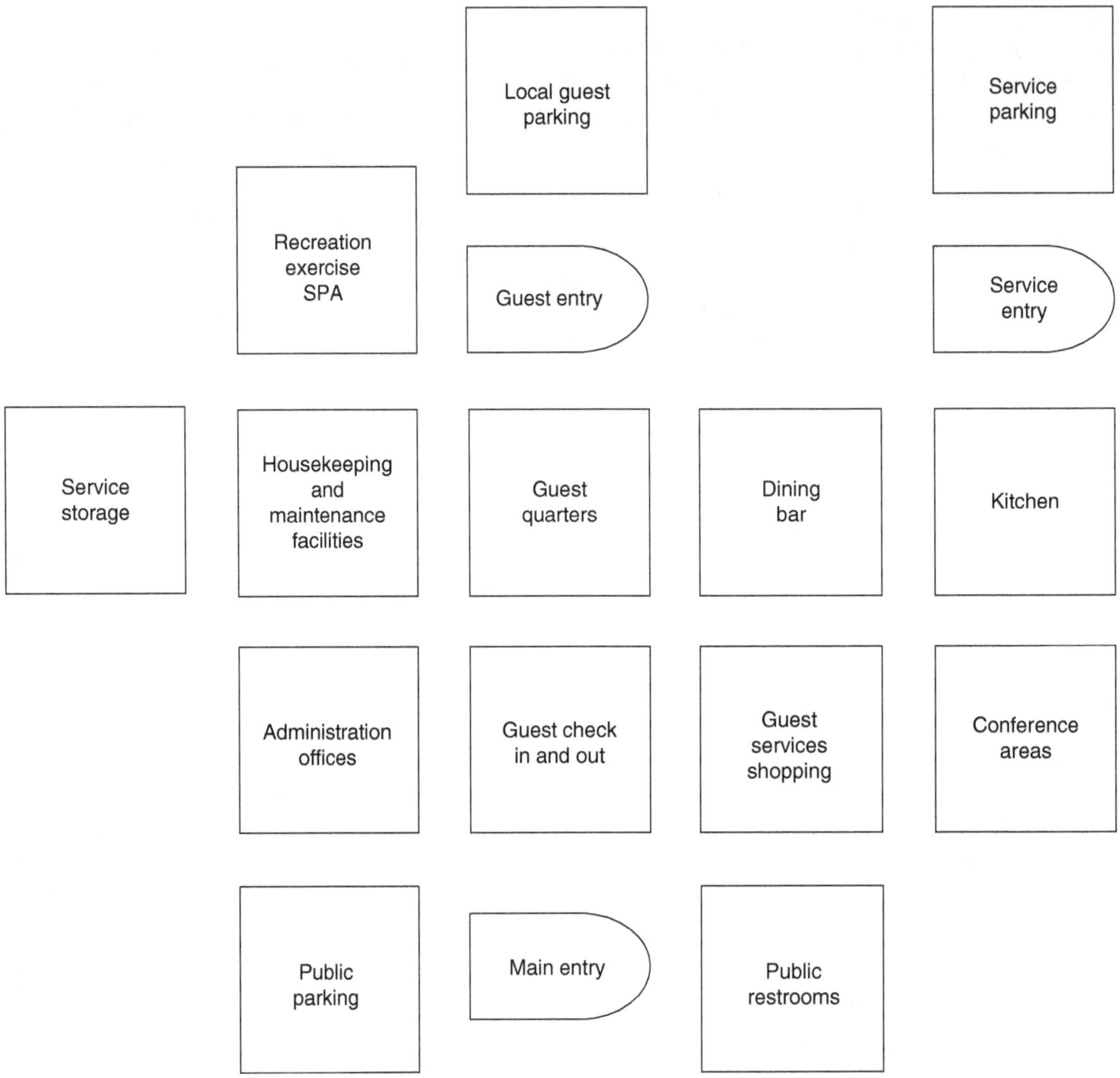

FIGURE 2.16 Public accommodations planning model.

guests still will not be happy with the system and thus probably will not patronize the complex again. The following checklists, guidelines, and other information should help the planner develop the best overall plan to meet the needs of all participants in the public accommodations system.

The size and complexity of public accommodations facilities obviously will vary from one concept to another; however, because of the typically large number of people and activities, special attention should be given to each of the more significant activity–space interface interactions. Figure 2.18 illustrates a typical array of interacting activity and space elements.

General Human Factors Priniciples to Consider During Conceptual Planning and Design

Guest Accommodation

1. First priority should be given to making all guest accommodations comfortable, convenient, and safe. Depending on the type of facility, a variety of guests should be considered: overnight guests; nonovernight guests who may utilize dining, meeting, or convention facilities; and interim guests of house guests. The more important considerations are as follows:

 a. Ease of locating and identifying the facility in terms of its location (i.e., with respect to highways and streets), signing, and approach system geometry.

 b. Ease of locating the entry to the site, parking, and identification of, and convenience to, the main facility entrance (i.e., not only should these elements be easily recognizable in terms of their location, arrangement, and physical appearance, but also proper signing should leave no doubt in the minds of guests as to where they should go to gain entrance into the facility). Appropriate external illumination must be provided in order for this to occur. The distance between parking and buildings should be minimal.

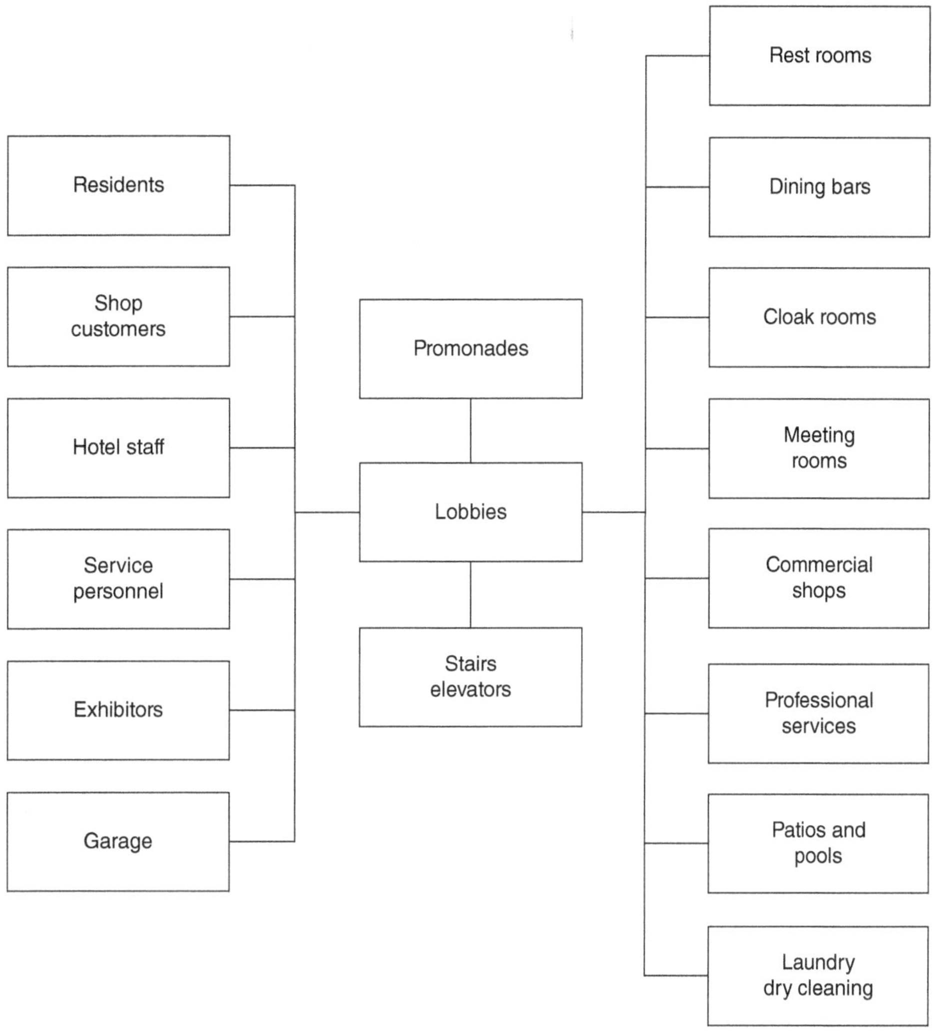

FIGURE 2.17 Sample arrangement of public accommodation functional elements.

c. Ease of locating the checkin area once the guest is inside. The checkin desk should be conspicuous from the point at which the guest first enters the facility. Appropriate, visible, and legible signing is an important aspect of the confidence that a guest requires upon entering.

d. Ease and convenience of the front desk to the guest quarters. This should be considered a prime prerequisite. Avoid arrangements that are confusing or that require considerable walking by guests. Although most hotels provide baggage service (which includes guidance to the guest quarters), a design objective should be to create a system that is simple enough so that guests can find their own rooms easily by themselves.

e. Convenience between the guest quarters and typical hotel facilities that guests will use during their stay (dining facilities, a bar, a barber or beauty shop, dry-cleaning facilities, a drugstore, etc.) and particularly the recreational facilities provided (swimming pool, game rooms, sporting facilities, etc.). Consider also special localized facilities that should be especially convenient to guests' rooms, e.g., soft-drink and ice dispensing machines.

f. Adequate guest-room size, configuration, furnishing, lighting, security, air conditioning, and sound isolation. Treat the guest quarters as a "home away from home." Provide adequate internal and external communication and control of the thermal environment.

g. For motel or motor hotel systems, provide adequate and convenient local parking for each guest room. For all systems, provide adequate and convenient visitor parking, typically separate from overnight guest parking.

2. Accommodation of Administrative and Maintenance Personnel: The myriad of interactions between the hotel or motel management and the service personnel

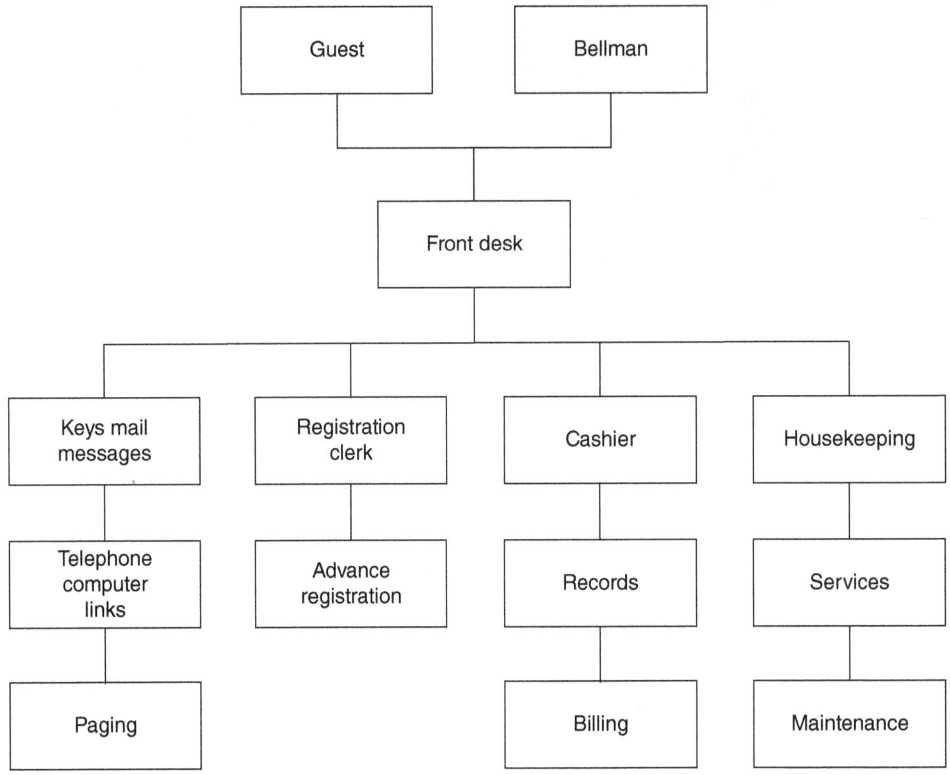

FIGURE 2.18 Typical activity interactions with the front desk.

should be analyzed carefully in order to provide a basis for identifying, specifying, and arranging special as well as general accommodations for their respective tasks. Individual performance efficiency with minimum interference of guests should be of prime concern, and time is perhaps the critical factor. Consider the following:

a. General Management: The location of the facility manager should be chosen both to maximize his or her convenience to guest contacts and to facilitate immediate supervision of the administrative staff (desk clerks, the cashier, the business agent, etc.).

b. Front-Desk Activity: The size and shape of the front-desk facilities should be based on a combination of numbers of clerical persons required (for peak loads) and their special furnishings and equipment. As indicated, this particular facility must also have proper interfacing characteristics with guests; i.e., it should be easily identified, not prone to traffic jams at peak arrival or departure times, etc.

c. Housekeeping Activity: First priority should be given to the time- and energy-saving convenience of those responsible for the service of the guest quarters. This activity should be addressed not only in terms of the overall facility layout but also in terms of the layout of individual guest quarters. In addition, certain architectural features, such as clearance beneath fixtures and cabinets, coved baseboards, and pullman counters, have been found to facilitate cleaning tasks. Although generally of secondary priority, housekeeping convenience of general and common facilities is equally important in terms of maintaining an acceptable appearance. Especially important to housekeeping effectiveness are the provision and proper location of housekeeping supplies and equipment in order to reduce the time spent by housekeepers and the number of steps they must take.

d. Maintenance Activities: Both internal and external maintenance considerations should be taken into account in terms of the general layout of the facility, special work-area and storage-area requirements, and, last but not least, the type and amount of landscaping that may be planned. In addition, special consideration should be given to the accessibility of the service area or areas with respect to the use of large vehicles for deliveries or debris disposal.

e. Food Service Activities: The dining room, coffee shop, bar, or other food service systems associated with hotels or motels should be designed according to good restaurant guidelines. Whether they are housed within the general hotel structural complex or developed as separate units, consider the guest convenience to and from the main hotel complex, especially during inclement weather.

f. Special Entertainment, Convention, and Exhibit Facility Activity: Such facilities should normally be considered adjuncts to the main hostelry mission of the hotel or motel; i.e., their placement and traffic patterns should be such as to minimize interference with guests' activities (many guests may not participate in events occurring within the special facilities and may become annoyed by heavy traffic through regular guest areas or by the excessive noise that such facilities tend to generate). Special attention should be given to the food service problems associated with these special facilities and to the unique problems associated with exhibitors and entertainers and their equipment. Added storage is a major consideration because these special activities require a considerable amount of portable furnishings, including chairs, tables, and stages. A special loading dock may also be required for delivery and removal of exhibits, musical instruments, and the like.

g. Other Factors: The following considerations should also be considered:

i. Fire safety (exits and emergency lighting)

ii. Security (including vandalism and theft)

iii. Accessibility to various delivery functions (mail, vendors, etc.)

iv. Accessibility to emergency crews and equipment (fire fighters, police, etc.)

v. Environmental control and illumination for common-use areas

vi. Signing and marking

vii. General communications (administrative and public address)

viii. Public restrooms (associated with the lobby, restaurant and bar, convention area, and swimming pool and with such activities as golf or tennis, which may require consideration of a locker room)

ix. Audio-visual aids (power, screens, and storage)

x. Access to critical areas by the handicapped

Guest Quarters. Although many alternative designs will provide satisfactory guest-room accommodation, Figs. 2.19–2.21 illustrate a number of desirable features to consider in planning such facilities. Among the more important points that Fig. 2.19 illustrates are the following:

1. Double doors associated with the entry vestibule provide extra noise protection from activities that go on in the corridor.

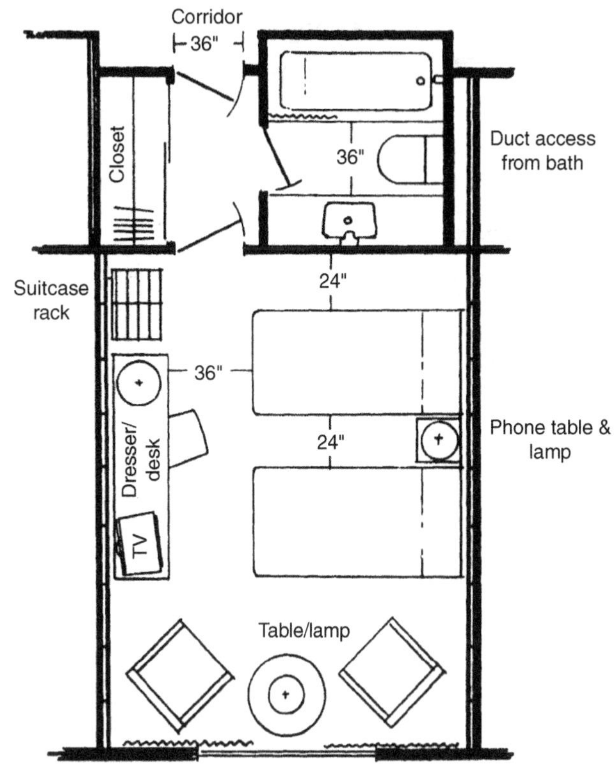

Figure 2.19 Guest quarters—desirable features.

2. Certain minimum clearances are observed, which ensures adequate passage through doorways and between furnishings.

3. Adequate clearance in front of the commode prevents the person who is opening the door from being caught between the door and the commode.

4. Clearance around the beds facilitates makeup.

5. A number of quarters should be specially designed to accommodate handicapped guests. These will require additional clearance around beds and clearance and handrails in the bathroom.

When floor space has to be limited for one reason or another, care should be taken to seek out special furniture arrangements in order not to crowd guests, especially during the daytime. Typical approaches have been to create special furniture that can be moved from one position to another, as in the accompanying example. Figure 2.20 illustrates a special technique for providing linen and bedding storage within the headboard unit. Such a device further provides the alternative daytime davenport for visiting with other guests.

More expanded quarters can take on any form, of course. Refer to the guidelines for apartments in the section on multifamily residential systems. Figure 2.21 illustrates a logical and convenient arrangement that includes a kitchenette and separate sleeping area, making such quarters complete enough for longer-duration occupancy.

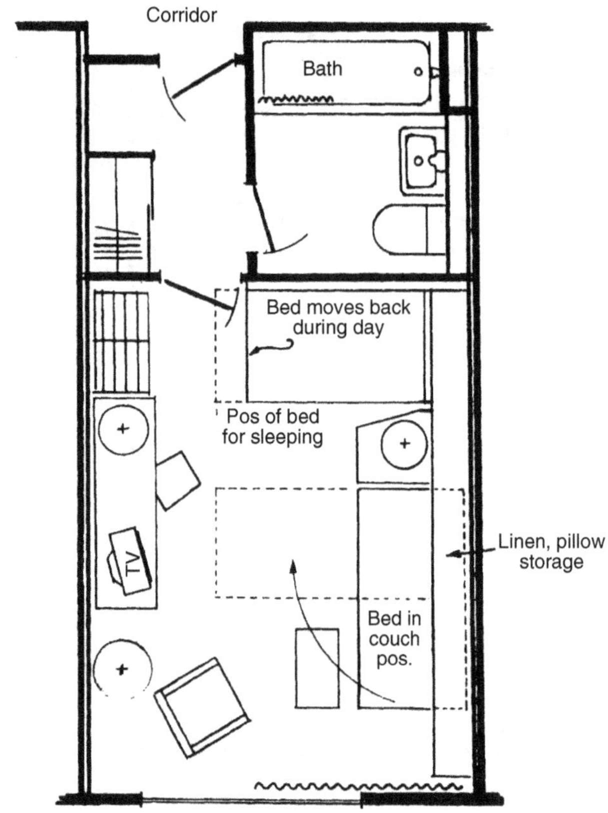

FIGURE 2.20 Guest quarters—limited floor space.

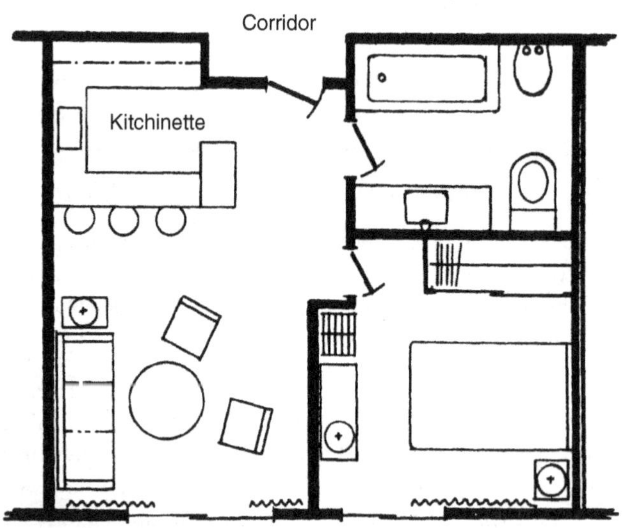

FIGURE 2.21 Guest quarters—expanded quarters.

The bathroom area in the example was enlarged, reflecting the desirability of more convenience for the guest who is staying longer.

Environmental Considerations. The following is a general discussion of environmental considerations that must be made when designing public accommodations. Specific environmental standards and guidelines are defined in Part 4, "Human Factors Data."

1. Noise and Vibration Control: Because of the nature of hotel and motel activity and the unique requirements for special air-conditioning and other noise-producing equipment, attention must be given to the problem of noise and vibration control. The following points provide a rationale for this concern and indicate some of the planning features that are significant in providing proper acoustic control:

   a. Traffic Noise: Although traffic may subside to some extent in some areas where public accommodations are located, traffic noise should be an important consideration in terms of site planning and facility organization. Guest sleeping areas obviously should be isolated as much as possible from traffic.

   b. General Operating Noise: Primary noise-producing equipment should be isolated not only from guest sleeping areas but also from certain conference areas. It is easier to minimize noise problems at the source than to create barriers between the source and an area that requires quiet.

   c. Guest-Produced Noise: An inevitable problem associated with hotel and motel operation is that some guests try to sleep while others come and go, entertain, or otherwise create commotion and noise. By means of proper arrangement of hallways, elevators, stairs, etc., the basic guest accommodations plan should provide as much separation as possible between these and sleeping quarters.

   d. Entertainment Noise: A common error is to place an entertainment facility right in the midst of the complex, with the result that the noise produced by a band or other entertainment tends to permeate the entire facility—obviously an undesirable approach to planning.

   e. Systems Vibration: Vibration (and often an attendant noise) is particularly annoying to guests who wish to sleep or rest. In planning, be especially cognizant of the possible production of vibration within and through the facility by passing traffic, low-flying aircraft, air-conditioning systems, etc.

2. Heating, Cooling, and Ventilating: Both natural ventilation and artificial air conditioning should be provided for all but the few uniquely blessed public accommodations because of the round-the-clock, year-round operating conditions typically required by these facilities. The following special considerations are germane at the planning stage:

   a. Provide enough independent control so that all areas can adjust the immediate area temperature and ventilation to match the individual activity need.

Guests require different temperatures and airflow for active versus rest cycles, public areas require adjustment to match small versus large crowds, and so on.

b. Provide humidity control as well as temperature and ventilation and consider the use of electronic dust-collecting systems to aid in preventing the spread of contagious diseases and in minimizing cleaning and refurbishment requirements.

c. Avoid the semiclosed air-conditioned concept, wherein fresh air cannot be introduced; this is especially important in the event of a breakdown in the air-conditioning system.

3. Illumination: Although decorative illumination is a desirable feature of the public accommodation facility, special care should be taken during planning not to let aesthetic lighting schemes interfere with ordinary seeing tasks. This includes consideration of both natural and artificial lighting features and systems. The principal areas that are commonly poorly executed in terms of illumination are the following:

a. Parking areas

b. Entryways

c. Restaurants and dining areas

d. Stairways and balconies

e. Hallways in guest-room areas

Especially important is the consideration of emergency lights that illuminate in the event of a power failure and the sudden extinguishing of normal lighting.

Hotel and Motel Safety Considerations.[1] For the following reasons, safety is particularly critical for hotel and motel operations (see Table 2.6):

TABLE 2.6 Maximum Exit Distance Travel for Escape

	Without Sprinkling System		With Sprinkling System	
	Meters	Feet	Meters	Feet
Places of assembly, convention rooms, restaurants, ballrooms	45	150	60	200
Areas of high fire risk (kitchens, boiler houses, etc.)	23	75	30	100
Apartments	30	100	45	150
Hotel guest rooms	30	100	45	150

Note: Distribution of room contents must not increase the distance more than 1½ times.

[1]*Life Safety Code,* NFPA no. 101, National Fire Protection Association, Boston, Mass.; and United Kingdom Home Office, *Guide to Fire Precautions Act, 1971,* H. M. Stationery Office, London. 1971.

1. The majority of the occupants are transient, are unfamiliar with the facility, and in many instances are too tired and/or too preoccupied with other concerns to exercise normal safety-conscious behavior.

2. Large numbers of people who may be sleeping in many separate rooms require separate warning and a variety of evacuation solutions.

3. Large groups of people congregated together in public areas create severe crowd control and evacuation problems.

4. It is difficult to devise a single disaster plan for mixtures of adults, children, the disabled, and the aged.

5. The fire loading of the typical hotel or motel is often substantial because of the large amount of furnishings that such a facility contains.

During conceptual planning, consider at least the following safety features:

1. Materials flammability
2. Fire warning systems
3. Water supplies for firefighting
4. Emergency exit doors
5. Sprinkler systems
6. Escape routes and distances
7. Fire extinguishers
8. Access for firefighting vehicles and equipment
9. Barriers to control the spread of fire (including building separation)
10. Warning and exit signs, markers, and instructions

In addition, consider the following specific design features:

1. Nonslip walkways and stair treads
2. Appropriate handrails, balcony and staircase guards, and pool guards
3. Electrical system grounding and outlet child-proofing
4. Elimination of single-step changes in walkways and room thresholds
5. Adequate illumination, including emergency lights
6. Elimination of tripping hazards, sharp corners, and edges or protrusions on structural or built-in cabinetry, railings, etc.
7. Exclusion of guests from hazardous operating machinery
8. Avoidance of floor-to-ceiling glass windows or doors through which individuals might walk

9. Minimization of escape-route travel distance and maximization of the number and identifiability of exits; escape routes that are direct, i.e., that do not require the user to make many changes of direction, traverse many stairs, or pass through many doors; and avoidance of dead ends

OFFICE BUILDINGS

Public and Private Office Buildings

Although an infinite variety of office building types could be addressed, this section deals principally with the large office building system. However, many of the planning guidelines presented here are equally applicable to smaller office complexes. The guidelines in this section outline some of the more important human factors considerations that should be addressed during initial office building planning.

Functional Design. First, focus on designing for the functions that must occur in the building. Understand the needs of the users. As a starting point, Table 2.7 lists some of the basic functions that might be associated with various office buildings.

TABLE 2.7 Functional Requirements Checklist for Office Building Planning and Development

Activity	Example
Accounting	Private and group account processing, client counseling, bookkeeping, and financial planning
Banking	Central or branch bank, teller service, accounts, and record keeping, safety deposit, loans, and financial advisory services
Commercial	Wholesale distributor sales center, private detective agency, employment agency, travel agency, secretarial services agency, mailing services, real estate agency, etc.
Governmental	Public records; offices of city, county, state, and federal elected official administrative, clerical, and engineering functions; issuance of legal permits and license; employment and welfare counseling and disbursement; courts, detention facilities; etc.
Legal	Attorneys, paralegal assistants, private counseling, research, and record keeping
Medical and dental	Doctor, nurse, and patient reception; diagnostic services; minor treatment and recovery area; laboratory testing; pharmacy; etc.

Building-Site Configuration. Since the typical large office building causes a concentration of people and vehicles within a fairly restricted area and since it often produces fairly high-density traffic loads at regular intervals, it is critical that approaches to, and departures from, the site be laid out to minimize traffic conflicts; that adequate and convenient parking be provided to maximize the ease of transfer from vehicles to and from the building or buildings; and that the approaches, parking, sidewalks, and building entrances be not only arranged but also signed for ease of understanding by the people who will be using the building, especially the transient visitor who is arriving for the first time.

Interior Layout of Office Spaces. Unlike many other large architectural systems, the office building generally is subject to almost continual rearrangement as tenant replacements occur. This requires careful consideration not only of the shape, size, and placement of the fixed features of the building's interior (to provide adequate flexibility for future refurbishment) but also of the basic geometric concept as it may provide an apparent logic for tenants and visitors. (See Fig. 2.22.)

1. The Office Complex as Part of a Larger Architectural System: Often the office complex may be an integral part of a larger system, e.g., a factory, a laboratory, or a governmental complex. In this case, the office portion should not be allowed to become "lost" within the overall architectural plan; i.e., its exterior appearance should provide a clear indication to the visitor of what it is (e.g., an office building, as opposed to a manufacturing facility), how to get to it, where to park, etc.

2. The Windowless Office Building: Although there are obvious advantages for the designer in utilizing a windowless building concept for office space (i.e., it simplifies many environmental control solutions), most office workers need visual contact with the outside world. Even though most modern office buildings are air-conditioned, the value of visual contact with what is going on outside cannot be overemphasized. On the other hand, "solid glass" concepts, popular with some designers, are equally inhuman in that they create many visual, environmental, and physical arrangement problems for the office workers and their clients and particularly for the individuals who must maintain the facility.

3. Key Subsystems Absolutely Necessary for All Office Buildings
    a. Restrooms for tenants, visitors, building management and maintenance personnel, and special tenants who may supply unique services such as security, food service, or other catering
    b. Building management and maintenance areas
    c. Building engineering and utilities areas
    d. Service and delivery docks
    e. Corridors, stairs, and elevators
    f. Entrances and exits for tenants, visitors, services, and emergency escape
    g. Parking for tenants, visitors, and services

4. Special Subsystems That May Be Desirable for Common Usage
    a. Cafeteria and kitchen
    b. Inside parking

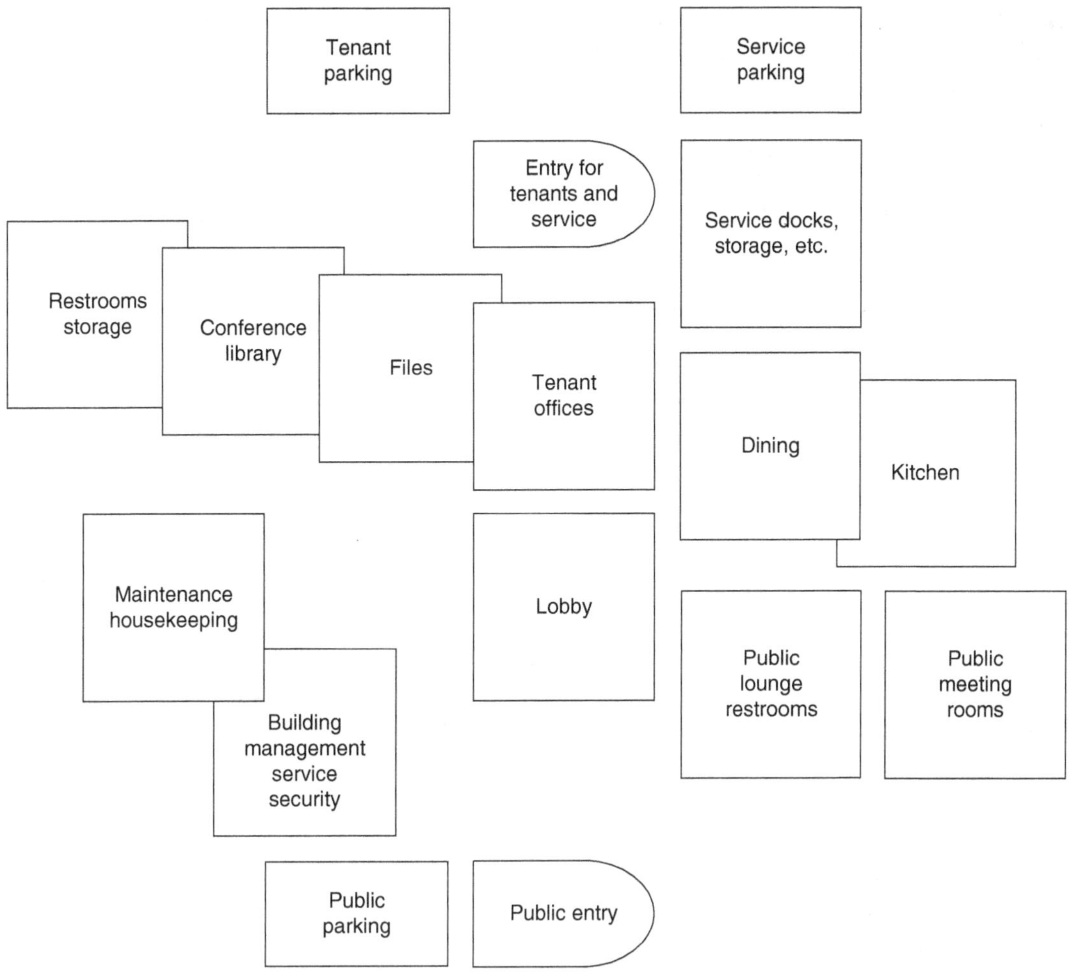

Figure 2.22 Public office building planning model.

c. Mail room

d. Tenant recreation (lounges, gymnasium, locker rooms, etc.)

e. Limited catering and shopping facilities (barber and beauty shop, drugstore, pastry shop, etc.)

f. Branch bank

5. Other Features Critical to the Initial Planning

a. Safety features (sprinkler systems, fire alarm system, fire hoses, hand extinguishers, emergency exits, and fire escapes)

b. Occupant conveniences (drinking fountains, public telephones, vending machines, etc.)

c. Environmental control (illumination, temperature, ventilation, humidity, noise, etc.)

d. Services (electric power, trash collection, building and office housekeeping, maintenance, security, etc.)

6. Among the More Important Considerations

a. Initial visitor and client convenience to the main entrance, to the office of the person the visitor or client wishes to visit, to public restrooms, and possibly to public meeting rooms.

b. Tenant ease of ingress and egress to the building and to the tenant's office and convenience of his or her office layout for each task to be performed.

c. Ease and convenience of building management and maintenance.

Individual Tenant Space Requirements

Wherever possible, plan the basic internal floor-space requirements per tenant on the basis of the tenant's needs (as opposed to creating the external walls and letting the tenant use what space is left inside as best he or she can). Some of the typical tenant needs that should be considered in developing floor-space envelopes are the following.

Reception Area. Although some tenants may be able to manage without a client or visitor reception area, this is not desirable. The reception area should provide enough space for a receptionist and his or her desk and chair, possibly a switchboard and file, and suitable waiting-room furnishings, e.g., a davenport, chairs, a coffee table, an end table, and magazine rack. Although the reception area can be made larger or smaller according to tenant needs, it should be considered from the standpoint of its location with respect to how the rest of the space might be partitioned. That is, the reception space acts as a lobby from whence visitors will proceed into the inner office complex.

Storage. Avoid the temptation to consider the internal space requirements of the tenant only in terms of his or her potential furnishings, e.g., desks, chairs, and file cabinets. Generally, most tenants will have to add portable storage cabinets or build in such cabinets, thus reducing the originally planned personnel and furnishings space. This factor alone suggests that the so-called minimum floor-space criteria found in many architectural handbooks should not be used; i.e., plan on the generous side wherever possible.

Lighting, Ventilation, and Utilities Service. Avoid the general approach of permanently installing light fixtures, ventilating outlets, electrical outlets, and/or telephone outlets. Because all tenants invariably place their equipment where they want it, there are no successful methods for providing acceptable locations for these items. Use modern, flexible techniques that provide basic service capacity in ceiling or floor recesses and employ flexible, modular techniques for removing and replacing interface materials and hardware, e.g., ceiling panels, ducts and vents, and electrical and telephone outlets.

Built-in Cabinetry and Modular Partitions. Develop modular components (or select off-the-shelf components) that allow for easy change and/or addition of landlord-provided cabinetry or partitions. That is, either provide no such amenities as part of the tenant package or plan a system that allows these elements to be added, changed, relocated, or removed with a minimum requirement for refurbishment as a result of damage to the basic structure.

Consider the needs for control flexibility when planning lighting and air conditioning. It should be possible for tenants to adjust temperature and ventilation at various points within their particular space, not at some single point that may not represent the specific conditions in some other part of the space (note that when tenants partition their space, they may destroy well-intended, general ventilation-flow patterns that the architect devised). Similarly, tenants should have the capability of selectively turning various light fixtures on or off within their space. However, it is also desirable to consider an added overall system switch so that tenants can turn off all lights (and possibly electric power) as they leave, without having to check out various parts of their own space to make sure that a light has not been left on in an inner office.

1. Other Tenant Space Services: Although these depend on the basic objectives of the specific office builder and/or the client, several other service needs of special tenants should be considered.

    a. Private restroom or restrooms

    b. Kitchenette (with appropriate water, power, sink, etc.)

    c. Provisions for a wet bar or bars

    d. Private shower or showers

    e. Private elevator

2. Individual Tenant Safety: Avoid the tendency to consider safety only from the point of view of overall tenant safety. The crux of any safety solution should always be to make sure that all occupants of the building have an equal opportunity to be safe. This requires specific attention to the hazard possibilities within each basic tenant space. Of particular importance is the planning of alternative escape possibilities, regardless of how an individual space may be laid out or partitioned by its tenant. Although a significant aspect of this problem may reside in the building management's rules and policies, even the best rules are of little help if the original structural plan does not provide adequate alternatives.

Other significant factors are the selection of nonflammable materials, proper execution of the utility system layout and installation, and adequate accessory items such as sprinkler systems, fire and smoke detection equipment, communications systems, and emergency power and illumination. Although planners cannot prevent tenants from introducing hazards such as flammable materials, they can make sure that floor coverings, drapes, and other permanent or management-supplied items are not dangerous.

Planning Considerations for Specific Architectural Features

Certain basic features, such as window systems, doors, staircases, and elevators or escalators, may be selected quite early in architectural planning. Although, in most cases, certain local ordinances must be observed in making these selection decisions, the following key human factors considerations are important to remember.

Door Concepts. Recent door concepts include the floor-to-ceiling configuration illustrated in Fig. 2.23. These concepts are not recommended from the human factors standpoint for two important reasons:

1. Such doors are heavier and are particularly hard for children, older persons, and handicapped individuals to handle.

2. These doors have an undesirable psychological impact; i.e., they are out of scale in terms of their relation to human body size. Although some people accept them eventually, others may feel dwarfed and threatened by them.

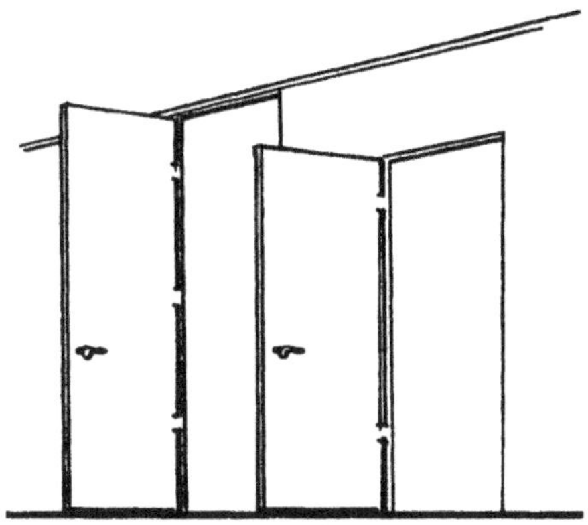

Figure 2.23 Smaller door preferred.

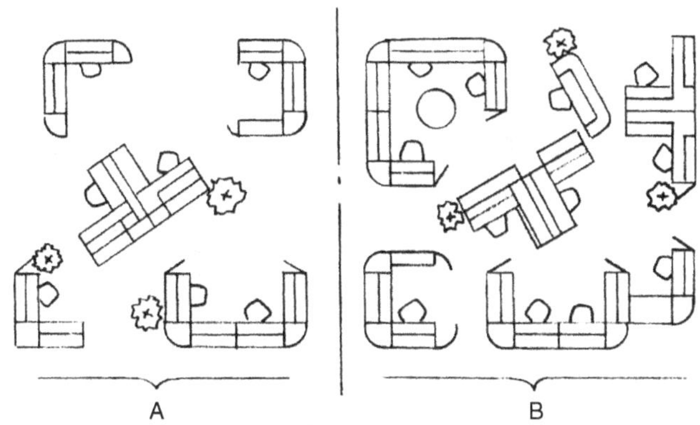

Figure 2.25 Plan views.

Doorway Dimensions. Clearance for the transfer of furniture is as important as clearance for the passage of people. It is undesirable to make movers stand a typical desk on end, twist it around the door jamb, or remove the door in order to get it in or out of an office.

Door and Window Hardware. Avoid leaving the choice of opening and closing hardware to someone else. The size and shape of handles, the forces required, the direction of motion, etc., should be compatible with human expectations and capabilities, especially those of the infirm and handicapped. (See Fig. 2.24; see also Part 3, "Component and Product Design.")

Figure 2.24 Width of door opening.

Arrangement Flexibility through Space Systems and Modular Furnishings.[2] Figure 2.25 shows how one arrangement (A) is easily modified to another arrangement (B) in a matter of a few hours by using some of the modular furnishings available today.

Note: Trends have been to the open office—elimination of cubicles that can be barriers to communications. In the past, this was referred to as "office landscaping." More recently, the term "war rooms" has been used. Depending on the project, open offices have shown to increase productivity over closed cubicles. This was found true for software development. Concentration or confidential conversations may need more privacy, however. As true with all human interface designs, it is best to assess the functions and then design to accommodate them. A combination of private space retreats and open office is the best solution. Also, it is recommended that some rectilinear patterning of clusters of work spaces be maintained in order that (1) visitors can find their way around and (2) orderly pathways are available for emergency exiting. (See Fig. 2.26.)

SCHOOLS

General

Philosophies concerning the creation of an optimum learning environment are constantly changing, and this has resulted in considerable experimentation in educational architecture. This section deals not with these philosophical aspects but rather with the fairly constant physical interface needs of students, faculty members, administrators, and others as they utilize facility subsystems in their everyday work, play, or recreation on the school site. Although many of the same types of building elements found in other systems are required on the school campus, the school presents numerous rather unique facility requirements because of both the nature of the educational processes and the variety of users involved.

The following are among the general, yet unique, considerations to be addressed during the conceptual stage of planning. (See Fig. 2.27.)

Site Location. School sites should be selected so that they are as convenient to the potential student population as practical and yet isolated from such disruptive influences as heavy traffic, high-density commercial and industrial areas, and establishments such as liquor stores.

Size of Site. To avoid the necessity of splitting and separating facilities and thus causing special problems for both students and faculty, sufficient property should be acquired to provide

[2]Action Office, Herman Miller Inc., Zeeland, Mich.

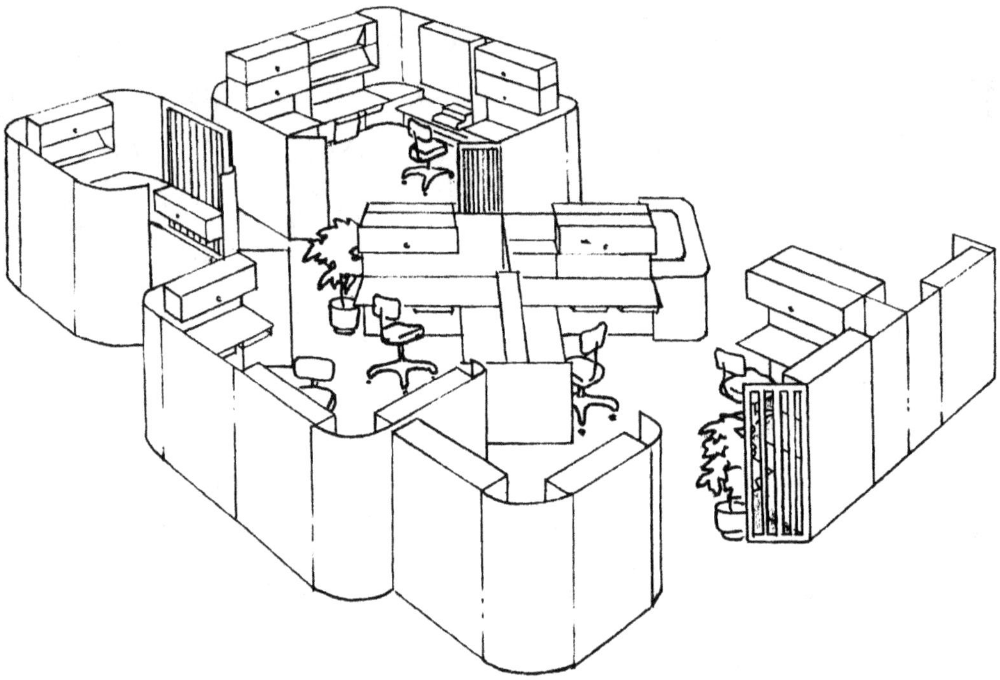

FIGURE 2.26 Perspective view of arrangement B.

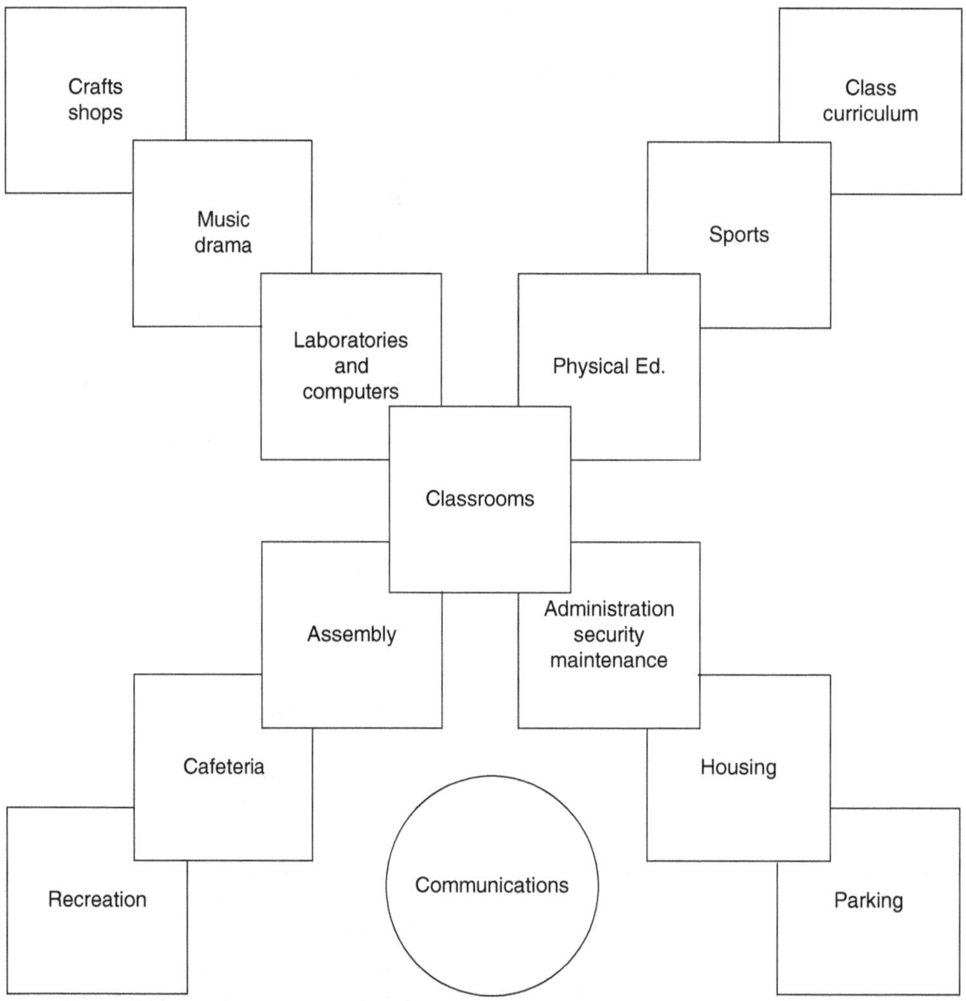

FIGURE 2.27 School facilities planning model.

for expected future expansion. Special consideration should be given to the parking requirements of secondary and college systems, since the foreseeable trend is for more and more students to have their own cars and for educational institutions to sponsor an increased number of public-oriented sporting events, thus adding to the parking needs of the campus.

Accessibility for the Handicapped. The requirement for equal educational opportunity for the handicapped has forced all school planners to give special attention to designing out architectural barriers. This makes it important to consider these problems before some aesthetic concept precludes an effective solution to the access problem of the handicapped student and other handicapped persons.

Among the more important planning considerations are the following:

1. Students: Optimum learning environment and safety
2. Faculty: Optimum teaching environment and access to students
3. Administration: Optimum work environment and communications
4. Maintenance: Ease of maintenance and freedom from vandalism

Accommodate Information Technology Advances. Advances in information exchange and storage systems can improve the effectiveness and efficiency of schools. Planning of all schools (from elementary through universities) should consider personal computers, multimedia demonstrations (including immersive and multisensory), training in the use of computers, and instructors and students remote from the campus.

Security and Resistance of Facilities to Vandalism and Acts of Violence. Presently, and for the foreseeable future, schools seem to be prime targets for deviant behavior, not only by a few of the students within the particular institution, but also by outsiders bent on disruptive objectives. This has raised many questions regarding general architectural concepts that have fostered a more open and friendly campus atmosphere. Although the facilities obviously should not give the appearance of a jail, they should minimize the opportunity for crime and should make it possible to provide proper on-campus disciplinary control.

Important Planning Considerations

The following are among the more important planning considerations:

1. Students: Optimum learning environment and safety
2. Faculty: Optimum teaching environment and access to students
3. Administration: Optimum work environment and communications
4. Maintenance: Ease of maintenance and freedom from vandalism

Special Planning Emphasis by School Type

Elementary Schools

1. The physical compatibility of facilities for use by children and adults is an obvious requirement that architects try to meet. However, the unique furnishings requirements of children often make the tasks of the adult faculty extremely difficult. This is true not only of the classroom furnishings but also of stairs, bathroom facilities, and many other jointly used features. By the same token, children's tasks may be made difficult when administrative facilities are designed primarily for adult staff members.

2. Play facilities are a unique requirement of the lower grades, and special attention should be given not only to the various types of play activities and devices provided for the children but also to the problems of faculty monitoring and/or participation.

3. The convenience and safety of student arrival and departure are a unique consideration of the elementary school facility plan. Students may walk to and from school, or they may travel by bicycle, bus, or private automobile. This requires consideration of site layout in terms of public transportation, special school busing systems, and parking for vehicles and bicycles.

High Schools

1. The school library, study halls, assembly halls, etc., are an important feature of the high school facility system, requiring anticipation of the special space and convenience needs of students at this level.

2. The gymnasium, swimming pool, track and field, tennis courts, etc., are an important part of high school planning. Sporting events require consideration of additional public parking and unique equipment storage.

3. Shops, laboratories, and craft facilities are typically introduced at the high school level, requiring consideration of the special facilities for storage, display, experimentation, etc., associated with these activities and also consideration of the unique safety requirements associated with use of toxic, combustible, and flammable substances.

4. Culturally oriented activities require consideration of special practices as well as storage and audience-participation facilities. Special note should be made of the potential theft and vandalism problems associated with musical instruments.

Colleges and Universities

1. On-campus housing is a unique consideration of college campus planning, particularly if the site is somewhat isolated and/or if administrative policies require first-year students to live on campus. The modern college campus often must accommodate both married and

single students, and the majority of students will have private cars, which create obvious parking needs.

2. The field house, stadium, athletic fields, etc., are an important part of the modern college campus, from both the educational and the recreational points of view. In many cases, indoor as well as outdoor sporting events draw exceptionally large public attendance, requiring that facilities be oriented not only to student needs but also to public needs (including especially the consideration of parking and traffic accommodation and control).

3. Student unions are an accepted part of college facility requirements and should be considered during initial planning; they should not be thought of as something that can be added later, when the economic situation is more favorable. The student union has become the center not only of student relaxation but also of student self-government and sociopolitical activity.

4. Technically oriented facilities (supercomputers, nuclear reactors, engineering test laboratories and field facilities, radio and TV stations, etc.) have become an important part of advanced educational philosophy. Because of the space and safety implications of these facilities, they must be addressed during the initial planning stage.

5. Public-oriented cultural and academic interface requires the careful consideration of both student and public use of such community-related facilities as Theaters, auditoriums, and libraries.

6. Other special facilities should also be considered during the planning stage at the college level. Some of these might include (1) educational training schools that are open to parents who would like their children to participate in advanced or experimental educational programs and (2) special research institutes created to study specific scientific or engineering areas for government or industry. In such cases, the planning should consider the physical advantages and disadvantages of combined use of the university facility versus separate facilities, though they may be located on the same site.

TABLE 2.8 Checklist for Other College and University Facilities[a]

The Following Checklist Identifies Some of the Facilities That Should Be Considered During College or University Campus Planning
1. Offices (administration, faculty, maintenance, and security)
2. Eating facilities (student, faculty, and guest cafeterias, dining rooms, coffee shop, and snack bars
3. Restrooms and lounges (students, administration, faculty, guests, and maintenance personnel)
4. Post office and mail collection
5. Communications (computer-based, telephone, intercom, public address system, radio, and TV)
6. Publications (campus news, yearbook, and technical publications)
7. Book and stationary store
8. Student union and alumni offices and facilities
9. Chapel and social services center
10. Visitors' information center
11. Outdoor Theaters
12. Housing (guests, administration, and faculty)
13. Fire station
14. Ground maintenance, landscape equipment, and nursery
15. Branch bank
16. Furnishings warehouse and shipping and receiving center
17. Agricultural experiment areas
18. Visual aids development facility
19. Campus radio and TV station
20. Health clinic and first-aid center or centers
21. Accessibility for handicapped
22. Walkways and roadways

Note: The campus walkway and roadway system should be an integral part of planning the above as well as planning the standard educational facilities, for the effectiveness with which this system interconnects the various facilities for both campus and itinerant pedestrians, bicyclists, and motor vehicle operators is of utmost concern to the campus administration. A key factor to be addressed is: How can we affect an optimum compromise between maintaining a semiprivate system and providing for public access?

[a]For specialized school facilities (e.g., medical and dental facilities), refer to J. De Chiara and J. H. Callender, *Time-Saver Standards for Building Types*, McGraw-Hill Book Company, New York, 1973.

School Site and Facilities Layout Considerations: Single, Large Structures versus Smaller, Independent Structures

Smaller, independent structures are more desirable because they make it possible to provide natural window light from opposite sides of a room. However, certain types of facilities benefit more from being relatively windowless and combined into a larger structure, e.g., a music building with practice rooms of various sizes or a gymnasium combining one large playing area for basketball, etc., with separate locker rooms, a swimming pool, and other physical education areas. (See Table 2.8.)

1. Indoor versus Outdoor Corridors: In milder climates, outdoor corridors are desirable because they give students an opportunity to get out of doors periodically as they go from one classroom to another. Outdoor corridors should have covers to protect students from rain.

2. Facilities Adjacency: In general, the following campus facilities should be located adjacent to each other:

a. Facilities for administrators and visitors

b. Facilities for faculty and students

c. Facilities for students and the primary assembly facility

d. The physical education facility and outdoor sports facilities

e. Classrooms and the bus loading area

f. The public parking area and the outdoor sports area and auditorium

3. Sound Isolation: Certain facilities generate loud sounds and thus should be spatially isolated from other facilities; e.g., music practice facilities and physical education facilities should be isolated from classrooms and libraries.

4. Centralization: Certain facilities should be centralized with respect to the overall campus layout in order to maximize the convenience of these facilities to all potential users.

a. The cafeteria

b. Restrooms

c. The student assembly or auditorium

5. Parking: Single, mass parking areas are least desirable. Several smaller and strategically located parking areas not only provide shorter travel distances but also help segregate traffic categories, i.e., vehicles of students, faculty and administrators, visitors, and service personnel.

6. Terrain Factors: Although some aesthetic benefits are obtained by selecting a hilly site and positioning certain buildings so as to provide inspiring views, important practical benefits accrue from a relatively level site. Perhaps the most important benefit is to the handicapped, for which vertical movement is considerably stressful and time-consuming. When the terrain is not reasonably flat, care should be taken to design and arrange buildings so the separate facilities to be used by given groups of users are on approximately the same level.

7. Maintenance: The layout and organization of buildings and other facilities on the campus are extremely important to those who must perform janitorial and landscape maintenance. In spite of the desirability of providing adequate open space between buildings and/or of placing buildings close together for economic reasons, either of these approaches can work to the disadvantage of maintainers. In the case of janitorial personnel, great travel distance increases the time required to perform duties; in the case of those who perform landscape maintenance, crowded conditions make it difficult to manipulate service vehicles among the buildings.

8. Visitor Orientation: An organized (more formal) rectilinear layout of campus facilities provides visitors a better chance to maintain their bearings and to follow directions for getting from one place to another. Curvilinear roads and pathways (although aesthetically pleasing and perhaps less boring) create confusion. Equally important is the planning of roadways on the large campus with a minimum of dead ends.

Structural Configuration Considerations

A wide variety of building configurations is, of course, acceptable for most school facilities. However, several important guidelines are noted within the accompanying figures.

1. Cross Ventilation: Although air conditioning may be planned, it is desirable to introduce a proper flow of natural air when temperatures warrant. (See Fig. 2.28.)

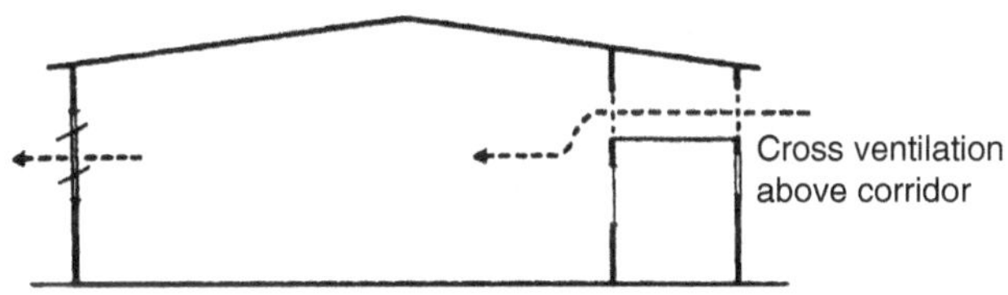

Figure 2.28 Structural consideration guidelines—Example 1.

2. Even Light Distribution: The structural configuration should provide evenly distributed natural light wherever possible, even though a satisfactory artificial illumination system is being proposed. However, special attention should be given to control of glare. (See Fig. 2.29.)

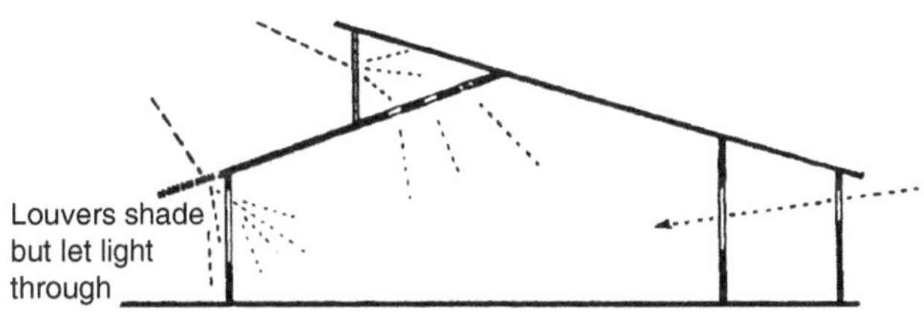

Figure 2.29 Structural consideration guidelines—Example 2.

3. Separated classrooms provide natural light and ventilation from two sides. The outdoor corridor allows students to enjoy a moment of fresh air as they go from one classroom to another. (See Fig. 2.30.)

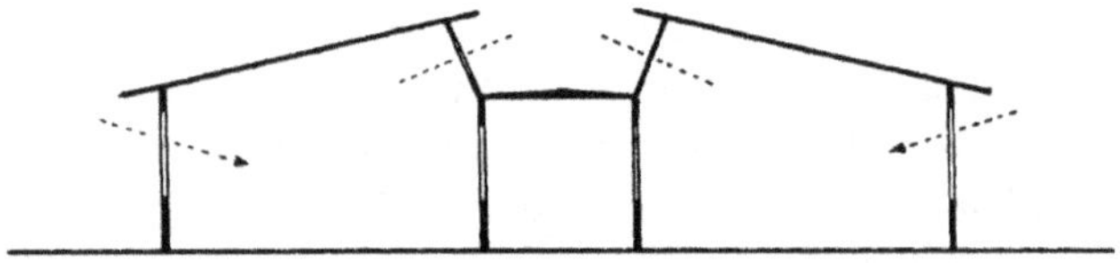

FIGURE 2.30 Structural consideration guidelines—Example 3.

PUBLIC LIBRARIES: CONCEPT DEVELOPMENT

The public library generally is no longer simply a place where books are stored and read or where materials are checked out to be taken home and read. The modern library has become a much broader system; in addition to performing the typical document storage and handling functions, it is a place in which to exhibit, provide entertainment, educate, and develop craft skills. Library specialists sometimes differ in their personal approach to the public library; some feel that they are primarily guardians of the materials they have been given (and thus are more concerned with keeping the public's hands off their treasured volumes), while others firmly believe that books should be in use as constantly as possible (and thus are more concerned with making it easy for the public to obtain materials). The obviously desirable architectural objective should, of course, be to provide a balance between effective safeguard of the library contents and maximum exposure of these contents to the public. From a purely human factors point of view, the public should probably be given first priority; i.e., the most important considerations should be ease of getting into the facility, of finding one's way around the facility, of locating the book one desires and finding a good place to use it while in the facility, and of checking out materials to take home.

Many of the library–user interfaces are similar to those found in other public office buildings, and the reader should refer to the sections dealing with office building planning. In addition, however, the following unique planning considerations relate particularly to library management and use functions.

Library Staff Functions

1. Administrative
2. Materials acquisition
3. Materials cataloging and filing and record keeping
4. Materials distribution
5. Overdue follow-up
6. Bookbinding and book repair
7. Research
8. Display and exhibit development
9. Materials checkout, return, and redistribution
10. Building maintenance

Patron Functions

1. Materials identification
2. Browsing
3. Reading and study
4. Copying
5. Listening
6. Checkout and return

Ancillary Functions

1. Lectures
2. Recitals and drama
3. Special materials exhibits
4. Education and crafts

Common Functions

1. Parking (staff, patrons, and delivery people)
2. Personal hygiene

Typical Library Spaces

1. Book stacks (reference and lending)
2. Computer workstations
3. Periodical, newspaper, and map racks
4. Checkout desk area
5. Card catalog file area
6. Reading areas
7. Audio-visual studies areas
8. Recital and lecture halls
9. Meeting rooms
10. Restrooms
11. Administrative offices
12. Workrooms (staff and public)
13. Archival rooms and vaults
14. Receiving and shipping
15. Storage (materials and equipment, maintenance supplies, and special vehicles, e.g., a bookmobile)

Environmental Factors

1. Nominal thermal regulation for habitability of staff and patrons
2. Low noise levels
3. High-quality illumination

4. Special environments for certain materials (archival books and other documents, film, etc.) and control of temperature, humidity, dust, etc.

5. Special acoustic qualities for recital and lecture spaces

6. Special lighting effects for exhibits, drama, etc.

Security

As noted earlier, philosophies differ with regard to public access to library materials and the importance of the loss of books and other materials. Special consideration, however, should be given to materials whose loss could be serious. These include such items as archival, one-of-a-kind volumes and special art objects that would be extremely expensive to replace or which could never be replaced. It is general practice to place such items in special areas, where access is controlled by the library staff. The planning should anticipate this need, but consider techniques that minimize potential loss and yet provide reasonable access—without which, the public is not being served properly.

Safety

The library, like other public buildings, is subject to hazards requiring careful consideration of traffic flow for emergency escape in the event of fire, properly designed stairways, etc. One special hazard exists where films are stored, requiring consideration of techniques to control such possibilities as fire in the film vault.

CHURCHES

Church Facilities Concept Development

Since philosophies vary regarding church facility design, it is vital that designers explore with their clients both these philosophical concepts and the basic functional needs typical of all building design. This section deals only with the latter aspects of church planning. Among the functions to be addressed for any church project are the following:

1. Worship service procedure as it pertains to the tasks of ministerial and lay persons, special furnishings, music, lighting and communication, and, finally, the congregation itself

2. Group instructional activities for both children and adults

3. Social activities involving large and small groups, within buildings and/or outside, and dining and kitchen facilities

4. Administrative functions including clerical and supervisory tasks, record keeping, publication, and accounting,

5. Ministerial activities including sermon preparation, counseling, and staff conferences

6. Lay program work, committee meetings, activity planning, program implementation, etc.

7. Music program activities including rehearsal, concerts, and special events

8. Housekeeping and ground maintenance, storage, shops, etc.

9. Special events (weddings, funerals, etc.)

10. Communications

11. Playgrounds, nursery, and sports facilities

12. Parking

13. Restrooms

The major facilities generally required for the complete church project are the following (see Fig. 2.31):

1. Sanctuary
2. Administration building(s)
3. Education building(s)
4. Maintenance facilities
5. Social hall
6. Ministerial housing (if on site)

Important planning considerations include the following:

1. Congregation: Effective worship, education, and social facilities

2. Staff: Efficient work space and convenience among facilities

Sanctuary Planning Considerations

Church buildings are a prime target for experimenting with unusual design features. Planners and designers are cautioned against letting their enthusiasm for aesthetic experimentation introduce unnecessary problems for the eventual user, including difficulties in seeing, hearing, and negotiating stairs and pathways; uncomfortable seating; and, most important, difficulties in maintaining the facility.

Entrances and Exits. Convenient entrances and exits must be considered for a variety of users (the ministers, the choir, the choir director, the organist, the ushers, and the congregation). Both exterior and interior entrances must be provided as necessary to allow people to move from one end of the sanctuary to the other without being seen by the congregation or exposed to inclement weather.

Seating. Seating should be provided not only for the congregation but also for the ministers and their assistants, the choir and its director, the ushers, and the audio-visual equipment operators.

1. Lighting: Both general and supplementary lighting must be provided both for adequate seeing tasks (e.g., reading and walking safely) and for special effects (illuminating specific decorative items, stage lighting effects, etc.).

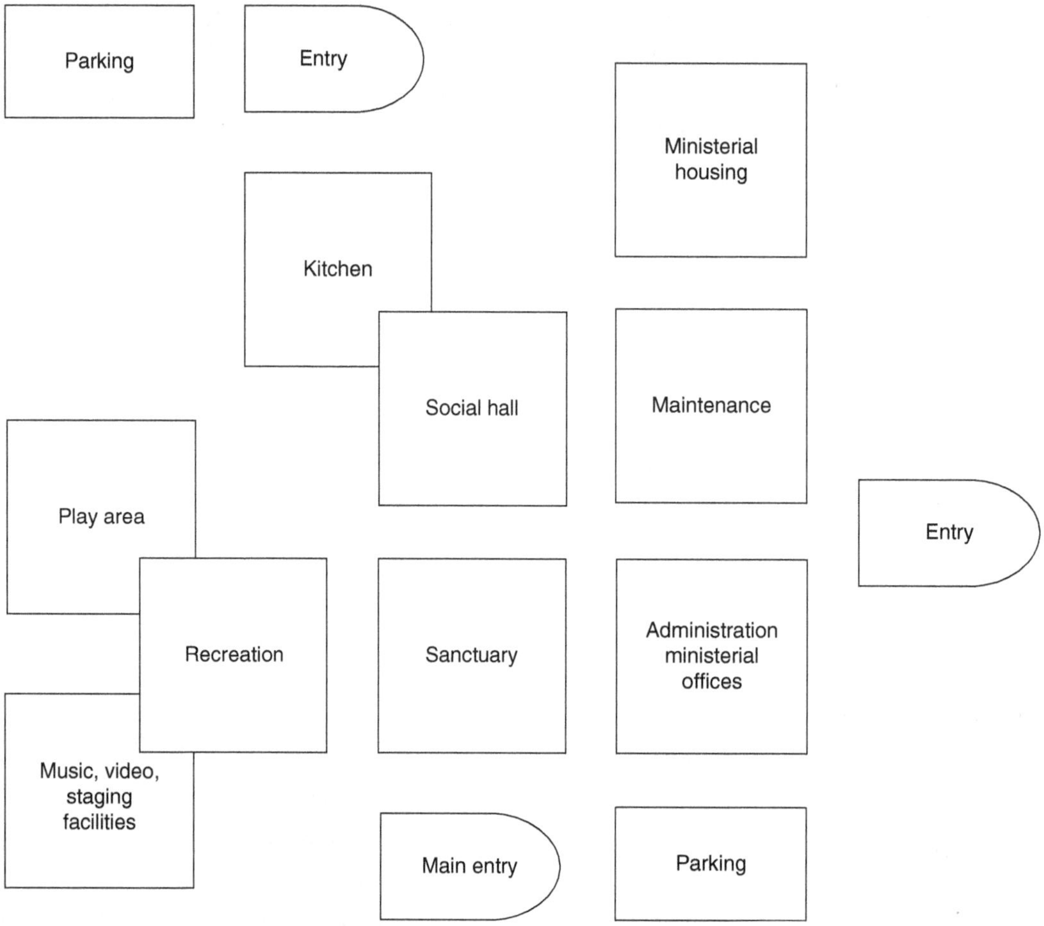

FIGURE 2.31 Church facilities planning model.

2. Communications: Depending on the size of the sanctuary, sound-amplification systems should be considered. In addition, consideration should be given to other aids such as special equipment for the hard of hearing, a recording system, and audio signaling between key participants, e.g., the minister, the organist, and the ushers.

3. Acoustics: In order to preclude interruption of a service, it is important to consider techniques for isolating the main sanctuary (nave) from the adjacent areas where people tend to generate noise, i.e., the entry vestibule, anterooms where the choir may gather, and areas where parents with small babies sit so that they can leave if a child becomes noisy and restless.

4. Special Areas: Several special areas often included as part of the sanctuary building require consideration during the planning stage in order to optimize their functional relationship with the main nave and chancel areas. These include the following:

a. Sacristy for communion preparation

b. Minister's study

c. Choir assembly area

d. Organ chambers

e. Choir and organ loft

f. Baptismal chamber

g. Audio-visual system monitoring booth

h. Bride's room

i. Ushers' materials facility (should include air-conditioning system controls, lighting controls, etc.)

5. Line of Sight: In spite of certain traditions in sanctuary layout, it is important that certain participants see each other. The more important visual links to consider are as follows:

a. The minister should be able to see the ushers, the organist, and the choir director.

b. The organist should be able to see the minister, the choir director, and the ushers.

c. The ushers should be able to see the minister and the organist.

d. The congregation should be able to see the minister, the choir, and special performers when the sanctuary is used for special events.

e. The audio-visual monitors should be able to see the general house lights, special lighting effects, and the exit lights.

6. Accommodations for the Elderly and the Handicapped: More than any other public facility, the church should be accessible to the elderly and the handicapped, since these people are typically among the most regular attenders. The key features to be considered are as follows:

a. Wheelchair ramps and a minimal requirement to negotiate stairs

b. Better-than-average illumination

c. Special aids for the hard of hearing

d. Restrooms specially equipped for the handicapped

e. Parking for the handicapped

7. Special Safety Considerations: In addition to the typical safety considerations for any public building, special attention should be given to exterior safety problems that are somewhat unique to the church situation, especially the problem of the safety of children with respect to vehicular traffic around and within the church site. A typical problem occurs when church buildings are built directly adjacent to sidewalks and streets. Children often are unattended for short periods between Sunday school classes, and they tend to roam and play on the sidewalks, often inadvertently running into the street in front of cars that are picking up passengers.

Sanctuary Planning Example

Figure 2.32, although not intended to represent any particular facility, provides several important ideas to be considered in the planning of any new sanctuary. Note the following:[3]

1. Convenience of the main entrance to the street

2. Parking for the general congregation and the staff

3. Large narthex (which is used as a gathering place, in spite of the best intentions of the ushers)

[3]For other church facilities (e.g., churches, auditoriums, offices, and schools), see J. De Chiara and J. H. Callender, *Time-Saver Standards for Building Types,* McGraw-Hill Book Company, New York, 1973.

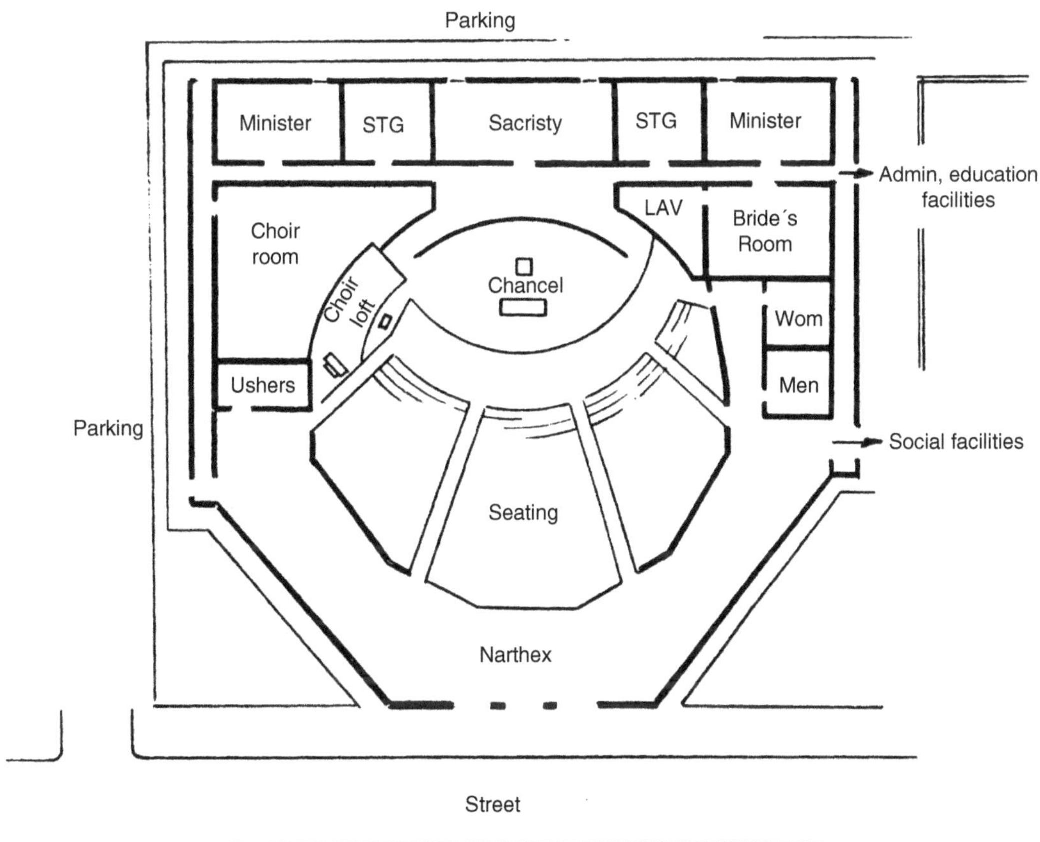

Figure 2.32 Example sanctuary.

4. Sound barrier between the main sanctuary and any outside walls

5. Reasonable viewing distance between any seat in the sanctuary and the pulpit

6. A choir position that allows sound to focus fairly well on all parts of the congregational seating and also allows choir members to view the pulpit

7. Complete inside access around the main sanctuary

8. Good sight lines between the minister, choir, organist, and ushers

9. Good staff access to other church facilities and good congregational access to social gathering areas

10. Restrooms for both staff and others

HOSPITALS

Hospital System Planning Model

The prime planning objective in hospital design should be to seek a balanced compromise between maximizing hospital staff efficiency and maintaining adequate respect for the personal needs, both physical and psychological, of the patients. Foremost among those needs (other than medical and therapeutic care) are the minimization of fear, boredom, and the feeling that they have lost complete control over their lives. (See Fig. 2.33.)

Hospital Facilities Concept Development

In conceptualizing and planning either the general or the specialized hospital facility, much depends on the nature and scope of the services required by the public or private client.

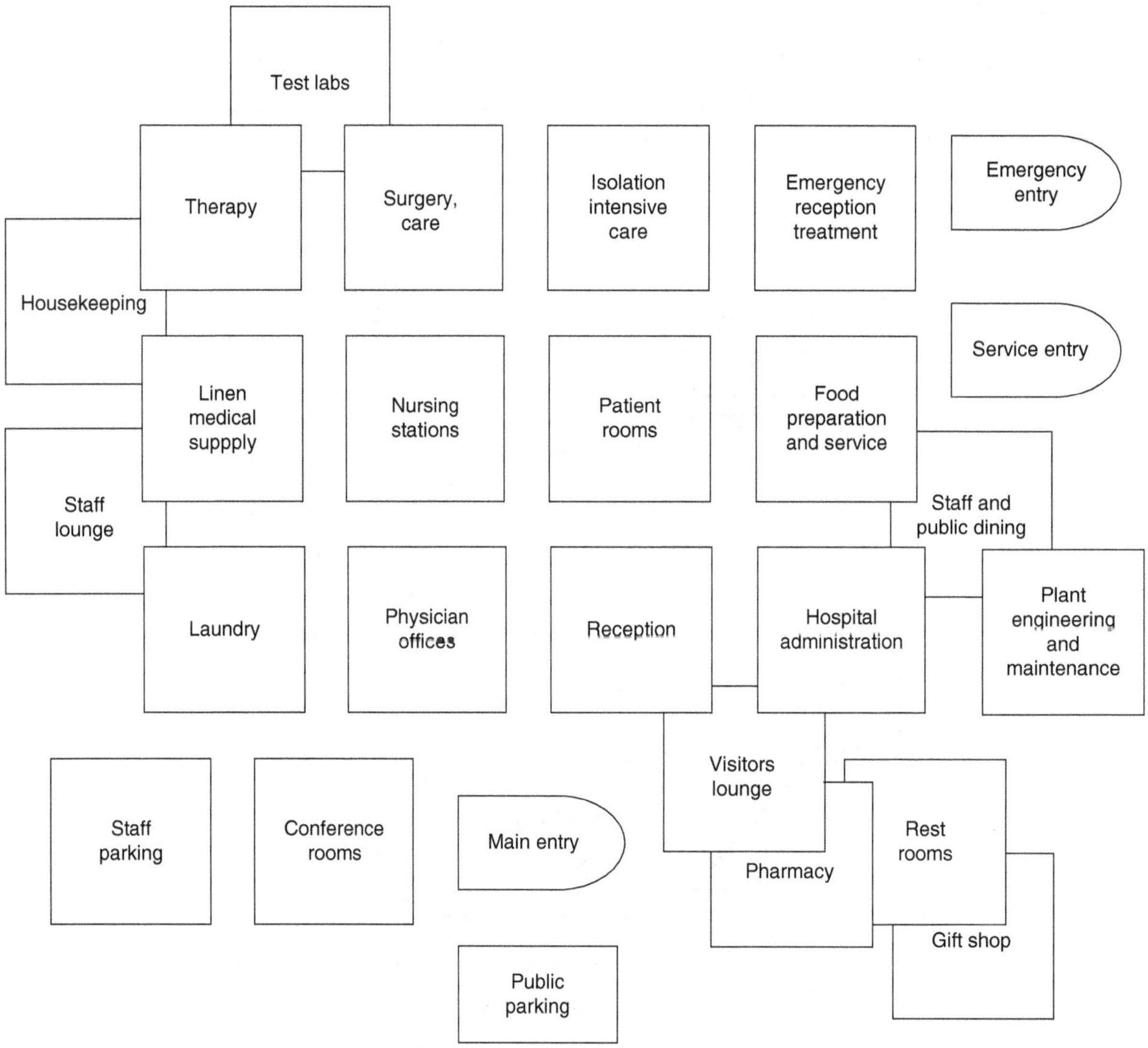

FIGURE 2.33 Hospital system planning model.

However, all hospital and convalescent facilities have many common human interface needs and design requirements, which is the subject of this section. It is perhaps important in the beginning, however, to remind the planner and designer that the primary purpose of any hospital facility is to serve the patient, a fact that is often overlooked in the rush to make the jobs of the hospital staff and the doctors easier. High priority should therefore be given to making the patient's task of getting to the hospital, checking into the hospital, receiving a proper diagnosis and treatment, having an acceptable recovery environment, and finally getting discharged with a minimum of strain as pleasant and convenient as possible. Obviously, it should be possible for the hospital staff and the doctors to perform their tasks as efficiently as possible, but not at the expense of the patient. *Note: Guidelines for the design of hospitals can be found in Design and Construction of General Hospitals by Public Health Service, U.S. Department of Health Education, and Welfare (1953) and the updated Guidelines for Design and Construction of Hospitals and Health Care Facilities, American Institute of Architects Academy of Architecture for Health (2001).*

Some of the observed deficiencies noted as follows should serve to emphasize the need for careful consideration of priorities.

Patients

1. Patients cannot find a parking place.
2. Patients cannot find the entrance to the hospital building.
3. Patients cannot locate the checkin or information desk.
4. Patients cannot find a place to sit while waiting to be admitted or to see a doctor.
5. The rooms are institutional-looking and often frightening.
6. The rooms lack privacy and freedom from sounds made by other patients or their visitors.
7. The room facilities are not designed for ease of use from the bed.
8. There is lack of space to receive visitors.

Nurses and Aides

1. The space layout requires considerable walking.
2. The equipment is awkward to operate.
3. Work spaces are not conveniently located or laid out for maximum efficiency.
4. Relaxation facilities are inadequate.

Doctors

1. The space layout requires considerable walking and is time-consuming.
2. Work spaces (operating rooms, etc.) are not laid out for maximum efficiency.
3. Relaxation facilities are inadequate.

Visitors

1. Confusing floor plans make it hard to find one's way around.
2. There is not enough parking.

Hospital Facilities Shopping List

It is doubtful that any list of possible facilities would include all possible considerations. Table 2.9, however, should provide the planner with a good start.

General Conceptual Considerations

The following considerations should be addressed early in the planning stage of any type of hospital facility. The planner or designer who is not acquainted with hospital design should obtain additional information from such sources as the American Hospital Association in Chicago and the U.S. Public Health Service in Washington, DC.

Hospital Location. Whenever possible, locate the hospital facility in a quiet neighborhood where it is away from heavily traveled, noisy highways or streets and yet adequately accessible to and from such routes. The access route should be direct and well marked so that emergency vehicle operators do not waste time trying to locate the facility.

Site Availability to Public Transit. The site should be chosen with the potentiality of public transportation access in mind, since both staff members and visitors often need to use public transportation to get to the hospital.

Planning for Emergency Evacuation Possibilities. Although one hopes that a disaster requiring emergency evacuation of hospital patients and staff members never occurs, this is a critical consideration in terms of both site selection and site and building layout.

Site Layout. Hospital grounds should be laid out to maximize the logic of the plan for visitors, especially those trying to bring an injured or very sick patient to the hospital for the first time. Although an attractive site is important, a confusing one may mean the difference between life and death.

Building Layout. The layout of buildings and building interiors should also evidence a logic that visitors can readily understand from their first-time, limited viewpoint. Although the staff may eventually learn to navigate from one place to another, the first-time visitor is easily confused by irregular pathways, curved or angular geometries, irregular elevations, etc.

Identification and Instructional Signing. Signing requirements should be addressed at the same time that grounds and building layouts are under consideration. Do not expect a sign to make up for a poor layout.

Removal of Architectural Barriers. The hospital, more than any other facility, is subject to constant use by those who are handicapped. Do not assume that all handicapped persons are being assisted by a nonhandicapped person; many are

TABLE 2.9 Hospital Facilities Shopping List

Administrative Personnel	Spaces Required
Hospital director and assistants	Offices
Business manager and assistants	Conference rooms
Secretarial and clerical personnel	Admissions and checkout center or centers
Bookkeeper	Staff lounges and restrooms
Engineering staff	Kitchen and dining facilities
Housekeeping staff	Mechanical and maintenance spaces
Maintenance staff	Laundry
Ambulance staff	Pharmacy
General pharmacist	Storage (records, housekeeping materials and equipment, linens and supplies, drugs, furniture, etc.)
Dietician, kitchen and dining room staff	Vault
	Parking
Medical Personnel	**Spaces Required**
Medical director and assistants	Offices and conference rooms
Staff doctors	Nursing stations
Itinerant doctors	Examination rooms
Director of nursing staff	Operating and treatment rooms
Nurses and aides	Laboratories
Dieticians	Nursery
Food preparers and aides	Emergency receiving rooms
Laboratory technicians	Kitchen and dining facilities
Physical therapists	Staff lounges, restrooms, lockers, and dressing spaces
Pharmacists	Parking
	Records, database, and communication facilities
Patients	**Spaces Required**
Emergency patients	Private and multipatient rooms
Patients requiring minor first aid	Lavatories and bathrooms
Patients requiring intensive care	Physical and mental therapy spaces
Patients requiring major surgery	Isolation wards and intensive care areas
Maternity patients	Nursery and play areas
Patients with communicable diseases	Day rooms and solariums
Psychiatric patients	Personal storage space
Children and infants	Emergency parking
Cancer patients	Temporary morgue
Others	
Visitors and Others	**Spaces Required**
Patients' visitors	Lobby, waiting rooms, and public restrooms
Social workers	Information center
Vendors	Personal computer and communications accommodations
Delivery personnel	Public eating facilities (kitchen)
Security personnel	Shop selling gifts, newspapers, and flowers
Funeral parlor personnel	Mail room
	Private sleeping rooms for overnight visitors
	Parking

not. Plan for the handicapped in the basic design so that modifications or add-ons are not required; e.g., a basic ramp entry should be the primary entry, not an auxiliary one.

Protection from Inclement Weather. Staff members, patients, and visitors should never be required to walk in the rain from building to building, be buffeted by wind on an outdoor balcony or roof solarium, or be unable to utilize special therapy facilities outside. These conditions should be considered in the initial concept.

Site Planning. Several important site planning considerations are as follows:

1. Central core (includes elevators, information center, admissions and checkout, administration, etc.) equally accessible from both the main and the emergency entrances
2. Ample parking for both the public and staff members arranged so that no one has to cross busy streets or roadways
3. Auxiliary lifts and stairs at each end of the building
4. Covered, drive-through entrances
5. Short distance from entries to the information point

THEATERS

Theater Concept Development

Whether a proposed Theater is restricted to a single type of theatrical performance or is to accommodate a variety of performances (sometimes simultaneously), similar user activity categories should be addressed in order to ensure that each user group is properly accommodated. The principal user categories include the following:

1. Patrons
2. Performers
3. Performance support personnel
4. Patron support personnel
5. Theater management
6. Facilities operation, housekeeping, and maintenance

The typical subsystems required in most Theaters include the following:

1. Parking (for each of the above categories)
2. Entrances and exits (separate customer, performer, service, and other entrances and exits)
3. Basic patron facilities (lobby, restrooms, cloakrooms, refreshment booths, stairs, elevators, and escalators)
4. Basic Theater management facilities (offices, restrooms, cash vault, plant engineering and equipment spaces, housekeeping and maintenance shops and storage, employee lounge, etc.)

5. Auditorium (including seating, lighting, air conditioning, special projection, and special-effect light and sound booths)

6. Main stage and backstage (including stage props management, revolving and elevating stage elements, stage lighting, and performer preparation spaces, e.g., dressing, instrument storage, and rehearsal)

7. Other performance support facilities (shops and storage for stage props, lighting and sound equipment, costumes, furniture storage, etc.)

Figure 2.34 illustrates the major Theater subsystems that impact on human users, categories include the following:

1. Human Users
   a. Patrons
   b. Performers
   c. Performance support personnel
   d. Administrative and managerial personnel
2. Subsystems
   a. Entrance and foyer
   b. Audience seating
   c. Stage
   d. Stage support and performer preparation area

Key Human Factors Considerations by User Activity Requirements

Visibility

1. Patrons need to see in order to get to and from their seats, often in semidarkness.

2. All patrons should be provided approximately equal visual access to all activity onstage (but not be exposed to what is going on offstage).

3. Ushers must guide patrons to their seats, often in semidarkness.

4. Performers and stagehands must move quickly and safely both backstage and onstage, often in the dark.

5. Stage lighting operators must be able to monitor the effect of their light control manipulations (preferably from the patrons' viewpoint).

6. There should be intervening, preparatory areas that allow patrons and others to adapt visually to a darkened auditorium or a bright external sunlit environment.

Audibility

1. Patrons should have equal auditory access to the important sounds emanating from a performance, but they should be isolated from unwanted sounds coming either from within or from outside the auditorium.

2. The acoustic characteristics onstage should fit performers' requirements to hear one another.

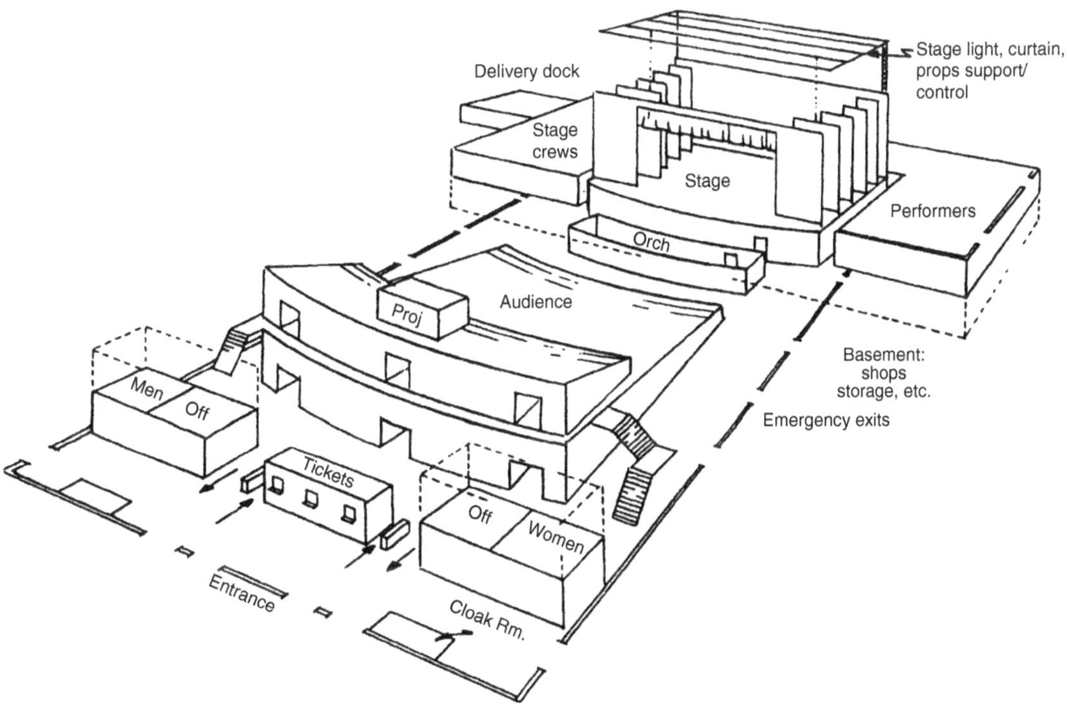

Figure 2.34 Conventional community Theater.

3. The acoustic characteristics within the auditorium should be designed to minimize distortion of desired sounds from the stage and to minimize possible patron-generated noise.

Thermal Comfort. Proper temperature, humidity, and ventilation control should be provided (particularly within the auditorium) to ensure patron comfort and freedom from drafts, undesirable odors, etc.

General Safety. The overall facility concept should be examined carefully from the standpoint of the safety of all Theater user categories, but especially in terms of the following:

1. Emergency escape
2. Nonflammable materials
3. Emergency lighting
4. Number, location, and operation of emergency exits
5. Disaster control (fire-extinguishing systems and equipment)

Other Considerations. Special attention should be given to the layout of the Theater in terms of the following:

1. Patron queuing both outside and inside the Theater to ease the flow of both entering and exiting customers
2. Isolation of patrons from backstage activities to avoid unnecessary interference with performers and/or others who must perform time-constrained tasks and to preclude the possibility of damage to stage equipment, instruments, etc.
3. General layout clarity and convenience, especially in terms of patrons' needs to locate their seats, find restrooms, etc.
4. External considerations, including isolation of patrons', Theater operators' and performers' parking and entrances; covered walkways to protect patrons who may have to wait outside in inclement weather; and sufficient space outside for people to make an emergency exit without having to run directly into an adjacent street, where they may be exposed to heavy traffic

Special Considerations for Each Theater Subsystem

Entrance. A sufficient number of entrances should be provided for the anticipated exchange of patrons entering and leaving the Theater. An entry vestibule large enough to accommodate purchasers of tickets (out of the rain) should be provided. The entrance and foyer area should be arranged so that patrons' eyes have time to adapt to bright sunlight following exposure to the typically low light levels of the Theater. Similar visibility problems occur for persons working backstage, where scenery must be changed in semidarkness and where performers must move from unlit to brightly lit areas, and vice versa.

Foyer. A large foyer should be provided (for each floor) because this must accommodate patrons both before the performance and during intermissions. Restrooms, lounges, and administrative and management offices should be provided off the foyer. Refreshment centers are also appropriate here, in addition to ticket booths and a cloakroom.

Audience Seating Area or Areas. The audience seating area should be designed for optimum viewing and listening and should have both comfortable seating and convenient and safe access to seats. A sloping floor with staggered seating can provide optimum viewing.

Orchestra Pit. A large orchestra pit should be provided when this facility is planned. Instrumentalists must have room to maneuver and must be able to see the conductor. Access to and from the pit (without exposure to the audience) is necessary.

Stage. A generous stage should be considered. It is simple to contract the stage size, but it is extremely difficult to enlarge it for special performances (i.e., ones for which the chorus and orchestra may be positioned on the stage together).

Backstage. The behind-the-scenes activity of typical stage productions requires considerable space, including access from one side to the other for moving scenery and performers. Included among the backstage facilities are scenery storage and preparation, performers' dressing rooms and restrooms, costume storage, curtain and light controls, and rehearsal studios. A shipping dock should be provided.

Special Features. Among the special features to be considered are a projection booth for motion pictures and sound control, mounting and control of special lighting effects, elevators for patrons and for staging, acoustic control features (sound reflection and absorption), and illumination for housekeeping and maintenance. A basement is desirable for many purposes, among which are storage, maintenance shops, and art studios for creating advertising materials.

Restrooms. Intermissions provided a limited period for patrons to use the restrooms. It is important that sufficient number and conveniently located facilities are provided. See Chap. 18, "Bathrooms, Restrooms, and Locker Rooms," for guidelines to select the proper number of restroom facilities. Also, consider traffic flow so that people entering the restrooms do not interfere with those exiting.

RESTAURANTS AND BARS

Restaurant Concept Development

Although restaurant concepts will vary depending on the size and extent of service planned, certain basic human interface features remain relatively constant, and the problems that patrons or staff members will have depend on how well these interfaces have been designed to fit human needs and constraints.

The principal functions to be addressed include the following:

1. Parking and public transportation accessibility
2. Entrance and exit convenience
3. Reception and/or access to the dining area (including waiting area, cloakroom, restrooms, etc.)

4. Dining and customer service

5. Food preparation and storage (including delivery and trash and waste disposal)

6. Dishwashing and garbage disposal

7. Storage for dishes, flatware, linens, utensils, etc.

8. Administrative and management areas (including staff lounge, locker, and restrooms)

9. Housekeeping and maintenance areas (including storage)

10. Other (performing stage, dance floor, dressing rooms, audio-visual system control and storage, etc.)

Each of the following architectural subsystems requires special individual as well as collective consideration:

1. Parking area or areas both for patrons' and staff members' vehicles and for service vehicles

2. Foyer, waiting room, lounge and restrooms, and cloakroom

3. Special lounge and bar (storage)

4. Main dining room or rooms

5. Intermediate service centers

6. Kitchen (food, utensil, and equipment storage)

7. Service staff lounge, restrooms, lockers, etc.

8. Manager's office

9. Special performer areas (instrument rooms and dressing rooms)

10. Sound and lighting control booths

11. Engineering spaces (environmental control)

12. Emergency exits

The important planning considerations are as follows (see Fig. 2.35):

1. Ample, well-laid-out, easily identifiable parking that will not cause vehicles to queue up in the street

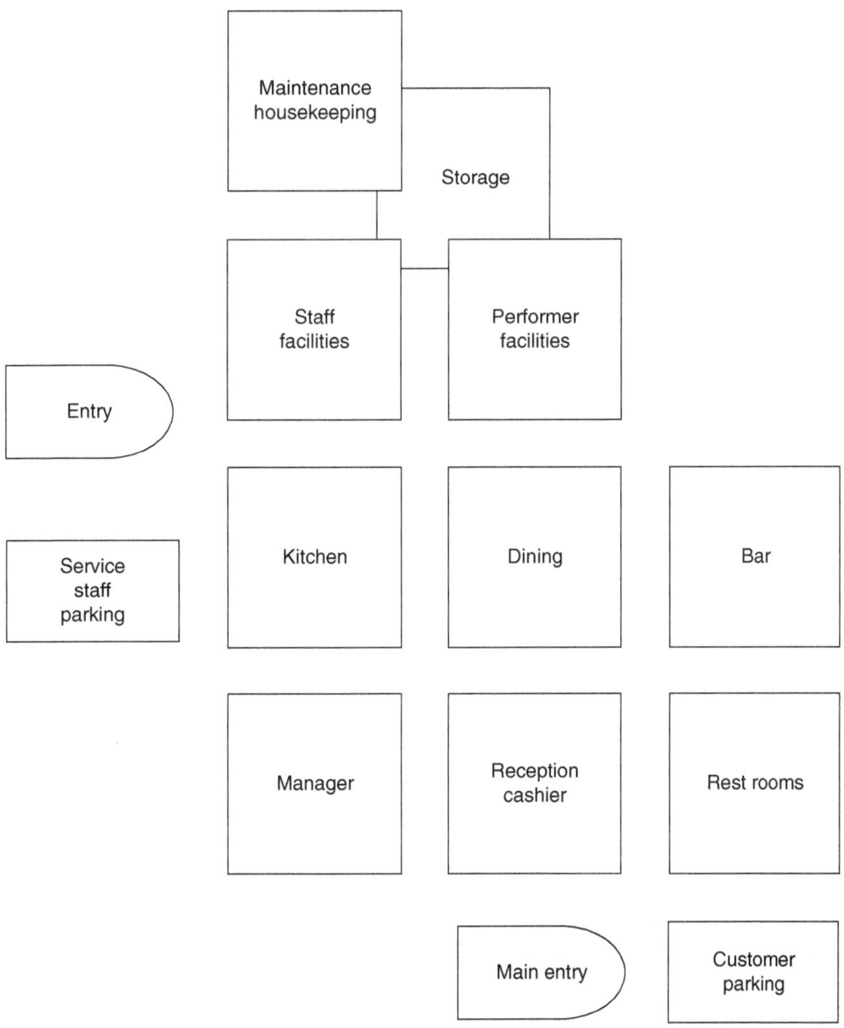

Figure 2.35 Restaurant and facility planning model.

2. An ample reception and waiting area with minimum conflict between incoming and departing guests

3. An ample, quiet, and pleasant dining area that provides reasonable privacy for individual diners and groups of diners

4. Efficient kitchen and service arrangements

Special Human Factors Problem Areas

Human factors problems typically occur in at least three primary areas: (1) the transitional space, where customers enter the restaurant and transfer to the dining area; (2) the dining area, where both customers and service staff must be provided uniquely convenient features; and (3) the kitchen area, which obviously requires special attention to convenience details to make the kitchen staff's output as efficient as possible.

General Considerations in Planning the Dining Area

From the Diner's Point of View. Upon entering the dining area, diners need to have a good preview of the total area so that (in case they are not being ushered by a maître d') they can choose the area and/or table that they believe meets their particular requirements and decide which is the best way to get there. Although there may be something intriguing about "hiding" various tables and creating devious pathways and nooks and crannies (which also are of some value from the standpoint of providing privacy and freedom from noise), care should be taken not to confuse patrons completely.

Fairly direct, clear, and generous pathways should be provided for patrons going to and from tables and/or dining areas. Not only is it annoying to bump into waiters and other patrons, but also some safety hazards are created when the layout forces crowding.

Reasonable privacy is desired by most diners. Plans that are based on "hub-to-hub" tables and seats do not provide the privacy most people want. In addition, most diners want to be free from physical disturbance, and therefore the plan should provide adequate space around tables and chairs so that waiters and waitresses can have free access without bumping their serving carts into the backs of diners' chairs.

The plan should provide pleasant views for the diners of the outdoors or of attractive decorative elements in the room. Diners should not have to look at the kitchen or serving activities.

Special attention should be given to the ventilation and acoustic aspects of the dining room. No one likes to sit in a draft or listen to the clatter of dishes in the kitchen.

From the Waiter's or Waitress's Point of View. Convenience to and from the kitchen and/or strategically placed service centers is, of course, of prime concern to waiters, waitresses, and busboys. Pathways to and among diners' tables are equally important in terms of clearance, directness, and sufficient illumination so that restaurant employees can see where they are going. Particularly important to waiters and waitresses is elimination of "blind spots," i.e., possible traffic conflicts at doorways, partition thresholds, stairs, etc.

General Considerations in Planning the Transitional Space of the Restaurant

Entrance. Remember that the main entrance to the restaurant is also the main exit for customers who have finished dining. Since people tend to approach the door (from either inside or outside) as though they were the only ones using it, avoid the temptation to minimize the entryway width (both inside and outside). In addition, remember that people often are in the process of putting on outer garments or adjusting umbrellas, and they are generally busy with small talk and jostling, unaware that they may interfere with others who are going or coming. It is recommended that the main entrance consist of separate entry and exit doors, clearly marked to advise the customer which is which. (Although not all people will read and follow instructions, most will, and this reduces the potential traffic conflict.)

The area just outside the entry should be planned so that customers are protected; that is, they should not have to step directly into the rain, onto snow-covered sidewalks, into bright sunlight, or into passing pedestrian traffic. The area just inside the entry should be large enough so that customers can step aside to reconnoiter, fold umbrellas, remove outer garments, and adjust their eyes (if the interior illumination is considerably darker than the out-of-door entry illumination).

Foyer. A fairly spacious foyer should be considered. This should be arranged so that new patrons can immediately survey and plan their next step. That is, patrons will be looking for a place to put their hats and coats, perhaps the restrooms, a maître d' or hostess or the pathway to the dining area, or a place to sit down and wait. Again, remember that the foyer will also be used by exiting patrons, and the layout should provide a clear delineation of the preferred paths for incoming as opposed to exiting patrons.

Depending on the type of restaurant, there may be the requirement for a cashier's stand in, or adjacent to, the foyer. Often this can be used as a natural divider between incoming and exiting patrons.

Dining-Area Entry. The transitional point between the foyer and the dining area should be clearly delineated so that it is obvious to incoming patrons. Avoid single step-downs or step-ups from the foyer to the dining area (a pet architectural device that leads to many accidents). This transition point should be well illuminated, not treated as a secret of the management. If the levels between the foyer and the dining area are to be different, create at least two steps and also provide a ramp for customers who cannot safely negotiate steps.

It is suggested that consideration be given to the problem of noise that may emanate from the foyer, which often is disturbing to diners. Obviously, certain acoustic treatments are available. However, a change in direction can also help ameliorate the problem. A typical plan incorporating some of the features discussed is shown in Fig. 2.36.

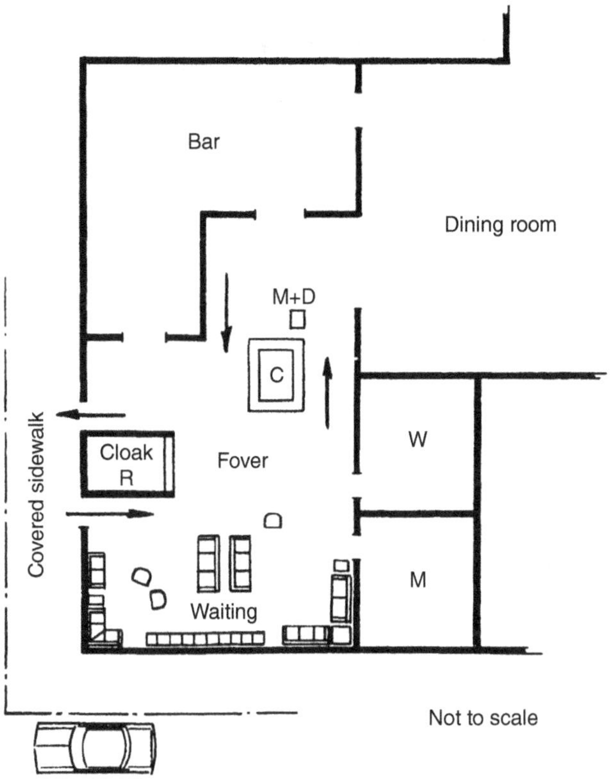

Figure 2.36 Example restaurant transitional.

Dining-Room Examples (See Fig. 2.37)

1. Initial view of possible booths and tables; good clearances and organization enhance traffic flow; service center or centers reduce number of steps that waiters or waitresses must take; and booths provide reasonable privacy and reduce general noise level.

2. Clusters of tables separated by partitions minimize noise transfer even more.

3. Wide angle of view can be provided even for the fast-food cafeteria.

Common Human Factors Design Mistakes in Fast Food Restaurants

The following are mistakes often found in the subsystems of fast food eating establishments. Figure 2.38 illustrates these mistakes. The fast food consumer is often in a hurry and has different human factors needs than a consumer in a more formal restaurant.

1. Slippery flooring—Because many of these establishments are open 24 hours, floor cleaning must be done while customers are using the premises. The wet soapy floors, the smooth easy-clean surfaces, and spilled food all lead to slipping accidents.

2. Additionally, high in/out traffic flow in rain or snow can make the floor wet and slippery at the entrances.

3. There is no waiting area or it is placed in an area that obstructs traffic around the order or pickup counter. People often must wait for an order and feel reluctant to occupy a table.

4. There should be enough room in front of the ordering and pickup counters for people to line up during rush periods. Often planners do not plan for peak periods and people must block aisles to stand in line. This creates many hazards because customers are carrying hot foods. Additionally, restaurant employees are cleaning tables and carrying trays through this congested traffic area. Often many of the customers in the aisles are small children who are difficult for an adult to see over a tray of food.

5. Employee exits from the kitchen side of the order counters are in areas that interfere with customer lines. These exits make it difficult for both the employees and customers to stay out of each other's way.

6. Tables reserved for smokers placed across from the ordering counters and salad bar create a health hazard for both employees and waiting customers. Because these restaurants are small they do not always have a physical barrier between the smoking and nonsmoking sections.

7. In Fig. 2.38 of a fast food floor plan, the salad bar interferes with other activities in the restaurant. Customers at the salad bar must select their salad, avoid slippery floors (possibly spilled food), avoid the line at the ordering counter, avoid employees cleaning, and avoid other customers carrying food to their tables or out of the restaurant.

8. Video arcades should be sectioned off from the rest of the restaurant. If these areas are not removed from other restaurant subsystems, they can create high noise levels and often can add to the traffic pattern problems. Carrying food through individuals who are concentrating on a video game can create accidents.

9. No restrooms for customers inside establishment.

Calculating Dining-Area Requirements

From the patron's point of view, clearance is perhaps the most important factor. Although the particular furnishings selected also must be considered in arriving at the final clearance and area dimensions, consider especially the points illustrated in Figs. 2.39–2.43:

1. In order for people to sit comfortably at a table, there must be clearance for their feet. This suggests a minimum 40-in. (102-cm) diameter for a table to seat four people. (See Fig. 2.39.)

2. In order for people sitting back to back to extricate themselves from the table, there must be at least 47 in. (119 cm) between tables. This can be reduced by about 5 in. (13 cm) if the chairs are arranged so that they are not back to back; however, this creates a clearance problem

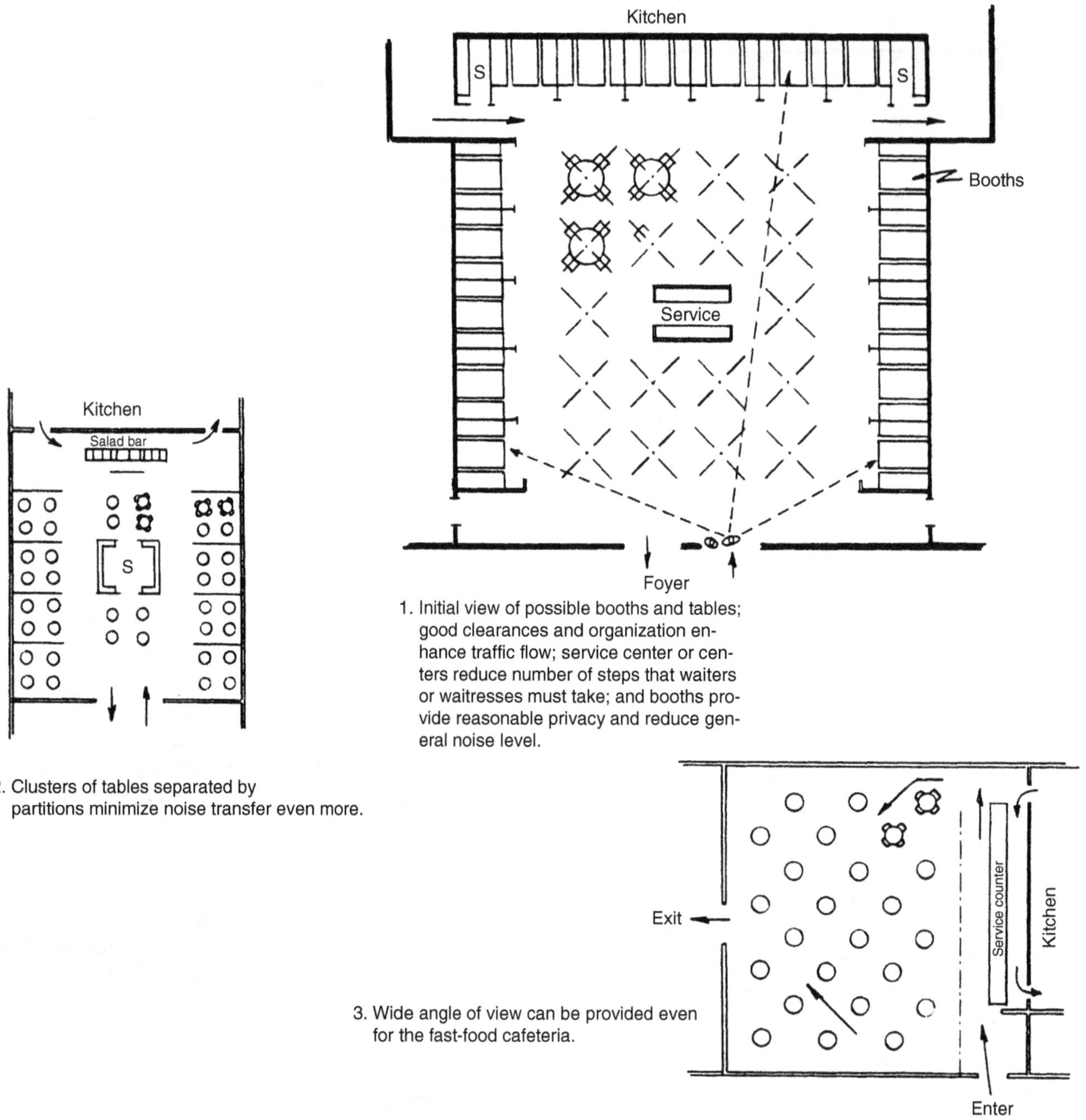

1. Initial view of possible booths and tables; good clearances and organization enhance traffic flow; service center or centers reduce number of steps that waiters or waitresses must take; and booths provide reasonable privacy and reduce general noise level.

2. Clusters of tables separated by partitions minimize noise transfer even more.

3. Wide angle of view can be provided even for the fast-food cafeteria.

FIGURE 2.37 Dining-room examples.

for the waiter or waitress who is trying to pass between chairs. (See Fig. 2.40.)

3. Provide at least 24 in. (61 cm) between seat centers so that people sitting side by side do not have to touch each other. The table should be at least 30 in. (76 cm) wide to provide space for dishes and clearance for feet. (See Fig. 2.41.)

4. A counter top must obviously be deep enough to accommodate the patron's dishes, but not so deep that the waiter or waitress cannot reach across it to clean off the counter. Patrons at a counter must be seated further apart so that they can get in and out (the centerline separation should be at least 27 in., 69 cm). (See Fig. 2.42.)

5. Eye heights between patron and waiter behind the counter should be relatively equal, but people (total strangers) sitting across the counter from each other should not have to be stared at. Separate them by at least 15 ft (4.5 m) or more. (See Fig. 2.43.)

General Considerations in Planning the Restaurant Kitchen

Worker Efficiency. Individual worker efficiency is, of course, the most important aspect of kitchen operation and

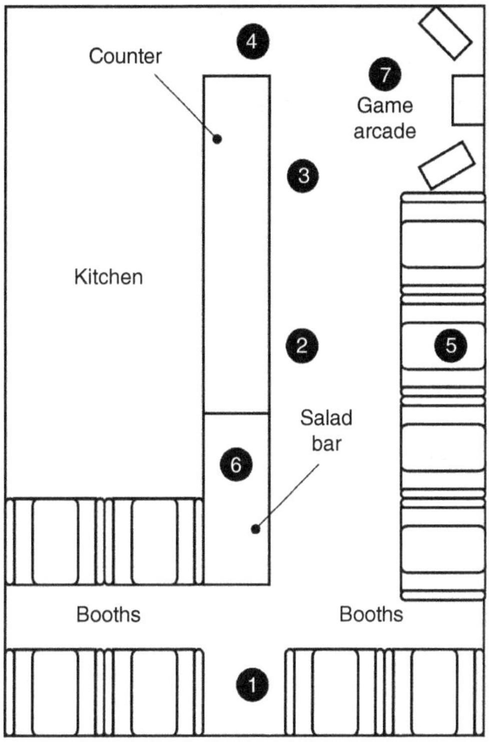

Figure 2.38 Dining-room mistakes.

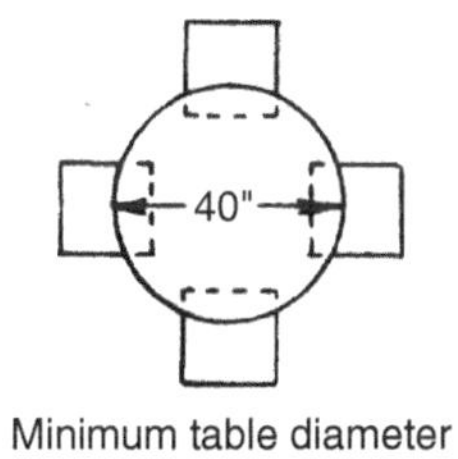

Figure 2.39 Dining-area requirements—Example 1.

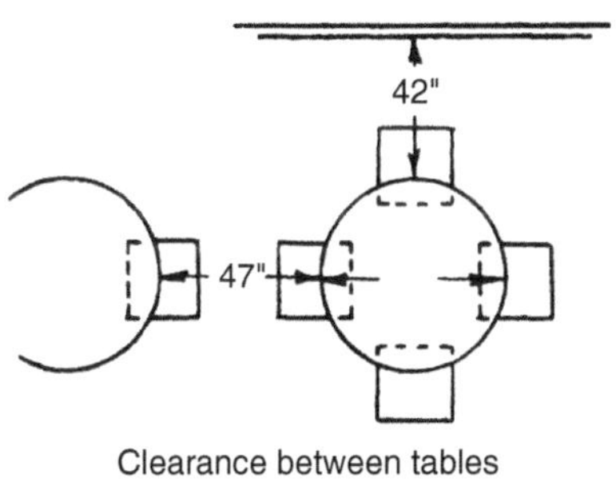

Figure 2.40 Dining-area requirements—Example 2.

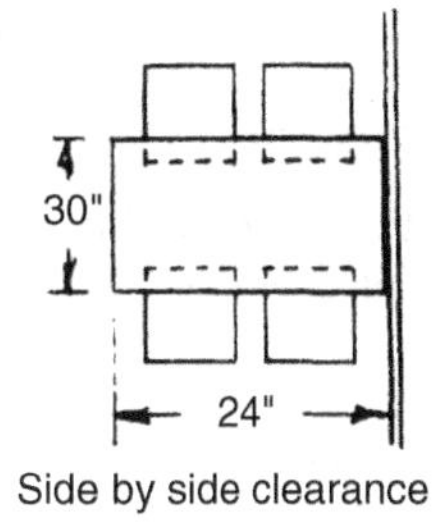

Figure 2.41 Dining-area requirements—Example 3.

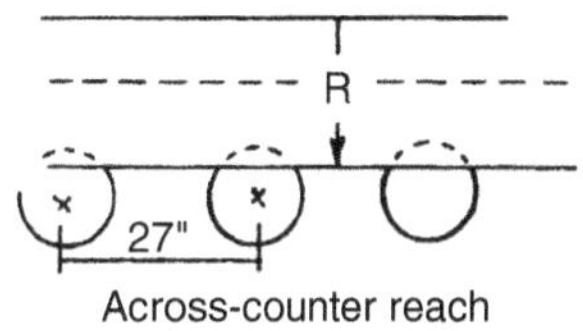

Figure 2.42 Dining-area requirements—Example 4.

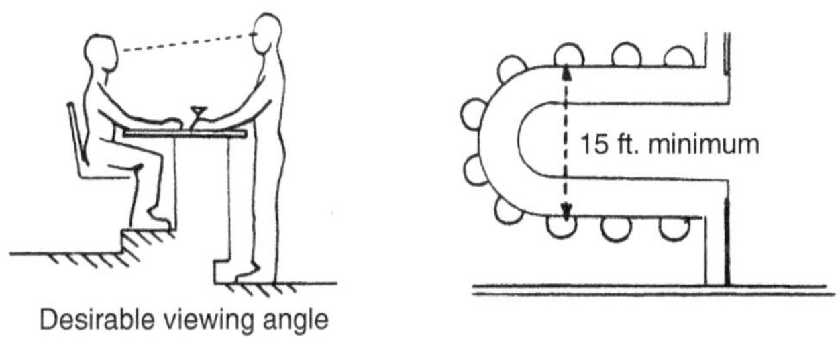

Figure 2.43 Dining area requirements—Example 5.

therefore should be uppermost in the mind of the planner and designer. Difficulty arises because of the variety of operations that all must be efficient. They include the following:

1. Receiving (all supplies)
2. Storage and retrieval (dry foodstuffs, produce, meats, refrigerated and frozen products, etc.)
3. Food preparation
   a. Precooking and dispensing stage (cutting, trimming, etc.)
   b. Cooking, baking, frying, etc. (involves acquisition of utensils and use of ranges, ovens, fryers, and other equipment)
   c. Dispensing (involves acquisition of dishes, flatware, crystal, trays, etc., and critical interaction between cooks and waiters and waitresses)

4. Cleanup, disposal, and storage of cleaned elements

   a. Garbage disposal

   b. Preparation of dirty dishes for washing and disposal of soiled linens

   c. Dishwashing and storage of clean dishes, utensils, flatware, cutlery, crystal, and special food-preparation equipment

5. Housekeeping (maintaining the kitchen facilities, i.e., cleaning counters, worktables, griddles, ovens, ranges, floors, venting hoods, etc.)

Safety. Consult local ordinances regarding the required safety features for commercial kitchens. In addition, however, consider particularly the following hazards, since they imply special planning considerations:

1. Splatter of hot grease and potential ignition of same
2. Breakage of crockery and glass and attendant cuts
3. Falls due to liquid spilled on floors
4. Cuts suffered while using special meat-cutting equipment
5. Staff collision injuries due to crowding, swinging doors, and blind intersections
6. Lighting to reduce accidents (See Fig. 2.44.)

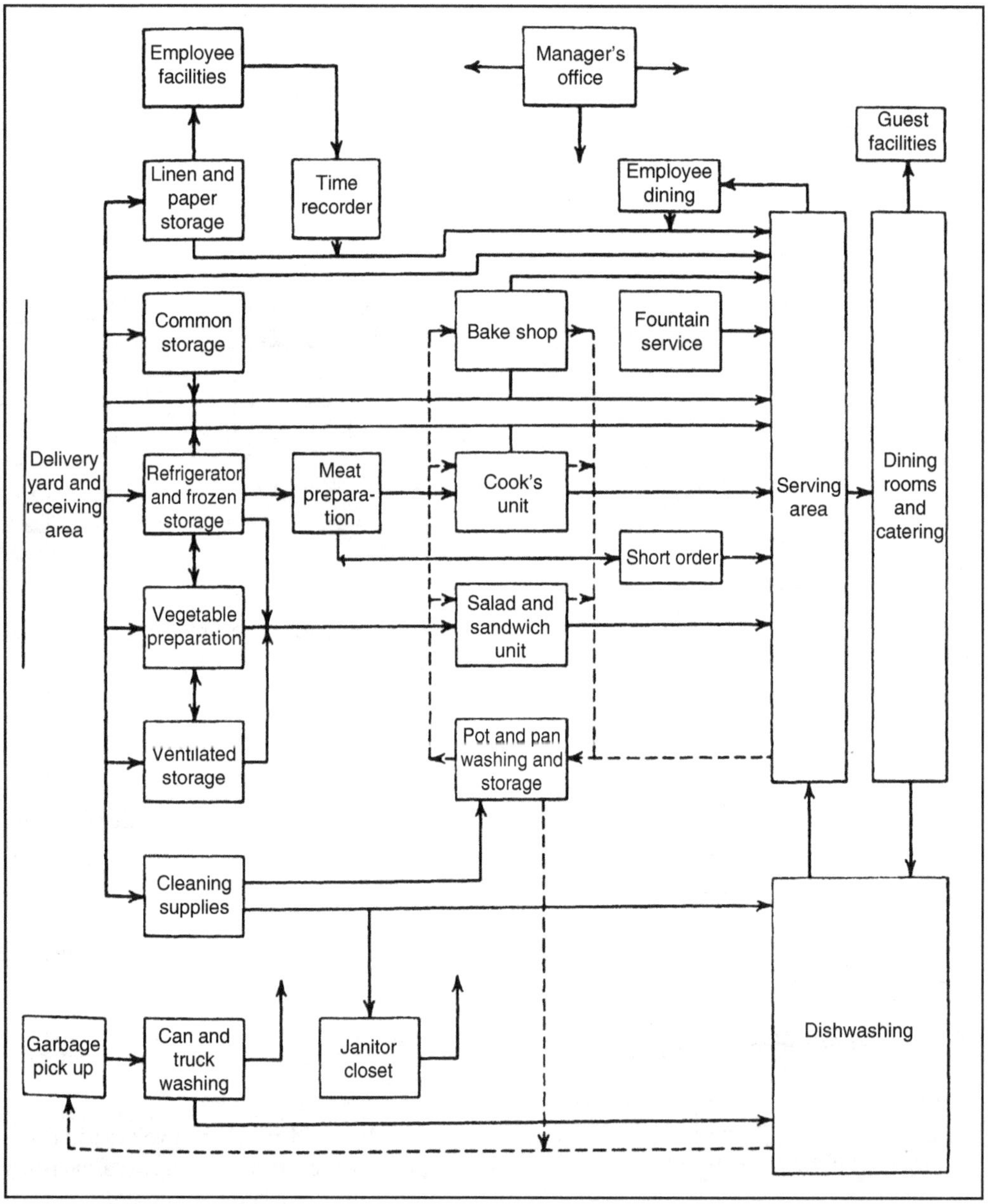

FIGURE 2.44 Functional flow diagram for restaurants.

Additional Support Facilities for Restaurant and Kitchen Employees

1. Restrooms
2. Locker and lounge areas
3. Showers
4. Staff dining area
5. Staff parking
6. Manager's office

FACTORIES AND WAREHOUSES

General

The primary objective in planning factory and warehousing facilities should be to expedite the flow of materials into, through and out of the facility. However, people are also involved, and they either enhance or degrade this objective, depending on how well their welfare and task interface designs are formulated in the initial planning stage. The following general functional requirements are common to almost all factories and should be addressed both individually and collectively during the planning stage:

1. Materials, equipment, and components receiving
2. Checking, sorting, and preliminary storage
3. Materials preparation
4. Manufacturing (cutting, forming, treating, drilling, etc.)
5. Assembly (electrical, mechanical, welding, soldering, etc.)
6. Finishing (anodizing, painting, etc.)
7. Inspection and testing
8. Packing
9. Predelivery storage
10. Delivery
11. Waste disposal
12. Management and supervision
13. Inventory control
14. Communications
15. Maintenance and housekeeping
16. Employee services
17. Security

Among the more important planning considerations are the following:

1. Flexibility for reorganizing spaces to accommodate product manufacturing shifts and expansion of plant facilities
2. Hazard monitoring and control
3. Efficient work flow
4. A nonstressful working environment for workers
5. Security
6. Minimal pollution of the surrounding environment (See Fig. 2.45.)

Special Human Factors Problem Areas

As indicated by the development of the Occupational Safety and Health Administration of the U.S. government (OSHA), personal health and safety are a prime concern. (See Table 2.10.) Whether the factory manufactures heavy equipment, hazardous materials, or toys, the planner should anticipate as many of the hidden as well as the human-precipitated hazards as possible and should plan the facility to prevent these hazards from occurring or to mitigate their possible effects. Equally important, however, is the planning of an efficient working environment so that the workers can do the best job of which they are capable.

Checklist for Identifying Typical Factory Hazards

1. Toxic fumes, dusts, and other substances
2. Chemical contaminants
3. Explosives and explosive material
4. Scalding liquids and steam
5. Falling objects, flying particles, and sparks
6. Moving machinery (including robotic systems), vehicles and parts and swinging cables
7. Cutting tools, machines, and sharp materials and parts
8. Presses and crushers
9. High-voltage electrical equipment and lines
10. Radiation
11. Irregular floors and walkways
12. High noise levels
13. Falls from scaffolds, ladders, and lifts

See also the section on safety in Part 3 of this book.

Checklist for Identifying Worker Efficiency Considerations

1. Visual: Provide proper illumination for each task.
2. Auditory: Provide protection from ear damage, communication interference, and undue annoyance and psychological stress.
3. Biomechanical: Ensure that human-facility interface features are designed to be used within human mobility and strength limits. Special attention should be given to the removal of architectural barriers for the handicapped worker during the initial planning stage.

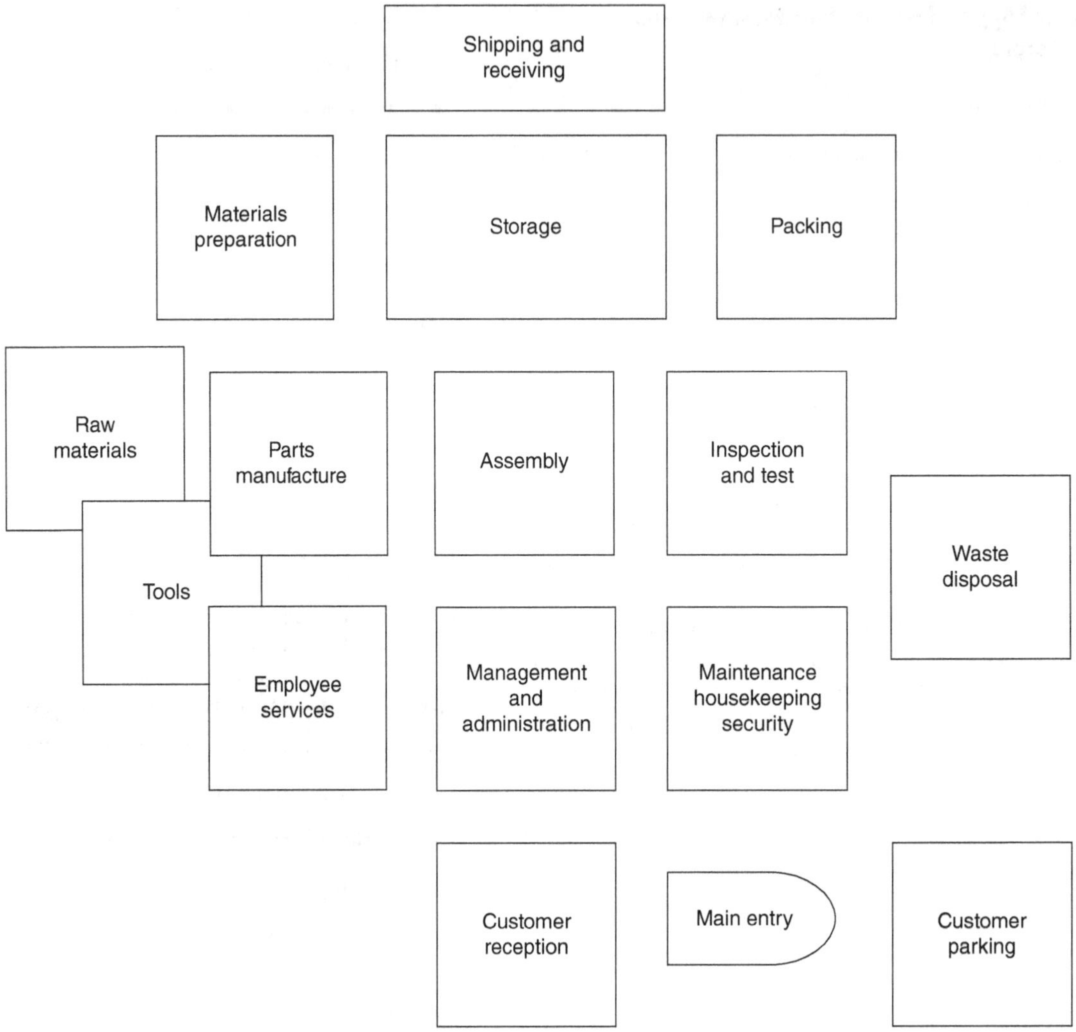

FIGURE 2.45 Factory and warehousing facilities planning model.

TABLE 2.10 Topics Covered by Occupational Safety and Health Act (OSHA)

Part 1910 Subparts:
D - Walking-Working Surfaces
E - Means of Egress
F - Powered Platforms, Manlifts, and Vehicle-Mounted Work Platforms
G- Occupational Health and Environmental Control
    Ventilation
    Noise
    Non-Ionizing Radiation
H - Hazardous Materials
I - Personal Protective Equipment
J - General Environmental Controls
K - Medical and First Aid
L - Fire Protection
M - Compressed Gas and Compressed Air Equipment
N - Materials Handling and Storage
O - Machinery and Machine Guarding
P - Hand and Portable Powered Tools and Other Hand-Held Equipment
Q Welding, Cutting and Brazing
R - Special Industries
    Pulp, Paper, and Paperboard Mills
    Textiles
    Bakery Equipment
    Laundry Machinery and Operations
    Sawmills
    Logging Operations
    Telecommunications
    Electric Power Generation, Transmission, and Distribution
    Grain Handling Facilities
S - Electrical
T - Commercial Diving Operations
Z - Toxic and Hazardous Substances

From Code of Federal Regulations, Title 29, Part 1910, Occupational Safety and Health Standards.

4. Cognitive: Ensure that the physical features of the facility are compatible with typical human expectancies and mental capacities for comprehension.

5. Energy Consumption: Create a plan that conserves human energy.

6. Working Environment: Provide an environment that will allow workers to operate within their most productive range.

SPORTS FACILITIES

Sports Facilities Planning Considerations[4]

Although the sports fan might be considered the primary user in terms of planning, numerous other groups of people have equally important needs. The major user categories to be addressed in planning the sports facility include the following:

1. Fans
2. Special guests (VIPs)
3. Officials
4. Athletes, coaches, and staffs
5. Facility management staff
6. Vendors
7. Security staff
8. Maintenance and housekeeping staff

Key subsystems that ordinarily will be associated with any major sporting facility include the following:

1. Parking lot
2. Ticket booths
3. Main stadium or arena structure, corridors, stairways, lifts, restrooms, lounges, dressing rooms, press box, score and information board, offices, meeting rooms, utility and engineering spaces, and vendor and storage spaces
4. Playing area (may require special, alternative facilities for different events, e.g., basketball, hockey, tennis, football, swimming, or track)

Important planning considerations are as follows (see Fig. 2.46):

1. Public access, traffic control, and parking
2. Public seating accessibility, comfort, and visibility of the playing area
3. Participant accommodations
4. Facility housekeeping and maintenance
5. Security

[4] J. De Chiara and J. H. Callender, *Time-Saver Standards for Building Types*, McGraw-Hill Book Company, New York, 1973.

Special Human Factors Problem Areas

1. Access to and from the facility parking area often is a major problem, not only for the sports fans, but also for the planner, who may not have control over related community thoroughfares and property. Failure to consider this problem early in the planning stage will invariably result in either considerable cost to modify existing roadways or both expense and disgust on the part of fans when a poorly planned access creates long and time-consuming traffic jams. An important objective should be to create multiple approaches to reduce the traffic load at any one entry and to lay out the site-roadway configuration to prevent traffic conflicts near the site.

2. Parking sites for major facilities generally have to be quite large to accommodate the typical vehicle load. If the parking pattern is not clear to drivers, considerable congestion can result, and many drivers will become lost out on the periphery. An objective should be to create a plan in which traffic flows logically and cannot be circumvented by the motorist who is anxious to get parked first.

3. Access from the parking area to the stadium or arena entrance and ticket windows is often extremely confusing to the new fan, who may have difficulty seeing the entrance from the point where his or her car is parked. An obvious objective, of course, should be to make these distances as short as practical. However, an equally important consideration is to provide obvious pedestrian pathways to the entrance (a convenience missing from most plans). Finally, the stadium, parking area, and pedestrian pathways should be well signed, not only so that fans can find the entrance, but, equally as important, so that they can find their cars again. A good addition is to slope the parking area upward toward the stadium, which gives people who are out on the periphery a clear view of the facility and of any instructional signs that have been placed at various locations for their benefit.

4. Seat location can become confusing and difficult for the fan since the tendency is to make all corridors, ramps, stairs, and doorways look alike. Consider the problem of clear signing during the planning stage, since this forces one to see the relationship between building configuration and how to make it clear to fans what they are supposed to look for and do to find their proper location. Color coding is a useful technique to consider.

5. A good view of the playing area is, of course, very important to the fan. Also consider the view of the scoreboard, which can provide a significant amount of information as well as entertainment. Early consideration of viewpoints can eliminate the problem of the unfortunate fan who has to sit behind a pillar or who cannot see over the head of the person in front. Seating comfort can be provided by following prescribed criteria for a good seat design. Too many planners fail to consider seat design

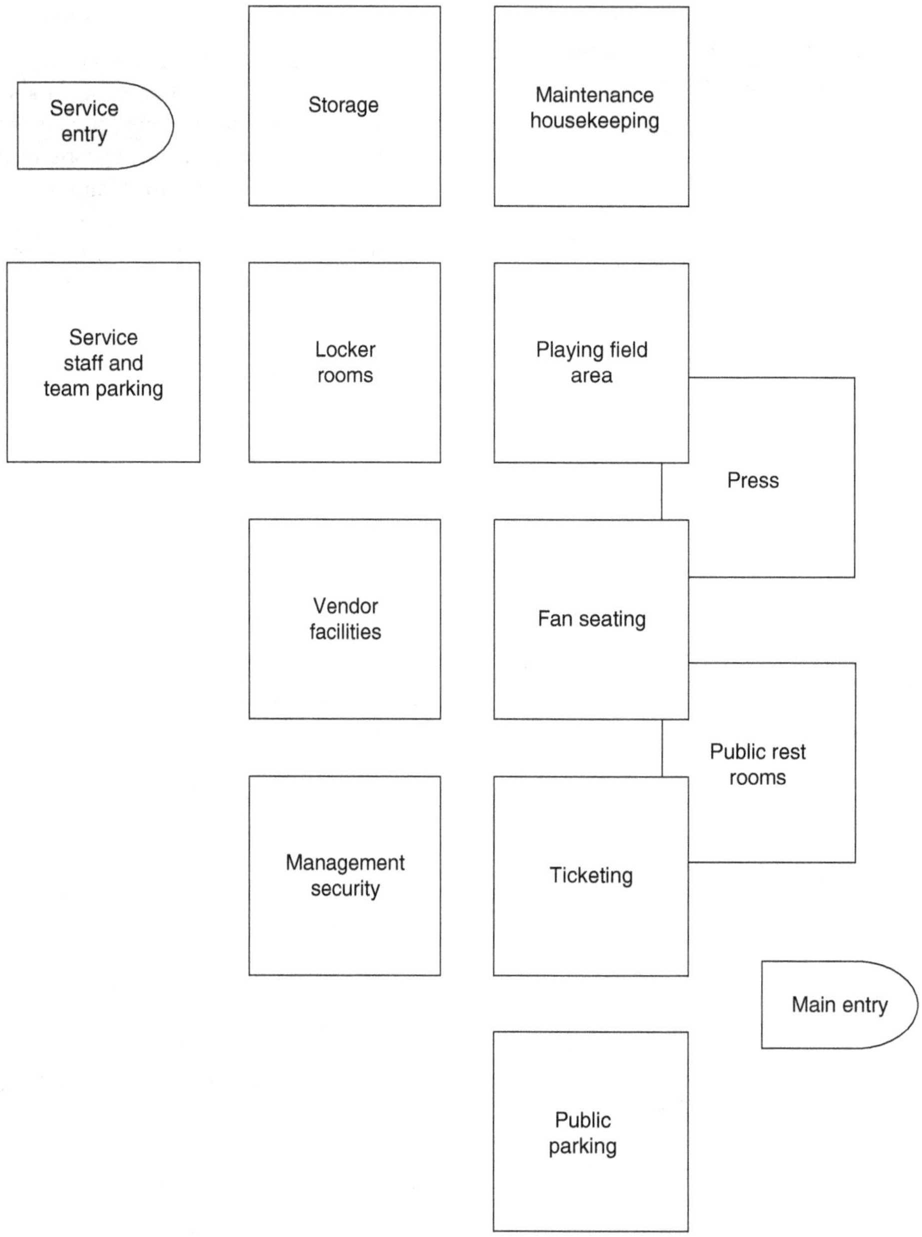

FIGURE 2.46 Sports facility planning model.

and purchase seats purely on the basis of low cost and ease of maintenance (also important issues). In seeking effective seat-view configurations, avoid creating steep stairways; these can become hazardous, especially for the excited fan who insists on running up and down them from time to time.

6. Special attention should be given to the handicapped fan who cannot use stairways or sit in general seating areas. Provide elevators, ramps, special wheelchair areas, and, of course, restrooms designed for the handicapped.

Vehicular and Pedestrian Flow Planning Examples

1. Street traffic does not tie up because of crossovers.

2. Parking and aisle system lends itself to easy identification and marking.

3. Customer parking area designations are compatible with seating sections.

4. Basic stadium and arena shape is rectilinear and oriented with north–south reference; useful for seating identification system.

5. A-sections at main entrance keep fans oriented.

6. Parking for private vehicle of fans is separate from parking for buses, parking for administrative personnel, and other special parking.

Figures 2.47 and 2.48 show how vehicular traffic can be controlled in terms of being directed to specific areas with minimum loss of control. In addition, specific pedestrian pathways (sidewalks) are provided to minimize the hazards created by fans who walk among parked and moving

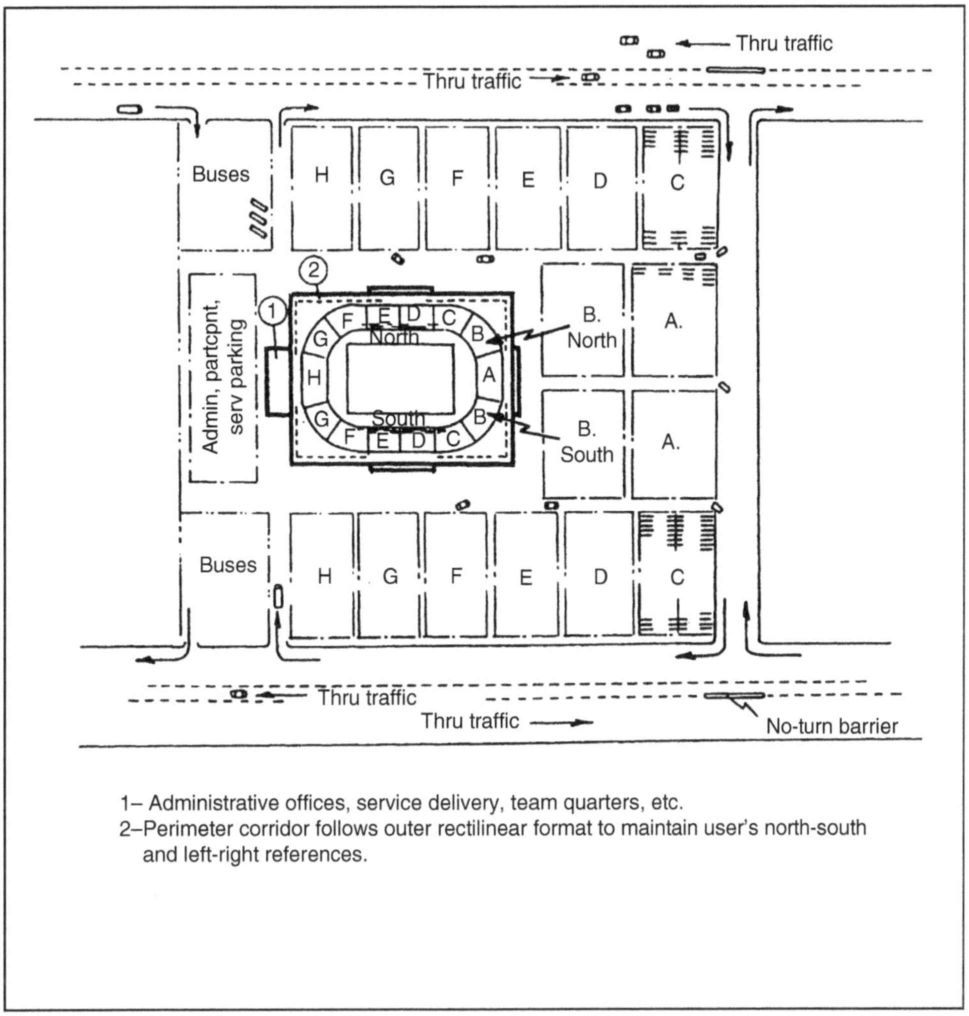

FIGURE 2.47 Sports facility traffic and parking concept.

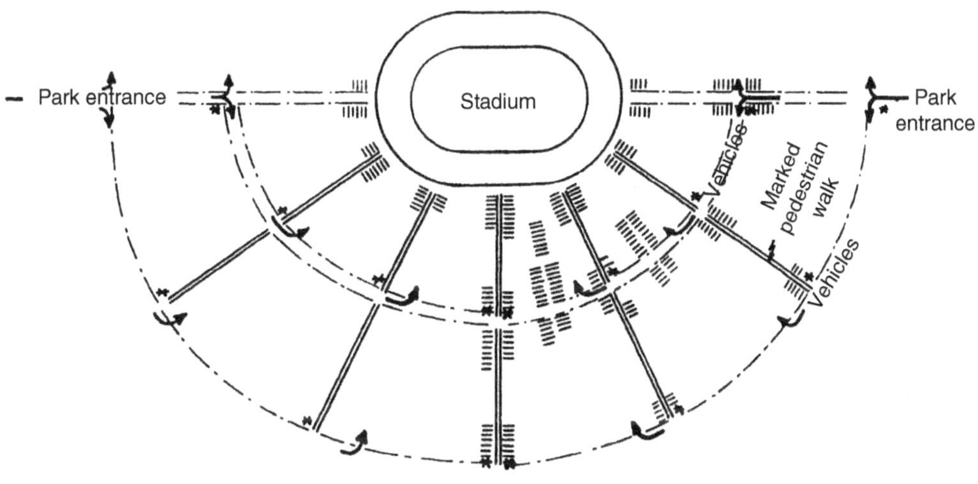

FIGURE 2.48 Sports facility traffic flow concept.

vehicles. Note that the vehicular–pedestrian geometry provides a natural flow to and from the stadium.

Temporary barriers can be placed at the points marked with an asterisk to prevent drivers from straying beyond the intended parking area being filled at a particular moment. Special traffic, such as press, participants, and vendors, can be routed on through these points at the discretion of parking-lot attendants.

Special Storage Considerations for the Sports Facility

Since many sports facilities must be designed for multiple uses (sports, conventions, circuses, musical shows, etc.), it is important to consider the storage requirements of all these events during the planning stage. The following are some of the items that must be not only stored but also handled readily during performances and other events:

1. Personal clothing and articles of participants
2. Athletic clothing and equipment
3. Special costumes
4. Musical instruments and music stands
5. Scenery, curtains, and modular booth accessories
6. Extra seats and chairs and special furniture
7. Special wheeled equipment (floor cleaners and ice sweepers)
8. Special field and court equipment (basketball goals, field goals, and equestrian equipment)
9. Special sound equipment, scoreboards, TV equipment, etc.
10. Records and money
11. Vendors' supplies
12. Advertising materials and equipment
13. Electrical accessories and lighting equipment
14. Internal and external maintenance materials, tools, and equipment
15. Special parking-lot accessories (signs, markers, and stanchions)
16. Shop materials (lumber and other materials used to make special stage props, field and court barriers, etc.)

Lighting

Consult appropriate lighting handbooks and/or architectural guides for lighting requirements for the various elements of the sports facility, as follows. Visual task and lighting principles are also discussed in more detail in Part 4 of this book.

1. Parking
2. Building exteriors
3. Stairs and corridors
4. Locker rooms and restrooms
5. Sporting events

Safety

General safety considerations for all categories of sports facilities include poorly designed and/or improperly lighted corridors, stairways, ramps, and escalators, which may cause people to trip or fall; lack of proper separation of pedestrian and vehicular traffic, which may lead to pedestrians' being run down by a vehicle; and crowding, which may lead to personal entrapment and/or injury during a panic exit due to the layout of escape routes, exit doors, etc.

Special consideration should, however, be given to the following possible safety hazards as they apply to specific sports facility users.

Spectators

1. Steeply angled bleachers
2. Railings with sharp corners and/or lack of handrails
3. Lack of protective barriers to keep players and other objects from entering the bleacher area
4. Narrow aisles between seats, which make it difficult to walk past someone who is sitting along the row of seats
5. Lack of adequate illumination on bleacher stairways

Sports Participants

1. Sharp or unpadded field or court features into which players may run or fall
2. Lack of barriers between players and sideline participants or onlooker
3. Improper court or field illumination, e.g., bright light that may cause a player to run into another player or a hazardous object
4. Slippery locker-room flooring

All Parking-Lot Drivers and Pedestrians

1. Lack of properly defined vehicle and pedestrian separation in terms of fences, walls, curbs, etc.
2. Guy wires across the path of a pedestrian who may be blinded by stray headlights or field lights
3. Lack of adequate parking-area illumination, which may cause pedestrians to trip over irregular surfaces, berms, etc.
4. Lack of illuminated signs to guide both pedestrians and drivers so that they are less apt to cross each other's path

First-Aid and Medical Facilities. All public sports facilities should have an appropriate number of strategically located first-aid stations to provide assistance to both spectators and sports participants.

AMUSEMENT CENTERS

Amusement Center Planning Considerations

Although the variety of amusement center concepts is almost limitless, certain critical user functions are fairly common and should be addressed as seriously as the primary theme of the center. The following are among the most important considerations:

1. Site access for both private and public transportation

2. Adequate and well-planned parking for both patrons and center operations personnel

3. Entrance, ticketing, and information to accommodate maximum patron load

4. A site layout that provides sufficient organizational clarity to ease patrons' problem of keeping themselves oriented throughout their stay at the center

5. An adequate number of easily located patron support services such as restrooms, dining facilities, personal storage facilities, drinking fountains, rest seating (with shade), telephones, a lost-and-found center, and informational signs

6. Adequate security for both patrons and center personnel (including their facilities, equipment, etc.) so that patrons cannot "wander behind the scenes"

7. Accessibility for handicapped patrons

8. Safety (especially where special exhibits, rides, or animals could create physical hazards)

9. Damage-resistant as well as wear-resistant design throughout

10. Site organization and structural and materials design that will minimize the time necessary to maintain the facility

11. Examination of queuing strategies for various anticipated exhibits, rides, entertainment, etc., during the initial planning stages. Consider the use of electronic paging of small queue groups.

12. Consideration of possible natural or man-produced disasters and the requirements for rapid egress of both patrons and center personnel, the use of disaster control equipment, and the problem of crowd control

13. Consideration of the effect the center may have on adjoining communities, i.e., noise, glare from lights, traffic, etc.

14. Consideration of potential expansion possibilities so that none of the above factors become distorted by the add-ons. (See Fig. 2.49.)

The more important considerations include the following:

1. Site access and parking

2. A logical layout to ease personal location, crowd control, housekeeping, and maintenance

3. Personal security and safety

4. Crowd control

Design for Safety

Although architectural planners may not have control over all the special amusement center equipment, such as special rides and exhibits, they should have control over, and therefore planning responsibility for, permanent features that relate to the hazards that may be associated with some of these special equipment features. Consider the following:

1. Wherever animal exhibits are part of the center plan, provide appropriate barriers to keep customers away from the animals and, obviously, to keep the animals from escaping.

2. Whenever special rides such as a merry-go-round, a Ferris wheel, or similar devices are expected, provide appropriate barriers to keep children from getting too near the moving elements and possibly being struck by the device.

3. Plan for permanent, safe electrical power transfer for rides, concessions, and other center exhibits so that concessionaires do not lay loose cables where patrons can trip over them and so that children cannot come in contact with high-voltage boxes or connections.

4. Provide smooth walking surfaces throughout the center. These should be permanent (i.e., paved), nonskid, and well drained.

5. Provide separate pathways for center maintenance and delivery vehicles so that they do not mingle with the customer traffic, either on the main grounds or on approaches to the grounds.

6. Provide adequate illumination in all areas, for both customers and operating personnel, so that they do not run into barriers, guy wires, or other objects in their path. Since most youngsters tend to stray from the normal pathways, other areas should be made inaccessible or should be illuminated.

7. If water features are planned (e.g., ponds, lakes, or waterfalls), lay these out so that small children (who are apt to stray from their parents) cannot fall into the water.

8. Provide strategically located and easily identifiable first-aid facilities for emergency treatment of both customers and operating personnel.

9. When laying out the center, consider the possibility of an emergency disaster requiring the grounds to be evacuated quickly and safely. This may require consideration not only of customers but also of other personnel, animals, etc.

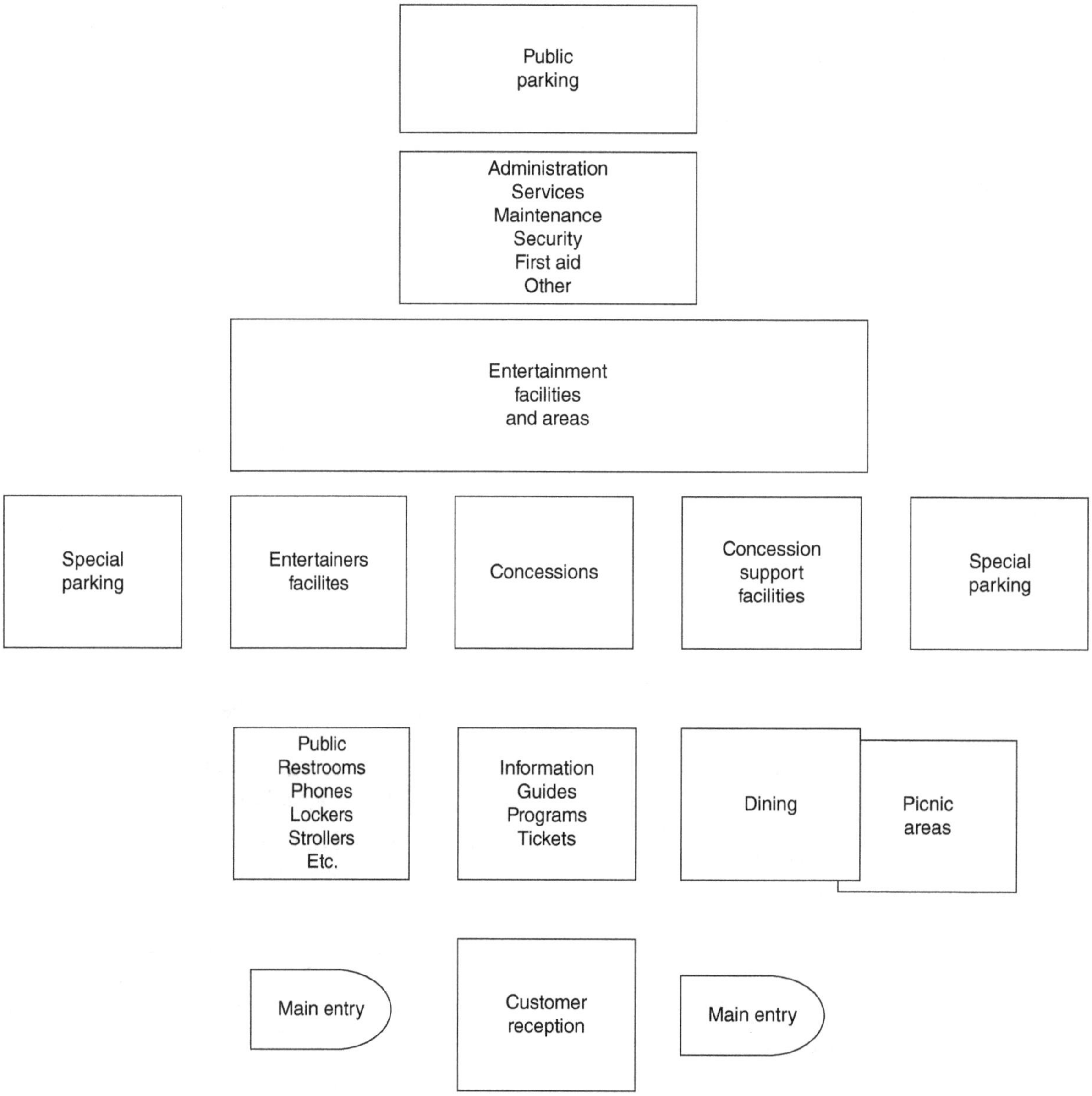

FIGURE 2.49 Amusement center planning model.

PARKS

Public Park Facilities Planning Considerations

The public park, of course, varies from the small urban or residential community recreational space to the large, semi-wild or wilderness space where hunting and camping can be accommodated. This section treats some of the human factor considerations common to all out-of-door recreational facilities.

Entrances and Exits. Typically, public park users would like complete freedom to enter or leave the park whenever and wherever they please. Unfortunately, the planner must seek a compromise between letting users have as much of this freedom as possible and exercising sufficient control to minimize such problems as protecting park facilities and natural features from destruction or damage, protecting park users against traffic hazards, and minimizing difficulties associated with park surveillance and maintenance. Thus, it is desirable to define park boundaries either by the natural barriers available or by means of man-made barriers and appropriately designed entrances and exits. Even the small neighborhood park benefits from a decorative fence that will prevent children from inadvertently running into the street and which will keep stray animals from injuring children inside the park. Larger parks should have guarded entrances and gates in order to control who and what goes into the park and what is taken out. The very large wilderness park is, of course, another matter. Although primary guarded entrances are still appropriate, their benefits depend entirely on the capacity of the natural barriers to encourage use of the designated entrances.

Roadways and Parking Areas. The two major problems regarding most park roadways relate to the adequacy of roadways to allow for passage of vehicles, especially large recreational vehicles, and to maintenance of safe speeds on these roadways (a major problem with the advent of dune buggies, motor bikes, etc., which, the riders or drivers believe, are meant only for racing). In the latter case, consideration must be given to the use of speed barriers (berms placed across the roadways at periodic intervals and/or regular police patrol). It is also suggested that, in planning the entrances and exits and the inner park roadways, the geometric relationships between the exterior roads or streets and the immediate inner roadway be such that people will not be tempted to use the park as a thoroughfare, i.e., to travel through it only to get from one side to the other, with no intention of enjoying the park as a facility.

Restroom facilities. With the exception of the wilderness park, most park facilities should be provided with an appropriate number of properly designed restroom facilities. Such facilities should be designed to require a minimum of maintenance and should be constructed and equipped to minimize the possibilities of misuse and vandalism. Additionally, restrooms should be accessible to the handicapped; i.e., they should have smooth approaches and large doorways, and there should be appropriate fixtures and handrails.

Lighting. In spite of the desire to leave parks as natural as possible, they are used by people who need to see. In addition, the urban park benefits from good lighting in terms of reduced vandalism and criminal activity. Of course, it is also important that the method for lighting itself be as vandal-proof as possible.

Signage. The question of signage should be addressed during the early stages of park planning so that the signage system can successfully explain or identify areas, roadways, etc., for the individual who may not know how the park is laid out or what facilities are available and where they are. Although it is desirable to design signs that are compatible with the rustic environment, this should not preclude their being legible and understandable to the observer. Consider the entire signage system at the beginning, including instructions, identification, and guidance information.

Park Service and Maintenance

The following list of possible park services should be reviewed during the planning stage in order to approach the accommodation of such service in the most cost-effective manner.

Park Patron Services

1. Picnic tables
2. Fire pits and barbecue pits or stoves (including gas)
3. Trash receptacles
4. Water outlets
5. Emergency communications and first-aid stations
6. General provisions store
7. Park benches (neighborhood park)
8. Prepared hiking or walking paths, bridges, stairways, etc.
9. Shelters for use during inclement weather
10. Prepared campsites (including flat areas for tent, vehicle, cooking equipment, and dining table and seats)

For large parks where considerable camping is planned, it is recommended that a special area be set aside for the handicapped user. This area should be as flat as possible, and/or appropriate ramps and pads should be provided so that persons using wheelchairs can get about easily within each campsite and can go to and from nearby restroom facilities.

Park Management Services. A variety of management services should be considered for the large park or recreational area, including the following:

1. Park ranger residence
2. Park ranger entrance and exit station
3. Service, fire, and inspection roads (not generally used by the public)
4. Park housekeeping (weed cutting, tree trimming and removal, trash pickup and disposal, etc.)
5. Facilities maintenance (road repair, fence repair, utilities repair, etc.)
6. Security (traffic control, crime prevention, and animal control)
7. Fire control, including provision of appropriate fire-fighting vehicles, equipment, and storage
8. First-aid station, equipment, and provisions
9. Landscape maintenance (watering, trimming, etc.)

Special attention should be given to the design of park facilities in terms of the ability of the materials selected to withstand frequent and continued use with minimum wear, damage due to misuse or vandalism, weather effects, and, of course, flammability. In addition, there is no reason why park facilities should not also be functionally convenient and comfortable. Avoid the tendency to believe that a rustic design cannot be convenient or comfortable. Also do not assume that most people want to preserve the beauty of your park and therefore will respect the natural and man-made facilities provided. In fact, most users will be inconsiderate, if not disrespectful, and careless at one time or another. It is these latter considerations that must establish the basic design goals even though they may compromise the more desirable functional goals of providing the most natural and enjoyable environment for those park patrons who will normally use the park as it was intended to be used.

TERMINALS

Terminal Facilities Concept Development

Although each type of transportation requires slightly different facilities and site conditions in order to accommodate the particular prime vehicular system (airplanes, buses, trains, ships, etc.), certain basic requirements should be considered for all terminal concepts.

For Passengers and Their Guests

1. Site access (by foot, private automobile, taxi, or bus)
2. Public parking for private vehicles (both short- and long term)
3. Vehicle-to-terminal transfer (from the curbside or from a parking lot)
4. Ticketing and baggage check
5. Security checks to prevent sabotage
6. Waiting (i.e., seating, temporary storage for personal articles, restrooms, dining or snacking facilities, information centers, and arrival and departure communications)
7. Transfer from terminal to prime transportation vehicular system (gates, ramps, mobile lounge, etc.)
8. Passenger assistance (porters, baggage carts, strollers, wheelchairs and carts, telephones, and public address system)

For Terminal Operations Personnel

1. Site access
2. Private employee parking and terminal access
3. Lounges and storage for personal articles
4. Administrative offices
5. Housekeeping and maintenance offices, shops, storage, and delivery facilities and parking

For Terminal Tenant Operations Personnel. Includes prime vehicular system operators, on-site government operations, rental-car agencies, social service agencies, public transportation agencies (e.g., a local bus or taxi), and concessionaires (a bank, barber shop, gift shop, restaurant, etc.)

1. Site access (service)
2. Special parking (personnel and service vehicles)
3. Primary public interface facility (internal and external)
4. Offices, storage, delivery facility, kitchen, etc.
5. Special prime vehicle services (docking, refueling, baggage handling, maintenance and repair, equipment storage, etc.)
6. Personnel facilities (lounges, locker rooms, toilet, conference rooms, etc.)
7. Dispatch and shipping facilities
8. Systems control (air traffic control tower and center, weather office, pilots' ready rooms, etc.)

Critical Considerations. The following are among the most critical considerations:

1. Efficient movement of passengers, baggage, and cargo
2. Rapid and efficient turnaround for the primary vehicular system
3. Effective 24-hour terminal facilities maintenance (See Fig. 2.50.)

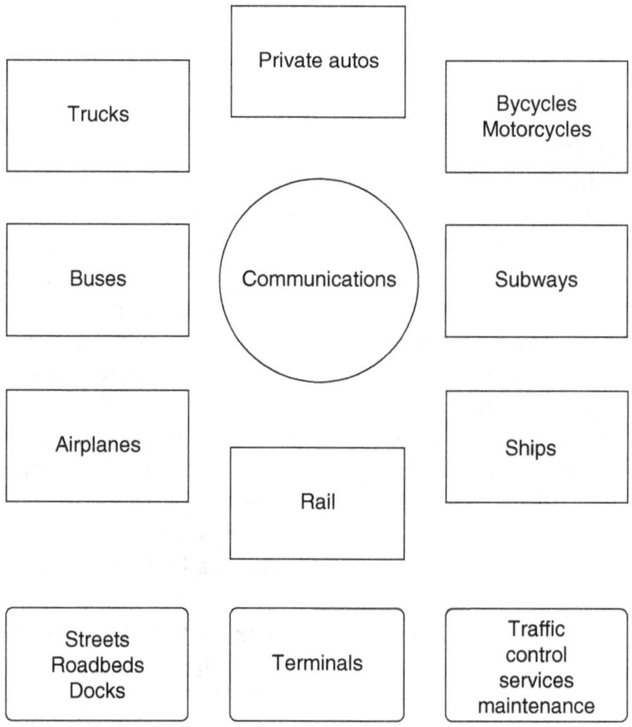

FIGURE 2.50 Transportation systems terminal facilities model.

Site Planning

The key human factors considerations to be addressed in site selection and planning are the following.

Accessibility. The site should be as nearly equally accessible to all expected users as is practical. This requires careful assessment of the points from which potential passengers will emanate and how they will probably get to the proposed site, i.e., via private or public vehicle, over what streets or roadways, etc. One should consider not only passengers but also other users, such as shippers of materials and equipment and operators who service the terminal.

Site Configuration and Size. The site shape and size must accommodate the particular prime vehicular system operator's needs in terms of the nature of the equipment operation. This requires careful assessment of the operating characteristics of the operator's equipment; e.g., airplanes need runways, docking ramps, etc.; railroads need tracks and switching and docking facilities; and bus and truck operators need parking, docking, and maneuvering facilities. However, in addition to the obvious concern for the prime vehicular system operator's needs, careful attention must also be given to the other side of the system: passengers, shipping clients, and terminal operators. For example, one should be careful not to allow concern for an airline operator's problems of maneuvering and docking airplanes create impossibly long and difficult transfer and walking problems for passengers.

Community Interaction. The location of a terminal site should be compatible with the general environmental needs of the community in terms of traffic congestion, noise, safety, and air pollution. Consideration should also be given to the interactions between the proposed terminal and other terminals in terms of roadway interconnections, local public transportation, and the effects that time and distance have on the interchange of passengers between terminals.

Terminal Facilities Layout. The proposed site should create the least amount of constraint on the layout of terminal facilities in terms of spacing buildings and orienting roadways, track, or runways. A common problem of many large terminal sites is that the odd shape of the original site creates a confusing maze of pathways and building arrangements. A great many of the eventual site users will be unfamiliar with the site and may have time constraints, and there may be environmental conditions that will further add to their possible misorientation when site layouts are not obvious and logical to them.

Terminal User Facility Orientation and Identification

The most important human factors considerations in terminal concept formation relate to the ability of the terminal user to identify elements of the facility and remain oriented as he or she approaches, enters, moves about, and departs. Although this would appear an obviously easy set of requirements to fulfill, a majority of terminals fail miserably in this regard. That is, the design creates confusion. The following scenarios illustrate where and when the need to identify elements of the terminal and become oriented is important to the first-time or infrequent visitor to a transportation terminal.

Approaching the Site. The critical user is usually one who is driving to the terminal site. Although this user probably has identified the general site location on a road map with respect to some major highway, and thus can reach the general terminal area easily, unless the street or highway markings are clear and/or the terminal building itself is visible from the highway, he or she may have considerable difficulty getting from the highway to the terminal building area. Site planning should include a direct and well-signed terminal-site roadway from the principal highway or street that terminal visitors will use.

Going from the Terminal-Site Entrance to the Terminal Building and Other Facilities. Again the critical user is driving his or her own car or a rental car to the terminal. The route from the terminal-site entrance to the terminal building, a public parking area, or a rental-car return area should be as direct and uncluttered by intervening interruptions (intersections, other buildings, other signs, other parking areas, etc.) as possible. In other words, the layout and signing for this portion of the user's terminal approach should leave as little doubt in the user's mind as possible as to whether he or she is on the right track. It is desirable to create a site layout that makes it possible for individuals to keep the terminal building in sight throughout the approach and in which the roadway does not gyrate through many curves or turns.

Going from the Parking Facility to the Terminal Building. The critical user is one who is pressed for time. Thus, it should be possible for this user to keep track of which way to go, after leaving his or her car in a parking lot. Either plan a parking transportation system that will take the user directly to the terminal building entrance or locate the parking facility so that the user can keep the terminal building in sight. The distance to the terminal building is obviously important if the user is to walk (carrying luggage). If an enclosed parking garage concept is to be used, careful attention should be given to the marking and signing so that the user can easily locate the signs. In deciding on a parking facility concept (i.e., whether it is to be a lot arranged in a certain way or a parking structure that has various levels and ramps), consider carefully the method by which the facility is to be encoded and marked. Avoid the use of abstract alphanumerics, the logic of which generally is not readily apparent to the visiting observer.

Becoming Oriented to the Terminal Building. Terminal buildings, especially large airport terminals, have a way of becoming very confusing to the casual user. This is often due to the fact that, in laying out the building, the designer gave priority to the space demands of parked aircraft. When aesthetic whims are added, such buildings become almost impossible to perceive as a logical and systematic geometric pattern. The key orientation and identification considerations for building conceptualization should be the following:

1. Entrances should be readily identifiable, not only by the signing, but also by the geometric features. All pathways should lead directly to these entrances; users should not have to take paths that turn, jog, circle, or curve.

2. Entrances should open directly to the things the user expects to see; i.e., as the user enters the building, he or she should have a direct view of the ticket counters, the information desk, indications of where the gates are, the restrooms, etc.

3. There should also be obvious placement of those interior features that a boarding passenger will need if he or she has to wait for a flight, i.e., the waiting-room area or areas, dining room, bar, snack bar, newsstand, etc. Concessionaire stands can be located in a variety of places since passengers who have a longer wait can look for these at their leisure.

4. Disembarking passengers have the same needs as those of boarding passengers, except in reverse. The most critical scenario is that of disembarking passengers who must make a quick connection between the plane they left and the one they have to catch. The arrangement concept should minimize the problems of getting out of the gate area of one airline and into the gate area of the other—preferably without having to go all the way back to the originating ticket-counter area of the terminal. An especially important layout consideration is the separation of areas for incoming passengers and reception areas for people who are meeting debarking passengers. Traffic jams occur when the design allows these people to block the path of passengers who are in a hurry to get to the next departure gate.

Terminal-Site Planning (Air Terminal)

1. Regardless of required runway axis (due to prevailing winds), plan the layout of terminal buildings and the approach to be "squared" with a principal highway. (See Fig. 2.51.)

2. Roadways into the terminal area should be organized in rectilinear patterns, and incoming traffic and departing traffic should not be forced to cross each other. (See Fig. 2.53.)

3. Plan for possible expansion. Do not allow industrial or business facilities to encroach on the approach areas in such a way as to preclude later terminal expansion or obstruct the view of the terminal for patrons who are approaching it. (See Fig. 2.52.)

Terminal-Site Planning (Bus and Rail)

Unlike air terminal sites, the bus or rail terminal is usually located within the city and is often constrained by adjacent street and building patterns. However, the following basic considerations are still important to the ease with which passengers and visitors can find their way around:

1. Provide sufficient terminal building setback so that site visitors are able to get off the street while they reconnoiter to determine where to deposit or pick up passengers at the curbside and/or where to locate the public parking lot.

2. The site drive should be laid out so that entering and departing vehicles do not conflict with one another.

3. Separate busline vehicle entry and departure drives should be provided so that these vehicles do not conflict with patron or visitor vehicles, city buses, or cabs.

4. Provide for taxi waiting lines.

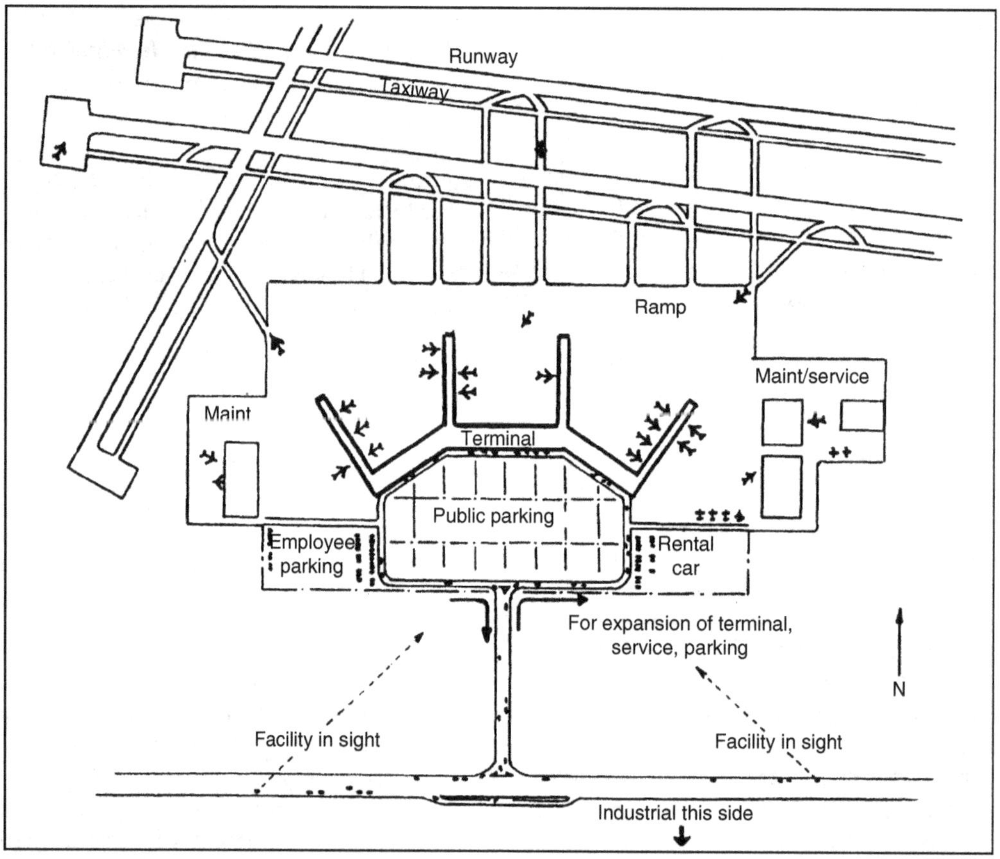

FIGURE 2.51 Example of good terminal site planning.

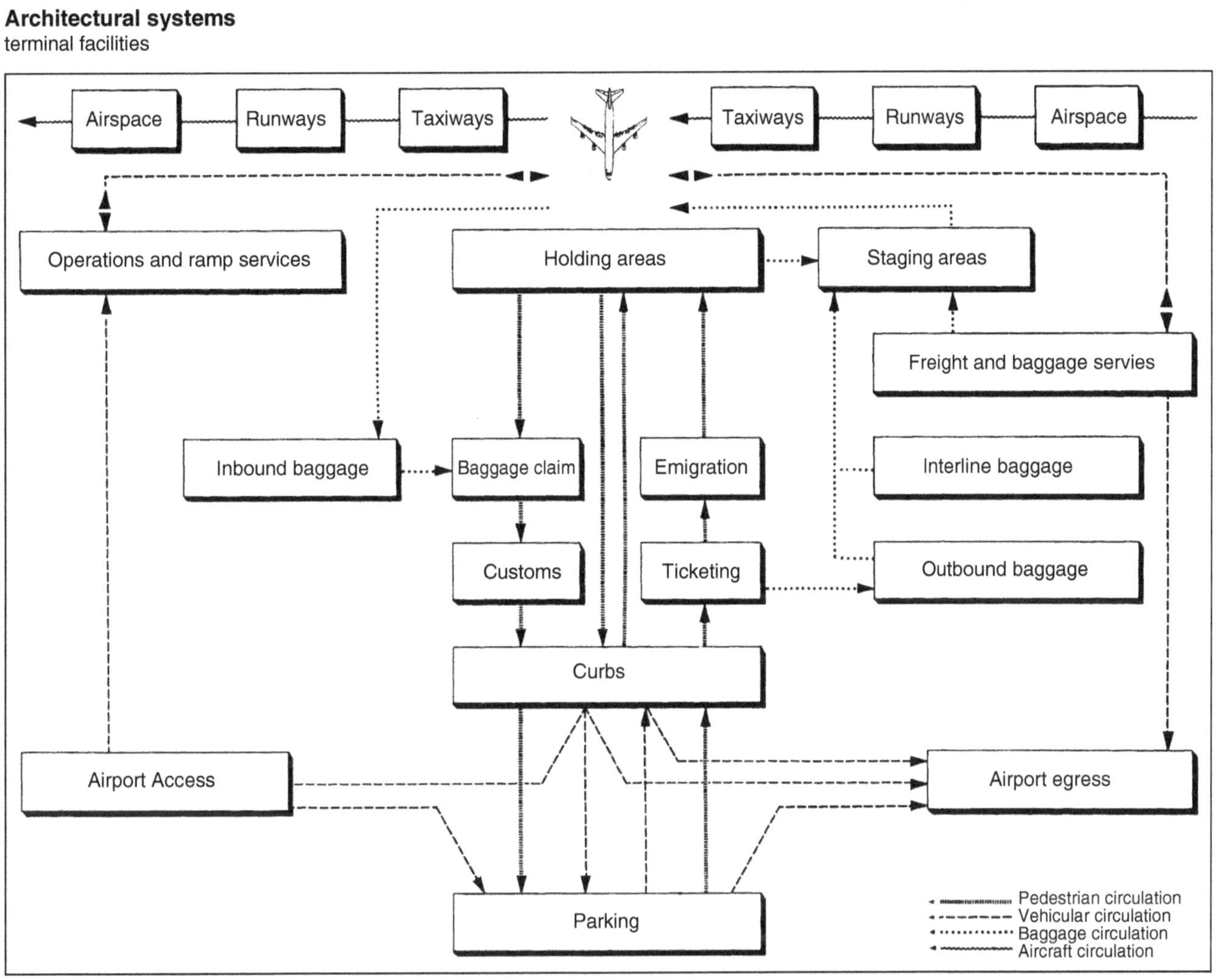

Figure 2.52 Air terminal activity circulation model.

5. Provide separate roadways and parking for service and freight vehicles so that these vehicles do not intermix with patrons' vehicles more than necessary. Whenever possible, do not locate freight facilities so that service trucks enter the site from the same side as vehicles proceeding to the passenger terminal building.

Selected User Interface Considerations for Initial Terminal Facility Conceptualization[5]

A diagonal ticket-counter concept helps minimize interference between passenger queues and other persons who are walking, standing, or milling about in the ticketing area. (See Fig. 2.54.)

Entrances, Exits, Corridors, and Risers

1. A sufficiently broad sidewalk is required outside main entrances in order to accommodate incoming passengers, baggage, baggage carts, passengers who are waiting for transportation, and people who are walking from one entrance to another or to a waiting vehicle. A minimum of 15 ft (4.6 m) should be provided. An overhanging roof should protect passengers from rain as they transfer from a vehicle to the sidewalk.

2. Extra-wide double doors should be provided so that arriving and departing passengers do not conflict with one another. An automatic tread-actuated door system that opens both doors at once is recommended, with a clear opening of at least 10 ft (3.1 m). It is recommended that more than one entrance be provided to help distribute passengers among the various ticket booths inside the terminal building.

3. When a two-level structural concept is practical, consider separating arriving and departing passengers so that one group transfers on one level and the other group on the other level.

[5]For a more detailed discussion, see J. De Chiara and J. H. Callender, *Time-Saver Standards for Building Types*, McGraw-Hill Book Company, New York, 1973.

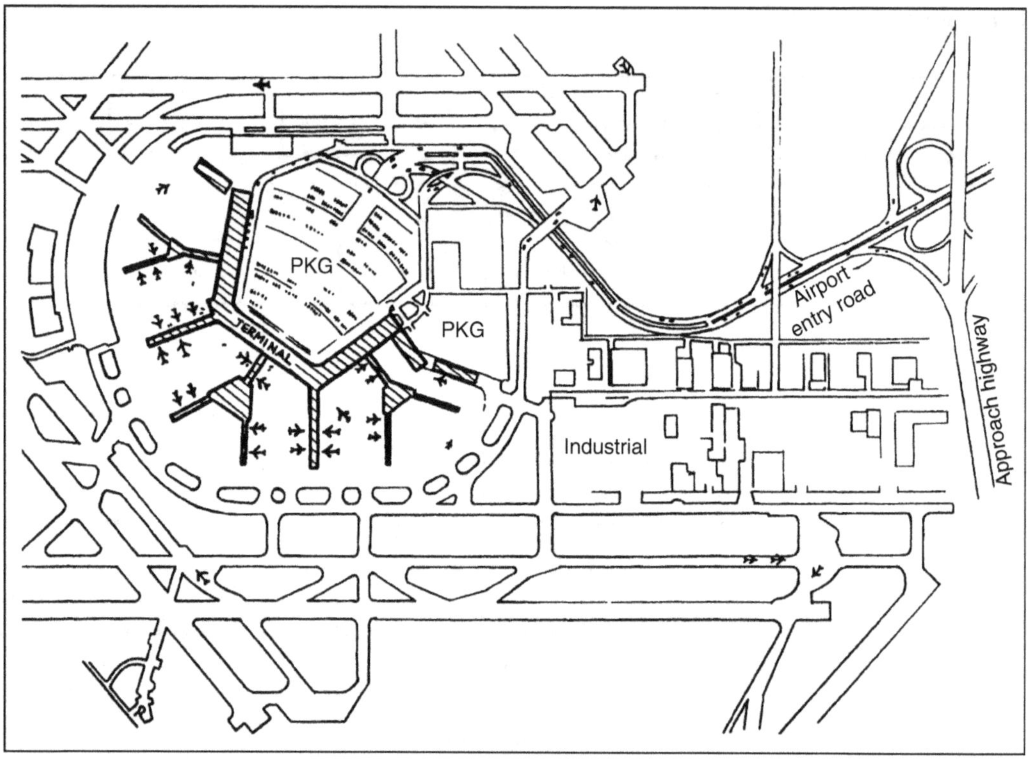

Figure 2.53 Example of poor terminal site planning. This terminal-site layout creates utter confusion for the traveler who must approach or leave the area for the first time. Note especially the lack of north-south geographic reference, the frequent turns from the highway to the terminal area, and the confusing industrial area on the way into the terminal area.

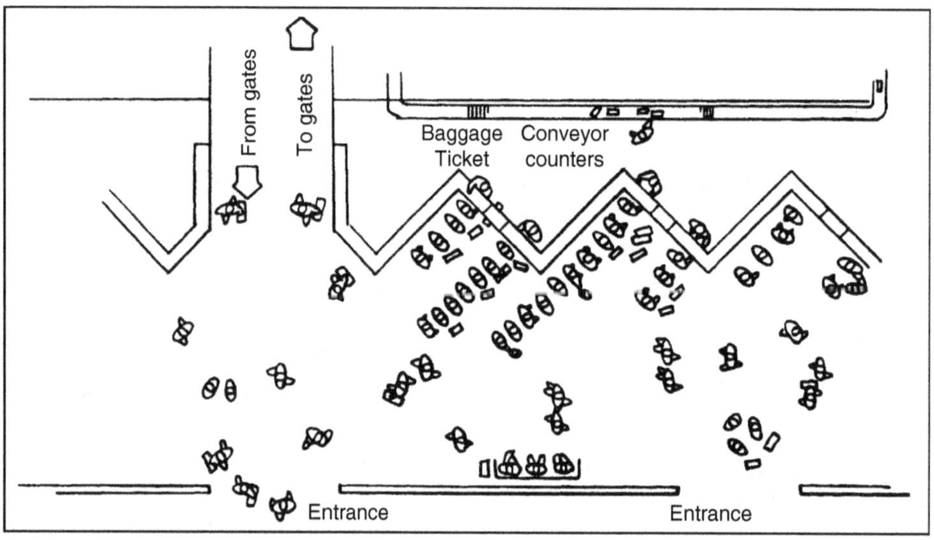

A diagonal ticket-counter concept helps minimize interference between passenger queues and other persons who are walking, standing, or milling about in the ticketing area.

Figure 2.54 Recommended ticket counter.

4. Entrance configuration should be obvious in terms of general appearance (i.e., avoid the temptation to camouflage an entrance for aesthetic reasons, such as by using an uninterrupted glass facade). Provide identifying signs to indicate which airline, bus, or railroad system is nearest a particular entrance. Orient the signs so that they are readable from the viewpoint of vehicle drivers and passengers (i.e., not as though drivers and passengers were facing the building), and make sure that the sign lettering is readable from expected viewing distances.

5. Provide flat entry pathways and thresholds so that people in wheelchairs can negotiate them safely and easily.

6. A double entry with an intervening vestibule should be considered for terminal buildings located in cold climates; this helps keep cold winds from entering the main building.

Main Ticketing and Baggage Check-in Area

1. The size and shape of this area should be based on the anticipated number of airline, bus, or train ticketing booths; the expected peak passenger queuing loads; and the obvious milling around created by nonpassenger as well as passenger patrons.

2. Information signs should be centrally located so that they are immediately visible to the incoming or transferring passenger and/or to the person who wants to know what plane, bus, or train is expected at what time and at which gate. If the area is large, consider several such installations, spaced appropriately along the ticketing-area corridor. Signs should be oriented so that they are readable from the typical viewing angle of the observer (i.e., observers should not be required to walk to a position in front of the sign before they can read it).

3. Several special facilities should be located within, or directly adjacent to, the main ticketing area:

   a. Information desk

   b. Traveler's aid desk

   c. Restrooms

   d. Telephones

   e. Seating

Passenger Waiting and Services Area or Areas. One or more passenger waiting and services areas should be provided for the terminal in which considerable waiting may occur between services. The following should be considered: *(When separate immigration facilities are required, the following should also be considered within the restricted areas used by passengers awaiting departures to foreign countries.)*

1. Restrooms, shoeshine stand, and parcel lockers

2. Dining, snack, and cocktail concessions and drinking fountains

3. Bank and money exchange facility

4. Drugstore, gift shop, flower shop, duty-free shops, liquor store, haberdasher, women's clothing store, newsstand, etc.

5. Insurance vendor's desk

6. Special airline lounges

7. Flight, bus, and train schedule board

8. Traveler's aids (baggage carts, strollers, etc.)

9. Clocks with updated departure and arrival information

10. Computing and digital communications facilities: WIFI, charging stations, electrical outlets, laptop work surfaces

Terminal Administrative Facilities. The following are among the important terminal facilities that will be needed both by the terminal management and by other nonpassenger service persons (e.g., government transportation agents):

1. General administrative and clerical offices

2. Employee lounges, restrooms, locker rooms, dressing rooms, private dining rooms, etc.

3. Communications centers (for security, public address systems, transportation systems control, etc.)

4. Maintenance and housekeeping facilities, including shops, equipment storage, and service docks and parking

5. First-aid station

Baggage-Claim and Local Transportation Area. Most modern transportation terminal concepts separate the baggage-claim and local transportation functions from the previously outlined functions. Among the more typical facility requirements are the following:

1. Incoming baggage collectors and dispensers (diverters, carousels, racetrack conveyors, etc.)

2. Temporary parcel storage (lockers and/or room)

3. Rental-car concessionaire booths

4. Telephones

5. Taxi, limousine, and city bus stands

6. Restrooms

Other Amenities

1. In-terminal hotel

2. Theater

3. Stairs, ramps, escalators, and elevators

4. Environmental control (illumination, temperature, etc.)

5. Observation deck (airport terminal)
6. Nursery, playrooms, TV areas, etc.

Air terminal buildings must be planned to ameliorate the detrimental effects of high-level jet aircraft noise, both for passengers and for all personnel associated with terminal operation or use. Additionally, all transportation terminal facilities should be accessible to the handicapped traveler.

Rapid-Transit Commuter System Terminals

Terminal facilities provided for local exchange of transit system passengers (including those using buses, rail vehicles, suspended rail vehicles, and air-cushion vehicles) often either are neglected in terms of patrons' needs or are designed in such a way that they are attractive but not necessarily serviceable. The following are some of the basic local terminal facility considerations that should be addressed during the planning stages of any transit system development.

Patron Parking. Fringe-area parking at key transit terminals on the perimeters of a city not only should be sufficiently large to accommodate expected user vehicles, but also should provide reasonable security for the vehicles left temporarily in these facilities.

Waiting Area and Shelter. A sheltered waiting structure should be provided to protect waiting passengers from inclement weather conditions. The facility should have heat for cold weather and good lighting both inside and out. Comfortable seating should be provided. A system map and schedules should be prominently posted. These should be designed for easy understanding by the average user (avoid art maps and complex street maps and use clear identification code techniques). A telephone should be provided, and if the station is at the far end of the line, there should be restrooms and a drinking fountain. The shelter and loading platform should be accessible for wheelchair users and the schedule and map data should be designed for use by the visually handicapped.

Safety. Special attention should be given to the potential hazards that may be associated with the terminal facility and the equipment and area around the facility, especially with regard to people walking or falling on the tracks.

Vending Machines. When self-service ticket and/or change vending machines are provided, these should be easy to use. The machines should also be vandal-proof.

Downtown Bus Stops

Downtown bus stops, although not strictly terminals in the sense of the facilities previously discussed, also have critical human factors considerations. Among the most important are the following.

Safety. Although midstreet island stops are appropriate in a few cases, the curbside stop is preferable because it prevents people from walking across traffic to get to the island. It is suggested that a guardrail be placed between waiting passengers and the street. It should be designed to force patrons to queue up behind it.

Signing and Illumination. Bus-stop identification, schedules, and other pertinent information should be legible not only to those with normal eyesight but also to the visually handicapped.

CHAPTER 3

TRANSPORTATION SYSTEMS

GENERAL

This chapter covers human factors considerations that should be addressed during the initial conceptual phases of any transportation system development. The key elements addressed include the primary transporter or vehicle, the roadway upon which the transporter moves, and the special terminal interfaces where various people and transporter systems interact.

The chapter begins with a series of general guidelines applicable to all elements of the transportation system; these are followed by separate sections dealing with the more specific questions to be addressed for each of the system elements.

Transportation Systems Planning Model

Although a particular transportation system may be of primary interest, it is generally the case that it must be made compatible with any or all of the total system elements indicated in Fig. 3.1. The key to this compatibility lies in a satisfactory solution to users' problems of interfacing with any or all of these elements; i.e., users usually do not confine their travel to a single element, even though that may be the major interaction for most of the trip.

Considerations for International Travel

Because of the opportunities that modern transportation system technology offers for international travel, it is important to address human factors on a global basis. That is, although a specific type of system may be proposed for fairly local use, it is probable that the system will at some time be used by individuals with widely differing backgrounds in terms of experience, language, and understanding. (See Table 3.1.) Thus, a local urban transit system might be used by individuals from many parts of the world as they travel on business or for personal pleasure.

1. Allow simple transition from one transportation system to another.

2. Provide systems to transfer packages and luggage between transportation systems.

3. Provide safe and clean terminals for basic human needs (bathrooms, rest, and food).

Among the typical system features to be considered from this international point of view are the following:

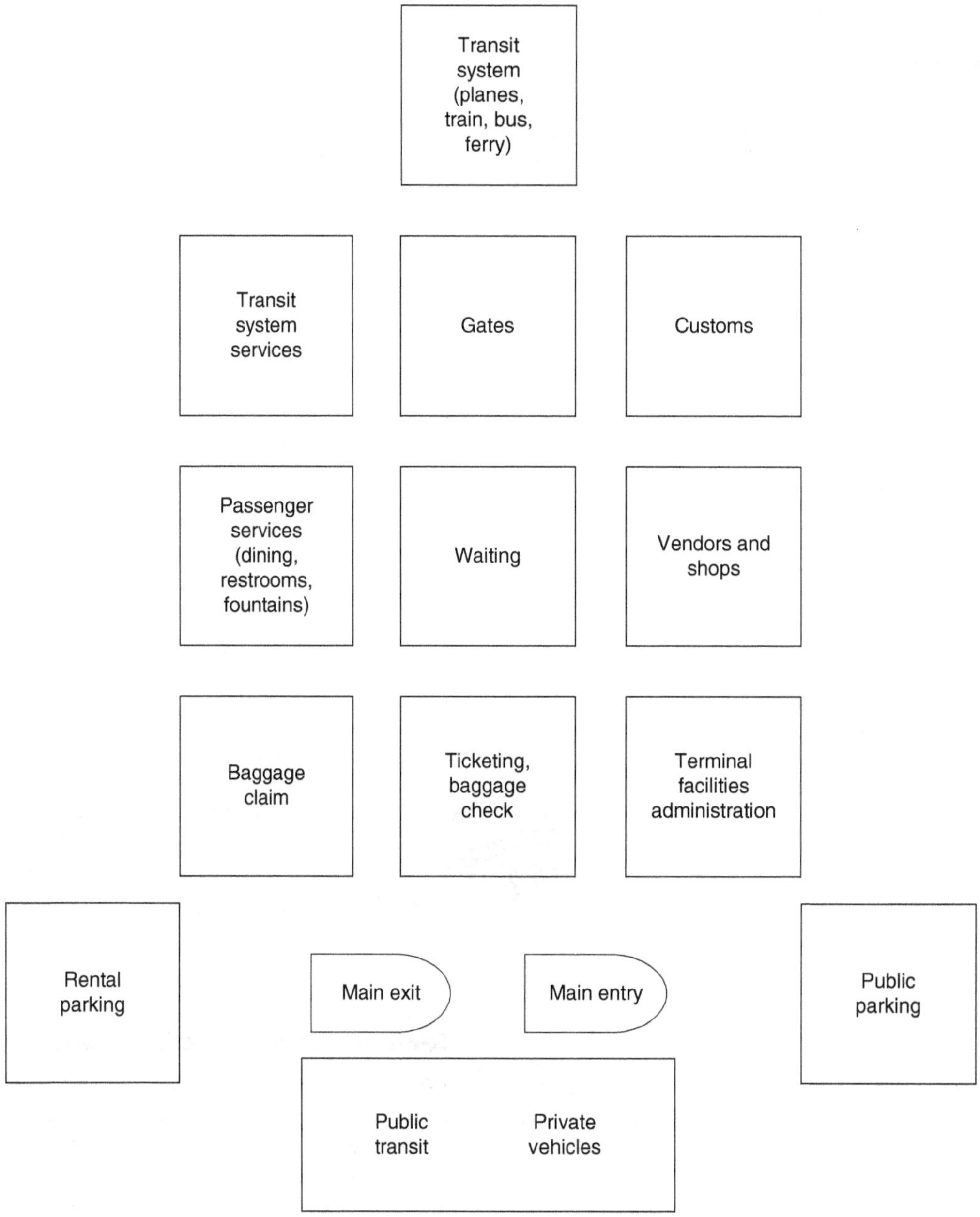

FIGURE 3.1 Transportation systems planning model.

1. Standardization of the driver controls in automobiles that may be rented by individuals of various nationalities so that the vehicles can be operated effectively and safely

2. Standardization of roadway patterns, marking, signing, and signaling devices so that drivers can "mesh" effectively and safely with the local traffic

3. Standardization of special terminal facilities so that travelers do not become confused or lost or cause special problems for terminal staff or vehicular systems operators

4. Standardization of facilities, equipment, and communications used by the operators of vehicular systems that may be acquired from a variety of equipment manufacturers (e.g., aircraft purchased by a carrier in one country from a manufacturer in another country and communication protocol and language used by various air, land, or sea traffic control organizations) so that the mix of foreign carriers does not become snarled at terminal areas, etc.

5. Standardization of equipment and facilities so that maintenance can be accomplished by a variety of technicians with varying backgrounds

TABLE 3.1 User Population Considerations

Characteristics of the User Population

The user population for transportation systems is perhaps broader and more complex than the user population for any other system; this makes it especially important to consider the individual differences of each user group not only during initial planning but also at each stage of each subsystem design and development. Among the most important characteristics to consider are the following:

Differences in Human Body Size

Any transportation system must accommodate persons of all sizes, from tiny babies to the largest adult. Although there are certain conventions (e.g., designing to fit a range from the 5th-percentile female to the 95th-percentile male), the traveler who does not fall within this arbitrary design range still must use the system.

Male–Female Differences

Recognition of certain male and female characteristics is required for public transportation, especially with respect to separate facilities.

Ethnic Differences

Differences in terms of language, experience, custom, etc., should be considered wherever the convenience and safety of the traveler are critical.

Age Differences

Although the middle-aged group typically establishes the norm with respect to the planning and design of a transportation system, consideration must be given to the unique requirements of children in terms of their limited understanding and to the needs of the elderly in terms of their lack of mobility, their slower responses, and their visual, aural, and other limitations.

The Handicapped

The physically handicapped use public transportation. Therefore, special attention should be given to their problems during the initial system planning phase. Corrections are difficult to make once basic barriers are established by the initial concept.

Behavioral Attitudes

Users of transportation systems represent a unique blend of the broadest possible range of attitudes. That is, users' attitudes will range from considerable respect for equipment and facilities to complete disrespect for either these elements or their fellow travelers. Anticipation of this is particularly important in order to preclude both operational and service problems.

System Use Patterns

A significant factor in system planning is rate of use of a proposed system so that the most cost-effective solution to the ebb and flow of travel patterns can be arrived at. This applies not only to the overall system, where roadway capacity and terminal capacity are of concern, but also to the individual vehicle size in terms of numbers of passengers and amount of baggage.

Special attention should also be given to world travelers who have special problems, e.g., the elderly, the handicapped, children traveling alone, and people traveling with small children.

The user population to be considered for transportation systems includes the following key individuals (see Table 3.2):

1. Vehicle operators
2. Passengers
3. Vehicle maintenance technicians
4. Terminal operators (management, ticketing and baggage-handling personnel, housekeeping and maintenance personnel, concessionaires, carrier service attendants, security personnel, etc.)
5. Visitors
6. Taxi drivers
7. Public transit drivers
8. Traffic controllers
9. Fire and disaster crews
10. Other supporting personnel (travelers' aids, medical staff, porters, immigration staff, etc.)

TABLE 3.2 General Classification of Human Considerations

Area of Consideration	Principal Human Considerations
Vehicle and terminal	Space and seating, ingress and egress (entrances and exits) Environmental stress (acceleration, deceleration, noise, vibration, temperature, atmospheric pressure, ventilation, etc.)
Convenience and mobility	Time: total trip time, waiting, walking, transfer, terminal organization, and special equipment operation. Operations: scheduling, frequency, and reliability of service. Intelligibility of information. Accessibility for elderly and handicapped persons
Safety and security	Emergency provisions; alarms, restraints, procedures, and first aid Security provisions: crime and vandalism control. Accident prevention
Social factors	Personal privacy. Crowding. Freedom to choose traveling companions
Psychological factors	Comfort. Freedom from anxiety and fear. Confidence with respect to what to do and where to go. Pleasurable features, i.e., features that are aesthetically pleasing, interesting, and nonmonotonous
System environment	Minimal congestion and delay. Minimal noise, air, and visual pollution and minimal physical interference
Maintenance	Reliability. Ease of maintenance. Resistance to wear, abuse, and vandalism

Human Factors Design Objectives by Vehicle User Category

All Occupants

1. Ease and safety of entry and exit (especially emergency escape) require a properly sized and shaped door opening and an optimum physical relationship between the opening and the seats, plus elimination of intervening impediments (especially snag-producing elements). Special care should also be taken to design doors and door-manipulating hardware so that doors are secure when closed and also are not confusing or difficult to lock, unlock, or open. The adjustable windows should be equally easy to operate (sometimes windows are the only escape route in an emergency).

2. Body, limb, and head clearance should be sufficient to minimize crowding and inadvertent bumping when the vehicle sways or goes over a bump.

3. All interior surfaces should be made as noninjurious as possible (e.g., there should be no sharp protrusions, edges, or corners, and appropriate padding should be provided) so that if an occupant strikes a surface when the car is moving, when getting in or out of the car, or during a crash, he or she will not be cut, bruised, pinched, or otherwise injured.

4. All seating should be designed to fully support occupants in a nonstressful position so that unnecessary fatigue is minimized during long-duration occupancies.

5. All interior materials used in the vehicle should be nonflammable.

6. Shatterproof glazing should be used in all windows.

7. Equally effective crash-restraint systems should be provided for all passengers.

Drivers

1. The driver station should be laid out so that all the driver's visual and manipulative tasks can be performed as easily and efficiently as possible (i.e., as correctly and quickly as necessary to provide safe response to the driving environment).

2. The controls and displays provided for driving should be designed so that they will provide the necessary input and output for proper vehicle control and also will not be confusing and cause the driver to make errors or take his or her attention away from the external information needed to cope with traffic.

3. The vehicle dynamics and vehicle control systems should be designed so that they allow not only an expert driver but also an inexperienced driver to cope with all driving conditions (especially emergency conditions associated with skids, rough surfaces, etc.) and are compatible with all drivers' sensorimotor response capabilities and limitations (i.e., strength, response time, coordination, and dexterity).

Maintenance and Service Personnel

1. Servicing features (fuel, oil, water, and battery) should be placed where they are easily accessible (with minimum hazard) and should have easily manipulated intervening features (caps, covers, etc.).

2. Typical service and maintenance features (battery replacement, engine timing, fan-belt adjustment or replacement, etc.) should not be designed only for convenience and ease of operation; special consideration should also be given to the related hazards created by the design and/or location of adjacent, intervening elements (e.g., components within an engine compartment that someone who is servicing the engine might touch or get caught in).

3. Consider the potential hazards of tire changing, lubricating the vehicle on a grease rack, etc., in terms of designing the basic vehicle and its components as well as the special supportive devices that have to be used in conjunction with these services.

Pedestrian Interfaces. All external surfaces of vehicles should be made as noninjurious as possible so that if a pedestrian is inadvertently struck by a vehicle or walks or runs into it, personal injury is as minimal as possible.

Special Users: Children and the Handicapped. All of the previous objectives should be evaluated in terms of possible use of a vehicle by handicapped persons. Although this is most critical in the case of handicapped passengers, one should also consider the possibility of service and maintenance personnel with certain types of handicaps.

Children require special attention as passengers, not only because of their size, but also because they may create safety hazards (e.g., small children may try to operate vehicle controls, windows, doors, etc., when they should not).

Other Human Factors Considerations. Other key human factors considerations should be addressed during the initial concept development stage. These include the following.

Riding and Handling Quality. Obviously, the automobile should not possess qualities that would make it difficult to control, especially under unusual conditions, i.e., on rough or slick roads, when it is heavily loaded, or when unusual maneuvers are necessary, such as making a sudden turn to avoid another vehicle or a pedestrian or object in the road. Equally important is a ride quality that will not create unreasonable stresses on the passengers. The latter consideration is secondary to the former since people vary in their interpretations of what is a satisfactory ride; e.g., a sports-car enthusiast wants to "feel the road," while a Wall Street executive probably does not.

Reliability. Obviously, most consumers appreciate a vehicle that is reliable and does not require continual repair. Most important, however, is a design that will not let the driver down in a pinch, i.e., an automobile whose brakes, steering, motor, tires, etc., will not fail.

Ease of Maintenance. Much of the cost of maintenance is related to the difficulty and time required to determine where a malfunction is and to the difficulty of getting to a part, removing it, and replacing it. Ease of maintenance can be increased primarily during the conceptual stage, before difficulties have been created by placing things in awkward places.

Design-Induced Driver Problems. A great many of the driver's problems are often created because a designer has chosen to use the driver interface as a prime target for artistic or aesthetic experimentation. That is, instead of recognizing that, to make the driver's tasks easier, controls and displays should be designed, located, and labeled in a certain way, the designer chooses to experiment with various methods and techniques for mechanizing these displays and controls, arranging them to create a special effect or adding decorative chrome that reflects in the windshield or into the driver's eyes, for example. In addition, designers often ignore good seating design criteria, thus creating difficulties in postural control or leg manipulation for the driver (e.g., soft cushions may allow the heavy driver to sink too low to see out or may cause clothing to bunch, and incorrect cushion dimensions may create pressures on the back of the driver's legs).

THE AUTOMOBILE

Planning Considerations for Automobiles

Although many other factors must also be considered in conceptualizing a new automobile, human factors should be addressed first. (See Fig. 3.2.)

Related Government and Industry Standards

The following government and industry standards relating to the human factors aspects of automobile design should be consulted during conceptual development. Although some of the information may not necessarily coincide with some human engineering design principles, the objectives are common and therefore provide a sound basis for conceptual analysis.

Federal Motor Vehicle Safety Standards (FMVSS)

101 Control location, identification, and illumination—Passenger cars, multipurpose passenger vehicles, trucks, and buses

102 Transmission shift lever sequence, starter interlock, and transmission braking effect—Passenger cars, multipurpose passenger vehicles, trucks, and buses

103 Windshield defrosting and defogging systems—Passenger cars, multipurpose passenger vehicles, trucks, and buses

104 Windshield wiping and washing systems—Passenger cars, multipurpose passenger vehicles, trucks, and buses

105 Hydraulic brake system—Passenger cars, multipurpose passenger vehicles, trucks, and buses

108 Lamps, reflective devices, and associated equipment—Passenger cars, multipurpose passenger vehicles, trucks, trailers, buses, and motorcycles

111 Rearview mirrors—Passenger cars and multipurpose passenger vehicles

113 Hood latch systems—Passenger vehicles, trucks, and buses

114 Theft protection—Passenger cars

118 Power-operated window systems—Passenger cars and multipurpose passenger vehicles

124 Accelerator control systems—Passenger cars, multipurpose passenger vehicles, trucks, and buses

125 Warning devices

201 Occupant protection in interior impact—Passenger cars

202 Head restraints—Passenger cars

203 Impact protection for the driver from the steering control system—Passenger cars

204 Steering control rearward displacement—Passenger cars

205 Glazing materials—Passenger cars, multipurpose passenger vehicles, motorcycles, trucks, and buses

206 Door locks and door retention components—Passenger cars, multipurpose passenger vehicles, and trucks

207 Seating systems—Passenger cars, multipurpose passenger vehicles, trucks, and buses

208 Occupant crash protection

209 Seat-belt assemblies—Passenger cars, multipurpose passenger vehicles, trucks, and buses

210 Seat-belt assembly anchorages—Passenger cars, multipurpose passenger vehicles, trucks, and buses

212 Windshield mounting—Passenger cars

213 Child seating systems

214 Side-door strength—Passenger cars

216 Roof crush resistance—Passenger cars

219 Windshield intrusion zone

302 Flammability of interior materials—Passenger cars, multipurpose passenger vehicles, trucks, and buses

572 Anthropomorphic test devices

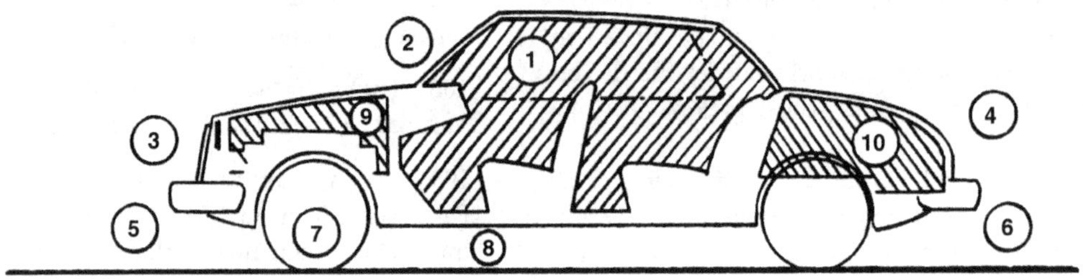

1. *Passenger Space,* i.e., a space that is large enough and geometrically appropriate for the number of passengers the automobile is planned to accommodate. Do not "short-change" passengers who are expected to be riding in the vehicle only once in a while.
   Passenger Seating, i.e., properly designed seats and restraints for *all* passengers. Do not force some passengers to sit in less-than-adequate seats or be less protected because these are used infrequently.
   Driver Control, Display, and Visibility, i.e., create an optimum driving configuration so that the driver has the best possible chance of operating the vehicle safely and efficiently in all expected environments.
   Environmental Control, i.e., adequate heating, cooling, ventilation, and noise and vibration control to minimize stress on passengers and especially on the driver.
   Passenger Compartment Structural Integrity, i.e., the compartment envelope should be able to withstand expected crash loads, possible penetrations, etc.
   Interior Safety, i.e., adequate passenger restraint and freedom from structural contact injury should be provided to minimize the possibility of injury or death during a crash.
   Ingress and Egress, i.e., ease of getting into and out of the vehicle should be a prime consideration, under both normal and emergency circumstances.
2. *Visibility,* i.e., provide the necessary visibility for the driver not only in terms of angle of view but also in terms of keeping that visibility under typical environmental disruptions, e.g., rain, snow, ice, or fog.
   Crash Protection, i.e., provide glazing that will not splinter or shatter and possibly cut passengers during a crash.
3. *Road Illumination,* i.e., provide a properly controllable headlight subsystem to ensure that the driver can see where he or she is going and can observe other vehicles, pedestrians, or obstructions.
   Identifying lights, i.e., provide auxiliary light subsystems that indicate when the driver plans to turn, which is the front or rear of the vehicle, and when the driver is going to stop.
   Headlight Beam Control, i.e., provide control of the headlight beam direction so that the driver can reduce the potential visual disturbance that the headlights may cause for an oncoming motorist.
4. *Rear-Collision Protection,* i.e., provide proper energy absorption and isolation of the fuel tank so that, in the event of a crash, the fuel tank is protected and passengers are not injured.
5. *Minimization of Injuries to Pedestrians,* i.e., provide a front-end configuration that is minimally injurious in the event the vehicle runs into a pedestrian.
   Crash Energy Absorption, i.e., provide a front-end design that will reduce the amount of energy transferred to the passenger compartment and will limit the possibility of penetration of the compartment.
6. *Rear-Impact Energy Absorption,* i.e., provide a design that will reduce structural distortion and the possibility of penetration of the passenger compartment.
7. *Ease and Safety of Tire Replacement,* i.e., adequate consideration should be given to the problems of replacing a flat or blown tire, including the safety aspects of jacking up the vehicle.
8. *Side-Impact Energy Absorption,* i.e., provide adequate internal padding and a design that will reduce structural distortion and the possibility of penetration of the passenger compartment.
9. *Ease and Safety of Maintenance and Service,* i.e., provide proper access to frequently serviced components; ease of refueling and of refurbishing, removing, and replacing parts; and protection of the service technician and the owner from maintenance hazards, such as electric shock, hot surfaces, moving parts, and fire.
10. *Baggage,* i.e., provide reasonable storage capacity and ease of loading and unloading, removing and replacing the spare tire, etc. Consider all user sizes and strengths.

FIGURE 3.2 Planning considerations for automobiles.

TRUCKS

Planning Considerations for Truck Design Conceptualization

Depending on the purpose of the truck, the following functional considerations should be addressed:

1. Driver and assistant workstation
2. Sleeping accommodation
3. Cab ingress and egress
4. Cargo loading and unloading provisions and package tie-down
5. Tractor-trailer connect and disconnect and trailer parking
6. Recreational vehicle (RV) camper module loading and unloading
7. Access for maintenance (cab over engine)
8. Dump-truck operation
9. Special cargo compartment environmental control (refrigeration, liquid filling, and dispensing, etc.)

Driver Workstation. Because of the necessity to place the cab in a relatively high position and because of the obvious rear-vision obstruction created by the cargo component, large truck configurations, especially, create several unique human factors problems over and above those associated with the automobile. Refer to the section of the automobile for basic driver dimensional configurations.

Although some truck configurations will require additional driver controls, it is important to recognize that truck drivers are also automobile drivers, and thus the general arrangement of controls and displays common to both types of vehicles should be similar; otherwise, truck drivers may make errors when they transfer from one type of vehicle to the other. The special truck driver workstation considerations to address are the following:

1. Cab Ingress and Egress: When the cab is placed high above the ground, special steps and handrails, accessible door handles and locks, etc., must be examined in terms of both the mounting and the dismounting processes; i.e., it is not sufficient to place these devices wherever there is a convenient attachment point. Consider the order in which they will be used and the best location to put them in order to provide the driver or passenger the best advantage for supporting and controlling body weight and movement. And make sure that the steps are not so far from the ground (or so far apart) that the act of mounting creates too much strain on the operator or offers the operator an opportunity to take a wrong step or lose his or her grip on a handhold or rail.

2. Frontal Visibility: When the cab is unusually wide and high, consider the problem of the driver's ability to see the front corners of the cab in order to clear adjacent vehicles or objects. This suggests a seating arrangement that places the driver close to the windshield and high enough to provide a good down angle of view. Consider the possibility of additional viewing windows placed in the lower front portions of the cab.

3. Rear Visibility: Because the typical cargo compartment occludes the rear viewing potential, appropriately designed and positioned cameras, sensors, or exterior mirror systems are usually necessary.

4. Cab Environment: Trucks that are expected to perform long-duration (long-haul) operations should always provide an acceptable driver environment (in terms of temperature, ventilation, noise, and vibration) in order to ensure minimal fatigue and optimum driver performance.

5. Service and Maintenance: Examine the overall service and maintenance requirements of the proposed truck system as early in the conceptual phase as possible in order to preclude awkward working arrangements that often are the result of "jury-rigging" to fit the constraints of a preconceived cab configuration. This is especially critical for very large rigs, in which the height of components from the ground makes it difficult to reach these elements without using special aids. Remember that maintenance often occurs out on the road, where such aids are not available.

6. Cargo Loading and Unloading: For large and/or heavy cargo-handling systems, consider self-contained loading and unloading subsystems that are suitable for operation by one person. In approaching the cargo-handling problem, examine all potential loading-site situations, including special docking interfaces. The analysis of the loading and unloading sequence should include the following as a minimum: (a) ground-up cargo manipulation, (b) dock-truck exchange, and (c) onboard cargo movement and tie-down.

7. Special loading and unloading systems are required for the transfer of liquid products. Consider these in terms of spillage and especially in terms of safety aspects when chemicals or other toxic substances are involved.

8. Special Operating Problems: Control of a vehicle both when it is loaded and when it is unloaded is an especially critical consideration in developing the truck system concept. Although truck drivers are usually well trained, it is still important to create a system that does not perform considerably differently when it is loaded from the way it performs when it is unloaded. Be especially cognizant of the hazards created by smaller trucks when they are empty (the rear is much lighter than the front) because these types of vehicles often are driven by less skilled drivers who are unaware of the drastic change in vehicle controllability.

Driver Viewing Position and Cab Concept. As Fig. 3.3 shows, the cab concept has considerable influence on the viewing position of the driver. Whether the truck is small or

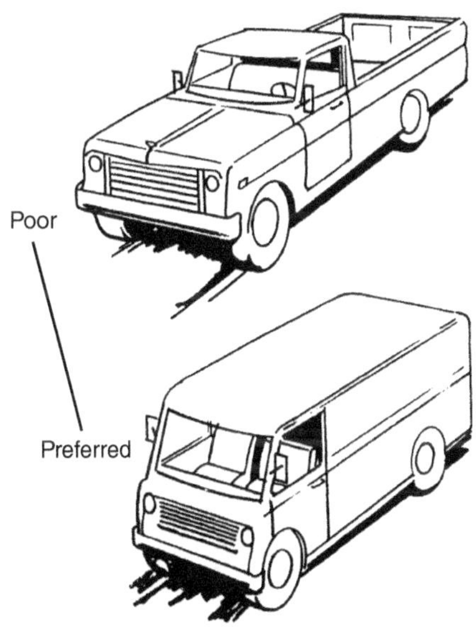

Figure 3.3 Fontal visibility.

large, the cab-over-engine configuration offers increased opportunity to place the driver where he or she can see more readily what may be directly in front of the vehicle.

As can be seen, the primary problem is the large, bulky hood, which obstructs the driver's view of the area around the front of the vehicle.

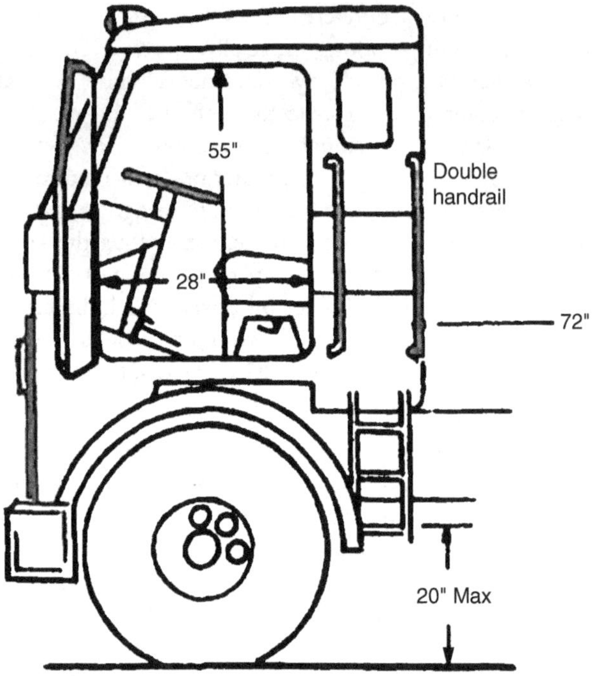

FIGURE 3.4 Tractor-trailer cab considerations—Example 1.

Tractor-Trailer Cab Considerations

Long-haul tractor trailers impose considerable stress on drivers, and consideration should be given to minimizing these stresses, especially in terms of avoiding cramped spaces, providing ease of ingress and egress, and controlling noise and vibration.

Access. Figure 3.4 shows some of the critical dimensional considerations for cab design. Special attention is directed to the boarding ladder, steps, and handrails. Too often these are located wherever there appears to be a suitable structural connection or surface. The step–handrail relationship should be such that the user is not thrown off balance as he or she mounts the steps and/or proceeds through successive steps to the cab. Note the fairly large threshold step, or platform, provided at the point where the user may have to change his or her body orientation for final cab entry and/or egress at the beginning of dismounting.

Sleeper Cabs. Sleeper cabs are commonly used in cross-country rigs, in which case it is important to provide a sleeping cubicle that is sufficiently large to ensure the user's rest. Figure 3.5 shows preferred interior clearance dimensions. The interior walls should have some type of padding to prevent bruises if the occupant is rolled about. Acoustic treatment and adequate temperature control and ventilation should be provided. Note also the suggestion for storage of personal items, e.g., things that the user might want to take out of his or her pockets before sleeping.

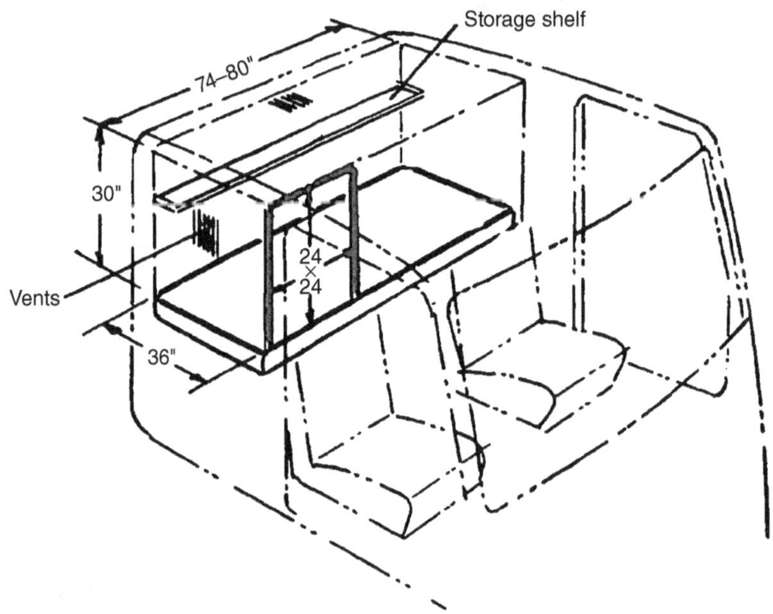

FIGURE 3.5 Tractor-trailer cab considerations—Example 2.

RECREATIONAL VEHICLES AND MOTOR-HOME VEHICLES

Although intended for a distinctly different purpose, recreational vehicles and motor homes often begin with a truck chassis and therefore are considered special-purpose trucks here. Since the variety of possible accommodations for such vehicles is almost limitless, the conceptual considerations below are limited to the most common human factors problem areas. (See Table 3.3.)

Modular Camper Concepts

1. Ease of camper shell installation or removal is particularly important to the owner who uses the pickup for other purposes. Provide a simple, safe method by which the owner can back the truck accurately underneath the camper unit and then lower the unit for securing it to the truck. (See Fig. 3.6.)

2. Provide external access to replenishment and disposal components and to external power, gas, and water hookups.

    a. Provide internal storage systems to prevent articles from being dislodged during travel.

    b. Provide at least two exits so that occupants can escape if the primary one is blocked by fire. Use only nonflammable materials in construction.

Motor-Home Concepts

Since motor-home operation is quite similar to camper operation, all the previous suggestions apply. The major additional consideration relates to the fact that the rear of the motor home may be occupied while the vehicle is under way. Adequate occupant restraints should be provided for seats that may be used, and all interior surfaces should be made as safe as possible; i.e., rounded corners and edges, padding, and appropriately positioned handrails should be used to keep passengers from falling in the event they are walking about while the vehicle is moving. (See Fig. 3.7.)

FIGURE 3.6 Modular camper concept.

TABLE 3.3 Recreational Vehicles General Accommodations Checklist

1. Toilet, sink, and shower
2. Range, oven, refrigerator, sink, and garbage disposal
3. Storage for food, dining and kitchen equipment and supplies, clothing, linens and housekeeping supplies and equipment, water (both hot and cold), and cooking gas
4. Seats, tables, and beds
5. Entertainment (radio, TV, video player, games)
6. Illumination and air conditioning
7. Emergency equipment (fire extinguishers, jacks and tire repair equipment, extra gasoline and oil, road flares, first-aid kit, etc.)

Other Operating Considerations

1. **Vehicle Leveling:** In order for certain systems to function properly, it may be necessary to level the vehicle.
2. **Vehicle Stability:** Recreational vehicles are characteristically affected by wind both when they are moving and when they are parked. Appropriate schemes should be provided either to minimize the possibility that the vehicle will become unstable and uncontrollable while traveling in a high wind or to reduce the effect of winds on the vehicle when it is parked and occupied as a temporary home.
3. **Resource Conservation:** A method for monitoring the use of certain resources is important so that stoves are not left on, supplies are not drained unnecessarily, etc.
4. **Security:** Owners of recreational vehicles often park in remote areas, which may expose them to the dangers of burglary, vandalism, or bodily harm. Windows and doors should be designed to minimize the possibility of wrongful entry.

FIGURE 3.7 Motor home concept.

PUBLIC TRANSIT BUS

Primary Human Factors Considerations for Public Transit Buses

1. Driver: Exterior visibility, control and display visibility and accessibility, visual monitoring of passengers, lack of reflections at night, and comfortable seat for long-duration occupancy

2. Onboard Passengers: Level floor, wide aisle, accessible handrails, good seats with knee and elbow clearance, visibility to look for stops (either seated or standing), illumination to read and move about safely, air conditioning, minimal noise, and reasonable ride quality

3. Boarding Passengers: Adequate bus identification that can be seen at a distance and nearby and convenient entry with handrails

4. Service Personnel: Convenient access to all components that frequently require servicing or maintenance

Any bus that will be used by the public should be configured so that handicapped persons and the elderly are not excluded from its use and/or put under stress because of the difficulties imposed. The principal features to consider are special wheelchair ramps or elevators, appropriately positioned handrails to aid the unsteady senior citizen, good illumination for the partially sighted, and an appropriate technique for securing wheelchairs aboard.

Nonflammable materials should be used throughout the public transit bus, and suitable window designs should be provided to allow passengers to exit from windows in an emergency.

Doorways should be wide enough for easy exit or entry. Note that the accompanying example provides two two-way doors as well as one one-way door in the rear for exit only. This makes for extra convenience and safety.

Vehicle Conceptualization from the Driver's Point of View

In addition to the primary task of controlling the vehicle, the bus driver is required to perform a wide variety of other tasks. Collectively, these establish the functional requirements for many bus features. The following functional requirements are important to consider.

Driving

1. External Visibility: City buses frequently operate under crowded traffic conditions; therefore, it is important that the driver be able to see the relationship between the bus exterior boundaries and adjacent obstacles. The driver must be able to see potential passengers waiting at the curbside, so as not to run into them, and to see a small child who may have wandered in front of the bus. In order for the driver to see clearly at night, steps must be taken to minimize the reflections that interior lighting might cause on the windshield.

2. Internal Visibility: The driver, of course, must be able to see the controls and instruments while driving. In addition, the driver needs to observe passengers who are standing or walking in the aisle, passengers who are creating a problem for other passengers, etc.

Passenger Service

1. Making Change: Although many system operators have gone to an exact-change or token operation because of frequent en-route robberies, the designer should be conscious of the driver change-making task, wherein the convenience of the coin machine, the need to see the fare being deposited, etc., are important design considerations.

2. Observing Boarding and Disembarking Passengers: The good driver needs to see that passengers are boarding safely so that he or she does not start the bus up suddenly and throw them off their feet. In addition, the good driver needs to see disembarking passengers, especially at rear doors, so that he or she knows the passengers are clear before closing a door and moving the bus.

3. Disembarking Signal Monitoring: The bus driver must be able to hear a disembarking passenger's signal so that he or she can stop the bus at the appropriate stop.

4. Controlling the Passenger Environment: The driver is responsible for providing appropriate illumination and temperature within the bus. This requires an appropriately selected and located set of controls so that the driver can find and operate them while driving.

Vehicle Conceptualization from the Passenger's Point of View

Entry Threshold. The initial step onto the bus should be reasonably convenient; i.e., the passenger should not be made to stretch to gain the first step up from the ground or curb. It is recommended that a ramp concept be considered for the main entrance in order to accommodate wheelchair users. There should be sufficient floor space directly adjacent to the driver and the fare box to accommodate the wheelchair user as he or she stops to make change. Appropriate handrails should be provided on both sides of the entry ramp or steps, and the steps should be well lighted at night. Nonskid treads should be provided so that passengers whose shoes are wet will not slip and fall.

Aisles. Aisles should be level so that they will not add to the possibility that standing or walking passengers may lose their footing. Aisles should be wide enough so that the standing or walking passenger does not bump seated passengers. Overhead as well as vertical and/or seat-mounted handholds should be provided so that standing passengers can steady themselves while the bus is moving. Aisles should be lighted at night (this is more appropriately accomplished when the fixtures are close to the floor, since overhead illumination may not reach the floor when the bus is crowded with standing passengers). Avoid single step-ups and step-downs; a single-step change is usually not observed, and people will trip and fall.

Seating. Fancy seats are not necessary on the typical public transit bus since the duration of occupancy is seldom very long. Seats should provide comfortable posture, however, and each seat should be wide enough to minimize interpassenger crowding. Equally important is the fore and aft spacing; room for both knees is important to the large passenger, and some space is needed to accommodate typical carry-on packages (unless an overhead package space is to be provided, which is not usually the case for public transit buses). Armrests should be provided along the aisle, not so much to serve as a place to rest one's arm, but to provide some passenger security and freedom from encroachment by someone standing beside the seat. No more than two seats should be placed side by side in transit buses, and no armrest should be placed between the two seats.

A handrail across the back of a seat pair is suggested; it aids passengers in getting out (especially the passenger who is seated by the window), and it acts as a steadying device in the event of sudden deceleration of the bus.

Some space and special securing facilities should be provided aboard the public transit bus for accommodating wheelchair users. Although some individuals could possibly transfer from a wheelchair to a regular seat, there is still the problem of the wheelchair to be dealt with.

Other Passenger Features. Passenger reading lights should be provided. These should be designed and positioned so that they serve their purpose (i.e., merely providing such lights does not guarantee that they will illuminate the passenger's reading material properly). Care should also be taken to minimize the possible reflection of these lights in the windshield, which makes the driver's external viewing task more difficult.

Passenger facilities for notifying the driver of a desire to get off at the next stop should be located so that they can be easily reached, by both seated and standing passengers.

Exit-door systems should be designed to preclude injury to the passenger; i.e., they should be designed so that a folding door will not strike a passenger who is standing too close, pinch a passenger's hand if it is in the wrong place, etc.

The passenger environment should be as pleasant as possible. Consideration should be given to providing adequate soundproofing from engine and traffic noise and adequate temperature, humidity, and ventilation control to combat unusual weather conditions. In addition, the ride quality should not be annoying; i.e., there should not be too much jolting, vibration, sidesway, or forward pitch during turning or stopping sequences.

Vandalism. Special consideration should be given to the problem of vandalism, especially in terms of materials selected for use in the bus.

External Features. The principal external feature of concern to the passenger is bus identification. Signs should be located both on the front and on the sides of the bus. The front sign should be positioned so that sun reflection will not obscure the information. All signs should have illumination so that they can be read at night.

Considerations for Handicapped Users[6]

Special provisions must be made to accommodate handicapped users who are confined to wheelchairs. The principal considerations relate to boarding and leaving the bus, maneuvering the wheelchair within the bus, securing the wheelchair against sudden stops and/or lateral forces occurring during turns, and providing special spaces for the wheelchair, both when the occupant remains in it and when he or she transfers from the wheelchair to a bus seat and keeps the folded wheelchair close by.

Figures 3.8–3.10 on wheelchair entry depict some of the possible boarding solutions that have been proposed by various bus manufacturers: (1) a ramp entry for a full-size bus at

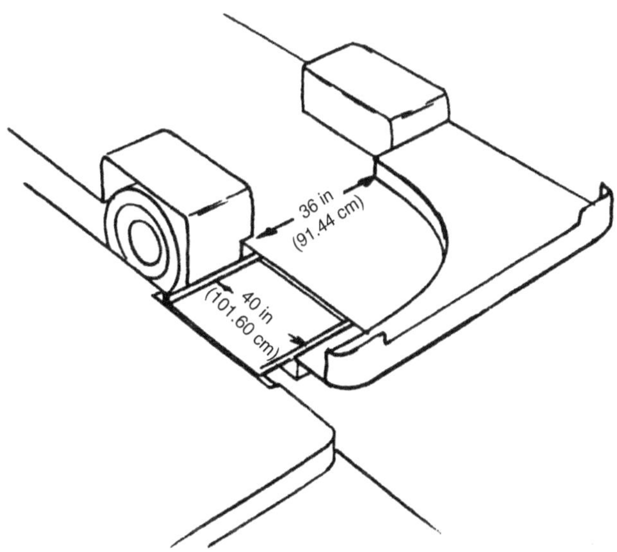

Figure 3.8 Wheelchair ramp entry for bus.

[6]T. L. Black and J. A. Mateyka, *Bus Design for the Elderly and the Handicapped,* SAE Paper 760082, Feb. 23, 1976.

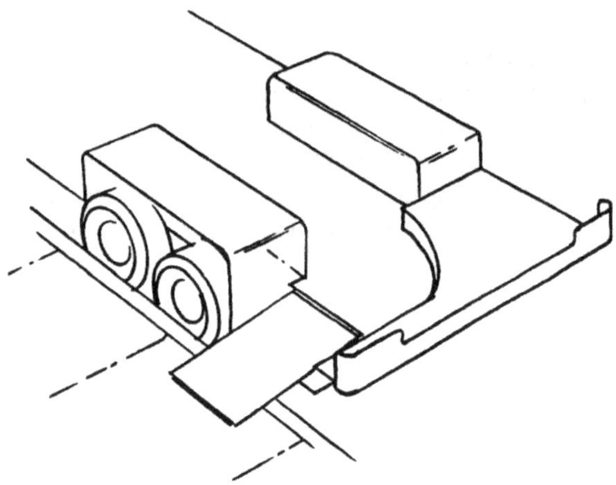

FIGURE 3.9 Kneeling bus wheelchair entry.

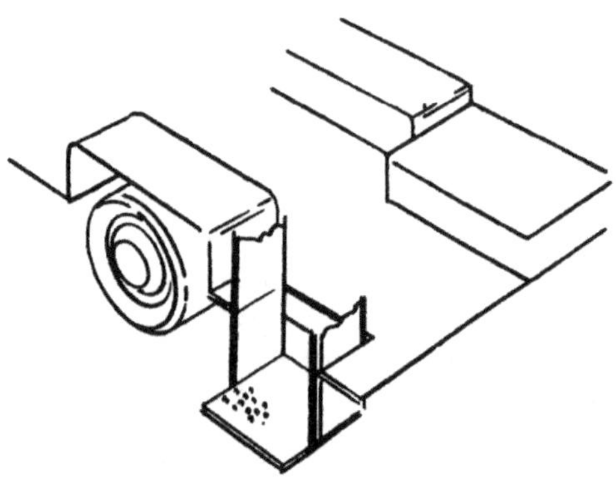

FIGURE 3.10 Wheelchair elevator for bus entry.

a standard curb, (2) a "kneeling" bus design in which a ramp provides for minor variations in curb- or street-height irregularities, (3) and an integral elevator configuration.

Note that in all these approaches, the floor section of the main aisle must be coincident with the ramp threshold level.

INTERCITY BUSES

Primary Human Factors Considerations for Intercity Buses

The major difference between conceptualizing an intercity bus configuration and conceptualizing a public transit bus configuration lies in the need to consider longer-duration occupancy and the special features required to reduce the effects of driver and passenger fatigue. It is also important to recognize that en-route speeds will be considerably higher than for the public transit bus system and special precautions are therefore required with respect to possible emergencies, including the possibility of accidents. The interior should be made as noninjurious as possible to protect the driver and the passengers in the event of a crash, and provisions should be made for emergency escape and for ease of rescue by fire and accident workers in the event of a major crash.

1. Driver: Exterior visibility, control and display visibility and accessibility, and long-duration seating comfort

2. Passengers: Long-duration seating comfort, including appropriate headrest and back recline, adequate leg room, above-average environmental control (especially ventilation), an onboard restroom, local garment and package storage, reading lights, adequate aisle space to bring carry-on baggage aboard, and above-average ride quality and noise suppression.

3. Remote baggage and package storage

4. Convenient access to all major subsystems that require quick turnaround maintenance

Seating for Transcontinental Buses

The basic dimensions should be approximately the same as for city buses; however, some important changes should be considered, namely, the addition of a headrest and recline capability, as noted in Fig. 3.11.

Although more fore and aft clearance between seats is desirable, 8 in. (20 cm) is the absolute minimum. Although a 45° recline is desirable, it is obvious that this requires more space.

A 30° recline provides a compromise that will be appreciated by passengers who want to lean back far enough to keep their heads from pitching forward as they attempt to sleep.

If a 45° recline is provided, the seat pan should adjust to a flat position, which is more compatible with the typical passenger's desire to stretch out his or her legs.

Avoid vertical contouring of the seat backrest and contouring of the seat pan for reclining seats. A slight concavity across the lateral axis of the backrest is desirable to reduce the effects of sidesway. The same caution applies to the design of the headrest portion of the seat back, i.e., the headrest should be the same width as the backrest, rather than tapering to a narrower section that requires the occupant to keep his or her head in exactly the center of the seat.

Other Amenities and Considerations

As shown in Figs. 3.12 and 3.13, it is sometimes desirable to place seats the equivalent of one step higher than the aisle (in order to keep the overhead lower). When such a configuration is used, provide aisle lighting so that passengers do not forget the step-down and thus stumble as they try to get out of their seats.

At least the following interior lighting should be provided:

1. Overhead general lighting for boarding

2. Aisle lighting

3. Reading lights that are selected and positioned so that one individual's light does not bother a fellow passenger who is trying to sleep

Overhead (captive) baggage compartments are desirable so that passengers can place garments and small luggage or

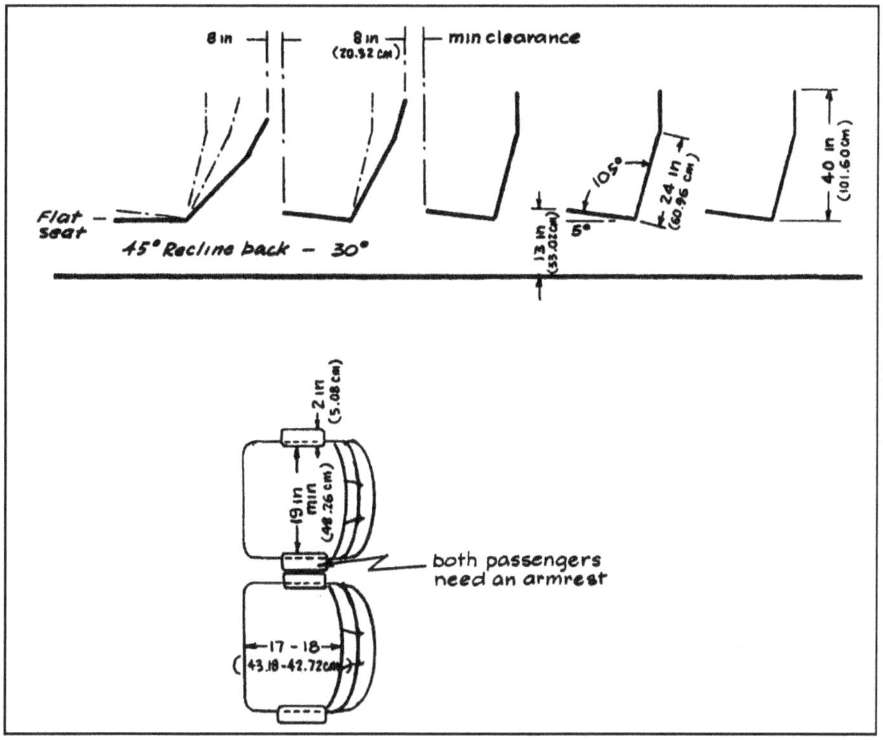

Figure 3.11 Recommended seating dimensions.

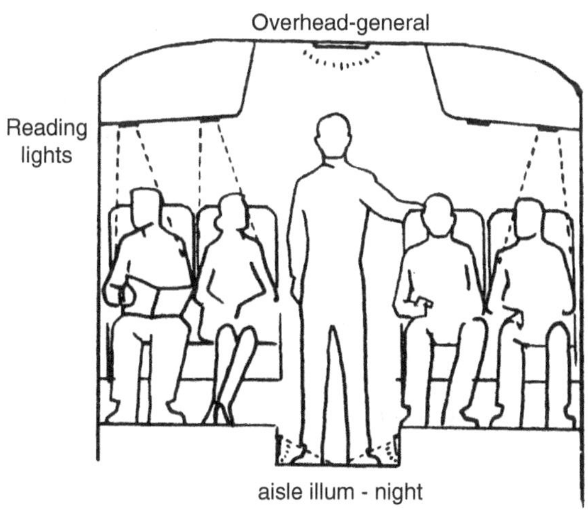

Figure 3.12 Intercity buses—recommended illumination.

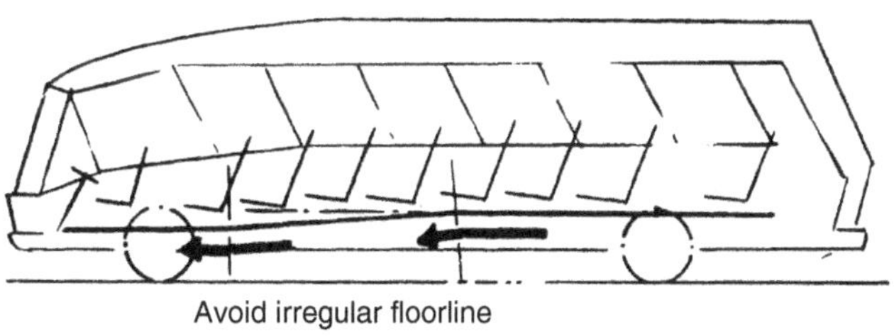

Figure 3.13 Intercity buses—irregular floorline.

parcels above their seats. Although open bins are common, these are not as effective as closed bins; the latter not only are easier to use (since open bin configurations must have some type of strap or elastic barrier to keep packages in place) but also prevent loose baggage from becoming a hazard in the event of a sudden stop or crash.

Dimensions for overhead bins should be determined through mockup evaluations because the primary problem of head clearance when a passenger is entering or leaving the seat will depend on several factors that can be evaluated only in the mockup.

Seat identification should be in a conspicuous place, i.e., one that can be viewed from the standing position, and should be located so that it is obvious which seat is being referenced.

Avoid vehicle configurations in which the aisle floor is not level; not only is a change in angle difficult for certain passengers (e.g., the aged) to negotiate, but also most people will not notice the change in angle and thus may fall.

The same caution should be exercised with respect to the aisle overhead; i.e., avoid any ceiling patterns that require a passenger to lower his or her head in order to avoid contact.

It is desirable to provide air conditioning for all buses, but especially transcontinental buses. Not only does air conditioning regulate temperature and ventilation, but properly designed systems also minimize the effects of exterior noise, generated either by the bus itself or by other traffic.

COMMERCIAL AIRCRAFT

Functional Requirements and Human Factors Considerations That Should Be Addressed at the Earliest Stages of Conceptual Design

Although it is obviously important that the aircraft fly properly and that it be economically feasible, the prime function of the machine is to carry people from one point to another; in order for the machine to do this effectively, various people have to perform certain tasks correctly, and the passengers must be made as comfortable as possible. The following functional requirements, by user and operator categories, should be addressed at the earliest stages of conceptual design in order to minimize the typical constraints that other engineering considerations may impose. Detailed aviation systems design standards can be found in the U.S. Federal Aviation Administration Human Factors Design Standard HF-STD-001. (See Table 3.4.)

1. Convenient ingress to the aircraft
2. Easy access to, and identification of, seats
3. Comfortable seats
4. Windows for external viewing
5. Safety belts
6. Reading lights, general cabin illumination, and aisle illumination
7. Flight attendant call buttons
8. Emergency oxygen masks, life preservers, cushions, and vests
9. Restrooms
10. Serving tables for eating
11. Entertainment outlets and controls
12. Storage for outer garments and personal articles
13. Storage for large baggage

TABLE 3.4 Typical Aircraft System Human–Machine Interface Areas

Flight Crew Workstation	**Emergency Escape, Survival, and Rescue**
Visibility	Exits, chutes, and rope
Display and control layout	Life rafts and flotation cushions
Seating restraint	Survival kits and first-aid supplies
Communications	Radio beacons
Cockpit lighting	**Exterior Lights**
Flight Attendant Stations	Navigation, landing, and anti-collision
Entrance door	In-flight inspection
Galleys	**Ground Support**
Garment storage	Refueling
Communications	Ground inspection
Seating restraints	Emergency repair
Food service equipment	Ramp towing
Passenger Accommodations	Baggage and cargo handling
Seating restraints	Lavatory service
Outside view	Galley service
Eating and entertainment	Overhaul and repair
Reading illumination	
Parcel storage	
Environmental controls	
Oxygen masks	
Restrooms	
Special lounges	

Passenger Service Personnel Needs

1. Seating, safety belts, oxygen masks, life preservers, and vests
2. Galley for preparing food and carts for serving
3. Communications with flight crew and passengers
4. Storage for personal articles
5. Controls for operating doors, emergency exit chutes, and entertainment equipment

Flight Crew Needs

1. Seats, safety belts, and oxygen masks (command pilot, copilot, flight engineer, and government inspector)
2. External viewing capability for taxiing, takeoff, in-flight observation, and landing
3. Controls and displays for aircraft operation, internal and external communication, and cabin environment control
4. Storage for personal articles
5. instrument panel and general cockpit illumination

In addition, the following must be provided for all persons aboard the aircraft:

1. Cabin pressurization
2. Temperature, humidity, and ventilation control
3. Life rafts for overseas configurations

Ground Crew Needs

1. Access for baggage and cargo, catering service, and housekeeping equipment loading and unloading
2. Access for refueling, engine inspection, and servicing
3. Access for overhaul and repair
4. Ground support equipment and communication connections
5. Means for towing

Other Needs

1. External lights (navigation, identification, landing, anti-collision, inspection, etc.)

2. External marking and instructions for rescue operations

3. Emergency exits, markings, and instructions

4. Onboard firefighting equipment and emergency lighting

Passenger Accommodation Considerations

The following are the most common areas of complaint by airline passengers and are due to improper planning in the earlier stages of design.

Seat Identification. Seat identification should be located so that it is readily seen, even when the aisle is crowded with passengers, and the identification label should provide an immediate and obvious indication of which seat is which (i.e., a window seat or an aisle seat).

Seat-to-Lavatory Distance. The longest distance a passenger should have to walk to a lavatory should not exceed about 50 ft (15 m). It should be possible for passengers to see the lavatory identification and occupied and unoccupied signs from a seated position.

Aisle Width. It should be possible for the largest expected passenger to walk down the aisle without bumping into the arms or shoulders of seated passengers.

Seat Width and Interpassenger Clearance. The distance between armrests should accommodate the largest expected passenger's hip width without the passenger's hips fitting tightly between the armrests. Armrests should be wide enough so that adjacent passengers can both rest their arms on the armrest (i.e., they should not have to compete for a place to put their arms).

Fore and Aft Seat Separation. There should be sufficient clearance between seats so that when a person reclines a seat back, the person behind still has plenty of knee clearance, room to put a good-sized magazine on his or her lap, and space to enjoy a meal without staring at the top of the front passenger's head.

Window-Seat Relationship. Window size and placement should be compatible with window use; i.e., the passenger directly adjacent to the window should not cover it up for his or her fellow passenger or have to "scrunch down" in order to see out.

Individual Passenger Service Modules. Overhead service modules should be located so that they can be reached easily by the smallest expected adult passenger. The service controls should be positioned so that each passenger can reach his or her own particular set of controls. Reading lights should be positioned and designed so that they illuminate only the specific passenger's reading material or dinner tray, not that of an adjacent passenger.

Lavatory Size and Layout. Adjacent lavatories should be arranged so that a person who is about to enter one lavatory is not struck by the door being opened by someone who is leaving the other lavatory. Lavatory doors should not face the general array of seated passengers (this is an embarrassment to many people). The interior of each lavatory should be large enough and/or arranged so that the largest expected adult passenger can maneuver inside it, including closing and opening the door. The space should also be compatible with the requirements of a mother who is trying to attend to a small child. At least one of the lavatories should be large enough (and properly equipped) for wheelchair users.

Seat-Belt and No-Smoking Signs. Such signs should be readily visible and readable from any passenger seat. The signs should also be visible and readable even when they are not illuminated so that passengers can identify them and know where to look when the captain or flight attendant indicates that they are in effect (i.e., avoid camouflaging important signs).

Seat-Reclining Controls. Place seat-reclining and other controls where the passenger can see them (positions inside the armrest or on the front, out of sight, are confusing).

Seat Design. Select seat designs that do not create special comfort problems for certain people, e.g., contours that fit only a few people's backs; seat-pan cushions that are too long for short-legged people; headrests that are too low for tall people and hit them in the shoulders, causing them to be uncomfortable; and cushions that are too soft, causing clothing to "bunch" and bind. Make sure that the upright position of the seat does not push the passenger forward into an extremely uncomfortable position. Seat arrangements should not be used that require an "inside" passenger to crawl across more than one person to get to the aisle and back again.

Flight Crew Accommodation Considerations

The following are the most common areas of complaint by airline flight crews and are due to improper planning in the earlier stages of system conceptualization.

Flight Compartment Overhead Clearance. Flight crew members should not have to bend over or watch out for low overhead features when they enter the flight deck compartment or try to get into their seats.

Pilot Seat Ingress and Egress. There should be sufficient clearance between any center console and each of the pilot's seats so that entry or exit is convenient; it should not be necessary to take special precautions to start with the proper foot or lift one's leg over the seat.

Joy-Stick (Wheel) Clearance. The main control device should be designed and/or mounted so that the pilot does not have to go through a series of contortions in order to get past the control column when getting into or out of his or her seat. The wheel should be designed so that the pilot can easily view any displays located directly forward of the wheel. No panel controls should be placed in such a way that the pilot has to reach through or around the control wheel to operate them.

Centrally Located Shared Controls and Displays. Either pilot should be able to reach shared controls without having to lean an appreciable distance. Instruments should be designed and/or positioned so that there are no parallax problems for either pilot (e.g., numbers, scales, or labels that are covered by bezels or other controls and pointer-scale relationships that are distorted because of the angle of view).

Instrument-Panel Illumination. All displays and labels should be equally readable during the day and at night, and the illumination system should be of even quality and brightness so that, if a pilot turns down the illumination level, one or more of the indicia do not disappear.

Window Reflections. By means of proper selection, positioning, etc., colors and lights used in the cockpit should be prevented from reflecting in the windshield or side windows and preventing the pilots from seeing out. The faces of instruments should be tilted so that the cover glass does not reflect the pilot's bright shirt front.

Crew Seat Design. All crew seats should be designed to minimize possible fatigue and stress due to long periods of use. Good posture control is vital; i.e., avoid soft cushions, sling backs, etc.

Adjustable Foot Pedals. Foot pedals should be adjustable so that the short-legged pilot can bring them within reach without having to move the seat forward; remote adjustment is desirable.

RAILWAY VEHICLES: PRIMARY HUMAN FACTORS CONSIDERATIONS

Although there is a wide range of variations in rail system concepts (intercontinental transportation, urban transit, etc.), certain common interfaces between the train crew, the passengers, the maintenance and service personnel, and the vehicle features remain constant for all systems. The following should be considered in the early stages of system development in order to ensure the efficiency, comfort, and safety of train crews, support personnel, and passengers. (See Table 3.5 and Fig. 3.14.)

Onboard, En-Route Train Crew Operations

1. Ingress and egress to the locomotive cab as well as to the principal power system compartment are critical in terms of both nominal and emergency operating scenarios. There should be direct access between the locomotive and other rail cars as well as between the typical external entry modes.

2. The layout of the engineer and assistant workstations should be designed for efficient monitoring of the external roadway, individual systems instrumentation, manipulation of appropriate controls, and the general convenience and comfort of operators.

3. The layout of the various onboard systems and equipment should be designed for ease of inspection, troubleshooting, servicing, and repair.

4. Adequate environmental control should be provided to protect crew members from the typical thermal, noise, and vibration stress produced by onboard systems and equipment.

5. Appropriate support amenities should be provided the crew (a lavatory, drinking water, storage for personal articles, etc.).

6. Appropriate communication systems should be provided both for onboard intercrew communications and for communication with external transportation system facilities.

7. Special consideration should be given to emergency- and safety-related scenarios, e.g., derailment, collision, loss of power for environmental systems, and engineer incapacitation.

Passenger Accommodation

Entryways, aisles, seating and/or special wheelchair tie-down spaces, restrooms, etc., should be given special attention in terms of the problems of passengers who may be in wheelchairs.

1. Exterior car identification should be clear and understandable, and signs and marking should be placed where boarding passengers can readily find them. Instructions should be designed to minimize the possible confusion created by a long series of cars

TABLE 3.5 Typical Train System Human Interface Considerations

TRAIN CREW LOCOMOTIVE ACCOMMODATION
Exterior inspection
Cab ingress and egress
Control stations
System compartment access
Environmenstal control
Exterior visibility
Forward roadbed illumination and exterior signal lights
Audio-visual warning systems
PASSENGER RAILCAR ACCOMMODATION
Exterior car identification
Ingress and egress
Intercar access
Aisleways
Seating
Personal article and baggage storage
Restrooms
Snack bar and dining car
Environmental control
Public address system
SERVICING PERSONNEL MAINTENANCE ACCESSIBILITY
Systems, equipment, and component access
Maintenance hazard protection
Housekeeping
Baggage handling
Coupling and yard handling

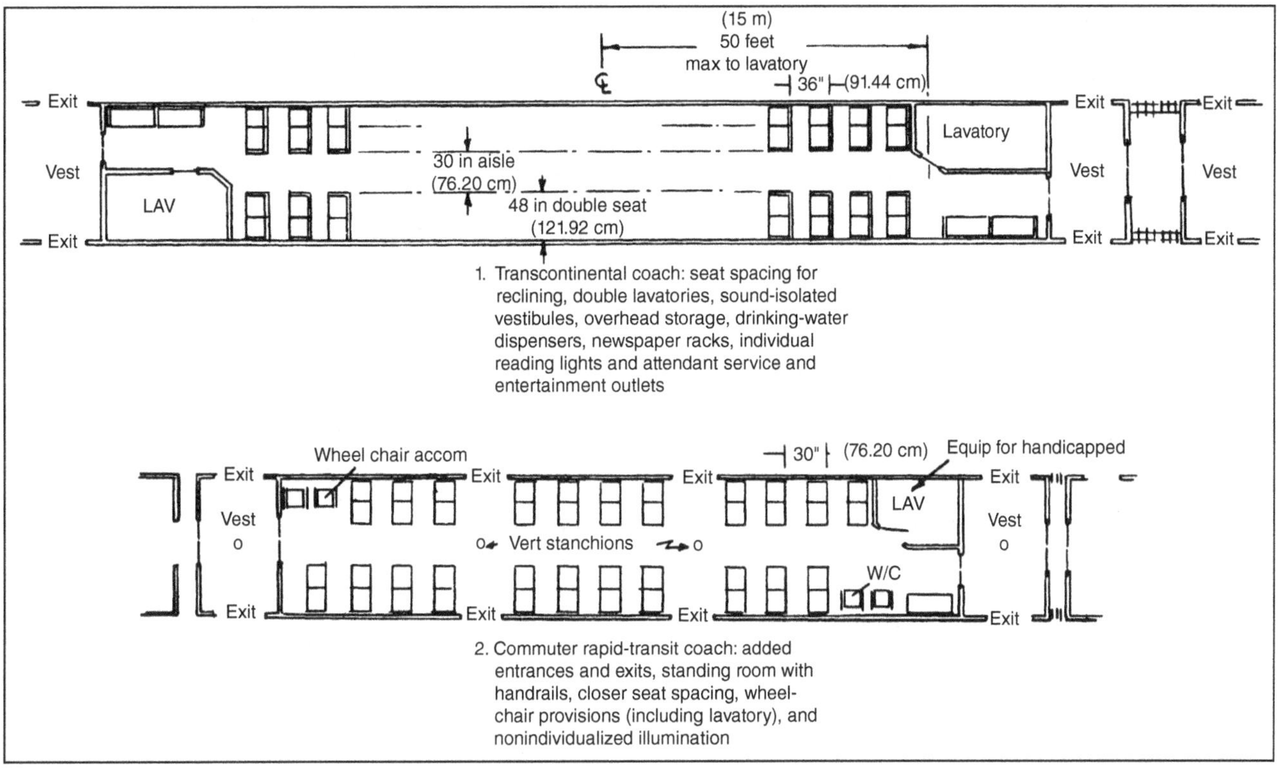

(The above are for illustrative purposes only and are not drawn to scale.)

Figure 3.14 Passenger coach conceptual suggestions.

when a passenger (limited by his or her observing position) must quickly locate a particular car. A typical problem occurs because signs cannot be read until the passenger is alongside the car or because signs are obscured by other passengers standing in front of them.

2. Entrances should be located and positioned for the ease of passengers carrying suitcases. They should provide a level entry for wheelchair use, and there should be appropriate handrails to assist the elderly and children. Entrances should be properly illuminated so that passengers can see where they are stepping. When steps are required to accommodate stations where there is no loading platform, these should be designed to minimize long stretches to reach the first step from the ground, and stair treads should be nonslip so that passengers who may have had to walk in rain or snow will not slip and fall climbing the steps. No unusual structural projections or sharp corners should protrude into the entryway; these could be a hazard for entering passengers. Handrails should be designed to accommodate both entry and exit (i.e., the mobility procedure is different for each situation).

3. Entry vestibules should be large enough to accommodate the passenger who is carrying suitcases, packages, or a small child. Vestibules should be well lighted, and appropriate signs should be provided so that the passenger knows which direction to go when reaching the vestibule. Doors leading from vestibules to the main car compartment should operate easily (i.e., they should not require considerable force to open, and they should not operate in a fashion that requires awkward manipulation of packages or handles).

4. Aisles within the passenger car should be wide enough to allow passengers to negotiate the baggage or parcels they may be carrying. There should be hand holds along the pathway, especially for the elderly, who may be unsteady on their feet. Aisles should be well lighted.

5. Seating should be designed for maximum comfort and convenience; i.e., the seat dimensions and the seat-pan and seat-back angles should provide good posture control. Seats used in trains that travel longer distances should have recline capability. There should be plenty of foot and leg space between seats (including the clearance behind a reclined seat). Independent reading lights should be provided for each passenger, and these should be positioned so that one passenger's light does not annoy an adjacent passenger who may want to sleep. Two-abreast seating should be maintained so that a

passenger who is seated by a window is not required to step around more than one person to get into the aisle. Fixed armrests should be provided on the aisle and window sides of seats, and a folding armrest should be provided between seats. The latter armrest should be wide enough so that both passengers can rest their arms at the same time. Seat-reclining controls should be provided on the fixed armrests. In addition, controls for individual reading lights, attendant call buttons, etc., should also be provided on these armrests. (Airlines have offered these special amenities for some time and, in order for rail travel to be attractive, similar conveniences should be provided rail travelers.) Consider the provision of removable trays at each seat so that food service or other activity is possible at the individual passenger's seat.

6. Whenever possible, restrooms should be provided at both ends of a passenger car so that some passengers are not required to negotiate the entire length of the car to reach these facilities. Restrooms should be properly signed so that passengers know where they are and when they are available.

7. Storage for personal baggage, garments, and parcels should be provided at each seat section. Such storage should be readily accessible for the shortest expected adult passenger. Consider parcel compartment designs that can be closed, which will keep the packages from being dislodged in the event the train suddenly decelerates.

8. Each passenger car should have the following special amenities:

   a. Drinking water

   b. Magazine and newspaper rack

   c. Trash receptacle

   d. Fire extinguisher or extinguishers

   e. Emergency lights (inertial actuated, self-powered)

   f. Emergency ventilation system (in the event the power for the main air conditioning or heating fails)

   g. First-aid kit

9. The entire interior of each passenger car should be made safe in terms of removal of sharp projections, corners, edges, etc.

10. Seat identification should be prominently and clearly displayed so that passengers can easily locate their seats.

11. Provide a public address system so that stations as well as other information or instructions can be announced to all the passengers simultaneously.

12. Provide an appropriate car environmental control panel so that the car attendant can adjust lighting, heat, cooling, and ventilation for each car. Apply appropriate techniques for the control of noise, and design car suspension to minimize the effects of vibration and sidesway.

Equipment Maintenance Features

Rail vehicles, like other public transportation system equipment, require daily servicing and maintenance that is time critical. The ease with which support personnel can accomplish their various tasks is directly related to how much attention is given to this aspect of system operation in the initial stages of conceptual design—to the physical arrangement of equipment and components, the type of enclosures that may be used around equipment, the types of materials used, etc. Special note should be taken of the problems that the activities of one crew may create for another, especially the following:

1. If maintenance technicians have to carry heavy, dirty, or oily equipment through passenger cars, there is a likelihood that this will add to the difficulties of housekeeping crews.

2. If equipment compartments do not have easily operated covers, there is a good chance that the interior trim and materials will be soiled or damaged and thus will have to be replaced to keep them looking decent.

3. If the interior decor, materials, etc., are not carefully selected and appropriately attached, typical cleaning and servicing procedures will soon result in a shabby-looking interior. The bulkhead, ceiling, and floor covering should be easily replaceable so that the basic car can be maintained in a relatively new-looking state. One of the most common complaints of rail travelers is that rail systems (unlike airplanes) are frequently dirty and unattractive.

4. The following critical exterior maintenance considerations should be addressed during initial conceptual planning:

   a. Car coupling and decoupling hazards

   b. Electrical grounding hazards

   c. Lack of illumination at key inspection and repair points

   d. Lack of appropriate safety interlocks to ensure that all maintenance personnel are aware of maintenance in progress (to prevent inadvertent car movement, power activation, etc.)

   e. Adequate provisions for replacement (with ease) of major equipment components (large power units, tanks, wiring cables, etc.)

   f. Adequately designed and positioned worker aids for climbing to and from, and for walking on top of, the locomotive and other rail cars

Typical Engineer and Assistant Operating Tasks to Be Considered in Developing a Locomotive Cab Concept

1. Ingress and egress
2. Starting and stopping the train
3. Controlling speed
4. Negotiating turnouts and crossovers and moving to a side track
5. Negotiating grades
6. Forming a train and/or locomotive consist
7. Detaching a locomotive consist
8. Passing trains and equipment
9. Obtaining clearance and receiving and transmitting messages
10. Managing auxiliary systems
11. Responding to malfunctions

Typical Locomotive Systems

1. Train brake system
2. Drive system
3. Propulsion system
4. Engine system
5. Pneumatic system
6. Dynamic brake system
7. Auxiliary systems
8. Internal environment systems
9. External environment systems
10. Locomotive brake system and track and train situation signal systems
11. Communications system or systems

Other Cab Features

1. Exterior visibility (windows, windshield wiper and washer, defogger, and defroster)
2. Lavatory
3. Drinking water and refrigerator
4. Fire extinguisher
5. Engineer and brakeman seats
6. Crashworthiness features

Note: Typically, two principal operator positions are required: one for the engineer and one for an assistant. The engineer's position is generally placed on the right side of the cab near a window so that the engineer can turn to see the rear. Where practical, the main headlight should be placed below the windshield so that light will not reflect on it, especially in foggy weather. Display control consoles should be laid out so that operators can easily see ahead of the locomotive at the same time they are monitoring and/or manipulating principal control devices. The displays should be positioned and/or hooded, however, so that they will not reflect in the windshield and obscure or confuse frontal external visibility.

ELEVATED AND SUSPENDED RAIL SYSTEM VEHICLES

Passenger cars used for roadbeds that are elevated or suspended present several unique problems, in addition to the interior problems and the problems of ingress and egress of typical rail systems.

Emergency Egress and Rescue

In the event such vehicles become stranded high above the ground because of a system malfunction, there are obvious hazards to passengers who might try to leave the vehicle by the normal exit doors. Conceptually, it is more desirable to confine the passengers until the car can be rescued with them safe inside.

Crash Protection

Typically, elevated or suspended cars need to be constructed of extremely light materials to save weight. This requires consideration of special concepts of energy absorption and passenger restraint to protect passengers in the event the car collides with another car.

Ventilation and Air Conditioning

Since passengers cannot get off at will in the event of a malfunction and since they must wait in the car for some time before rescue, emergency power must us provided to keep the car ventilated and cooled or heated.

Emergency Communication

People who are stranded temporarily may become frightened because of the close confinement high above the ground, and therefore communication with system authorities should be provided so that passengers will know that someone is aware of their plight and also so that they will know when they can expect to be rescued.

STREETS, ROADS, AND HIGHWAYS

Street, Road, and Highway Systems Concept Development

The primary objective of any street, road, or highway system is, of course, to provide vehicular and pedestrian links between geographic points. Since the main purpose is to make it possible for people and their vehicles to get from one point to another, any proposed system should be people oriented. From a user's point of view, any system should provide the following.

Convenience. There should be the opportunity for as rapid and direct movement as possible from where a person is to where that person wants to go. User convenience differs. For example, pedestrians need sidewalks, street crosswalks, and pedestrian overpasses and underpasses to cross high-speed expressways and freeways; local motorists need arterial streets between major shopping areas, restricted but connecting residential streets, and means for crossing or by-passing major expressways and freeways; and transient motorists need noninterrupted pathways around or across communities, methods for changing from one major highway to another, and periodic access to local services.

Orientation. The system organization, layout, and identification features should make it easy for motorists and pedestrians to know where they are, to know what path they should follow to get where they want to go, and to know they are in the right place when they get there.

A uniform street grid oriented to a north–south and east–west reference is easy to understand and follow. Odd and non-uniform patterns are hard to understand and retain in one's mind; angular, rather than perpendicular, intersections cause people to lose their direction; and traffic circles cause people to forget where they are. Streets that do not continue across an intersection leave one in doubt as to where the continuing path is. (See Figs. 3.15–3.17.)

Safety. The system should not include any built-in booby traps for the motorist or pedestrian, and any features that could be struck should be designed so that injury potential will be as minimal as possible. Typical built-in booby traps to watch for include the following:

1. For Motorists

   a. Blind corners, curves, and hilltops

   b. A sudden stop at the end of a sharp, blind curve

   c. Sudden narrowing of the road and disappearance of a lane

   d. Short preparation distance between an advisory sign and the decision and maneuver point

   e. Sharp, unbanked curves

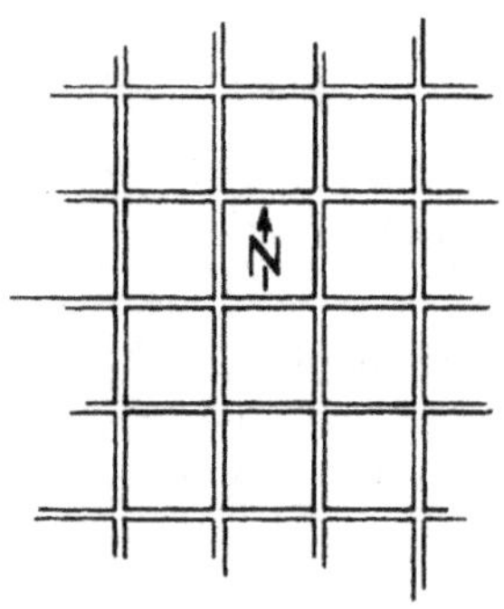

FIGURE 3.15 Road patterns—desirable patterns.

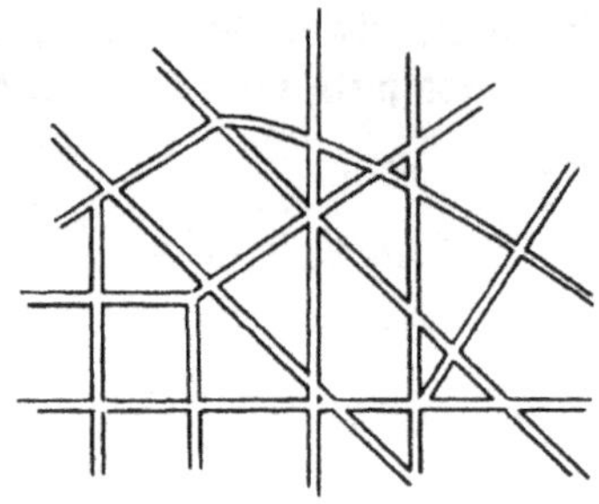

FIGURE 3.16 Road patterns—haphazard patterns.

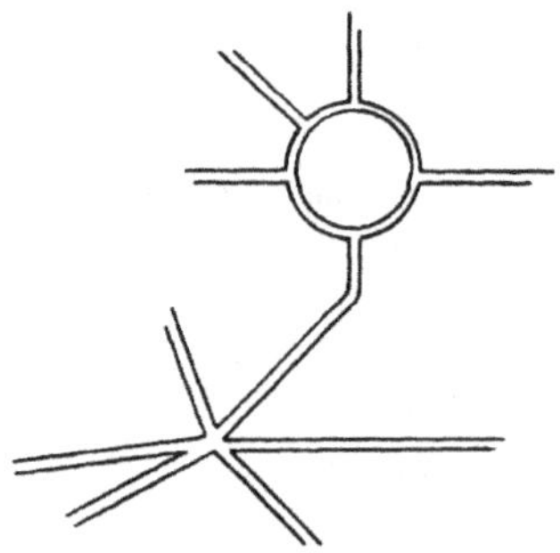

FIGURE 3.17 Road patterns—non-uniform intersections.

   f. Varying freeway interchange concepts (right versus left, under versus immediate approach to left overpass route, etc.)

   g. Nonuniform curvature (radius) of on- and off-ramp turn

   h. A common approach road used for both on and off ramps to and from a freeway

   i. Ill-defined road edges (lack of contrast)

   j. Lack of lane marking on multilaned street, frontage, or off-ramp roadways

   k. Advisory or warning signs not located on the side of the road on which motorists normally look for them

   l. Signs positioned so that other vehicles block motorists' view of them

   m. Overhead advisory lane arrows that do not coincide with lanes from the driver's viewing angle and lane designations that shift back and forth on succeeding signs

   n. Highway names that imply a location rather than a route (e.g., the name "Riverside" implies that the road goes to that city, even though motorists need to take the highway in the opposite direction)

o. Marked bike lanes that are also used for curbside parking

p. Deep ditches alongside the road without appropriate guardrails

q. Abrupt changes or irregularities in the road surface

r. Lack of a "slow truck" lane on steep hills

2. For Pedestrians

a. Crosswalks at blind curves

b. Automobile stop lines not set back from the crosswalk

c. Lack of sidewalk along the roadway

d. No place to walk along the freeway or cross the center divider in an emergency

e. No sidewalk on bridges

f. A sloping driveway that crosses a sidewalk

g. Lack of a wheelchair ramp at an intersection

h. Lack of intersection illumination

i. Light-signal duration that is too short to allow pedestrians to cross fully, and lack of a safety platform in the center

j. Lack of crosswalk safety marking and no tactile cues for the blind to use in following the crosswalk

3. For Bicycle and Tricycle Riders

a. No separate bike lanes and lanes that disappear at corners and on bridges

b. Curbside automobile parking allowed in bike lanes

c. Sidewalks that are directly adjacent to residential streets, making it possible for children to ride off the curb and into the street

4. Other safety considerations in planning street, road, and highway systems include the following:

a. Perpendicular drive and road entries to main highways or thoroughfares

b. Energy-absorbing barriers at bridge abutments and roadway divisions

c. Crash barriers at the ends of pedestrian center-island platforms

d. Guardrails between center-island pedestrian waiting platforms and adjacent vehicle lanes

e. Freeway emergency pull-off areas

f. Oncoming headlight glare barriers

g. Nonskid road surfaces

h. Rumble strips and raised lane markers for the nighttime warning of motorists

i. Illumination at all intersections, on bridges, and in tunnels

j. Following the Manual on Uniform Traffic Control Devices principles and standards for implementation of appropriate marking and signing.[7]

Other Considerations Pertinent to Conceptual Planning

Traffic-Noise Attenuation. The problem of potential traffic noise, especially in certain areas of a city, should be addressed very early in the planning stage, since there are various ways to avoid creating the problem in the first place. These include routing heavily traveled thoroughfares away from residential areas, hospitals, schools, etc., and taking advantage of natural noise barriers, such as hills and groves of trees, and of prevailing winds that will carry the sounds away from critical areas.

Landscaping. Advanced landscape planning is important not only from the standpoint of aesthetic objectives but also in terms of precluding visual sight problems for motorists and of providing a reasonable system for maintenance that will not introduce traffic bottlenecks, e.g., the interference that is often caused by highway landscape crews and equipment when they have to park along the highway or by sprinkler systems that spray water across a roadway, thus creating hazards such as wet roadways and spray on the motorist's windshield.

Other Built-in Hazards

1. Bridge rail designs that block the motorist's view of merging traffic because the designer was not aware of how important it is to create a railing pattern that the motorist can see through or past.

2. No pull-off space on the centerline side of a multilane freeway. In many in stances a motorist who is in trouble cannot get across the several lanes to the typical emergency shoulder.

3. No pull-off space on very long bridges or in tunnels.

4. Lack of provisions for reducing the blinding effects of oncoming headlight glare, especially at curves.

5. Introduction of sharp dips or humps in the roadway, which can cause a motorist who is driving too fast to lose control.

6. Use of alternating concrete–macadam (light and dark) merging. Although lane patterns are marked, this may confuse the motorist (especially at night), who tends to follow the color of the surface rather than the lane marking.

[7]U.S. Department of Commerce, Bureau of Public Roads, *Manual on Uniform Traffic Control Devices for Streets and Highways*, 2009.

Intersection Designs. In many areas of the world, traffic intersections are changing from direct road crossings (sometimes controlled by stop signs or lights) to roundabouts. In a roundabout, traffic is diverted to flow in a circular direction around a central hub. Traffic can then exit the circle onto one of the intersecting roads. Studies show the superiority of roundabouts in many situations. Roundabouts reduce the possibility of high-speed collisions between cross vehicles. Roundabouts can also speed traffic flow by eliminating delays caused by unnecessarily long traffic lights. Roundabouts are standard in some countries and are becoming more popular in the United States.

Street and Roadway Systems Planning Relative to Mobility of Handicapped Pedestrians

In most urban areas, there is a need to increase the mobility capability of certain handicapped persons. The principal user of concern is the person who is blind or partially sighted. Although it is important to provide sufficient smooth shoulder along unimproved residential roadways even for the pedestrian who is not handicapped, it is extremely important to plan for such pedestrian paths for the blind. Those who are totally blind or almost blind usually must find their way by using a cane. The pathways must provide a high contrast that can be detected by the partially blind who may be able to see light and shadow, but not detail. These pedestrian pathways should be designated and controlled in such a way that obstructions such as fire hydrants and telephone poles are set outside the path.

The urban signing system should be standardized as much as possible. That is, only certain kinds of posts or poles should be allowed at intersections, and these should be placed in the same relative positions so that blind persons can easily determine where they are at an intersection and so that they can know which poles have tactile identification devices on them.

In fully developed areas where sidewalks are built, these should be large enough for two pedestrians in wheelchairs to pass each other; they should also be wide enough to accommodate a fire hydrant or other utility within the sidewalk area without taking up the space needed by handicapped pedestrians. No poles should be supported by guy wires that could be run into by a blind person who is feeling along the surface with a cane. All standard street, bus, and other roadway signs must be placed above the heads of pedestrians.

Freeway Lane-Drop Concepts

Under certain circumstances it is necessary to reduce the number of lanes on a freeway. In many current designs, considerable confusion is created because the lane drops are not physically well defined; they are difficult to see because they tend to blend into the background, or they are hidden by topographic features, such as the crest of a grade. In addition, advance warning signs or traffic control devices are frequently misleading or obscure. The following guidelines will help eliminate some of the obvious confusion that motorists may experience:

1. The lane drop should be placed where the surface of the roadway remains continuously visible for a significant distance.

2. The lane drop should not be located near "attention-dividing" conditions, e.g., at ramps or where there is complicated directional signaling or signing.

3. The lane-drop taper should provide conspicuous visual clues that inform the driver that the lane is ending and which allow a smooth rather than an abrupt lane transition in the taper area.

4. The lane drop should be located on the side of the freeway that is best with respect to given traffic and geometric considerations.

5. The lane should "appear" to end on the same side of the freeway as the operational lane drop.

6. When a lane drops at an exit ramp, an escape area of appropriate dimensions should be provided to allow for smooth transition into through lanes.

7. When a lane is added at an on ramp and dropped at a nearby off ramp, entering drivers should be warned well in advance that the lane in which they are traveling is not a continuous lane for through traffic.

8. Traffic control devices used in conjunction with lane-drop configurations should be consistent and designed so that they can be "read" sufficiently ahead of decision-making points.

Uniform Traffic Control for U.S. Highways[8]

Definition of a Traffic Control Device. A sign, signal, marking, or other device used to regulate, warn, or guide traffic, placed on, over, or adjacent to a street, highway, private road open to public travel, pedestrian facility, or shared-use path by authority of a public agency or official having jurisdiction, or, in the case of a private road open to public travel, by authority of the private owner or private official having jurisdiction.

Definitions Relating to Signs

1. Sign—Any traffic control device that is intended to communicate specific information to road users through a word, symbol, and/or arrow legend. Signs do not include highway traffic signals, pavement markings, delineators, or channelization devices.

2. Guide sign—A sign that shows route designations, destinations, directions, distances, services, points of interest, or other geographical, recreational, or cultural information.

3. Lane-use control signal—A signal face displaying indications to permit or prohibit the use of specific lanes of a roadway or to indicate the impending prohibition of such use.

[8]U.S. Department of Commerce, Bureau of Public Roads, *Manual on Uniform Traffic Control Devices for Streets and Highways*, June 1988.

4. Sign legend—All word messages, logos, pictographs, and symbol and arrow designs that are intended to convey specific meanings. The border, if any, on a sign is not considered to be a part of the legend.

5. Regulatory sign—A sign that gives notice to road users of traffic laws or regulations.

6. Warning sign—A sign that gives notice to road users of a situation that might not be readily apparent.

Definitions Relating to Markings

1. Center line markings—The yellow pavement marking line(s) that delineates the separation of traffic lanes that have opposite directions of travel on a roadway. These markings need not be at the geometrical center of the pavement.

2. Channelizing line markings—A wide or double solid white line used to form islands where traffic in the same direction of travel is permitted on both sides of the island.

3. Delineator—A retro reflective device mounted on the roadway surface or at the side of the roadway in a series to indicate the alignment of the roadway, especially at night or in adverse weather.

4. Edge line markings—White or yellow pavement marking lines that delineate the right or left edge(s) of a traveled way.

5. Lane line markings—White pavement marking lines that delineate the separation of traffic lanes that have the same direction of travel on a roadway.

Traffic Sign Categories

1. Regulatory signs—Provide notice of traffic laws and regulations, which usually are unenforceable in the absence of such signs.

2. Warning signs—Call attention to conditions that are potentially hazardous to traffic operations and that otherwise may not be immediately apparent.

3. Guide signs—Show route designations, destinations, directions, distances, services, points of interest, and other geographical or cultural information. (See Fig. 3.18.)

Sign Shape Codes

1. Octagon is reserved exclusively for the stop sign.

2. Equilateral triangle with one point downward is reserved exclusively for the yield sign.

3. Round is reserved for the advance warning of a grade-crossing.

4. Pennant, an isosceles triangle with its longest axis horizontal, is used to warn of "no passing" zones.

5. Diamond is used only to warn.

6. Rectangle, ordinarily with the longer dimensions vertical, is used for regulatory signs—with the exception of stop and yield signs.

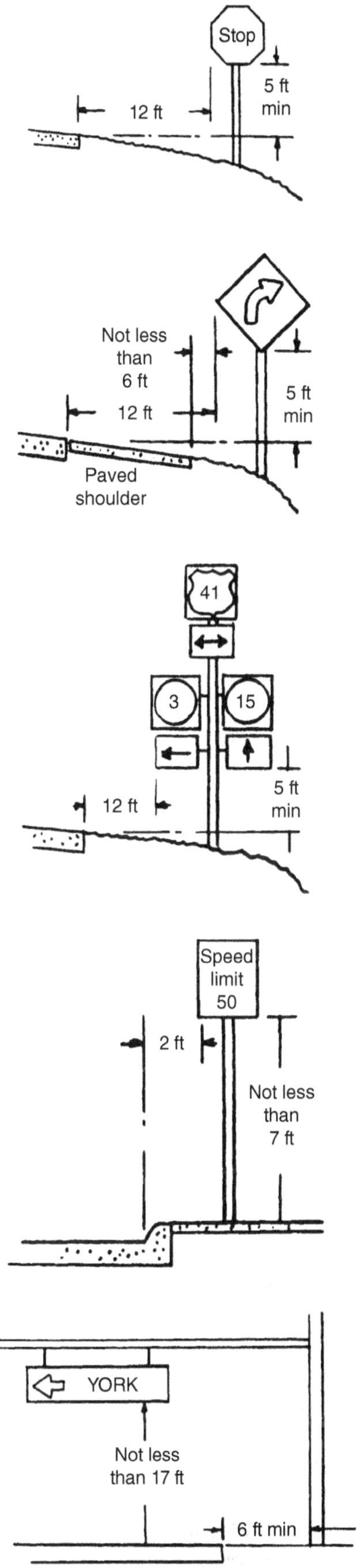

Figure 3.18 Traffic signs.

7. Trapezoid shape may be used for recreational-area guide signs.

8. Pentagon, with the point up, is used for school advance warning and school crossing signs.

Sign Color Codes

1. Red—Used as a background color for stop signs, multiway stop supplemental signs, *Do Not Enter* messages, *Wrong Way* signs, and on interstate highway shields; as a legend color for yield signs, parking prohibition signs, and the circular-outline and diagonal-bar prohibitory symbol.

2. Black—Used as a background on one-way signs, certain truck weigh-station signs, and night speed-limit signs specified in the Manual on Uniform Traffic Control Devices. Black is also used for messages on white, yellow, and orange signs.

3. White—Used as a background on route markers, guide signs, and regulatory signs (except stop signs) and for the legend on brown, green, blue, black, and red signs.

4. Orange—Used exclusively as a background color on highway construction and maintenance signs.

5. Yellow—Used as a background color for warning signs and for school signs.

6. Brown—Used as a background color for guide and information signs related to points of recreational or cultural interest.

7. Green—Used as a background color for guide signs (except those using brown or white) and mileposts, and as a legend color with background for permissive parking regulations.

8. Blue—Used as a background color for information signs related to motorist services and the evacuation route marker.

Roadway Sign Location. Street and road signs must be located where the motorist expects to find them, and they should be positioned far enough in advance of the motorist's proposed change of path so that he or she has time to prepare for the change.

1. Ordinarily, a motorist expects a street sign to be located on his or her side of the street. However; a motorist also needs to verify that he or she is still traveling on the right street. This type of location for street signs is applicable to all but the very widest streets. Signs used for city streets should be as large as necessary to be seen far enough in advance (i.e., lettering heights should be based on the average speed range of the street system) so that motorists do not have to slow down at every intersection to read the signs. (See Fig. 3.19.)

2. When very wide streets and broad intersections are involved, a large central sign in the center of the intersection is desirable because motorists may be too far from a curbside sign to see it above the tops of large vehicles or from the most distant lane in which they may be traveling or stopped. (See Fig. 3.20.)

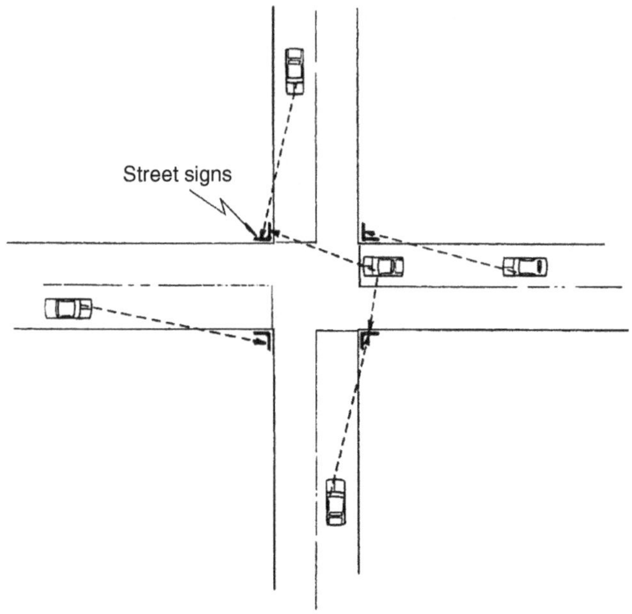

FIGURE 3.19 Typical street sign location.

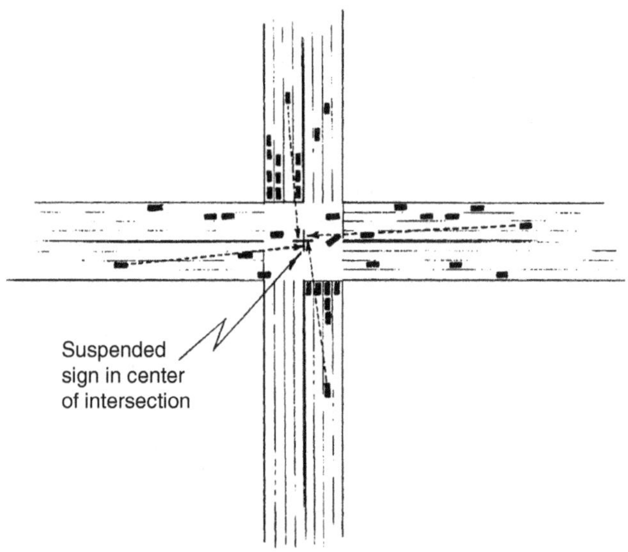

FIGURE 3.20 Sign location for wide streets.

3. A minimum of two sign locations should be considered for the high-speed highway turnoff situation. The first sign must be positioned far enough in advance so that the motorist can see and read it and still have time to prepare for the various maneuvers that he or she must make. A second sign should also be provided to reconfirm that the off ramp is the one the motorist desired.

Signal Lights[9]. Early planning should include the location of appropriate signal lights to ensure the most effective flow of

[9] U.S. Department of Commerce, Bureau of Public Roads, *Manual on Uniform Traffic Control Devices for Streets and Highways*, June 1988.

traffic in both urban and rural areas. A three-color system should be used: red for "stop," yellow for "caution," and green for "go ahead." Because there are many colorblind drivers, the colors selected for these basic hues should meet the requirements in *Vehicle Traffic Control Signal Heads and Traffic Signal Lamps,* published by the Institution of Transportation Engineers.

All traffic signal lights should be located consistently throughout a given system; i.e., they should be located on the corners nearest the lane of traffic being regulated. When very wide freeway configurations are subject to signal lights, it is desirable to place a second set of lights in the center of the roadway.

Additional early-warning signal lights in advance of an impending signaled intersection are desirable for high-speed roadways in order to give the motorist time to slow down for a red light that is about to be sequenced.

Requirements for signal placement are defined in the U.S. Department of Transportation's Manual on Uniform Traffic Control Devices. Signals should be located so that they are visible well before the stop line. Table 3.6 shows the minimum sight distance for signal visibility as a vehicle approaches the signal. The column in the table labeled "85th Percentile Speed" refers to the speed at or below which 85 percent of the motor vehicles travel.

As the vehicle approaches the signal, it should be within the normal visual range of the driver—20° from the forward vision axis. Figure 3.21 shows recommendations for lateral (side-to-side) and longitudinal (distance to the front) signal

TABLE 3.6 Signal Visibility

85th-Percentile Speed (mph)	Minimum Site Distance (ft)
20	175
25	215
30	270
35	325
40	390
45	460
50	540
55	625
60	715

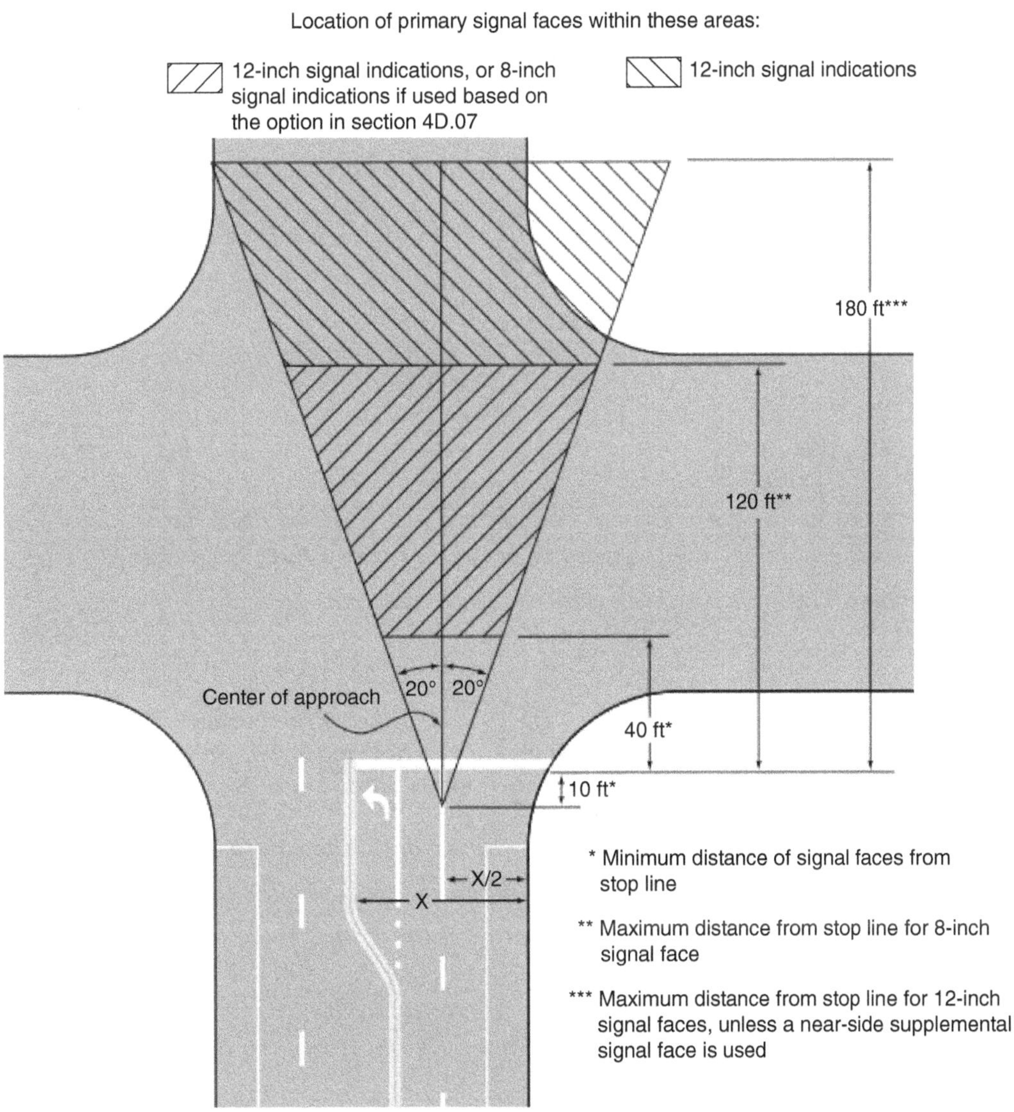

Figure 3.21 Recommendations for latitude and longitudinal signal placement.

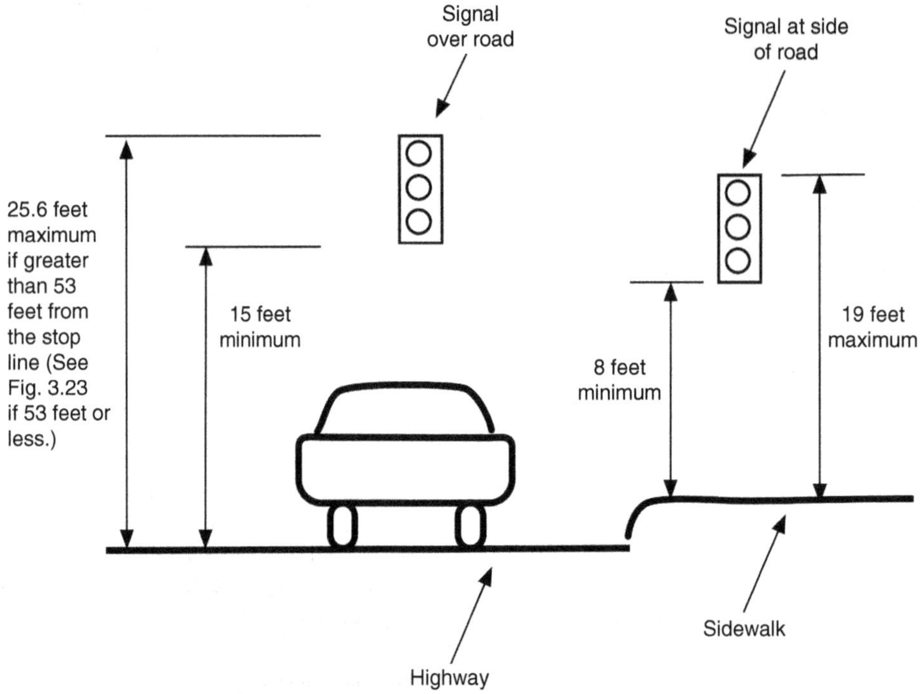

Figure 3.22 Recommendations for signal height.

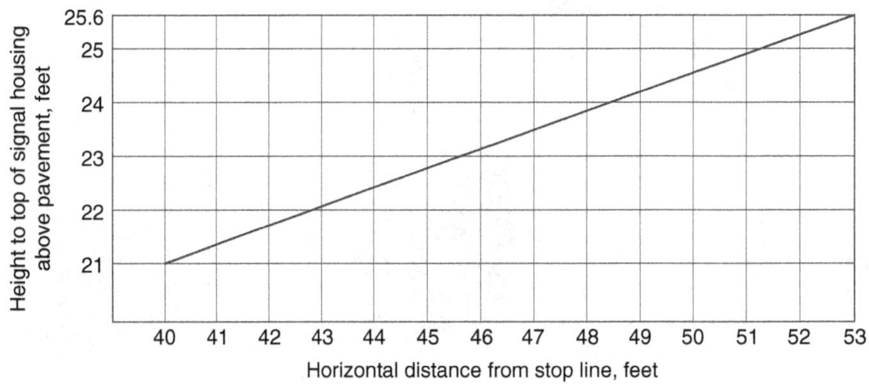

Figure 3.23 Maximum signal height at 53 ft or less from the stop line.

placement. The distances depend on the diameter of the light signal face (8 or 12 in.). The driver is assumed to be 10 ft in back of the intersection stop line.

Figure 3.22 shows that minimum heights are for clearance and maximum values keep the signal within the driver's visual range. The maximum height can vary with the distance from the driver. The signal must be lower if it is close to the driver so that it remains within the optimum visual cone. Figure 3.23 shows maximum signal heights for signals 53 ft or less from the stop line.

CHAPTER 4

MILITARY AND GOVERNMENT SYSTEMS

GENERAL

General Considerations in Development of Military Systems

Military missions demand maximum performance efficiency from both the hardware and the personnel elements of their systems. Thus, the military establishment places more strict requirements on the designer and manufacturer in terms of making sure that people can operate equipment with minimum error. In addition, the military establishment is particularly sensitive to the problem of creating equipment that is too sophisticated for the types of people it has available to operate its equipment, since this not only induces system failures but also imposes lengthy and costly training requirements on the services. In initiating any new military system concept, the designer and/or planner should be thoroughly familiar with at least two important documents:

1. *MIL-STD-46855 Human Engineering Requirements for Military Systems Equipment and Facilities:* This document outlines the responsibilities of both the procuring agency and the system manufacturer for creating a system in which system operators are assigned appropriate roles in equipment operation and maintenance, as well as a system designed to minimize operator and maintain or failures, excessive training demands, and unnecessary hazards when the equipment is delivered and put in operation. The document specifies the analysis, documentation, design review, and testing required to ensure that the final system will meet the needs of the procuring agency and the operating forces.

2. *MIL-STD-1472 Human Engineering Design Criteria for Military Systems, Equipment and Facilities:* This document provides an array of good human engineering design principals, criteria, and guidelines that are based on research and operating experience. Development contractors are expected to follow these guidelines in developing their system concepts and their final design of human–machine interface hardware. When they think they cannot, they are required to explain this in writing.

There are nearly 3 million active and reserve personnel in the United States military (based on 2011 estimates). Humans are both a critical and expensive component of military systems. It is therefore important that the developers of new systems consider all aspects of human involvement in the system. The military refers to this comprehensive view as "Human Systems Integration" or HSI. Human factors engineering addresses one aspect of HSI, focusing on the interface of humans with the hardware, software, and the environment of the system. Other aspects (or "domains") of HSI include the following:

1. *Manpower*—The number of military and civilian personnel required and available to operate, maintain, sustain, and provide training for systems.

2. *Personnel*—The cognitive and physical capabilities required to be able to train for, operate, maintain, and sustain material and information systems.

3. *Training*—The instruction or education and on-the-job or unit training required to provide personnel their essential job skills, knowledge, and attitudes.

4. *System Safety*—The design features and operating characteristics of a system that serve to minimize the potential for human or machine errors or failure that causes injurious accidents.

5. *Health Hazards*—Consideration in the design features and operating characteristics of a system that create significant risks of bodily injury or death; prominent sources of health hazards include acoustics energy, chemical substances, biological substances, temperature extremes, radiation energy, oxygen deficiency, shock (not electrical), trauma, and vibration.

6. *Soldier Survivability*—The characteristics of a system that can reduce fratricide, detectability, and probability of being attacked and minimize system damage, soldier injury, and cognitive and physical fatigue.

7. *Habitability*—The living and working conditions that are necessary to sustain the morale, safety, health, and comfort of the personnel.

Each of the United States military branches (Army, Air Force, and the Navy) has programs to address HSI.

These elements or domains of HSI all address the human in the system and they must integrate successfully to assure an effective, efficient, affordable, and safe system. On a system development project, human factors engineering personnel must coordinate their work with system designers as well as representatives from the other HSI domains. The findings and knowledge can be shared between the domains. System design solutions must incorporate a balanced address of the needs of all the HSI domains.

Figure 4.1 shows some of the information that might be exchanged between a human factors engineering program and the other HSI domains.

In general, military system users will fall within a well-defined age group, the range of their body sizes will be fairly well defined because of selection procedures, and their backgrounds, in terms of education and experience, will be fairly well defined by service entrance requirements. Note that this is considerably different from the situation that exists with regard to nonmilitary systems, which must be designed to fit a much broader range of users who are often ill defined because of lack of research on the general population. Many of the known facts about anthropometry, physiology, psychology, human–machine system interaction, etc., have been generated from military-sponsored studies. These data are available in the references previously noted, which provide a ready source of information that can be used very early in the system concept development stage. The following are among the key human factors considerations that will invariably be addressed in system conceptualization and concerning which the planner and designer can find suitable information available:

1. Military personnel anthropometric data

2. Military personnel biomechanical performance data

3. Military personnel information-processing performance data

4. Military personnel control response performance data

5. Military personnel physiological and psychological data

6. Environmental reaction data

Appropriate examination and utilization of these data can provide the planner of a new system concept with sound bases for determining when and how to use people in the proposed system (i.e., when a human should be required to perform a system function or when the function should be automated), the best type of interface for the particular manual function in order to ensure maximum human response effectiveness, and the kind of living or working environment that is required to preclude injury or maintain effective human performance.

Typical Functional Considerations for Military Systems

Hand-Held Weapons

1. Unpacking, assembly, and emplacement

2. Adjustment and calibration

3. Carrying and stowage

4. Loading, aiming, and firing

5. Disassembly and cleaning

6. Service and repair

7. Personnel hazards

8. Training and training devices

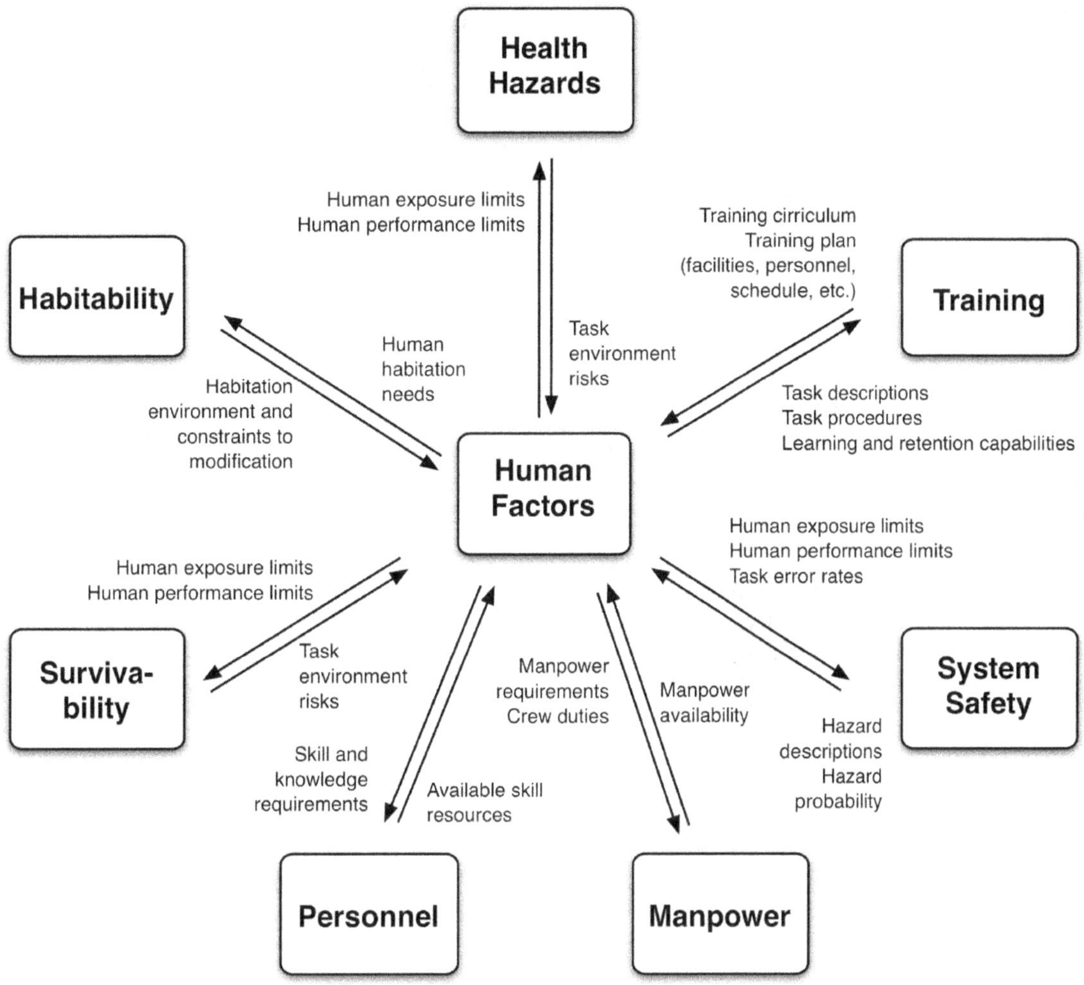

Figure 4.1 HSI domains.

Large Non-Hand-Held Weapons

1. Transport
2. Uncovering, erection, and emplacement
3. Adjustment and calibration
4. Loading, aiming, and firing
5. Disassembly and cleaning
6. Service and repair
7. Personnel hazards
8. Training and training devices

Vehicular Systems: Personnel and Weapons Carriers

1. Ingress and egress
2. System checkout
3. System operation
4. Personnel accommodations
5. Service, maintenance, and repair
6. Stowage
7. Personnel hazards
8. Training and training devices

Communications Command and Control

1. Installation, activation, adjustment, and calibration
2. Operation
3. Maintenance
4. Dismantling, loading, and transport
5. Personnel accommodation
6. Personnel hazards and working environment

Ground-Based Housing

1. Space accommodation and arrangement
2. Living and working environment
3. Security and safety
4. Housekeeping and maintenance
5. Assembly and transport (mobile)

Areas of Analysis

For all military systems, the following areas of analysis should be addressed from the very beginning and refined throughout the system development:

1. System Mission, Operational Requirements, and Function Definition: This information provides the basis on which all other concept and design decisions are made and justified.

2. Human versus Machine Function Allocation: This is a trade-off necessary to ensure that humans are assigned tasks that they are best capable of doing.

3. Operator and Maintainer Hardware Interface Selection and Design: This information ensures that the devices the human must use are appropriate to his or her physical and mental capabilities and do not make demands upon the human that may cause failure to perform as planned.

4. Human-Environment Compatibility: This information determines how to preclude and/or minimize health hazards and also defines habitability control limits that will ensure the maximum performance efficiency of operators, maintainers, and personnel in general.

5. Job and Task Descriptions: This information helps determine the kinds of skills that are required and the number of people that will be needed to operate and maintain the system. It also provides the basis for defining training requirements necessary to bring personnel skills to the level required to ensure satisfactory system operation.

6. Training Concepts and Plan: This information is necessary in order to determine training content and method, training aids, devices and/or simulation hardware support, and the scheduling required to design, develop, fabricate, and install all elements noted, not the least important of which is scheduling of personnel, instructors, and students.

7. Training Equipment and Facilities: This is a critical area in many instances because it involves fairly extensive hardware and/or facilities design and construction. It must be started early enough to ensure that classrooms, simulators, and other complex systems required for training are ready prior to the delivery of the basic system.

8. Operating and Training Publications: Operations, maintenance, and other technical publications are required to support both the training operations and the regular operations of the system. The length of time between the date on which sufficient information is available and the date on which the documents are needed generally is very short, and thus sensitive communication between the design and publications staffs is required on a continuing basis throughout system development.

9. Test and Evaluation: This is one of the most important requirements imposed by military system agencies, since it implies the need to "prove the system" (both human and machine) prior to delivering it for field use. Human engineering test and evaluation, however, begins early in any program and continues at successive points to the last and final field test. Plans for this program must dovetail with other engineering test plans in order to preclude duplication or omission.

Military agencies typically require documentation of the above activities. These documentation end items often are identified as follows:

1. Personnel and Equipment Data: A general file of all data generated during system development and relating to the human factors.

2. Human Engineering: Specific human–machine interface design results.

3. Manpower, Personnel, and Training Requirements information: Data on tasks, skill levels, job descriptions, and procedures suitable for defining training objectives, personnel selection, etc.

4. Training Concept and Training Plans: A detailed plan and schedule for preparing the training cadre and the students in time to operate and maintain the delivered system.

5. Training Equipment Development Plan: Hardware and facilities procurement specification and schedule.

6. Technical Orders and Manuals Development Plan: Preparation of all publications required to support training and the operation of the system.

7. Personnel Subsystem Test and Evaluation Plan: A plan integrated with general engineering test and evaluation to ensure proper involvement of human factors specialists in predesign, detailed design, engineering testing, mockup evaluations, system simulations, supporting human factors research, and predelivery and postdelivery field tests in which both people and hardware or software components are evaluated in real-time operation of the system.

8. Final Human Engineering Reports: A separate human factors report generally must accompany the final engineering report delivered by a contractor to the procuring agency.

Figure 4.2 illustrates how human factors activities should progress parallel with prime and supporting hardware development. Note that all three emanate from the same initial decisions concerning operational requirements, mission definition, and functional requirements and that all interact throughout the development so that the final human–machine system can provide a total compatibility among all elements.

AIRCRAFT

Human Factors Considerations in Military Aircraft Conceptual Design

Although many of the human factors considerations are the same in the case of both military aircraft and private or commercial aircraft, certain military operational requirements create unique problems that must be resolved in the design of military aircraft systems. The following are among these special problem areas.

High Altitudes. Certain types of military aircraft operate at extremely high altitudes, where the hazards of loss of pressurization require use of special pilot protective garments and consideration of unique emergency ejection systems.

High Performance. Unusual dynamic stress on the pilot during critical maneuvers requires giving special consideration to the pilot's protection, while at the same time ensuring that the pilot's performance as a controller remains intact.

Combat Hazards. Armor may be required to protect the pilot from gunfire or shrapnel, glare shield may be required to protect the pilot's eyes from nuclear flash blindness, and/or the pilot must be protected for emergency ditching, i.e., when the pilot is required to eject at high or low altitudes, when the aircraft is in difficult positions, or when the pilot has to survive in rough or frigid seas.

Typical Aircraft Design Areas Requiring Consideration of Human Factors

Airframe Design

1. Facilities for passage and movement of crew
2. Flooring
3. Headroom
4. Passageways
5. General illumination
6. Sharp protrusions
7. Interior doors and hatches
8. Exterior doors
9. Door checks and locks
10. Emergency escape provisions
11. Emergency landing provisions
12. Ditching structure
13. Installation of heavy equipment
14. Turnover structure
15. Openings for maintenance, loading, airdrop, etc.
16. Body pressurizing
17. Soundproofing
18. Upholstering and carpeting

Transparent Assemblies

1. Materials selection (e.g., impact, color, and shatter resistance)
2. Visual distortion
3. Angle of incidence
4. Degree of curvature
5. Light transmissibility and reflection
6. Sighting area
7. Scanning area
8. Sliding canopies and locks
9. Bullet-proofing and bird-proofing
10. Windshield clearing
11. Thermal barrier

Gear Retracting System

1. Manually operated
2. Power operated
3. Gear position indicator
4. Landing gear controls
5. Emergency control
6. Malfunction warning
7. Ground safety locks

Personnel Safety

1. Walkways
2. Steps, handgrips, and rails
3. Armor plate

Other

1. External maintenance access openings and covers
2. Cargo doors and opening and closing systems
3. Weapons attachment

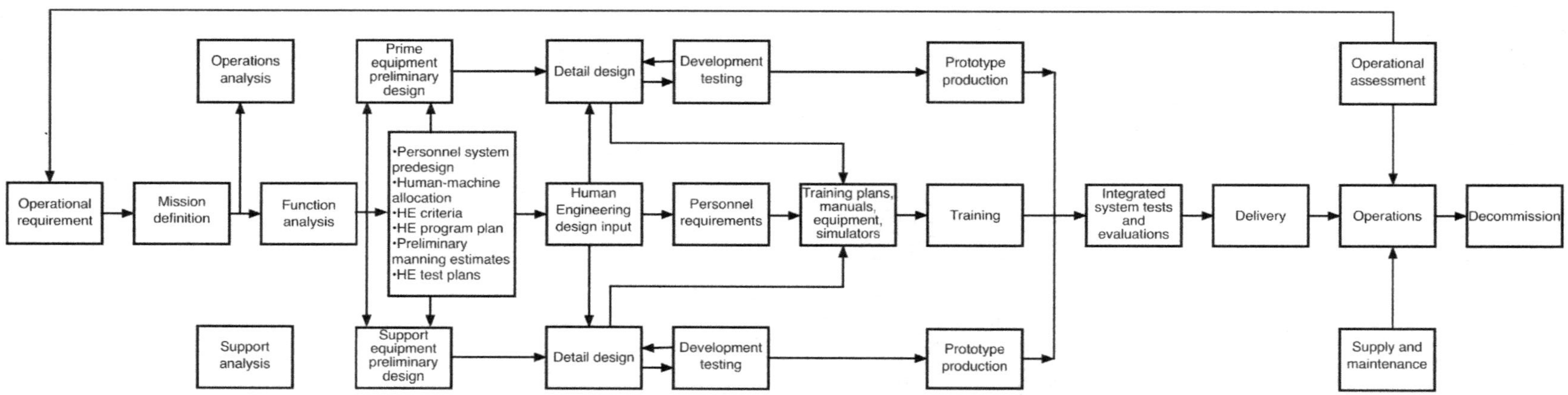

FIGURE 4.2 Typical military system development progression. (From D.W. Conover and W.E. Woodson, *Human Engineering Guide for Equipment Designers*, University of California Press, 1964.)

4. Soundproofing
5. Watertightness
6. Pressure-tightness

Aircrew Stations

1. Workplace arrangement
2. Vision (internal and external)
3. Seating (nominal and ejection)
4. Separable crew compartment
5. Flight controls and displays
6. Engineering controls and displays
7. Special operations controls and displays
8. Internal and external communication controls and displays
9. Weapons systems controls and displays
10. Personal protective systems connections, controls, and displays
11. Map stowage
12. Internal and external lighting systems controls and displays
13. Internal environmental systems controls and displays.

Unique Considerations Related to Combat Operations

Vision. Military pilots have unique visual problems that must be addressed during the conceptual stage of aircraft system development. Among the more significant are the following:

1. Formation Flying: Formation flying, especially at night, requires consideration of the tasks of both internal instrument viewing and external adjacent aircraft viewing. In combat, it is generally required that all telltale lights be extinguished that might give the aircraft's position away to the enemy. In order to see their associates' aircraft in silhouette, pilots must be completely dark-adapted.

2. Aerial Combat: It is vital that the pilot be able to see possible enemy aircraft approaching from any angle, which requires an excellent viewing angle from the side, top, and rear as well as in front of the aircraft. In addition, rearview mirrors are desirable so that the pilot can keep tabs on what is behind while visually attending to other forward viewing tasks.

3. Ground Reconnaissance: In many type of military missions, the pilot and/or another flight crew member is required to inspect the ground for targets. When this is done on an almost continuing basis, the aircraft must be designed to provide wide down-angle vision.

4. Night vision.

Control. In high-performance aircraft operations, not only the speed but also the dynamics of certain maneuvers create observational and control problems for the pilot and the crew members. In addition, combat injuries should be anticipated in terms of how a pilot will be able to control the aircraft for return to base. Typical considerations are the following:

1. High g Forces: High g forces may preclude a pilot from operating critical controls properly; thus, control designs should be made as insensitive to such a possibility as is practical.

2. Disabling Wound: The pilot is subject to loss of an arm or leg because of enemy gunfire; however, if control concepts provide a reasonable opportunity to use another limb, the pilot should be able to return to base.

3. Control Damage: Certain critical parts of the aircraft determine whether the machine is flyable. Whenever possible, control damage should not preclude a return to base.

MARINECRAFT

Human Factors Considerations in Marinecraft Concept Development

Environment. Both surface and undersea operational environments create numerous unique human factors problems that should be addressed early in the conceptual development of any seaborne vessel and the various pieces of equipment that may be placed aboard for special mission accomplishment. The principal factors to consider are the following:

1. Ship Motion (e.g., Pitch, Roll, Heave, and Buffet Caused by Rapid Crossing of Wave Crests): These motions impact on personnel not only in terms of potential sickness but also in terms of task interference and general safety (i.e., loss overboard).

2. Equipment Noise and Vibration: Because of the necessity to conserve space and minimize vessel size, personnel and equipment must necessarily be housed in close proximity. Adequate noise and vibration control is required in order to maintain crew performance at acceptable levels. Control must be exercised not only in working spaces but also in living spaces.

3. Weapons Systems Noise and Hazards: Guns, missiles, and aircraft operating aboard various vessels each create special noise and hazard conditions. Although priority is necessarily given to maximizing the effectiveness of each of these systems, it is possible to degrade such performance if insufficient consideration has been given to reducing the effects of weapon noise and certain failure modes on the crews that must control the weapons. Early consideration of the hazards to the crew usually provides more opportunity to minimize potential crew performance degradations or hazards to personnel in the event of system failures (explosion, fire, aircraft crashes, etc.).

4. Atmospheric Control: Military vessels (especially submersible vessels) present unusual atmospheric control requirements in order to maintain the crew in a healthy and efficient state. The idea that unfavorable environments are just part of the job is no longer valid within the context of the more sophisticated and demanding systems of today, in which human failure can no longer be tolerated. Air conditioning for crew living spaces is as important as it is for electronic equipment if the total human–machine system effectiveness is to be maintained. It is difficult to add atmospheric control if insufficient space and power were provided by the basic concept.

Space. Although onboard space is generally at a premium, lack of space is often the result of a failure to appreciate the way in which space impacts on crew performance. Traditional ship architecture tends to follow outmoded trends toward designing crew spaces according to "minimums," as opposed to creating spaces that will maximize crew effectiveness. The principal space-related features of vessels that make a considerable difference in crew performance effectiveness are the following:

1. Head Clearance: Traditionally, low overheads are created because between-deck dimensions are based on minimum head clearances; in many cases it is forgotten that pipes and ducts, which will consume part of the space, will be installed. The worst problems are created when multiple berthing is imposed and the berths have so little vertical separation that crew members cannot sit up in bed. This is more than a matter of convenience, as lack of head clearance often leads to unnecessary head injuries and/or to inefficiencies such as slowing the movement of personnel through passageways.

2. Cramped Living and Working Conditions: Lack of adequate planning in the initial conceptual phase creates spaces that interfere with job efficiency because of restrictions on mobility. More prevalent, however, is the restriction imposed on the general habitability of crew living quarters where there is insufficient space for storing personal belongings, for relaxing, or for performing personal hygiene tasks, such as toilet, showering, and dressing.

3. Passageway Clearance: Not only is it important to create passageways and hatches that are large enough for crew transfer, but it is also mandatory that such passageway clearance accommodate passage of crew members and the equipment they may have to transfer from one part of the ship to another.

4. Space Organization: Organization of ship space requires obvious compromises. However, human requirements for basic living and work convenience are the key to effective crew performance and eventual mission success. The crew-occupied space must be compatible not only with environmental needs (e.g., isolation from excess ship motion, vibration, and noise) but also with typical personnel traffic patterns as these relate to crew fatigue and time lost en route from one space to another. Traffic patterns should be thoroughly analyzed from both a nominal and an emergency standpoint.

Special Features. The following special features aboard ship are particularly important to crew efficiency and safety:

1. Lighting: Although electric power is always at a premium, good lighting both for general purposes and for special tasks is important to crew performance efficiency and safety. Adequate lighting does not necessarily imply tremendous power demands if the lighting plan is properly conceived and executed. Many lighting problems are created because too little attention was given in the initial stage of design conceptualization to the types of lighting fixtures or to the light reflection factors in selecting overhead, bulkhead, decking, furnishing, or equipment colors.

2. Ventilation: Even the most sophisticated air-conditioning system often fails to provide a cost-effective living and working environment if initial planning for proper ventilation is not taken into account. Not only personal efficiency but also health is affected when ventilation is not adequate to provide sufficient air exchange within closed compartments to maintain the proper oxygen content, remove toxic elements from the air, minimize undesirable odors, and so on.

3. Communications: Effective communication among the ship's personnel is obviously critical for efficient ship control and systems operation. Traditionally, ship communications systems have been almost universally poor in terms of intelligibility of voice transmissions. Intelligibility standards should be stressed during initial specifications preparation, and system design and/or selection should be monitored closely. Communications failures due to misinterpretation of low-fidelity voice transmission should not occur under today's technological state of the art.

4. Safe Design: Numerous features that traditionally present hazards to personnel are present throughout most vessels. These should be reviewed carefully during concept planning so that they do not recur as a result of default, i.e., previous design practices that are assumed to be acceptable. Among the key issues are sharp corners and edges; improperly designed ladders, stairs, railings, or handholds; slippery deck or stair surfaces; low overheads created by various equipment installations (booms, ducts, platforms, weapons, etc.); and special

equipment or devices that create hazards in the event of some failure during operation (cargo or fuel line transfer systems, cable failures in deck cargo-handling systems, carrier aircraft barrier systems, etc.).

5. Escape, Survival, and Rescue: Although considerable guidance is available in regard to this important aspect of ship design, each new configuration often presents unique problems that should be addressed as early as possible in conceptual development. Some of the typical emergencies that should be reviewed in detail are those related to entrapment below deck in the event of hull penetration and flooding, explosion and fire, nuclear reactor failure, and shifting of cargo.

6. Crew Protection in Combat: Protection of deck crews from typical enemy action during combat, although a familiar requirement in ship design must be given special consideration with each new vessel. This is particularly true as new deck-mounted weapons are introduced, since the proposed protective device (e.g., armor) must provide not only protection but also access for the operating crew and reasonable assurance that other crew members will not be exposed to unusual hazards created either by the weapon's operation or by the fact that it may be the target of enemy fire.

7. Special Equipment-Produced Hazards: Special hazards may be produced by radar, laser, or other exotic systems planned for a new ship system. Each must be examined in detail in order to create a proper preventive concept to mitigate hazards to the crew.

Crew Living-Quarter Concepts

Although a great deal of the emphasis has often been placed on the "cosmetic aspects" of these spaces, the planner should recognize that there are significant gains to be made by planning ahead in terms of the amount of space allocated and the manner in which this space is used or arranged. Some of the pertinent considerations are presented below.

Amount of Space. The amount of space allowed for each person aboard ship should be based not on rank but on need. Every man or woman requires approximately the same amount of space for sleeping, for storing clothing, and for dressing, showering, etc. Where needs differ in terms of space requirements, decisions should be based on the special tasks that various members of the crew have to perform. For example, top-level officers require private conference space, middle-management officers and noncoms require special space for bookwork, and others require special space for record keeping.

Space Configuration. Provision of private spaces should be based not on rank but on specific needs to separate persons or activities. The lowest-ranking crew member has the same basic psychological needs for privacy as the highest-ranking officer. However, privacy can be provided in different ways. The officer may need a suite of spaces in order to keep berthing, personal hygiene articles, and personal bookwork close to a conferencing space, but these require physical partitioning to avoid certain conflicts, e.g., visual and audio attenuation. However, it should be remembered that enlisted men and women also have needs that suggest reasonable separation or adjacency of spaces. For example, a sleeping crew member should not have to put up with the leisure-time noise or other disturbances of crew members who are on a different shift. There is a considerable rationale for creating shipboard living-quarter arrangements that would be similar to the domestic shore-based apartment that may be shared by several persons; i.e., the suite would provide separate sleeping quarters, a study area, and adjacent leisure-time and lavatory spaces.

Personal Storage. Every man or woman aboard ship has the same basic need to have adequate space to store his or her personal belongings in a secure and convenient manner. There can be no defensible rationale for forcing an enlisted man or woman to stuff garments and personal articles into a bag or under a mattress. Persons need (and are required) to look as neat as their commanding officers, and they should have the opportunity to do so. Similarly, there is little defense for allowing commanding officers to bring trinkets, pictures, or other personal items aboard and for providing special shelves or cabinets for display of these articles, while directing enlisted persons to forgo these important amenities, which add to satisfaction with one's personal space.

Appearance of Quarters. The argument that officers, as a general rule, are more used to attractive surroundings than many enlisted personnel because of their previous civilian experience is less valid today than in the past. Today, in schools, homes, hotels, restaurants, etc., practically everyone is exposed to the very finest in architecture, interior decor and furnishings, and other environmental features. Differences in cost are also often given as a reason for not providing attractive crew quarters. Modern technology, however, has eliminated most of these differences. Thus, the systems planner should recognize the importance of appearance in planning not only the officers' quarters but also the crew quarters. Some of the major features to consider in terms of appearance include the use of good lighting; attractive color schemes, carpeting, and deck, bulkhead, and overhead sheathing; and furnishings that do not look "institutional."

Two important considerations are pertinent to the question of appearance, however. First, materials must meet fireproofing requirements because of the seriousness of the fire problem aboard ship. Second, interior decor and furnishings must be appropriately sturdy, wear-resistant, and noninterfering in terms of task activities that will be performed; i.e., exotic colors and patterns, dark colors, poorly distributed illumination, etc., should not be allowed just to satisfy a single designer's or skipper's personal decorating whim.

Special Facilities. Although they need not be directly adjacent to crew quarters, several important amenities are worth considering during initial concept development because they require additional space. Among these are special lounges, snack bars, game rooms, and library and study areas. In considering these facilities, one other tradition must be confronted, i.e., the separation of spaces for officers, noncoms, and enlisted persons. Modern Navy planners are just beginning to recognize the fallacy of "off-limits" traditions. The modern naval crew should be considered a team. Although the members' specific responsibilities differ, the sum total of their efforts must result in the success of an operation. Thus, from a human factors point of view, most facilities aboard ship (other than private quarters) should be open to the entire crew. Not only does this increase crew morale, but it also provides an important variety of alternatives to reduce boredom during long sea-duty periods.

HUMAN FACTORS CONSIDERATIONS IN MISSILE SYSTEMS CONCEPT DEVELOPMENT

The primary human factors considerations in missile system development relate to (1) handling, (2) checkout, (3) launch, and (4) tracking. Each of these areas of concern varies depending on the particular system; e.g., missiles may be carried by hand or transported by ground vehicles, rail vehicles, aircraft, or marinecraft. Missiles may be launched from hand-held launchers, from vehicles, or from underground or aboveground launch pads. Missiles may also be tracked from any combination of ground, air, and seaborne stations. Finally, missile checkout may occur in a special facility far from the scene of combat, or it may have to be performed in the field within the range of enemy fire.

Hand-Held Missiles and Launcher Systems

The following are some of the principal human factors considerations in developing handheld systems:

1. Proper size, weight, and balance to ensure the efficient and safe handling of the weapon as the operator picks it up, carries it, places it into the aiming and firing position, and returns it to the carrying or storage position

2. Aiming and firing features that make it easy to hold the weapon steady, match the eye to the aiming system, and actuate the firing mechanism

3. Hazard control features to prevent injury to the operator and/or adjacent personnel (flash blindness, flash burn, toxic fume inhalation, etc.)

Note: It may be necessary to package the system in a modular fashion in order to meet the first requirement, including the possibility of transport of separate elements by more than one person. When this is done, assembly and disassembly become another important human factors consideration in terms of ease and accuracy of performing these tasks.

Missiles and Rockets Launched from a Mobile Platform

The following are some of the principal human factors considerations in developing these systems:

1. Loading the missile, i.e., the ease and safety aspects of picking up the missile, transferring it, and placing it on the launcher. The obvious hazards associated with dropping the missile during any stage of handling must be analyzed, and a suitable design must be defined that will minimize typical human errors of tripping, falling, losing hold, or inadvertently striking the missile against structural elements of the prime carrier or launcher device.

2. Control station configuration, i.e., the layout of the specific workplace or work places in which the principal operator (and/or the principal operator's assistants) will perform appropriate checkout, arming, aiming, firing, and other tasks. The safety implications of the system for other personnel and/or equipment that might ordinarily be in the vicinity of the system should also be of concern.

3. Control of special hazards such as possible burns or flash blinding from missile plume, inhalation of toxic gases, and injuries incurred when a missile explodes on the launcher.

Ground Launch Systems for Large Missiles

Because of the size of certain very large missile systems, the principal handling has to be performed by equally large, automated transporters, cranes, and erecting devices, and the actual checkout, countdown, firing, and control of the missile must occur within the protective confines of specially prepared, blast-proof facilities. Therefore, human factors considerations should be addressed in each of the following areas:

1. Transporter operation

2. Launch-pad facility operation

3. Launch control facility operation

4. Other support facilities (i.e., ground, sea, and/or airborne tracking stations)

A significant aspect of the larger and more complex missile systems is the coordination of large numbers of people, all of whom must contribute on a timely basis to the sometimes long and arduous series of steps leading to the final firing of the missile. Since many such weapons may never be fired until a very strategic moment, they may remain in a "ready" state for months or even years. In order to maintain such a state of readiness, system checkout and simulated firing exercises must be conducted on a regular basis. Thus, there are essentially two systems designed to work independently and/or together through certain stages of checkout and firing. Both the hardware and the personnel subsystems must be monitored and tested regularly. Since most such weapons are of a strategic nature and since inadvertent launching could contribute to a major confrontation between world powers, a key aspect of system design is to create preventive features and procedures to preclude unauthorized weapon firing.

The power and energy of the larger missile systems present a major safety problem, because of both the missile itself and the warhead, especially if it is nuclear; it is particularly important that definitive safety analyses be performed early in the conceptual stages, including scenarios taking into account the human failure that typically occurs when there is a large group of people. Special attention also should be given to the design of various launch-area and control-center facilities in terms of such special hazards as the following:

1. Falling from high platforms at the launch-pad site
2. Escape and rescue of personnel working at the higher levels of a launching platform or tower in the event of a premature explosion
3. Handling of, and/or protection from, high-pressure hose breaks
4. Launch-pad monitoring to make sure that personnel are "clear" prior to a missile launch

The launch control center is a major area for human factors analyses during concept formulation, whether it is land based or sea based. Particular attention should be given to the selection and conceptualization of the various individual and large-screen, team visual displays necessary to maintain continuing communication among the many operators, technicians, and commanders.

Habitability within the blast-proof control center should be of prime concern during the initial planning stage since such facilities often are extremely complex and costly and do not lend themselves to easy modification later on. Particularly important are such considerations as emergency power for air conditioning, illumination, and special filtering systems required in the event of nuclear attack and/or local disaster.

Finally, because of the high injury potential of many large systems, early consideration should be given to emergency treatment facilities within the launch and control complex, including the special equipment and procedures that should be provided for rescue and treatment of the injured in the event of a major disaster on the launch pad.

Although they are all of a slightly different nature, each of the various types of launch and control facilities requires approximately the same consideration of human–machine operations, housing, and maintenance; i.e., although an underground launch site is somewhat different in execution from an aboveground site, the basic living and working requirements are similar. Likewise, the seaborne launch and control (primarily submarine) system, although usually somewhat smaller, requires consideration of the total range of human-machine interactions and housing and protection requirements.

SPECIAL HUMAN FACTORS CONSIDERATIONS FOR ARMORED VEHICLE CONCEPTUALIZATION

Ingress and Egress

Although there are many constraints to contend with in designing an armored vehicle such as a tank (e.g., small and perhaps poorly located hatches), rapid and safe ingress and egress is often a key operational consideration, particularly in combat. Hatches should also be located to minimize the exposure of personnel who are either entering or departing because the typical top hatch places the user in a prime position for the enemy sniper's shot. Emergency escape from more than one hatch also is an important consideration in the event the tank crew is forced to leave and can pick the safe side from which to depart.

Interior Crew Positioning

In spite of the desirability of a low tank profile from the military operations point of view, crew members can perform better and withstand the sometimes tortuous dynamics of the tank's progress over rough terrain longer if they are positioned well, i.e., if a reasonably normal sitting posture is provided.

Exterior Visibility

A major problem to be resolved in tank design is that of seeing out when the tank is "buttoned up" for combat. Although hard to obtain, 360° visibility is obviously desirable. Equally important, however, are the vertical viewing-angle requirements necessary for some severe types of terrain operations. Optical systems are obvious methods for obtaining these objectives, but to be effective, they must be defined before the basic external package has precluded their potential effectiveness.

Internal Crew Environment

The typical armored vehicle concept creates tremendously severe environmental problems for the crew. Special attention early in the concept formation may, however, allow minimization of the effects of noise, vibration, pitch and roll, and temperature, humidity, and ventilation (especially the latter, since the "buttoned-up" condition implies a need for a semiclosed life support system). Safe interior design to prevent contact injuries requires special consideration for removing sharp corners and projections and for providing suitable restraints and handholds and comfortable, shock-absorbing seating.

Field Maintenance and Repair

Although the typical armored vehicle cannot be pushed to the nearest garage by the crew, simple field service and maintenance can be designed into the basic system if considered

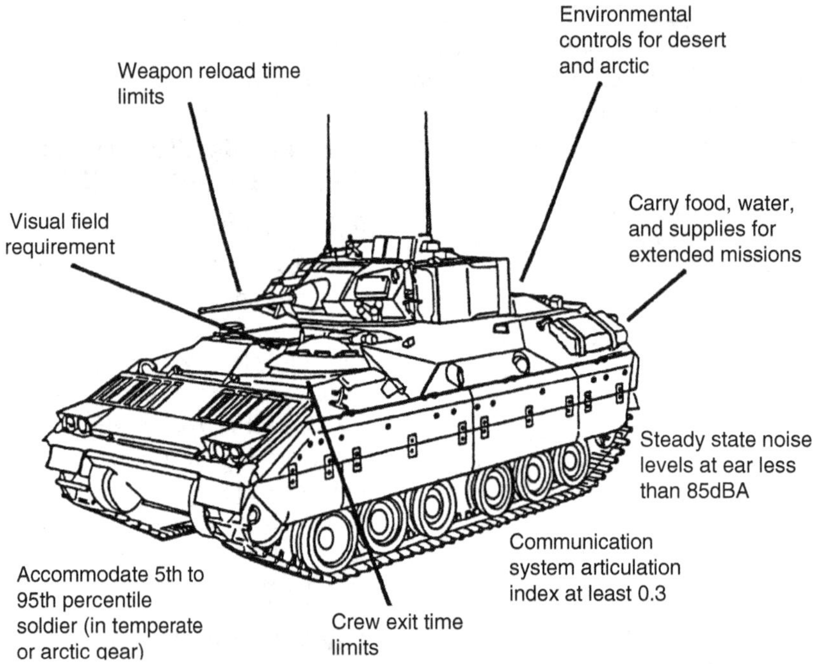

FIGURE 4.3 Typical human factors considerations for armored vehicles.

early in the planning stage. Such maintenance should not require special tools. (See Fig. 4.3.)

SPECIAL HUMAN FACTORS CONSIDERATIONS IN FIELD ARTILLERY WEAPON SYSTEM CONCEPTUALIZATION

Operations

1. Disengagement of the weapon from the prime mover
2. Manhandling the weapon into position
3. Site survey and final weapon positioning and securing
4. Prefiring weapon preparation
5. Loading
6. Aiming
7. Firing
8. Expended shell expulsion and disposal
9. Communications
10. Cleaning and field service
11. Securing the weapon
12. Reengagement of the weapon to the prime mover

Environmental Factors

1. Climate (cold, rain, snow, ice, etc.)
2. Darkness
3. Hazards associated with firing (noise, recoil, flash blindness and burns, flying particles, noxious and toxic gases and fumes, etc.)
4. Topographic deterrents (rough terrain, interfering vegetation, unstable surfaces, etc.)

Personal Equipment Constraints

1. Cumbersome and constraining winter garments, gloves, mittens, helmet, face mask, etc.
2. Special equipment interferences (back-pack, belt-hung articles, small arms, etc.)

Typical Problem Areas

Most field artillery weapons are heavy and awkward to handle, and they present serious personnel hazards both during their initial maneuvering and during firing sequences. As indicated by the above operating, environmental, and operator considerations, the basic packaging design must reckon with numerous, often conflicting objectives. Although it is difficult to anticipate and suggest specific design features for all future field artillery weapons concepts, several typical problem areas have been identified with most past and current systems. (See Fig. 4.4.)

1. Too little attention is paid to creating suitable handholds so that crew members can maintain a secure grip on the field piece as it is being manipulated.
2. Too little attention is paid to creating an efficient operator position so that a gunner can assume the proper posture for observing various displays, sighting, and manipulating various firing and sighting controls.
3. Too little attention is paid to creating an efficient ammunition loading system that will accommodate the sometimes heavy and awkward shells and guide them into the breach easily and safely without requiring unnecessarily great precision.

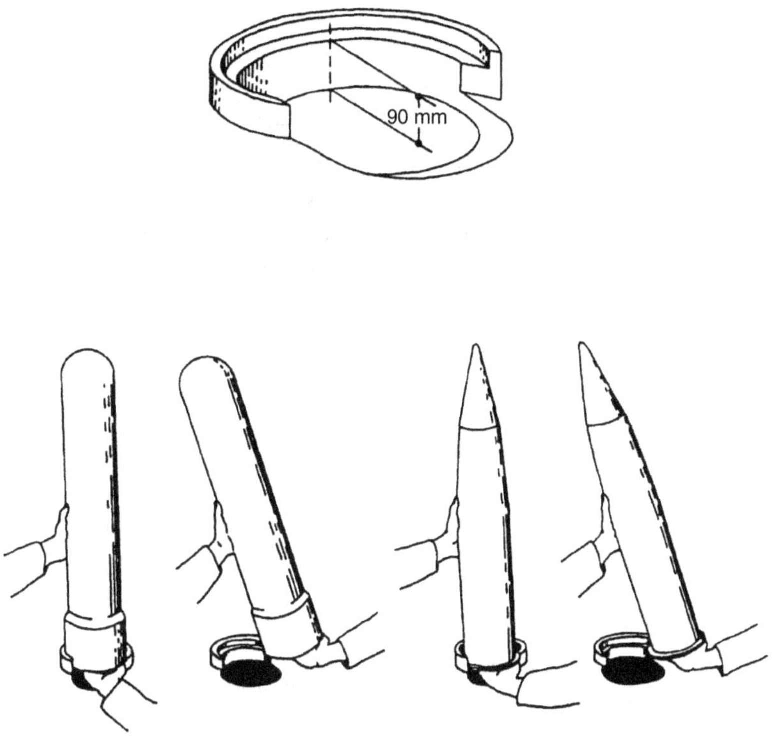

Figure 7-1. Retainers for floor mounted vertical ammunition and missile ready racks

7.1.3.3 *Upright mounts*. Upright-mounted ammunition weighing over 18.1 kg should have a floor retainer which has sufficient clearance to allow removal by the 95th percentile gloved hand.

MIL-HDBK-759A
30 June 1981

Figure 4.4 Ammunition handling design guidelines.

4. Too little attention is paid to designing system elements so that they will not require extra care when being cleaned, serviced, or replaced and to minimizing the need for special tools and other aids for dismantling and reassembling various parts of the fieldpiece.

5. Too little attention is paid to designing weapons that either cannot be jammed or can be unjammed simply and safely.

TYPICAL HUMAN FACTORS CONSIDERATIONS FOR CONCEPTUALIZATION OF HAND-CARRIED SMALL-ARMS WEAPONS

Weight

A shoulder-fired weapon should not weigh more than 7 lb (3.2 kg) (unless it rests on top of the operator's shoulder, as a bazooka-type launcher). A hand-held weapon (pistol) should weigh no more than 3.5 lb (1.6 kg).

Center of Gravity

Hand-held weapons should be designed so that when the operator holds the weapon in the normal firing position, its center of gravity is within 20 in. (51 cm) of the operator's body. If the center of gravity must be farther away, provide special support in the form of a tripod or similar device.

Size and Shape

Avoid very long weapon configurations, particularly because of the difficulties they create in manipulating the weapon, i.e., picking it up, carrying it without interference, maneuvering it for aiming in close quarters, etc. Critical dimensions relate to gripping (both hands), reaching a trigger, aligning the eye to sighting devices, and reaching specific elements for cocking, loading, adjusting sights, etc. (See Fig. 4.5.)

Manipulation and Carrying Aids

Consider the potential difficulties of rapid acquisition of the weapon from a holster, from a back-carry sling position, from the ground, or from another weapon holder. The prime objective should be a design that allows the operator to swing the weapon rapidly and directly into a firing position; designs that require the weapon to be repositioned, rotated, etc., should be avoided.

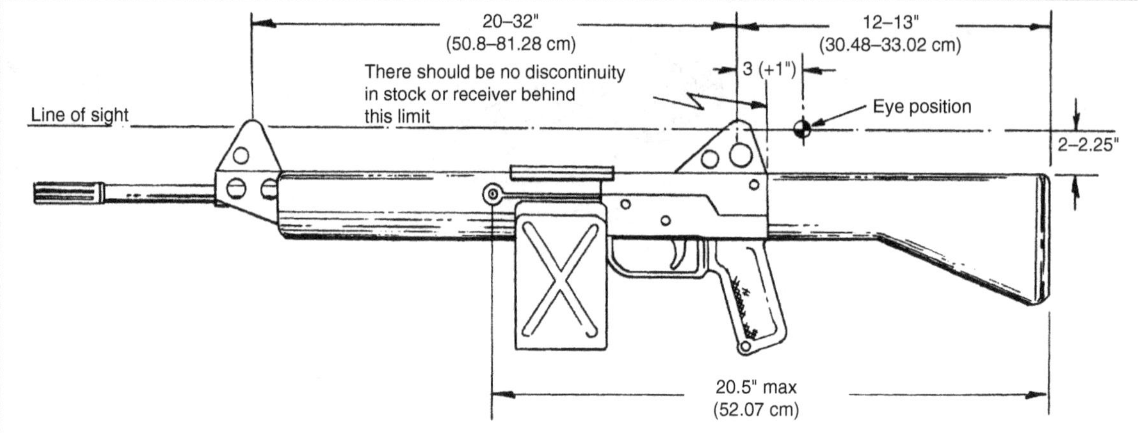

MIL-HDBK-759A, 30 June 1981

FIGURE 4.5 Recommended small arms dimensions.

Weapon Operation Interference

Design the weapon package, including grips, sights, etc., so that it can be used by an operator who may be encumbered by typical military clothing, a helmet, a gas mask, gloves, or mittens.

Safety Devices

Safety lock systems should be devised that will not seriously increase the time required to fire the weapon but which will provide protection against inadvertent operation while the weapon is being carried, cleaned, adjusted, or stored.

Firing Hazards

The general weapon design should minimize the potential for operator injury in terms of violent recoil effects (i.e., injury from the sight or other elements that might strike the face), flash burns, pinching by sliding parts, skin puncture by sharp protrusions, and cartridge expulsion after firing.

Sighting Features

The sighting system should be designed so that all components are positively retained, so that the rear aperture will not flutter during recoil, so that windage and elevation controls will not be disturbed by the recoil, etc. The main sighting concept should make it possible for gunners to see both the target and the necessary reference marks, the background environment, etc. Setting should be immediately obvious so that operators are not required to move their bodies or the weapon. Appropriate index-mark illumination for the night firing should be provided.

CHAPTER 5

SPACE SYSTEMS

GENERAL HUMAN FACTORS CONSIDERATIONS FOR SPACE SYSTEMS CONCEPT DEVELOPMENT

Human factors considerations permeate every aspect of space system operation. Although many of the subsystems may include much automation, the human remains a key element in space systems and equipment checkout, inspection, servicing, maintenance, and operation. The following are key areas within the total space complex that require a detailed examination of the human's potential roles and the design of suitable interfaces to allow these roles to be efficiently performed.

At the Launch Site

1. Booster and payload assembly, test, transfer from the assembly facility to the launch pad, inspection, adjustment, and fueling on the launch pad

2. Booster and payload launch control, emergency abort, in-flight communications and control, onboard data receipt, malfunction analysis, and in-flight maintenance instruction

3. Personnel hazards control at the launch site (noise, explosion, fire, electric shock, toxic contaminants, personnel falling from high platforms, objects being dropped from platforms on personnel below, etc.)

4. Astronaut holding facility, mission simulators, emergency medical facilities, special mission laboratories, and shop facilities

Within Manned Payload Vehicles

1. Ingress and egress (for both normal and emergency escape prior to launch and for in-space extravehicular and/or intervehicular transfer)

2. View ports

3. Crew accommodations and furnishings

4. Life support systems (food, water, and waste management)

5. Crew protective systems (space garments, radiation shielding, meteor shielding, reentry heat shield, and internal environmental control)

6. Crew monitoring (physiological and psychological)

7. Special performance aids (space tools, tethers and restraints, mobility aids, and remote manipulators)

8. Other (special foods and containers, in-space communicating and visual signaling devices, and astronaut extravehicular self-propulsion systems)

GENERAL CONSIDERATIONS WHEN DESIGNING SPACE CREW ACCOMMODATIONS

Space is a unique environment for humans. Important human factors considerations for designing for space are listed below:

1. Space anthropometrics
2. Human performance in space
3. Life and health support systems in space
4. Architectural considerations for reduced gravity environments
5. Spacesuit design
6. Artificial gravity

The most significant and unique change in space is the variation in gravity. In orbit or away from large masses (such as Earth or Moon) gravity is nearly zero (called microgravity). The human and all objects become weightless. Planets such as Mars or Moon are smaller than Earth and have less gravity. This is called partial gravity.

Space Anthropometrics

The effects of weightlessness on anthropometrics are shown in Table 5.1. Three basic things happen to the body:

1. Fluids normally pulled down by gravity raise to the upper body. The body interprets this as a fluid excess and promotes urination.

2. The spine elongates without gravity and the total body length increases about 3 percent.

3. The relaxed body in weightlessness also takes on a characteristic semistooped posture (the "neutral body posture"). Work and living facilities should accommodate this posture. Part 4 of this book shows the neutral body posture.

Human Performance in Space

Table 5.2 shows changes in human performance capabilities observed in space. This information is general, and no quantitative data are available. However, the designer should be aware of these changes and design to accommodate them.

TABLE 5.1 Anthropometric Changes in Weightlessness

	Anthropometric Change		
		Long-Term Mission (>14 days)	
Parameter	**Short-Term Mission (1–14 days)**	**Before vs. During Mission**	**Before vs. After Mission**
Height	Slight increase during first week (~1.3 cm or 0.5 in.). Height returns to normal R+0. Increases caused by spine lengthening.	Increases during first 2 weeks and then stabilizes at approximately 3% of premission baseline. Increases caused by spine lengthening.	Returns to normal on R+0.
Circumferences	Circumference changes in chest, waist, and limbs. Changes due primarily to fluid shifts.		
Mass	Postflight weight losses average about 3.4%; about 2/3 of the loss is due to water loss, the remainder due to loss of lean body mass and fat. Center of mass shifts headward approximately 3–4 cm (1–2 in.) See Paragraph 3.3.7.3.2.1 for details.	In-flight weight losses average 3–4% during first 5 days; thereafter, weight gradually declines for the remainder of the mission. Early in-flight losses are probably due to loss of fluids; later losses are metabolic. Center of mass shifts headward approximately 3–4 cm (1–2 in.)	Rapid weight gain during first 5 days postflight, mainly due to replenishment of fluids. Slower weight gain from R+5[a] to R+2 or 3 weeks.
Limb volume	In-flight leg volume decreases exponentially during the first mission day; thereafter, the rate of decrease declines until reaching a plateau within 3–5 days. Postflight decrements in leg volume up to 3%; rapid increase immediately postflight, followed by slower return to premission baseline.	Early in-flight period same as short missions. Leg volume may continue to decrease slightly throughout mission. Arm volume decreases slightly.	Rapid increase in leg volume immediately postflight, followed by slower return to premission baseline.
Posture	Immediate assumption of neutral body posture.	Immediate assumption of neutral body posture.	Rapid return to premission posture.

[a]Recovery day plus postmission days.

From NASA-STD-3000.

TABLE 5.2 The Effects of Space on Human Performance

Human Performance Capability	How Space Affects Human Performance
Vision	Visual field and acuity reduction during launch and reentry accelerations. Also visual reaction time and tracking may be degraded. Atmospheric light scatter does not exist. The loss of this cue can degrade perception of object shape, distance, location, and relative motion. The lack of a horizon, overhead sun, and fixed vertical in weightlessness can degrade recognition and perception of object shape, distance, location, and relative motion.
Hearing	No evidence of change in space.
Taste and smell	Slight decrease in sensitivity to taste and smell in space.
Vestibular system	Microgravity can cause vertigo and posture and movement illusions. Roughly half the astronauts experience "space motion sickness" that can cause severe nausea and vomiting. It may last 2–4 days. Pitch and roll head movements may increase the effects.
Kinesthetic system	In microgravity, humans are not able to estimate differences in mass as well as they can on Earth. Sensitivity decreases about one half.
Motor skills	Motor skills and response time are degraded when persons first encounters a change in gravity levels (e.g., Earth gravity to microgravity back to Earth). Adaptation may take weeks. Motion is limited in high acceleration.
Strength	Muscles that support the body in gravity (e.g., leg muscles) atrophy and lose strength in microgravity. Supplemental exercise is needed. In microgravity, people need restraints and counterforce surfaces to exert force. With proper support in microgravity, humans can move objects weighing hundreds of pounds.

From NASA-STD-3000.

Life and Health Support Systems in Space

In long-term missions, in reduced gravity the body will atrophy and possibly cause permanent disability on returning to Earth. Special facilities and procedures should be provided. Table 5.3 lists the zero-gravity effects and countermeasures.

TABLE 5.3 Effects of Zero Gravity on Crew Health and Countermeasures

Zero-Gravity Effects	Possible Countermeasures
Cardiovascular deconditioning	Low-resistance, high-frequency exercise of large muscle groups (aerobic exercise)
Fluid loss	Fluid loading prior to 1 G entry; pharmaceuticals
Bone mineral loss	Skeletal loading through low-frequency, high-resistance exercise (anaerobic exercise); pharmaceuticals
Disorientation; space adaptation syndrome; neuromuscular patterning not adapted to gravity environment	Psychomotor exercises; pharmaceuticals
Loss of muscle mass, strength, and endurance	Exercise of specific muscle groups (aerobic and anaerobic)

From NASA-STD-3000.

Architectural Considerations for Reduced Gravity Environments

Orientation. There is no "up" or "down" in weightlessness. Provide the following:

1. Consistent vertical orientation—"Up" should be consistent within a workstation or activity center.

2. Visual orientation cue—Provide a visual cue to allow a person to quickly orient to the local vertical.

3. Separation—Separate workstations and activity centers that have orientations differing by more than 90°. (See Fig. 5.1.)

Interior Volume. Provide sufficient area and volume for turning and passage in reduced gravity environments.

Restraints and Mobility Aids. The body "floats" in weightlessness. The designer must provide restraints and mobility aids for crew performance, comfort, and safety. Table 5.4 shows NASA requirements for restraints and mobility aids.

Spacesuit Design

Space has no atmosphere. People work in space in an enclosed, pressurized facility, or outside with a portable pressurized spacesuit. Working in a spacesuit is referred to as "extravehicular activity" or EVA. EVA living and working poses many safety and human factors problems. The spacesuit design requirements in NASA-STD-3001 help ensure the EVA worker's safety. (See Table 5.5.)

NASA-STD-3001, Space Flight Human-System Standard, is a two-volume set of National Aeronautics and Space Administration (NASA) standards directed at minimizing health and performance risks for flight crews in human space flight programs. Volume 1 of NASA-STD-3001, Crew Health, sets standards for fitness for duty, space flight permissible exposure limits, permissible outcome limits, levels of medical care, medical diagnosis, intervention, treatment and care, and countermeasures. Volume 2 of NASA-STD-3001, Human Factors, Habitability, and Environmental Health, focuses on human physical and cognitive capabilities, and limitations and defines standards for spacecraft (including orbiters, habitats, and suits), internal environments, facilities, payloads, and related equipment, hardware, and software systems with which the crew interfaces during space operations.

Volume 1 of NASA-STD-3001 considers human physiologic parameters as a system, much as one views the

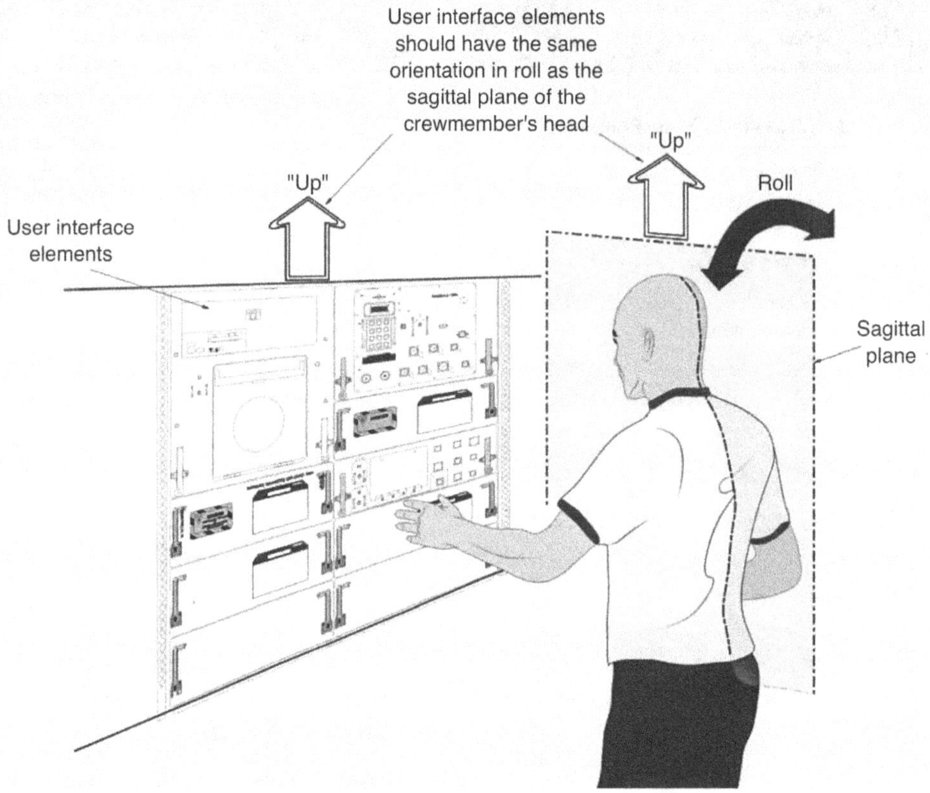

FIGURE 5.1 Equipment orientation requirements in microgravity. (From NASA-STD-3001.)

TABLE 5.4 Design Requirements for Restraints and Mobility Aids
Crew restraints shall be provided to assist in maintaining body position and location in reduced gravity conditions or during high accelerations.
Crew restraints shall be designed to accommodate the crewmember for the duration of the task.
Crew restraints to be used in microgravity applications shall be designed for compatibility with the neutral body position.
Crew restraints shall not interfere with crewmembers' performance of tasks.
Crew restraints shall provide for the operation of controls during reduced gravity, as well as during dynamic or multiaxis accelerations.
Mobility aids shall be standardized, clearly distinguishable, and located to aid the crew in starting or stopping movement, changing direction or speed, or translating equipment.
All fixed and portable IVA (Intravehicular) mobility aids shall be designed to withstand expected forces of the crew without failure or sustaining damage.
Mobility aids shall be provided for ingress, egress, and escape of suited crewmembers.
Where appropriate, mobility aids shall be provided for the crew to conduct IVA operations.

From NASA-STD-3001, Volume 2.

engineering and design of a mechanical device. Doing so allows the human system to be viewed as an integral part of the overall vehicle design process, as well as the mission reference design, treating the human system as one system along with the many other systems that work in concert to allow the nominal operation of a vehicle and successful completion of a mission. In Volume 2, the focus turns to human-system integration where the context is about how the human crew interacts with other systems, including the habitat and the environment. The focus is on performance issues during a mission—whether the human and the system can function together (within the environment an habitat) and accomplish the tasks necessary for mission success.

In addition to Volume 2 of NASA-STD-3001, NASA has published a supplementary document, NASA/SP-2010-3407, Human Integration Design Handbook (HIDH). HIDH is a compendium of human space flight history and knowledge. It is organized in the same sequence as NASA-STD-3001, Volume 2, and provides useful background information and research findings. While the HIDH is not a standard or a requirement, it can be a resource for preparing the program-specific requirements. The HIDH can be used not only in the preparations of requirements but as a useful tool for designers.

Standards and topics addressed in NASA-STD-3001, Volume 2, and HIDH are as follows.

1. Anthropometry, biomechanics, and strength—The physical size, shape, reach, range of motion, strength, and mass of crewmembers. HIDH explains how to determine the correct data for a project and how this information should be used to create a design that fits the crew.

2. Human performance capabilities—The physical, cognitive, and perceptual capabilities and limitations of humans in space. Topics covered include physical workload, visual and auditory perceptions, and cognitive workload.

3. Natural and induced environments—The habitable range for environmental factors (air, water, contamination,

TABLE 5.5 Spacesuit Design Requirements

Suited Donning and Doffing [V2 11001]
The system **shall** accommodate efficient and effective donning and doffing of spacesuits for both nominal and contingency operations.

Translation Paths for Suited Crewmembers [V2 11002]
The expected translation paths **shall** be large enough to accommodate a suited crewmember and the motions necessary for translation.

Mobility Aid Provision for Suited Operations [V2 11003]
Mobility aids **shall** be provided along expected translation paths of suited crewmembers.

EVA Mobility Aid Standardization [V2 11004]
The color of all EVA mobility aids **shall** be standardized with distinguishing characteristics such that they can be readily distinguished from items that are not to be used as mobility aids.

EVA Translation Path Hazard Avoidance [V2 11005]
EVA translation paths **shall** be designed to prevent exposure to hazards.

Suit Pressure Set-Points [V2 11006]
The suit **shall** provide the capability for the crew to select discrete suit pressure set-points within the suit operating pressure ranges.

Suit Equilibrium Pressure [V2 11007]
Suits **shall** maintain pressure within 0.1 psi (0.689 kPa) after the suit has achieved an equilibrium pressure for a set-point.

Suit Decompression Sickness Treatment Capability [V2 11008]
Suits **shall** accommodate DCS treatment.

Suited Noise Exposure Limit [V2 11009]
Suits **shall** limit noise (not including impulse noise) exposure at the ear to NC-50 or below without the use of hearing protection, unless such protection is included in the nominal suit configuration.

EVA Suit Radiation Monitoring [V2 11010]
EVA suits **shall** provide or accommodate radiation monitoring and alerting functions to allow the crew to take appropriate actions.

Suited Crewmember Heat Storage [V2 11011]
The system **shall** prevent the energy stored by each crewmember during nominal suited operations from exceeding the limits defined by the range 3.0 kJ/kg (1.3 Btu/lb.) > ∇Q stored > −1.9 kJ/kg (−0.8 Btu/lb.), where ∇Q stored is calculated using the 41 Node Man or Wissler model.

Restraints for Suited Operations—Provision [V2 11012]
Restraints **shall** be placed to ensure that optimum reach and work envelop of the suited crewmember and be adjustable to maximize the work envelop.

Suited Body Waste Management—Provision [V2 11013]
Suits **shall** provide for management of urine, feces, menses, and vomitus of suited crewmembers.

Suit Urine Collection [V2 11014]
Suits **shall** be capable of collecting a total urine volume of $V_u = 0.5 + 2t/24$ L throughout suited operations, where t is suited duration in hours.

Suit Urine Collection per Day [V2 11015]
For contingency suited operations lasting longer than 24 hours, suits **shall** be capable of collecting and containing 1 L (33.8 oz.) of urine per crewmember per day.

Suit Feces Collection per Day [V2 11016]
During contingency suited operations, suits **shall** be capable of collecting 75 g (0.15 lb.) (by mass) and 75 mL (2.5 oz.) (by volume) of fecal matter per crewmember per day.

Suit Isolation of Vomitus [V2 11017]
Suits **shall** provide a means for the isolation of vomitus from the EVA crewmember's face.

Suited Field of Regard [V2 11018]
Suits **shall** provide a field of regard sufficient to allow the crewmember to accomplish required suited tasks.

Suit Helmet Optical Quality [V2 11019]
Suit helmets **shall** have sufficient optical qualities to allow the crewmember to accomplish required suited tasks.

Suit Helmet Luminance Shielding [V2 11020]
Suit helmets **shall** provide protection to suited crewmembers from viewing objects with luminance that could prevent successful completion of required suited tasks.

Suit Helmet Visual Distortions [V2 11021]
Suit helmets **shall** prevent visual distortion.

Suit Helmet Displays [V2 11022]
Suit helmet field of regard **shall** be unencumbered if helmet- or head-mounted displays are provided.

Suit Information Management [V2 11023]
The system **shall** allow the crew to effectively input, store, receive, display, process, distribute, update, and dispose of information on consumable levels, suit status and alerts, and biomedical data to the suited crewmember, to other crewmembers, and, when feasible, to the ground controllers.

Ability to Work in Suits [V2 11024]
Suits **shall** provide mobility, dexterity, and tactility to enable the crewmember to accomplish suited tasks within acceptable physical workload and fatigue limits.

Suited Nutrition [V2 11025]
The system shall provide a means for crew nutrition while suited.

Suited Hydration [V2 11026]
The system **shall** provide a means for crew hydration while suited.

Suited Medication Administration [V2 11027]
The system **shall** provide a means for administration of medication to a suited crewmember.

From NASA-STD-3001, Volume 2.

acceleration, acoustics, vibration, radiation, and temperature) that will ensure that humans can perform safely and effectively.

4. Habitability functions—Considerations for the daily functions of the crew inside the spacecraft, including eating, sleep, hygiene, waste management, and other activities to ensure a habitable environment.

5. Architecture—Guidance for the development and integration of overall spacecraft size and configuration, and layout of location and orientation aids, traffic flow and translation paths, hatches and doors, windows, and lighting.

6. Hardware and equipment—Human factors standards and guidelines for the design of hardware and

equipment such as tools, drawers and racks, closures, mounting hardware, handles and grasp areas, restraints, mobility aids, fasteners, connectors, visual access, packaging, clothing, and crew personal equipment.

7. Crew interfaces—The design of interfaces through which information is exchanged between the crew and systems. Topics include visual displays, audio displays, controls, and labels.

8. Extravehicular activities—The human factors design guidelines and standard for EVAs performed by suited crewmembers outside the pressurized environment of a flight spacecraft (during space flight or on a destination surface). It also addresses off-nominal operations performed inside unpressurized spacecraft.

In addition, the HIDH contains general guidance on developing program-specific requirements and developing a human-system integration process throughout system design.

Artificial Gravity

Artificial gravity is a possible way to maintain body conditioning in weightlessness. Artificial gravity is created by rotating the space habitat and substituting centrifugal force for gravity.

The forces on the human body depend on the angular velocity and rotational radius. The relationship of these factors is shown in Fig. 5.2. The figure also shows a human occupancy "comfort zone." Undesirable artificial gravity byproducts are Coriolis forces. The figure shows that the movement of the crew affects the size and direction of the Coriolis forces they experience.

Figure 5.3 shows typical design considerations to overcome the deleterious effects of Coriolis.

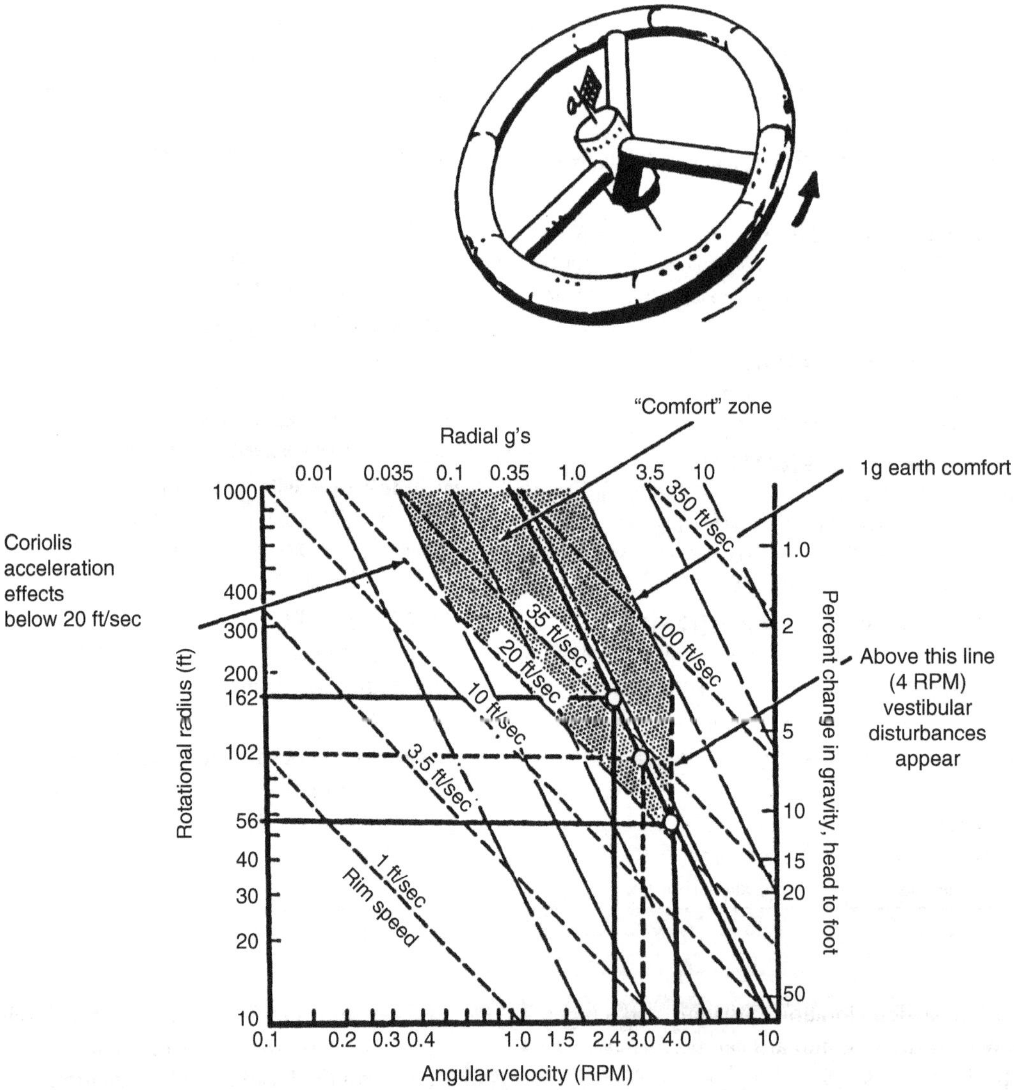

FIGURE 5.2 Effects of artificial gravity depend on rotational radius and angular velocity.

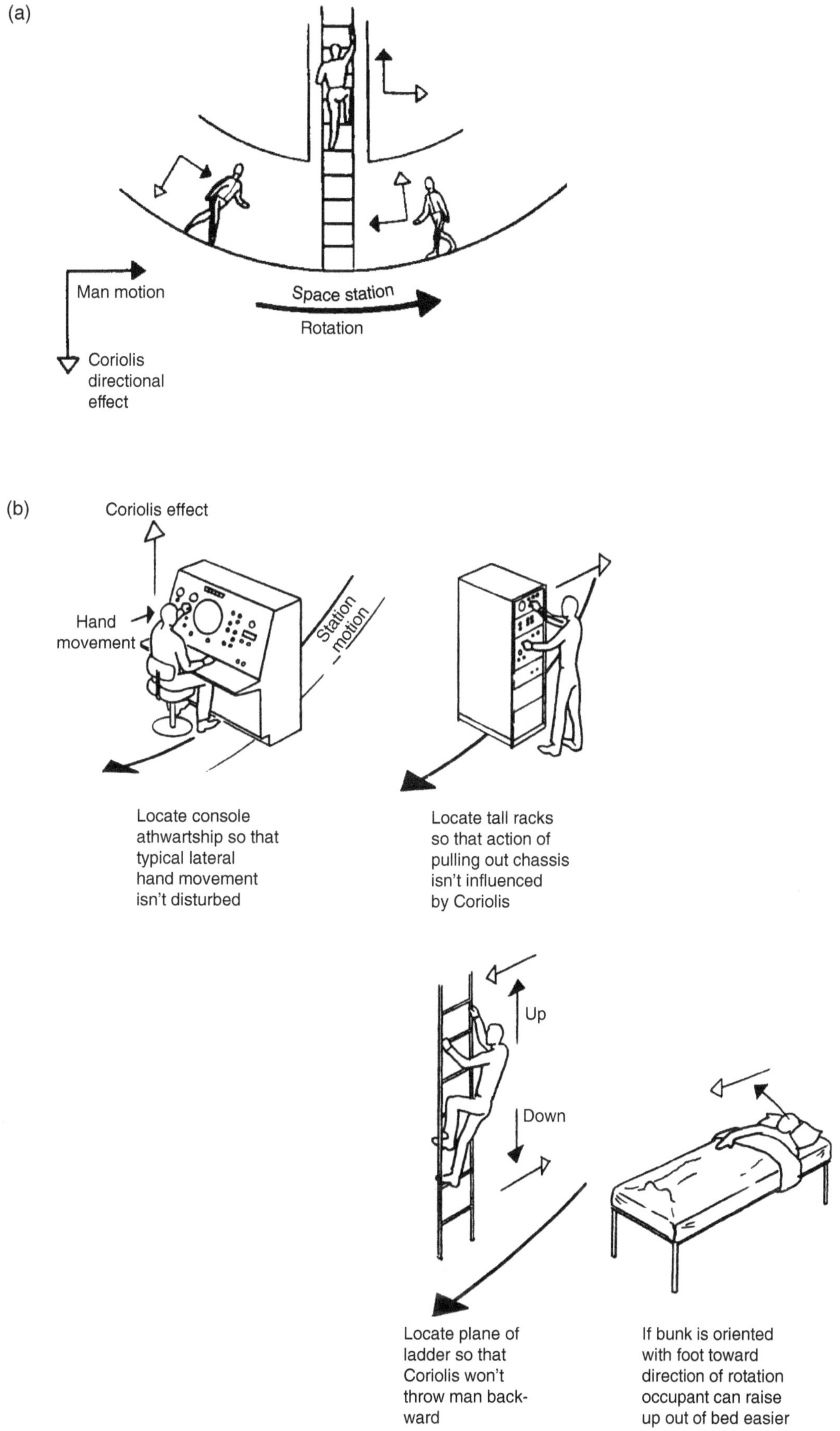

Figure 5.3 Coriolis forces depend on crew movement.

ADDITIONAL INFORMATION

NASA has published a human factors design standards and design handbook specifically for space. The standard is Human Factors, Habitability, and Environmental Health, NASA-STD-3001, Volume 2. The handbook is the Human Integration Design Handbook, NASA/SP-2010-3407. Both documents cover the above topics and other in great detail. An earlier standard, NASA-STD-3000 has background information that was not included in NASA-STD-3001. The information in the older standard is still valid and is included in this book to provide additional understanding.

CHAPTER 6

INDUSTRIAL SYSTEMS

GENERAL

General Considerations in Development of Industrial Systems

The success of industrial systems is generally measured in terms of productivity or output. Effective human performance is a key factor in producing a desired output in the factory, in the laboratory, or in the construction or agricultural industry. Past experience has shown, however, that many industrial systems are conceptualized around machinery, materials, and architectural features, with too little thought given to what workers should be doing, how they should be doing it, and where they should be doing it. As a result, these systems have tended to be designed more to accommodate the machines and materials than to make sure that workers can contribute maximally to the eventual industrial output. Fortunately, this bias toward the machine rather than the worker in the industrial system is changing through the efforts of ergonomics, or the human factors approach to the design of work systems.

Ergonomists consider that the work system must be designed to perform a particular work task and that it is constituted of the human supported by his or her work equipment and the surrounding environment. The key human factors issues in conceptualizing an effective work system include the following.

Role of the Worker. The role of the worker should be based on his or her perceptual, cognitive, physical, and psychological capabilities and limitations, not on what the technicalities of preconceived electromechanical or other purely physical features of the technologically oriented inventor create for the worker to do. Just as industrial system managers would not want a computer to make their decisions; industrial machines should not direct the work activities of the industrial workers. Each member of the industrial team should have a meaningful job to do.

Role of the Industrial Machine. The role of the industrial machine should be to support the worker in accomplishing his or her assigned task. The machine should be assigned functions to complement the worker's capabilities, relieve certain stresses, and perform activities that are needed but which are beyond human strength, speed, or perceptual capacities.

Role of the Industrial Environment. The role of the industrial environment (i.e., the architectural facility, temperature, noise, illumination, etc.) should be to provide working conditions that will ensure that both the worker and the machine can function at maximum capacity. It is often noted that planners are more concerned about machines than they are about workers; e.g., air conditioning may be provided for computers, while factory workers "sweat" on the assembly line.

Worker Safety. The design of the work process should be based not only on the objective of maximizing output but also on that of safeguarding the worker's well-being and health. Safety results only when both machine and human failures are anticipated and designed out of the system. Although safety training and supervisional procedures are important, hazards are created at the beginning by lack of proper consideration of the modes of machine, environmental, and human failure.

Human Factors Principles to Consider during the Conceptual Phase

The principal areas in which human factors principles are important considerations during the conceptual phase of industrial systems development are the following:

Design of the Workplace

1. Ingress and egress, and internal traffic flow
2. Working position, seating, and layout of materials and tools
3. Communications
4. Illumination
5. Temperature, ventilation, and humidity control
6. Noise and vibration control
7. Clearance and protection from hazards (materials and vehicular movement, electric shock, toxic contaminants, dust, smoke, fumes, fire, etc.)

Design of Machines, Tools, and Equipment

1. Visual displays, instructions, labels, and signs
2. Controls and handles
3. Protective features (guards, handrails, fail-safe switches, locks, warning devices, blast and noise barriers, etc.)
4. Ease of maintenance and service (i.e., inspection, adjustment, calibration, troubleshooting, removal and replacement, and portability)
5. Supplementary illumination for critical seeing tasks
6. Safety marking (i.e., the appropriate use and location of standard safety symbols, colors, marking, and signs)

Design of Supportive Implements and Devices

1. Vehicular equipment
2. Scaffolds, platforms, ladders, and stairs
3. Lifts, loaders, cranes, and materials pallets
4. Automated materials-handling devices (conveyor belts, elevators, remote manipulators, etc.)
5. Process-monitoring consoles
6. Cleaning and maintenance (vats, chambers, furnaces, etc.)
7. Machine controlled/robotic manufacturing systems

Specific, Critical Human Factors Considerations in Industrial System Workplace Design

Worker Posture. Whether workers are standing or seated at a workbench or machine or are riding on some type of industrial vehicle, dolly, or crane, their working posture is extremely important. That is, if the hardware forces workers to remain in an awkward position for a long time, they obviously will become fatigued and will be more apt to make mistakes or incur some type of physical disability over a period of time.

Worker Mobility and Balance. Tasks that involve considerable muscular activity and/or the continual maintenance of balance place considerable stress on workers, especially when the tasks continue for extended durations. Tasks should be designed in such a way that workers do not have to continually remind themselves of potential loss of balance, possible muscle overstrain, the potential for rupture, etc. The workplace within which tasks are performed should not be restrictive in the sense that workers cannot assume a normal posture and balance; use symmetrical, ballistic movements; or distribute the strain by using first one hand and then the other.

Weight, Force, and Precision Requirements. Tasks should be created that do not place excessive (and especially continuous) demands on workers in terms of lifting weights, applying force, and making precise movements.

Visibility. Task environments and the design of the various task elements should be based on the fact that people need to see what they are doing. They need to see where they are walking and where the perimeters of a package that they are carrying are with reference to the pathway obstructions. They need to see clearly the spatial and positional relationships between a tool and the material they are about to bend, cut, drill, or treat, and they need to see clearly the markings on tools, machine adjustment controls, and monitoring or inspection displays.

Environmental Conditions. Although the healthy worker can adapt quite readily to a considerable variation in temperature, humidity, or ventilation, the stress of compensating is sometimes cumulative and can lead to sudden collapse in the event of a sudden task stress (not to mention the potential for irritations and interemployee distress). Most serious, however, are the effects on worker performance and morale when noise, vibration, and atmospheric contaminants are not properly controlled. Uncontrolled noise not only leads to temporary and/or permanent hearing losses but may also interfere with crucial communications (e.g., hearing warning signals). Uncontrolled vibration may interfere with specific task performance, cause annoyance stress to the point of incurring employee aggression or absenteeism, and affect the employee's health after a time. Respiratory disabilities are perhaps the most insidious of the uncontrolled hazards, second only to the hazards associated with toxic substances where workers are exposed to toxic fumes, dusts, chemicals, etc. It is important to consider all the potential environmental hazards in terms of system design control, as opposed to discovering that such hazards exist after the system is designed. The latter condition is too often used as a guiding principle during system concept development; it is based on the assumption that it is easy to "fix" a problem if it occurs.

Human Energy and Efficiency, and Motion Economy

Time and motion specialists espouse the desirability of designing work systems and procedures to economize on worker motion and thus improve human output efficiency and reduce fatigue potential. Attention should be given to motion economy during the conceptual stage of industrial system design because it is here that modifications in hardware, procedure, and worker utilization are most easily made. As indicated earlier, once the machine is built and the architectural features of the factory are fixed, the human must do all the adapting. Various industrial handbooks present the principles of motion economy, but perhaps the best known is the one by Barnes.[10]

Human energy consumption, on the other hand, is a factor that is less well understood in terms of its overall effect on human performance; i.e., although excessive energy demands for long durations obviously lead to physical fatigue, mental fatigue may be the worst enemy of the worker in terms of how it contributes to serious task failures. A basic objective of work design should be to conserve the worker's energy, but not to the extent that boredom is introduced. Actually, boredom may contribute to the energy cost.

Ultimate worker efficiency is necessarily a combination of many variables. Although the design cannot be expected to compensate for all the human variability (much of which may have little to do with the work or the job), the job or task design should provide workers with the best possible chance to perform their assigned tasks as effectively as they can. The application of good human engineering and time and motion principles early in the conceptual stages of system development generally leads to energy conservation and output efficiency.

[10]M. Barnes, *Motion and Time Study*, John Wiley & Sons, Inc., New York, 1968.

PRINCIPLES OF WORKER EFFICIENCY AND MOTION ECONOMY

Plant Layout

1. Organize plant facilities so that workers can easily learn where things are located with respect to one another, i.e., so that they can recognize how materials, people, and products flow through the physical plant system.

2. Provide recognizable visual references so that workers know where they are and where they are going. This includes not only signs and other codes but also visually distinctive architectural features.

3. Provide convenience between internal and external features (employee parking and workstations, delivery vehicle access and shipping and receiving storage, customer parking and administrative stations, etc.).

4. Provide convenience between separate buildings and special testing and/or storage facilities, i.e., minimize travel distances between facilities that interact frequently.

5. Provide appropriate separation between facilities where serious noise or safety hazards may compromise worker efficiency or safety.

6. Arrange for internal functions that are related and interact frequently to be located close together.

7. Provide a sufficient number of conveniently located employee support services to minimize travel time (restrooms, cafeterias, tool cribs, employee personnel services, etc.).

Individual Workstation Planning

1. Provide a definite, fixed place for all tools and materials.

2. Locate tools, materials, and machine controls and displays close to the point of use.

3. Arrange for delivery of materials to a point close to the work site and the point of use. The worker should not have to go for parts or materials.

4. Design the height of the workplace and the seat so that alternate sitting and standing are possible.

5. Configure the workplace and seat arrangement to allow the operator to maintain the least fatiguing posture.

6. Design the work process so that both the worker's hands begin and complete their motions at approximately the same time.

7. Design the process so that both the worker's hands are not idle at the same time.

8. Lay out the work so that the worker's arm motions are generally in opposite and symmetrical directions and are approximately simultaneous.

9. Lay out the work so that momentum can be used to advantage—but not if it must be overcome by muscular effort.

10. Lay out the work so that smooth, continuous, curved motions of the arms and hands occur, as opposed to straight, irregular, suddenly changing motions.

11. Lay out the work so that ballistic motions are used, as opposed to restricted, precisely controlled movements.

12. Create work-flow patterns that are natural, rhythmic, and easy.

13. Lay out the workstation so that as few eye fixations, as close together as possible, are required.

14. Wherever a specific movement requires repetitive or precise control, assign the motion to the hand or digit that provides the required degree of strength or dexterity, but avoid overloading a particular hand or finger unnecessarily.

15. Provide jigs, fixtures, and foot-operated aids to relieve the hands of unnecessary and restrictive tasks.

16. Combine two or more tools wherever practicable.

17. Preposition tools and materials as much as possible.

18. Provide an auditory signal to indicate when long-duration machine operations are completed so that the operator is free to perform other tasks.

19. Place controls and displays where they require the least excess movement of torso, head, limbs, or eyes.

20. Provide lighting that is appropriate to the specific tasks, as opposed to general illumination for an entire area.

Interaction between Workstations

1. Arrange machines, workbenches, etc., so that work can be passed directly and naturally from one to the other, i.e., not across aisles, around corners, or past an intervening workstation.

2. Arrange a production line so that products finished at one station can be moved to an interim waiting point adjacent to the next operation; individual stations will then not be held up in accepting the next work package.

MATERIALS HANDLING

Human Factors Considerations in Materials-Handling Systems Conceptualization

Although a great deal of the handling of materials coming into, or being shipped from, the industrial complex, as well as the actual handling within the facility, may be accomplished by machines of one kind or another, humans interface with this process either directly or indirectly via machine controls and displays. The primary human factors considerations fall into two major categories, as follows.

Prevention of Materials Damage. Waste of materials due to damage is of considerable economic consequence to the manufacturer. Therefore, it is extremely important to examine the materials-handling concept in terms of the potential errors the human element could make with respect to movement of all materials in the receipt, manufacturing, and shipping phases of the proposed operation. This examination should include consideration not only of raw materials and substances but also of packaged items. The potential for human error often exists at the following points during materials handling:

1. Dock transfer to and from a transportation system (rail, truck, ship, etc.)

2. Shipping and storage-area transfer (i.e., to and from the loading dock and areas designated for temporary storage of either the preprocessed materials or the finished and packaged products)

3. Transfer within the plant (i.e., between processing, assembly, test, and packaging stations)

Prevention of Personal Injuries. Personal injuries also have a considerable economic impact on an industrial operation in terms of lost time and possible legal liability. More important from a human factors point of view, however, is the direct consequence to the worker and his or her family. The latter consideration should guide the planner because it assumes that most workers do not want to be injured and that it is therefore incumbent on the systems designer to create as safe a place to work as is technologically possible. The typical areas in materials handling where personal injury often occurs include the following (see Fig. 6.1):

1. Direct interface with raw materials that are heavy, sharp, rough, toxic, breakable, or explosive

2. Storage and stacking, where materials fall on people

3. Automatic material movement or manufacturing, where workers may be struck by the material or machine or may fall into or be caught by machine elements

4. Toxic liquids or chemicals, which may spill onto workers during transfer

Transport and Lifting Appliances

Transport and lifting appliances include: Narrow-gauge railway superseded by overhead trolleys or trackless electric trolleys with trailers; Long-distance transport by Lorries; Internal transport by traveling cranes of different types, fork lift trucks, etc.

Semiautomatic Warehouse Systems

Continuous Trolley Concepts. Tripping and collision hazards must be considered. Design of the trolley track should be such that personnel cannot step into the track slot

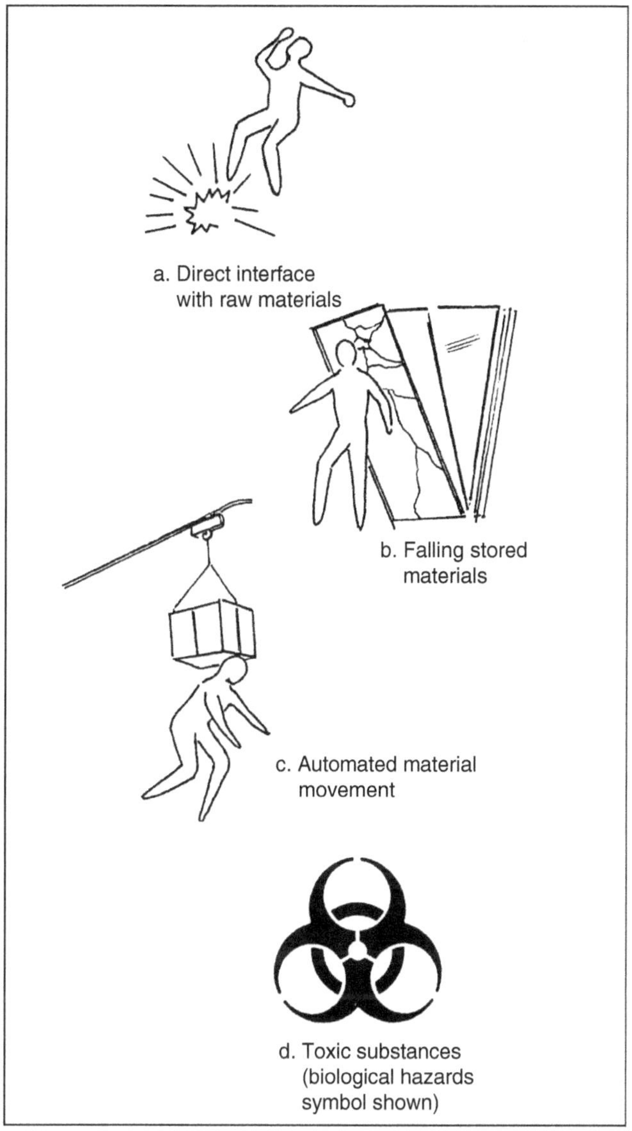

Figure 6.1 Causes of injuries in materials handling.

and turn an ankle. Forklifts should be painted with appropriate hazard markings, and audio alarms should make it possible for operators to warn other workers. (See Fig. 6.2.)

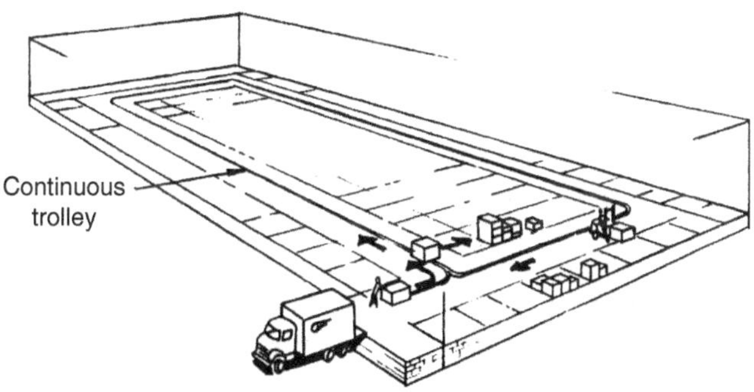

Figure 6.2 Continuous trolley.

Partially Computer-Controlled Trolley, Elevator, and Conveyor Concepts. Inadvertent activation while performing maintenance tasks and limited visibility in high-density stack areas are common hazards. Manual operation during system down periods requires special consideration of access into areas and positions that may not have been planned for human occupation, and audio alarms should be provided to warn workers that conveyors are moving. (See Fig. 6.3.)

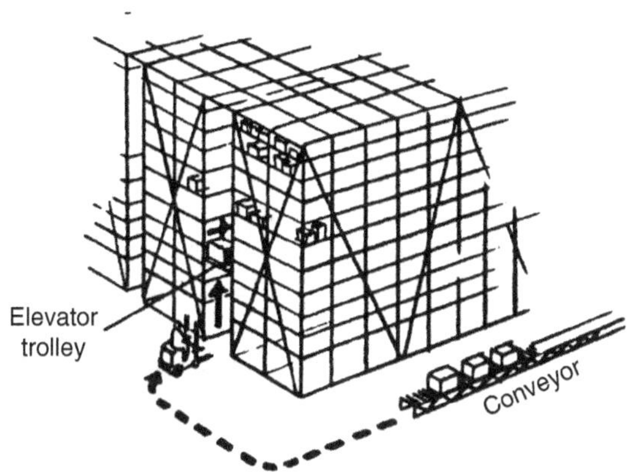

Figure 6.3 Elevator trolley.

Human Factors Considerations in Conceptual Design of Materials-Handling Vehicles

Special vehicles designed for the express purpose of handling materials within the industrial plant should be approached from the operator's point of view, i.e., in terms of what the operator has to see and operate in order to manipulate the vehicle and its load in an efficient and safe manner. Some of the key issues are the following.

Visual Clearance of Vehicle and Load. The vehicle operator should be positioned so that he or she can see the perimeters of the vehicle and its load with reference to pathway obstructions.

Visual Observation of Load Transfer. The vehicle operator should be positioned so that he or she can maintain continuous observation of the load as it is raised and lowered and otherwise moved into position, i.e., onto a pallet, into a storage area, into a truck, etc.

Vehicle Controls and Displays. Controls and displays required to operate the vehicle should be located and positioned so that the operator can reach or see them, while at the same time visually monitoring the vehicle and load clearance and load manipulation.

Operator Support and Restraint. Select the most efficient posture for the vehicle operator in terms of his or her ability to control the vehicle and load, considering alternate standing and seated modes. Consider the possible need to provide restraints to keep the operator from falling from the vehicle.

Operator Protection. Provide appropriate guards and/or screens to prevent materials or components being transported from toppling onto the operator. Consider weight and balance dynamics in terms of tipping over the vehicle. Provide a warning system to indicate when too heavy a load is being manipulated or when an impending load shift is approaching the imbalance point.

Fail-Safe Features. Consider the typical misuse behaviors of the average driver and provide fail-safe features to prevent the driver from leaving the vehicle insecured on an incline, leaving the vehicle in gear while he or she is away from the work position, driving too fast, etc.

HUMAN FACTORS CONSIDERATIONS IN PRODUCT MANUFACTURING SYSTEMS CONCEPTUALIZATION

Although the nature of product manufacturing varies considerably, depending on the particular product, and although many of the operations involve automatic manipulation of materials, humans interface with this process at many points, either directly or indirectly through machine controls and displays. The primary human factors considerations fall into the following main categories.

Prevention of Materials Waste

Waste of materials and damage of components and parts used in manufacturing the product are of considerable economic consequence. Therefore, it is extremely important to examine the entire manufacturing process concept in terms of potential human errors that could be made by the worker at the machine, at the assembly bench, in the test lab, or at the packing and shipping center. Human error often occurs at the following points during product manufacture:

1. Cutting and forming raw materials
2. Fastening materials together
3. Treating materials
4. Testing materials
5. Assembling components and parts
6. Testing assemblies
7. Packaging assemblies

Prevention of Tool and Machine Damage

Although many of the fabricating machines and tools are standard items procured from another manufacturer, these should be selected in part on the basis of the fact that they are minimally prone to misuse by the average factory worker. Tooling that is new should definitely be designed with this in mind. Some of the typical misuse potentials are as follows:

1. Putting the wrong kind of material into a machine
2. Putting material into a machine in the wrong way
3. Improperly securing material in a machine
4. Overtorquing material fasteners
5. Scarring critical sliding surfaces with hand tools
6. Burning out a motor by leaving it on too long
7. Failing to monitor a machine operation and allowing the machine to damage itself when the material in process fails
8. Inadvertently putting or dropping a hand tool into exposed gears or into a chain or belt system
9. Damaging a machine while moving it from one place to another

Prevention of Personal Injuries

Personal injuries on the job have considerable economic impact on an industrial operation, in terms of both the lost time and the liability potential. Again, however, the impact is greatest in terms of the potential effect on the employee and his or her family. Manufacturing processes should be conceived on the assumption that no employee wants to be injured and that it is therefore incumbent upon the planner to create a system that is as safe as is technologically possible. The following are typical of the injuries that can occur in the manufacturing process:

1. Entrapment in exposed moving machinery elements (gears, chains, belts, sliding devices, wheels, cranks, etc.)
2. Being hit by flying particles, broken materials, and sparks
3. Being struck by moving beams or materials or by machine components
4. Being crushed by stamps and dies
5. Being burned by hot machine surfaces, by hot materials being produced from a machine, or by a blast furnace
6. Being blinded by operations such as welding
7. Being electrocuted by electrical shorts or exposed high-voltage contact points
8. Being inadvertently exposed to electromagnetic or nuclear radiation

A critical incident analysis of personal injury should be performed early in the manufacturing concept formulation. This analysis should be done not only in terms of the manufacturing process personnel but also in terms of factory maintenance people, supervisory personnel, and visitors, who may be less familiar with the hazard potentials and safety rules of the installation.

In conjunction with this analysis, consider also the requirements for responding to the potential injuries that could occur in spite of a planner's best intentions and/or precautionary design. Consider alarms, special facilities such as contamination-removal showers, first-aid equipment, and a nearby first-aid station. Last, but not least, include in the basic layout plan for any fabrication facility easy access for outside emergency help and/or movement of emergency medical equipment to injured persons.

HUMAN FACTORS CONSIDERATIONS IN HEAVY CONSTRUCTION SYSTEMS CONCEPT DEVELOPMENT

Large earthmoving machines, construction cranes, heavy materials-loading devices, and the like generally dwarf the human operator because of their necessarily large size. Yet few of these systems can be made sufficiently automatic to eliminate either the machine operators or the support personnel who must work closely with the machine system. Most such systems present similar human factors problems that should be addressed as early as possible in the concept formation stage; otherwise, human tasks become extremely difficult, and the potential for personal injury is great. The following key human factors problem areas should be examined and resolved along with the basic machine design questions.

Operator Position and Visual Surveillance of Machine Operation

Because of the size of many large construction machines, it is difficult to find an appropriate position from which an operator can see all that is going on, i.e., the perimeters of the machine for maneuvering clearance and avoiding irregularities that could cause the machine to be upset, the juxtaposition of other machine features that the operator may be maneuvering to lift or dump materials, and/or personnel who may be working within the area near the machine.

Machine Upset

A serious problem in heavy construction machine operation is the potential for upsetting the machine as a result of either its movement across irregular terrain or inadvertent imbalancing due to improper loading or load mismanagement.

Inadvertent Contact with High-Voltage Wires, Obstructions, and/or Personnel

Incidents involving contact with high-voltage wires are reported frequently, indicating that too little thought is given to this hazard. Although operator's carelessness may explain many of these incidents, it is apparent that operators are often overloaded in terms of what they are required to watch and do. Similarly, many accidents in which machines have been run into obstructions and/or personnel have been due to the fact that the operators could not see where they were going or were too busy with other coincidental operating tasks.

Injuries Associated with Mounting and Dismounting and/or Being Thrown from a Machine

Personnel often are required to work at locations high above the ground without suitable guards, ladders, platforms, or tethers. The machines typically used for these operations are designed without due regard for sharp protrusions, corners, or edges, which personnel bump into or fall against. Ladder and step rungs and/or platform surfaces may be slippery (especially when the worker's shoes are covered with mud or snow) and thus contribute to falls.

Loss of Control of Cables and Loads

Machines that utilize cable-lifting systems often are sensitive to high wind disturbance; the hoisted load may become unmanageable and thus be allowed to drop or swing into adjacent structures or the crane itself, or even the total system may become unbalanced and upset.

Inappropriate Movement of Machine Elements

Occasionally the machine operator inadvertently moves the crane member, the hoisting cable, the scoop, or another device in the opposite direction from the one intended. Since even the most skilled operators do this from time to time, it is apparent that the basic control system is incompatible with the natural direction-of-motion expectancies of the operator.

Working Environment

Heavy construction operations are often subject to fairly stressful and inconveniencing environments, i.e., cold, heat, snow, rain, ice, high wind, dust, noise, and vibration. Too often, these conditions are accepted as hazards of the trade, and little concerted effort is made to design the system to minimize the effects of such environmental stress on operator performance.

Communications

The combination of distance between operating personnel and the typical noise generated by machinery creates a serious problem in terms of communications between operating personnel. In addition to warning systems, special voice communication systems should be considered an integral part of the total system package.

System Maintenance

Most heavy construction equipment must be serviced out of doors, in many instances at the actual construction site. Considerations for the ease and safety of field maintenance should be part of the early concept design formulation, as opposed to figuring out how to maintain the system after certain constraints have been built in. In addition to the important objective of accessibility, special consideration must be given to the typical problems of handling extremely large and heavy components. Much of this work must also be accomplished by personnel who are burdened by heavy winter garments.

OCCUPATIONAL SAFETY AND HEALTH ADMINISTRATION STANDARDS

Occupational Safety and Health Administration safety standards should be reviewed as a major consideration during initial industrial systems concept development. Some of the key standards for review include the following:

Part 1910

Subpart D: Walking—working surfaces

Subpart E: Means of egress

Subpart F: Powered platforms, personnel lifts, and vehicle-mounted work platforms

Subpart G: Occupational health and environmental control

Subpart H: Hazardous materials

Subpart I: Personal protective equipment

Subpart J: General environmental controls

Subpart K: Medical and first aid

Subpart M: Compressed gas and compressed air equipment

Subpart N: Materials handling and storage

Subpart O: Machinery and machine guarding

Note: For a more complete listing and description, refer to the *Federal Register,* Title 29-Labor, Part 1910.

CHAPTER 7

AGRICULTURAL SYSTEMS

GENERAL CONSIDERATIONS FOR AGRICULTURAL SYSTEMS CONCEPT DEVELOPMENT[11]

In conceptualizing the typical farm facility, human factors considerations should be addressed for the following:

1. Main and secondary site access
2. Living center
3. Service yard
4. Public parking
5. Family garden
6. Poultry center
7. Dairy center
8. Grain and feed center
9. Other animal centers
10. Machinery center
11. Bulk feed and grain storage
12. Utility systems
13. Security and lighting

Since the typical farm is often isolated, it requires considerable self-sufficiency, and many of the human factors considerations become more critical than they would be if the system were more accessible, in terms of time and distance, to typical urban amenities such as supplies, fire and police protection, doctors, and hospitals.

The farm system should be approached in much the same manner as any other system in terms of laying out facilities for convenience, functional relatedness, work flow or sequence of operations, and, most important, ease of handling emergencies.

Topographic and soil characteristics, of course, influence the placement of crops; thus the location of the principal farm structures must represent a suitable compromise between access to the main highways and access to the principal crop fields.

[11]J. De Chiara and J. H. Callender, *Time-Saver Standards for Building Types,* McGraw-Hill Book Company, New York, 1973.

Finally, natural environmental considerations necessarily influence not only the location but also the orientation of the principal farm structures. Principal environmental considerations include sun path and prevailing winds. The most suitable compromise between environmental conditions involves other natural features that are important to a farm system that is both effective and economical, i.e., natural foliage that may serve as a windbreak, natural drainage, and natural water resources.

Fires are a particularly critical hazard for isolated farmers and their families because of the difficulty of controlling them once they have started. Appropriate spacing of farm structures and positioning with respect to the possible spread of fire by prevailing winds are both important considerations.

Buildings and roads should be designed for maximum utility in terms of the kinds of machinery that will traverse or utilize these facilities. Consider "drive-through" structures and turnarounds so that the machine operator is not required to do a lot of backing.

Build up roadways, service areas, and buildings so that water drains away. Orient buildings and roadway relationships so that snow will not build up across roads or entrances. Provide gates and/or appropriate cattle guards so that the farmer can control the animals and yet traverse the farm without having to mount and dismount from the vehicle any more than is absolutely necessary. The entire farm should be fenced.

CONCEPTUAL DEVELOPMENT OF FARM MACHINES

Perhaps the most significant human factors consideration in developing farm machinery is the fact that these systems will often be operated by persons with minimum skill and/or understanding of mechanical systems. On small farms, not only will the farmer himself operate much of the machinery, but also his wife and children often will assist him both around the immediate farm and in the fields. This requires that farm machines be made simple to operate and as free of hazards as possible.

In addition, since the farmer will expect to perform the major share of the machinery servicing and maintenance, machines should be designed so that they can be maintained with a minimum of training and special skill and without the need for special tools and test equipment.

The following are some of the more important considerations for the design of farm machinery.

1. Mobile Machines (Tractors, Combines, Soil Conditioners, Planters, etc.)

   a. They should be easy to hook together.

   b. They should be easy to operate (start, steer, stop, back, adjust, monitor, etc.).

   c. They should be easy to mount and dismount.

   d. They should be difficult to operate improperly, turn over, or become dislodged from.

   e. They should be easy to service and inspect, and it should be possible to remove and replace parts using only common tools.

   f. They should have special features such as comfortable seats and closed, heated, cooled, and well-illuminated cabs.

2. Fixed Machines (Huskers, Conveyors, Pumps, Auxiliary Power Systems, Milkers, etc.)

   a. They should be simple to operate.

   b. They should be easy to load and unload.

   c. They should be simple and safe to unjam.

   d. They should be easy to service and maintain with common tools.

   e. They should have appropriate safety features, guards, automatic power cutoffs, proper electrical grounding, etc.

CHAPTER 8

COMMUNICATIONS SYSTEMS

GENERAL

Introduction

In conceptualizing a communications system, whether it is large and complex (such as a worldwide satellite system) or relatively simple and limited (such as an internal closed-circuit television system in a factory), the common denominator should always be the human need. That is, communications between people should be analyzed and planned on the basis of human task needs and thus should be designed to meet these needs in the most effective manner. This requires special consideration of human input–output characteristics, or, if you will, human channel capacities and sensory channel sensitivity. In developing a communications system concept, consider at least the following:

1. Communications mode (i.e., visual, auditory, tactile, and/or some combination)

2. Communicating environmental constraints (i.e., noise, environmental interference, and the need for security or privacy)

3. The time factor (i.e., how much information must be transmitted within some prescribed time frame)

4. Reliability requirements (i.e., whether you can afford to repeat or lose some of the information without degrading system performance objectives)

5. Convenience (i.e., whether immediate accessibility is critical)

6. Multiuser requirements (i.e., whether the same communication is important to several users simultaneously)

7. Personal innuendo (i.e., the value of face-to-face expression)

8. International language (i.e., whether it is critical that people with differing language backgrounds understand the communication)

9. Freedom from interruption (i.e., whether it is critical that a given communication not be interrupted prior to its complete transmission)

10. System cost, both the initial cost and the cost of maintenance

11. The effectiveness of the human–machine interface design (i.e., ease of use of handsets, microphones, earphones, controls, etc.), considering the environmental and other constraints involved in use of the interface (temperature, noise, vibration, clothing restrictions, etc.)

Considerations for Development of Communications System Concepts

The following discussion is limited primarily to visual and auditory communications between people within the working community, as opposed to communications for the purpose of recreation or entertainment. A primary objective should be to create a system that supports other human task activities, e.g., the exchange of information among individuals and the transmission of instructions, commands, warnings, orders, etc. Such systems should be based on the capabilities and limitations of the human as well as on the information transfer needs of the users. Consider the following:

1. Type of information (spoken messages, pictures, coded messages, digital data, etc.)
2. The time factor (i.e., how rapidly the information must be transmitted)
3. Security (i.e., whether the information is to be accessible to only one person or to anyone who might be within hearing or seeing distance of the receiving device)
4. Number of intended recipients (i.e., whether the message is for general broadcast for the benefit of a large group of recipients)
5. The receiving environment (i.e., whether certain interferences at the receiving station may mask the incoming communication)
6. Reliability requirements (i.e., whether the situation permits repetition of the message if it is not understood the first time)
7. Quality (i.e., whether the information requires high fidelity in order to be fully understood)
8. Language (i.e., whether it is critical that people of differing language backgrounds understand the communication)
9. Personal factors (i.e., how much value there is to seeing facial expressions or hearing voice inflections)
10. Directionality (i.e., how important it is to associate the transmission with a direction or source)
11. Message capture (i.e., whether it is necessary to have a record of the transmission)
12. Condition of the addressee (i.e., whether he or she is busy, is using his or her hands or eyes, or is under special physical stress, such as high g forces)

Communications hardware technology has reached the state at which the planner can prescribe almost any level of fidelity or reliability and almost any mode. Therefore, the choice is limited primarily by cost. However, since the primary purpose is to allow people to communicate with one another, systems concepts should generate from human needs and from basic human sensorimotor characteristics and limitations. In defining the characteristics of any proposed system, consider the following.

For auditory communications:

1. Loudness (intensity)
2. Pitch (frequency)
3. Timbre (signal and voice profile)
4. Signal-to-noise ratio
5. Signal pattern (duration, intermittency, and frequency shift)
6. Sound location
7. Public as opposed to private

For visual communications (including text and moving or still images):

1. Brightness
2. Color
3. Contrast
4. Duration
5. Pattern and shape
6. Alphanumeric legibility

The following are some of the important human factors objectives for communications systems:

1. Detectability: The intended receiver can sense the signal.
2. Recognizability: The intended receiver can tell what the signal is.
3. Intelligibility: The intended receiver can tell what the signal means.
4. Conspicuousness: The signal is attention-getting.

Although communications systems designers obviously are concerned with the basic transmission medium and technology, the human factors relate primarily to the immediate interface between the system and the sender, listener, or visual observer. Although tactile communication interfaces may have some application in isolated instances, the primary questions are whether the message originator should transmit by voice, encoded aural signal, or written or graphic means; whether the message will be received by aural or visual means; and whether there is increased value (e.g., reliability) if the message is transmitted using both visual and auditory modes. Table 8.1 summarizes guidelines for selecting auditory or visual alarms.

TABLE 8.1 Guidelines for Using Auditory vs. Visual Alarms

Auditory Alarm Preferred for	Visual Alarm Preferred for
■ Simple message	■ Complex message
■ Short message	■ Long message
■ Not referred to later	■ Referred to later
■ Requires immediate action	■ Does not require immediate action
■ When visual system overloaded	■ When auditory system overloaded
■ Moving persons	■ Stationary persons

From NASA Human Integration Design Handbook, 2012.

When to Use Auditory Communication

1. When the message is simple
2. When the message is short
3. When the message does not necessarily have to be referred to later
4. When the message deals with events in time
5. When the message calls for immediate action
6. When the visual system is already over loaded
7. When the intended receiver cannot look at a visual display or may have to move about continually
8. When the receiving environment is incompatible with seeing (e.g., when it is too bright or when it is necessary to avoid visual signals that could interfere with the intended receiver's adaptation to the dark)
9. When the intended receiver's visual capacity is impaired as a result of a dynamic stress (e.g., high g forces)
10. When the intended receiver's eyes may be closed (e.g., when he or she is asleep)
11. When a visual signal strength is borderline and the addition of a coincident auditory signal would provide redundant, reinforcing confirmation

When to Use Visual Communication

1. When the message is very complex
2. When the message is long
3. When the message needs to be referred to later
4. When the information deals with location in space
5. When the message does not necessarily call for immediate action (e.g., when the receiver needs to finish some other task)
6. When the auditory system of the receiver is overburdened
7. When the receiving environment is too noisy to ensure reliable receipt of an aural message
8. When the receiver will remain in a position where he or she can continue to watch the visual display
9. When the information characteristics cannot be reliably described by words
10. When face-to-face communication may provide added understanding between communicators

When to Use a Private Rather Than a Public Interface. An important decision in developing any communication system concerns whether, and under what circumstances, to use individual rather than group visual or auditory displays. Although there are no hard and fast rules, the following should be considered:

1. Use group display interfaces when:
   a. Several people should receive the same information simultaneously.
   b. Several people should receive the information in the same format.
   c. Some people require general information at the same time that they are busy sending or receiving individual communications.
   d. The transmitter may not know who the persons are who should receive the information.
   e. The individual channel that the transmitter wishes to use is temporarily unavailable.
   f. The receiving individual is temporarily away from his or her individual communication interface.
   g. Input to a general status display is being generated by several individuals on a continuing basis.
   h. It is inconvenient for the required number of listeners or viewers to gather around an individual display device.
2. Use individual display interfaces when:
   a. The transmission involves only two persons.
   b. The information need not be shared with any other persons.
   c. The transmission is or should be private.
   d. The information is being modified by the individual for his or her own benefit.
   e. The individual needs to record the information for future reference.

When to Use Handsets, Intercoms, or Public Address Systems. Although there are obvious overlaps in the manner in which handsets, intercoms, and public address systems can or should be used, the following are guidelines for making this decision:

1. Handsets should be used primarily when privacy is desired although the proliferation of mobile phones makes that medium viable for public announcements.

2. Intercoms should be used primarily when the individual is too busy to look up a number and dial it.

3. Public address systems should be used primarily for making public announcements.

Other Communicating Techniques and Rationales for Their Use

1. Colored and/or flashing light signals, horns, bells, claxons, sirens, etc., are particularly useful for issuing warnings and conditional communication because they can be made conspicuous without materially interfering with other communications.

2. Digitized communications can be used to reduce the amount of paper transfer.

3. Encoded signals can be used to maintain communication security.

4. Flag, light, and hand signals can be used to minimize the cost of more expensive systems.

ALARMS

A summary of design recommendations for auditory alarm and warning devices is given in Table 8.2.

TABLE 8.2 Summary of Design Recommendations for Auditory Alarm and Warning Devices

Conditions	Design Recommendations
If distance to listener is great	Use high intensities and avoid high frequencies
If sound must bend around obstacles and pass through partitions	Use low frequencies
If background noise is present	Select alarm frequency in regions where noise masking is minimal
To demand attention	Modulate signal to give intermittent "beeps" or modulate frequency to make pitch rise and fall at rate of about 1–3 cps
To acknowledge warning	Provide signal with manual shutoff so that it sounds continuously until action is taken

ARCHITECTURAL PLANNING FOR OPTIMUM COMMUNICATION

Communications systems requirements should be considered at the beginning of architectural concept development because the effectiveness of the final communication depends on the inherent characteristics of the architectural configuration; that is, avoid the tendency to try to add a communications system to an architectural plan that is frozen. Few, if any, communications systems are completely satisfactory unless the architectural and communications systems are optimized as an integrated concept.

Key factors in planning the architectural and communications systems include the following:

1. Plan specifically for the total array of expected communications, i.e., direct person-to-person speech (without benefit of intervening aids such as the telephone), private telephone conversations, public address systems, emergency alarms, special intercom connections, mobile communications closed-circuit television, signal lights, special visual displays, and directional and instructional signs.

2. Establish requirements concerning privacy and prevention of annoyance.

3. Determine both inside and outside communication interference possibilities and identify the exact nature of these in terms of loudness, frequency distribution, etc.

4. Examine and define the potential physical interactions between each type of communicating mode and the proposed architectural features that may produce undesirable effects on the communication link (reverberation, echoes, sound time lag, sound shadow, visual barriers, effect on sound localization, visual glare, seeing distance, illumination level, etc.).

5. Organize the facility in terms of ease of developing a logical and meaningful signing system.

Key architectural factors that should be considered relative to auditory communications systems include the following:

1. The shape and size of interior spaces and other occupied areas (e.g., hallways and stairwells)

2. The surface texture and shape of floors, walls, and ceiling and of furnishings that are expected to be relatively permanent

3. The physical positions of speakers (e.g., people) and listeners

4. The position of communications system hardware elements (e.g., loudspeakers, intercoms, and microphones)

5. The number of people who may occupy a space and thus will tend to absorb sound

Key architectural factors that should be considered relative to visual communications systems include the following:

1. Viewing distances

2. Ambient lighting effects

3. Artificial lighting effects (e.g., decorative lighting)

4. Day versus night ambient and artificial lighting effects and their influence on internal seeing conditions

5. Individual display and light control needs relative to interference among different facility users

Maintenance of both the communications system and the related architectural features of the planned facility should also be considered early and simultaneously. Like plumbing, many communications systems become inaccessible for maintenance because of the tendency to hide wiring and hardware. Although the aesthetic objective is laudable, all visual and audio communications systems hardware that can be expected to need service or replacement should be made accessible, especially so that wear and tear on architectural features (e.g., ceiling and wall panels) is minimized.

Finally, if a proposed communications system (for a public auditorium, church, campus, factory, etc.) cannot be fully implemented at the beginning or if there is any possibility of future system expansion, plan the architectural facility for ease of adding new wiring, controls, connections, etc.

CHAPTER 9

CONSUMER PRODUCT SYSTEMS

HUMAN FACTORS IN CONSUMER PRODUCT CONCEPTUALIZATION AND DESIGN

There are at least three distinct, but not necessarily mutually exclusive, areas in which human factors should be considered relative to the conceptualization and design of consumer products such as appliances, entertainment and recreational equipment, tools, and toys. These are discussed briefly as follows:

Safety

Products should not be designed so that they could fail and cause injury to the user or so that they could easily be misused in a manner that could lead to injury to the user, nor should they be made out of materials that might be toxic, poisonous, or otherwise hazardous to the health of the user. Lack of a proper safety analysis during the conceptual design stage typically results in the following:

1. Products in which toxic or flammable materials are used
2. Products that are made of breakable materials or that have sharp protrusions, corners, or edges
3. Products in which exposed moving parts cause pinches, cuts, or amputations
4. Products that can be used improperly to strike another person or to shatter glass
5. Products that can be swallowed
6. Products that can cause electric shock because of improper insulation or grounding
7. Products that people can easily fall from
8. Products that can produce injurious noise, extreme heat, or flying particles that could puncture a person's skin or eye
9. Products that could cause rupture or strain when lifted or injury to a foot if dropped
10. Products that can cause burns

Operability and Maintainability

In order for a product to be truly acceptable, it has to be easy to operate and maintain. Most consumers fail to appreciate this aspect of a product fully until after they have purchased it. There is considerable evidence that the typical consumer will not purchase another product of the same make once he or she has discovered that it is difficult to operate or maintain.

Typical features that show up in many consumer product designs are the following:

1. The product is hard to pick up, hold onto, manipulate, or maneuver.

2. The product is hard to open and close, and the latches and handles are difficult to manipulate.

3. The controls and displays are poorly labeled, hard to read, difficult to understand, and awkward to adjust or operate.

4. The product is difficult to load and unload and requires the user to stretch, reach, bend, etc.

5. The product requires too much force or too much precision to operate properly, resulting in frequent errors in use.

6. The product is packaged in such a way that it is difficult to get to parts that need adjustment or replacement.

7. A presumably movable product is not provided with appropriate handholds or casters so that it is easy to move to a new location.

8. The product contains hard-to-find parts, with the result that acquiring replacement parts later is impossible or time-consuming.

9. It is practically impossible to keep the product properly adjusted long enough for it to do its job.

10. The product wears the operator out because of its weight, the awkward way it has to be held, the noise it makes, or the way it vibrates.

Attractiveness

Although most manufacturers and designers are fully aware of the importance of product attractiveness to salability, and thus pay more attention to this aspect of product design than to any other, attractiveness is an elusive variable when one tries to define it. A critical comment is in order also: Although attractiveness is an admirable and desirable trait in product conceptualization, it is extremely important that, in an effort to make something attractive, the designer be careful not to compromise product safety, operability, or ease of maintenance.

Since attractiveness is truly a human factors consideration, it is important to recognize that it may be a transient phenomenon; i.e., what one considers attractive this year is not necessarily considered attractive several years from now. This point is made to refute the argument that the aesthetic function is the most important consideration in consumer product conceptualization; i.e., a product that works well can continue to do so over many years and through many different packaging treatments, but only if the initial conceptual objective was to make the product work well. It should also be emphasized that a product can be made attractive after the objectives of operability and maintainability have been developed. Some of the hazards of starting with the aesthetic objective are the following:

1. Aesthetic features may interfere with product recognition. For example, there is often a tendency to camouflage a product, i.e., to make it look like something that it is not.

2. Aesthetic features may make visually displayed information impossible to read, as when white labels are printed on bright gold metallic backgrounds.

3. Exotic-looking control knobs may interfere with the user's ability to adjust the control easily and precisely, or the fancy surface treatment or shape may actually hurt the user's hand or fingers.

4. A streamlined cover or case may make it difficult to pick up the device or may cause a person to drop the object because there is no way to get a good grip on it.

5. An attempt to completely hide all visible means of fastening a cover on a product may cause the owner to damage the cover while trying to get it off for maintenance or replacement of a part.

6. Ornaments placed on a product to make it attractive often turn out to be hazards, or at least annoyances, because the user is constantly catching his or her clothing or hands on the ornamentation.

7. Color used for aesthetic reasons often becomes "visual noise" around an important warning display, making the display insufficiently conspicuous to attract the user's attention.

8. Fancy logos are often much more conspicuous, readable, and intelligible than more important user instructions. In some cases, important labeling is actually left off because the designer felt the label would disturb the overall aesthetic appearance that he or she was trying to create.

9. Furniture and related products are often less than useful, operable, comfortable, or movable because of the interferences created by aesthetic objectives. For example, deep, puffy cushions do not provide needed posture control, and they make it impossible for the short-legged user to sit comfortably.

10. Products in which sharp edges and corners are typical styling features often injure the user.

Misuse Behavior as a Key Safety Consideration

Liability should be an important consideration in conceptualizing any consumer product; i.e., the possibility of consumer injury and a potential lawsuit should stimulate any planner to investigate thoroughly the possible failure modes of a proposed product design, including especially the misuse behaviors of expected consumers.

Avoid the excuse that "I can't worry about every moron who may buy or use my product." These words have come home to haunt many manufacturers and, more recently, many designers. Worrying about potential misuse of a product can become an advantage in that it generates product features that will make the product more salable, e.g., easier to operate without making mistakes.

The following are some of the typical misuse behaviors that consumer product planners should consider:

1. Children

   a. Children are extremely curious. As a result, they investigate and examine or try to play with products or product elements that they should not touch. If there is a potential hazard, design the product so that children cannot reach or operate the device.

   b. Small children investigate by touching things, putting their fingers or hands into openings, putting things into their mouths, or at least trying to put their tongues on things. Make sure that objects that are small enough to swallow cannot be removed, that holes are either too small to allow passage of fingers or arms or large enough so that fingers and arms can be pulled back out, and that surface finishes are not toxic and cannot flake off.

   c. Children like to climb or reach for something on top of a surface that they cannot see. Package products so that there are no climbing aids. Design hazardous products, like stoves, so that children cannot reach the controls or burners.

   d. In spite of a designer's best intentions, most children will use a toy in an entirely different manner from the way it was intended to be used; e.g., they will use it to pound with, stick it in their own or another child's eyes, or try to pull it apart to see what makes it work. Make sure that toys are sturdy enough so that they will not break or come apart; soft enough so that they will not break something else or injure another child; devoid of sharp edges, corners, and projections; and made of materials that are not toxic.

   e. Children are unsteady on their feet, and they often run so fast that they are unable to maintain their balance or control their bodies. As a result, they often fall or run into objects. Products around which children may play should be designed with the idea that a child might fall or run into them; i.e., there should be no sharp edges or corners, and they should be made of unbreakable materials.

2. Adults

   a. Adults often attempt to operate a product without looking at the instructions first. As a result, they may do something to create a hazard for themselves and/or for others around them. Products should be designed as nearly as possible to be understandable without requiring the user to read special instructions; i.e., the critical instructions and labels should tell how to start, adjust, and turn off the product.

   b. Adults are often absentminded or easily distracted, and they often forget to be careful. Products that have a hazard potential should have safety lockout features and/or added protective features that will minimize the hazard in the event the user forgets and turns a wrong knob, for example.

   c. Adults may be in a hurry, may bump into things in their haste, or may not look to see where they are putting their hands or feet. Products should be designed so that even if the user does tend to hurry, injury will not result.

   d. Adults tend to believe that they are smart enough to take shortcuts (even though there may be a warning to the contrary). Often these shortcuts end in tragedy. Products that have such hazard potential should be designed so that shortcuts are out of the question.

   e. Many adults are willing to take risks, even though they know there is a chance of making a mistake or a miscalculation, either because they think they can get away with it or because they truly believe their time is more important. Products should be designed so that they cannot be operated in any way other than the right way; i.e., there should be no opportunity for risk taking.

3. The Handicapped: Many handicapped consumers are forced to try to operate products that were not designed with their infirmities in mind. As a result, they are subject to hazards that other adults are not exposed to. Since the number of handicapped persons is increasing and since they are becoming a significant portion of the consumer product market, products should be designed so that people with handicaps can use them efficiently.

DESIGN CONCEPTUALIZATION FOR AN INTERNATIONAL CONSUMER MARKET

Designing a product that may be marketed around the world requires knowing something about the differences between consumers that are critical to the product's safe and efficient use and to its acceptability. The following are some of the human factors that are important to an understanding of a product's operation and acceptability in different cultures.

Anthropometric Differences

People of various nationalities differ in terms of their body dimensions. Some are taller, heavier, and stronger than others. Some have larger hands, longer reaches, and more strength than others. Products designed for large consumers might be too large or too high or have reach requirements that are too great. By the same token, certain products designed for the smaller individuals might not provide sufficient clearance.

Technological Experience

An ethnic population that is not used to many of the modern products that a more technically oriented population is used to will neither appreciate the technical complexities that may be built into a product nor understand how to operate or maintain the product. Thus, the complexity of a product should be geared to the expected user population. Even though a product may not seem as sophisticated and efficient as it could be, its final effectiveness and acceptance might be much higher if it were less complex.

Language Barriers

A major problem for designers of equipment that may be used in many parts of the world is the difficulty of labeling a product and/or creating appropriate instructions for each potential user. Translating instruction manuals is fairly simple; however, labeling to provide universal understanding presents a more difficult problem. As will be noted elsewhere in this handbook, using international symbols in labeling is one way to ensure that people who speak different languages can still understand and operate a product or device.

Ethnic Differences in Terms of Styling

Although of less importance than in past years, various styling features have considerable significance to members of some religious or ethnic groups. It should be noted that much of this can be treated in a cosmetic manner; i.e., after a basic product package has been developed for the total market, cosmetic features appropriate to particular user markets can be added to increase user acceptance.

Regulations

Foreign markets (particularly the European Union) may have regulations governing human factors and safety aspects of a product design. Products intended to be marketed in these areas should be designed and tested to meet foreign regulations. Depending on its type, products sold to the European Union market will have to meet a General Product Safety Directive and additional requirements defined by specific sector legislation. Products regulated under these rules include the following:

1. Chemicals
2. Toys
3. Personnel protective equipment
4. Cosmetics
5. Pharmaceuticals
6. Machinery
7. Recreational craft

SAFETY CONCEPTS BY PRODUCT CATEGORY

Although the following safety considerations by no means represent all the potential safety problems to be addressed for each of the product categories, they provide a point of departure for planning product design concepts.

Electrical Appliances

1. Electrical grounding
2. Overload fuses
3. Nonmetallic user interface hardware
4. Automatic power cutoff with cover removal
5. Adequate insulation
6. Warning and caution labels and instructions for operation and maintenance

High-Temperature-Producing Products

1. No direct contact between the internal heat source and the case
2. Insulation to attenuate radiant heat transfer to the exterior
3. Low-heat-transfer hardware to user interfaces (e.g., handles and controls)
4. Nonflammable materials and finishes

Mechanical Products with Moving Parts

1. Covers over moving gears, chains, belts, and activating levers
2. Captive spring-loaded parts
3. Fracture-, chip-, and distortion-resistant metal, composition, or plastic parts

Products with Explosive Components

1. Nontoxic substances
2. Minimal energy release commensurate with purpose
3. Structural integrity of the container
4. A replacement package and a method of replacement that preclude mishandling and other forms of misuse

Toys for Small Children

1. Nonflammable materials
2. Nontoxic materials
3. Unbreakable materials
4. Sufficient size to preclude swallowing the toy or putting it far enough into the mouth to strangle and no small parts that can be removed and put into the mouth
5. No sharp corners, edges, or protrusions

6. No cords or wires that could be swallowed or wrapped around the user or a playmate

7. No metal parts that could be inserted into an electric outlet

8. No articulating element in which the child could be pinched

Tools and Cutlery

1. Appropriate electrical and mechanical component grounding and guards

2. Guards to prevent contact by flying particles or sparks

3. Safety catches and covers

4. Heat insulation around heat-producing parts (nonflammable)

5. No cords, if possible; otherwise, quick release connectors

6. Appropriate warning labels and instructions on the product

7. Safety features associated with replacement of parts and cleaning or servicing

Furniture

1. Structural integrity for expected loads

2. Antitip construction

3. Nonflammable and nontoxic materials

4. No sharp corners or edges

5. Lockable casters (when used)

6. A foldable design that will not cut or pinch the user or collapse accidentally during use and a design that lets the user know when the product is fully locked in place

Other Products

1. Lockable medicine cabinet doors

2. Refrigerator doors that can be opened from inside in the event a child gets shut in accidentally

3. Stove burners and controls arranged so that a child cannot get to them

4. Space heaters designed so that children cannot be burned by hot exteriors

5. Ladders that will not collapse and which the user can be sure are locked in place

6. Stair rails designed so that small children cannot get their heads caught between the rail supports or fall between the sup ports

7. Baby furnishings designed to prevent the child from falling between the mattress and the rails, from lowering a rail or getting caught between the rail supports, and from being cut or pinched by rail-lowering hardware; nontoxic materials and finishes on parts of the product that a child might touch with his or her mouth

PART 2

SUBSYSTEMS DESIGN

POINTS TO CONSIDER

When working with subsystems, the following points may help to clarify communications between you as human factors professionals and designers.

1. Thinking in terms of systems and subsystems is handy from a design and operational point of view. All of the elements in a subsystem, including a human, coordinate together to form a functional system. Designers usually work at the subsystem level—systems engineers assure the integration of all of the subsystem designs. Humans usually function within a subsystem—their workstation and the displays and controls are all part of the subsystem. Some subsystems have very little human interface. The cockpit is a subsystem of an aircraft with a high level of human interface. The hydraulic subsystem, however, may have very little human interface. Installation and maintenance may be the only interface of some subsystems. Think of different systems—hospitals, spacecraft, or highways. What are some of the subsystems of each and which ones involve humans? Those are the subsystems that should be the focus of human factors specialists. Interfaces with other subsystems—that may be only maintenance—can be critical, however, and should not be overlooked.

2. Human factors considerations should be consistent across the subsystems within a system. For example, a yellow stripe in one area of a plant should mean the same in another area. User size assumptions should also be consistent. Can you think of other human factors considerations that must be consistent across subsystems?

3. The design process usually requires selecting between alternatives. The criteria for selection may include cost or weight or availability of components. Human factors criteria can include speed of operations, consequence of human failure (catastrophic versus forgiving), level of training required, or aesthetics. The importance, or weighting, of these criteria will vary with difference subsystems. For example, aesthetics might not be important for fire extinguishing, but speed of operation is. In doing quantitative trade-off evaluations to select a design, these criteria might be given different numerical weights according to their importance. Think of a subsystem and the human factors criteria for the design. How would you weight each of the criteria (say on a 1 to 5 scale) if you were to make a selection between different design configurations?

CHAPTER 10

SUBSYSTEMS INTRODUCTION

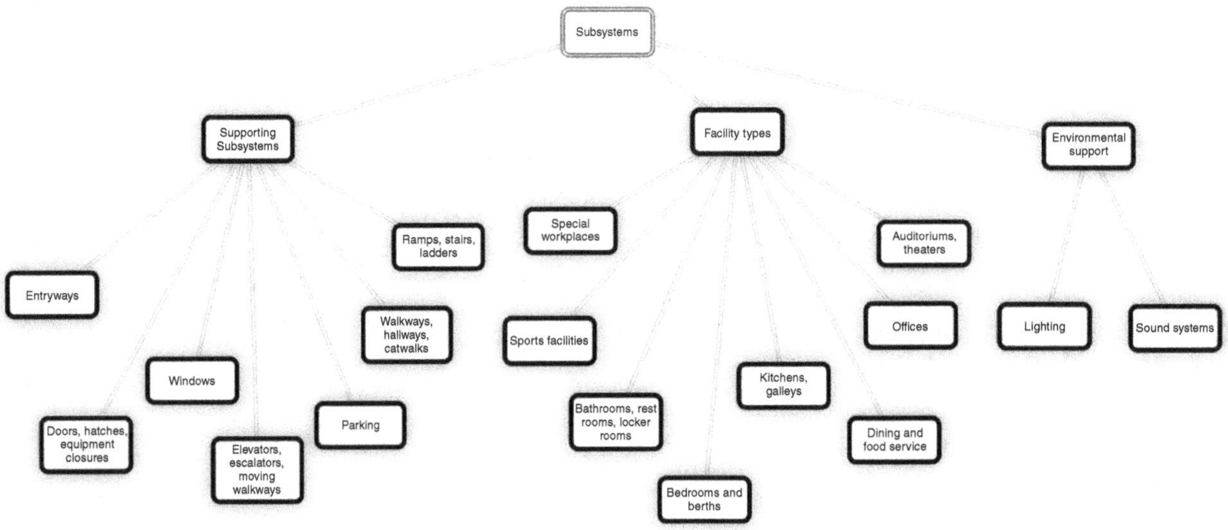

Although many of the subsystems discussed in Part 2 of this book are sufficiently complex to be referred to as "systems," they are included here because they generally occur as part of a much larger system. When they are addressed during the overall *system* conceptual planning phase, they often are merely referred to, and no specific recommendations are made concerning the way in which they should be designed.

Thus, a particular designer may be given the task of designing a subsystem without receiving much direction from the original overall system planners, which means that the designer is left with the job of analyzing the functional requirements and actually "conceiving" as well as designing the subsystem.

In this part of the book, subsystem considerations are discussed in more detail than the system conceptualization considerations in Part 1. It will also be noted that in many instances, considerable attention is directed to the typical problem of selecting from among alternative already designed and produced systems and components. Typically, architects, interior designers, and industrial designers should not create new designs, nor do they usually have the time to do so; rather, they must search out the system and component products that most effectively meet their immediate needs. Their job, then, is to select the right systems and components and put them together in the most user-effective manner.

INITIAL PLANNING

The following planning steps should be taken before designing any subsystem:

1. Determine the constraints placed on the subsystem by the overall system concept. These include both operational and cost constraints.

2. Make a detailed operational and functional requirements analysis for the subsystem itself, even though a preliminary analysis has already been made and a general specification has already been prepared.

3. Determine which functions should be performed by some type of equipment and which should be performed manually by a human and establish specifications for both types of functions.

4. Identify the equipment, displays and controls, furnishings, and other support items that are generally available off the shelf; i.e., avoid reinventing the wheel.

5. Develop grouping and traffic analyses to describe how the elements in the subsystem will probably be used and prepare a rough scenario of the various operations and/or the tasks that will be performed.

6. Establish criteria for searching out and evaluating off-the-shelf elements for the subsystem and perform a trade-off analysis to determine the best selection.

7. Begin putting elements together in alternative combinations and arrangements, utilizing appropriate analytic methods, such as time-line analysis, link analysis, and mockup evaluations.

8. Prepare a detailed specification for the subsystem, including performance requirements for each element in the system. Include salient human engineering requirements as well as engineering requirements.

DESIGN

The design process, whether new or off-the-shelf products or components are utilized, should always start with the user; i.e., start with the user at a task level and work outward, step by step. The following are typical steps:

1. Determine who the user or users will be: a child, a housewife, an office worker, an engineer, a technician, a client, an equipment operator, a deliveryman, a housekeeper, a maintenance worker, and so on. Then determine what functions the user is to perform, individually or in conjunction with others.

2. Determine how each user will perform the functions he or she is responsible for and what tools, equipment component interfaces, and furnishing support each will probably have to use.

3. Define the immediate "use space" required by the users and their related devices and tools and then define the interrelationships among several use spaces so that necessary communications and traffic flow are optimized.

4. Maximize interface details both for each immediate subsystem user and for multiuser interactive features; i.e., design and/or select displays, controls, consoles, seats, illumination, etc., which are well engineered from the human factors point of view and which will help assure that each user has the maximum opportunity to perform whatever task he or she is assigned as efficiently as possible.

5. Maximize the environmental conditions surrounding user spaces so as to minimize the possibility that noise, vibration, extreme temperature, or poor ventilation may degrade the user's performance, cause annoyance to the user, or affect his or her health and safety.

6. Throughout each of these steps, continue to evaluate, test, and reiterate in order to develop a balanced and successful trade-off between human and hardware priorities. Above all, keep in mind that the ultimate success of the system depends in most cases on how well the user performs, which is a combination of the user's appreciation, skill, and physical and psychological interaction with the system hardware.

TEST AND EVALUATION

It is particularly important in subsystem design to verify concept details at each step in the design process. Avoid the temptation to assume that because certain subsystems have traditionally been designed in a certain way, they have been successful in the past. Remain sensitive to the possibilities of improvement in user performance. In many cases this requires merely a more thorough search for improved products or components, i.e., ones that evidence good human engineering in their design.

One of the most important steps in evaluating a proposed subsystem concept is to construct mockups of alternative designs so that the concept can be evaluated from the point of view of both physical space and procedures. Avoid the mistake of believing that a three-dimensional task environment can be evaluated by looking at drawings alone. People work, live, and play in three-dimensional environments, and therefore it is vital that we confirm that a proposed space and human–equipment interface allows the user to perform his or her various tasks as efficiently as possible. Remember that a given design should adapt to the user, rather than the other way around. Although humans are obviously quite adaptable, they often are forced to pay a considerable penalty in terms of impaired performance and physical or psychological stress.

Finally, learn to evaluate the system from the standpoint of its performance rather than its appearance. Many seemingly attractive systems or products are nothing more than that. They are nice to look at, but you would not want to use them!

CHAPTER 11

ARCHITECTURAL ENTRYWAYS

ENTRYWAY DESIGN

General Principles for Architectural Entryway Planning and Design

1. Provide two or more entries for all facilities that are to be occupied by people (preferably located at opposite ends of the facility) so that there will always be an alternative emergency escape route and so that the various user functions will not interfere with one another more than necessary.

2. Locate entryways so that internal and external functional relationships and associated activities are as convenient as possible; e.g., the travel distances should be as short as practicable between the family car or delivery vehicle and the kitchen, between the kitchen and the outdoor patio, between the swimming pool and a bathroom, and between the laundry room and the outdoor clothesline.

3. Do not hide or camouflage main entries; place them where they can be seen from the typical visitor's approach view. A hidden entry provides ideal camouflage for a burglar who is trying to force open the door.

4. Create an easily negotiated approach to the entry, especially when there may be a significant grade change; i.e., provide properly designed ramp, stairs, handrails, path surfacing, and illumination.

5. Create an appropriate docking interface for service entrances that must provide special vehicle interfacing. Keep in mind the special pedestrian relationships, especially where safety may be concerned.

6. Create an appropriate external threshold (a porch, platform, deck, balcony, etc.) that will safely accommodate expected pedestrians and provide sufficient clearance to manipulate packages, furniture, equipment, wheelchairs, etc.; to clear swinging doors; and to avoid conflicts between entering and exiting traffic.

7. Create a similarly appropriate internal threshold (e.g., a hall, lobby, or reception area), keeping in mind the potential conflicts created by putting on and removing outer clothing, clearing swinging doors, and manipulating furniture or equipment to other parts of a facility.

8. Provide proper illumination for both the external and the internal entry thresholds so that people can see where they are going, what they are doing, and with whom they may be conversing. If identification displays are required, make sure that these also are illuminated.

9. Provide an appropriate communications system, such as a buzzer, bell, intercom, or visual identification.

10. Provide protection from inclement weather.

Residential Entrances

There should always be one entry that is primarily "guest-oriented" and one or more others that are for family use only and/or for infrequent use by service people. The following considerations are important:

1. The guest entrance should be readily visible and identifiable from the normal approach by both social guests and others such as the mail carrier and the paper carrier.

2. Family and service entrances need not be readily visible or identifiable by guests, but they should be located for maximum convenience by family members (including access between the family car and the kitchen area for grocery transfer, access to and from rear yard facilities and internal facilities such as the kitchen and bathroom, etc.).

3. All entrances should have an adequate threshold area so that the user is not inconvenienced as he or she opens the door. Such external thresholds should have adequate cover to protect the user from rain, and there should be proper lighting so that the user can find a lock, a doorknob, etc. The threshold to the guest entrance should be large enough so that several guests can gather on the threshold, under cover, without being knocked over when the door is opened. Lighting should be provided so that the guests can be seen and recognized. (Many porch lights are positioned so that guests' faces are in shadow or so that their shadow makes it difficult for them to see where they are stepping.)

4. All entry doors should be provided with vandal-proof locks, and any windows in doors should be located so that the glass cannot be broken for easy access to an inside locking bolt or door handle.

5. A sufficient number of doors should be provided and located to ensure that emergency escape is possible from any part of the house in case of fire. If windows are designated as alternative escape routes, they should be accessible from inside and should open to a plausible and safe escape route on the outside.

6. At least one entrance should be accessible for use by a person who is confined to a wheelchair; i.e., the threshold between the walkway and the porch should be properly ramped.

7. Internal entryway or foyer space should be provided so that there is adequate room for guests to collect inside and remove their outer clothing without being jostled by the closing of the door. A cloak closet should be located in this area so that wet garments will not drip onto the floor of a living-room area. Avoid combining the entryway with a living room, thus requiring people to track through the living room to get to other parts of the house.

Entryway Design to Provide Protection from Inclement Weather. Provide cover at all entryway thresholds, not only to reduce the discomfort of users caused by rain, snow, wind, etc., but also to minimize interference or hazards caused when snow, ice, or rain collects on the entry threshold.

Avoid styling features, such as the one shown in Fig. 11.1, which obviously neither provide protection nor prevent rain, snow, or ice from collecting on the threshold platform.

Figure 11.1 No eaves—no protection.

Figures 11.2 and 11.3 illustrate various approaches to providing cover at the entrance:

1. The inset type provides protection not only from rain but also from wind.

Figure 11.2 Covered entry—example 1.

2. The covered entryway is suitable for locations where rain or wind is less frequent or severe.

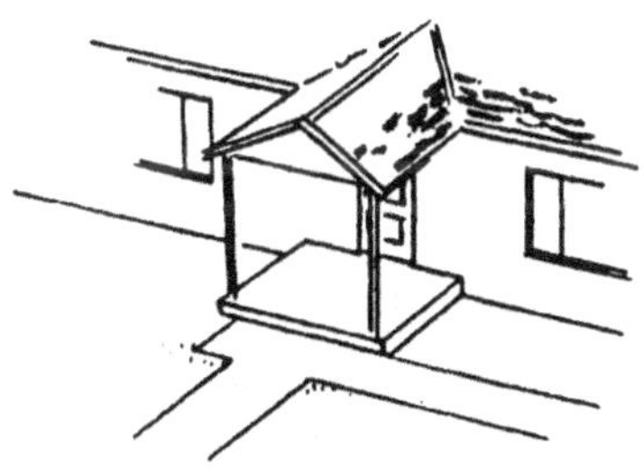

Figure 11.3 Covered entry—example 2.

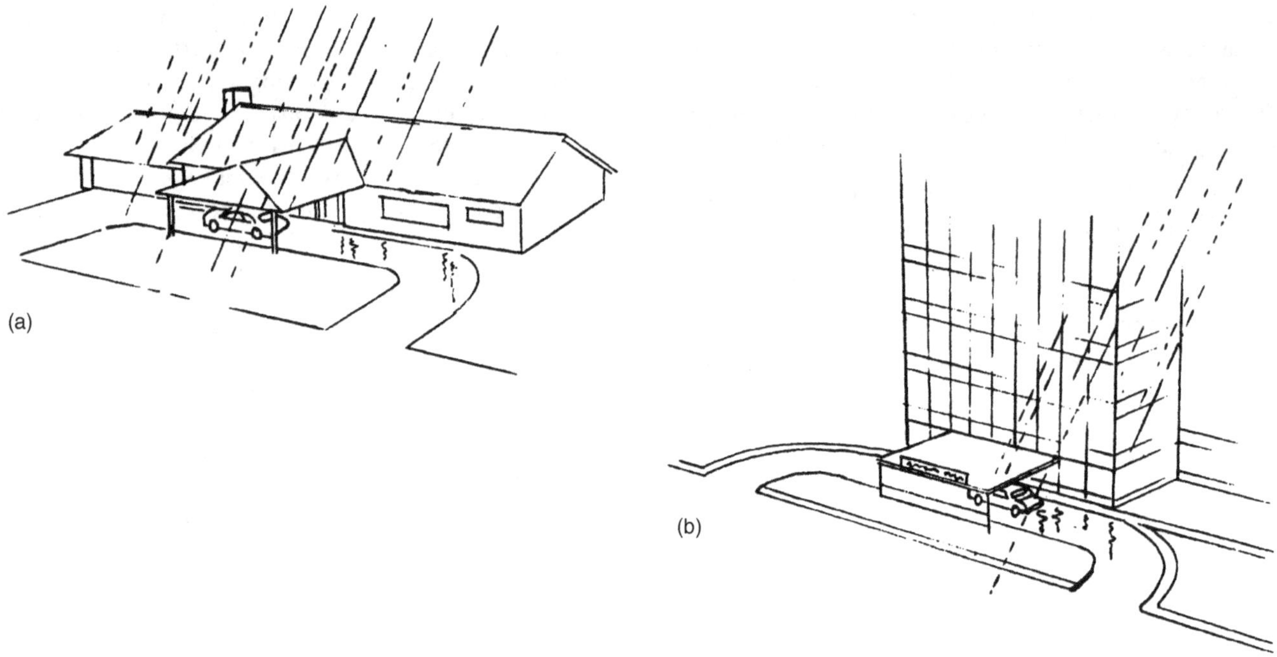

FIGURE 11.4 Complete protection: (a) example 1 and (b) example 2.

3. Complete protection from the vehicle to the residence or building is desirable in order to eliminate the need to run or use an umbrella. (See Fig. 11.4.)

Entry and Exit Locations

1. The main entry should be convenient for visitors from their point of arrival. (See Fig. 11.5.)

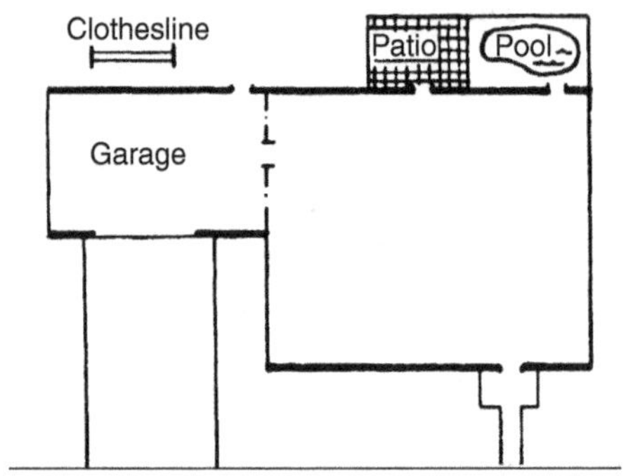

FIGURE 11.5 Entry and exit location.

2. There must be more than one exit in case of fire. Exits from opposite sides of the house are desirable in case one is blocked. Exit through a garage is unacceptable because of the high possibility that a fire may be associated with an automobile.

3. Other convenient entries should be located relative to external and internal positional relationships.

Multifamily Dwellings Entry and Exit Location

Tradition has allowed apartments, motels, and hotels to be designed with only one entrance and exit. From a safety standpoint this is unacceptable. Multiple doors should be provided. (See Fig. 11.6.) Equally unacceptable is the assumption that occupants can escape through windows or from balconies. Children, the elderly, and handicapped persons are usually unable to use the latter methods of exiting without serious injury.

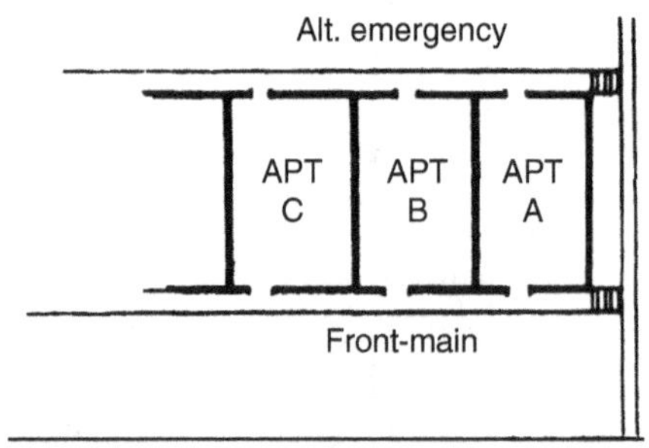

FIGURE 11.6 Multifamily dwellings entry and exit location.

COMMERCIAL AND INDUSTRIAL BUILDINGS

Entry and Exit Locations

Entry and exit locations should be designed for visitors' convenience, to minimize traffic-flow conflicts, and to reduce problems associated with entering and exiting the building. (See Fig. 11.7.)

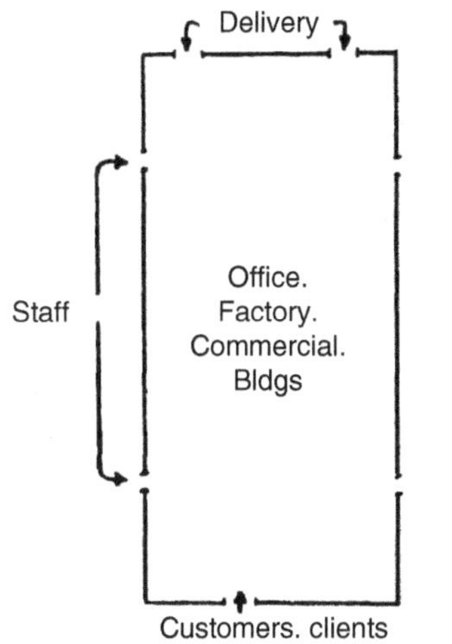

Figure 11.7 Commercial entry and exit location.

1. Main entries should be readily obvious to visitors, clients, and other infrequent users, and such entries should be convenient to their point of arrival and/or planned parking area.

2. Separation of entries by user function is desirable to minimize traffic-flow conflicts and bottlenecks due to internal activities such as loading and unloading and to provide privacy and security.

3. Provide multiple entries for building occupants when there is a considerable influx or departure problem.

4. Emergency exists, including alternative escape routes, should be accessible to blind persons and to persons in wheelchairs in accordance with the U.S. Department of Justice 2010 ADA Standards for Accessible Design.

Exterior Configurational Implications Relative to Entrance Recognition

It is desirable to select a building configuration that provides an immediate indication of where the main entrance is located. Although tenants soon learn to find less conspicuous entrances, the visitor can easily become confused by certain exterior features of the office building.

1. Figure 11.8 shows that people generally expect a main entrance to be on the broad side of a building, as opposed to being on one end.

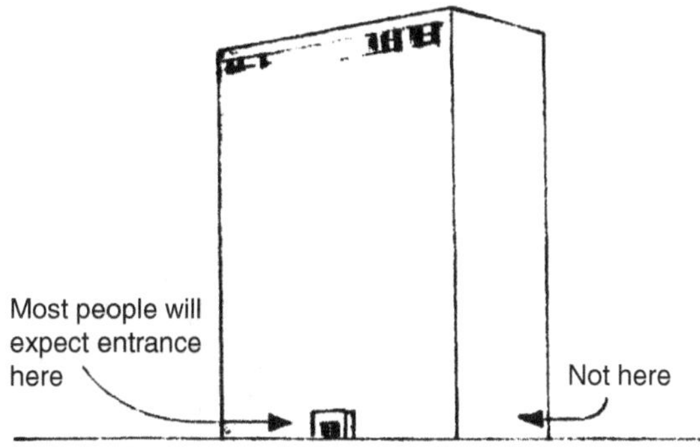

Figure 11.8 Broadside main entrance—example 1.

2. Symmetrical configurations such as the one shown in Fig. 11.10 create confusion because there are essentially equal choices. If one entrance is to be used as the main entrance, it should have a special size, shape, or other emphasis, and it should be clearly marked.

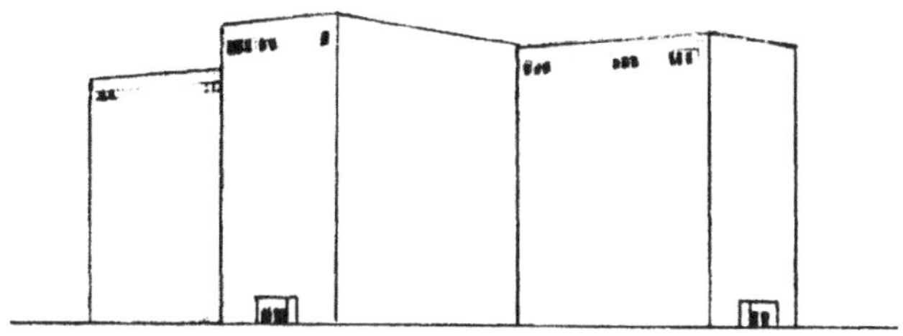

Figure 11.9 Clearly marked main entrance—example 2.

3. When an office section of a larger complex is provided, it is important that the exterior appearance make it obvious to the observer which part of the overall complex is the office section, rather than some other section, such as the apparent factory section shown in Fig. 11.10.

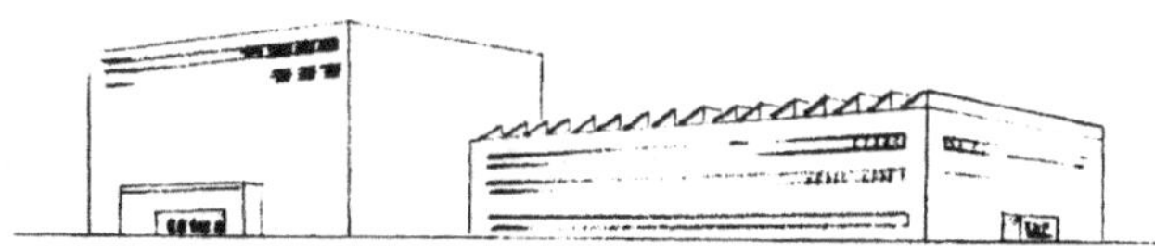

Figure 11.10 Define factory section from office section—example 3.

MAJOR GRADE APPROACH CONSIDERATIONS

Whenever feasible, the approach to all facilities should be as level as possible in order to ensure minimum inconvenience and maximum accessibility for the elderly and the handicapped. Figures 11.11 and 11.12 emphasize some of the most common problems created by lack of attention to the approach to entryways:

Avoid long, unbroken stair patterns. When the natural grade precludes the use of ramps or other methods for providing level entry approaches, provide rest platforms. If the building is for public use, provide a series of ramps or elevators or escalators.

FIGURE 11.11 Stair patterns.

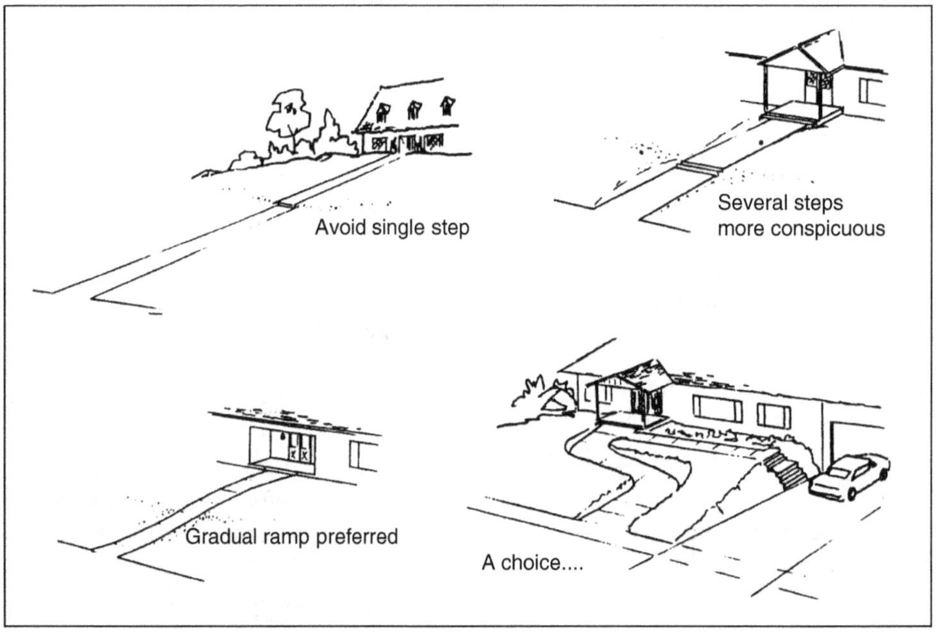

FIGURE 11.12 Preferred methods for gradual approach grade.

VEHICULAR ENTRYWAYS

Separate Vehicular and Pedestrian Entryways

Avoid combining pedestrian and vehicular entryways, especially in industrial, public, and similar facilities. A separate pedestrian entryway, such as shown in Fig. 11.13, is desirable to avoid hazardous conflict between pedestrians and vehicles. Separate entryways alone will not suffice, however. Appropriate marking on surfaces both inside and outside should also be provided.

Separation of the two entryways also helps prevent pedestrians from using the vehicular entryway and from wandering into the path of a vehicle, once they are outside.

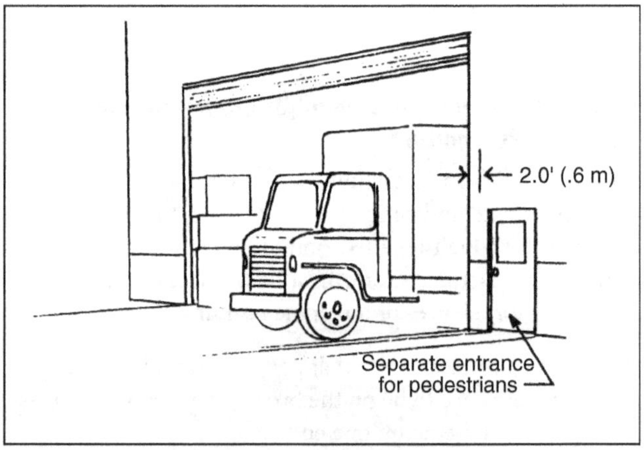

FIGURE 11.13 Avoid pedestrian/vehicle conflict.

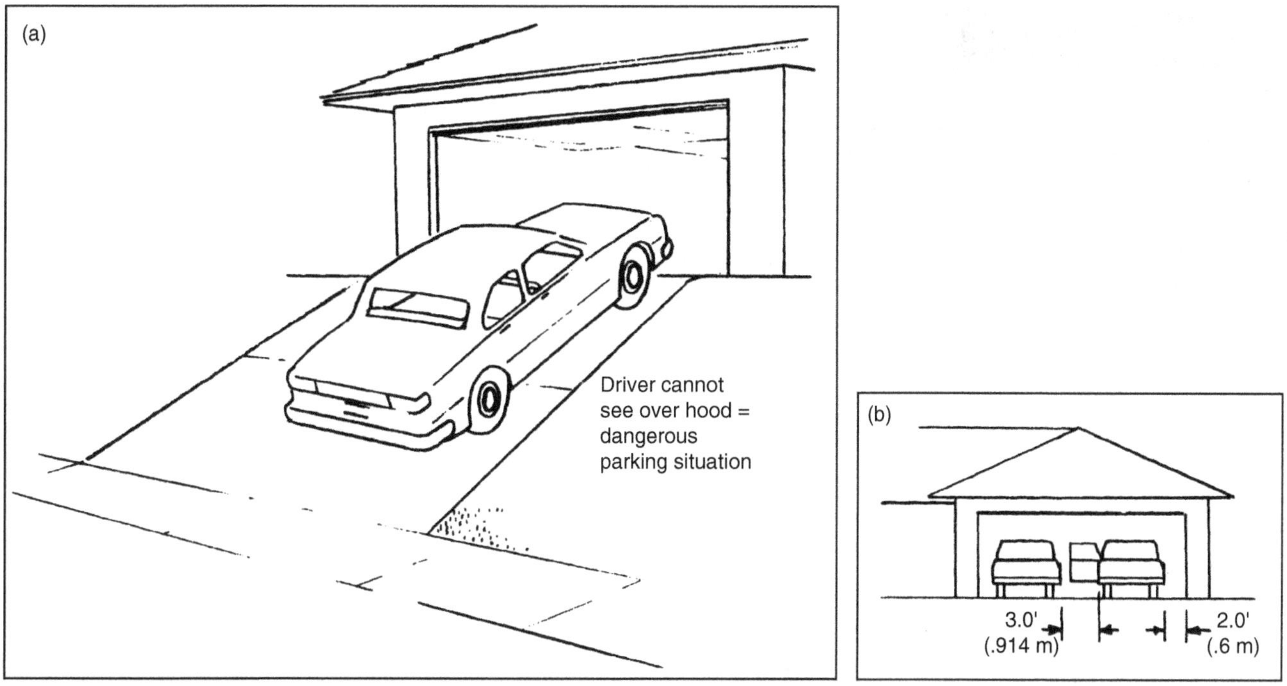

Figure 11.14 (a) Cars should be level prior to entering garage. (b) Recommended car door clearance.

Entry Approach Inclines

Too steep an approach to the vehicular entryway prevents the driver from seeing required clearance needs. Cars should be level just prior to entering. (See Fig. 11.14.)

CHAPTER 12

DOORS, HATCHES, AND EQUIPMENT CLOSURES

GENERAL PRINCIPLES

The primary functions provided by doors, hatches, and equipment closures include the following:

1. Access from one side of an enclosure to another
2. A security barrier
3. A privacy shield
4. An environmental barrier

In considering doors for a given project, one needs to address several user interface variables. The following is a useful checklist for the designer:

1. Doorway clearance needs
2. The type of door or doors (material; door opening geometry; door manipulating hardware, latching, and/or locking system; see-through requirements; door position retention; the type of threshold; etc.)
3. Automatic versus manual door articulation
4. Identification
5. Sealing against drafts, extreme pressure differentials, water leakage, dirt and dust, etc.

From a user's point of view, several dos and don'ts should be noted:

1. Do not use revolving doors or all-glass frameless doors; these are apt to be misused and cause personal injury.
2. Avoid the use of extremely large and/or heavy doors if they are to be manipulated by hand.
3. Provide view ports on free-swinging doors.
4. Do not combine vehicle and pedestrian doors; i.e., provide separate entryways for pedestrians and vehicles.

5. Provide ample doorway and door clearances; i.e., avoid crowding a person who is approaching or passing through a doorway or past an opening door.

6. Avoid using fancy "decorator" door handles that are hard to grasp or manipulate without hurting one's hand.

7. Place door handles where they provide the best advantage for gripping and manipulating the door; there should also be sufficient clearance between the handle and the adjacent structure to minimize the possibility that a person will scrape or pinch his or her hand or fingers.

8. When required, select door latching or locking systems that are easy to operate and locate them where visual guidance for a key is reasonably near the operator's eye level and/or viewing perspective.

9. Select door hardware that does not have sharp corners or edges that could cause injury if a person fell or bumped against the device.

10. All metal hatches and equipment doors or covers should be designed with radiused corners and edges so that users will not get injured by the sharp features. This applies to doors on vehicular equipment as well as to equipment closures.

11. Use materials for doors that are not affected by changing weather conditions so that the doors will not stick when wet or rattle when dry.

12. Select and/or design doorway and door thresholds that are relatively smooth (i.e., level with adjoining floor surfaces) so that pedestrians will not trip, wheelchairs will not be tipped over, and vehicles or equipment cars will not be suddenly jolted by the abruptness of the threshold.

13. Passenger doors for automobiles should be reinforced to minimize possible penetration by side impact from another vehicle, and doorframe and door latching systems should be designed to preclude inadvertent unlatching of the door in the event of a crash.

14. Door handles on automobiles (both inside and outside) should be inset so that neither a pedestrian (outside) nor a passenger (inside) can be injured by contact with an exposed handle.

15. Place door or other access handles on vehicles where they can be seen and easily manipulated; i.e., avoid "hiding" such devices.

16. Design opening and closing hardware for airline hatch doors so that it can be easily manipulated by flight attendants as well as by the crew.

17. Design hardware for airline emergency door releases so that the average passenger can easily locate and operate the device without assistance from a crew member (who may be incapacitated) and make sure that the instructions for operating the device are easy to follow.

ARCHITECTURAL DOORS

The dimensions indicated in Fig. 12.1 are recommended for most pedestrian doors. This clearance envelope and tread design will admit all adults and a person in a wheelchair. Smaller openings not only preclude the wheelchair user but also limit the passage of large pieces of furniture. (Note that the width includes any encroachment of the door and door hardware.) Avoid larger doors than shown; they become difficult for children and wheelchair users to manipulate, and when the door is of the swinging type, considerable clearance is required for the door swing.

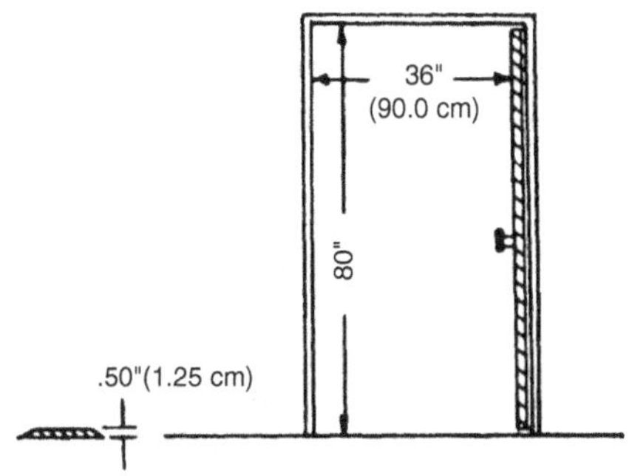

FIGURE 12.1 Minimum door clearance envelope for adults and wheelchairs.

Sliding doors have advantages over swinging doors in terms of the swing clearance noted above; in addition, an open swinging door takes up valuable wall space that could otherwise be used for furniture or equipment. When a sliding door is used, however, it is vital that the door handle remain easily accessible and that the door be designed so that fingers cannot be crushed accidentally when a person opens the door. (See Fig. 12.2.)

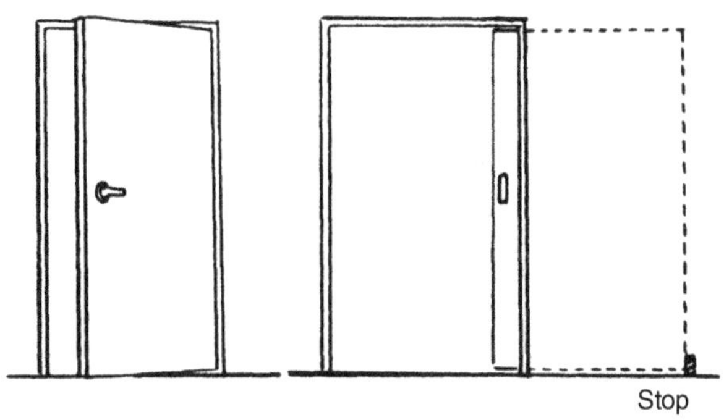

FIGURE 12.2 Sliding doors eliminate swing clearance requirements.

Double doors are useful in many applications because only one door can be used to admit people, thus preventing wind, rain, or snow from entering. When it is necessary to move extra-large furnishings or equipment through the doorway, both doors provide this convenience with clearance to spare. Exterior main entry doors should have a visual access of some kind so that the host can determine who is outside. Note the security grille in Fig. 12.3.

FIGURE 12.3 Double doors provide added clearance when necessary.

Generally, exterior swinging doors should open outward so that people are minimally hindered in the event they are making an emergency escape.

As shown in Fig. 12.4, door hardware should indicate to the user whether to push, pull, or turn a handle to open a door. Avoid the temptation to camouflage these features.

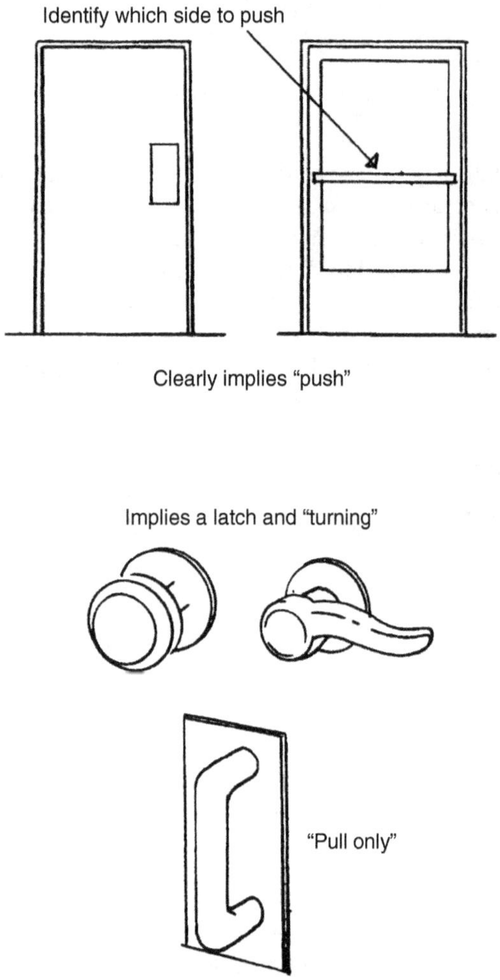

FIGURE 12.4 Hardware indicates action required to open door.

Avoid double door configurations that have a narrow divider between them, as shown in Fig. 12.5. Experience has demonstrated that people tend not to observe the divider and to run into it.

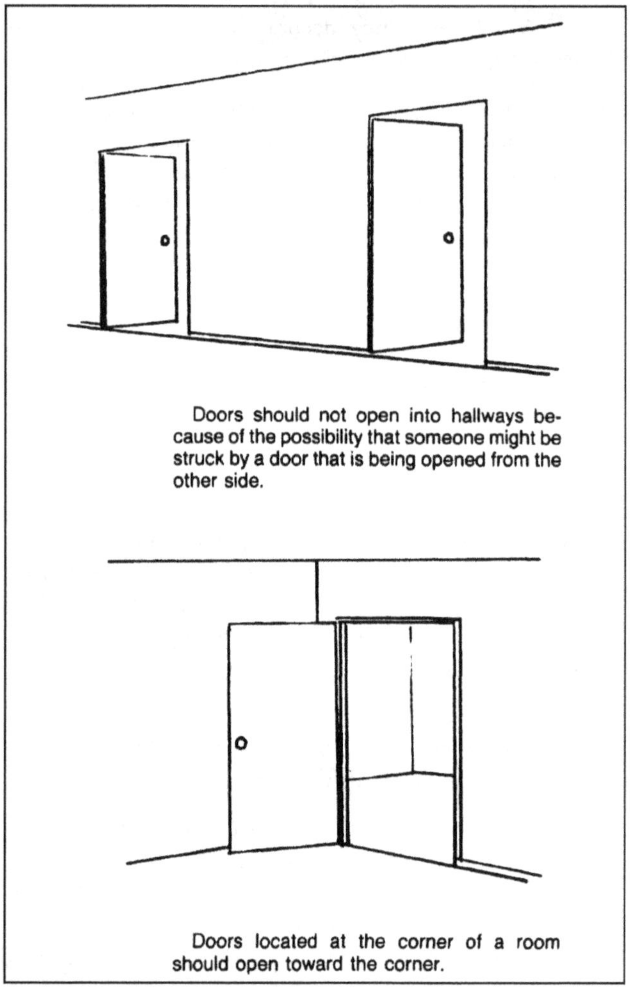

FIGURE 12.5 Recommended door swing direction.

Decorative doors are generally acceptable unless they have a built-in hazard, as in the case of the Dutch door, shown in Fig. 12.6. People tend not to give themselves adequate clearance around the upper half when it is swinging clear, as shown in Fig. 12.7. Children often bump their heads on the upper half. Such doors are also hard to control in a sudden wind gust when the two halves are not fastened together.

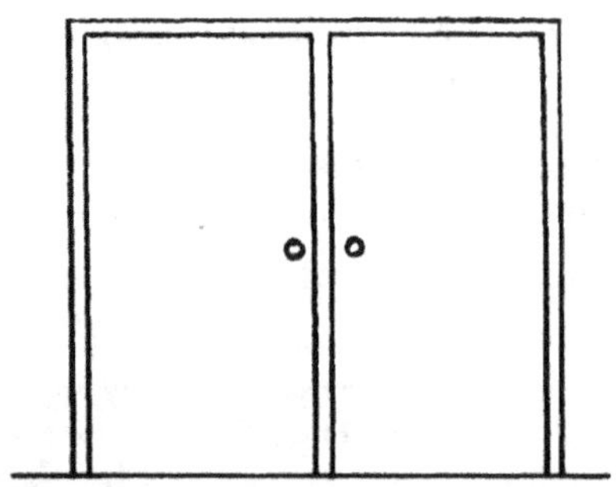

FIGURE 12.6 Post is a hazard.

Figure 12.7 Upper part of door can be a clearance hazard.

Accordion doors provide some useful advantages such as less weight, less loss of wall space, and little, if any, encroachment into the room area. (See Fig. 12.8.) However, consider the specific accordion design in terms of potential injury, namely, pinching of fingers that are carelessly placed near the folds.

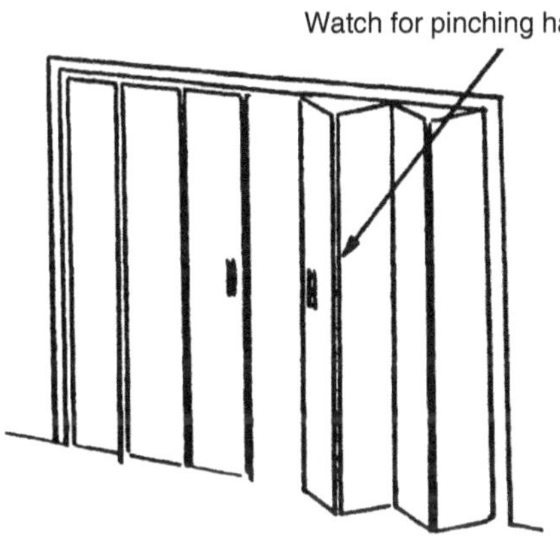

Figure 12.8 Accordion doors eliminate door swing clearance requirements.

Selection and Location of Door Manipulation Devices

1. Shape and Size of Door Handles: Although most adults can manipulate a door fairly well using a variety of handles, children and certain handicapped users have difficulty with some types. The best type for all users is the L-shaped handle, as shown in Fig. 12.9. This type of handle does not have to be squeezed, which is an extremely difficult task for some users, and its appearance gives an immediate indication of what one is to do with it. Avoid oddly shaped handles that are hard to get hold of and fancy decorator handles that have sharp patterns that cut into the user's hand or fingers.

Care should be taken in relating door handle selection and the planned installation. Figure 12.9 illustrates a typical problem: not enough finger clearance remains once the door jamb is installed.

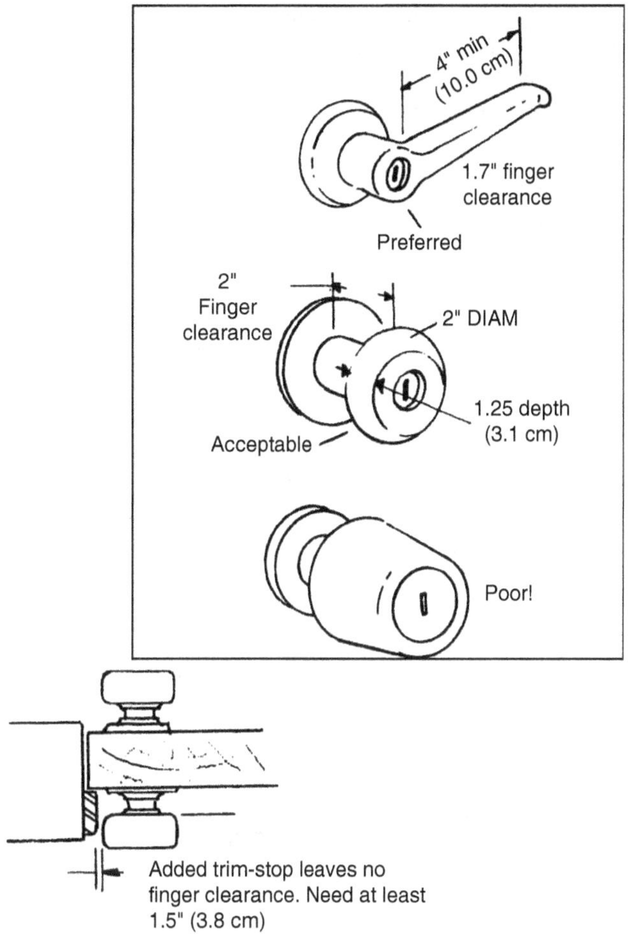

Figure 12.9 Recommended door handle design.

2. Door Locks: Door lock systems should be selected and located so that users can see what they are doing when they attempt to find the keyhole. Combination handles and locks are not recommended because when the handle is placed for easy use, the lock is too low. A separate lock placed near eye level is preferred.

All external architectural doors should be provided with dead-bolt-type burglar-resistant locks. (See Fig. 12.10.)

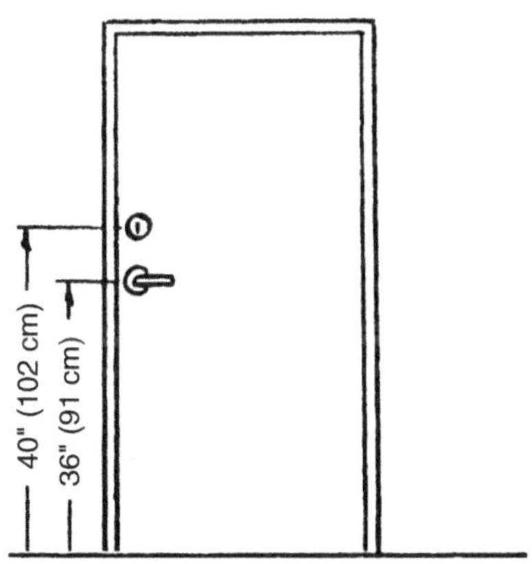

Figure 12.10 Door handle and lock installation.

3. Finger Latches and Door Pulls: The principal considerations in the design and/or selection of door latch and pull devices include finger clearance and force requirements either to unlatch the device or to pull the door open.

Although many different latch configurations are available or possible, Fig. 12.11 shows the clearance needs of the average user. Note that more finger clearance is needed for an outdoor application because the user may be wearing gloves. The depth of the door pull is especially important because children, elderly persons, and those with orthopedic handicaps do not have sufficient fingertip strength to move a heavy door. Although the minimum depth shown is generally acceptable, a deeper finger stall is desirable; i.e., ¾-in. (1.9-cm) depth allows people with weaker fingers to manipulate a door more easily.

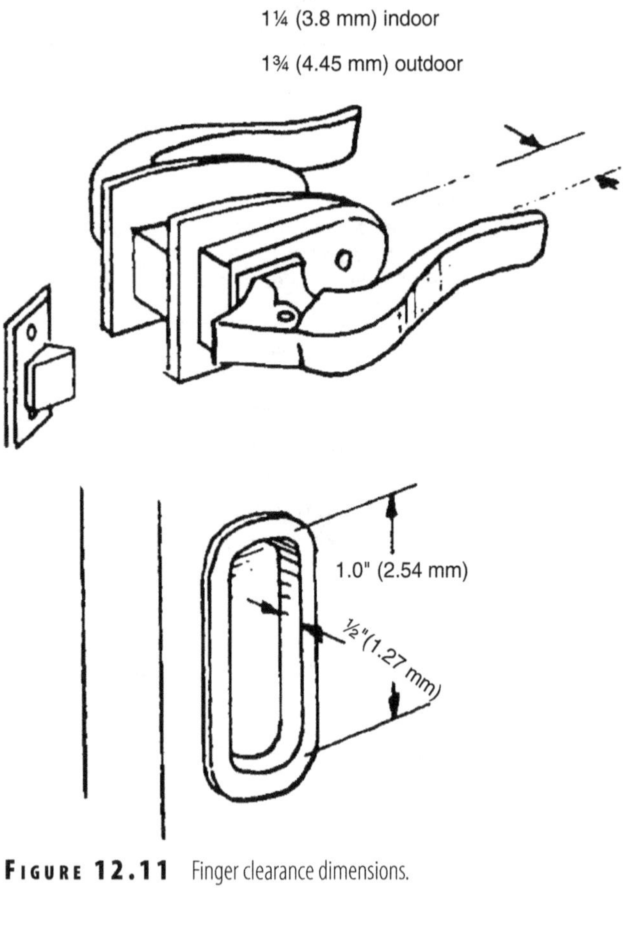

FIGURE 12.11 Finger clearance dimensions.

Safety and Emergency Door Features

A typical "panic" door installation (see Fig. 12.12) provides emergency exit without requiring the user to manipulate locks or latches. The door must obviously open out, or with the flow of traffic. The force required to unlatch the system should be sufficiently low so that the weakest member of the user population can unlatch the door (the door must also not be so heavy that such persons cannot push it all the way open). A door "catch" should be located outside so that the first person opening the door can secure it in an open position.

The standard height for door latches used by most architects is 36 in. (91 cm) above the floor level. However, this is too high for a person in a wheelchair, and therefore the lower height of 34 in. (86 cm) is recommended.

The latch shown in the figure should automatically retract when the door opens so that if someone lets the door swing closed, it will not relock. If such doors are not meant to be used except in an emergency, provide an auditory alarm to indicate when the door has been opened. If the door is a fire door, place a sign on the door identifying it as such.

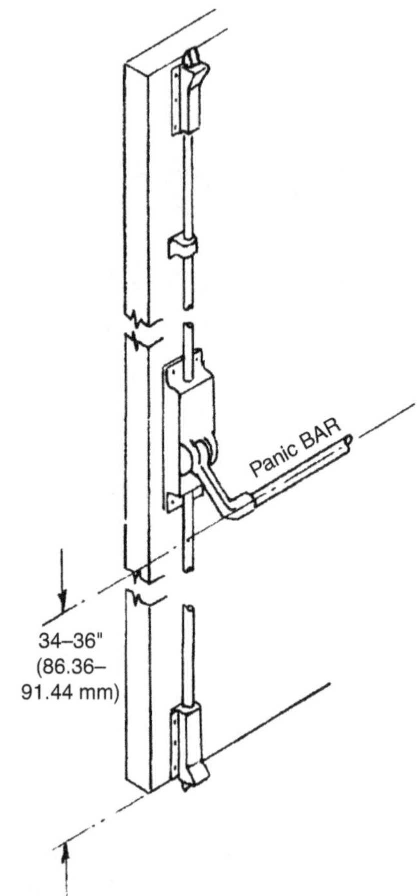

FIGURE 12.12 Recommended height for emergency door latch.

Open Elevator Doors. Commercial elevator systems (see Fig. 12.13) should always be provided with a guard gate or bars, as appropriate. There should also be an auditory alarm to indicate either that the sidewalk doors are being opened or that the guard gate is being lowered in preparation for the elevator arrival.

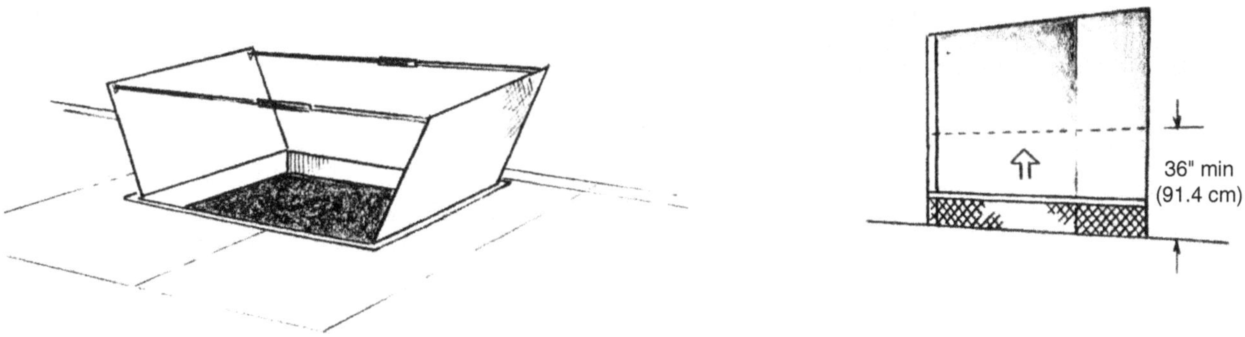

Figure 12.13 Provide guards and audible alarms as necessary for elevator doors.

EQUIPMENT AND CABINET DOORS

Cabinet doors and doors for gaining access to equipment and home appliances should be designed and located to simplify the user's problem of getting the doors open and holding them open while he or she is working or putting something inside, and they should be planned to "fit" the constraints of the expected range of installations. Figure 12.14 illustrates a few of the common problems associated with typical door designs.

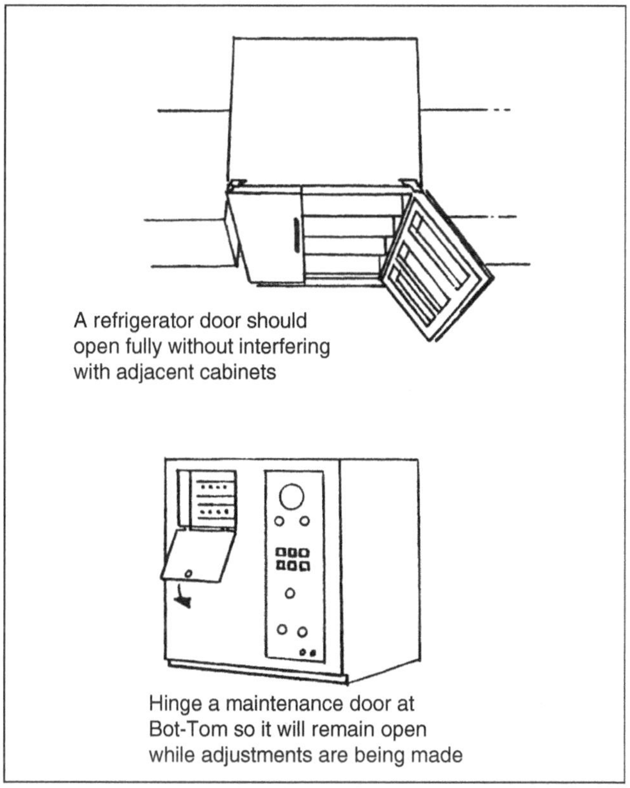

Figure 12.14 Door design problems.

Injury-Causing Door Features

The following common errors in design may cause considerable difficulty for the user.

1. Incorrect Door Hinge Pattern: Equipment or cabinet doors should not be mounted so that the user's hand might be injured if he or she opens the door too far. In some instances the user may be applying considerable force in order to overcome the restriction of a door clasp; when the clasp suddenly lets go, the user's knuckles may be cracked on an adjacent structure before he or she can reduce the amount of force being applied. Never locate the handles of adjacent doors so that they could coincide during an opening procedure. (See Fig. 12.15.)

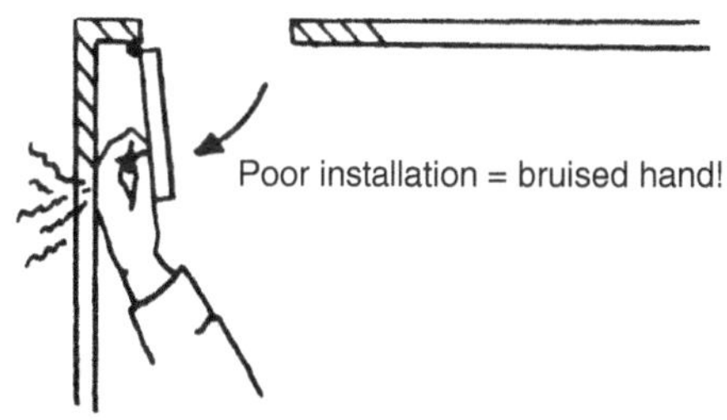

Figure 12.15 Hinge doors to allow clearances when opening and closing.

2. Sharp Corners: Although the cost of eliminating sharp corners may not seem warranted to the manufacturer, the typical equipment user is not very sympathetic to the so-called cost-effective sharp corner that may just have punctured his or her hand. Be especially alert for sharp corners that people might inadvertently hit with their heads. (See Fig. 12.16.)

3. Sliding Doors: Provide stops so that people will not pinch their fingers as they slide a door against another part of the cabinet. At least 1½ in. (3.8 cm) of clearance is needed to provide finger protection. (See Fig. 12.17.)

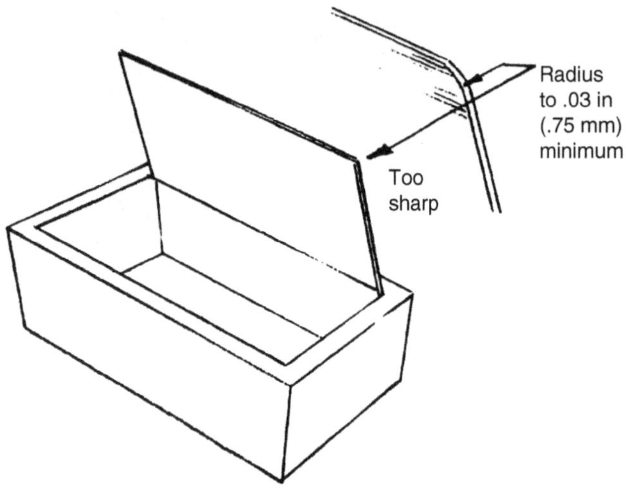

FIGURE 12.16 Radius sharp corners and edges.

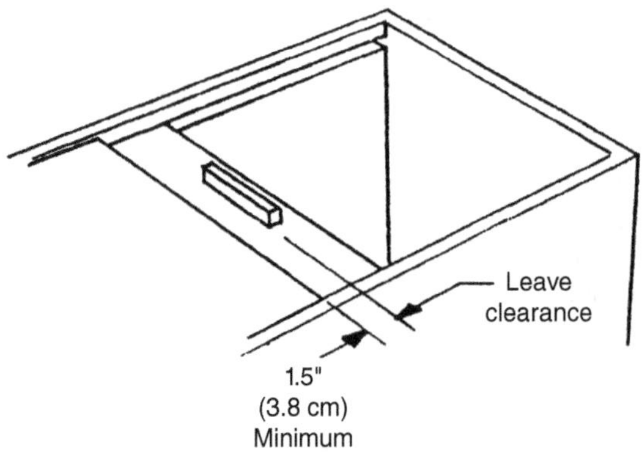

FIGURE 12.17 Provide door stops to protect fingers.

Equipment Door Size

Equipment door size obviously relates to the specific activities that may take place as technicians put their hands and/or an object through the opening. The door size should permit the following:

1. Using a common screwdriver with freedom to turn the hand through 180° (see Fig. 12.18)

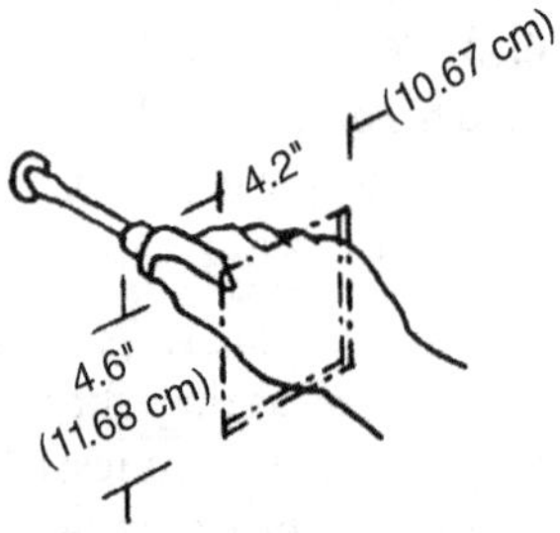

FIGURE 12.18 Equipment door size—example 1.

2. Using pliers and similar tools that require gripping (see Fig. 12.19)

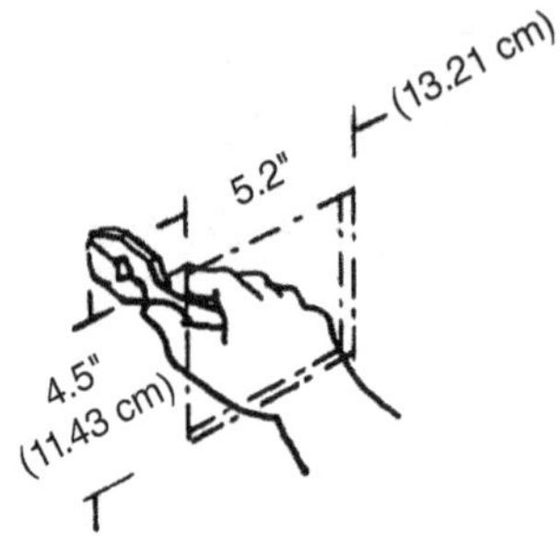

FIGURE 12.19 Equipment door size—example 2.

3. Using a T-handle wrench with freedom to turn the tool and hand through 180° (see Fig. 12.20)

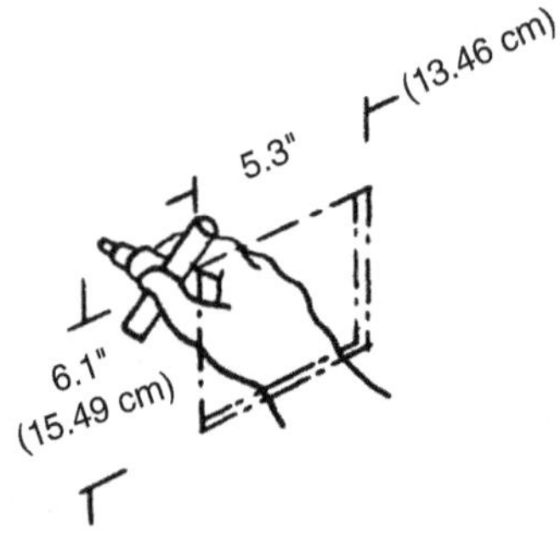

FIGURE 12.20 Equipment door size—example 3.

4. Using an open-end or box-end wrench with freedom to turn the wrench through at least 60° (see Fig. 12.21)

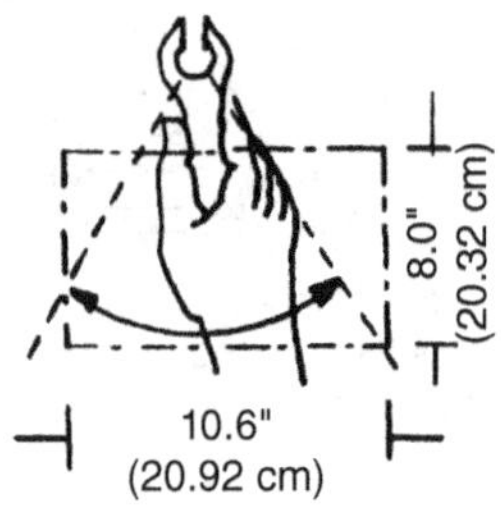

FIGURE 12.21 Equipment door size—example 4.

5. Grasping and manipulating small objects [up to 2¼ in. (5.7 cm) wide] with one hand (see Fig. 12.22)

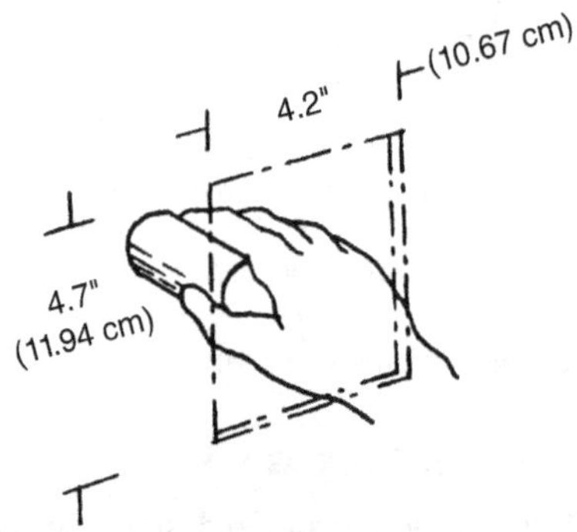

FIGURE 12.22 Equipment door size—example 5.

6. Grasping large objects with one hand (see Fig. 12.23)

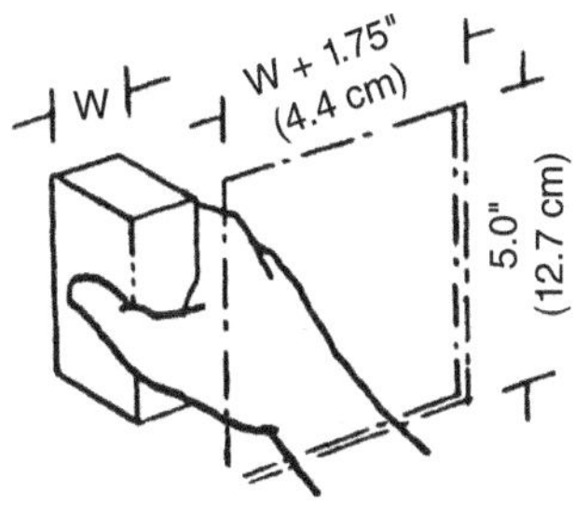

Figure 12.23 Equipment door size—example 6.

7. Grasping large objects with two hands, with the hands extended through the opening up to the length of the fingers (see Fig. 12.24)

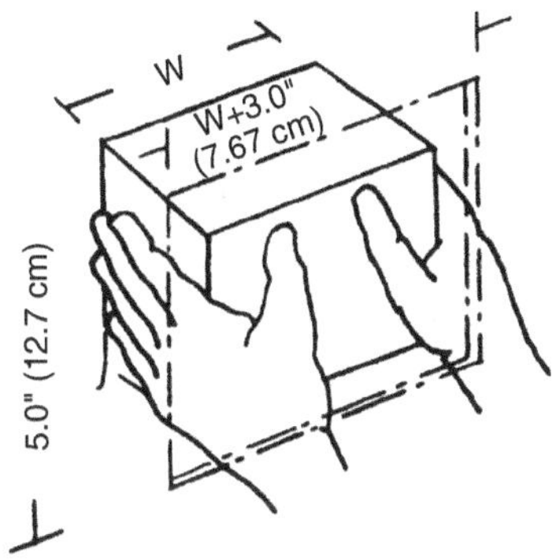

Figure 12.24 Equipment door size—example 7.

8. Grasping large objects with two hands, with the arms extended through the opening (see Fig. 12.25)

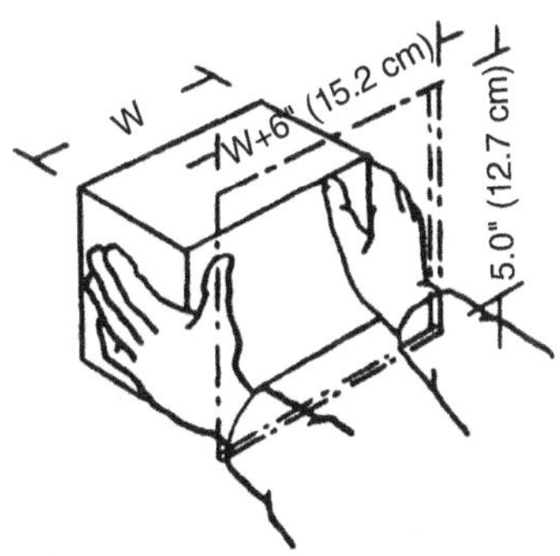

Figure 12.25 Equipment door size—example 8.

VEHICULAR DOORS

General

Actually, door shape, size, and location should always be considered in the initial phase of vehicular profile or concept development. When these overall profiles are initially established on the basis of styling alone, door design options are already so limited that the door's ultimate utility, as far as the user is concerned, is only barely acceptable. Key issues that should be addressed early in the vehicular concept stage are shown in Figs. 12.26–12.29:

1. Doors should be attached so that they can be opened at least 90°. Intermediate detented positions should be provided so that the door will remain in place. (See Fig. 12.26.)

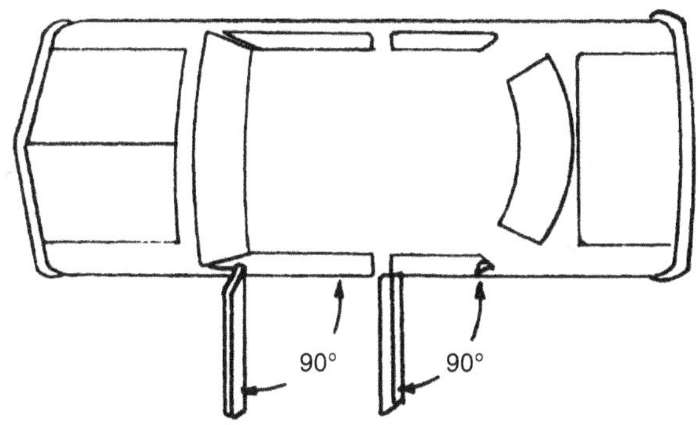

Figure 12.26 Vehicular doors—open forward.

2. All exposed door corners that face the user as he or she gets in or out should be radiused to prevent injury. (See Fig. 12.27.)

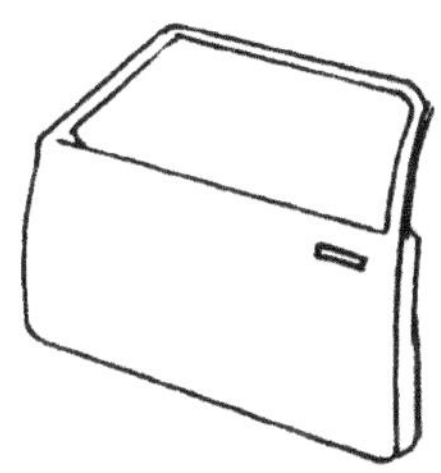

Figure 12.27 Vehicular doors—radius export door corners.

3. Engine and trunk lids should have a spring-assist and positioning lock. (See Fig. 12.28.)

Figure 12.28 Vehicular doors—provide spring assist and position.

4. The interior deck of a trunk should be flush with the bumper so that the owner does not have to lift packages over the trunk lip or bumper. (See Fig. 12.29.)

The same precautions noted for the design of exterior door hardware should be taken with respect to interior door hardware.

FIGURE 12.29 Vehicular doors—deck should be flush with bumper.

Because of requirements that interior vehicle hardware be made as noninjurious as possible, so that occupants will not be thrown against sharp protrusions during a crash, many designers have tended to hide door handles. However, it is confusing to the user under normal operations to have to hunt for a door handle, and in an emergency there may not be time to do so. Door handle location and operation should be obvious. It is quite feasible to do this and still retain a noninjurious design.

Similar precautions apply to the standard mechanical window crank.

To avoid confusing the user, electrically powered window controls should be arranged in one of the layouts shown in Fig. 12.30. Labels should be provided, however, so that the user will know that these are the window and/or lock controls.

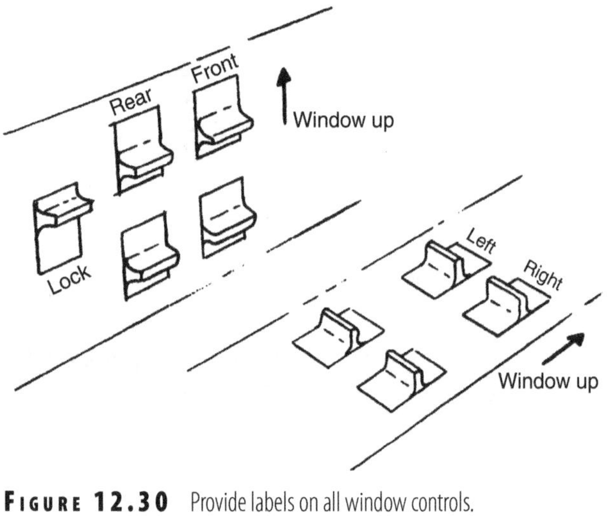

FIGURE 12.30 Provide labels on all window controls.

Any manual door lock controls should be positioned as indicated in Fig. 12.31 so that the user can reach them. Avoid the error of locating them near the aft doorframe merely because that is where they are typically found; i.e., when the occupant's seat is considerably forward of the rear edge of the door, the plunger should be moved to coincide with the operator's reach.

FIGURE 12.31 Position manual door locks within easy reach.

Mass-Transit Vehicle Doors

Although all door systems used in mass-transit vehicle systems should be openable by both operators and passengers, the primary door opening systems should be designed to be under the control of the onboard operator to prevent passengers from opening doors either intentionally or unintentionally.

1. Select door system configurations that do not pose hazards to a boarding passenger who is standing too close to the vehicle when the door is being opened. As Fig. 12.32 indicates, a sliding or folding door system generally provides a slim profile, whereas swinging doors are obviously a hazard.

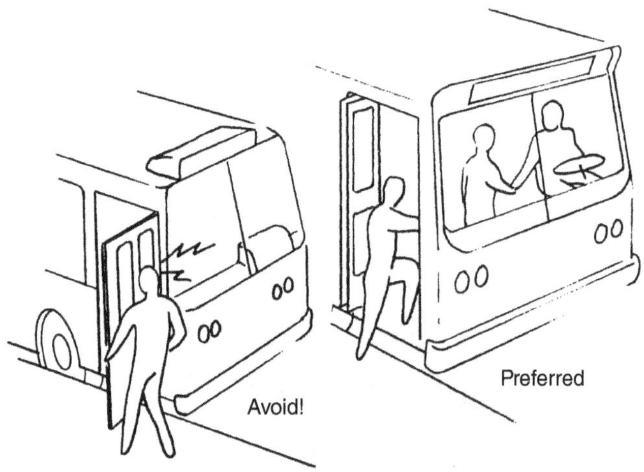

FIGURE 12.32 Multitransit vehicle doors—sliding or folding door preferred.

2. The most common design failure in passenger entry doors for commercial aircraft is insufficient head clearance for taller passengers. The problem exists primarily in smaller aircraft. Consider conspicuous warning signs when head clearance is inadequate. (See Fig. 12.33.)

FIGURE 12.33 Multitransit vehicle doors—provide head clearance.

3. Sliding clamshell-type doors are generally preferred for rail cars used in rapid transit, e.g., subways and other urban systems. Emergency bumper stops are vital to prevent passenger entrapment. (See Fig. 12.34.)

Figure 12.34 Multitransit vehicle doors—sliding doors preferred for rapid transit.

VEHICULAR HATCHES

General

Since many vehicular hatches have an emergency escape function, they are considered auxiliary doors. Nevertheless, they must be large enough to make escape easy and sure, and they should be free from elements upon which people could become snagged. The dimensions shown in Fig. 12.35 represent the best guidelines for various typical hatch-use functions.

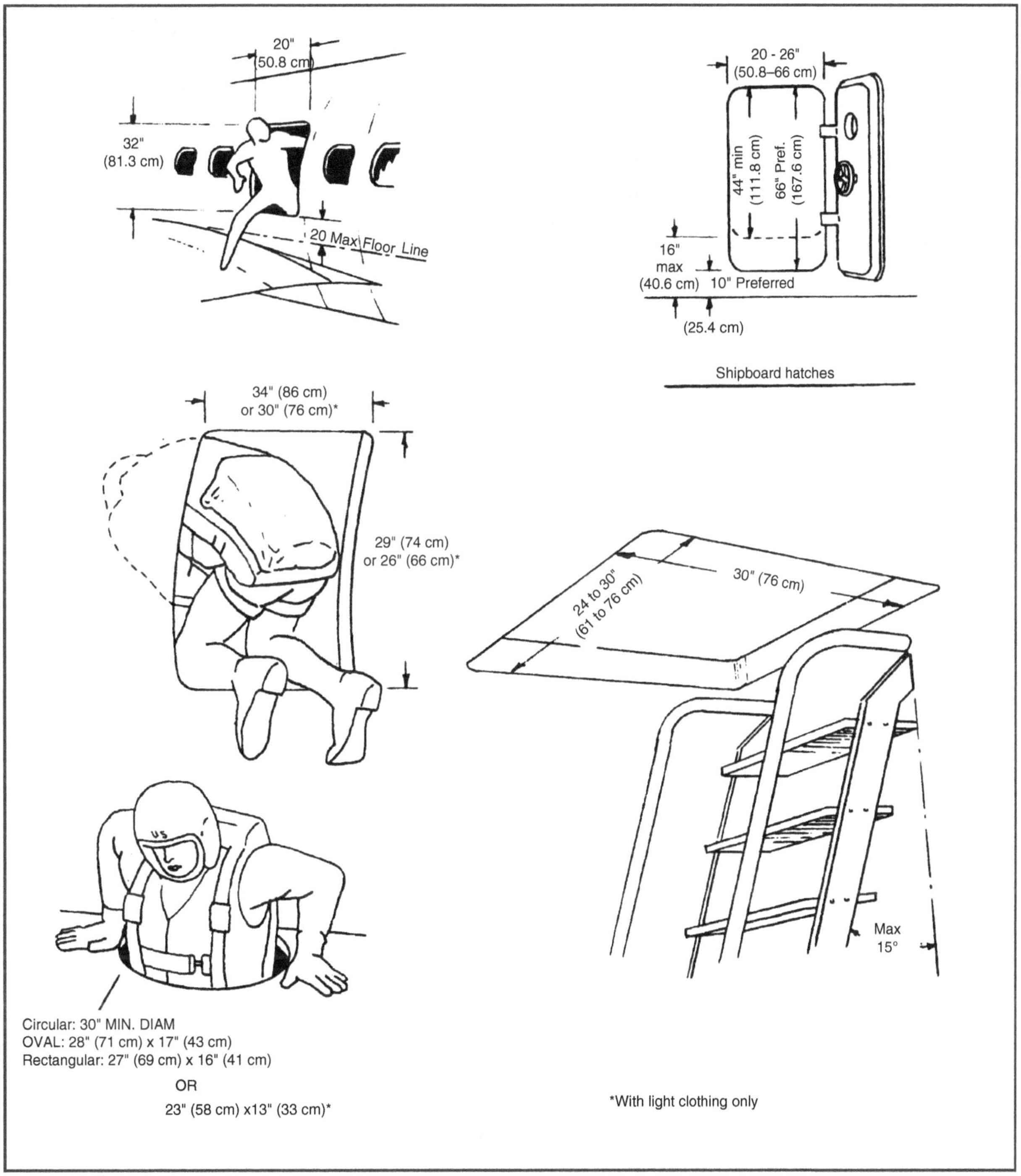

Figure 12.35 Minimum hatch dimensions.

Aircraft Canopy Hatch

Beyond the obvious requirement to design the fighter-type canopy so that it permits the pilot or pilots to have clear, unobstructed ingress and egress to their pilot station, special consideration must be given to rapid and safe removal of the canopy in the event of emergency abandonment of the aircraft. (See Fig. 12.36.) One must address at least the following emergency operational problems in defining the nature of the canopy configuration and removal system:

1. Ditching at zero speed, at ground level, from under water, at high altitude, or at high speed
2. Alternative ditching modes, i.e., by using the ejection seat and/or by going over the side
3. Control of an ejected canopy to prevent its contacting personnel or aircraft before it is clear
4. A manual backup canopy release in addition to automatic canopy ejection, which is required in most emergencies

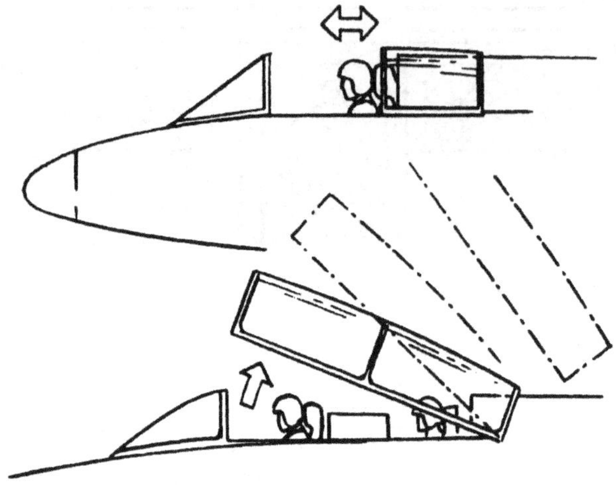

FIGURE 12.36 Aircraft canopy hatch.

CHAPTER 13

WINDOWS

GENERAL

Window Functions

The basic function of a window subsystem should be identified before a custom window for any particular application is selected or designed. Note that the aesthetic function should be considered only after the other functional needs are completely satisfied. In many cases, the earlier functional needs can be satisfied, while at the same time making the window design quite attractive. The primary functions of windows are as follows:

1. To provide a view of the external world from inside an enclosed architectural facility, vehicle, or other occupied system
2. To provide natural light for the interior of an occupied facility
3. To provide ventilation
4. To provide an auxiliary emergency escape route
5. To provide a visual communication link between two sides of an otherwise solid wall
6. To provide an aesthetically desirable focal point for an architectural system

The secondary or auxiliary functions of windows may also include the following:

1. To provide natural light filtering
2. To provide a thermal and/or acoustic barrier
3. To provide a moisture and/or dust barrier
4. To provide one-way viewing for covert observational requirements

Special Window Design and Selection Considerations

1. Location with respect to functional needs
2. Size with respect to ventilation, escape, and maintenance requirements
3. Security configuration (locks, guards, etc.)
4. Access and ease of opening and closing, cleaning, etc.

5. Location and/or materials relative to possible breakage

6. Screening the open window to prevent insect penetration

7. Glare control

8. Supplemental windows (storm windows, draft shields, etc.)

Typical User Problems

A brief review of the most common problems people have with windows provides further insight into the features that good window design should have. The following are among the most frequent complaints made by homeowners, office workers, and others:

1. Ventilation versus Draft: Poor window design and/or location makes it practically impossible to obtain the desired ventilation without introducing undesirable drafts, rain, dirt, or dust.

2. Observation: People cannot see out from their typical working or resting position, cannot observe children playing outside, or cannot observe arriving guests or an unknown visitor at the door.

3. Visual Task Efficiency: People cannot perform required visual tasks efficiently or without becoming fatigued because of direct glare from the window and/or because of the extreme contrasts created between the bright window and the interior workplace arrangement.

4. Window Operating Difficulties: Windows are difficult to open and close because of their location and/or because of the particular opening and closing hardware system selected.

5. Lack of Security: With a few exceptions, most commonly produced window systems have extremely poor locking systems and/or guard systems or have no such systems.

6. Window Installation Methods: Many window systems are installed with little consideration for the additional window-related features that may have to be added; i.e., there may be a lack of wall space and/or a solid surface on which to mount blinds or drapes.

7. Window Maintenance: Many window configurations create extremely difficult cleaning problems and replacement problems, such as removal and replacement of broken glass or obsolete hardware.

8. Special Hazards: Hazards are created by windows that have no frames, which extend to the floor, which are very large and/or very high, which require the use of a ladder for cleaning, or which are located so that they open out across an adjacent path. Sometimes a window is the last remaining escape in an emergency, and if it is too small or is located out of reach, this final route is unavailable.

9. Expense: Some window designs are extremely costly, not only at the time of purchase, but also on a continuing basis as maintenance and replacement requirements occur. The "all-glass" concept, which comes and goes in terms of popularity, creates severe glare control problems, adds to the cost of heating and cooling, is a continual cleaning problem, and generally results in considerable replacement expense for the owner.

Windows versus Windowless Architecture

Although there are certain times when a windowless environment is important for the work being carried on within a facility, such as a photographic laboratory or a light and vision experimental laboratory, people tend to fare better psychologically when they have visual access to the outside world. Therefore, most living and working spaces should have windows. The objectives for windows include the following:

1. Specific observation of outside activities, such as when a mother watches her children at play; when a plant supervisor checks a specific operation that is under way; when an air controller observes aircraft landings, taxiing, and takeoff; or when a railroad switchman observes rail yard movements.

2. Observation and/or inspection of what is happening inside a facility, such as when a person looks inside a restaurant before entering, when a shopper examines articles in a retail store window display, or when a patrol officer makes a security inspection.

3. Observation of the out-of-doors is very important in many situations, such as schools, factories, hospitals, and convalescent homes. A moment of relaxed observation of the outside world often gives the office or industrial worker a "second wind" during a long, hard work period.

4. Although there are conflicting opinions about the value of a sunlit living and working environment in relation to general health, most experts are of the opinion that people who work or live in windowless environments are physiologically and psychologically less well off than those who have the benefit of sunlit living and working environments.

5. Emergencies will always arise when power failures cause loss of illumination, ventilation, or air conditioning. Therefore, from a cost-effectiveness point of view, the windowless architectural facility may lose its primary advantage of ease of interior environmental control.

ARCHITECTURAL WINDOWS

User-Oriented Considerations in the Selection and Use of Windows in Architectural Systems

Decisions concerning the location, number, size, and type of windows to be used in architectural systems should be based upon a user-oriented philosophy. That is, one should approach these decisions from the point of view of those who will be inside the occupied spaces of the home, office, factory, school, hospital, or church, rather than from the point of view of creating architectural effects for the benefit of passersby. Decisions concerning the selection and use of windows should be based on questions such as the following:

Where Should Windows Be Located?

1. They should be located so that they will maximize desired exterior viewing angles from key visual reference points.
2. They should be located so that they will distribute incoming, natural light effectively for visual tasks and for the enhancement of the internal psychological environment.
3. They should be located so that good cross ventilation can be obtained by opening various windows.
4. They should not be located where they present unnecessary hazards.
5. They should be located so that they are accessible for cleaning and repair.

How Many Windows Should There Be, and What Size Should They Be?

1. There should be enough windows and/or sufficiently large windows to permit the desired natural light to enter where it is needed. However, the windows should not be so numerous or so large that there is a problem of light control or of cleaning and maintenance.
2. There should not be so many windows that there is not enough wall space to arrange furnishings.
3. There should be enough properly located windows to provide secondary emergency escape.
4. There should not be so many windows or so much glass that a thermal or noise insulation burden is placed on the owner of the facility.

What Types of Windows Should Be Selected?

1. Select windows that permit the desired exterior viewing.
2. Select windows that permit maximum ventilation control.
3. Select windows that provide desired light control.
4. Select windows that provide both security from burglary and ease of use for emergency escape.
5. Select windows that are easy to clean from the inside.

What Decorative Features Should Be Selected?

1. Select windows that are compatible with the basic architectural style of the proposed facility.
2. Select windows that convey a desired symbolism.
3. Select windows that will not be impossible to clean, repair, or replace.

An important aspect of window selection is the particular hardware that goes with the window, i.e., devices that will be used to open and close, lock, or adjusts the window that will be used to replace broken glass.

Consider the type of glazing that best fits each application, i.e., glass that provides thermal insulation, semi reflecting glass, colored glass, or plastic safety material.

Consider the type of window framing material with regard to weathering, rust, corrosion, need for frequent painting, etc.

Screens. Except in unusual climates (i.e., areas where there is never a problem with insects), window types should be selected that will accept screens. Screen removal and replacement should be simple and such that potential damage to the screens is minimal. The method of attachment should be such that the screens are accessible from inside the building unless the windows are positioned so that the screens can be reached and removed or replaced by a person who is on ground level or on a solid surface.

Note: The size, shape, and weight of removable windows and screens must always be compatible with the reach and strength capabilities of the persons who are normally involved in cleaning them; e.g., residential windows are smaller than industrial and commercial windows.

General Characteristics of Window Types[1]

1. Double-hung: Not apt to sag; easy to install screens or storm windows; only 50 percent openable; does not protect from rain when open; hard to operate over an obstruction; inflowing air cannot be diverted up or downward; horizontal members obstruct view. (See Fig. 13.1.)

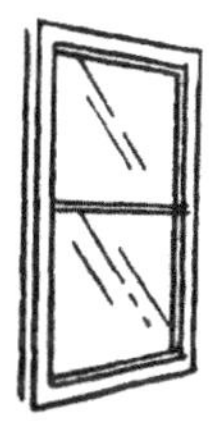

FIGURE 13.1 Double-hung.

[1]Reprinted from *Selection of Windows*, Technical Reprint No. 5, Building Research Institute.

2. Double-hung (reversed): The same as above but easier to clean.

3. Casement (out): 100 percent vent opening; will deflect drafts; does not protect from rain; a hazard close to walkways; vertical members obstruct view; may sag if not structurally strong; inflowing air cannot be diverted downward; interferes with drapes or furniture. (See Fig. 13.2.)

Figure 13.2 Casement out.

4. Casement (in): The same in many respects as the above but easier to clean. (See Fig. 13.3.)

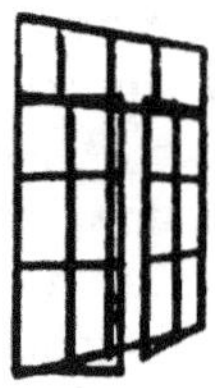

Figure 13.3 Casement in.

5. Awning, canopy: Not apt to sag; screens easy to install; 100 percent opening; will deflect drafts; offers rain protection when partly open; diverts inflowing air upward; vertical members obstruct view; inflowing air cannot be diverted be cleaned from inside. (See Fig. 13.4.)

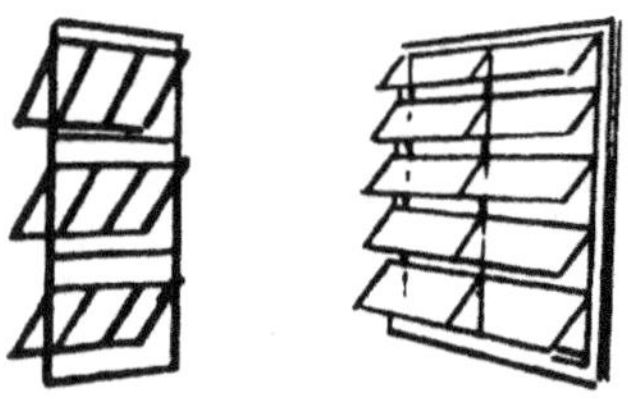

Figure 13.4 Awning canopy.

6. Pivoted (vertical): Not apt to sag; 100 percent opening; easy to clean; will deflect drafts; no protection from rain when open; hazard when located close to walkway; inflowing air cannot be diverted downward; interferes with drapes and furniture; screens or storm windows difficult to provide; fairly easy to clean. (See Fig. 13.5.)

Figure 13.5 Pivoted vertical.

7. Pivoted (horizontal): The same as the above, except offers rain protection when partly open; diverts inflowing air upward; inconvenient operation when over an obstruction; horizontal members obstruct view. (See Fig. 13.6.)

Figure 13.6 Pivoted horizontal.

8. Top hinged (out): Not apt to sag; screen and storm windows easy to install; provides 100 percent opening; offers rain protection when partly open; inconvenient operation when over an obstruction; hazard when close to walkway; inflowing air cannot be diverted downward; sash has to be removed for washing. (See Fig. 13.7.)

Figure 13.7 Top hinged out.

9. Bottom hinged (in): Not apt to sag; 100 percent opening; easy to clean; will deflect drafts; offers rain protection when partly open; diverts inflowing air upward; interferes with drapes and furniture. (See Fig. 13.8.)

Figure 13.8 Bottom hinged in.

10. Fixed sash: Not apt to sag; large size practical; sash has to be removed for cleaning. (See Fig. 13.9.)

FIGURE 13.9 Fixed sash.

11. Jalousie: Provides 100 percent opening; will deflect drafts; offers rain protection when partly open; diverts inflowing air upward; horizontal members obstruct view; excessive air leakage; hard to clean. (See Fig. 13.10.)

FIGURE 13.10 Jalousie.

12. Monitor: Not apt to sag; 100 percent opening; will deflect drafts; offers rain protection when partly open; diverts inflowing air upward; hazard when located close to walkway; horizontal members obstruct view; sash has to be removed for cleaning. (See Fig. 13.11.)

FIGURE 13.11 Monitor.

13. Projected: Offers rain protection when partly open; diverts inflowing air upward; hazard when located close to walkways; horizontal members obstruct view; somewhat difficult to clean from inside; not apt to sag. (See Fig. 13.12.)

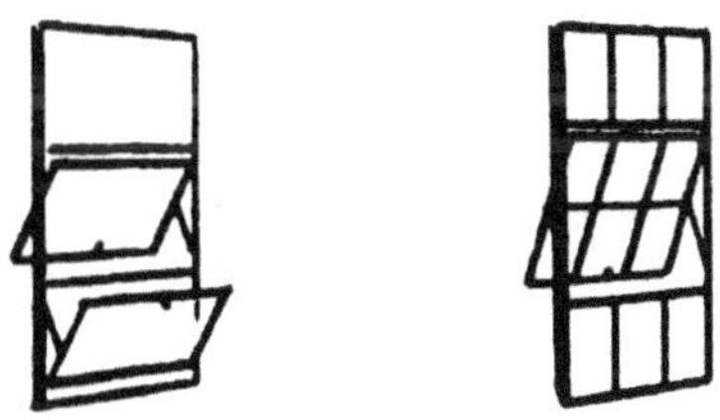

FIGURE 13.12 Projected.

14. Monitor: Not apt to sag; screen and storm windows easy to install; large sizes practical; only 50 percent openable; does not protect from rain when open (although screen helps); inconvenient when located over obstruction; vertical members may obstruct view; inflowing air cannot be directed or diverted; storm windows difficult to provide; sash has to be removed for cleaning (large size windows difficult to handle). (See Fig. 13.13.)

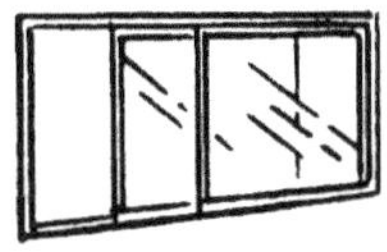

FIGURE 13.13 Horizontal sliding.

Window Type and Hardware Considerations

Key considerations from the user's point of view should be the following:

1. Ease of opening and closing
2. Ease of latching and locking
3. Ease of cleaning

For the following reasons, select sliding window configurations whenever appropriate and practical:

1. They do not encroach upon interior space, nor do they project outside, possibly into the path of someone who is walking past an open window.
2. The opening operation is direct and obvious.
3. Screening is usually simpler.
4. They generally can be cleaned from inside.

The following precautions should be considered:

1. Select windows that are easy to slide, i.e., that do not bind or hang up.
2. Select windows with manipulating hardware that is easy to get hold of.

The next most preferred window types are those that utilize a mechanical hand crank (as opposed to those that are opened or closed without some device to hold them in place). The examples shown in Figs. 13.14–13.17 are typical. Note the suggested hardware dimensions; these are desirable so that the user can manipulate the crank or lock easily.

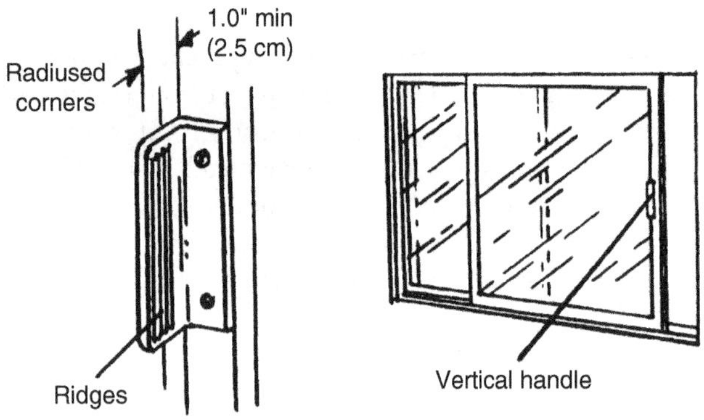

FIGURE 13.14 Window handle dimensions for horizontal slider.

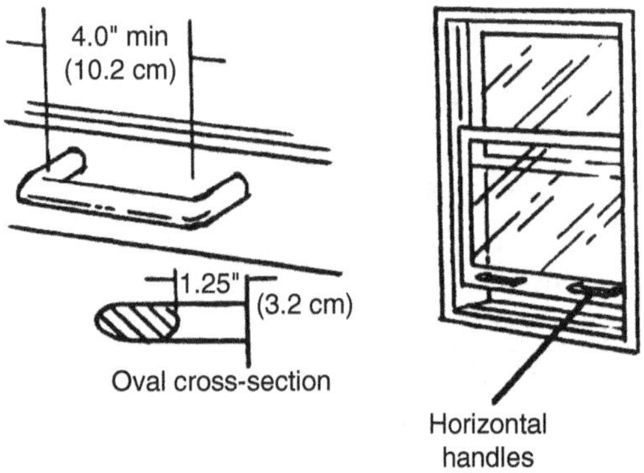

FIGURE 13.15 Window handle dimensions for double-hung window.

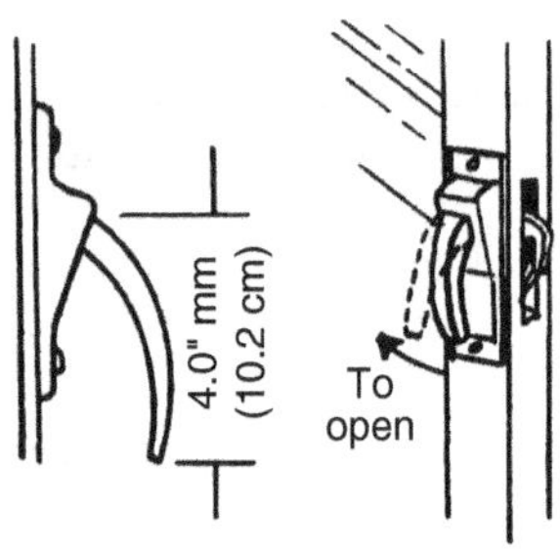

FIGURE 13.16 Latch dimensions.

Avoiding User-Induced Injuries

Window concepts and window placement should be analyzed carefully to determine what possible user-induced injuries might occur. Many hazards that could lead to injury can be minimized by proper design. Figures 13.18–13.25 illustrate several common design-related hazards that often occur in the home environment:

1. Use safety or tempered glass when a window is lower than 18" from the finished floor. See local building codes for other requirements. (See Fig. 13.18.)

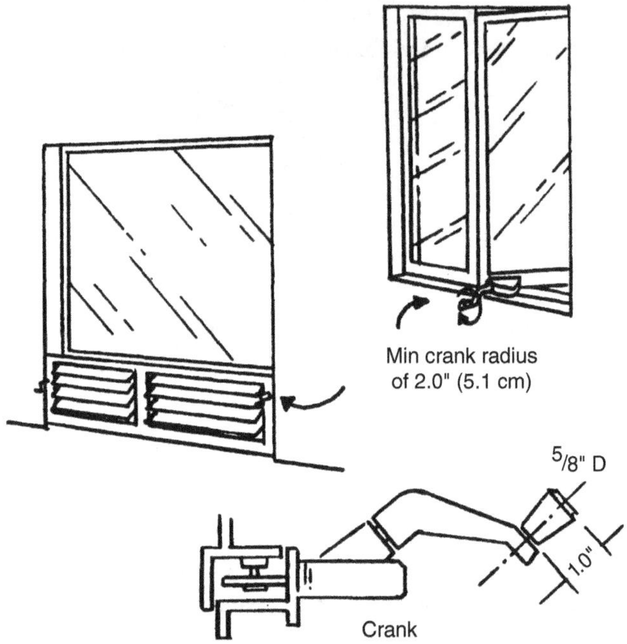

FIGURE 13.17 Crank dimensions for casement and jalousie windows.

2. Avoid placing large glass windows where furnishings might inadvertently be pushed into them. (See Fig. 13.19.)

3. Avoid placing large windows at the bottom of narrow stairs. (See Fig. 13.20.)

4. Provide a method for getting to the outside of large glass windows so that they can be safely cleaned. (See Fig. 13.21.)

FIGURE 13.18 Use safety glass.

FIGURE 13.19 Consider possible furniture placement.

FIGURE 13.20 Consider possible falls.

FIGURE 13.21 Provide methods to clean windows.

5. Placing windows where it is difficult to reach them either to open or close them or to clean them is a gross design error. As shown in Fig. 13.22, not only may this cause an injury, but it may also encourage people to take chances that could lead to even more serious injuries.

FIGURE 13.22 Select correct window design and placement.

6. Care in the total planning is also required to avoid placing certain kinds of windows so that they can lead to the kind of human error (and obvious injury) that Fig. 13.23 indicates is about to occur.

FIGURE 13.23 Don't invite misuse!

7. When a window is obviously the only alternative emergency escape route, select a window that a person can get out of quickly and safely and locate the window where it can be reached easily. Consider not only access from inside but also a safe escape route once the person is through the window. Building codes should be checked for egress requirements. (See Fig. 13.24.)

8. "Jut-out" windows such as those illustrated in Fig. 13.25 are extremely hazardous to both sighted and blind persons. Sighted individuals tend to turn from their position in front of a store window that is flush with the sidewalk and to walk directly into the protruding window before checking to see whether the coast is clear. This occurs most often if the window that juts out

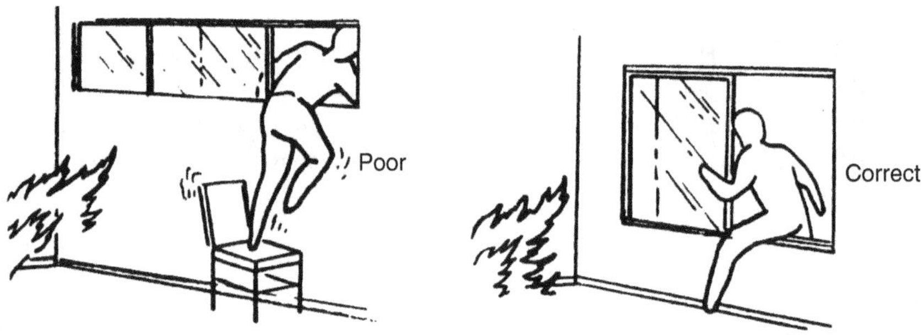

FIGURE 13.24 Consider emergency escapes.

has no frame around it. The blind person who is using a cane to trail the storefront wall often runs directly into the jut-out window before the cane makes contact, especially if the jut-out window ledge is above knee level.

FIGURE 13.25 People move before they look—avoid creation of booby traps.

Implications of Building Configuration and Window Patterns

The primary purpose of windows in an office building is to provide outside viewing for occupants. In spite of this obvious purpose, there is often a tendency to consider windows and window patterns as a key aesthetic feature of the building during initial conceptualization. In their enthusiasm for this latter objective, planners may forget, or at least may downplay, the primary visual function, failing to take problems of glare and maintenance into account. Figures 13.26–13.29 emphasize several important human factors considerations.

1. Window Location: Windows should be located so that they will provide some benefit to the occupants, not merely symmetrical exterior geometric patterns. (See Fig. 13.26.)

2. Window Size: Windows should be large enough to allow for viewing and natural illumination, but not so large as to increase glare and make cleaning difficult and time-consuming. (See Fig. 13.27.)

3. Glare: Although interior techniques can be used, consider glare control techniques in the initial structure configuration. (See Fig. 13.28.)

4. Cleaning: Window cleaning from the outside is hazardous at best. Consider ways to minimize the necessity for suspending window cleaners outside the building (balconies, windows that can be cleaned from the inside, etc.). (See Fig. 13.29.)

FIGURE 13.26 Glare cleaning problems.

FIGURE 13.27 Reduce glare, expand view.

FIGURE 13.28 Hazardous, hard-to-control glare.

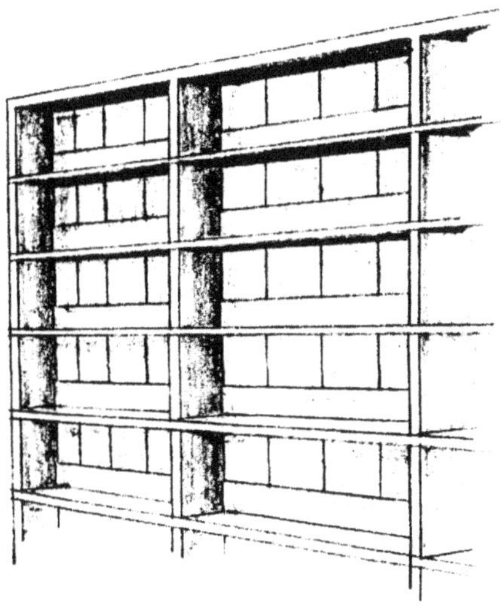

FIGURE 13.29 Glare control, easy to clean.

VEHICULAR WINDOWS

General Considerations

Driver Visibility. From the human factors point of view, the most important consideration relates to the ability of the driver to see everything outside the vehicle that he or she should see. The size, shape, and transmissibility characteristics of the windows should be such that there is minimal obstruction of the driver's exterior view as a result of roof support structure, reflections on either the interior or the exterior surface of the window glass, optical distortions in the glass itself, or the inability to keep windows clear of fog, mist, rain, snow, or ice.

1. Angle of View: Essentially, the driver should be able to see typical traffic features (e.g., other vehicles, the roadway, pedestrians, and signs or signals) in all directions, either directly or by means of appropriately positioned rearview mirrors. The driver should also be able to see key referencing elements of the vehicle, i.e., the front and rear corners or fender limits.
2. Glare and Reflection: Windows should be configured (along with other elements of the vehicle, such as instruments and interior reflecting surfaces) to minimize the amount of direct sun glare and/or reflected patterns on the glazing that might interfere with the driver's ability to see clearly.

Passenger Visibility. Although of lesser priority, it is important that passengers also be able to see out of a vehicle, whether it is an automobile, a bus, a transport aircraft, a train, or another type of vehicular system. Although the things that passengers have to see are perhaps less critical in most cases than those that the driver must see, passengers of public transit vehicles need to be able to observe signs that help them determine where they are or when they should get off. This is particularly important for the passengers of urban transit buses, inter-urban transit buses, and subway trains. For this reason, passenger windows in these vehicles should be designed so that passengers can accomplish these seeing tasks when either sitting or standing.

Ventilation. Vehicular windows should be designed for ease in controlling direct ventilation within the vehicle (except for special cases, such as the air-conditioned train or bus).

Emergency Escape. Windows often may be the only means for emergency escape and/or rescue. Window opening shape and dimensions, as well as opening methods, should be made compatible with the physical and mental capabilities and limitations of the driver or operator and the passengers.

Crashworthiness. Shatterproof window glazing is required in all vehicles that may be subject to crash conditions. It may also be desirable to consider "pop-out" windshields for private or smaller highway vehicles where it is possible that passengers could be thrown into the windshield.

Window Clearing. Most modern vehicles are now equipped with reasonably effective windshield wiping, washing, defogging, and/or defrosting systems. Some are being equipped with similar systems to clear the rear and side windows. It is recommended that all vehicles have systems to perform these clearing functions for all windows that a driver may, at one time or another, need to look out of in order to drive as safely as possible.

Pressurized Windows. All window systems in aircraft that fly above 20,000 ft (6100 m) should be designed to maintain cabin pressure under any condition of flight.

Aircraft Windshield Penetration. The forward windshield and/or the windows in any high-performance aircraft should be designed to preclude the penetration of the glazing by foreign objects, e.g., birds. Refer to FAA regulations pertaining to windshield "bird-proofing."

Distortion. Windshield glazing materials should always be of sufficiently high quality and configuration so as not to alter the appearance of objects viewed through the window or seriously distort the color of important traffic control devices such as signal lights, color-coded signs, and runway or taxiway markings. Select glazing materials that will not deteriorate as a result of weather, sun, heat, cleaning compounds, etc.

Safety Considerations

Important safety considerations are described as follows and illustrated in Fig. 13.30.

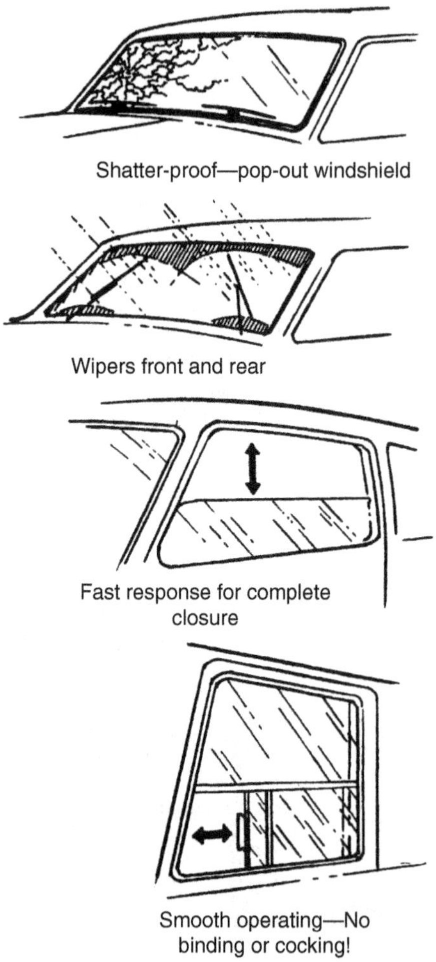

Figure 13.30 Vehicular windows.

1. All glazing in vehicular windows, whether for automobiles, trucks, buses, aircraft, or marine-craft, should be shatterproof.

2. There should be some means for clearing rain, snow, ice, mud, or collected dirt from outer surfaces and for clearing moisture condensation from inner surfaces.

3. Objectives for clearing windows should be based on the critical operator viewing needs, i.e., the viewing angles the operator requires in order to see ahead, to the side, and to the rear of the vehicle.

4. External surface washing capability should be provided for the critical windows, i.e., the windshield and the rear window.

5. Window clearing controls should be located where they can be reached quickly, and they should operate in a simple, direct manner so that the operator is not delayed while trying to understand complex labels or instructions. These labels must be visible at night as well as during the day.

(Refer to SAE standard practices for a complete description of window design requirements.)

Window Operating Convenience

Window controls should be located conveniently, and they should operate smoothly and easily. A window crank with a 2- to 3-in. (5.1- to 7.6-cm) radius is preferred, with a maximum torque of no more than 40 in. lb (4.5 m N). No more than two to three turns should be required to raise or lower a window. Electrically powered windows should include an emergency lockout so that a child will not accidentally crush a hand or choke. The driver should be able to prevent passenger window controls from being operated.

Avoid booby traps that can cause injuries and/or inconveniences for the vehicle user, particularly the following:

1. Sharp Corners: Sharp corners can become a hazard to the person who is hurrying to get into a car. This applies not only to the upper rear corner of a doorframe or frameless window glass, but also to special features such as the window wings frequently provided to control ventilation.

2. Window Opening Size and Shape: Windows that may have to be used for emergency egress must be of a sufficient size and an appropriate shape so that a person can get his or her body through the opening. Preferably, the opening system should cause the window glass to disappear completely, leaving the opening clear of glazing.

3. Powered System Malfunction: Powered systems should not fail with the window closed if the particular window is intended as an auxiliary emergency escape route.

CHAPTER 14
PARKING

AVERAGE AUTOMOBILE DIMENSIONAL GUIDELINES

Very large cars take approximately 26-ft (7.8-m) turning radius. The maximum or conservative dimension should be used whenever possible. (See Table 14.1.)

Although it has been standard practice for some time to specify garage and parking-lot spaces using considerably smaller values than shown in Fig. 14.1, such standards do not account for the difficulties that less skilled drivers have in maneuvering a vehicle in and out of tight quarters. In addition, current standards, because of the ever present objective of putting as many cars into as little space as possible, are based on mixes of large and small cars and assume that most people can get in and out of a car without fully opening the doors and/or that all drivers can park parallel to surface markings provided.

Note especially that the spaces in parking areas set aside for persons who must manipulate a wheelchair in and out of a car must be wider than the other spaces. (See Fig. 14.1.)

TABLE 14.1 Average Automobile Dimensional Guidelines

		Small Cars	Large Cars
L	Overall length	15.0 ft (4.5 m)	20.25 ft (6.1 m)
W	Overall width	–5.9 ft (1.8 m)	6.81 ft (2.0 m)
H	Overall height	–5.0 ft (1.5 m)	5.4 ft (1.6 m)
WD	Width plus one door	8.7 ft (2.6 m)	10.49 ft (3.2 m)
W2D	Width plus two doors	11.48 ft (3.4 m)	14.18 ft (4.3 m)
WB	Wheelbase	8.82 ft (2.7 m)	12.48 ft (3.7 m)
T	Tread C/C	4.68 ft (1.4 m)	5.23 ft (1.6 m)
RC	Road clearance	4.7 in. (11.9 mm)	6.5 in. (16.5 mm)
OHF	Overhang, front	2.36 in. (7.7 mm)	3.38 in. (11 mm)
OHR	Overhang, rear	3.63 in. (11.9 mm)	5.7 in. (16.6 mm)
AA	Approach angle	14.4"	29.71"
AD	Departure angle	9.6"	21.21"
AR	Ramp breakover	7.3"	16.21"

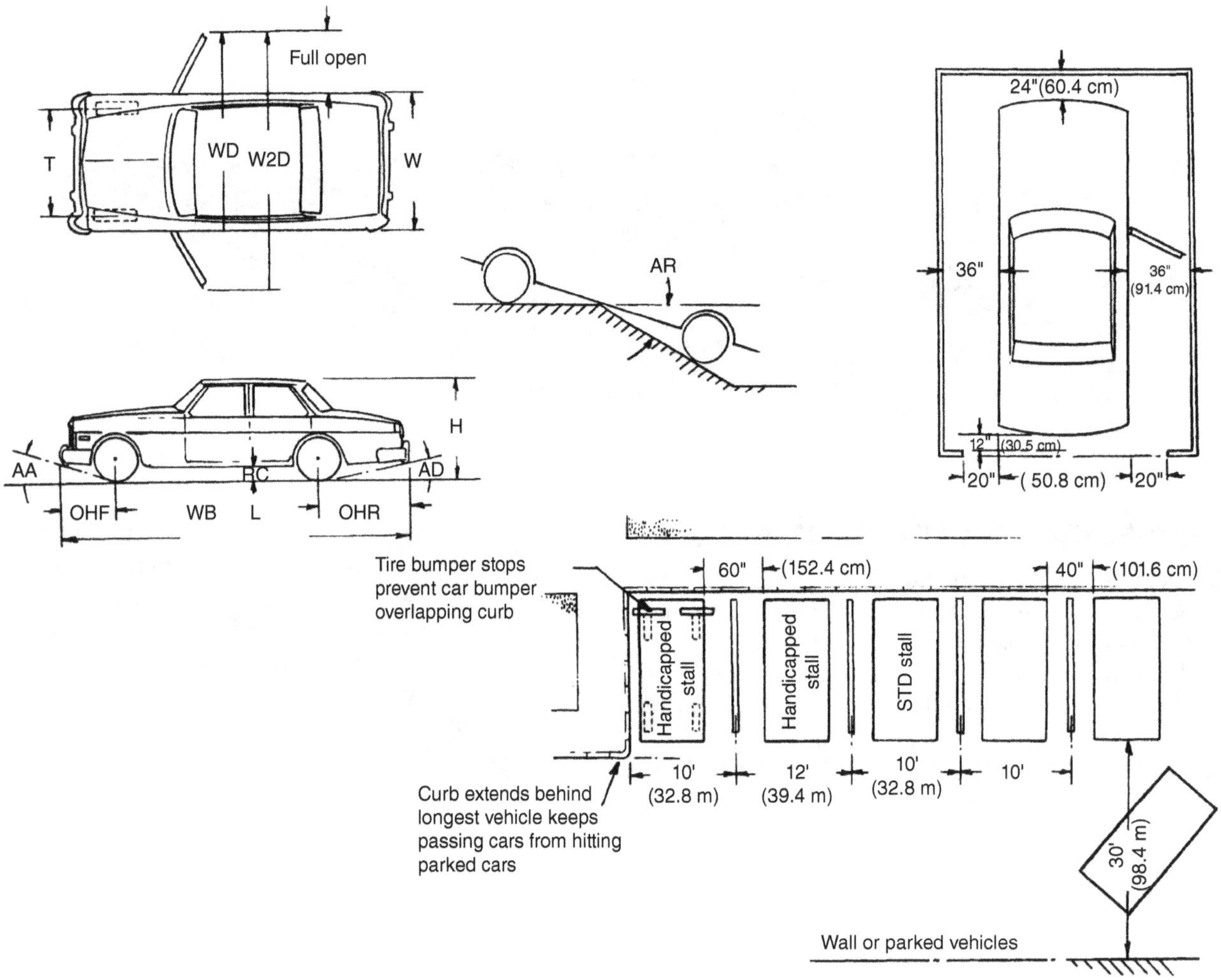

Figure 14.1 Auto parking facility design guidelines.

MULTIVEHICLE PARKING

Preferred Arrangement for Parking

The preferred arrangement for multivehicle parking is shown in Fig. 14.2. Although perpendicular parking will accommodate more cars in less space, many drivers have considerable difficulty manipulating their cars so that they are not skewed within the marked stalls. Also, it is preferred that the arrangement allow drivers to turn left into the stall because drivers are positioned on the left side of a car and can see what they are doing better than when they have to look across a large expanse of hood to the far corner of a car.

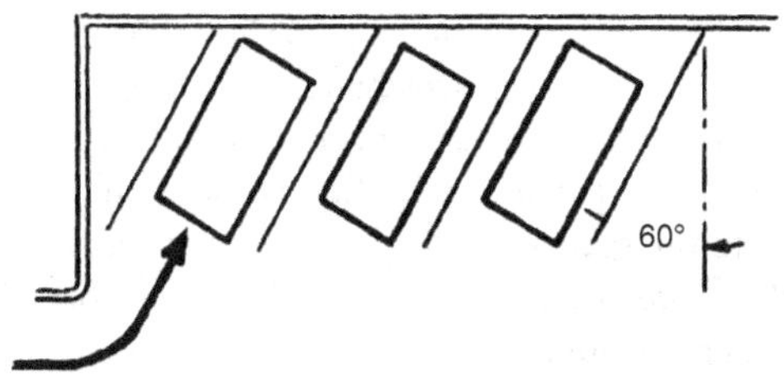

FIGURE 14.2 Left turns into stall make it easier for drivers.

Preferred Back-Out Clearance and Circulation Pattern

The preferred back-out clearance and circulation pattern for the parking garage or parking lot are shown in Fig. 14.3. This clearance permits the worst combination of two long vehicles, side by side, and the clearance will be sufficient so that even the poorest driver can be sure that his or her right front fender and bumper have cleared the rear fender and bumper of the car to the right. Although it is recognized that this circulation pattern takes more space than current parking pattern standards, it precludes some of the typical rear-end collisions that occur when two people try to back out at the same time, and incoming traffic and departing traffic are kept separated.

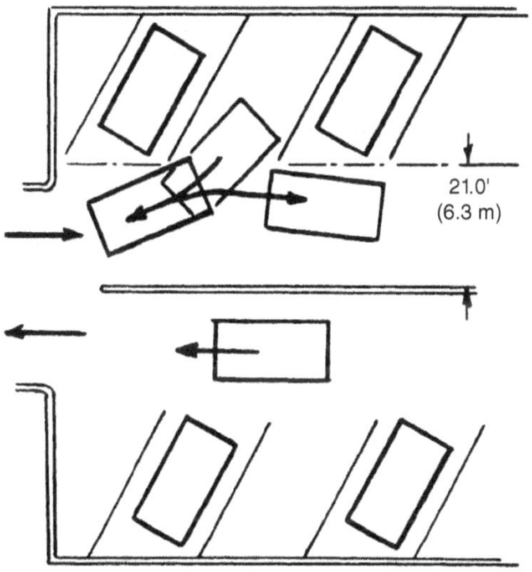

FIGURE 14.3 Back-out clearances.

Complete separation of the entrance and exit is even more desirable, as indicated in the Fig. 14.4.

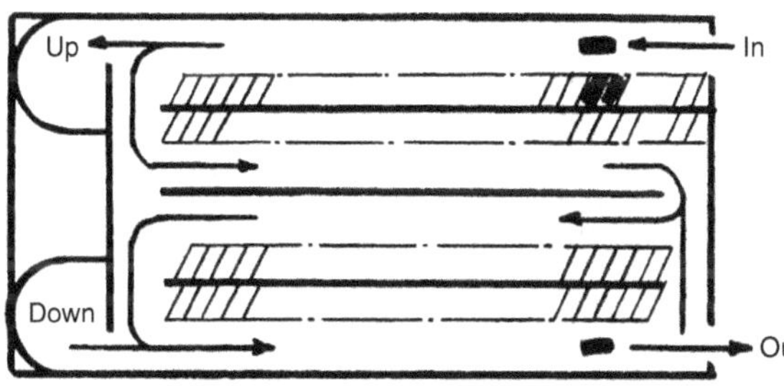

FIGURE 14.4 Separate entrances and exits.

The current trend toward smaller cars makes it desirable to consider separate sections for larger or standard-size vehicles and for smaller ones; this provides an opportunity to save considerable space in the multivehicle parking lot or parking garage. When a separate small-car area is set aside, the spacing for each car can be reduced from the recommended 10 ft (3.1 m) centerline to centerline, and the rear backing clearance can be reduced to 21 ft (6.4 m) for perpendicular stalls and 18 ft (5.5 m) for 60° stalls. This reduction is practical not only because there is a basic reduction in the width of the smaller vehicles but also because drivers can see the perimeters of the vehicle better and thus can be more precise in their maneuvering. The spacing for drivers who are in wheelchairs, however, should still be about 12 ft (3.7 m).

COVERED GARAGE PARKING

Pedestrian Access

Especially in multistory parking buildings, one should pay particular attention to the matter of how persons who have parked in the proposed structure are to get to and from their cars. Although part of this question involves the physical aspect, i.e., by elevator and/or by stairs, another and equally important part relates to identification of the owner's vehicle position so that he or she can find it again.

The multilevel parking garage should be laid out so that the relationship between floors and pedestrian departure and return features is coincident; i.e., avoid some of the typical approaches in which, for example, an elevator entrance is placed at one level, and a stairway for that same level is placed at a point that appears to the parking client to be on the next higher or lower floor or level. The preferred arrangement is to place both the elevators and the stairs in the same area or in adjacent areas.

Illumination

Often multilevel parking garages are designed with less than desirable illumination. This occurs most often with the "open-sided" designs because designers believe there will be enough natural light for motorists to see. Unfortunately, this often works in reverse; i.e., the glare contrasts against a rather dark interior during the day, making it difficult for motorists to see. Although it is generally not necessary to have as extensive artificial illumination during the day as at night, in the case of the multilevel parking garage the opposite may be true. For the same reason, the interiors of the open-sided parking garage should also be painted a light color, thereby taking advantage of "bounce" light and thus reducing the necessity for extra daytime artificial lighting.

Parking-Space Allowances

The number of parking spaces to be provided for various types of buildings will, of course, vary with the type of building and with the availability of public transportation. The following guidelines provide a good starting point for planning purposes:

Residential

1. Single-family dwelling (separate house)

2. Multifamily dwelling (efficiency apartment) 1.0 per dwelling unit

3. One- or two-bedroom apartment 1.5 per dwelling unit

4. Apartment with three or more bedrooms 2.0 per dwelling unit

Commercial

1. Offices 3.3 per 1000 ft²
2. General retail 4.0 per 1000 ft²
3. Shopping centers 5.5 per 1000 ft²
4. Restaurant 0.3 per seat
5. Motor hotel 1.0 per guest room 0.5 per staff member

Industrial Facilities 0.6 per employee

Churches 0.3 per seat per staff member

Hospitals 1.2 per bed

Note: There should be a minimum of three stalls for handicapped persons per 100 cars for any public facility. These special parking places should be located the minimum distance possible from main entrances.

Parking-Area Lighting Fixtures

Needless to say, parking facilities should be well lighted at night. Use of fluorescent-type lighting is preferred in order to ensure maximum diffusion and minimum shadow. Avoid the temptation to use a few fixtures at relatively high positions above the ground in order to reduce the initial cost. Generally this requires higher foot-candle levels, and in the long run these fixtures may cost just as much to maintain as more lower-powered fixtures spaced closer together. A minimum of about 3 (maintained) fc measured at ground level is required for people to find their way around in the typical parking lot at night.

Parking-Lot Signing

A logical signing system should be developed at the same time that the parking arrangement is being considered. All signs and markings should be visible (not obscured by the tops of cars) and readable, i.e., well illuminated at night, large enough for the typical viewing distances, and legible (appropriate contrast and alphanumeric font, etc.).

Other Features

1. Curbs and Tire Bumpers: In spite of the higher initial cost, curbs or tire bumpers are desirable to keep order in the average parking lot. In addition, such devices may prevent damage to vehicles and thus possible liability for the parking-lot proprietor.

2. Speed Control Berms: Raised ridges in the parking-lot surface at strategic points are useful to keep drivers from driving too fast. These should always be painted to contrast with the primary surface, however; otherwise, pedestrians may trip over them.

3. Wheelchair Facilities: There should be special ramps to allow persons in wheelchairs to get from their cars (at ground level) to whatever primary walkways are associated with the particular facility. These should be located close to the special parking spaces provided, and the route between these spaces and the ramps should not cause handicapped persons to be exposed any more than is absolutely necessary to moving vehicles as they go from their cars to the special ramps.

4. Marking of Pedestrian Paths: When very large parking lots are involved, consider the addition of special walkways to various parts of the lot and/or plan to mark safe paths to these areas on the lot surface.

Exits and Entrances

Care should be exercised in planning garage exits so that departing motorists can see ahead. Figures 14.5–14.7 depict several critical considerations (lateral view and view over the hood and/or beneath the upper windshield brow):

Provide the motorist a good peripheral view before he or she crosses the pedestrian right-of-way. (See Fig. 14.5.)

Avoid inclines that are so steep that the car's hood prevents the driver from seeing where he or she is going. (See Fig. 14.6.)

Allow the car to travel over a sufficiently large flat space prior to crossing the sidewalk and entering the street.

In residential design, avoid creating short driveways on inclines; i.e., provide a level space between the garage and the sidewalk or street on which a car can sit without being in danger of rolling out of the driveway in the event a parking brake fails. (See Fig. 14.7.)

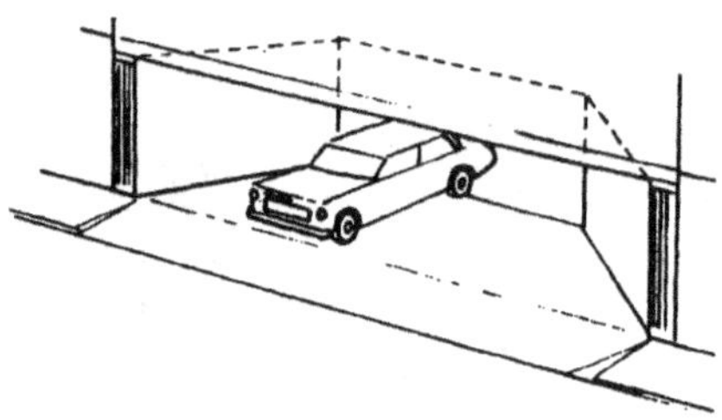

Figure 14.5 Provide driver with a clear view of sidewalk.

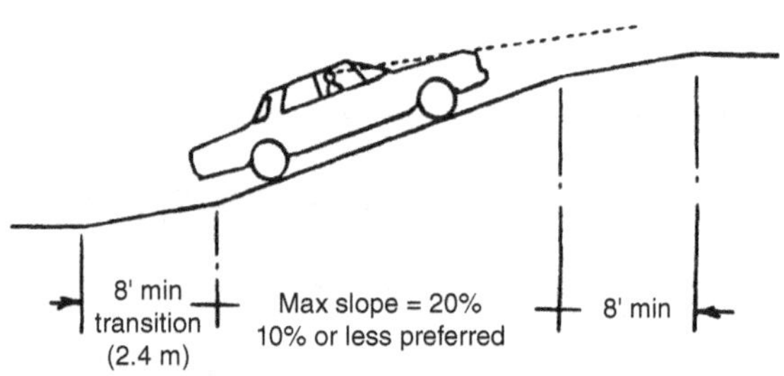

Figure 14.6 Incline requirements.

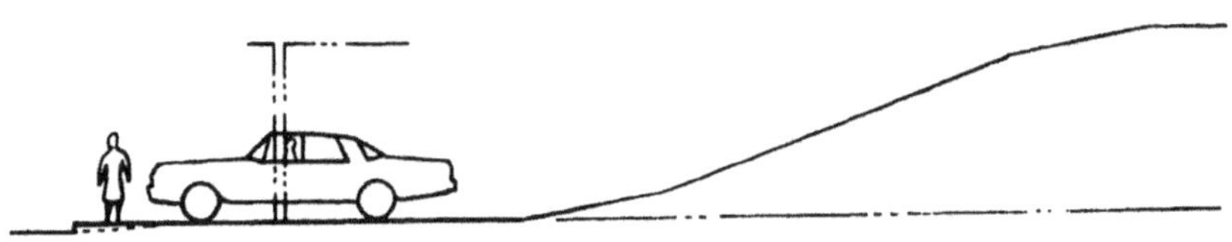

Figure 14.7 Provide level area before crossing sidewalk and street.

CHAPTER 15
WALKWAYS, HALLWAYS, AND CATWALKS

EXTERIOR SIDEWALKS

General

Exterior sidewalks should be wide enough so that persons who are traveling along the walkway do not feel as if they are walking a tightrope. The guidelines illustrated in Figs. 15.1–15.3 take into account elderly persons' unsteadiness as well as the need to feel that one can walk rapidly and easily without scraping walls or falling off a curb:

1. The minimum width for a sidewalk that normally would be used by only one person at a time is about 30 in. (76 cm) [36 in. (91 cm) is preferred because individuals are usually carrying something]. (See Fig. 15.1.)

2. Lightly traveled sidewalks should be at least 4 ft (1.2 m) wide so that two adults can pass each other without bumping. (See Fig. 15.2.)

3. Moderately traveled sidewalks in suburban areas should be at least 7 ft (2.1 m) wide in order to accommodate the most common passing situation, in which two people pass a single person going in the opposite direction, without crowding. (See Fig. 15.3.)

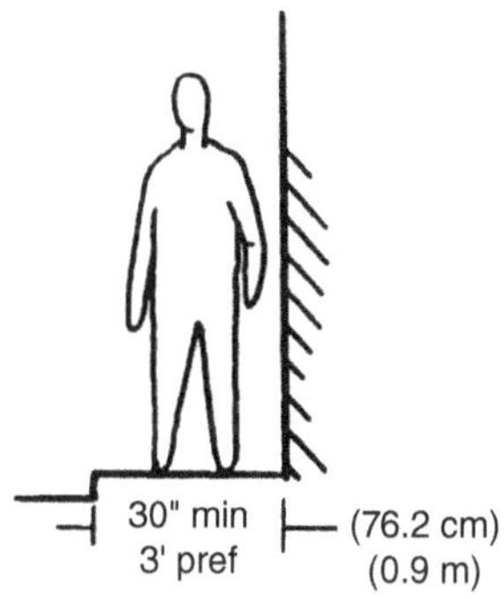

FIGURE 15.1 Exterior sidewalks—minimum width.

4. The busy sidewalk, especially in commercial urban environments, should be at least 12 ft (3.7 m) wide. Figure 15.4 illustrates some of the needs that dictate ample space for widely varying sidewalk activities.

Design Specification for Wheelchair Ramps between Sidewalk and Street

The criteria shown in Fig. 15.5 will provide not only optimum passage of the wheelchair but also tactile cues for blind

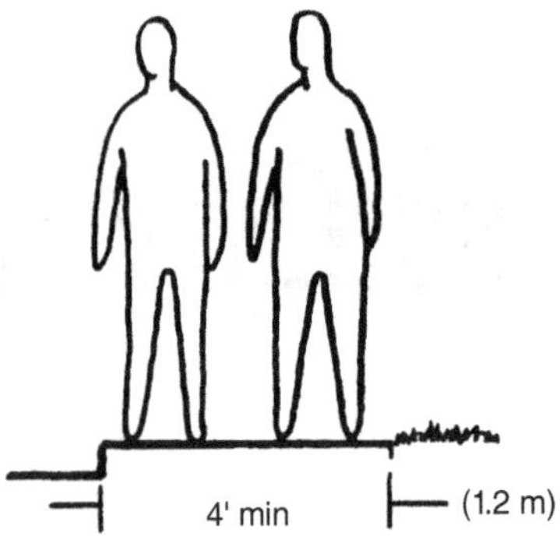

FIGURE 15.2 Exterior sidewalks—lightly traveled.

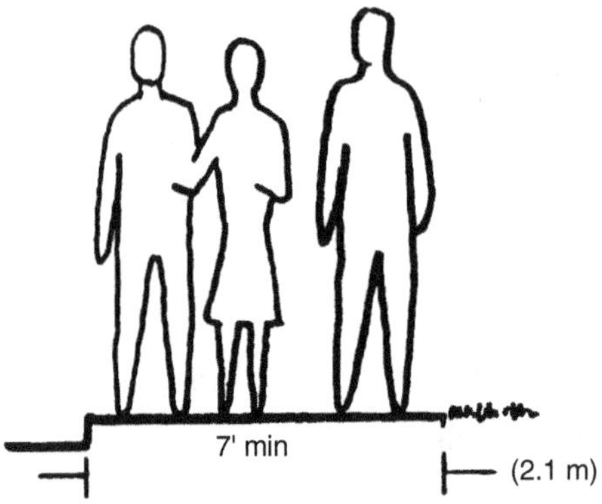

FIGURE 15.3 Exterior sidewalks—moderately traveled.

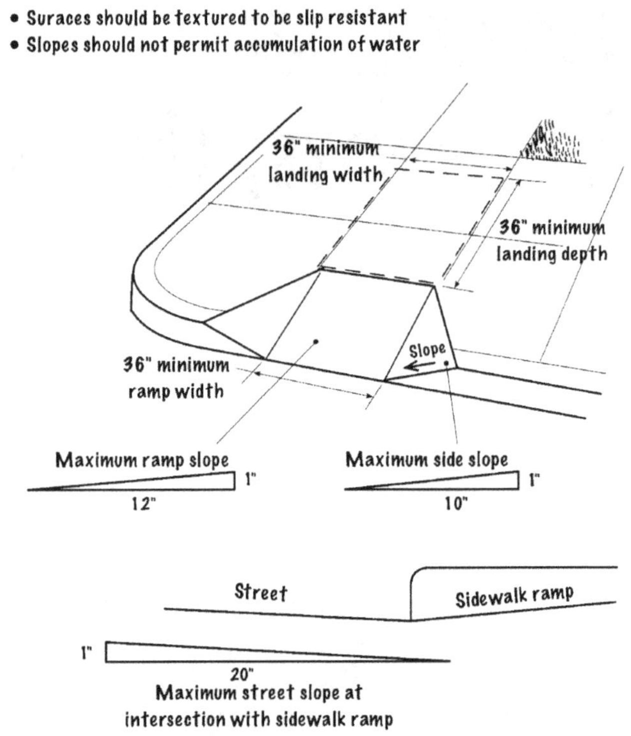

Source: 2010 ADA Standards for Accessible Design, US Department of Justice

FIGURE 15.5 Recommended wheelchair ramp design.

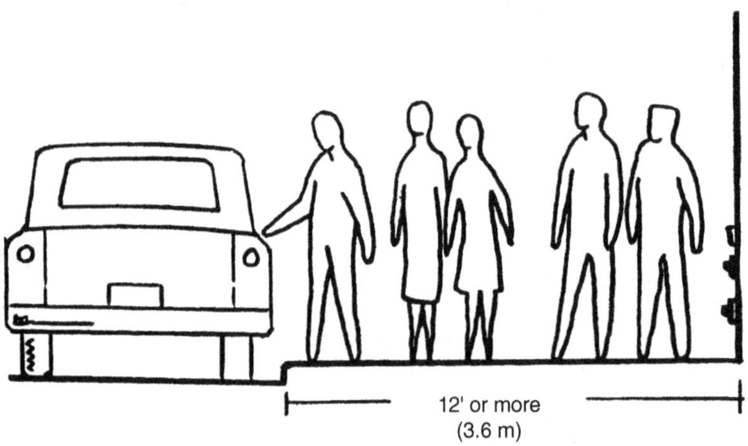

FIGURE 15.4 Exterior sidewalks—busy.

pedestrians and visual cues that will help prevent other pedestrians who do not notice sidewalk irregularities from stumbling.

Each of the special features has been derived from a number of attempts to create the optimum compromise for all users.

INTERIOR HALLWAYS AND AISLES

Corridor- and Stair-Width Guidelines

Although people can pass by each other in fairly cramped quarters if necessary, they do not like to, and in certain cases it is hazardous to crowd people, such as during an emergency exit situation. Figures 15.6–15.11 and dimensional guidelines are representative of the desirable widths for the following office building situations.

1. Main Corridors for Ground-Floor Level: People often walk in pairs, and generally one couple dislikes touching another couple as they pass. (See Fig. 15.6.)

2. Primary Corridors for Upper Levels: Typically, a couple needs to pass at least a single individual going in the opposite direction. (See Fig. 15.7.)

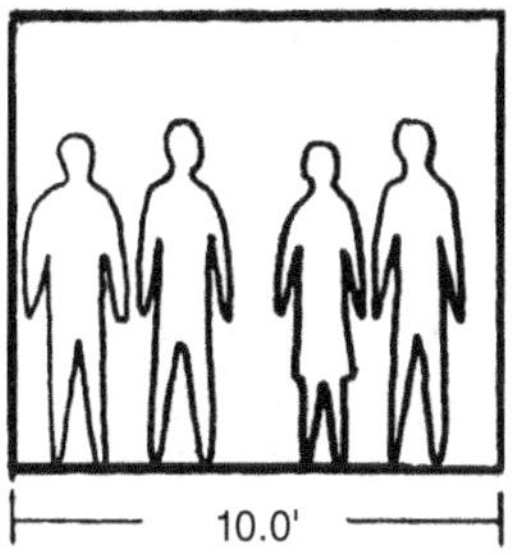

Figure 15.6 Corridor stair-width guidelines—main corridor.

Figure 15.7 Corridor stair-width guidelines—primary corridors for upper levels.

3. Secondary Corridors, Any Level: People generally are not walking in pairs, and/or traffic is light and local. (See Fig. 15.8.)

Figure 15.8 Corridor stair-width guidelines—secondary corridors for any level.

4. Within-Office Areas: The aisle is typically bordered by half partitions (at least on one side). (See Fig. 15.9.)

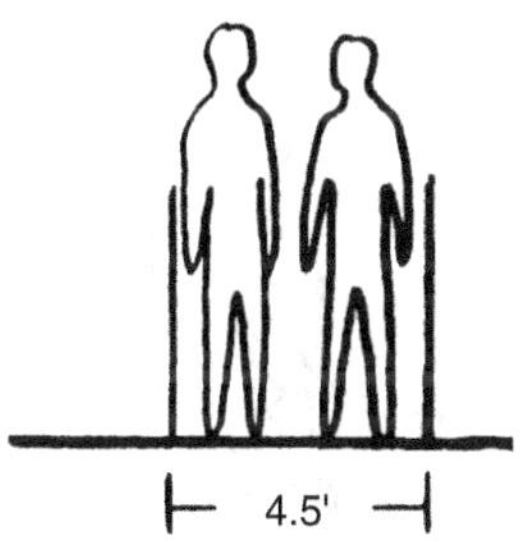

Figure 15.9 Corridor stair-width guidelines—within office areas.

5. Elevator Foyer: Although a larger space is desirable at ground levels for very large office buildings, the dimensions shown in Fig. 15.10 should be about minimum for any floor level; 25 percent less is generally adequate when elevators are on only one side of the corridor.

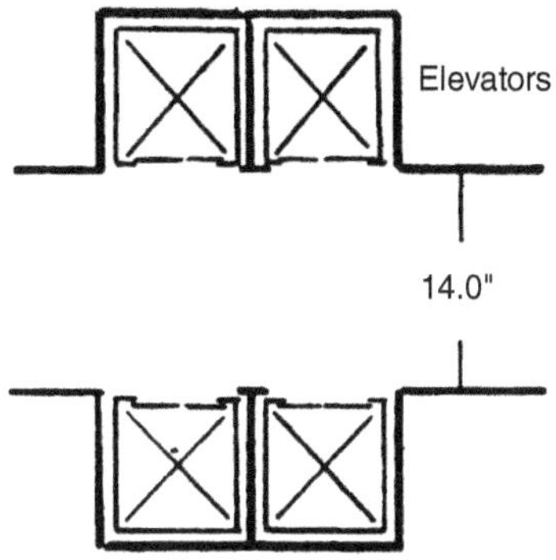

Figure 15.10 Corridor stair-width guidelines—elevator foyer.

6. Stair Width: The width between dual handrails shown in Fig. 15.11 should be the minimum for any stairwell. Consult local ordinances regarding standards for corridors and stairs.

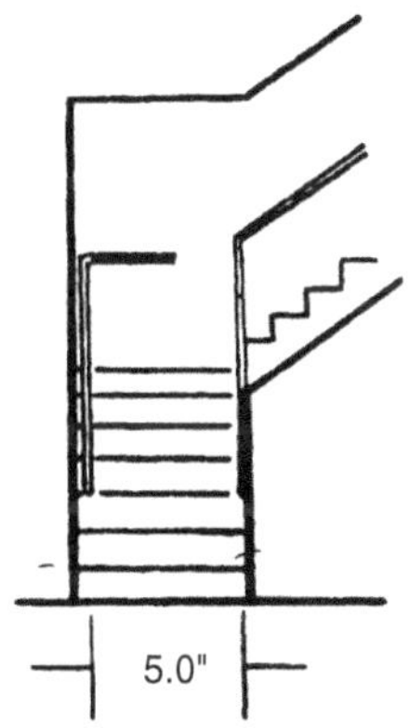

Figure 15.11 Stair width.

General Considerations

Interior hallways, usually being defined by walls on both sides, should be wide enough so that people do not have to walk carefully in order to keep from bumping into a wall, into other people, or into equipment that is attached to the wall or is being carried or moved on wheels. Although this may sound like an unnecessary caution, most standards for hallway widths do not account for all these factors. For example, a hallway standard is often applied without consideration of the fact that a drinking fountain, a public telephone, or a fire extinguisher will later be hung on one of the walls.

Figures 15.12–15.15 reflect preferred dimensions to ensure lack of conflict during passage of persons and/or equipment through typical hallway configurations:

1. Although a minimum of 30 in. (76 cm) is wide enough for one person, designers should always plan for the person in a wheelchair situation, which requires more space. The extra width will also accommodate a person on crutches. (See Fig. 15.12.)

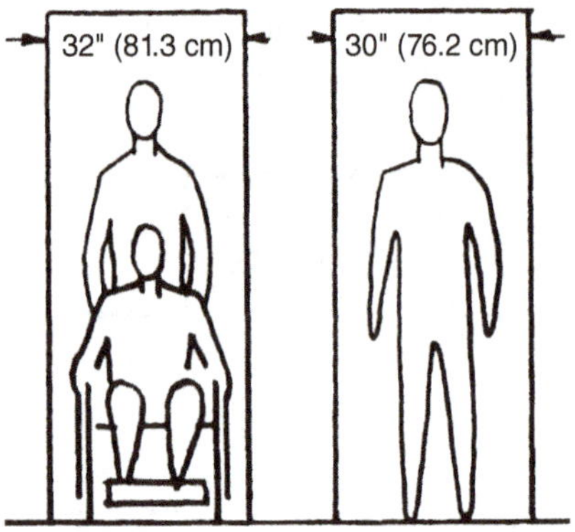

FIGURE 15.12 Interior hallways and aisles—minimum widths.

2. A person should not have to step aside or crowd against a wall to let another person pass. (See Fig. 15.13.)

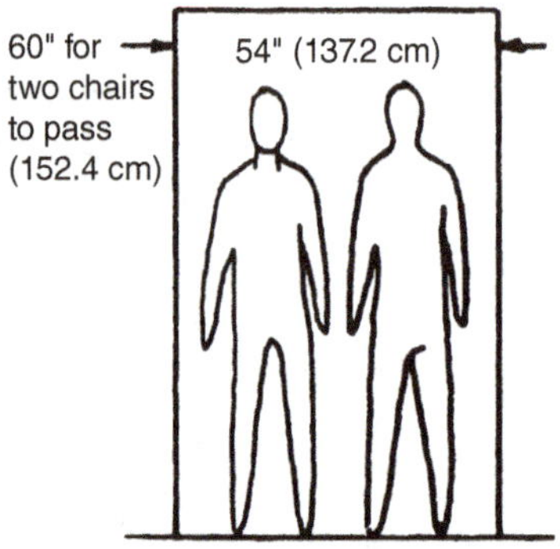

FIGURE 15.13 Interior hallways and aisles—passing.

3. When mobile equipment can be identified, the additional space required for moving such equipment past people who are using the hallway should be determined and designed for accordingly. (See Fig. 15.14.)

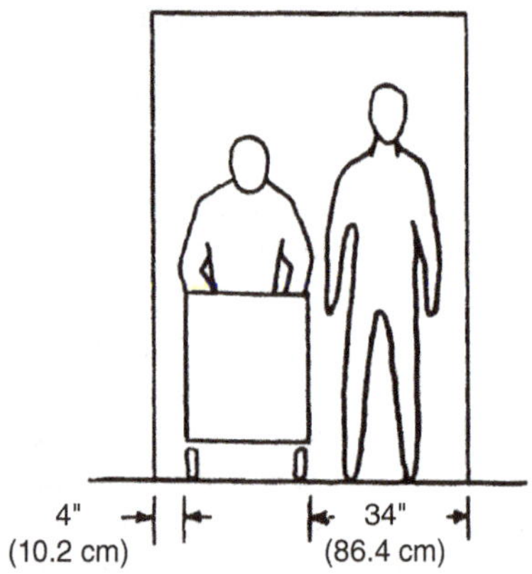

FIGURE 15.14 Interior hallways and aisles—moving equipment.

4. Aisleway widths for public transportation systems should not be narrowed to the point where people who are using the aisle bump into and annoy fellow passengers. If it is expected that the transportation vehicle will be accommodating persons in wheelchairs, be sure that the aisle is widened so that wheelchairs can pass freely, or lay out the vehicle so that persons in wheelchairs can park their wheelchairs near the door (e.g., in a public transit bus). (See Fig. 15.15.)

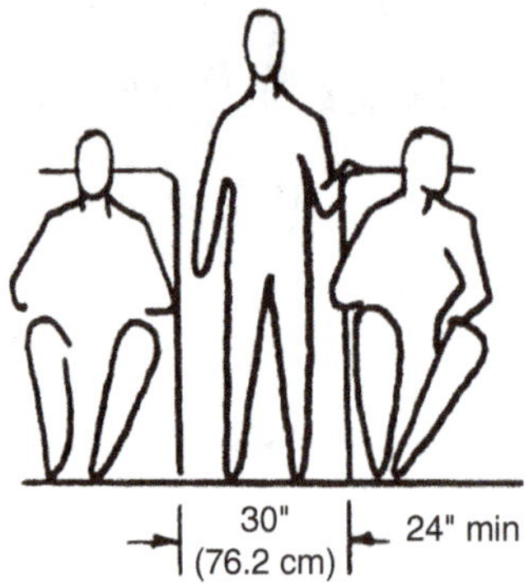

FIGURE 15.15 Interior hallways and aisles—for public transportation systems.

Hallway Path Definition

A defined pathway limit slightly apart from the walls of a hallway tends to keep people from wandering or absent-mindedly cutting the corner at a change of hallway direction. Avoid the use of large-radius cove design at the juncture of the wall and floor; people tend to trip or step on the radiused area and fall. (See Fig. 15.16.)

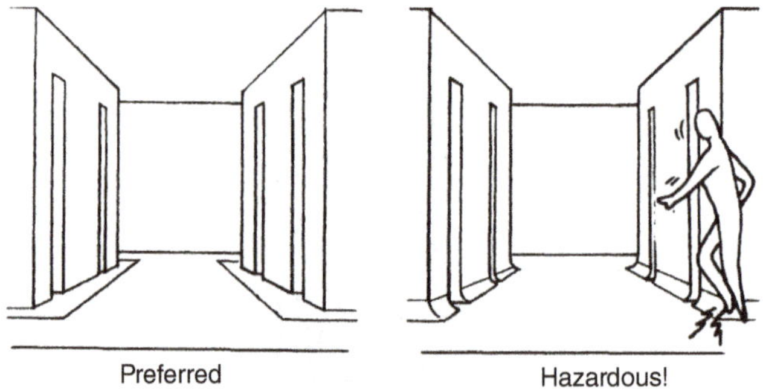

FIGURE 15.16 Hallway paths—avoid cove design.

In laying out hallways, do not place doorways directly adjacent to corners where the hallway junctures with another hall and/or the single hallway merely turns the corner. The doorway should be at least 5 ft (1.5 m) from the corner; more distance is preferred. (See Fig. 15.17.)

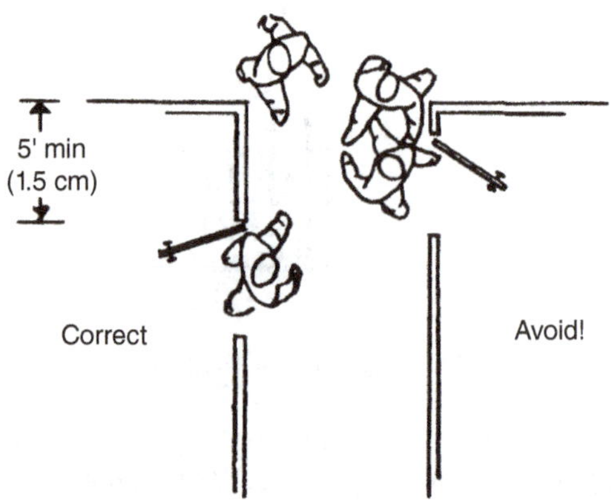

FIGURE 15.17 Hallway paths—doors should be 5¢ from corners.

Make sure that hallways are adequately lighted both during the day and at night. Natural light during the day should come from the side of a hallway, not from the end, where it tends to contrast with the less light exterior and blind the person who is facing the bright sunlight. (See Fig. 15.18.)

If a building is not ordinarily occupied at night except by security personnel, plan the hallway lighting circuits so that only some of the fixtures will burn.

SPECIAL WALKWAYS AND PASSAGEWAYS

Industrial or Military Applications

Figures 15.19 and 15.20 illustrate a few of the special types of walkways and passageways that often are associated with industrial or military applications. Figure 15.19 shows the minimum passage envelope for the adult male, walking at a slow and careful pace. Passageways of this type should be laid

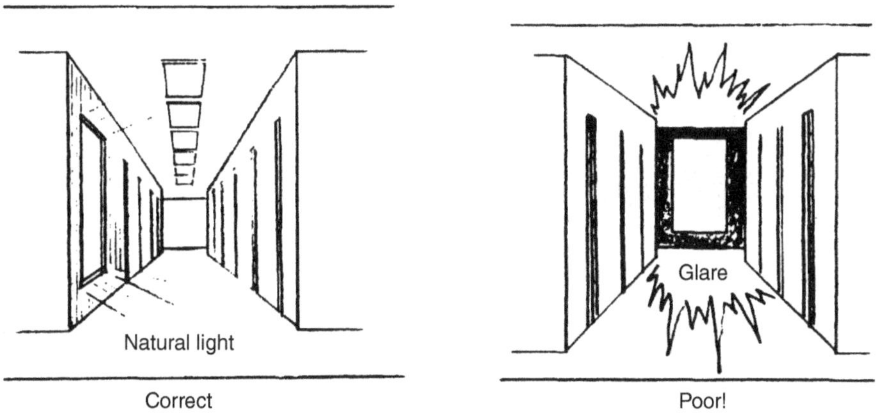

Figure 15.18 Hallway paths—avoid natural light entering only at ends of hallways.

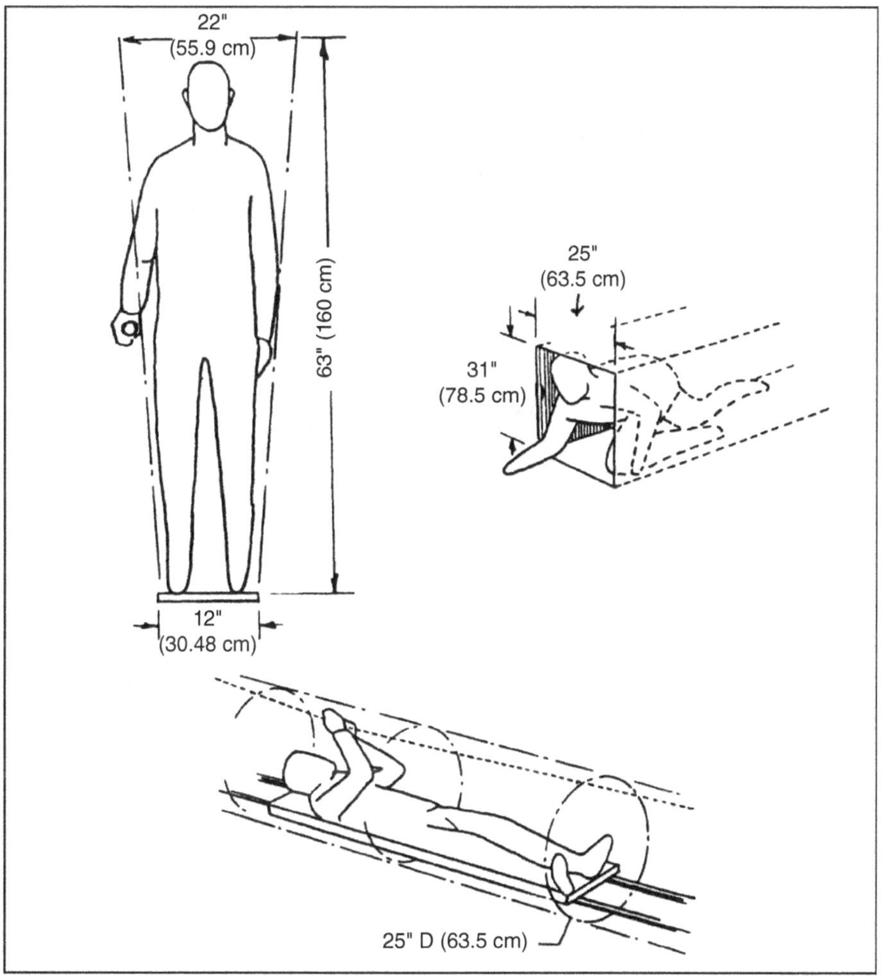

Figure 15.19 Special hallways—minimum passageway dimensions. (From MIL-STD-1472G.)

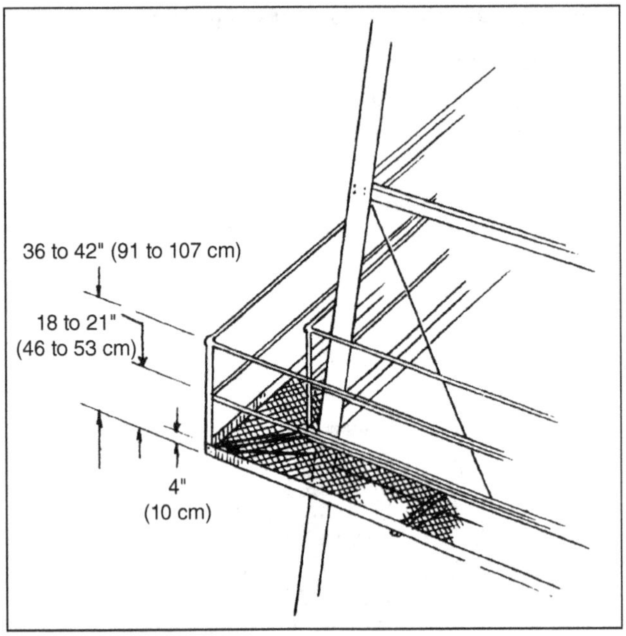

FIGURE 15.20 Special hallways—recommended dimensions for elevated platforms.

out in a straight line whenever possible, with no intervening obstacles which people have to step over, or which might snag their clothing.

Walkways and catwalks should have nonskid surfaces, especially when they are outdoors and when they are extremely high above the ground-floor level indoors.

Catwalks should have handrails on both sides, a midrail, and a toe board for safety.

Rough edges, bolt ends, and other protrusions must be avoided. The interior surfaces of crawl spaces and tunnels should be painted a light color.

Space Vehicles

Space vehicle passageway size depends on the following:

1. Gravity level.
2. Need to carry packages and equipment.
3. The size range of the crew and the type of clothing worn—People need to wear bulky space suits when outside a pressurized habitat.
4. Number of persons using passageway. (Figure 15.21 shows passageway size requirements for a specific case. More detailed information is in the NASA Human Integration Design Handbook.)
5. Body orientation—For a minimum passageway, a person can slip through in a prone position. Turning corners or performing other tasks will require other body positions and more room.

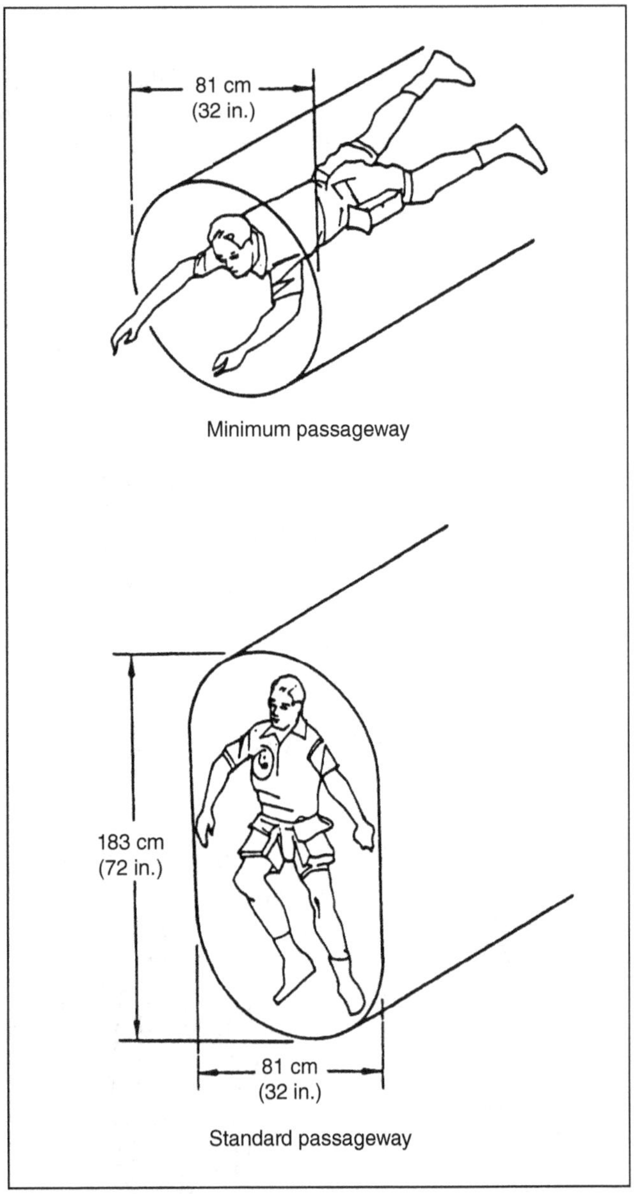

FIGURE 15.21 Space vehicle passageway sizes for one person in light clothing in zero-gravity environment.

CHAPTER 16

RAMPS, STAIRS, AND LADDERS

Figure 16.1 gives guidelines for selecting when to use ramps, stairs, or ladders based on the incline to be climbed and material handling requirements. Ladders should not be selected when equipment must be hand carried since both hands should be free to grasp the ladder. Where the equipment is bulky or in excess of 13 kg (29 lb), use ramps instead of stairs and steps.

RAMPS

Ramp Guidelines

The primary purpose of a ramp is to provide a smooth change of elevation for wheeled vehicles, mobile equipment, and wheelchairs. Unless a ramp is quite shallow, it is not preferred over a properly designed staircase for many applications. The principal considerations for deciding on the use of a ramp as opposed to a staircase are the following:

1. Wheeled vehicles, mobile equipment, or wheelchairs must be accommodated.
2. There is sufficient space to meet the ramp angle guidelines.
3. The ramp can be positioned where opening windows or doors will not interfere with it.

It should be noted in considering ramps that, although more energy may be required to go up a ramp, the major hazards occur when people are going down a ramp. Although healthy, nonhandicapped individuals can negotiate a fairly steep ramp (up to about 15°), the main criteria for designing most ramps should be based on the needs of the elderly or handicapped user. The guidelines shown in Figs. 16.2–16.5 are generally accepted by those who design for the handicapped person, who must travel by wheelchair.

A slope of 5 percent is the maximum allowed for a ramp before it is required that a handrail be added. When the slope becomes greater, it is very taxing for most wheelchair users to "pull the hill." Thus this upper limit is a preferred slope guideline for all ramps, whether they are for indoor or outdoor applications.

No ramp should exceed about 8 percent slope under any circumstances.

If a ramp slope is created that is over the 5 percent grade, handrails should be provided to assist individuals who are weaker and/or less steady on their feet. There should be a separate railing for both sides of such ramps.

Nonskid surfaces should be provided on all ramps.

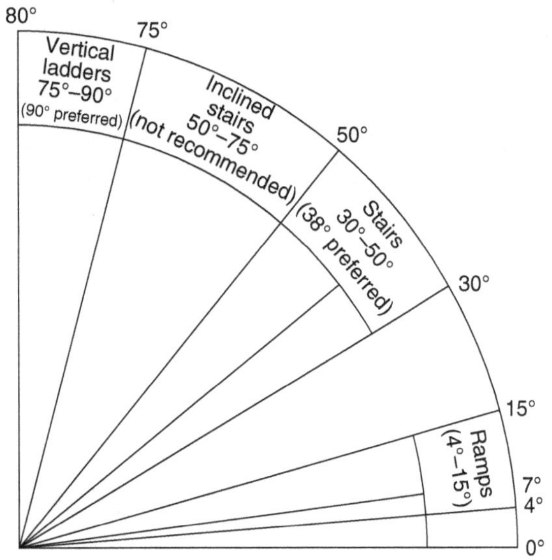

Type	Angle of inclination (in degrees)	Preferred angle of inclination (in degrees)
Vertical ladders	75–90	90
Inclined stairs	50–75	Not recommended
Stairs	30–50	38
Ramps for personnel	7–15	7–8
Ramps for material handling	4–7	4

FIGURE 16.1 When to use ramps, stairs, and ladders.

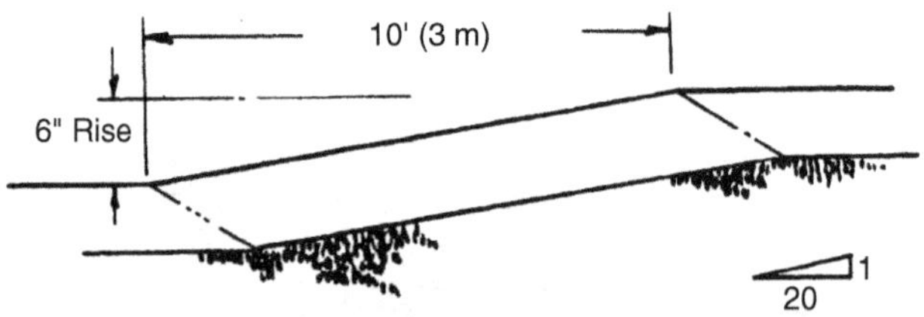

FIGURE 16.2 Ramps—maximum allowed for handicapped users.

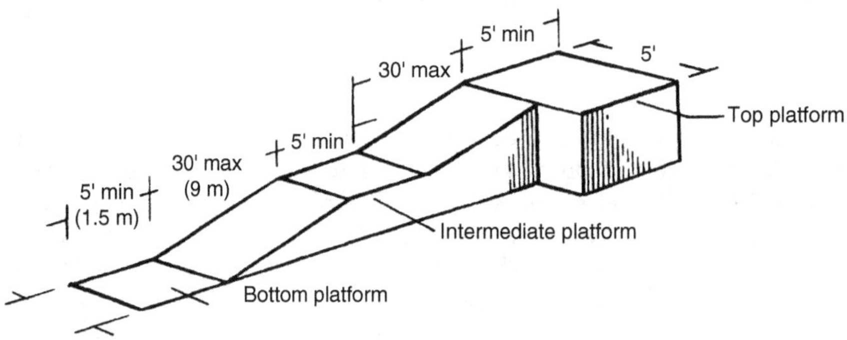

FIGURE 16.3 Ramps—straight run ramp.

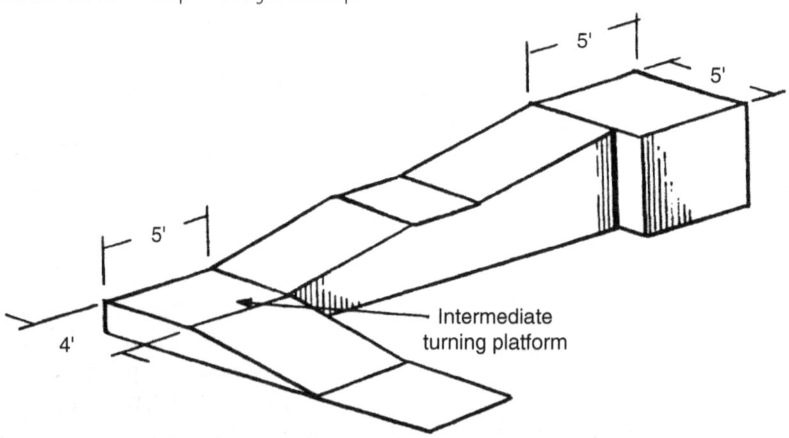

FIGURE 16.4 Ramps—ramp with turning platform.

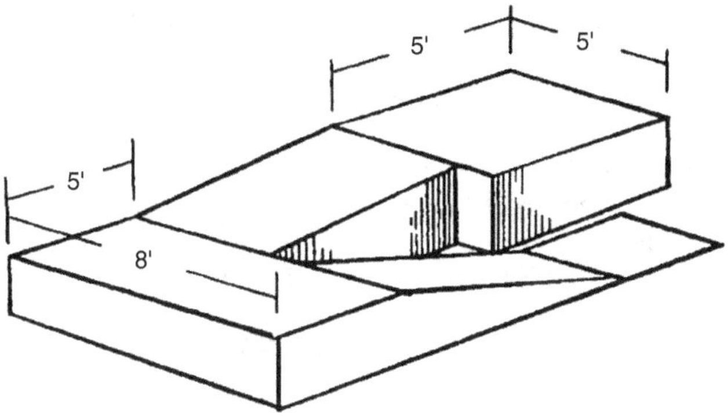

Figure 16.5 Ramps—ramp with intermediate switch-back platform.

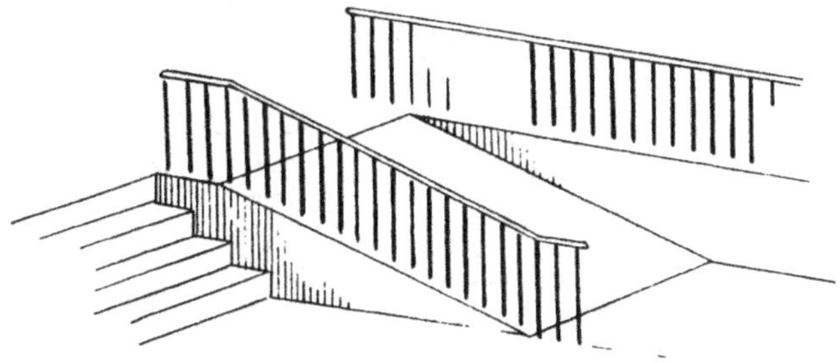

Figure 16.6 Ramps and stairs—combination.

Ramps should not extend more than 30 ft (9.2 m) before an intervening level "resting platform" is provided. There should also be similar flat platforms both at the bottom of the ramp and at any point at which the ramp system changes direction. The guidelines shown in the figures are best for wheelchair users but are equally valid for all users.

Ramps and Stairs

1. In order to improve traffic flow in busy public buildings, consider a combination ramp and staircase system, which allows the nonhandicapped to "move through" without waiting for someone who is using the ramp (who is usually slower). This scheme prevents a handicapped person from trying to hurry to get out of the way, which could cause an accident. (See Fig. 16.6.)

2. Handrails: Handrails should be provided on all ramps or platforms that are more than 2 ft (0.6 m) above grade. Note the preparatory horizontal element at both the bottom and the top of the ramp handrail shown in Fig. 16.7. This helps the user adjust to the change from a horizontal to a ramped surface.

3. Stairs: Since stairs may at some time be used by both elderly persons and handicapped persons not confined to wheelchairs, stairs should be designed so that they are negotiable without stubbing a toe. Elderly and handicapped persons often are not aware that they are not picking up their feet high enough to clear the next tread. (See Fig. 16.8.)

STAIRS

Optimizing Stair Design

Studies have shown that a staircase slope of approximately 30° requires the least expenditure of energy on the part of the

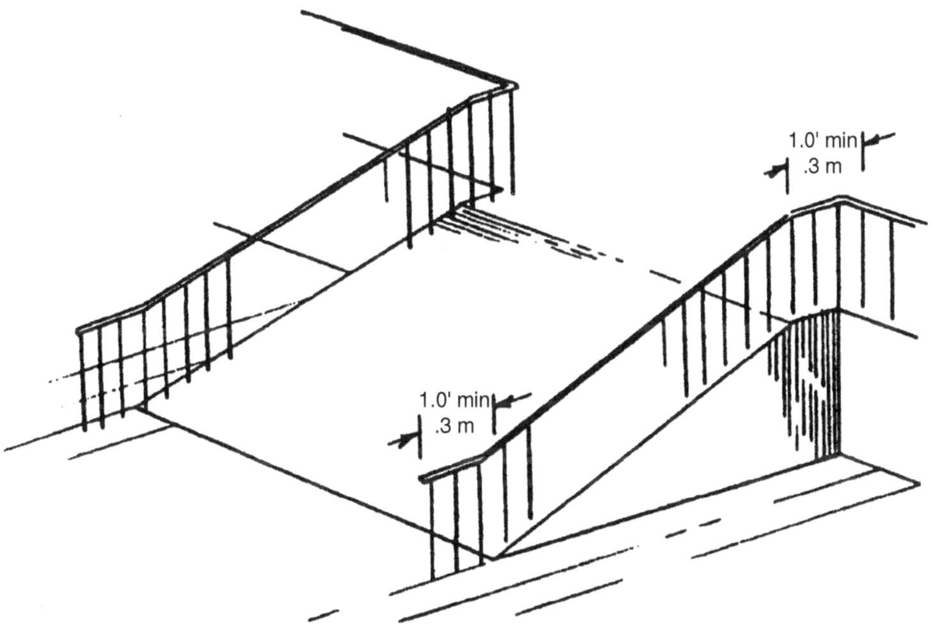

Figure 16.7 Handrails.

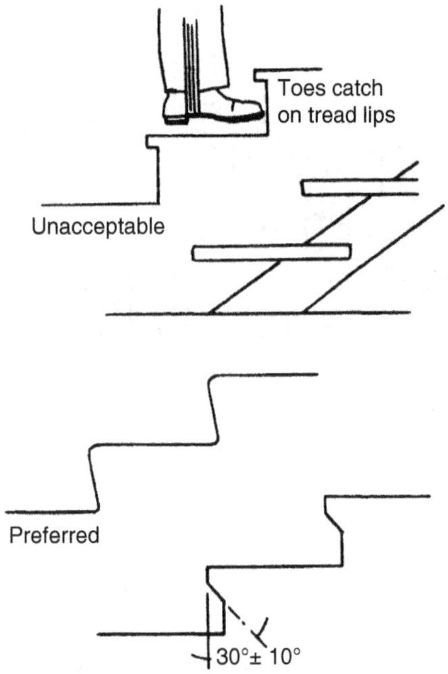

FIGURE 16.8 Stairs.

average adult. Such a slope requires a tread depth of approximately 11 in. (28 cm) and a riser height of about 6.5 in. (16.5 cm). (See Fig. 16.9.)

This, of course, does not adequately take into consideration the two groups who have perhaps more stairway accidents than any other, i.e., children and the elderly. Unfortunately, it is probably impractical to design separate staircases for everyone (except in special cases that might be identified as custom installations).

Other important optimizing features include, obviously, the proper width (a minimum width between handrails of 30 in., 76 cm, for one person at a time and of 52 in., 132 cm, for two persons at a time), the overhead clearance, and the height of the handrail (all stairs should have at least one handrail, and preferably two). See Fig. 16.10.

Stair Design in Relation to Equipment and Furniture Handling

The recommended spacing between stair rails for safely moving the typical large piece of furniture in the home is shown in Fig. 16.11. This dimension will also accommodate the movement of other large items such as a grand piano and a standard pool table (with the legs removed, of course).

It should also be noted that windows located at staircase landings are vulnerable when furnishings are being manipulated within the landing; therefore, the designer should consider the possibility of eliminating such windows.

Special Safety Considerations in Stair Design

Spiral staircases should not be used as a general rule. The varying tread depth (the inside and the outside dimensions differ) often leads to missteps and accidents.

All main stairways within a single structure or dwelling should be alike in terms of tread-riser configuration. (See Fig. 16.12.)

Do not leave the backside of a staircase open, or people (especially the blind) may walk into the stair structure.

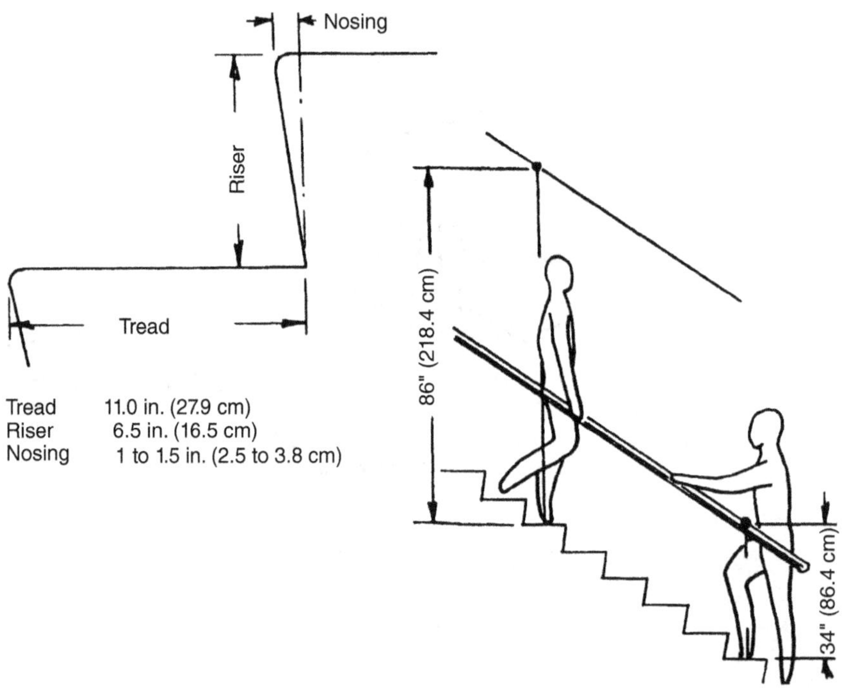

FIGURE 16.9 Preferred design.

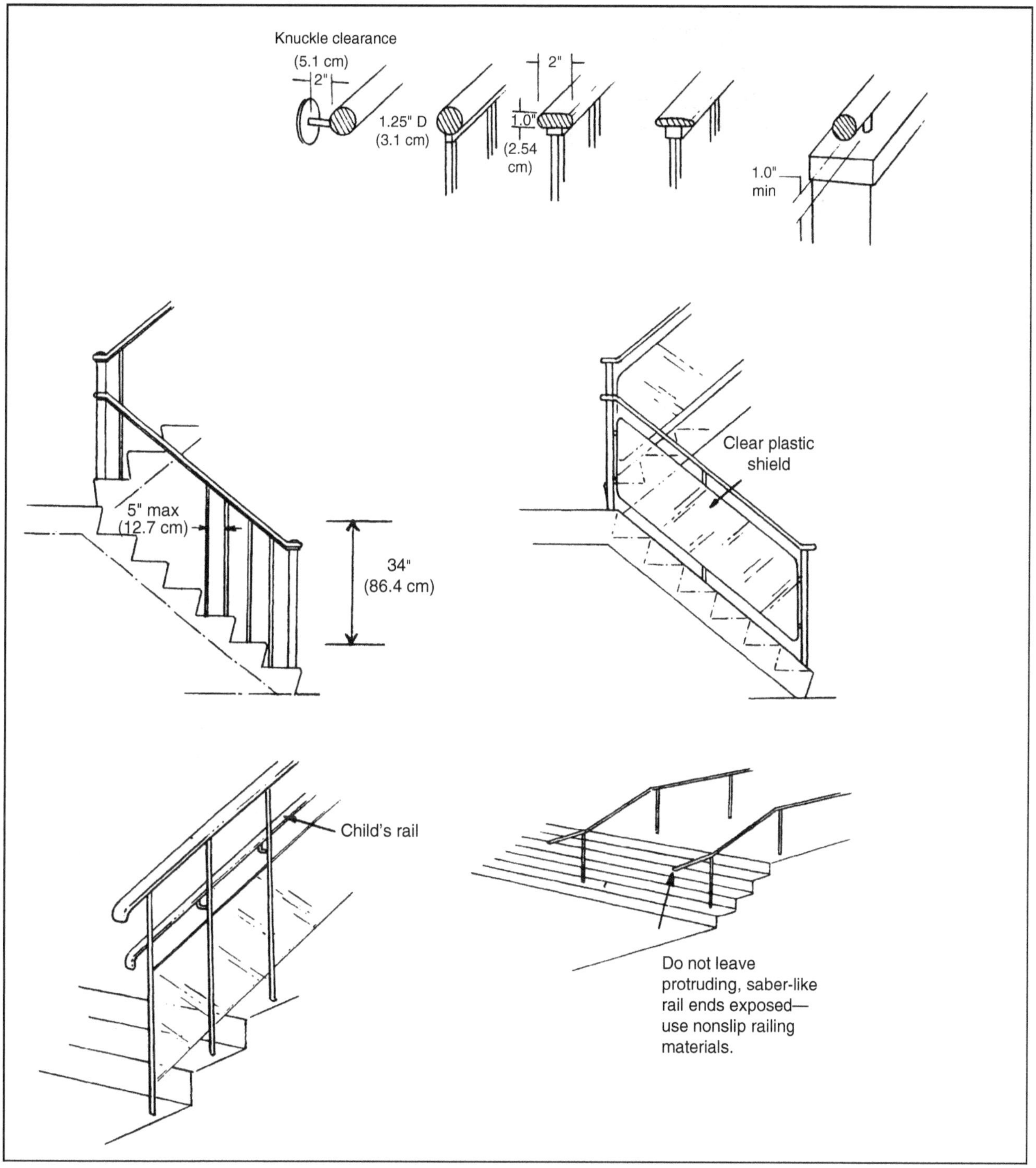

Figure 16.10 Special safety considerations.

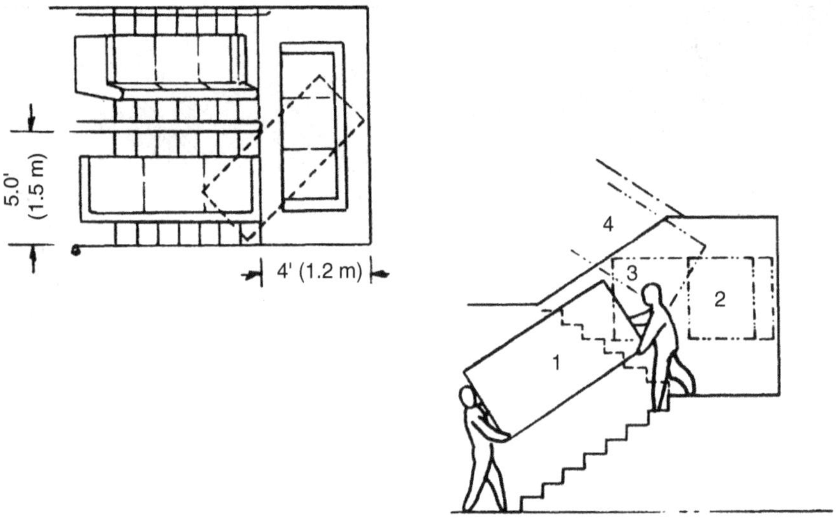

FIGURE 16.11 Railing design recommendations.

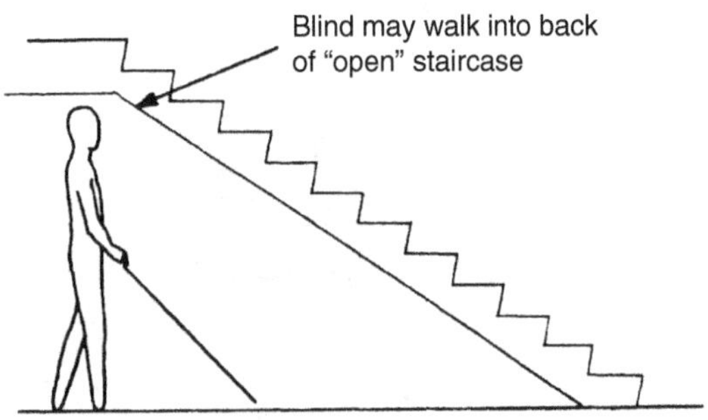

FIGURE 16.12 Furniture handling requirements.

Special Industrial Stairs

When specialized stair applications require some compromise of the preferred slope, tread and riser, and other previously noted features, one should still consider the possible hazards that may be associated with stair use. (See Fig. 16.13.)

LADDERS

Ladders and Stair-Ladders

Figures 16.14–16.16 illustrate Occupational Safety and Health Administration requirements for fixed ladders and the military design requirements for fixed ladders and stair-ladders.

Boarding Ladders

Special step and handrail design requirements are often associated with transportation system vehicles and equipment and with industrial, construction, and agricultural vehicles and equipment. Planning for the ingress and egress or mounting of these special devices should always occur before basic configurations make it impossible to create a suitable ladder, step, and/or handrail subsystem. In most of these cases, the relationship between the ladder or step element and the associated handholds or handrails is particularly critical to the safety of the person using them. Figure 16.17 on p. 198 illustrates an obvious concern by the designer to integrate the ladder handrail into the overall machine concept early enough so that the driver will have a reasonably safe approach to the driver's compartment and back again to the ground. Although not quite so obvious, the short ladder and platform provided at the rear also demonstrate the designer's concern for the person who has to service the engine.

In spite of the apparent logic of planning steps and handrails from the very beginning, it is common for large machinery to be designed and packaged almost completely before any thought is given to where steps and handrails for entering or leaving the machine should be placed. The following principles should be observed in considering the design of these special types of entry devices:

1. The first step from the ground must be reachable by the shortest expected user, and at least two handholds must be accessible to this person while he or she is still on the ground.

2. The steps or rungs must accept the user's shoe, with the shoe being located for a firm step; i.e., the midpoint of the shoe, not just the toe, must rest solidly on the step. Each step or rung should be sufficiently wide so that the person can stop and rest both feet on it.

3. The position of each succeeding step and its associated handhold must be planned so that the user's final entry into the machine or vehicle will be compatible with sitting in the seat. Remember that when a person climbs a ladder, the hand and leg that are making the next move are on opposite sides of the body. If handholds and steps are not planned to conform to this natural "climb pattern," the person will more than likely end up with the

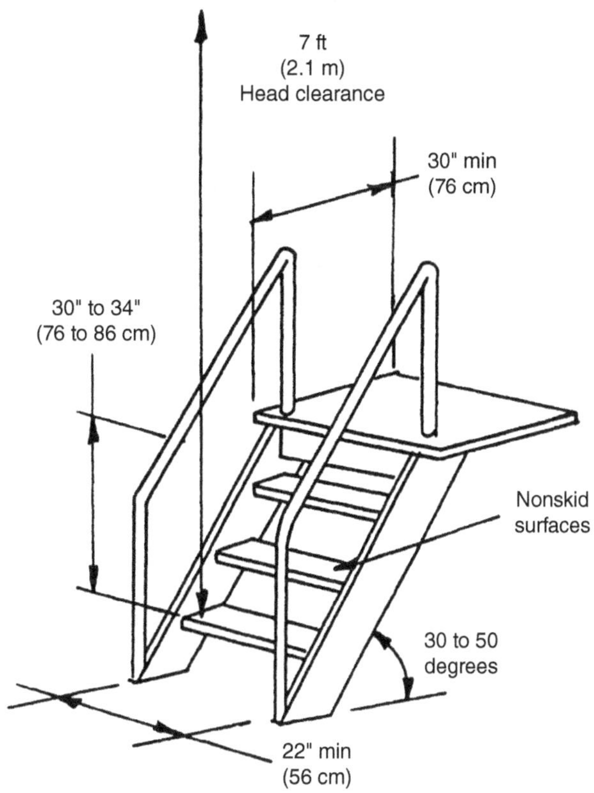

Possible rise and run combinations

Angle to horizontal	Rise (in.)	Tread run (in.)
30°35'	$6^1/_2$	11
32°08'	$6^3/_4$	$10^3/_4$
33°41'	7	$10^1/_2$
35°16'	$7^1/_4$	$10^1/_4$
36°52'	$7^1/_2$	10
38°29'	$7^3/_4$	$9^3/_4$
40°08'	8	$9^1/_2$
41°44'	$8^1/_4$	$9^1/_4$
43°22'	$8^1/_2$	9
45°00'	$8^3/_4$	$8^3/_4$
46°38'	9	$8^1/_2$
48°16'	$9^1/_4$	$8^1/_4$
49°54'	$9^1/_2$	8

Figure 16.13 Fixed industrial stair design requirements.

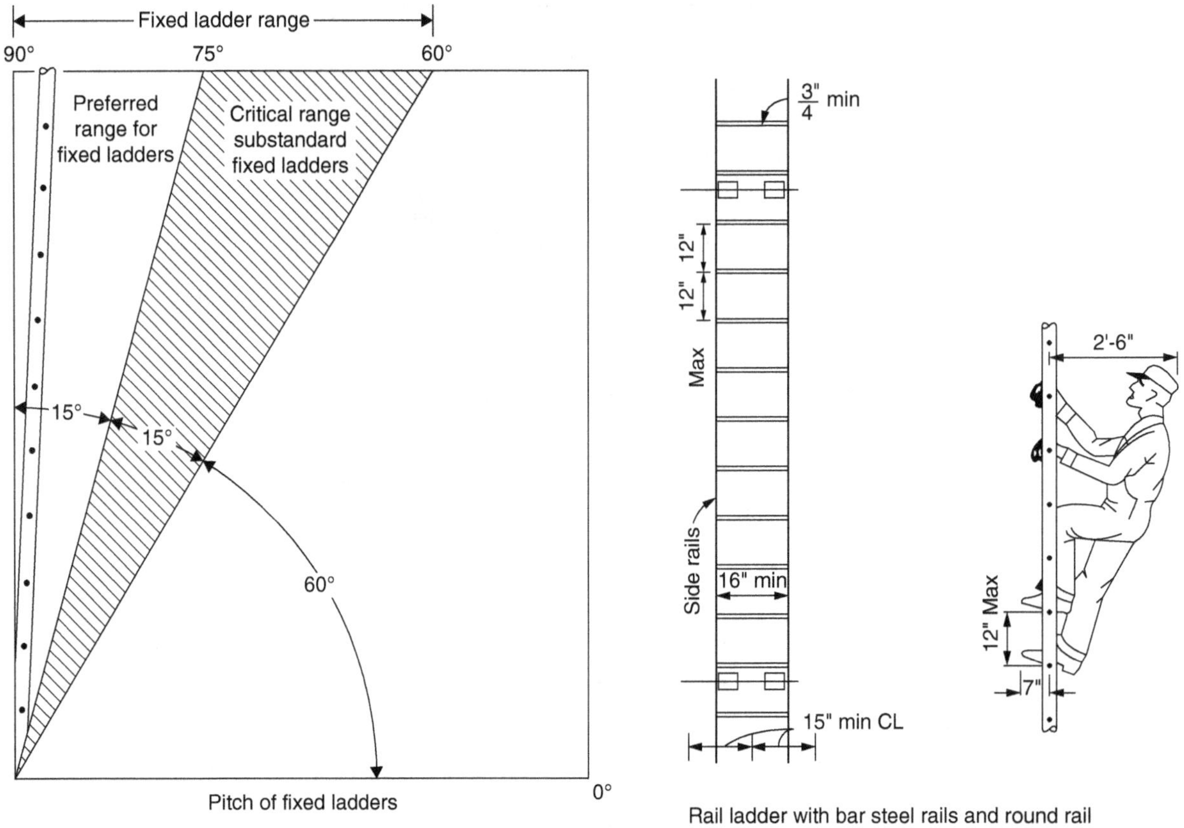

Rail ladder with bar steel rails and round rail
(From OSHA, 29 CFR 1901.27, 1987.)

FIGURE 16.14 Fixed ladder design requirements.

Dimension	Minimum	Maximum	Recommended
A Rung thickness:			
Wood	32 mm (1-1/8 in.)	38 mm (1-1/2 in.)	35 mm (1-3/8 in.)
Protected metal	19 mm (3/4 in.)	38 mm (1-1/2 in.)	35 mm (1-3/8 in.)
Corrosive metal	25 mm (1 in.)	38 mm (1-1/2 in.)	35 mm (1-3/8 in.)
B Rung spacing	230 mm (9 in.)	380 mm (15 in.)	300 mm (12 in.)
C Height, rung to landing	150 mm (6 in.)	380 mm (15 in.)	380 mm (15 in.)
D Width between stringers	300 mm (12 in.)		460–530 mm (18–21 in.)
E Climbing clearance width	610 mm (24 in.)		760 mm (30 in.)
Clearance depth:			
F In back of ladder	150 mm (6 in.)		200 mm (8 in.)
G On climbing side (range)	910 mm (36 in.) 75° to 760 mm (30 in.) for 90°		
H Height of stringer above landing	840 mm (33 in.)		910 mm (36 in.)
J Height from lower elevation to bottom rung		380 mm (15 in.)	

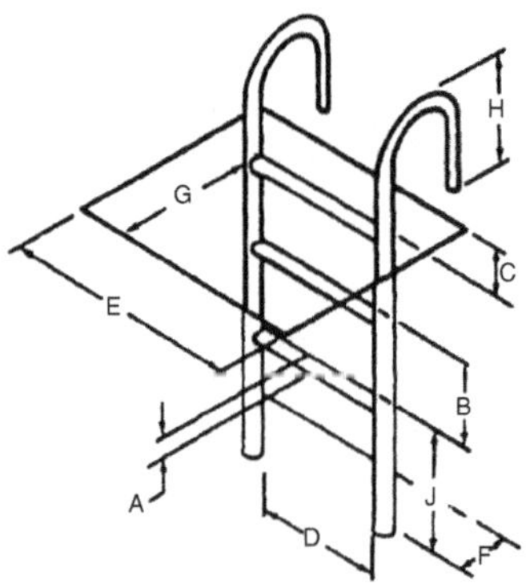

FIGURE 16.15 Fixed ladder design requirements. (From MIL Standard—1472 D.)

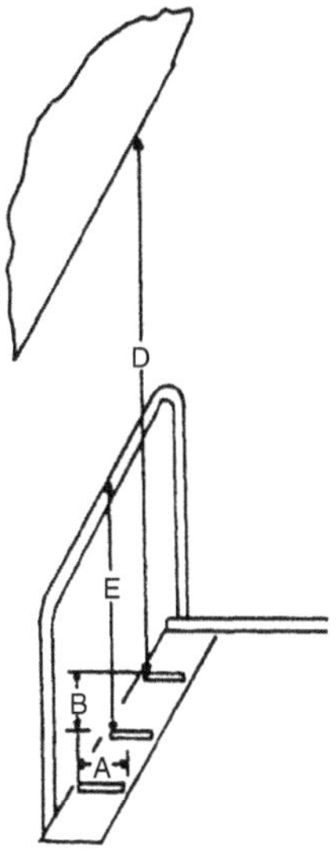

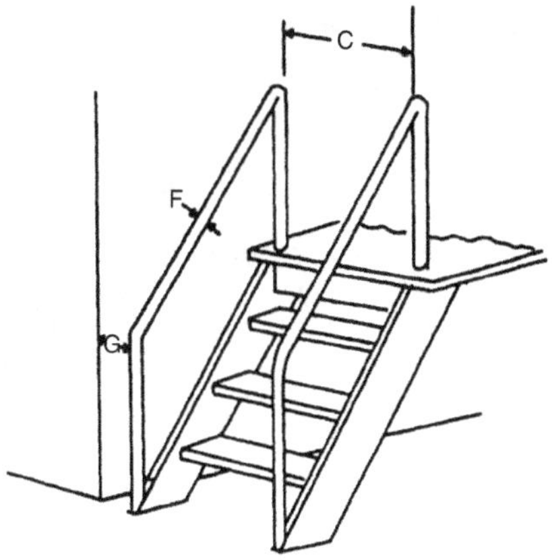

Dimension	Minimum	Maximum	Recommended
A Tread depth range:			
For 50° rise	150 mm (6 in.)	250 mm (10 in.)	215 mm (8-1/2 in.)
For 75° rise (open ladders only)	75 mm (3 in.)	140 mm (5-1/2 in.)	100 mm (4 in.)
B Riser height	180 mm (7 in.)	300 mm (12 in.)	230 mm (9 in.)
C Width (handrail to handrail)	530 mm (21 in.)	610 mm (24 in.)	560 mm (22 in.)
D Overhead clearance	1730 mm* (68 in.)	…	1930 mm (78 in.)
E Height of handrail (from leading edge of tread)	860 mm (34 in.)	940 mm (37 in.)	890 mm (35 in.)
F Handrail diameter	32 mm (1-1/4 in.)	75 mm (3 in.)	38 mm (1-1/2 in.)
G Rail clearance from wall	50 mm (2 in.)	…	75 mm (3 in.)

*Whenever the distance D is less than 1.880 m (74 in.), the overhead obstruction should be painted with yellow and black stripes.

FIGURE 16.16 Stair-ladder design requirements. (From MIL Standard—1472 D.)

wrong foot ready to enter the vehicle; i.e., usually a person cannot hold and step from the same side without swinging. Do not create a system that forces the user to step on a hubcap, tire, or other irregular surface.

4. The specific contour of handrails must follow the gripping, pulling, and supporting pattern associated with both entry and exit; entry and exit are generally quite different and in some cases require more than one handrail geometry.

5. Both ladder rungs or steps and handrails must have nonskid or nonslip surfaces. Handrails should be placed on both sides.

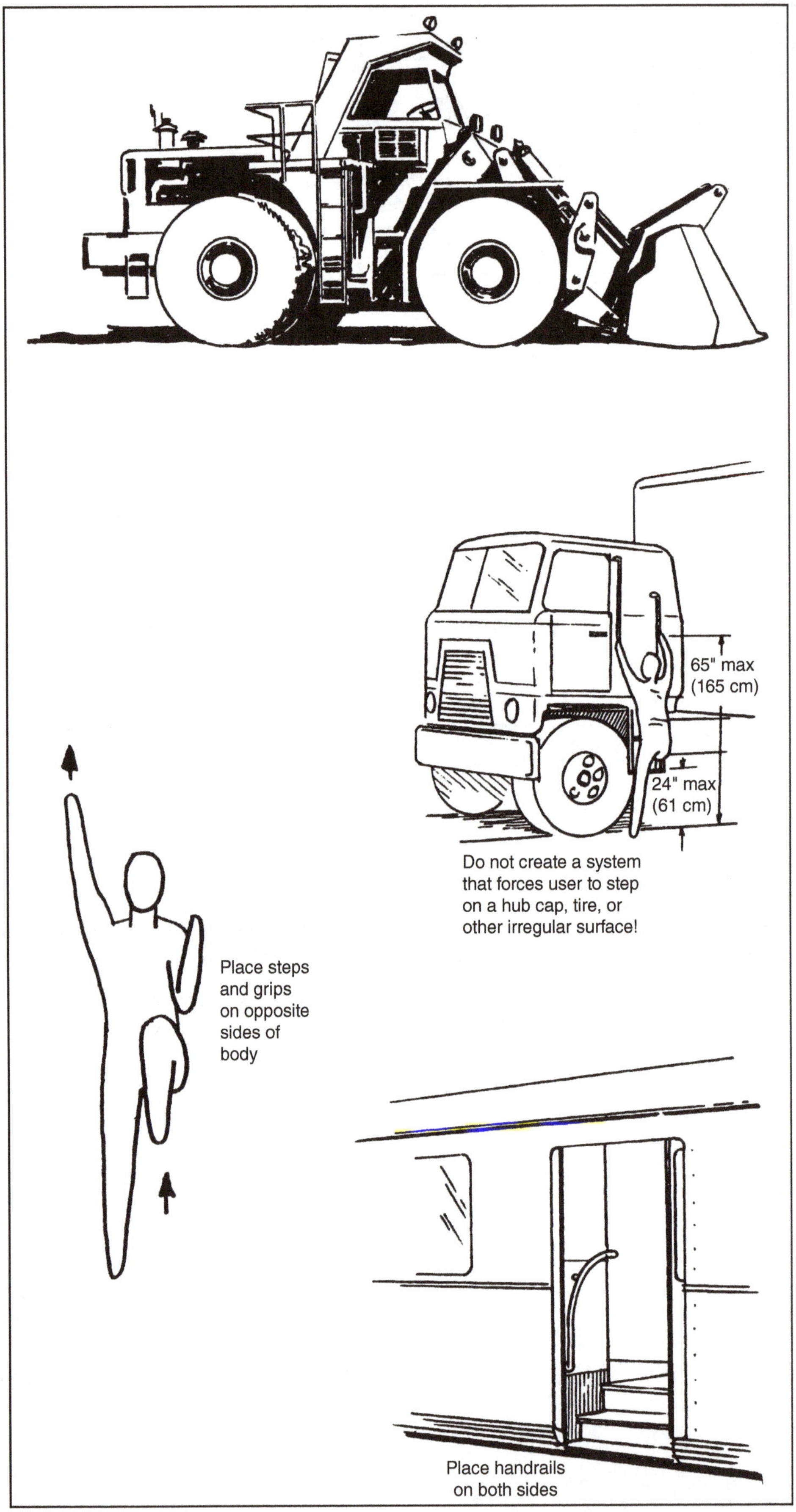

FIGURE 16.17 Boarding ladder design considerations.

CHAPTER 17

ELEVATORS, ESCALATORS, AND MOVING WALKWAYS

ELEVATORS

Critical Variables

Critical variables to consider are the following (see Fig. 17.1):

1. Capacity
2. Velocity
3. Acceleration and deceleration
4. Normal and emergency ingress and egress
5. External and internal controls
6. Emergency communication
7. Illumination
8. Ventilation
9. Door operating variables (e.g., speed and safety bumper)
10. Entrance and threshold matching
11. Handrails

Elevator standards generally are established by local ordinances, which should be consulted. It is also suggested that the American Standard Safety Code for Elevators (A17.1-1965) be used to establish basic mechanical requirements.

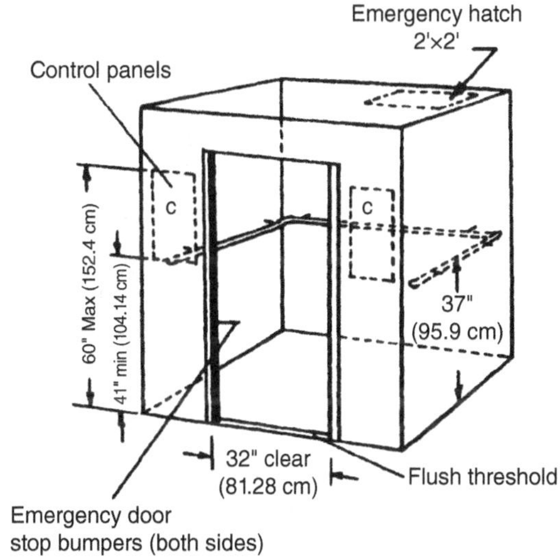

FIGURE 17.1 Elevator dimension.

In addition, consideration should be given to the problem of passenger escape and rescue in the event of a power failure.

General Guide for Determining Number and Type of Elevators[2]

Elevators should be provided for any residence facility that is more than one story high because of the possibility of use by persons in wheelchairs.

Elevators should be provided for residences and office buildings that are four or more stories high (public buildings that are frequented by handicapped persons should have elevators if they are more than one story high and if there are no other ramped approaches to the second story). (See Fig. 17.2.) The accompanying table (Fig. 17.3) provides guidelines for determining the number of elevators required in a facility.[3]

External and internal control panels should be located and arranged so that orthopedically handicapped persons can locate and reach the controls. There should be an audio signal for all passengers, and, in addition, the tonal signal should provide a cue for blind persons as to which way the elevator is expected to move, i.e., a higher-pitched signal for "up," and a lower-pitched signal for "down."

Visual signs, floor indicator displays, and panel controls should be arranged and labeled so that they are visible, legible, and understandable. Avoid special abbreviations that are known only to the tenants. The floor-selection panel shown in Fig. 17.2 illustrates desirable control-button identification; i.e., the vertical arrangement provides vertically distributed numbers that relate to the vertical arrangement of the floors. When there are several subterranean floors, separate the buttons for these floors from upper-floor controls by an intervening space and a general panel label that identifies the lower button array as "basement."

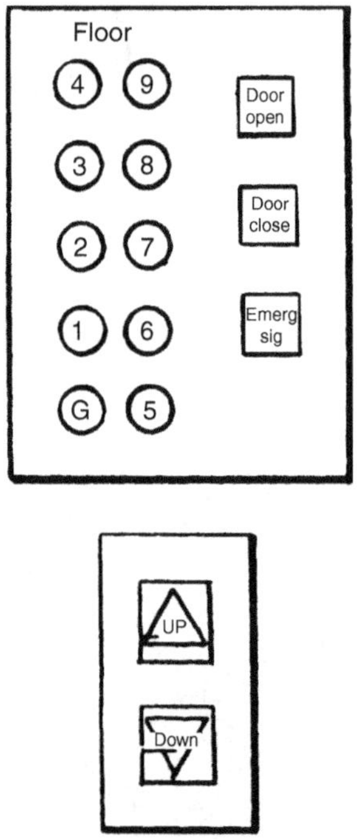

FIGURE 17.2 Elevator control layout.

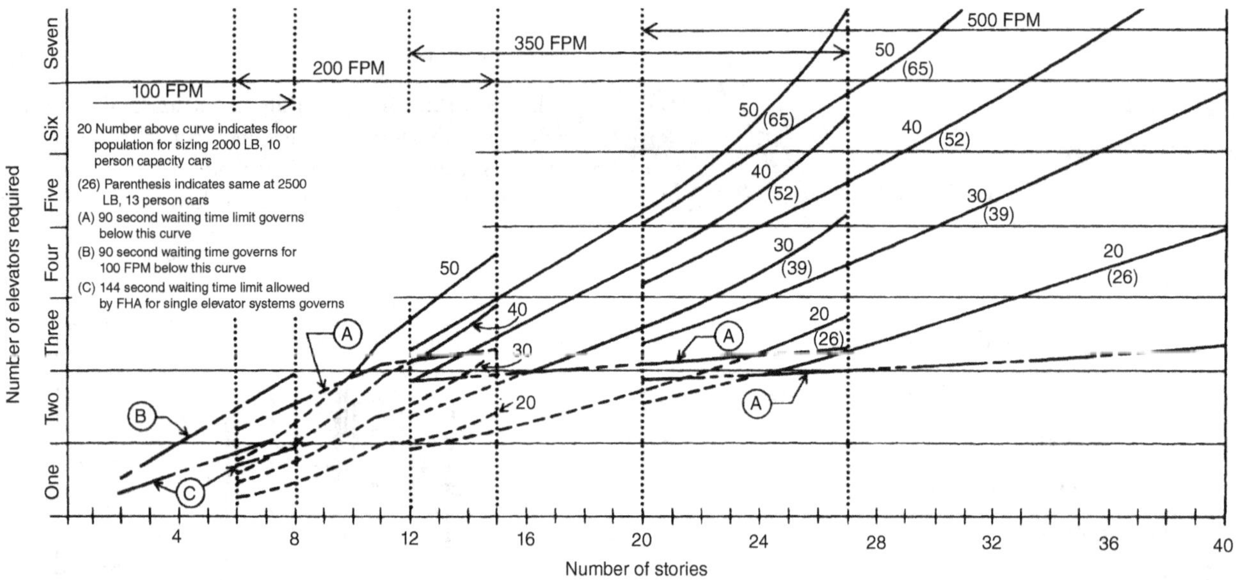

FIGURE 17.3 Number of elevators required.

Push buttons, as well as floor-identification indicators, should illuminate when a control button is actuated and/or when a particular floor is reached.

Outside control buttons should be shape-coded as shown in Fig. 17.3, and the translucent button should illuminate in the letters.

---

[2] J. De Chiara and J. H. Callender, *Time-Saver Standards for Building Types*, McGraw-Hill Book Company, New York, 1973, p. 83.

[3] J. De Chiara and J. H. Callender, *Time-Saver Standards for Building Types*, McGraw-Hill Book Company, New York, 1980, p. 77.

Interiors of elevators should be properly illuminated so that deep shadows are not cast by passengers. It is recommended that only light colors for walls, ceiling, and floor be used in order to help distribute interior illumination. There should be an emergency light source in the elevator that is not dependent on system power.

Note: It is suggested that elevators in facilities frequented by the blind have control buttons that can be factually identified. Consider also an audio signal so that the blind person can "count" floors as they are passed.

ESCALATORS

The escalator generally is an efficient method of moving people in certain department store and lower office building interfloor transfer. The escalator industry has developed a fairly consistent set of standards which it follows to ensure the safety of users. In addition, the following special human factors considerations are important in the selection and installation of such facilities (see Fig. 17.4):

1. The angle of incline should be approximately 30°.
2. Speed should not exceed about 120 ft/min (37 m/min).
3. The width between handrails should be approximately 32 in. (81 cm).

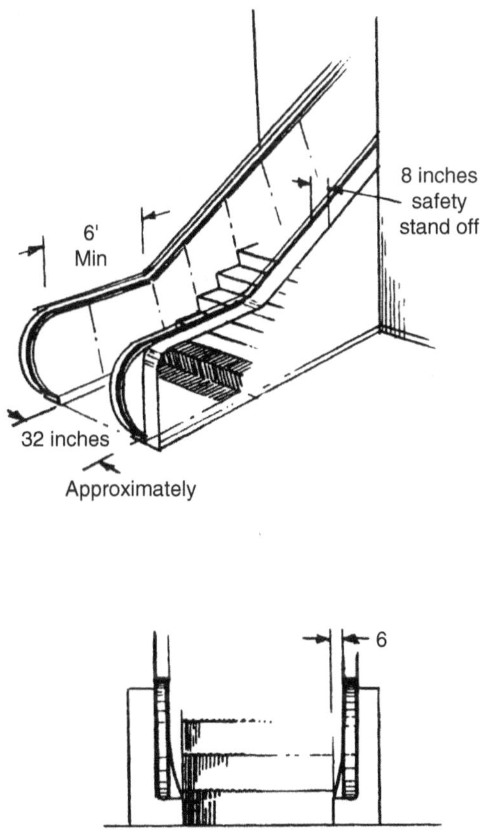

Figure 17.4 Escalator dimensions.

4. Users should be able to grasp the handrail at least 6 ft (1.8 m) prior to the first rise of the moving steps so that they can be prepared for the rate of stair movement.
5. The handrail and step movement should be synchronous so that users do not have to continually change their hand grip.
6. The inside, fixed panel should curve toward the handrail so that users' legs or clothing will not drag.
7. There should be at least 8 in. (20 cm) of clearance between the handrail and any wall alongside the railing so that users cannot inadvertently run their hands or arms into any wall or adjacent structure.
8. The handrail should be approximately 3–4 in. (7.6–10.2 cm) wide and have a curved surface that is easy to grasp.
9. The tread depth should not be less than about 13 in. (33 cm) so that even the largest shoe can rest solidly on the tread while the escalator is in transit.
10. There should always be good illumination at entry and departure points.
11. The escalator and/or the escalator well installation should be designed so that users cannot catch their hands or arms between the moving railing and the adjacent structure.
12. Steps should be suitably marked so that the user can easily see the edges of the step and therefore avoid stepping on parts of two steps, which could upset the individual as the steps change from a common level to elevated levels.
13. There should be no sharp edges, corners, or gaps that could cause injury if a person fell down on the steps.
14. Appropriate signing should be provided so that people can easily identify where the escalator is, which direction it is going, and what the rules are concerning use (no shopping carts, no children without parental supervision, etc.).
15. All escalators should be clustered together (i.e., up and down assemblies) so that people do not have to hunt for the escalator to the next level.

MOVING WALKWAYS

The moving walkway may be a useful device for minimizing some of the difficulties that elderly persons and persons carrying heavy baggage experience in airports, large shopping malls, and other public facilities where people must walk long distances. (See Fig. 17.5.) Some of the more important human factors considerations are the following:

1. Speed should never exceed about 100 ft/min (31 m/min), and preferably should not exceed 90 ft/min

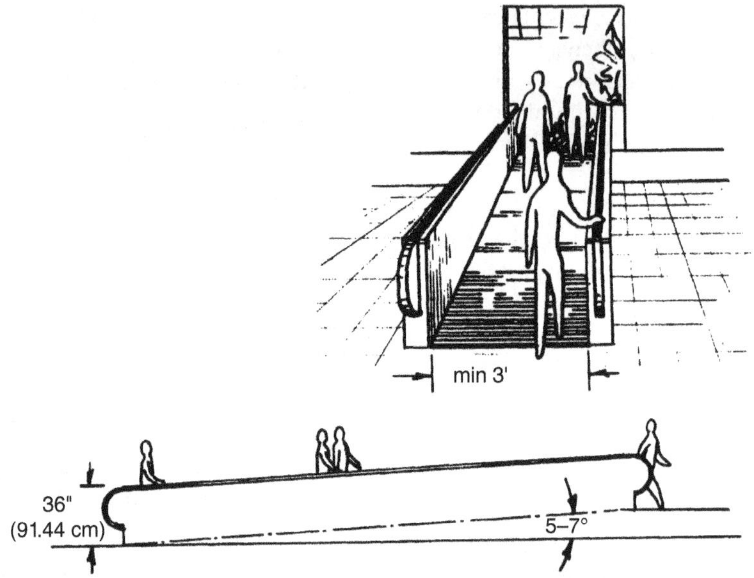

FIGURE 17.5 Moving walkway dimensions.

(28 m/min). An alternative walkway should be provided for those who wish to walk faster, and/or the moving walkway should be made wide enough so that fast walkers can pass those who are holding their position and "riding" the system.

2. The width of the walkway should not be less than 32 in. (81 cm) so that persons in wheelchairs can also use the walkway. Widths for more than three persons are undesirable.

3. The moving walkway can be ramped as much as 7°–10° and still be acceptable to older passengers. Although some safety codes permit up to 15°, this is too steep for older persons who are unsteady on their feet or who have difficulty seeing. Good illumination should be provided not only at the entrance and exit points but also along the walkway if it is wide enough for persons to "walk the ramp."

4. The ramp or walkway surface must be as smooth as practical; i.e., there should be no "bumps" created by the walkway propelling or supporting system. Moving walkway surfaces typically are ribbed to provide good nonslip characteristics. The rib pattern should be clearly visible so that persons who are entering the walkway can observe the approximate speed of the walkway before stepping onto it.

5. Systems should be designed so that, in case of malfunction, the walkway "coasts" to a stop rather than stopping abruptly.

CHAPTER 18

BATHROOMS, RESTROOMS, AND LOCKER ROOMS

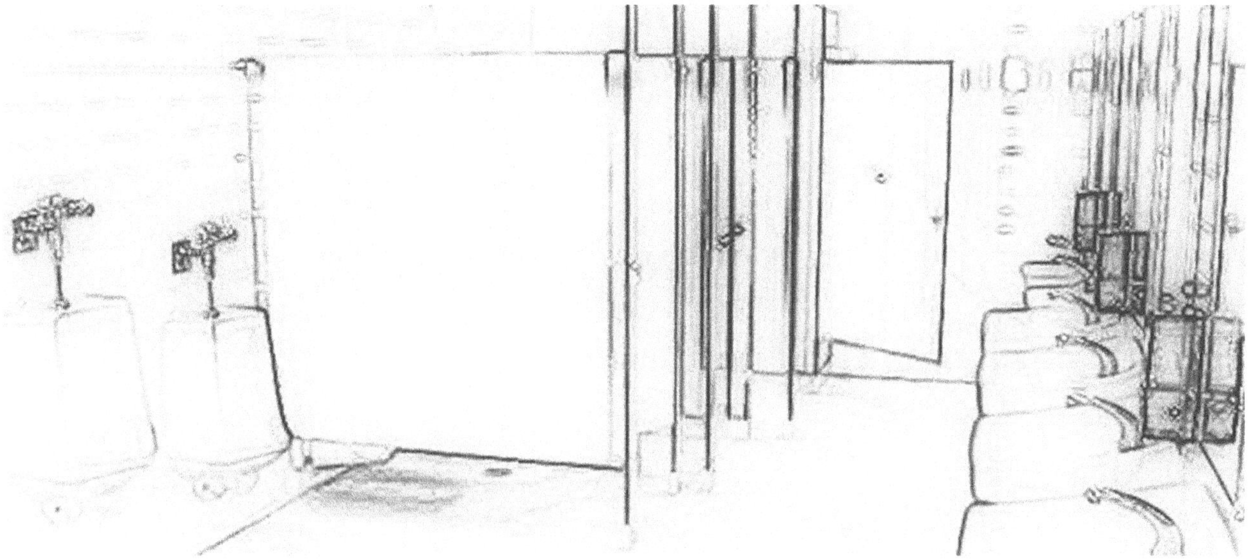

BATHROOMS

Residential Bathrooms

The residential bathroom should be conceived as a series of the following related but independent activity centers.

Head and Hand Grooming Center. This activity center typically is oriented around a washbasin and mirror, where the user performs one or more of the following tasks:

1. Washing hands and face
2. Brushing teeth
3. Washing, drying, brushing, and combing hair
4. Shaving
5. Applying cosmetics
6. Manicuring
7. Performing minor first aid

Bathing Center. This activity center typically involves either partial or full immersion of the body in water, either in a tub or in a shower, and must provide access to soap and other bath items, washcloths, and towels.

Toilet Center. This activity center typically involves use of a commode and, in some cases, a bidet, and it must provide access to toilet tissue.

Storage Center or Centers. The typical residential bathroom should provide storage for articles used in each of the above activity centers plus bathroom housekeeping items, including at least the following:

1. Washcloths, hand towels, and bath towels
2. Personal grooming equipment and cosmetics
3. Toilet tissue and feminine hygiene supplies
4. Medicine and first-aid supplies
5. Housekeeping supplies and equipment

The selection and arrangement of bathroom fixtures and cabinetry should be based on the premise that the facility will be used by a variety of people, including adults, children, and elderly and handicapped persons, and that in some cases more than one person may use the facility simultaneously. It is especially important to consider bathroom layout from the standpoint of the dynamic movement envelopes of people as they utilize various bathroom features. Avoid the temptation to crowd cabinets and fixtures merely to fit a preconceived space.

The Half Bath. The half bath, or "powder room," does not contain a full body wash (such as a shower or tub). Although it obviously requires less space than other types of bathrooms, care must be taken not to make it so small and crowded that the user is inconvenienced, either while utilizing the two primary fixtures or when entering, exiting, or manipulating the door.

As long as there are other major bathroom facilities within the residence or office, many arrangements are possible. Figure 18.1 illustrates two arrangements and shows the critical clearance dimensions that must be observed. A common design error occurs when it is assumed because the room is smaller, smaller fixtures can be located closer together. Remember that the size of the user does not change just because the fixtures are smaller.

Although one can dispense with some of the normal storage elements of the primary bathroom, the half bath should always include an ample towel-rack fixture, an appropriately accessible toilet-tissue dispenser, a mirror, and space for a wastebasket.

Powder rooms are often located near the front part of a house or apartment so that they are handy for guests. Do not neglect consideration of the acoustic problem; it is not only annoying but also extremely embarrassing when the sounds of the toilet flushing can be heard in nearby guest areas. Similarly, provide good ventilation for such facilities, since it is equally embarrassing when odors drift from the facility to nearby guest-occupied areas.

The Three-Quarter Bathroom. The three-quarter bathroom may be used as either an auxiliary or a primary facility. Preferably, it should not be used as a primary facility except in hotels, motels, and minimal-efficiency apartments.

Although many arrangements for the three-quarter bath concept are possible, Fig. 18.2 illustrates the critical dimensions to be considered.

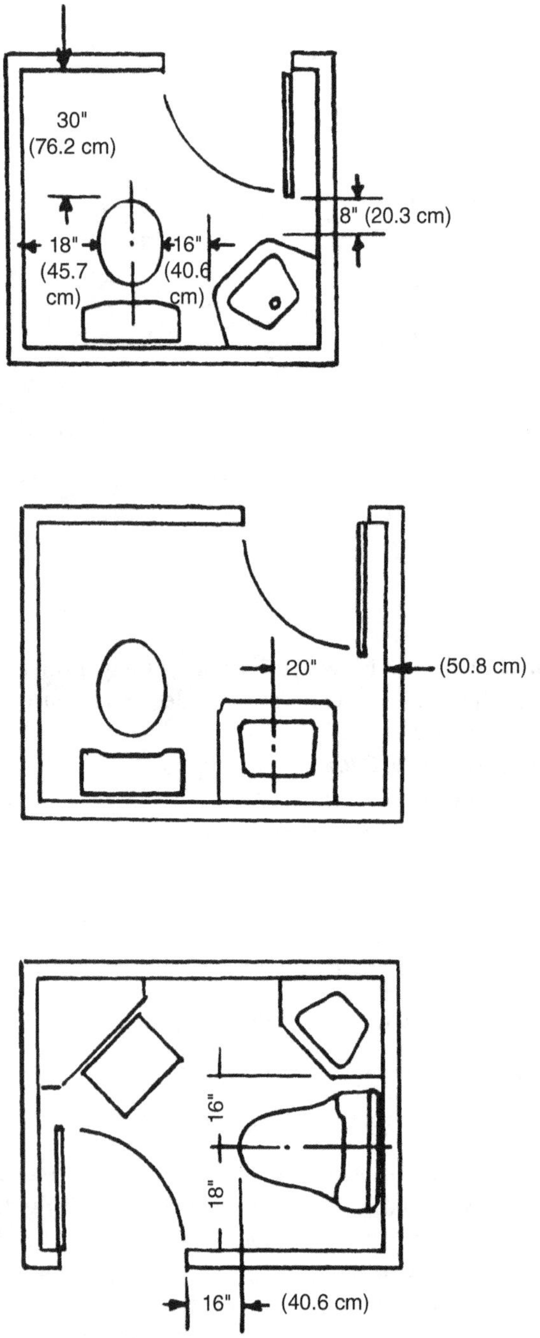

FIGURE 18.1 Examples of half baths.

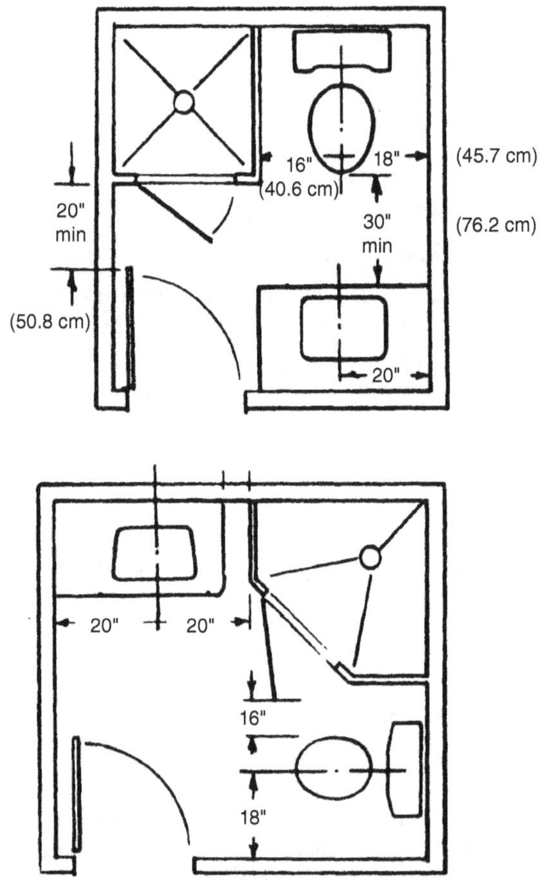

FIGURE 18.2 Dimensions for three-quarter baths.

Whenever the three-quarter bath becomes the primary facility, it should be designed so that it is accessible for persons who are permanently or temporarily confined to a wheelchair. Figure 18.2 provides important dimensional guidelines for making such a facility maximally accessible.

Although most home buyers probably will not want to accept some of the less attractive, special fixtures that have been created for wheelchair users, minimal inconvenience still depends on the ability of the wheelchair user to get as close to the fixtures as possible; thus one must be particularly careful not to design cabinets under which the chair cannot pass. (See Fig. 18.3.) Handrails can be designed so that they are not too unattractive, and they can be helpful even to those who are not handicapped.

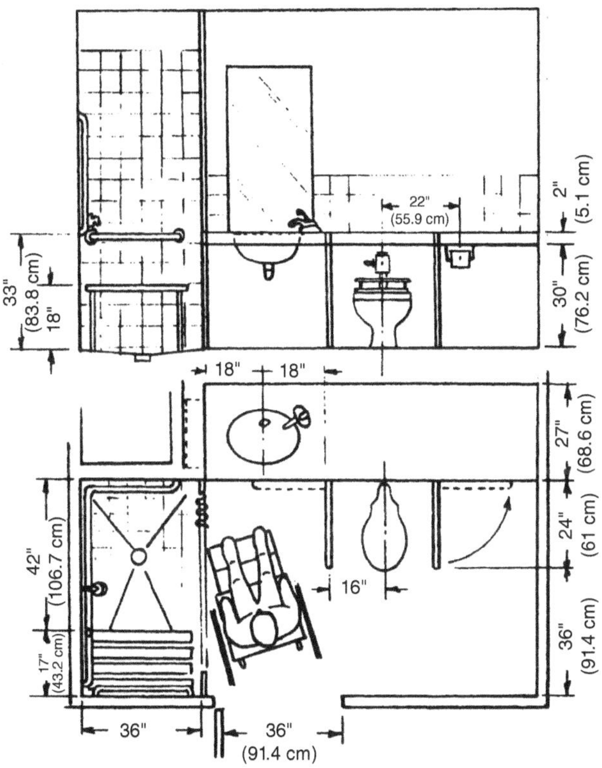

Figure 18.3 Dimension required for a wheelchair user.

The Full Bathroom. The full bathroom contains at least one commode, one washbasin, and a combination tub and shower. Many variations and combinations are acceptable as long as the typical clearance dimensions shown in Figs. 18.4 and 18.5 are maintained.

1. In Fig. 18.4, the architect has provided separation of the commode and bathtub from the washbasin grooming center, thus allowing the use of the facility by two people at the same time.
2. An additional convenience, as shown in Fig. 18.5, is simultaneous use of the commode and bathtub.

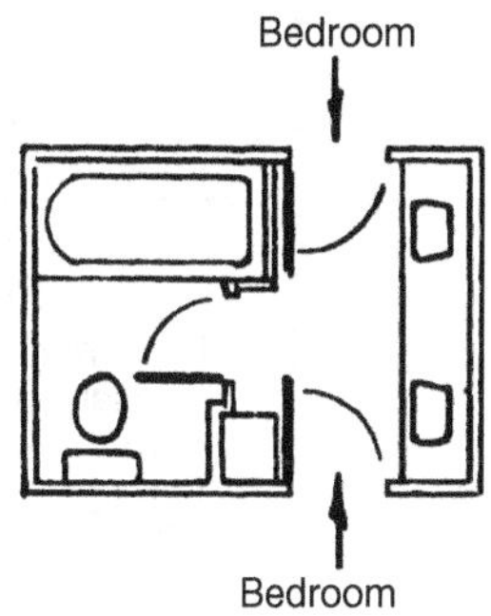

Figure 18.4 Separation of a commode from a bathtub and wash basin.

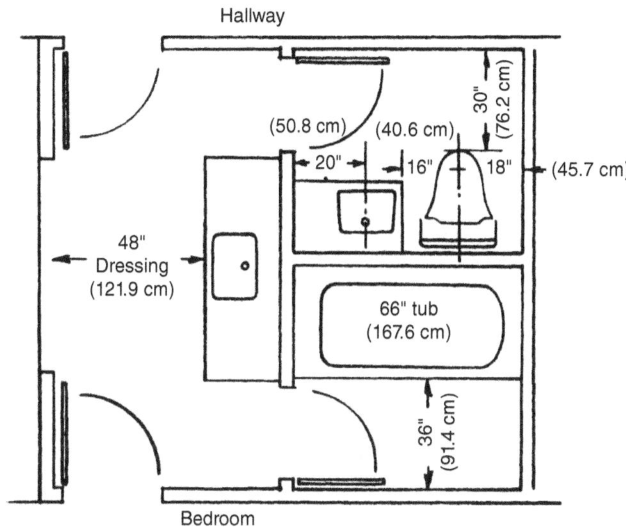

Figure 18.5 Separation of a commode from a bathtub—simultaneous use.

Guidelines for Selection and Installation of Bathing Facility Fixtures

Although the typical tub and shower configuration is generally created using off-the-shelf fixtures, a certain amount of customizing (actual positioning) is left up to the designer. (See Fig. 18.6.) The following guidelines contain suggestions for making things safer and more convenient for the user:

1. Place the shower and bath controls where they can be reached easily from both the showering and the bathing position and also from outside the tub.

2. Provide insets for soap dispensers and for shampoo and ointment containers. Note that the shower control is recessed within the inset so that if the bather should slip, he or she will not accidentally strike the control.

3. Handrails should be provided to aid the bather in getting up and down. The high rail is helpful if the bather loses his or her balance.

4. Select a tub that has a slight lip around the edge; this prevents water from running onto the floor.

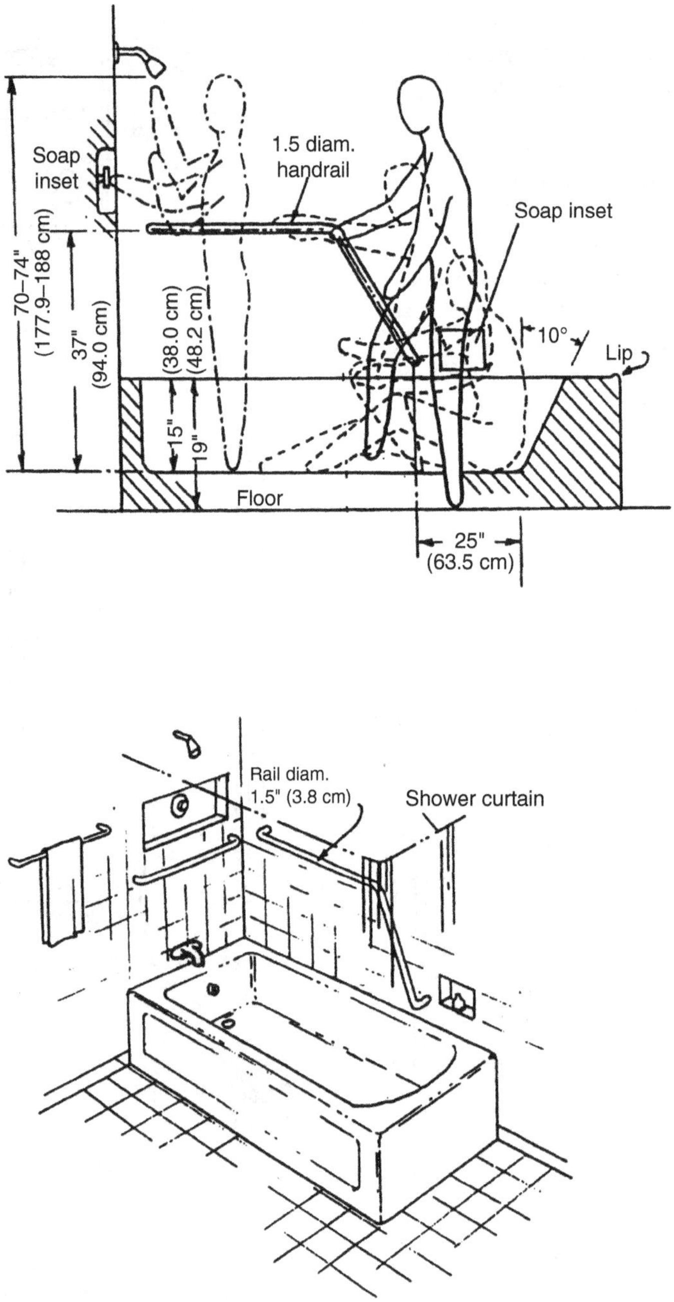

Figure 18.6 Recommended tub-shower configuration.

5. Select a tub whose edge is not so high that it is difficult to step over.

6. Place at least one towel rack close at hand so that the bather can reach a towel without dripping water all over the floor.

7. Use nonslip material for standing surfaces.

Shower Stalls. Shower installations that are not intended for use by wheelchair users should be designed taking into account the considerations and dimensional guidelines illustrated in Fig. 18.7.

An idealized shower is illustrated in Fig. 18.7. Note the following:

1. A seat is included so that the user can sit down to scrub his or her feet. The controls are accessible from outside as well as from inside (and are mounted on a module to minimize injuries).

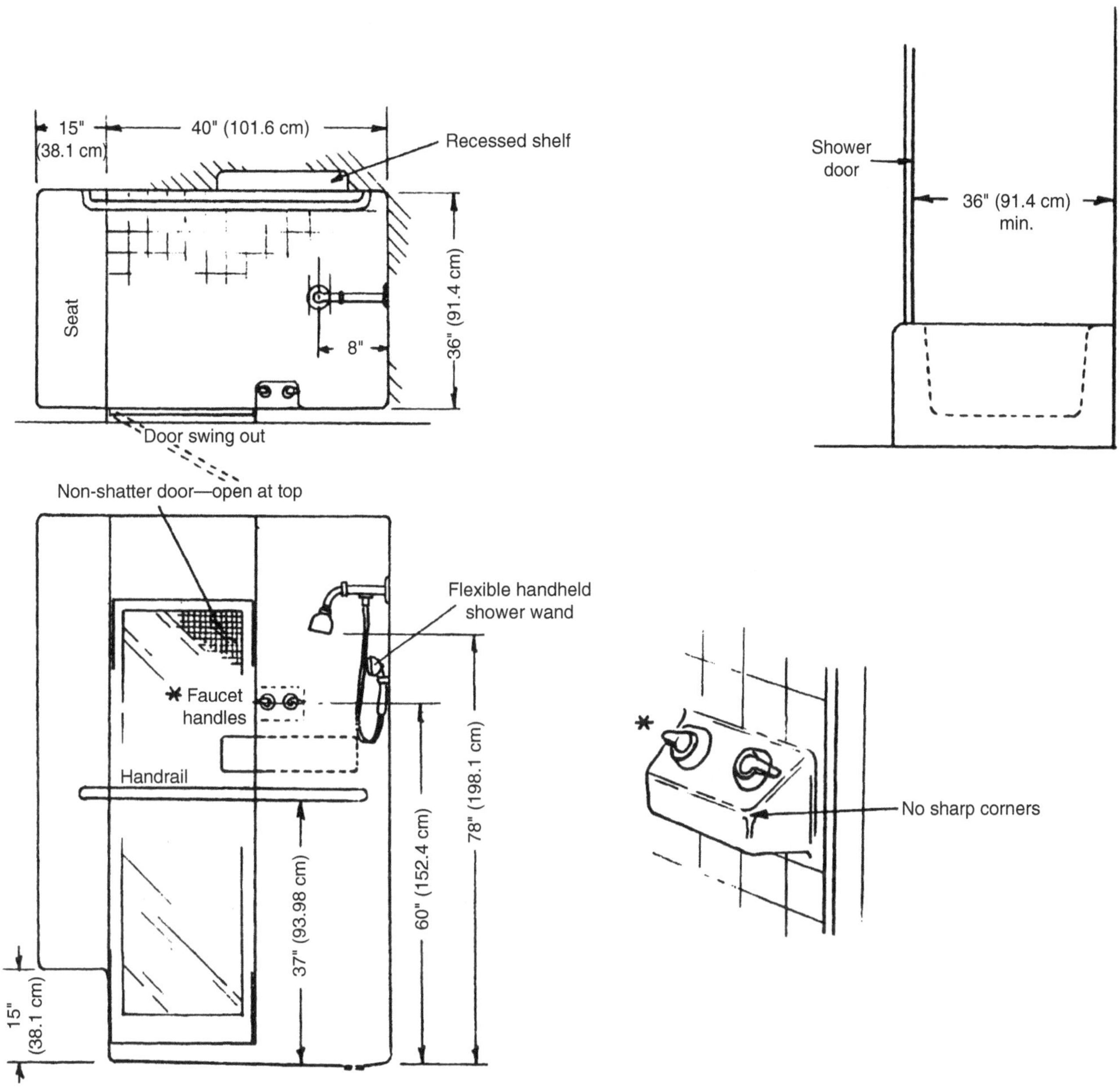

Figure 18.7 Shower dimensions.

2. A portable shower head as well as the typical fixed shower head is provided.

3. A handrail is provided.

4. A recessed toiletry and soap shelf is provided.

5. The drain is at the far end of the stall so that the user does not have to stand on it.

6. A shower partition and door should not be used with a tub unless the minimum interior clearance is at least 36 in. (91 mm).

Washbasin and Fixture Selection. The desire to create unusual aesthetic effects in bathrooms has led to fixture designs that are sometimes poorly engineered from the human factors point of view. A washbasin can, however, be both attractive and convenient and easy to use if one is careful to look for certain features that have a significant impact on the utility of the basin or its hardware, i.e., faucets and control devices. The following guidelines are suggested:

1. Basin Shape and Faucet Configuration: The basin shape and faucet configuration should combine to capture water rather than splatter it all over the user and/or the floor. (See Fig. 18.8)

2. Basin Size and Shape: The basin must be large enough so that users can get their hands and faces inside

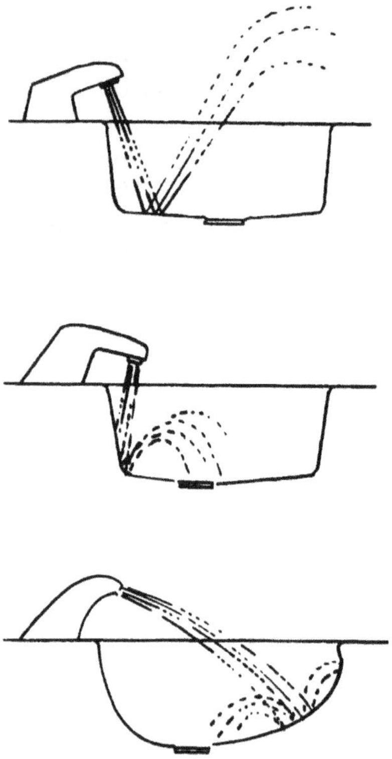

FIGURE 18.8 Basin and faucet should capture water and not splatter.

it without bumping their heads on the faucet or having water run down their arms.

3. Control Handles: The shape of control handles should be compatible with the most efficient gripping and handle manipulation objectives. Oddly shaped handles may be attractive, but they are often very difficult to manipulate. The handles illustrated in Fig. 18.9 show poor versus preferred configurations.

4. Sharp Corners: Sharp edges or corners should be avoided on basins and hardware.

5. Separate Controls versus Single Combined Controls: Separate controls for hot and cold water are preferred in many applications because it is easier to manipulate them more precisely and because there is less confusion about which is which and about how the controls operate.

Hot water should always be on the left, and cold water on the right. The handles should rotate as shown in Fig. 18.10 (note that this is somewhat contrary to the operation of single valve controls).

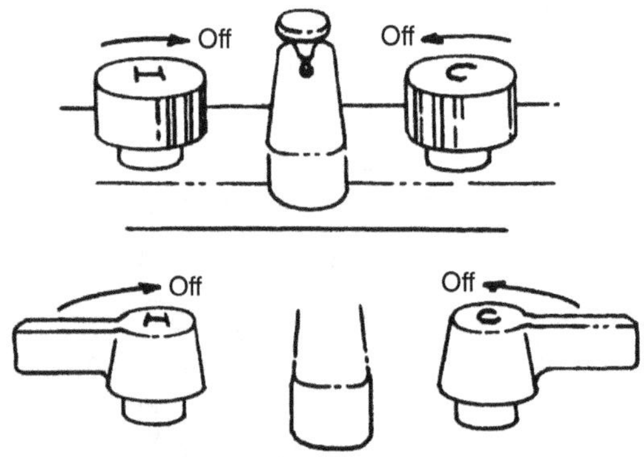

FIGURE 18.10 Typical separate controls.

A single control is often desired because it allows the user to use one hand to operate the control and the other hand to test the water temperature as it is being adjusted.

The common type of control shown in Fig. 18.11, although theoretically very handy, is difficult to operate precisely because the pulling action required to open and close the valve is not the best direction of motion as far as human capabilities are concerned.

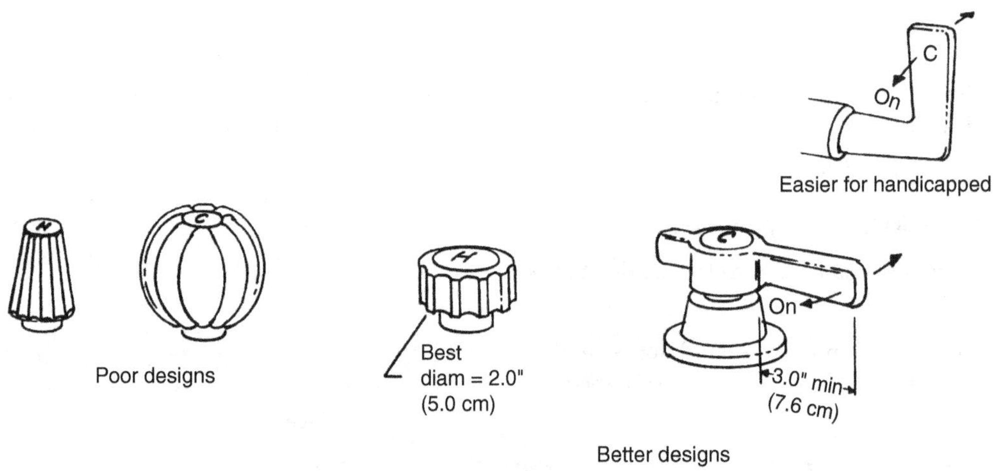

FIGURE 18.9 Faucet handles should fit hand.

Figure 18.11 Awkward single control.

A better single control configuration is illustrated in Fig. 18.12 (a). Not only is the configuration shown more compatible in terms of precise manipulation, but also the handle shape indicates readily whether cold or hot water will probably predominate when the handle is pulled forward to turn the water on.

The handle shown in Fig. 18.12 (b) is also good, and in addition it allows the user to change the position of the spout.

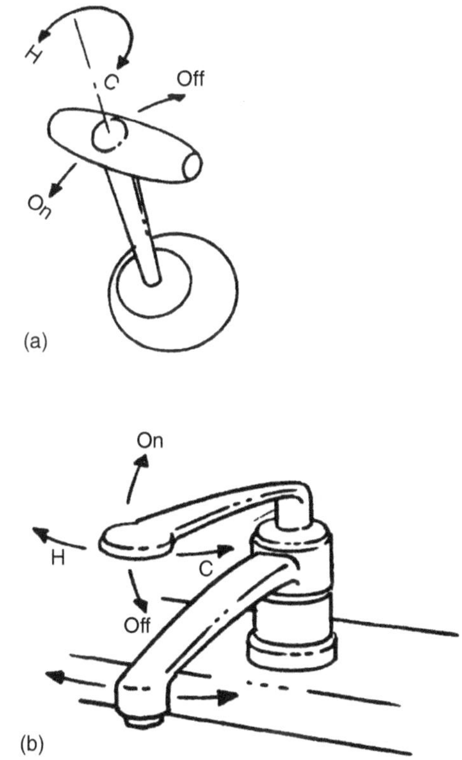

Figure 18.12 **(a) and (b)** Better single control designs.

Studies regarding bathroom usage by Dr. Alexander Kira[4] have produced some interesting suggestions regarding washbasin shape and size. Figure 18.13 illustrates some of the principal features suggested by Kira.

[4]Alexander Kira, *The Bathroom.* The Viking Press, Inc. New York, 1976.

As can be observed, the general shape of the basin shown in Fig. 18.13 allows users to wash their hair without fear of bumping their heads, it places the stream of water where it is most effective for rinsing, and it provides sufficient space so that users' elbows do not drip water all over the floor.

When looking for a suitable basin to propose for a particular architectural client, avoid the tendency to select very small basins just because they are cheaper or because insufficient space has been provided. In order that the necessary space accommodation can be built into the plan, space considerations should not be frozen until an appropriate array of fixtures have been evaluated. Although a fixture as small as 15 in. (38 cm) in diameter can be used successfully for, say, a powder room, nothing smaller than about 18 in. by 20 in. (46 cm by 51 cm) should be considered acceptable for three-quarter or full bathroom concepts.

Although various studies have shown that people of differing stature prefer washbasins at differing heights, the 30-in. (76-cm) height shown in Fig. 18.13 is required so that

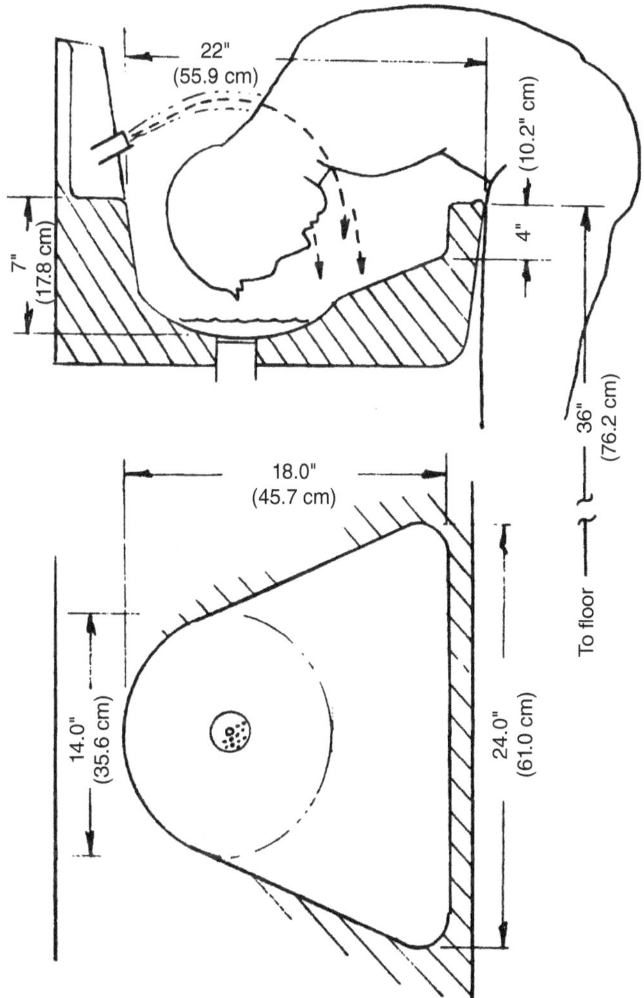

Figure 18.13 Suggested wash basin dimensions.

an adult *can* bend and place his or her head in the basin for hair washing (i.e., without having water run off onto the floor). In addition, this is the maximum height for children to be able to reach into the basin.

Of course, it would be preferable to create adjustable-height basins (which, incidentally, is feasible with the advent of flexible piping). This would allow a sink to be lowered for use by small children and raised so that a wheelchair could fit beneath it or so that it would be more convenient for very tall persons.

Special Industrial Washbasins. In many industrial settings, it is desirable to select a special washbasin design that allows several workers to wash up at the same time or which permits workers whose hands are covered with dirt and grease to turn on the water with their feet. (See Fig. 18.14.)

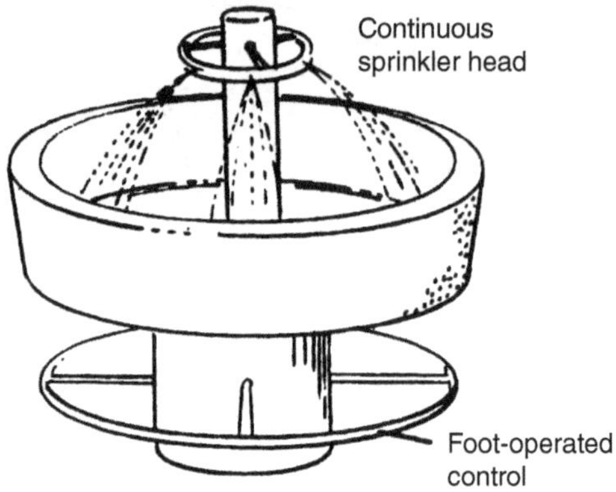

FIGURE 18.14 Special hands-free wash basins.

Special basin and control systems should also be considered for other applications, such as locker rooms. Although the locker-room application may seem unnecessary, the advantages should not be overlooked; i.e., not only can several persons use the basin simultaneously, but also the single configuration takes less time to clean than an equivalent number of separate basins.

Industrial facilities in which potentially hazardous materials are used should also have special safety basins and/or showers permitting workers who have been accidentally sprayed by a chemical to be rushed to the device and to have water sprayed over their bodies and faces as quickly as possible. The designer should check with the manufacturers of special safety equipment and/or plumbing to determine which device is best for a particular application.

Integration of the Head and Hand Grooming Center. A fully satisfactory head and hand grooming center requires more than a washbasin. Figure 18.15 illustrates a number of the special integrating features that should be considered in developing the residential bathroom.

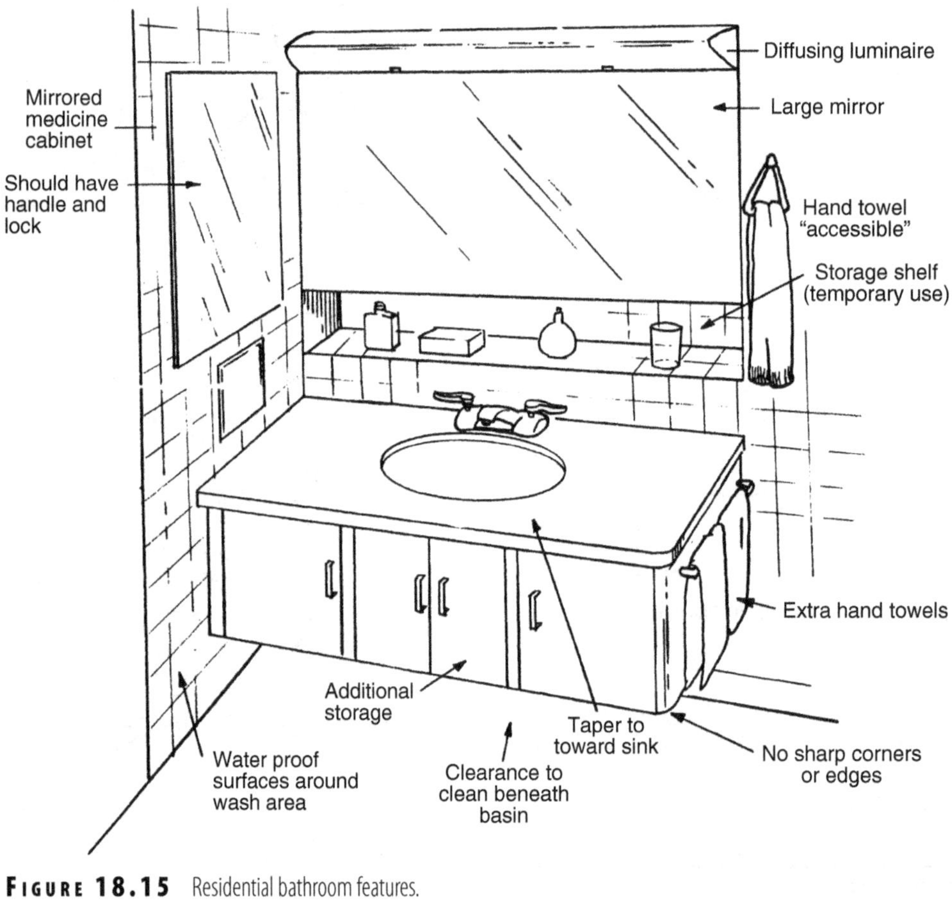

FIGURE 18.15 Residential bathroom features.

Note: The cabinetry and pullman top should have radiused corners so that people will not be injured if they accidentally bump into the cabinet.

Special Bathroom Features for the Handicapped

General. Although in certain cases a bathroom must be given special treatment in order to accommodate a particular handicapped user, in many instances preplanning can make it easier to adapt these features to the bathroom that already exists.

Figure 18.16 illustrates how a pullman type of washbasin configuration can be made attractive and yet usable by persons in wheelchairs. Some of the special basins that have been designed to accommodate the handicapped are as attractive as the standard residential basins.

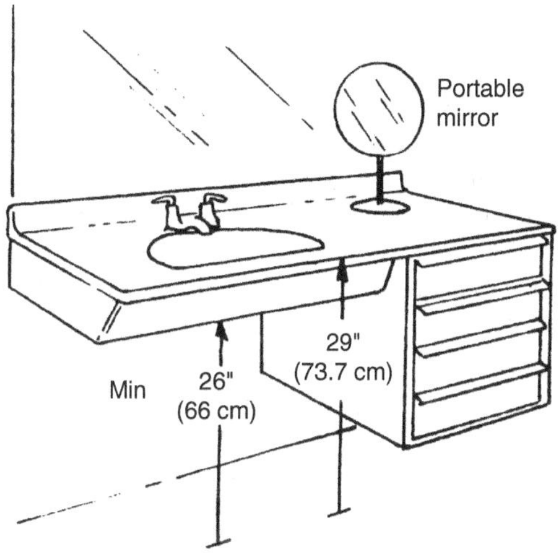

Figure 18.16 Wash basin wheelchair clearances.

Transfer from a wheelchair to a commode is often difficult for the handicapped person. One solution is illustrated in Fig. 18.17. The swinging arms can be moved out of the way while the wheelchair is placed alongside the commode. After the person has transferred to the commode, the handrails can be used by folding them out next to the commode. This sketch also illustrates how an adjustable seat can be set on top of the regular seat to make it easier for the handicapped person to transfer to and from the wheelchair.

Figure 18.18 illustrates a more permanent installation for a person who is not confined to a wheelchair but who has difficulty getting up and down from a commode.

Note: The normal height of the toilet seat above the floor should be about 15 in. (38 cm).

Tub and Shower Fixture Guidelines for Facilities Used by the Orthopedically Handicapped. There obviously must be enough room for the wheelchair user to maneuver about within the bathroom. In addition, a sliding rather than a hinged door should be provided. Minimum dimensional guidelines are noted in Fig. 18.19.

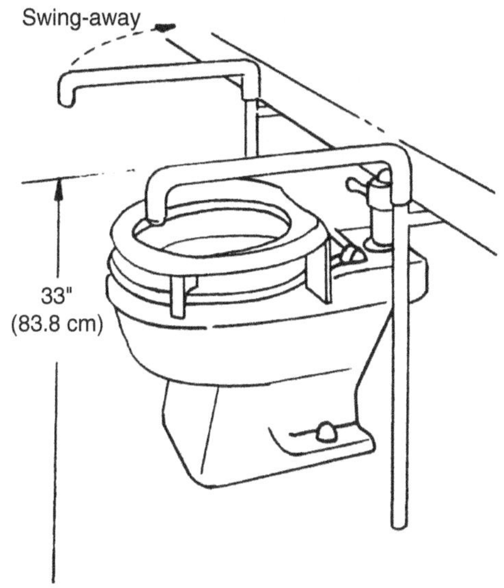

Figure 18.17 Design to accommodate wheelchair users.

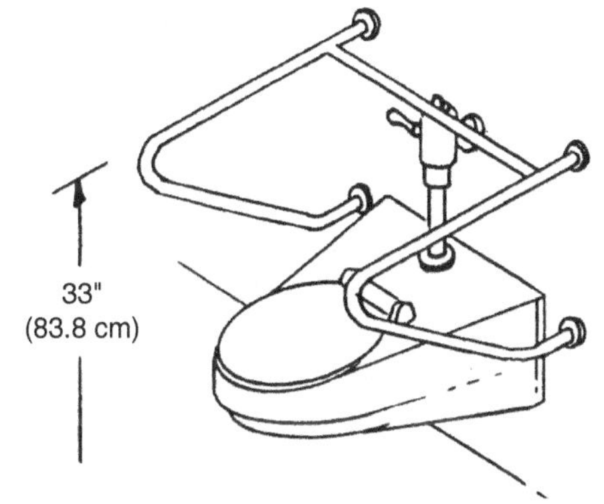

Figure 18.18 Design to accommodate users not confined to a wheelchair.

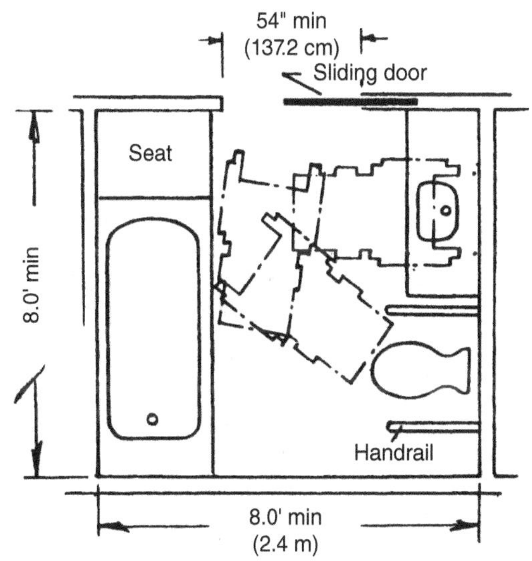

Figure 18.19 Provide room to maneuver wheelchair.

Other special considerations include the following:

1. A seat placed at the end of the tub and designed so that the handicapped person can easily shift back and forth from the wheelchair to the tub. (See Fig. 18.20.)

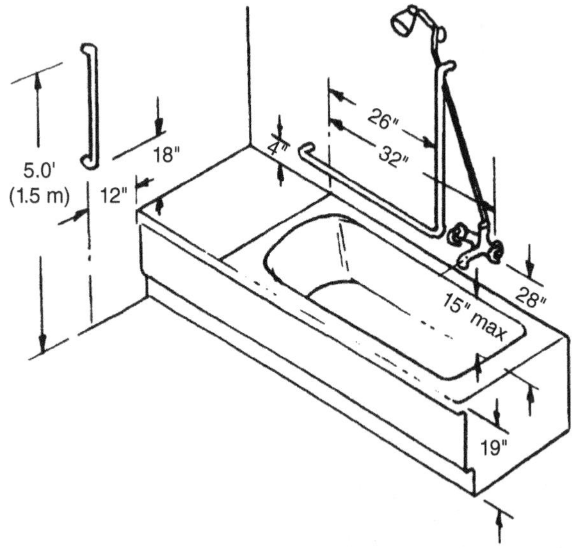

FIGURE 18.20 Considerations for wheelchair users.

2. A handrail positioned so that the user can obtain maximum support. Also look for a tub that has a grip on the outside. This allows certain individuals to support their full body weight as they lower or raise themselves to and from the bottom of the tub. (See Fig. 18.21.)

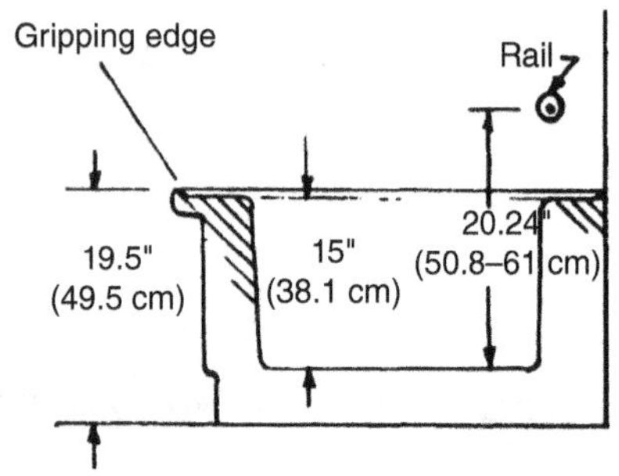

FIGURE 18.21 Handrail and gripping edges.

3. Look for specially designed shower extenders to make manipulation of the shower head as easy as possible while the user is supporting himself or herself from the handrail.

4. Look for faucet handles that are easily pushed, rather than gripped and turned.

Note: Although the above suggestions provide generally useful aids for the wheelchair user, the designer should investigate more fully the range of user problems of a particular client. These suggestions assume that the handicapped users still have considerable use of their arms and hands and sufficient arm strength to move themselves with both arms. Some persons, however, may be unable to use their arms and hands.

PUBLIC RESTROOMS

Public Restroom Guidelines

Except for private or very small offices, separate rest-room facilities should be provided for men and women. Single restrooms jointly used in the smaller office should be provided with an inside lock to ensure the privacy of the occupant.

It is desirable in most public buildings to locate restrooms for men and women fairly close together since most people expect them to be in the same area. This is particularly important in public buildings, where the facilities are used primarily by visitors who will not be familiar with the building layout.

Public restrooms should be located near the entrance to a building. The tradition of "hiding" the restroom because it should not be too conspicuous is contrary to human needs. Restrooms used primarily by the building staff and its occupants should be centrally located on each floor so that users have approximately equal distances to travel to the facility.

Although the number of restrooms obviously must vary according to the basic layout of a particular building, the number of fixtures provided should be determined by the number of people who are expected to use them. Table 18.1 provides general guidelines for determining the number of toilets (or "water closets"). Lavatories should also be available. The lavatory should have hot and cold water, hand soap, and hand drying facilities.

Special Requirements for Public Restrooms

1. Individual stalls should be provided for all toilets. These should have doors that can be secured from the inside.

2. Individual privacy panels should be provided all urinals.

3. The following accessory fixtures should be provided in all restrooms:

a. Paper towel dispensers (rolling cloth towel dispensers are not recommended) and/or hot-air hand dryers and soap dispensers

b. Sanitary napkin dispensers (women's restrooms) and disposal containers

c. Mirrors above all washbasins

d. Shelves above all washbasins

e. Toilet paper dispensers (all toilet stalls)

f. Garment racks or hooks

TABLE 18.1 Bathroom Fixtures in Various Facilities

Table - Recommended number of bathroom fixtures for various facilities (adapted from OSHA regulations in 29 CFR 1910.141 and the Illinois State Plumbing Code)

<table>
<tr><th></th><th colspan="5">Bathroom Fixtures</th></tr>
<tr><th>Facility</th><th>Water Closet (WC)/Toilet</th><th>Urinals (Note 1)</th><th>Lavatory (Lav)</th><th>Drinking Fountains</th><th>Bathtubs, Showers</th></tr>
<tr><td>Workplace (per OSHA, 29 CFR 1910.141)</td><td><table><tr><td>Number of employees</td><td>Water closets</td></tr><tr><td>1 to 15</td><td>1</td></tr><tr><td>16 to 35</td><td>2</td></tr><tr><td>36 to 55</td><td>3</td></tr><tr><td>56 to 80</td><td>4</td></tr><tr><td>81 to 110</td><td>5</td></tr><tr><td>111 to 150</td><td>6</td></tr><tr><td>Over 150</td><td>Additional fixture for every 40 employees</td></tr></table></td><td>Where toilet facilities will not be used by women, urinals may be provided instead of water closets, except that the number of water closets in such cases must not be reduced to less than 2/3 of the minimum specified.</td><td><table><tr><td>Number of employees</td><td>Lavs</td></tr><tr><td>1 to 15</td><td>1</td></tr><tr><td>16 to 35</td><td>2</td></tr><tr><td>36 to 60</td><td>3</td></tr><tr><td>Over 60</td><td>Additional fixture for every 45 employees</td></tr></table></td><td>1 per 75 employees</td><td>None</td></tr>
<tr><td>Dormitories</td><td>1 per 10 males plus 1 for each additional 25 males over 10
1 per 8 females plus 1 for each additional 20 females over 8</td><td>1 per 25 males.
Over 150, add 1 urinal for every additional 50 males.
Over 400, add 1 urinal for every additional 200 males.</td><td>1 per 12 people.
Over12, add 1 lav for every additional 20 males, and 1 lav for every additional 15 females</td><td>1 per 75 persons</td><td>1 per 8 persons (for females, add 1 bathtub for every 30 females and add 1 bathtub for every 50 for over 150 females)</td></tr>
<tr><td>Assembly Places: Sports Arenas, Stadiums, Convention Halls, Airports</td><td><table><tr><td>Num of people</td><td>Num WCs for males</td><td>Num WCs for females</td></tr><tr><td>1–100</td><td>1</td><td>2</td></tr><tr><td>101–150</td><td>2</td><td>3</td></tr><tr><td>151–200</td><td>2</td><td>4</td></tr><tr><td>201–300</td><td>3</td><td>5</td></tr><tr><td>301–400</td><td>3</td><td>6</td></tr><tr><td>401–500</td><td>4</td><td>7</td></tr><tr><td>501–650</td><td>4</td><td>8</td></tr><tr><td>651–800</td><td>4</td><td>9</td></tr><tr><td>Over 800</td><td>1 additional WC for every 700 males</td><td>1 additional WC for every 200 females</td></tr></table></td><td><table><tr><td>Num of males</td><td>Num of urinals</td></tr><tr><td>1–100</td><td>1</td></tr><tr><td>101–200</td><td>2</td></tr><tr><td>201–400</td><td>3</td></tr><tr><td>401–600</td><td>4</td></tr><tr><td>Over 600</td><td>1 additional urinal for every 250 males</td></tr></table></td><td><table><tr><td>Num of people</td><td>Num of lavs</td></tr><tr><td>1–200</td><td>1</td></tr><tr><td>201–400</td><td>2</td></tr><tr><td>401–750</td><td>3</td></tr><tr><td>Over 750</td><td>1 additional lav for every 400 people</td></tr></table></td><td>1 per each set of male and female public restrooms</td><td>None</td></tr>
</table>

(Continued)

TABLE 18.1 Bathroom Fixtures in Various Facilities (*Continued*)

Table - Recommended number of bathroom fixtures for various facilities (adapted from OSHA regulations in 29 CFR 1910.141 and the Illinois State Plumbing Code)

Facility	Bathroom Fixtures: Water Closet (WC)/Toilet			Urinals (Note 1)		Lavatory (Lav)		Drinking Fountains		Bathtubs, Showers
	Num of people	Num WCs for males	Num WCs for females	Num of males	Num of urinals	Num of people	Num of lavs	Num of people	Num of drink founts.	
Assembly Places: Theaters, Auditoriums, Other Facilities for Spectator Events	1–100	1	2	1–100	1	1–200	1	1–100	1	None
	101–150	2	3	101–200	2	201–400	2	101–500	2	
	151–200	2	4	201–400	3	401–750	3	501–1000	3	
	201–300	3	5	401–600	4	Over 750	1 additional lav for every 400 people	1001–5000	1 additional fount for every 500 people	
	301–400	3	6	Over 600	1 additional urinal for every 300 males			Over 5000	1 additional fount for every 1000 people	
	401–500	4	7							
	501–650	4	8							
	651–800	4	9							
	Over 800	1 additional WC for every 400 males	1 additional WC for every 170 females							
Malls, Stores, Mercantile Units	Num of people	Num WCs for males	Num WCs for females	Num of males	Num of urinals	Num of people	Num of lavs	Num of people	Num of drink founts.	None
	1–100	1	2	51–200	1	1–200	1	1–100	1	
	101–150	2	3	201–400	2	201–400	2	101–500	2	
	151–200	2	4	401–600	3	401–750	3	501–1000	3	
	201–300	3	5	601–800	4	Over 750	1 additional lav for every 350 people	1001–5000	1 additional fount for every 500 people	
	301–400	3	6	Over 800	1 additional urinal for every 300 males			Over 5000	1 additional fount for every 1000 people	
	401–500	4	7							
	501–650	4	8							
	651–800	4	9							
	Over 800	1 additional WC for every 500 males	1 additional WC for every 175 females							

<table>
<tr><th></th><th></th><th></th><th></th><th></th><th></th></tr>
<tr><td>Restaurants, Pubs, Lounges (Places Serving Food or Beverages to be Consumed on the Premises)</td><td><table><tr><td>Num of people</td><td>Num WCs for males</td><td>Num WCs for females</td></tr><tr><td>1–50</td><td>1</td><td>1</td></tr><tr><td>51–100</td><td>1</td><td>2</td></tr><tr><td>101–150</td><td>2</td><td>3</td></tr><tr><td>151–300</td><td>2</td><td>4</td></tr><tr><td>Over 300</td><td>1 additional WC for every 200 males</td><td>1 additional WC for every 100 females</td></tr></table></td><td>1 urinal for every 150 males.</td><td><table><tr><td>Num of people</td><td>Num of lavs</td></tr><tr><td>1–100</td><td>1</td></tr><tr><td>101–200</td><td>2</td></tr><tr><td>201–400</td><td>3</td></tr><tr><td>Over 400</td><td>1 additional lav for every 200 people</td></tr></table></td><td>None</td><td>None</td></tr>
<tr><td>Worship Paces and Funeral Homes</td><td>1 WC for every 250 males
1 WC for every 125 females</td><td>1 urinal for every 250 males</td><td>1 lav for every 125 persons</td><td>None</td><td>None</td></tr>
<tr><td>Elementary School - Student Use</td><td><table><tr><td>Num of students</td><td>Num of WC</td></tr><tr><td>1–20</td><td>1</td></tr><tr><td>21–50</td><td>2</td></tr><tr><td>Over 50</td><td>1 additional WC for every 50 students</td></tr></table></td><td>See Note 1</td><td><table><tr><td>Num of students</td><td>Num of lavs</td></tr><tr><td>1–25</td><td>1</td></tr><tr><td>26–50</td><td>2</td></tr><tr><td>Over 50</td><td>1 additional lav per restroom for every 50 students</td></tr></table></td><td>1 for every 75 students</td><td>None</td></tr>
<tr><td>Secondary Schools, Colleges, Universities</td><td>1 WC for every 40 males
1 WC for every 20 females</td><td>1 urinal for every 40 males</td><td>1 lav for every 40 students</td><td>1 for every 75 students</td><td>None</td></tr>
<tr><td>Day Care, Nursery, and Preschool Centers (excludes infants 0-15 months)</td><td><table><tr><td>Num of students</td><td>Num of WC</td></tr><tr><td>1–10</td><td>1</td></tr><tr><td>11–175</td><td>1 additional WC for every 25 students</td></tr><tr><td>Over 175</td><td>1 additional WC for every 50 students</td></tr></table></td><td>See Note 1</td><td><table><tr><td>Num of students</td><td>Num of lav</td></tr><tr><td>1–10</td><td>1</td></tr><tr><td>11–175</td><td>1 additional lav for every 25 students</td></tr><tr><td>Over 175</td><td>1 additional lav for every 50 students</td></tr></table></td><td>1 for every 75 students</td><td>None</td></tr>
</table>

(*Continued*)

TABLE 18.1 Bathroom Fixtures in Various Facilities (*Continued*)

Table - Recommended number of bathroom fixtures for various facilities (adapted from OSHA regulations in 29 CFR 1910.141 and the Illinois State Plumbing Code)

	Bathroom Fixtures				
Facility	**Water Closet (WC)/Toilet**	**Urinals (Note 1)**	**Lavatory (Lav)**	**Drinking Fountains**	**Bathtubs, Showers**
Hospital Individual Room	1 per room	None	1 per room	None	1 per room
Hospital Ward Room	1 per 8 patients	None	1 per 8 patients	1 for every 75 patients	1 for every 8 patients
Institutional Other than Hospital or Penal	1 WC for every 25 males 1 WC for every 20 females	1 urinal for every 50 males	1 lav for every 10 persons	1 for every 75 persons	1 for every 8 persons
Penal Institution For Inmate Use	1 per cell 1 for every 8 inmates in a dormitory	None	1 per cell 1 for every 8 inmates in a dormitory	1 for every 75 inmates	1 for every 8 inmates

Note 1: Urinals may be substituted for water closets, not to exceed ½ of the required total number of water closets.

4. All large restrooms should have a special maintenance area that includes a special sink for rinsing mops and storage space for cleaning implements and supplies and for paper supplies.

5. Adequate illumination should be provided, including luminaires above each mirror for grooming and several general luminaires for overall mobility and maintenance illumination.

6. Adequate air conditioning and/or natural ventilation should be provided, depending upon the climate and/or seasonal temperature conditions.

7. All restrooms should be laid out in such a way that there is an intervening antechamber and/or privacy screen so that persons using the facility cannot be seen from outside as someone enters or leaves the restroom. Care must be taken to make this separating space large enough so that people who are coming and going can pass without interference. It is recommended that separate entry and exit doors be provided.

8. There should always be at least one toilet stall, one urinal, and one washbasin designed especially for handicapped persons (see the guidelines for the removal of architectural barriers).

9. Rest-room signs should be provided whenever the public may use the restrooms. *(Restrooms in dormitories, aboard ship, etc., should have additional facilities such as showers, dressing areas with seats, storage facilities for personal belongings, clothing hooks, and racks for towels, the latter being located for maximum convenience to washbasins and showers.)*

Separate Restrooms in Public Buildings. Use terms such as "Boys" and "Girls," or "Men" and "Women" and/or international symbols, such as those shown in Fig. 18.22. If facilities for the handicapped are provided within the men's or the women's restrooms, also provide the international symbol shown in the figure. Both printed and symbolic information should also be made tactually identifiable; i.e., the print or symbol should be raised, and braille code symbols should be placed below each letter or symbol so that partially sighted and blind persons can also identify the restroom door.

The Single Lavatory. When only a single lavatory that may be used by the public is provided, use the words "lavatory," "toilet," or "water closet," plus the international symbol denoting that it is for use by the handicapped. If the lavatory may be used by blind persons, provide raised lettering and/or symbols. Avoid the use of artistic or humorous graphics. Although these may seem aesthetically pleasing or amusing to the designer, they are often merely confusing to the person who needs to find and use the facility.

Occupied and Unoccupied Signs. Such signs should not be considered a substitution for the primary restroom or lavatory sign. One or the other of the signs should be illuminated when that particular condition prevails. Such signs should be legible from the position of the observer who is farthest away, e.g., an airliner passenger seated in the most remote position from the lavatory. The primary lavatory sign, like the OCCUPIED or UNOCCUPIED sign in an airplane, rail car, or bus, should always be visible by the observer who is farthest away. Do not use a primary sign that is not illuminated either internally or by reflected light.

FIGURE 18.22 Provide international symbols.

Restroom Dimensional Guidelines

Figure 18.23 illustrates some of the key fixture spacing considerations in planning any public restroom facility.

1. There should be at least 36 in. (91 cm) of clearance behind any washbasin or urinal fixture so that an

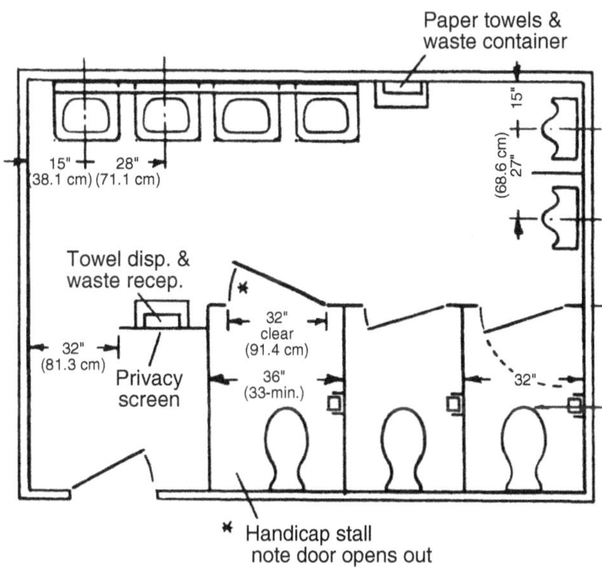

FIGURE 18.23 Public restroom clearance guidelines.

individual can pass other people without disturbing them. An additional 15 in. (38 cm) should be added where it is necessary for a wheelchair to pass. (See Figs. 18.24 and 18.25.)

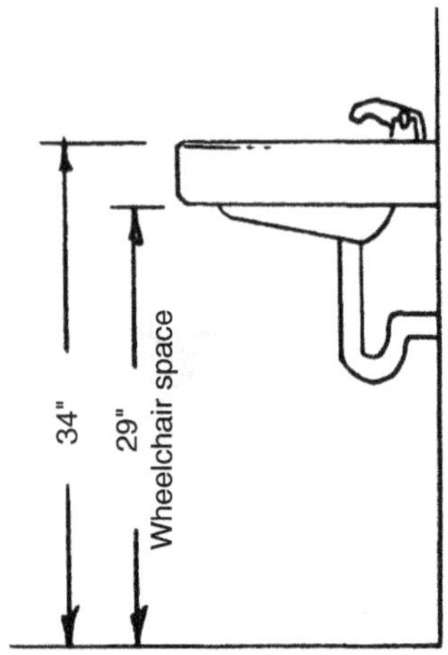

FIGURE 18.24 Sink clearances.

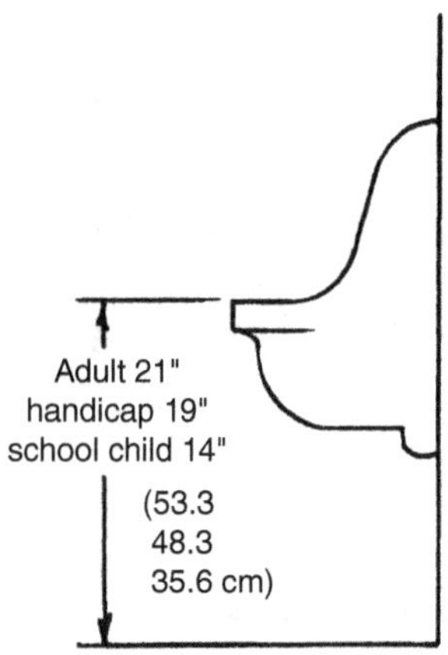

FIGURE 18.25 Urinal clearances.

2. Regular toilet stalls should be at least 32 in. (81 cm) wide, and there should be rigid monitoring of the installation of toilets and partitions to make sure that the fixtures are centered within the stall envelope. Many poor installations cause the toilet user to be crowded against one of the stall partitions.

3. The guidelines shown in Fig. 18.26 pertain to the special requirements of the wheelchair user. Thus the reader should refer to other sections for appropriate washbasin heights for the nonhandicapped user and/or should place other fixtures according to normal architectural practice.

4. Fixtures and partitions for restrooms should be suspended so that it is easy to clean beneath them. (See Fig. 18.27.)

5. For the sake of safety, use a nonskid floor covering. For ease of cleaning, provide a slight slope to a floor drain placed near one corner of the restroom. Use wall material that resists damage and is easy to clean.

LOCKER ROOMS

Locker-Room Guidelines[5]

Figures 18.28 and 18.29 show two examples of locker-room arrangement and critical dimensions. Key considerations should also include the following:

1. Good ventilation in toweling and dressing areas
2. Water control curbs
3. Nonskid floor surfaces throughout
4. Good lighting

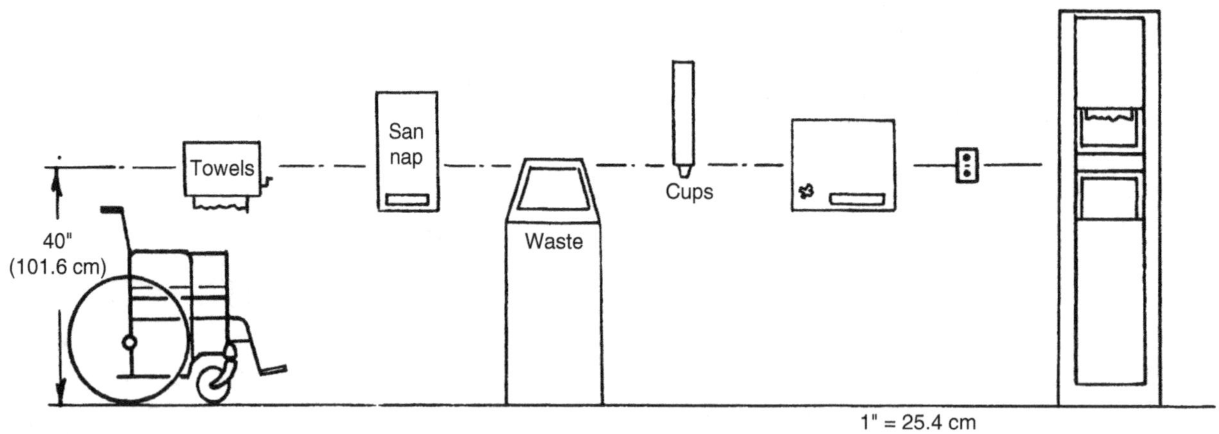

FIGURE 18.26 Wheelchair requirements.

[5]From J. De Chiara and J.H. Callender, *Time-Saver Standards for Building Types*, McGraw-Hill Book Company, New York, 1973.

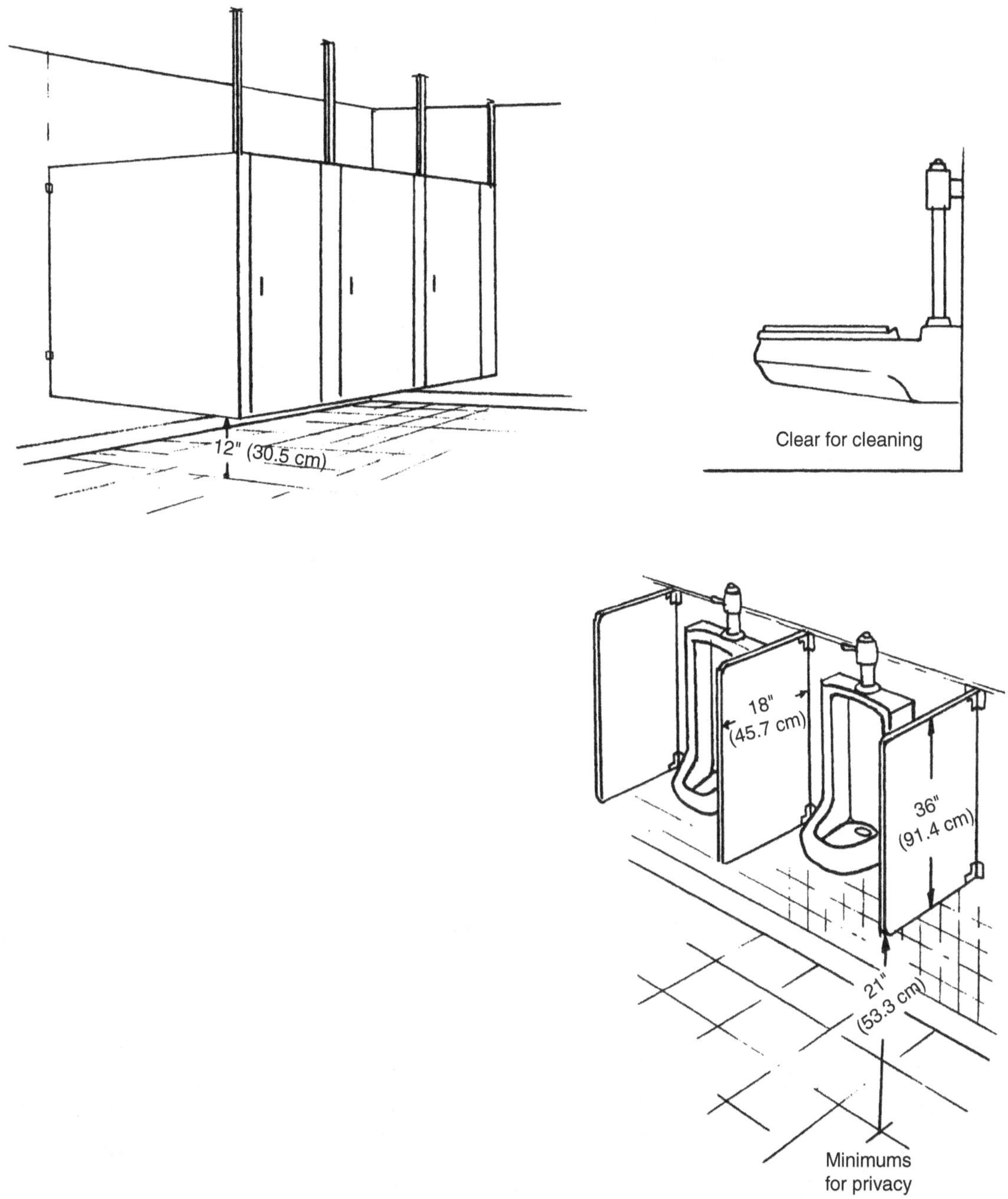

Figure 18.27 Suspended fixtures for cleaning.

5. Ample accessories such as mirrors, clothing and towel racks and hooks, waste receptacles, drinking fountains, and lockable, ventilated lockers

6. A first-aid room directly adjacent to the locker-room facilities

Note: In communal shower rooms, adjustable-height, non-clogging shower heads at least 4 ft (1.2 m) apart should be provided. There should be both individual and master controls.

See Figs.18.30 to 18.33 for examples of physical education facilities.

Special Safety Considerations for Bathroom, Restroom, and Locker-Room Designs

Following are some of the unsafe features most commonly found in bathroom designs:

1. Improper door arrangements. As Fig. 18.34 shows, the process of opening the door may force the user to back into the tub, or a person who is entering the restroom may walk directly into the door being opened by a person who is leaving the restroom.

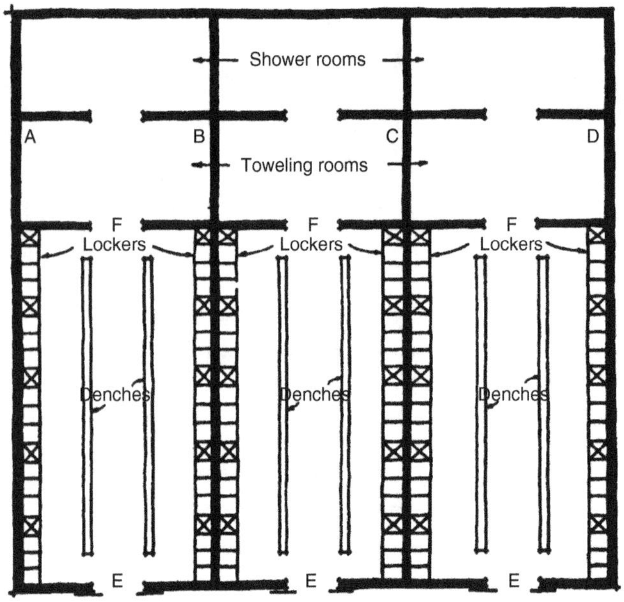

FIGURE 18.28 Locker-room layout.

2. Light switches and electric outlets located where they can be reached while a person is sitting or standing in water.

3. Hot appliances (heaters) located where they are easily backed into or touched inadvertently.

4. Nonstructural towel racks located where they are used as a supporting handrail.

5. Fixtures, cabinets and controls, spouts, and towel racks that have sharp corners or edges that can cause injury if a person falls or bumps against them.

6. Windows located over tubs. This tempts people to stand up in the tub or on the edge of the tub to open and close the window, which typically re-suits in slipping and falling.

7. Bath or shower stall enclosure materials that break easily if someone inadvertently falls against the enclosure or door.

8. Use of slippery floor materials in an effort to provide a shiny, attractive surface.

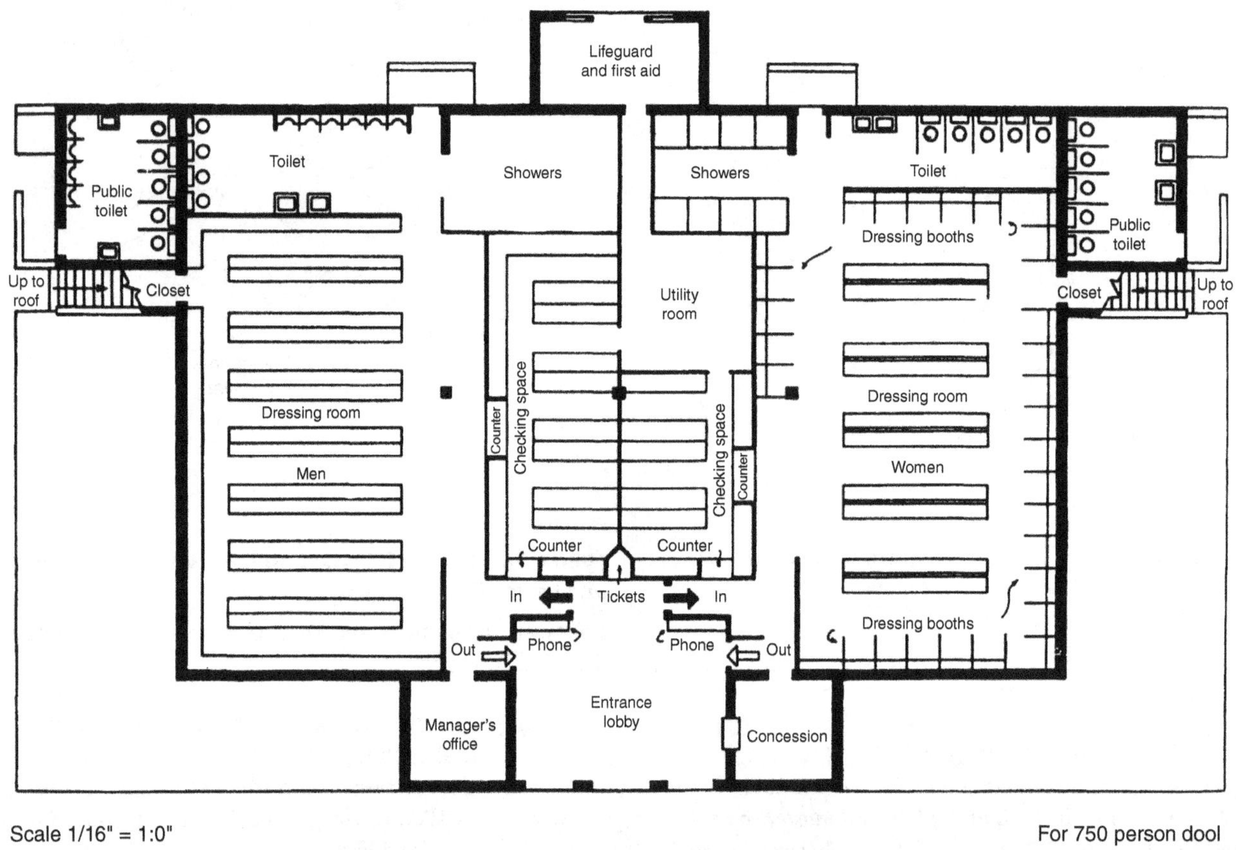

FIGURE 18.29 Plan of typical bath house.

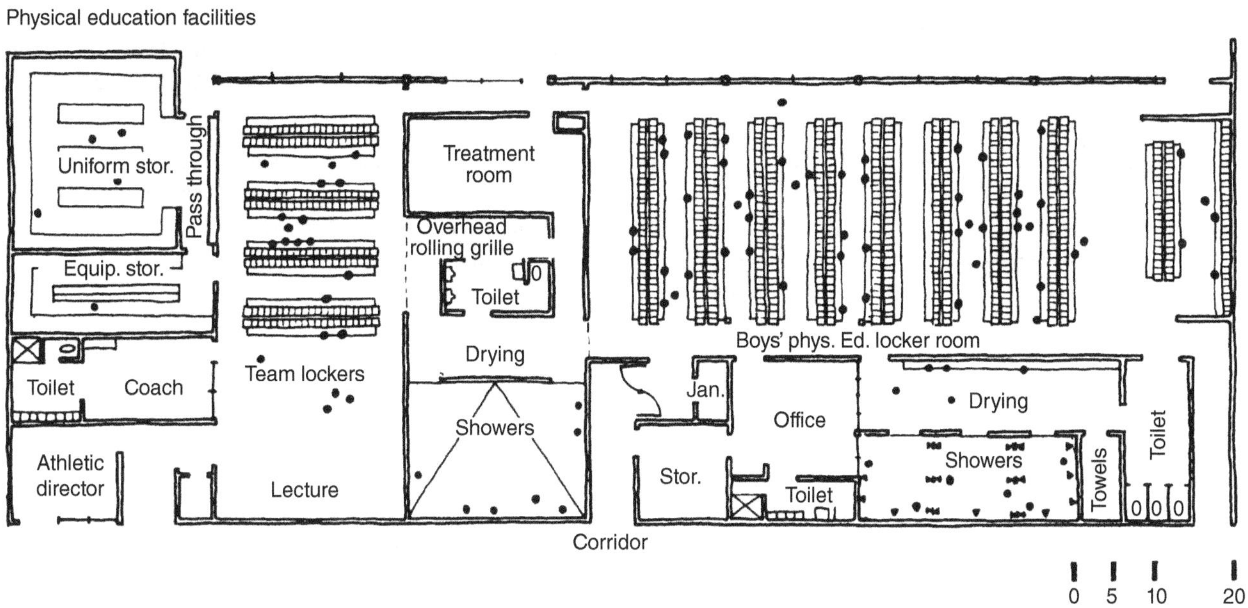

Figure 18.30 Boy's physical education and team locker rooms with dual user showers and toilets.

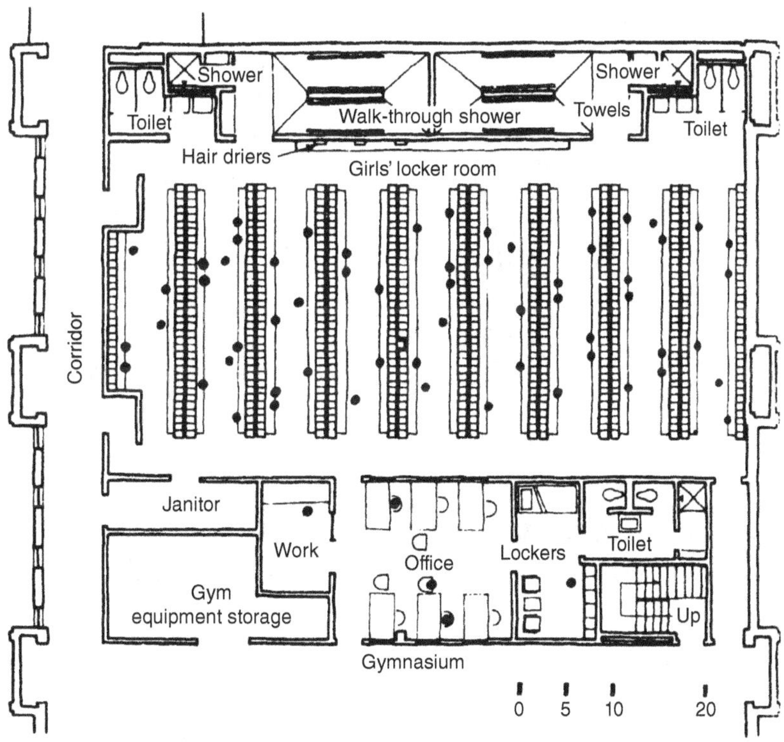

Figure 18.31 High school girl's locker room.

9. Medicine cabinets positioned so that if people leave the door open while bending over the washbasin, they almost invariably bump their heads on the open door.

10. Lack of a convenient shelf, with the result that a person may put an electric appliance down on a wet surface.

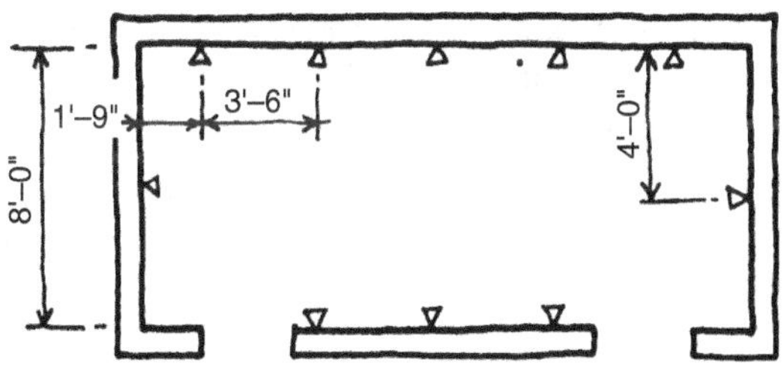

Figure 18.32 Conventional shower (in-wall piping).

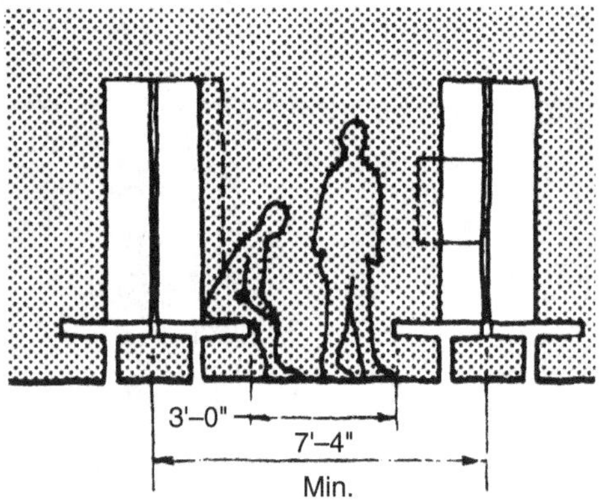

- Combined wider aisle and dressing area
- Seat for each locker bank less congestion
- Seat provides aisle protection from door swing
- Positive through-locker ventilation
- Economical use of floor area

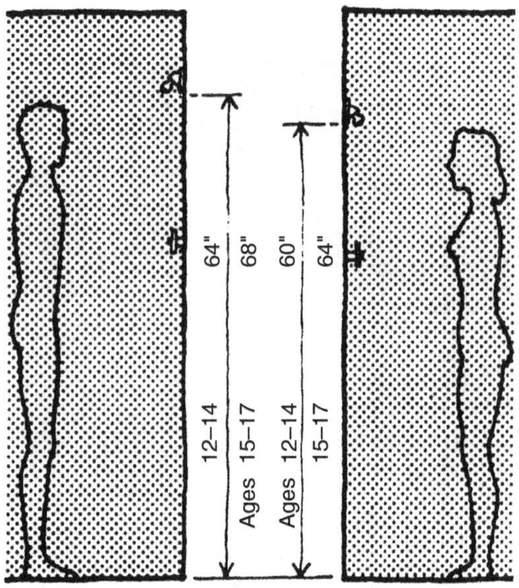

FIGURE 18.33 Recommended shower head height.

Maintenance and Housekeeping Considerations

1. Seamless joints and surfaces are extremely important in the design and installation of cabinets and fixtures in the bathroom since such seams trap mixtures of water, dust, soap, and cosmetic substances.

2. Smooth hardware designs are important; fancy filigree entraps water, soap, and dirt and is extremely hard to clean.

3. Suspended cabinets and fixtures are preferred because floor-mounted features tend to collect dirt at the base; this is hard to remove by the typical methods of sweeping or mopping.

4. Plenty of clearance must be provided around such fixtures as the toilet tank if it is necessary to remove the tank lid in order to adjust, service, or repair tank components.

5. Intersections between the floor and wall and pullman-type sink tops should be radiused for ease of cleaning.

6. Standard practice invariably locates important plumbing connections and pipes within unaccessible walls. Care should be taken during initial planning to provide quick access to the points which may leak or which require frequent replacement of seals or parts.

7. Toilet fixtures should be selected that provide easy access for replacing toilet-seat covers (which often become loose and are difficult to replace because the fasteners are hidden inside the tank). These fasteners invariably rust unless they are made of nonrusting materials, and it becomes impossible to loosen the rusted nuts; then they have to be cut off, which often mars the porcelain.

8. So that they will not stick, doors, windows, and the frames for these should be made of materials that are not affected by the moisture that is typically produced within the bathroom.

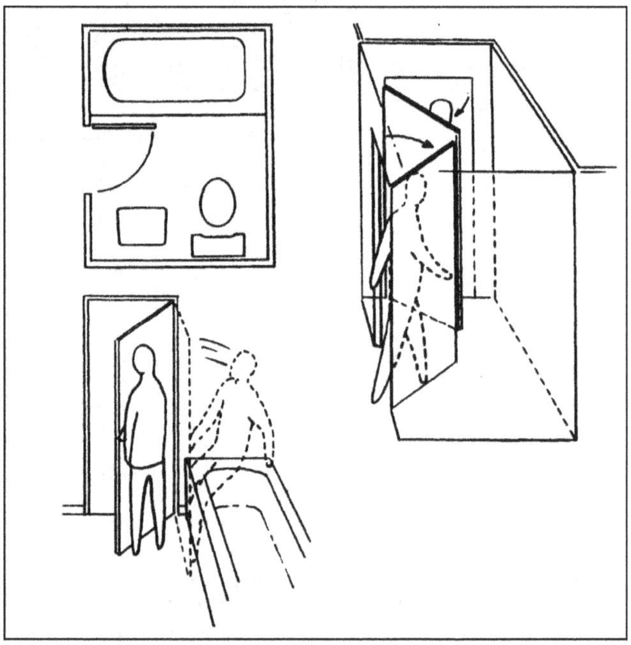

FIGURE 18.34 Improper door arrangements.

CHAPTER 19

BEDROOMS AND BERTHING SUBSYSTEMS

BEDROOMS

Figure 19.1 illustrates the three standard types of bedrooms: (1) the bedroom with a single bed for one occupant, (2) the bedroom with a double bed for two occupants, and (3) the bedroom with twin beds for two occupants.

Space should always be provided for standard furniture elements and for the necessary clearance to move around them for normal use and housekeeping, and space should also be provided for garment storage. Note also the desirability of providing for nightstands.

Since many variations are not only possible but also desirable to satisfy varying client desires, the illustrations do not include room dimensions, nor do they necessarily represent any preferred room shape. For initial planning purposes, the general space guidelines shown in Table 19.1 are provided. However, one should always derive the final floor-area requirements on the basis of the type of furniture currently in vogue.

Bedroom Planning

Bedroom size and shape should be planned around the use of standard furnishings that people can purchase at a local furniture store. (See Fig. 19.2.)

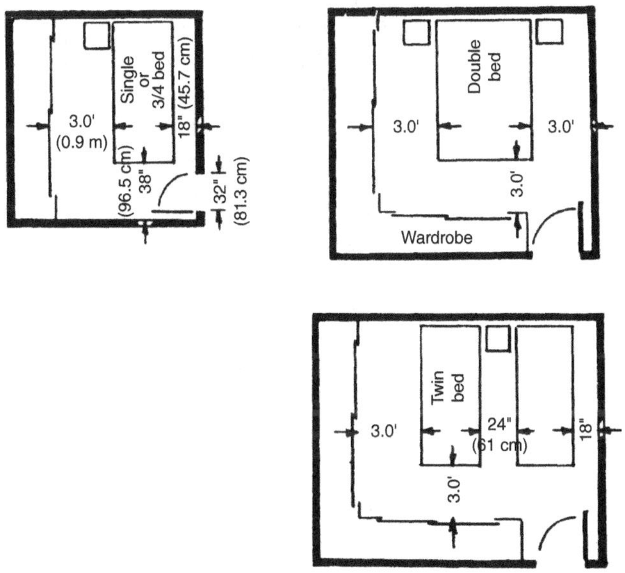

FIGURE 19.1 Three standard types of bedrooms.

TABLE 19.1 Typical Minimum Bedroom Square Footage (Not Counting Closets)

DIMENSIONS	APPLICATION
9 × 10 ft	Single adult; one child nursery (single bed or crib)
10 × 12 ft	Singe adult; one youth (single bed)
12 × 16 ft	Couple (standard double bed)
16 × 16 ft	Couple (twin beds)
16 × 18 ft	Couple (king-size bed)

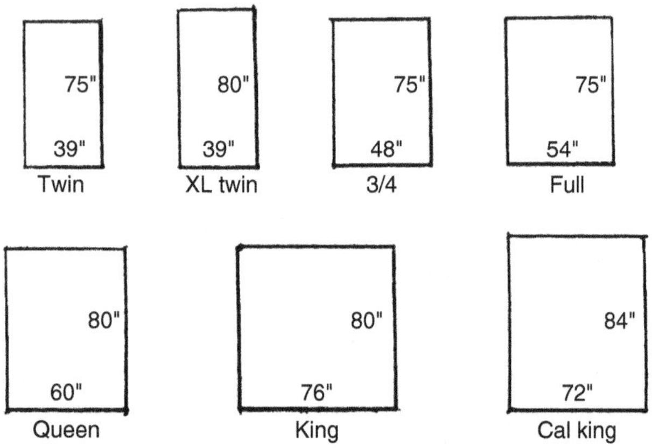

FIGURE 19.2 Typical bed dimensions.

In considering the various furnishings, keep in mind the following arrangement constraints:

1. The head of a bed should be against a clear wall, not against a window or a built-in closet door.

2. Beds should be located in such a way that there are no drafts across the head of the bed.

3. High chests and dressers with mirrors should be backed against a clear wall, not a window.

4. Plenty of clear dressing space must be provided in front of dressers and chests.

5. Remember that furnishings may interfere with window manipulation and curtain operation.

6. Plan double-occupancy rooms and furnishings so that there is an effective separation of the dressing activities of the two occupants—i.e., "his" closet and dresser and "her" closet and dresser—and the associated dressing space required by each person.

Built-in Closets for Bedrooms

Figures 19.3–19.5[6] present general guidelines for providing closet space and various built-in conveniences in bedroom-related closets.

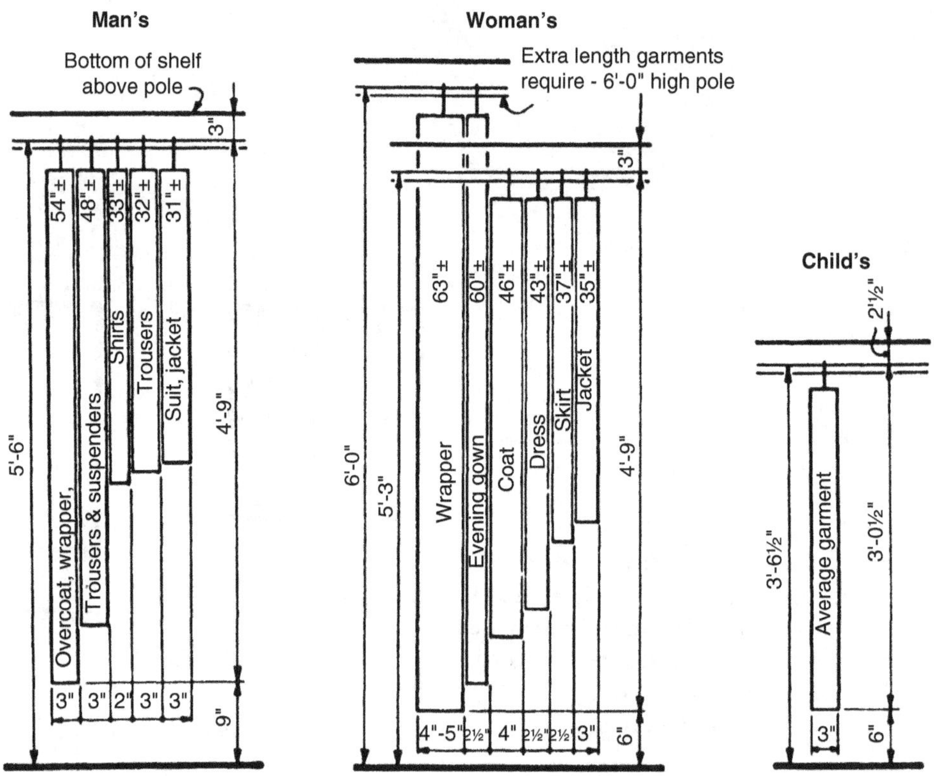

FIGURE 19.3 Sizes of clothes hung in closet.

[6]From J. De Chiara and J.H. Callender, *Time-Saver Standards for Building Types*, McGraw-Hill Book Company, New York, 1973.

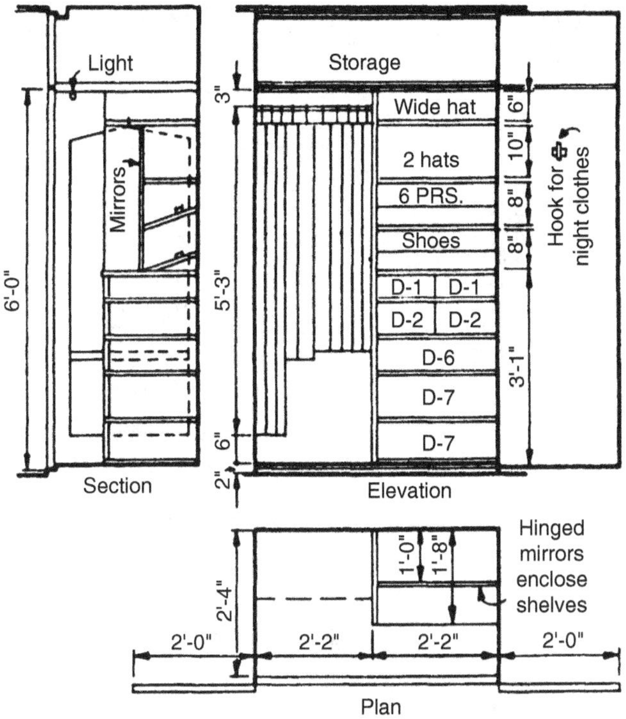

FIGURE 19.4 A four-foot closet combining hanging and shelf space with drawers.

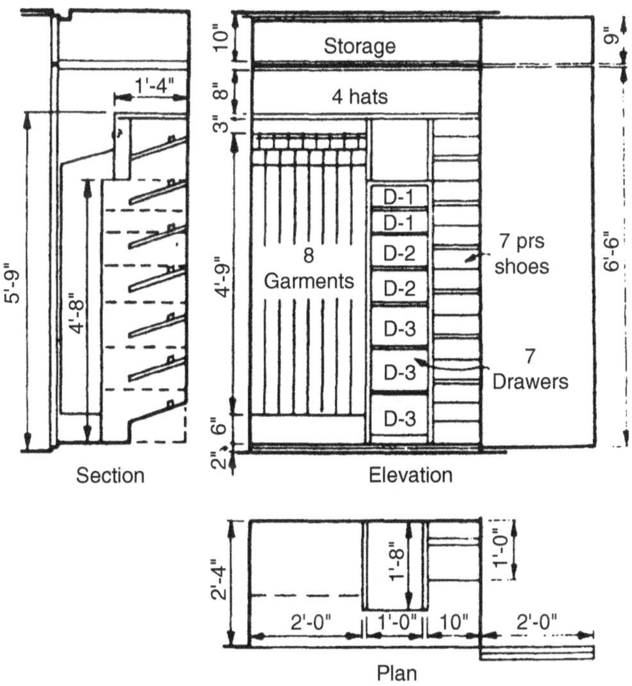

FIGURE 19.5 A four-foot closet with seven drawers.

Note that there are several basic storage categories listed as follows:

1. Storage of hanging garments, both long and short
2. Shelf or drawer storage of garments that normally should be folded and should lie flat
3. Shoe storage
4. Hat storage
5. Miscellaneous storage, which includes storage of ties, grooming items, luggage, etc.

Closets for Children[6]**.** Closet for infants up to about 5 years old should have low hanging pole shelves and drawers permitting habits of care and orderliness to be developed at an early age. (See Fig. 19.6.) Upper part would be used by adults. Note two sets of doors.

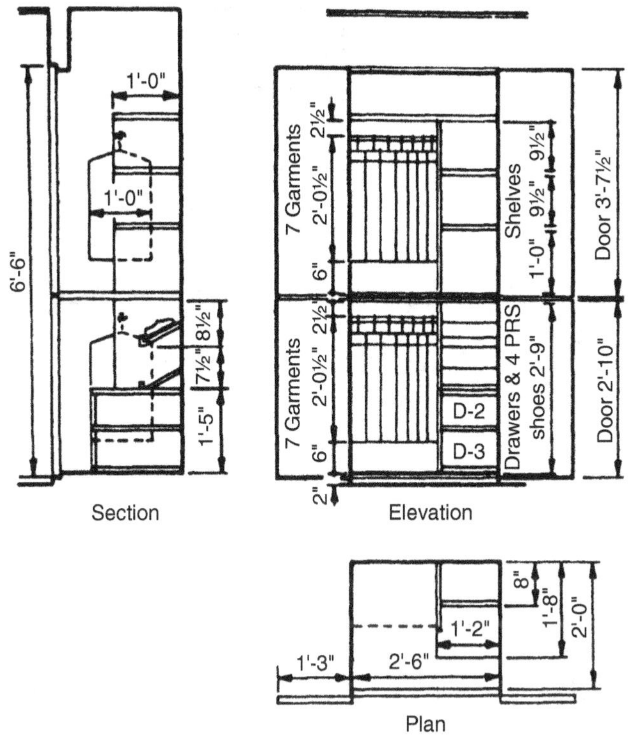

FIGURE 19.6 Closets for infants and toddlers (up to 5 years old).

Alternate, and larger, closet for an infant up to 5 years of age. (See Fig. 19.7.) Trays or drawers for folded garments at an upper level for adult use. Hanging space, drawers, and shelf available to child using the lower doors.

Small closet designed for a child of 6 to 10 years. (See Fig. 19.8.) Pole at higher but easily reached level. Drawers and shoe racks at convenient heights. Ample shelf room provided above for the storage of possessions.

Closet for youngster up to 10 years old providing greater length of hanging pole and different shoe arrangement, trays instead of cleat racks. (See Fig. 19.9.) A large shelf for hats, toys, or "collections" available to child.

Walk-in Closets. The chief considerations for walk-in closet design relate to accessibility and clearance. (See Fig. 19.10.) They include the following:

1. A minimum 30-in. (76-cm) doorway width (32 in., 81 cm, if the closet is to be used by a person in a wheelchair) should be provided.

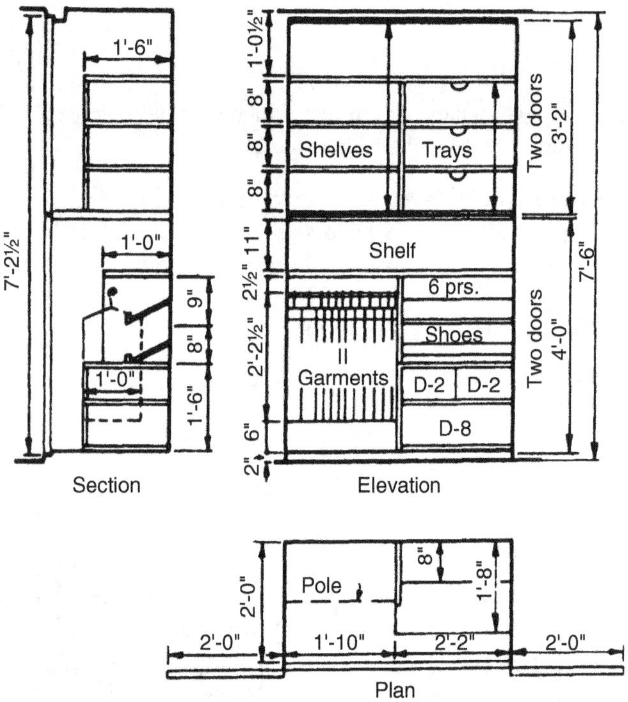

FIGURE 19.7 Closets for infants and toddlers (up to 5 years old)—alternate.

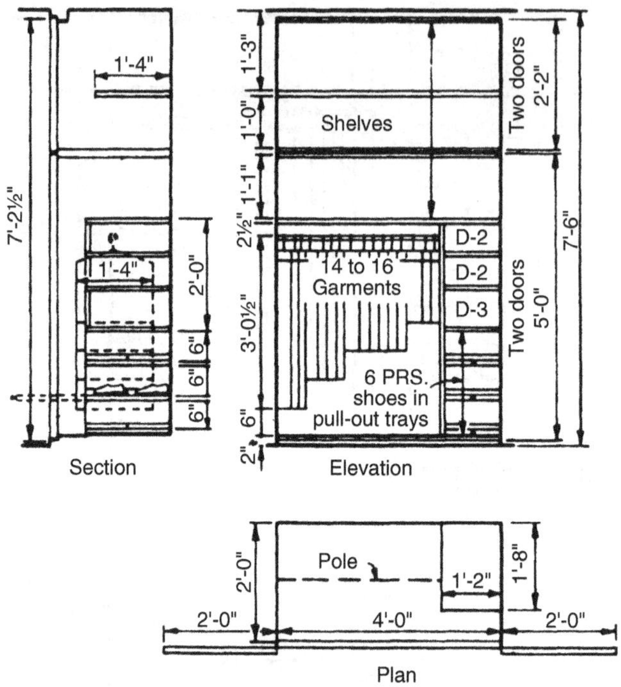

FIGURE 19.9 Closets for children up to 10 years old.

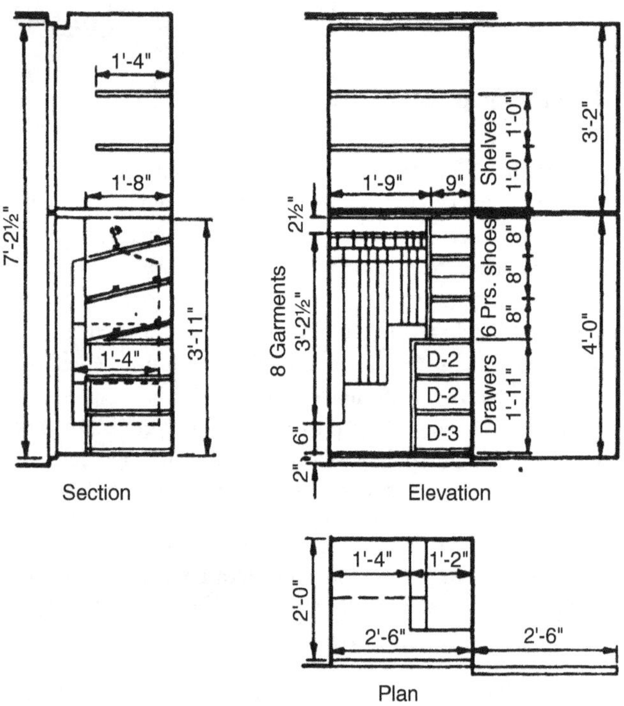

FIGURE 19.8 Closets for children 6–10 years.

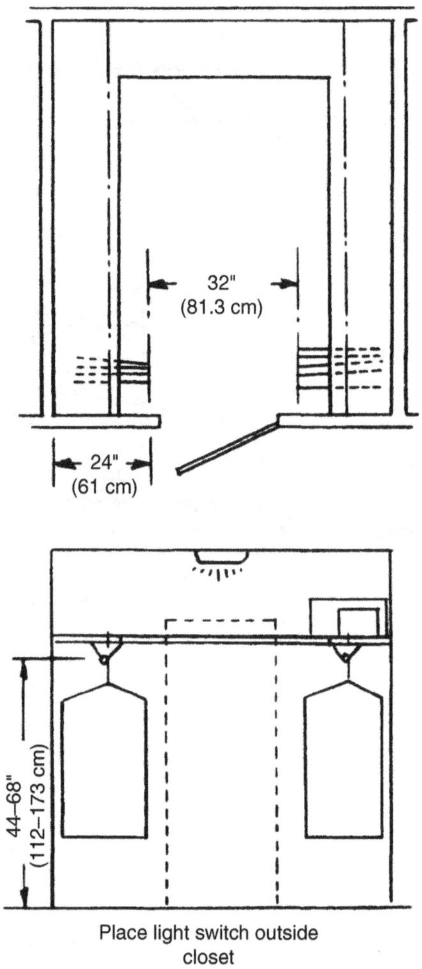

FIGURE 19.10 Dimensions for walk-in closets.

2. There should be a minimum of 32 in. (81 cm) between garments hung opposite each other.

3. There should be at least 1 in. (2.5 cm) of clearance between the garments and the adjacent wall.

4. A garment rack or pole should be provided that is high enough so that clothes will not drag on the floor; about 44 in. (112 cm) is required for children's clothes (on hangers), and about 68 in. (173 cm) for women's dresses.

5. There should be good lighting inside the closet so that colors are visible.

6. The light switch for the closet should be placed on the wall outside the closet—not inside, where it can be hidden by clothing.

7. Where possible, good ventilation should be provided inside the closet.

8. If an overhead shelf is provided, as shown in the figure, it should not extend out so far as to keep the light from shining on the clothing.

SHIPBOARD BERTHING

Guidelines

Contrary to past practice aboard U.S. Navy ships, berths (beds) should be at least 84 in. (213 cm) long and 36 in. (91 cm) wide so that they will fit the largest members of the Navy's shipboard crews.

There should be at least 32 in. (81 cm) of vertical clearance between stacked bunks so that crew members can get in and out easily, 84 in. (213 cm) long and 36 in. (91 cm) wide so that can sit on the edge of the bunk to put on shoes and socks, and can sit up (partially) in bed to read a book.

The crew member's bunks (unlike the situation in an officer's quarters) are their "home away from home." They can call their bunks their own, and it is here that they find some privacy and store their few personal belongings. Therefore, berthing should make it possible to shut out some of the everyday sights and sounds by means of appropriately soundproofed partitions and privacy curtains. The curtains should be designed so that they can be snapped to the bunk structure and thus keep the crew member from rolling out of bed when the ship rolls.

Figure 19.11 illustrates some important features that were developed in a recent shipboard habitability study.

Berthing for officers where private staterooms are provided should follow good bedroom practice, but with the special requirement that things must be tied down to keep them in place whenever the ship pitches or rolls.

Special attention should be given to the selection of pleasant colors for shipboard berthing areas (as opposed to the standard Navy gray).

Ventilation within stacked bunks should be controllable by the occupant when the curtains are closed.

Shipboard Wardrobes

Although certain wardrobes can be designed with objectives similar to those for closets in residential facilities (primarily

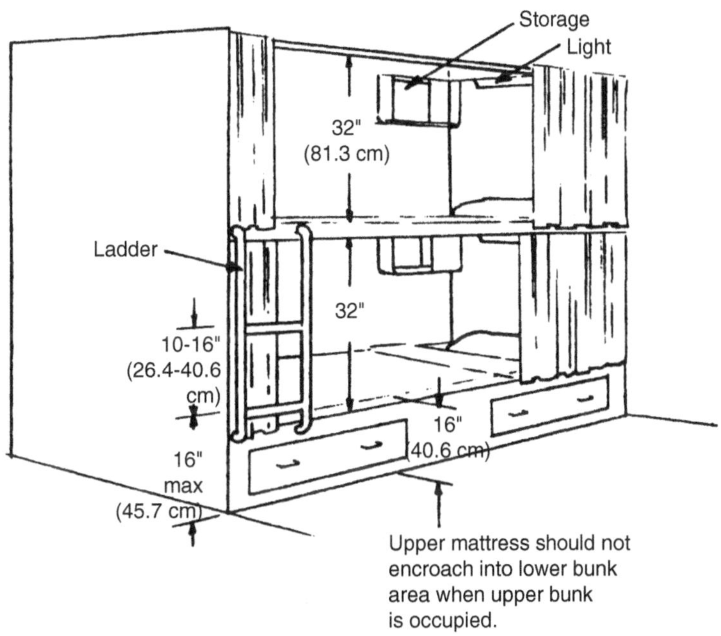

Figure 19.11 Crewmember bunks.

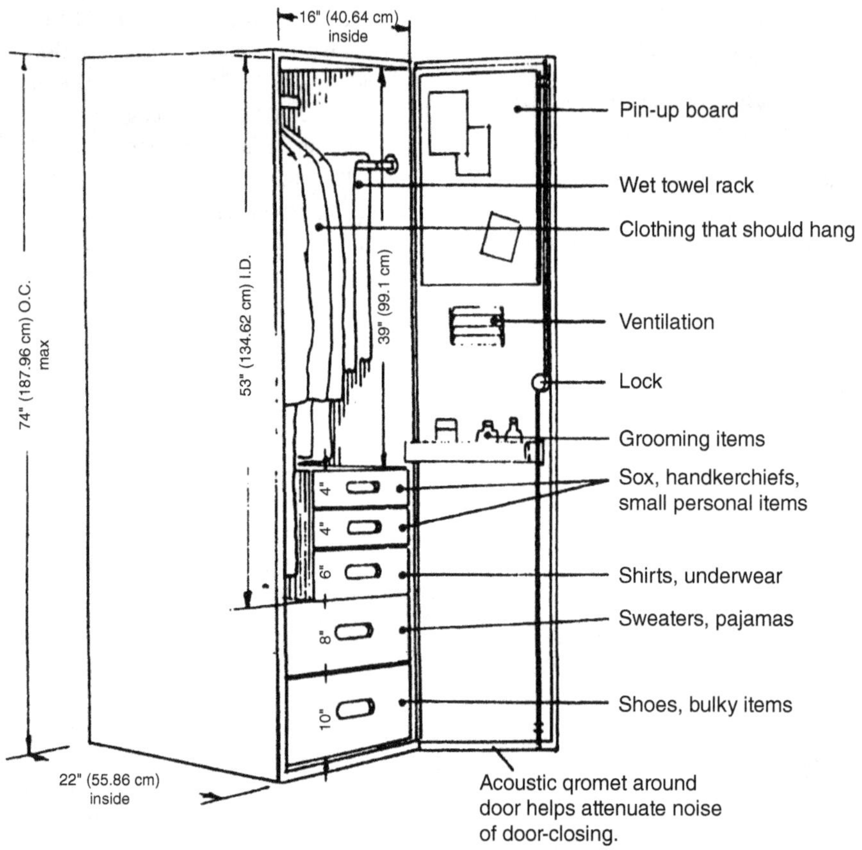

Figure 19.12 Crewmember wardrobe—general requirements.

for key officers aboard ship), more austere facilities must be designed for the major share of a ship's crew. Figure 19.12 shows general requirements for minimum wardrobe design that are based on the personal belongings that each crew member normally has aboard.

An acoustic grommet or bumper is required to prevent the door from making noise when it is being closed. Expanded versions of this design are desirable when space allows.

Berthing System Integration

Adapting the berthing subsystem to oddly shaped spaces is often difficult. Avoid the tendency to cramp an individual berth just to "squeeze it in."

An objective should be to arrange related crew-member functional groups in the same area so as to include not only their bunks but also their lockers, sanitary facilities, and any lounging space in an integrated manner. Figure 19.13 shows an example of "fitting" one crew's berthing aboard a particular ship.

Key issues addressed here were the semi-isolation of the actual sleeping area from the noisier dressing area.

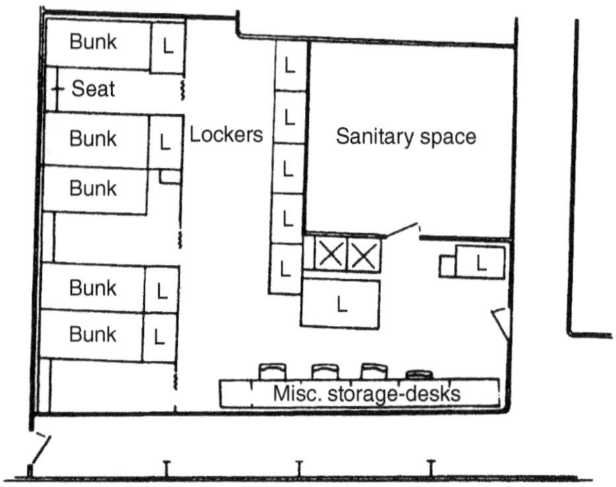

Figure 19.13 Example of berthing.

(See Fig. 19.14.) Note that acoustic treatment consisted of carpeting and of individual curtains on each bunk and on doorways to sleeping compartments. (All special materials must be fireproof.)

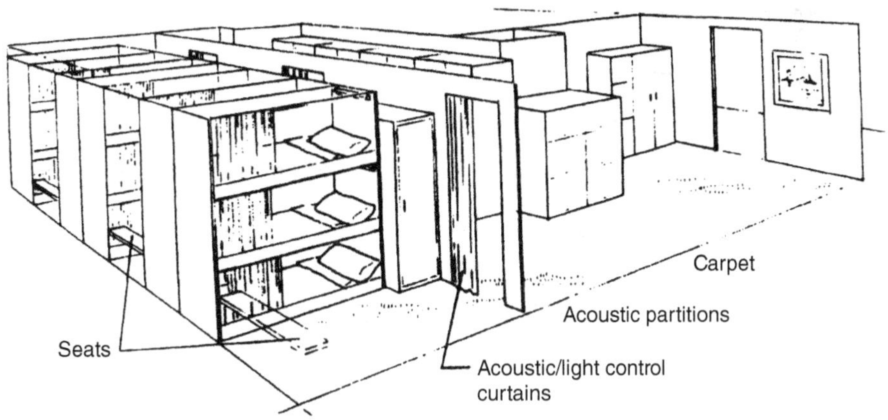

Figure 19.14 Semi-isolation of the sleeping area.

COLLEGE DORMITORY PLANNING

Guidelines

According to a study on university residential planning, the single dormitory room must accommodate at least a bed, a desk, a wardrobe, a bookcase, a soft chair, and the clearance space required by the entry door, *plus* the additional "use space" required around each element so that it can be used appropriately.

Figures 19.15–19.17 provide general element-use space requirements for each of the identified furniture elements.

It is suggested, however, that several other furnishing elements also be considered in the development of room sizes and shapes because these are equally necessary to meet the needs of each student:

1. A wastebasket
2. A hamper for dirty clothes
3. A place to hang outer garments that may be wet
4. A bulletin board on which to mount memos, pictures, schedules, etc.
5. A stand or shelf for an alarm clock, radio, or personal mementos
6. A cabinet for storage of personal hygiene and grooming aids

Obviously, a considerable variety of room layouts could be developed, depending on the size and shape of the spaces being considered. Usually, planners are of necessity interested in the most cost-effective layout. General objectives in developing any layout are the following:

1. To provide a well-defined resting, dressing, and grooming area
2. To provide a well-defined study area
3. To provide an adequate visiting area that does not require a visitor or visitors to occupy the host's sleeping area

When two students are to occupy the same general space, it should be arranged for minimum interference by either student of the other's critical activities, i.e., sleeping and studying. Consider at this level a common visiting area that can be used by the students and their friends. A few examples are shown in Figs. 19.16 and 19.17 (which are not drawn to scale).

Single-Occupancy Rooms

Although it is possible to squeeze the student into less space, approximately 120 ft^2 (10.8 m^2) is recommended as a minimum. (See Fig. 19.15.)

Note that in both examples, the bed has been positioned as far from the corridor doorway as possible to minimize interference with sleep.

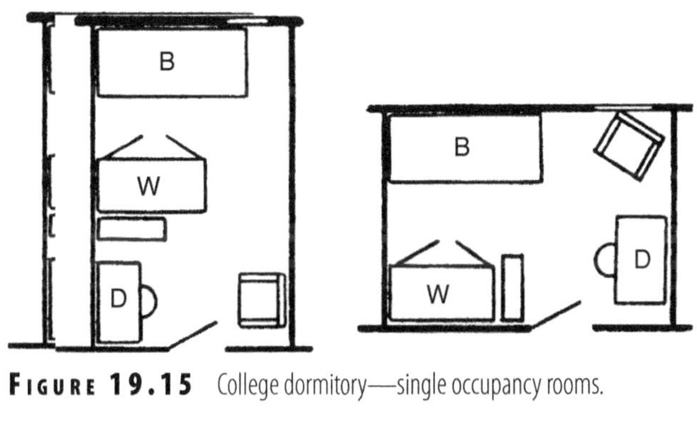

Figure 19.15 College dormitory—single occupancy rooms.

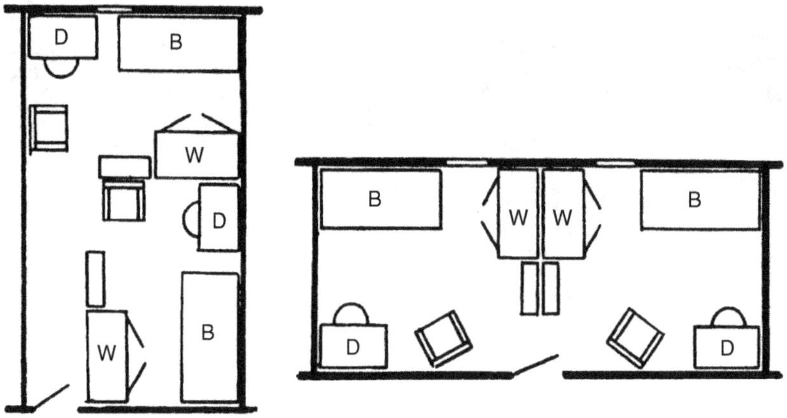

FIGURE 19.16 College dormitory—double occupancy rooms.

Double-Occupancy Rooms

It is important in planning double-occupancy layouts to maintain as much equality as possible between the two spaces, not only in terms of actual floor space, but also in terms of the apparent convenience of each space. (See Fig. 19.16.)

Other alternatives are important to consider when planning dormitory space, including the suite concept. A minimal example is illustrated in Fig. 19.17, where a common social area has been added to the double-occupancy space. Also, this example illustrates an important addition of extra storage space adjacent to the student's quarters (as opposed to central storage at some distant point).

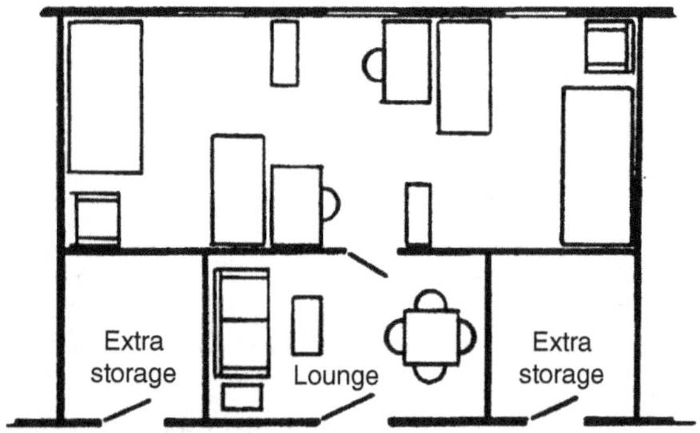

FIGURE 19.17 College dormitory—other alternatives.

CHAPTER 20

KITCHENS AND GALLEYS

RESIDENTIAL KITCHEN PLANNING

The residential kitchen should be planned in terms of work centers. The following are typical:

1. Food Preparation: One or more locations where raw materials are cleaned, cut, mixed, etc., prior to cooking or serving.

2. Cooking and Baking: One or more locations where food is cooked or baked, e.g., on a stove, in an oven (conventional and microwave), or in a special electric cooking vessel or device.

3. Food Storage: One or more locations where raw or already prepared materials and food are stored before and after meals (refrigerator, freezer, pantry, cupboards, etc.).

4. Culinary Utensil Storage: One or more locations where pots, pans, cutlery, special electric mixing devices, etc., are stored.

5. Dish and Flatware Storage: One or more locations where serving dishes, glassware, and flatware are stored.

6. Washing and Cleanup: One or more locations where dishes, pots and pans, and other implements used for cooking are emptied, rinsed, stacked, and either hand-washed or deposited in an automatic dishwasher.

7. Miscellaneous Storage: One or more locations where soaps, detergents, cleaning materials and devices, and other special serving devices such as trays and cutting boards, are stored.

Primary kitchen arrangement objectives should be the following:

1. Locate stored items close to the primary task involving these items.

2. Arrange the work centers so that activity flows logically from one point to another in the order in which tasks usually are performed.

3. Arrange the work centers so that there is adequate clearance to accommodate the task mobility requirements.

4. Arrange the general kitchen so that extraneous traffic will not interfere with the in-kitchen tasks.

5. Arrange the kitchen to accommodate other individuals who may join in preparing and serving food.

6. Arrange the windows so that a person who is working in the kitchen can look outside.

Traditional Kitchen Layout Examples

Single-Wall Layout. This is suitable for small "efficiency" apartment kitchens where space is limited and there is typically a single user. (See Fig. 20.1.)

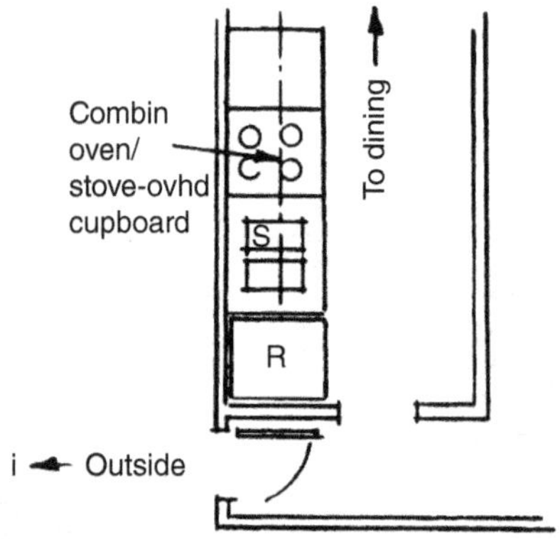

FIGURE 20.1 Single-wall layout.

Pullman Layout. This is suitable for a smaller residence. A parallel cabinet and appliance arrangement minimizes the necessity to string out the work flow. (See Fig. 20.2.)

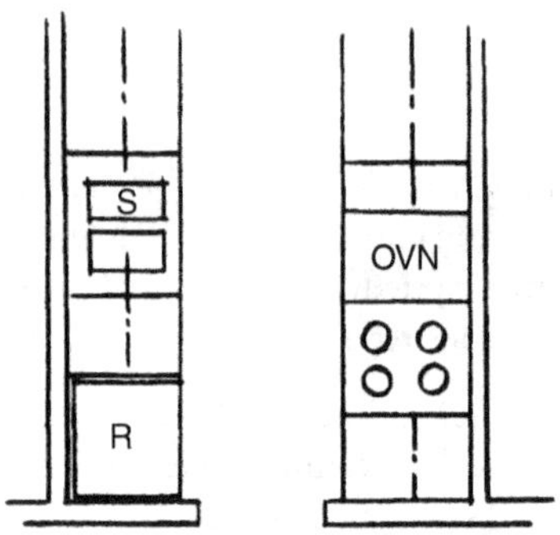

FIGURE 20.2 Pullman layout.

L-Shaped Layout. This is suitable for a small- to medium-sized residence. It allows combining work-center relationships and permits the addition of extra storage space. Its most important virtue is the reduction of traffic congestion and the ease with which additional kitchen help is accommodated. (See Fig. 20.3.)

U-Shaped Layout. This is suitable for a medium-sized to large residence where enough space can be allowed between rows of cabinets. It can be varied in many ways, including the addition of an "island counter," so that several people can work in the kitchen simultaneously with minimum traffic conflict. The island layout also tends to minimize the walking distance between common work centers such as the stove, refrigerator, and sink. (See Fig. 20.4.)

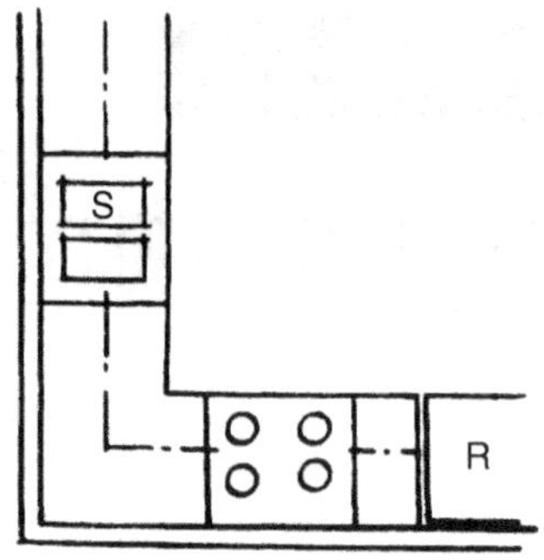

FIGURE 20.3 L-shaped layout.

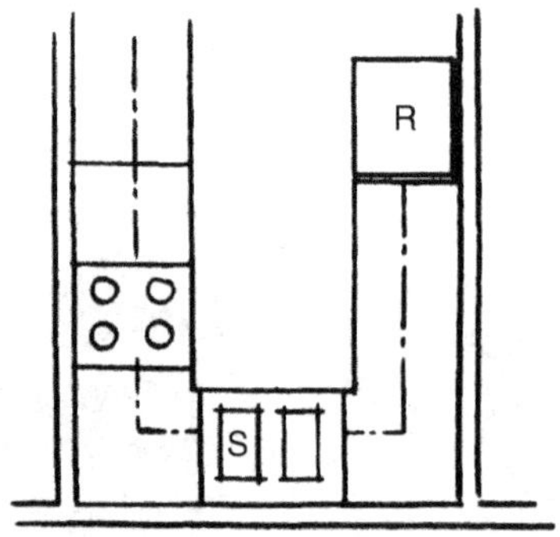

FIGURE 20.4 U-shaped layout.

Island Concept. Perhaps the greatest benefit provided by the island concept is that a primary work center, such as the stove or sink, can be equally convenient to many other work centers. This is important to the person who cooks for many guests.

When an island concept is used, consider the addition of a small utility counter for mixing, cutting, or temporary storage of cooking utensils. Also take advantage of the island to provide additional storage space below the range or sink counter. (See Fig. 20.5.)

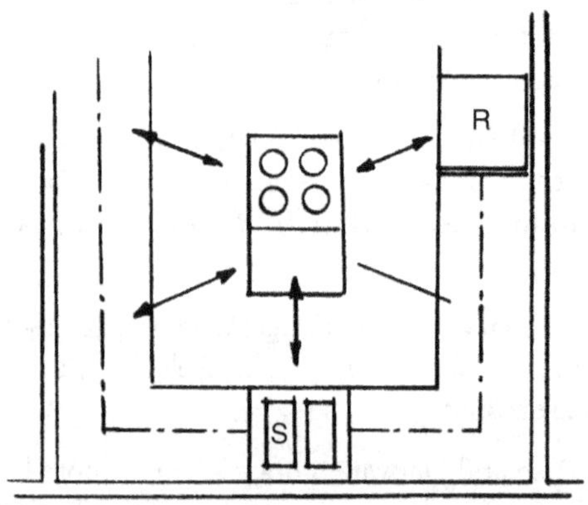

FIGURE 20.5 Island concept.

Work Triangle Concept. The basis of good kitchen layout has for many years been considered to be the so-called work triangle. This consists of an arrangement that limits the walking distances between the three key work centers to no more than 22 ft (6.6 m). The key centers, as shown in Fig. 20.6, include the sink, range, and refrigerator. Obviously, storage requirements are associated with each work center, i.e., food storage, utensil storage, dish storage, etc.

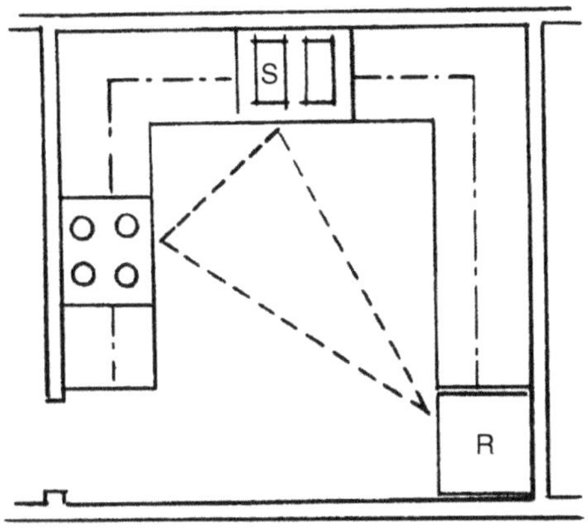

FIGURE 20.6 Work triangle (22 ft).

Work Flow. Equally important in kitchen arrangement is the work-flow principle. An attempt should be made to arrange elements in such a way that the necessity to randomly retrace steps as one works in the kitchen is minimized. (See Fig. 20.7.) The basic sequence typically is as follows:

1. Acquiring and cleaning raw materials
2. Mixing (often adding water)
3. Cooking or baking
4. Dishing up
5. Serving
6. Gather dishes and food
7. Store food
8. Clean dishes
9. Clean kitchen
10. Store dishes

General Traffic Considerations

1. Lay out the kitchen so that extraneous traffic does not normally flow through the work triangle. (See Fig. 20.8.)

2. Provide clear space between facing cabinets or appliances of at least 48 in. (122 cm). When such fixtures are at right angles to each other and are separated by a passageway, the fixtures should be at least 30 in. (76 cm) apart.

3. In L-shaped or U-shaped layouts, the minimum edge distance between an appliance and an adjacent corner should be 9 in. (23 cm) for the edge of a sink, 16 in. (41 cm) for the refrigerator, and 14 in. (36 cm) from the edge of the nearest range burner.

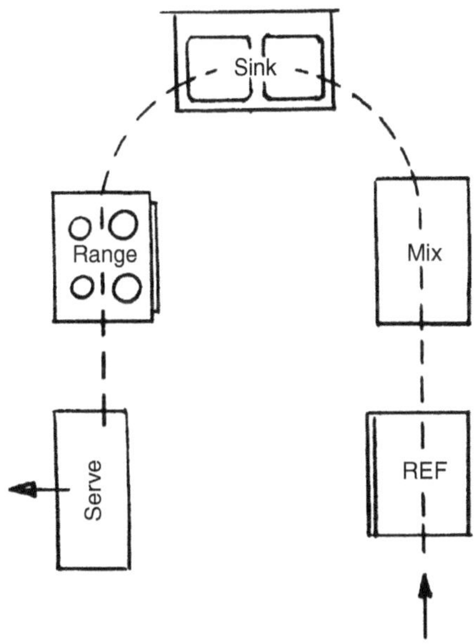

FIGURE 20.7 Partial work flow example.

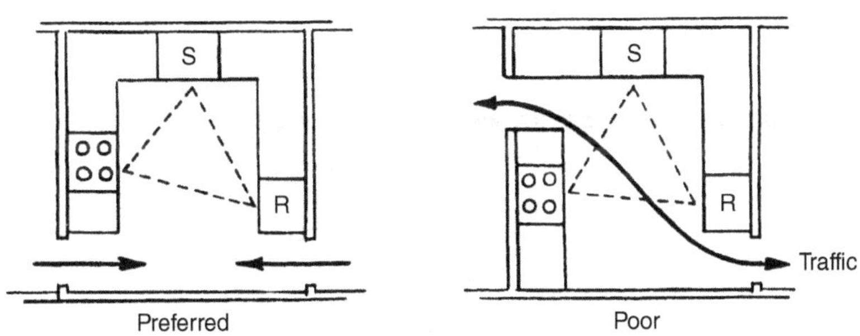

FIGURE 20.8 Lay out work triangle to avoid extraneous traffic.

Key Kitchen Workplace Dimensions

Basic requirements for kitchen cabinets, established by the U.S. Department of Housing and Urban Development (HUD), are shown in Table 20.1 and Fig. 20.9.

Other related specifications are as follows (see Figs. 20.10 and 20.11):

1. Work centers may be combined; kitchen multiuse space shall equal or exceed the largest frontage of any one of the work centers being combined, plus 6 in. (15 cm).

TABLE 20.1 Kitchen Cabinet Requirements

	Minimum Width (in.)		
Work Centers	Two Bedrooms	Three Bedrooms	Four or More Bedrooms
Sink	24	32	32
Counter and base cab at each side	20	24	30
Range	24	30	30
Counter and base cab at one side	20	24	30
Refrigerator (space)	36	36	36
Counter at latch side	15	15	18
Mixing (base and wall cabinet)	36	36	42

2. Provide at least 9 in. (23 cm) from the edge of the sink or range to any adjacent corner cabinet and 16 in. (41 cm) from the latch side of the refrigerator.

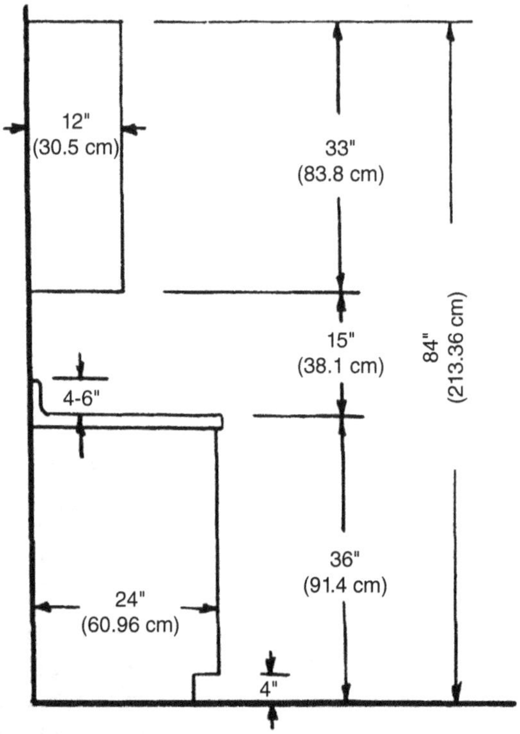

FIGURE 20.9 Basic requirement for kitchen cabinets.

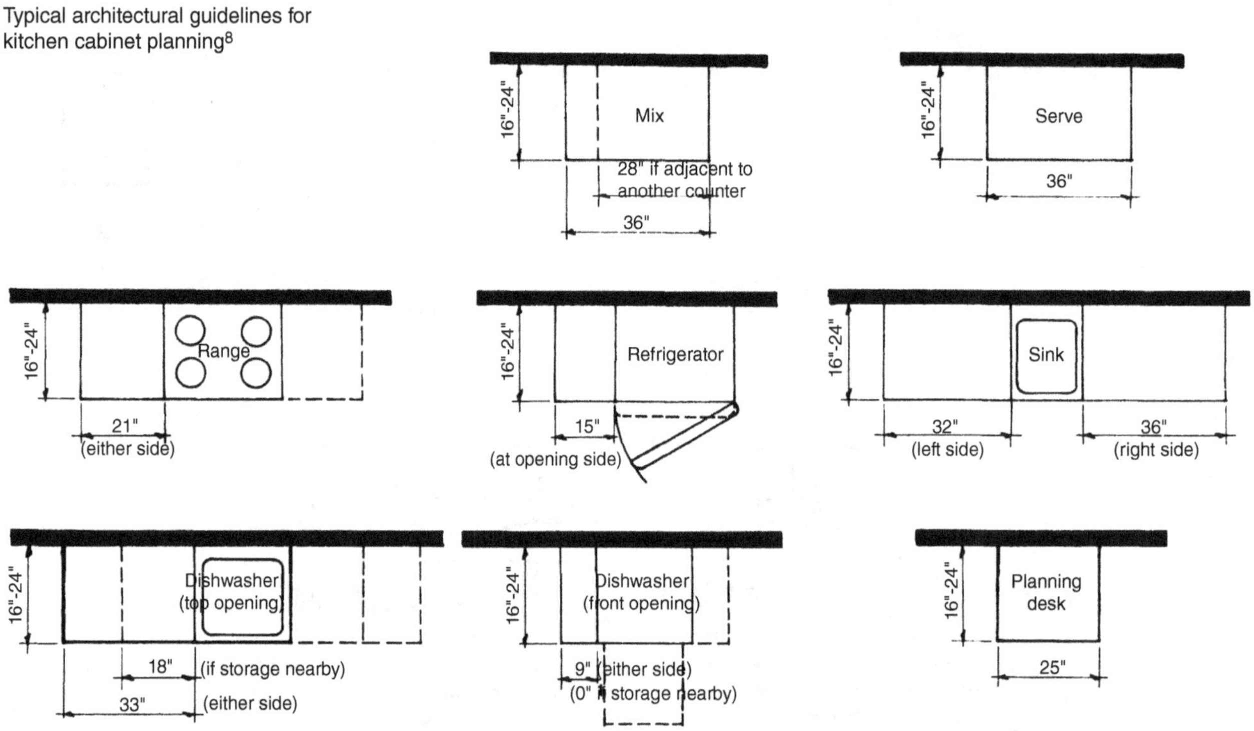

FIGURE 20.10 Minimum counter-width dimensions.

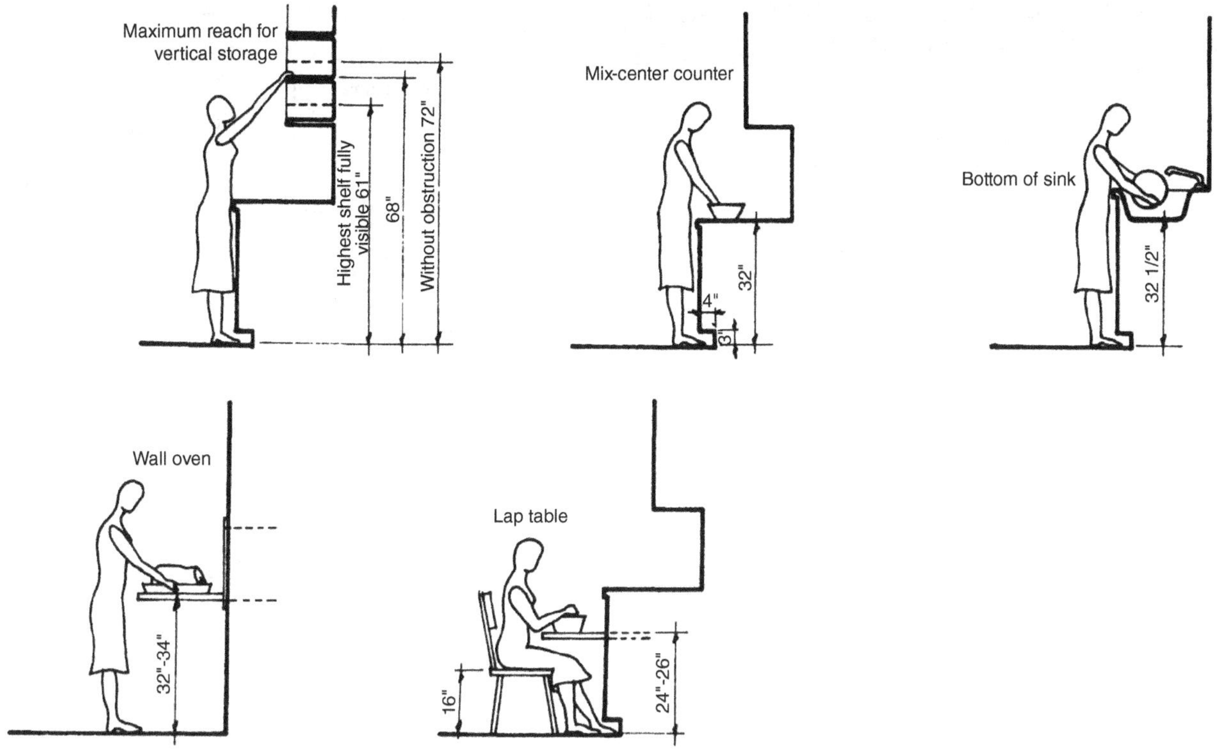

Figure 20.11 Comfortable working heights.

3. 24-in. (61-cm) sinks are permissible when dishwashers are provided.
4. Counter-space minimums are as follows:
   a. 52 in. (132 cm) for one- or two-bedroom units
   b. 60 in. (152 cm) for three-bedroom units
   c. 72 in. (183 cm) for four-bedroom units
5. Base cabinet width minimums are as follows:
   a. 68 in. (173 cm) for one- or two-bedroom units
   b. 72 in. (183 cm) for three-bedroom units
   c. 84 in. (213 cm) for four-bedroom units
6. Wall cabinet width minimums are as follows:
   a. 68 in. (173 cm) for one- or two-bedroom units
   b. 72 in. (183 cm) for three-bedroom units
   c. 84 in. (213 cm) for four-bedroom units

Typical Architectural Guidelines for Kitchen Cabinet Planning[7]

Planned Storage. Avoid the tendency to make storage an afterthought. That is, consider the arrangement and size of a kitchen not only in terms of where you plan to place the major fixtures and appliances but also in terms of the supporting storage spaces for each of the work centers. Also, avoid thinking of storage only in terms of a "bin" into which things are thrown. Each type of storage requirement should be addressed in terms of what is to be stored, the shapes and sizes of articles to be stored, how the items are normally positioned and stacked, and, obviously, approximately how many of each item there will be. Certain items are stored best in drawers; others, in bins; and still others, on shelves.

Consider the problems of weight. For instance, a large drawer full of flatware or heavy steel cookware will be sufficiently heavy so that it may require special drawer slides.

Drawers should be used that cannot be pulled all the way out, possibly leading to the dumping of the drawer's contents. Doors on cabinets should swing in the direction most convenient for the person who is trying to get into the cupboard or cabinet. Avoid very large swinging doors because they get in the way when opened. Do not place a door where the user might inadvertently mash his or her hand against an adjacent structure or appliance when opening the door.

The first step in analyzing storage needs should be to determine how much of each item must be stored. Table 20.2 can be used as a guide.

Activity-Storage Relationships. Although few people agree with the designer's decisions concerning where things should be stored in the kitchen, some consideration must be given to likely relationships that will occur so that the shape of the storage (drawer depth, shelf spacing, etc.) can accommodate

[7] From J. De Chiara and J.H. Callender, *Time-Saver Standards for Building Types*, McGraw-Hill Book Company, New York, 1973.

TABLE 20.2 Equipment and Food Supplies Stored at Range Center

	Number Stored		Storage Space per Item (in.)*		
Item	Limited	Liberal	Side to Side	Front to Back	Height
Equipment					
Potato masher	1	1	3½	13	4½
Knives, forks, spoons	3	3	3½	13	3
Frying pan, 10½-in.	1	1	11	17½	5½
Frying pan, 9-in.	1	2	9½	16	5
Frying pan, 6-in.	0	1	6	12	5
Pot lids	2	4	10½	10½	1
Potholders	4	8	7	7	2**
Food supplies					
Rice, 1-lb pkg.	1	1	2½	4	6½
Spaghetti, 1-lb pkg.	1	1	2½	11½	6
Coffee, 1-lb can	1	1	5½	5½	4
Oatmeal, 3-lb box	1	1	6	6	11
Macaroni, 1-lb pkg.	1	1	2	5½	9
Tea, 8-oz pkg.	1	1	2½	4½	7

*Dimension of the item (including lid, if any) plus clearance for handling.
**Provides for stack of six potholders.

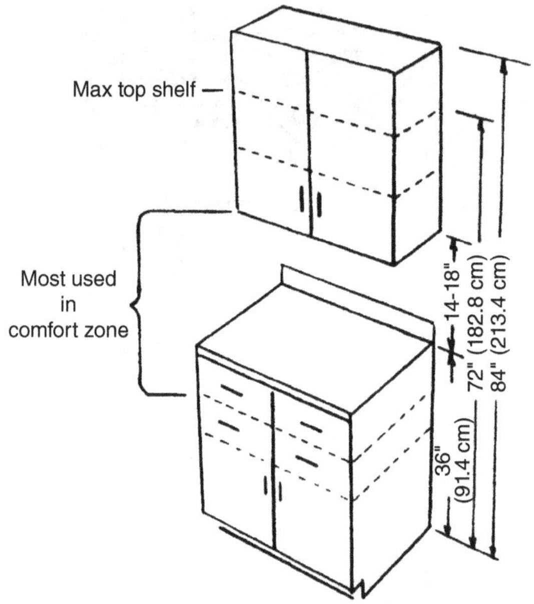

Figure 20.12 General storage.

what is expected at each location. The generalizations in Figs. 20.12–20.15 are meant to assist in developing an appreciation for activity—storage relationships.

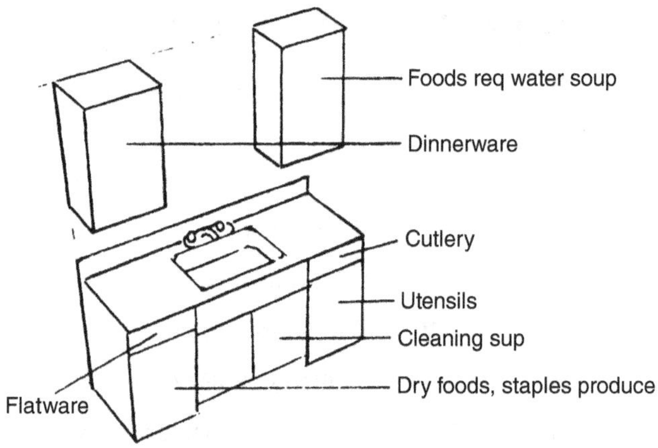

Figure 20.13 Sink area.

Design for Health and Safety

Circuits and Switches. Use ground fault circuit interrupters for all kitchen circuits and locate switches far enough from the sink so that no one whose hands are in water or who is touching a sink fixture can reach them. (See Fig. 20.16.)

Lighting. Good-quality illumination is required throughout the kitchen, not only for inspecting food and washing dishes but also for utilizing cutlery in a safe manner. Key points to note—i.e., the stove top, underneath cabinets, and around the sink—are illustrated in Fig. 20.17.

Appliance Selection and Installation. Select cooking appliances that have controls that cannot be operated by small children and which do not require the person who is using them to reach across hot burners or utensils. (See Fig. 20.18.)

Mount stove hoods where they are not easily bumped when the person who is using the stove bends over the stove top. The hood should have radiused corners.

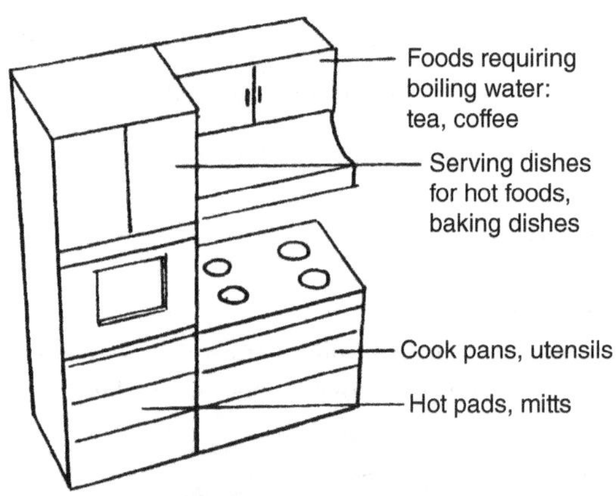

Figure 20.14 Cooking area.

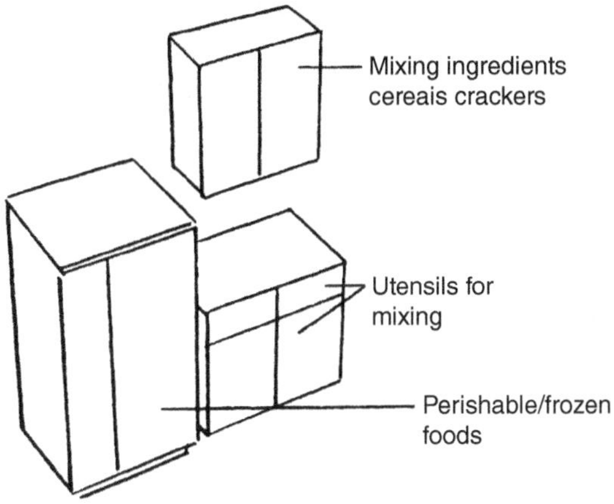

Figure 20.15 Refrigerator area.

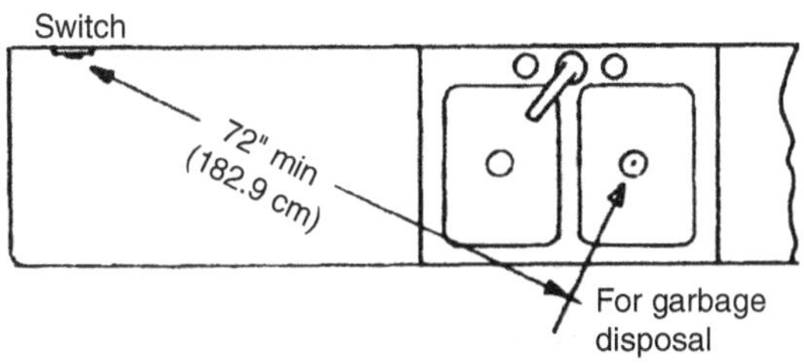

Figure 20.16 Locate electrical switches away from the sink.

Figure 20.17 Provide good quality illumination.

Figure 20.18 Add child-proof interlocks or safety devices.

Make sure that the stove or oven selected is well insulated so that the exterior surface will not become hot when the appliance is in operation. (See Fig. 20.19.)

Figure 20.19 Do not locate stove controls where users can get burned.

Special Designs for the Handicapped Homemaker

Design for the handicapped homemaker should include the following (see Fig. 20.20):

1. Leave frequently used pots and pans exposed and within reach.
2. Select a range that has controls near the front.
3. Consider a mobile oven rather than the built-in type because such ovens are more accessible.
4. Provide several alternative work shelves positioned at a convenient height for mixing and provide a hole in which the mixing bowl is captured so that it does not have to be held.
5. Provide clearance beneath the sink for wheelchair armrests.
6. Provide shallow, roll-out shelves for storage of small items.
7. Provide clearance for wheelchair foot-rests.
8. If the handicapped person can stand, provide strategically placed handholds along the edges of the cabinets so that the person can hold on and/or steady himself or herself while working.
9. Locate a telephone in the kitchen.
10. Where possible, provide power assisted adjustability for countertop and appliance heights.

Special Considerations for the Blind or Partially Sighted Homemaker

1. Provide high-level (a minimum of 50 fc at the working surface) shadowless lighting throughout the kitchen.
2. Use contrasting colors to help the partially sighted homemaker differentiate dishes and pots from work surfaces, special controls from their immediate background, electric cords and outlets from work surfaces or the splashboard, etc.

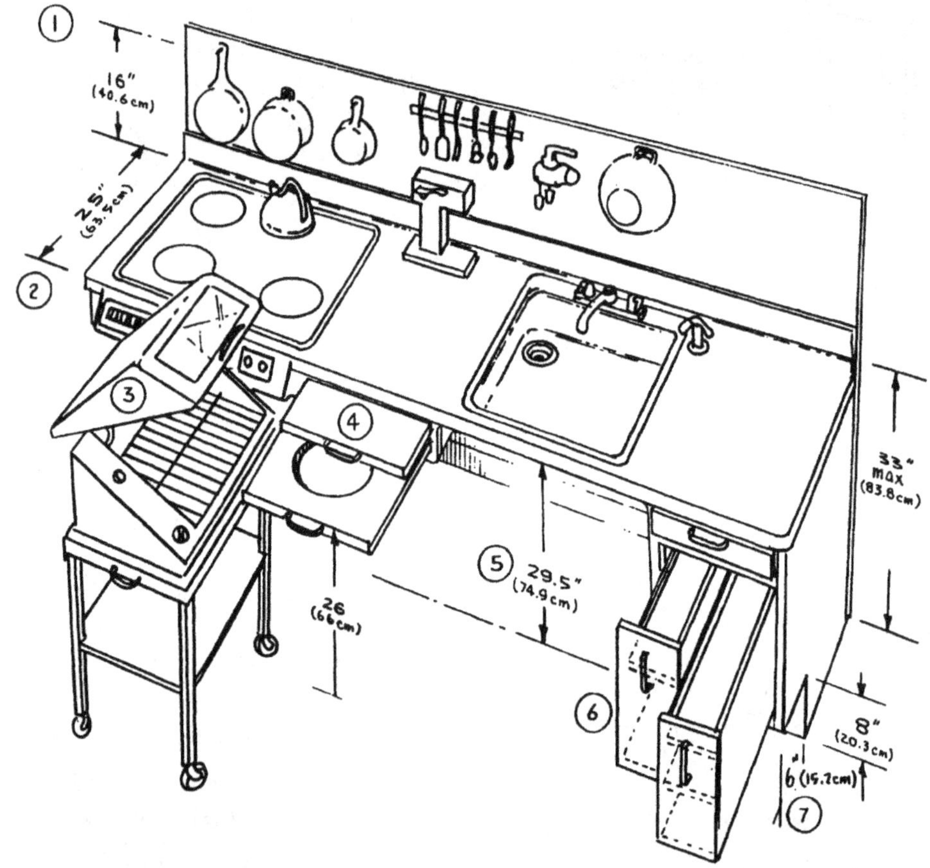

Figure 20.20 Optimizing convenience for the wheelchair user.

3. Provide tactile coding on range and oven controls and labels.

4. Do not include any cabinetry that is suspended in the path that the homemaker is expected to take.

5. Select appliances that have auditory cues to indicate when cooking is completed.

6. Make sure that there are no sharp corners on cabinets or appliances.

7. Provide a lip around counter tops where water is used or where mixing is done, so that liquids or objects will not drain or roll off onto the floor.

8. Consider providing a tactile coding system for identifying storage areas, i.e., a braille or other tactile code on the outside of cabinet doors or drawers to indicate what is stored within.

GALLEYS

The galley typically provided aboard commercial aircraft, commercial and private ships and boats, etc., requires considerable ingenuity not only in terms of finding sufficient space for everything and fitting this space to irregular contours but, most important, in terms of creating a "captive design" that will keep everything in place when the motions of the airplane or boat are sufficient to cause objects to shift position, doors to fly open, etc. (See Fig. 20.21.)

In spite of a typical lack of space, one should still attempt to apply the same organizational principles in arranging the galley that were presented in the section on kitchens. Although there is usually less urgency aboard private craft, time and efficiency are extremely critical for commercial systems. An important objective should be to design storage systems so that they serve both pre-serving and post-serving functions. Since each new galley must be customized for the particular vehicle, there is little point in suggesting specific arrangements. However, the following guidelines are applicable to all galley installations:

1. Everything should be storable in a captive state, i.e., by means of latching doors, secured trays, shelf rails, or special brackets.

2. Counter tops must have "spill lips" to prevent liquids from spilling over the edge.

3. Devices that must remain in service, such as coffee pots, should be deep-set within retaining burners.

4. Plug-in warmers or cookers should plug in and not require electrical cord lead-ins.

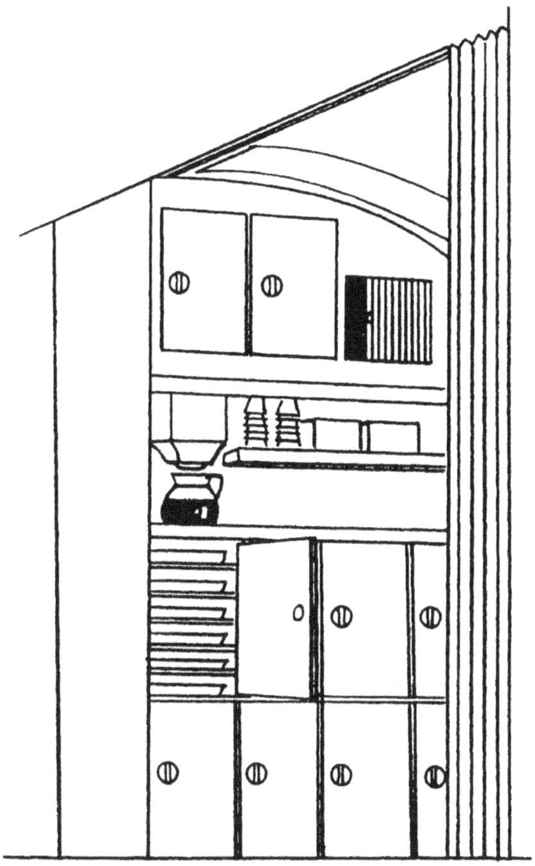

Figure 20.21 Commercial aircraft galley.

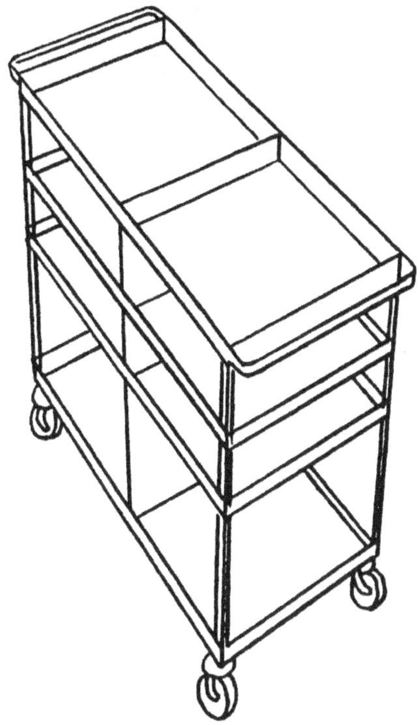

Figure 20.22 Serving carts must have locks on wheels.

5. Serving carts should be designed to complement and extend the galley subsystem whenever possible. Such carts should also be designed to "capture" all items carried, and, in addition, they must be provided with wheel locks to prevent the cart from being displaced if extensive vehicular movement is experienced. (See Fig. 20.22.)

6. Galley lighting should be of high quality, although the intensity level need not be more than about 30 fc at the working surface.

7. Galley cabinets should be designed so that spilled liquids cannot drip into electrical wiring or control switches. Cabinets should have as few crevices as possible since these are difficult to clean, especially when the work load of attendants is heavy.

8. It is always desirable to provide some method for closing off the galley from the rest of a passenger cabin area. This helps prevent unnecessary interference by passengers and also keeps them from seeing the galley area during cleanup time.

9. Galleys should be located so that they are convenient to servicing and loading activities.

10. Above all, storage of waste materials must be provided for. Discarded packaging materials tend to become space-consuming.

11. All galley cabinets should be designed so that there are no sharp edges or corners against which attendants can suddenly be thrown and injured. Provide strategically located handholds so that attendants can support themselves against excessive vehicular motions.

CHAPTER 21
DINING AND FOOD SERVICE

DINING FACILITIES

When the combination service and dining configuration includes an overhead storage cabinet and/or a stove ventilating hood, care should be taken to ensure that the cabinet does not create a hazard; i.e., it should not be likely that either the person doing the serving or the diner will lean forward and bump his or her head on the cabinet. The approximate dimensions shown in Figs. 21.1–21.5 can be used as a guideline. Note also that the counter must not be too deep; otherwise, the person doing the serving cannot reach to the far edge to serve diners or wipe off the counter.

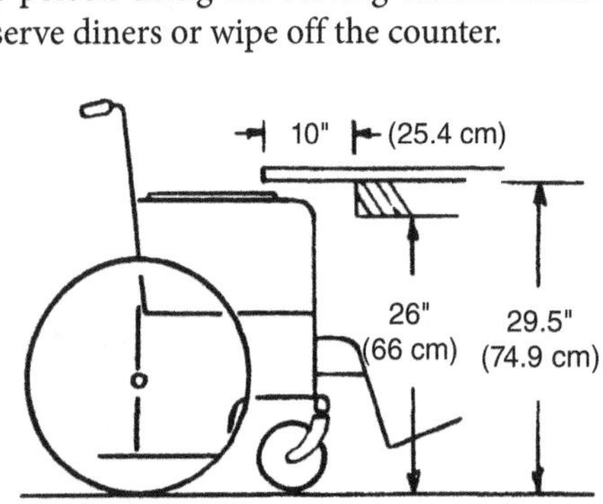

FIGURE 21.1 Dining to accommodate the wheelchair.

The 36-in. (91-cm) counter height is recommended when the counter contains a range top because this is a comfortable height for stirring food in cooking utensils on the stove.

Combination situations in the typical restaurant require an additional consideration, i.e., the frequent communications between the customer and the waiter or waitress.

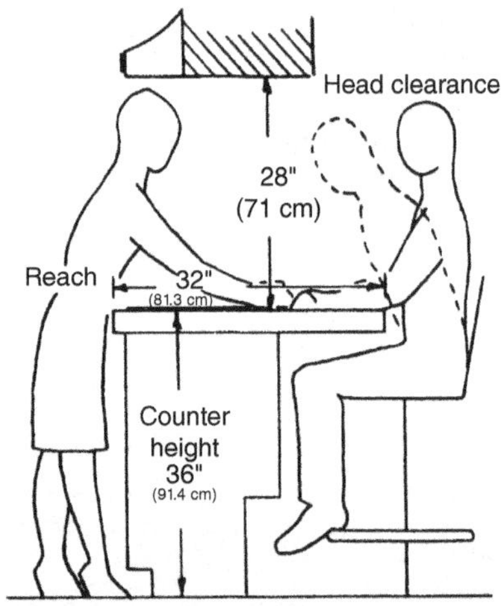

FIGURE 21.2 Overhead clearance.

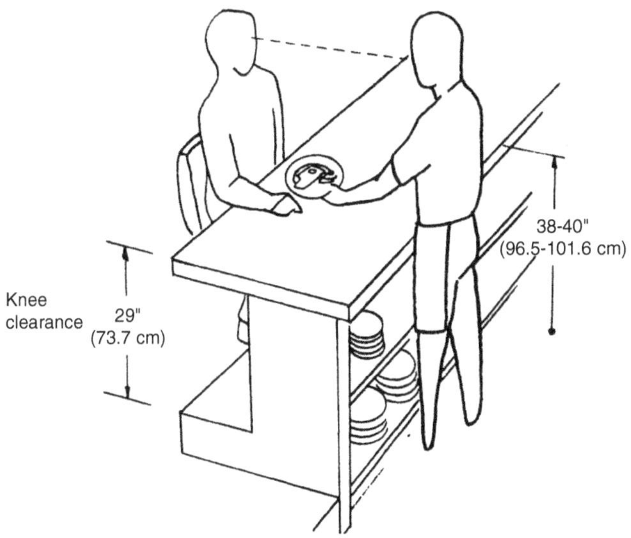

Figure 21.3 Serving counter.

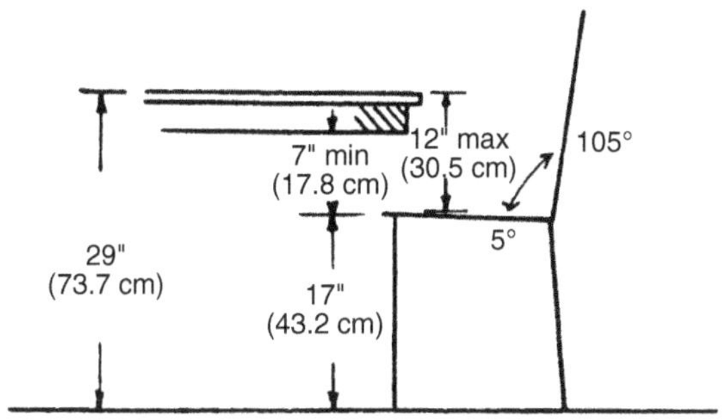

Figure 21.4 Chair seating.

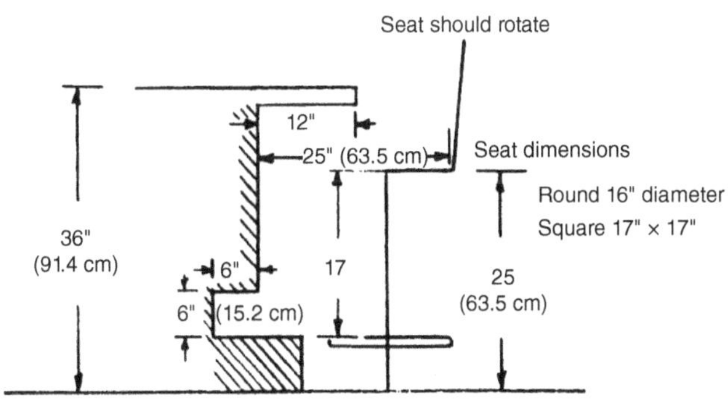

Figure 21.5 Stool seating.

An objective should be to establish an approximately horizontal face-to-face relationship. The approximate dimensions shown in the figures will provide a good and comfortable compromise between the several geometric relationships:

1. Eye-to-eye level
2. Comfortable dining-table height
3. Comfortable counter-top serving height
4. Reasonable storage access behind the counter
5. Reach limits of the waiter or waitress who must clean the counter.

Lateral spacing for comfortable food service and noninterference for the single or multiple diner situations requires proper accommodation of typical dishes, glassware, and flatware. The important considerations include the following:

1. Sufficient space for a full table setting should be provided. (See Fig. 21.6.)

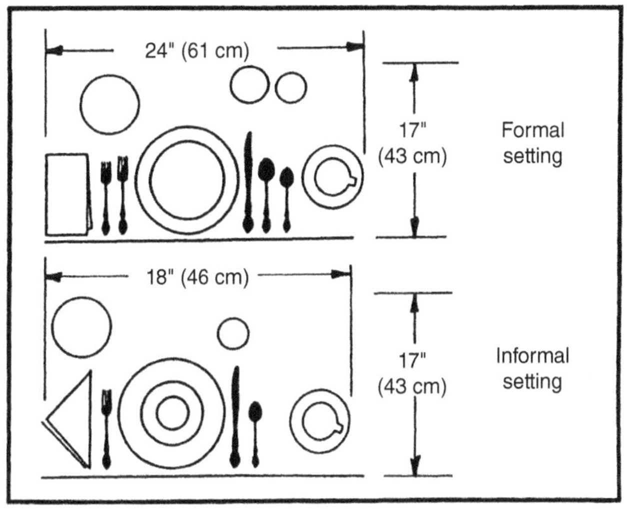

Figure 21.6 Comfortable table setting envelope.

2. Comfortable fore and aft positioning of the diner should be provided. (See Fig. 21.7.)

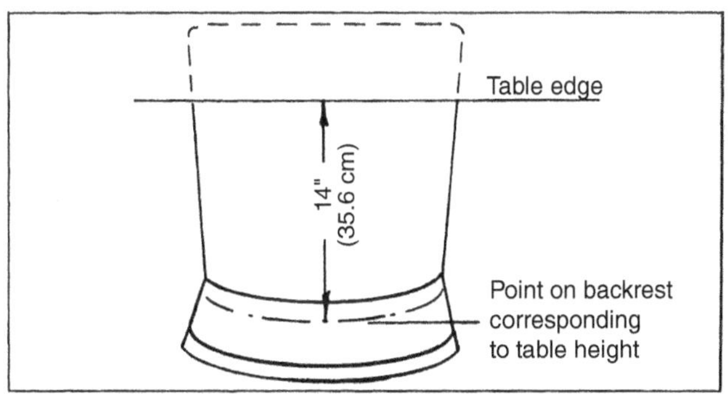

Figure 21.7 Comfortable eating distance from table (adults).

3. There should be comfortable spacing between adjacent diners to minimize interference while eating. (See Fig. 21.8.)

4. In addition to adequate center-to-center spacing of adjacent diners to accommodate them while they are eating, there should be sufficient space between the chairs so that people can get into the chairs with minimum disturbance of the other diners. If fixed seats are involved, it is preferred that the seats be round and that they revolve

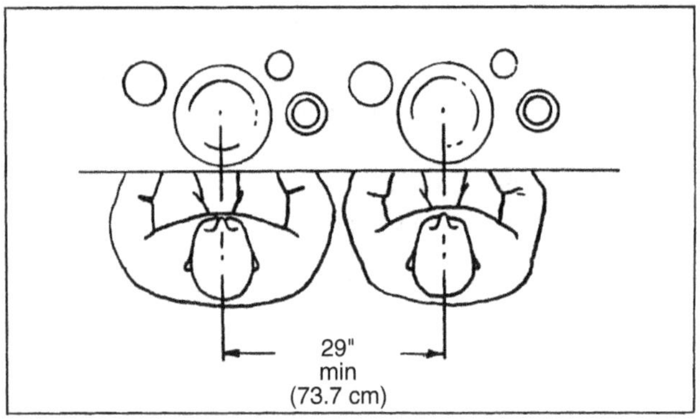

FIGURE 21.8 Minimum separation for eating (adults).

so that the person getting in and out will have less problem clearing his or her legs between the table and/or the adjacent seat. (See Fig. 21.9.)

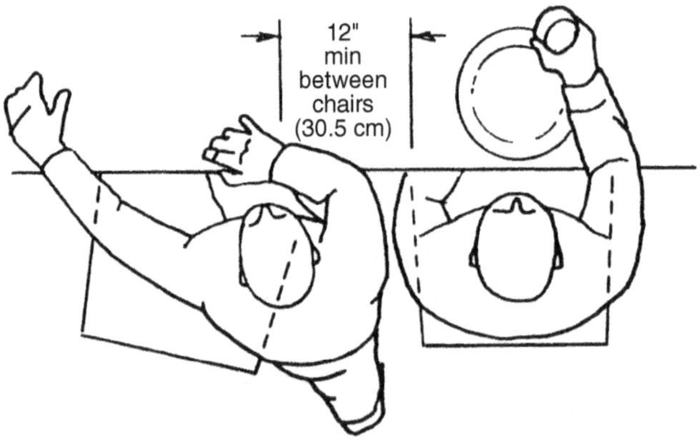

FIGURE 21.9 Provide clearance between diners.

5. Foot and knee interference between facing diners should be minimized. Toe-to-toe clearance is particularly important in booth design, where there is no leeway to move seats. (See Fig. 21.10.)

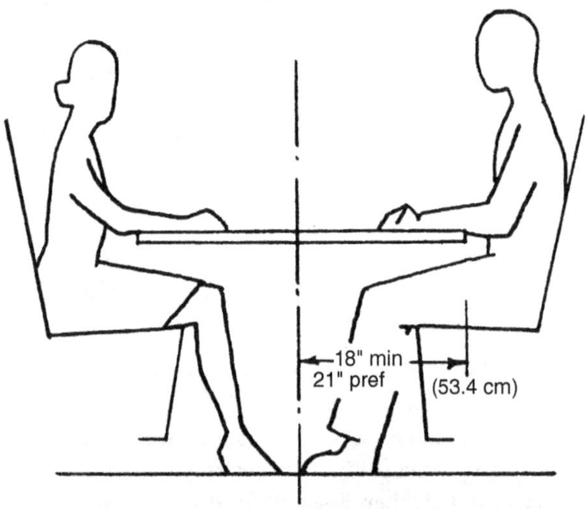

FIGURE 21.10 Provide foot and knee clearance at corners.

6. Similar knee and foot clearance is needed for diners who are sitting at the corner of a table or counter, i.e., 90° to one another. The minimum clearance for facing or corner-positioned wheelchairs is 21 in. (53 cm). (See Fig. 21.11.)

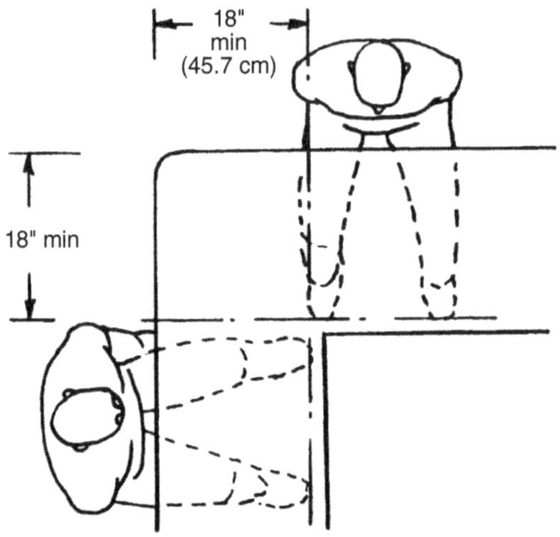

FIGURE 21.11 Foot and knee clearance at corners.

7. Clearance around tables should be adequate to permit other persons to pass by seated diners and to allow a diner to get up suddenly from the table. The minimum for a typical home environment would be as shown in Fig. 21.12; i.e., this would prescribe the envelope within which no other furniture encroaches.

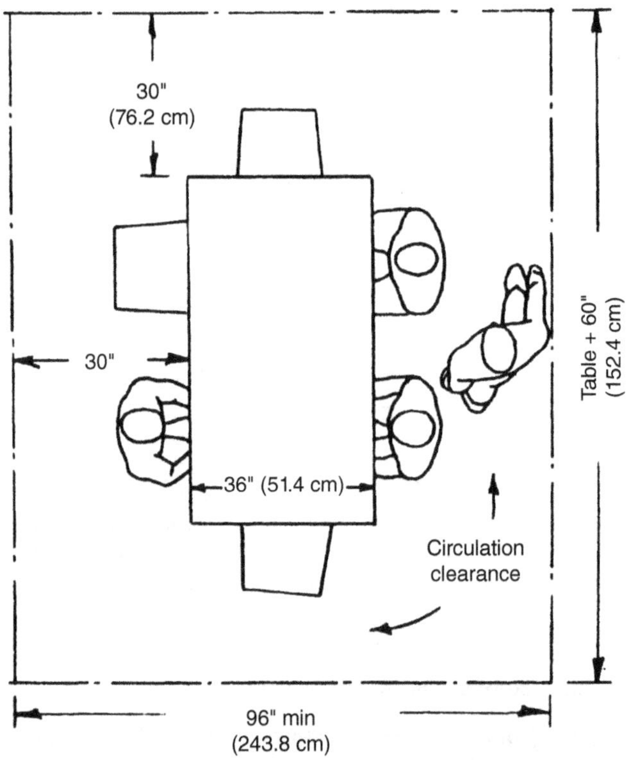

FIGURE 21.12 Table clearance.

8. In planning the commercial dining room, the best table-chair orientation is 48 in. to the general traffic flow, as shown in Fig. 21.13. This set of dimensions should be considered minimum; more clearance is desirable.

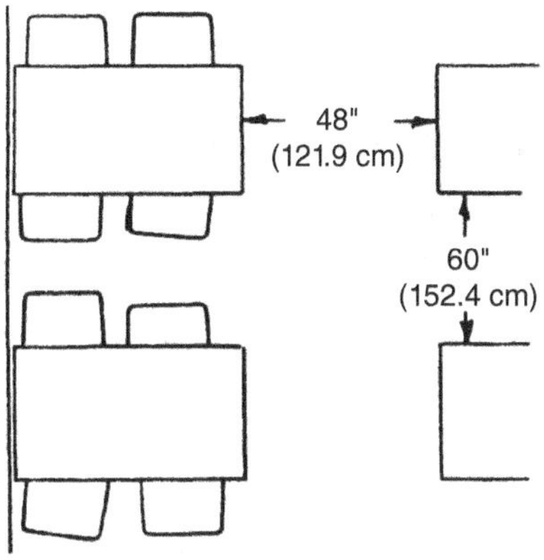

Figure 21.13 Table orientation.

9. The high-efficiency dining-room or cafeteria arrangement shown in Fig. 21.14 illustrates suggested clearance dimensions for that type of facility.

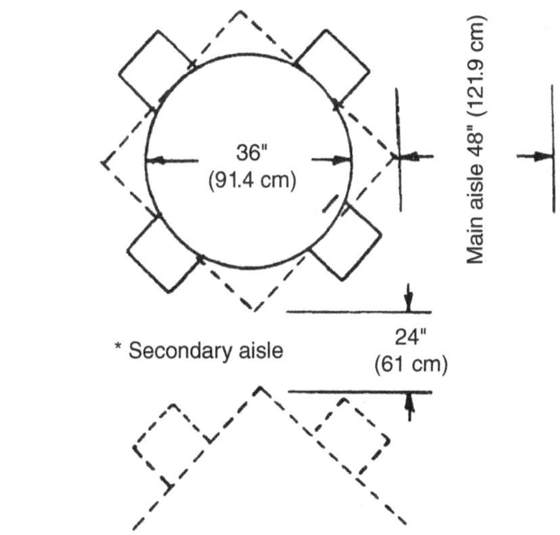

Figure 21.14 High-efficiency clearance.

The important point is that the space envelope for any dining area should be determined on the basis of the furniture, people, and traffic flow.

COMMON HUMAN FACTORS DESIGN MISTAKES IN FAST FOOD RESTAURANTS

The following are mistakes often found in the subsystems of fast food eating establishments. (See Fig. 21.15.) The fast food consumer is often in a hurry and has different human factors needs than a consumer in a more formal restaurant.

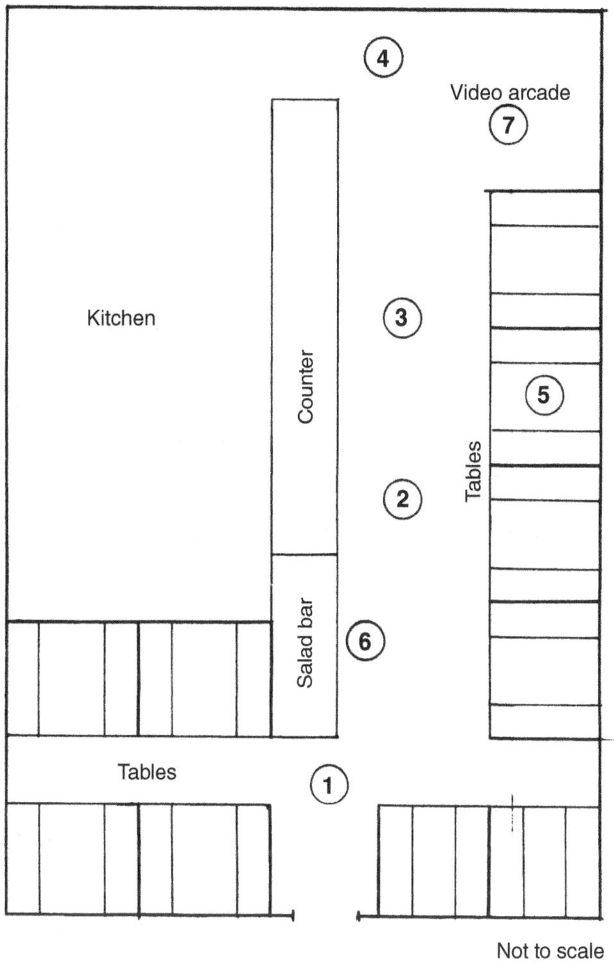

Figure 21.15 Common human factors design mistakes in fast food restaurants.

1. Slippery flooring—Because many of these establishments are open 24 hours, floor cleaning must be done while customers are using the premises. The wet soapy floors, the smooth easy-clean surfaces, and spilled food all lead to slipping accidents. Additionally, high in/out traffic flow in rain or snow can make the floor wet and slippery at the entrances.

2. There is no waiting area or it is placed in an area that obstructs traffic around the order or pickup counter. People often must wait for an order and feel reluctant to occupy a table.

3. There should be enough room in front of the ordering and pickup counters for people to line up during rush periods. Often planners do not plan for peak periods and people must block aisles to stand in line. This creates many hazards because customers are

carrying hot foods. Additionally, restaurant employees are cleaning tables and carrying trays through this congested traffic area. Often many of the customers in the aisles are small children who are difficult for an adult to see over a tray of food.

4. Employee exits from the kitchen side of the order counters are in areas that interfere with customer lines. These exits make it difficult for both the employees and customers to stay out of each other's way.

5. Tables reserved for smokers placed across from the ordering counters and salad bar create a health hazard for both employees and waiting customers. Because these restaurants are small they do not always have a physical barrier between the smoking and nonsmoking sections.

6. In the figure of a fast food floor plan, the salad bar interferes with other activities in the restaurant. Customers at the salad bar must select their salad, avoid slippery floors (possibly spilled food), avoid the line at the ordering counter, avoid employees cleaning, and avoid other customers carrying food to their tables or out of the restaurant.

7. Video arcades should be sectioned off from the rest of the restaurant. If these areas are not removed from other restaurant subsystems they can create high noise levels and often can add to the traffic pattern problems. Carrying food through individuals who are concentrating on a video game can create accidents.

8. No restrooms for customers inside establishment.

CAFETERIA FOOD SERVICE FIXTURES

A wide variety of standard as well as custom food service fixtures are generally available to the designer. However, some of these are less convenient than others. The following are some of the important considerations in selecting an appropriate fixture (see Fig. 21.16):

1. Tray slide racks should be high enough to allow a wheelchair armrest to pass underneath; i.e., there should be at least 30 in. (76 cm) of clearance, but the tray height should not exceed 34 in. (86 cm).

2. There should be at least 10 in. (2.5 cm) of clearance to pass dishes through the glass sanitary shield opening.

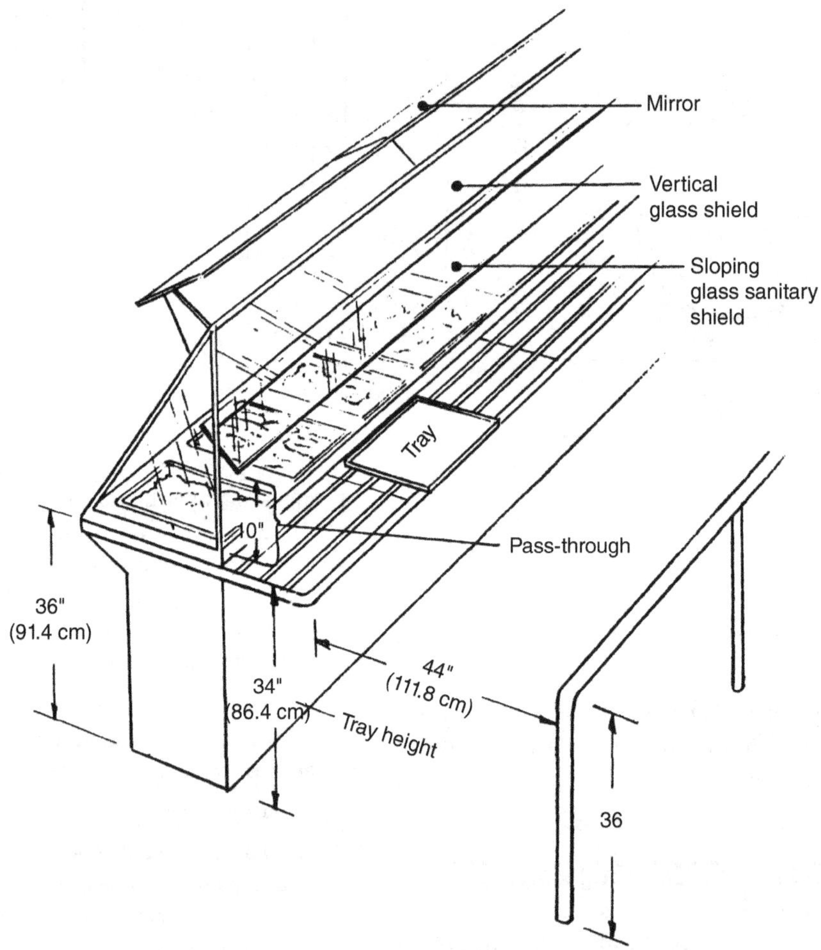

Figure 21.16 Considerations for self-service cafeteria fixtures.

3. There should be overhead mirrors so that the patron in a wheelchair can see what food is available.

4. If there is to be a guardrail to control traffic, it should provide an aisle at least 44 in. (112 cm) wide.

SPACE REQUIREMENTS[8]

A "scramble system" is often best for school cafeterias because it provides fast service on a daily basis. Duplication of service modules minimizes cross-traffic interference. (See Fig. 21.17 and Table 21.1.)

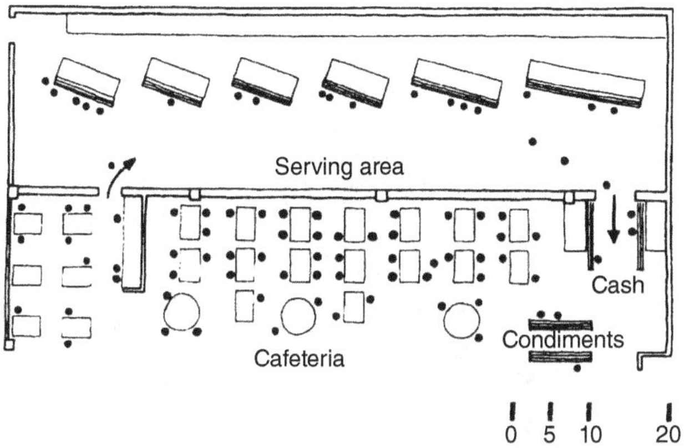

Figure 21.17 Scrabble system.

TABLE 21.1 Food Service Space Requirement Chart for School, College, and Commercial Lunch Programs

Planned Enrollment or Patrons	Seats Required	Area Designation, sq ft: Kitchen*	Area Designation, sq ft: Serving	Number of Counters**
400	170	1500	700	1
500	210	1650	800	
600	250	1800	1540	2
700	290	1950	1540	
800	335	2100	1920	
900	375	2250	1920	
1000	420	2400	2310	3
1100	460	2550	2310	
1200	500	2700	2690	
1300	540	2850	2690	
1400	585	3000	2690	
1500	625	3150	3080	4
1600	670	3300	3080	
1700	710	3450	3460	
1800	750	3600	3460	
1900	790	3750	3460	
2000	835	3900	3850	5
2100	875	4050	3850	
2200	920	4200	4230	
2300	960	4350	4230	
2400	1000	4500	4620	6
2500	1040	4650	4620	
2600	1085	4800	5000	
2700	1125	4950	5000	
2800	1170	5100	5000	
2900	1210	5250	5390	7
3000	1250	5400	5390	
3100	1290	5550	5770	
3200	1335	5700	5770	
3300	1375	5850	5770	
3400	1420	6000	6160	8
3500	1460	6150	6160	
3600	1500	6300	6540	

*Kitchen space:
150–650 students = 3 and 4 sq ft per student
650–2000 students = 2 and 2¼ sq ft per student
2000–6000 students = 1½ and 1¾ sq ft per student
** Counter = 35 to 40 lineal feet of serving equipment.

[8]From J. De Chiara and J. H. Callender, *Time-Saver Standards for Building Types*, McGraw-Hill Book Company, New York, 1973.

CHAPTER 22

OFFICES

GENERAL GUIDELINES FOR CREATING AN EFFECTIVE OFFICE

1. Provide a private office (i.e., a floor-to-ceiling partition or walls) only when necessary. A private office is required where professional or supervisory counseling or confidential discussions are a regular occurrence.

2. Provide open but semi partitioned and/or landscaped offices when the conditions warranting a private office are not present. The open office provides for better interoffice communication, less feeling of being crowded, easier traffic flow (because of the lack of doors that must be closed), and better ventilation.

3. Provide transparent but sound-absorbing partitions where supervisory visual access is required; most workers need to have protection from the noise emanating from neighboring offices and also need a sense of "place," which the semi partition concept provides.

4. Do not put more than two persons in one office or partitioned space. Provide general workrooms and/or conference spaces for those times when workers need to get together for a common activity.

5. Do not separate workers (by means of walls or partitions) who must share items of equipment on a continuing basis.

6. Do not place noisy equipment in the same work area with workers who need the equipment only infrequently; i.e., copiers, sorters, etc., should be separated from clerks, typists, bookkeepers, etc.

7. Use modular office partitioning and furnishing concepts wherever frequent changes in work effort or activities occur; permanent walls generally waste space when an activity changes and/or when undesirable crowding occurs.

8. Determine office location on the basis of the convenience of customers and clients, staff members, support personnel, and/or housekeeping and service personnel. Priority should be based on expected frequency of contact.

9. Centralize library, copy service, filing, rest-room, and other commonly used facilities. Convenience should be equalized among all workers. This also applies to office-exit relationships.

10. Do not isolate some workers from window exposure. All workers like to have a view of the outside.

11. Keep frequently interacting functions on the same floor.

12. Separate customer and worker spaces, except where there is a necessary relationship. Consider private avenues between the main entrance and the special customer-related activities requiring confidentiality.

13. Organize the office space so that there are straight-line pathways, i.e., avoid making workers or customers take devious, winding paths.

14. Provide space and facilities for storage of personal items: coat, umbrella, purse, briefcase, coffee cups, photographs, posters.

THE PRIVATE OFFICE

1. The smallest acceptable private office space is about 7.5 ft × 7.5 ft (2.3 m × 2.3 m). This can accommodate a small table or desk and a maximum of four side chairs.

2. The private office should be designed and arranged so that no one has to face a window while working or conversing with another person.

3. If the office does not have a window of its own, it is recommended that the floor-to-ceiling partition be transparent or divided so that the upper half is transparent. Curtains can be used if the occupant needs to have visual privacy for brief periods.

4. Each private office should have good general illumination. If the office is less than about 15 ft (4.6 m) square, the ceiling and walls should be painted a light color to provide adequate light distribution.

5. The ventilation system should be such that independent control can be accomplished within the office, even though the door may be shut. Do not place a general temperature-sensing control for a suite of offices in one office.

THE LANDSCAPED OFFICE

The landscaped office concept can be accomplished either through the use of special furnishings (i.e., the furniture arrangement can be used to separate various workers visually) or by means of partial partitions. When using partitioning it is particularly important not to create unnecessary dead ends. Not only are these confusing to the person who is trying to locate someone, but they may also be a safety consideration in the event of an emergency escape. In designing or selecting a modular partitioning or furniture system, choose one which is sturdy and self-supporting, which does not have "feet" that can become tripping hazards, which provides good acoustic characteristics, and which wears well and is easily cleaned.

OFFICE SUPPORT REQUIREMENTS CHECKLIST

1. Electric outlets
2. Telephone and digital communication outlets
3. General materials storage
4. Drinking fountain
5. Restrooms
6. Security systems, alarms, fire extinguishers, etc.
7. Housekeeping equipment storage

FACILITIES FOR THE HANDICAPPED

Most offices will be occupied by either handicapped workers or handicapped clients at one time or another. Therefore, every office should be examined in terms of removing those barriers that make it either difficult or impossible for a handicapped person to enter and use the office. The principal considerations are the following:

1. Wide doorways and hallways
2. Ramps and elevators
3. Tactile cues for the blind
4. Special rest-room fixtures
5. Special furniture to accommodate wheelchairs

OFFICE IDENTIFICATION AND SIGNING

From a human factors point of view, proper office identification and signing are one of the most important requirements of office planning. Too often this requirement either is forgotten or is purposely overlooked for aesthetic reasons. *Do not ignore this requirement!* There is no reason to create confusion purposely; in addition, it is possible to make the signing scheme both useful and attractive.

ENVIRONMENTAL CONTROLS

Good lighting and comfortable air conditioning are prerequisites of efficient office work. Refer to the sections on environmental control and on lighting. These controls, as well as acoustic planning, should be accomplished as part of the initial office facility planning. When this has not been done, utilize special consultants in each of these areas to assist in devising the best systems to overcome any faults the basic building may already possess.

"OPEN OFFICE" MODULAR SYSTEMS

The popularity of open office systems is due to the flexibility that such systems offer, plus the fact that they considerably reduce the amount of time and labor required to rearrange an office space when a change in workplace layout is required. (See Fig. 22.1.)

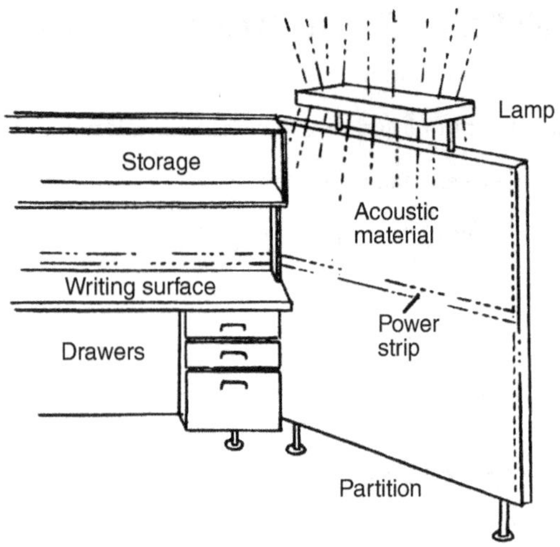

FIGURE 22.1 Modular systems example.

In choosing a particular system, the following should be considered:

1. Partitioning modules can be made to stand sturdily without being secured to the primary building structure. Partitions should be designed so that other modules can be attached at any level.

2. Partitions should provide reasonable nonpenetration of sound and should permit the application of sound-absorbing material when necessary.

3. Partitions should contain wiring or cables for desired electric power, plus communication lines and ambient lighting module additions. Cable connections should be accessible for both overhead and floor-level attachment.

4. Partitions should have adjustable legs so that the partition can be either lowered to the floor or raised to allow ventilation beneath the panel.

5. Hung-type modules (which are more desirable from the standpoint of ease of cleaning beneath them) should include the following:

   a. Writing and drawing surfaces

   b. Office machine support surfaces (typewriters, etc.)

   c. Drawers

   d. Bookcases

   e. Cabinets (with sliding, tambour, or folding doors)

   f. Special supports for telephones

   g. Supplementary and overhead ambient light fixtures

6. Ease of assembly and disassembly should be considered; i.e., avoid systems that can be safely erected or changed only by factory specialists or that can be wired only by licensed electricians.

EXTERIOR BUILDING CONFIGURATION AND IDENTIFICATION

Although business identification is an important aspect of all architectural planning, identification of the large office building is perhaps even more critical because of the nature of its transient user needs. Many visitors may visit the building only once or infrequently, and when they do, they often are in a hurry. Figures 22.2–22.4 emphasize several important aspects of building identification:

1. Identification from a distance helps the person plan ahead and become prepared to look for specific streets, numbers, and/or special directional signs to parking and driveways. (See Fig. 22.2.)

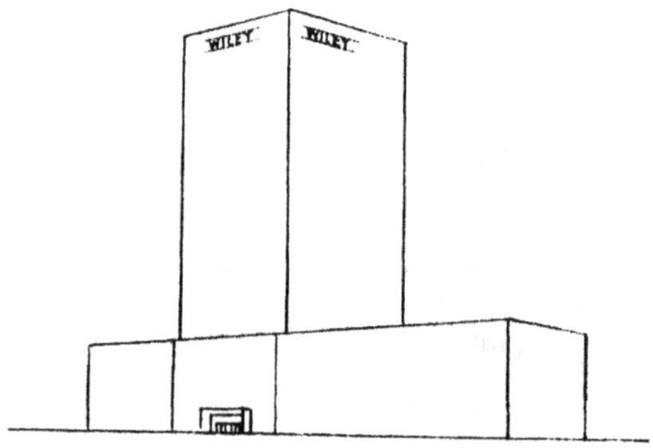

FIGURE 22.2 Exterior building configuration—identification from a distance.

2. Identification near the main building site is important, especially when there are a number of similar office buildings located side by side. Thus, as Fig. 22.3 illustrates, the identification (e.g., the building name and the street number) must be placed lower, or nearer the observer's normal line of sight. Note that the size of identification signs and/or numbers must be commensurate with the viewing distance considered appropriate for the observer's planning needs.

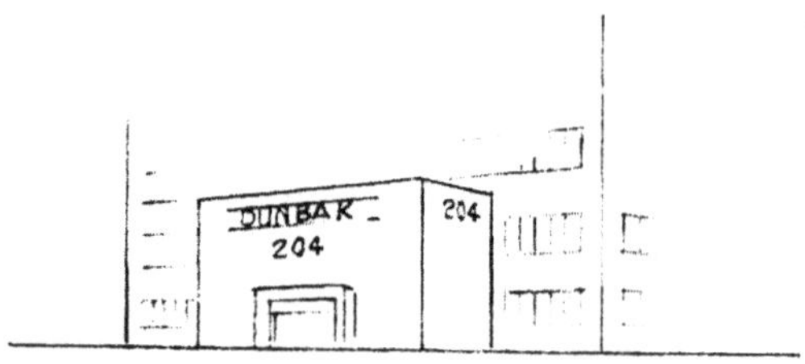

FIGURE 22.3 Exterior building configuration—identification near the main building site.

3. Identification of parking entrances is especially critical when the office building is located in a high-density area. The location and size of the sign must be

such that it can be seen in time for the person who is driving to plan any lane changes, and it should not be blocked from view by other cars or pedestrians that may be located between the driver and the sign. (See Fig. 22.4.)

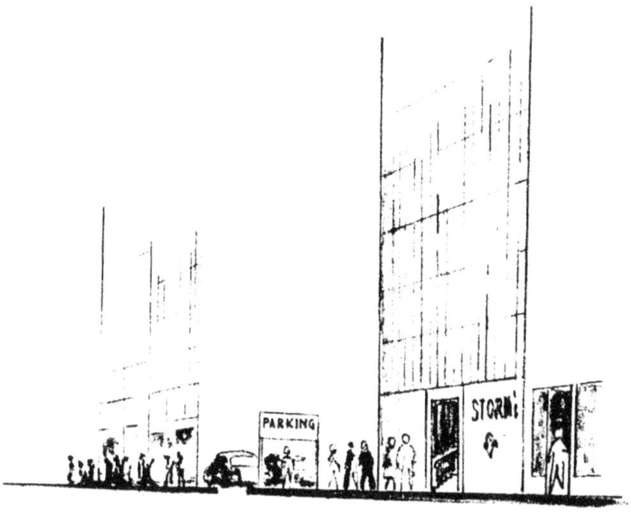

Figure 22.4 Exterior building configuration—identification of parking entrances.

Avoid the temptation to leave identification problems until later. In many cases the shape of the proposed structure and/or special ideas about decorative symbolism make it impossible to add an adequate identification system.

INTERIOR PLANNING

Interior Office Building Planning Considerations

An important planning question is: What functions do I locate on which floor of the multistory office complex? The following checklist provides a number of guidelines.

Ground Floor (Street Level)

1. Lobby—information directory or personnel station
2. Stairs and elevators
3. Building management offices
4. Security office or offices
5. Special tenant facilities (i.e., a bank, a restaurant, a travel agency, and commercial shops that require walk-in traffic exposure and convenience)

Upper Floors

1. Foyer (adjacent to elevators and stairs)
2. Restrooms
3. Professional tenant offices
4. Emergency stairs

Special Areas

1. Underground parking level or levels, with stair, elevator, and ramp access.
2. Tenant cafeteria (preferred at midlevel for essentially equal convenience to all upper floors). This requires a special service elevator.
3. Sun deck (preferred at the upper level for tenant use only). The location should ensure adequate sun exposure during the times it would normally be used.
4. Combined public and tenant patio (preferred at ground level, probably associated with a ground-floor restaurant establishment).
5. Shopping arcade (preferred at ground level for easy access for walk-in traffic).

The next planning question is: How do I plan the interior space on specific floors to provide the maximum utility for various tenants? In the case of a special long-term tenant for whom the floor space is to be customized, a detailed analysis of that customer's needs and unique desires is, of course, necessary. For general planning purposes, however, the suggestions on the following pages provide useful guidance. However, these suggestions are intended primarily to ensure adequate consideration of the human factors.[9]

General Interior Planning Guidelines

Although there are considerable differences between various building floor plans, key human factors considerations should be kept in mind. Figure 22.5 emphasizes some of these factors.

Travel Distances

1. Locate stairs and elevators so as to reduce the distances tenants must travel and to make more than one emergency escape pattern possible.
2. Locate restrooms so as to reduce the distances tenants must travel; also provide visitor facilities off the lobby near the entrance (i.e., avoid hiding restrooms from visitors). *(The restroom is the most important facility within any public building. The past practice of trying to make such facilities inconspicuous and/or private is not consistent with human need.)*
3. Locate maintenance closets so that the building maintenance personnel do not have to carry equipment long distances.
4. Locate utility rooms or closets so that technicians do not have to travel considerable distances between these and office spaces.
5. Provide a service elevator that is independent of the normal passenger elevators and that connects with a service entrance rather than with normal tenant or visitor entrances.

---

[9]From J. De Chiara and J. H. Callender, *Time-Saver Standards for Building Types*. McGraw-Hill Book Company, 1973.

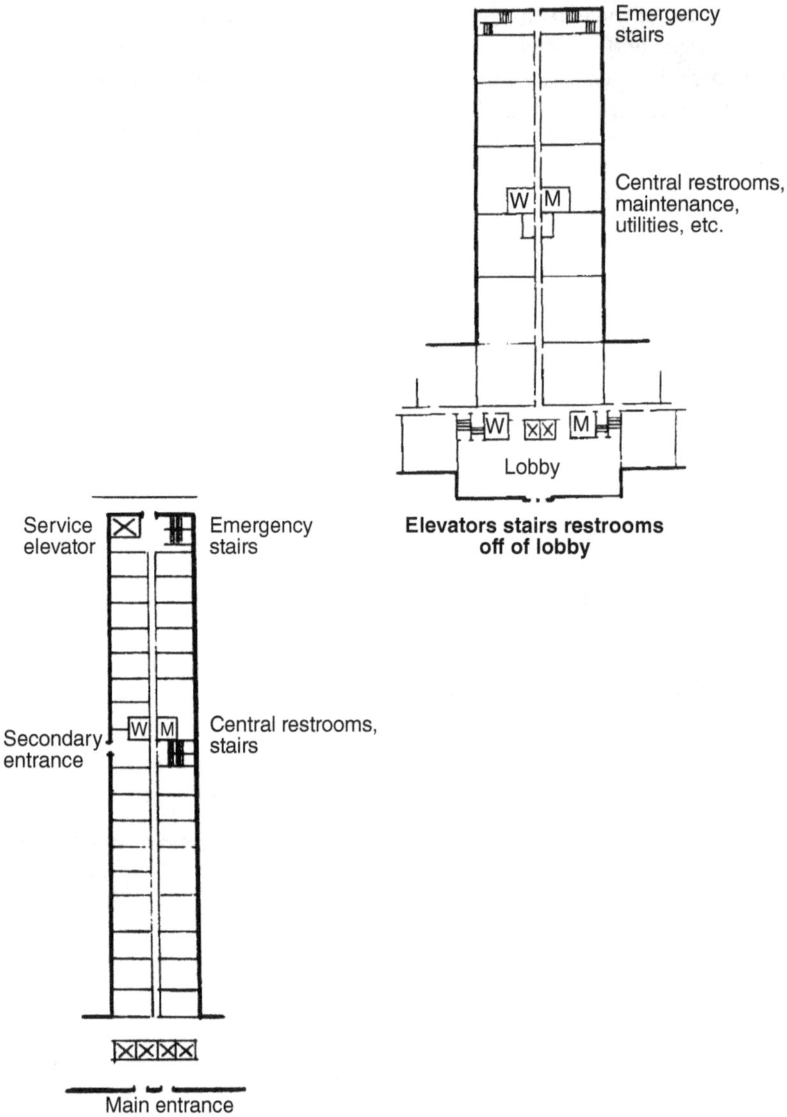

Figure 22.5 Office building interior examples.

6. Plan emergency escape facilities that minimize the distances that the person in a wheelchair must travel to an elevator or to an exit. Many such special facilities are placed where the person in a wheelchair has to travel further than anyone else.

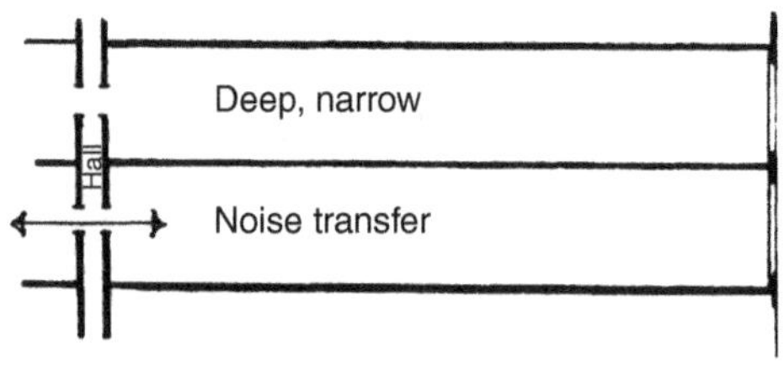

Figure 22.6 Avoid narrow offices with doors placed opposite each other.

GENERAL PERMANENT SPACE COMPARTMENTALIZATION

When permanent walls between general tenant office spaces are planned, consider the following guidelines:

1. Avoid deep, narrow spaces. People need to have visual access to the outside. (See Fig. 22.6.)

2. The equal-sided space is generally more easily utilized by the individual tenant for sub partitioning and/or "open-space planning." In addition, it offers more opportunity to provide window and light opportunities for more people.

3. Door positions should be staggered, not placed opposite each other, as shown in the Fig. 22.7. Not only does this concept introduce more noise transfer, but it also creates potential traffic interference.

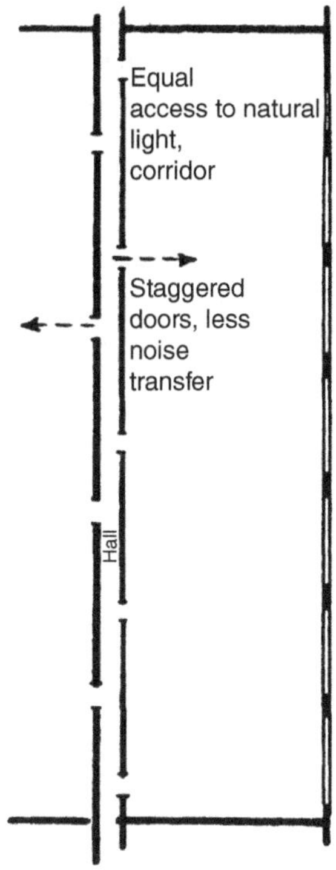

Figure 22.7 More desirable office layout.

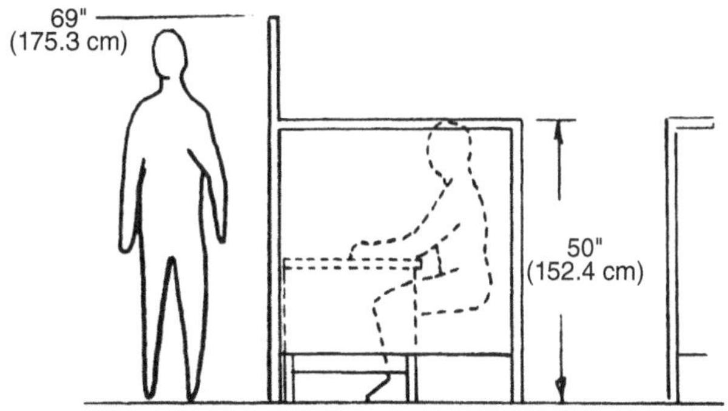

Figure 22.8 Partitioning for semiprivacy—line of sight minimized from the main corridor.

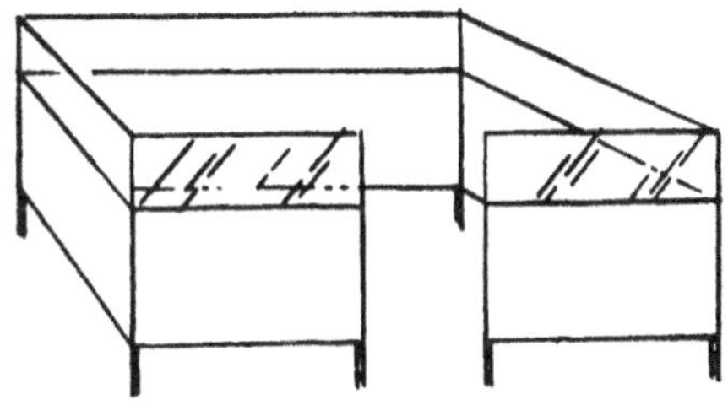

Figure 22.9 Partitioning for semiprivacy—transparent section at the top of partition.

4. Where feasible, the space configuration shown in Fig. 22.7 is more desirable, for several reasons:

   a. There are equal window and light opportunities for all tenant staff members.

   b. There is more direct access to any inner offices that the tenant may have established by means of partial partitions.

   c. There is more direct access by all tenant staff members to the corridor in an emergency.

PARTITIONING FOR SEMIPRIVACY

Many modular partitioning systems are available to the designer. Figures 22.8–22.10 illustrate several points to consider in choosing a system:

1. Provide a high partition where the line of sight should be minimized from a main corridor but where visual communication may be desirable within an area. (See Fig. 22.8.)

2. By using a transparent section at the top of the partition, it is possible to provide an internal line of sight and still reduce the transfer of unwanted sounds. (See Fig. 22.9.)

3. Combine high and low partitions, but with air circulation at foot level. (See Fig. 22.10.)

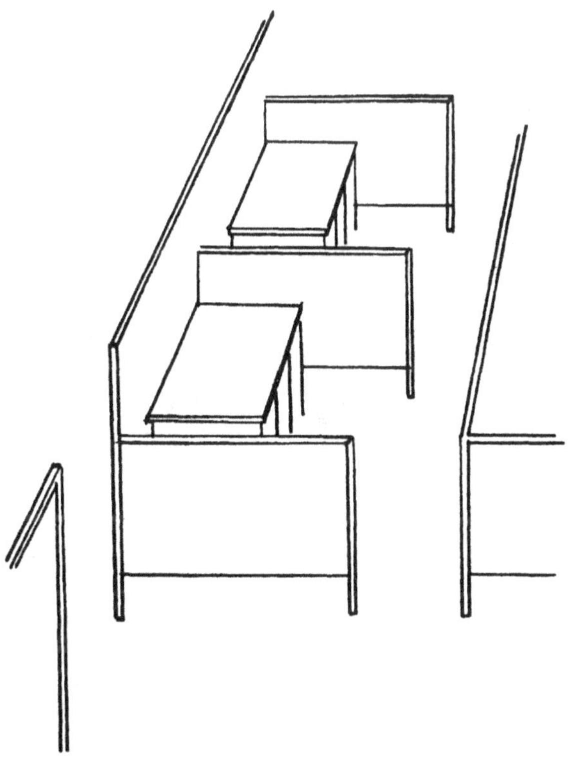

Figure 22.10 Partitioning for semiprivacy—combination of high and low partitions.

OFFICE BUILDING LAYOUTS[9]

Various types of office building layouts are shown in Figs. 22.11–22.13.

ESTIMATING FLOOR SPACE

Hints for Estimating Floor-Space Requirements

To estimate floor-space requirements in offices, consider both the area occupied by furnishings and that occupied by office personnel under all anticipated dynamic movement patterns (i.e., moving a chair away from a desk, swinging around to a file cabinet, reaching for the wastebasket, getting out of a chair, and passing by furniture and/or other people). Figures 22.14–22.20 provide general dimensional suggestions for typical situations.

1. The typical L-shaped secretarial configuration requires a floor area as indicated in Fig. 22.14. If the secretary greets visitors on either side of the desk, additional space is required as indicated. Generally there

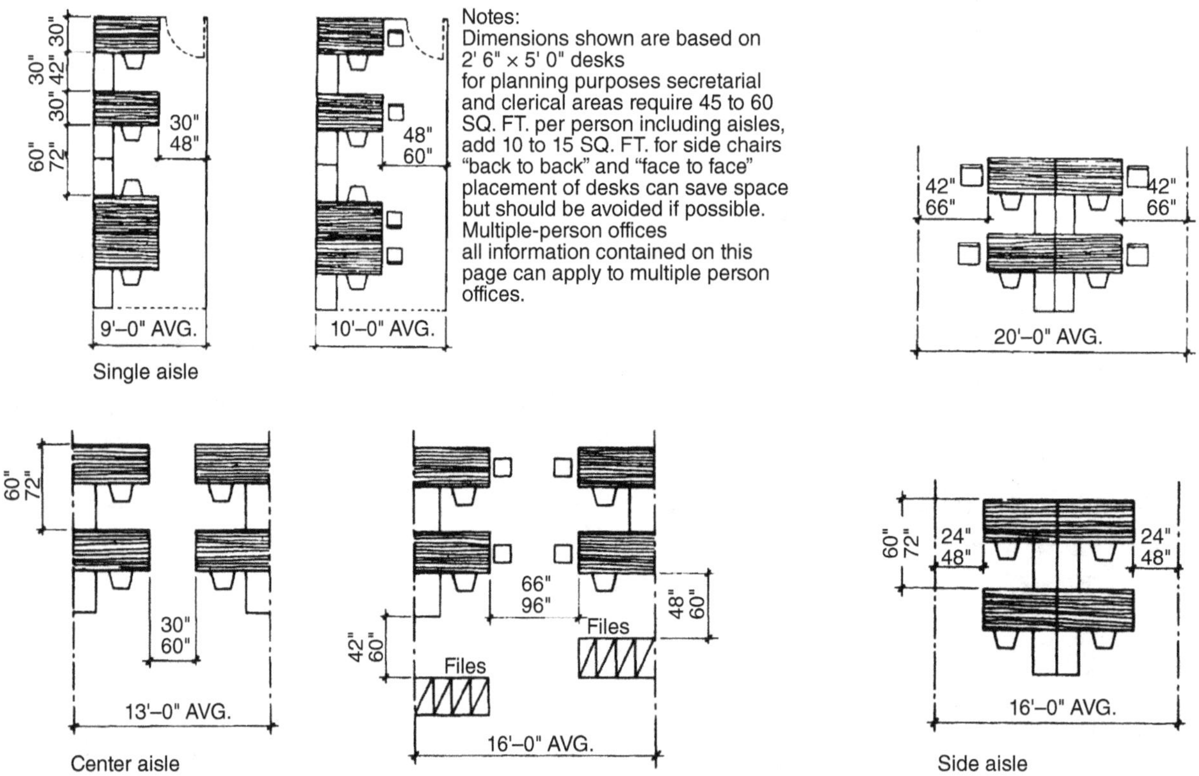

FIGURE 22.11 Conventional office layouts.

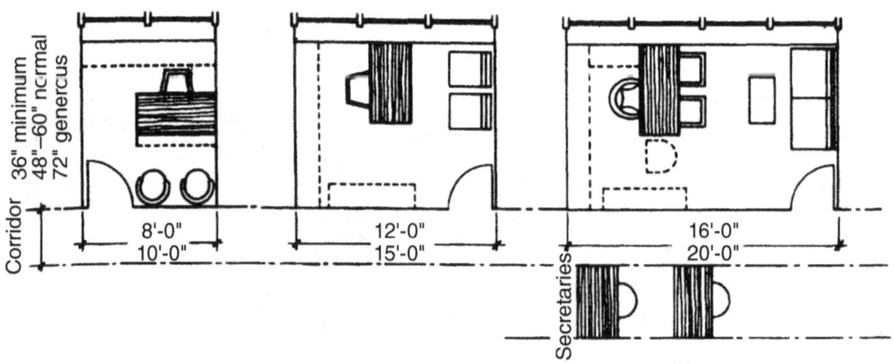

FIGURE 22.12 General clerical work spaces.

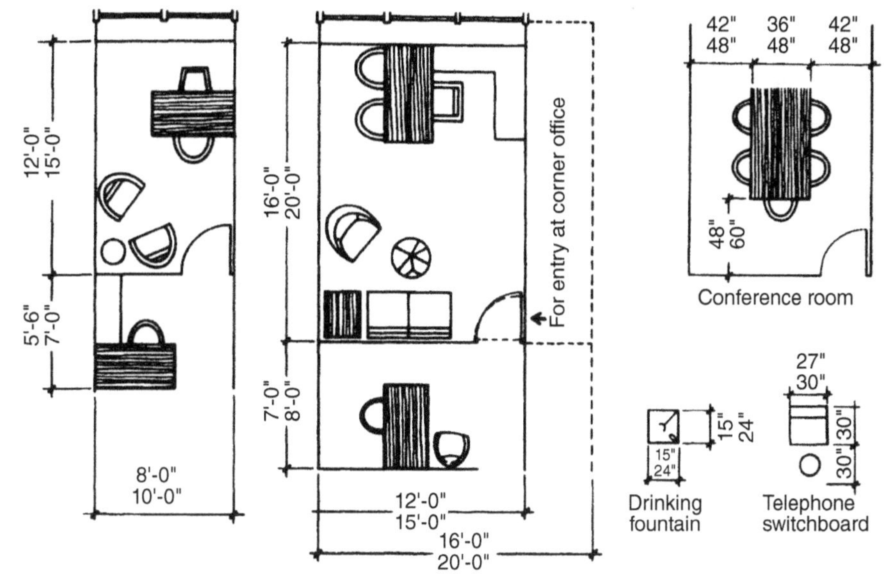

Figure 22.13 Private offices.

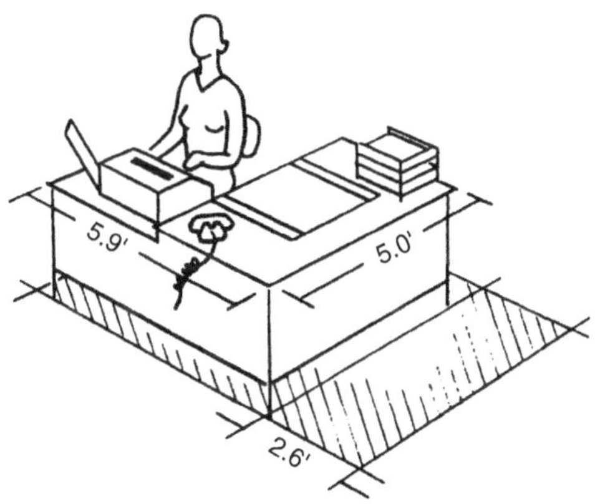

Figure 22.14 L-shaped secretarial configuration.

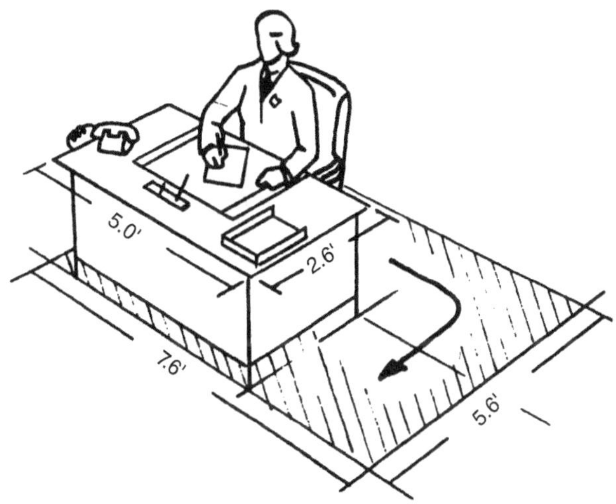

Figure 22.15 Minimum office desk size.

is enough space beneath the desk for a small wastebasket. The secretary usually leaves the desk toward the rear; the space required for this is included within the total area shown.

2. The minimum office desk size that is acceptable for any kind of office work is shown in Fig. 22.15, along with the additional floor area required so that the occupant can move away from and around his or her desk. This assumes that such a worker does not ordinarily greet or communicate with others at the desk, and thus no added space has been indicated for visitor floor space.

3. The more appropriate standard executive desk (conference type) is 6 ft × 3 ft (1.8 m × 0.9 m), as shown in Fig. 22.16. The minimum working floor space (which includes room to move away from and around the desk,

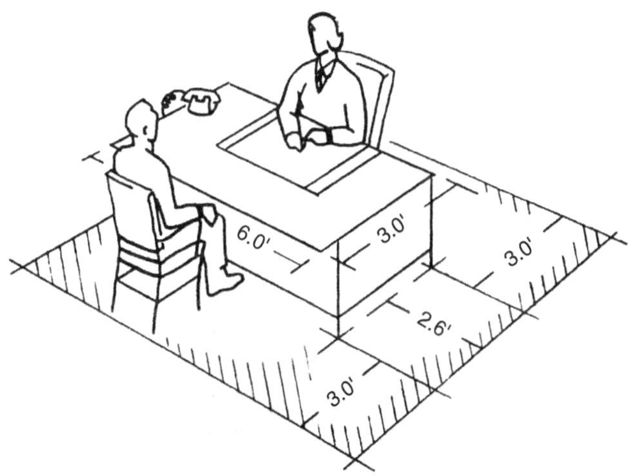

Figure 22.16 Executive desk.

plus space for one or two visitors sitting opposite the primary occupant) is indicated in the figure. This area should be considered minimum for such activities as counseling clients and discussing sales, but it is too small for extensive work that may require adjacent storage of reference materials.

4. A preferred working space for the executive who not only meets a few clients or counselees but also does a considerable amount of work requiring additional file storage space is shown in Fig. 22.17. Note the additional clearance for walking past a visitor (i.e., it should be possible to walk past a visitor while he or she sits in the side chair undisturbed).

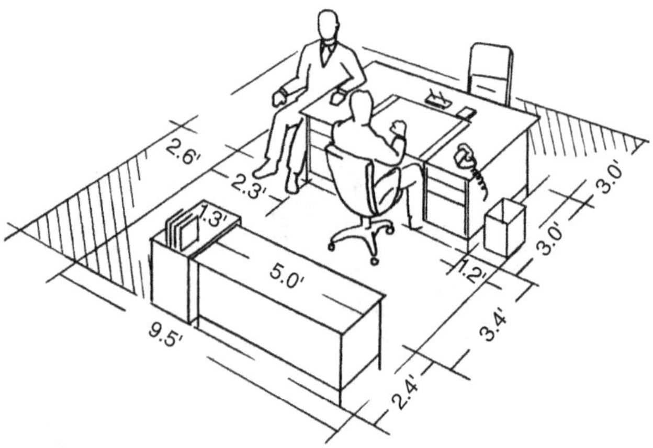

FIGURE 22.17 Working space for executives.

5. Minimum clearance in front of the standard metal storage cabinet is shown in Fig. 22.18. If, however, it is expected that other people may have to pass by while the secretary or other office person is working at the cabinet, an additional 24 in. (61 cm) of clearance is necessary to avoid conflicts. Even more clearance is required if the individual passing is pushing a portable cart or carrying heavy packages. Therefore, a good rule of thumb is always to provide at least 30 in. (76 cm) behind anyone who is working from a standing (dynamic) position.

FIGURE 22.18 Clearance for a metal storage cabinet.

6. Minimum clearance in front of the typical metal file cabinet is shown in Fig. 22.19. Once again, these dimensions assume that no one is passing behind the person. If this is expected, use the 30-in. (76-cm) clearance rule of thumb.

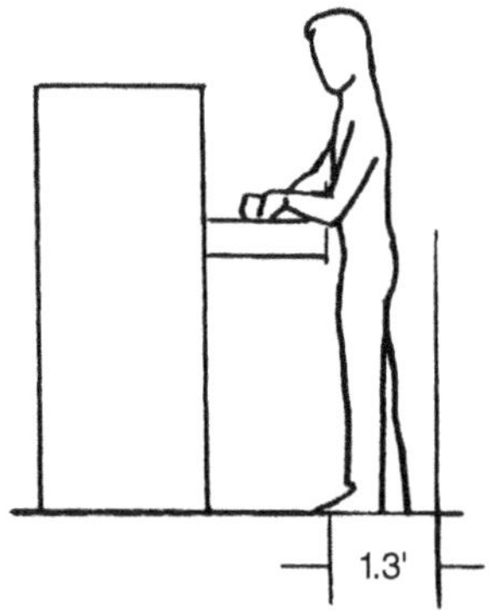

FIGURE 22.19 Minimum clearance for a metal file cabinet.

7. Various other working situations call for consideration of the specific dimensions of the equipment being used. As shown in Fig. 22.20, certain basic clearances are required just so that the operator (in this case, a draftsman) can move the stool around to more convenient positions. In addition, note the suggested clearance behind the board to allow large drawings to drape across the board without being creased or torn.

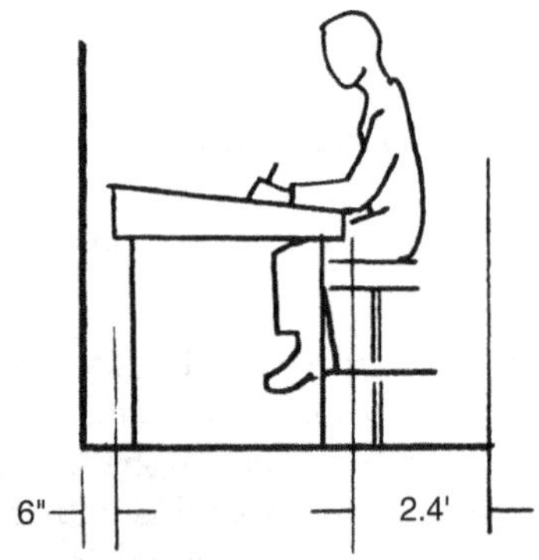

FIGURE 22.20 Other basic clearances.

Typical Dimensional Characteristics of Office Furniture[8]

For estimating floor space, it is recommended that the following guidelines be used:

Desks: Secretary—60 in. × 30 in. (152 cm × 76 cm); executive secretary—72 in. × 36 in. (183 cm × 91 cm); supervisory personnel—72 in. × 36 in. (183 cm × 91 cm); other executives and managers—72 in. × 36 in. (183 cm × 91 cm); or larger. (See Tables 22.1 and 22.2 and Fig. 22.21.)

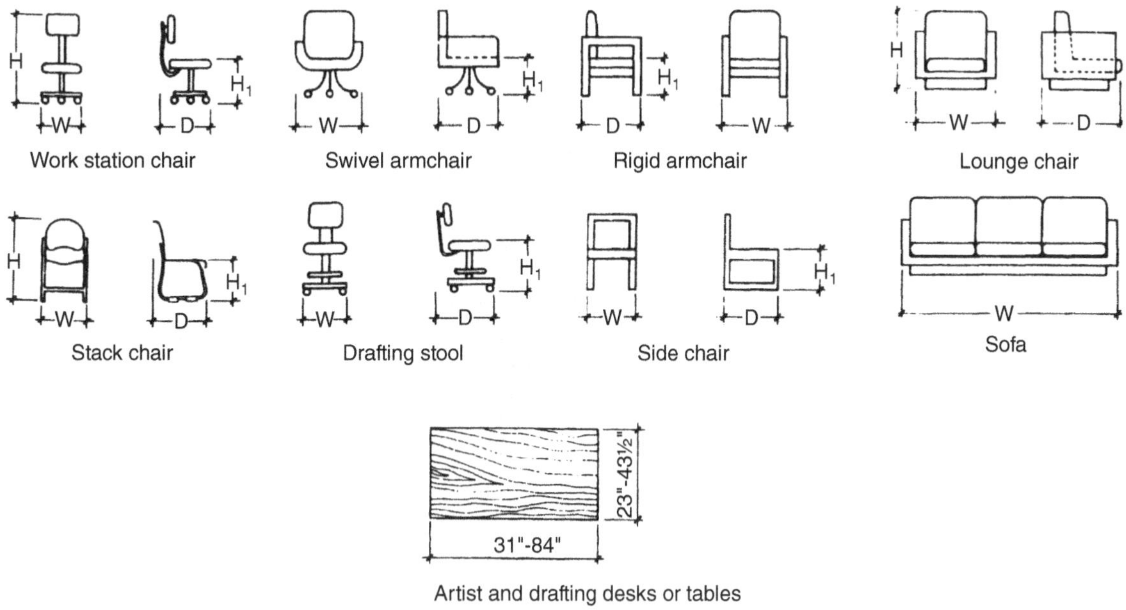

FIGURE 22.21 Furniture dimensions to determine floor space.

TABLE 22.1 Chair Dimensions

	Secretarial		Swivel Armchair		Rigid Armchair		Stack Chair		Rigid and Adjustable Drafting Stool		Side Chair	
	Std	Range	Std	Range	Std	Range	Std	Range	Std	Range	Std	Range
W	1'5"	1'4"–1'8"	2'4"	1'8"–2'6"	1'10"	1'6"–2'3"	1'9"	1'6"–1'11"	1'6"	1'5"–2'0"	1'8"	1'4"–2'0"
D	1'-7½"	1'6"–2'0"	2'3"	1'8"–2'6"	1'10"	1'7"–2'8"	1'9"	1'7"–1'10"	1'8"	1'6"–2'0"	1'10"	1'6"–2'8"
H	2'6"	2'5"–2'10"	2'9"	2'6"–3'0"	2'6"	2'4"–2'10"	2'6"	2'4"–2'9"	3'0"	2'11"–3'6"	2'6"	2'4"–2'10"
H_1	1'5"	1'4"–1'8"	1'5"	1'4"–1'10"	1'6"	1'4"–1'7"	1'5"	1'5"–1'6"	2'4"	1'5"–2'10"	1'6"	1'5"–1'7"

TABLE 22.2 Lounge Chair and Sofa Dimensions

	Lounge Chair		Sofa
	Std	Range	
W	2'6"	2'6"–3'4"	D, H, and H_1 similar 2 seats 5'0"–6'7" 3 seats 6'0"–7'6" 4 seats 7'8"–9'0"
D	2'7"	2'2"–3'4"	
H	2'6"	2'1"–3'4"	
H_1	1'3"	1'0"–1'6"	

CHAPTER 23

AUDITORIUMS AND THEATRES

AUDITORIUM DESIGN GUIDELINES

Patron Seating Arrangement

Primary attention should be given to the positions of the worst seats in the house in terms of viewing distances and angles. Consider the following:

1. No patron should be seated so close to the stage that he or she has to look up more than about 15° continuously (a maximum of 30°) in order to observe actors, singers, or other performers on the stage or screen. (See Fig. 23.1.)

2. No patron should have to keep his or her head turned continuously to the right or left more than about 15° (a maximum of 30°) in order to watch a performance. (See Fig. 23.1.)

3. Surveys of where people sit indicate that they invariably avoid seats that are outside the angular range illustrated in Fig. 23.2, even though they could probably still see adequately from these seats.

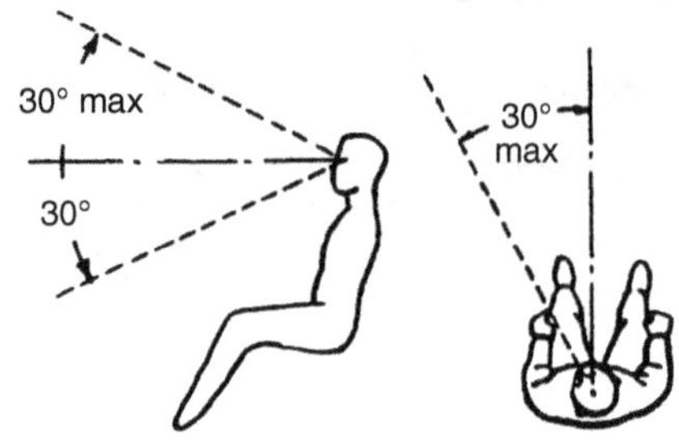

FIGURE 23.1 Maximum viewing angles for patrons.

4. Patrons seated in balconies should not have to look down more than about 30° in order to see the action at the leading edge of the stage. (See Fig. 23.1.)

5. Patrons in the last row of balcony seats should be able to see as much of the upper portion of scenery backdrops as required to appreciate its relationship to the stage action.

6. Patrons in front seats on the main floor should be able to see the stage surface.

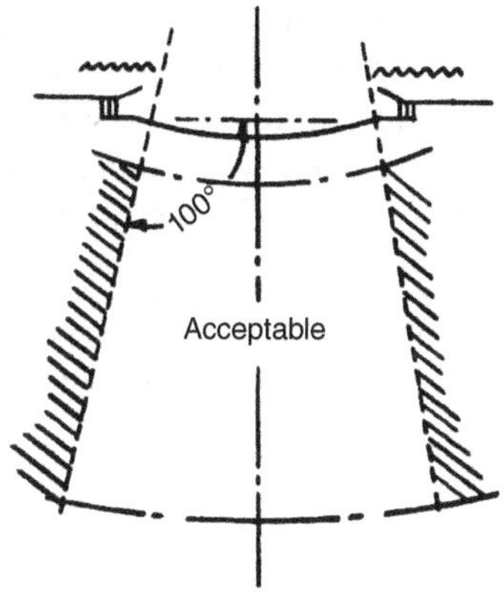

FIGURE 23.2 People avoid seats outside the acceptable viewing angle.

7. The angle and distance for viewing a screen from the front row should be as shown in Fig. 23.3.

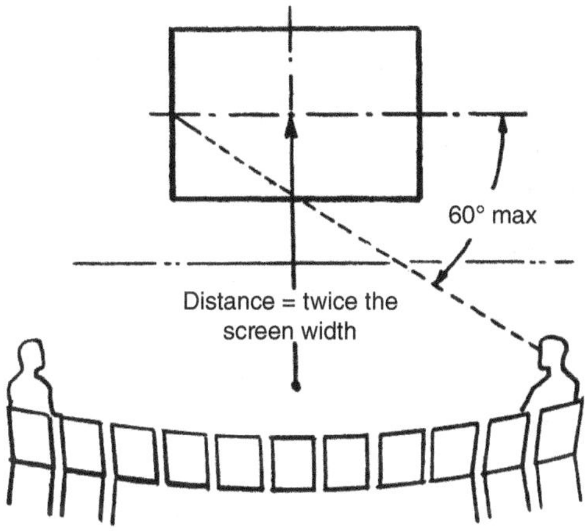

FIGURE 23.3 Viewing angle and distance for front row seats.

8. The maximum viewing distance to recognize a performer's facial expression is about 75 ft (23 m); gross gestures, about 125 ft (38 m); and gross movements, about 150 ft (46 m).

9. The relative arrangement of seating and stage appurtenances (e.g., scenery and curtains) should be such that patrons see only what they are intended to see.

Minimize patron discomfort (i.e., having to keep the head in awkward positions for long periods), consider the suggestions illustrated in Figs. 23.4 and 23.5: arrange seats so that patrons can see past patrons in front of them.

Use a sloping floor rather than a step configuration wherever possible, but do not make it so steep that people fall down or so that persons in wheelchairs cannot negotiate the aisles. (See Fig. 23.4.)

There should be enough space between seats so that patrons can get in and out with minimum annoyance to other patrons. (See Fig. 23.5.)

Tilt a movie screen so that both main-floor and balcony observers have an approximately equal chance to observe a good screen image. (See Fig. 23.6.)

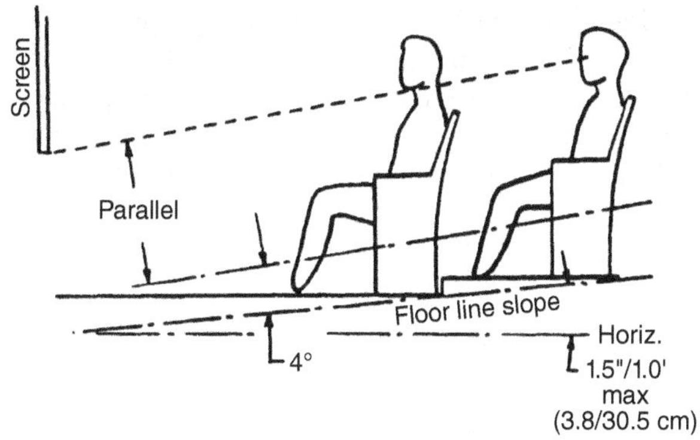

FIGURE 23.4 Sloping floor rather than step configuration.

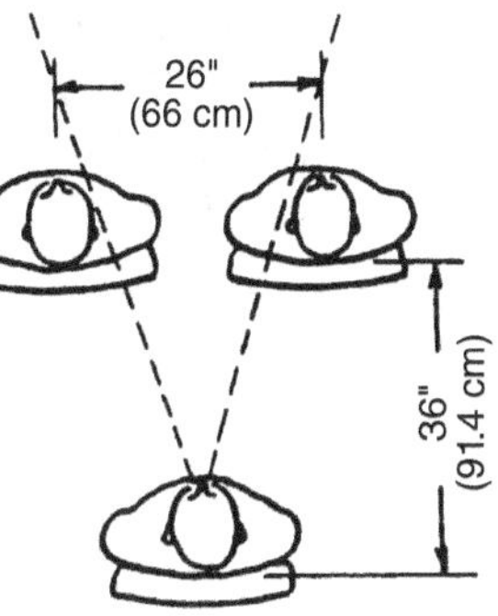

FIGURE 23.5 Space between seats.

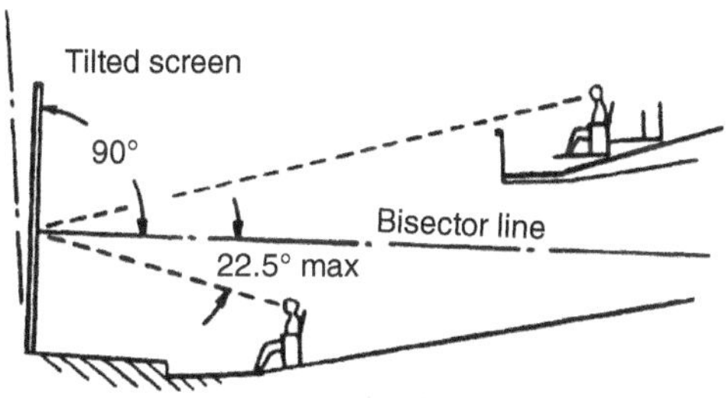

FIGURE 23.6 Tilted movie screen.

Although a generally "fan-shaped" seating arrangement is desirable, it requires that considerable individual seat adjustment be made in order to create straight aisles and not to include hazardous pockets where patrons could become entrapped during an emergency evacuation. (See Fig. 23.7.)

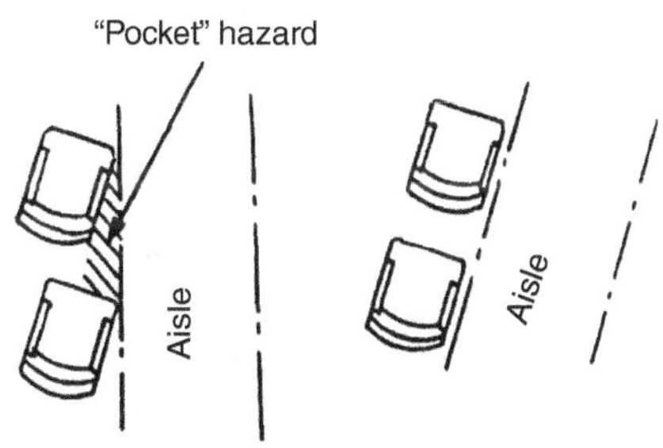

FIGURE 23.7 Fan-shaped seating.

Aisles should be large enough for two people to pass each other comfortably in the darkened theatre. Because the central section of any seating arrangement contains the best viewing positions, it is suggested that a central aisle be avoided and that a configuration such as that shown in Fig. 23.8 be used.

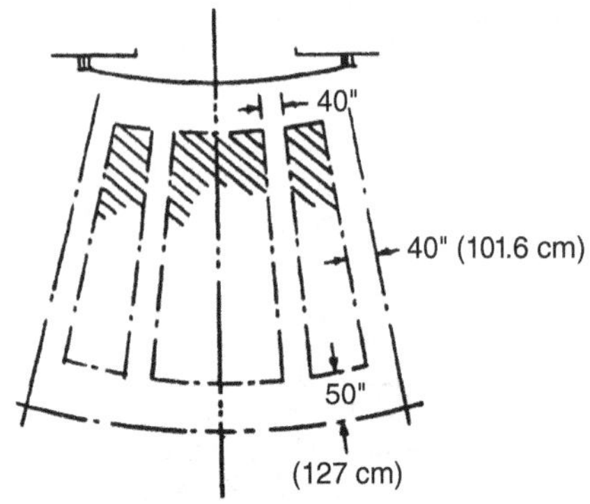

FIGURE 23.8 Aisle size.

Max of seven seats "to" an aisle. Avoid very long rows of seats. It should not be necessary for the patron in the center seat of any row to have to pass over more than about seven seats to get to one aisle or another. (See Fig. 23.9.)

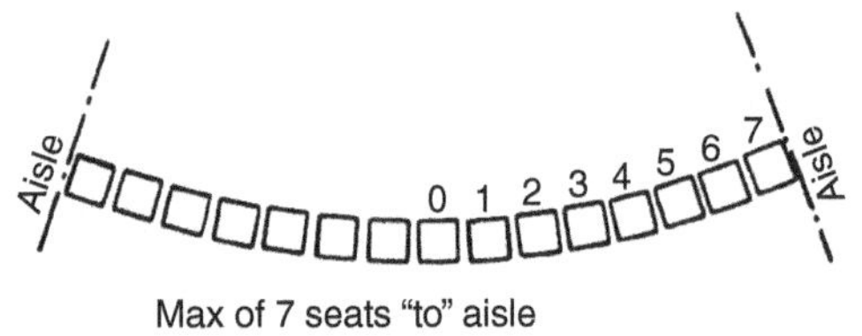

FIGURE 23.9 Max seats to an aisle.

GENERAL REQUIREMENTS FOR AUDITORIUM SOUND, LIGHT, AND TEMPERATURE CONTROL

Sound Systems

Sound systems should be designed in conjunction with the architectural plan, rather than added on after the architectural features and constraints are already fixed. The acoustic features of the auditorium and the location of the microphones and loudspeakers must combine to create speech or music that sounds as natural as possible (i.e., there should be minimum distortion in terms of frequency, intensity, and/or identification of sound source direction). Among the important sound system characteristics to consider are the following:

1. Audience: The speech or music should appear to come from the speakers, actors, or musicians when they can be seen by the audience. Microphones and loudspeakers should be positioned so that individual sounds from actors on opposite sides of a stage or sounds from instrumental groups on opposite sides of an orchestra appear to emanate from the portions of the stage where the actors or orchestra members are being observed. A sound control operator should be positioned approximately in the center of the audience area so that he or she can monitor and balance the sound from the audience's point of view.

2. Performers: The sound system should be designed so that a single lecturer or speaker, stage actors, and/or musicians can hear themselves and their fellow performers. Means should be provided for "covert communication" between stage performers, stage or orchestra directors, and sound booth operators.

Lighting Systems

Separately controllable lighting systems should be provided for the following:

1. General Ambient Illumination: Even distributed, general ambient illumination should be provided for audience areas (at least 30 fc), the general storage area, and backstage areas to facilitate movement of people and equipment and to allow the audience to read programs.

2. Auditorium aisle Illumination: Aisleways should be illuminated when the main house lights are dimmed.

3. Stage Lighting: A variety of stage lighting systems should be considered, depending on the types of performances anticipated for the auditorium. The following are among the more important considerations:

   a. Footlights, floodlights, and spotlights for illuminating the performers must be provided.

   b. Overhead stage lights for illuminating major portions of a stage (e.g., illuminating music for an orchestra) should be provided.

   c. Special-effects stage lights should be provided to create various color and brightness conditions to represent day and night changes during a play or opera or to create special psycho logical effects, as when red is used to indicate terror.

   d. Emergency lights must be provided throughout the auditorium so that, in the event of a power failure, the audience, the performers, and the theatre staff personnel can find their way to safety.

   e. Independent control of each of the above lighting systems should be provided. Dimmer circuits should be provided for each system. Lighting control stations should be located so that operators have an optimum position for observing the effects of their light system adjustment.

4. Orchestra Pit lighting: An appropriate number of conveniently located electric outlets should be provided within an orchestra pit so that musicians' music stand

lights can be connected with a minimum of loose extension cables entwined across the pit (an extreme hazard when orchestra members have to find their way to and from their positions in the dark).

5. Speaker lectern light control: A supplementary lectern luminaire should be provided so that lecturers and speakers can see their notes. In addition, it is desirable to provide on the lectern controls for house lights so that the speaker can turn off house lights, turn on projectors, raise and lower a movie or slide screen, etc.

Temperature Control Systems

It is generally desirable to provide an automatically controlled temperature and ventilation system that adjusts to the conditions of the auditorium. A separate air-conditioning system may be desirable for the stage areas, where heat from special lighting often makes the performance areas almost unbearably hot. Care should be taken during the design of the auditorium stage areas and during the air-conditioning system installation not to introduce undesirable noises that may interfere with satisfactory hearing conditions, i.e., noises made by equipment or airflow. Similarly, ventilation within the audience area must not introduce drafts that will make patrons uncomfortable. Locate system sensors where the audience sits, not in out-of-the-way places such as hallways, ceilings, or vestibules.

Safety

Refer to local ordinances for safety requirements. In addition, consider at least the following:

1. At least one emergency exit per 100 patrons and a minimum of two per theatre, located at opposite ends of the theatre.
2. Emergency lights (self-powered and self-actuated)
3. Fire-retardant carpet and upholstery materials
4. Sprinkler system plus strategically located fire extinguishers.

THEATRES

Tables 23.1–23.3 and Fig. 23.10 present the spatial requirements for theatres.

TABLE 23.1 Spatial Requirements for Various Types of Theatrical Productions

	Pageant and Symphonic Drama	Grand Opera	Vaudeville, Review
General characteristics	Dramatic episodes, processions, marches, dances, and crowd scenes. Masses of performers engaged in simple but expansive movements before very large audiences	Large numbers of performers on the acting area at one time; often more than one hundred in big scenes and finales. Movement is martial processions and group dances and the costumes are elaborate. Soloists perform downstage center, close to the footlights but within the bounds of the conventional proscenium, principals play twosome and group scenes in the area near the audience, and choruses and supernumeraries require space upstage. The ballet and the chorus of soldiers, pilgrims, peasants, or what not, sometimes fill the entire acting area. The performance is viewed objectively by the audience and does not benefit by intimate contact between performance and audience	Vaudeville and revue emphasize the human scale. Although the vaudevillian keys his performance to the last row in the gallery, the form is characterized by intimate direct relationship between performer and audience: monologues strait to the front, confidential asides to the front row, and audience participation in illusions. Other acts (acrobatics, etc.) are played across the line of audience vision for maximum effect.
Acting area size	From 2000 to 5000 sq ft depending on the scale of the pageant.	Minimum: 1000 sq ft Usual: About 2500 sq ft Reasonable maximum: 4000 sq ft	Minimum: 350 sq ft Usual: About 450 sq ft Reasonable maximum: 700 sq ft
Shape	Rectangular with aspect ratio between 1 to 3 and 2 to 3.	Quadrilateral with an aspect ratio between 1 to 2 and 2 to 3. Sides converge toward the back of the stage, following the sight lines from the extreme lateral positions.	Rhomboid with aspect ratio about 1 to 3. Sides converge toward back of stage following the site lines from the extreme lateral seats.

(Continued)

TABLE 23.1 Spatial Requirements for Various Types of Theatrical Productions (*Continued*)

	Pageant and Symphonic Drama	Grand Opera	Vaudeville, Review
Arrangement	Long dimension of acting area perpendicular to general sight line. Audience entirely on one side, elevated to perceive two-dimensional movement. Large openings at ends and in side opposite audience for processions, group entrances, and exits. Some elevation of portion of acting area opposite audience, purely for compositional reasons.	Long dimension perpendicular to the general sight line. Audience elevated to perceive two-dimensional movement.	Long axes of the acting area perpendicular to the optimum sight line. Audience grouped as close as possible to the optimum sight line. The forestage is an essential part of the acting area; steps, ramps, and runways into the house are useful.
Proscenium	Either no proscenium with performers entering the "pageant field" from beyond the lateral sight lines, or structural or natural barriers to delineate the side limits of the acting area and conceal back stage apparatus and activity. "Curtains" of sliding panels, lights or fountains for concealing the acting area; often the concealment is by blackout only.	Width equal to the long dimension of the acting area.	Width equal to the long dimension of the acting area. Flexibility is to some advantage in review but of little value in vaudeville.
Orchestra	Space for 100 musicians between audience and acting area. Conductor must see performance.	Pit for 60 to 80 musicians. Conductor must have good view of action.	Music and music cues closely integrated with both vaudeville and review performances. Pit space for from 15 to 30 musicians. Conductor and percussionist must have good view of the action.
Comment	Primarily an outdoor form, it is often staged in makeshift or adapted theatres, utilizing athletic fields and stands or natural amphitheaters. A few permanent pageant theatres have been built.	Movement in two dimensions in acting area is a significant visual component, predicating elevation of the seating area to make this movement visible.	Most of the visual components of vaudeville and review are such that they are perceived best in the conventional audience-performance relationship. The comic monologist who must confront their audience is defeated by the open stage and arena arrangements.

	Dance	Musical: Folk Opera, Operetta, Musical Comedy, Musical Drama	Legitimate Drama
General characteristics	Graceful and expressive movements of human figures in designed patterns, chiefly in two dimensions but with the third dimension introduced by leaps and carries. Occasional elevation of parts of the stage floor. Singles, duets, trios, quartets, groups. The movement demands maximal clear stage space.	These forms embody on a smaller scale the production elements of grand opera, plus a certain freedom and a quest for novelty which encourage the development of new performance devices. Close audience contact of soloist and specialists is borrowed from vaudeville and revue. Big scenes involve many dances, singers, and show girls, often with space-filling costume and movement. Fifty people on stage at one time is not unusual.	Of all production types, legitimate drama places the greatest emphasis upon the scale of the human actor. The importance of the individual actor requires that stage space and scenery do not dwarf him. Dominance of plot, locale, and characterization requires verisimilitude in the size and relationship of scenic objectives. Too small an acting area crowds actors and furniture, hampers stage action, and detracts from the dramatic effect which is the sole aim of the performance. Too large an acting area diminishes the actor in scale and renders his performance ineffective by weakening the effect of his gestures and movement.
Acting area size	Anything under 700 sq ft is constricting. Reasonable maximum: 1200 sq ft	Minimum: 600 sq ft Usual: About 1200 sq ft Reasonable Maximum: 1800 sq ft	Minimum: 240 sq ft (12 by 20 ft) Usual: About 525 sq ft (15 by 35 ft) Reasonable maximum: 1000 sq ft (25 by 40 ft)

(*Continued*)

TABLE 23.1 Spatial Requirements for Various Types of Theatrical Productions (*Continued*)

	Dance	Musical: Folk Opera, Operetta, Musical Comedy, Musical Drama	Legitimate Drama
Shape	Rhomboid with aspect ratio about 3 to 4. May project into and be surrounded by audience (open stage or arena) since frontal aspect of performers has minimal and space-filling quality has maximal significance.	Proscenium: Rhomboid with aspect ratio between 1 to 2 and 2 to 3. Sides converge toward the back of the stage following the sight lines from the extreme lateral seats. Arena: Circle, square, or rectangle (3 by 4 aspect ratio) or ellipse (3 by 4 aspect ratio).	Proscenium: Quadrilateral with an aspect ratio about 1 to 2. Sides converge toward the back of the stage following the sight line from the extreme lateral seats. Open Stage: Semicircle, quadrilateral, or polygon projecting from a proscenium or from an architectural façade. Arena: Circle, square, rectangle, polygon, or ellipse with about 3 to 4 aspect ratio. Entrances from diagonal corners and in middle of one or both long sides.
Arrangement	Nearly square acting area so that dance patterns may be arranged in depth and movement may be in many directions including along the diagonals. Many dance figures require circular movement. Many entrances desirable, especially from the sides of the acting area.	Proscenium: Long axis of acting area perpendicular to the optimum sight line. Mechanized mobility of structural parts to produce changes in acting area arrangement are desirable. Forestages, sidestages, acting area elevators. Arena: Numerous wide entrances for actors and stage hands via the aisles or through tunnels under the seating banks. Ramps preferable to stairs or steps. Experimentation possible in rendering stage flexible by lifts, and in development of flying systems over the acting area.	The realistic style of dramatic production confines the performance to an acting area entirely inside the proscenium. The apron is not used. Most historic styles and much modern dramatic theory demand more freedom of audience-performance relationship than the realistic style and call for the projection of the performance toward, into, and around the audience. For this, projecting aprons, forestages, sidestages, runways, steps, and ramps and aisles are all to some degree useful. To meet the demands of different styles and stylists, the acting area for drama must be capable of assuming many shapes. To confine it within the proscenium opening is adequate for the realistic style but inadequate for the others; to project it toward, into, or around the audience in any rigidly unalterable form is likewise adequate for one style but inadequate for others.
Proscenium	Proscenium not really necessary; though useful as concealment for lighting instruments and dancers awaiting entrances, other devices such as pylons, moveable panels, and curtains may be substituted.	Usually as wide as the acting area, but should be adaptable to changes in arrangement of the acting area described in the preceding column. Arena: None	Width equal to long dimension of the acting area. Moving panels to vary width, openings in proscenium splay to form side-stages, movable pylons or columns by which the opening may be subdivided are all desirable. Flexibility and mobility are increasingly desirable. The application of motive power under remote control to the movement of structural parts to produce different arrangements appears desirable but is costly. Manually alterable parts, particularly forestage proscenium panels and sections of the stage floor, if not unwieldy, are reasonable substitutes.
Orchestra	Music almost always accompanies dancers. For dance, as part of opera or musical show, orchestra is in pit. For dance as specific performance, as in ballet, orchestra may be in remote location and music piped in. Maximum orchestra for dance: 60 musicians in pit for classical ballet. Minimum: 1 drummer	Music an integral auditory component, sometimes integral visually. Elevating orchestra pit to accommodate from 20 to 40 musicians. Arena: Orchestra pit beside the acting area parallel to long axis and opposite principal entrance. This unavoidably imparts a performer orientation toward the orchestra and favors the seats in that general direction.	Orchestral music is sometimes an integral visual part of the performance, but most generally it is a purely auditory component. It is not generally necessary for the orchestra to be seen by the audience, but because cueing of music is so extracting, the conductor must see the action. It is reasonable to provide a pit for 15 to 30 musicians, but the flexibility cited above must be provided, either by portable pit covers, steps, and platforms, or by mechanized orchestra lifts. There is opportunity for originality of arrangement.
Comment	Dance in its various manifestations is the performance form best suited to the open stage or arena since it possess the least amount of facial-expression significance and the greatest amount of movement and pattern in two or three dimensions. Elevation of the audience to perceive best the patterns of dance is desirable.	The assumption by ballet of a greater share in the performance of musical comedy indicates the need for a high general sight line from the audience.	The various forms of theatre used by legitimate drama are discussed fully earlier in this table.

TABLE 23.2 Community Theaters—Typical Space Requirements

Spaces	Areas* (sq ft)	Remarks
Vestibule and gallery	1200	Less area would hamper use of space as gallery and meeting place, area may be increased in proportion as auditorium capacity exceeds 800. Good lighting is necessary.
Checkroom	240	Minimum unless checkroom does not serve auditorium or unless patrons do not check overcoats.
Lobby	1000	See vestibule; mechanical ventilation needed here.
Ticket office	50	Minimum; for larger houses additional administration office (50 to 80 sq ft) is required. Ticket windows (2) and wall space (approx. 4 ft × 8 ft) are necessary.
Lounge – Rehearsal room	750	Minimum size, equal to acting area of stage; mechanical ventilation needed.
Administrative	350	Minimum; area varies. Outside light and air needed.
Men's and Women's toilets	250 each	Consult codes; areas ample for the capacity of 800; either mechanically ventilated or outside light and air needed.
Auditorium	5600	Minimum for conventional seating; may increase to 7000–8000 sq ft for aisleless seating. Area includes forestage (removable seats). Outside light undesirable.
Radio studio	300	Can be reduced to 200 sq ft; no outside light; mechanical ventilation needed.
Control room	70	Minimum; mechanical ventilation needed.
Director's room	20	Minimum, but adequate.
Quiet room	30	Acts as sound insulation between circulation and radio unit.
Project room	200	Ample, includes toilet and lavatory; consult code requirements.
Spotlight booth	400	Area may be divided into three booths: one on center with stage, one at each side of auditorium.
Stage	3500	Ample; 2,800 sq ft minimum, 3,500 usual average except for encircling stage. Air conditioning in conjunction with auditorium desirable; no outside light; top of stage house louvered (consult codes); if conventional stage, minimum height, floor to grid, 70 feet.
Stage workshop	1500	Sometimes reduced to 1,200 sq ft Outside light, if clear glass, preferable from north; if obscure, orientation unimportant.
Scene storage	1000	Minimum; larger if possible.
Costume workshop	420	May reduce to 300 sq ft; north light desirable.
Costume storage	210	Minimum; no outside light; preferably ventilated; must be dry.
Costume dyeing	80	Minimum; no outside light required; unless outside air provided, must be mechanically ventilated.
Six dressing rooms**	680	Each room required access to two lavatories; size not changed with size of building; stars' dressing rooms each need private toilet and shower; all preferably air conditioned.
Makeup room**	130	Minimum; Used also for dressing, requires two lavatories; preferably air conditioned.
Two chorus rooms**	440	Reasonable minimum; three lavatories needed in each; preferably air conditioned.
Two bathrooms	300	Reasonable minimum.
Stage manager	150	Minimum.
Discussion room	750	Can be used for rehearsal; area determined by acting area.

*Based on auditorium capacity of 800.

**Dressing, chorus, make-up rooms require mirrors, preferable three-sided type, movable; and overhead lighting, mirror-lighting equipment.

TABLE 23.3 Proscenium Widths, in Feet, for Kinds of Theatrical Production

	Minimum	Usual	Reasonable Maximum
Drama	26	30–35	40
Vaudeville, revue	30	35	45
Musical comedy, operetta	30	40	50
Presentation, opera	40	60	80

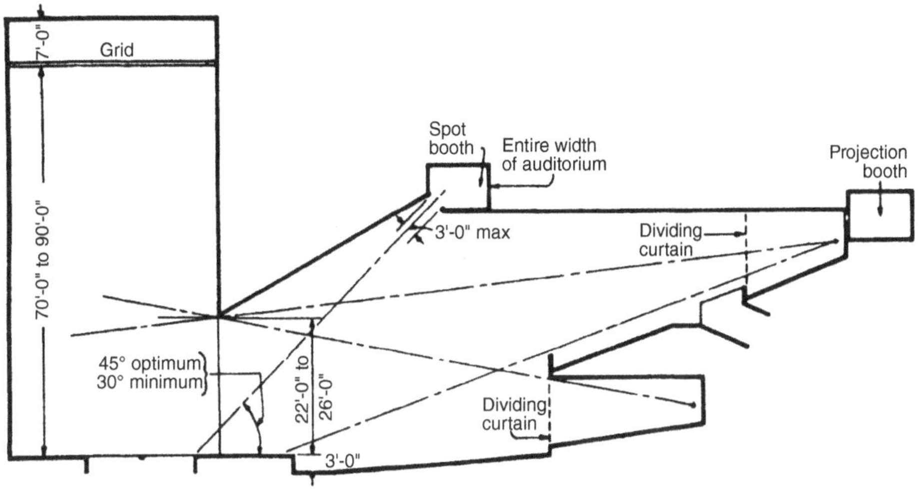

Figure 23.10 Theatre design guidelines.[10]

[10]From J. De Chiara and J. H. Callender, *Time-Saver Standards for Building Types,* McGraw-Hill Book Company, 1973.

CHAPTER 24

SPECTATOR REQUIREMENTS FOR SPORTS FACILITIES[11]

Photo courtesy of Kate Kubicek.

SEATING

Most arenas now being built are employing theater-type upholstered seats. The minimum recommended width is 19 in., and they should vary up to 23 in. center to center for the prime areas. In the wedge-shaped sections in curved rows, a mix of widths is usually used to make the ends of each row come out as flush as possible. Where risers are 5 in. or over, seat stanchions should be riser-mounted to facilitate cleaning. Seats should be self-rising with perforated acoustical treatment on the seat bottoms.

CONCESSIONS

Concession stands for food, beverages, souvenirs, and coat checking should be provided at convenient locations in the promenade areas. Counter areas should be as long as practical and if possible recessed in alcoves to prevent backup of patrons into circulation spaces. Storage space should be provided immediately adjacent to each counter area so that food items can be restocked to the sales area during a game or performance without a trip to the central supply point. If it is contemplated by the arena management that an outside concession firm be brought in to run the operation, it should be selected as early as possible and participate in the planning process. If the firm is an experienced national operation, it will have strong points of view on counter locations, size, visibility, and utility requirements among others. If not built into the original building, the concessionaire's desires will likely prevail at a later date and unsightly and costly additions result.

TOILET ROOMS

Sets of men's and women's restrooms should be provided at one or more locations on each public level. Their layout must provide for peak loads during the 15- to 20-min intermission periods when hundreds of patrons will pass through each room. It is ideal if a one-way traffic flow can be developed with an in and out doorway separated by some distance. Within, the space should be divided with the water closets and urinals located near the entrance and the lavatories in a space near the exit. Also it is desirable if possible to design each toilet room so that half of the space can be closed off by some means during events of small attendance. This will save

[11]From J. De Chiara and J. H. Callender, *Time-Saver Standards for Building Types*, McGraw-Hill Book Company, 1973.

a good deal of operating cost for cleaning. Plumbing-line capacity should be studied carefully for peak use and generous pipe spaces with good access provided.

GRAPHICS

A good graphics and signing control program is important not only for an attractive appearance, but for controlling and expediting crowd movement. Signing can help establish a clear pattern of movement that can easily be comprehended by the patrons. Seat colors in the arena can be keyed to ticket colors to identify the various areas or categories of seating. This can be done on a horizontal basis with rings of seats changing color as they change from one price category to another. Or the arena can be divided into quadrants each with its own color key. In cases where the arena sits within a large parking field, this color system may even extend to the exterior and guide patrons to the proper entrance as they park and approach the building. Within the seating area, signs designating sections, rows, seats, etc., should be large, clear, and located in easily read places. Signs for restrooms, concessions, telephones, etc., should also be of good size and clear and consistent in style. In the lobby ticketing area, space must be provided for coming attraction signs, current-event pricing, and seating plans for various event setups. It has proved successful also to have a scale model of the arena seating including colors and section identification within the ticket sales area to assist patrons with their ticket purchases.

CHAPTER 25

SPECIAL WORKPLACES

GENERAL PRINCIPLES FOR CONCEPTUALIZING SPECIAL WORKPLACES

Working Position

Workers or operators should stand when (see Fig. 25.1):

1. The task they must perform requires reaching beyond a point that typically can be reached comfortably while seated.
2. They have to move frequently from one type of task or workplace to another.
3. They frequently interface with others who are standing.
4. Several persons have to work jointly at a large display area such as a map.

Workers or operators should be provided an alternative sit–stand configuration when (see Fig. 25.2):

1. They normally remain at a task more than 30 min, but others frequently must observe their operation.
2. They need the advantages that seating provides for their major task, but often have to move to another position for a temporary standing task.

FIGURE 25.1 Standing tasks.

Workers or operators should sit when (see Fig. 25.3):

1. They must work at the same task for an extended time period (30 min or more).

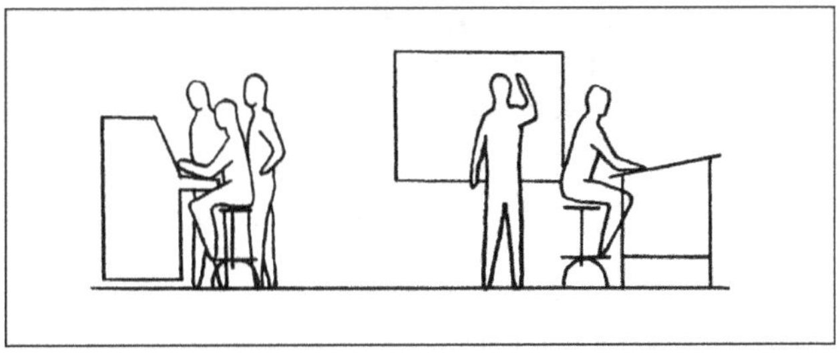

Figure 25.2 Sit–stand tasks.

2. Task precision demands steadiness both for them and for the devices being used.

3. They need to be restrained to prevent their displacement by dynamic environmental forces.

4. They have to use their feet to operate controls.

5. They are required to apply maximum force to operate a control.

6. They have to perform extensive writing tasks.

General Worker and Operator Positioning Considerations

Avoiding Awkward Positioning

1. Minimize the necessity for operators to lean; i.e., arrange the workplace so that operators can maintain a more or less normal alignment of their bodies, especially when standing.

2. Operators should not have to use their maximum reach capability; i.e., they should be able to shift their bodies to a better position.

3. Operators should not have to sit or stand for long periods with their head, torso, or limb positions skewed; i.e., they should be able to keep their heads, necks, torsos, and limbs in a symmetrical relationship as much as possible.

4. Operators should not be forced to work frequently or for long periods with their hands and arms above normal elbow level.

5. Avoid positioning operators in supine or prone positions.

6. If the limbs must be extended for long periods, provide some kind of limb support.

7. Configure the workplace so that operators can see what they are doing without having to assume an awkward head or body position to see around their hands or the control device.

Creating Arrangements That Are Compatible with Normal Limb Movement Patterns and Reach Limitations

1. Minimize the need for operators to move their torsos during continuous control operation, except where a ballistic advantage is desirable.

Figure 25.3 Examples of sitting tasks.

2. Repetitive arm and leg motions should be in the direction that limbs articulate normally about typical shoulder, elbow, wrist, hip, knee, or ankle pivot points.

3. When both hands or both feet are involved simultaneously, create a configuration in which the motions are opposite rather than in the same direction; i.e., the right hand should rotate in the opposite direction to the left, and the right foot should push forward while the left relaxes in the aft direction, as in aircraft rudder pedal movement.

4. Minimize the necessity for operators to shift their position in a seat in order to look at a display or operate a control. In certain situations, however, it may be desirable to provide seat movement, i.e., fore and aft, laterally, or rotating.

5. Avoid workplace arrangements that force operators to stand too close to, or lean across, a hazardous element, such as a moving belt or chain or a hot component.

6. Take into account the extra clearance requirements and/or reduced mobility caused by special apparel that operators may be required to wear and make sure that there are no features within the mobility envelope that could prevent proper limb movement, i.e., components that might be bumped, which limit full limb movement, or upon which clothing might become snagged.

Force Application

1. Place manipulable objects or controls in positions that are compatible with the best geometric relationship for applying pushing, pulling, or rotating force by operators whenever such forces cannot be minimized.

2. Provide an appropriately positioned support against which operators can minimize counterforce effects, e.g., an armrest, handrest, or backrest.

3. Arrange force-demanding controls in positions where operators can apply the necessary force without disturbing their normal body position, especially when this might interfere with a primary visual or control activity and/or with necessary body-referencing posture.

Minimizing Fatigue-Producing Workplace Arrangements

1. Provide backrests for seated operators.

2. Provide armrests.

3. Provide handrests when operators are using a continuously operated controller, such as a desk-mounted joy stick or roller-ball controller.

4. Provide footrests. (Refer also to the discussion on Awkward Positioning.)

Minimizing Potential Safety Hazards

1. The operator's posture should be such that his or her body is in a position that can accommodate dynamic force inputs, such as excessive spinal loads during deceleration, the impact of a sudden landing, or the buffeting of a vehicle as it impacts against rough terrain or ocean waves.

2. Provide body restraints for vehicle operators that will protect them from impacting the interior structure or equipment and which will not allow them to slip into a position in which the restraint itself, such as a seat belt, becomes a hazard.

3. Make the workplace as safe as possible in terms of locating and designing components so that they will not penetrate an operator's skin upon contact, i.e., they will spread the energy rather than concentrate it.

Key Factors to Consider for Each Type of Operator Working Position

1. Standing Operators: A smooth, level surface should be provided. There should be sufficient surface area for operators to establish an adequate spread of their feet, to move when necessary, and to brace their feet when required. A nonslip surface should be provided if a platform is subject to movement, as in a vehicular environment. A resilient surface should be provided if operators must stand all day. Visually observed workplace elements should be arranged so that they can be seen without excessive movement, so that there is no parallax for reading instruments, and so that reading distance and size of display are compatible with readout precision. Manipulative tasks should be arranged and/or oriented to be compatible with respect to reach convenience, required motion patterns and excursions, force application requirements, precision demands, and response speed. Special attention should be given to the implications of eye-hand coordination.

2. Sit–Stand Operators: The eye reference for both seated and standing operators should be approximately the same. The arrangement of visual displays that have to be monitored by both seated and standing operators should be such that there is the same level of readout accuracy from both positions; i.e., operators should not make reading errors because of parallax. If both seated and standing operators may have to use the same control device, place this in a position that will minimize interoperator interference. Provide a footrest for seated operators.

3. Seated Operators: Provide a seat that ensures optimum working posture for the tasks being performed. Arrange visual displays and controls so that an operator's hand will not cover critical displays and so that the displays are normal to the expected viewing line of sight. If the primary viewing task is directed outside the workplace (vehicle control), position the most important internal displays so that they can be viewed without excessive eye movement from the nominal exterior line

of sight. Position all controls so that, in manipulating them, operators do not appreciably move their nominal eye reference and possibly miss seeing important events occurring outside or on the principal internal display.

4. Miscellaneous Considerations:

   a. Organizational Standardization versus Individual Adjustment: When similar work places are to be repeated, a compromise should be sought between the benefits of a standardized workplace (which permits operators to shift from one workplace to another without confusion) and the benefits that accrue from allowing individual operators to modify their workplace to fit each type of activity more conveniently (i.e., to rearrange tools, materials, or other aids so that they are more accessible).

   b. Illumination: Do not assume that a generalized type of illumination will satisfy all task-seeing requirements; i.e., the type and location of light fixtures should be determined according to the varying needs of the operator. Especially important is the creation of seeing conditions that minimize glare and reflection problems.

   c. Storage: Almost all workplaces require the provision of storage space for books, files, tools, or writing and drawing equipment and materials. Equally important is the provision of storage space for communications devices, discarded items, and personal articles such as glasses or eye shields.

   d. Service and Maintenance: Careful consideration should be given to ensuring that, in arranging the workplace and/or packaging of equipment, access to critical elements for service or other maintenance is provided.

   e. Design for Handicapped Operators: Whenever a workplace may be used by a person in a wheelchair, a blind person, or a deaf person, special care must be taken not only to dimension the workplace properly but also to provide those special features necessary for the handicapped operator to perform more effectively within his or her limitations (e.g., tactile features for the blind and special visual signals for the deaf).

   f. Environmental Considerations: Be aware of the possibility that a particular individual workplace may reduce the effectiveness of general ambient thermal, ventilation, and noise control systems, thus requiring additional features to ensure that the operator does not become overheated or is not subjected to respiratory contamination or to interfering or annoying noises that may be produced within his or her particular workplace.

   g. Manipulative Clearance for Large Materials: Often large materials must be maneuvered into place before they can be worked on. Provide adequate clearance so that such materials can be maneuvered into place easily.

BASIC TASKS IN RELATION TO WORKPLACE DIMENSIONING (SEE FIG. 25.4)

1. Writing: The writing surface should be at approximately elbow level and should be relatively horizontal (maximum 5° slope); greater slopes make it difficult to keep a pen or pencil from rolling off the surface.

2. Typing: The center of the keyboard should be at about elbow level.

3. Data Entry Keyboard: The center of the keyboard should be at about elbow level. An associated visual display should be positioned so that the operator's line of sight is perpendicular to the face of the display.

4. Drawing: A drawing board used for precision work should slope approximately 3° to 4°; this provides the best compromise between reaching, viewing, and precise instrument manipulation. For less precise drawing, sketching, or artwork, the drawing surface should be adjustable, so that the artist can match the size of the work with the best position for sketching, painting, etc.

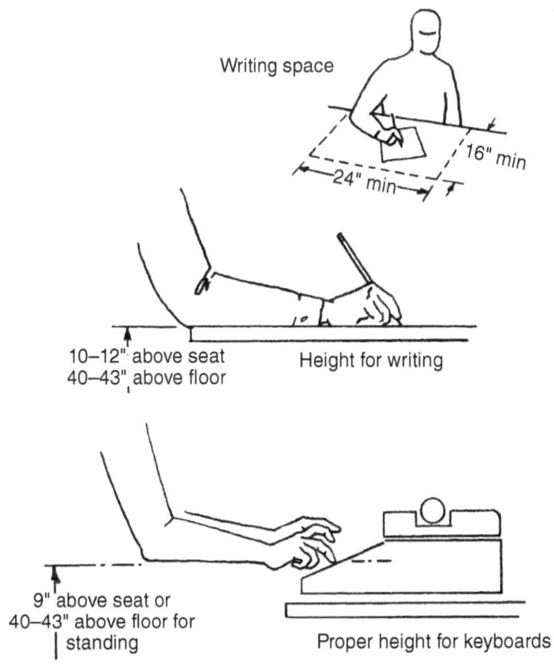

FIGURE 25.4 Nominal guideline dimensions. Although the general dimensional suggestions are recommended to approximate the optimum, one should be careful not to assume that they necessarily apply exactly to a specific problem. Use of a test mockup to verify these is recommended.

A sit–stand arrangement is most desirable for average mechanical drafting work since it allows the worker to reach all areas of a large drawing easily.

5. Storing and Retrieving Shelves: Design and/or select storage shelving that allows the worker to reach and grasp objects to be stored or retrieved easily and without fear of dropping them. Avoid deep, high shelving since it is easy for small objects to be pushed to the rear of a shelf, where they are neither visible nor manually accessible.

6. Office Machine Use: Either select machines that have been properly designed (including special stands) or select or build tables that assure proper positioning of the principal work level of the machine. The work level should be at approximately the operator's elbow height.

7. Tool Operation: Various tools require that the work level be carefully chosen, especially if there is a requirement for precision and/or the application of controlled force.

COMPUTERS

Human–Computer Interface

Figure 25.5 illustrates elements of the human–computer interface:

1. Input hardware
2. Display hardware
3. Computer software
4. Workplace layout

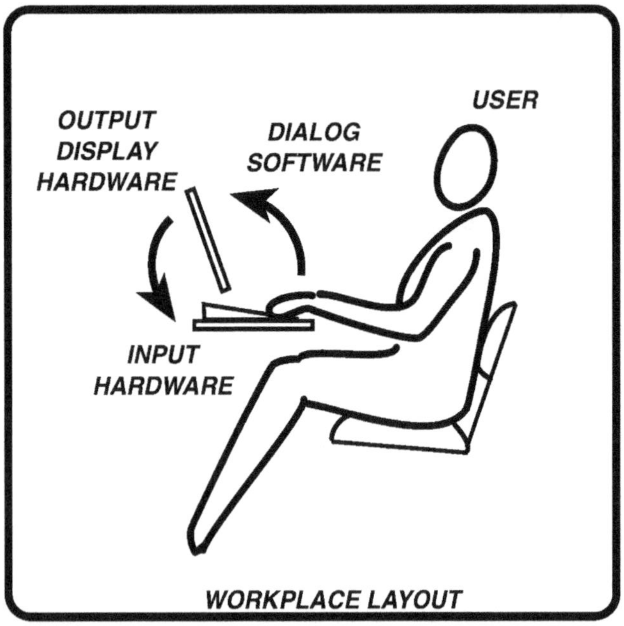

FIGURE 25.5 Elements of the human computer interface.

Design requirements for each element will be defined in the next sections. Details and more extensive discussion are in the following resources:

1. ANSI/HFS 100-2007: American National Standard for Human Factors Engineering of Visual Display Terminal Workstations, 1988
2. MIL-STD-1472G: Human Engineering Design Criteria for Military Systems, Equipment and Facilities, Department of Defense, 2012
3. MIL-HDBK-761A: Human Engineering Guidelines for Management Information Systems, Department of Defense, 1989
4. Guidelines for Designing User Interface Software, S. Smith and J. Mosier, 1986

Input Hardware

Table 25.1 lists computer input hardware devices, when they should be used, and their operating and design requirements.

Computer Dialog Software

Computer dialog software integrates the user, the input, and the display devices. Figure 25.6 shows a simple computer operation model. Tables 25.2–25.4 list important software design requirements for each computer operation. More detailed software requirements are in Part 3 of this book.

Computer Workstation

Guidelines for designing a computer workstation are shown in Fig. 25.7. The guidelines are in the "American National Standard for Human Factors Engineering of Visual Display Terminal Workstations," ANSI/HFS 100-1988, February 1988.

ELECTRONIC CONSOLES AND RACK-MOUNTED PANELS

Seated Operator Consoles

Figure 25.8 shows recommended dimensions for standard seated consoles. The figures are from military standard MIL-STD-1472G. The console accommodates 5th percentile females to 95th percentile size males.

Standing Operator Consoles

Figure 25.9 shows recommended dimensions for standard standing consoles. The dimensions are from military standard MIL-STD-1472G. The console accommodates 5th percentile females to 95th percentile size males.

TABLE 25.1 Computer Input Devices Operating and Design Requirement

Input Device	Use	Operating Requirements	Design Requirements
Keyboard	Entry of alphabetic, numeric, or special function information into a system.	1. Keys can be fixed function or variable function: *Fixed Function Keys*—A fixed function key should consistently accomplish a single function. It should require only a single actuation to accomplish the function. *Variable Function Keys*—Computer software can change the function of a variable function key. The functional status of the key should be displayed on the keyboard or on the computer display. 2. The operator should be able to repeat characters and symbols by holding the key down for longer than 0.5 seconds.	Standard layouts for alphanumeric keyboards are in ANSI (X4.23-1982) and MIL-STD-1280. See Part 3 for additional design requirements.
Mouse	1. Used on flat services to control the position of the cursor on an associated display. 2. May use for data pickoff or entry of coordinate values. 3. Not for generation of free-drawn graphics.	1. Zero-order control (movement of the cursor proportional to the displacement of the mouse) 2. Moveable in any direction without a change in hand grasp. Cursor to move smoothly in the same direction + or − 10°. 3. Operable with either the left or right hand. 4. It should be possible to move the cursor through the full range without exceeding the full mouse movement range. 5. Provide operator with aid to bring the cursor back if driven off the screen.	See Part 3, "Controls."
Touch screen	Direct user interface with controls presented on a data display screen.	Provide positive indication of touch screen control actuation.	See Part 3, "Controls."
Speech input	Consider only when data entry cannot be accomplished through more reliable methods (keyed entry, for example).	1. Minimize input vocabulary. 2. Allow easy error correction. 3. Allow alternate entry words. 4. Allow user to pause input and continue after interruption.	Information on auditory systems is in Part 3
Light pen	1. Two-axis controller (similar to grid and stylus) that works at the display screen. 2. Use primarily for noncritical, imprecise work such as item selection. 3. May be used for free-drawn graphics.	1. Provide feedback for: (a) position of the light pen on the display, and (b) system activation. 2. The cursor should be at the coordinates selected by the light pen. 3. If used for free-drawn graphics, make the refresher rate high to give the appearance of a continuous track. 4. Movement of the light pen should result in smooth movement of the cursor in the same direction.	1. Dimensions are in Part 3. 2. Provide a clip near the screen for hanging the light pen when not in use. 3. Pushbutton switch on pen top should have a resistance of 2 to 5 ounces (0.5 to 1.4 N).
Grid and stylus	1. Data pickoff from display 2. Entry of points on a display 3. Generation of free-drawn graphics. 4. Zero-order control only.	1. Cursor should be at coordinates on the display selected by the stylus. 2. If used for free-drawn graphics, make the refresher rate high to give the appearance of a continuous track. 3. Movement of the light pen should result in smooth movement of the cursor in the same direction.	The grid may be on the display or remote from the display. Grids displaced from the display should be: 1. Approximately the display size. 2. Mounted directly below the display. 3. In the same orientation as the display.

(*Continued*)

TABLE 25.1 Computer Input Devices Operating and Design Requirement (*Continued*)

Input Device	Use	Operating Requirements	Design Requirements
Ball control	1. Positioning of the cursor on a display. 2. Do not use when you require cursor autoreturn to an origin. 3. Zero-order control only. 4. Allows unlimited accumulative travel in any direction. This can be useful for navigation. 5. Not for free-drawn graphics.	1. Provide operator with aide to bring the cursor back if driven off the screen. 2. Allow rapid gross positioning and precise fine positioning. 3. Permit movement in all directions. 4. Eliminate backlash and cross-coupling.	1. Dimensions and resistances are in Part 3. 2. Provide the operator with wrist or forearm support.
Joystick—stiff (isometric)	1. Positioning of the cursor on an associated display. 2. Faster but less accurate than an isotonic joystick. 3. Good if precise return to origin is required.	1. Provide operator with aid to bring the cursor back if driven off the screen.	1. Dimensions and resistances are in Part 3. 2. Provide the operator with wrist or forearm support.
Joystick—displacement (isotonic)	1. Positioning of the cursor on an associated display. 2. Generation of free-drawn graphics. 3. More accurate than isometric joystick. 4. Better for longer operating periods than isometric joystick (less force required). 5. Can use stick as mounting platform for secondary controls (thumb and finger switches).	1. Allow rapid gross positioning and precise fine positioning. 2. If used for free-drawn graphics, make the refresher rate high to give the appearance of a continuous track. 3. The delay between control displacement and cursor movement should be as small as possible and less than 0.1 second.	1. Dimensions and resistances are in Part 3. 2. Provide the operator with wrist or forearm support.

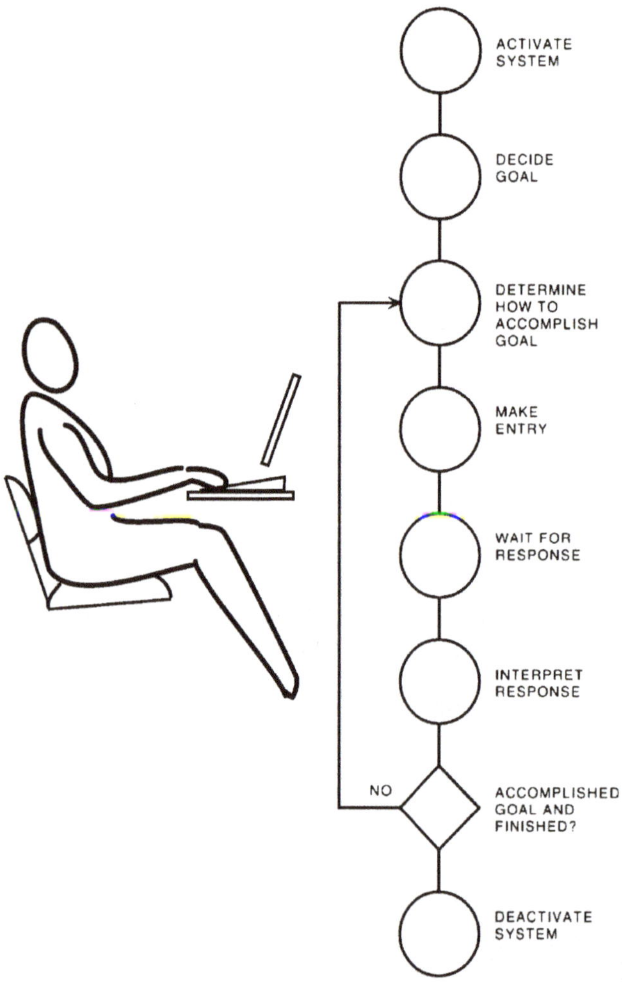

Figure 25.6 Computer operating model.

See-Over Consoles

If the operator needs to see over the console to monitor other displays, reduce the console height so that the operator is not required to stretch his or her torso or neck. (See Fig. 25.10.)

Large Plot Subsystems

Figure 25.11 illustrates general dimensional guidelines for large plot subsystems; these are based on (a) limitations created by knee clearance if the operator sits and (b) convenient use area when the operator stands.

Horizontal versus Vertical Arrangement Other Considerations

As a general rule, panels for either seated or standing operation are easier to use when the displays are laid out for horizontal, sequential scanning. This is due to the fact that an operator's eye motions and head movements are less tiring in the horizontal direction. Studies also have shown that operators can respond slightly faster when they attempt to read instruments or other visual information from horizontally arrayed data than when this is presented in vertical arrays. On the other hand, standing operators, because they are free to walk backward and to the side, can more easily adjust to control and display arrangements that are organized in vertical patterns. (See Fig. 25.12.)

Arranging for Added Convenience

Although the typical console or equipment operator has considerable reach capability, it is desirable to minimize the necessity for reaching, especially if it may be required on a frequent basis. The so-called wraparound concept is helpful

TABLE 25.2 Computer Software Design Guidelines

Computer Operation	Software Design Guideline	Reference
Activate system	1. Separate and distinct step. 2. Rapid as possible—if not possible, advise user of status. 3. Automate whenever possible. 4. Do not require user knowledge of internal system mechanisms.	MIL-HDBK-761A
Decide goal	Provide users with task oriented guidance. Guidance should describe system capabilities and available procedures. Provide guidance online within the program.	"Guidelines for Design User Interface Software"
Determine how to accomplish the goal	Design a dialog between the user and the computer that is compatible with the user and the task requirements. Table 25.3 shows possible dialogs and the characteristics of each dialog.	MIL-HDBK-761A
Make entry	1. Every input should produce a consistent perceptible output. 2. Be consistent with the data entry method and the method for changing or canceling data entries. 3. Reduce workload with minimized input actions and required use memory. 4. The user, rather than the computer system, should pace data entry. 5. Always provide an explicit ENTER action to initiate processing of the entered data. 6. Provide feedback of the acceptance or rejection of a data entry. 7. Provide a way to correct erroneous entries without reentering correct information. 8. For spelling or other common errors, provide feedback of errors, recommendations, and option for acceptance of the recommendation. 9. Delete or cancel an entry with an explicit action and require user confirmation before making a permanent deletion.	"Guidelines for Designing User Interface Software" MIL-STD-1472G
Wait for system response	Provide the user feedback for all data entry. Table 25.4 gives maximum acceptable response times. If a response time is longer than 1.0 second, provide a message that the computer is responding. Notify user that the response is complete.	MIL-STD-1472G
Interpret system response	1. Display data in a usable form and consistent with standards and conventions familiar to the user. 2. Minimize abbreviations in data display. 3. Avoid lengthy text displays; use printed matter instead, if possible. Reading lengthy text on a computer screen may be 20 to 30 percent slower than reading printed material. 4. Display text in wide columns (at least 50 characters per line). 5. Separate paragraphs by at least one line. 6. Allow the user to zoom in graphic displays to see greater details. 7. Use codes to help interpret display: a. When coding displays, adopt meaningful or familiar codes, rather than arbitrary codes. b. Define display codes that have special meaning. c. Use hatching and simple patterns to code graphic areas. d. Consider coding with geometric shapes or symbols. People can readily distinguish approximately 15 different shapes. e. Limit the use of size coding to two or three sizes. f. Limit the use of brightness coding to two values. One value should be at least twice as bright as the other. g. Use color for coding but only after addressing possible use in monochrome format. h. Do not overuse color so that the display is cluttered. i. Consider flash coding to attract attention to critical items. Flash at a rate of three to five flashes per second. Limit flash rates to 2. The second rate should be less than 2 per second. 8. Consider auditory displays to supplement visual displays. 9. If display is larger than a single frame, provide a simple method to page or pan/scroll through the display. a. Be consistent about panning (moving window, fixed data) vs. scrolling (fixed window, moving data). b. Provide the user with an overview of position in a panned or expanded display. 10. For windows. a. Be consistent with window control and control of items in a window. b. Label windows. c. Indicate active windows. d. Overlay windows should not destroy data beneath them. 11. Interpret an error. a. Provide feedback before the user enters an error into the system. b. Display the error and any message until correction. c. For stacked commands, show the sequence with the error highlighted.	"Guidelines for Designing User Interface Software" MIL-STD-1472G
Deactivate system	1. Inform the user if data will be lost on log-off. Allow the user to save open files under some defined file name. 2. Provide an audible warning signal before automatic log-off.	MIL-HDBK-761A

TABLE 25.3 Guidelines for Selecting Computer Dialog

Question and Answer	Form Filling	Menu Selection	Function Keys	Command Language	Query Language	Constrained Natural Language	
				X			Arbitrary control/data entry sequences
						X	Poorly defined/broad interface definition
					X	X	Unpredictable information retrieval
				X			Wide range of control entries
			X				Frequent control/transactions
		X	X				Small/constrained command choice set
	X						Complex control
		X					Large command set
X							Routine data entry
X							Entry order constrained
	X						Data entry flexibility needed
		X					Little arbitrary data input
	X						Slow computer response time
X		X					Fast computer response time
				X			Highly trained users
	X				X		Moderately trained users
X		X				X	Little training

From MIL-HDBK-761A.

TABLE 25.4 Maximum Accep Computer Response Times

System Interpretation	Response Time Definition	Maximum Acceptable Response Time (s)
Key response	Key depression until positive response; for example, "click"	0.1
Key print	Key depression until appearance of character	0.2
Page turn	End of request until first few lines are visible	1.0
Page scan	End of request until text begins to scroll	0.5
XY entry	From selection of field until visual verification	0.2
Pointing	From input of point to display point	0.2
Sketching	From input of point to display of line	0.2
Local update	Change to image using local data base; for example, new menu list from display buffer	0.5
Host update	Change where data is at host in readily accessible form; for example, a scale change of existing image	2.0
File update	Image update requires an access to a host file	10.0
Inquiry (simple)	From command until display of a commonly used message	2.0
Inquiry (complex)	Response message requires seldom used calculations in graphic form.	10.0
Error feedback	From entry of input until error message appears	0.2

From MIL-HDBK-761A.

as long as care is taken to ensure that unusual interferences are not introduced.

The wraparound console is especially helpful for the seated operator, as shown in Fig. 25.13 (top). If the seat has to be secured to the floor (as it does aboard ship), a careful analysis of reach limits must be made. If a movable seat is acceptable, the arrangement can be expanded, taking advantage of a chair with casters.

A sit–stand operation (see Fig. 25.13, bottom) that includes both worktable and panel rack operations can be accommodated best if the rack can be made movable.

Multioperator Stations

When designing multioperator stations (see Fig. 25.14):

1. Start with basic single-operator requirements.
2. Expand dimensions for additional operators to prevent accidental infringement.
3. Accommodate commonly used equipment.

EVALUATING THE SITE LINE IN LARGE GROUP DISPLAY AND OBSERVER ARRANGEMENT

When a number of operators must view a large visual display in conjunction with console operation, the large group display should be arranged so that the various observers can see the display from their normal working positions without having to peer around their neighbors or their neighbors' equipment. Many alternative arrangements are, of course, possible, and sometimes a compromise must be made because of facility space constraints.

Figures 25.15–25.18 and guidelines are helpful in making sure that all the sight-line requirements have been evaluated.

1. As Fig. 25.15 illustrates, the angle of view may be improved by seating the operators further apart.

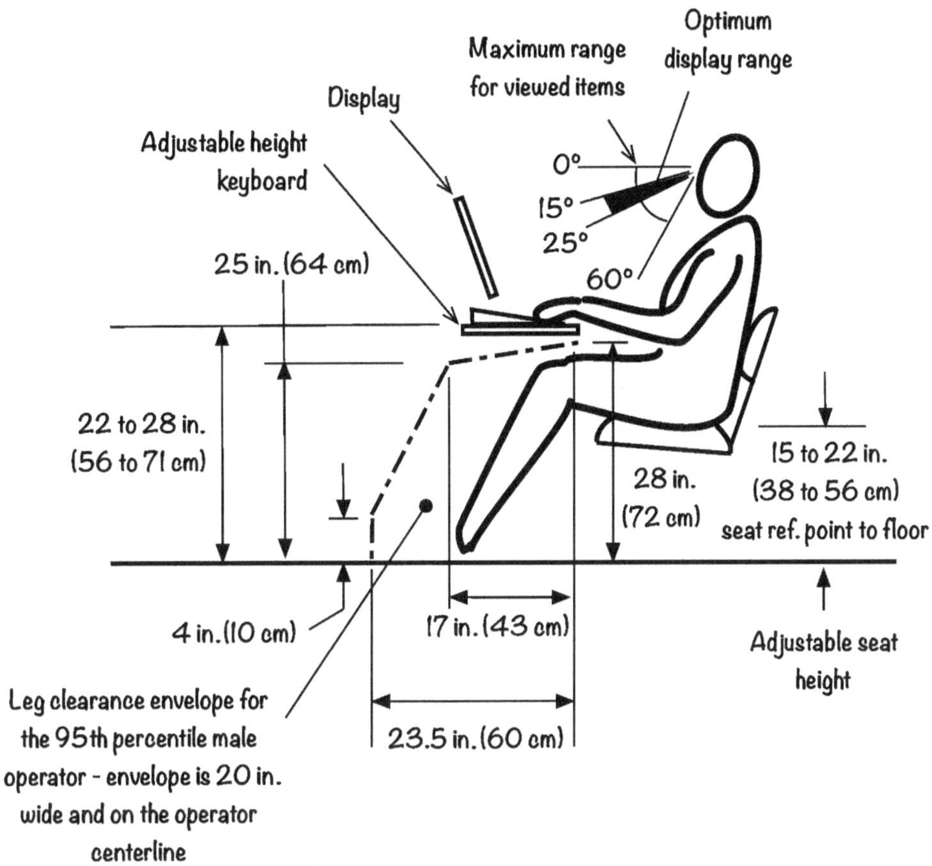

Figure 25.7 Computer workstation dimensions.

2. Operators and their consoles can be staggered in order to maintain the recommended screen-observer line-of-sight limit of 60°, as shown in Fig. 25.16.

3. When the recommended clear line of sight cannot be accomplished by lateral spacing, consider placing the operators in the rear on platforms so that they can see over the heads of those in front. (See Fig. 25.17.)

4. If, because of space constraints, the large display cannot be placed far enough away from a group of observers, consider breaking up the display so that the edges can be rotated for a more suitable angle of viewing. (See Fig. 25.18.)

GUIDELINES FOR DRAWING AND PLOTTING

Figure 25.19 shows the guidelines for drawing and plotting.

CONFINED WORKPLACES

As a general rule, avoid creating situations in which workers have to work in close quarters and/or in the awkward positions illustrated in Fig. 25.20. Additional dimensions are in standards MIL-STD-1472.

When this cannot be avoided, consider the general limiting dimensions provided in these illustrations. Critical factors are the following:

1. There must be sufficient vertical clearance for the worker to enter and leave the environment.

2. There must be sufficient horizontal and lateral clearance for the worker to perform the task (considering the nature, size, and configuration of any tool being used).

3. In the supine position, it is critical that the worker's eyes do not have to be too close to the specific work point; otherwise, the worker's eyes cannot adjust for clear viewing.

4. Openings must be large enough to accommodate both the tool package and the operator's hand, finger, or arms, as shown in Fig. 25.20.

Note: These opening dimensions are for shirt-sleeve conditions. If heavier clothing may be worn, the openings have to be at least 1 in. (2.5 cm) larger on all sides [1.5 in. (3.8 cm) is better].

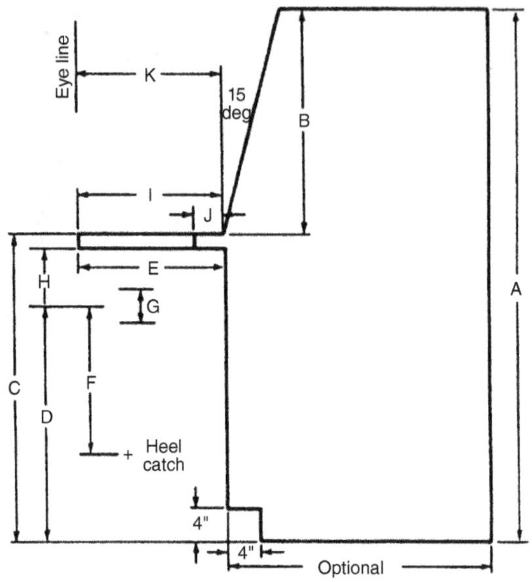

Key	Dimensions	1 mm	(in.)
A	Maximum total console height from standing surface	See table XX	See table XX
B	Suggested vertical dimension of panel, incl sills		
C	Writing surface: Shelf height from standing surface		
D	Seat height from standing surface at midpoint of "G"		
E*	Minimum knee clearance	(460)	18
F*	Foot support to sitting surface**	(460)	18
G*	Seat adjustability	(150)	6
H*	Minimum thigh clearance at midpoint of "G"	(190)	7.5
I	Writing surface depth including shelf	(400)	16
J	Minimum shelf depth	(100)	4
K	Eye line-to-console front distance	(400)	16

*Not applicable to console types 4 and 5 of table XX.
**Since this dimension must not be exceeded, A heel catch must be added to the chair if "D" exceeds 460 mm (18 in.).

Note: A shelf thickness of 25 mm (1 in.) is assumed. For other shelf thickness, suitable adjustments should be made.

Table XX. standard console dimensions

Type of console	Maximum total console height from standing surface	Suggested vertical dimension of panel (including sills)	Writing surface: shelf height from standing surface	Seat height from standing surface at midpoint of G	Maximum console width (not shown)
	A	**B**	**C**	**D**	
1. Sit (W/vision over top)*	1.170 m (46 in.)	520 mm (20.5 in.)	650 mm (25.5 in.)	435 mm (17 in.)	1.120 m (44 in.)
	1.335 m (52.5 in.)	520 mm (20.5 in.)	810 mm (32 in.)	595 mm (23.5 in.)	1.120 m (44 in.)
	1.435 m (56.5 in.)	520 mm (20.5 in.)	910 mm (36 in.)	695 mm (27.5 in.)	1.120 m (44 in.)
2. Sit (W/O vision over top)	1.310 m (51.5 in.)	660 mm (26 in.)	650 mm (25.5 in.)	435 mm (17 in.)	910 mm (36 in.)
	1.470 m (58.0 in.)	660 mm (26 in.)	810 mm (32 in.)	595 mm (23.5 in.)	910 mm (36 in.)
	1.570 m (62.0 in.)	660 mm (26 in.)	910 mm (36 in.)	695 mm (27.5 in.)	910 mm (36 in.)
3. Sit-stand (W/standing vision over top)	1.535 m (60.5 in.)	620 mm (24.5 in.)	910 mm (36 in.)	695 mm (27.5 in.)	910 mm (36 in.)
4. Stand (W/vision over top)	1.535 m (60.5 in.)	620 mm (24.5 in.)	910 mm (36 in.)	NA	1.120 m (44 in.)
5. Stand (W/O vision over top)	1.830 m (72. in.)	910 mm (36 in.)	910 mm (36 in.)	NA	910 mm (36 in.)

*The range in "A" is provided to allow latitude in the volume of the lower part of the console; note relationship to "C" and "D".

FIGURE 25.8 Standard Console Dimensions. (From MIL-SD-1472G.)

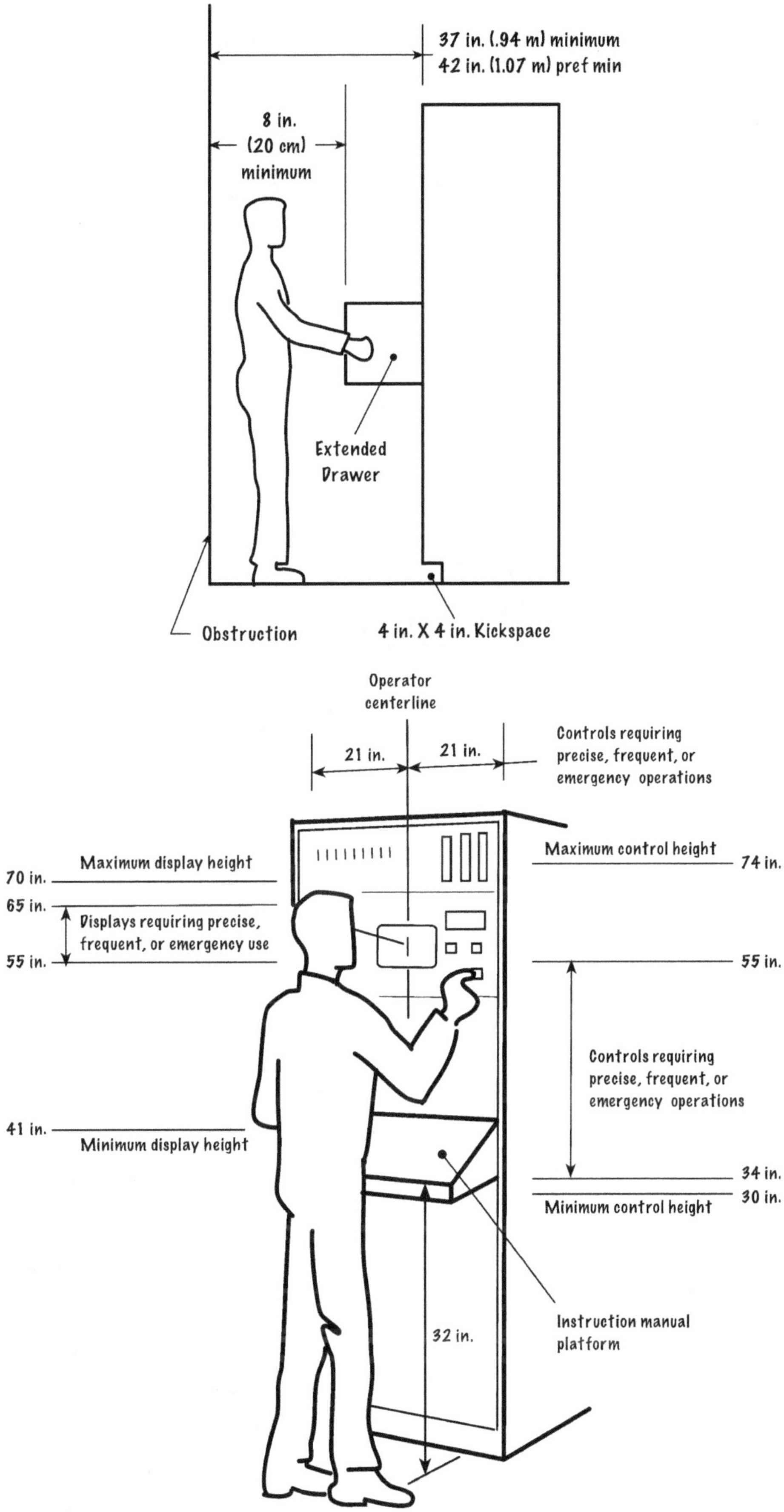

Figure 25.9 Workspace dimensions for a standing console figure changes.

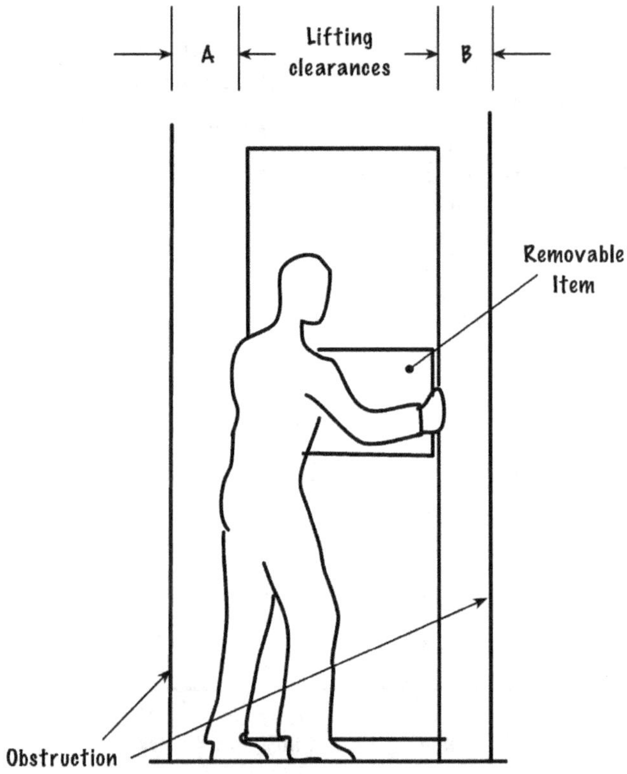

Weight of Removable Item	Minimum Clearances A	B
Less than 44 pounds (20 kg)	18 in. (46 cm)	4 in. (10 cm)
More than 44 pounds	18 in. (46 cm)	18 in. (46 cm)

FIGURE 25.9 *(Continued)*

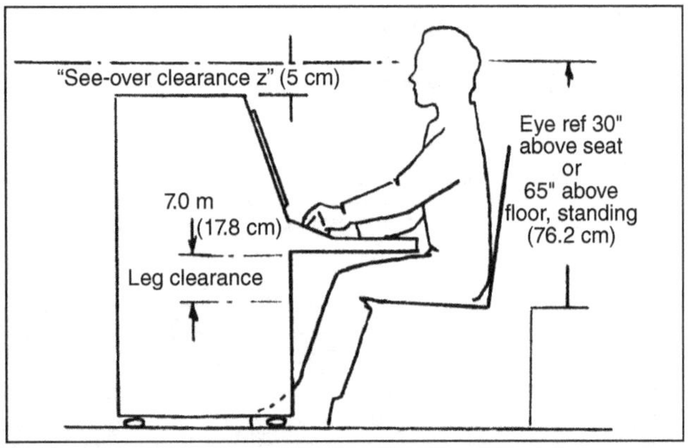

FIGURE 25.10 Guidelines for a "see-over" console.

LIBRARY FACILITIES

General

Dimensional guidelines for key adult library workplaces are illustrated in Figs. 25.21 and 25.22. These dimensions should be altered to fit young people as follows:

1. For teenagers:
    a. Top shelf height 62 in. (158 cm)
    b. Reading table height 27 in. (69 cm)
    c. Counter height 35 in. (89 cm)
    d. Catalog worktable height 35–38 in. (89–97 cm)
    e. Map and book table height 34 in. (86 cm)
    f. Top of monitor height 43 in. (109 cm)
    g. Keyboard height 24–28 in. (61–71 cm)

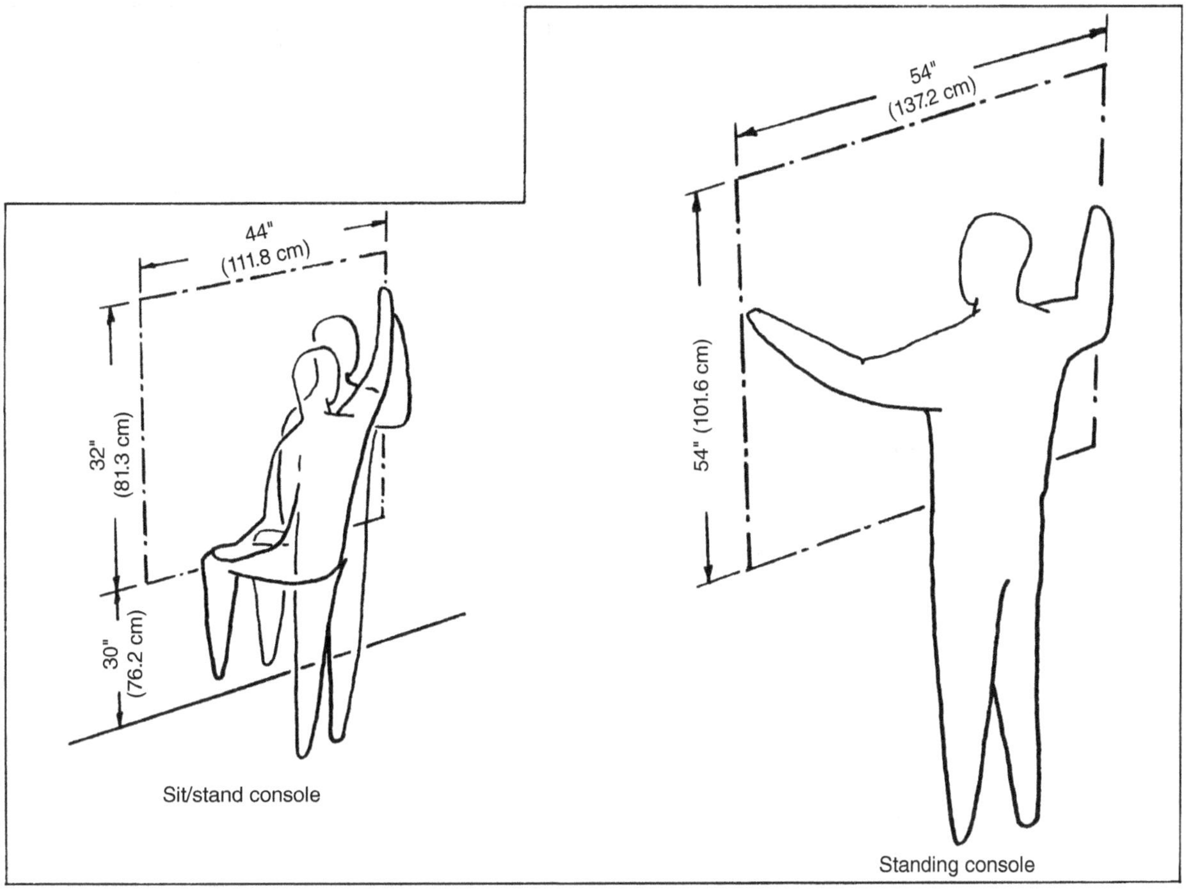

Figure 25.11 Large plot console reach ranges.

2. For grade school children:
   a. Top shelf height 48 in. (122 cm)
   b. Reading table height 21 in. (53 cm)
   c. Counter height 30 in. (76 cm)
   d. Map and book table height 30 in. (76 cm)
   e. Top of monitor height 37 in. (94 cm)
   f. Keyboard height 16–22 in. (41–56 cm)

Reading and Study Carrels

Visual and auditory isolation is often desirable for library patrons. Figures 25.23–25.25 illustrate configurations as well as dimensional guidelines for designing and/or selecting the various types of carrels.

FACTORY WORKPLACES

Workplaces in which fabrication and assembly operations are performed should be laid out not only in terms of placing the critical visual and manual interfaces so that they are compatible with efficient manipulation of materials, parts, and tools but also in terms of work flow.

The accompanying examples represent typical layouts that have been used from time to time to increase the overall efficiency of factory workers.

The safety aspects of the factory workplace are also important, as shown in Fig. 25.26. All moving parts of machines should be guarded and equipped with fail-safe emergency cutoff switches.

MAINTENANCE WORKPLACES

Although equipment maintenance often is not considered as important as the operator's interface, design of the maintenance interlace is important to the maintenance technician and therefore should be considered at the same time that operator interfaces are being designed. Several typical examples of maintenance problems, created by lack of proper attention, are illustrated in Figs. 25.27 and 25.28.

Different equipment packaging concepts either aid or hinder the maintenance technician, as illustrated in Figs. 25.29 and 25.30.

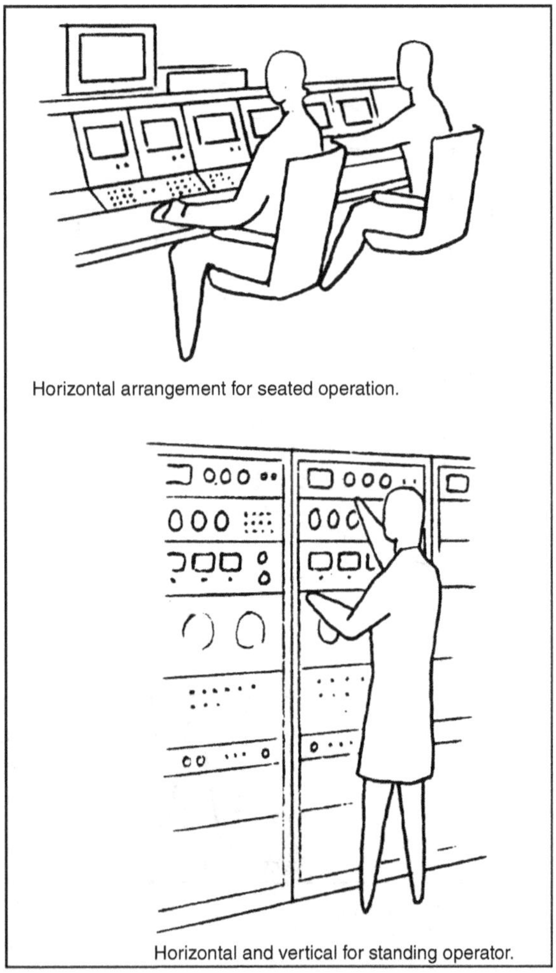

FIGURE 25.12 Horizontal and vertical arrangements.

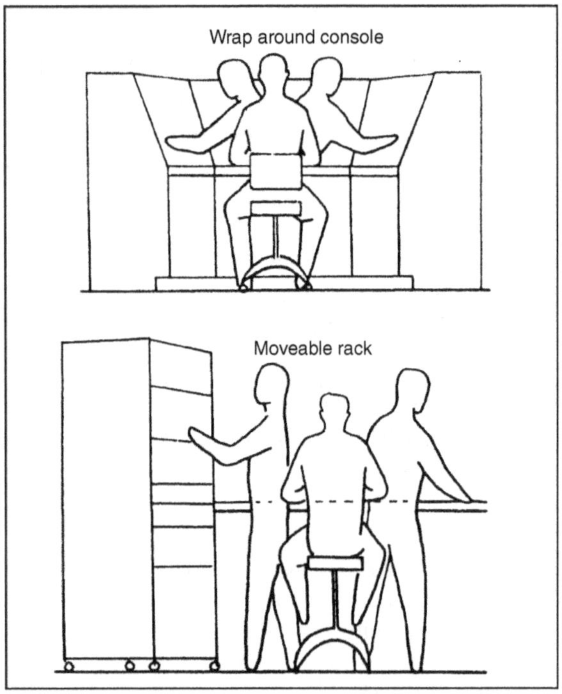

FIGURE 25.13 Ways to improve convenience.

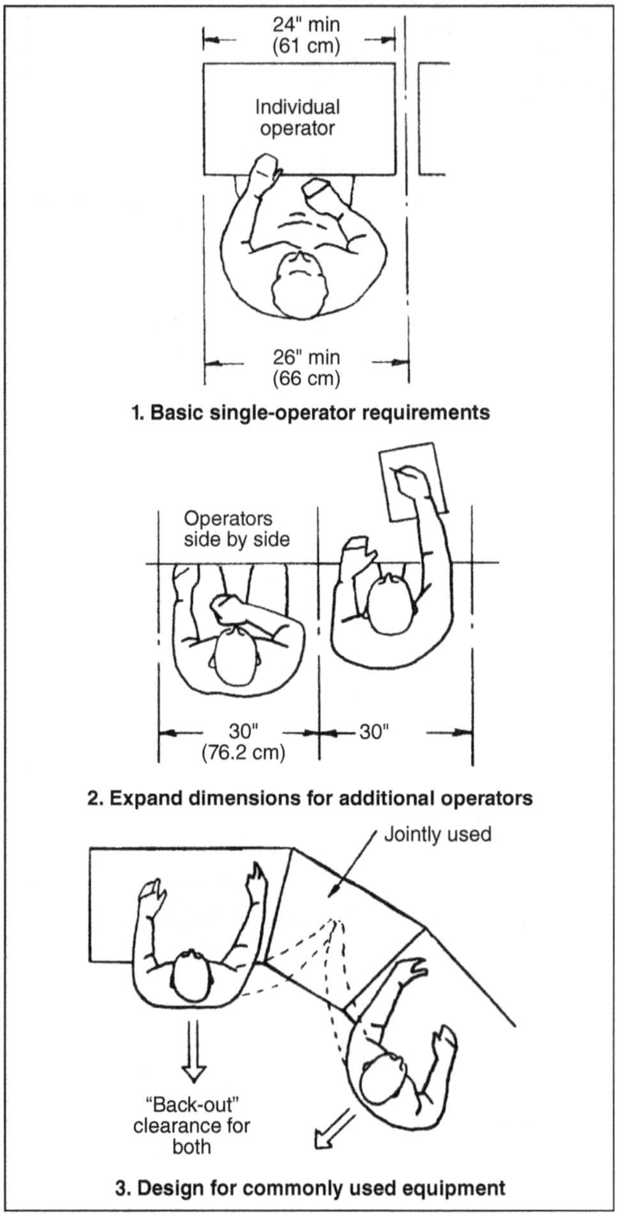

FIGURE 25.14 Designing multioperator stations.

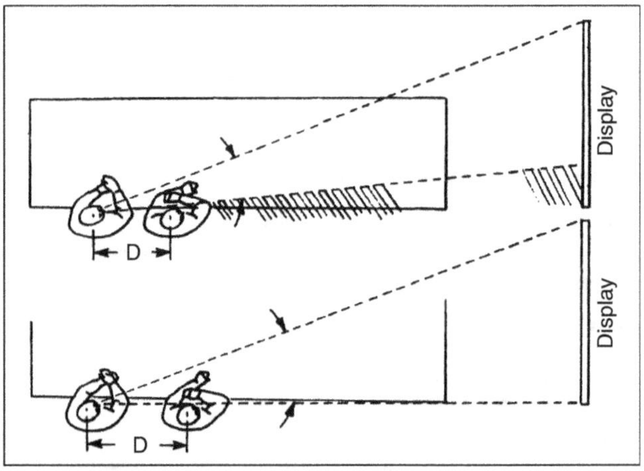

FIGURE 25.15 Increase spacing (D) to improve display visibility.

Access is better if the case can be lifted off the chassis. More freedom to work is provided, and there is no cover to prop up.

Avoid designs in which the chassis has to be lifted out of the case or in which a cover has to be removed or propped out of the way.

Design covers so that the minimum number of fasteners have to be removed.

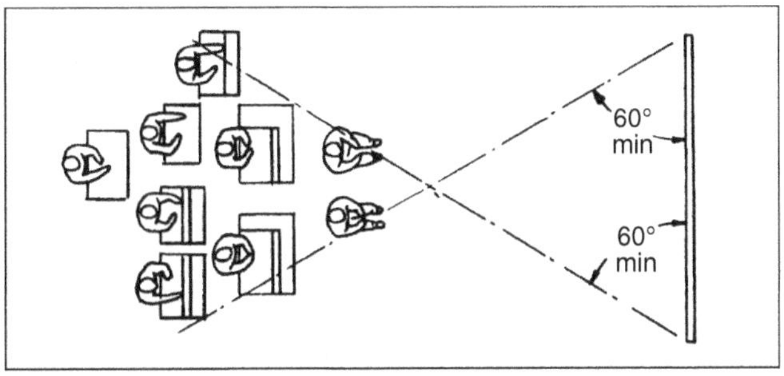

Figure 25.16 Stagger observers.

Fasteners should be captive so that they do not become mislaid or lost.

When high voltages are involved, removal of the case or cover should automatically shut the power off. A secondary switch or jumper should be used to reactivate the power when maintenance is being performed.

OPERATOR CONTROL PANEL LAYOUT

Basic Principles

Four basic principles should be considered in planning the layout of an operator workplace or, more specifically, the equipment control interface (usually a panel containing controls and displays). These are discussed below.

Sequence of Operations. The order in which operators will normally use controls and displays is important in terms of their being able to find what they want and being able to go through the sequence with a minimum possibility of missing a step and also in terms of reducing unnecessary motion due

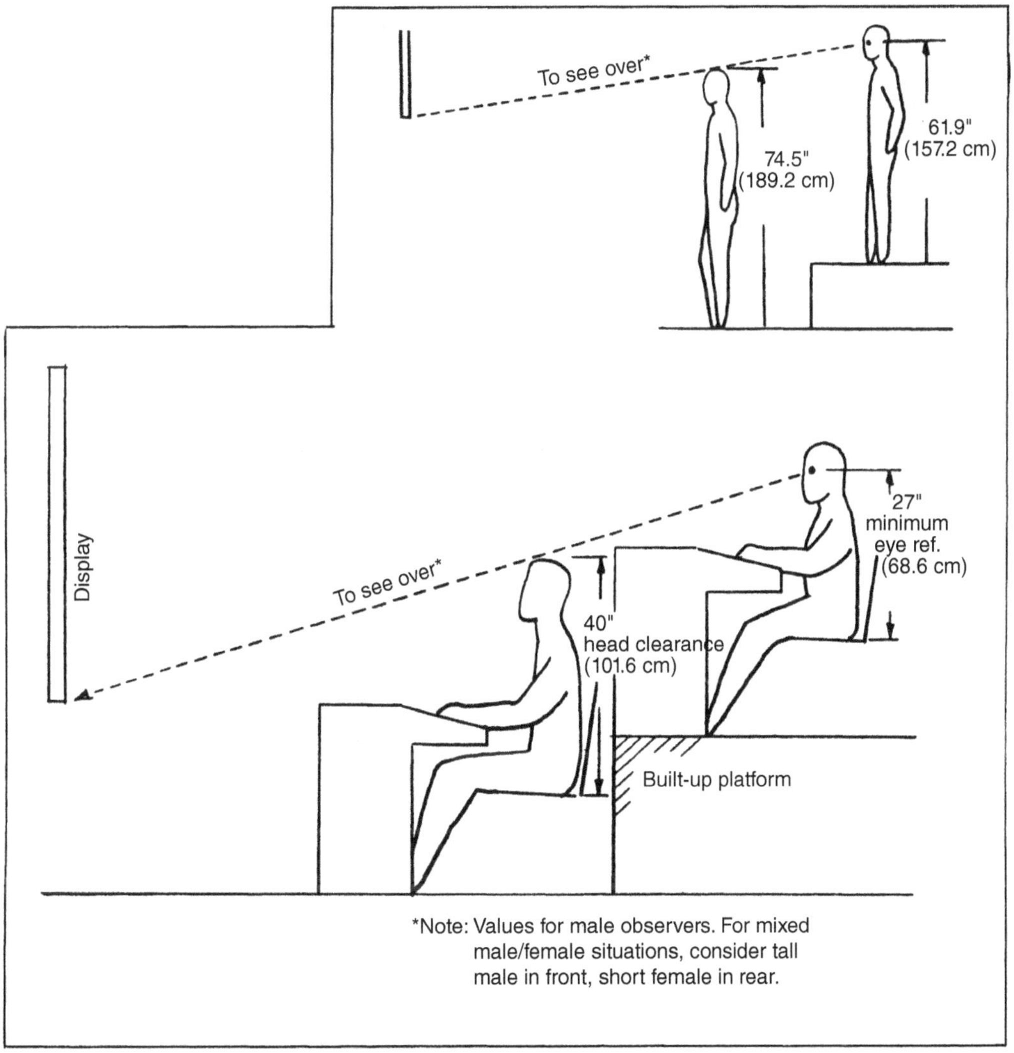

Figure 25.17 Use platforms.

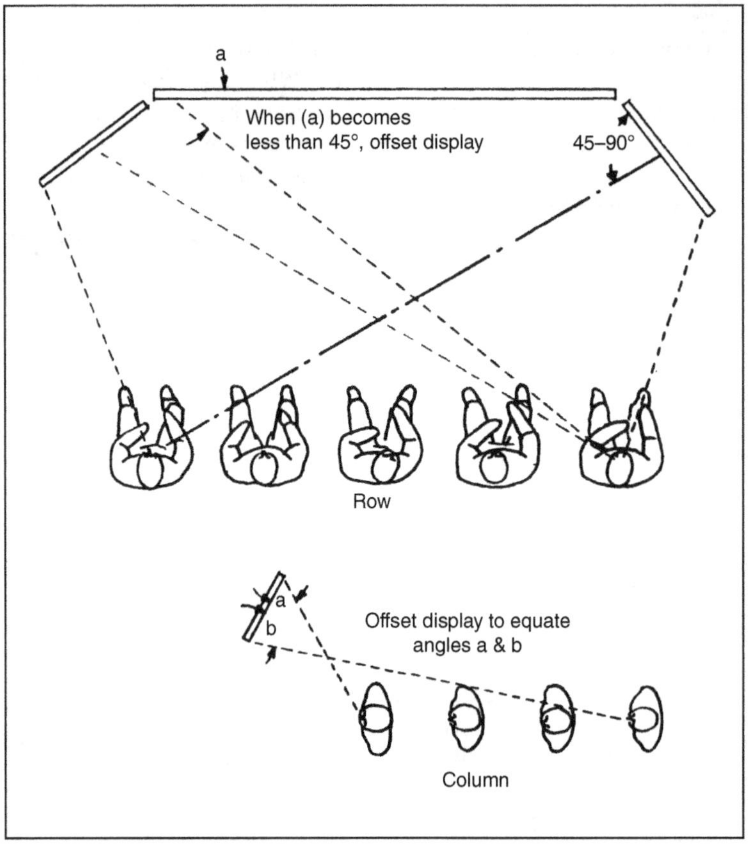

FIGURE 25.18 Rotate edges of display.

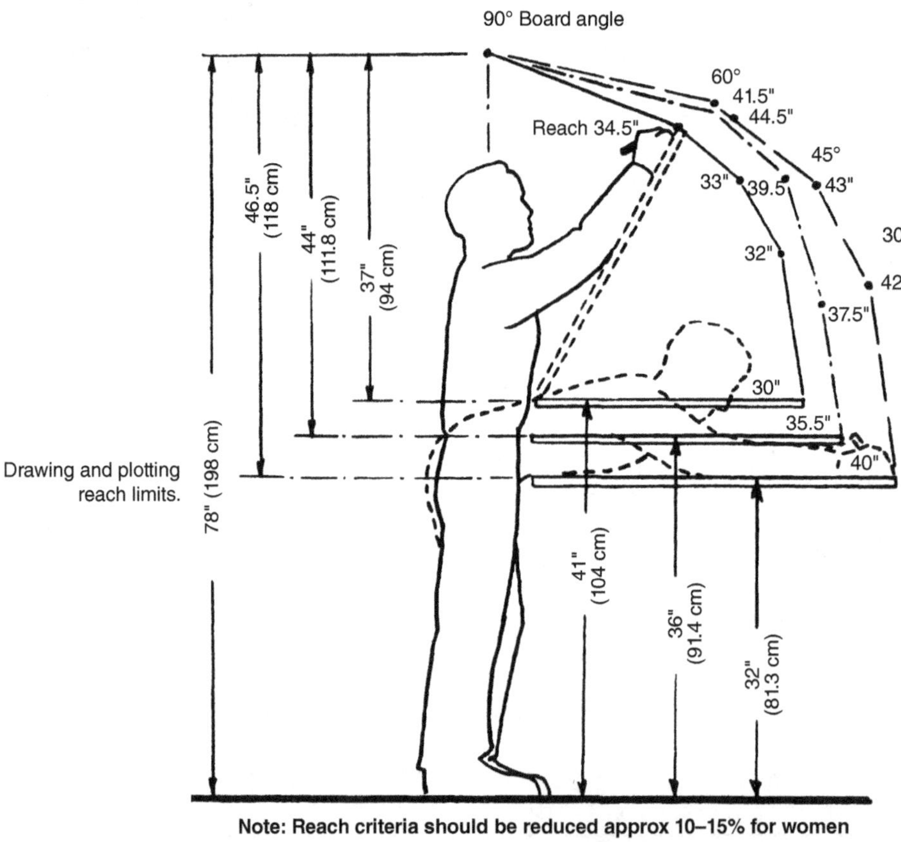

FIGURE 25.19 Guidelines for drawing and plotting.

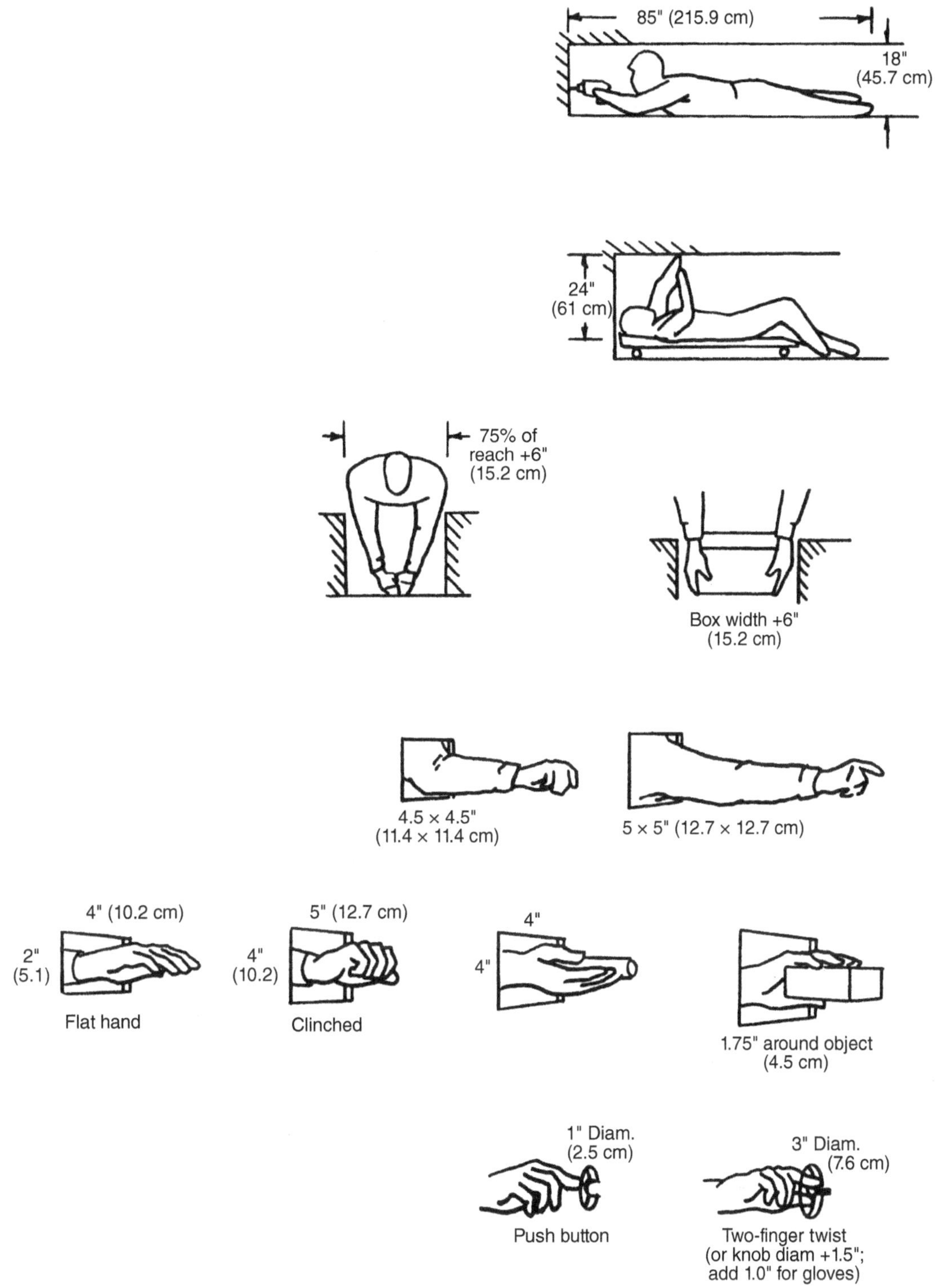

Figure 25.20 Limiting dimensions.

to "backtracking." Although some operations may not have a single sequential pattern, there are usually one or two elements that proceed according to a fairly standard sequence of steps. The sequential arrangement should be from left to right and top to bottom.

Functional Grouping. Certain functions within almost any operation are related to each other, and/or certain controls are related to certain displays. Related functions should be arranged together so that their association is readily apparent to the operator. When such functions are not grouped, operators tend to get lost and to wander from one point to another trying to find what they want. Although there may be a few elements within the workplace that cannot be placed together for obvious reasons (i.e., a foot pedal obviously

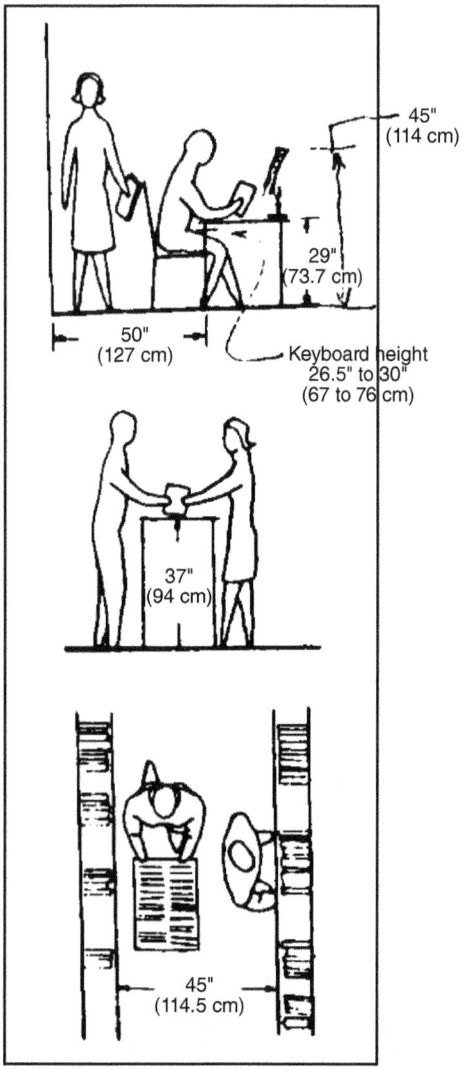

FIGURE 25.21 Dimensional guidelines for adults.

FIGURE 25.23 Carrels with temporary book storage shelves.

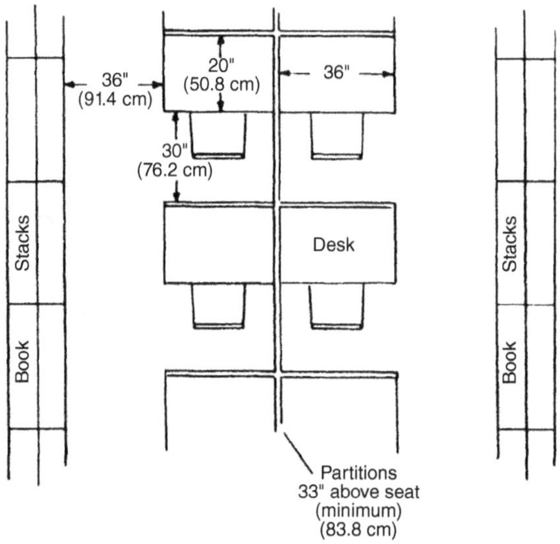

FIGURE 25.24 Simple, partitioned reading stalls placed within the stack area to enhance convenience to source materials.

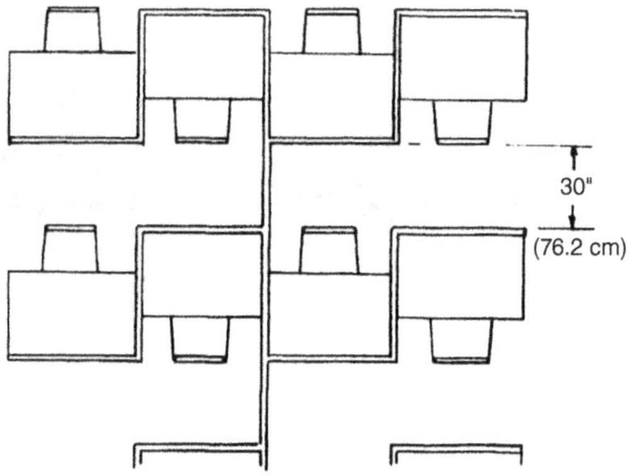

FIGURE 25.25 Minimum carrel configuration to make maximally effective use of limited space.

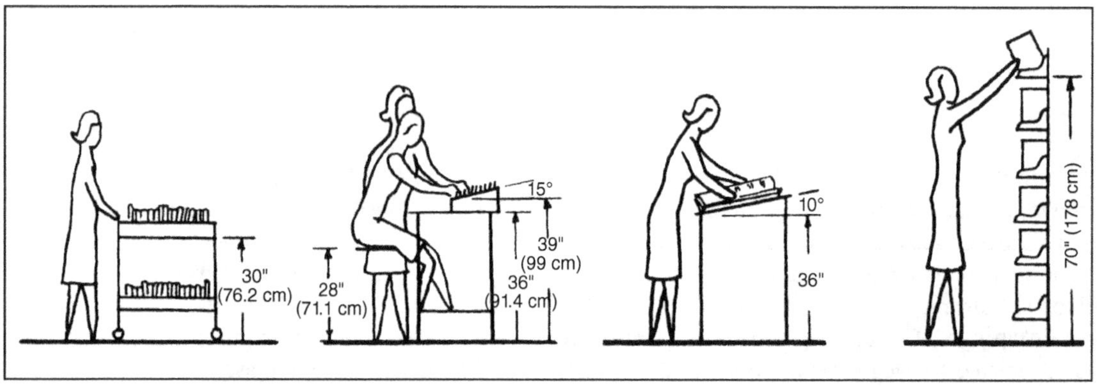

FIGURE 25.22 Dimensional guidelines for adults.

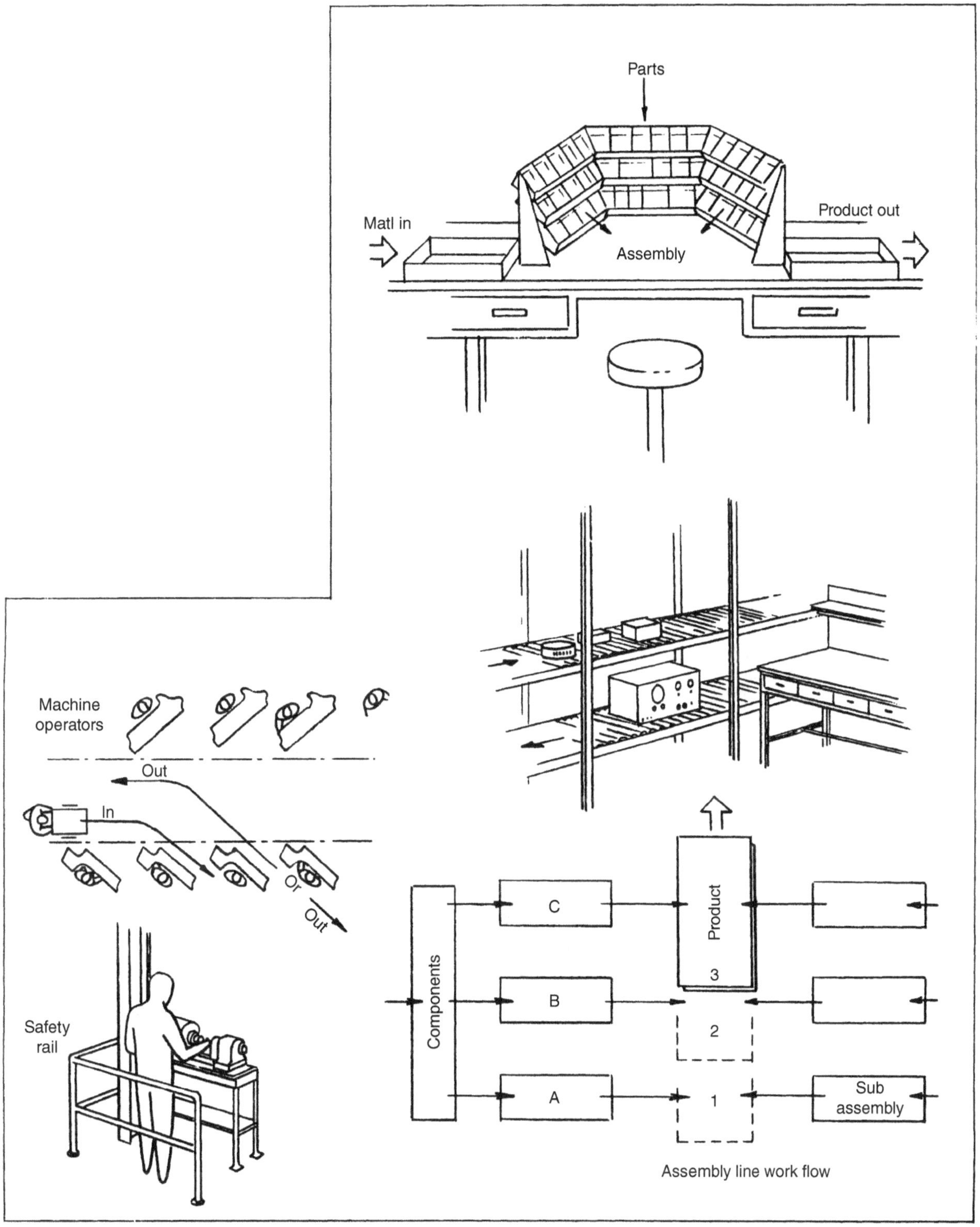

Figure 25.26 Typical factory workplace layouts.

cannot be placed where an associated visual display would be), an attempt should be made to avoid scattering controls and displays anywhere on the panel without regard to functional associations.

Frequency of Use. Some controls or displays will probably be used more frequently than others. These should be located so that they are maximally convenient; i.e., visual displays should be near the normal sight line, and controls should be near the hand when the arm is in a normal position. For example, a frequently used visual display should not be located where the operator must continually turn his or her head or body in order to see it, and a control should not be

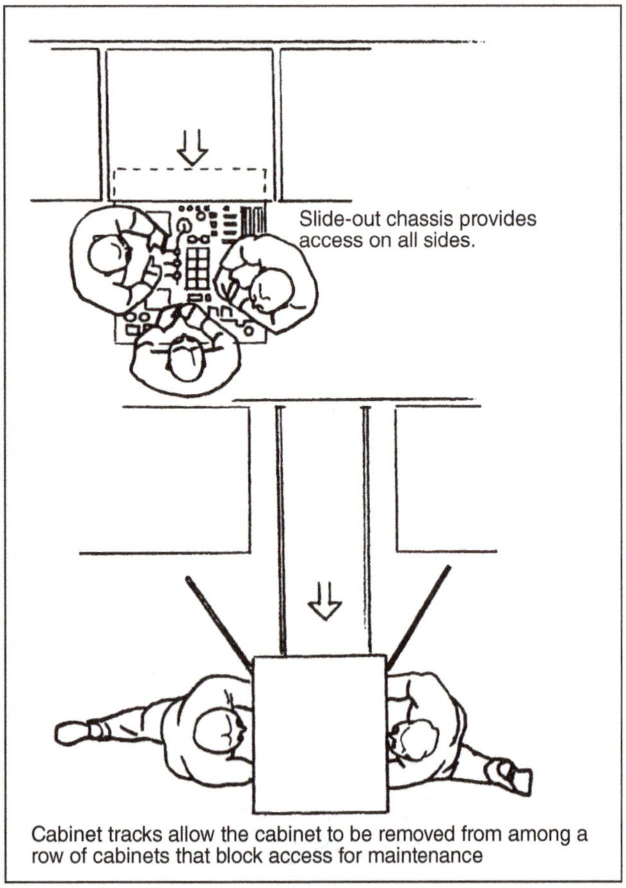

Figure 25.27 Provide access by building in slides and rails.

placed where the operator continually has to reach a long distance for it.

Emergency Use. Once in a while there will be a control that is critical under emergency situations; i.e., it may be vital that the operator get to it and operate it successfully in the shortest possible time. Obviously, if such a control is hidden, out of reach, or otherwise difficult to get to and operate, the operator may not be able to reach it in time.

Note: Each of the above principles must be considered in light of a particular operator interface problem; there is no set priority that can be all-prevailing.

Primary Visual Reference Principle. When it has been determined that an operator is expected to visually monitor a primary display for a great share of the time, that display should be located for maximum comfort, usually directly in front of the operator and approximately at eye level or slightly below. In addition, the things that the operator should be able to see fairly regularly in conjunction with the principal viewing tasks should be within about a 30° cone around the principal line-of-sight line. This includes visual displays and indicators and the controls that the operator may have to look at while he or she is adjusting them.

A typical console situation in which the visual reference principle must be applied is the large electronic display requiring the operator to continually watch for and track sonar or radar targets. The center of the display should coincide with the comfortable line of site. If the operator has to share more than one line of sight, seek a compromise reference, as shown in Fig. 25.31.

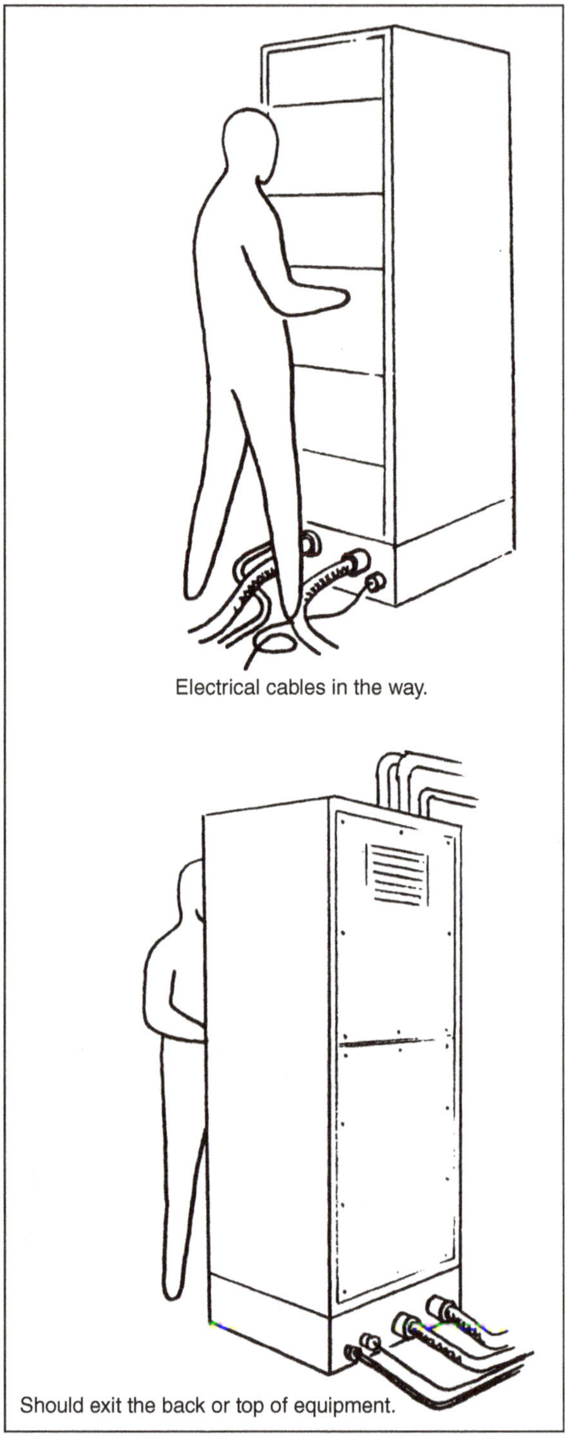

Figure 25.28 Route cables out of the way.

Positioning Panel Elements Relative to Line of Sight

Visual displays, including indicator lights, instruments, panel and control labels, etc., are useless to the operator if he

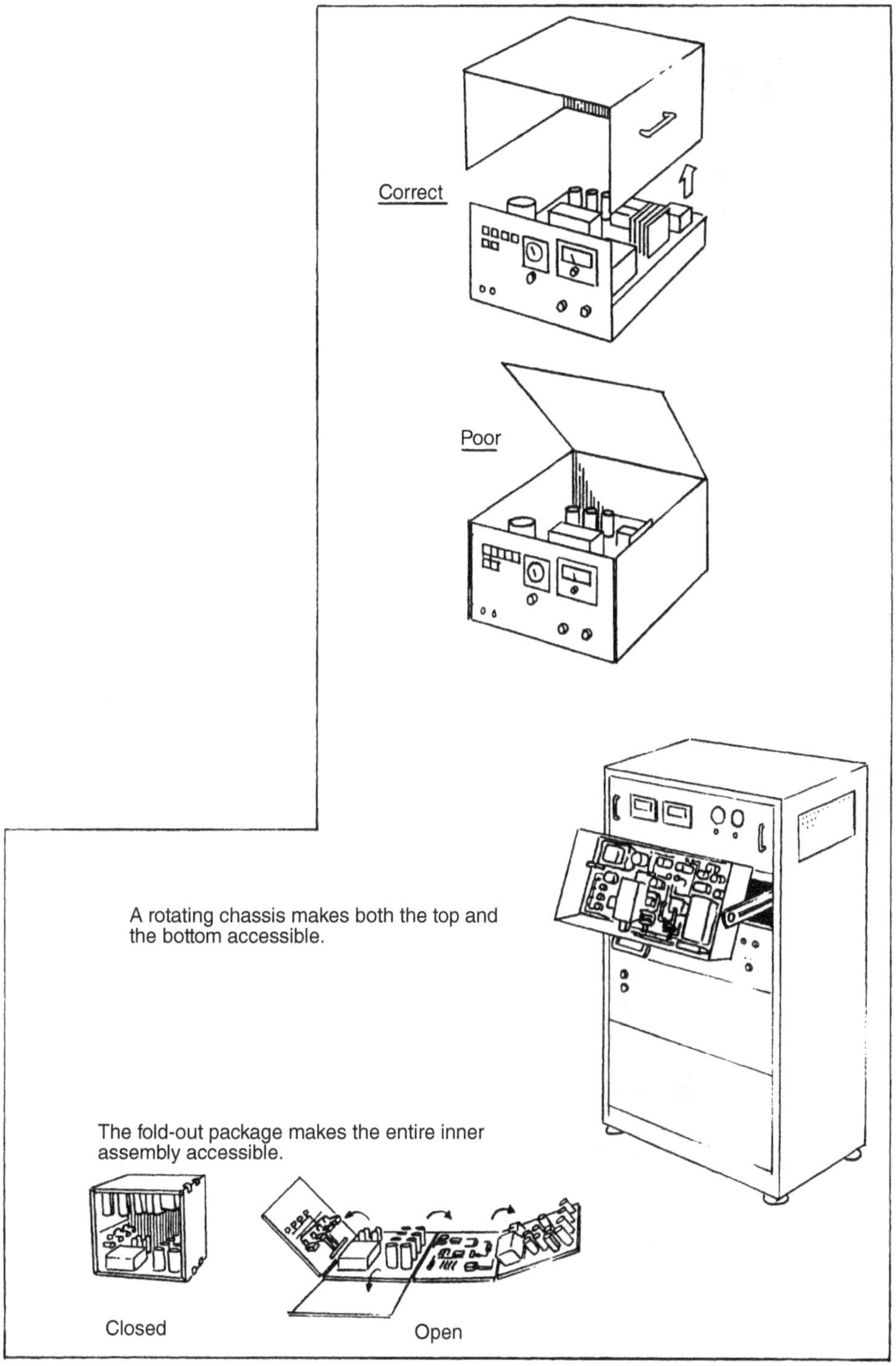

Figure 25.29 Design equipment packaging to improve maintenance access.

or she cannot see them. This requires that no control (joy stick, steering wheel, etc.) be placed between the observer's eye and other displays. In addition, it requires that care be taken in the placement of control and display labels so that when critical, the operator's hand will not cover up an important label that may be used in adjusting the control.

As a general principle, labels should be located in a consistent manner with respect to being above or below a control or display; i.e., avoid having some labels above a control, while others are below a control—at least on the same panel.

However, the choice of label position should also reflect consideration of the expected operator's eye reference; i.e., if the control is above the eye reference position, the label is more likely to be visible than if it were placed above the control. (See Fig. 25.32.)

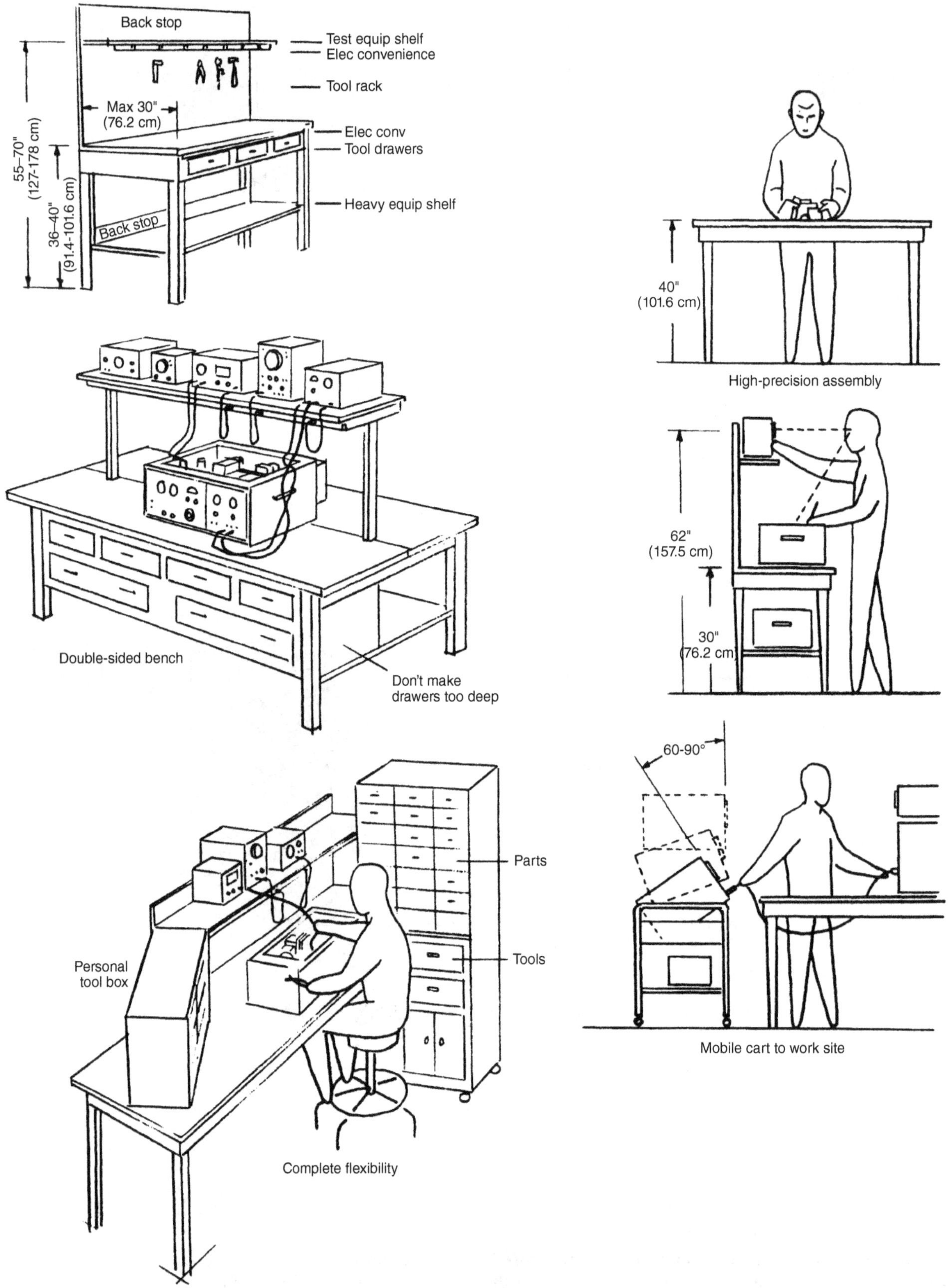

FIGURE 25.30 Shop workplaces should be designed for maximum flexibility so that a variety of tasks can be accomplished in the most convenient manner.

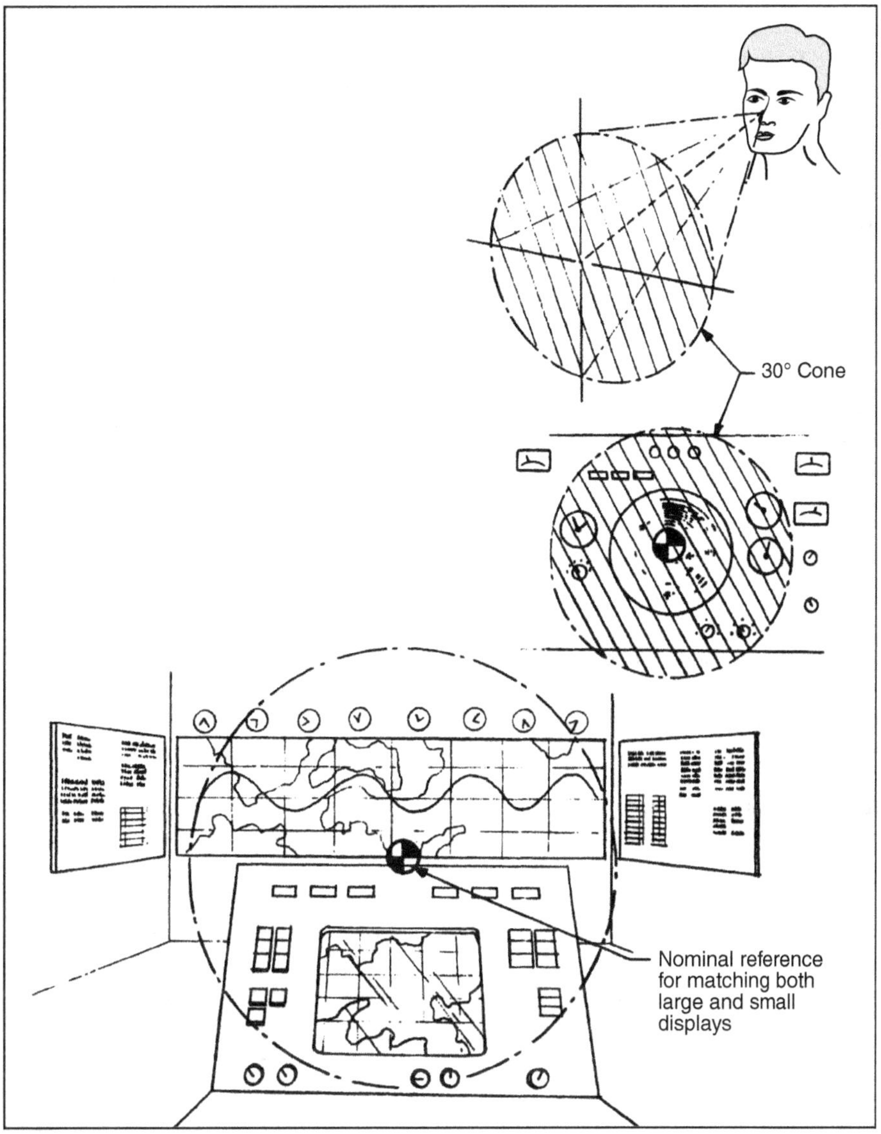

Figure 25.31 Place displays in the primary visual area.

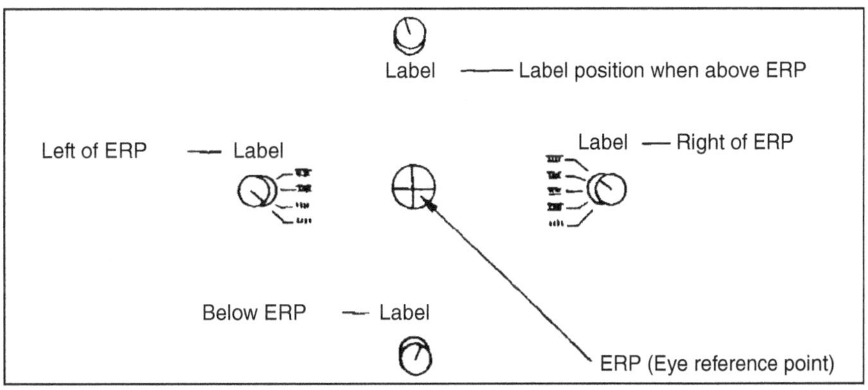

Figure 25.32 The eye reference point (erp) defines the relative location of controls and labels.

Display Arrangement

When the circumstance includes an array of displays that must be monitored on a regular basis to determine whether conditions are approximately "normal," the check-reading task is made easier for the operator if the displays are arranged so that their indications (e.g., pointer positions) generally reside in the same relative place on the instrument (see Fig. 25.33). Arrange the normal pointer reading at the 12 o'clock position when the array is vertical and at the nine o'clock or three o'clock position when the array is horizontal.

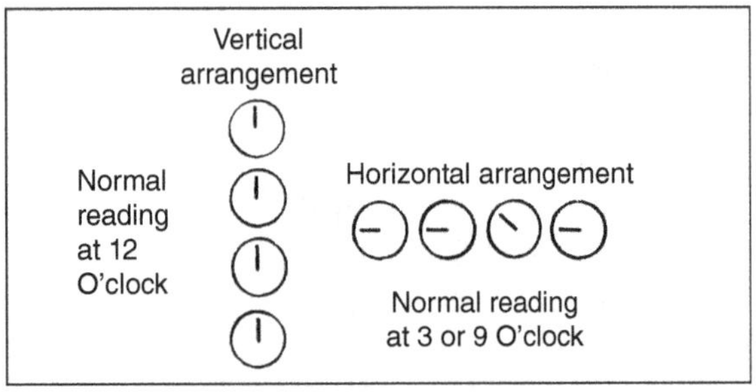

FIGURE 25.33 Simplify dial check reading.

If displays are ordinarily read in sequence, arrange them in order, either horizontally or vertically. (See Fig. 25.34.)

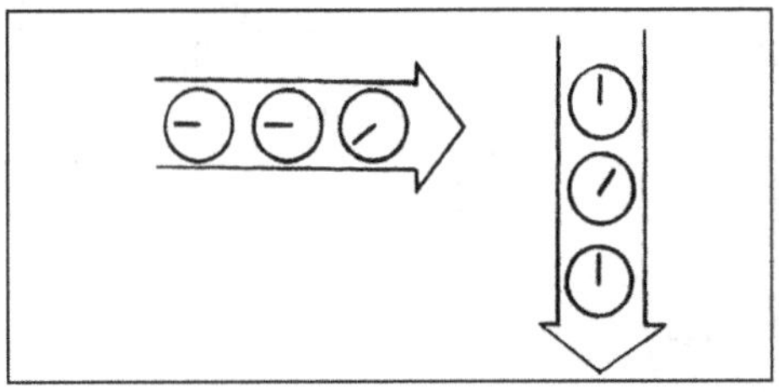

FIGURE 25.34 Arrange displays in sequence of use.

When a display is adjustable by means of an associated control, the control should be located as close to the display as practical. However, since there may be a directional relationship, as shown in Fig. 25.35, care must be taken to position the control relative to the display so that there is no doubt in the operator's mind as to the direction in which the control should be moved in order to cause the displayed element to move in the right direction. Controls that are placed below their respective displays are less confusing than those that are placed to the right or left of the displays.

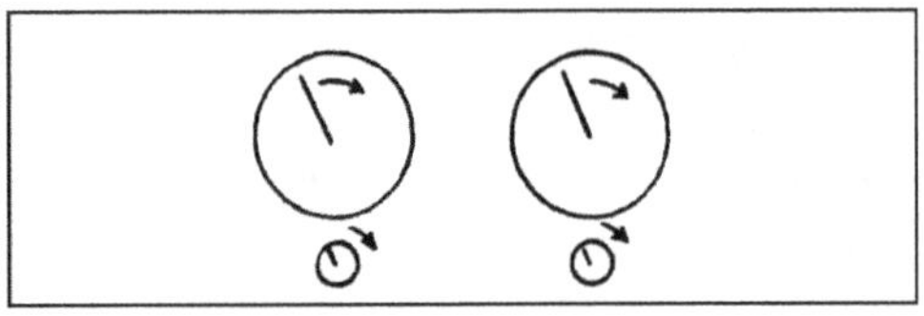

FIGURE 25.35 Place controls and associated displays together.

When it is impractical to place controls close to the displays they affect, try to arrange both the displays and the controls in similar patterns (see Fig. 25.36) so that there is a similarity between the two.

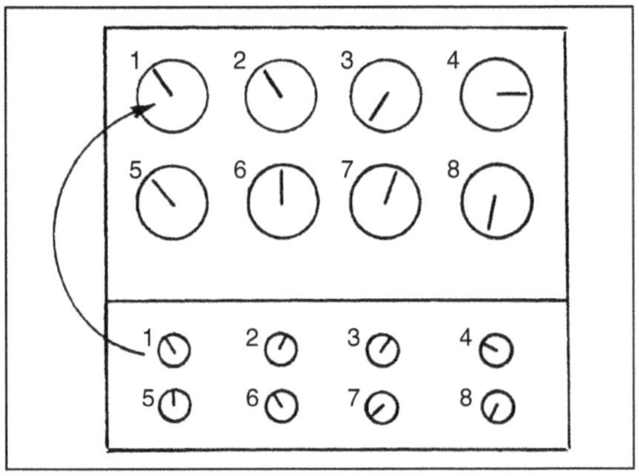

FIGURE 25.36 Arrange controls and displays in similar patterns if not possible to locate them together.

When a control causes something to move in space outside the console (e.g., an antenna), control-element directions of motion should also be similar, i.e., clockwise, counterclockwise, left to right, up and down, etc.

Although not highly recommended (except where panel space is limited), relationships between concentric-shaft knobs and their display counterparts should be defined as shown in Fig. 25.37; i.e., the top knob is associated with the first display on the left, and the bottom knob is associated with the last display on the right.

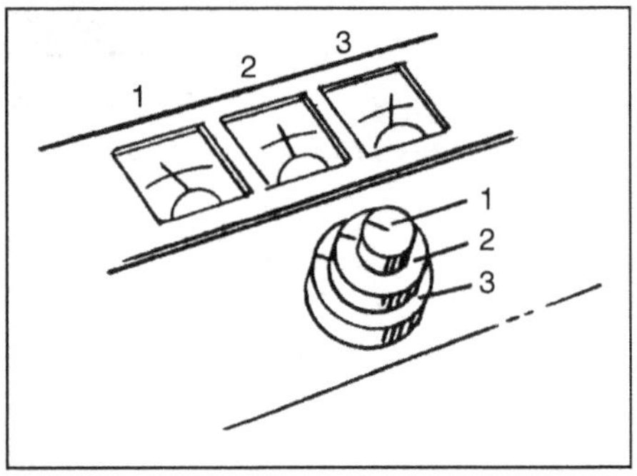

FIGURE 25.37 Recommended arrangement of displays associated with ganged controls.

In certain situations aboard an aircraft (namely, the engineer's or other auxiliary crew member's workplace), relationships between controls on the panel and external counterparts (e.g., the engines) should be as shown in Fig. 25.38. The general principle to be followed is that operators face the panel as though the engines were in front of them, depending on the direction in which they are facing. It is highly desirable, however, to face all operators "forward" with respect to the direction in which the vehicle is traveling.

An important objective in panel layout should be to make the functional grouping of controls and displays readily apparent to the operator. There are various effective methods for doing this, some of which are illustrated in Fig. 25.40:

1. Spatial separation is effective when there is sufficient panel area. This is the preferred method.

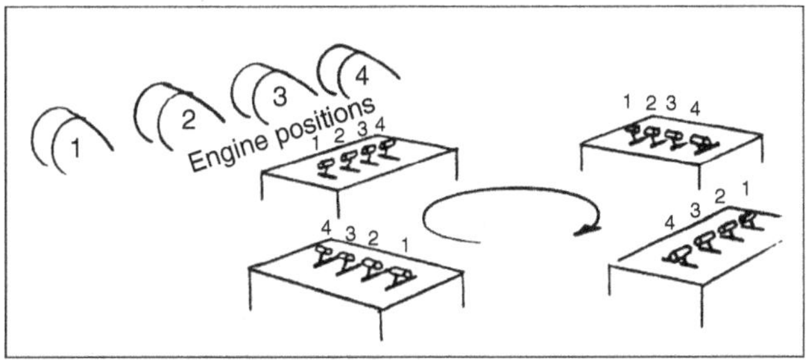

Figure 25.38 Recommended control arrangements when operator faces away from displays or "forward" direction.

2. When the panel is necessarily space-limited, one can use lines or borders around the various functions. This is particularly effective if the functional grouping is irregular.

3. If one functional group is especially important for the operator, it can be made more conspicuous by using a different color for the entire background for the group from that which is used for the rest of the panel.

4. Subpanels—either the protruding or the inset type, as shown in Fig. 25.39—are effective for setting off the various functional groupings.

5. In some cases, a separate package can be attached to the main package, especially if the package is small but needs a special physical orientation to make it easier to use.

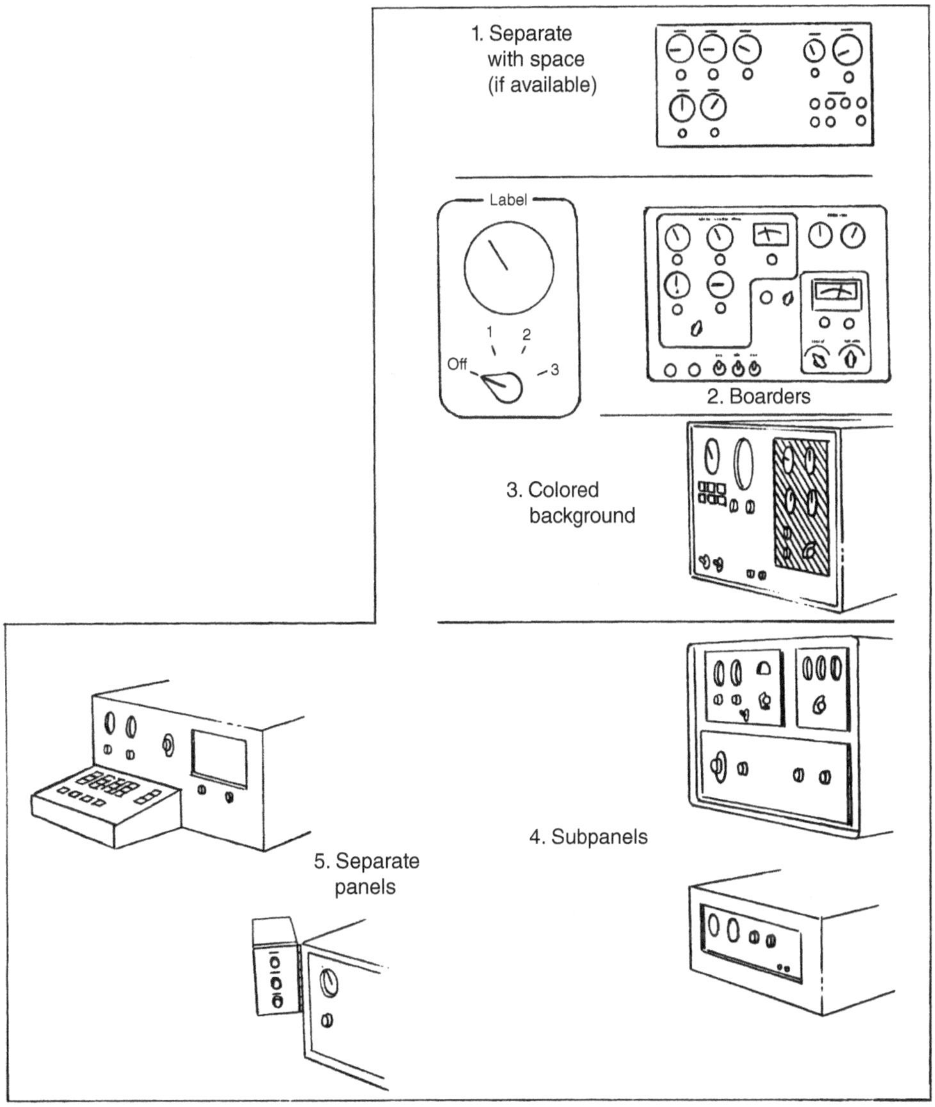

Figure 25.39 Ways to separate functional groups of controls and displays.

Panel Arrangement According to Sequence and Frequency of Expected Use

Whenever possible, try to determine the order in which an operator will probably use each display or control on the proposed panel and to estimate the frequency of the various sequential links. In this way, it is possible to layout the instruments and/or controls so that those that are used frequently are closer together, thus requiring less eye and hand movement and travel distance. (See Fig. 25.40.)

Generally, the equipment designer knows how he or she expects equipment to be operated. Where in doubt, the designer should search for someone who may have experience operating similar equipment. Together they usually can determine the best layout for the new equipment. (See Fig. 25.41.)

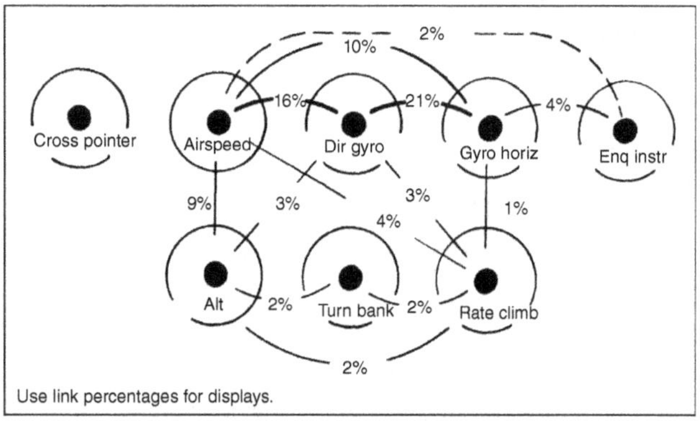

Figure 25.40 Group controls and displays that the operator uses together.

By tracing the operating steps through a proposed panel layout, it will become apparent when there are too many crossing or reversing pathways.

While this process is being accomplished, observe any situations in which the operator's hand seems to get in the way of seeing the next visual display.

Also remember that, as a general rule, visual displays should be located at the top of the panel, and controls should be located at the bottom (the natural positional relationship between the eye and the hand).

VEHICLE OPERATOR WORKSTATIONS

As shown in Fig. 25.42, the vehicle operator's position relative to external and internal viewing requirements depends on the size and shape of the proposed vehicle.

External Visibility Objectives

The proper reference eye position, plus the vehicle body and window configuration, should be combined to meet the following requirements (see Fig. 25.43):

1. To see the roadway surface to within 10 ft (3.1 m) of the front of the vehicle
2. To see an overhead signal tight when the vehicle is stopped at a stop line
3. To see the foremost left and right fender excursions (mobility clearance)

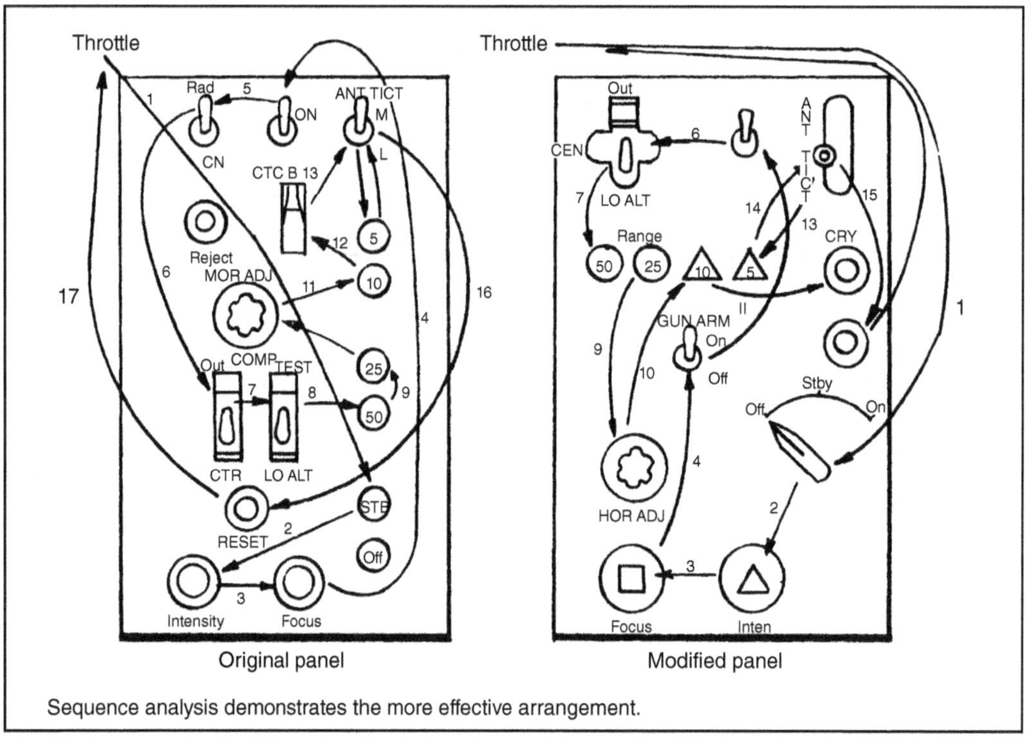

Figure 25.41 Arrange controls and displays by sequence of use.

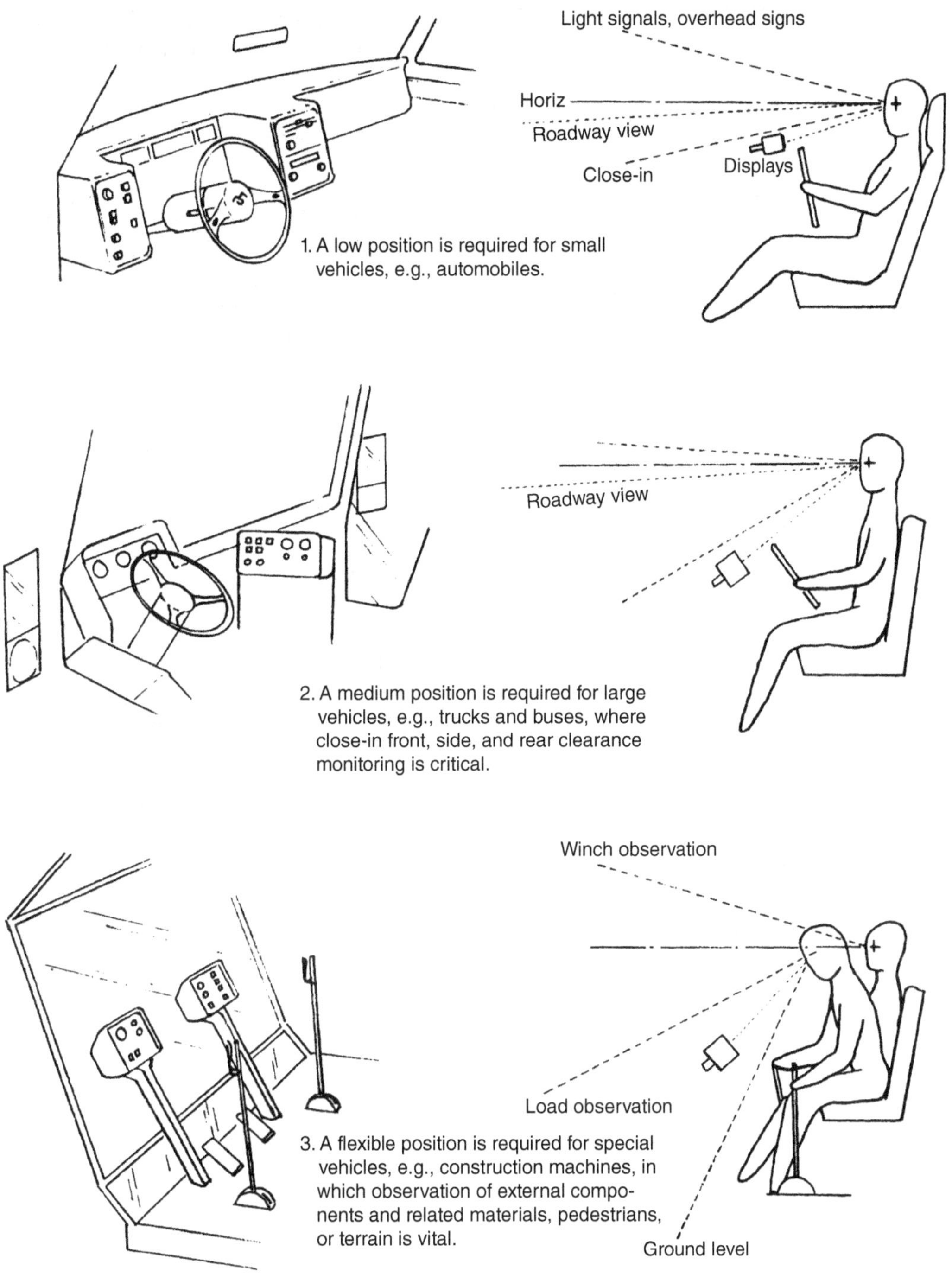

Figure 25.42 Vehicle operator's position.

4. To see a pedestrian who is about to cross in front of the vehicle (A—pillar occlusion angle)

5. To observe a rear-approaching motorcyclist in the left or right lane (B—pillar occlusion angle)

6. To observe the aftmost left and right fender excursions (backing mobility clearance)

Although rearview mirrors should also be provided (i.e., an interior mirror plus right and left rearview mirrors), everything possible should be done to arrange the driver's eye position and the vehicle's window and exterior body contours (especially the fender lines) so that direct external viewing requirements can be met.

These viewing needs should be met for all drivers whenever possible (at least 90 percent of the population), considering the expected variations in seat adjustment.

Placement of rearview mirrors should not incur viewing occlusions, nor should seat belts and hardware or headrests

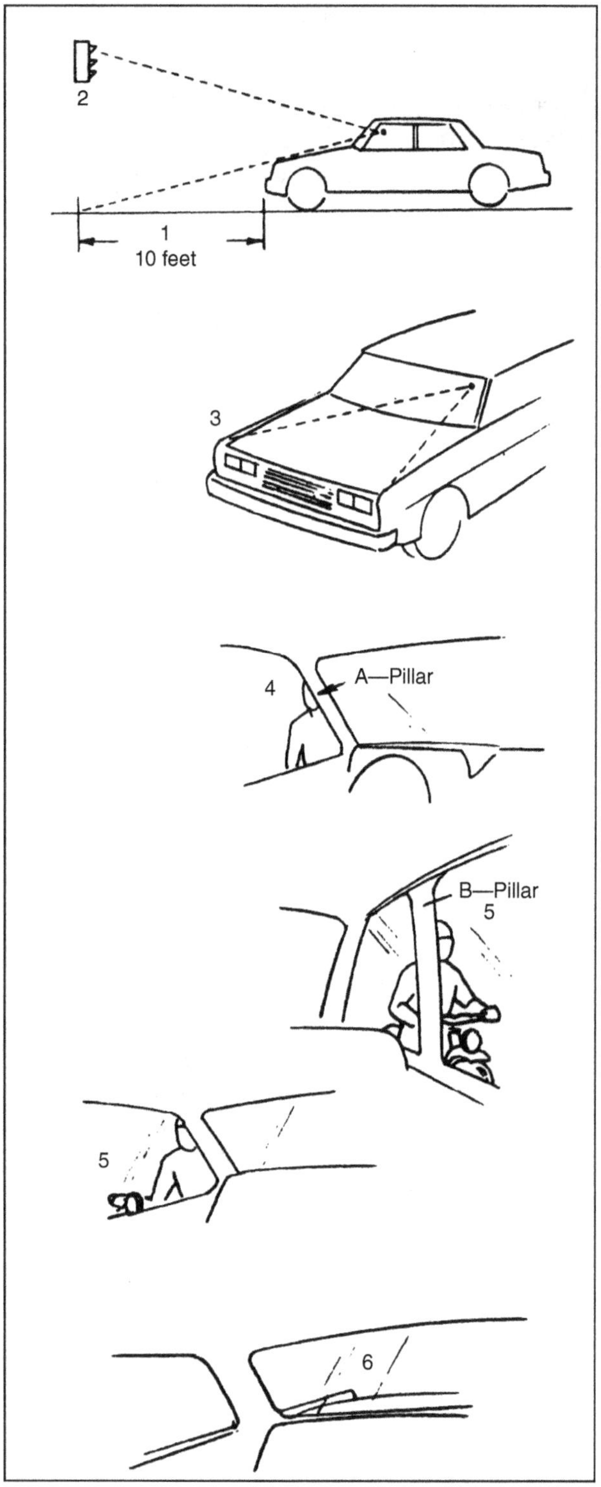

Figure 25.43 Visibility objectives in ground vehicle design.

be designed or positioned so as to block any of the above viewing requirements.

It should be possible to adjust rearview mirrors when sitting with the eyes in the normal driving position; i.e., the driver should not have to lean to make such adjustments.

Exterior mirrors located on both sides of the cab should be remotely controllable by the driver. It is unsatisfactory to depend on manual positioning because most operators forget from time to time to position these mirrors until after they are on the road. (See Figs. 25.44–25.47.)

Layout of Automobile Driving Controls and Displays

The automobile control and display location and arrangement should be standard among all automobiles. This is because a great many drivers transfer among a variety of automobile makes and models. If the driver station layout varies considerably, drivers can become confused and may select the wrong control, look at the wrong display, or become so distracted trying to find the right control or display that they fail to watch the road. The following suggestions concerning standardization are based on a number of studies that have attempted to establish the most common driver expectancies in terms of locating and operating controls and displays in automobiles.

As indicated in Fig. 25.48, primary visual displays often used in conjunction with observing the road ahead are positioned so that they can be seen above the steering wheel rim. Other controls and displays are placed on panels to the side of the wheel, on wheel spokes, or on levers mounted on the steering column so that the controls are within fingertip reach (within 6 in., 15 cm, of the wheel rim). All controls and displays should be properly labeled for both day and night observation.

Table 25.5 summarizes the SAE J1138 recommendations for vehicle hand controls. The table identifies priority for reach location (high or low), priority for visibility (high or low), how the control should be labeled (identification, setting, or both identification and setting), and location (right or left side).

Application Guidelines for Location of Automobile Driver Controls

Basic Functional Categories

1. Vehicle control (steering, accelerating, braking, using the clutch, etc.)
2. Visibility control (internal and external lights, window clearing, and rear viewing)
3. Warning systems (turning signals and horn)
4. Environmental systems control (fresh air, airflow direction, airflow rate, heating, and cooling)
5. Driver information, communications, audio entertainment, etc.
6. Driver seat adjustment
7. Exit systems (door latches and locks)
8. Convenience systems (window lifts, tailgate actuator, etc.)

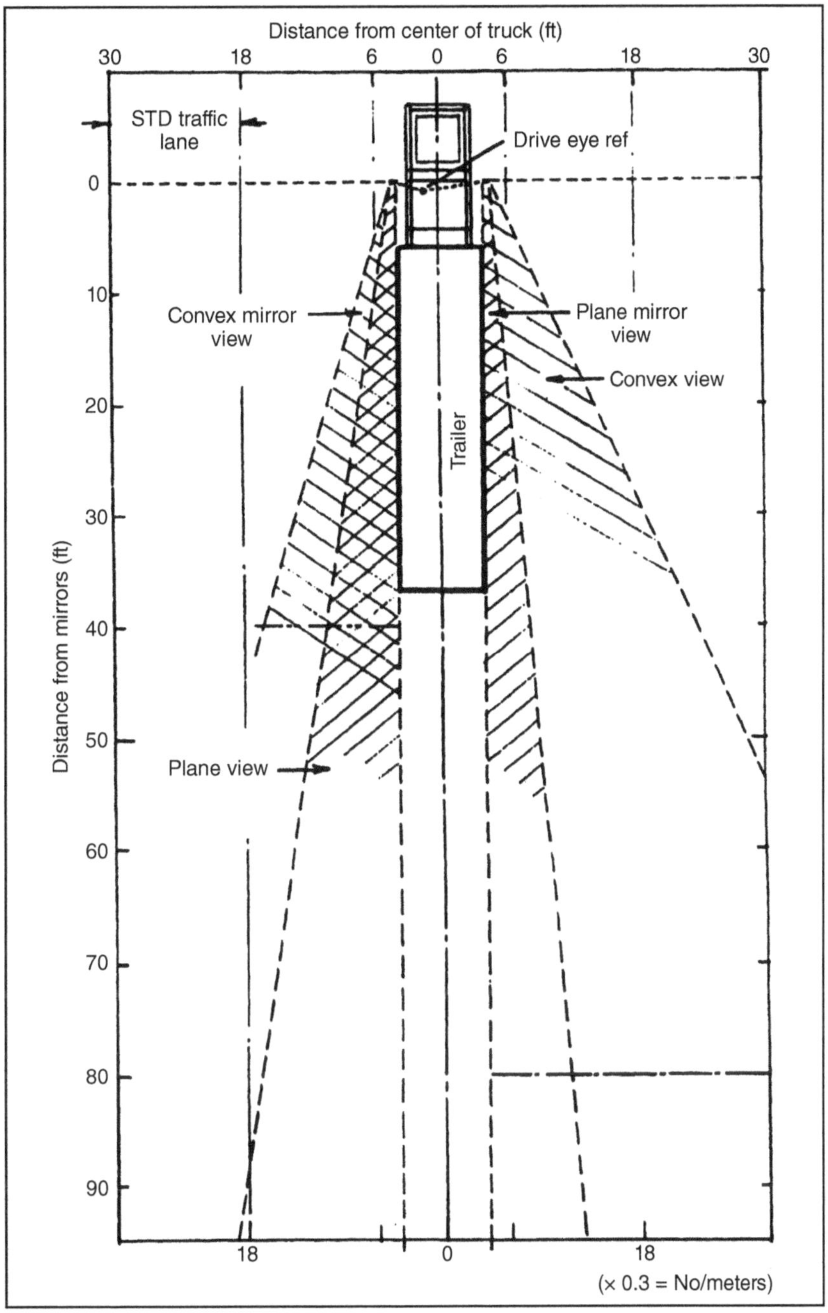

Figure 25.44 Recommended horizontal field of view with rearview mirrors on tractor-trailer trucks.

Use Priorities

1. Priority 1: Controls that require (or may require) immediate error-free response while the vehicle is under way. Such controls should be located so that the hand travels a minimum distance away from the normal "hand-on-the-wheel" reference position.

2. Priority 2: Controls that, although reach time is not as critical, should be placed where the driver can see them peripherally while watching the road (without excessive head or torso movement). Foot controls obviously are exceptions to these first two priorities.

3. Foot controls obviously are exceptions to these first two priorities.

4. Priority 3: Controls that are used only when the vehicle is stopped. Although they do not have to be reached quickly or be visible while the driver watches the road, they should not be hidden from view, nor should they be out of reach from the driver's normal position.

Priority 1 controls are often referred to as "fingertip reach controls"; priority 2 controls should be within the 30° vision cone for driving. Figures 25.49 and 25.50 further illustrate the various priority conditions.

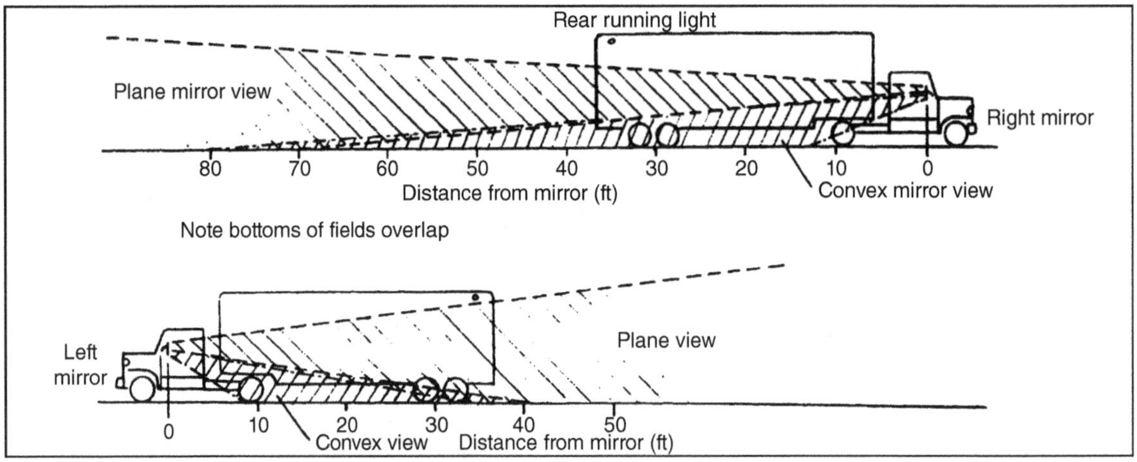

FIGURE 25.45 Recommended vertical fields of view with rearview mirror systems mounted on tractor-trailer.

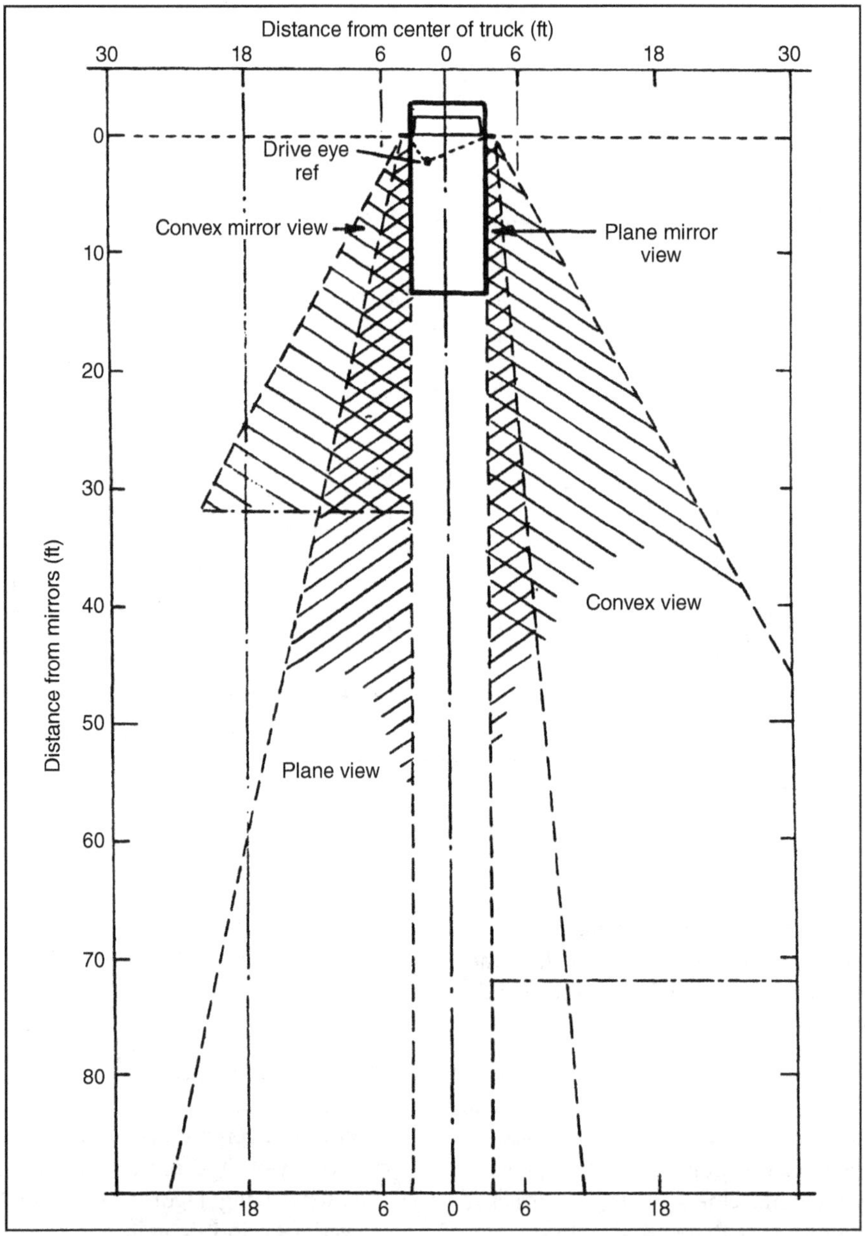

FIGURE 25.46 Recommended horizontal field of view for the straight truck mirror system.

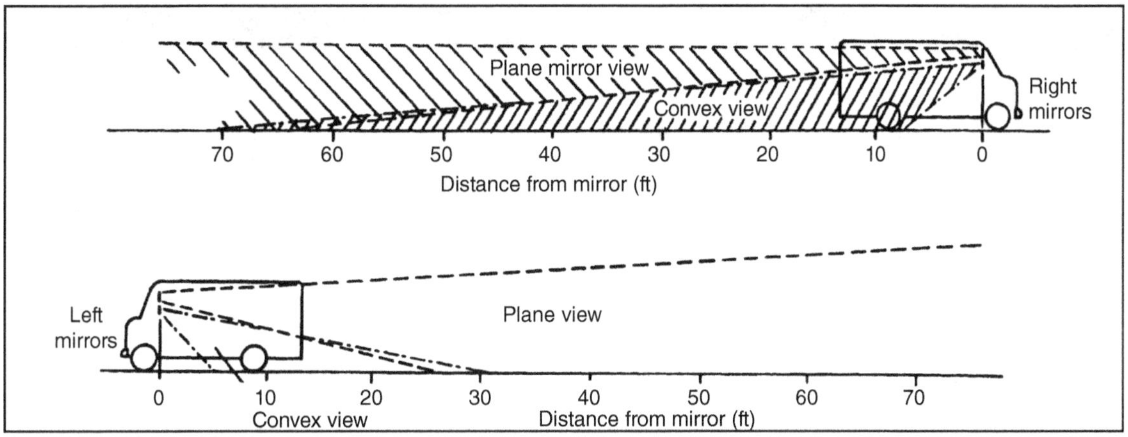

Figure 25.47 Recommended vertical field of view afforded by rearview mirror systems for straight truck.

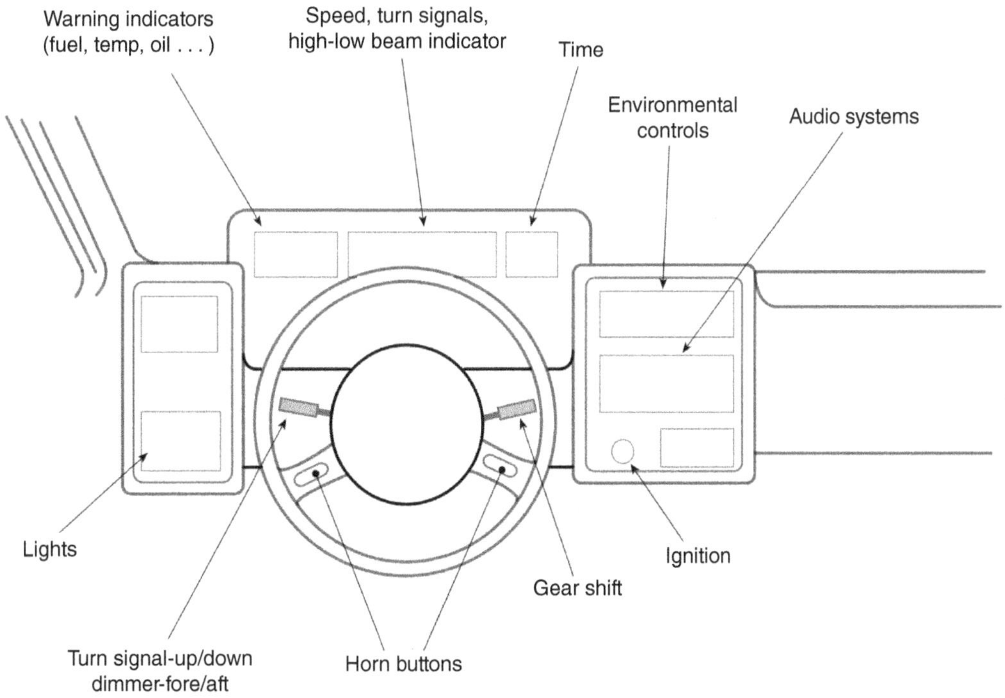

Figure 25.48 Standardized layout of automobile controls and displays.

TABLE 25.5 Automobile Controls

	Reach Location Priority (Note 1)	Viewing Priority (Note 1)	Labeling (Note 2)	Location (Note 3)
2- to 4-wheel drive transfer	L	L	S	R
Airbag disable	L	L	B	R
Ashtray	L	L		R
Audible horn	H	H		
Audio (i.e., volume, selection devices)	H	H	S	R
Cigarette lighter	L	L		R
Climate control air distribution			S	
Climate control fan			B	
Climate controls	H	H	B	R
Clock	L	L	S	R
Cruise controls	H	H	S	
Day/night mirror	H			R
Dome/map lights	L	L	S	
Electronic transmission shift point	L	L	B	R
Fog/driving lamps	L	L	B	
Gearshift	H			R
Hand brake	H			R
Hand release for foot-operated parking brake			I	L
Hazard flasher	H	H	I	
Headlamp	H	H		L
Headlamp dimmer	H			L
Hood release	L		I	L
Ignition	H			R
Manual gearshift pattern		H		
Master lighting control			B	
Panel illumination intensity	L	L	B	
Gear selection display		H	S	
Overdrive on/off		L	B	R
Power door lock	H	H	S	L
Power mirror	H	H	B	L
Power seat adjusters	L		I	L
Power windows	H	H	S	L
Rear defroster	H	H	B	
Rear washer/wiper	H	H	B	
Remote fuel door release	L		I	L
Remote trunk release	L		I	L
Steering	H			
Steering wheel tilt/telescope	L			
Telephone handset	L			R
Traction assist	L	L	B	
Trip odometer	L	L		
Turn signal	H			L
Windshield defroster	H	H	I	R
Windshield washer/wiper	H	H	B	

Note 1: H = high, L = low
Note 2: I = identification, S = setting, B = identification and setting
Note 3: R = right, L = left.

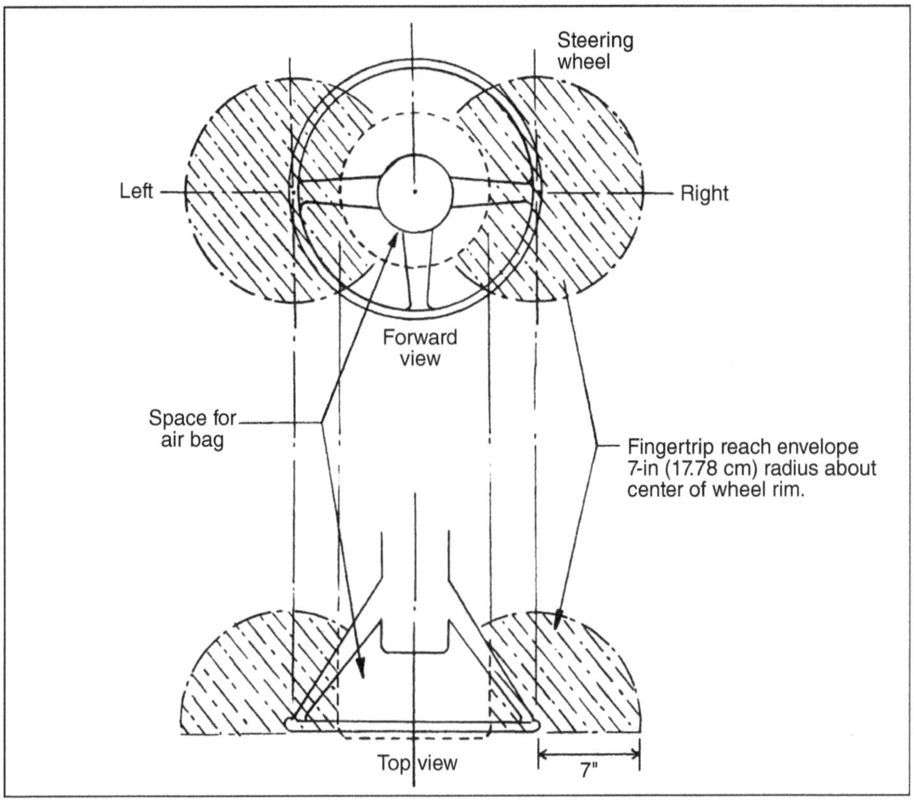

Figure 25.49 Recommended location for priority 1 controls—fingertip reach areas.

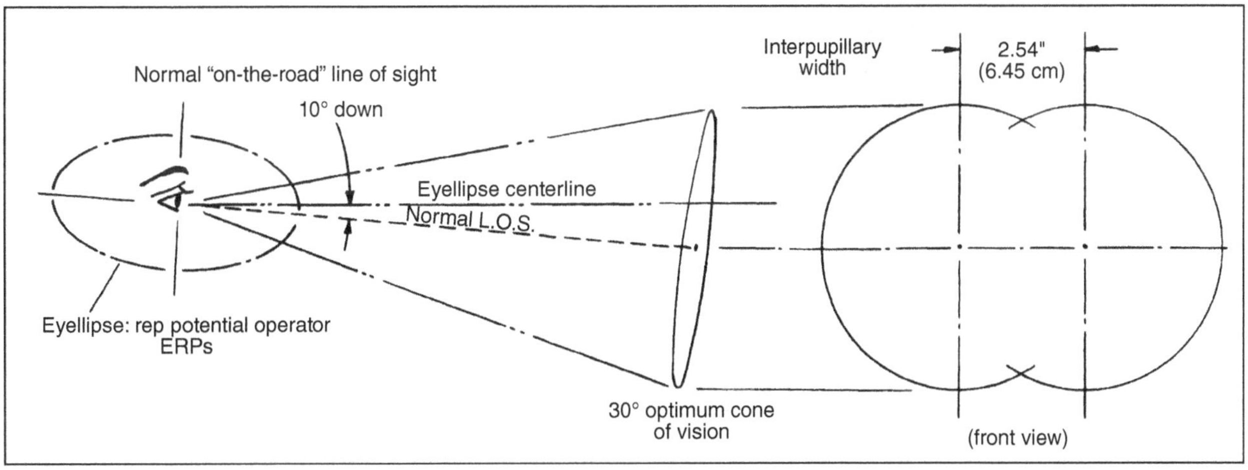

Figure 25.50 Recommended location for priority 2 controls—30° visual cone.

Optimized Driver Station Profile[12]

These specifications provide optimum positioning for the range of American drivers from the 95th-percentile female through the 95th-percentile male to ensure the best posture, external and internal visibility, and access to all controls without requiring the driver to lean or reach and thus risk losing visual contact with the road. (See Fig. 25.51.) Approximation at these dimensions will ensure the maximum efficiency and comfort for all drivers.

[12]W. E. Woodson, et al., *Instrument and Control Location, Accessibility and Identification,* Report MFI 69-106. U.S. Department of Transportation, NHTSA Contract FH-II-6907, 1969.

General Checklist for Initial Concept Development (See Fig. 25.52)

Driver Positioning for Safe External Viewing

For several reasons, listed below, it is most critical to seat the driver of an automobile (and of any other highway vehicle, for that matter) relatively high so that he or she can see the road and the hood of the vehicle. (See Fig. 25.53.) It is equally important to minimize any serious obstructions to lateral viewing and/or viewing out the rear.

The driver in a low position cannot see over the steering wheel, dashboard, or hood and has little information upon which to base his or her steering or speed.

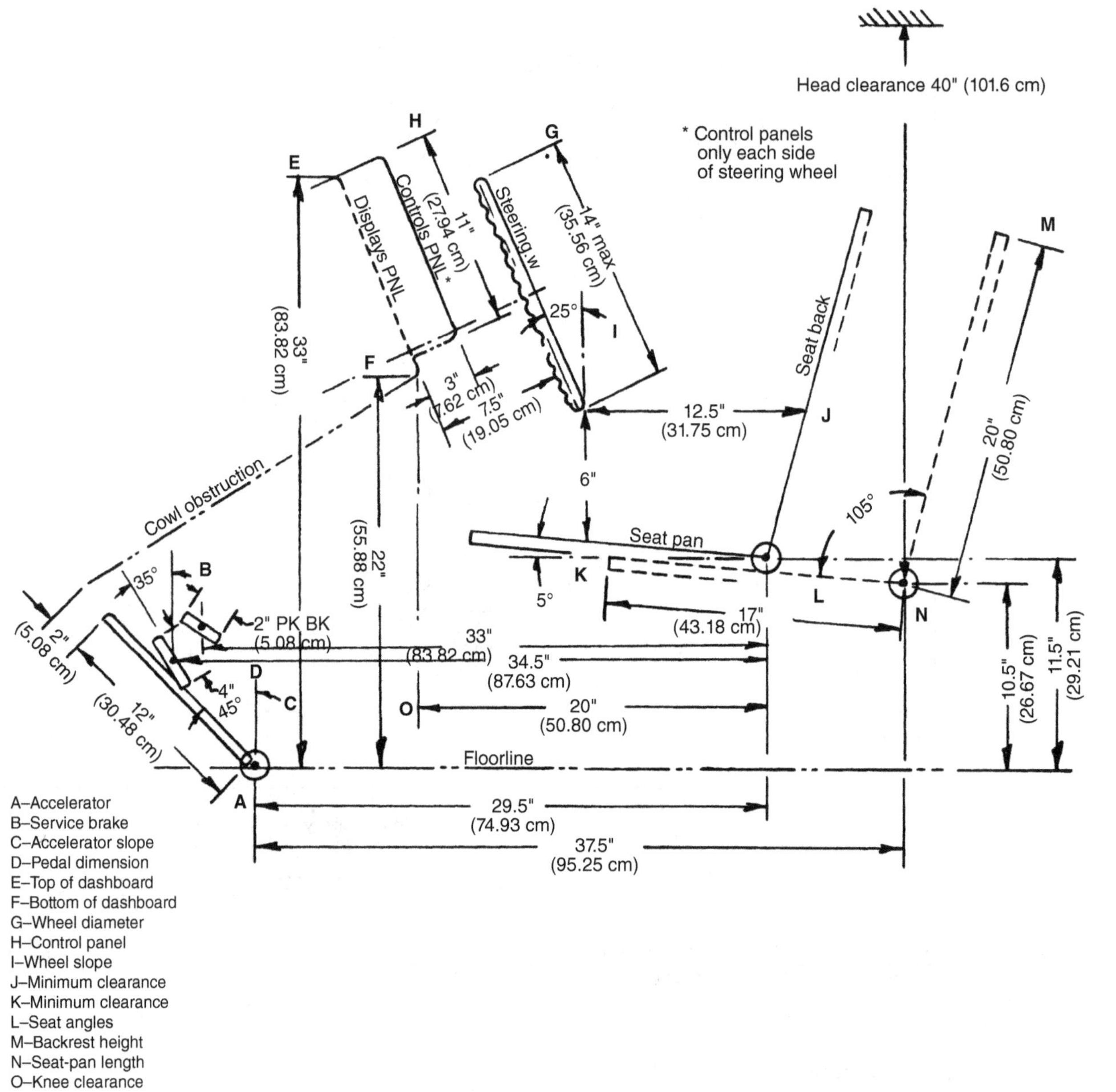

FIGURE 25.51 Optimized automobile driver station. (From W.E. Woodson, et al., instrument and Control Location, Accessibility and Identification, Report MFI 69-106. U.S. Department of Transportation, NHTSA contract FH-11-6907, 1969.)

Ingress and Egress Clearance

As automobiles have become lower and lower, principally for styling reasons, the difficulties for the larger individual with respect to entry and exit have become considerably greater, to the point where the probability of striking one's head is quite high. Although many injuries are undoubtedly sustained as a result of such styling, the consumer continues to accept the inconvenience (and the sore head). Good practice should follow the dimensional criteria shown here. (See Fig. 25.54.)

In addition to the doorway opening clearance, adequate space between the rim of the steering wheel and the seat should be provided. This minimum clearance (e.g., 7 in., 18 cm) should be measured when the seat is in the full forward position. A seat-adjustment range of at least 8.0 in. (20 cm) is required in order for the seating arrangement to accommodate the full range of American drivers (i.e., a 5th-percentile female and a 95th-percentile male). (See Fig. 25.55.)

Other Human Factors Considerations

Other key human factors considerations should be addressed during the initial concept development stage. These include the following.

Riding and Handling Quality. Obviously, the automobile should not possess Qualities that would make it difficult to control, especially under unusual conditions, i.e., on rough

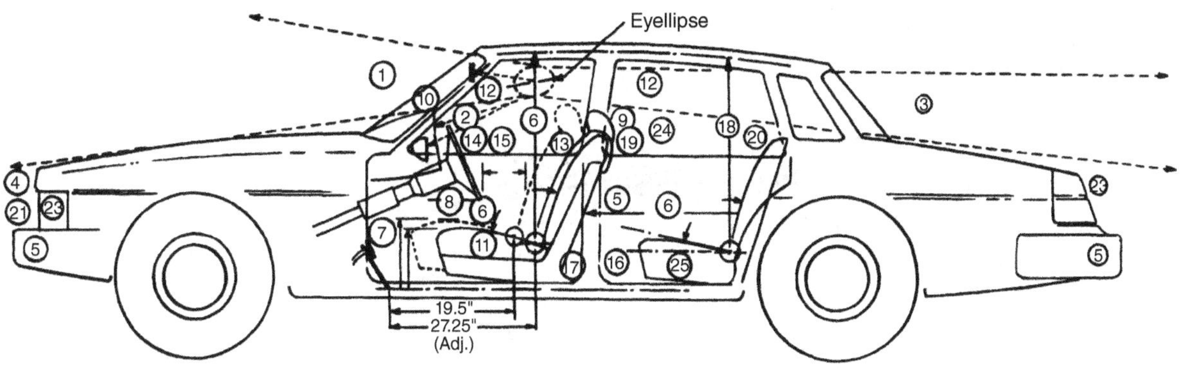

1. Exterior visibility (signs and signals, over the hood and dashboard, and vehicles in front and to the side)
2. Interior visibility (dash displays and hand controls)
3. Exterior visibility (vehicles in the rear and to the side)
4. Noninjurious exterior pedestrian interface
5. Energy-absorbing front, rear, and side; cabin envelope integrity in crash
6. Driver position (reach controls and clear head)
7. Ease of ingress and egress
8. Clearance for legs and knees around dashboard and steering wheel
9. Headrest—whiplash protection
10. Energy-absorbing surfaces, and buried or noninjurious controls
11. Posture-control seat (adjustable fore and aft and up and down)
12. Glare and reflection control
13. Crash restraint system
14. Display and control illumination
15. Labels, instructions, and warnings
16. Ease of ingress and egress
17. Knee and foot clearance
18. Headroom
19. Noninjurious front-seat back
20. Crash restraint system
21. Roadway illumination
22. Interior illumination
23. Running and signal lights
24. Environmental control (temperature, vibration, ventilation, and noise)
25. Fireproof interior materials

Figure 25.52 General checklist of initial concept development.

or slick roads, when it is heavily loaded, or when unusual maneuvers are necessary, such as making a sudden turn to avoid another vehicle or a pedestrian or object in the road. Equally important is a ride quality that will not create unreasonable stresses on the passengers. This latter consideration is secondary to the former since people vary in their interpretations of what is a satisfactory ride; e.g., a sports-car enthusiast wants to "feel the road."

Reliability. Obviously, most consumers appreciate a vehicle that is reliable and does not require continual repair. Most important, however, is a design that will not let the driver down in a pinch, i.e., an automobile whose brakes, steering, motor, tires, etc., will not fail.

Ease of Maintenance. Much of the cost of maintenance is related to the difficulty and time required to determine where a malfunction is and to the difficulty of getting to a part, removing it, and replacing it. Ease of maintenance can be increased primarily during the conceptual stage, before difficulties have been created by placing things in awkward places.

Design-Induced Driver Problems. A great many of the driver's problems are often created because a designer has chosen to use the driver interface as a prime target for artistic or aesthetic experimentation. That is, instead of recognizing that, to make the driver's tasks easier, controls and displays should be designed, located, and labeled in a certain way, the designer chooses to experiment with various methods and techniques for mechanizing these displays and controls, arranging them to create a special effect or adding decorative chrome that reflects in the windshield or into the driver's eyes, for example. Designers often ignore good seating design criteria, thus creating difficulties in postural control or leg manipulation for the driver (e.g., soft cushions may allow the heavy driver to sink too low to see out or may cause clothing to bunch, and incorrect cushion dimensions may create pressures on the back of the driver's legs). (See Fig. 25.56.)

Fore and aft crowding not only is uncomfortable for the rear-seat passenger but also may be the cause of unnecessary injury during a crash. The recommended clearance is for the case in which the front seat is adjusted to its aftmost position. (See Fig. 25.57.)

Common Control Layout Mistakes

Poor planning often results in the accident illustrated in Fig. 25.58; i.e., moving the visor has displaced the mirror, which could present a hazard. The time and attention required to readjust the mirror may be just enough to cause the driver not to notice an impending hazard on the roadway.

Similar problems occur relative to steering when levers are placed too close to the wheel rim.

It is poor layout when drivers are forced to reach (and get their eye reference out of position) to adjust a rearview mirror; i.e., drivers should be able to adjust the mirror from their normal eye reference position since that is where they will be viewing the mirror while driving. It is unforgivable, then, to provide an inside remote mirror control which also requires reaching. (See Fig. 25.59.)

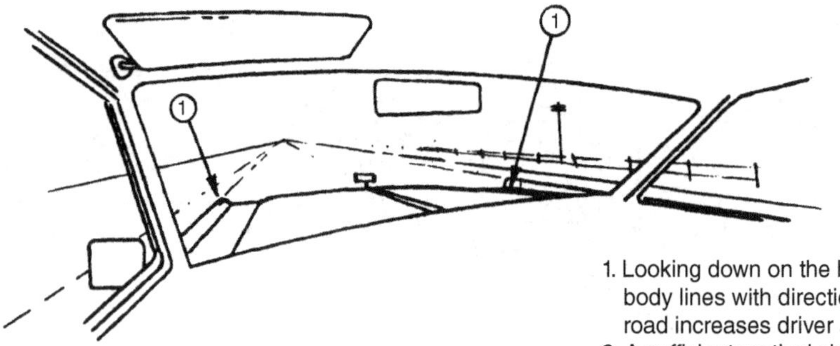

1. Looking down on the hood and being able to relate body lines with directional references at the side of the road increases driver steering accuracy.
2. A sufficient vertical viewing angle and a high driver position are necessary so that the driver can see a small child in front of the car.
3. See traffic signals, and see overhead signs.

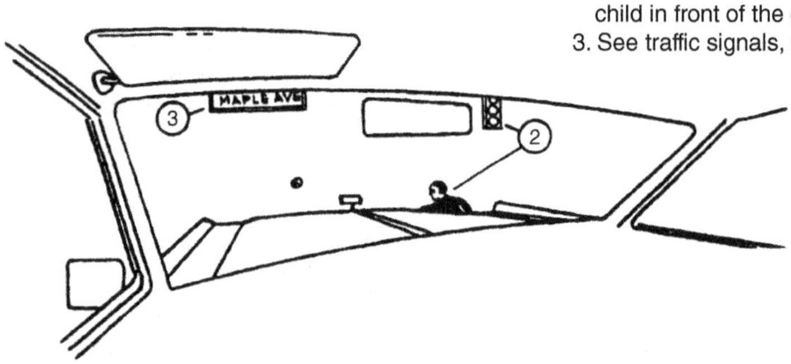

The driver in a low position cannot see.

FIGURES 25.53 Driver positioning for safety external viewing.

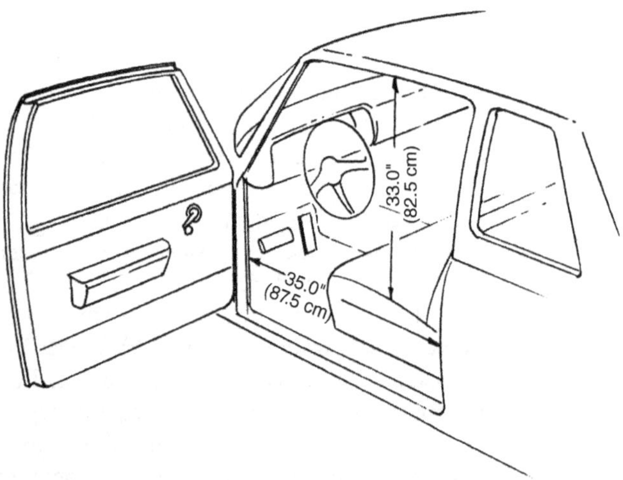

FIGURE 25.54 Automobile ingress and egress clearances.

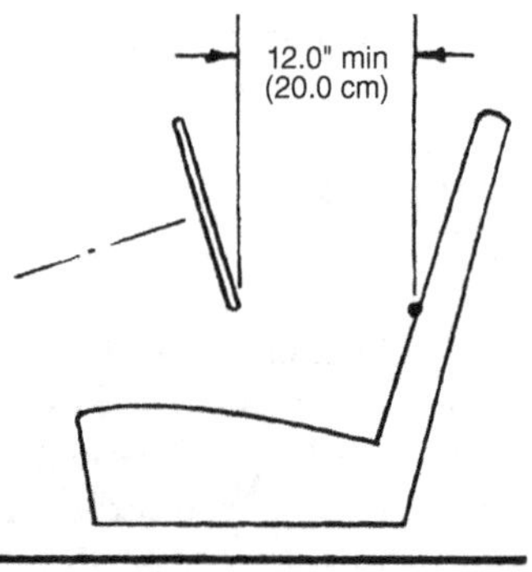

FIGURE 25.55 Steering wheel clearance.

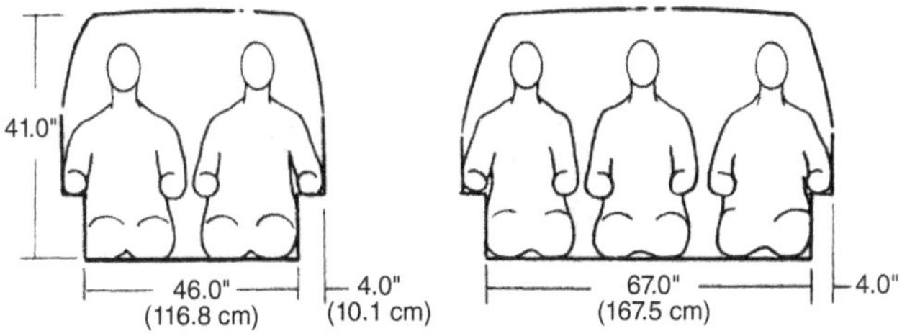

Figure 25.56 Passenger space requirements.

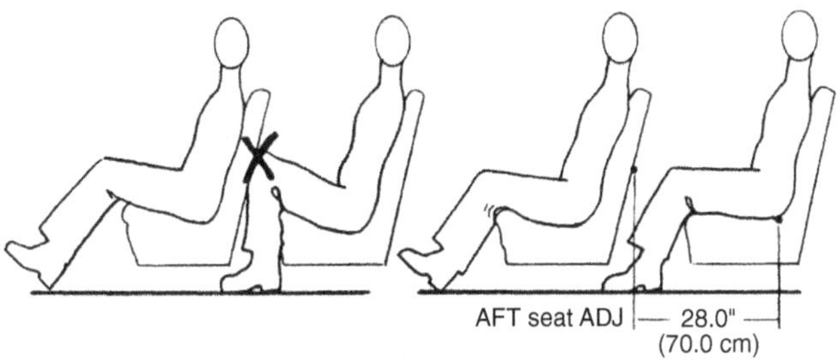

Figure 25.57 Fore and aft seating requirements.

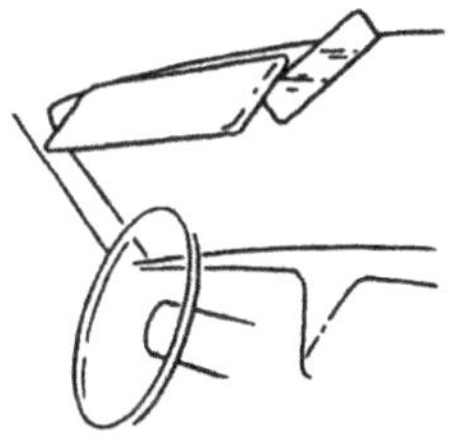

Figure 25.58 Poor planning resulting in operator distraction.

Figure 25.59 Driver should not be forced to reach and remove their eyes from the road.

Use of options such as automatic window controls and door locks should be based on their effect on driver performance and safety. Drivers frequently find that a window on the opposite side of the car needs to be closed while they are underway. Instead of stopping, many drivers will try to reach across the car to operate the control. If the driver's attention is diverted too long from the road, there may be false inputs to the steering wheel or accelerator pedal. Automatic controls should be used for a performance reason. (See Fig. 25.60.)

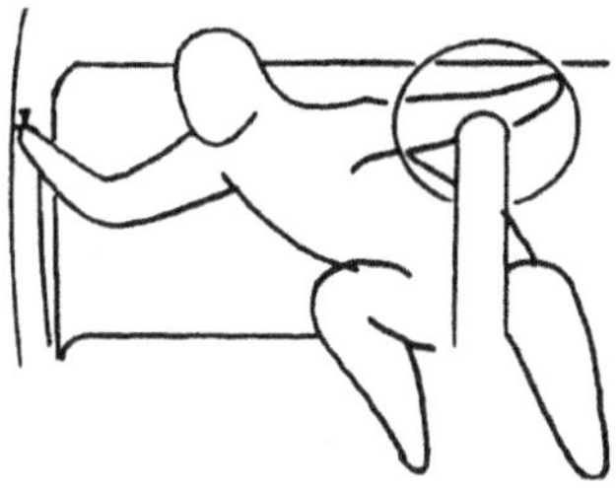

Figure 25.60 Automatic controls should be considered for performance reasons.

Door handles should be located where they are visible and quickly accessible. In years past, door handles were very accessible and visible. But they were also a hazard since one could strike them and incur injury or inadvertently actuate them and open a door while the car was under way. Overzealous attempts to correct these deficiencies often result in "hidden" and confusing door handle configurations, such as those illustrated in Fig. 25.61. Looking down at the armrest, the observer seldom can see the handle. Although the vehicle owner may get used to this, a guest may not be able to find the handle in time to get out in an emergency.

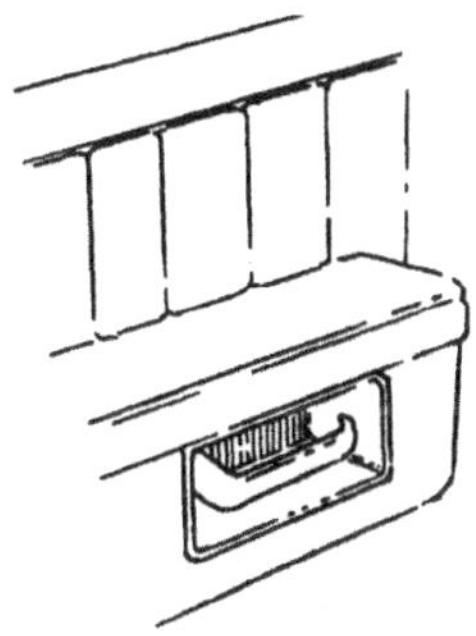

Figure 25.61 Hidden door handles should be avoided.

Handles should be shaped so that the hand slips normally into position for operating the handle. Many handles are made so that users must twist their hands or wrists into unusual positions in order to get hold of them.

Foot pedals should be at approximately the same level to each other so the driver does not have to raise his or her foot high to keep from catching a toe on the adjacent pedal. (See Fig. 25.62.)

Figure 25.62 Foot pedals should be placed at the same heights.

Illumination and Glare

Vehicle drivers are confronted with a unique visibility problem in that they must operate their vehicles under many different ambient illumination conditions. Observe the following while planning the new vehicle and its driver control and display environment:

1. Avoid using any highly reflective material in the general forward viewing area; i.e., no glossy paints or chrome trim should be used on the instrument panel, window frames, mirrors, hood, or fenders.

2. Artificial lighting should be provided for all critical control labels so that they can be read during the day and at night.

3. Artificial lighting should be provided for all critical visual displays so that they can be read during the day and at night. (Dimming capability should be provided for instrument control lighting so that the driver can adjust the levels for the best seeing conditions.)

4. Provide adjustable glare shields mounted on the header above the windshield.

5. Provide an adjustable headlight glare control for the interior rearview mirror.

6. Provide a slightly tinted (neutral color) windshield glazing (see the appropriate federal motor vehicle safety standard).

7. Provide a glare shield for the instrument control panel and/or an inset design to prevent external ambient light from obscuring instruments and label detail and/or to prevent these details from reflecting on the windshield at night. Use dark colors and a matte finish on any interior surface (windowsills, steering wheel or other control devices, upholstery, etc.) that might reflect on the windshield during the day or at night.

8. Slope the cover glass on instruments to minimize reflections of the sun and the driver on the glass.

Making the Driver Station and Passenger Compartment Safe

Because of the potential hazards associated with moving vehicles (i.e., sudden changes of direction and deceleration), the interiors of all vehicles should be examined closely to make sure that they contain no special "contact hazards." Figures 25.63–25.68 illustrate some of the typical areas in automobiles requiring special consideration:

1. Steering Wheel: The hub should be broad and set back from the rim, and levers should not protrude in the driver's direction. Steering wheels should adjust in and out as well as up and down to accommodate different arm lengths. (See Fig. 25.63.)

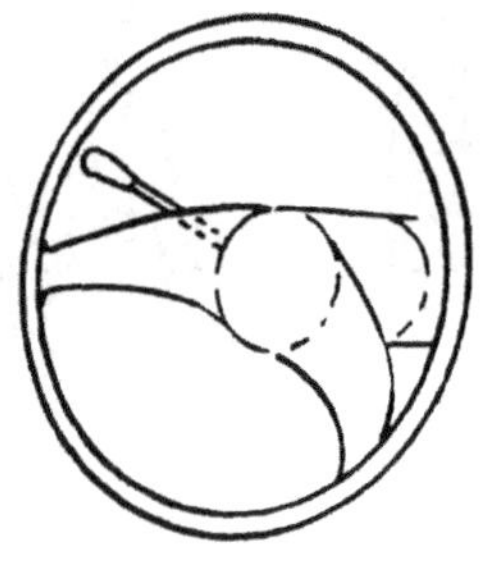

Figure 25.63 Driver station and passenger compartment safety—steering wheel.

2. Frontal Areas: Energy-absorbing padding should be provided on all dash surfaces, and all controls should be inset and collapsible and/or should present a large contact surface. (See Fig. 25.64.)

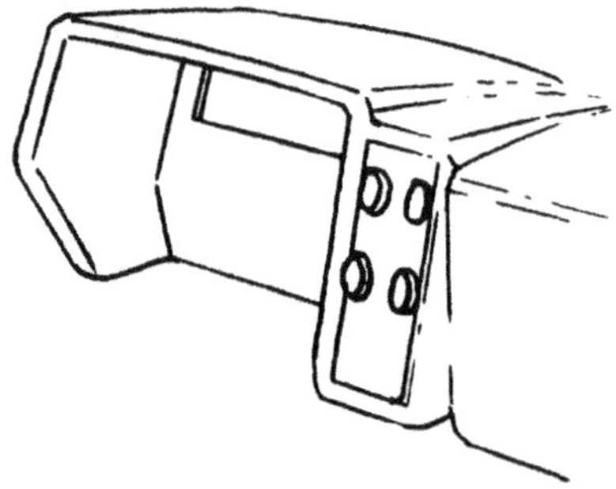

Figure 25.64 Driver station and passenger compartment safety—frontal areas.

3. Glare Shields: These should be made of frangible materials, and the hardware that holds them should not protrude into the expected head movement envelope. (See Fig. 25.65.)

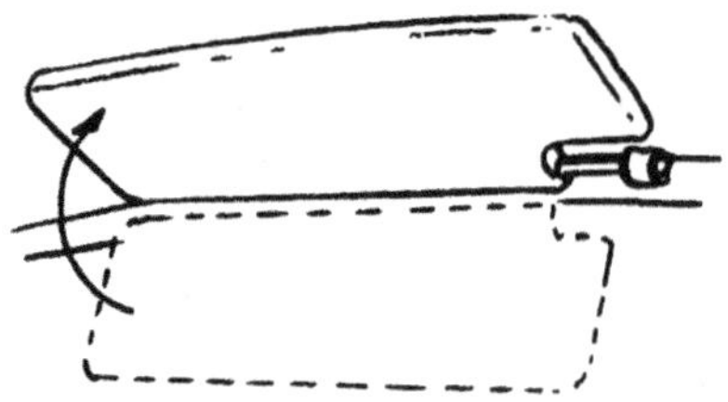

Figure 25.65 Driver station and passenger compartment safety—glare shields.

4. Interior Rearview Mirror: The mirror should have a padded frame and be mounted so that it is easily broken away in the event the driver's head is thrown into the mirror during a crash. (See Fig. 25.66.)

5. Door and Window Handles: All door-mounted hardware should be designed and positioned to minimize the possibility of an occupant's being impaled on the device during a crash. On the other hand, the door

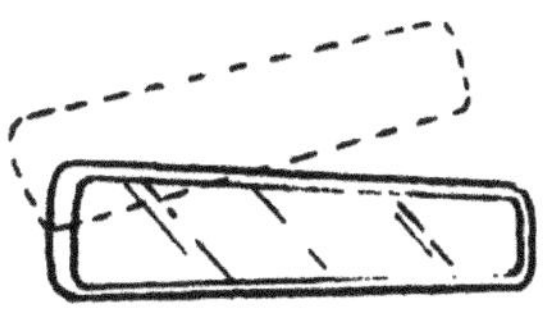

Figure 25.66 Driver station and passenger compartment safety—interior rearview mirror.

handle must not be hidden or made so obscure that the occupant cannot find it quickly and operate it easily in an emergency. (See Fig. 25.67.)

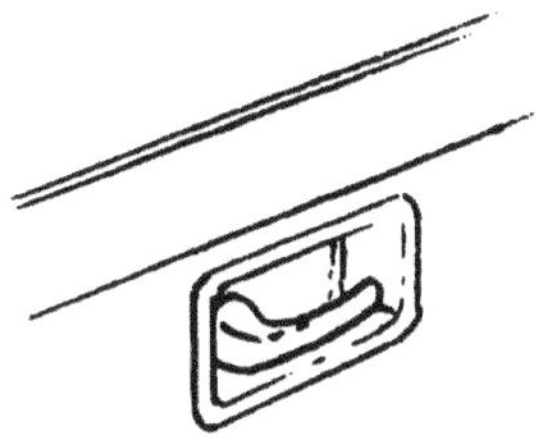

Figure 25.67 Driver station and passenger compartment safety—door and window handles.

6. Seat Security: All seats (and adjusting systems) must be fait-safe; i.e., the seat and/or adjusting system must not suddenly let go and allow the occupant to be thrown forward during a crash. (See Fig. 25.68.)

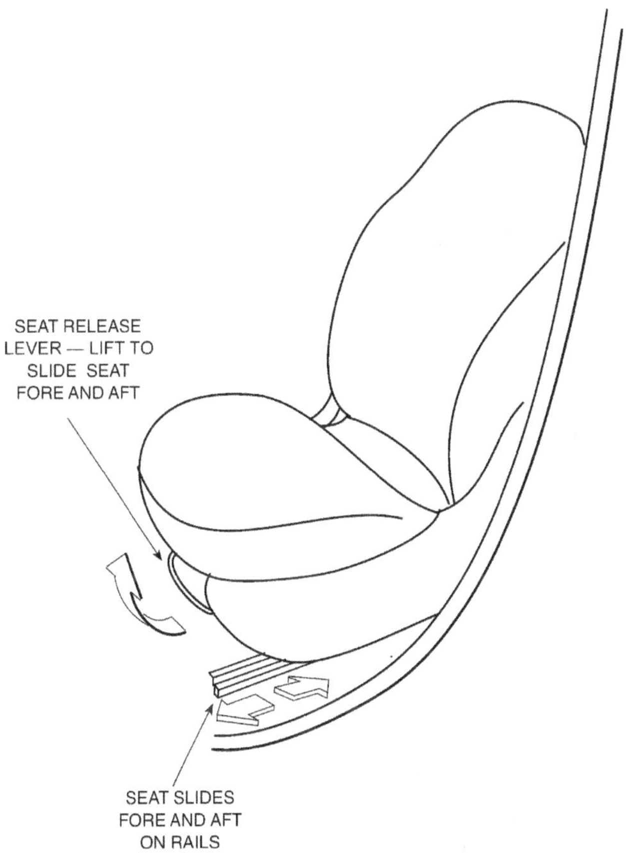

Figure 25.68 Driver station and passenger compartment safety—seat security.

Vehicular Design to Accommodate the Handicapped

Specific design guidelines and requirements for accommodation of people with disabilities are spelled out in ADA (American with Disabilities Act) Standards for Accessible Design published by the United States Department of Justice. (See Fig. 25.69.)

1. Wheelchair Users: Accommodation of the wheelchair user, whether he or she is a driver or a passenger, involves primarily allowance for movement of the wheelchair in entering or leaving the vehicle, maneuvering within certain public conveyances, and securing the wheelchair once it and its occupant are aboard. Wheelchair dimensions alone are not sufficient to define the clearance needs at doorways; generally, additional space is required for turning, folding, and parking the wheelchair, and special attention must be given to threshold and floor characteristics (how level, whether there are gaps between the vehicle threshold and the loading surface, etc.) in order to minimize the possibility that the chair will be upset. A unique and critical requirement for vehicular systems (e.g., public transit systems) is the necessity of securing the wheelchair against dynamic forces when the vehicle is under way and, more importantly, in the event of a crash. (Note that wheelchairs themselves are not designed to be crashworthy.) In addition to providing the extra clearances for the wheelchair itself, special attention must be given to the accessibility of special vehicular controls and other features that the handicapped person may have to reach and operate from the wheelchair and/or operate as a driver with less-than-normal use of limbs, hands, or fingers.

2. The Blind and Partially Sighted: Attention should be given to critical visual interfaces that the partially blind have to see or read, such as displays, signs, and markings. This includes not only the legibility of words and symbols, but also the selection of colors that can be recognized by the color-deficient, illumination levels and contrasts that make up for the reduced visual response sensitivity of the elderly, and/or elimination at extreme glare, which adversely affects even the reduced visual capacity of certain visually handicapped persons.

3. The Deaf and Partially Deaf: Care should be taken in developing certain critical communication systems to recognize that many people may not hear warning signals or advisory communications by transit personnel or be able to understand verbal messages that are minimally satisfactory (i.e., those with poor transmission quality).

4. Persons with Other Handicaps: Some transit system users, although ambulatory, may lack full use of their hands, may be too weak to apply normal force to door handles or controls, or may not be able to push heavy doors. Similarly, many people cannot reach overhead handrails, grip and turn doorknobs, or bend sufficiently to keep from striking their heads on low overheads.

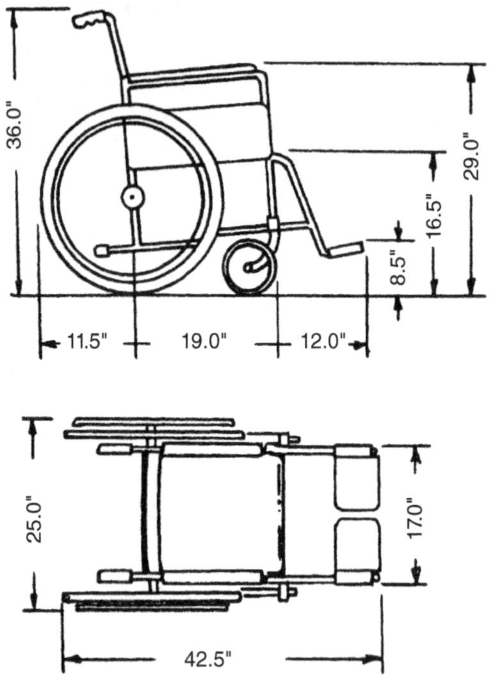

Collapsible wheelchair dimensions

Length 42 in. (107 cm)
Width (open) 25 in. (64 cm)
Seat height 19.5 in. (49.5 cm)
Armrest and floor height 29 in. (74 cm)
Push-handle height 36 in. (91 cm)
Width (folded) 11 in. (28 cm)
Fixed turn radius 18 in. (46 cm) (wheel to wheel)
Fixed turn radius 31.5 in. (80 cm) (front to rear)

Desirable turn space 63 × 56 in. (160 × 142 cm)
Passing (two wheelchairs) 60 in. (152 cm)

Adult occupant limits
Vertical reach (unilateral) 60 in. (152 cm)
Average horizontal reach (table height) 30 in. 76 cm)
Average diagonal to wall (height above floor) 48 in. (122 cm)

FIGURE 25.69 Design criteria for the handicapped.

Motor Vehicle Safety Standards

The National Highway Traffic Safety Administration in the U.S. Department of Transportation publishes motor vehicle safety design standards. Table 25.6 list standards that specifically address the human interface.

Bus Driver Workstation Geometry

General. Figure 25.70 shows the key geometric and dimensional features required to provide a workstation that is compatible with the range of adult drivers from a 5th-percentile female driver to a 95th-percentile male driver. The criteria represent a more or less ideal design that would not be restricted because of other factors such as structural thickness and curvature. Because one cannot attain these ideal geometries, these criteria should be looked upon as a guide; i.e., one should develop an adjustable mockup of the proposed workstation using these data, but adjusting them as necessary to fit within the existing constraints.

Consider these factors:

1. Continuous, extending reaching increases driver fatigue.

2. A pedal may not be pressed as hard, or far, if it is at the limits of the driver's leg-foot reach.

3. Holding one's foot at an awkward angle (as on the accelerator) tires the driver overall, not just his or her foot or ankle.

4. A steering wheel that is too big or which is positioned in such a way that the driver has to go through drastic torso and arm motions to turn corners not only tires the driver but may also encourage "sloppy" steering.

5. An optimally designed seat, even one that appears to be too upright or too firm, may in the long run minimize fatigue with long usage periods. A seat that is too soft often adds to fatigue because it fails to provide support

TABLE 25.6 U.S. Federal Motor Vehicle Safety Standards

101	Control location, identification, and illumination—passenger cars, multipurpose passenger vehicles, trucks, and buses
102	Transmission shift lever sequence, starter interlock, and transmission braking effect—passenger cars, multipurpose passenger vehicles, trucks, and buses
104	Windshield wiping and washing systems—passenger cars, multipurpose passenger vehicles, trucks, and buses
108	Lamps, reflective devices, and associated equipment—passenger cars, multipurpose passenger vehicles, trucks, trailers, buses, and motorcycles
111	Rearview mirrors—passenger cars and multipurpose passenger vehicles
118	Power-oriented window systems—passenger cars and multipurpose passenger vehicles
124	Accelerator control systems—passenger cars, multipurpose passenger vehicles, trucks, and buses
125	Warning devices
201	Occupant protection in interior impact—passenger cars
202	Head restraints—passenger cars
203	Impact protection for the driver from steering control system—passenger cars
204	Steering control rearward displacement—passenger cars
205	Glazing materials—passenger cars, multipurpose passenger vehicles, motorcycles, trucks, and buses
206	Door locks and door retention components—passenger cars, multipurpose passenger vehicles, and trucks
207	Seating systems—passenger cars, multipurpose passenger vehicles, trucks, and buses
208	Occupant crash protection
209	Seat-belt assemblies—passenger cars, multipurpose passenger vehicles, trucks, and buses
210	Seat-belt assembly anchorages—passenger cars, multipurpose passenger vehicles, trucks, and buses
213	Child seating systems
214	Side-impact protection—passenger cars
216	Roof crush resistance—passenger cars
302	Flammability of interior materials—passenger cars, multipurpose passenger vehicles, trucks, and buses
572	Anthropomorphic test devices

and because of the "bunching" it causes. Proper seat-pan angle and seat-pan-to-backrest angle are the most important contributors to fatigue minimization. A straight, rather than contoured, seat pan and back cushion are preferred. The backrest should, however, have a slight concavity in order to counteract the effects of sidesway.

A lap belt is provided for the driver and the belt is anchored to the seat in such a way that the belt geometry will stay constant regardless of where a particular driver adjusts his or her seat.

Control and Display Layout. Although various bus configurations vary in terms of the subsystems that have to be monitored and controlled by the driver, all layouts can and should be organized by system function. As noted in Fig. 25.71, electrical and environmental systems should be located on the left of the driver because these are the first systems he or she activates and checks before moving the bus; i.e., since one normally proceeds sequentially from left to right, the first elements to be used are placed on the left. The central panel contains those instruments (and related controls) that are monitored most often as the driver is maneuvering the bus; i.e., although they are not the second step in pre-checkout, they occupy a preferred position relative to the nominal in-motion driving task. The other subsystems, including engine, brakes, fuel, etc., are located on the right-hand panel. Finally, the fare box is placed at the most obvious driver-customer interface point.

Pertinent Federal Aviation Administration Regulations[13]

Before beginning the design of any commercial aircraft, the designer should review both the FAA, Human Factors Design Standard document and the FAA requirements pertaining to crew and passenger accommodation, such as the following. (See Fig. 25.72.)

Criteria for Determining Minimum Flight Crew. The following are considered by the FAA in determining the minimum flight crew under section 25.1523:

1. Basic Workload Functions: The following basic workload functions are considered:
    a. Flight path control.
    b. Collision avoidance.
    c. Navigation.
    d. Communications.
    e. Operation and monitoring of aircraft engines and systems.
    f. Command decisions.
2. Workload Factors: The following work load factors are considered significant when analyzing and demonstrating workload for minimum flight crew determination:

[13]Code of Federal Regulations, Title 14, Aeronautics and Space.

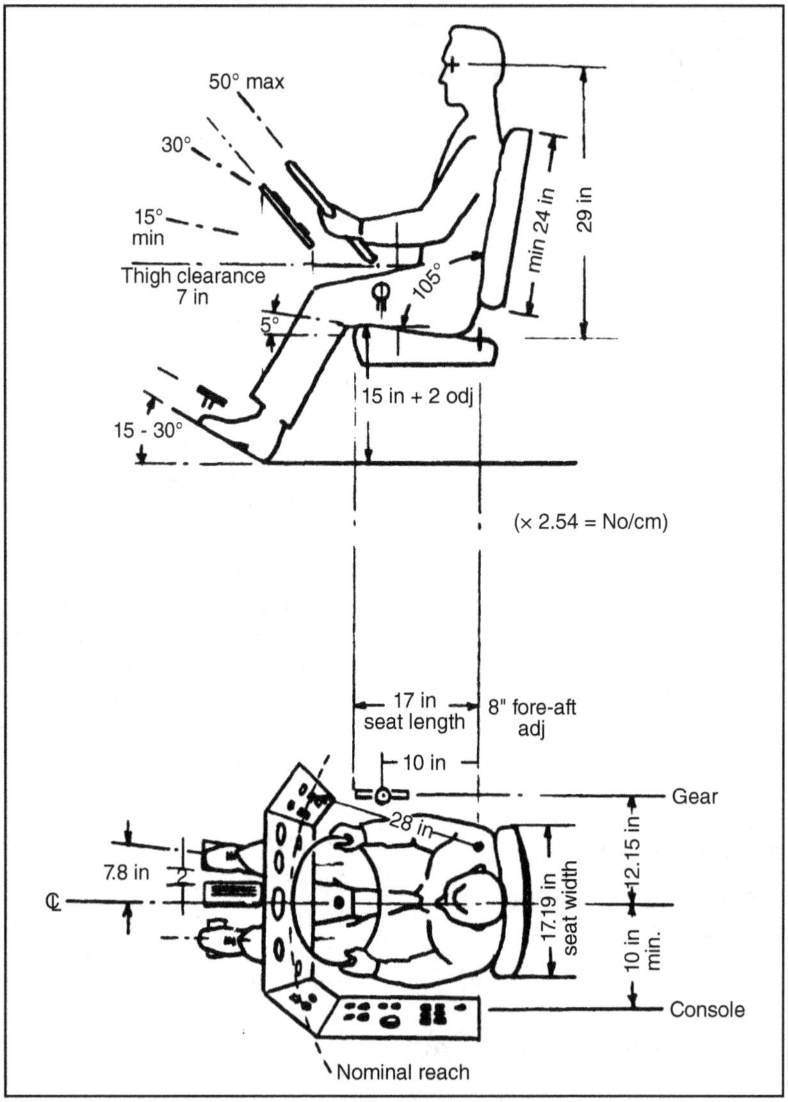

FIGURE 25.70 Guidelines for bus driver station layout.

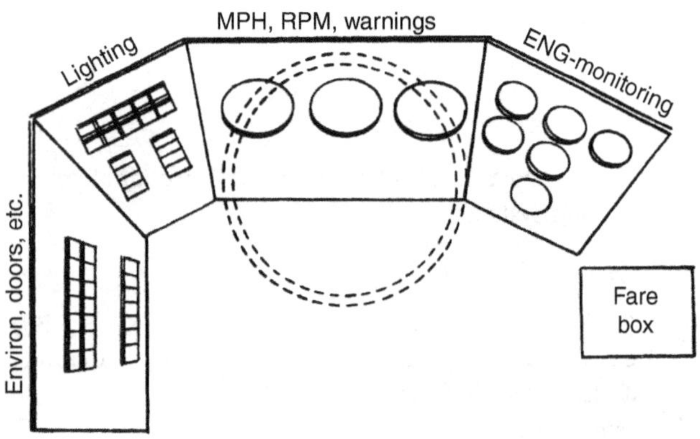

FIGURE 25.71 Bus control and display layout.

a. The accessibility, ease and simplicity of operation of all necessary flight, power, and equipment controls, including emergency fuel shutoff valves, electrical controls, electronic controls, pressurization system controls, and engine controls.

b. The accessibility and conspicuity of all necessary instruments and failure warning devices such as fire warning, electrical system malfunction, and other failure or caution indicators. The extent to which such instruments or devices direct the proper corrective action is also considered.

c. The number, urgency, and complexity of operating procedures with particular consideration given to the specific fuel management schedule imposed by center of gravity, structural or other

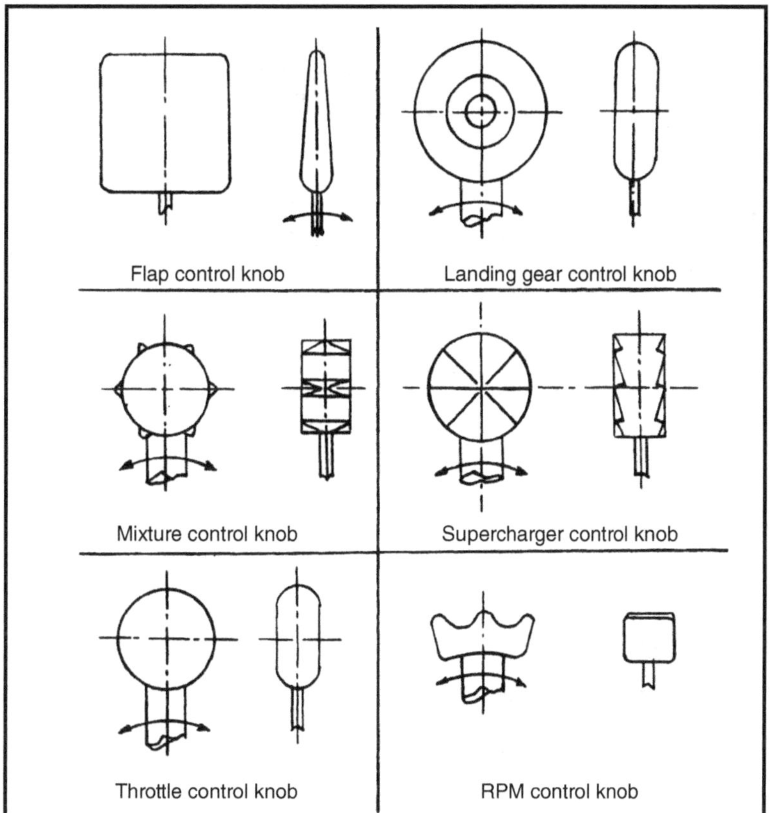

§ 25.781 Cockpit control knob shape.
Cockpit control knobs must conform to the general shapes (but not necessarily the exact sizes or specific proportions) in the accompanying figure.

Figure 25.72 Example of federal aviation regulations.

considerations of an air-worthiness nature, and to the ability of each engine to operate at all times from a single tank or source that is automatically replenished if fuel is also stored in other tanks.

d. The degree and duration of concentrated mental and physical effort involved in normal operation and in diagnosing and coping with malfunctions and emergencies.

e. The extent of required monitoring of the fuel, hydraulic, pressurization, electrical, electronic, deicing, and other systems while en route.

f. The actions requiring a crewmember to be unavailable at his assigned duty station, including: observation of systems, emergency operation of any control, and emergencies in any compartment.

g. The degree of automation provided in the aircraft systems to afford (after failures or malfunctions) automatic crossover or isolation of difficulties to minimize the need for flight crew action to guard against loss of hydraulic or electric power to flight controls or to other essential systems.

h. The communications and navigation workload.

i. The possibility of increased workload associated with any emergency that may lead to other emergencies.

j. Incapacitation of a flight crewmember whenever the applicable operating rule requires a minimum flight crew of at least two pilots.

3. Kind of operation authorized requires consideration of the operating rules under which the airplane will be operated. Unless an applicant desires approval for a more limited kind of operation, it is assumed that each airplane certificated under this Part will operate under IFR conditions.

Cockpit Standardization

Military agencies are particularly adamant about cockpit standardization in order to minimize pilot confusion when transferring from one aircraft to another. Following are selected examples of U.S. Air Force standardization guidelines.[14]

[14] *Crew Stations and Passenger Accommodations,* AFSC DH 2-2. Design Handbook Series 2-0. Aeronautical Systems Headquarters. Air Force Systems Command. Wright-Patterson Air Force Base. Ohio. Also see MIL-STD-1333, Aircrew Station Geometry for Military Aircraft.

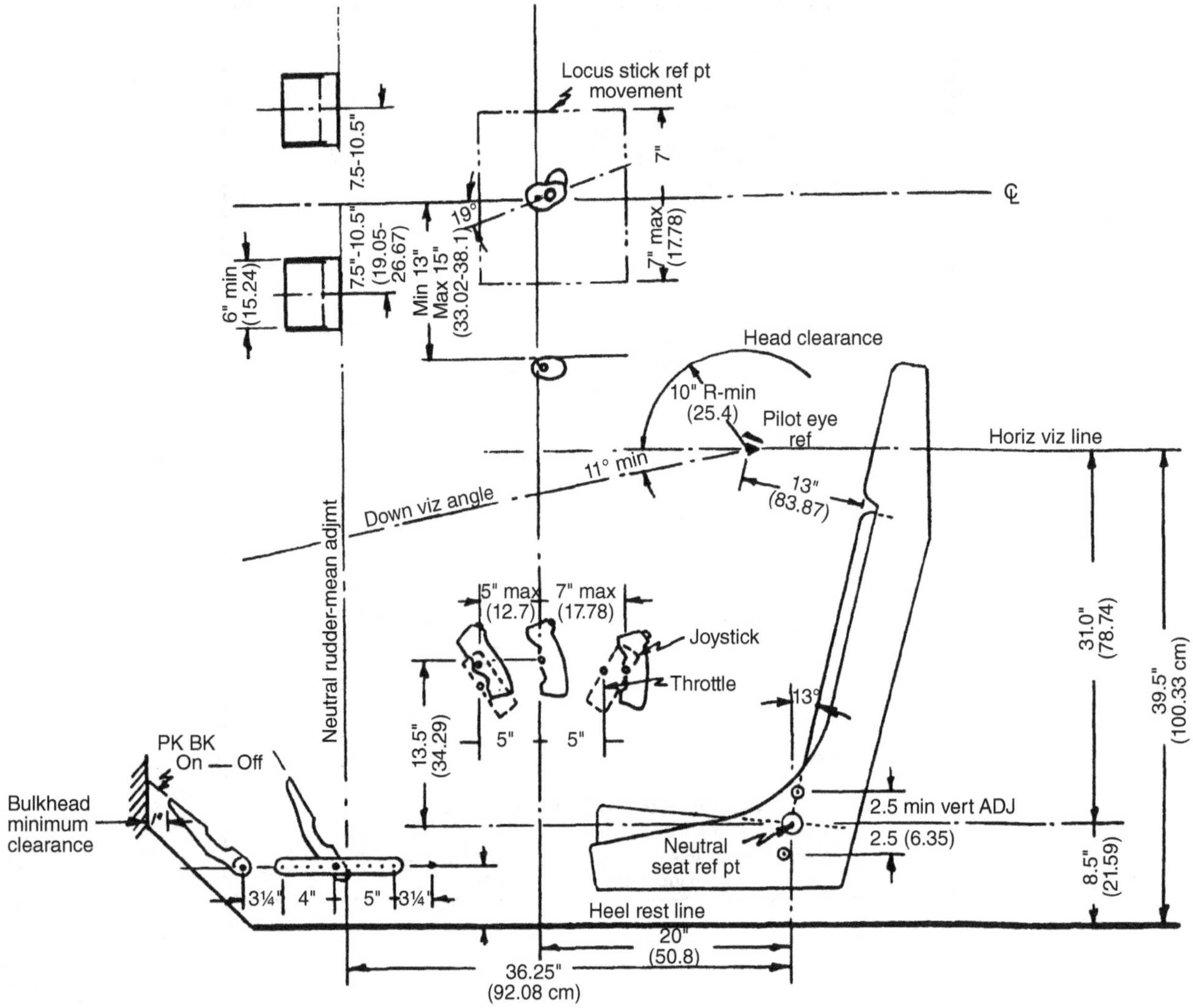

FIGURE 25.73 Cockpit: basic dimensions—fixed wing.

Cockpit: Basic Dimensions—Fixed Wing (See Fig. 25.73)

1. The pedal "brake off" angle, with respect to the vertical shall remain constant throughout the entire rudder pedal travel.

2. Three inches of rudder pedal adjustment is acceptable when seats provided with ± 2.500 in. of vertical and 1.500 in. of fore and aft adjustments are utilized.

3. Rudder pedal adjustment shall be in increments of 1 in. or less.

4. The stick and throttle reference point is defined as the point at which the pilot's second finger is in contact with the forward face of the control.

5. Reference spec MIL-B-8584 for brake pedal angles and dimensions.

6. All measurements are based upon the seat reference point at the centerline of the seat in the neutral position.

Cockpit: Basic Dimensions—Helicopter (See Fig. 25.74)

1. Adequate clearance shall be provided to allow unrestricted operation of the rudder pedal throughout the full range of travel.

2. The pedal "brake off" angle with respect to the vertical shall remain constant throughout the entire rudder pedal travel.

3. The collective pitch stick travel shall not exceed 12 in. from pull-down position of the grip.

4. The seat shall be provided with fore and aft and vertical adjustment independent fore and aft and vertical adjustment is desirable. However, diagonal adjustment is acceptable.

5. Rudder pedal adjustment shall be in increments of 1 in. or less.

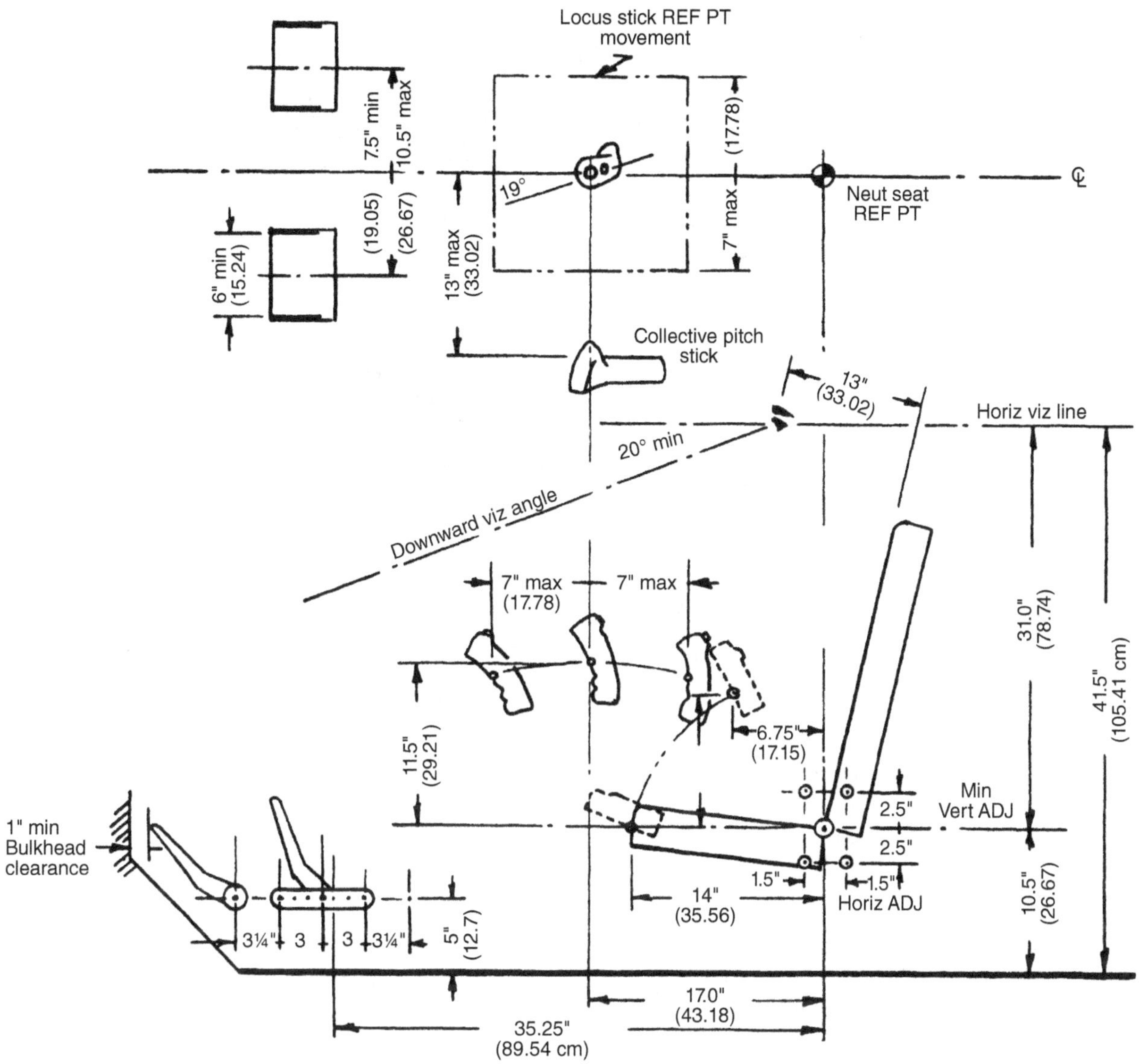

Figure 25.74 Cockpit: basic dimensions—helicopter.

6. The stick reference point is defined as the point at which the pilot's second finger is in contact with the forward face of the stick grip.

7. Reference spec MIL-B-8584 for brake pedal angles and dimensions.

8. All measurements are based upon the seat reference paint at the centerline of the seat in the neutral position.

Cockpit: Clearance Dimensions—Ejection Seat (See Fig. 25.75)

1. There shall be no projections such as the throttle, landing gear control, instrument panel, etc., into the ejection seat envelope that would interfere with seat ejection.

2. All measurements are based upon the seat reference point at the centerline of the seat in the neutral position.

3. The 30-in. minimum ejection clearance line (parallel to the ejection path and measured perpendicularly to the plane of the seat back) shall be provided from the seat reference point. For airplanes not requiring ejection seats the minimum cockpit opening shall be 24 in. by 24 in.

4. The seat shall be provided with vertical adjustment as shown.

Cockpit: Basic Dimensions—Cargo and Bomber (See Fig. 25.76)

1. When ejection seats are required, the vertical seat adjustment shall be parallel to the ejection rails.

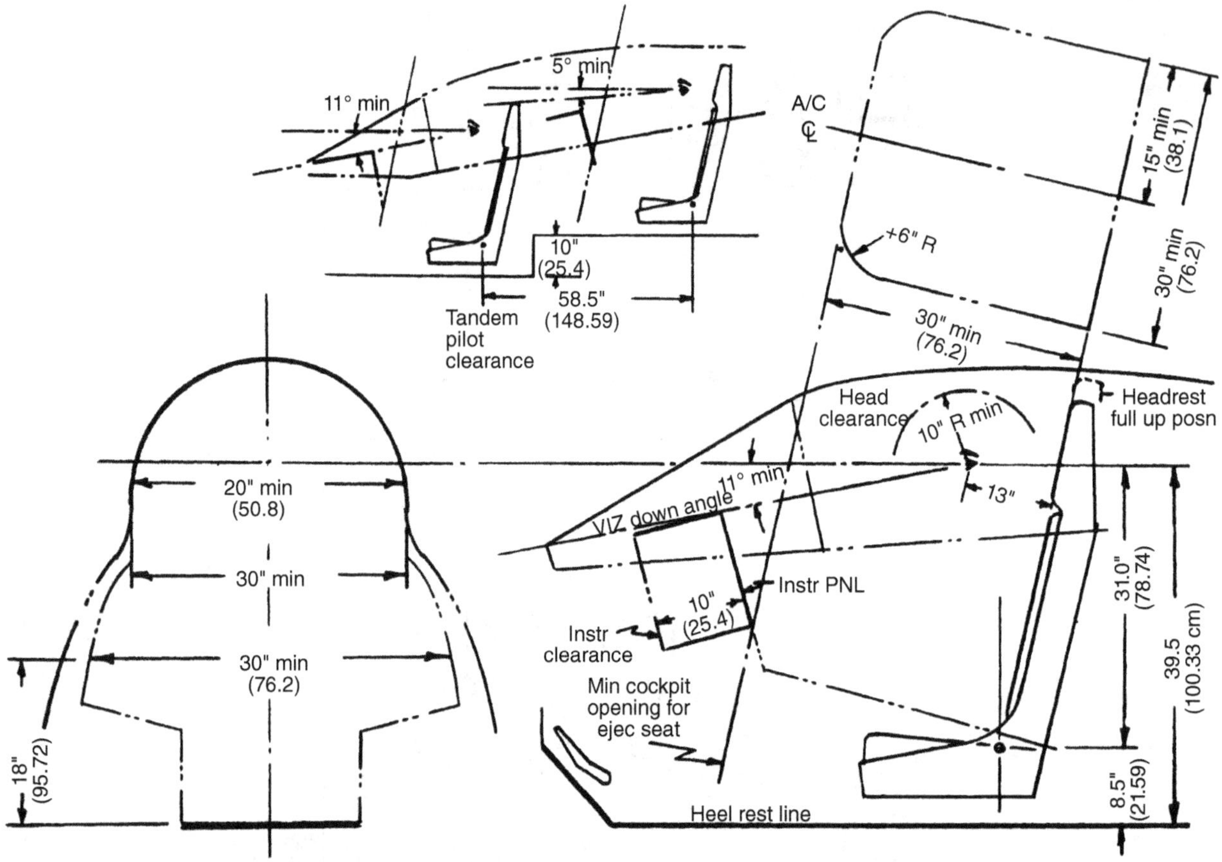

Figure 25.75 Cockpit: clearance dimensions—ejection seat.

2. The rudder pedal "brake off" angle with respect to the vertical shall remain constant throughout the entire rudder pedal travel.

3. Rudder pedal adjustment shall be in increments of 1 in. or less.

4. The wheel reference point is defined as the point at which the pilot's second finger is in contact with the forward face of the control wheel.

5. The seat shall be provided with fore and aft and vertical adjustment. Independent fore and aft and vertical adjustment is desirable. However, diagonal adjustment, as shown, is acceptable in lieu of vertical adjustment.

6. Reference spec MIL-B-8584 for brake pedal angles and dimensions.

7. All measurements are based upon the seat reference point at the centerline of the seat in the neutral position.

Aircrew Station Controls and Displays

1. Dividing lines shall be 0.062 ± 0.016 wide. (The above markings shall be finished with luminescent fluorescent material conforming to MIL-L-25142.)

2. Main borders shall be 0.156 ± 0.031 wide.

3. Numerals designating engine numbers shall be 0.250 minimum height.

Military fixed-wing aircraft controls and displays must be located and arranged as defined in MIL-STD-203, Aircrew Station Controls and Displays. Figures 25.77–25.80 are from MIL-STD-203G.

RESTRAINT SYSTEMS

Guidelines

1. Some type of restraint against crash forces should be provided for all automobile passengers. (See Fig. 25.78.)

2. Restraint systems should be made and installed in such a manner that they do not interfere with the driver's tasks while he or she is driving.

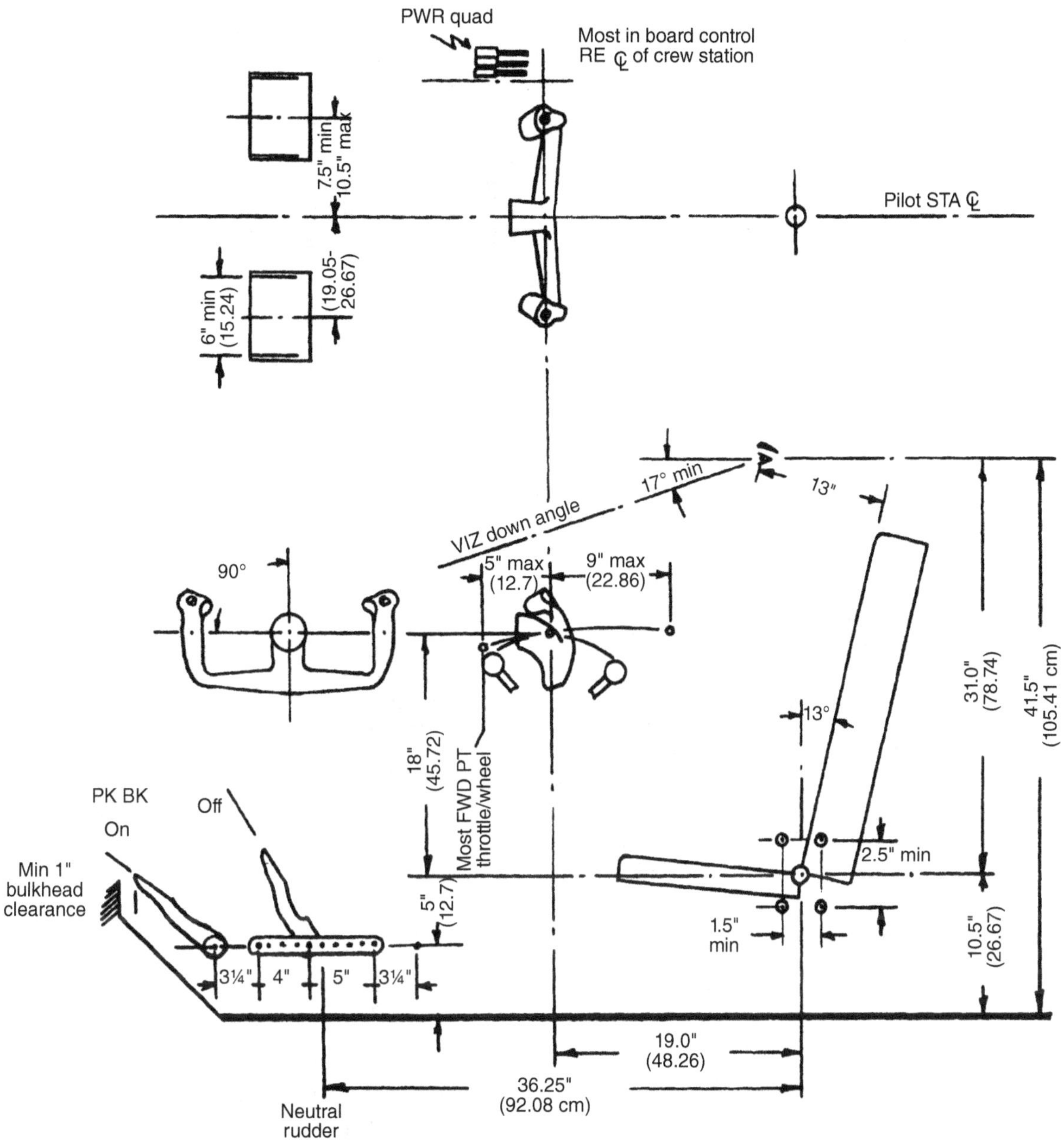

Figure 25.76 Cockpit: basic dimensions—cargo and bomber.

3. If a seat-belt system is used, the seat belts must not incur injury during a crash, especially in terms of cutting across the occupant's neck or abdomen.

4. The seat-belt geometry should be comfortable; i.e., the belts should not rub on the occupant's neck or shoulder, cross or press uncomfortably on a female driver's breasts, or make it difficult for the driver to lean forward (i.e., against the retractor forces of a self-adjusting belt system).

5. The seat-belt webbing should retract into a position that makes it easy for the occupant to get in and out of the car, and it should not fall loosely so that it can be trapped by the closing door and/or fall behind the seat so that the next user cannot locate the latch plate that he or she expects to use in fastening the belt. *Refer to U.S. Department of Transportation Standards FMVSS 208 and 209 for specific requirements in the development, use, and testing of motor vehicle seat restraint systems.*

6. Passive restraint systems are generally of two types: passive belt systems and air bags. Passive belt systems

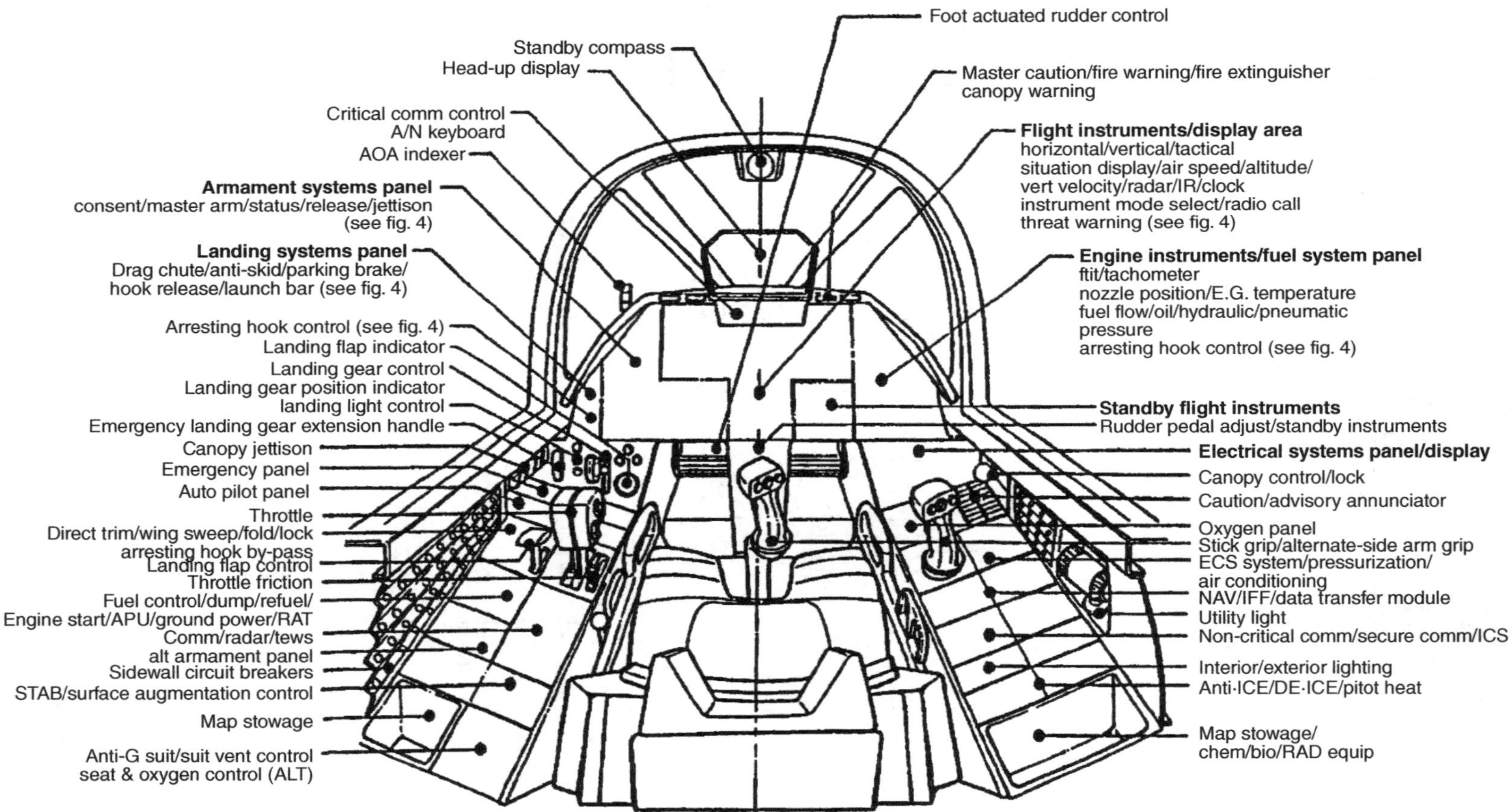

Cockpit arrangement attack fighter, single ejection seat (from MIL-STD-203G, aircrew stations controls and displays)

(a)

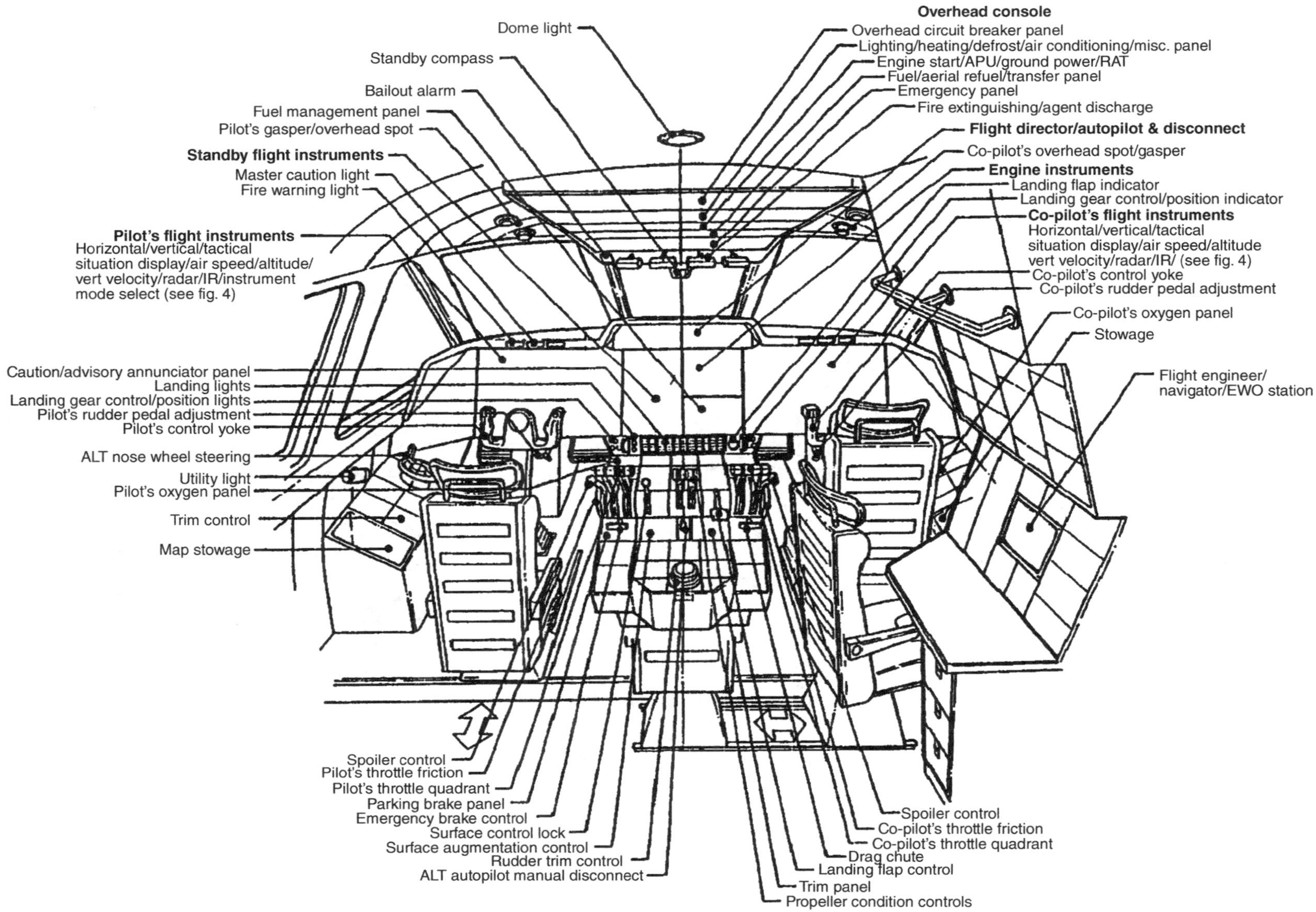

Cockpit arrangement
patrol/cargo/transport/AEW, side-by-side, multi-flight crew, non-ejection seat
(from MIL-STD-203G, aircrew stations controls and displays)

(b)

FIGURE 25.77 Aircrew station controls and displays.

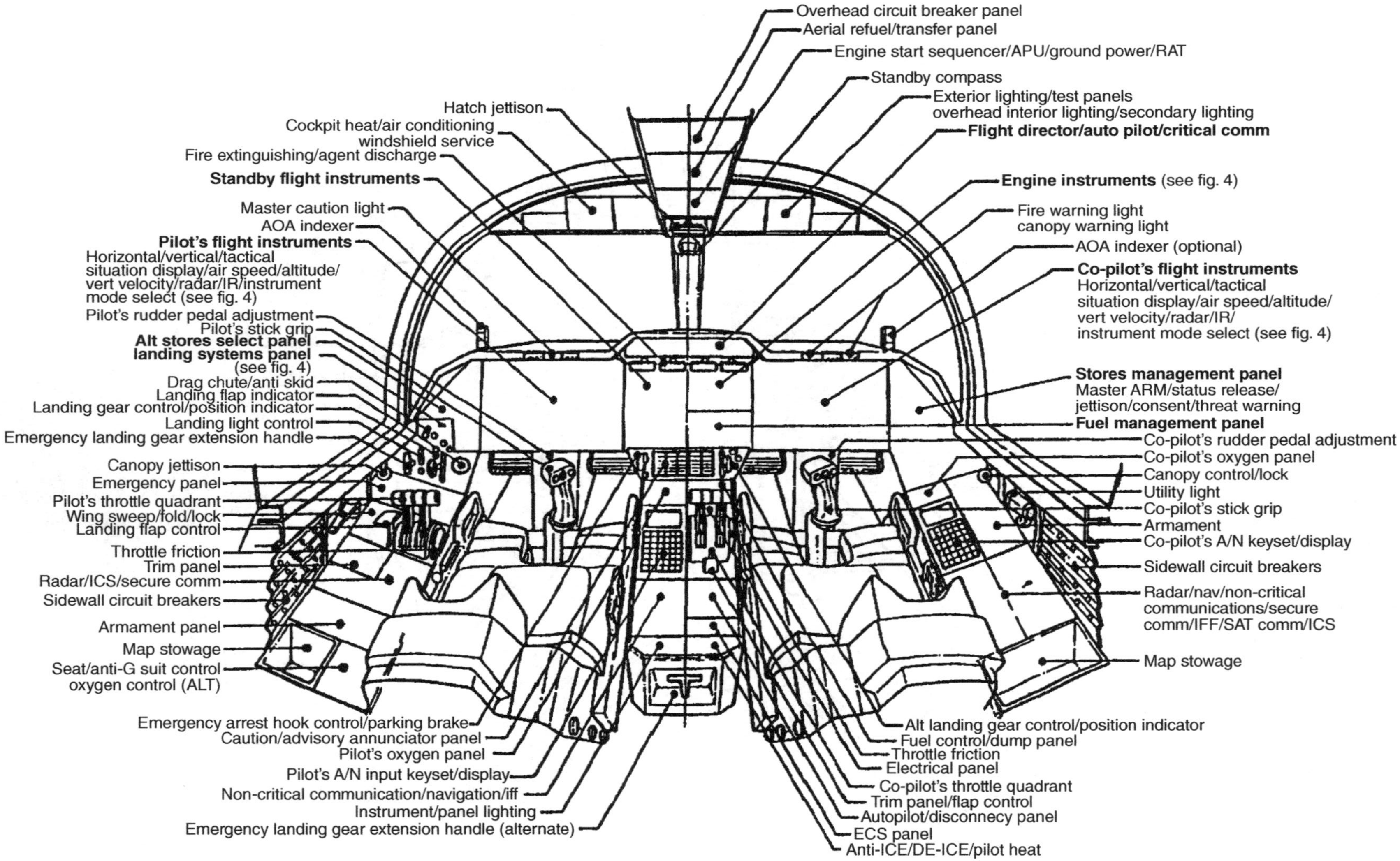

Cockpit arrangement
attack bomber, ASW, side-by-side, ejection seats
(from MIL-STD-203G, aircrew stations controls and displays)

(c)

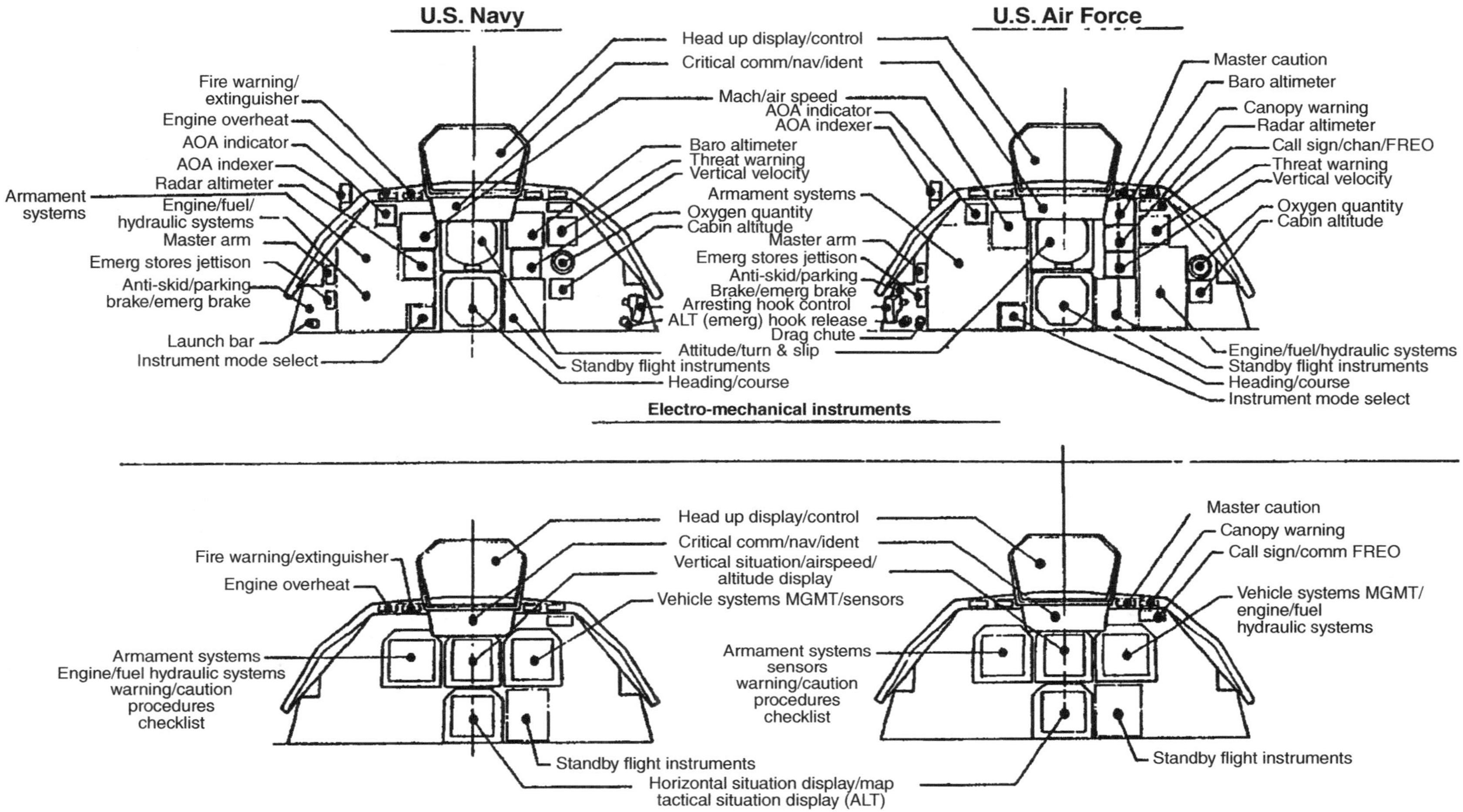

Instrument panel variations attack fighter, single ejection seat (from MIL-STD-203G, aircrew stations controls and displays)

(d)

FIGURE 25.77 *(Continued)*

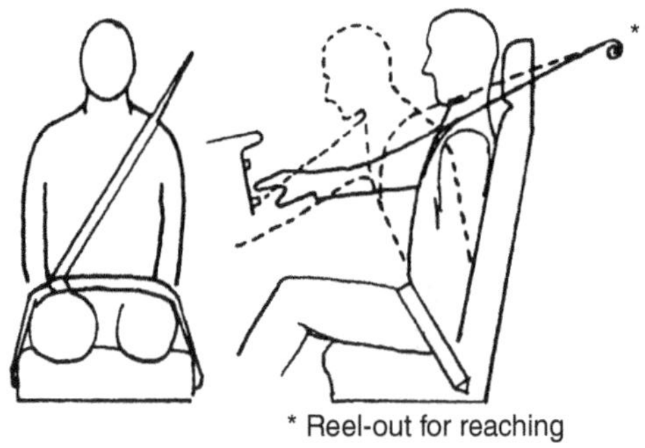

FIGURE 25.78 Restraint systems.

should be designed with the above suggestions in mind. Air-bag systems should be designed so that they effectively restrain the occupant's upper torso and head and also prevent the buttocks from sliding forward. This requires carefully planning the location of the stowed air bag in relation to the position and varying sizes of occupants. The air bag should inflate upon the sudden deceleration of the vehicle under typical crash conditions (but must not inflate under normal vehicle decelerations). The bag should deflate shortly thereafter so that the occupant can escape if necessary.

7. Special restraint systems should be provided for children since the above systems are effective only for adult occupants (see current accepted child restraint system products).

Safety Belts[15]

Not only must safety belts be designed to restrain a vehicle occupant properly during a crash, but they also should be comfortable to wear; otherwise (when they have a choice), people will try to avoid wearing them.

The sketches in Fig. 25.79 illustrate the most common complaints about seat belts (the chief complaints are associated with the torso belt). In the first sketch, the belt rides against the wearer's neck. In the second sketch, the belt rides atop the wearer's shoulder tip. The third sketch illustrates where the belt should fall, i.e., across the chest, midway between the breasts. Figure 25.80 illustrates the optimized belt-crossing pattern for a torso belt to ensure that it is comfortable for a range of adult wearers from the 5th-percentile female to the 95th-percentile male (U.S. population).

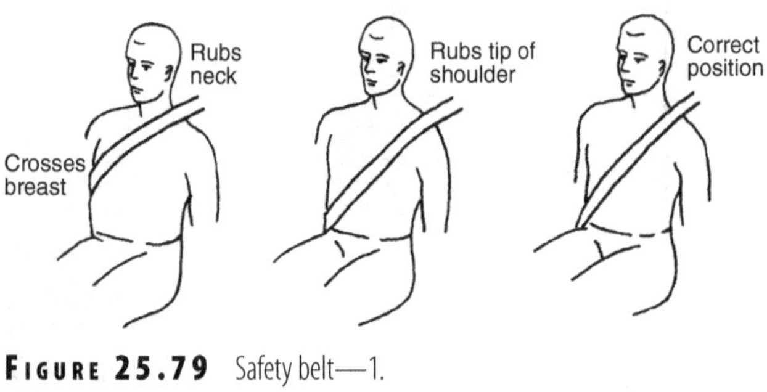

FIGURE 25.79 Safety belt—1.

[15]W.E. Woodson et al. *Development of Specifications for Passive Belt Systems.* MFI Report 78-109 (DOT Contract DOT-HS-7-01617), 1978.

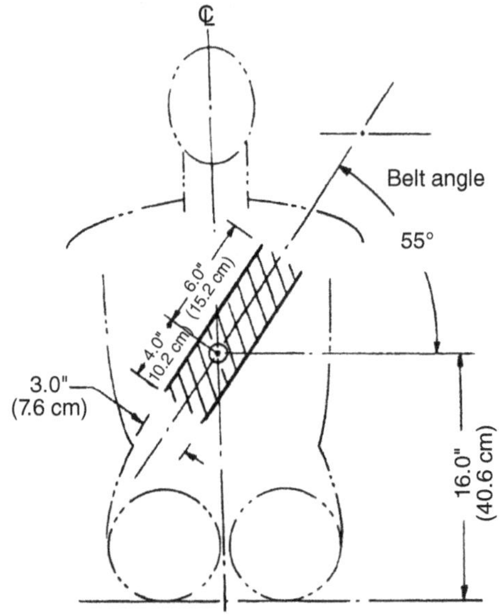

FIGURE 25.80 Safety belt—2.

These specifications are related to a standard 50th-percentile anthropomorphic dummy, seated properly in the mid-adjustment position of an adjustable front automobile seat. The cross-hatched envelope represents a compliance envelope used to determine whether a particular seat belt meets the comfort criteria.

Seat-Belt Design Guidelines

Although a lap belt alone provides some measure of protection for the vehicle passenger, a combination lap belt and shoulder harness is more effective because the shoulder harness minimizes the possibility that the occupant's head will be thrown against a structural barrier. (See Fig. 25.81.) The combination integrated shoulder harness and lap belt is considered an active three point system; i.e., it is secured at three places, and the occupant manually dons the system.

The important considerations are as follows:

1. The latch plate should be positioned so that it is easily accessible; i.e., it should not be hidden behind the seat or be inaccessible because of an interfering door-mounted armrest. It is preferred that the latch plate and webbing be anchored to a vehicle-sensitive emergency-locking retractor so that the webbing can be pulled smoothly and easily across the lap.

2. The buckle should extend on a semi flexible stem from the opposite side of the occupant so that it, too, is easily accessible; i.e., it should not be buried between seat cushions, where it is hard to get hold of.

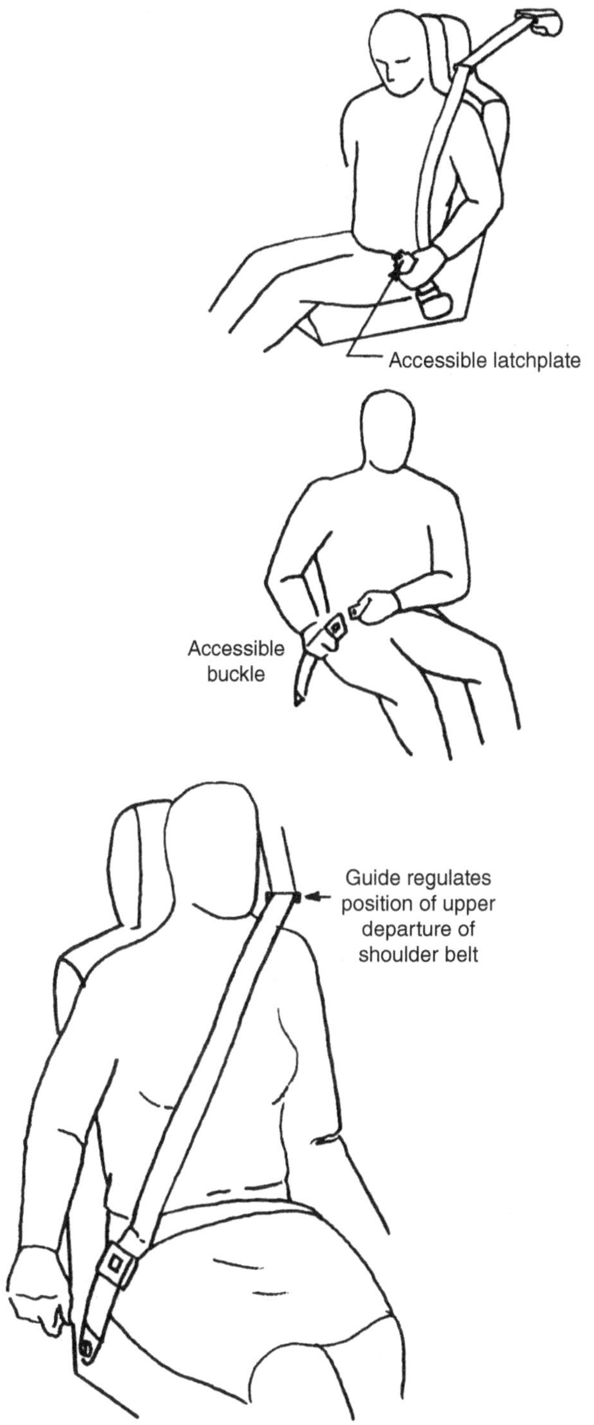

Figure 25.81 Seat-belt guidelines.

3. The shoulder harness and lap-belt webbing elements should be sewn together so that the two elements are both donned with the simple movement of only one element, i.e., the latch plate. The lower juncture between shoulder and lap webbing should occur approximately on top of the occupant's pelvic arch. The upper departure of the shoulder webbing should be slightly above and behind the shoulder to ensure that the webbing will cross the shoulder midway between the neck and shoulder tip. If these upper and lower departure points are properly maintained, the webbing will cross the occupant's chest at approximately the sternum point, which is the most comfortable and safe wearing position. The shoulder harness can be lethal in an accident if it stretches across the occupant's neck. Up and down adjustment of the upper departure point will help safely position the belt for vehicle occupants ranging from children to large adults.

Seat-Belt Geometry for Maximum Comfort

People are more apt to wear seat belts if they are comfortable. To obtain this comfort, use the guidelines provided in Fig. 25.82. The key issues are positioning the departure points for the upper and lower ends of the torso belt so that the belt crosses the wearer's chest approximately at the sternum (midway between the breasts), does not rub against the neck, and does not "pull down" on the shoulder. The dimensions shown provide this desired geometry for at least 90 percent of the driver population as long as the departure points (±1.0 in., 2.5 cm, at the lower point; ±½ in., 1.3 cm, at the upper point) are maintained. Provide departure point adjustments to fit the remaining 10 percent.

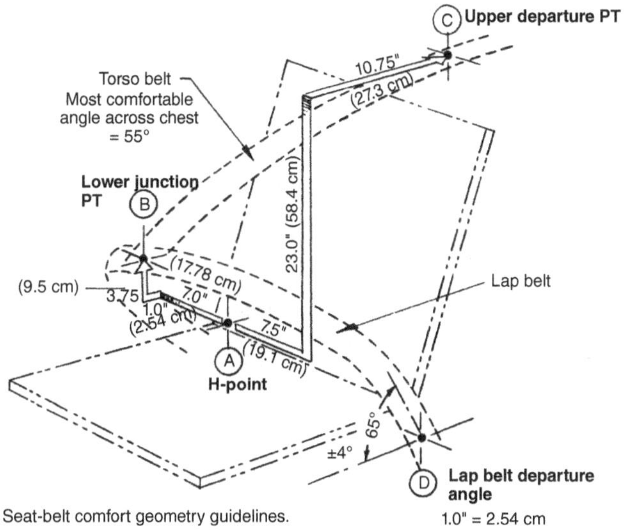

Figure 25.82 Seat-belt comfort geometry guidelines.

The lap-belt approach angle shown is necessary to cause the belt to stay on the wearer's lap and not creep up on his or her abdomen, which could result in injury during a crash.

Note: Points B and C are referenced to the standard automotive industry reference, e.g., the H-point (A). For a definition of the H-point, see SAE J826, Society of Automotive Engineers.[16]

[16] *Devices for Use in Defining and Measuring Vehicle Seating Accommodation*, SAE J826 Society of Automotive Engineers, Inc., Warrendale, Pa.

It should be noted that the recommended geometry must remain constant while seat position is adjusted, preferably by attaching the webbing retractor belt guide to the seat itself and/or by adding seat-mounted guides to maintain the lap-belt anchor points.

Passive Seat Belts

The passive seat-belt concept provides automatic donning and doffing; i.e., when the door is opened (and/or when some motorized device is provided to pull the webbing away from the occupant), the seat belt is pulled away from the occupant's body so that he or she can get out of the car without manually manipulating the buckles or webbing. When the person gets back into the car and closes the door, the webbing once again envelopes his or her body and provide restraint.

Two-Point System. Typically this consists of a single shoulder belt plus a knee bolster on the lower edge of the dash. This bolster prevents the occupant from sliding forward during a crash, and the torso webbing keeps the occupant's upper torso from slamming forward into the dashboard. A typical configuration is shown in Fig. 25.83. Although the retractor can be placed in the inside position, it is preferred in the outside position so that the webbing is let out as the driver or passenger leans forward to reach the controls.

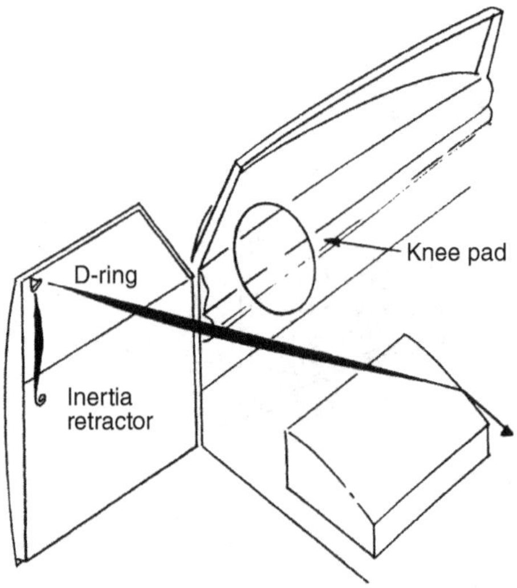

FIGURE 25.83 Two-point system.

Three-Point System. This includes both a lap and an upper torso belt. The system operates much like the two-point system in that it is actuated by opening and closing the door (or with motorized pullers). Added ingress and egress may be obtained by means of a special hook located as shown in Fig. 25.84. This hook, however, must automatically dump the webbing before the vehicle can be driven off. Otherwise, the hook could be used to defeat the purpose of the system.

Note: Both systems should be laid out as nearly as possible according to the seat-belt comfort geometry guidelines noted previously.

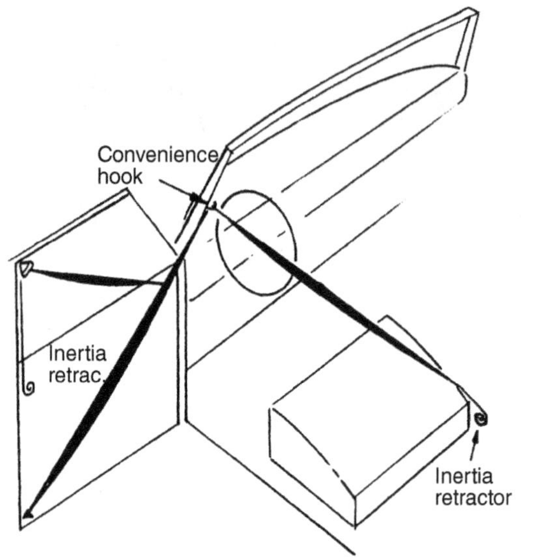

FIGURE 25.84 Three-point system.

Other Passive Seat-Belt System Considerations

1. Manually Operated Convenience Hook: Care should be taken to place the hook device where it is within easy reach but where the occupant's head will not strike it in the event of an accident. The hook should not have sharp edges or corners, but it should hold the belt securely so that it does not slip of (easily) until the ignition is actuated, which "dumps" the webbing so that it will fall around the occupant.

2. Interference of Webbing with Door Manipulation: When special motorized track systems are installed, they should not cause the in-transit webbing to "capture" the occupant's arms as he or she is opening or closing the door. For instance, avoid placing the track above a door handle because that is where the person's hand will be when the belt begins to articulate.

3. Emergency Release: It is vital to provide a means for belt release in an emergency. If a release buckle system is used, place it in the inside position because that is where the occupant will expect to find it. Also use a design that makes the release button visible from the occupant's viewpoint. Making the button red also helps identify it as an emergency item.

4. Motorized Retraction Rates: If the belt retracts too slowly, people become impatient; if it retracts too fast, they may be startled. A good compromise is to have the cycle fall between 1.5 and 3 s.

5. Bench-Seat Application: Passive belt systems can be used with bench seats as long as it is recognized that there is no way to accommodate a middle-position occupant; i.e., it is practically impossible for a person to get

past a system that is positioned on the outside in order to get to the middle position of a bench seat.

6. It is equally important to recognize that it is not practical to utilize a two-point passive system in a bench-seat configuration—at least on the front passenger's side of the vehicle—because if a tall driver moves the bench all the way aft for comfort, a small front passenger will probably be too far away from the protective knee pad for it to be effective. Therefore, if a two-point passive system is used with a bench seat on the driver's side of the car, a three-point passive system should be provided for the front passenger.

Other Active Seat-Belt Considerations

In order for an active seat belt to be donned easily, care must be taken in the design of the various hardware components.

1. The latch-plate element must be designed so that it is possible to grasp it firmly and yet leave the end of the plate exposed sufficiently to insert it into the buckle.
2. Although the buckle end of the belt does not have to be pulled out, it is necessary to hold it in a position that makes latch-plate insertion possible. As in the case of the latch-plate, there must be sufficient gripping area behind the buckle so that the buckle will be exposed.
3. The torso webbing pressure on the occupant's shoulder should not exceed 4 lb (28 kN). If it becomes necessary to increase this amount to cause proper belt retraction, some type of tension-relieving system should be added.
4. The latch plate should always return to a common, accessible position, either beside the seat or near the shoulder.

Special Considerations for Seat-Belt System Design and Installation

Ideally, design of the seat-belt system should take place concurrently with development of the seat and vehicle body designs since there are significant physical relationships that should be maintained. First, the geometry of the seat belt and torso harness should remain constant regardless of the size or weight of the occupant. (See Fig. 25.85.) If very soft adjustable seats are used, it is difficult to maintain this constant geometric configuration; i.e., the angle of the seat belt will become more shallow (if the belt is anchored to the floor) as the seat is moved forward to accommodate a shorter occupant. (See Fig. 25.86.) When the seat cushion is soft and this angle becomes too shallow, the occupant tends to depress the cushion during a crash deceleration, his or her pelvis rotates, and the seat belt rides up on the abdomen, often resulting in severe internal injuries. If the shoulder harness pulls down on the shoulder because of the positioning of the torso belt anchoring, this pelvic rotation is increased even more.

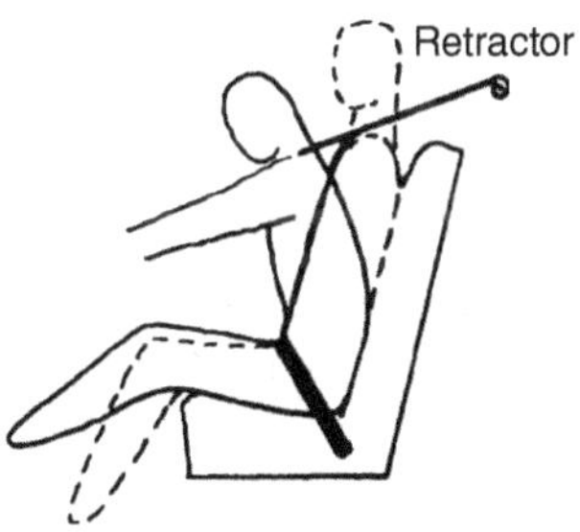

Figure 25.85 Seat-belt considerations—correct geometry.

Figure 25.86 Seat-belt considerations—incorrect geometry.

Although some forward motion of the occupant's upper torso is acceptable, the buttocks should remain as securely against the seat back as possible. The two-point system, using a knee bolster, as shown in Fig. 25.87, is one method for minimizing the forward motion of the buttocks.

Figure 25.87 Seat-belt considerations—two-point harness with knee bolster.

It is obvious that the proper geometric conditions could be maintained more effectively if seat cushions were relatively firm and the seat was not adjustable so that the belts could be anchored to the vehicle structure. This could be accomplished for all seats except the driver's seat, which drivers of various sizes must be able to adjust to fit their control interfaces. (See Fig. 25.86.)

Whatever the system, the following loads on the occupant should not be exceeded:

Forward head deceleration 80 g/3 ms
Rotational head deceleration 7000 to 10,000 rad/s^2
Horizontal chest deceleration 60 g/3 ms
Horizontal pelvis deceleration 50 to 80 g

It may be necessary to consider special load-limiting devices in order to keep loads within these limits.

Whiplash Prevention

When a vehicle is impacted from the rear, the occupants may be subject to severe whiplash injuries because their heads have no support. Headrests are one solution. However, an adjustable headrest does not ensure injury prevention because many people fail to adjust a headrest properly. A permanent headrest is preferred. It should be designed to provide minimum clearance between the headrest and the occupant's head during normal operation, but not so much that the head and neck can rotate too far (a maximum clearance of about 4 in., 10.2 cm, is suggested).

Since a permanent headrest may create a visual obstruction for the driver, it is suggested that the headrest be made from a transparent material so that the driver can see through it. A successful design is shown in Fig. 25.88.

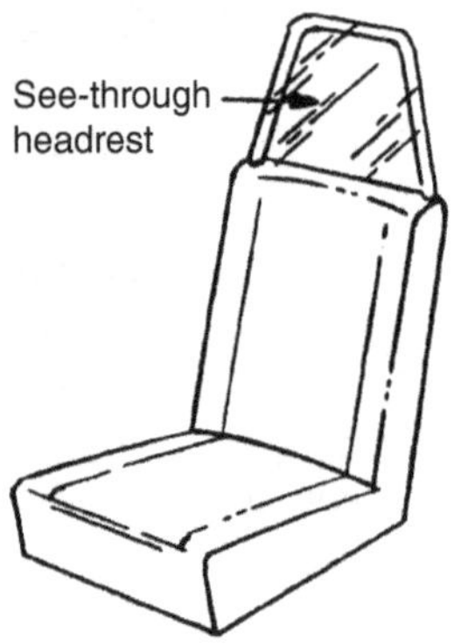

Figure 25.88 Seat-belt considerations—see-through headrest.

Webbing Retractors

The "inertial reel" type of retractor is recommended, as opposed to others, such as the common belt-sensitive systems. The latter systems tend to "lock" as the user attempts to buckle up.

Any retractor should be of the emergency locking type, however; i.e., the system should be free under gradual movements of the wearer, but under specified decelerative loads it should immediately lock out the belts from further webbing extension.

Other Restraint Systems

Vehicle upset is a major hazard for many agricultural and construction machines because of the rough terrain on which they operate. Roll-over guards such as those shown in Fig. 25.89 should be considered mandatory.

In addition to roll-over protection, the partial or full cab cover also provides protection from rocks or other debris that could fall on the driver.

Seat belts are suggested for such machines not only to protect the operator from being thrown from the vehicle but also to help control the operator's position and thus provide a steadier base from which to observe and manipulate. Such belts should, however, have a quick-release buckle so that the operator is not detained in the event the vehicle must be abandoned quickly.

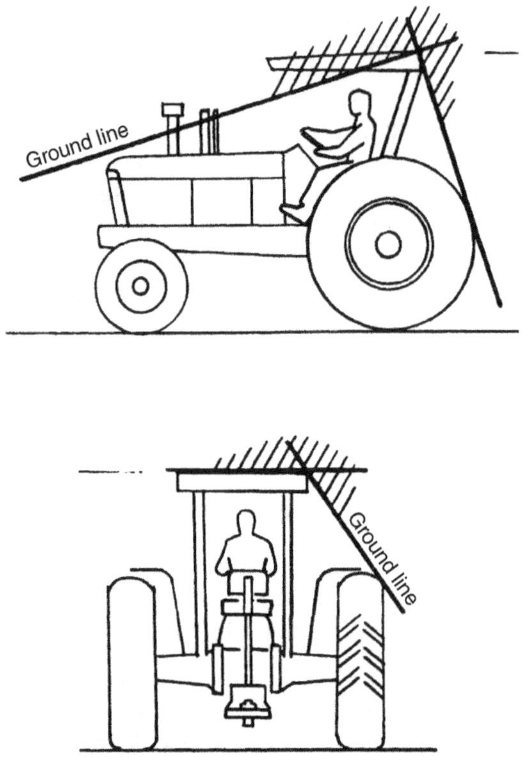

Figure 25.89 Roll-over protection envelope for off-road vehicles.

Handholds for Public Transit Buses

Because city buses start and stop frequently (and sometimes with little warning), handholds are particularly important for passenger safety. (See Fig. 25.90.) Also, because of the wide variety of passenger sizes and infirmities, it is important to provide a variety of handholds, including those on the backs of seats, handholds on vertical poles, and overhead handrails. (See Fig. 25.91.) Since specific bus configurations cannot be anticipated, specific recommendations cannot be provided. The particular configuration should be mocked up in full scale, and various handholds and rails should be placed experimentally to determine the type and placement that will provide maximum passenger access without interference. Handrail diameter should be approximately 1½ in. (3.8 cm).

AGRICULTURAL AND CONSTRUCTION EQUIPMENT

Principal Human Factors Problem Areas in Agricultural and Construction Machine Design

1. Ingress and egress (steps, handrails, doors, and door opening devices)
2. Ingress and egress to operator position (clearance)

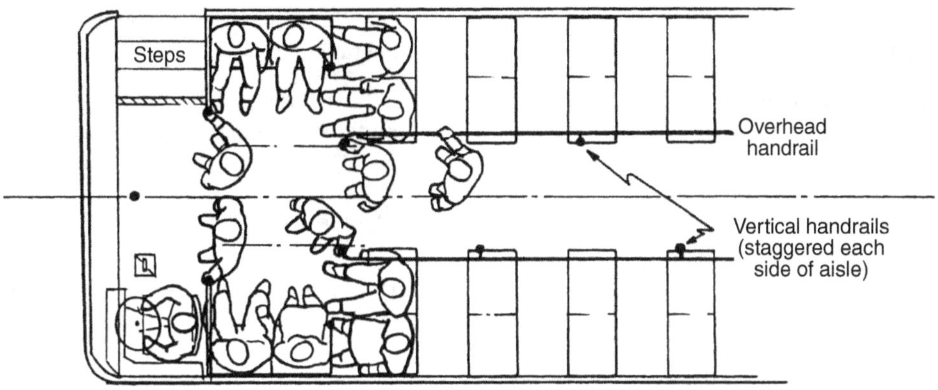

Figure 25.90 Handholds for passenger safety.

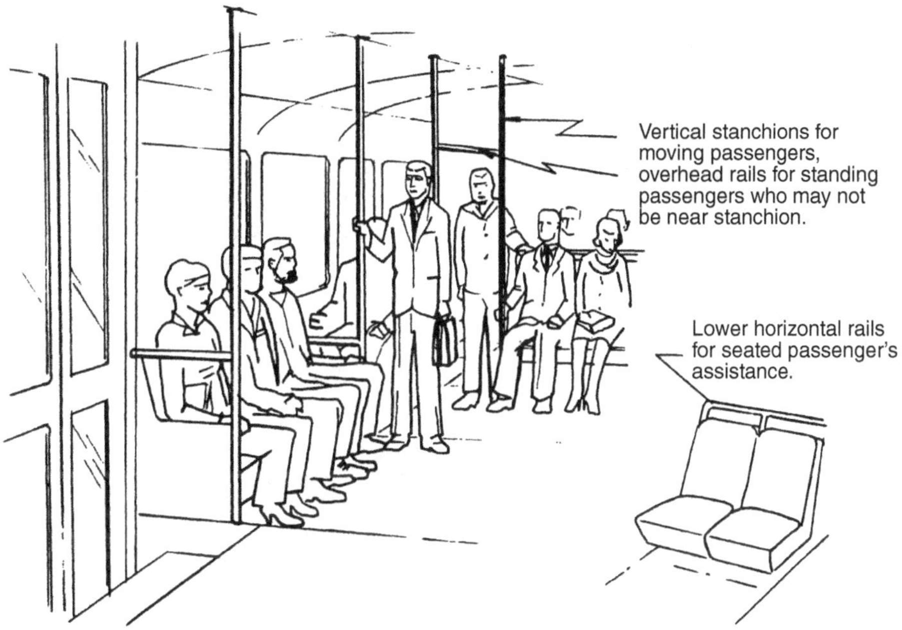

Figure 25.91 Provide a variety of handrails.

3. Arrangement of seat controls and displays (visibility and reach)

4. Exterior viewing (critical operating areas and components)

5. Lighting (interior and exterior for safe operation and identification)

6. Operator habitability (control of temperature, ventilation, vibration, and noise)

7. Maintenance (inspection, servicing, troubleshooting, and removal and replacement of parts)

8. Safety (impact, roll-over, electric shock, burns, toxic materials and substances, falls, and accidental contact with sharp elements)

Priorities during Design Conceptualization

1. Locate the operator position for maximum task surveillance.

2. Arrange task controls and displays for maximum monitoring and manipulative effectiveness and minimum strain and fatigue.

3. Provide adequate control over potentially degrading environmental factors to ensure operator performance efficiency.

4. Provide necessary safety features to preclude injury to other persons who may be working in the vicinity.

5. Provide adequate features to facilitate the servicing and maintenance functions required to keep the machine operating effectively.

6. Position the operator for the best view for steering and monitoring key machine operations.

7. Provide good seating to maintain posture and security during a rough ride.

8. Arrange controls and displays for easy monitoring and reach.

9. Provide appropriate exterior lights for both steering and equipment monitoring at night.

10. Provide a safety cab to protect the operator from noise, dust, and weather.

11. Provide good ingress and egress, including handrails and platforms.

12. Eliminate all sharp corners, protrusions, and edges.

13. Provide adequate access for inspection, servicing, and repairs.

Operator Controls for Industrial Equipment[17]

1. The brake pedal should be actuated by the right fool. When traction control is not foot actuated, brakes may be operated by either foot.

2. When separate brake pedals are provided for independent right-hand and left-hand braking, it should be possible to obtain combined and/or equalized control.

3. When a foot pedal traction control is used it should be positioned for left-foot operation.

4. If the control is hand operated, the control should move aft for disengagement.

5. When a combination clutch and brake control is used, it should be positioned for left-foot operation; forward and/or downward for clutch disengagement and brake engagement.

6. When a hand-operated clutch control is used, its movement should be rearward and/or downward for disengagement.

7. An engine speed control should move generally forward or upward to increase speed.

8. When the speed control is foot operated, it should move forward and/or downward for speed increase.

9. When a hand-operated forward-reverse lever is used, forward motion should coincide with forward vehicle movement, and vice versa.

10. Lift control levers should be located for right hand operation.

11. If a foot pedal is used for lift control, downward motion of the toe should lower the lift and vice versa.

12. If a single lever (other than tiller type) is used for steering, right control motion should turn the vehicle to the right, and vice versa.

13. When a tiller control is used, a right lateral movement should result in a left vehicle movement.

14. When two levers are provided for steering, the right lever should control the right-hand element, the left lever the left element. Forward lever movements should cause that particular element being controlled to move forward and increase in speed, and vice versa.

Note: It should be noted that the several direction-of-motion relationships listed above and in other machine guidelines prepared by the SAE are drawn, for the most part, from traditional practice, although some parallel results of human engineering research (see the section on control design in Part 3). It is recommended that, when a new control function is required, the specific type of control and its movement patterns be based on the human engineering recommendations presented in Part 3. For guidelines for motor grader controls, see SAE J1071.

Construction and Industrial Equipment Instrument Guidelines[18]

Figure 25.92 presents general guidelines for the design of instrument faces and their arrangement in construction and industrial equipment, as recommended by the Society of Automotive Engineers.

In general, instrument faces should be designed according to good human engineering practices, as discussed in Chap. 28, "Visual Displays." Since the figure represents only a few of the possible instrument requirements for many types of equipment and machines, it behooves the designer to establish sound functional requirements before attempting to select a particular visual display. Then the designer should create a display that provides the necessary information in the simplest, clearest, and most legible format possible. The panel arrangement should be based on an iterative layout study in which the several criteria are exercised for the best trade-off between the following:

1. Sequential operation
2. Frequency of use
3. Functional grouping
4. Emergency accessibility

[17]*Reprinted from Operator Controls on Industrial Equipment—SAE Recommended Practice.* SAE J297. Society of Automotive Engineers. Inc., Warrendale, PA.

[18]*Instrument Face Design and Location for Construction and Industrial Equipment—Recommended Practice,* SAE J209, Society of Automotive Engineers. Inc., Warrendale, PA.

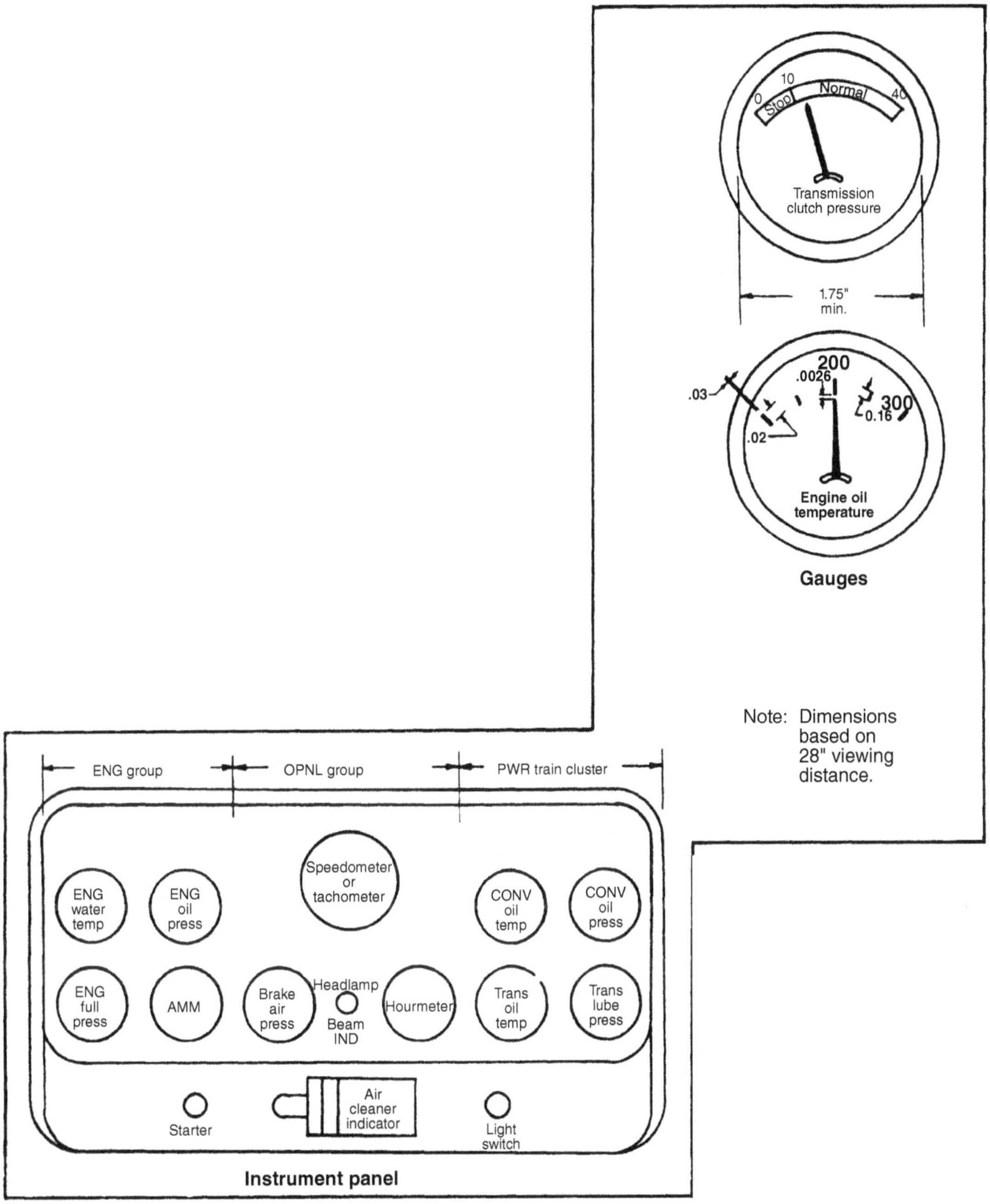

Figure 25.92 Recommended control and display layout.

Industrial and Agricultural Machine Enclosure Guidelines[19]

Figure 25.93 shows the operator cab clearance guidelines were generated by the automotive and machinery industry to guide designers. The space envelope occupied by an operator of such equipment must be sufficiently generous to minimize the possibility that the operator will strike the structure or other elements of the cab, especially when the vehicle moves over rough terrain. These guidelines should be used only as the first step, however. A full-scale mockup should be created to provide a more specific evaluation of each specific operator workstation, including all the controls and other elements required in the cab enclosure, each of which can become a contact hazard.

[19] *Operator Enclosures (Cabs)—Human Factor Design Considerations,* SAE J154a, Society of Automotive Engineers, Inc., Warrendale, PA. See also ISO 3411:2007—Earth-moving machinery—Physical dimensions of operators and minimum operator space envelope-moving machinery—Physical dimensions of operators and minimum operator space envelope for operator dimensional reference.

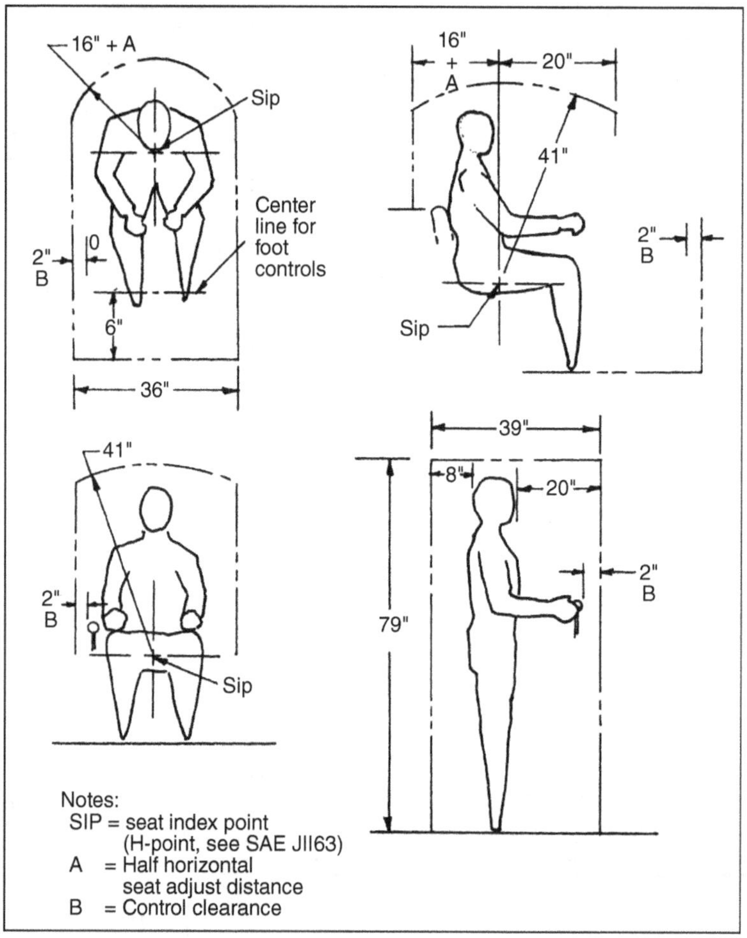

Figure 25.93 Enclosure guidelines.

Location of Operator Mounting Steps and Handrails or Handholds

In planning the location of steps and handholds to help the operator mount or dismount, consider the interaction between the operator's hands and feet at each discrete pause position and also the intervening maneuvers, which are dynamic. The operator requires not only a sure footing that does not exceed the limits of comfortable leg and foot manipulation but also strategic handhold locations that are "natural." As shown in Fig. 25.94, continuous handrails are preferred to individual handholds because they make it possible for the operator's hands to follow through without having to "search." It is also important that dual handrails be provided when the "path" is skewed, as shown, because the operator's body progresses through a series of axial and center-of-gravity shifts that could cause the operator to lose his or her grip or footing with only one hand for support. Note also that the suggested handrail configuration provides both "pulling" and static gripping capability.

Miscellaneous Dynamic Task and Design Considerations Associated with Vehicular Systems

Steering. A wide variety of vehicles, of course, require steering wheels. Whenever possible, avoid using wheels with very large diameters. As indicated in Fig. 25.95, the large, flat wheel requires frequent awkward (and tiring) body motions. A smaller wheel designed to minimize operator reach problems could be used in vehicles that have power steering. Unfortunately, the traditional large wheel is often used because it is believed necessary in order for the driver to apply enough force.

Lever Operations. Avoid the tendency to place levers so that they are awkward for the operator to reach. Typically, awkward location is dictated by the mechanical linkage. When this is the case, the linkage should be rearranged. (See Fig. 25.96.)

Severe Motion. When it is anticipated that the vehicle system will be subject to severe motion conditions, provide a means for the operator to "hang on." (See Fig. 25.97.)

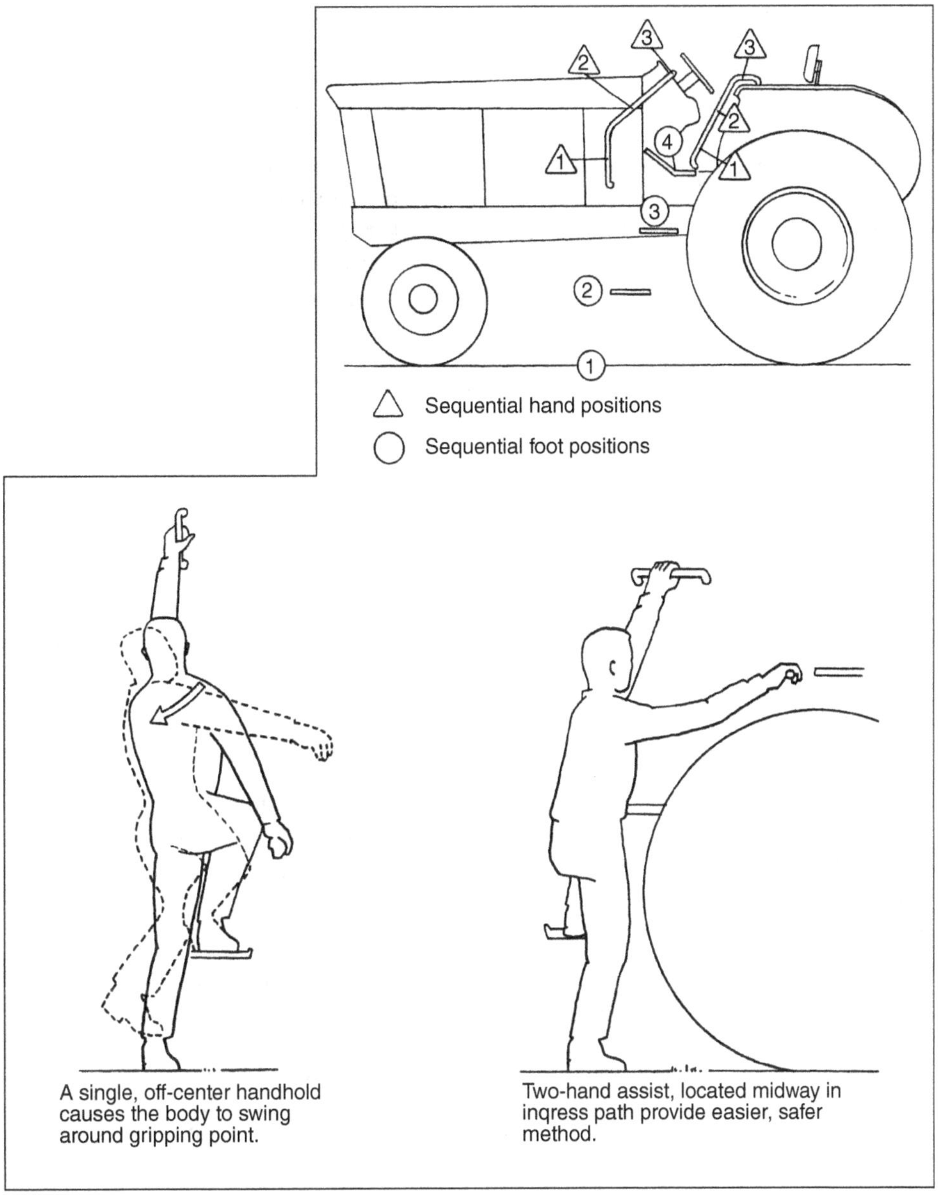

Figure 25.94 Example analysis and design of steps and handholds.

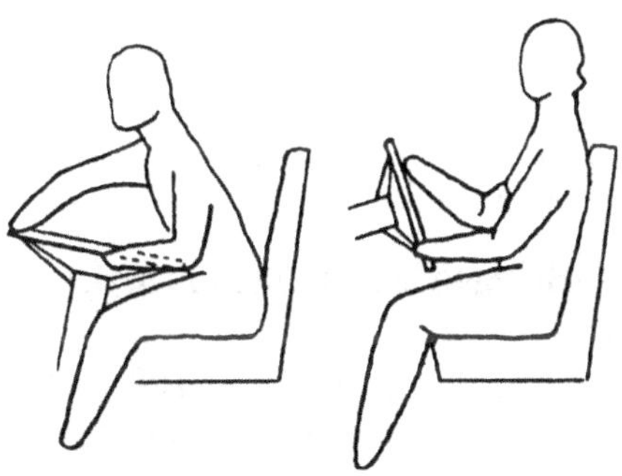

Figure 25.95 Avoid wheels more than 16 in. diameter.

Figure 25.96 Avoid awkward reach requirements.

FIGURE 25.97 Provide handrails.

Visibility. Visibility for backing or observing other machines, workers, or objects should be planned. (See Fig. 25.98.)

FIGURE 25.98 Provide visibility.

STORAGE SPACE

Storage Space as a Special Workplace

The location and design of a storage space should receive the same systematic appraisal and attention as other workstations, especially when the storage space is utilized frequently. Above all, avoid the temptation to treat the problem of storage space as something to be taken care of after all the more important operator work-station problems are solved. Design of the storage subsystem should consist of more than "making do" with whatever leftover space one can find in the facility or vehicle.

General Principles

1. Provide storage space where it is needed, not at some distant out-of-the-way, or inaccessible location.

2. Design the space to "fit" the items to be stored; do not design just a generalized volume of space or shelf area.

3. Provide enough space not only for the predicted use but also for unexpected or future expansion.

Content Considerations

1. Size, shape, and weight

2. Special storing conditions, e.g., package position and protective and/or safety precautions to prevent deterioration and damage Label exposure, i.e., the necessity to have identification labels immediately accessible as the storage container or storeroom door is opened and the contents are viewed.

Special Feature Considerations

Content separation, securing, and preserving.

Content adjustment as needs change.

Ease of ingress and of storage and retrieval.

Ease of cleaning the storage area (dirt and spills.)

Content security, i.e., inaccessibility to unauthorized persons.

Illumination so that the user can move about safely and/or can see to identify and manipulate stored items, make space adjustments, or clean.

Environmental control, i.e., control of heating, cooling, ventilation, humidity, shock, and vibration, etc.

Key Element Design Considerations

1. Location

2. Shelving and partitioning

3. Doors and covers

4. Handles, latches, and locks

5. Content securing

Storage Type and Location in Relation to Order and Frequency of Use. In many situations storage cabinets are built with little thought given to what type of storage is needed at what locations, e.g., a kitchen in the typical home. Although this is not always the case, kitchen cabinets may be created merely to fill in the spaces between appliances. The cabinetmaker may be told to provide a variety of drawers and cabinets to fill specified dimensions. As a result, the user often finds that things cannot be stored next to the area in which they are needed because the storage space provided does not accommodate them. Examine the use sequence and/or the frequency for each given storage requirement and make sure that the storage plan has some direct relationship to the storage and retrieval needs of the eventual user.

Storage Shelf Depth in Relation to Shelf Height and Package Weight

Avoid the temptation to provide general shelf storage space without regard to human limitations in terms of reaching and lifting. Obviously, people cannot reach very far into a shelf that is very high or very low. In addition, people cannot support or manipulate as heavy a package when a shelf is either very high or very low as they can when the shelf is somewhere between waist and shoulder height.

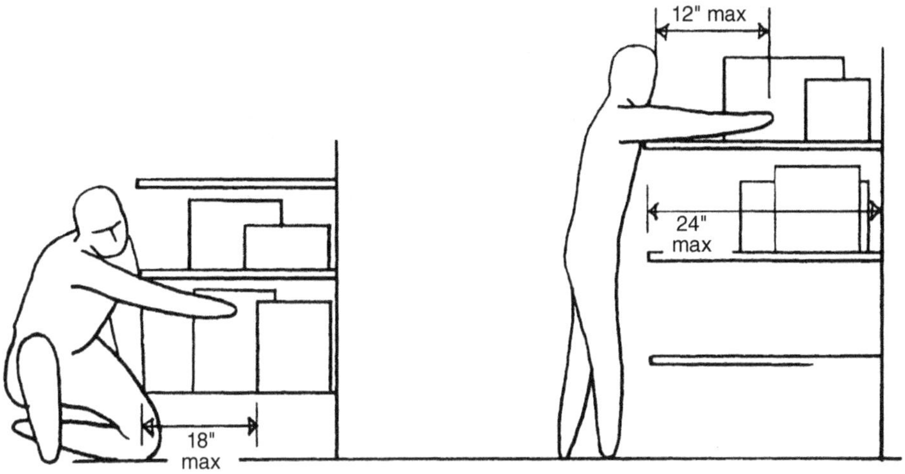

Figure 25.99 Recommended shelf depth.

See Chap. 35, "Human Strength," for carrying and lifting limits. Figure 20.99 illustrates the difficulties created by extremely deep shelving. The following are general guidelines for specifying shelf depth:

1. Below waist height, the maximum shelf depth should not exceed about 18 in. (46 cm).

2. Above shoulder height, the maximum shelf depth should not exceed about 12 in. (31 cm).

3. At waist to shoulder height, the maximum shelf depth should be about 24 in. (61 cm).

Storage Cabinet Drawer and Door Design

Although there are some aesthetic as well as constructional advantages to certain types of cabinets, these sometimes create problems for the user. Figures 25.100–25.102 provide some guidelines for designers, whether they are selecting ready-made units or designing new cabinets:

1. Lap-front drawers or doors are not preferred because they make it easier to catch and pinch one's fingers and because any liquids that are spilled down the front of the cabinet tend to enter the drawer or cabinet. (See Fig. 25.100.)

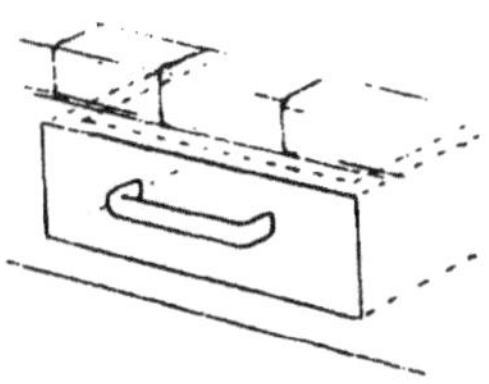

Figure 25.100 Flush fronts preferred.

2. Handles should be on the front of drawers and doors so that the user knows where they are. (See Fig. 25.101.)

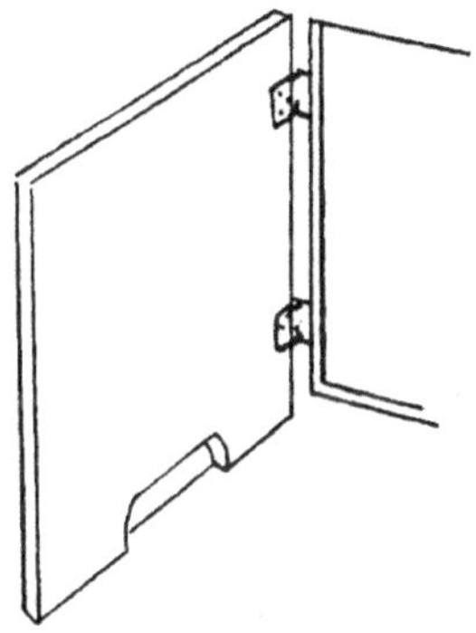

Figure 25.101 Hidden handles undesirable!

3. Corner adjacent drawers should be avoided, and similar door arrangements should be planned so that when two adjacent doors swing open, their handles do not coincide. (See Fig. 25.102.)

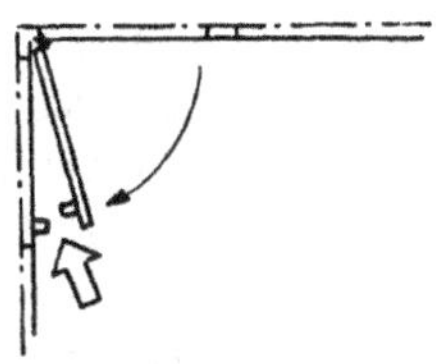

Figure 25.102 Potential "pinch" point.

Magnet or spring-return types of door closures are preferred over latching types (except for those cases where dynamic forces might cause the door to fly open, as in vehicular applications).

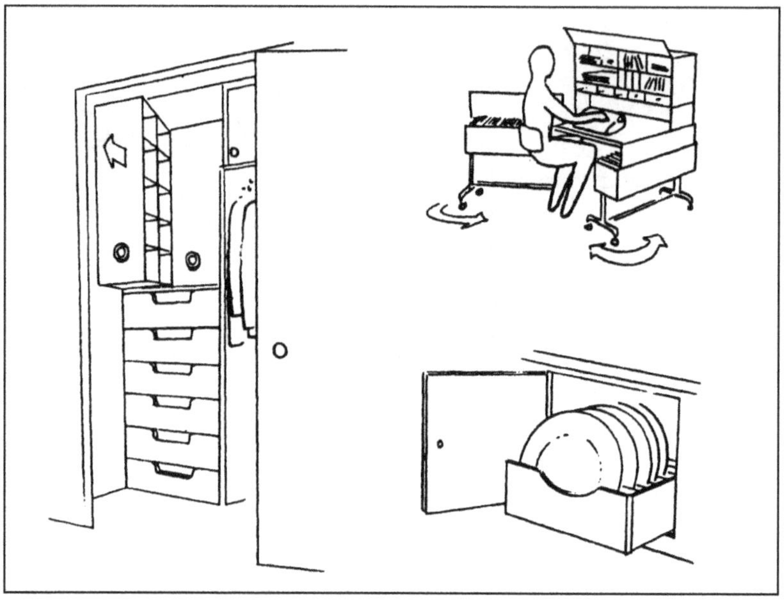

FIGURE 25.103 Bring the stored item to the user.

A desirable design objective should be to "bring the stored item to the user." That is, instead of designing fixed shelves, bins, etc., consider the possibility of making these elements slide out, rotate, or move into more convenient positions for the user to store or retrieve articles. Figure 20.103 illustrates some of the schemes that various designers have used successfully.

In executing such designs as shown here, it is extremely important that the mobility devices operate smoothly and easily and that there be proper means for securing shelves or drawers so that they stay where the user wants them. Equally important is ensuring that a drawer, for instance, cannot inadvertently be pulled all the way out, causing its contents to be spilled on the floor.

Always provide handles that are visible and easy to grasp on all such mobility systems. Heavy-duty, smoothly functioning slide or motion systems should be provided, especially if the shelf or drawer supports heavy articles.

Storage Requirements in Moving Vehicles

In addition to the basic requirements discussed above regarding planning for the more stable, fixed storage place, when planning storage aboard various vehicles (whether for private or commercial use), the designer must consider the added difficulties caused by the dynamic environment. The critical issue involves making sure that cargo stays put! Although most designers are aware of this need in commercial applications and/or are forced to comply with specifications that define cargo tie-down requirements, a similar concern is often not displayed with respect to private vehicles.

Figures 25.104–25.106, although not representative of all possible situations, provide at least three basic approaches for cargo storage capture: (1) place stored items inside a closed compartment; (2) provide railings or insets to minimize lateral shifting; and/or (3) provide tie-down features so that cargo can be strapped in place.

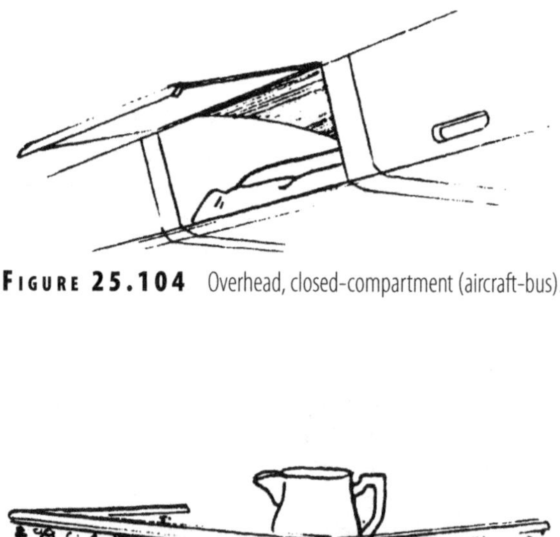

FIGURE 25.104 Overhead, closed-compartment (aircraft-bus).

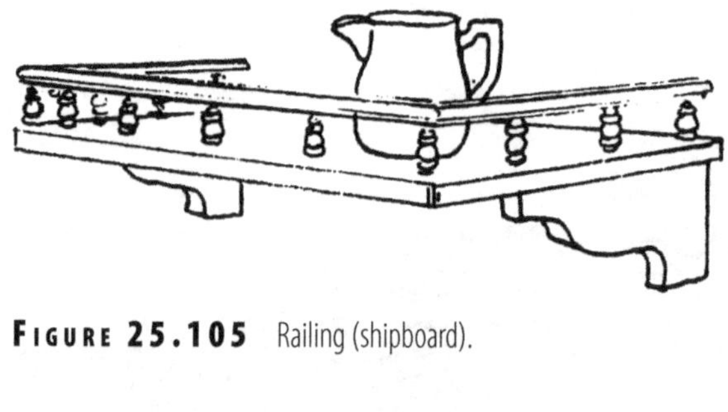

FIGURE 25.105 Railing (shipboard).

The family station wagon shown in Fig. 25.106 illustrates a problem that is typically ignored. Many a passenger has become a statistic because some loose cargo became a missile during a crash. Something as small and light as a tape cassette has been known to kill a passenger because it was left unsecured on the back window ledge of the family automobile.

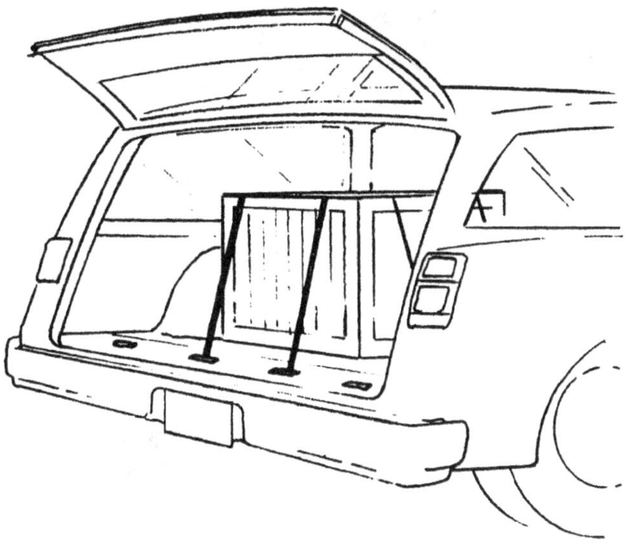

Figure 25.106 Tie-downs!

CHAPTER 26

LIGHTING

LIGHTING SYSTEM DESIGN

Purpose

Lighting is typically provided for two main purposes: (a) to enable people to see what they need to see in order to perform various tasks and (b) to create decorative effects. It is important to differentiate between these purposes because the objectives of each are different. In some cases, decorative lighting schemes may interfere with task illumination unless there is a constant awareness that the observer, to see well, must have enough of the proper type of light, placed where it does the most good, and that there should be no extraneous light or reflections that might interfere with the observer's seeing process.

Seeing tasks are not always the same; thus, the amount and type of lighting and the method for providing illumination should be tailored to fit the particular seeing task. A few examples will serve to illustrate how different tasks and lighting system requirements can be:

1. Reading a book or newspaper
2. Reading instruments or video displays in a darkened room or cockpit
3. Performing surgery or sewing on dark material
4. Inspecting for flaws in material or metal surfaces
5. Matching colors for material selection
6. Developing film in a darkroom

Types of Lighting

1. Natural Light: The proper manipulation and use of natural light is very desirable for many applications, especially because it tends to make things we look at seem more "natural."

2. Artificial Light: Artificial light is used when there is not enough natural light available for the desired seeing task and/or when it may be necessary to control the color of the light for a special seeing task.

Lighting Methods

The three basic types of illumination either for illuminating seeing tasks or for creating decorative lighting effects are (1) general illumination, (2) task illumination, and (3) accent illumination.

1. General Illumination: General illumination should be used when the seeing requirements are broad and varied, including seeing to move about safely, observing other people or general features within the surrounding area, performing housekeeping chores, moving materials or equipment, and performing general maintenance tasks.

2. Task Illumination: Task illumination is required when general illumination is not available and/or when the nature of the task is such that the general illumination is insufficient, is not appropriately directed, or is of such a quality or color as to make specific seeing tasks difficult to perform. Supplementary illumination also is generally used for decorative effects since it can be precisely controlled in terms of color, direction, and amount.

3. Accent: Accent illumination is used to draw the eye to a specific item or area. It can be used for decoration or to provide information.

Lighting Techniques

1. Floodlighting: As the term implies, this type of lighting is usually external to the object or objects that are to be illuminated, literally flooding them with light. Floodlighting is generally used to illuminate large areas, either indoors or outdoors, but it can also be used in more concentrated, localized situations such as for illuminating a control panel, a doorway or platform, a theatre stage, a parking lot, or an airport apron. Although some control can be exercised over the flood pattern, it should be noted that the illuminated area covered by a floodlighting system will vary from the beam center to its periphery. Care must also be taken in using the floodlighting technique to consider the problems that might be associated with shadows, which are a by-product of the floodlighting technique.

2. Internal Lighting: A wide variety of internally generated lighting techniques are available to make objects or visual displays "self-luminous." Typical applications for which internal lighting is appropriate include signs, panel labels, vehicle or operator console instruments and displays, shadow boards, and marker and signal lights. Internal lighting can be created in many ways, including the use of luminous phosphors, back-lighted screens, edge-lighted plastics, and end-lighted plastic "light pipes." The primary advantage of internal lighting is that more precise control can be obtained over where the light will appear.

Light Level

Avoid the temptation to believe that merely providing higher and higher illumination levels will result in improved seeing conditions. Although the level or intensity of light is obviously important, light can be too bright as well as too dim for the seeing task. Also remember that as light levels increase, glare problems also increase. There is usually a point below which seeing is difficult and another point above which seeing either does not improve or becomes more difficult for each type of seeing task. Adequate illumination levels depend on the age of the viewers. As we age, we require several times more illumination. Table 26.1 shows the minimum illumination changes with age.

TABLE 26.1 Age and Minimum Illumination Levels

Age of Viewer	Minimum Illumination Level
Less than 25 years old	X
Between 25 and 65	2 times X
Over 65 years old	4 times X

Color

The color of the light used (i.e., for general illumination, supplementary illumination, or signal lighting) is important from several standpoints:

1. A broad-spectrum light (white light) is desirable to make the colors of objects or people seem natural. Natural light is used as the baseline reference since it is by this reference that we observe most things during everyday living. Several types of artificial light have been created to simulate natural light, but some are better than others for certain purposes. Care must be taken to select the artificial light color that is most appropriate for the task or seeing conditions being created. Incandescent luminaires typically provide more reds and yellows, which impart a warm appearance to viewed objects. Fluorescent lights, on the other hand (depending on the specific luminaire), tend to accentuate the blue. Often, a colored material selected under fluorescent light will not look the same when the material is viewed under natural sunlight. Light-emitting diode, LED, technology allow designers' flexibility of a wide range of colors. LEDs can replicate incandescent lighting, provide monochrome colors, or the white light of fluorescent lighting.

2. Monochromatic light (i.e., red, yellow, blue, and green) may be desirable for certain types of seeing tasks and/or for decorative purposes. Such colors are not, however, desirable for general seeing conditions; they make surface colors appear abnormal and/or impossible to recognize. Monochromatic light is particularly useful for creating color-coded displays, e.g., traffic signal lights, airport runway and taxiway lights, and warning displays.

Light Fixture Type and Usage

An extensive array of lighting fixtures, lamps, and materials is available for creating almost any type of lighting that is required. Some precautions need to be taken, however, in selecting commercially available fixtures. Some provide effective task illumination, but others are merely attractive to look at. When looking for appropriate fixtures off the shelf, designers should analyze the seeing task fully so that they can evaluate the merits (or, more often, the demerits) of the fixtures they are contemplating for a particular application.

An important factor to consider is the location of lighting fixtures, for even a good fixture's value can be lessened if it is improperly positioned relative to the seeing task. In addition, it is important to consider the light control requirements in terms of the possible need for dimming and/or selectively illuminating only certain fixtures at certain times during an operation.

Fixture Servicing

The ease with which light fixtures can be cleaned and relamped is perhaps as important as the overall lighting system design concept. It has been estimated that some lighting systems output is reduced by 25 percent six weeks following its installation. In selecting fixtures for any lighting system, consider the following servicing features:

1. The fixture should be readily accessible using only the simplest stepladder. Avoid placing fixtures at extreme heights or in positions where it becomes necessary to employ special devices and/or where there is access only through difficult crawl spaces.

2. It should be easy to remove intervening light control panels without damaging them, and the method of fastening and releasing these devices should be obvious and not require special tools. Many current schemes that are offered by manufacturers or designed on a custom basis seem to purposely camouflage the method for removing a ceiling panel, egg-crate screen, or translucent filter panel. Removable screens or panels should not be so large that they are unwieldy to handle efficiently, and they should not be made of material that is too fragile and easily damaged.

3. Light-filtering or light-distributing panels or covers should be made of materials that do not discolor with age.

4. Fixture housings should be designed so that there is sufficient clearance to get hold of the lamps and manipulate them in and out of sockets.

5. Choose fixtures that either allow dirt or dust to drop through naturally (open egg-crate screen) or that are semi sealed against dust and dirt collection.

6. Preferably, use fluorescent fixtures that contain lamps no longer than 40 in. (102 cm). Longer tubes are difficult to handle safely.

LIGHTING TERMINOLOGY

Lighting terms and units are illustrated in Fig. 26.1. Two sets of units are commonly used for lighting: radiometric and photometric. Radiometric units are used to describe the distribution of energy within spectral boundaries without regard to human visual characteristics. Photometric units are used to describe the distribution of energy within the visual spectrum. Photometric units are weighted to reflect the variation of human sensitivity to light.

Process for Designing Lighting Systems

Figure 26.2 (from NASA Human Interface Design Handbook) shows the process for designing lighting systems. Designers should first analyze tasks and create a task lighting system. Through modeling (with computers or physical mockups), the designer should then assess the general lighting to see if it is adequate for "general" tasks such as transit between work areas. Inadequately illuminated areas should then be supplemented with additional lights.

ILLUMINATION LEVEL

General Illumination Level Guidelines

The Illuminating Engineering Society of North America (IESNA) has detailed lists of illuminance recommendations. They determine lighting requirements by categorizing tasks into three categories—with subcategories defined by task and visual demands. (See Tables 26.2 and 26.3.)

Table 26.4 shows the illumination level guidelines for maintaining proper brightness contrast between visual tasks and background.

Computer Workplace Illumination

ANSI/HFS Standard Number 100-2008 recommends illumination levels for computer workstations. Ambient illumination should be sufficient to support the visual tasks that will be done at the computer workstation but should not degrade the image quality on the display. For emissive computer workstation screens, high illuminance (above 500 lux) can reduce character contrast and impair visual performance. Nonemissive display technology, on the other hand, benefits from high ambient illumination.

Commercial and Industrial Illumination

The illumination levels in Tables 26.5 and 26.6 are typical minimum illuminance values for home, commercial, and industrial environments. An excellent source for illuminance standards is the IES Lighting Handbook published by the IESNA. The handbook is updated regularly and has extensive illuminance recommendation lists. Other illuminance standards are published by the Department of Defense

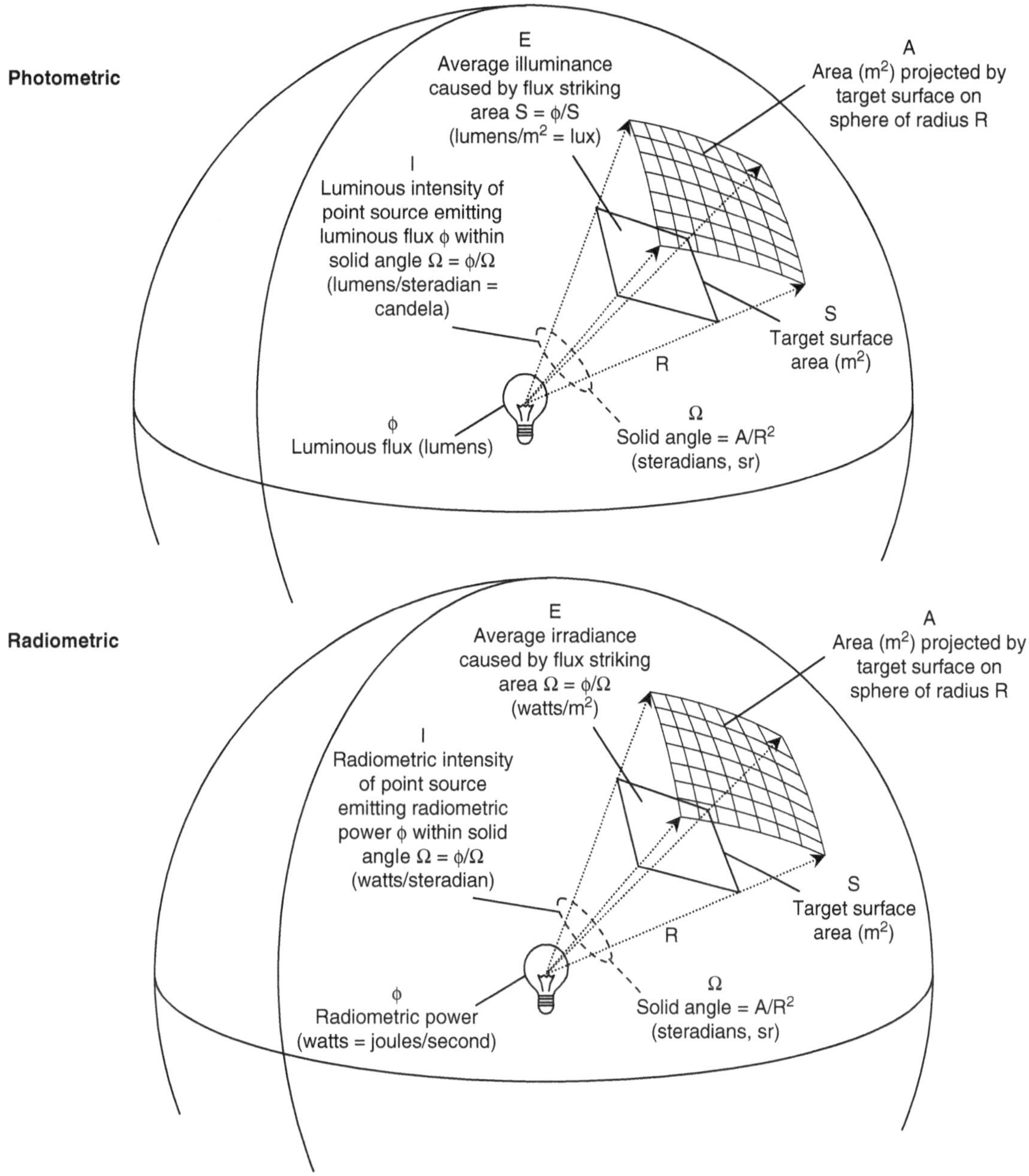

Figure 26.1 Lighting terms and units.

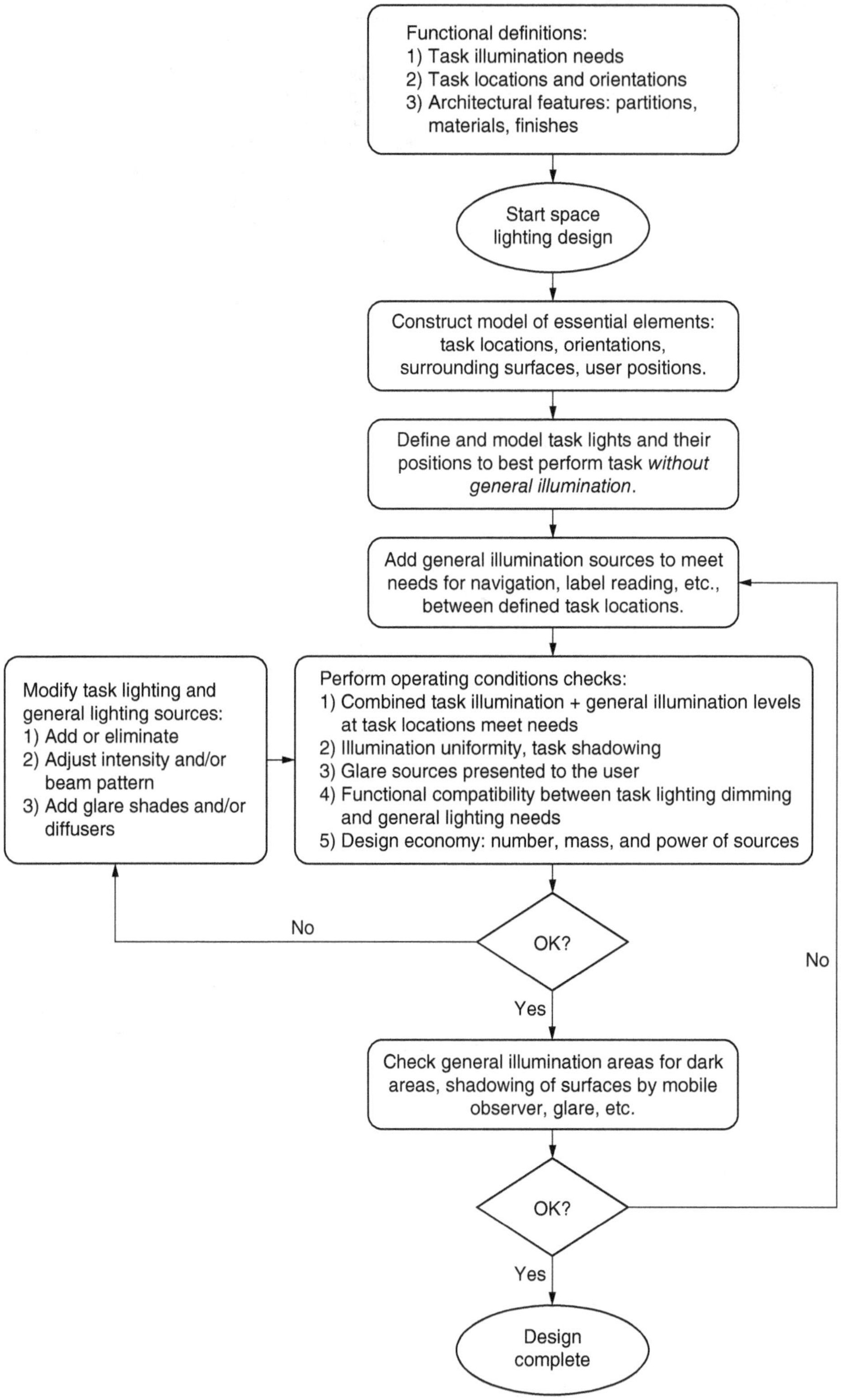

Figure 26.2 Lighting system design process.

TABLE 26.2 General Illumination Level Guidelines

Task Category	Subcategory	Example	Illuminance Recommendation
Orientation and simple visual tasks	A. Public spaces	Ballrooms, social events	30 lx (3 fc)
	B. Simple orientation for short visits	Shopping mall concourse	50 lx (5 fc)
	C. Simple visual tasks	Hotel room general	100 lx (10 fc)
Common visual tasks	D. Tasks with high contrast and large size	Library stacks	300 lx (30 fc)
	E. Tasks with low contrast/large size or high contrast/small size	Auto maintenance shop	500 lx (50 fc)
	F. Tasks with low contrast and small size	Fine woodworking	1000 lx (100 fc)
Special visual tasks	G. Tasks with near threshold vision demands	Operating room	3000 to 10,000 lx (300 to 1000 fc)

TABLE 26.3 Special Situations where Visual Dark Adaptation is Required

Work Situations	Light Level (fc)	Type of Illumination
Darkened control rooms and cockpits at night	0.01–0.4 0.03–0.10 0.10–1.00	Illumination on work surfaces Black and red instrument face detail (scale marks, numerals, and labels) Indicator lights (not to exceed 1 in. 2.5 cm square)
Ready rooms where crew members are required to maintain some dark adaptation prior to a mission at night	0.50–2.00	Illumination at tabletop
Move theater during film running	0.50–2.00	Illumination in aisle and immediate vestibule

Note: Red illumination is required in military workplaces and ready rooms. Low-level white light is satisfactory in theaters.

TABLE 26.4 Limiting Brightness Ratios for Glare Reduction

Comparison	Environmental Classification* A	B	C
Between tasks and adjacent darker surroundings	3:1	3:1	5:1
Between tasks and adjacent lighter surroundings	1:3	1:3	1:5
Between tasks and more remote darker surfaces	10:1	20:1	**
Between tasks and more remote lighter surfaces	1:10	1:20	**
Between luminaires and adjacent surfaces	20:1	**	**
Between the immediate work area and the rest of the environment	40:1	**	**

Note: Direct glare arises from a light source within the visual work field. It should be control by:

1. *Avoiding bright light sources with 60° of the center of the visual field. Since most visual work is at or below the eye's horizontal position, placing luminaires high above the work area minimized direct glare*
2. *Using indirect lighting*
3. *Using more relatively dim light sources, rather than a few very bright ones*
4. *Using polarized light, shields, hoods, or visors to block the glare in confined areas*

**A)—Interior areas where reflectances of entire space can be controlled for optimum visual conditions; B) Areas where reflectances of immediate work area can be controlled, but there is only limited control over remove surroundings; C) Areas (indoors and outdoors) where it is completely impractical to control reflectances and difficult to alter environmental conditions.*

***Brightness-ratio control not practical.*

From MIL-HDBK-759A.

TABLE 26.5 Recommended Illuminance Values for Home and Commercial Environments

Facility	Light Level (fc)
Art galleries:	
General	30
Paintings	30–60
Sculpture	30–100
Auditoriums:	
Lobbies	30
Seating area (nonperformance)	10–30
Seating area (during performance, in aisle)	1–3
Orchestra (music stands)	15–20
Banks:	
Lobby	30
Officer platform area	30–50
Teller line	50–75
Rear posting and keypunch area (work surface)	75–100
Depots and terminals:	
Waiting rooms	30
Ticket counter	50–75
Baggage room	50–75
Platform	30
Restrooms	30
Hospitals	
Lobbies and hallways	50–75
Patient rooms	3–50
Laboratory and examination rooms	75–100
Surgery and delivery	Up to 2500

(*Continued*)

TABLE 26.5 Recommended Illuminance Values for Home and Commercial Environments (*Continued*)

FACILITY	LIGHT LEVEL (FC)
Hotels:	
Lobbies	30–75
Front office	50–75
Halls and elevators	20–30
Restrooms and guest rooms	20–50 plus reading lights to 75
Kitchen	100
Laundry	50–75
Office Buildings:	
Lobbies, halls, and elevators	20–30
Offices	75–100
Residences:	
Entryway and porch	10
Hallways and stairs	10–25
Kitchen	50–75
Laundry	50–70
Library	50–100
Sewing room	30
Bathroom—general sink and mirror	50
Workshop	70
Restaurants:	
Quick-service counter	50
Intimate dining area	15–30
Kitchen	70
Cashier area	50
Schools:	
Study areas and classrooms	50–100
Drafting room	100
Offices	50
Restrooms	30
Retail Stores:	
Circulation area	30
Showcase	100–150
Sports:	
Archery	10
Badminton	30
Baseball—seating	3–10
Baseball—field	100–150
Basketball	50
Gymnasium, general exercise room	30
Gymnasium, locker room	20
Hockey	20
Boxing	200
Football	100
Swimming pool, general overhead	10
Swimming pool, under water	100
Tennis	30–50
Volleyball	20

TABLE 26.6 Recommended Illuminance Values for Industrial Environments

FACILITY	LIGHT LEVEL (FC)
Assembly:	
Rough seeing	30
Rough and difficult seeing	50
Medium-difficult seeing	100
Fine seeing	500
Auto Body:	
Frame assembly	50
Chassis assembly	100
Final inspection	200
Bakery:	
General	30–50
Hand decoration	100
Bookbinding:	
General	70
Embossing	200
Chemical Laboratory	30–50
Cleaning and Pressing:	
General inspection	50
Spotting	500
Hand pressing	150
Repair and alteration	200
Electrical Equipment Manufacturing:	
General	50
Coil winding	100
Foundries:	
General	30
Molding and pouring	50
Grinding and chipping	100
Fine inspection	500
Glassworks:	
General	30–50
Fine grinding and polishing	100
Engraving	200
Iron and Steel Manufacturing	20–30
Machine Shops:	
General	50
Fine bench work	100
Paint Manufacturing:	
General	30
Color matching	200
Printing Industries:	
Pressing	70
Composing	100
Proofing	150
Sheet Metal Works:	
General cutting and bending	50
Scribing and inspection	200

(*Continued*)

TABLE 26.6 Recommended Illuminance Values for Industrial Environments (*Continued*)

Facility	Light Level (fc)
Shoe Manufacturing	200–300
Textile Mills:	
General	30
Carding, spooling, and spinning	50–150 (depending on color)
Inspecting and Weaving	100
Welding:	
General	50
Precision arc	1000
Woodworking:	
General	30
Finishing	100

Note: Where color matching or working on darker colors is involved, higher illumination levels are required. In order to avoid spectral flare, when the work involves close visual inspection of materials with a high-gloss finish, light levels should not be too high. Although the above standards appear to be very specific, the reader is cautioned not to assume that the specified illumination levels are either adequate or necessary. Each visual task should be examined in detail to determine seeing difficulty. A lighting mockup is desirable to test the proposed lighting scheme as well as the proposed levels.

(MIL-STD-1472) and the National Aeronautics and Space Administration (NASA-STD-3001), and the Human Integration Design Handbook.

SURFACE COLORS

Table 26.7 shows the appearance ratings of typical surface colors when viewed under various artificial light sources.

TABLE 26.7 Appearance Ratings of Typical Surface Colors When Viewed Under Various Artificial Light Sources

Color	Daylight	Standard Cool White	Deluxe Cool White	Standard Warm White	Deluxe Warm White	Incandescent Lamps
Maroon	Dull	Dull	Dull	Dull	Fair	Good
Red	Fair	Dull	Dull	Fair	Good	Good
Pink	Fair	Fair	Fair	Fair	Good	Good
Rust	Dull	Fair	Fair	Fair	Fair	Good
Orange	Dull	Dull	Fair	Fair	Fair	Good
Brown	Dull	Fair	Good	Good	Fair	Good
Tan	Dull	Fair	Good	Good	Fair	Good
Golden yellow	Dull	Fair	Fair	Good	Fair	Good
Yellow	Dull	Fair	Good	Good	Dull	Fair
Olive	Good	Fair	Fair	Fair	Brown	Brown
Chartreuse	Good	Good	Good	Good	Yellowed	Yellowed
Dark green	Good	Good	Good	Fair	Dull	Dull
Light green	Good	Good	Good	Fair	Dull	Dull
Peacock blue	Good	Good	Dull	Dull	Dull	Dull
Turquoise	Good	Fair	Dull	Dull	Dull	Dull
Royal blue	Good	Fair	Dull	Dull	Dull	Dull
Light blue	Good	Fair	Dull	Dull	Dull	Dull
Purple	Good	Fair	Dull	Dull	Good	Dull
Lavender	Good	Good	Dull	Dull	Good	Dull
Magenta	Good	Good	Fair	Dull	Good	Dull
Gray	Good	Good	Fair	Soft	Soft	Dull

Good—color appears most nearly as it would under an ideal white-light source, such as north skylight
Fair—color appears about as it would under an ideal white-light source, but is less vivid
Dull—color appears less vivid
Brown—color appears to be brown because of small amount of blue light emitted by lamp
Yellowed—color appears yellowed because of small amount of blue light emitted by lamp
Soft—pinkish cast because of red light emitted by lamp

GLARE AVOIDANCE AND SHADOW PREVENTION

Shadows and glare are the most commonly encountered problems in lighting system designs. The three most serious conditions are illustrated in Figs. 26.3–26.5.

Shadows are cast by the observer because of improper luminaire positioning. Obviously, the light source should not be behind the observer. Typically, the luminaire should be located slightly to the right or left so that the light comes from over the observer's shoulder. General practice suggests that the light come from over the left shoulder because most people are right handed. A lighting mockup is recommended for determining the best location. (See Fig. 26.3.)

Figure 26.3 Light source from behind created undesirable shadows.

The light source should not be visible to the observer; i.e., the light bulb or lamp should be screened either by some type of baffle or by a distributing screen. Once again, since layout errors have a way of slipping by, use a lighting mockup to make sure that the observer cannot see the light source directly from his or her various working positions. (See Fig. 26.4.)

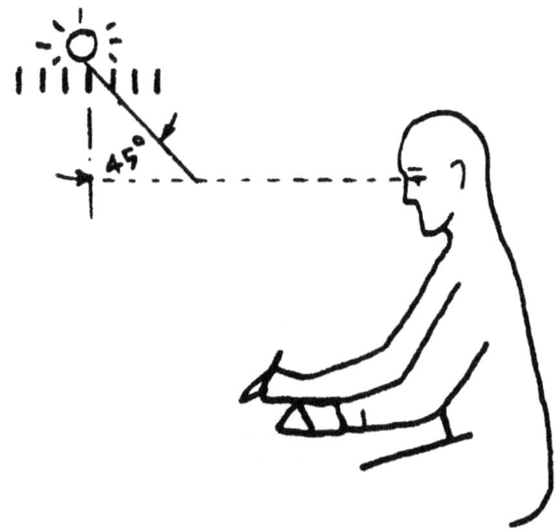

FIGURE 26.4 The light source should be screened.

Ambient glare from frontal or peripheral windows and/or similarly large and bright artificial sources not only makes seeing difficult but also produces general worker fatigue. Many methods are available for dealing with this problem, including reorienting the worker, putting shades over the windows, and adding light-filtering materials to the glazing. (See Fig. 26.5.)

FIGURE 26.5 Frontal or peripheral light can create glare.

Although this is not always the prerogative of the lighting system designer, try to eliminate glossy surfaces in front of the observer.

Diffusing light sources (fluorescent) are recommended to minimize glare and shadow effects.

INTEGRAL LUMINOUS CEILING LIGHTING SYSTEMS

A popular concept has been the ceiling that provides a combination of overhead general lighting, acoustic treatment, and heating and ventilation, plus the basic supporting structure (as shown in Fig. 26.6). When used properly, this concept provides excellent seeing conditions for large-area installations. A common failure, however, occurs when the architect or interior designer paints the surfaces of the fixture insets a dark color, which absorbs much of the light from each fixture. This also makes the contrast between lighting fixtures and support structure extreme, which causes the lights to glare rather than create a pleasing atmosphere.

Many other overhead lighting systems fail to provide the desired seeing conditions for the same reason. Always minimize the contrast between the lighted fixture and its surrounding support structure or decorative trim.

MODULARIZED, COMBINATION LIGHTING SYSTEMS

Energy conservation often can be obtained in conjunction with good lighting, as illustrated in Fig. 26.7. The objective here is to create a combination ambient and task lighting system that can "move with the workplace." Instead of trying to create a general ceiling lighting system that will be appropriate for all possible seeing situations, the designer should consider the modular approach, which illuminates only the areas that need it; in addition, since the luminaires are tailored to the modular workplace, the lighting will always be right for that workplace. This is important to note because, after a general lighting system has been installed, many seeing problems are created by the placement of furnishings in such a way that they block the intended light rays and the general light distribution.

Systems such as that shown in Fig. 26.7 must be designed so that the luminaires on top of the modular cabinets are above the heads of persons passing by. In addition, to be most effective, such overhead lighting requires a normal-height (rather than a double-height) ceiling of light color.

LIGHT-LEVEL CONTROL

Light-level adjustment is desirable for a wide variety of applications because it enables the operator to optimize his or her seeing conditions. The principal applications are the following:

1. Instruments and Control Panels: The ability of operators to adjust the light levels of a vehicle or of console instruments and/or the illumination of control panel labels is extremely important when they are confronted with changing ambient lighting conditions and/or operating needs; i.e., operators should be able to balance brightness levels among various visual displays, reduce interfering glare when they must maintain lower visual adaptation levels, and/or optimize special display viewing conditions (e.g., display reference scales).

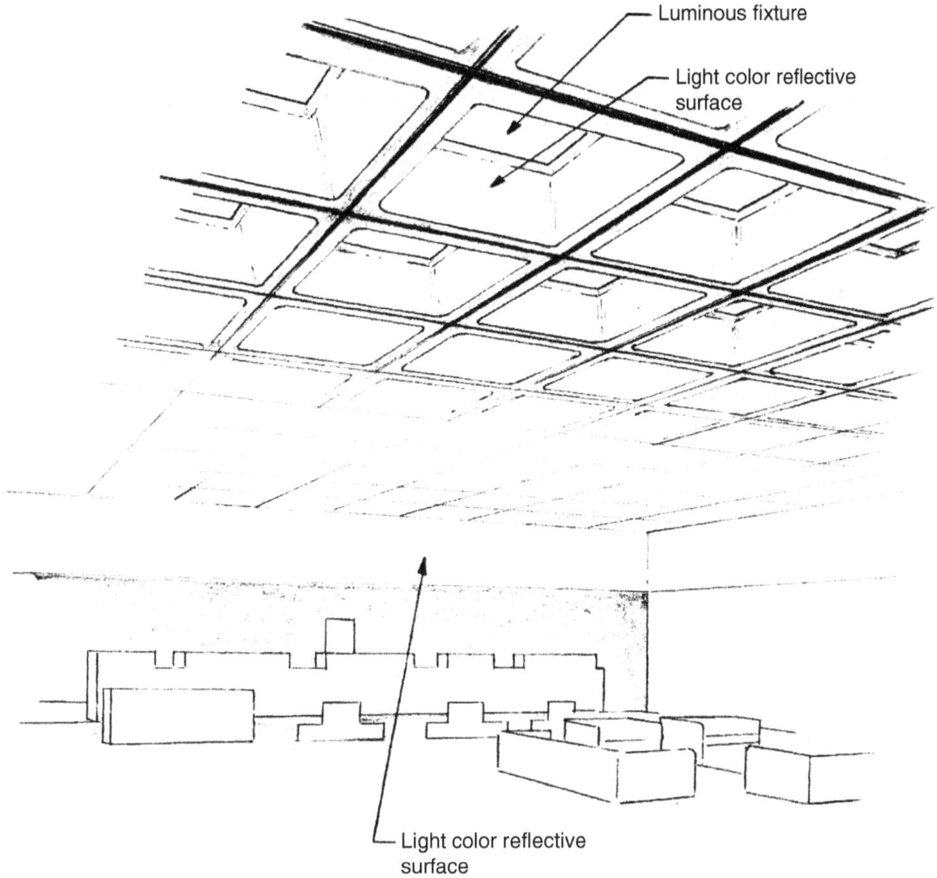

FIGURE 26.6 Lighting system example.

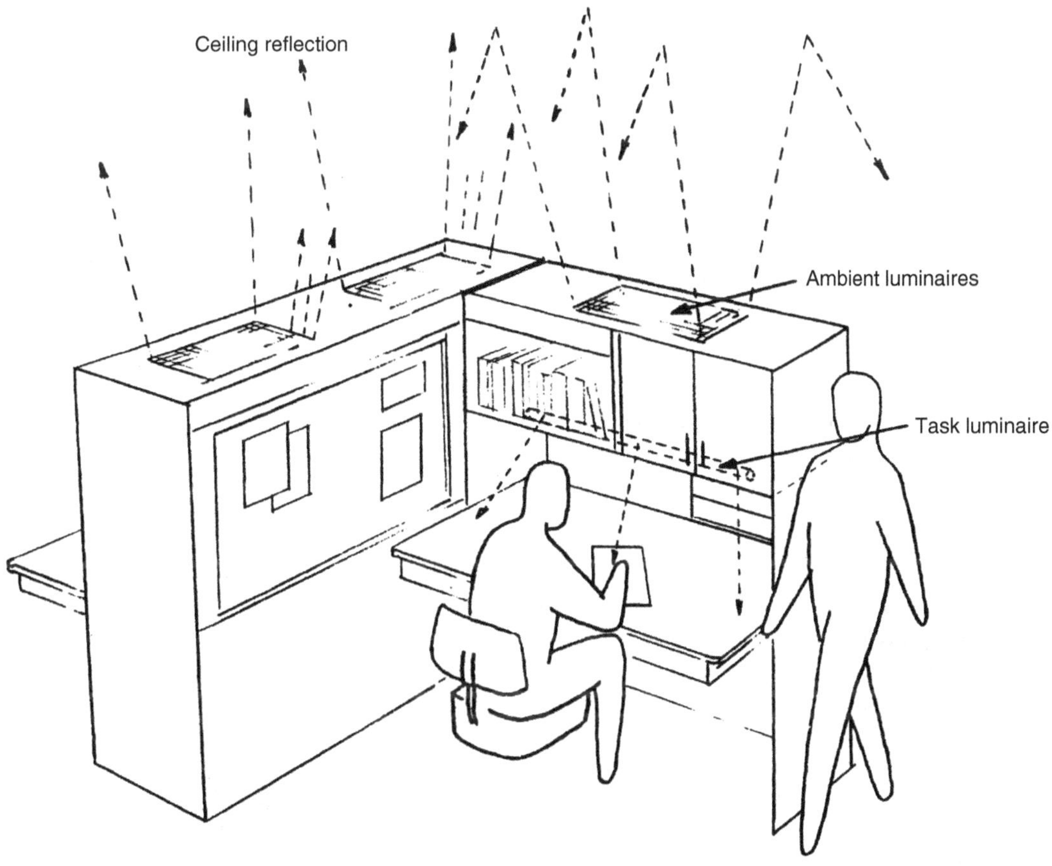

FIGURE 26.7 Combination ambient and task lighting example.

2. Ambient Illumination Level: It is often desirable for operators to be able to adjust ambient room lighting in order to improve their ability to see special visual displays, e.g., to detect targets on a display, to view projected images, or to view a TV program. In addition, the ability to modify the ambient lighting level of various other architectural spaces may be important for psychological reasons, such as when lower lighting is desired for an intimate dining environment.

Note: It is sometimes desirable that a dimming system be designed so that the light cannot be turned completely off because, if a system has been previously turned off, it may create a problem for the succeeding operator or user, who may come into a dark room or sit at a darkened control panel and not be able to find a light switch or recognize obstacles while trying to locate the switch.

ENERGY SAVING

Because of the need to save energy, it often becomes necessary to use fewer lights and/or to guarantee that lights are turned off when they are not being used. Consider the following:

1. Use selective circuit control concepts so that it is possible to turn off various luminaires. Care must be taken in creating such system control capability not to destroy the general acceptability of seeing conditions; i.e., the light distribution (although less intense) should continue to be balanced for a given visual operating area.

2. Provide central control for large-area lighting systems so that the person who leaves last can turn off all remaining lights.

3. Consider the use of automatic light control (timing and sensing systems) so that certain lights are turned on or off to match operating needs. Similar automatic control of special exterior light control shutters should be considered; i.e., as long as natural light can be used, it is foolish to waste artificial light.

4. Technological advances have created more efficient lighting. For example, incandescent bulbs are not being replaced by florescent and LED lamps. Not only are the new lights more efficient, they have a greater variety of color choices, and they last longer.

Using a dimmable luminous ceiling lighting system is a better energy-saving approach than selectively turning out every other luminaire because the latter approach creates light and dark spots, and often a dark spot occurs directly over a desk or other operator position.

The same consideration applies to hallway illumination; i.e., it is better to turn off one lamp within a fixture than to turn off every other luminaire along the hallway.

An automatic "user-presence" sensing system can be used effectively to turn out lights in unoccupied spaces.

GUIDELINES FOR LIGHTING FIXTURES

To provide the best seeing conditions as well as the most pleasant seeing conditions (i.e., minimum annoyance), consider the *luminous ceiling*. The intensity of the light will not have to be as high, and the distribution will be even and free from extreme brightness contrasts. (See Fig. 26.8.)

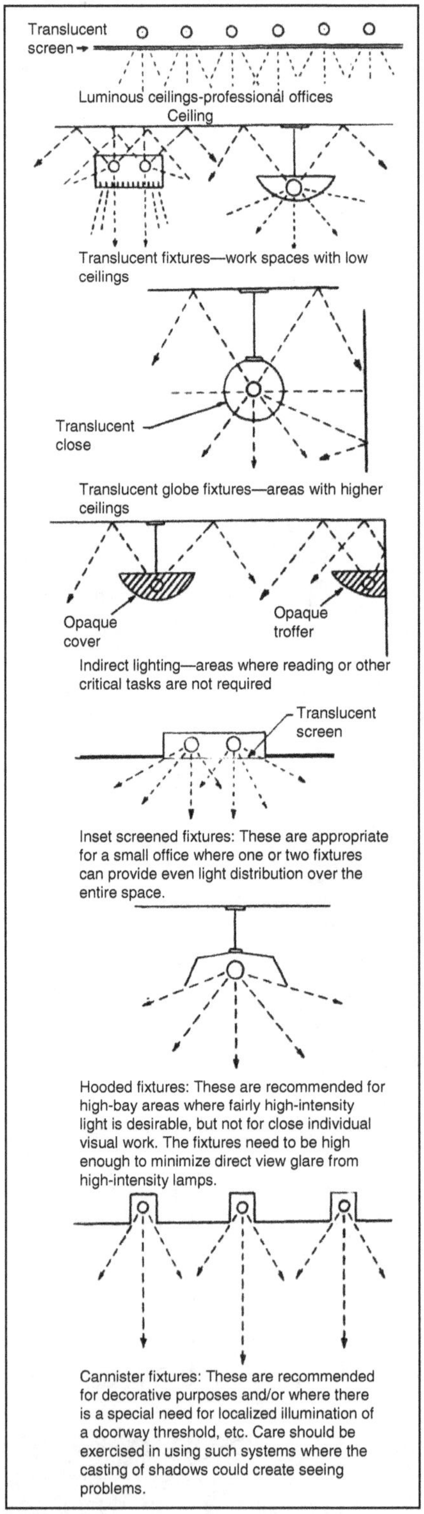

FIGURE 26.8 Fixture examples.

The next best system consists of *translucent fixtures*, which reflect light from the ceiling and also direct light through the fixture screen. The egg-crate configuration has the advantage of not collecting dirt or insects, which eventually cut down the amount of light obtained.

Translucent globe fixtures provide excellent general light distribution and minimize the collection of dirt, which can reduce the lighting effectiveness of the fixture.

Note: Suspended fixtures require sufficient ceiling height in order to prevent direct glare from the fixture.

Indirect lighting systems rely on "bounce light," i.e., the light is never direct and thus annoying to the observer. This type of lighting is useful when there is no requirement for a high degree of brightness on the visual task.

The *inset screened luminaire* is less desirable than previously discussed fixtures because it does not provide an opportunity for the light to reflect from the ceiling and thus improve general, even distribution. It can be effective, however, for "over-the-task" illumination as long as the fixture and task positional relationships remain constant.

The *hooded fixture* provides high efficiency for high-bay factory or sports applications.

The *canister fixture*, although popular for decorative lighting, is perhaps the least desirable luminaire system for good seeing. It tends to create a "spotlighting" condition in which extreme shadow is created whenever anyone walks beneath the light. Although it can be used as a general illumination system, an inordinately large number of fixtures are required in order to provide reasonable light distribution.

In selecting any of the above, one should pay special attention to the manner in which bulb or lamp replacement and filter removal and cleaning can be accomplished.

SUPPLEMENTARY LIGHTING FOR THE INDIVIDUAL WORKPLACE

Floodlighting techniques for individual work place illumination should be designed so that the operator can select or adjust the best position for his or her seeing task; sometimes this changes.

The key consideration is to place the luminaire so that it does not cause light rays to reflect from the work surface into the operator's eyes and so that the light source (lamp) is not directly visible to the operator.

In general, the best lamp position is behind and to the side of the operator. If the operator is right-handed, the lamp should be to his or her left; if the operator is left-handed, it should be to his or her right. Thus it is necessary to provide a flexible positioning system. (See Fig. 26.9.)

Figure 26.10 illustrates a typical flexible-arm luminaire that provides a wide range of positioning possibilities.

For some special tasks, such as engraving, the light source must be placed so that the light rays actually cast shadows because these provide important guidance cues for the worker.

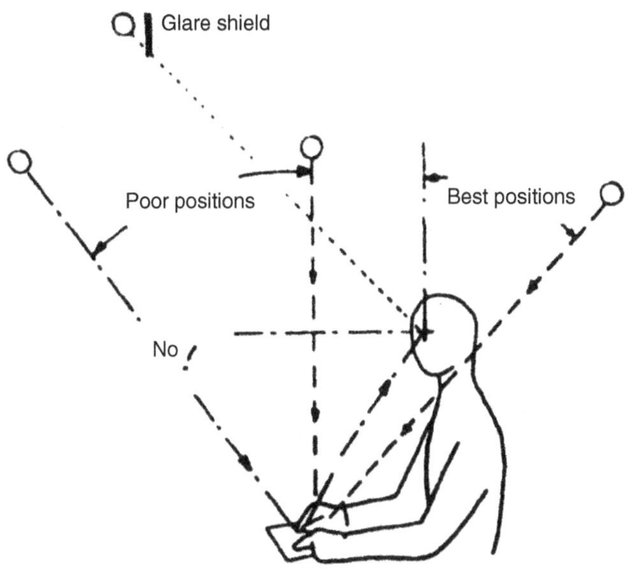

FIGURE 26.9 Light source position.

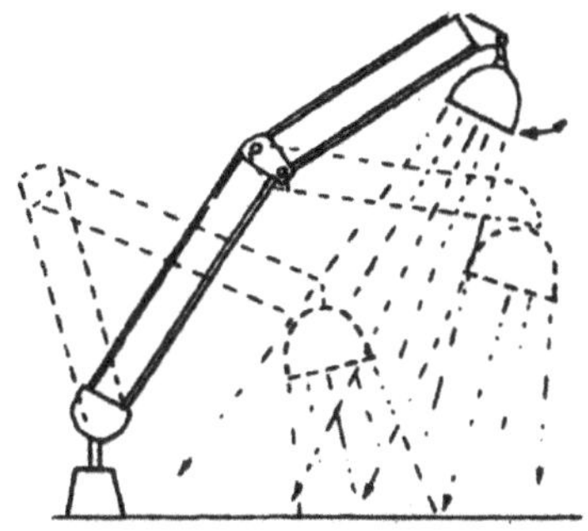

FIGURE 26.10 Wide range of position desirable.

A low position for the light source, as shown in Fig. 26.11, is typical of the special situations in which shadow casting is advantageous.

FIGURE 26.11 Some tasks require low light position.

Illumination for personal grooming tasks requires careful fixture selection and placement. The lamp sources and adjacent surface colors used should provide light that is ample and properly distributed and which is flattering to the complexion. (See Fig. 26.12.)

FIGURE 26.12 Provide ample illumination.

Figure 26.13 illustrates an arrangement of fixtures in a pattern that illuminates both sides of the user's face, plus his or her hair. The illumination level at the person's face should not be less than 50 fc.

FIGURE 26.13 Provide illumination that lights both sides.

The wall lamps shown in Fig. 26.14 are more typical of a decorative approach and tend not to provide as good illumination as one would like. When this type of lighting is used, it is extremely important that the lamps be placed at approximately "face height," or about 28 in. (71 cm) above the seat. Lampshades should be as transparent as possible, and adjacent wall areas should be light in color. Warm colors are required since these are flattering to the complexion.

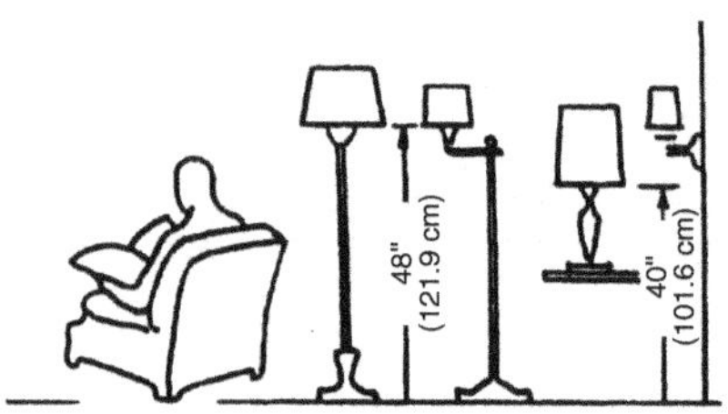

FIGURE 26.14 Lamps should be placed at about 28 in.

Flexible, pull-down fixtures are effective when shades are transparent.

Lampshades for all other lamps should also be transparent.

Various lamps are available that permit adjusting the position of the light for the best illumination.

ILLUMINATING FLOORS, WALKWAYS, AND STAIRS

An overhead floodlighting system is not adequate for illuminating floor, walkway, or stair surfaces at night or in darkened rooms. The light is blocked by the pedestrian's body, causing dangerous shadows.

An entryway such as that shown in Fig. 26.15 should be provided with two types of luminaires: one to illuminate the faces of the guests standing at the threshold of the doorway and the other to illuminate the steps and the porch surface.

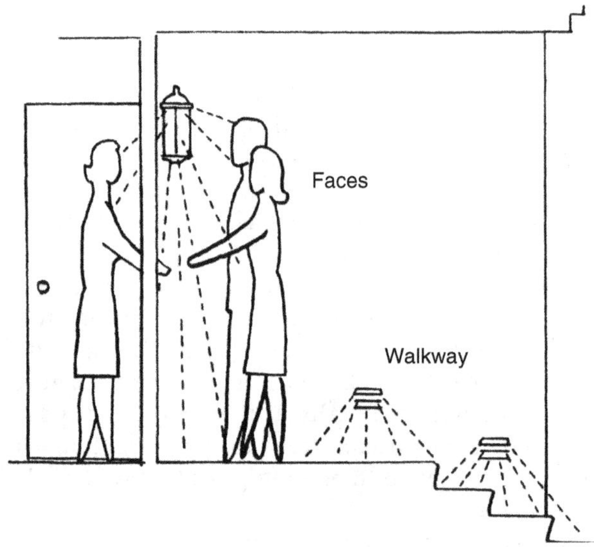

FIGURE 26.15 Porch lighting.

Although standard practice is to provide a minimum of about 3 fc, it is desirable to have at least 10 fc (for those whose eyesight is poor).

Luminaires should be placed so that the light patterns overlap whenever possible. If for some reason this is not possible, it is critical that good illumination be provided at the threshold of the stairs. (See Figs. 26.16 and 26.17.)

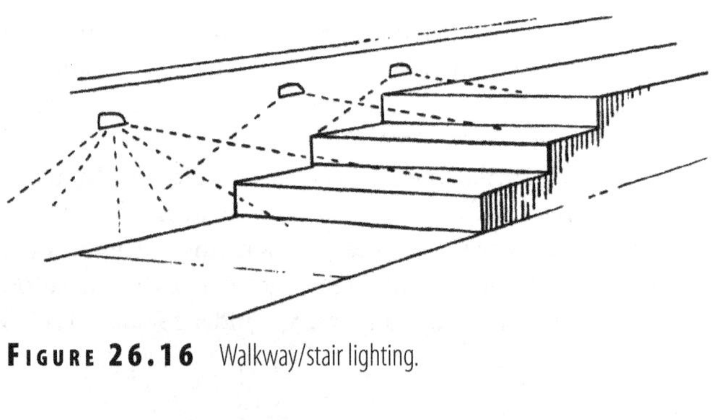

FIGURE 26.16 Walkway/stair lighting.

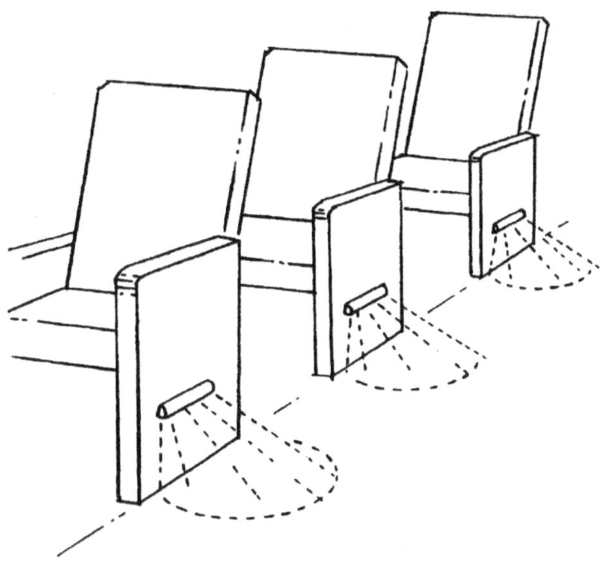

FIGURE 26.17 Theatre floor lighting.

Aircraft passengers should not have to step out of a well-lit interior into the dark night. (See Fig. 26.18.)

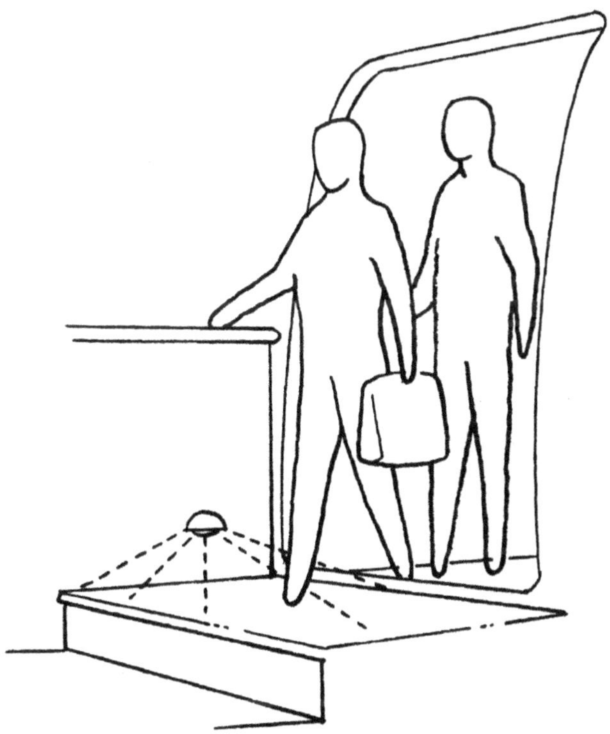

FIGURE 26.18 De-planning passenger should not be required to step from well-lighted cabin into the dark.

LIGHTING GRAPHICS ON A WALL

Place graphics at the approximate eye height of the viewers (H). (See Fig. 26-19.) Then mount lights so that they shine down onto the midline of the graphics at an angle of approximately 30 degrees from the wall. If there are many graphics, it is best to illuminate the entire wall, rather than each one.

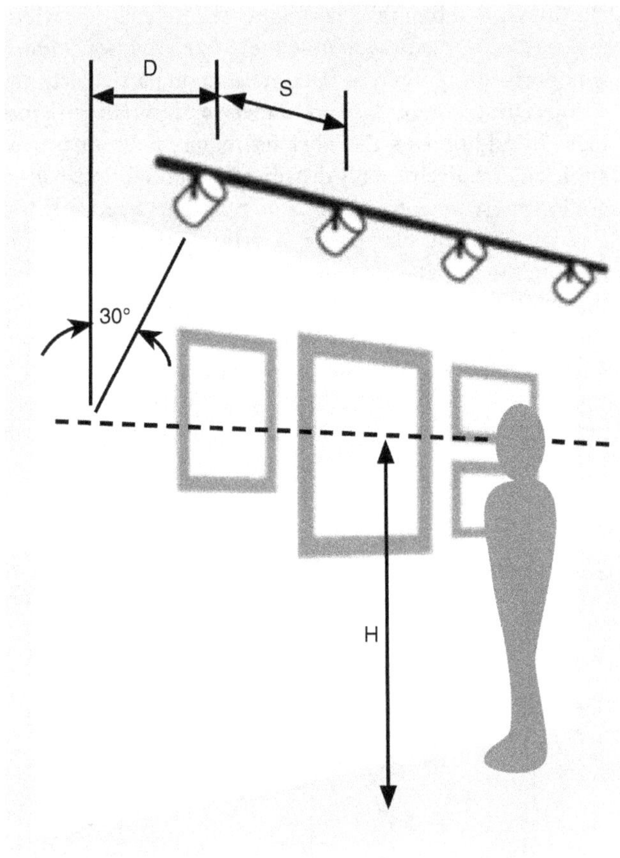

FIGURE 26.19 Lighting graphics on a wall.

The distance between multiple lights (S) should be less than twice the distance from the wall ($S < 2 \times D$).

OUTDOOR LIGHTING

The principal objectives of outdoor floodlighting include the following:

1. Illumination of roadways, walkways, parking areas, runways, and taxiways
2. Illumination of obstacles, including people, vehicles, obstructions, and stairways
3. Illumination of signs and pathway markers
4. Decorative illumination of buildings, sculpture, etc.
5. Illumination of external work and storage areas
6. Illumination for the purpose of crime prevention

Roadway Lighting[20]

Traffic safety requires proper illumination of the roadway, vehicles, and pedestrians. In addition, consideration should be given to unusual conditions that make visibility important, such as road work that is in progress.

[20]For roadway lighting standards, see *IES Lighting Handbook*, Illuminating Engineering Society. New York.

Although automobile headlights theoretically provide a major share of roadway illumination, fixed roadway lighting is important, especially at intersections, on particularly hazardous curves, on bridges, in tunnels, and at railroad crossings. In addition to the obvious requirement to provide sufficient, relatively evenly distributed floodlighting, consideration must be given to the potential interference factors that floodlighting may create. A primary objective of good floodlighting practice should always be to put the light on the object to be seen, not in the eyes of the observer.

As indicated in Fig. 26.20, roadway floodlights should be high enough to minimize glare when the fixture (light source) is visible to the motorist. The position of the streetlight should be such that it will maximally illuminate the pedestrian who is crossing the roadway.

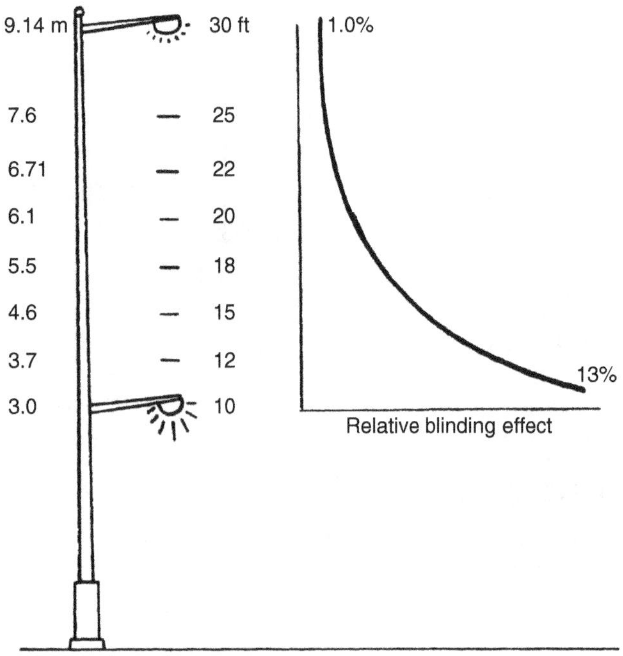

FIGURE 26.20 Lights must be high enough to minimize glare.

Light sources used to illuminate roadway signs must be designed and or positioned so that the light does not shine into the eyes of motorists. Roadway lights must not create confusing shadows across the roadway.

Parking-Lot Illumination

Evenly distributed, nonglaring light should be provided for parking lots. Many good lighting systems are available to the designer, but some are more decorative than practical from the standpoint of effective seeing.

In certain situations, it is critical that floodlights do not interfere with airborne traffic; pilots have enough problems with the many confusing lights around an airport without having a poorly conceived parking-lot lighting system direct light into their eyes while they are trying to land. (See Fig. 26.21.)

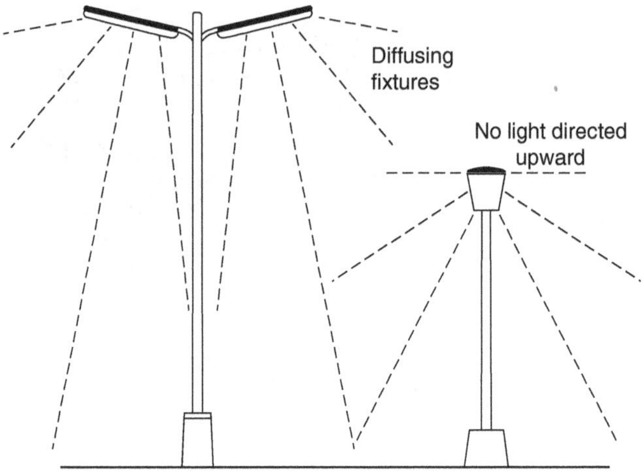

FIGURE 26.21 Parking-lot illumination.

Urban Parkway Lighting

Heavily traveled parkways should be illuminated throughout their peak load areas and especially at multiroad exchange points. (See Fig. 26.22.)

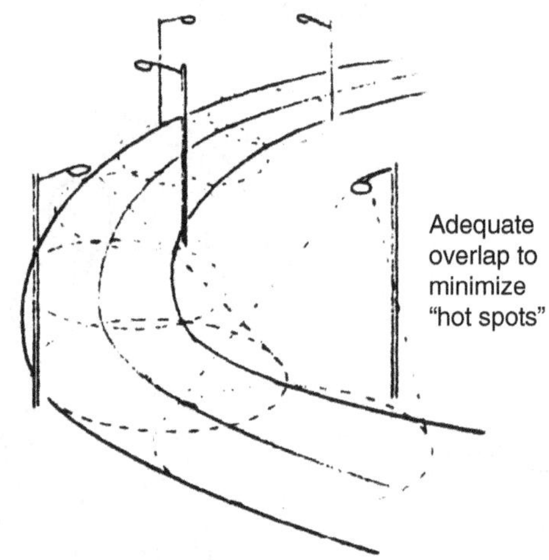

FIGURE 26.22 Parkway lighting.

Decorative lighting of buildings must not shine in the eyes of passing motorists or pedestrians who may be approaching the building. (See Fig. 26.23.)

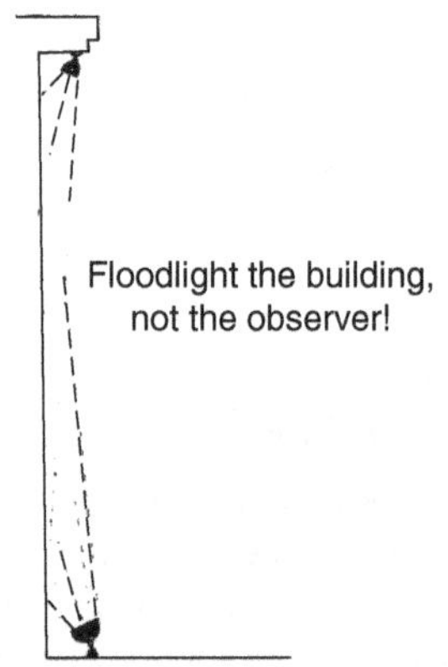

FIGURE 26.23 Building lighting.

Illuminated signs should be designed so that the letters are self-luminous rather than being floodlit; deep shadows make such signs difficult to read. (See Fig. 26.24.)

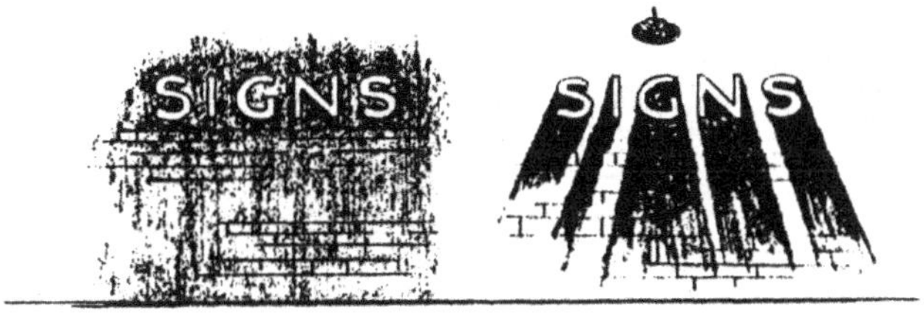

Preferred Undesirable

FIGURE 26.24 Illuminated signs.

Security Lighting

Typical objectives for security lighting systems should be to illuminate the following:

1. All entrances to buildings
2. All gates
3. Inner fenced-in areas
4. Perimeter security fences
5. Immediate areas outside fences
6. Parking lots

Luminaires should be selected and located to provide as even and overlapping illumination as practicable. However, light sources should not be positioned so that they interfere with the visual tasks of persons who are still working at night or of passing motorists or special operators such as pilots, who may be landing or taking off at a nearby airport. (See Fig. 26.25.)

SPECIAL CONSIDERATIONS FOR INTERIOR VEHICLE LIGHTING

General

Vehicles that are operated at night present special problems of reflections on windshields emanating from the lighted instruments, the illuminated structure, the seats, and the passengers. Figures 26.26–26.28 point up some of these problems.

Although each vehicle configuration presents a special case, the best way to control these undesirable reflections is to use the following techniques:

1. Keep illumination levels no higher than is required to provide adequate seeing. (See Fig. 26.26.)

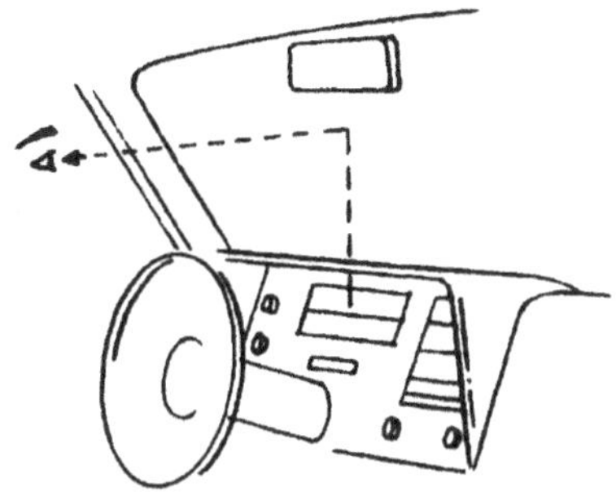

FIGURE 26.26 Keep illumination levels low.

2. Locate luminaires so that they cannot be seen in the windshield.

3. Use glare shields over instrument panels. (See Fig. 26.27.)

4. Recess instruments that are to be illuminated.

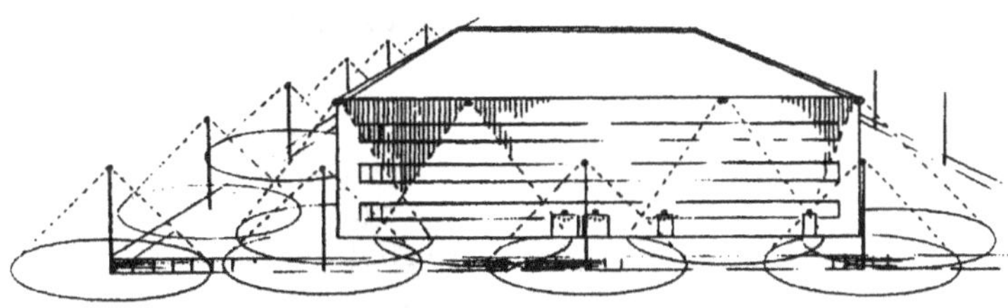

Illumination guidelines for specific points within and around the secured area are as follows:

Building entrances	10 to 30 fc
Areas adjacent to buildings	10 to 15 fc
General inner storage yard	5 to 10 fc
Perimeter fence	3 to 5 fc
Outer parking lot	3 to 10 fc

FIGURE 26.25 Security lighting examples.

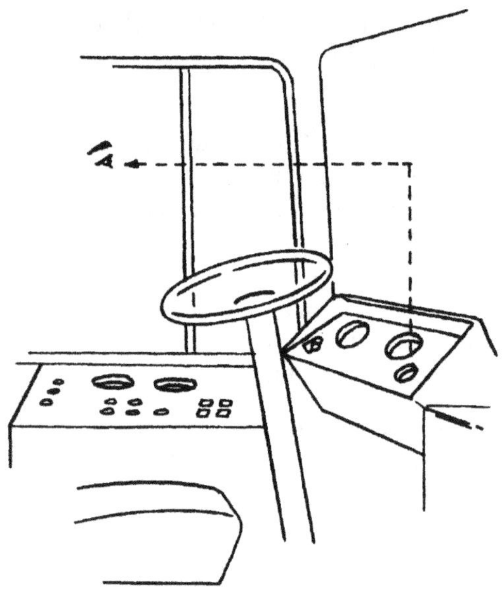

FIGURE 26.27 Use glare shields.

5. Provide curtains between the passenger and driver compartments. (See Fig. 26.28.)

FIGURE 26.28 Provide curtains.

6. Control the slope of the windshield so that reflections (if they cannot be prevented) do not occur on the parts of the windshield that are critical for driving.

7. Use low-reflectance surfaces on structures that could be reflected in the windshield.

8. Provide continuously controllable illumination at the command of the driver.

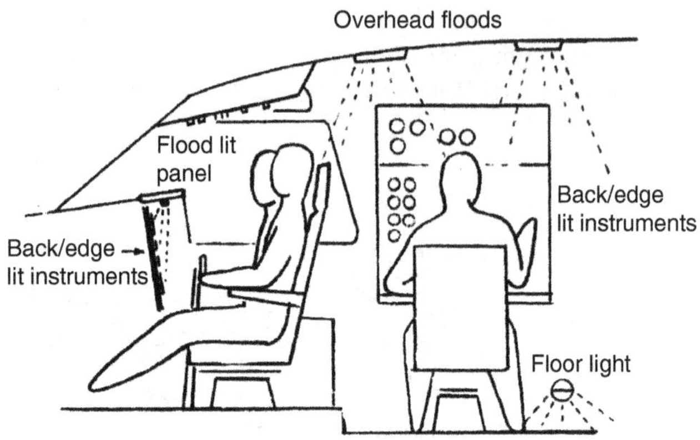

FIGURE 26.29 Flight deck lighting requirements.

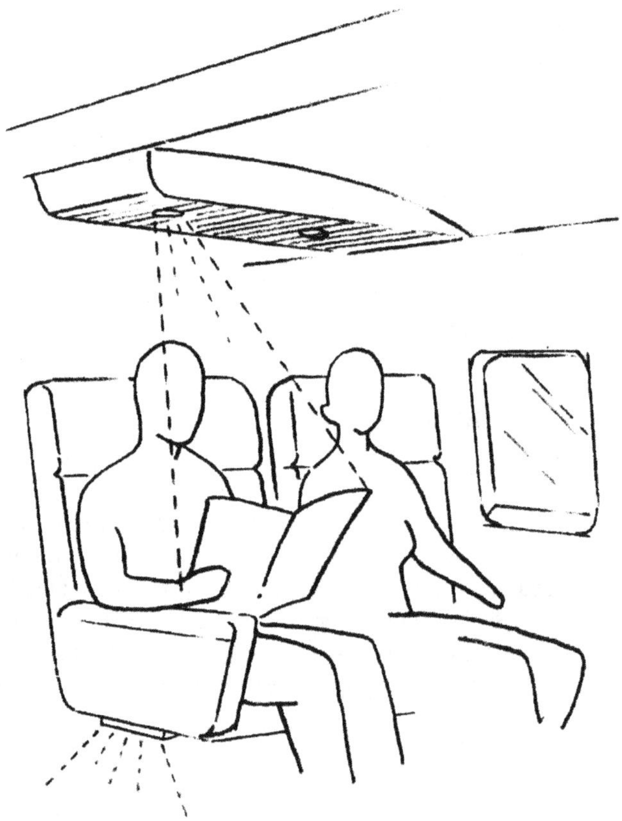

FIGURE 26.30 Provide both aisle and task lighting.

Techniques

Although Figs. 26.29 and 26.30 are of aircraft, the principles apply to all vehicles that are used at night.

For Passengers. Provide both general overhead ambient light for moving in aisles and performing tasks in restrooms. At seats, provide reading lights that are positioned so that passengers can control their own light, i.e., so that one passenger can see to read while his or her fellow passenger sleeps. Provide floor illumination so that passengers can find their way around the cabin even though the overhead lights or reading lights are extinguished. Provide emergency lights that illuminate automatically in the event of a crash (and/or can be turned on by a crew member for noncrash situations) and which are not dependent on the primary electric power source of the vehicle.

For Flight Decks. Provide general overhead lights for nonflight maintenance, internal instrument, and panel lights for normal flight operation, panel floodlights for use during thunderstorms or at other times when the panel lighting is not sufficient, and floor or deck illumination for moving about the compartment. Instrument lighting should be red to maintain dark adaptation. Both red and white instrument panel lights should be continuously controllable by the crew, and a sufficient number of circuits should be provided to allow different parts of the panel to be controlled separately. Instrument lights must not reflect on any critical windscreen or window surface.

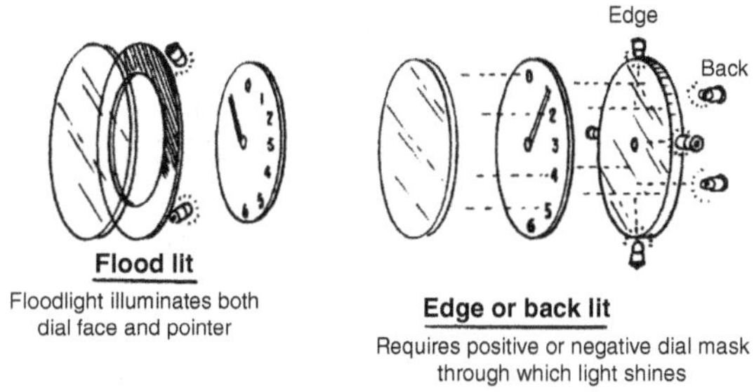

Figure 26.32 Instrument light techniques.

Floodlighting Techniques. Adjustable-position luminaires are preferred over fixed fixtures because they allow the operator to place the fixture in a position that best suits his or her needs. Especially important is the operator's ability to illuminate panel control and display labels without causing light to fall on self-illuminating monitors.

As shown in Fig. 26.31(a), an egg-crate baffle is useful to control the direction of the light rays. It is also possible to inset a monitor so that it is in shadow, as shown in Fig. 26.31(b). The flexible arm support shown in Fig. 26.31(c) allows the operator to minimize shadows that may be caused by control knobs and switches.

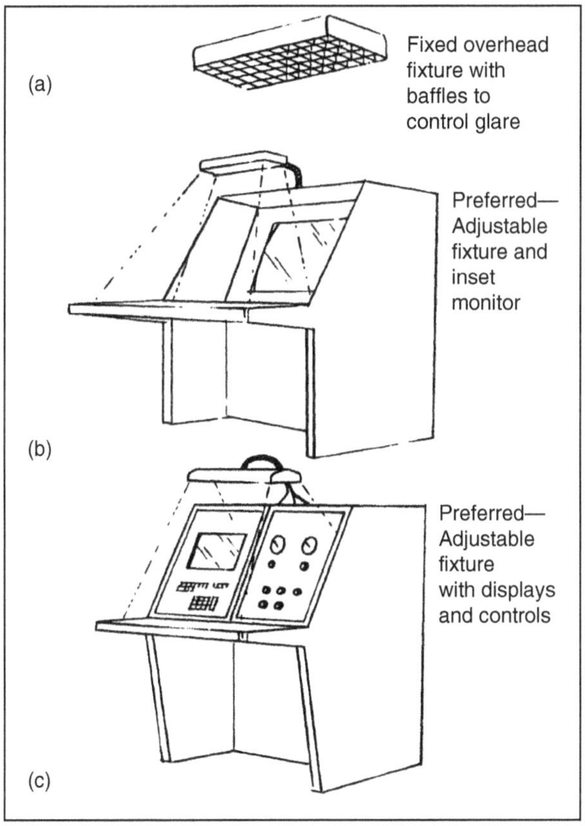

Figure 26.31 Flood lighting techniques.

Care must be taken to make the contrast between labels and panel background sufficient so that the floodlighting will make the labels legible.

When there are several deep-set instruments on the console panel, it is desirable to create an internal display illumination system. Although this can be done by means of a floodlight within the instrument case (d) (in Fig. 26.32, note the baffle, which keeps light sources from being seen by the operator), an edge-lighting or back-lighting technique is preferred.

Transillumination Techniques. A transparent plastic core provides an effective means for transferring light from a lamp to screen print or engraved labels on a panel. The light from the lamp may be introduced either from the back or from the edge of the plastic core, light rays bounce back and forth inside this core and break out wherever the plastic surface has an engraving or contact with another material.

As illustrated in Fig. 26.33, a two-layer sandwich of white, covered with black material, is sealed to the transparent

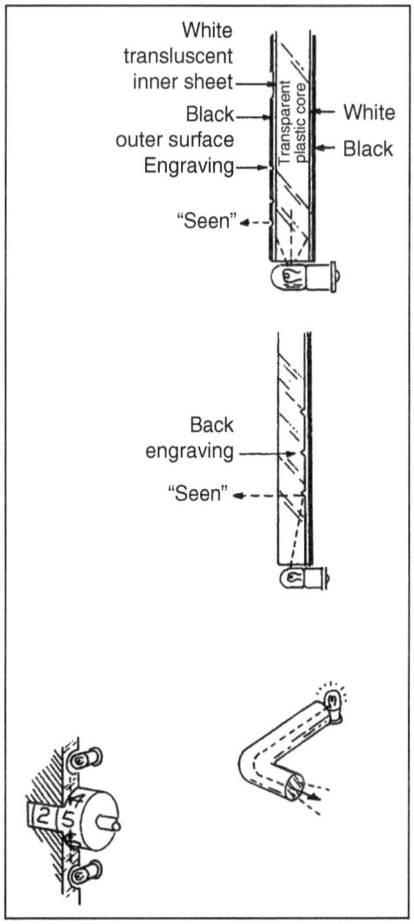

Figure 26.33 Ways to use transparent plastic to illuminate instruments.

light-transmitting core. Labels are engraved through the black outer material (which is translucent to let the light pass). Wherever these engravings occur, the tight will show through, making the character visible both at night, when the panel is lighted, and during the day, as the ambient tight reflects from the white character layer.

Figure 26.33 shows how back-engraved characters can be illuminated by transillumination, leaving the front surface of the core uncovered. In this case, a black material is placed behind the core so that the lighted characters are contrasted against this black background.

Light piping is also a useful technique in which transillumination is employed (as shown in Fig. 26.33). As long as the clear plastic rod is not in contact with anything, the light will travel from one end to the other and break out at the far end when it is in contact with some other material.

Illumination for Military Systems

Table 26.8 gives the minimum illumination levels for various tasks and task locations. When night vision is required, low-level white lighting (with the capability to dim to zero) shall be used.

TABLE 26.8 Specific Task Illumination Requirements

	Illumination Level*	
Work Area for Type of Task	**Preferred Lux (fc)**	**Minimum Lux (fc)**
Assembly, bench, and inspection		
Coarse work detail	540 (50)	325 (30)
Medium work detail	810 (75)	540 (50)
Fine work detail	1075 (100)	810 (75)
Precise work detail	3230 (300)	2155 (200)
Assembly (missile component)	1075 (100)	540 (50)
Bakery	540 (50)	320 (30)
Battery room	300 (28)	200 (18)
Bomb shelters and mobile shelters	20 (2.0)	10 (1.0)
Bridge	540 (50)	325 (30)
Business machine operation	1075 (100)	540 (50)
Cargo handling (weather decks)	215 (20)	150 (14)
Chart room		
General lighting	540 (50)	150 (14)
On chart table	1075 (100)	810 (75)
Circuit diagram	1075 (100)	540 (50)
Control rooms		
General lighting	540 (50)	325 (30)
Computer work	300 (28)	300 (28)
Computer workstation	540 (50)	300 (28)
Consoles (front)	540 (50)	325 (30)
Consoles (rear)	325 (30)	110 (10)
Compressor/pump/generator	300 (28)	200 (18)
Crane cab	540 (50)	400 (37)
Dials/gauges	540 (50)	325 (30)
Electrical equipment testing	540 (50)	325 (30)
Mess room (cafeteria)	540 (50)	300 (28)
Muster station	540 (50)	200 (18)
Medical space		
Waiting area	755 (70)	540 (50)
Exam/treatment room	1075 (100)	810 (75)
Missile facilities		
General inspection	540 (50)	325 (30)
Repair and service	1075 (100)	540 (50)
Storage areas	215 (20)	215 (20)
Offices		
General	755 (70)	540 (50)
Computer workstation	540 (50)	300 (28)
Ordinary visual tasks	540 (50)	325 (30)
Paint room		
General	540 (50)	325 (30)
Color matching	2155 (200)	2155 (200)
Passageways (walkways, hallways)	215 (20)	150 (14)
Pump room	340 (32)	225 (20)
Pump rows, valves	200 (18)	150 (14)
Purifier room (water or oil)	500 (46)	215 (20)
Radar room	540 (50)	200 (18)
Radio Room	540 (50)	540 (50)
Reading		
Large print	325 (30)	110 (10)
News print	540 (50)	325 (30)
Small type (prolonged reading)	755 (70)	540 (50)

(Continued)

TABLE 26.8 Specific Task Illumination Requirements (*Continued*)

Work Area for Type of Task	Illumination Level* Preferred Lux (fc)	Minimum Lux (fc)
Elevators	325 (30)	215 (20)
Emergency lighting	55 (5.0)	30 (3.0)
Emergency generator room	340 (32)	225 (20)
Escape trunks	55 (5.0)	30 (3.0)
Fan room	200 (18)	150 (14)
Fire pump room	300 (28)	200 (18)
Food storage		
Nonrefrigerated	540 (50)	200 (18)
Refrigerated	215 (20)	100 (19)
Galley	755 (70)	540 (50)
Gyro room	300 (28)	200 (18)
HVAC room	300 (28)	200 (18)
Instrument shop	540 (50)	325 (30)
Laundry		
General	540 (50)	540 (50)
Hand pressing	1395 (150)	1395 (150)
Lay down areas	540 (50)	200 (18)
Laboratory	540 (50)	360 (33)
Library	755 (70)	540 (50)
Loading/off-loading areas	320 (32)	150 (14)
Lounges	540 (50)	300 (28)
Manifold area	215 (20)	100 (9.0)
Machine operation (automatic)	540 (50)	325 (30)
Machine shop		
General lighting	540 (50)	325 (30)
Fine bench work	1075 (100)	745 (80)
Maintenance platform	540 (50)	200 (18)
Machinery room	540 (50)	200 (18)

Work Area for Type of Task	Illumination Level Preferred Lux (fc)	Minimum Lux (fc)
Recording	755 (70)	540 (50)
Recreation rooms	540 (50)	300 (28)
Repair rooms		
General lighting	540 (50)	325 (30)
Instrument repair	2155 (200)	1071 (100)
Sanitary spaces		
General	540 (50)	325 (30)
Sinks and mirrors	540 (50)	540 (50)
Toilets	540 (50)	150 (14)
Scales	540 (50)	325 (30)
Screw fastening	540 (50)	325 (30)
Service areas (general)	215 (20)	110 (10)
Shaft alley	215 (20)	110 (10)
Snack and coffee bar	540 (50)	500 (46)
Stairs and ladders	215 (20)	110 (10)
Storage areas		
General warehouse	110 (10)	55 (5.0)
Large parts	110 (10)	55 (5.0)
Small parts	300 (28)	300 (28)
Fine parts	540 (50)	540 (50)
Steering gear room	540 (50)	300 (28)
Switchboards	540 (50)	325 (30)
Tanks	215 (20)	215 (20)
Testing (see Assembly)		
Welding/mechanical shop	500 (46)	500 (46)
Windlass room	215 (20)	100 (9.0)
Work and repair areas	700 (65)	540 (50)

From MIL-STD-1472G.

*As measured at the task object or 76 cm (30 in.) above the floor.

Note: Task lighting inside vehicles shall be designed to accommodate the specific tasks to be performed within the vehicle while still ensuring that the lighting will not distract or impair the driver's performance, nor create an unacceptable visual signature. This may include continuously variable intensity controls.

CHAPTER 27

GUIDELINES FOR SOUND SYSTEMS

The quality of a sound system will be only as good as the quality of the poorest element within the system, e.g., the microphone, amplifier, loudspeaker, or headset. However, the ultimate satisfaction of the listener also depends on the proper handling of the acoustic environment and on the location of the input and output elements of the system. (See Fig. 27.1.)

First, select quality components; the microphone, amplifier, and speakers or headsets should faithfully capture and transmit the speech or music without introducing extraneous noise or distortion.

Second, determine the environmental conditions within which the sound must originate and within which the output must be heard and consider where and how to place and/or use the input and output components.

SIGNAL CHARACTERISTICS

Signal Type	Frequency Range, Hz	Dynamic Range, dB
Minimum speech recognition	75–7,000	50
Voice recognition	50–11,000	60
Music and vocalists	30–12,000	80
Orchestral instruments	15–16,000	125

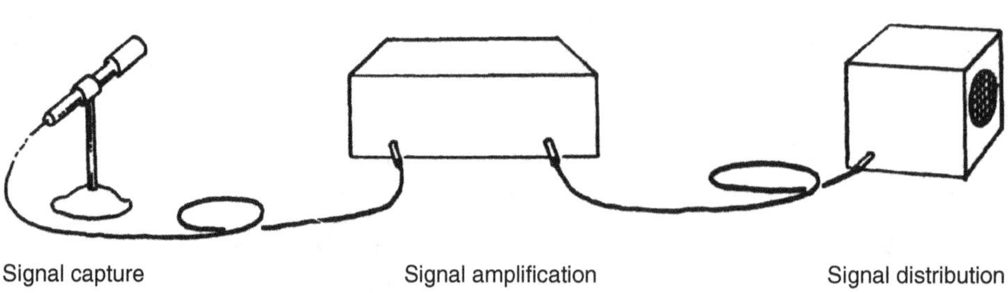

FIGURE 27.1 Signal characteristics.

MICROPHONES

Consider the following:

1. Single versus multiple inputs
2. Mobility of speaker, vocalist, or instrumentalist
3. Speaker's manual constraint (e.g., to hold the microphone)
4. Ambient noise
5. Special speaker constraints (headgear, oxygen mask, etc.)

When special microphone-use situations require that the microphone be contained in a mask, make sure that the mask is designed so that the speech is not distorted by the enclosure; i.e., it should not be muffled or sound as though the speaker is talking in a barrel, and breathing sounds should not be sufficient to mask the speech. (See Fig. 27.2.)

Console operator's hands are so completely occupied that he or she must use a headset microphone placed as close to the user's mouth as practical. If the ambient noise level is high, consider using a highly directional and/or noise-canceling microphone. (See Fig. 27.3.)

Lectern microphones should not be highly directional because speakers tend to wander or turn away. Either mount the microphone on an adjustable arm or use a lapel microphone so that the speaker can move about without increasing the distance between his or her lips and the microphone. (See Fig. 27.4.)

Stage performers generally use tiny head-mounted microphones. Professional performers, such as singers, may prefer to use highly directional microphones (wired or wireless) quite well, keeping it close to their lips and aligned for the best sound capture. It is, however, desirable to provide a microphone stand so that the performer can easily retrieve and/or store the microphone between numbers. (See Fig. 27.5.)

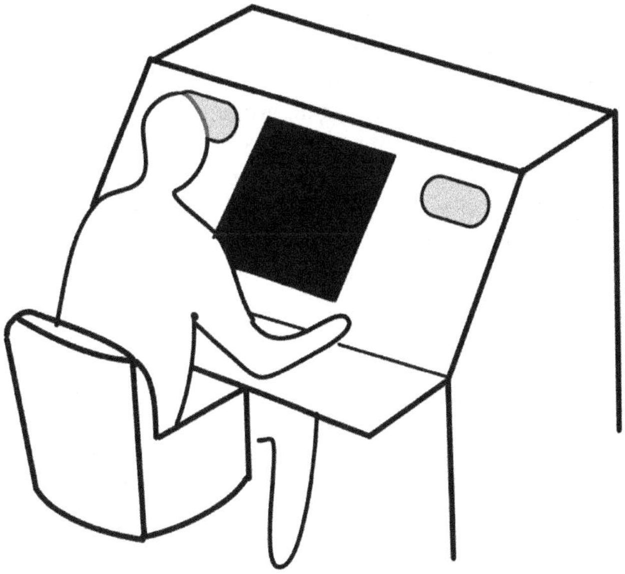

FIGURE 27.3 Hands-free microphone.

FIGURE 27.4 Lectern microphone.

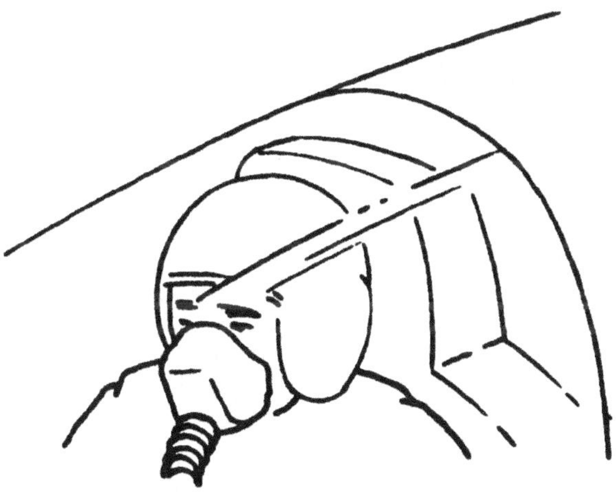

FIGURE 27.2 Microphone in mask.

FIGURE 27.5 Performer's microphone.

Select conference phones with omnidirectional microphones when it is necessary to pick up the voices of several people. Omnidirectional microphones are effective for small groups of people who are sitting around a desk or conference table. The average distance of talkers should be calculated, and more than one microphone should be provided, depending on the effective pick-up distance of the microphone selected. If ambient noise levels are high, however, consider individual noise-canceling microphones. (See Fig. 27.6.)

FIGURE 27.6 Conference microphone.

When vocal or instrumental groups need to have a combination of group and or individual amplification, use a combination of omnidirectional and directional microphones so that the sound control engineer can selectively choose overall and/or individual sounds, i.e., pick up an individual soloist or instrumental choir and yet maintain sufficient accompaniment to provide proper solo-accompaniment balance. (See Fig. 27.7.)

Performances on large stages often require the use of a number of microphones strategically positioned so that the individual voices of actors positioned at different points on the stage can be balanced in terms of average loudness. If the stage is shallow, this usually can be accomplished with a row of microphones arranged across the front of the stage. (See Fig. 27.8.)

When the stage is deep, several microphones usually have to be suspended both laterally and in depth throughout the stage area.

The most desirable approach is to use cordless microphones so that an individual actor will have complete mobility and yet not be able to "walk away" from the microphone.

FIGURE 27.7 Microphone for groups.

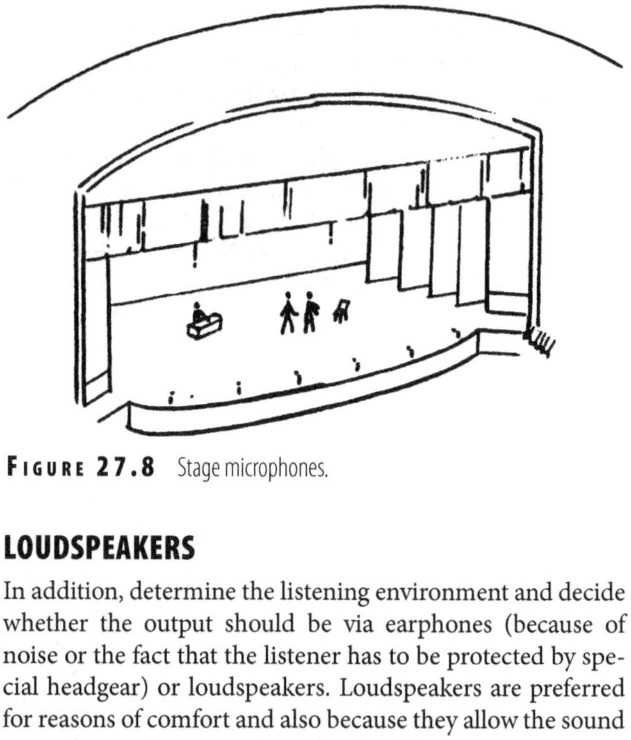

FIGURE 27.8 Stage microphones.

LOUDSPEAKERS

In addition, determine the listening environment and decide whether the output should be via earphones (because of noise or the fact that the listener has to be protected by special headgear) or loudspeakers. Loudspeakers are preferred for reasons of comfort and also because they allow the sound to reach more listeners. Consider the following:

1. As a general rule, loudspeakers should be positioned so that the sound is directed "at" the listener. Although speakers can be positioned so that the sound reflects off hard surfaces before reaching the listener, this should be done only when music is being amplified, since the listener is concerned not with speech intelligibility but

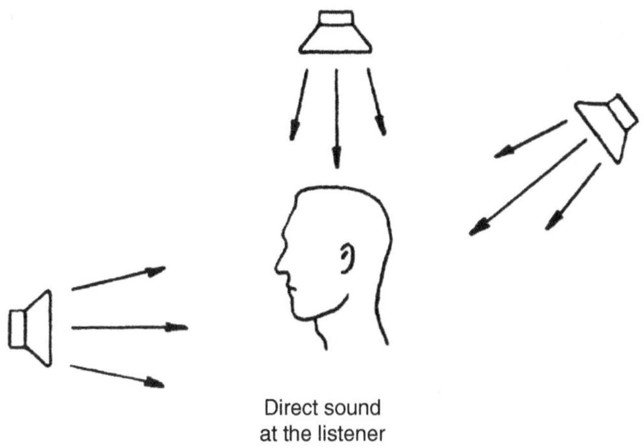

Figure 27.9 Loudspeakers direction.

with blended sounds from an instrumental group. (See Fig. 27.9.)

2. When the sound is intended primarily for an individual listener, place the loudspeaker close to the listener at approximately ear level. Use several loudspeakers when the sound is meant to reach many people at the same time. This reduces the necessity to overdrive individual loudspeakers. (See Fig. 27.10.)

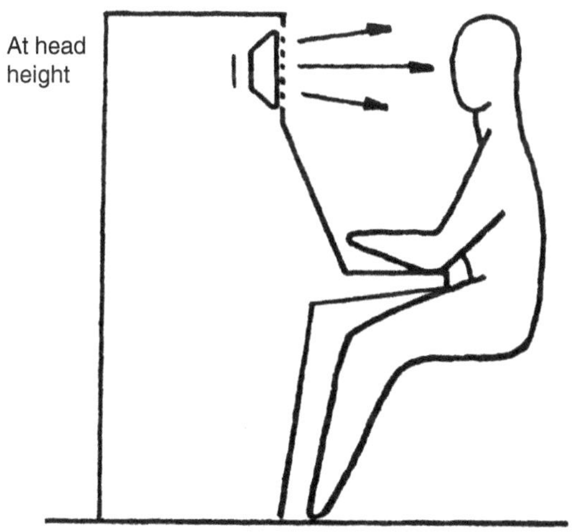

Figure 27.10 Loudspeaker proximity to ear.

3. Several loudspeakers should be used for large stage situations where it is desirable to maintain audio-visual positional integrity with what the listener sees on the stage; i.e., what is happening on the left side of the stage is "miked" and distributed through a speaker on the left side of the stage, and the opposite is accomplished on the other side of the stage. The center speaker should provide enough of both left and right sounds to fulfill the needs of the most laterally distant listeners. (See Fig. 27.11.)

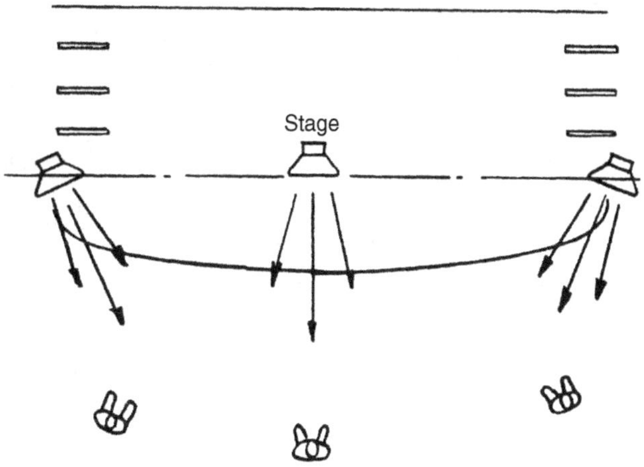

Figure 27.11 Loudspeakers for audiences.

4. Even when the speech of a single lecturer is being amplified by a sound system, it is important that the voice of the lecturer seem to be originating from where he or she is standing. The loudspeaker should be located either in the lectern (directed toward the audience) or above the lecturer's head. Avoid locating the loudspeaker too far to one side and never place it behind the audience. If the auditorium is too large or if the ceiling is too low for the voice of a single speaker to reach all listeners in the back of the room, use speakers (in addition to the front speakers) at several points on the ceiling in order to provide adequate sound levels so that the people in the rear can hear. (See Fig. 27.12.)

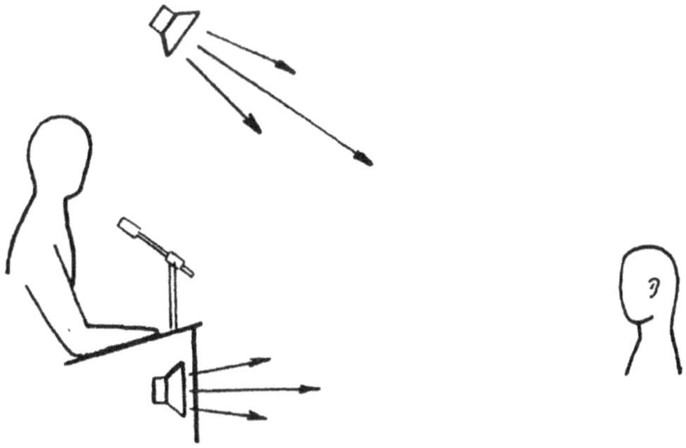

Figure 27.12 Loudspeakers from lecturer.

5. The preferred speaker location in very large auditoriums is shown in Fig. 27.13. Use a number of speakers (some of which should be designed especially for higher frequencies, and others for low frequencies) in order to provide adequate sound levels without introducing distortion.

Sound amplification for motion pictures requires speaker location based on the same principles used for live

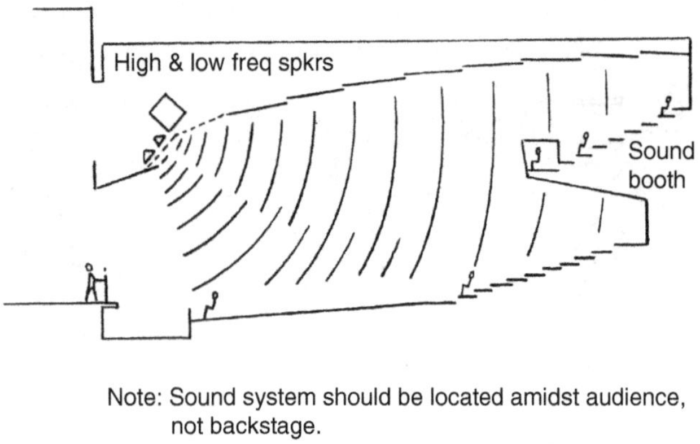

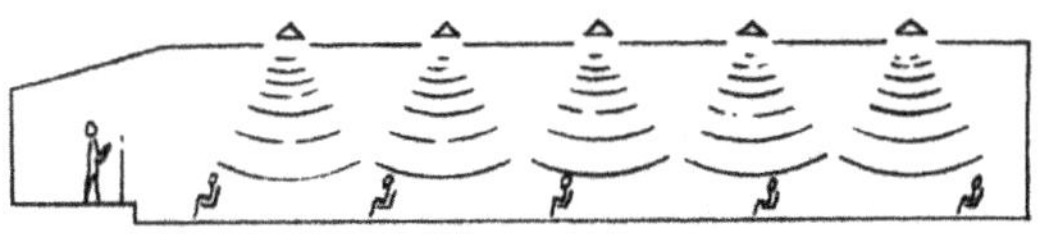

Figure 27.13 Loudspeakers for large auditoriums.

performances; i.e., the sound should seem to come from the visual source (screen area). In the case of smaller screens, for which it is not necessary to create stereophonic relationships between opposite sides of the screen, a single overhead speaker is adequate.

6. For "wide-screen" movies, it is helpful to provide a number of speakers in order to create the audio-visual spatial impact necessary to provide total audio-visual fidelity. The preferred method is to utilize a sound-transparent screen through which the sound is projected. (See Fig. 27.14.)

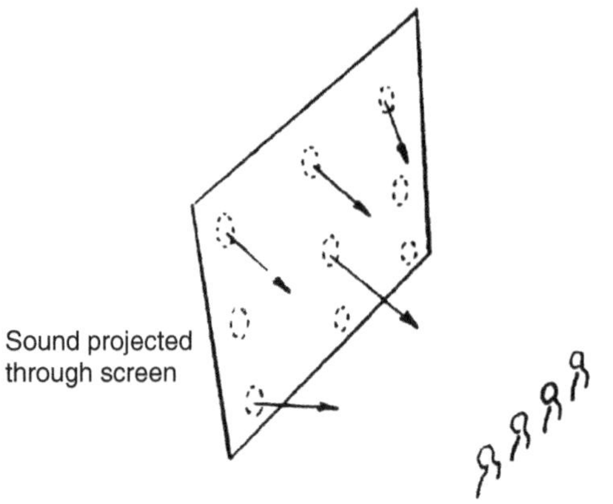

Figure 27.14 Loudspeaker placement to aid visual localization.

7. For the most realistic stereophonic response, loudspeakers should be placed so that the distances A, B, and C shown in Fig. 27.15 are equal.

8. As long as no visual-audio association is required, it is often more dramatic to surround the listener with sound, as shown in Fig. 27.16. If the performance is stereophonic, the left front and left rear speakers are always coupled, and the right front and right rear speakers are coupled.

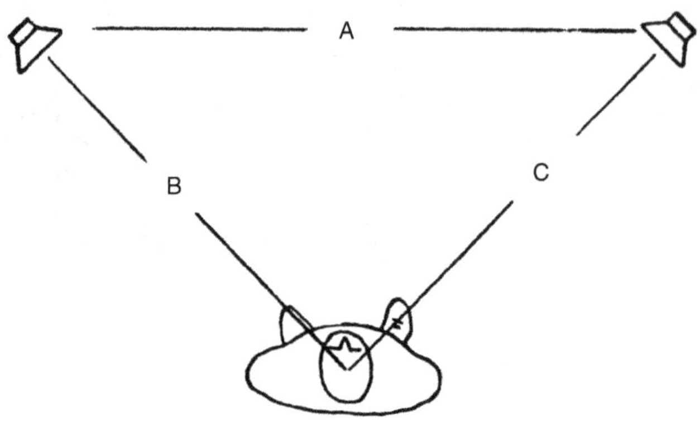

Figure 27.15 Loudspeaker placement for stereophonic effect.

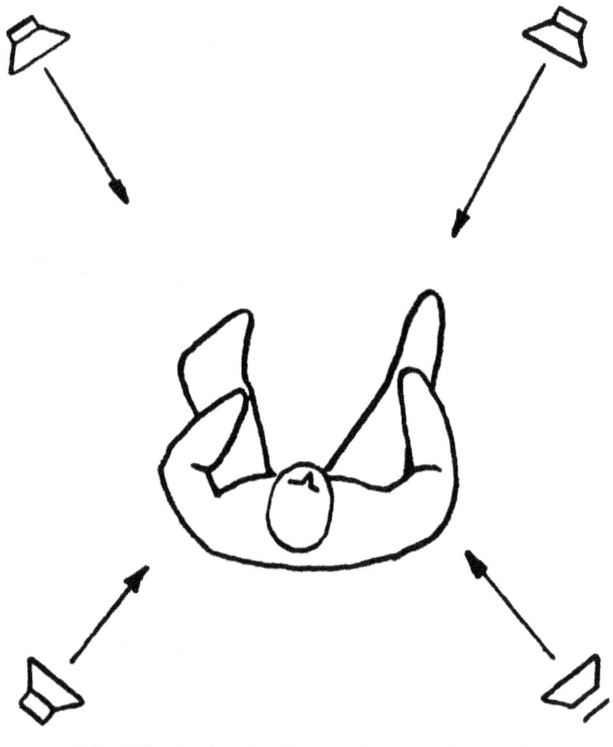

Figure 27.16 Loudspeaker placement for surrounding sound.

Note: Whenever practical, as with individual listeners or small groups of listeners, it is wise to provide controls to adjust stereo relationships.

9. Dynamic sound movement can further enhance the listener's enjoyment, especially in the case of motion picture presentations. As shown in Fig. 27.17, a series of speakers are arranged laterally across the audience's viewing-listening axis, and speakers are successively reinforced from left to right (or fore to aft for nonvisual

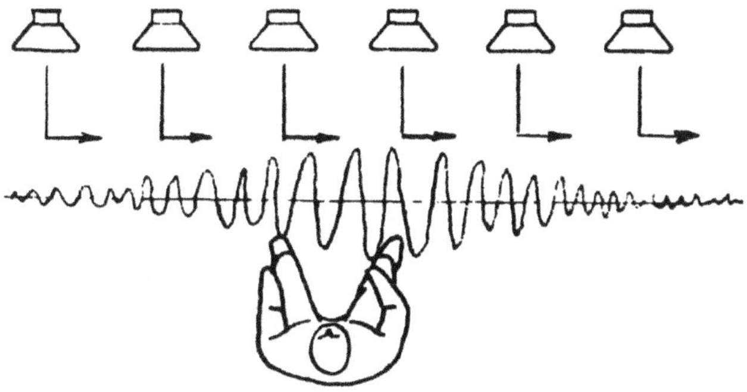

Figure 27.17 Loudspeaker placement for dynamic sound.

sound dynamics). Such dynamic motion simulation can also be used with earphones since the process is accomplished electronically.

10. Multichannel listening via a single loudspeaker is generally less satisfactory than listening by means of several speakers (each channel is piped through its own independent speaker). To be most effective, speakers (and channels) must be arranged at least 10° apart in order for the listener to differentiate reliably (i.e., to localize the sound from an individual speaker).

It is desirable to provide the operator with a separate volume control for each speaker and also to give the operator the alternative of selecting an automatic gain control when signal strength varies significantly. (See Fig. 27.18.)

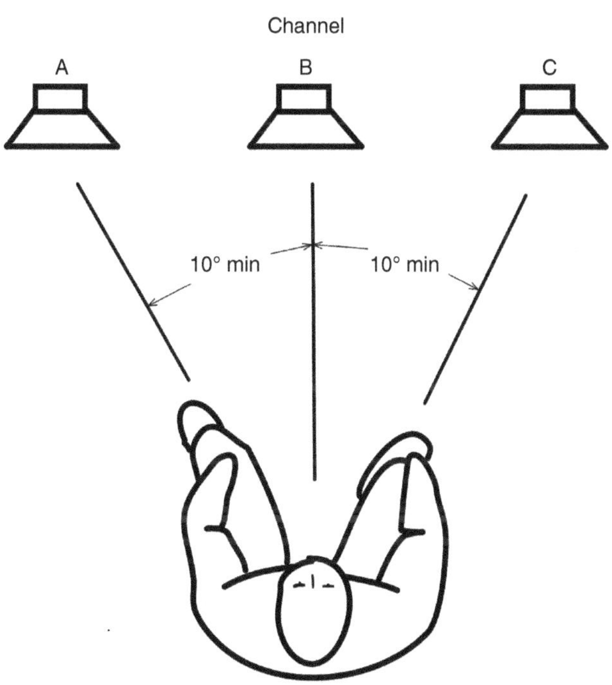

Figure 27.18 Loudspeaker placement for localization.

EARPHONES AND HEADSETS

Selection of the proper headset and or earphones requires careful consideration of the following:

1. Fit is perhaps the most important consideration as far as the user is concerned. The headset should be light in weight and adjustable.

2. Earphones should be cushioned so that they feel soft as they press on the head. The cushion should be large enough so that it contacts the user's head and cheek, but not the outer ear.

3. The earphone and headset support should be adjustable so that the bracket assembly does not have to touch the top of the user's head and so that the tension that holds the earphones snugly against the head will not create too much pressure.

4. The cavity inside the earphone should not be so large that the sound-generating element is too far away from the ear canal.

5. Headsets and earphones designed for maximum ambient noise occlusion include an ear insert. Preferably, this should be custom molded to fit each ear. If not, the ear insert must be extremely soft and pliable so that it can be adjusted for a tight fit without creating undue pressure. Ear inserts must be made of nontoxic materials.

6. Sounds can be transmitted to the ear through head bones.

SOUND SYSTEM OPERATING CONTROLS

Whether the system is complex or simple, it should have a control station that does not require a sound system engineer to operate it. The following considerations relate to the design of the system operator's interface:

1. Provide a separate on/off switch with a light to indicate when the system is powered.

2. Provide a separate control over the primary system features that the nontechnical person would expect: speaker or channel selection, microphone selection, volume adjustment, and bass and treble adjustment.

3. Apply good human engineering principles to panel layout (functional grouping, sequence of operation, proper labeling, etc.).

4. Provide easily read and understood visual displays (meters, telelights, etc.).

5. Use appropriate control types for selection and continuous adjustment functions. Provide feedback to keep the operator continuously advised. Use a computer interface or push buttons, toggle switches, and rotary switches. These are easier and less confusing to operate than slide and thumb-wheel controls.

PART 3

COMPONENT AND PRODUCT DESIGN

INTRODUCTION

Part 3 of this book covers items that contact directly: controls, displays, etc. These are the design details—the components of systems and subsystems. Although a system may be perfectly designed (accommodations for the correct number of properly trained personnel), a system will fail if the components are inoperable. The guidelines in this book will help ensure that humans are able to operate and maintain the detailed system components.

Designing For The Extremes Versus The Average

No phase of system development is more important than knowing whether to design for the average user population or for the extremes of the population. For example, designing a work seat for everyone when the requirements of the task require that each worker be positioned exactly right in order to do the task properly and with minimum stress; designing a display for average visual or auditory capabilities when it is obviously possible that users with subnormal perceptual capacities could make costly mistakes; and creating a tool that people are expected to know how to use when the consumers who will buy the tool have little or no skills relative to tool use.

Behavioral Expectancies

Sometimes a new idea seems to have all the ingredients necessary to improve user efficiency, but because the immediate interface is so radically different from what people expect, we tend to induce operator error and confusion. Test each idea on a sufficiently large and representative sample of the expected user population to demonstrate either that few people seem to have problems or that appropriate indoctrination and training will overcome negative expectancies and that subsequent reversals are not probable.

Expected Payoff

What are the expected payoffs that accrue from being more sensitive to component detail? First, it is generally at the immediate component interface that people make mistakes, many of which may result in product damage or personal injury. Therefore, from the standpoint of product liability, we should try to prevent human error.

Second, it is at the immediate interface that a great many of the confusions or stresses occur that make product users dissatisfied enough with a product not to purchase it ever again. This potential loss of future sales should stimulate the designer to try to provide the most effective interface possible the first time.

Third, in this day of proliferating technology and hardware, making a product available to the largest market is highly competitive. If a product requires special user skill to overcome basic component interface deficiencies, that product may end up only in the hands of a small number of consumers because the rest of the potential buyers are looking for something they can use easily and correctly without having to become experts.

Overview

The topics and examples selected for inclusion in Part 3 represent components and/or product items that most often occur in almost any system or subsystem. For example, there is generally some kind of visual display, auditory display, control device, handle, fastener, tool, or furnishing item required in order for a system or subsystem to be operable. Although these components can be generally listed during concept development and although some of the detailed specifications for the components can be described in sufficient detail to make sure that key operator issues have been properly identified, the specific details at the component level need serious attention to ensure that the final system can take advantage of all the capability of its intended user.

Part 3 provides information dealing with the following component and product elements:

1. Visual displays
2. Auditory displays
3. Controls
4. Fasteners
5. Tools
6. Furniture and appliances

The suggestions and guidelines represent basic human interface requirements and generally apply to any type of system or subsystem; i.e., whether one is concerned with a military, commercial, architectural, or vehicular product or whether the system application deals with specialized populations (e.g., engineers) or general populations (i.e., typical consumers).

It should be noted, however, that many of the suggestions and guidelines are general in nature. Some of the dimensional guidelines are merely beginning references. The reader is cautioned to examine each suggestion in light of his or her specific operating need, to use common sense in applying the suggestions, and, most importantly, to test and evaluate each specific proposed design in a simulated operational setting, using a representative number of expected user subjects to confirm that the design is actually compatible under use conditions.

POINTS TO CONSIDER

In applying human factors engineering to detail designs, the following points are useful to consider.

1. Humans interact with items in three broad modes: intellectually, perceptually, and physically. In assessing the human–component interactions, it is often helpful to look at these broad categories of task demands. Physically demanding components may require that workers are properly rested, adequate in size and strength, and have unrestricted movement. Intellectual tasks often

require uninterrupted focus and proper references and training. Perceptually intensive tasks may depend on careful lighting and noise control. Think of various system components (keyboards, hammers, chairs, etc.) and general types of demands (physical, intellectual, and perceptive) associated with each of them.

2. Humans rarely interface with one item at a time—a keyboard is used with a desk, a monitor, a light, and a chair, for example. This means the chair, desk, light, monitor, and keyboard all have to be located to be used together and accommodate the size of the user. A perfect chair for a living room will probably not work. Components cannot be designed in isolation. This is why task analysis (see Chap. 56) is such an important part of the application of human factors engineering. Think of one component. Now think of all the other components that a person might be using with that item (glasses, gloves, clothing, and lights). Do they affect the human factors considerations and the design?

3. Human factors engineering is concerned with usability—however, it may also focus on nonusability. For example, stove controls and pill bottles should not be usable by small children. Protective measures guard against inadvertently starting power tools. Levers and handles must be designed to prevent the user from putting equipment in an unsafe mode. And controls must be strong enough not to break if the user tries excessive force. Hot surfaces, sharp edges, and moving parts all need user protections. Review all tasks and designs to ensure that usability does not lead to safety hazards.

Chapter 28

Visual Displays

GENERAL GUIDELINES (SEE TABLE 28.1)

1. Use the simplest display concept compatible with the information transfer needs of the operator or observer. The more complex the display, the more time it takes to read and interpret the information provided by the display, and the more apt the observer or operator is to misinterpret the information or fail to use it correctly. Consideration should also be given to creating displays that are easy to read and interpret, as opposed to displays that are merely attractive and/or visually engaging. Avoid the temptation to make displays decorative at the expense of their being readable and interpretable.

2. Use the least precise display format that is compatible with the readout accuracy actually required and/or the true accuracy that can be generated by the display-generating equipment. Requiring operators to be more precise than necessary only increases their response time, adds to their fatigue or mental stress, and ultimately causes them to make unnecessary errors.

3. Use the most natural or expected display format sufficient for the type of information or interpretive response requirements. Unfamiliar formats require additional time to become accustomed to them, and they encourage errors in reading and interpretation as a result of unfamiliarity and interference with habit patterns. When new and unusual formats seem to be needed, consider experimental tests to determine whether such formats are compatible with basic operator capabilities and limitations and/or whether the new format does in fact result in the required performance level.

4. Use the most effective display technique for the expected viewing environment and operator viewing conditions (lighting, acceleration, vibration, operator position, mobility restrictions, etc.). Match the display technique to the operator's constraints; do not make the operator match the display.

5. Optimize the following display features:

   a. Visibility (the viewing distance in relation to size, viewing angle, absence of parallax and visual

TABLE 28.1 Display Design Checklist

Overall Configuration	
Display Identifying Features—Displays have identifying features (such as location, size, shape, and color) that allow the operator to correctly navigate, locate, and identify the display in a timely manner.	*Display characteristics need to make displays easy to identify to prevent confusion and loss of situational awareness.*
Display Area—The system provides sufficient display area to present all critical task information within the operator's visual field that is covered by eye movement only.	*To ensure that critical tasks can be performed quickly, easily, and accurately, especially during critical mission phases, it will be important to avoid scrolling or switching among several display pages, and to avoid excessive head or body movement by the crewmember to view several displays.*
Display Content	
Display Interpretation—Displays are accurately interpretable within the time required to meet mission needs.	*Displays must be interpreted efficiently. That is, the information displayed, information organization, and other elements support easy identification and interpretation.*
Visual Display Legibility—Displays are legible in the viewing conditions expected during task performance.	*Legibility includes both text elements and meaningful graphic elements such as symbols, icons, and maps. Legibility will depend upon display properties, such as resolution and contrast, as well as text properties, such as font contrast, color, and size, background color and texture, as well the environment. Significant environmental factors include viewing angles and distances, ambient illumination (level, glare, reflections), and movement (vibration) of the reader and the display. The visual capabilities of the reader must also be considered.*
Display Readability—Displays are readable.	*The characters, symbols, and language in the display must be familiar to the reader.*
Display Information—Information displayed is sufficient to allow the operator to make decisions and perform the intended actions.	*Displays must contain the information necessary to perform the task at hand. Too little or too much information will decrease efficiency or will make task completion impossible.*
Display Information Relevancy—Information displayed is relevant to the current task.	*Operators must have access to all necessary and sufficient information to accomplish their task. If there is information displayed that is not important for the current task, it may slow down task completion by causing confusion.*
Display Information Flow—Information displayed follows a logical flow (e.g., from top to bottom, left to right).	*Design requirements need to ensure that the information displayed supports the task flow; it is intuitive and easy to follow. This will ensure that the task is accomplished in a timely manner.*
Display Navigation—Navigation within and between displays is intuitive.	*Design requirements need to ensure that the operator can navigate within and between displays without loss of situational awareness and in a timely manner. Unnecessary steps in navigation may increase task time and may reduce situational awareness of the operator. In general, it is recommended to have a shallow navigational structure for navigation instead of a deep structure to make navigation more transparent.*
Display Nomenclature—The nomenclature for each item or process is self-explanatory and directs the operator to the function or usage of the item.	*Design requirements need to ensure that item and process names are easy to understand and to remember. This limits the time spent on recognizing and understanding names. Also, this limits the training needed to understand the nomenclature of the items.*
Display Coding Redundancy—For critical information and critical tasks, when color is used to convey meaning, the system provides an additional cue.	*Design requirements need to ensure that in case of critical tasks there is more than one way to identify relevant items. If for some reason (lighting, distance, or other factor) one characteristic is difficult to discern, there is still another redundant cue to help with identification.*
Measurement Units—Units of measure are displayed with their corresponding values.	*Design requirements need to ensure that units can be identified with the correct magnitude and scale. This ensures correct decision making when comparing or using these units in some other way.*

From NASA-STD-3001, Volume 2.

occlusion, visual contrast, minimal interference from glare, and adequate illumination): Consideration should be given to possible visual anomalies of the expected user population, i.e., lack of normal visual acuity, color deficiencies, and nearsightedness and farsightedness. These user limitations should be considered not only in terms of selecting the proper display concept but also in terms of executing the specific features that make up the display.

b. Conspicuousness (the ability to attract attention and distinguishability from background interference and distraction): Design or select dynamic displays (instruments that have moving elements, warning indicators, etc.) that let the observer know if the display is malfunctioning. If it is not obvious (e.g., a pointer that suddenly drops to zero when it should be indicating some value), consider a special indication of malfunctioning, such as a warning light or flag.

c. Legibility (pattern discrimination, color and brightness contrast, size, shape, distortion, and illusory aspects).

d. Interpretability (meaningfulness to the intended observer within the viewing environment; requirements for interpretation, extrapolation, special learning, and training; and general reliability in terms of retention of meaning).

VISUAL DISPLAY SELECTION (SEE TABLE 28.2)

USE OF LIGHTS TO DISPLAY INFORMATION

Changes in display status shall signify changes in functional status rather than results of control actuation alone. The absence or loss of a signal or visual indication must not be used to denote a "malfunction," "no-go," or "out-of-tolerance" condition, or to indicate a "ready" or "in tolerance" condition. This is only acceptable if the status or caution light filament and its associated circuitry can be easily tested by the user and user perception of such events is not time critical (MIL-STD-1472G, 2012).

Simple Indicator Lights

Simple indicator lights are addressed as follows, with examples of how simple light or sets of lights can be used to convey information by the fact that the light is on or off, by the color of the light, by the position and/or shape of the light, and/or by whether the light is steading or flashing. *(Legend lights are displays that combine lights and letters, numbers, or symbols. Legend lights are usually preferred over simple indicator lights because they display more information.)*

Power ON Lights. These are used to indicate that electrical power is available to the equipment. (See Fig. 28.1.) Such indicators should be limited to one per piece of equipment or workstation since several lights will confuse the operator unless each pilot light is labeled. Pilot lights should not be brighter than necessary to determine that they are either lit or not lit (depending on the prevailing ambient illumination). Otherwise, they may interfere with other visual tasks since they tend to cause glare if the general ambient light conditions are fairly dim.

Function Active Lights. These are used to indicate that a particular function is occurring, as in the case of the handheld recorder shown in Fig. 28.2; i.e., the pilot light advises the operator that the recorder is in the "dictate mode" and that the tape reel is running.

Special Hazard Condition Lights. Some equipment, products, or tools present potential hazards that the operator may not be aware of unless a warning light is provided. In Fig. 28.3, the soldering iron will be hot when power is fed to the soldering element and the operator forgets and becomes careless about touching the soldering element and/or lays the tool down before it has cooled. The pilot light is helpful in this case to remind the operator to be careful.

Operating Mode Lights. A single push-button control switch is often used on some equipment in order to reduce the number of switches. This is practical because the push button can be back-illuminated, not only with a single light, but, as in the case of the copying machine, also with both a white and a red back-lighting source; the red light indicates that the machine is in a warm-up mode, and the white light indicates that the machine is ready for copying. (See Fig. 28.4.)

Localizer Lights. Simple pilot lights can be effectively used to help identity one of a series of systems or devices. The first sketch in Fig. 28.5 illustrates how a pilot light could be activated to indicate to the operator which speaker he or she is hearing. The second sketch shows a similar application relative to identifying, say, a hatch that is still open prior to a submarine dive. The possible applications of this positional light cue are many, and the designer has only to remember that the light indicators and the devices they represent are spatially related.

Malfunction Lights. A simple light can be used as an initial cue that some malfunction has occurred or is about to occur. If there is more than one such light at the same workstation, however, appropriate labeling is required to ensure that the operator does not become confused. In some situations it may be more appropriate to use only the malfunction light, as opposed to pilot lights that indicate that everything is functioning properly. This approach is particularly relevant if there is a requirement to keep the general ambient light level very low so that operators are able to maintain their dark adaptation. A particularly good example of the use of a malfunction caution or warning light is shown in Fig. 28.6. The light of the automobile gas gauge comes on to tell the driver that the vehicle is almost out of gas. This warning light may prevent the driver who does not watch the gauges on a regular basis from running out of gas.

TABLE 28.2 Visual Display Selection

Display Requirements	Type of Display	Rationale	Example
Go/no-go, start/stop, ON/OFF	Light Mechanical flag	Easy to tell when the light is on or off Easily detected; economical space use	
Identification	Light or lights Symbol or symbols	Easy to detect; can be coded by means of location, number, color, repetition, or on-off duration	
Instructional	Labels and printed words Arrows Pictorial symbols	Single word or multiword message clarifies meaning; arrow implies direction; pictorial illustrates feature	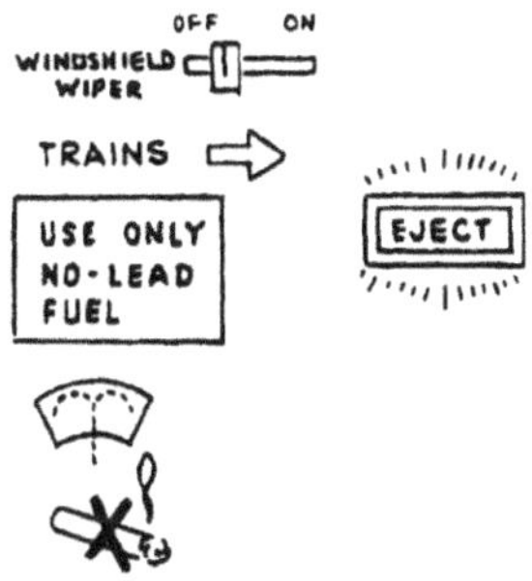
Exact quantity	Digital readout Printer	Single number value requires no further interpolation or extrapolation	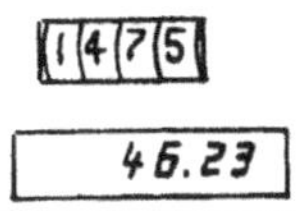
Approximate quantity or positional relationship	Moving pointer and fixed scale with numbers; moving scale and numbers with fixed pointer	General pointer orientation gives quick cue to scale-pointer relationship; rate of change; reference to scale limits	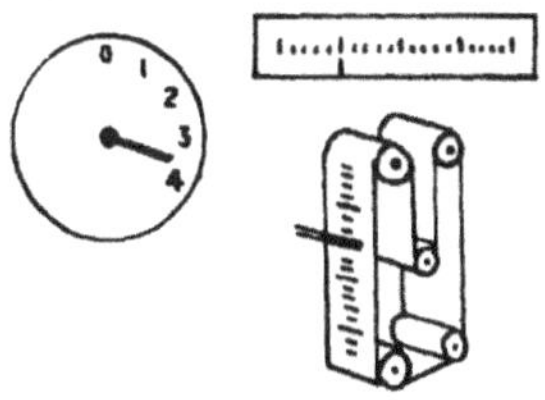
Data entry	Key set Pointer knob Counters	Alphanumeric keys provide immediate data selection reference Easy to position pointer adjacent to desired scale mark or number or label Most accurate display of numerical setting	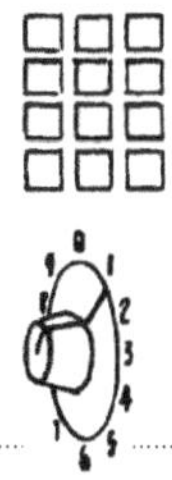
Tracking and hooking targets	CRT with associated data input and pick-off controller (joy stick, mouse, track ball, pantograph, etc.); indexing marks introduced on CRT with continuous capability for repositioning marker and adding alphanumerics or reference symbology	Provides convenient means for "pointing" to target and adding orientation reference marks or alphanumerics	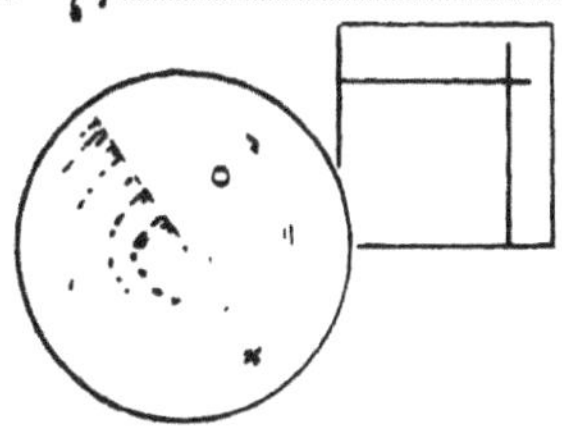

(*Continued*)

TABLE 28.2 Visual Display Selection (*Continued*)

DISPLAY REQUIREMENTS	TYPE OF DISPLAY	RATIONALE	EXAMPLE
Geographic position	Plan position reference analog: map or electronic map and grid display	Shows direct comparison of operator's position with geographic area and features; can show superimposed targets of interest	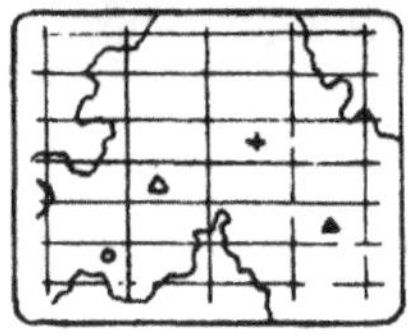 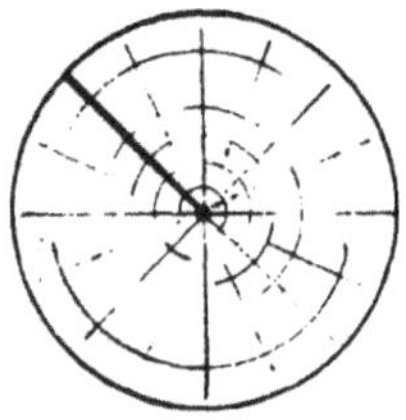
Aircraft attitude	Electromechanical or electronic display of position of aircraft relative to pitch and roll	Shows direct comparison of "own vehicle position" relative to horizon	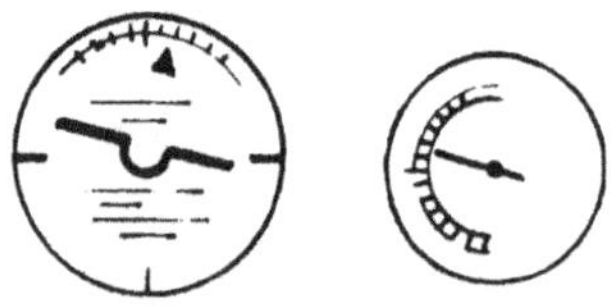
Command guidance	Electronic or mechanical analog of predicted position, path, and energy consumption; surrogate picture of external view (heads-up display)	Allows observer to anticipate what will occur (e.g., predicts) and provides substitution for external view when conditions limit direct view	
Equipment performance analysis	Meters CRT Pen recorder	Single parameter is easy to interpret Multiparameter relationships are easy to see Permanent record for later analysis	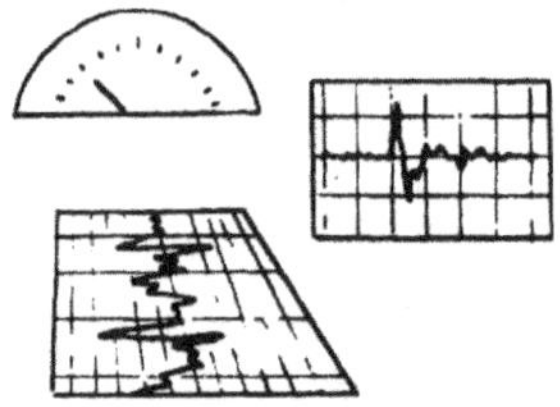
Surface traffic control	Lights, signals, signs, and markers	Multiple observer usage requires displays on or juxtapositioned with roadways, runways, and taxiways	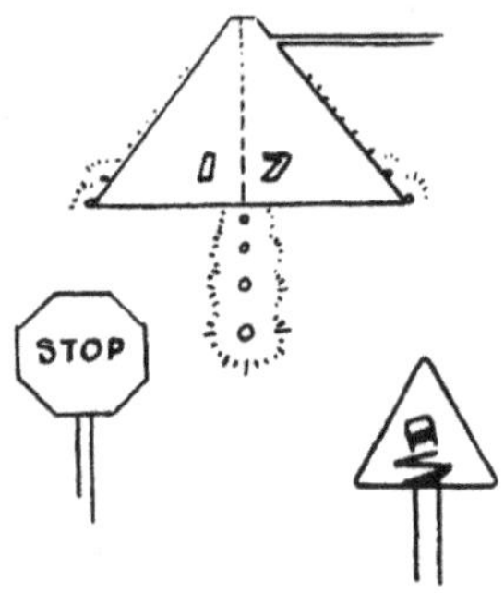

(*Continued*)

TABLE 28.2 Visual Display Selection (*Continued*)

Display Requirements	Type of Display	Rationale	Example
Large-group monitoring and signboard and scoreboard displays	Static, painted, back-lighted, or multilight signs (lights effective only at night)	Least expensive to install and maintain; permanent	 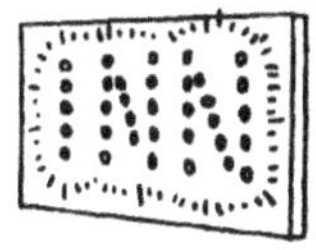
	Slide projection (front and rear: effective only in darkened room)	Static information: temporary exposure of successive pictorial or data items	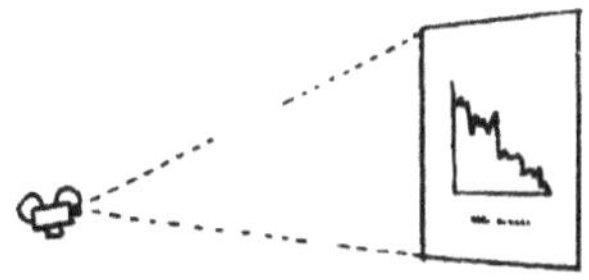
	Matrix type, moving message, or pictorial sign (multilight)	Effective at night for advertising or sports scoreboards	
	Mechanical character-changing message board	Useful for message status boards in transportation terminals	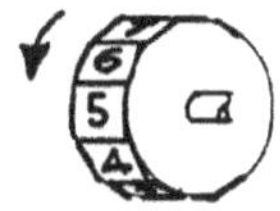
	CRT and TV	Easy to change data; multi-color possible; static or dynamic displays	
	Motion pictures	Primary use for movies, although has potential for training simulation	
	Edge-lighted plotting displays (darkened rooms only)	Command and control operation centers-plotting by hand on rear side	

FIGURE 28.1 Power on light.

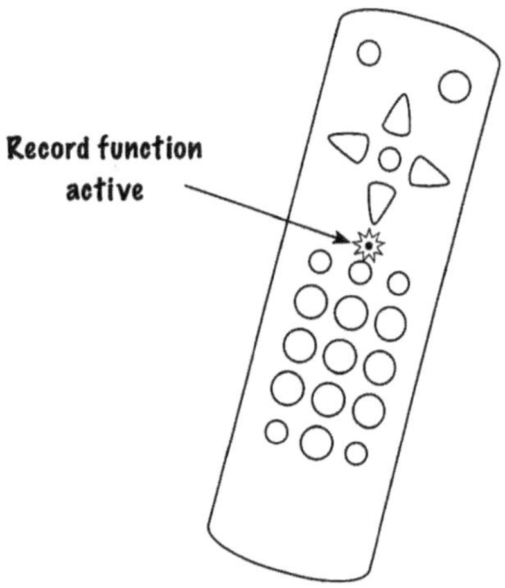

FIGURE 28.2 Function active light.

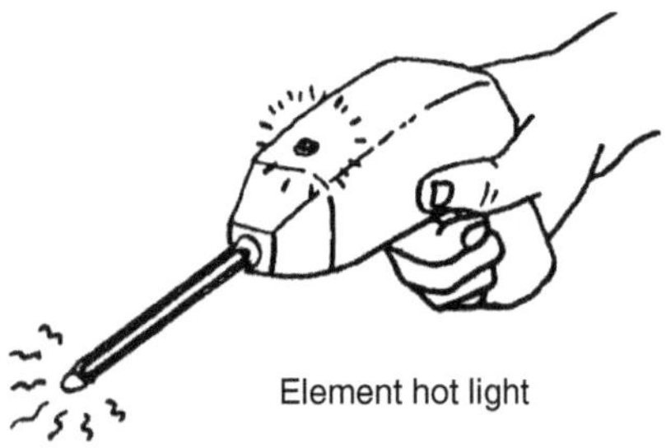

FIGURE 28.3 Hazard light.

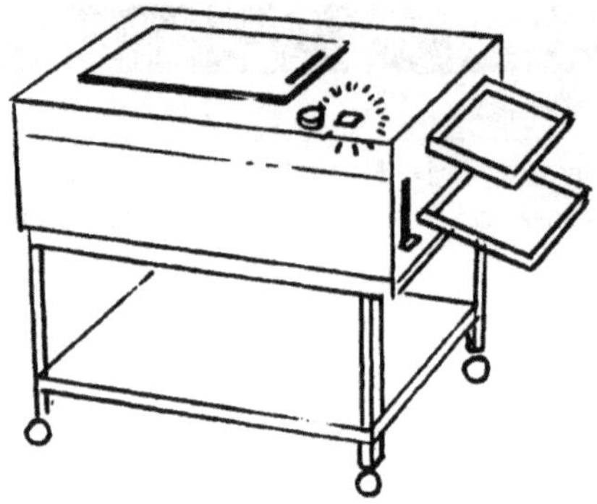

FIGURE 28.4 Operating mode light.

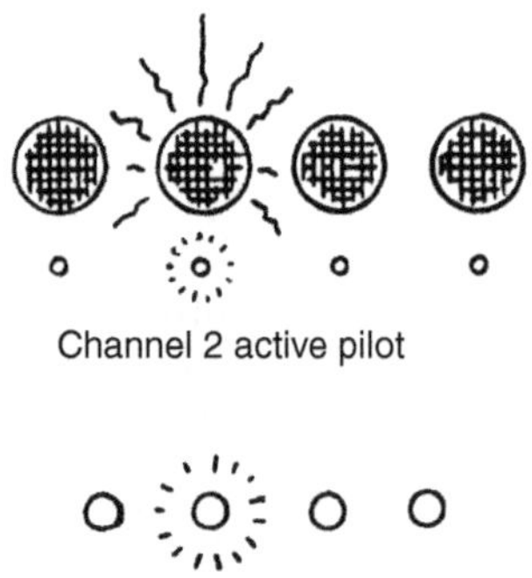

FIGURE 28.5 Localizer lights.

FIGURE 28.6 Malfunction lights.

Outdoor Signal Lights. Outdoor signal lights are widely used for controlling traffic, warning observers about obstacles, etc. Figure 28.7, which shows fairly common signal lights, is meant to remind readers of the possibilities for simple light signaling and information transfer. The single flashing signal light shown in the figure not only attracts attention but, because it is amber in color, also implies the need for caution. The most effective flash rate is between 3 and 10 per second.

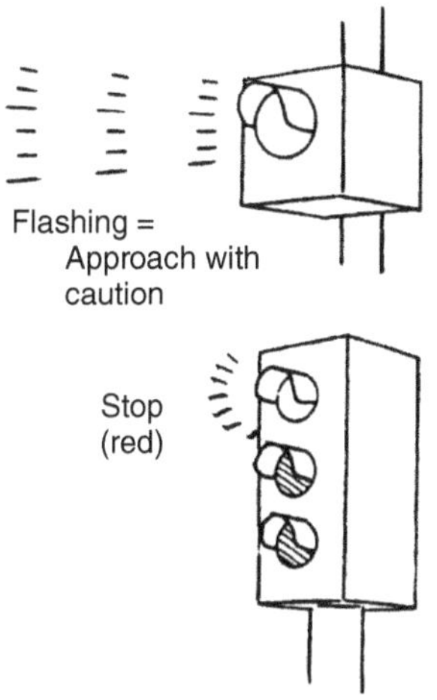

Figure 28.7 Outdoor signal lights.

Light Display Variables

Viewing Angle. In most situations it is desirable that the light display be visible from the widest viewing angle possible. An extended lens is suggested to increase the effective viewing angle. Avoid burying the light below the surface of an instrument panel or other structural surface except for special cases where it may be desirable to limit the viewing angle so that only certain observers can see the light. When it is desirable to make the light visible only to certain observers, use louvers, special directional lenses, or turn the display toward the intended observer. The limited view is appropriate for such devices as a stop light, which is intended only for one group of motorists and which would be confusing for other motorists.

Colored Lights. Certain color-coding conventions are fairly well established and should be followed. Unless it can be demonstrated that changing the color code would enhance performance, do not change colors that already have significant meaning to the majority of observers. The following are some of the conventions:

1. Red Lights: Use for danger, warning, fire, taillights for highway vehicles, and portside navigation lights.
2. Yellow or Amber Lights: Use for caution, slow, power on, turn signals for highway vehicles, and marine-craft identification.
3. Green Lights: Use for go, ready, functioning correctly, and interior turn signal indicator for highway vehicles and starboard side navigation lights.
4. Blue Lights: Use for high-beam headlight indicator for highway vehicles and taxiway lights for airfields.
5. White Lights: Use for airfield runway lights.

Size and Color Coding. Table 28.3 explains the use of size and color codes to help differentiate among the several functions and urgency categories. It must be recognized, of course, that the value of such encoding lies in the fact that the devices are repeated in several places, thus allowing operators to learn to recognize the size differences.

Color as an Indicator of Operational Condition

1. Red should be used to alert an operator to the fact that some situation makes the system inoperative, e.g., an error, a failure, or a no-go or life-endangering condition.
2. Yellow should alert an operator to a situation in which caution, recheck, or delay is necessary, i.e., a condition that, if not attended to, could lead to a dangerous situation.

TABLE 28.3 Coding Simple Indicator Lights

	Color				
Size and Type	Red	Yellow	Green	White	Blue
≥25 mm (1 in.), flashing (3–5 s)	Emergency condition (impending personnel or equipment disaster)				
≥25 mm (1 in.), steady	Master summation (system or subsystem)	Extreme caution (impending danger)	Master summation (system or subsystem)		
≤13 mm (0.5 in.), steady	Malfunction; action stopped; failure; stop action	Delay; check; recheck	Go ahead; in tolerance; acceptable; ready	Functional or physical position; action in progress	Advisory

From MIL-STD-1472G.

3. Green should indicate that equipment is operating satisfactorily or that one can proceed. It can be used to indicate the successful completion of steps within a process, thereby establishing a basis for continuing with the next step.

4. White should indicate items such as status, alternative functions and selection modes, a test in progress, or similar items that imply neither success nor failure of system conditions.

5. Preferential use of blue should generally be avoided because it has no standard meaning, except when it has been assigned a special significance within a given operating system; e.g., a blue reference light has been used by some automobile designers to indicate that headlights are in the high-beam condition.

A flashing red light should be used when an extreme danger is present; i.e., the flashing light indicates a situation of more importance than a continuous red light. A flash rate of between three and five flashes per second, with approximately equal on and off times, is recommended.

The flash-coding technique is also suggested for other colored lights when it is desirable to increase the conspicuousness of the signal. Typical examples include a flashing amber traffic signal in advance of a regular stop-lighted intersection and an alerting signal on a communications status board.

Additional general information on the use of color codes can be found in MIL-STD-1472G, paragraph 5.2.2.3.2.

Example Warning, Caution, and Advisory Light Colors. In military equipment installations, it is required that master warning and/or caution light indicators be placed directly in front of operators to ensure that they will not miss seeing the light when it illuminates. Illumination of such a light indicates to operators that they should immediately look for specific indicators that will tell them what is wrong. The latter indicators are typically located where they can be functionally grouped. Some of the types of warning, caution, and advisory functions that are identifiable by means of annunciator light displays are listed below:

1. Warning Indicators (Red Light)
   - a. Bailout
   - b. Cabin pressure failure
   - c. Canopy unlocked
   - d. Emergency fuel failure
   - e. Engine overheated
   - f. Engine on fire
   - g. Fuel system failure
   - h. Landing gear lower failure
2. Caution Indicators (Yellow Light)
   - a. Aileron trim failure
   - b. Antiskid-off
   - c. Automatic pilot failure
   - d. Constant speed drive failure
   - e. Defroster failure
   - f. Door open
   - g. Engine icing
   - h. Generator no.—Inoperative
   - i. Oil pressure low
3. Advisory Indicators (White, Green, or Blue Light)
   - a. Alternate current
   - b. De-ice
   - c. Direct current
   - d. External power on
   - e. Marker beacon
   - f. Radar on
   - g. Trimmed for takeoff

Military Standards for Indicator Colors. Because of their special importance to the military, the colors of indicator lights of various types are specified and are rigorously controlled to ensure that all observers (including those who may have some inherent visual color deficiency) will recognize the color they are observing. For color-coding specifications, refer to MIL-C-25050, Colors, Aeronautical Lights, and Lighting Equipment General Requirements.

Flashing Lights. The main benefit of the flashing light is that it is more conspicuous. On the other hand, too many flashing lights easily destroy the conspicuousness of any one signal. The same is true of color. The use of flashing lights must be minimized. Flashing lights may be used only when it is necessary to call the operator's attention to a mission critical condition requiring immediate attention. No more than two flash rates shall be used. The two flash rates should differ by not less than 2.0 Hz. The higher flash rate shall not be greater than 5.0 Hz. The slower flash rate shall be not less than 0.8 Hz. The higher flash rate shall reflect more critical information. The flash cycle should have approximately equal amounts of ON and OFF time. Flashing lights that could be simultaneously active should have synchronized flashes.

Flashing lights are recommended wherever it can be anticipated that an observer needs reminding that a situation exists or is about to occur; in certain cases this need arises when there is a high probability that the observer may be visually occupied and needs the benefit of a more conspicuous visual signal. The typical applications and conventions are as follows:

1. A preparatory flashing amber light in advance of a traffic signal.

2. An anticollision flashing light on aircraft because of the special alerting qualities of a flashing (and/or rotating) light. Although the basic anticollision light is typically red, high-intensity white flashing lights have been shown to be conspicuous at greater distances.

3. Flashing lights (both red and blue) for police and emergency vehicle identification and alerting.

4. A white/green flashing light to aid identification of an airfield.

Brightness. The luminance of transilluminated displays shall be compatible with the expected ambient illuminance level, and shall be at least 10 percent greater than the surrounding luminance. Where glare must be reduced, the luminance of transilluminated displays should not exceed 300 percent of the surrounding luminance (MIL-STD-1472G, 2012).

The desired brightness level of any signal light depends on the ambient lighting conditions, the brightness of other lights in the field of view, and the need to identify a change in situation (i.e., a given signal may be made to glow brighter when a functional change has occurred).

Often it is helpful to increase the conspicuousness of a conditional change by combining a brighter light with flashing. Care should always be exercised, however, not to blind an observer at a critical moment by increasing the brightness of a light. A typical case in point is the change in brightness of the taillight of a car when the brakes are applied. Although the objective is to make the change in brightness sufficient to let the observer in the rear know for sure that the brake has been applied, the observer may be blinded by an extremely bright light and make precarious evasive maneuvers that are not warranted.

To be effective, of course, a signal or pilot light has to be in a spot where the observer is most likely to be looking. Other positional considerations are also important, such as the following:

1. A power pilot light should be located in the place where the observer first looks at the equipment (typically on the upper part of the equipment).

2. A high-beam indicator light should be located at the top of the vehicle instrument panel near the observer's typical line of sight as he or she watches the road. The same applies to turn indicator lights.

3. Perimeter-defining lights on vehicles should be as near to the edges of the vehicle envelope as is practical; otherwise, the observer may come too close.

4. Traffic signal lights should be positioned high enough so that they can be seen over the tops of other vehicles, but not so high that when the driver stops, he or she cannot see the signal because of the limits of the vehicle's upper windshield frame.

5. Warning lights (and/or lighted indicator lights) should be located within about 15° of the principal line of sight of an operator in order to make sure that the operator will notice a warning while concentrating on something else (e.g., watching the road ahead, in the case of a motorist, or watching a monitor).

6. Stop lights used on the rear of automobiles should be placed at a height above the road approximating the eye level of the driver of the car behind (or slightly above this level). This has been shown to reduce the incidence of rear-end collisions.

Shape. It is important for an operator to be able to tell quickly which lighted indicator is a pilot light, an annunciator, or a telelight and which is a lighted push button. Avoid the temptation to select hardware components that all look alike, e.g., use a round fixture for a pilot light, a rectangular fixture for an annunciator or telelight, and a square fixture for a push button. Figure 28.8 shows how much easier it is to tell the difference between components when each has a different shape.

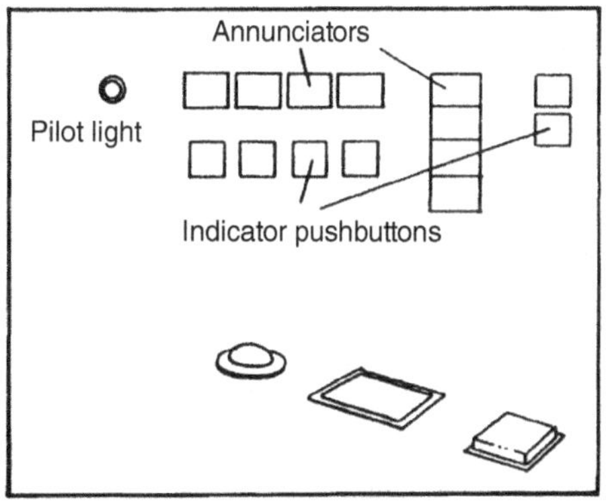

FIGURE 28.8 Shape coding.

It is also desirable to select push-button hardware in which the button extends about ⅛ in. (0.32 cm) out from the panel and annunciator hardware that is more or less flush with the panel. The best pilot light is one that also extends from the panel so that an operator can tell from an angle when the light is on.

USE OF WORDS TO DISPLAY INFORMATION

The use of words as a display device adds another dimension for increasing information transfer. However, it may also add complexity and increase the time an observer requires to recognize and understand the meaning of the information one is trying to convey. The most common word display is perhaps the sign, which can be used for identification, for warning or caution, for guidance, and for instruction.

Graphics specialists are often given the task of producing signs. In some instances, they are concerned primarily with the attractiveness of the sign rather than with its readability. Refer your graphics specialist to the guidelines presented in this book if you are not sure that he or she is familiar with what is required to make a sign readable.

Sign Standards

Certain types of signing have been standardized in order that the size, shape, color, and wording of particular signs will be the same wherever they may occur. For example, the U.S. Department of Commerce, Bureau of Public Roads, issues a Manual on Uniform Traffic Control Devices for Streets and Highways. Similar standards have been devised for various industrial safety signing. It is important to acquire and use these standards rather than to invent new signs.

Traffic signs fall into several distinct categories: (1) regulatory signs, (2) warning signs, (3) guide signs, and (4) pavement and curb markings. Key issues for signing are the following:

1. Legibility
2. Size in relation to viewing distance
3. Illumination
4. Clarity of meaning
5. Location relative to how and when the sign is to be used by the observer
6. Sign shape

Sign Colors

Color can be used to encode and/or categorize various signs. For example, highway signs are typically red, yellow, black, or white, with other colors such as green and blue interspersed for less critical information. The important factors to consider in deciding to use special color codes are the following:

1. Adhere to certain color codes that are already standardized: red—danger or stop; yellow—caution or be prepared; green—OK or go ahead.
2. Select colors or color combinations that enhance the legibility of the sign lettering or symbols.
3. Limit the number of colors selected to those which the observer can differentiate reliably; i.e., those which the observer can verbalize and recognize independently of a comparison with another color close by. The average individual can differentiate only a few colors in terms of faithful verbalization: red, yellow, orange, blue, green, brown, purple, white, and black.
4. In general, select saturated colors, as opposed to pale colors when you want the colors to have special significance.
5. Use luminous or fluorescing color to enhance color discrimination under marginal lighting or viewing conditions.
6. Illuminate colored signs with essentially white light in order to ensure reliability of color identification.
7. Use aviation red, yellow, or blue lights to ensure that color-deficient observers also can recognize these particular colors.
8. Do not alter critical colors to make them match an aesthetic color scheme.

Orientation Signs

Designers often fail to recognize the importance of providing signs to help people identify where they are and where they are going. Typical of these are signs that identify an entrance, a parking area, a bus stop, and an elevator. Once again, such signs should be legible and intelligible, and they should be located so that the observer can easily associate the sign with the physical features pertinent to the situation.

Emergency Signs

Identification of features that may be needed in an emergency is especially important. The typical exit sign is generally prescribed by local ordinance. Other, similar signs include those indicating an emergency escape route, a fire extinguisher, and a first-aid station.

Company Logos

Although one might argue with a specific logo used by a particular company, it is important to the company (and probably to most of its customers) that the logo be retained in any signing created for that company.

Caution Signs

In addition to the caution signs, one generally associates with large outdoor installations (e.g., traffic control), similar signing is applicable to other situations, such as the sign posted on an automobile gas filler cover indicating that unleaded fuel is to be used.

Operating Control Labels

The push buttons for elevator call are typical of this type of signing. Usually, the button that is pushed also illuminates to indicate that the request has been made.

Mechanical Flags

The mechanical flag has been used successfully for indicating the functional condition of a parking meter and/or a control panel instrument. In a sense, this provides a go/no-go signal using words rather than a light.

ARROW AND POINTER GRAPHICS

Design or select directional pointers that are recognizable as pointers and that are sufficiently conspicuous to stand out among an assortment of other geometric designs and backgrounds. (See Fig. 28.9.) The idealized pointer proportions shown in Fig. 28.10 provide proper geometric emphasis to maximize recognition.

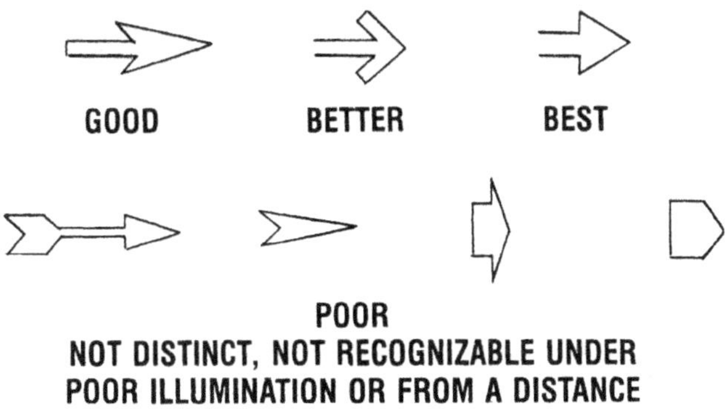

FIGURE 28.9 Directional pointer examples.

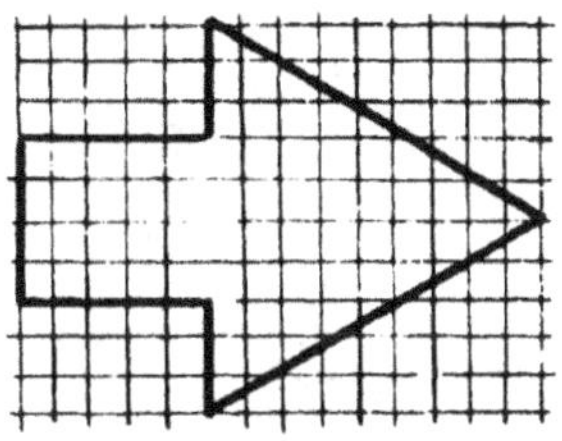

FIGURE 28.10 Ideal pointer proportions for directional signs.

Although many people are aware of the value of the arrow symbol as a directional aid for highway signs and other architectural guidance signs, not many designers have taken advantage of this special graphic device to help the consumer in other situations where guidance is necessary. The following guidelines may be helpful in deciding when to use the arrow symbol and how to use it effectively.

1. The arrow tends to attract attention since it implies "Follow this." (See Fig. 28.11.)

FIGURE 28.11 Directional aid—example 1.

2. The use of several arrows properly aligned above separate aisles or highway lanes provides immediate identification of guidance information related to the specific aisles or lanes. (See Fig. 28.12.)

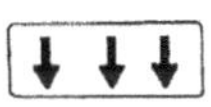

FIGURE 28.12 Directional aid—example 2.

3. An arrow with a tail that obviously changes direction provides immediate recognition of a directional change. (See Fig. 28.13.) In a series of arrows, it is immediately recognized when one or more of the arrows point to another direction, as opposed to continuing in the direction the observer might be going.

FIGURE 28.13 Directional aid—example 3.

4. A single arrow accompanied by simple place names immediately implies to the observer that these places may be found by following in the direction in which the arrow points. (See Fig. 28.14.)

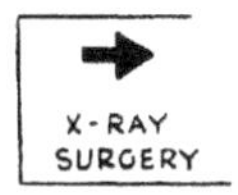

FIGURE 28.14 Directional aid—example 4.

5. Arrows are very useful on complex machine panels (especially those which may be used by the public, such as a vending machine). (See Fig. 28.15.) Observers generally will follow the arrows, while they may often ignore the best instructional material. The addition of "step numbers" almost assures that observers will follow instructions.

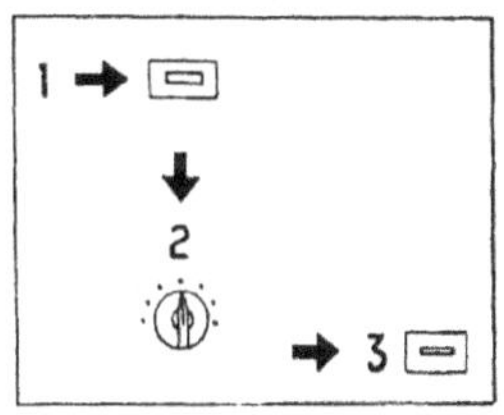

FIGURE 28.15 Directional aid—example 5.

GUIDELINES FOR OPTIMIZING LETTERS AND NUMBERS TO BE USED ON SIGNS, CONTROL PANELS, AND INSTRUMENT FACES

In spite of some controversy over the best style of character for use in signing, labeling, and instrument marking, at least the following parameters are agreed upon as important to consider in selecting a given character style for a specific application (see Fig. 28.16):

1. Proper height/width ratio, e.g., character proportions

2. Proper stroke width, e.g., width of the lines making up the character

3. Maximum figure-ground contrast

4. Adequate overall size for the viewing distance expected

5. A simple rather than an elaborate style, i.e., generally a block style without serifs

6. Upright rather than slanted (italic) characters

Stroke width	A, B Lead	A÷B = 0.14 to 0.17 dark characters on light background 0.13 to 0.14 light characters on dark background 0.10 trans-illuminated characters
Height to width ratio	B, A E	B÷A = 0.6 to 0.8 letters B÷A = 0.6 numbers (except = 0.8 for 4 and = 0.1 for 1)
Letter spacing within word	A, B Ship	Letter spacing (A) = stroke width (B)
Letter spacing between words	A and for B	Letter spacing (A) = stroke width (B) × 3 +/– 1/2 B
Spacing between lines of words	B Words spaced A	Capital letters minimum A = 1/2 B

Figure 28.16 Guidelines for character size and spacing.

7. Adequate intercharacter spacing
8. All capital letters for short labels and signs (as opposed to capital and lowercase letters for extended, instructional materials)

Choosing Correct Type Styles

When selecting a type style, one should look for certain key letters or numbers that are the most difficult to read. The several comparison characters shown in Fig. 28.17 are especially important to examine because they are most easily confused.

Figure 28.17 Examples of typically poor type styles.

Simple block styles should be used for control panel and instrument labeling applications. Typefaces to be avoided are those which:

1. Have uneven stroke widths
2. Have extended serifs
3. Include internal patterns or stripes
4. Are italicized
5. Are stenciled
6. Appear like handwritten script
7. Are shadowed or made to look three dimensional
8. Look like Old English script
9. Are distorted to look tall and thin or wide and fat

Creating Fonts

Each alphanumeric character's recognizability depends on the proper emphasis of specific features such as diagonals, flat tops, and easily discernible loops. Some type fonts tend to obscure these features. Some examples of how certain characters can be distorted to the point of confusion are shown in Fig. 28.18.

Figure 28.18 Variations in character forms (underlined are preferred).

The following points are important:

1. The letter A must have a clearly delineated space above the horizontal stroke. The letter B should have approximately equal loops at the top and bottom.
2. Cs and Gs are easily confused with each other or with Os if the break is not clearly discernible, if the horizontal stroke of the G is not long enough, or if the O tends to appear too square.
3. The E must not have its horizontal strokes too close together.
4. The M and the W must have a sufficiently long center section.
5. The P must have a large enough loop.

6. The S and the 5 are easily confused if the S is too square or if the horizontal top of the 5 is not long enough.

7. The 1 and the I are easily confused with each other unless they are made to look different. In this instance the serif is perhaps most helpful.

8. The U and the V are easily confused unless the vertical sides of the U are maintained, as opposed to the slanting sides of the V.

9. The Y needs a long tail to be discernible from the V. However, the V-shaped top also must be definite to keep the Y from being mistaken for a T.

10. When the 6 and the 9 have loops that are too large and tails that are too curved, they are not easily recognized. The loops must be readily apparent, but the tails need to be straight.

More guidelines for the style of label characters are in SAE-AS18012, Markings for Aircrew Station.

Brightness and Color

To ensure adequate legibility, there must be adequate visual contrast between alphanumeric characters and the background against which they are viewed. (See Fig. 28.19 and Table 28.4.) The following should be considered:

1. Under normal illumination conditions (i.e., where the observer does not have to be dark-adapted), use either dark characters against a light background or light characters on a dark background. The reading rate is the same.[1]

2. Where the observer must maintain a dark-adapted condition, use light characters on a dark background.

3. The character should be at least twice as light (or dark) as the background. Greater contrast is preferred and is especially important if the expected observer population includes persons with impaired vision, if the viewing conditions will be less than optimal, or if physiological stress factors are anticipated, such as oscillation vibration, motion, or high g forces.

4. Avoid the use of glossy or highly reflective metallic materials for either the lettering or the background (except where the lettering must be illuminated remotely, e.g., by automobile headlights). Spectral unevenness usually causes portions of the letter or numeral to become illegible as a result of the variation in light reflection.

TABLE 28.4 Suggested Color Contrast Selections in Order of Expected Visual Efficiency

CONDITIONS	CHARACTERS	BACKGROUND
Average or higher levels and quality of illumination	Black	White
	Black	Yellow
	White	Black
	Dark Blue	White
	White	Dark Red, Green, and Brown
	Black	Orange
	Dark Green and Red	White
	White	Dark Gray
	Black	Light Gray
Poor level and quality of illumination	Black	White
	White	Black
	Black	Yellow
	Dark Blue	White
	Black	Orange
	Dark Red and Green	White
Dark adaptation required	White	Black
	Yellow	Black
	Orange	Black
	Red*	Black
	Blue and Green	Black

*Low-level red light is required to maintain the greatest dark adaption level.

Note: When (0.07–0.34 cd/m^2) lettering is internally illuminated so that the letter is always bright against an essentially black background, almost any color is satisfactory as long as the brightness is above the minimal threshold, i.e., as long as the light is visible. Opaque letters must be illuminated with an essentially white light; otherwise, the colored illumination may decrease the contrast between the character and its background.

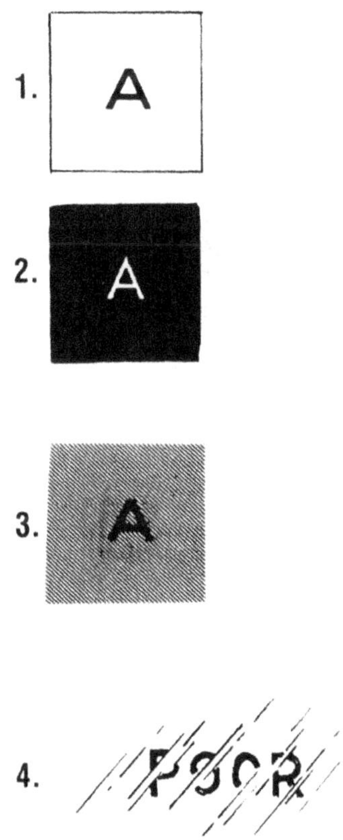

FIGURE 28.19 Brightness and color.

[1]Legge, G. E., Pelli, D. G., Rubin, G. S., and Schleske, M. M. (1985). Psychophysics of reading—I. Normal vision. *Vision Res*, 25(2), 239–252.

Character Size

All things being generally equal (legibility factors, illumination, atmospheric interference, etc.), the larger the letters and numbers are, the better they are seen and recognized up to the point where the observer would be unable to see an entire character at a single glance. (See Fig. 28.20.) Many people make a common error, however, in applying the above generalization; i.e., they believe that they should also make the character bolder as it is made larger for more distant viewing. As noted in the discussion of stroke width, to maintain legibility while increasing the size of a character, one must also maintain the proper stroke–width relationship.

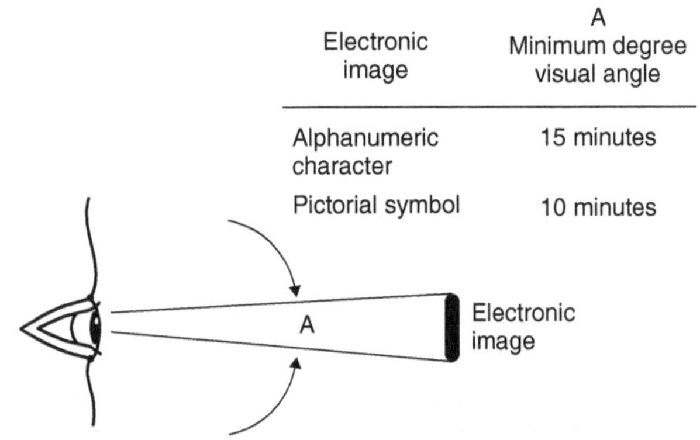

FIGURE 28.20 Minimum electronic image height.

Tables 28.5 and 28.6 provide guidelines for determining how high a letter or numeral has to be for the given viewing distances.

TABLE 28.5 Character Height vs Luminance and Viewing Distance

	HEIGHT*	
MARKINGS	LOW LIGHTING, 3.5 CD/M² (1 FL) OR BELOW	NORMAL LIGHTING, ABOVE 3.5 CD/M² (1 FL)
Critical markings with position variable (e.g., numerals on counters)	5.0–8.0 mm (0.20–0.31 in.)	3.0–5.0 mm (0.12–0.20 in.)
Critical markings with position fixed (e.g., numerals on fixed scales, controls and switch markings, or emergency instructions)	4.0–8.0 mm (0.16–0.31 in.)	2.5–5.0 mm (0.10–0.20 in.)
Noncritical markings (e.g., identification labels, routine instructions, or markings required only for familiarization)	2.5–5.0 mm (0.10–0.20 in.)	2.5–5.0 mm (0.10–0.20 in.)

From MIL-STD-1472G.

Note: Values assume a 710 mm (28 in.) viewing distance (D). For other distances, multiply the above values by D/710 mm (D/28 in.).

TABLE 28.6 Character Size Requirements in Pixels

WIDTH	HEIGHT	APPLICATION
5	7	Uppercase, numerals
4	5	Subscripts
	+2	Lowercase
	+2	Diacrits
7	9	Reading

From NASA Human Interface Design Handbook.

MESSAGE INTELLIGIBILITY

To be effective, words in labels and instructional materials have to be selected carefully so that their meaning is perfectly clear to the observer.

1. Use common terms that originate from typical language usage and/or from standard lists of terms for special fields or groups of users (pilots military personnel, etc.).

2. Use whole words rather than abbreviations wherever space permits. If abbreviations are required, refer to the list of standard abbreviations created by the appropriate "user" group (military, Department of Transportation, etc.). However, check these lists for duplications; it is undesirable to use the same abbreviation for two different functions. Acronyms may be used sparingly if they have been well established over a long period. They are least desirable for consumer products because of the wide variation in user experience.

3. To identify instruments, use terms that indicate what the instrument measures rather than the name of the device: OHMS not OHMETER; RPM not SPEEDOMETER; HRS/MIN not TIMER.

4. Avoid the use of words that may be interpreted one time as a noun and another time as a verb whenever both may occur at the same workplace: FIRE = a fire has occurred (warning); FIRE = to fire a weapon (command).

5. Always use capital letters for labels and short instructions because they can be read at a greater distance than capital and lowercase letters. Capital and lowercase letters should be used only for extended sentence messages where it is necessary to provide punctuation.

6. Generally, instruction labels should be as brief as possible within the limits of clarity. If an instruction implies several sequential steps, arrange each step on a separate line, preferably with a dot at the beginning of each line.

7. Avoid the use of words that may have a negative psychological effect; e.g., the word "speed" itself implies traveling at a fast rate and therefore may encourage a motorist to drive fast.

8. Avoid using the number 1 in combination with letters when encoding such things as airport terminal gates. For example, if you are planning to encode the gates with letter–number combinations, such as K-I, K-2, K-3, etc., do not include the letter I because it is easily confused with the number 1; i.e., 1-I might look like an H from a distance.

9. When special precautionary words are required, select ones that provide a sense of urgency, hazard, or danger. Avoid words that seem to leave it up to the observer to decide whether there is a danger involved, since he or she may not have the necessary experience to make such a judgment.

10. The lettering for all labels should be oriented so that they read from left to right, not around corners, on the side, or up and down.

11. A manufacturer's label must not overshadow the primary display labels and markings. Place the manufacturer's label in an inconspicuous place, in small print.

12. Control and display labels should be located consistently, either below or above the panel component, especially on the same panel. The decision to locate the label above or below components should be based on the eye position of the observer; i.e., if the panel labels are above the eye reference level, the controls may extend out so far that they obscure a label placed above the control, or the opposite might be true if the control panel is considerably below the eye reference level. Random placement of labels on the same panels makes them difficult to find and associate with the pertinent function.

13. Avoid running adjacent labels together. This often happens because control elements were spotted on the panel without considering where the labels would go or how much space they might require.

14. Utilize size and/or color coding to help the observer differentiate between levels of importance on your panel; i.e., the system label should be slightly larger than subsystem labels, and these should be larger than component labels. The character height of the smallest label must be sufficient for the maximum viewing conditions expected. Avoid the tendency to settle for some arbitrary policy set by your engineering department for all equipment turned out by the company; this may be completely unsatisfactory.

15. As a general rule, all labels should be made as permanent as possible; i.e., they should be wear and damage resistant. If labels are to be read under low light levels, they must be illuminated by means of ambient room lighting, special supplementary floodlighting, back or edge lighting, and/or luminescent materials or substances. If raised letters are provided (i.e., cast into a panel or package housing), there must be additional coloring to make the letters contrast with the background. Do not assume that a raised letter that is the same color as the basic background will be readable just because it is raised. Engraved or depressed lettering should also be of a color that contrasts sharply with the background. In addition, the engraving should always be filled level with the panel surface so that it will not collect dirt and eventually become unreadable. Avoid glossy inks or paints and plastic or glass dust covers, unless these are specially treated to minimize spectral reflectivity.

PANEL LABELING

Start with the "smallest" lettering and make the letter height minimum size for the viewing distance and conditions. Each next larger lettering size (see Figs. 28.21–28.23) should be at least 25 percent larger for differentiation.

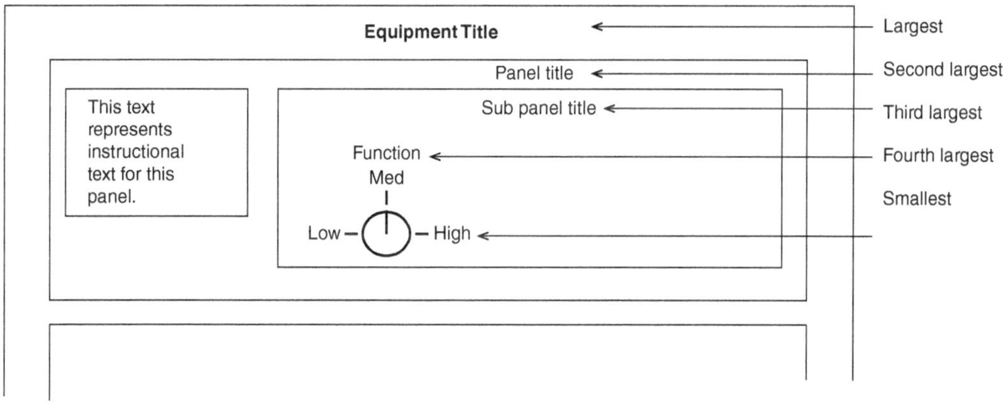

FIGURE 28.21 Relative sizes of panel labels. (From MIL-STD-1472G.)

Panel Labeling Standards		
Label Designation	**Letter Size**	**Location**
1. Panel title	18 pt (0.187 in.)	Centered; ¼ in. from top edge of panel
2. Panel subsection	14 pt (0.156 in.)	Centered at top of subsection; ¾ in. from top edge of panel
3. Subtitle	12 pt (0.125 in.)	¼ in. above components or 1/8 in. above labels of individual components
4. Toggle switch	10 pt (0.093 in.)	¼ in. above and below standard switch
5. Single component	12 pt (0.125 in.)	¼ in. above component
6. Rotary switch positions	10 pt (0.093 in.)	¼ in. from apex of pointer, line from pointer to label

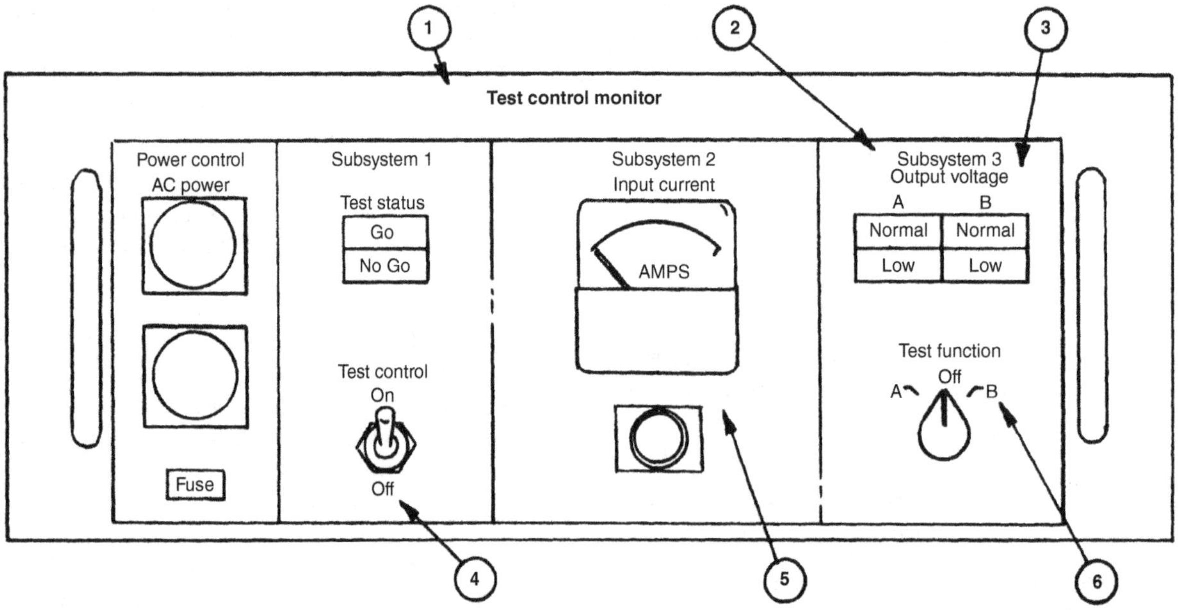

Figure 28.22 Panel labeling standards.

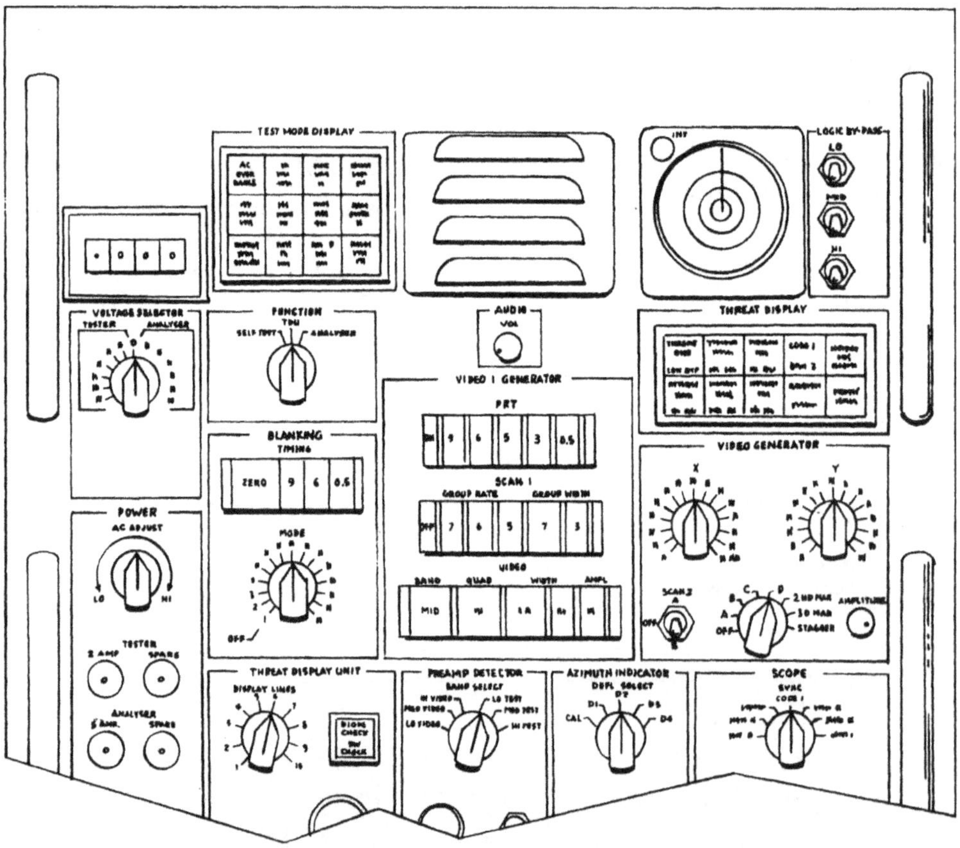

Figure 28.23 Example of a relatively well-executed labeling program for a fairly complex panel.

WORD DISPLAYS IN CONJUNCTION WITH CONTROLS

Words used directly on certain types of control devices increase the utility and efficiency of control identification and operation by combining the identifying variables of control position, light, and words. (See Fig. 28.24.) See Chap. 30, "Controls," for a more complete discussion of this application of word and light combinations.

Figure 28.24 Examples of word labels.

Several precautions are in order relative to placing word labels directly on the control device.

The control must not rotate so that the label may become unreadable. If the control surface area is too small for a fully legible label, place the label adjacent to the control; i.e., do not feel that you have taken care of your obligation just because a label is there—it has to be readable. Push buttons should extend out from a control panel, as compared with a telelight (i.e., not a control); otherwise, an operator will be unable to tell which similar-looking word-light displays are really controls and which are merely advisory displays.

TACTILE SIGNS FOR THE BLIND AND PARTIALLY SIGHTED

Public signing plans should consider the problems of blind and partially sighted users. Typical signs that should be provided for the blind include the following:

1. Street signs
2. Bus-stop signs
3. Intersection signal control signs
4. Exterior building signs (especially for restroom buildings)

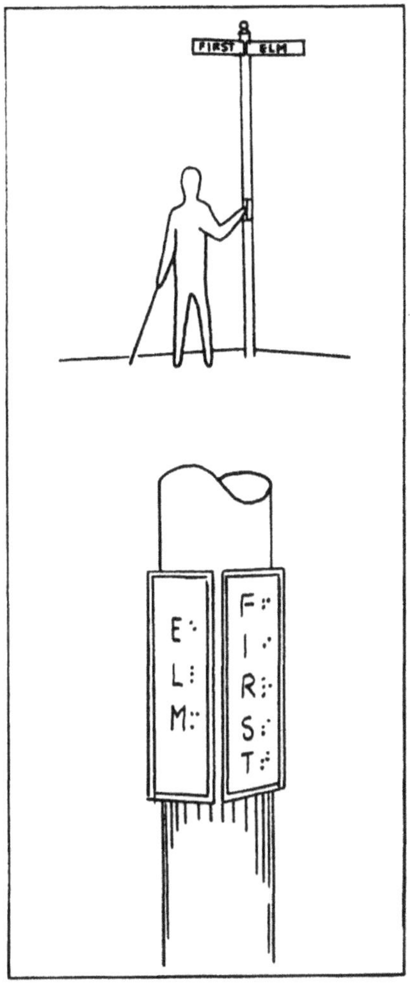

Figure 28.25 Tactile sign example.

5. Interior office and public facilities signs (restrooms, elevators, etc.)

Figure 28.25 illustrates a suggested approach to providing tactile sheet signing.

1. The tactile sign must face the street so that the person knows with which street to associate it.
2. The sign should have raised letters for the partially sighted and nonbraille user, plus braille code for individuals who are trained in the use of this device. Consider a system such as an audio signal to alert the blind user to the sign.

PICTORIAL SYMBOLS

Pictorial symbols make it possible for persons with different language backgrounds to recognize and understand a single symbol. (See Fig. 28.26.) Unfortunately, it is not always possible to pictorialize all informational conditions since some things would not be recognized by the observer because he or she has never seen the actual object. An example is the

FIGURE 28.26 Pictorial symbols.

choke system used on some vehicles. A symbol that is significant to engineers because they have seen the actual system or device not only is meaningless to average drivers but also may encourage them to believe that the symbol implies something entirely different; e.g., the typical hourglass pictorial suggested by the automotive designer could be construed to mean "time" to the lay person. Some principles to consider in deciding whether a pictorial signing method should be used include the following.

1. Pictorial Comprehensibility: If a particular situation includes items that cannot be completely pictorialized; i.e., if some items are easy to pictorialize, but others are not, it is better to stick to a word labeling system;

observers will only be confused if they find some pictorials and some word labels on an operator panel.

2. Combined Pictorial and Word Labeling: Whenever there is any doubt as to whether some observers will be able to understand the meaning of a pictorial, use both pictorial and word labels.

3. Size–Distance Compatibility: Some pictorial patterns may be effective only when the viewing distance and lighting conditions are optimum; be sure that a particular pictorial pattern does not lose its identity when it becomes smaller or more distant and/or when the ambient lighting and/or atmospheric conditions are not good.

4. Orientation: Do not place pictorials on components that may be reoriented, i.e., turned or moved so that the pictorial may not appear right side up. Although the observer might figure out what the pictorial is after considerable study, a prime purpose of the single pictorial is to elicit a quick response with a single symbol, as opposed to several words in a label.

5. Sufficient Detail: In creating a pictorial to represent some actual visual element, provide just enough detail to make the symbol recognizable, and no more. Fine detail often cannot be seen under certain distance and lighting conditions and may serve only to distort the impression the symbol creates. On the other hand, be extremely careful not to over stylize pictorial symbols just to create an artistic rendition. Such stylization often makes all symbols begin to look alike.

6. Isolation: Some type of border should always be used around a pictorial; otherwise, it may blend with background images.

See ISO 7000 (1989), Graphical symbols for use on equipment, and ISO 7001 (1991), Public information symbols, for more symbols.

Examples of Pictorial Signs for Public Accommodations

With careful thought, signs and symbols can be used in ways that will make them more effective. Some pictorials may be used for slightly different purposes, e.g., the baggage symbol. By adding slightly to the pictorial, the observer can be informed as to whether the location is for baggage check-in, inspection, or storage. (See Fig. 28.27.) Similar clarification for public transportation stops is helpful; in addition to showing a picture of the vehicle, providing a stop sign or platform helps clarify that the location is specific, i.e., that this is where one should stand to wait for a bus or train. (See Figs. 28.28 and 28.29.) Individual pictorials should be tested to determine the appropriate size for the expected viewing distance.

With a white sign background, outline the field definition and black figure. This is the only condition in which the symbol outline should be delineated. (See Fig. 28.30.)

Figure 28.27 Baggage check-in.

Figure 28.28 Bus stop.

Figure 28.29 Tram (rail) stop.

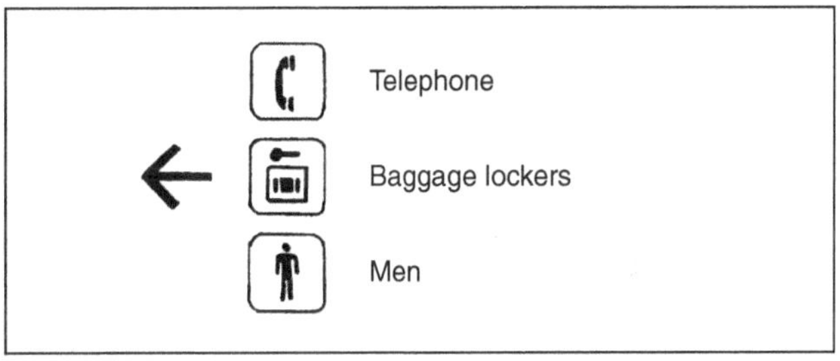

Figure 28.30

Legibility Criteria[2]

Figure 28.31 illustrates the results of pragmatic testing of several symbols (Ticket Purchase, Elevator, and Taxi), and represents a rough guide to size/distance relationships. For the purpose of this illustration, "legibility" was defined as the recognition of the various elements that make the symbol understandable without the aid of wording or preconditioning. "Recognition" of the symbols after they are learned is another matter, and we feel that it cannot be meaningfully tested at this time except in the case of those few well-known symbols such as First Aid, Men, Women, etc. The testing was done in daylight using symbols with black "figures" on white "symbol fields" displayed on a black sign background. The figure shows the result of the testing of the Ticket Purchase

[2]U.S. Department of Transportation, Symbol Signs: Development of Passenger/Pedestrian Oriented Symbols for Use in Transportation-related Facilities, DOT-OS-40192 (NTIS PB-239-352), 1974.

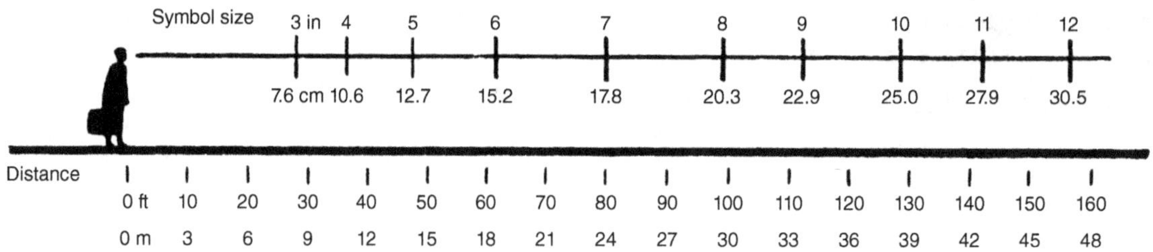

FIGURE 28.31 Symbol size for legibility.

symbol. The distances from which the Taxi symbol was legible were 10 percent greater, and for the Elevator symbol, 30 percent less.

Suspended Signs

The suspended sign is preferred when there is sufficient ceiling height because the sign is less easily confused with the facility structure. (See Fig. 28.32.) Suspended signs should not extend below about 7 ft (2.1 m), however, for they then become obstructions that are easily bumped into either by very tall people or by things that people may be carrying.

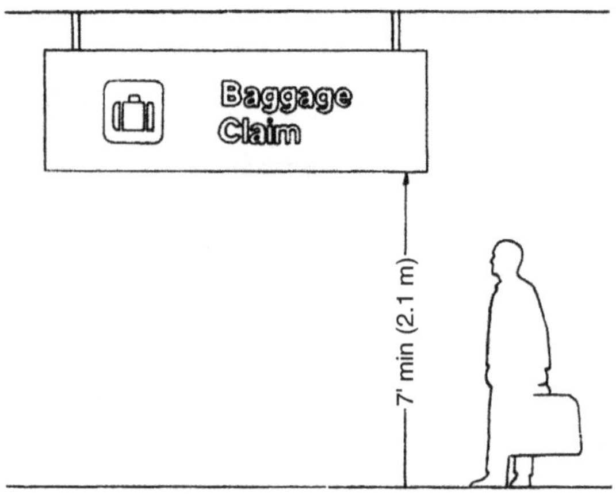

FIGURE 28.32 Suspended sign example.

Multidirectional Signs

Where several directions are indicated on one sign it is best to keep the arrow in a constant position relative to the symbols, regardless of arrow direction. (See Fig. 28.33.)

ROAD SIGNS

The functions of signs are to provide regulations, warnings, and guidance information for road users. Words, symbols, and arrows are used to convey the messages. Road signs are defined by their function as follows:

1. Regulatory signs give notice of traffic laws or regulations.

2. Warning signs give notice of a situation that might not be readily apparent.

3. Guide signs show route designations, destinations, directions, distances, services, points of interest, and other geographical, recreational, or cultural information.

Regulatory and warning signs should be used conservatively because these signs, if used to excess, tend to lose their effectiveness. Conversely, route signs and directional guide signs should be used frequently because their use promotes efficient operations by keeping road users informed of their location.

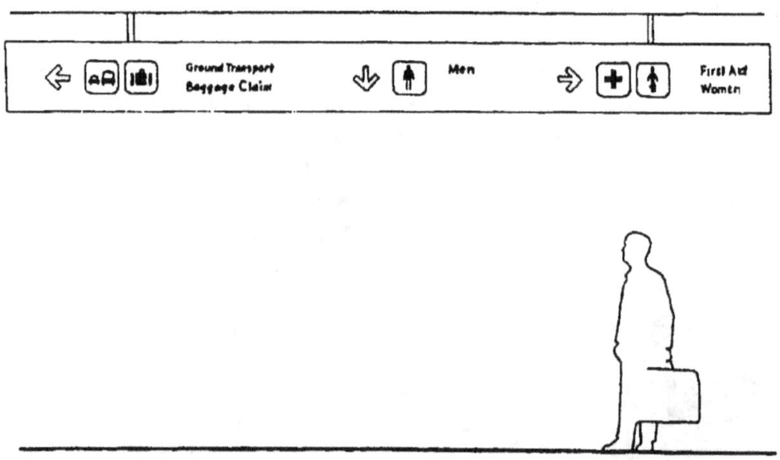

FIGURE 28.33 Keep arrows in constant position relative to symbols.

Road signs are defined primarily by their:

1. Shape
2. Color
3. Content—graphics and text

The Federal Highway Administration of the U.S. Department of Transportation has standardized these road sign characteristics to make signs easier and faster to recognize and use. These standards are published in the Manual on Uniform Traffic Control Devices and Standard Highway Signs. Both are available online from the Federal Highway Administration.

Sign Shapes

Shapes shown in Fig. 28.34 should be used exclusively for specific signs or series of signs.

Sign Colors

The Federal Highway Administration has assigned meanings for sign background colors. (See Fig. 28.35.)

Sign Content

Wording. Sign word messages should be:

- As brief as possible
- Uppercase letters (except names of places, streets, highways—then first letter cap then lowercase)
- Large enough to be seen (at least 1 in. high per 30 ft of viewing distance)

Borders. Signs should have borders. Borders help the reader focus on the content and to quickly identify start and end points.

Symbols. Some of the sign symbols are shown in Fig. 28.36.

Guidelines for street markings

So that markings placed on roadway surfaces will remain legible while the motorist is moving toward the markings, such as elements as the letters, numerals, and arrows must be elongated as indicated in Fig. 28.37.

Visual Background Interference

It is easy to forget that the visual background within which a sign or signal must be observed may interfere with successful location of the sign and/or with easy reading of the sign or recognition that a signal light is illuminated or not. (See Fig. 28.38.) Two major background interference factors should be considered: (1) glare from the sun or from very bright artificial lights and (2) general visual clutter from objects, other signs, crowds of people, or vehicles.

When such interference is anticipated, it is important to provide a sufficient border area around the significant elements of the sign to ensure that the observer can focus on and read the symbols or words and/or recognize that a signal light is on.

For symbolic or alphanumeric signs, a good rule of thumb for the border area is to provide approximately the same amount of clear border area as is required for the symbology (including the clear space between the lines and the lettering).

Shape	Signs
Octagon	Stop*
Equilateral triangle (1 point down)	Yield*
Circle	Grade crossing advance warning*
Pennant shape/isosceles triangle (longer axis horizontal)	No passing*
Pentagon (pointed up)	School advance warning sign (squared bottom corners)* County route sign (tapered bottom corners)*
Crossbuck (two rectangles in an "X" configuration)	Grade crossing*
Diamond	Warning series
Rectangle (including square)	Regulatory series Guide series** Warning series
Trapezoid	Recreational and cultural interest area series national forest route sign

* This sign shall be exclusively the shape shown.

** Guide series includes general service, specific service, tourist-oriented directional, general information, recreational and cultural interest area, and emergency management signs.

Figure 28.34 Use of sign shapes. (From Manual on Uniform Traffic Control Devices, U.S. Department of Transportation, 2009.)

Type of sign	Legend								Background										
	Black	Green	Red	White	Yellow	Orange	Fluorescent yellow-green	Fluorescent pink	Black	Blue	Brown	Green	Orange*	Red*	White	Yellow*	Purple	Fluorescent yellow-green	Fluorescent pink
Regulatory	X		X	X					X					X	X				
Prohibitive			X	X										X	X				
Permissive		X													X				
Warning	X															X			
Pedestrian	X															X		X	
Bicycle	X															X		X	
Guide				X								X							
Interstate route				X						X				X					
State route	X														X				
U.S. route	X														X				
County route					X					X									
Forest route				X							X								
Street name				X								X							
Destination				X								X							
Reference location				X								X							
Information				X						X		X							
Evacuation route				X						X									
Road user service				X						X									
Recreational				X							X	X							
Temporary traffic control	X												X						
Incident management	X												X						X
School	X																	X	
ETC-account only	X																X****		
Changeable message signs																			
Regulatory			X***	X					X										
Warning					X				X										
Temporary traffic control					X	X			X										
Guide				X					X			X**							
Motorist services				X					X	X**									
Incident management					X			X	X										
School, pedestrian, bicycle					X		X		X										

* Fluorescent versions of these background colors may also be used.

** These alternative background colors would be provided by blue or green lighted pixels such that the entire CMS would be lighted, not just the legend.

*** Red is used only for the circle and slash or other red elements of a similar static regulatory sign.

**** The use of the color purple on signs is restricted per the provisions of paragraph 1 of section 2F.03.

FIGURE 28.35 Use of background sign colors. (From Manual on Uniform Traffic Control Devices, U.S. Department of Transportation, 2009.)

Figure 28.36 Recreation and cultural interest area signs. (From Standard Highway Signs, U.S. Department of Transportation, 2004.)

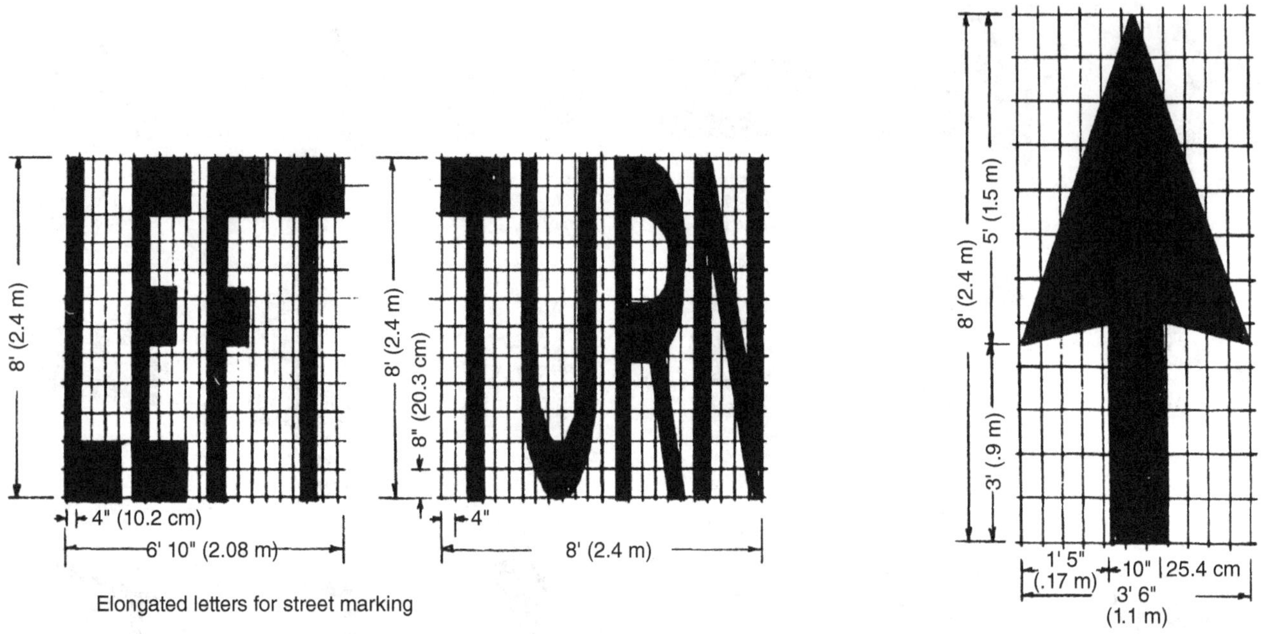

Figure 28.37 Guidelines for street marking.

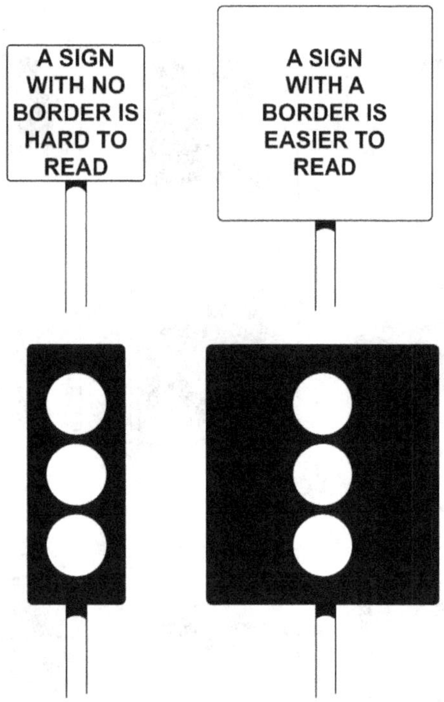

Figure 28.38 Consider sign background.

For outdoor traffic-light signals, when glare from the sun is a common condition for signal viewing (i.e., when the signal light has to be located so that a driver looks into the setting sun on a daily basis), a shield should be provided around the lights. Such shields should be painted matte black to accentuate the contrast between the illuminated signals and the adjacent shield.

As a rule of thumb, the shield border around the lights should be about two to three times the diameter of the lights.

Combining Words and Lights

As noted earlier, providing both a light and a word expands the possibilities for signing visual information displays. The addition of the words increases the probability that the meaning of the sign will always be clear, and the addition of the light provides a means for emphasizing the particular sign or signal mode and/or allows one to use the same display unit to present alternate information. One of the most common devices is the lighted pedestrian sign that switches from a "walk" to a "don't walk" instruction. Figure 28.39 shows a typical design for a pedestrian signal.

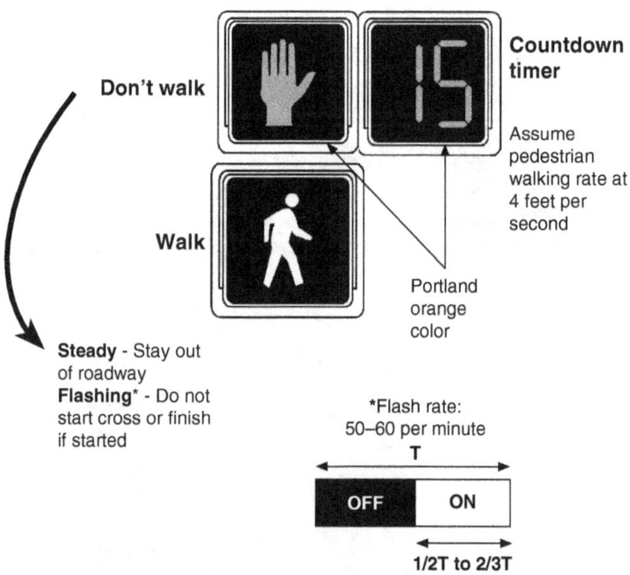

Figure 28.39 Typical design for a pedestrian crossing signal.

Combination light-word signal devices have long been successfully used for vehicle and equipment control monitoring. These are variously referred to as "telelights," "annunciator lights," etc.

One has the choice of using dark printing with a lighted background or the reverse, i.e., illuminated printing with a dark background. The general rule is to reserve the lighted background for emergency or warning displays, since this makes the display much more conspicuous and attention demanding.

The question is invariably raised as to whether the lettering of an inactive display should be visible. As a rule, it should be visible because it is important for the operator to have previous knowledge of what display comes up, and where.

PICTORIAL SYMBOLS FOR VEHICLE CONTROLS AND DISPLAYS—FEDERAL STANDARDS

Certain controls in automobiles are required to have international pictorial symbols, as indicated in Fig. 28.40, which is taken from Motor Vehicle Safety Standard, no. 101, Federal Controls and Displays. Note, however, that from a good human engineering standpoint, all labels, whether they consist of words or symbols, should always be illuminated for night use. Other important principles include the following:

1. If a control is rotatable, the symbol or word should be placed on the panel next to the control rather than on the control so that the label will always appear upright.
2. A word label should accompany any pictorial symbol in order to ensure symbol intelligibility.

For a more complete description, see Henry Dreyfuss, *Symbol Sourcebook: An authoritative Guide to International Guide Symbols*, John Wiley & Sons, New York, 1972.

PICTORIAL SYMBOLS FOR AGRICULTURAL EQUIPMENT

The various pictorial symbols shown in Fig. 28.41 were developed by the Tractor Technical Committee of the Society of Automotive Engineers. Standards for symbols are in ISO/CD 3767, Tractors, machinery for agriculture and forestry, powered lawn and garden equipment—symbols for operator controls and other displays.

The purpose of these symbols is to provide a more universal type of communication that overcomes language barriers when a particular piece of equipment may be used in countries where operators speak different languages.

The color combinations and the sizes of these symbols require adjustment to each particular application (e.g., viewing distance, ambient lighting conditions, or the necessity to indicate the urgency of action by the use of color, such as red for warning or danger and yellow for caution).

Note: From a human engineering point of view, it is desirable also to include word labels with each of the pictorials; although the pictorial may be generally meaningful, a word label provides additional assurance that the user will understand the meaning upon first seeing it. Obviously, the words must be in the language of the particular user group.

OTHER SYMBOLIC AND PICTORIAL GRAPHICS

Properly designed and/or selected symbolic or pictorial graphics can be used to reduce an observer's "reading time," but they must be chosen carefully, either because they are already standardized or on the basis of experimental testing.

Abstract Symbols

Abstract symbols should not be used for identification purposes where the user is not familiar with the symbols. Although the various engineering professions have developed standardized symbology to aid them in the preparation and rapid reading of schematics, such symbology is unfamiliar to the average user of control panels or products. Quasi-abstract analog symbology can be used to a limited degree for control panels and product labeling. Examples are shown in Fig. 28.42. Other common man-made as well as natural features are frequently presented in graphic form, as shown in the figure.

For standardized symbology recommended for various engineering disciplines, refer to the American National Standards Institute at http://www.ansi.org/.

Map Graphics

Although the science of cartography provides considerable guidance in the development and use of map graphics, which should be used as a basic reference, certain key factors should be kept in mind in creating special geographic displays. (See Fig. 28.43.) The symbols can be printed on paper maps but are also very likely to be viewed on computer monitors or smaller, portable navigation screens (including cell phones).

SAFETY SIGNS

Regardless of whether the proposed sign falls within the specific classification of a warning or caution sign or signal or a hazard advisory or instruction, the primary objective should always be to make the device noticeable, recognizable, and understandable. Eight of the key factors follow.

1. Conspicuousness

The sign should stand out among adjacent visual distractions because of its size, shape, color, or symbology or because it is moving or flashing. The sign or signal must be located where it is expected that most people will be looking; it should not be hidden behind a post or an operator control or below the heads of other observers or the tops of equipment, parked cars, etc. Above all do not camouflage the sign or signal to keep it from spoiling the decor planned for the site.

2. Emphasis

The most prominent feature of a warning sign should be words or symbols that imply danger. Include such words as "danger," "hazard," "caution," and "warning." Use standardized

Table 1
Identification and illustration of controls

Column 1	Column 2	Column 3	Column 4
Hand operated controls	*Identifying words or abbreviation*	*Identifying symbol*	*Illumination*
Master lighting switch	Lights	 5	———
Headlamps and tail lamps	(Manufacturer option)[2]	(Manufacturer option)[2]	———
Horn	Horn	 4	———
Turn signal	———	 3 5	———
Hazard warning signal	Hazard	 5	Yes
Windshield wiping system	Wiper or wipe		Yes
Windshield washing system	Washer or wash		Yes
Windshield washing and wiping combined	Wash-wipe or washer-wiper		Yes
Heating and or air conditioning fan	Fan	 Or 	Yes
Windshield defrosting and defogging system	Defrost, defog or def.		Yes
Rear window defrosting and defogging system	Rear Defrost, rear Defog. rear Def., or R-Def.		Yes
Identification, side marker and or clearance lamps	Marker lamps or MK Lps	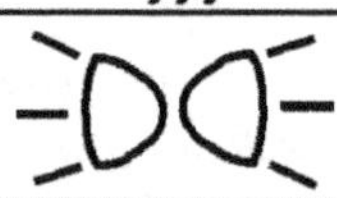 2 5	Yes
Manual choke	Choke	———	———
Engine start	Engine start[1]	———	———
Engine stop	Engine stop[1]	———	Yes
Hand throttle	Throttle	———	———
Automatic vehicle speed	(Manufacturer option)	———	Yes
Heating and air conditioning system	(Manufacturer option)	(Manufacturer option)	Yes

1. Use when engine control is separate from the key locking system.
2. Separate identification not required if controlled by master lighting switch.
3. The pair of arrows is a single symbol. When the controls for left and right turn operate independently, however, the two arrows may be considered separate symbols and be spaced accordingly.
4. Identification not required for vehicles with a GVWR greater than 4536 kg; or for narrow ring-type controls.
5. Framed areas may be filled.

FIGURE 28.40 Vehicle controls.

Table 2
Identification and illustration of displays

Column 1	Column 2	Column 3	Column 4	Column 5
Display	*Telltale color*	*Identifying words or abbreviation*	*Identifying symbol*	*Illumination*
Turn signal telltale	Green	Also see FMVSS 108	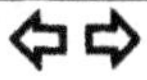 1, 5	————
Hazard warning telltale		Also see FMVSS 108	 2, 5	————
Seat belt telltale	————4	Fasten belts or fasten seat belts also see FMVSS 208	 Or 	————
Fuel level telltale		Fuel	 Or 	————
Gauge	————			Yes
Oil pressure telltale		Oil		————
Gauge	————			Yes
Coolant temperature telltale		Temp		————
Gauge	————			Yes
Electrical charge telltale		Volts, charge or Amp	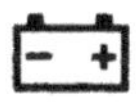	————
Gauge	————			Yes
Highbeam telltale	Blue or green 3	Also see FMVSS 108	 5	————
Brake system 8	Red 3	Brake, also see FMVSS 105 and 135	————	————

1. The pair of arrows is a single symbol. When the indicator for left and right turn operate independently, however, the two arrows will be considered separate symbols and may be spaced accordingly.
2. Not required when arrows of turn signal telltales that otherwise operate independently flash simultaneously as hazard warning telltale.
3. Red can be red-orange. Blue can be blue-green.
4. The color of the telltale required by S4.5.3.3. of standard no. 208 is red; the color of the telltale required by S7.3 of standard no. 208 is not specified.
5. Framed areas may be filled.
8. In the case where a signal telltale indicates more than one brake system condition, the word for brake system shall be used.

Figure 28.40 *(Continued)*

Table 2 *(continued)*

Column 1	**Column 2**	**Column 3**	**Column 4**	**Column 5**
Display	*Telltale color*	*Identifying words or abbreviation*	*Identifying symbol*	*Illumination*
Malfunction in anti-lock or	Yellow	Anti, anti-lock or ABS. also see FMVSS 105 and 135	———	———
Variable brake proportioning system 8	Yellow	Brake proportioning also see FMVSS 135	———	———
Parking brake applied 8	Red 3	Park or parking brake, also see FMVSS 105 and 135	———	———
Malfunction in anti-lock	Yellow	ABS, or antilock; trailer ABS, or trailer antilock, also see FMVSS 121	———	———
Brake air pressure position telltale	———	Brake air, also see FMVSS 121	———	———
Speedometer	———	MPH, or MPH and km/h 7	———	Yes
Odometer	———	——— 6	———	———
Automatic gear position	———	Also see FMVSS 102	———	Yes
Low tire pressure telltale (that does not identify which tire has low pressure)	Yellow	Low tire. Also see FMVSS 138		———
Low tire pressure telltale (that identifies which tire has low pressure)	Yellow	Low tire. Also see FMVSS 138		———
Tire pressure monitoring system Malfunction telltale 9	Yellow	TPMS	———	———

3. Red can be red-orange. Blue can be blue-green.
6. If the odometer indicates kilometers, then "KILOMETERS" or "km" shall appear, otherwise, no identification is required.
7. If the speedometer is graduated in miles per hour and in kilometers per hour, the identifying words or abbreviations shall be "MPH and km/h" in any combination of upper or lower case letters.
8. In the case where a single telltale indicates more than one brake system condition, the word for brake system shall be used.
9. Alternatively, either low tire pressure telltale may be used to indicate a TPMS malfunction. See FMVSS no. 138.

FIGURE 28.40 *(Continued)*

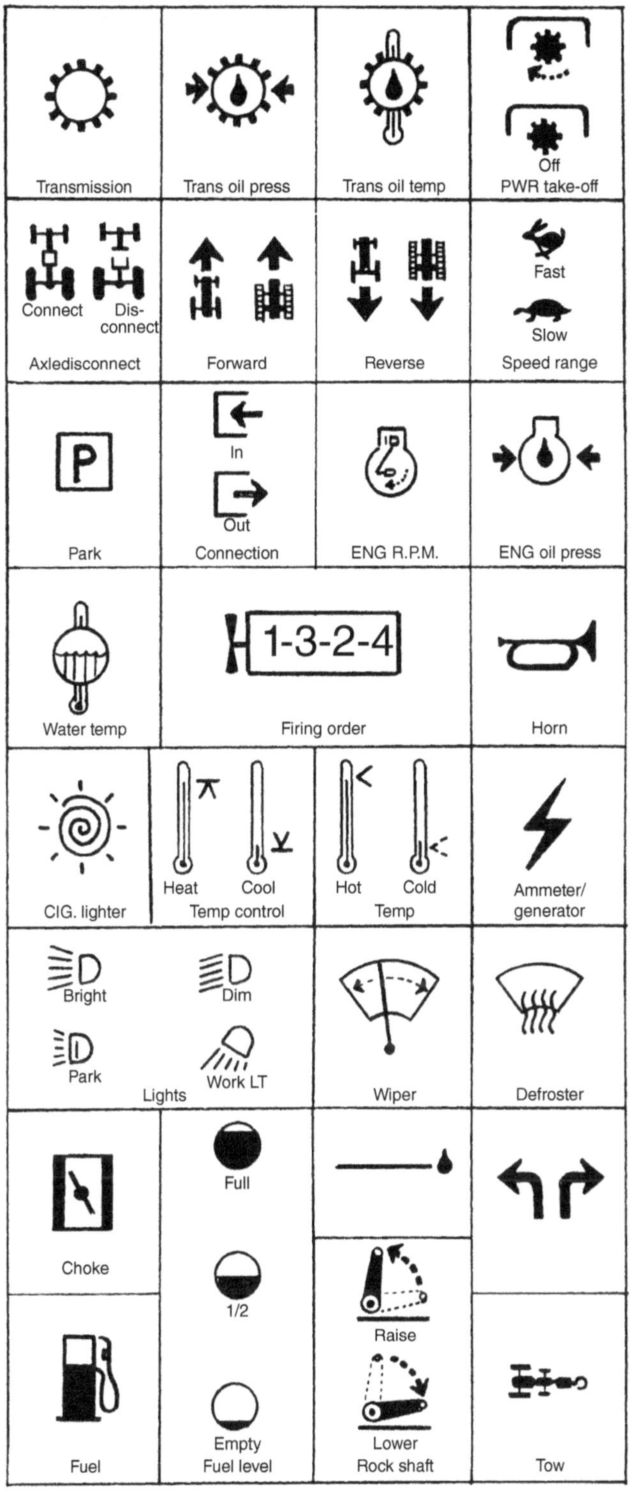

Figure 28.41 Standard symbols for agricultural equipment.

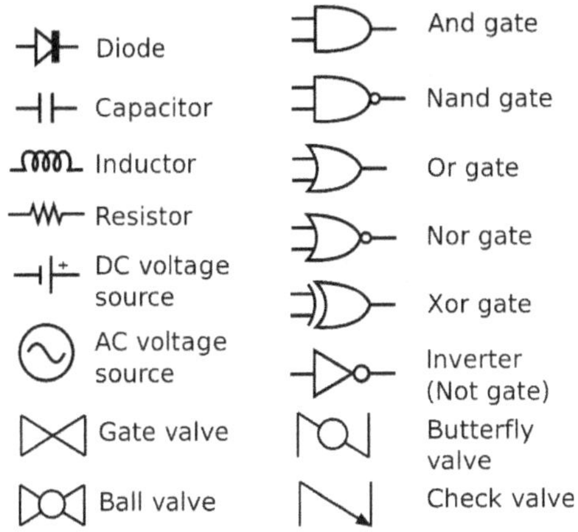

Figure 28.42 Standard engineering symbols.

symbology that by convention immediately indicates the nature of the hazard, e.g., a skull and crossbones for poison, the "fan" symbol for radiation, and the slow vehicle triangle for a roadway hazard. Add unique borders around the sign, such as yellow and black stripes or a checkerboard design.

3. Legibility

When words and messages are used, make sure that they are legible by selecting the proper letter style and size and by making the contrast between the letters and the background sufficient for whatever lighting condition may exist. Provide a sufficient border around the lettering so that other background influences, such as other signs, lights, or patterns, will not cause confusion.

4. Simplicity

Use as few words as possible; i.e., avoid long, complex advisory information or instructions that take too much time to read. Tell the observer exactly what to do or not to do. Avoid acronyms and abbreviations.

5. Intelligibility

Make it clear what the hazard is and what may happen if the warning is ignored. Especially avoid contradictions or implications that imply some doubt as to the actuality or seriousness of the hazard.

6. Visibility

Make sure the sign is visible under all expected viewing conditions, i.e., during the day and at night, if the artificial light fails, under marginal atmospheric conditions, in bright sunlight, etc.

7. Permanence

Choose devices and/or sign materials that are resistant against aging, wear, soil, and deterioration due to sunlight or cleaning methods, and use techniques that are resistant to vandalism.

8. Standardization

General. Use standard signing words, symbols, etc., wherever they already exist, such as those recommended by Occupational Safety and Health Administration (OSHA), the Federal Housing Authority (FHA), the Department of

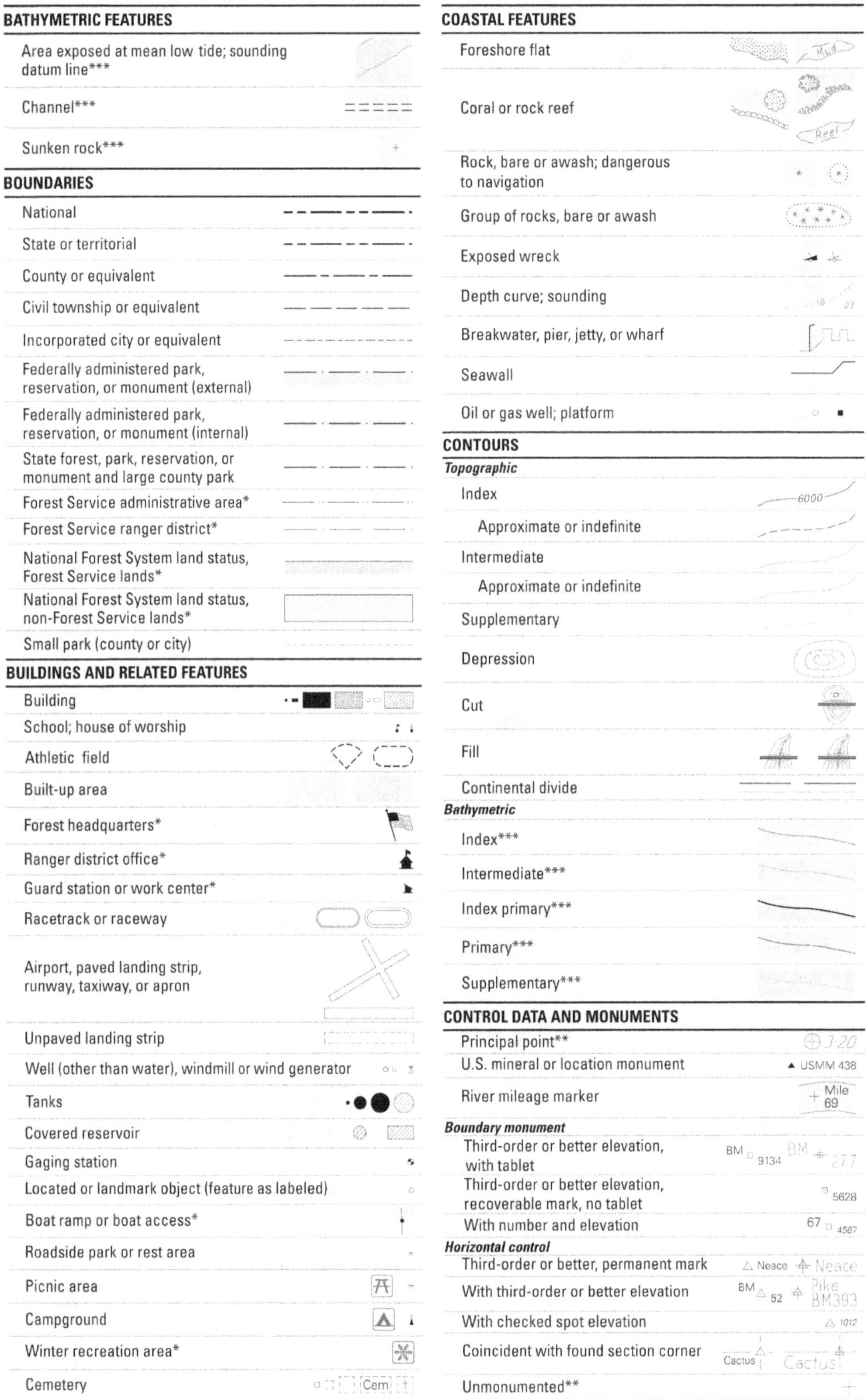

FIGURE 28.43 Topographical map symbols. (Published by the US Department of Interior.)

Commerce, and the military. (See Fig. 28.44.) Although some of these standards may not be in complete accord with the recommendations suggested here (i.e., they may not represent the best human engineering), they are usually so well established that changes to comply with human engineering principles might confuse the observer. When no such standard exists, however, apply the above principles and guidelines in creating any new signing.

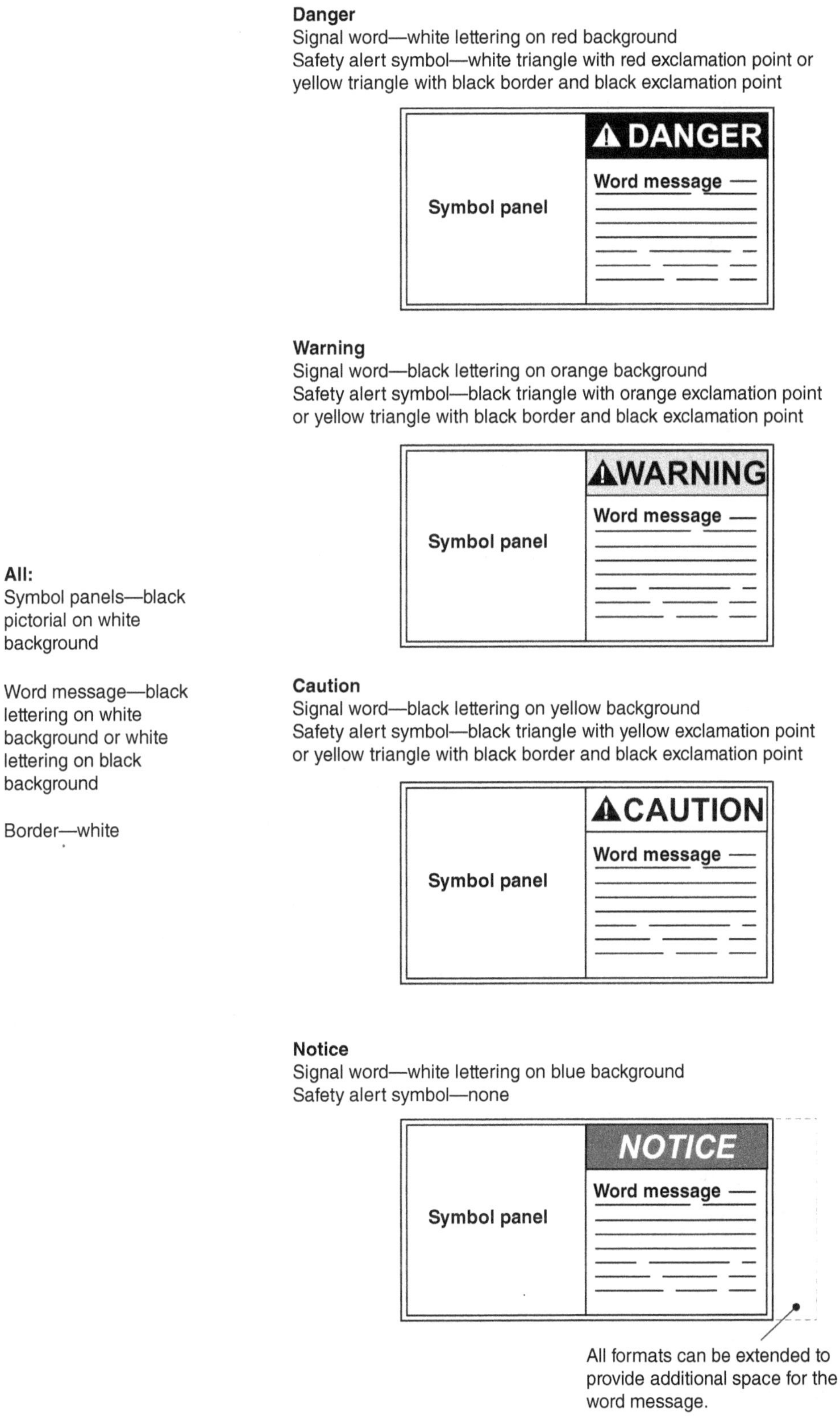

Figure 28.44 Horizontal, 3-panel format, product safety sign standards. (From ANSI Z535.4, 2007.)

The American National Standards Institute (ANSI) has developed a series of standards for identification of safety hazards. The standards are as follows:

ANSI Z535.1: Safety Color Code

ANSI Z535.2: Environmental and facility safety signs

ANSI Z535.3: Criteria for safety symbols

ANSI Z535.4: Product safety signs and labels

ANSI Z535.5: Safety tags and barricade tapes (for temporary hazards)

ANSI Z535.6: Product Safety Information in Product Manuals, Instructions, and Other Collateral Materials

ANSI Z535 series correspond (not fully) with ISO 3864 that specifies international standards for safety symbols.

The ANSI standards address four types of safety messages:

1. Environmental and facility safety signs (ANSI Z535.2)—These are generally permanent, large signs to be seen at a safe viewing distance.

2. Product safety signs and labels (ANSI Z535.4)—These signs are smaller and are usually affixed to the product. They often contain more detailed information than the facility safety signs.

3. Safety tags and barricade tapes (ANSI Z535.5)—Safety tags and barricade tapes are used to identify temporary hazards only until they are eliminated or the hazardous operation is complete.

4. Product Safety Information in Product Manuals, Instructions, and Other Collateral Materials (ANSI Z535.6)—These messages may be manual, on a single sheet of paper, or in an electronic document. They can contain detailed safety information spanning several pages.

The safety color codes in Z535.1 and the safety symbols in Z535.3 apply as guidelines to all four types of safety messages.

Product Safety Signs. A brief description of the requirements in ANSI Z535.4: Product safety signs and labels is as follows:

1. Layout—Product safety signs normally consist of a signal work, a word message, and a symbol panel. Figure 28.45 shows a typical layout of a product safety sign.

2. Signal Words—Safety signs use four signal words to indicate the potential for personal injury or property damage (DANGER, WARNING, CAUTION, NOTICE). The signal words are written in all capital letters and their height should be at least 50 percent greater than the height of the lettering in the word message.

If there is a potential for personal injury, then the safety alert symbol—the exclamation mark inside a triangle—is placed before the signal word. If the hazard consequence is property damage only (not personal injury), then "NOTICE" is the preferred signal word. The safety alert symbol should not be used with "NOTICE." The "CAUTION" without the safety alert symbol is a nonpreferred alternative to "NOTICE."

Figure 28.46 is a guide for the selection of the proper signal word.

3. Word Message—The word message is written in simple format that can be quickly understood by all potential product users. It should contain the following information: hazard identification, how to avoid the hazard, and what could happen if the hazard is not avoided.

4. Symbol Panel—Safety symbols are graphic representations of the word message. ANSI Z535.3 provides criteria for safety symbols and gives examples. Safety symbols may replace a word message (describe the type of hazard, potential consequences of the hazard, or evasive/avoidance actions to be taken) or used to clarify, supplement, or substitute for just a portion of the word message. The symbol must be comprehended by uses and ANSI Z535.3 describes methods for demonstrating this.

Examples of Safety Signs

ANSI and OSHA are standardizing the general format and design of caution and danger signs for use in industrial establishments and the home so people will recognize them quickly and understand what they mean. (See Fig. 28.47.) Books are available to help product manufacturers design their own labels.

Radioactive warning signs should contain the standardized pictorial symbol shown in Fig. 28.48.

Safety-Related Advisory Signs. The examples shown in Fig. 28.49 represent typical advisory signs that are safety related but which do not necessarily fall into the warning or caution categories. Such signs are also available from the typical safety equipment manufacturer and are designed to meet the OSHA standards. Do not try to create a new type of advisory sign before checking to see whether a similar accepted sign is already available. When it has been determined that a desired sign has not been standardized by some governmental agency, then consider the following in developing a new advisory sign:

1. Use as few words as possible to convey the message clearly.

2. Use a pictorial whenever it can be made clearly obvious to the expected observers, considering their background and experience. Do not invent vague symbols that have no relationship to something real, i.e., something that people have seen and will recognize.

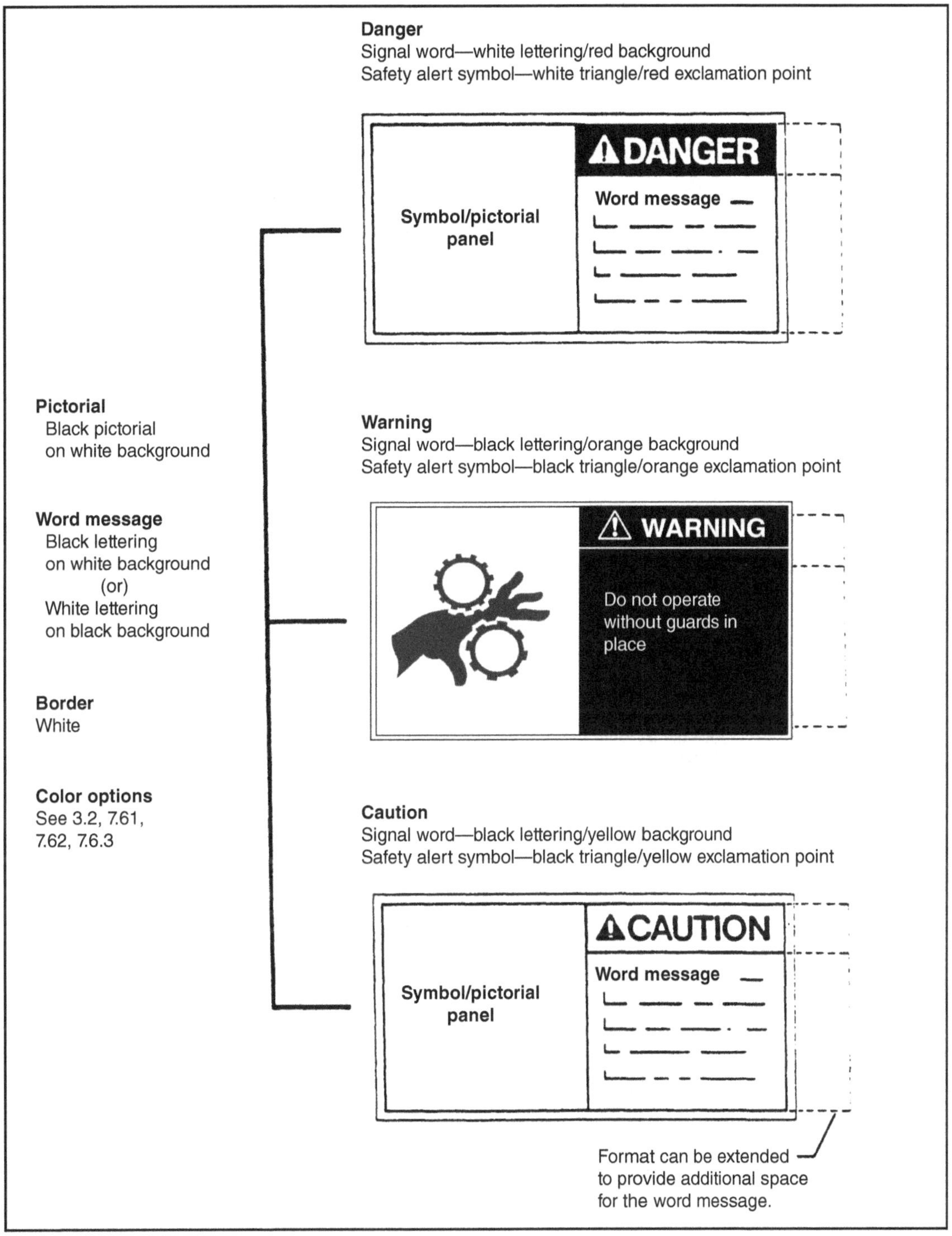

Figure 28.45 Product safety sign.

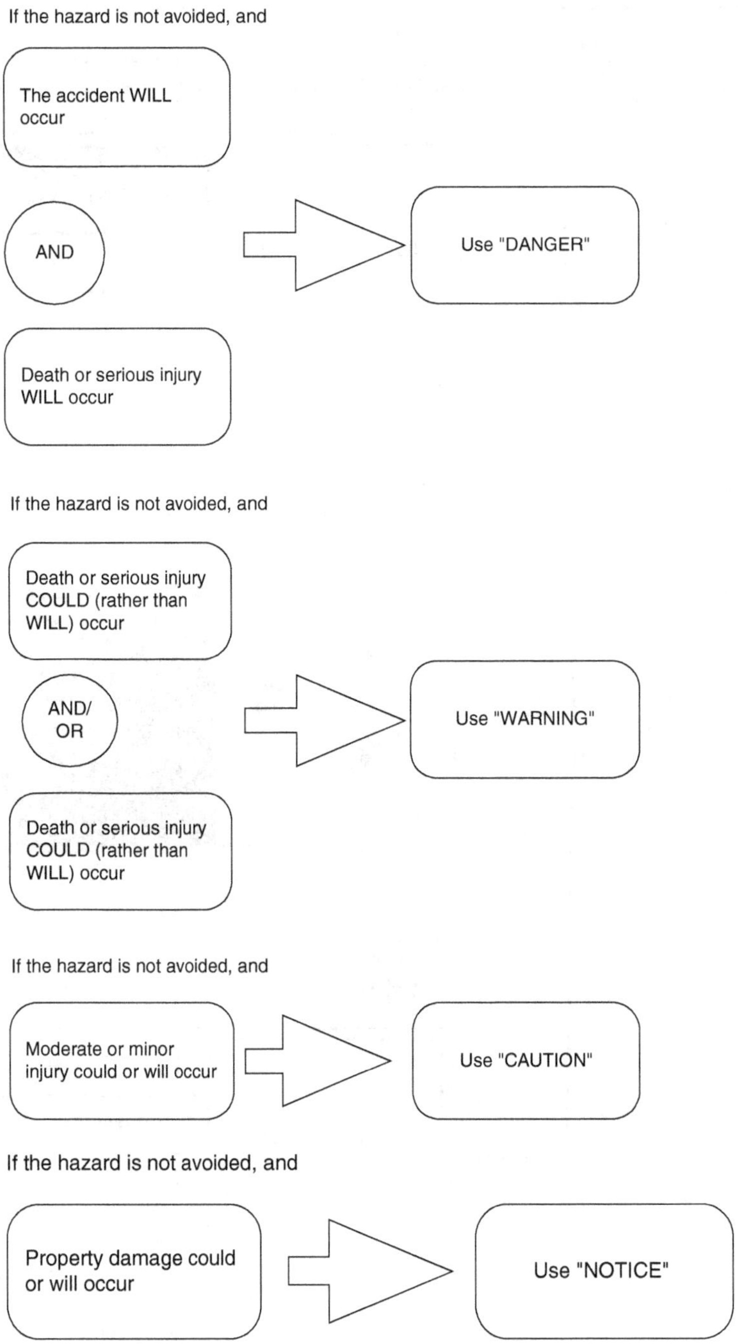

Word message
The word message is written in simple format that can be quickly understood by all potential product users. It should contain the following information:
- Hazard identification
- How to avoid the hazard
- What could happen if the hazard is not avoided

Symbol panel
Safety symbols are graphic representations of the word message. ANSI Z535.3 provides criteria for safety symbols and gives examples. Safety symbols may replace a word message (describe the type of hazard, potential consequences of the hazard, or evasive/avoidance actions to be taken) or used to clarify, supplement or substitute for just a portion of the word message. The symbol must be comprehended by the users and ANSI Z535.3 describes methods for demonstrating this.

FIGURE 28.46 Flowchart.

Figure 28.47 Example symbols for product safety signs developed by FMC Corporation.

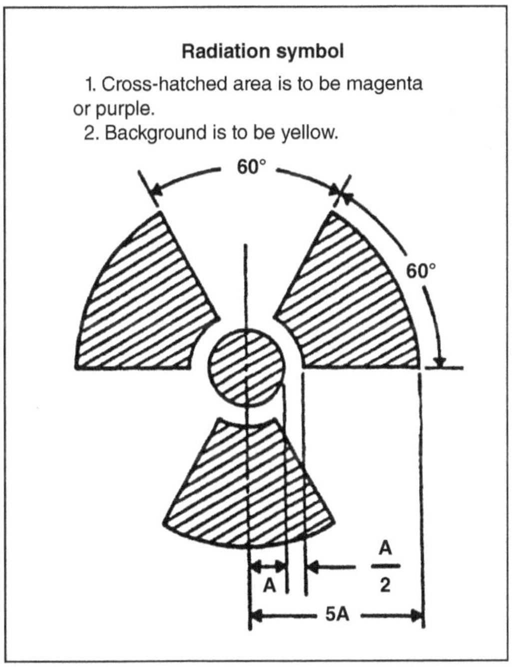

Figure 28.48 Radiation symbol required.

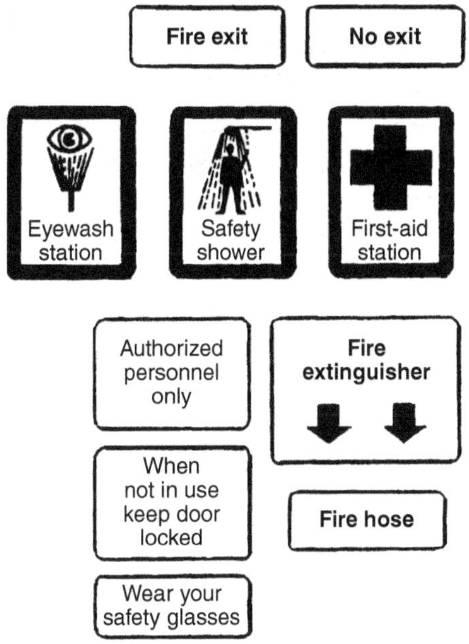

Figure 28.49 Example safety advisory signs.

3. Use maximum figure-ground contrasts and avoid the use of "pretty" colors just to be different. The best approach is to utilize the same color combinations that are used for standardized signs: black on white or yellow; white on black or red. When in doubt about colors to use on a facility that is already color-coded, such as a fire box or circuit box, provide a rectangular white background on which words or symbols are painted in black.

OSHA Hazard Communication Standard. OSHA has adopted hazardous chemical labeling requirements as part of the Hazard Communication Standard, 29 CFR 1910.1200 (HCS). This brings it into alignment with the United Nations' Globally Harmonized System of Classification and Labelling of Chemicals (GHS).

The HCS requires chemical manufacturers, importers, or distributors to ensure that each container of hazardous chemicals leaving the workplace is labeled, tagged, or marked with the following information (see Fig. 28.50):

1. *Name, Address, and Telephone Number* of the chemical manufacturer, importer, or other responsible party.

2. *Product Identifier* is how the hazardous chemical is identified. This can be (but is not limited to) the chemical name, code number, or batch number.

3. *Signal Words* are used to indicate the relative level of severity of the hazard and alert the reader to a potential hazard on the label. There are only two words used as signal words, "Danger" and "Warning." Within a specific hazard class, "Danger" is used for the more severe hazards and "Warning" is used for the less severe hazards.

4. *Hazard Statements* describe the nature of the hazard(s) of a chemical, including, where appropriate, the degree of hazard. For example: "Causes damage to

HCS pictograms and hazards

Health hazard	Flame	Exclamation mark

• Carcinogen • Mutagenicity • Reproductive toxicity • Respiratory sensitizer • Target organ toxicity • Aspiration toxicity	• Flammables • Pyrophorics • Self-heating • Emits flammable gas • Self-reactives • Organic peroxides	• Irritant (skin and eye) • Skin sensitizer • Acute toxicity • Narcotic effects • Respiratory tract irritant • Hazardous to ozone layer (non-mandatory)
Gas cylinder	**Corrosion**	**Exploding bomb**

• Gases under pressure	• Skin corrosion/burns • Eye damage • Corrosive to metals	• Explosives • Self-reactives • Organic peroxides
Flame over circle	**Environment** (Non-mandatory)	**Skull and crossbones**

• Oxidizers	• Aquatic toxicity	• Acute toxicity (fatal or toxic)

FIGURE 28.50 HCS pictograms and hazards.

kidneys through prolonged or repeated exposure when absorbed through the skin."

5. *Precautionary Statements* describe recommended measures that should be taken to minimize or prevent adverse effects resulting from exposure to the hazardous chemical or improper storage or handling. There are four types of precautionary statements: prevention (to minimize exposure); response (in case of accidental spillage or exposure emergency response, and first-aid); storage; and disposal.

6. *Pictograms* are graphic symbols used to communicate specific information about the hazards of a chemical. The GHS uses a total of nine pictograms. Each pictogram consists of a symbol on a white background framed within a red border and represents a distinct hazard(s). The pictogram on the label is determined by the chemical hazard classification.

CODING

Display Coding

Coding can be used to:

1. Facilitate discrimination between individual displays
2. Identify functionally related displays
3. Recognize the relationship between displays
4. Identify critical information within a display (unusual values, changed items, items to be changed, high priority messages, special areas of the display, errors in entry, criticality of command entry, and targets)

Consistent meaningful codes should be used. Coding must not be used if it reduces legibility or transmission time. All coding within the system must uniform. The following techniques are used to code displays:

1. Color
2. Brightness
3. Flash
4. Size
5. Pattern
6. Location
7. Underlining
8. Symbol
9. Shape

Color Coding. Display color coding can be employed to differentiate between classes of information in complex, dense, or critical displays. Color should not be used for gaining attention outside the optimum visual field. Color coding must be used consistently within a display and across displays of other systems used by the same users. The following are guidelines for using color coding.

When the user must recognize categories of information, use no more than seven colors.

To maximize discriminability, use colors having the dominant wavelengths and color values (using the International Commission on Illumination, CIE, system) listed in Table 28.7.

TABLE 28.7 Dominant Wavelengths and Color Values

Color Name	Nanometers (nm)	CIE Value (x, y, Y)
Red	700	0.6078, 0.3441, 31.05
Orange	600	
Yellow	570	0.4209, 0.5040, 111.4
Yellow-green	535	
Green	500	
Blue green	493	
Blue	470	0.1566, 0.0808, 13.33

When using color to group elements of a display together (i.e., to show parts are related or different) and the color has no meaning beyond a grouping function, up to 30 colors may be used.

1. Hue saturation can be used to indicate relative intensity (e.g., best, hottest, wettest, safest, deepest). Saturation coding is best done in grayscale, except when showcasing the following: hotter to cooler (where saturation changes from red to blue are used) and dangerous to less dangerous areas (where intensities of red are used).

2. Colors used should concur with design standards and common color association meanings for the user population (see the next section).

3. When color coding is used, luminance must be more than 10 cd/m^2.

4. If the user population includes color-blind users, color coding must not be the only means of coding information. Wavelengths above 650 nm should not be used and the color values given in Table 28.8 must be used to ensure discrimination of red and green. Also consider the effects of protective eyesore, visors, or other filters.

5. Colored symbols must differ significantly from their background and among each other. Methods for calculating the contrast and difference using the CIE color system are detailed in paragraph 5.2.2.3.2 of MIL-STD-1472G.

TABLE 28.8 Values for Color-Blind Users

Color	CIE Value (x, y, Y)
Red	0.6078, 0.3441, 31.05
Green	0.2564, 0.4372, 39.56
Yellow	0.4209, 0.5040, 111.4
Blue	0.1566, 0.0808, 13.33

6. Color-coded symbols must be adequate size for accurate color perception: major dimension of isolated large symbols should subtend 45 minutes of visual angle but not less than 30 minutes. The height of small characters and symbols must subtend at least 20 minutes of visual angle.

7. Symbols must be color-filled rather than simply outlined in color.

Flash Coding. Flash coding should only be employed to call the user's attention to mission critical events. The following are guidelines for using flash coding:

1. The percentage of "on" time must be equal to but not less than the percentage of "off" time; a 50 percent duty cycle is preferred.

2. No more than two flash rates should be used. The two flash rates should differ by not less than 2.0 Hz, with higher flash rate not be greater than 5.0 Hz and the slower flash rate not less than 0.8 Hz. The higher flash rate should reflect more critical information.

3. Characters that must be read must not flash; an adjacent flashing symbol or flashing background may be used to add emphasis to text.

4. Event acknowledgment or flash suppression control should be provided.

5. Flash only a small area of a display at any time.

Display Symbol Size Coding. Use no more than three size levels. The major dimensions of the larger size must not be less than 150 percent of the major dimension of the smaller.

Pattern and Location Coding. Pattern and location coding can be used to reduce user search time by restricting the area to be searched to prescribed segments.

Underlining Coding. Underlining may be employed to indicate unusual values, errors in entry, changed items, or items to be changed.

Symbol Coding. Symbol coding may be used to enhance information assimilation from data displays, and must be in accordance with the following:

1. Symbols should be analogs of the event or system element they represent or be in general use and well known to the expected users.

2. When special symbols are used to signal critical conditions, they must be used for only that purpose.

3. When a special symbol is used to mark a word, the symbol must be separated from the beginning of the word by one space.

Additional Coding Guidelines

1. The number of discrete symbols should be held to a minimum.

2. Symbol meanings should be standardized.

3. Symbols should be recognizable without reference to some comparison standard.

4. If combination codes are used, the respective meanings should be decodable separately without confusion.

5. Two codes maximum combination.

6. Symbol shapes should reflect a natural relationship to the event or type of target the symbol is supposed to represent.

7. Codes should be easy to learn and retain.

8. Symbols should differ widely in shape: avoid variations of a single form, e.g., a circle and an ellipse.

9. Redundant cues (i.e., those which associate a symbol with a particular size) increase coding efficiency.

10. External modifiers surrounding a symbol should be avoided.

Use of Colors for Signing and Marking Safety Hazards

Table 28.9 lists safety colors commonly used and their values. Table 28.10 summarizes the guidelines for application of equipment safety colors.

TABLE 28.9 Values for Commonly Used Safety Colors

Color Name	Dominant Wavelength (in nanometers)	Munsell Code
Red	610	5.0R/3.9/15.4
Yellow	582	3.3Y/8.0/143
Green	515	3.2G/4.9/11.1
Light Blue (cyan)	494	2.7GB/7.9/6.0
Dark Blue	476	2.9PB/4.1/10.4
Magenta	430	6.5P/4.3/9.2
White		2.5PB/9.5/0.2
Black		N/0.8

NUREG-0700, Human Systems Interface Design Review Guidelines, Nuclear Regulatory Commission, 2002.

TABLE 28.10 Equipment Color-Coding Guidelines

Color	Meaning	Typical Use
Red	Fire protection equipment and apparatus	Fire alarm boxes; fire exit signs; fire extinguishers, buckets, and pails; fire hose locations; fire blanket boxes; fire hydrants, pumps, and sirens; fire sprinkler piping
	Danger	Danger signs; safety cans or other portable containers of flammable liquids having a flash point at or below 80°F (26.9°C) (excluding shipping containers); red lights on temporary obstructions or construction
	Stop	Stop button used for emergency stopping of machinery; emergency stop bars on hazardous machinery
Orange	Dangerous parts of machines or energized equipment which may cut, crush, shock, or otherwise injure; purpose is to emphasize danger when enclosure doors or guards are open and safety hazard exists	Safety starting buttons; exposed edges of pulleys, gears, cutting devices, rollers, power jaws, etc.
Yellow	Physical hazards, such as stumbling, tripping, falling, or striking against an object	Handrails, guardrails, or top and bottom treads of stairways where caution is needed; lower pulley blocks and cranes; pillars, posts, and columns which might be struck; materials-handling equipment, such as industrial tractors, trucks, trailers, forklifts, conveyors, or gantry cranes; piping systems containing dangerous materials
	Caution	Caution signs
Green	Safety	Safety bulletin boards
	Location of first-aid equipment	Gas masks; first-aid kits; stretchers; safety deluge showers; safety signs
Blue	Caution against starting, using or moving equipment under repair or in use	Scaffolding and ladders; electrical controls; valves
Purple	Radiation hazards (used in combination with yellow for tags, labels, signs, and floor markers)	Containers of radioactive materials; disposal cans for contaminated materials; signal lights to indicate when radiation-producing machines are in operation
Black and/or white	Traffic and housekeeping markings	Directional signs; dead ends of aisles or passageways

1. Red should be used to identify:

    a. Fire protection equipment, fire apparatus, and fire alarm boxes

    b. Fire blanket boxes

    c. Fire buckets or pails

    d. Fire exit signs

    e. Fire extinguishers

    f. Industrial fire hydrants

    g. Fire pumps

    h. Fire sirens

    i. Post indicator valves for sprinkler systems

    j. Stop bars on hazardous machines

2. Orange should be used to identify:

    a. Dangerous parts of machinery energized equipment that could cut, crush, shock, or otherwise injure people. The inside of a cabinet door to indicate when the door is open, when guards around moving parts are removed, or an electrical circuit box is open, exposing high-voltage potential

3. Yellow should be used to identify:

    a. Possible physical hazards such as being struck, falling, stumbling, tripping, or being caught between moving and stationary devices. Yellow and black striping around the hazard is recommended

4. Green should be used to identify:

    a. Safety equipment

    b. First-aid equipment

5. Blue should be used to identify:

    a. Situations in which equipment should not be started or moved

    b. Situations in which equipment is being repaired or worked on

    c. Electrical circuit boxes

6. Purple should be used to identify:

    a. Radiation hazards including x-rays, alpha rays, beta rays, gamma rays, neutrons, protons, deuterons, and mesons. Yellow should be used in combination with purple for markers such as tags, labels, signs, and floor marking

7. Black and white should be used as basic colors for designating traffic and housekeeping markings. Black and white striping and/or checkerboard borders and solid areas may be used for added conspicuousness.

For additional information, refer to ANSI Z53.1, Safety Color Code.

Electrical and Hydraulic Conductor Color Coding. Wires should be color coded by (in order of preference): solid-color insulation, solid-color insulation with colored-strip tracer, or, colored braid insulation with woven tracer. Every wire in a cable should be color coded over its entire length. Wires should be coded using discriminably different patterns of solid colors and solids with striped tracers (see Tables 28.11 and 28.12). For more than 21 wires, see MIL-STD-686. If a wire's color may be obscured, wires may be coded with numbered metal tags. The length of wires should be long enough to allow for at least two repairs.

Fluid Line Color Coding. Color coding standards for marking fluid lines are defined in ASME A13.1-2007, Scheme for the Identification of Piping Systems, and are shown in Table 28.13.

Shipped Material Color Coding. Hazardous materials shipped between states must be labeled according to the rules set forth in the U.S. Code of Federal Regulations 29 CFR 172. Table 28.14 shows the required labeling background colors.

Marking of Military Equipment

Following are some of the military specifications that should be referred to with respect to the color and marking of military equipment:

MIL-C-3702: *Cable, Power, Electrical, Ignition, High Tension*

MIL-M-8555: *Missiles, Guided-Design and Construction*

MIL-M-11745: *Marking and Labeling of U.S. Army Marine Craft*

MIL-M-19590: *Marking of Commodities and Containers of Radioactive Materials*

MIL-HDBK-1473: *Color and Marking of Army Material*

Refer to the following military standards:

MIL-STD-101: *Color Code for Pipelines and Compressed Gas Cylinders*

MIL-STD-129: *Marking for Shipment and Storage*

MIL-STD-161: *Identification of Bulk Petroleum Products and Missile Fuels*

MIL-STD-172: *Color Code for Containers of Liquid Propellants*

MIL-STD-195: *Marking of Connections for Electric Assemblies*

TABLE 28.11 Electrical Conductor Color Coding

Instructions:

1. Find the number of the conductor to be color coded.
2. The colors at the right of the number are the appropriate combination for that conductor. For example, if a cable consists of 12 conductors, the twelfth color combination would be black with white tracer. The eight color combination would be red with black tracer. The fifth color combination would be orange without tracer, and so on.

No. of Conductors	Basic Color	Tracer
1	Black	None
2	White	None
3	Red	None
4	Green	None
5	Orange	None
6	Blue	None
7	White	Black
8	Red	Black
9	Green	Black
10	Orange	Black
11	Blue	Black
12	Black	White
13	Red	White
14	Green	White
15	Blue	White
16	Black	Red
17	White	Red
18	Orange	Red
19	Blue	Red
20	Red	Green
21	Orange	Green

Note: If a cable has concentrically laid conductors, the first combination or color applies to the center conductor. If a cable contains various sizes of conductor, the first color applies to the largest, continuing in order of conductor size.

From MIL-STD-1472G, 2012.

TABLE 28.12 Hydraulic Conductor Color Coding

Function	Color	Function Definition
Intensified pressure	Black	Pressure in excess of supply pressure induced by a booster or intensifier
Supply pressure	Red	Pressure of power-actuating fluid
Charging pressure	Intermittent red	Pump inlet pressure higher than atmospheric pressure
Reduced pressure	Intermittent red	Auxiliary pressure lower than supply pressure
Metered flow	Yellow	Fluid at a controlled flow rate (other than pump delivery)
Exhaust	Blue	Return of power-actuating fluid to reservoir
Intake	Green	Subatmospheric pressure, usually on intake side of pump
Drain	Green	Return of leakage of control-actuating fluid to reservoir
Inactive	Blank	Fluid within the circuit but not serving a functional purpose during phase being represented.

From MIL-STD-1472G, 2012.

TABLE 28.13 Fluid Line Color Coding

Fluid	Color Coding
Fire quenching fluids	Red
Toxic and corrosive fluids	Orange
Flammable fluids	Yellow
Combustible fluids	Brown
Portable, cooling, boiler feed, and other water	Green
Compressed air	Blue

TABLE 28.14 Shipped Material Color Coding

Material	Required Shipping Label Background Color
Explosive	Orange
Nonflammable gas	Green
Flammable gas	Red
Poisonous gas	Brown
Flammable solid	Red/white stripes
Spontaneously combustible	Red and white
Dangerous when wet	Blue
Oxidizer	Yellow
Inhalation poison	White
Infectious	White
Radioactive	White and yellow
Fissle materials	White
Corrosive materials	White and black

MIL-STD-450: *Signs for Contaminated or Dangerous Areas*

MIL-STD-686: *Cable and Cord, Electrical, Identification Marking and Color Coding of it*

MIL-STD-709: *Ammunition Color Coding*

MIL-STD-1458: *Radioactive Materials Marking*

MIL-STD-13231: *Marking of Electronic Items*

Note: The reader is advised to check current specifications, standards, and local ordinances and regulations as they may apply, since such regulatory materials continue to undergo modification and updating.

QUANTITATIVE READOUT DIGITAL DISPLAYS

The digital readout generally provides the most rapid and accurate method for presenting purely quantitative information. In selecting or designing such displays, consider the following (see Fig. 28.51):

1. The numeral characters should be as legible as possible in terms of character style, height/width ratio, stroke width, and figure contrast.

2. Mechanical counter readouts (drum type) should increase in value by an upward movement of the drum. The face of each drum (and its numerals) should not be buried deeply in the counter window. Individual drums that separate numerals are not desirable because they tend to obscure the numbers and make the total set of numbers harder to read. Wide separators between drums are similarly undesirable. However, such separators are useful for imprinting decimal points and/or commas when the numerical value contains many characters, e.g., 1,000,370.

3. Numbers should always read left to right; vertical arrays are undesirable.

4. When it is necessary to use other types because of a requirement for more rapid response, dot and line matrices are permissible. Although sloping matrix characters have been used extensively, these are less desirable. The slope should never exceed about 11°. Electrically pulsed counter rates should not exceed 50 per second. For design criteria of electronically generated displays, see Electronic Displays section in this chapter.

5. Electronically generated digital characters should be as free of brightness variations, jitter, and flicker as possible.

6. Although various colors are acceptable when they have special meaning, the preferred colors are black on white and white on black for general use.

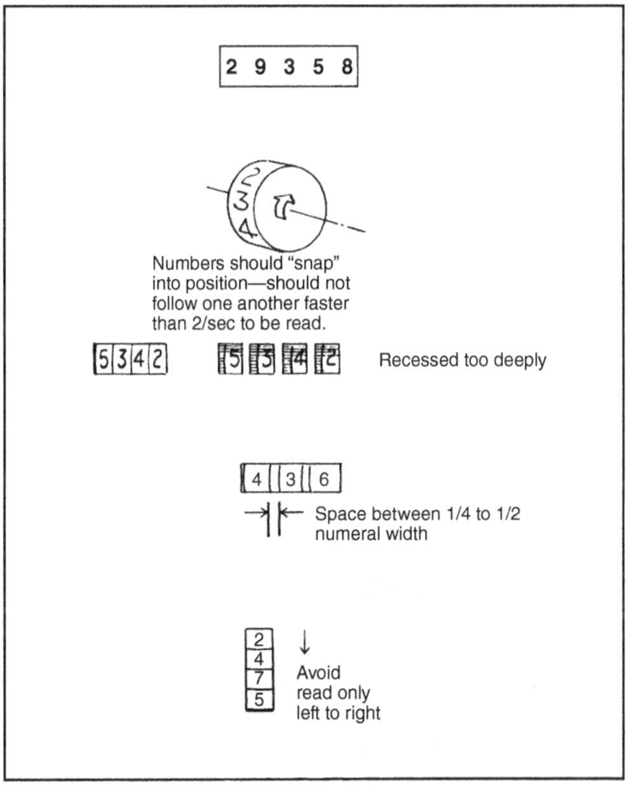

Figure 28.51 Design guidelines for mechanical counters.

DIALS AND GAUGES

Dials and gauges are used primarily to provide quantitative information. However, as opposed to a digital display (which provides extremely precise quantitative information), the dial or gauge also provides some additional information in the form of advance warning, rate of change, and/or opportunity to make "cross-dial" extrapolation; this is due to the fact that the pointer position and motion act as an additional qualitative cue as to what is happening.

First, and perhaps foremost, whenever possible avoid the use of dials and/or gauges in which double pointers or scales are used. A single scale and pointer format is best because it provides the least reading error.

As shown in Fig. 28.52, both round (or half-round) and linear or rectangular formats are possible. It has not been shown that either format results in any difference in readout effectiveness.

Place the pointer at the bottom or to the right of the scale. (See Fig. 28.53.) When the pointer is on the right or bottom of the scale, the direction of the pointer movement for increase and decrease is "natural." However, the increase and decrease direction of motion is reversed when the pointer is to the left or top of the scale.

When several round dials have to be clustered for ease in check readings (i.e., when an operator must glance at them quickly to ascertain that things are functioning properly), the instruments can be oriented so that all the pointers are aligned alike when the instruments are indicating normal

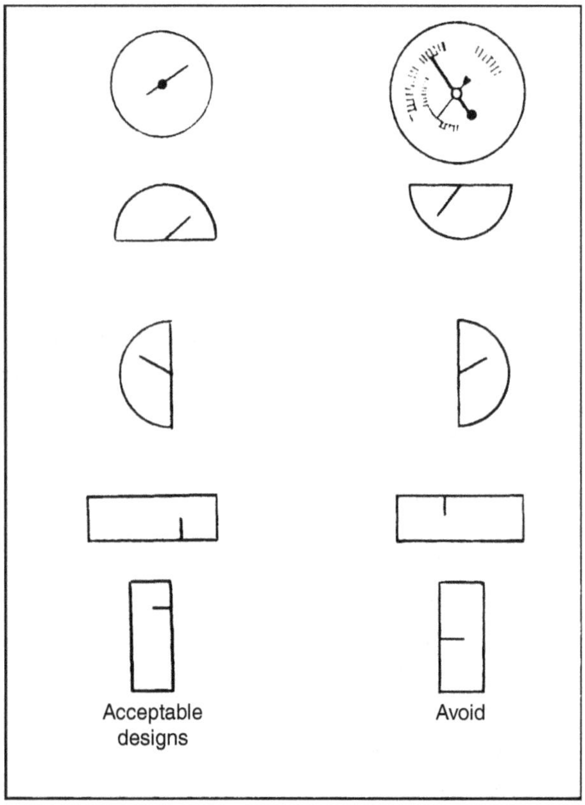

FIGURE 28.52 Dial and gauge design recommendations.

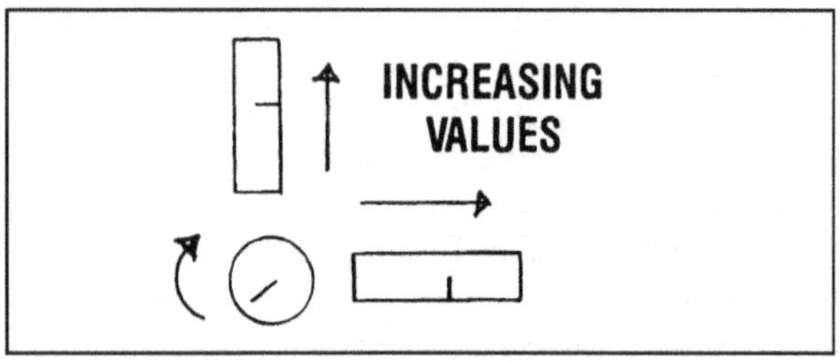

FIGURE 28.53 Recommended direction of needle movement.

operation. This makes a single pointer misalignment easy to see.

Vertical arrays should be oriented so that normal pointer positions are vertical; horizontal arrays should be oriented so that normal pointer positions are horizontal (see Fig. 28.54).

Design Considerations

Dial Size. The size of a dial or rectangular gauge normally should be determined by the number of scale graduation marks required (which is a matter of the inherent precision required). Instrument manufacturers have tended to standardize their instrument package sizes for the sake of production convenience and therefore have sometimes put more markings on the instrument face than can be accurately read. The spacing of scale graduation marks must be great enough so that the observer can discriminate between one mark and another, and also see the relation between the marks and the pointer, without taking an inordinate amount of time peering at the instrument.

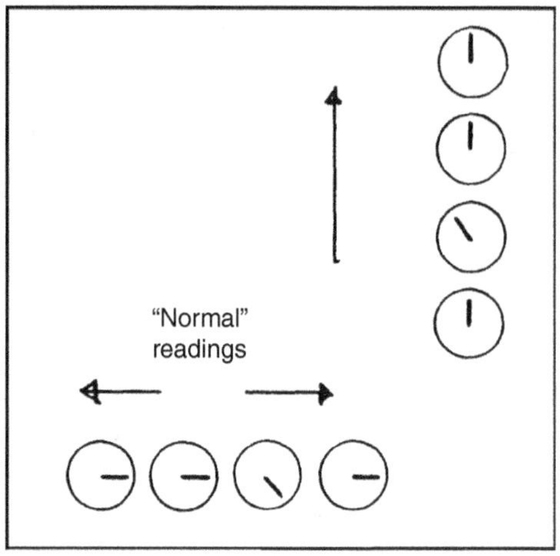

FIGURE 28.54 Recommended "normal" needle position in clustered displays.

In addition, the size of the dial depends on how deeply inset the face is below the bezel. (See Fig. 28.55.) When the manufacturer's case design includes a deeply inset dial face, the bezel tends to limit the viewing angle, with the result that (depending on the viewing angle) some of the numbers and/or scale marks may not be visible. Some designers have created poor formats to overcome this problem, such as placing the numbers "inside" the scale to make sure that they are not obscured. As will be noted later, this format is counter to good pointer-scale relationships and results in a pointer that partially covers up the scale numbers.

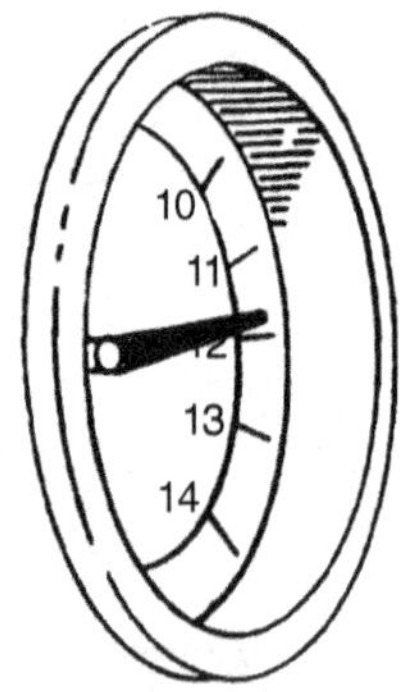

FIGURE 28.55 A deep set bezel limits viewing angles.

Dial Size in Relation to Required Number of Scale Marks. Too many scale marks, too close together, make dial reading difficult and error prone. In fact, it is impractical to have a dial that is so small that the inner annulus of the scale is less than 1 in. (2.5 cm) in diameter. Use Table 28.15 as a guide in determining the size of dial necessary to accommodate the required number of scale marks. For example, if the viewing distance is approximately 12 ft (3.7 m) and if

TABLE 28.15 Minimum Diameter of Inner Annulus at Various Viewing Distances

No of Scale Marks	20 in.	3 ft	6 ft	12 ft	20 ft
50		1.3 in.	2.6 in.	5.0 in.	9.0 in.
100	1.4 in.	2.6 in.	5.0 in.	10.0 in.	17. in.
150	2.0 in.	3.9 in.	8.0 in.	15.0 in.	26.0 in.
200	2.9 in.	5.0 in.	1.0 in.	21.0 in.	34.0 in.
250	3.5 in.	6.4 in.	13.0 in.	26.0 in.	43.0 in.
300	4.0 in.	7.7 in.	15.0 in.	31.0 in.	51.0 in.
350	5.0 in.	9.0 in.	18.0 in.	36.0 in.	60.0 in.

50 scale graduation marks are required, the inner annulus of the scale must be at least 5 in. (13 cm) in diameter.

A common design error occurs when the designer believes that a specified level of accuracy must be provided and therefore crowds the scale marking in order to get a certain number of marks within the constraints of the dial-face diameter. The result is that the observer cannot differentiate between the marks and not only makes errors in reading but also spends more time reading; thus, the basic objective is lost.

The designer should either reconsider the need for the level of accuracy thought to be required (i.e., the instrument itself may not provide the level of accuracy it was assumed to) or use no more than nine marks, since the observer can probably interpolate the pointer position between fewer, properly spaced marks more accurately and reliably than he or she would be able to do with the poorly marked dial. (See Fig. 28.56.)

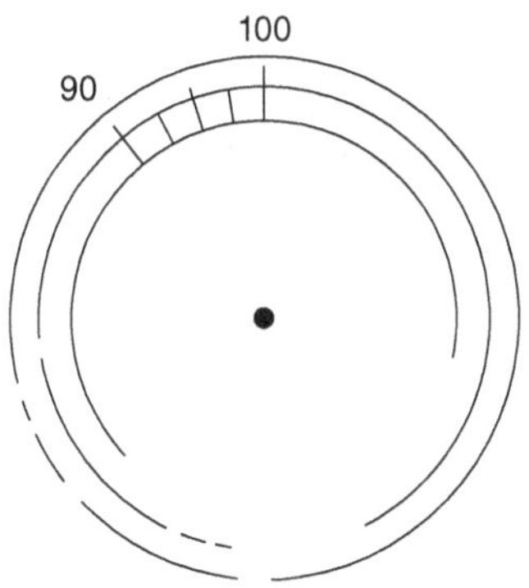

FIGURE 28.56 Use no more than nine marks between numbers.

Scale-Marking Design Errors. The following are some typical scale-marking design errors to be avoided (see Fig. 28.57):

1. Dots instead of lines
2. Thick marks
3. Marks joined by a heavy base line
4. Long marks spaced close together

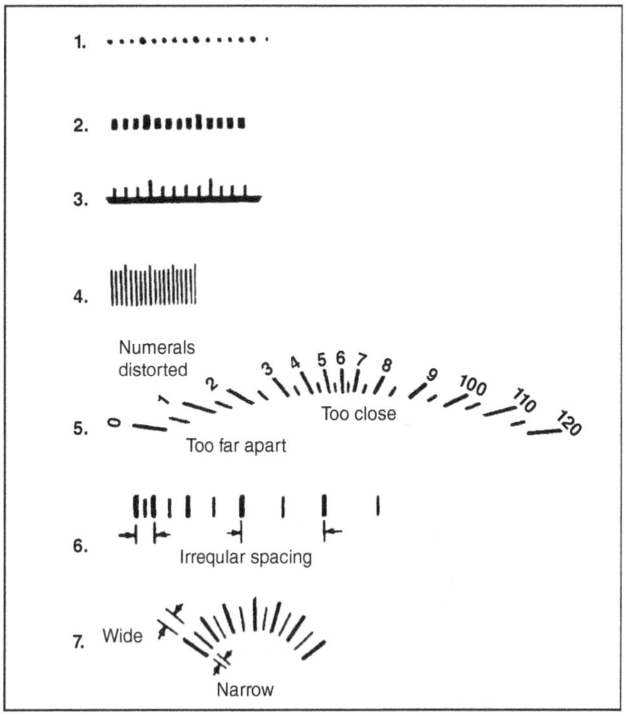

FIGURE 28.57 Avoid these scale marking errors.

5. Irregularly spaced marks about an arc, which cause the observer to shift his or her spatial reference
6. Uneven spacing, which leaves the observer in doubt as to the value of each mark
7. Significant spacing variation between the ends of the marks, which, because of the tight radius of the dial, makes it hard for the eye to follow along the scale-pointer axis

Pointer-Scale Relationships

1. For ease of reading, the scale-pointer relationship should be as shown in Fig. 28.58. The point tip should be the same width as the scale mark, and it should fall just short of the mark. Placing the pointer just inside the scale annulus ensures that it will never be confused with a mark.

2. The plane of the scale and the pointer also should be the same; otherwise, the observer may misread the scale-point association because he or she was looking at the instrument from an angle (parallax). The figure illustrates the proper arrangement for pointer and scale plane.

3. Finally, the numbers associated with the scale should be outside the scale ring so that the pointer will not obscure the number. If space is limited (for curved or arc scales) and only gross readings are required, numbers may be placed on the inside of the scale. This can avoid undue scale constriction.

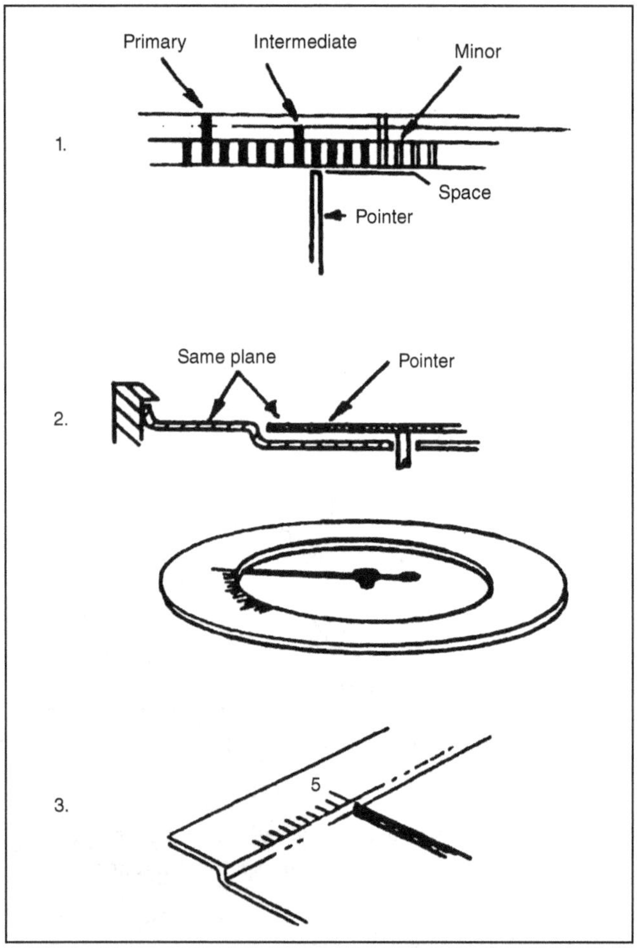

FIGURE 28.58 Correct relationship between pointer and scale.

Instrument Scale-Marking Guidelines. Experiments by military psychologists produced the scale-marking specifications shown in Fig. 28.59. Scales were read with fewer errors and faster response when these dimensional characteristics were present. The 28-in. (71-cm) viewing distance is typical of the aircraft cockpit situation. For other viewing distances, the proportions should be maintained, even though values are increased or decreased; i.e., multiply each dimension times the viewing distance, or

$$\text{Dimension at 28 in.} \times \frac{x \text{ in.}}{28}$$

Direction of Pointer Movement. Scale numbering should reflect the expected direction of pointer movement; i.e., pointers should move to the right, up, or clockwise to indicate a value increase. (See Fig. 28.60.)

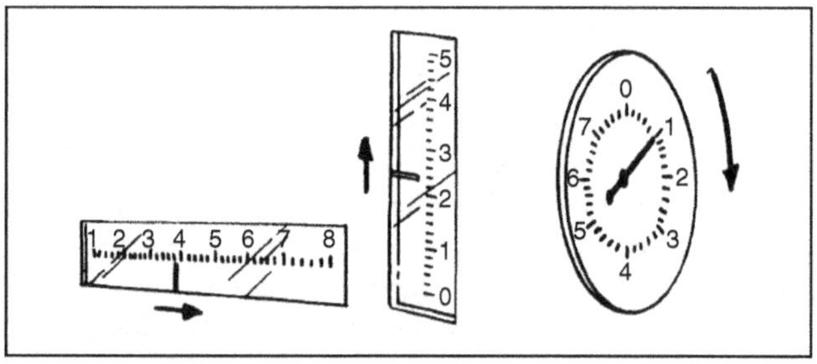

FIGURE 28.60 Pointer movement to indicate value increase.

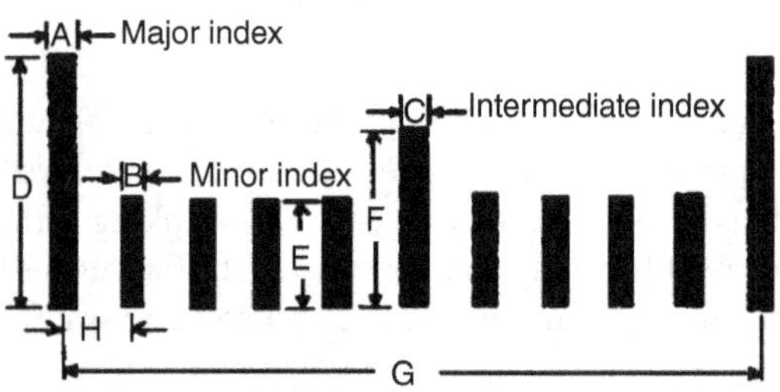

Dimensions of dark markers on light background, visual angle [1]	
A Width of major scale index	1.16 mrad (4 min) [2]
B Width of minor scale index	0.87 mrad (3 min) [2]
C Width of intermediate scale index	1.16 mrad (4 min) [2]
D Length of major scale index	7.86 mrad (27 min)
E Length of minor scale index	3.49 mrad (12 min)
F Length of intermediate scale index	5.82 mrad (20 min)
G Width of gap between major scale index	25.02 mrad (86 min)
H Width of gap between minor scale index	2.62 mrad (9 min)

Notes:

[1] For most applications with a dark graduation mark on a light background, the width of the minor graduation mark can be used for major and intermediate graduation marks as well. Use of this strategy allows the width of the pointer tip to be the same as all of the graduation marks. Visual angles are for longest anticipated viewing distance.

[2] 4.36 mrad (15 min) for light markers on dark background.

FIGURE 28.59 Recommended scale mark dimensions. (From MIL-STD-1472G.)

Scale Interval. For most people, scale intervals of 1 through 10 (or multiples thereof) are easiest to interpret without error. (See Fig. 28.61.) The next easiest are intervals of 2, 4, 6, 8, and 10. However, intervals such as 3, 6, and 9 (or 4, 8, and 12) can cause confusion.

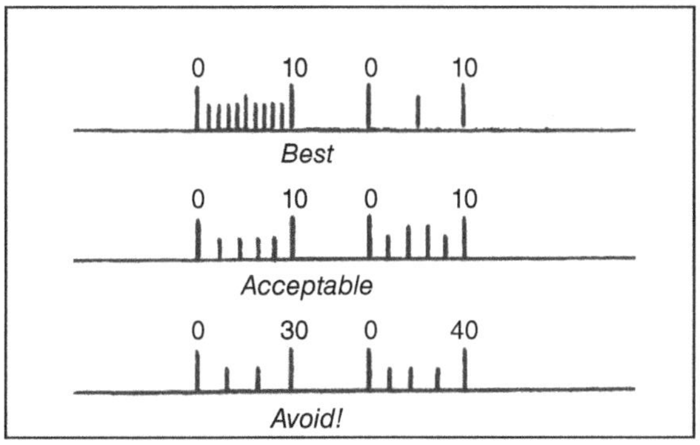

Figure 28.61 Scale intervals should be multiples of 10.

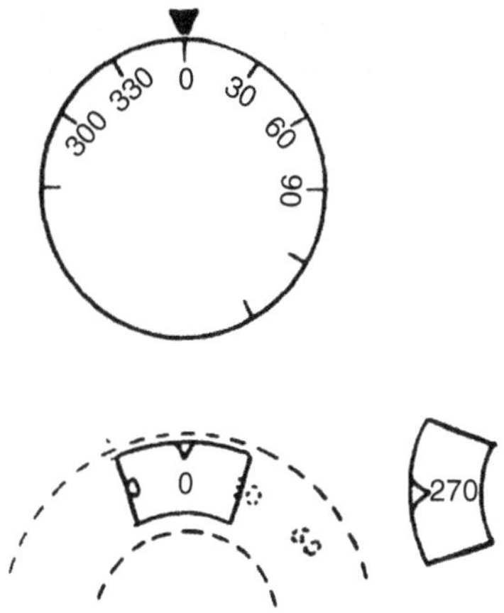

Figure 28.62 Readings should be upright.

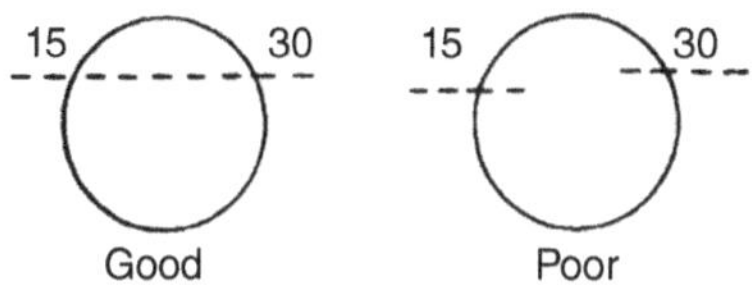

Figure 28.63 Make scale symmetrical.

Other Design Considerations

1. Make sure that numerals are oriented upright rather than slanted and turned on their side. (See Fig. 28.62.) The exception to this rule is the situation in which the dial may rotate, in which cases it is desirable to orient each numeral so that its base is toward the center of the dial, which makes the numeral appear upright when it rotates to an indexing position, preferably at the top of the dial. Alternative orientations may be necessary if the display exposes only one-half of the dial and the display is oriented to emphasize an up-and-down direction of motion.

2. If a scale covers only part of a circumference, attempt to balance the number positions so that they present a symmetrical appearance (see Fig. 28.63).

3. Consider the use of patterned or colored range strips when the primary readout is to determine when a system is performing within specified ranges, i.e., when it is less important to read specific values. (See Fig. 28.64.)

4. Although it is always desirable to optimize the size and shape of numerals used on dial displays, circumstances may dictate a smaller instrument face. It is better to use smaller numerals and thus avoid the clutter and illegibility that result when characters of a so-called optimum size are crowded together. (See Fig. 28.65.)

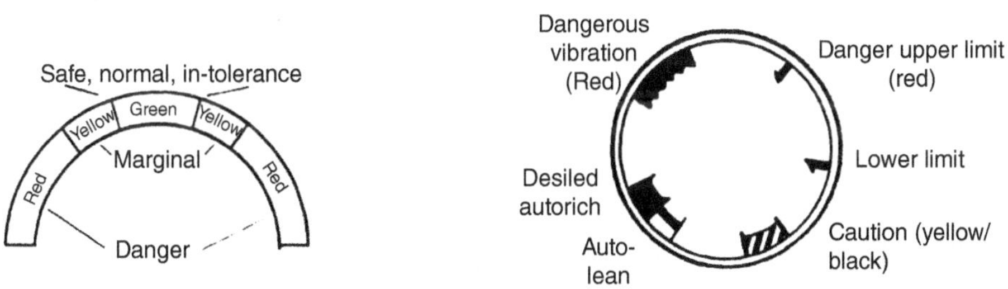

Figure 28.64 Shape and color-coding ranges. (From MIL-HDBK-1472G.)

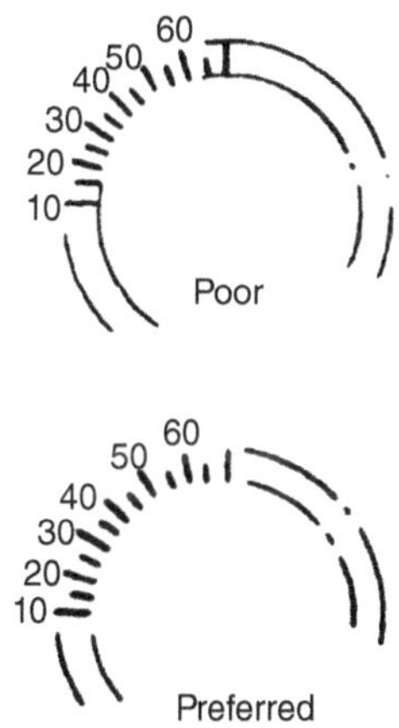

Figure 28.65 Do not crowd numbers.

5. Whenever the dial scale is of finite length, be sure to leave a sufficient break between the beginning and end of the scale so that observers do not mistakenly assume that the scale continues and/or that they should continue reading around the scale, but with some special extrapolation. (See Fig. 28.66.)

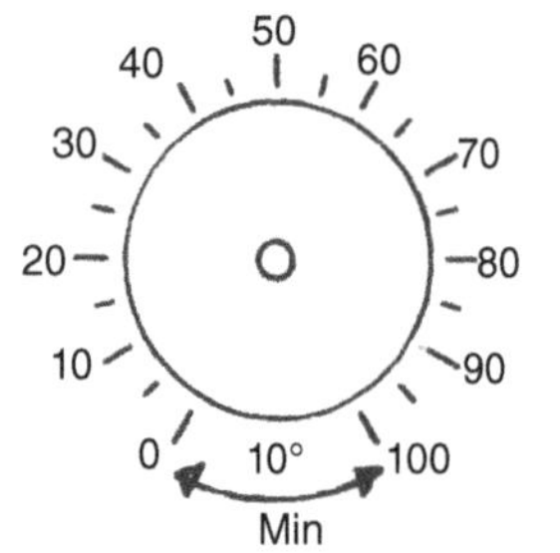

FIGURE 28.66 Have break between beginning and end of finite scale.

6. When several dial or rectangular, scaled instruments or gauges are mounted within a single workstation or on the same instrument panel, make sure that the scale breakdown is alike on all the instruments; i.e., do not mix two different scale breakdowns, as shown in Fig. 28.67.

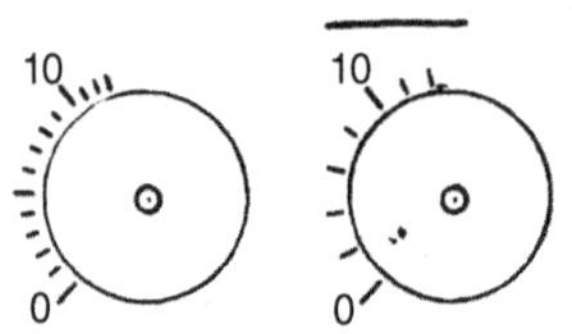

FIGURE 28.67 Dissimilar scale values on same panel = errors.

7. Instrument faces for use when night or dark adaptation is not critical are more easily seen if the markings are black on a light-colored face. Even though the instrument may have to be illuminated at night (e.g., an automobile speedometer), the lighter background can still be dimly lit and provide a more readable display. In the case of the automobile speedometer, the display is also easier to see when sunlight reflects on the cover. For blackout conditions, however (airplane cockpits, etc.), light characters and marks on a black background are mandatory to help the observer maintain proper dark adaptation.

Design of Pointers for Dial and Gauge Instruments

For maximum reading accuracy, a pointer should be of equal width throughout so that the parallel sides visually project to match the parallel-sided scale marks to which the pointer must be associated.

The width of the pointer should be exactly the same width as the scale marks. That is why, in a previous discussion of scale marking, it was noted that all scale marks should be the same width. In many applications, however, the thin pointer is not conspicuous enough, especially when the user has to depend on a brief, quick glance to see how a particular system is performing. Thus, it is necessary to increase the width of the pointer, i.e., to make it bolder and more obvious during a quick scanning look. In such cases, the pointer should be designed as shown in Fig. 28.68. Note, however, that although the main stem of the pointer is enlarged, it is still tapered at the tip so that the width of the tip still matches the scale-mark width.

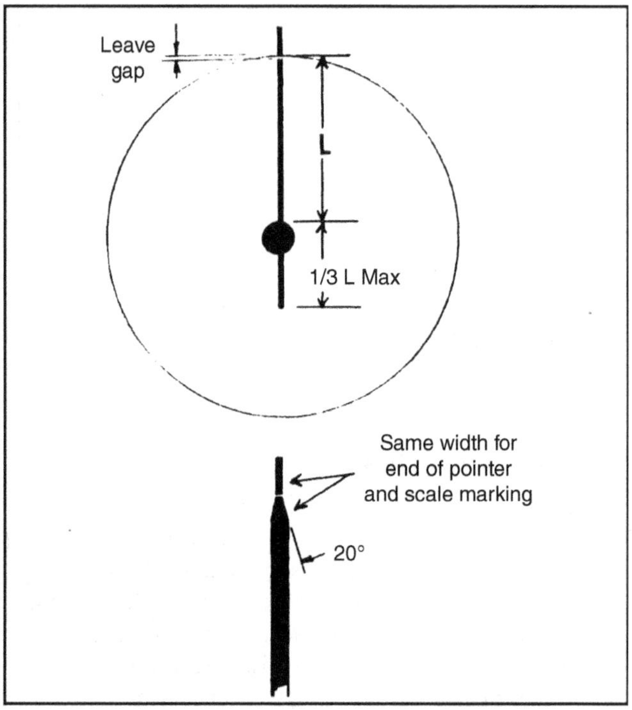

FIGURE 28.68 Correct pointer design.

Generally speaking, if a pointer has a "tail" on it, the observer can judge slight movements of the pointer better than if the pointer is tailless. The tail should not be more than one-third the length of the pointing segment, however; otherwise, there may be confusion as to which end is the pointing end.

Above all, avoid artistic pointer designs. (See Fig. 28.69.) Pointer illumination is often difficult. A translucent dial face

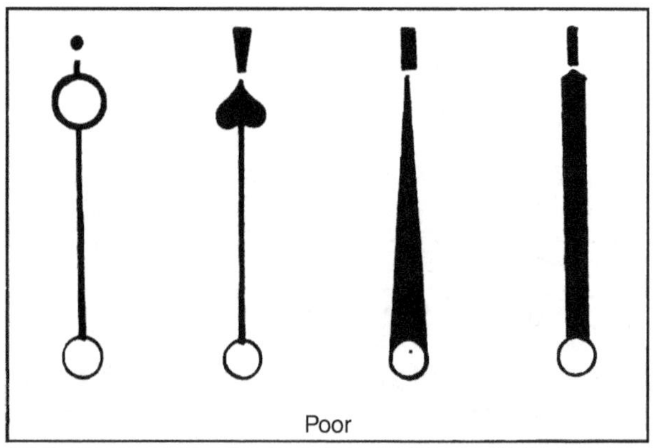

FIGURE 28.69 Poor pointer design.

provides a good background for silhouetting a black pointer. Another useful technique, shown in Fig. 28.70, consists of a plastic pointer ring inside a scale ring. Both can be either back- or edge-lighted with considerable success.

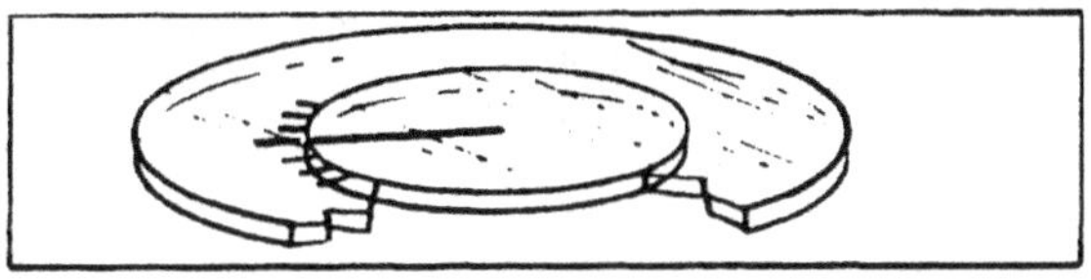

FIGURE 28.70 Design idea for transilluminating dial.

Multiple Pointer Display Considerations

A problem of pointer visibility and discrimination can occur when two or more pointers are necessary. (Three pointers are to be avoided if at all possible.) Several typical design errors are shown in Fig. 28.71. The two pointers are the same color, the shorter pointer tip is too far from the line of marking, and the tails of both pointers are the same length.

FIGURE 28.71 Poor design.

Figure 28.72 shows several ways to improve the visibility and discriminability of the dual pointer system. The color contrast and the variation in the length of pointer tails help clarify pointer identification. The length of the shorter pointer has been increased so that the tip of the pointing end is not so far from the line of marking, and a point has replaced the blunt tip, illustrated in the first example. Other ideas for aiding discrimination between pointers are shown in Fig. 28.72.

When two pointers are required on the same instrument dial, the upper one is generally too far from the surface on which its corresponding scale marking is located. One way to solve this problem, as illustrated in Fig. 28.73, is to bend the upper point in order to bring the pointer tip closer to the dial surface. Another method is to place the scale marks on a raised plane so that the marks are about midway between the two pointers.

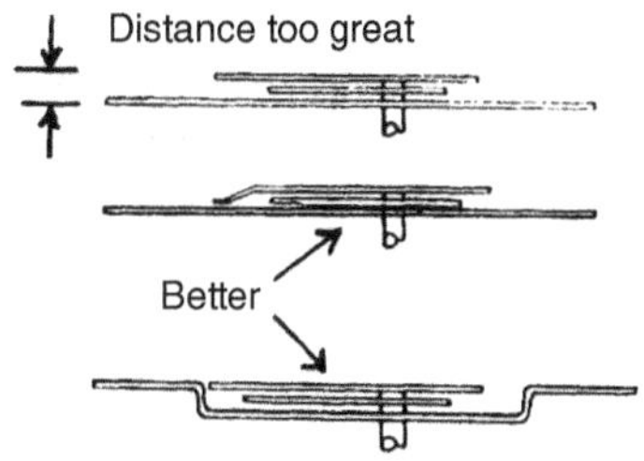

FIGURE 28.73 Double pointers on same dial.

When the particular instrument requires use of a "moving scale" that must be referenced against an indexing mark or pointer, it is usually necessary to make the index marker or pointer more conspicuous than an ordinary pointer because of the lack of length and pointer or index motion. Figures 28.74 and 28.75 illustrate acceptable design practice. Special attention is called to the double scaled instrument shown in Fig. 28.74. Here, making the central pointer pictorial has made it more conspicuous and identifiable; i.e., it represents the ship on which the operator is working. In other words, the operator knows immediately that the ship symbol represents the ship's heading, while the diamond pointer represents the "true" heading.

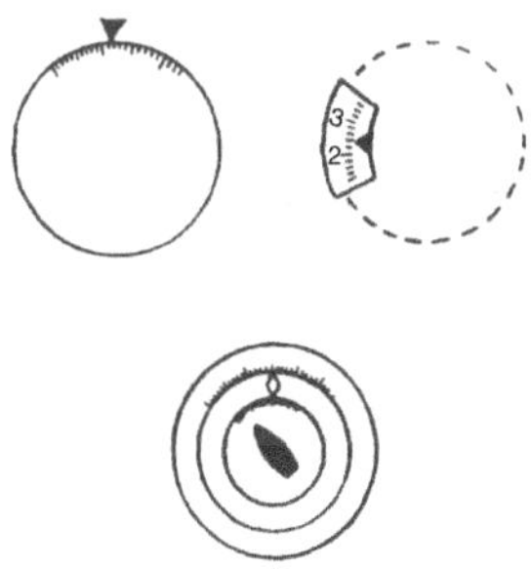

FIGURE 28.74 Acceptable designs for a moving scale.

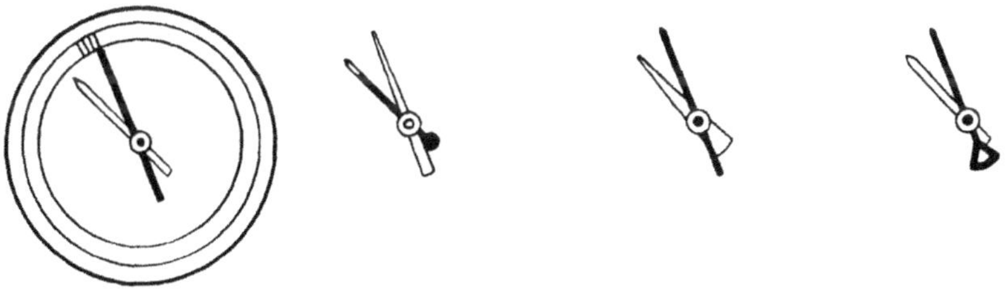

FIGURE 28.72 Better designs.

When a moving-tape type of gauge is used in order to provide an expanded scale, the indexing pointer can be designed either as a pointer positioned at the side or as an indexing or "lubber" line across the tape. (See Fig. 28.75.) In some cases two pointers may be required: one to indicate where the observer's position is now and the other to tell the observer which way to go.

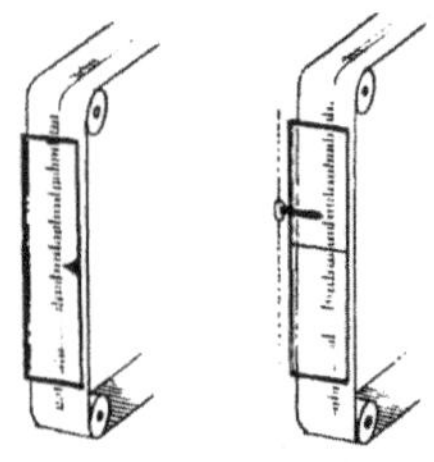

FIGURE 28.75 Acceptable designs for moving-tape gauge.

Many other pointer concepts and designs are possible, as Figs. 28.76 and 28.77 demonstrate.

In the case of the inset dial with its own separate pointer, care must be taken not to make the smaller display so small that it cannot be read. (See Fig. 28.76.) One should begin with the smaller one and make it legible; it then wilt dictate how large the primary display must be.

FIGURE 28.76 Size of smaller dial dictates size of larger dial.

Various navigation instruments require an entirely different approach to pointer design, as illustrated in Fig. 28.77. The cross-pointers move independently to indicate the positional relationship between where the pilot is and where he or she should steer to obtain the proper flight path. The double scale system provides accurate guidance information.

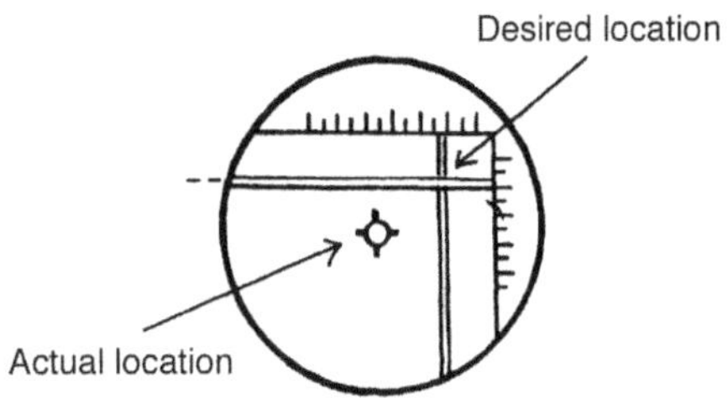

FIGURE 28.77 Flight path cross-pointer.

Example of Wristwatch Dial Design

Although the design shown in Fig. 28.78 shows some artistic license, it also illustrates how one designer managed to apply

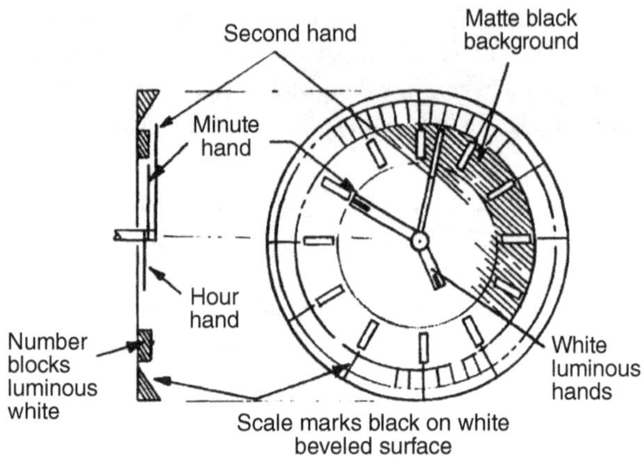

FIGURE 28.78 Example of wristwatch dial design.

a number of the principles discussed above. This watch, incidentally, can be read under the most severe sunlight as well as nighttime lighting conditions. Note these features: coplanar number marks and hands, a primary minute hand that does not cover the blocks, a second hand that is adjacent to the minute and second marks, extremely good contrast, and excellent differentiation between the hands due to size and shape.

ORIENTING AND POSITIONING DISPLAYS

General

Orienting and positioning oneself or one's equipment requires a special kind of display. Although a quantitative reference often is needed, it is important to display some pictorially represented reference for ease of interpretation.

The two displays shown in Fig. 28.79[3] illustrate how pictorial as well as quantitative information can be combined. Figure 28.79 (a) shows a realistic representation of a submarine in a dive. Figure 28.79 (b) shows a simpler yet somewhat pictorial representation.

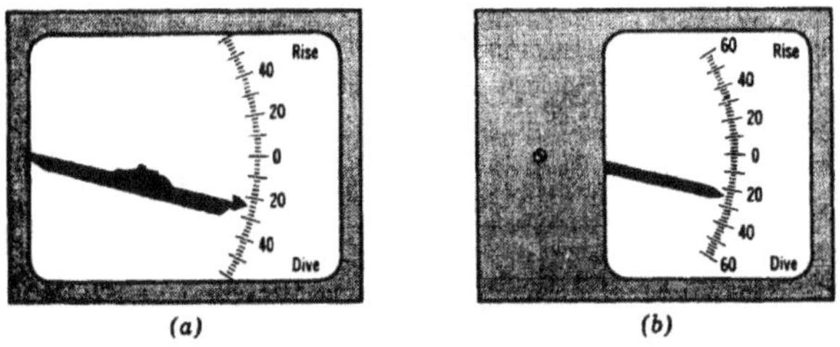

FIGURE 28.79 Pictorial and quantitative information combines.

Aircraft orientation is a typical system requiring special pictorial representation of the current attitude of the aircraft.

[3]Illustrations after C. R. Kelley, *Manual and Automatic Control*, John Wiley & Sons, Inc., New York, 1968, p. 93.

As shown in Fig. 28.80,[4] two alternatives are potentially available. In the first sketch (a), a picture-window view of how the earth's horizon tilts relative to the aircraft is provided. The second sketch (b) shows a different perspective in that a model represents the banking of the aircraft. The tilting horizon is often referred to as an "inside-out" display, and the tilting model as an "outside-in" display.

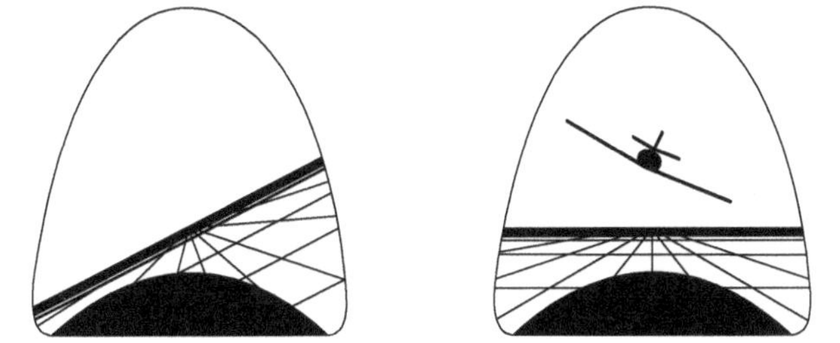

(a) (b)

Figure 28.80 Aircraft attitude.

An aircraft or similar vehicle also needs a display of its geographic position and directional attitude. A typical plan position indicator (PPI) would provide a picture of the aircraft on a simulated map, as illustrated in Fig. 28.81.[5] Several possibilities exist, of course. The first sketch (a) shows a north-oriented map with a moving and turning aircraft symbol; the second sketch (b) shows a north-oriented moving map with a turning but not-moving aircraft symbol; the third sketch (c) shows a turning map with a moving but not-turning symbol; and the fourth sketch (d) shows a turning map with a stationary, not-moving aircraft symbol.

Although each format has good and poor features (the symbol is moving in the wrong direction, the map is upside down, etc.), the one shown in the fourth sketch is more realistic.

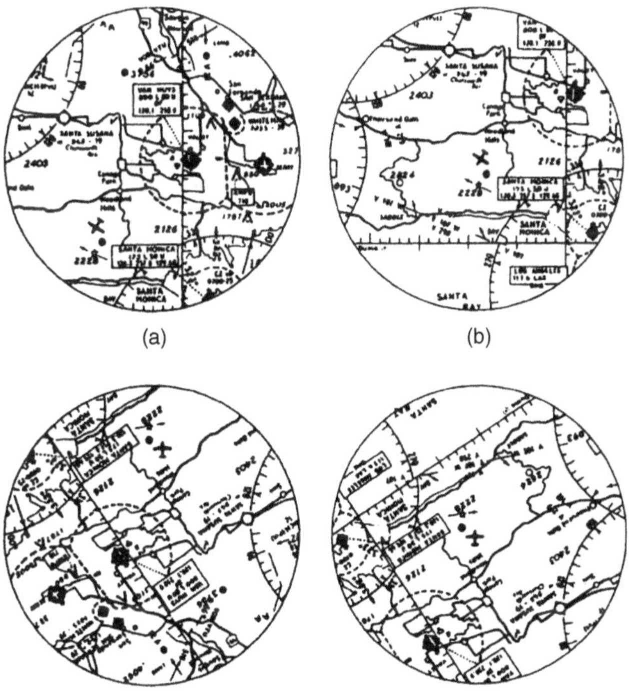

(c) (d)

Figure 28.81 Plan position indicators.

Head-Up Display Concept

Because pilots may intermittently fly from an instrument flight rules to a visual flight rules environment when flying in and out of clouds, they need to be prepared to shift instantly from their attitude instrumentation to the actual "outside" view. The head-up display concept was created to provide the instrumented display at approximately the same position on the windscreen at which outside reference objects would appear. Figure 28.82 illustrates attitude reference marks projected onto the windscreen. Note that they are of the inside-out format, coinciding precisely with what the pilot actually sees when it is possible to see out of the windshield.

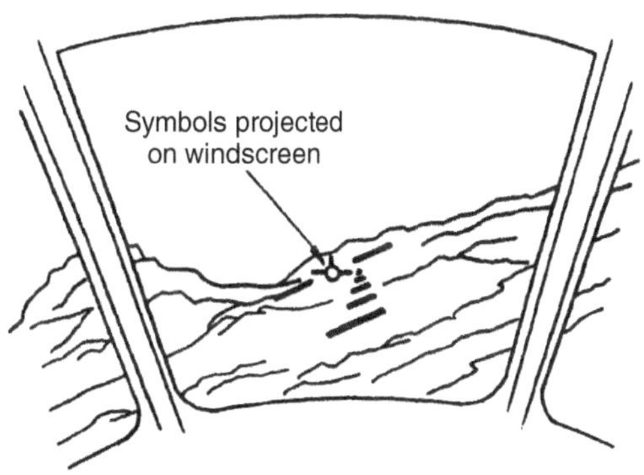

Figure 28.82 Head-up displays.

ELECTRONIC DISPLAYS

General

A broad range of display technologies are now available and under development. Figure 28.83 illustrates the range of currently available display technologies in hierarchical organization. The displays are arranged according to viewing mode (projection, direct-view, and virtual) and whether the technology intrinsically generates light (emissive) or uses light from a separate external or internal source (nonemissive).

Table 28.16 describes and rates (from low to high –1 to 6) the attributes of some of these current display technologies.

Electronic Display Design Requirements

Table 28.17 provides an overview of design features required for electronic displays (data source: NASA-STD-3001, Volume 2). Further discussion of these details follows.

[4]Illustrations after C. R. Kelley, *Manual and Automatic Control*, John Wiley & Sons, Inc., New York, 1968, p. 117.

[5]Illustrations after C. R. Kelley, *Manual and Automatic Control*, John Wiley & Sons, Inc., New York, 1968, p. 97.

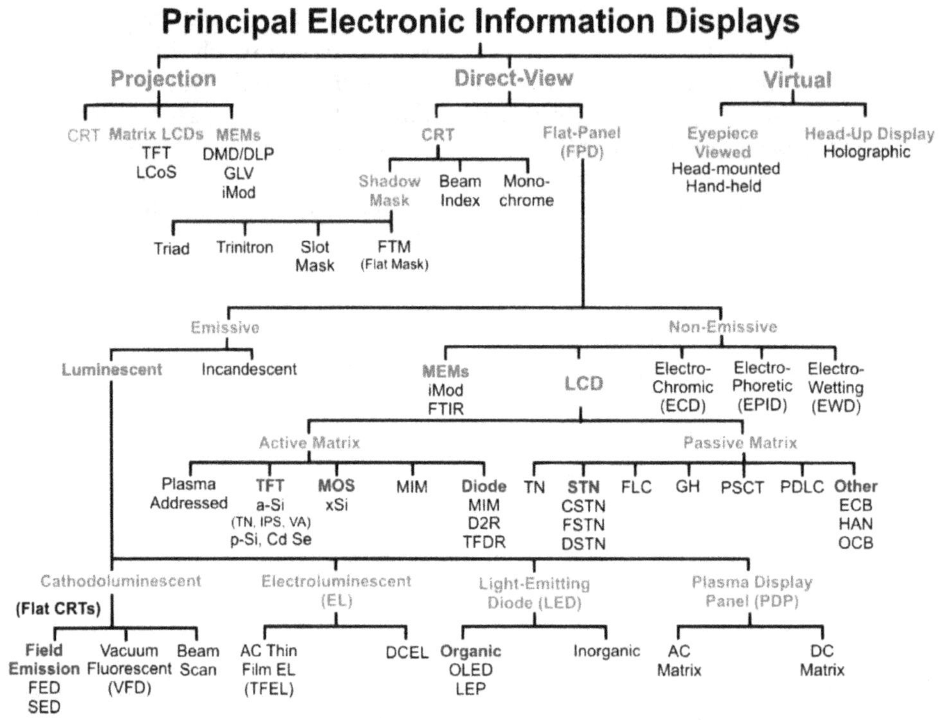

Figure 28.83 Principal electronic information displays.

Luminance and Contrast. Luminance is the intensity of the light emitted by a display. It must be sufficient for the display to be seen in the use environment. And yet the luminance must not be too great as to cause glare or interfere with night vision needs. The following are display luminance requirements:

- The brighter characters must have a luminance of at least 35 cd/m^2. Luminance should be adjustable—at least through a range not less than 50:1.
- The contrast ratios between the lightest and darkest areas or between a task area and its surroundings should not be less than the ratios specified in Table 28.18.
- Contrast between characters and a background should be 6:1 or greater. The preferred contrast is 10:1 or greater. Provide a control to vary the contrast between the characters and the background.

Viewing Angle and Display Distance. Visual displays should be normal to the line of sight. Displays should never be placed so that the viewing angle is less than 45° from the normal line of sight. (See Fig. 28.84.) Ideally, display viewing distance should be greater than 16 in. to accommodate the typical viewers resting focus. Anything less might cause eye strain.

Resolution. Visual resolution is the sharpness or fineness of detail. The visual resolution of a display is defined as the number of pixels (horizontal or vertical) per degree of visual angle. The more pixels there are in a visual degree, the greater the resolution.

A pixel is the smallest imaging element of a display capable of reproducing a full range of luminance colors.

Pixel pitch is defined as the distance between corresponding points (e.g., centers) on adjacent pixels. Pitch is usually specified in millimeters (mm), where different, separate horizontal and vertical pitches may be specified. The pitch of current (2007) LCDs can vary by as much as a factor of 6, depending on application.

Other terms such as "size" or "definition" are also sometimes used to describe the number of pixels in the display matrix. However, the display size in pixels does not say anything about the resolution—only the number of pixels and not their density. A number of sizes have become industry standards with particular names. Table 28.19 shows a number of the more common standard sizes and their names.

Ideally, displays should have a minimum resolution of 32 pixels per degree. Figure 28.85 shows what that means and some example resolutions.

Table 28.20 shows the resolution for six representative displays at a number of viewing distances.

In the table, the viewing distance H is the height of the display. H can be calculated by multiplying pixel pitch by the number of pixels across the display height. Also, the cell phone listed in the table is an older model with a screen approximately 1.5 in. high.

If the pixel pitch and the viewing distance are known, then visual resolution V can be computed using the formula shown in Fig. 28.86.

In the equation, D_p is the viewing distance in pixels (distance in millimeters divided by pixel pitch), and the arctangent is computed in degrees. Over the range of interest (D_p > 100), this can be accurately approximated by the equation shown in Fig. 28.87.

TABLE 28.16 Current Display Technologies and Ratings of Their Attributes (Low to High—1 to 6)

Display Type	Spatial Addressability	Spatial Resolution	Temporal Response	Luminance	Contrast	Ambient Contrast	Grayscale Performance	Viewing Angle	Color Gamut	Ambient Color Gamut	Physical Package	Application Flexibility	Other Attributes
CRT-shadow mask	6	5	5	2	5	2	6	6	5	2	1	1	■ Subject to various geometric distortions such as barrel, pincushion, and trapezoidal ■ Subject to flicker due to impulse response—requires high refresh rates
Plasma display	5	5	3	4	5	2	3	6	5	2	3	1	■ Subject to contouring artifacts at low gray levels ■ Subject to perceptible motion artifacts when viewing dynamic imagery
Light-emitting diodes and organic light-emitting diodes	5	5	4	4	5	2	6	6	5	2	6	6	■ Subject to internal specular reflections under ambient lighting ■ Circular polarizer reduces luminance ■ Differential aging of color materials can cause color shifts over time ■ Active-matrix configurations subject to motion blur with dynamic imagery
Liquid crystal displays	6	6	3	4	4	2	5	4	5	3	6	6	■ Subject to motion blur with dynamic imagery due to AM sample-and-hold behavior and LC response time
CRT-based projection display	6	4	5	2	4	2	6	2	5	2	2	2	■ Subject to various geometric distortions such as barrel, pincushion, and trapezoidal ■ Subject to misregistration of primary color images ■ Subject to flicker due to impulse response—requires high refresh rates
LC imager-based projection display	6	4	4	4	4	2	5	2	5	3	3	2	■ Transmissive LC light valves subject to motion blur with dynamic imagery due to AM sample-and-hold behavior and LC response time ■ Transmissive LC light valves subject to "screen-door effect" due to low aperture ratio ■ LCoS only minimally susceptible to motion blur and "screen-door effect"
MEMS imager-based projection display	6	4	4	5	4	2	3	2	5	3	3	2	■ Subject to contouring artifacts at low gray levels ■ Subject to perceptible motion artifacts when viewing dynamic imagery ■ Field-sequential color configurations subject to color breakup and flicker

Image Polarity. If the ambient illumination in the vicinity of the display is 540 lux or greater, dark characters and symbols on a light background should be used rather than light characters on a dark background.

Segment and Dot Matrix Characters. Electronically generated characters can be formed by segments or dots. Figures 28.88 and 28.89 show an example of each. Generally, dot matrix characters are more readable than segmented characters because of the greater capability to produce apparent curved lines. Increasing the number of dots or segments improves character readability. Figure 28.90 contains recommendations for character formation.

TABLE 28.17 Electronic Display Design Requirements

Metric	Minimum	Maximum	Context	Notes
Luminance (peak white luminance)	25 cd/m²	—	Emissive displays	≥100 cd/ m² preferred
Contrast ratio (ambient contrast ratio)	10	—		Includes ambient illumination
Color gamut area	0.17	—	Color displays	Fraction of CIE 1976 u′v′ chromaticity space
Viewing angle	−45deg	+45 deg		Four-point viewing angle (left, right, up, down), contrast and color gamut criteria met
Resolution (spatial resolution)	32 pixels/deg	—	Image and video displays	
Frame rate	60 Hz	—	Video displays	
Moving edge blur	—	15 ms		Using metric GET (preferable) or BET; use average of five equal lightness levels, including white and black
Number of colors	224 (1,627,716)	—	Image and video displays	
	212 (4,096)	—	Text and graphics displays	
Number of gray levels	28 (256)	—	B/W image and video displays	

Notes: Except where noted, metrics are defined as in International Committee on Display Metrology (ICDM), Display Measurement Standard (DMS 1.0), or Video Electronics Standards Association (VESA) Flat Panel Display Measurements (FPDM 2.0). Further details on metrics may be found in Chapter 10, Crew Interfaces, of the NASA, Human Integration Design Handbook—HIDH.

From NASA-STD-3001, Volume 2.

TABLE 28.18 Contrast Ratios

Comparisons	Environmental Classification		
	A[1/]	B[2/]	C[3/]
Between lighter surfaces and darker surfaces within the task	5:1	5:1	5:1
Between tasks and adjacent surroundings	3:1	3:1	5:1
Between tasks and more remote surfaces	10:1	20:1	[4/]
Between luminaries and adjacent surfaces	20:1	[4/]	[4/]
Between the immediate work area and the rest of the environment	40:1	[4/]	[4/]

Notes:
From MIL-STD-1472G.
[1/]A = Interior areas where reflection off entire space can be controlled for optimum visual conditions.
[2/]B = Areas where reflection off immediate work area can be controlled, but there is only limited control over remote surroundings.
[3/]C = Areas (indoor and outdoor) where it is impractical to control reflection and difficult to alter environmental conditions.
[4/]Contrast ratio control not practical

TABLE 28.19 Selected Standard Matrix Display Dimensions in Pixels

Name	Width × Height
VGA	640 × 480
SD	640 × 480
HD	1366 × 768
Full HD	1920 × 1080
XGA	1024 × 768
UGA	1600 × 1200
SXGA	1280 × 1024
UXGA	1600 × 1200
WUXGA	1920 × 1200
WQXGA	2560 × 1600
QSXGA	2560 × 2048

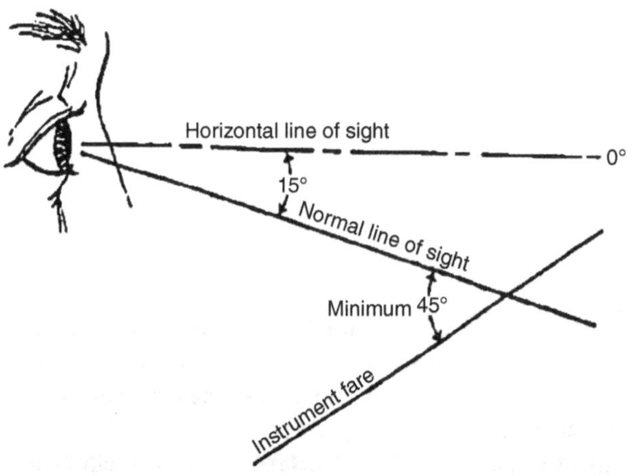

FIGURE 28.84 Viewing angle. (From MIL-STD-1472G.)

Suggested Electronic Character Optimized Set Using a 7 X 9 Matrix. The suggested character formation shown in Fig. 28.91 contains most of the alphanumerics and supporting symbols required for data processing. They are designed to minimize recognition errors, or at least possible confusion between one character and another. Although lowercase letters have been designed, they are not recommended because it is difficult to create lowercase features that are not

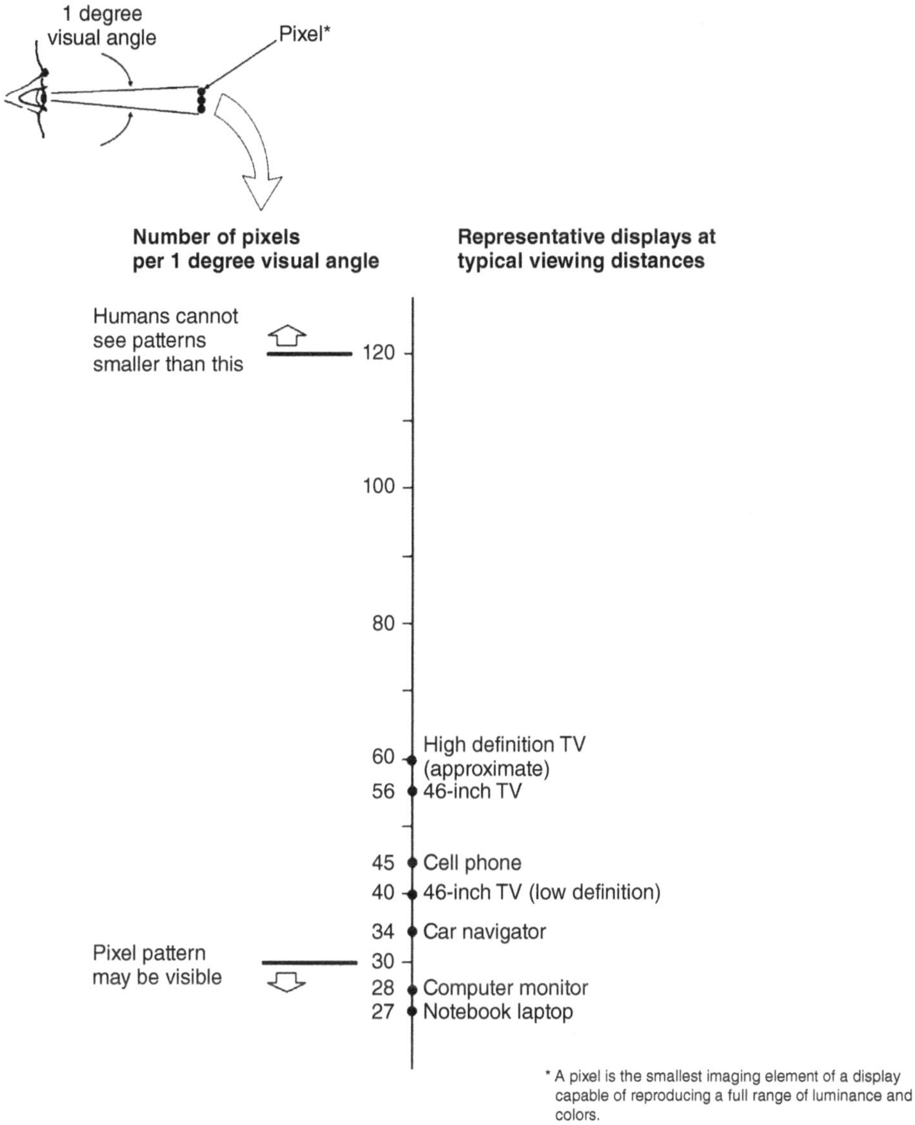

Figure 28.85 Electronic image resolution.

TABLE 28.20 Example of Displays and Viewing Distances

Pixels (vertical)	Application	Pixel Pitch (mm)	Viewing Distance, H (cm)					
			1H	2H	3H	4H	6H	8H
320	Cell phone	0.126	5.6	11.2	16.8	22.3	33.5	44.7
480	Car navigation	0.190	8.4	16.8	25.1	33.5	50.3	67.0
768	46" TV	0.746	13.4	26.8	40.2	53.6	80.4	107.2
1080	46" TV	0.530	18.8	37.7	56.6	75.4	113.1	150.8
768	Notebook	0.297	13.4	26.8	40.2	53.6	80.4	107.2
1600	Monitor	0.251	27.9	55.9	83.8	111.7	167.6	223.4

A typical viewing distance is indicated in shaded cells.

$$V = \frac{1}{\left(\tan^{-1}\left(\frac{1}{Dp}\right)\right)}$$

FIGURE 28.86 Equation 1.

$$V = \frac{Dp}{57.2957}$$

FIGURE 28.87 Equation 2.

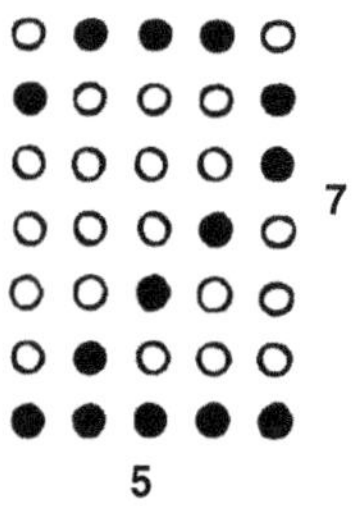

FIGURE 28.88 Example of dot matrix character (5 × 7 matrix).

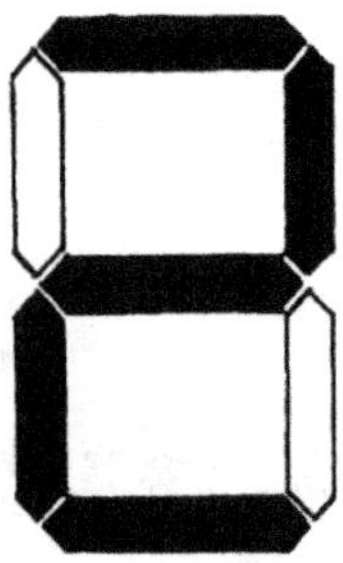

FIGURE 28.89 Example segmented character (7 segments).

Character type	Recommended use
Segmented displays	
7 Segments	Numerical information only
14 Segments	Preferred for general applications
16 Segments	
Dot matrix displays	
5x7 Matrix	Minimally acceptable
7x9 Matrix	Preferred
8x11 Matrix	Minimum if symbols are rotated
15x21 Matrix	Preferred if symbols are rotated

FIGURE 28.90 Recommendations for segmented and dot matrix character designs. (From MIL-STD-1472D.)

easily confused, especially if the observer has to read information rapidly.

Electronic Display Format (Radar and Sonar). Electronic displays used for detecting, identifying, and tracking target information usually are presented in either rectangular or polar-coordinate form, i.e., an A-scan or PPI format. (See Fig. 28.92.) For obvious reasons, the A-scan display face should generally appear as a rectangularly shaped display, and the PPI should appear circular.

Regardless of the type of scope presentation, certain features, such as the following, should be designed according to good human engineering design principles.

1. Display Size: For a single operator involved in a target search task, use a relatively small display (7–14 in., 18–36 cm) to reduce the area for visual scanning. For tracking targets, use 10–17 in. (25–43 cm). If more than one person must work at the display station, consider diameters of 24–30 in. (61–76 cm).

2. Display Cursors: Cursors help the operator relate target position to actual range and bearing values. They should always be accompanied by indexing scales laid over or around the display face and/or electrically connected to digital panel readouts. Electronic cursors are preferred because they do not produce the parallax problems that mechanical cursors often create.

3. Grid Overlays: Target position can be interpreted much faster if a grid overlay is supplied. The more accurate the reading requirement, however, the more elaborate the grid structure must become. Grids are more confusing than useful on displays that are less than about 14 in. (36 cm) in diameter. Even here it may be desirable to create a grid that will not obscure small target traces. For example, the minimum spacing between range rings on a polar display should be on the order of 1° (36 min visual angle subtended at the eye), or about 0.50 in. (1.27 cm) at an average 18-in. (46-cm) viewing distance. A mixture of solid and dashed grid lines or range rings helps reduce the potential for obscuring targets and also aids in identifying major range and bearing elements (numbered).

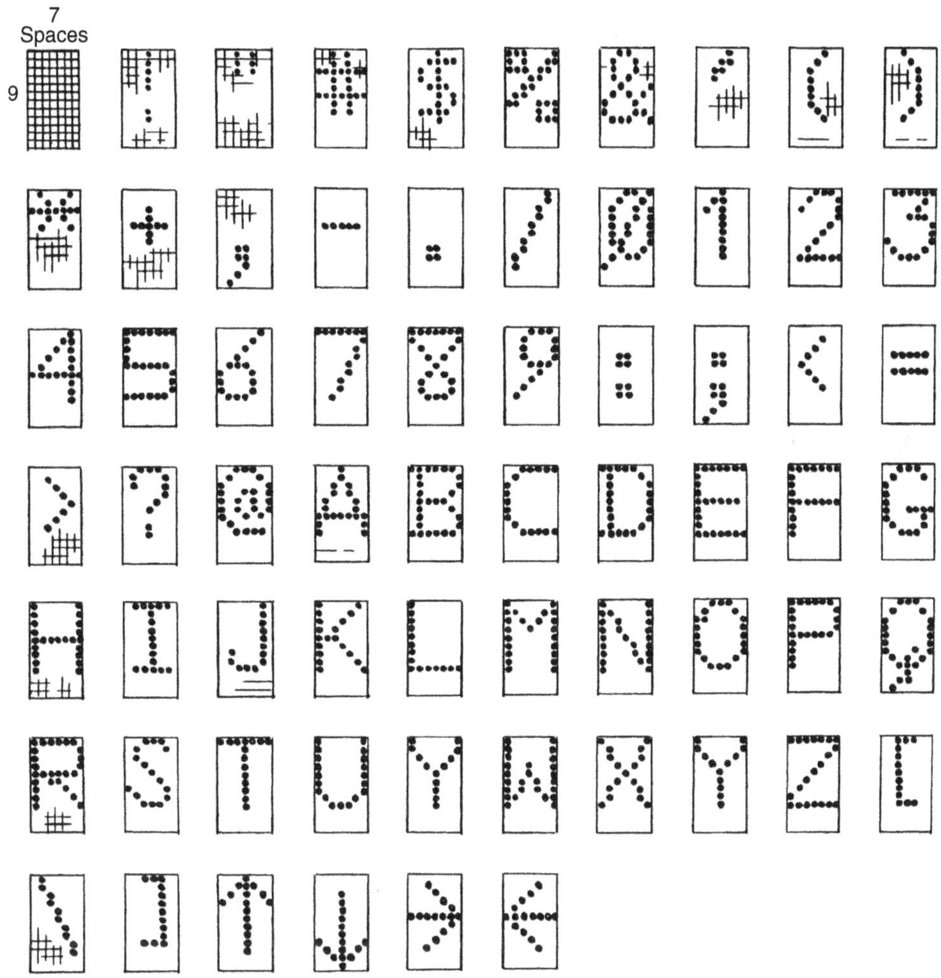

Figure 28.91 Suggested character formation for data processing.

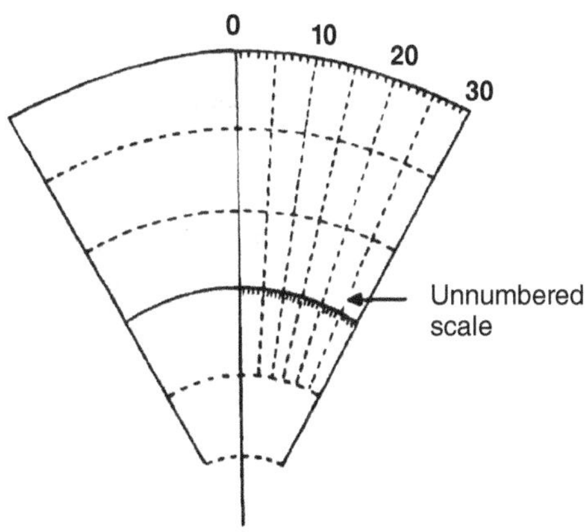

Figure 28.92 Example polar coordinate grid.

GROUP-VIEW DISPLAYS

Group-view displays are increasingly used to operate complex systems, such as nuclear plants, space operations, missile launches, and offshore drilling platforms. (See Fig. 28.93.) Advances in technology have improved the quality, size, and flexibility of these displays. Group-view display systems can have the following benefits.

- Provide an overview of system status
- Help to identify areas where additional information is required
- Operators are aware of each other's activities
- Allow for operator exchange and coordination during problem resolution

Developers of group-view systems must consider the following when designing the controls and displays.

Display Types

Display types include centrally located large screen, individual redundant display devices, and multiple walkup displays. A walkup display is smaller and not located in the operator's immediate area. Table 28.21 shows the appropriate use for these types of group-view displays.

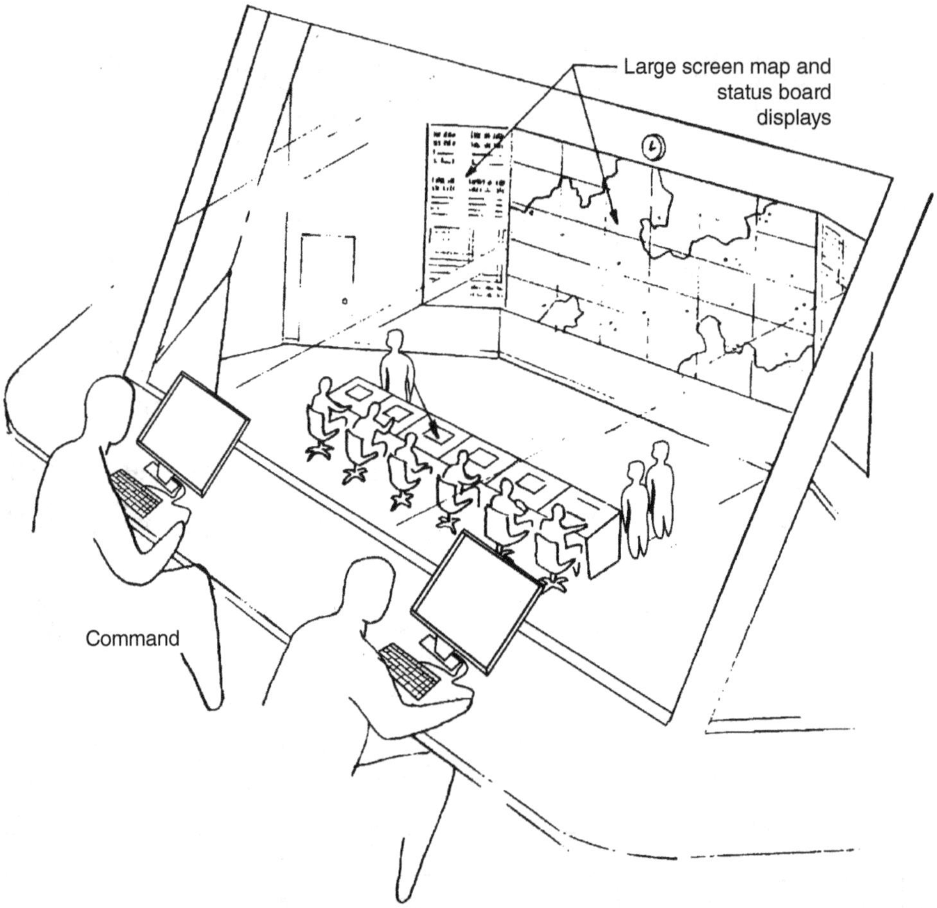

FIGURE 28.93 Group displays.

TABLE 28.21 Appropriate Use of Group-View Displays	
DISPLAY DEVICE	**CONDITION WHEN DEVICE IS PREFERRED**
Large-screen displays and redundant, small screen displays	■ Immediate access is required ■ Operator must be able to view information from multiple locations in the control room ■ Operator often cannot leave usual work area to go to a walkup display ■ Crew interaction requirements (e.g., verbal, gesturing) are low, or computer-based communication is provided.
Walkup display	■ Immediate access is not required ■ Operator does not need to view information from multiple locations in the control room ■ Operator can leave work area to go to a walkup display

The following are guidelines for the design of large-screen displays.

- It should not be used to modify or delete critical information inadvertently or arbitrarily.
- Screen dimensions and viewing distance must be based on an interrelated analysis of the number of operators, their information requirements, and their locations.
- The display should not be closer than half the display width or height (whichever is larger).
- Off-angle viewing decisions must account for distortions of dimension, color, and brightness.
- Seating and traffic must not obstruct vision.
- Character and symbol sizes must be based on maximum viewing distances.

Controls

1. Controls can be conventional knobs and dials, computer controls, and soft/automated control systems.

2. Rules for control interactions must be defined: can multiple operators perform the same function at the same time or will operators control functions one at a time and in sequence?

3. Should controls be at individual workstations, portable, or at group-view displays?

4. In analyzing and configuring control interactions, focus on the following operating sequence:

   a. Select the variables or components to control

   b. Identify the physical control or control input field (for a computer-based control)

   c. Input data or adjustments

   d. Monitor the system feedback

Communications

The system must be configured to support the appropriate communication systems:

1. Headset and microphones

2. Direct communication

3. Portable transceivers

4. Public address enunciators

5. Computer/human communications

6. Emergency and backup communication systems (see Fig. 28.94)

OPTICALLY PROJECTED DISPLAYS

1. The symbol size should be approximately 10 to 15 min of arc resolved at the observer's eye.

2. The aspect ratio should be approximately 1.33:1.48.

3. The symbol stroke should be about one-sixth to one-tenth the character height.

4. The viewing distance should be approximately four times the image width.

5. The viewing angle should be between 20° and 30° from the centerline of the display. Image luminance should be at least 10 fL.

6. The direction of light and dark contrast is not important for legibility. The ambient light impinging on the screen should not exceed about 0.02 fL.

7. The contrast ratio should be about 500:1, measured with no film in the projector.

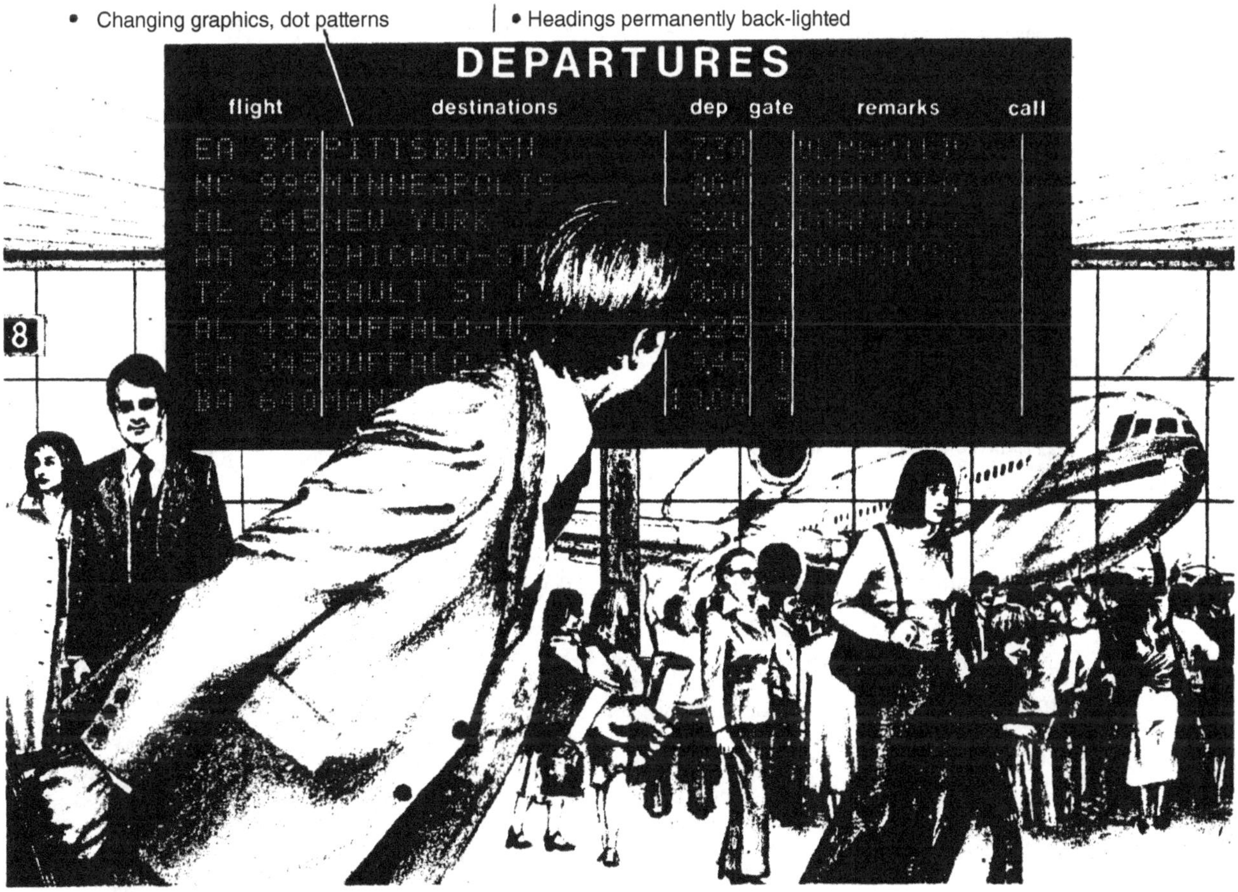

FIGURE 28.94 Lighted status board. (Illustration by permission of Ferranti-Packard Limited. Mississauge. Ontario, Canada.)

8. Matte screens are generally recommended because screen brightnesses are essentially equal at all viewing angles. However, they are only about 85 to 90 percent efficient. When light output from the projector is limited, a lenticular screen should be used (for approximately 1.5 to 2.0 gain).

Projectors

1. Provide an easily understandable system for setting up the equipment, i.e., for erecting supports for movie reels; attaching the reels; threading the film (preferably automatic); inserting slides (right side up); attaching the slide case, tray, or a carousel; adjusting the projection angle, focus, and brightness; etc.

2. Provide an easily understandable set of controls, properly labeled so that they can be seen in a dimly tit room.

3. Provide sufficient lengths of power cord, audio cord, and remote control cord, color coded so that they are not easily confused.

4. Use low-noise motors and fan systems and/or insulate so that the projector noise will not interfere with audio communication.

5. Design the package so that it does not weigh over 25 lb (11.3 kg) and is small enough for one-person transport.

6. Design for ease of lamp replacement and regular servicing.

To determine the required projector light output, take the projection surface area and divide this by the screen gain. Multiply this by the recommended screen brightness (foot-lamberts) shown below:

1. For motion pictures:
   a. 5 fL Minimum (marginal for some observers)
   b. 10 fL Satisfactory
   c. 15 fL Excellent
   d. 20 fL Maximum (flicker threshold for some observers)
2. For slides:
   a. 1 fL Absolute minimum (difficult to distinguish color from black and white)
   b. 1.5 fL Minimum for gross images
   c. 5 fL Minimum for detailed slide information
   d. 10 fL Satisfactory
   e. 20 fL Excellent
3. For projected TV:
   a. 2 fL Gross black-and-white images
   b. 10 fL Maximum (to avoid flicker thresholds)

Black- versus White-Background Slide Presentation. When slides are presented in a relatively dark room for extended periods of time, it is recommended that the slide backgrounds be dark with white graphics. This allows the observer's eyes to become dark-adapted and remain so until the presentation is completed. If, on the other hand, only one or two slides are presented at a time, with intermittent raising of the ambient room illumination, and/or if slides are presented in a medium to fairly high ambient illumination environment, it is best to present black graphics on a light background.

Colored Backgrounds. Colored backgrounds provide useful relief and stimulation in slide presentations; in addition, colors can be used to present information. Colored background should, however, be of a light shade so that black graphics are maximally legible. More saturated color should be reserved only for graphics themselves (i.e., without overlayed graphics or alphanumerics).

Slide Preparation. The proportion of lantern slides should be approximately 7:10. The ratio of alphanumeric character height on the screen to the most distant observer should be about 1:300, i.e., 1.0-in. (2.5-cm) characters for a 25-ft (7.5-cm) viewing distance. As a general rule, no character should be less than about 0.45 in. (1.1 cm) high, with a stroke width of approximately 0.006 in. (0.015 cm). In preparing a chart for projection, enlarge the copy about three times so that characters are about 0.125 in. (0.318 cm) high.

Making Slides for Opaque Projection. Alphanumeric characters typed in all capital letters will be legible when converted to a 2 in. by 2 in. (5.1 cm × 5.1 cm) slide when the material fits within a 3.0 in. × 4.5 in. (7.6 cm × 11.4 cm) area. Illustrations should fit the horizontal format of the screen; i.e., the width should be about 1½ times the height. (See Fig. 28.95.)

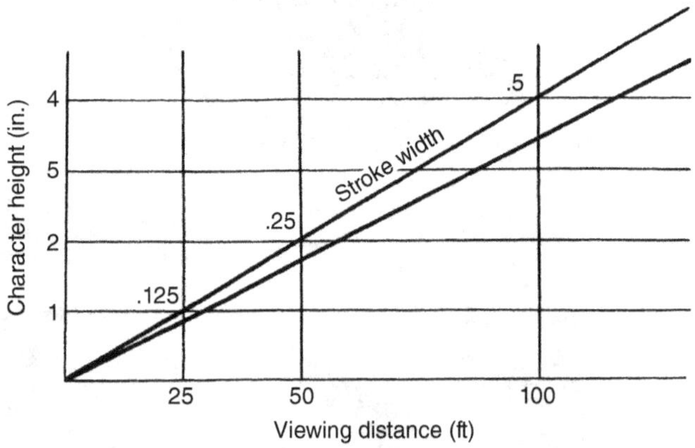

FIGURE 28.95 Required character height and stroke width.

As a general rule, limit line copy to about 10 lines, with five to seven words per line. The space between rows of characters should be at least one-half the height of the characters, and preferably a full character separation.

Hand-Held Optical Aids

Optical aids are used primarily to offset reduction in visibility due to extreme viewing distance. As a general rule, viewing with two eyes is preferred over single-eye viewing because it preserves a certain amount of depth perception.

Optical aids are most useful for object recognition and inspection, as opposed to detection. The latter is difficult because as object size is magnified, so too is the relative motion of the optical image. For identification of objects, the size of an airplane up to a distance of 3000 yd (2742 m), a 2× magnification is appropriate. Increases in magnification make it difficult to shift from naked eye to optically aided viewing. If higher magnifications are necessary, some means should be provided for steadying both the optical device and the observer.

Visibility through an optical device depends on the amount of light transmitted. Thus, the more separate glass surfaces there are in an optical system, the greater the loss of light. All such surfaces should be antireflection coated.

When an optical device is to be used for night viewing, provide reticle illumination with variable intensity control from 0 to about 0.02 fL. Binocular devices should have means for adjusting the interocular distance independently from individual eye viewing distance. (See Table. 28.22.)

TABLE 28.22 General Magnification Criteria

Distance	Magnification
1000 yd (914 m)	1–1.5x
3000 yd (2742 m)	2–3x
5000 yd (4570 m)	3–4x
Above 5000 yd (should be tripod mounted)	6–8x

Direct-View Optical Instrument Design[6]

1. User Visual Considerations

   a. Visual spectrum: 400 nm (400×10^{-9} m) through 700 nm

   b. Visual acuity: 60 s of arc for the emmetropic eye

   c. Maximum tolerable astigmatism: ¼ diopter Vernier acuity: 10 s of arc

   d. Stereoscopic acuity: 12 s of arc

   e. Extraocular limits for onset of blurred vision and/or possible fusion loss:

      i. Visual spectrum: 400 nm (400×10^{-9} m) through 700 nm

      ii. Visual acuity: 60 s of arc for the emmetropic eye

      iii. Maximum tolerable astigmatism: ¼ diopter Vernier acuity: 10 s of arc

   f. Stereoscopic acuity: 12 s of arc

   g. Extraocular limits for onset of blurred vision and/or possible fusion loss:

      i. Abduction: 15 min

      ii. Adduction: 30 min

      iii. Supraduction: 15 min

   h. Stimulation-intensity range:

      i. Smallest detectable (threshold) using rod vision: 10^{-6} mL; cone vision: 1 mL

      ii. Largest tolerable cone vision: 104 mL

2. Optical Parameters

   a. Magnification: between 1 and 20 power

      i. Rifle, pistol sights: 4 power

      ii. Monoculars or binoculars: 8 power

Note: If more than one magnification is required, use discrete magnifications rather than a "zoom" technique, unless accuracy is not important.

3. Field of View: The field of view should be compatible with the intended use, such as search and detection, recognition, identification, or aiming.

4. Entrance Pupil: The entrance pupil should be equal to the product of the magnification and the exit pupil diameter and is therefore defined by these.

5. Exit Pupil: The exit pupil should be consistent with the intended use.

   a. Daylight use: not less than 3.0 mm

   b. Twilight and darker use: not less than 7.0 mm

6. Eye Relief: A long eye relief (25 mm) is desirable on vehicular-mounted sights to protect the user from gun recoil, observing on the move, etc.; 15 mm permits the user to wear glasses.

7. Eyepiece Adjustment: Fixed-focus eyepieces set between 0.50 and 1.00 diopter may be used for 4 power or less. Eyepiece dioptric adjustments and a scale (± 3.0 diopters, although 4.0 diopters is more desirable—in 0.5-diopter increments) should be provided on all instruments over 4 power.

8. Axial Resolution: Axial resolution should be equal to, or better than, 60 s of arc divided by the magnification in order to create an "eye-limited" instrument.

9. Aberrations: Aberrations should be according to standard optical practice.

[6]Human Factors Engineering Design for Army Material, MIL-HDBK-759, March 1975.

10. Luminous Transmission: Luminous transmissions should be greater than 50 percent. The elements should be antireflection-coated, except for the focal plane components.

11. Parallax: The reticle should be focused to the target range of primary interest.

12. Reticles: Reticle lines should subtend 2.0 min at the eye. (Lines are preferred to dots.) A ring should be used, as opposed to a spot and/or a small crosshair. Reticle illumination should be provided for instruments to be used during twilight or lower ambient conditions (with dimming capability).

13. Binocular and Biocular Considerations

   a. Provide interpupillary adjustment scaled from 58 to 73 mm (1-mm intervals). Magnification differences between two barrels should not exceed 2 percent, and luminous transmission difference should not exceed 5 percent.

   b. Collimation, i.e., alignment of the binoculars, should not exceed the following at the eye: 15.0 to 40.0 min/arc divergence, or 25.0 min/arc convergence.

14. Filters: Consider the use of neutral density filters to reduce glare from the sun or other high-level light sources.

15. Eyecups: Use appropriate eyecups to cushion the user's head from the device and to keep stray light from entering the eyepiece and/or the observer's eye. The radius in the figure defines a surface of revolution within which an eyepiece and cup must be designed to avoid interference with facial features.

Reticle Design Suggestions (See Fig. 28.96)

1. Use the minimum number of reticle lines or circles necessary to provide adequate reference; too many lines interfere with the view of the target.

2. Provide a stable frame of reference by means of vertical and horizontal reticle lines. Avoid too many intervening angular lines, except when the viewing task does not require vertical and horizontal reference; concentric circles allow the observer to judge the size of an object that may be changing direction, e.g., an aircraft target.

3. Use broken reticle lines when too many lines tend to obscure the target.

4. If a scale reference is required, consider placing the scale to one edge of the reticle plate, thus minimizing the potential interference of the scale with the primary viewing area.

5. Optical aids that may be used where there is the possibility of extremely bright flashes (e.g., an atomic flash) should be equipped with an automatic filter system, since the hazard of flash blindness is increased by the optical system.

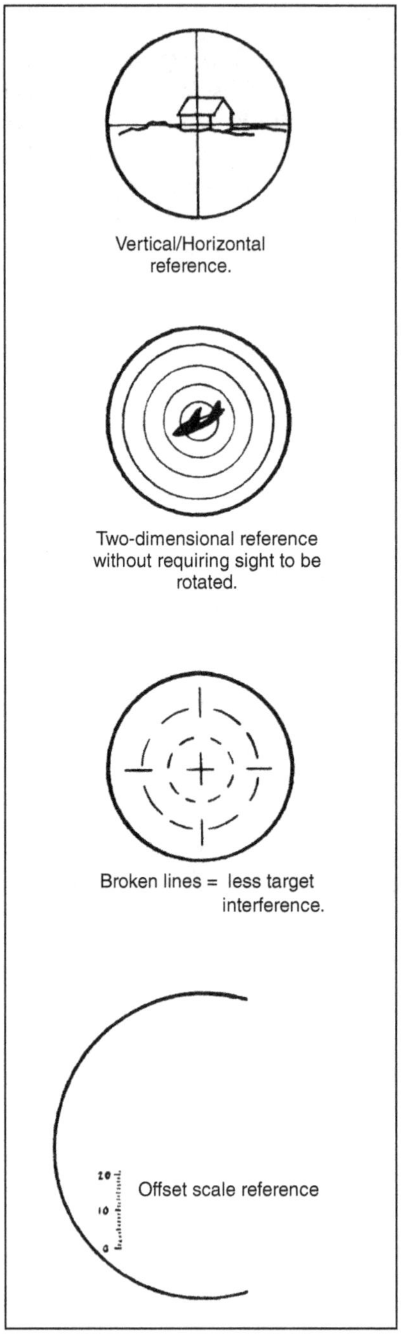

Figure 28.96 Reticle design recommendations.

Telescopic Sight Reticle Visibility. A collimator sight permits observation of a reticle with the eye focused at infinity, or with the same eye accommodation as required to view a target, and does not require that the eye be precisely positioned. Light from an illuminated reticle passes through a collimating lens placed one focal length from the reticle, rendering the light rays leaving the reticle parallel. Collimated light from the target scene by use of a coated glass plate (beamsplitter), permitting light both to pass straight through from one face and to reflect light striking the opposite face. By using dichroic filters for beam splitters, color discrimination can be made between the target and the reticle; i.e., dichroic

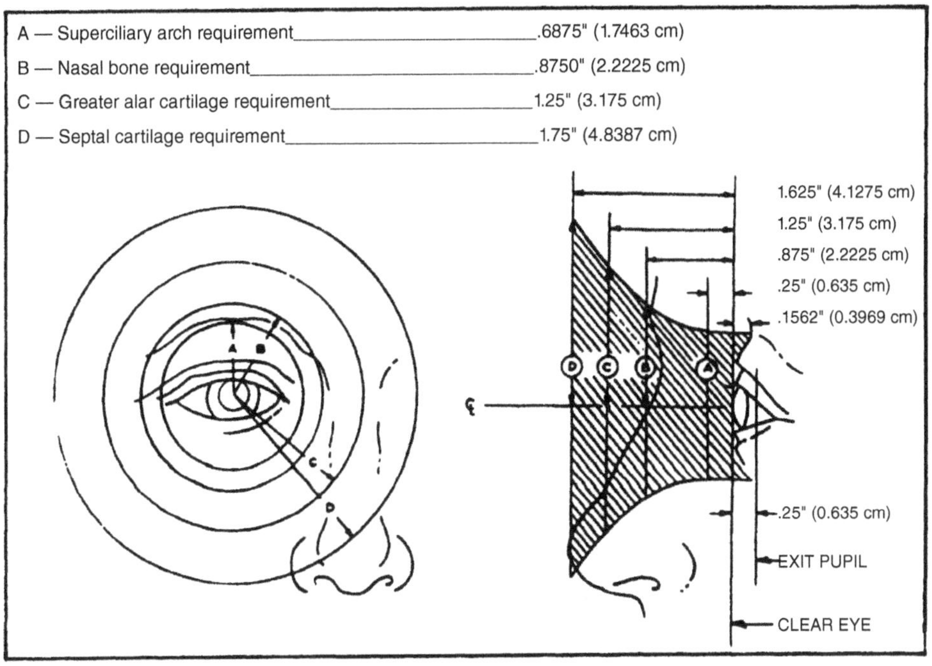

Figure 28.97 Anatomical limits on axially symmetrical ocular metal parts. (MIL-STD-1472G.)

filters reflect what they do not transmit, and the ratio of reflectance to transmittance can be made to vary with wavelength. (See Figs. 28.97 and 28.98.)

Use of Optical Instruments. Both eye alignment and control manipulation must be considered in determining the proper working height for optical instruments. Because of the multiple points of articulation in an operator's back and neck, it is difficult to anticipate exactly how operators may have to position their bodies and heads in order to align their eyes with the eyepiece of a particular instrument. The most practical approach is to obtain the actual instrument that is to be used and then mock up a mounting geometry that has sufficient adjustment for operators of various sizes to try out. In some cases it will not be possible to "fix" the work height so that it is acceptable for all operators; an adjustable worktable must then be provided.

Figure 28.99[7] illustrates the approximate eye-positioning path of a person sitting in a fixed seat and using an articulating optical device. (The designer is cautioned not to consider this diagram a final recommendation; rather, it is useful only in terms of preparing a mockup test bed for further evaluation with an expected range of users.)

WEBSITES

Websites are a means of rapidly and inexpensively communicating vast amounts of information. They are interactive and users can tailor them to get the information they need. Navigating through the website information can be a challenge, however. It is vital that website designers focus on the abilities and limitations of the users.

Website developers must consider all aspects of the user–system interface. Figure 28.100 shows a simple representation of a website–user interface:

1. The user first searches for information on the Web.
2. Once found, the user reviews and examines the information.
3. Finally, the user may manipulate useful information by saving it or sending it on to others.

For a webpage to properly meet its goals, developers need to consider all these potential events in the interface process. The following pages provide recommendations for designing successful user interfaces. The recommendations address each of the user–system interface events. The recommendations also include guidelines for the website development process itself. The information was extracted and reformatted from the website posted by the U.S. Department of Health and Human Services, www.usability.gov.

Website Design Process

1. **Website Discovery**—The website should be in the top 30 found from a major search engine.
2. **Website Goals**—Define what websites will do and the desired level of usability.
3. **Content**—Based on user requirements.
4. **Applets, Plugins, or Other Applications Required to Interpret Page Content**—Test to ensure that they can be used by assistive technologies.

[7] F. E. Randall et al. Human Body Size in Military Aircraft and Personal Equipment, Army Air Forces Technical Report no. 5501, June 10, 1946.

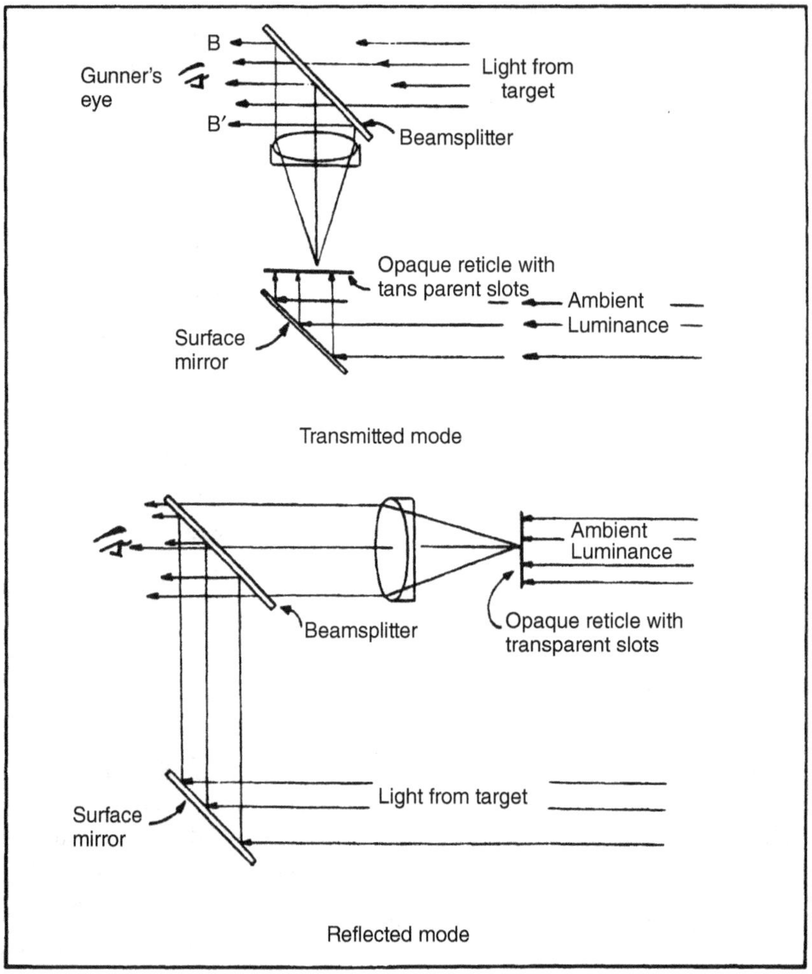

Figure 28.98 Schematic of a reflex-collimating sight.

5. **Accessibility to All**—Meet the Rehabilitation Act, Section 508.

6. **Diverse Audiences**—Provide information in multiple formats if necessary to address diverse audiences.

7. **Scripting Language Display Information**—Ensure it can be read by assistive technology.

8. **Client-Side Image Maps**—Provide client-side image maps instead of server side to improve accessibility.

9. **Multimedia Elements**—Provide synchronized equivalent alternatives to ensure accessibility.

10. **Style Sheet**—Organize documents so they are readable without requiring a style sheet.

11. **Text-Only Pages**—Provide equivalent information and functionality if compliance with accessibility provisions cannot be accomplished in any other way.

12. **Compatibility with User Computing Systems**—Account and design for the computer systems (browsers, operating systems, connection speeds, screen resolution), that are the most popular within the user population.

13. **Online Forms**—Ensure that users using assistive technology can complete and submit.

14. **User Errors**—Anticipate possible user errors, and program to identify these errors and prevent when possible.

15. **Fluid Layouts**—Use formatting for fluid layouts so that page content automatically adjusts to various monitor sizes and resolution settings.

16. **Iterative Design Process**—Develop and test prototypes through an iterative design approach.

17. **Test Participant Comments**—Solicit test participant comments during user interface development and evaluation.

18. **Number of Usability Test Subjects**—Select the right number of test subjects: inspection—minimum two to five specialists (more is better); performance (early design testing)—six users minimum; performance assessment—determine number to meet desired statistical power.

19. **Pre and Post Revision Testing**—Evaluate websites before and after usability changes.

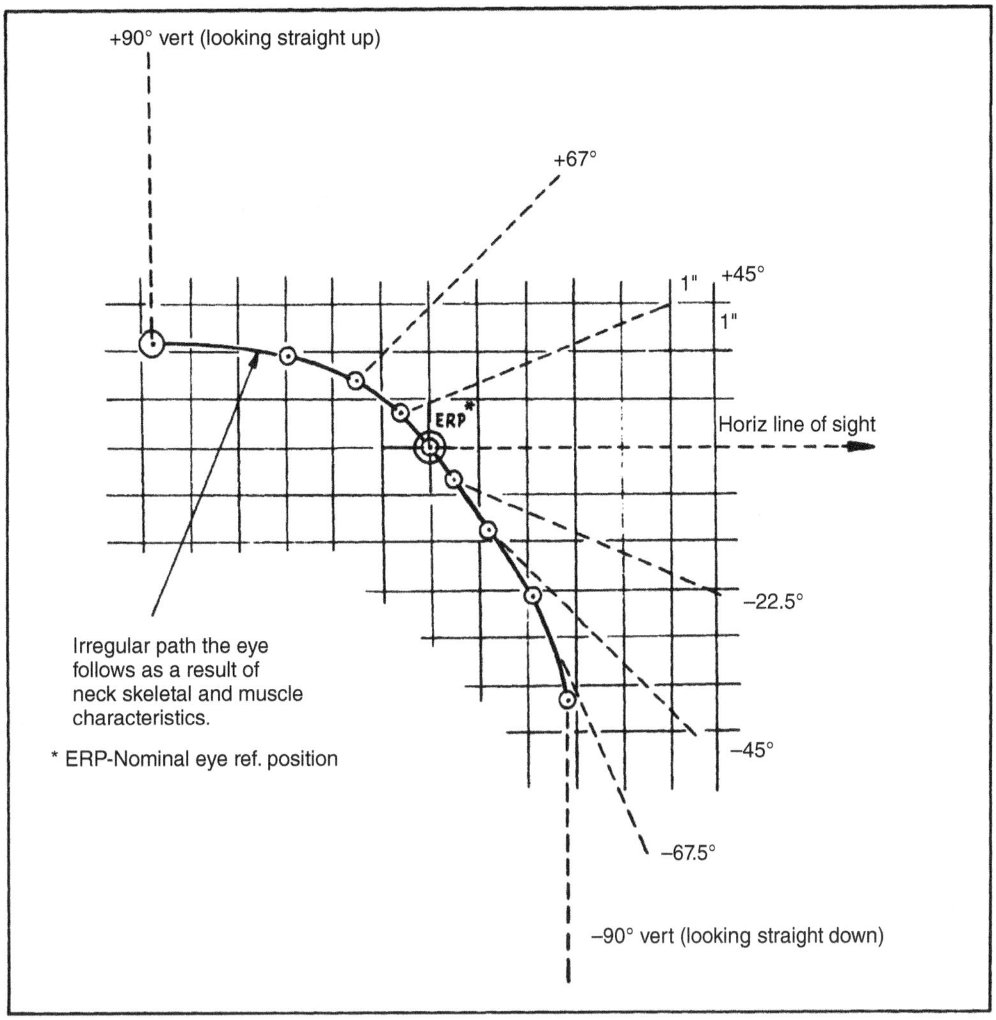

FIGURE 28.99 Eye-positioning path for aligning on an optical device.

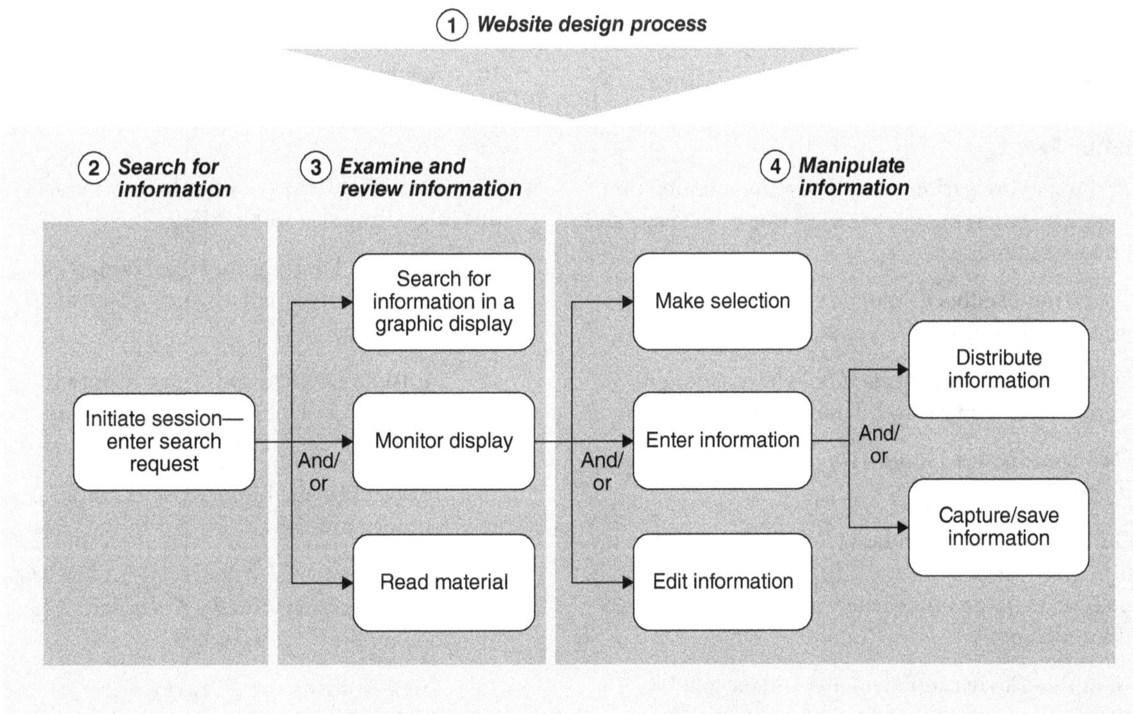

FIGURE 28.100 User–system interface.

20. **Choice of Prototyping Tools**—Select the best tools based on the phase of the design, the required fidelity of the prototype, and skill of the person creating the prototype.

21. **Prioritize Development Activities Carefully**—Don't put off addressing interfaces simply because it seems they will easy for the user to understand and use. Attack the most severe problems before the less severe, more frequent problem. Remember that expert judgment of problem severity is often wrong and severity is better evaluated by usability tests. And, focus on user performance first; things like color and graphics can be secondary.

22. **Use Inspection Evaluation Results with Caution**—Know that inspection evaluations are often very wrong. Inspection evaluations should not be used directly but may best be used to identify potential usability issues to evaluate during usability testing. (Inspection evaluations include heuristic evaluations, expert reviews, and cognitive walkthroughs.)

23. **Download Speeds**—Take measures to prevent images from slowing page download speeds.

24. **Screen Flicker**—Avoid flicker frequencies 2–55 Hz.

25. **Changes**—Announce major changes to a website on the homepage.

26. **Widget Selection**—Use familiar widgets and use them in commonly used manner.

27. **Download or Wait Times**—Minimize and warn of long times (0 to 10 s—show hourglass; longer than 10 s—show a progress indicator).

28. **Timeouts and Session Expirations**—Warn about.

Information Search

1. **Primary Navigation Menu Location**—Locate the primary navigation menu on the left side of the page and group lower menus.

2. **Location Feedback**—Provide feedback to let users know where they are in the website.

3. **Link Labels**—Link labels should be meaningful, understandable, and easily differentiated by users.

4. **Navigation Dead Ends**—Do not create navigation dead ends.

5. **Links to Contents on Long Pages**—Provide quick links to avoid lengthy scrolls. Allow users to see overview of the content, to go immediately to a section, and then return to the top.

6. **Graphic Navigation Elements**—Make graphic navigation elements to look like actual items and controls (for example, file folder tabs or "start" buttons).

7. **Search Engines**—Search engines should search the entire site, or clearly communicate which part of the site will be searched.

8. **Search Capabilities**—Provide search capabilities where appropriate—don't make the user go back to a home page.

9. **Search Terms Used**—A search engine should respond to users' terminology.

10. **Search Help**—When necessary, provide information to help improve searches. Help the user with short tip and hints rather than long, complex instructions.

11. **Use Appropriate Menu Types**—Use "sequentially displayed" submenus for simple forward-moving tasks, and use "simultaneous" menus (where all menus and submenus are displayed simultaneously) for more complex trial and error tasks.

12. **Navigation Options**—Always provide the user a way to navigate from a page (for example, do not drop the *Back* button function).

13. **Navigation Elements**—Clearly differentiate navigation elements from one another, but group and place them in a consistent and easy to find location on each page.

14. **Navigation Page Prose Text Length**—Be concise and limit the number of words on navigation pages.

15. **Simple Searches**—Accommodate simple searches with a small number of words. (This is how most people research.)

16. **Related Content**—Provide links to other pages in the website with related or explanatory and supportive content.

17. **Link Hierarchy**—If links have a hierarchy, show this in lettering size and bolding.

18. **Link and Destination Page Names**—Make the link name consistent with the title or headings on the destination page.

19. **Multiple Links to Important Content**—Ensure that important content can be accessed from more than one link.

20. **Text and Image Links**—Use text links in preference to image links.

21. **Used Links**—Use color changes to indicate to users when a link has been visited. (A standard is blue for unused and purple for used.)

22. **Click-Ability Cues**—Provide sufficient and consistent cues for graphic and text items that are clickable. Cues can include combinations of underline, blue color,

location, arrows, bullets, and appearance (push buttons or tabs). Items that are not clickable should not look like they are.

23. **Text Links**—When using text links (links embedded within text), the link text should accurately describe the link's destination.

24. **Pointing-and-Clicking**—"Pointing-and-clicking" is preferred over "mouseover" (where display change occurs simply by moving the mouse pointer to a spot) when selecting menu items from a cascading menu structure.

25. **Text Link Lengths**—Make text link lengths long enough to be understood but try to keep in one line.

26. **Internal and External Links**—Indicate to users when a link will move them to a different location in the same website or to a different website.

27. **Search Types**—Ensure that user is aware of all types of searches available.

28. **Upper and Lowercase Fonts for Search Terms**—Search terms should not be case sensitive.

29. **Search Results**—Search result content and format should be what user wants and expects.

30. **Navigation Links**—Allow users to skip repetitive links.

Information Examination and Review

1. **First Impression**—Use the homepage to create a good impression of the quality of the site and the information.

2. **Logos**—Place an identifying logo in a consistent place on every page.

3. **Banner Ads and Decorations**—Graphics should not look like banner ads or decorations or else they will be ignored.

4. **Look of Homepage**—Ensure the homepage *looks* like the homepage.

5. **Homepage Options**—Present all major options on the homepage.

6. **Purpose of Site**—Clearly and prominently communicate the intent of the website on the homepage.

7. **Order**—Arrange items in lists for the most efficient and successful user performance—usually with the most important items at the top of the list. Where no obvious order applies, organize lists alphabetically or numerically.

8. **List Numbers or Bullets**—Use bullets items in a list are of equal importance; use numbers if items have an order.

9. **Headings, Titles, and Category Labels**—Use many headings and category labels. They should be understandable by novice users, descriptive of the information, consistent, unique, and not too general. This also applies to column and row headings in tables and page titles.

10. **Subcategories**—Use subcategories and visually distinguish category levels.

11. **Highlights**—Visually highlight important information (using color, bolding, size, placement, underlines, etc.), but do not clutter the page.

12. **Frame Titles**—Provide frame titles to facilitate frame identification and navigation.

13. **Nontext Element that Conveys Information**—Provide a text equivalent.

14. **Tab Labels**—Make tab labels clearly descriptive of their function or destination.

15. **Layout for Comparison**—If users need to compare items, structure pages so that this can be easily done.

16. **Layout for Importance**—Use a page layout that show the level of importance of content and clickable items—from most to least important. The top center is a prime location.

17. **Consistent Page Layout**—Locate most and least important and clickable items in a consistent place on the pages.

18. **Alignment**—Align elements to make page to make it easier to scan for information.

19. **Frames**—Use fixed frames when certain functions must remain visible.

20. **Page Layout for Scrolling Feedback**—Ensure that the location of headings and other page elements does not create the illusion that users have reached the top or bottom of a page when they have not.

21. **Site Maps**—Provide site maps for websites with many pages.

22. **Page Lengths**—In general, use shorter pages for home pages, navigation pages, and pages that need to be quickly browsed and/or read online.

23. **Text Line Lengths**—If reading speed is most important, use longer text line lengths (75–100 characters per line). If acceptance of the website is most important, use shorter text line lengths (50 characters per line).

24. **Animation, Video, and Audio**—Use animation video and audio meaningfully—avoid unnecessary distraction to users or long download times.

25. **Large images**—Do not put large images that fill the first screen, information following may be ignored.

26. **Use of Graphics**—Limit the use of graphics to only important images convey the intended message to the user, otherwise they may be unnecessarily distracting and take a long time to download.

27. **Data and Graphics**—Include actual data with data graphics.

28. **Monitoring**—Use graphics to display information that changes and must be monitored.

29. **Animation Introductions**—Provide text introductions to animations.

30. **Thumbnail Images**—If large images are not critical to understanding the website, use thumbnails to preview them.

31. **Logical Organization**—Organize information logically so that it meets the users' needs and the website goals.

32. **Scanning**—Design content pages to facilitate user understanding by just scanning the page.

33. **Necessary Information**—Be sure that all information necessary to fulfill the website goal (but no more) is displayed when and where it is needed.

34. **Grouping**—Group all related information and functions.

35. **Path to Information**—Minimize the number of actions and pages necessary for the user to get to the desired information.

36. **Quantitative Information**—Format quantitative information to minimize the time to understand it (using charts, tables, and graphs as necessary).

37. **Visual Appearance**—The visual appearance of the text on the pages (size and spacing of characters; the colors used for labels, fonts, and backgrounds; and the locations of labels, etc.) should be consistent within a site.

38. **Attention Attracting Features**—Use attention attracting features (movement, font size, font style, bolding, graphics, bright color, etc.) when appropriate for important information, but avoid clutter and confusion.

39. **Font Types**—Use commonly used and familiar fonts for best reading speed and understanding.

40. **Font Size**—Use at least 12 point font on all webpages.

41. **Color**—Use color (but not color alone) for grouping or separating information.

42. **Color Backup**—Ensure that all information conveyed with color is also available without color.

43. **Color Coding**—When using color coding to communicate information, make sure there are instructions that identify what the colors mean.

44. **Prose Text Colors**—Prose text is most easily read and understood when it is black on a plain, high-contrast background.

45. **Text Formatting Consistency**—Be consistent with text formatting such as paragraph indents, time formats, telephone number formats, etc.

46. **Upper- and Lowercase Fonts for Prose Text**—For prose text, use the mix of upper and lowercase fonts rather than all lower or all uppercase.

47. **Images**—Use images and graphics to facilitate learning.

48. **Horizontal Scrolling**—Use a page layout to eliminate the need for horizontal scrolling.

49. **Writing Style for Comprehension**—To improve comprehension, use an active voice, keep instructions affirmative (what to do instead of what not to do), and limit text length. (A sentence should not contain more than 20 words and a paragraph should not contain more than six sentences.) Also, the first sentence of a paragraph should introduce the paragraph contents.

50. **Size of Homepage**—Limit the homepage to one screenful, if possible. Limit the amount of prose text on the homepage.

51. **Information Density and Clutter**—Information density should be low enough so that pages are not cluttered and important search targets are clearly available.

52. **Moderate Use of White Space**—Overuse of white space does not improve search capabilities and can cause unnecessary scrolling.

53. **Long Data Sets**—Partition long data sets into shorter subgroups, to help read and memorize the data and detect entry errors.

54. **List Length**—Show all or as many items in a list as possible without requiring the user to scroll.

55. **Home Page Access**—Enable users to access the homepage from any other page on the website.

56. **Unsolicited Windows or Graphics**—Do not display.

Information Manipulation

1. **Data Entry Method**—Users should be able to stay with one entry method as long as possible (instead of alternating between a mouse and keyboard, for example).

2. **Check Boxes**—Use check boxes to allow users to select one or more items from a list.

3. **Measurement Units**—For data entry fields, specify the desired measurement units in the field labels rather than requiring users to enter them.

4. **Data Entry Field Label Location**—Place data entry field labels near the associated fields to avoid confusion.

5. **Visibility of Entered Data—**Make data entry fields large enough to show the user all of their entered data no or minimal scrolling.

6. **Radio Buttons**—Use radio buttons for mutually exclusive selections ("pick just one"). Never use just one radio button.

7. **Default Values**—When choice is likely, use this selection as a default value.

8. **Cursor Placement for Data Entry**—Place the cursor at the first data entry point and then when possible provide the option for auto tabbing from data field to data field.

9. **Lists of Items to be Selected**—Use open list in preference to a drop-down list for displaying items to be selected.

10. **Data Entry Methods**—Data entry through typing is usually preferred and faster than selection from a list.

11. **Shift Key**—Minimize the need to use the shift key for data entry.

12. **Sequence of Tasks or Actions**—If tasks or actions must be performed in a specific order, make the order logical, clear, and consistent.

13. **Required and Optional Data Entry Fields**—Distinguish clearly and consistently between required and optional data entry fields.

14. **Push-Button Labels**—Clearly label push buttons to describe what will happen if they are clicked.

15. **Help Materials—**Use user's terminology.

16. **Workload**—Reduce user workload: (a) Design to user memory limits; (b) Standardize task sequences; and (c) Display information in a directly usable format.

17. **Data Entry Field Label Consistency**—Make data entry field characteristics consistent—such as length (one word versus a sentence), vocabulary (noun versus a verb), and terminology.

18. **Data Entry Field Label Clarity**—Label data entry fields so that users understand the desired entries (avoid unfamiliar, vague, or jargon terminology).

19. **Duplicate Data Entry**—Avoid users having to enter the same data more than once.

20. **Double Selection**—Ensure that doubling selection gestures (clicking or tapping) does not cause problems.

21. **Printing**—Format to allow.

CHAPTER 29

AUDITORY DISPLAYS

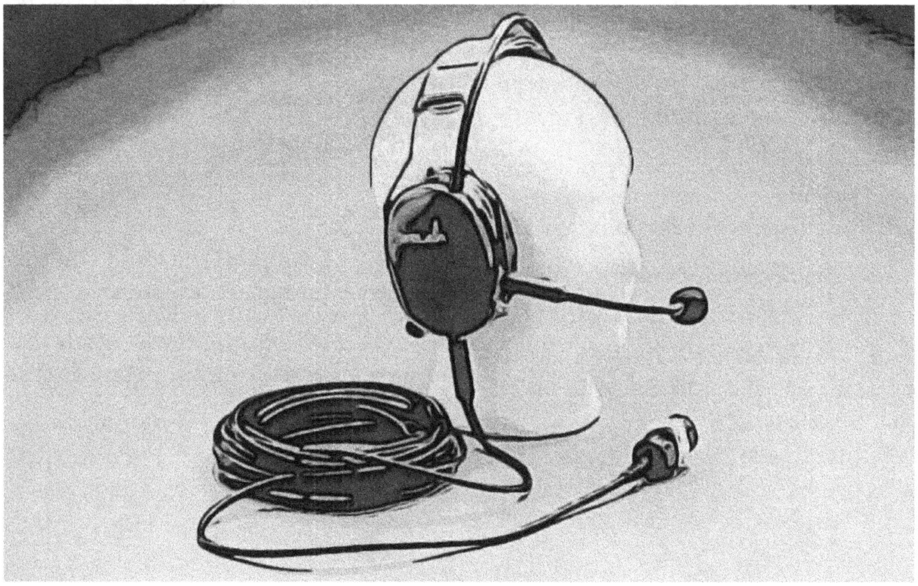

SELECTION GUIDELINES

General

General guidelines for selection of auditory displays are given in Table 29.1. Although the choice of system may be obvious in most cases, there are specific operational reasons for selecting one type of communications system over another. The following guidelines are intended to help the planner or designer decide which type of system to use for a particular operational environment:

1. Public Address System: Use for nonspecific emergency announcements for all diversely located listeners. Do not use for routine messages intended for one individual if the announcement may interfere with localized communication.

2. Intercom System: Use for small group-specific, two-way communication and/or for individual communication between people who are physically isolated from one another, where telephone dialing would create a considerable waste of time (e.g., for communications between a shop supervisor and personnel at factory workstations). Do not use where conversational privacy is necessary.

3. Telephone: Use for situations requiring individual conversational privacy.

4. Radio: Use for situations where one or more of the conversants are moving from one location to another (e.g., a security guard on foot or a police officer in a patrol car).

5. Audio plus Video Conferencing: Use where face-to-face visual reference and/or reference to the object under discussion adds considerably to communication reliability.

Type of Communicating Interface

1. Handset (Telephone): Use when at least one hand can always be free to hold the handset.

2. Headset (Earphone or Earphones and Microphone): Use when both hands are often occupied.

3. Wireless Microphone: Use when a fixed microphone or hand-held microphone with cable would restrict the user's freedom to move about.

TABLE 29.1 Auditory Display Selection Guide

Display Configuration	Use Criteria	Human Engineering Considerations
Steady continuous tone 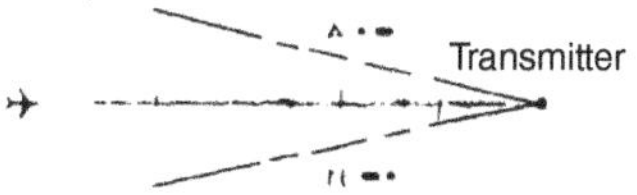	Use in both ears for tracking, i.e., both ears equally loud for "on target," shifting left or right when "off target."	Select the frequency which is least subject to other nonverbal auditory signals or typical vehicle system noises.
Prerecorded spoken message *Now hear this!* *"this is your captain speaking" calling Doctor Brown . . . come to emergency!*	Use when communication flexibility is necessary. Use when it is necessary to identify the source of a message. Use when a stressful situation might cause the listener to forget the meaning of a coded signal. Use when a simple coded signal cannot adequately give directions or instructions to the listener. Use when ambient masking noise characteristics obviate the use of simple tonal signals. Use when other complex tonal signal possibilities have already been exhausted, i.e., have been assigned and cannot be duplicated.	Utilize the most concise message possible that still retains maximum intelligibility and minimum ambiguity.
Intermittent tone 	Use for "start and stop" timing. Use for continuous information where rate of change of input is low.	Avoid a "train" of signals which might be confused with similar electrical system noise.
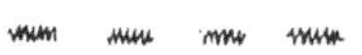	Use for irregularly occurring signals, i.e., alarms.	
  	Use as indicator of speed, i.e., a change in rate of signal occurrence. (Faster signal repetition indicates an increase in vehicle speed.) Use for continuous-wave message transmission, i.e., Morse code.	
Warble and undulating tones 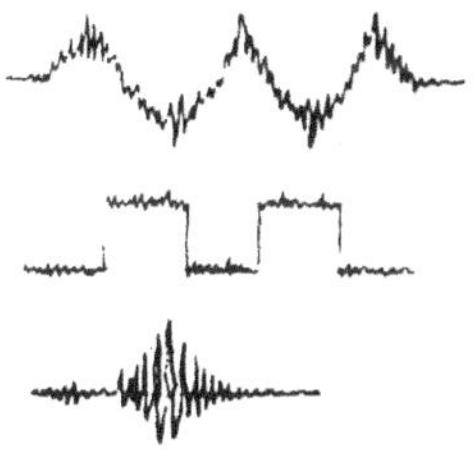	Use when masking noise characteristics are unknown or cover a broad frequency spectrum in random pattern. Use pitch differences to represent "up and down" relationships. Use for locating a null position such as in orienting a radio beam.	Avoid a signal that sounds like any characteristic navigation signal or radio signal such as might occur when two carriers present a "beat frequency" effect. Where distraction is a critical factor, abrupt onset of the audio signal should be avoided, as well as apparent movement of the signal in space.
Bell, buzzer, horn, siren, and whistle	Use bell for fire alarm. Use horn for emergency warning, equipment malfunction, pressure leak, hatch open, etc. Use whistle (typical of "Bosun's whistle") for "All Hands Alert" (followed by further communication). Use buzzer for individual operator alert to receive further communication. Use Claxon for "Report to Duty Stations."	The signal should be "distinctive" and unlikely to be obscured by other noises. Where crew members' attention may be highly concentrated on a task, use signals with relatively high alerting capacity. Signals should be at least 10 dB above the ambient noise level. They should be repeated so as not to be missed. Caution signals should be provided with reset and volume controls. Consider a multistage warning or alarm system, i.e., an initial tone signal followed by the critical signal, followed by speech where appropriate, to indicate the specific nature of the hazard. Alarm signals should focus attention on the first 0.5 s, with all essential information within the first 2.5 s. Use "sudden" onset. Concentrate signal energy between 250 and 2500 Hz, with signal readily identifiable on the basis of components below 2000 Hz.

(*Continued*)

TABLE 29.1 Auditory Display Selection Guide (*Continued*)

Display Configuration	Use Criteria	Human Engineering Considerations
Microphones 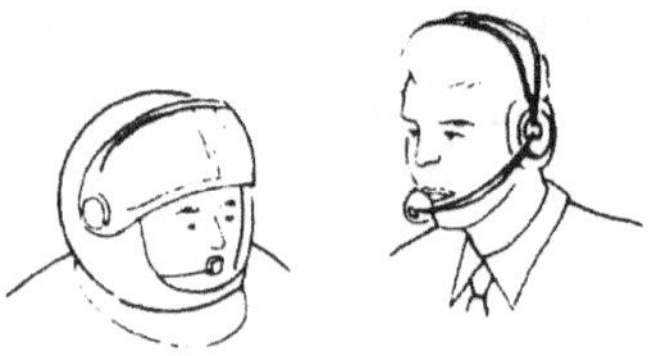	Provide microphone installation in all pressure helmets. Provide microphones for each operator position for shirt-sleeve operation. Microphones should be head-mounted for all situations where the operator will have both hands occupied.	Provide maximum convenience and minimum discomfort.
Telephone handsets 	Use at operator stations where the operator does not perform communications as a primary activity and can pick up the handset with essentially no interference with his or her regular duties. Use in situations where it is desirable to reduce the fatigue caused by wearing a headset.	
Earphones 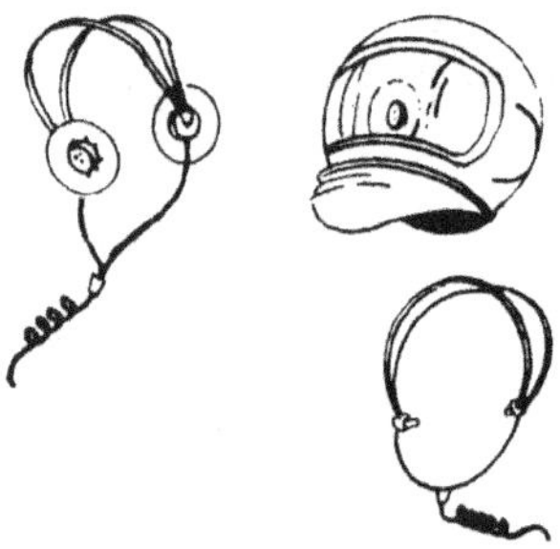	Provide earphones in all pressure helmets. Provide two portable head-mounted earphones for each operator position for shirt-sleeve operation.	The comfort and convenience of the operator as well as audio quality must be considered so that the operator will use the equipment as designed.
Loudspeaker 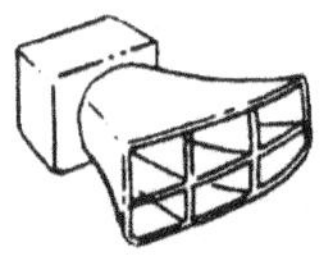	Use loudspeakers only when it is desirable to communicate with more than one person at the same time and when it is anticipated that one or more of the parties concerned will not be near a phone or will be wearing headsets or a space helmet (i.e., a public address system).	

Selection of Signal

Select the signal according to the function. Use Table 29.2 as a guide.

GENERAL PRINCIPLES FOR AUDITORY DISPLAY DESIGN

Situation Compatibility

The effectiveness of any auditory display depends on proper consideration of the sound environment within which the display must operate; i.e., a spoken message is easily obscured by other spoken messages, a particular tonal signal is easily masked by other similar tonal signals, and any auditory signal or message may be masked by frequent, extremely loud bursts of noise such as would occur during a battlefield bombardment or the noise of a jet aircraft afterburner. Table 29.3 gives guidelines for selecting audio signals based on their intended function.

Common Signal Conventions

Certain types of signals (sirens, bells, claxons, etc.) are recognized and associated with certain activities, such as those of fire fighters and the police. Such characteristic signals should not be used for other purposes when the situation is such that the more common convention is in use. Other natural relationships, such as high frequencies being associated with "up" and low frequencies being associated with "down," should be taken into account when considering the use of pitch as an auditory code.

Standardization

Once a particular auditory signal code is established for a given operating situation, the same signal should not be designated for some other display.

Listener Sensitivity

The signal frequency of auditory displays should be compatible with the midrange of the ear's response curve for both

TABLE 29.2 Selection of Audio Display Signal Based on Function

Function	Types of Signals: Tones (Periodic)	Complex Sounds (Nonperiodic)	Speech
Quantitative indication	*Poor* Maximum of 5 to 6 tones absolutely recognizable	*Poor* Interpolation between signals inaccurate	*Good* Minimum time and error in obtaining exact value in terms compatible with response
Qualitative indication	*Poor to fair* Difficult to judge approximate value and direction of deviation from null setting unless presented in close temporal sequence	*Poor* Difficult to judge approximate deviation from desired value	*Good* Information concerning displacement, direction, and rate presented in form compatible with required response
Status indication	*Good* Start and stop timing. Continuous information where rate of change of input is low	*Good* Especially suitable for irregularly occurring signals (e.g., alarm signals)	*Poor* Inefficient; more easily masked; problem of repeatability
Tracking	*Fair* Null position easily monitored; problem of signal-response compatibility	*Poor* Required qualitative indications difficult to provide	*Good* Meaning intrinsic in signal
General	Good for automatic communication of limited information. Meaning must be learned. Easily generated	Some sounds available with common meaning (e.g., fire bell). Easily generated	Most effective for rapid (but not automatic) communication of complex, multidimensional information. Meaning intrinsic in signal and context when standardized. Minimum of new learning required

From MIL-STD-1472G.

TABLE 29.3 Design of Auditory Warning Displays

Operational Objective	Guidelines
Audibility in high-noise; level environment	■ The signal energy should be concentrated within a narrow frequency band that is lowest in the background noise spectrum. A signal-to-noise ratio of 5:1 should be maintained if at all possible. The audio signal power at the listener's ear should be approximately 300 mW. ■ If the ambient noise is such that the warning signal would have to reach intensity levels that could be painful or injurious, provide the listener with a headset or a combination headset within an acoustic helmet. ■ If the nature of the ambient noise is such that it is loud across most of the audible frequency range, consider the use of a repeating signal (with shutoff control) and provide a noise-operated gain control to maintain a compatible signal-to-noise ratio as the background noise changes. ■ Use a noise-canceling microphone so that background noise will not be picked up along with the warning signal or the speaker's voice. Consider delaying the signal to one of the listener's ear (by 0.01 s) or reversing the phase of one earphone of the headset.
Fast operator response	Make the signal brief and concentrate the signal energy at the beginning. If a canned voice signal is used, provide a short alerting signal (0.5 s) just before the speech message to prepare the listener.
Operator distraction	Provide the operator with a temporary "silencing control" so that he or she can prevent the warning signal from interfering with a particularly critical ongoing task. The shutoff control must, however, have an automatic reset feature so that the warning system is not permanently disabled. Do not provide the operator with a volume control to quiet an interfering warning signal; the operator will generally turn it down to the point where he or she may never be alerted to an emergency.
Interference of warning signal with other important signals	Deliver the warning signal to one earphone via a headset and the other signals to the other earphone. Alternate the warning and general signal input positions every 2 to 3 s. Avoid frequency bands normally used by common communication or navigation signals.
Reduction of signal confusion	Avoid selecting warning signals whose particular characteristics are easily confused with common signals such as navigation signals, Morse code, sonar returns, radio or electronic countermeasure signals, and typical electrical interference.
Transmission mode	Where practicable, send warning signals via a different system from those being used for speech communications.

pitch and loudness; i.e., avoid the use of signals at the extreme ends of the sensitivity curves, where response reliability is more easily masked.

Dual Mode

When the noise environment is unknown or suspected of being difficult to penetrate, consider one or both of the following: (1) using a shifting frequency signal that passes through the entire noise spectrum and (2) combining the auditory signal with a visual signal.

Quantifiable Signals

When the tonal signal information is to be quantifiable, provide a reference tone, e.g., a base-line loudness or pitch against which the primary signal can be compared.

Privacy

If a signal must occur in an area in which only certain personnel should be privy to its purpose and others are not to be unduly annoyed, select a simple bell tone that can be recognized among ambient speech sounds without being loud.

Select Listeners

Consider a simple repetition code signal to indicate who is to respond, e.g., to answer a phone or to respond to a previously established set of task events.

Annoyance

Avoid the use of extremely loud signals that may startle listeners, and add to the overall noise levels, or interfere with local speech activity. When the ambient noise level gets too high, use earphones.

NONVERBAL AND ALARM SIGNALS

Two-Phase Signals

Attention should be focused on the first 0.5 s of the alerting signal and then upon the first 2.0 s of the identifying or action signal. Alerting signals should precede action signals by a time interval of 2.0 to 8.0 s for isolated signals and of 0.3 to 2.0 s for action signals occurring in sequence. When the alerting signal consists of a "glissando," the audible signal should spend at least 0.1 s in each octave band from lowest to highest.

Where masking is a critical factor, warning signals should be concentrated in frequency bands that are unused or little used by primary communication (command, navigation, advisory, etc.). Major concentration of energy should be between 250 and 2500 Hz/s and the signal should be readily identifiable on the basis of sound components below 2000 Hz. Where possible, avoid the following:

1. Trains of impulses that resemble electrical interference
2. Simple glissandi that might be confused with sounds made by carriers, i.e., beat frequency and oscillator effects
3. Periodic impulses similar to radar signals
4. Random noise similar to that generated by such equipment as an air conditioner
5. Musical sounds
6. Noise similar to static or sporadic radio signals
7. Modulated or interrupted tones resembling navigation signals or coded radio transmissions

Where warning signals delivered to headsets might mask other essential auditory signals (e.g., navigation and communication signals), separate channels may be provided to direct the warning signal to one ear and the background signal to the other ear. Presentation of warning signals may also be dichotic and alternating with the dichotic presentation of communication and navigation signals.

Where earphones cover both ears of the operator during normal system operation, the warning signal should be directed to the operator's headset as well as externally.

Audio warning signals should be at least 10 dB above the loudest expected ambient noise level. Such signals should not exceed a maximum intensity of 110 dB in areas (near the sound source) that must occasionally be occupied by unprotected personnel for as long as 10 min duration. The signal energy should be concentrated in the frequency band in which background noise is lowest.

Single-Phase Signals

All essential information should be contained in the first 0.5 s of the signal. Considerations regarding masking and signal strength are the same as for the two-phase signal.

False Alarms

Audio warning systems should be designed in such a way that any possibility of false alarms is minimized.

System Independence

The circuitry of audio warning systems should be independent of the system in which they are to provide warning information so as to preclude warning system failure in the event of primary system failure.

Operator Control

All audio warning systems or devices should be provided with a test button, switch, or other means of verifying the audio aspects of the signal at any time. Whether audio warning signals are designed to terminate automatically, by manual control, or both, an automatic "reset function" should be provided. The reset function should be controlled by a sensing mechanism that should recycle the signal system to a specified condition as a function of time or the state of the signaling system. The signal duration should be at least 0.5 s and should continue until an appropriate response is made by the operator. The completion of the proper corrective action should automatically terminate the warning signal.

A shutoff switch should be provided that is controllable by the operator, the sensing mechanism, or both, depending on careful consideration of the operational situation and personnel safety factors. (Caution signals should be provided with manual reset and volume controls.) Shutoff switches should not prevent the presentation of other warning signals and should reset automatically after an interval of time.

Steady, Continuous Tonal Signals

Signal frequency should be confined between 400 and 1500 Hz. According to NASA-STD-3001, the sound pressure level of a signal must meet at least one of the following criteria.

a. Using measurements of A-weighted sound levels [ISO 7731:2003(E), Ergonomics—danger signals for public and work areas—Auditory danger signals, method (a) in 5.2.2.1], the difference between the two A-weighted sound pressure level (SPLs) of the signal and the ambient noise is greater than 15 dBA (LS,A to LN,A > 15 dBA).

b. Using measurements of octave band SPLs [ISO 7731:2003(E), method (b) in 5.2.3.1], the SPL of the signal in one or more octave bands is greater than the effective masked threshold by at least 10 dB in the frequency range from 250 to 4000 Hz (LSi,oct to Lti,oct > 10 dB).

c. Using measurements of 1/3 octave band SPLs [ISO 7731:2003(E), method (c) in 5.2.3.2], the SPL of the signal in one or more 1/3 octave bands is greater than the effective masked threshold by 13 dB in the frequency range from 250 to 4000 Hz (LSi,1/3oct to LTi,1/3oct>13 dB).

Intermittent Tonal Signals

Signal frequency should be confined between 400 and 1500 Hz. The signal should exceed its masked threshold by at least 15 dB. Do not exceed a sound power level of 135 dB. Set the signal level 60 dB or more above the absolute threshold. Select a signal midway between the masked threshold (noise) and about 110 dB. Provide automatic volume control when the dynamic range is 20–30 dB or less.

Warble or Undulating Tonal Signals

The signal frequency should be confined between 500 and 1000 Hz. Use a pitch rise and fall rate of about 1–3 Hz. Where an alerting signal consists of a glissando, the audible signal should remain at least 0.1 s in each octave band from the lowest to the highest frequency. Where the operator's task is to detect frequency change, set the signal at least 30 dB above the absolute threshold.

VARIOUS SIGNAL DEVICES AND CHARACTERISTICS (SEE TABLE 29.4)

Buzzers

Usually of low-intensity ranges, low frequency (150–400 Hz). Generally suitable for relatively quiet situations where the distinctive sound commands attention but does not cause "alarm."

TABLE 29.4 Characteristic Intensity Ranges and Predominant Frequencies for Specific Signal Devices

Signal	Average Intensity (dBA) at 10 ft	Average Intensity (dBA) at 3 ft	Predominant Audible Frequency (Hz)
For Large Areas (High Intensity)			
4-in. bell	65–77	75–83	1000
6-in. bell	74–83	84–94	600
10-in. bell	85–90	95–100	300
Horn	90–100	100–110	5000
Siren	100–110	110–121	7000
For Small Areas (Low Intensity)			
Heavy-duty buzzer	50–60	70	200
Light-duty buzzer	60–70	70–80	400–1000
1-in. bell	60	70	1100
2-in. bell	62	72	1000
3-in. bell	63	73	650
Chime	69	78	500–1000

Bells

Normally of higher intensity and frequency than buzzers. Suitable for situations where the ambient noise is higher (although they can also be used at lower intensity in fairly quiet environments such as a bank or department store, where they provide communication to operating personnel without disturbing customers).

Horns

Usually of high intensity (90–100 dB) and middle to lower frequency. However, horns can be used with higher intensity and higher frequency up to 3000–4000 Hz. Because they generally are of longer duration, they are not suitable for quiet environments, but rather for high-noise-level situations where urgent emergency warning is desired.

Chimes

Similar to quiet bells, typically in the frequency range of 500–1000 Hz. Suitable primarily for situations where the resident personnel require signaling without disturbing customers or visitors. Excellent for hospitals and similar operating situations.

Sirens

Can be made to penetrate almost any noise environment. The combination of high intensity and constantly changing frequency is not easily masked by spurious or continuous ambient noise conditions. Primarily for emergency signaling by police, fire fighters, etc.

VERBAL (SPEECH) SIGNALS

Word Selection

In selecting words for use in auditory warning signals, priority should be given to aptness, conciseness, and intelligibility (in that order). When a set of standard words are chosen, i.e., a phonetic alphabet (Able = A, Baker = B, Charlie = C, etc.), select words of two or more syllables rather than one-syllable words and test to make sure that no two words in the set sound similar.

Talker Characteristics

Voice communication systems should be designed to make maximum use of features that will preserve and enhance the intelligibility of speech produced by the "average talker" with a generalized "American accent" (for the United States), average intensity level (approximately 70–75 dB SPL), and average speech rate. When a message is to be conveyed in another language, select a talker whose basic language training and experience are in the particular language; e.g., extreme care should be taken in selecting an English-speaking translator to provide messages in German, Spanish, or French to make sure that his or her pronunciations will not be misconstrued by the listeners because of improper emphasis, colloquial discrepancies, etc.

Presentation

Verbal signals that occur frequently (e.g., status and caution) should be presented in a fairly formal manner, as opposed to an impersonal one. Although emphasis is important in circumstances where a sense of urgency must be conveyed, care should be taken to train the talker not to convey emotional stress or anxiety. Less frequent and more urgent signals may be less formal and delivered in a more personal manner to instill confidence.

Signal Characteristics

1. Speech signals should fall within the range of 200–6100 Hz.

2. The audio signal power should be approximately 300 mW at the listener's ear.

3. The signal-to-noise ratio should be at least 5:1.

4. Provide distortion-free signals and minimize masking effects that may occur as a result of nonuniform frequency response or a lack of uniform amplification.

5. Maintain intelligibility nearly equal to that available from a given system at sea level by using a pressure-sensitive device to adjust the gain of the amplifier (as pressure changes, e.g., at higher altitudes).

6. Provide microphones and earphones that have been designed to give uniform frequency response at all altitudes.

7. Verbal warning signals should be "processed" only when necessary to increase or preserve intelligibility, i.e., by increasing the strength of consonant sounds relative to vowel strength. Where speech signals must be relatively intense because of high ambient noise, use "peak clipping" to protect the listener against auditory overload.

8. Where speech communication signals are presented dichotically via earphones, one of the following methods should be used for the recorded warning signals:

   a. Dichotic presentation (individual input to each ear)

   b. Distinctive, nonmasking, nonspeech, component to identify the speech as being part of the warning signal, i.e., superimposed on a speech component

   c. A very distinctive voice for recording warning signals, e.g., a female voice rather than the typical male voice

Where confusion might exist, avoid scrambled speech effects that might be confused with "monkey chatter" from adjacent channels.

Verbal Warning Devices

1. Initial alerting signal (nonspeech) to attract the listener's attention and to designate the general problem.

2. A brief, standardized speech signal (message) that identifies the specific condition and suggests the appropriate action that should follow.

3. Verbal alarms for critical functions should be at least 20 dB above the speech interference level (SIL) at the operating position of the intended listener. SIL describes the effectiveness of noise in masking speech. It is the average (in decibels) of the sound levels of masking noise in the 600–1200, 1200–2400, and 2400–4800 octave bands.

4. The voice used in recording verbal warning signals should be distinct and mature. The warning message should always be presented in a calm manner, with the intent of eliciting a rational reaction from the listener.

TELEPHONE HANDSET PACKAGING AND DESIGN

The key factors in packaging the telephone handset are the following:

1. The earphone and mouthpiece should be accommodated to the user's head so that regardless of the size of the head or its unique shape, the user can hold the earphone close to the ear, and the mouthpiece close to the lips (without breaking contact with the ear). Ideally, the axis of the speech projection should coincide with the receiver axis, as shown in Fig. 29.1. Because of differences in head size, this is not possible without an adjustable handle. An acceptable compromise is shown in Fig. 29.2.

2. The handle should fit the hand comfortably. The figure indicates the maximum acceptable dimensions. Avoid square cross-sectional shapes and edges.

3. When a "talk switch" is provided on the handset, place it as shown in Fig. 29.2 so that the handset can be used with either hand.

4. Use coiled or retractable cords with handsets to minimize cord tangles. Handset cords that are used in fixed operator stations are easily knocked on the floor. Unless there is a need for a longer cord, shorten the cord so that it prevents the handset from ever striking the floor and causing damage to the handset.

5. The total weight of the handset should not exceed 10 to 11 oz. (0.28 to 0.31 kg).

6. Phones used in vehicles should be provided with "securing cradles" and/or other devices that will prevent the movement of the vehicle from unseating the handset. Consider speakerphones or other hands-free operations.

7. Handset bases should be designed so that they are not easily displaced, especially if the base has a dialing device. Consider the use of rubber feet, suction cups, and/or other special holding devices.

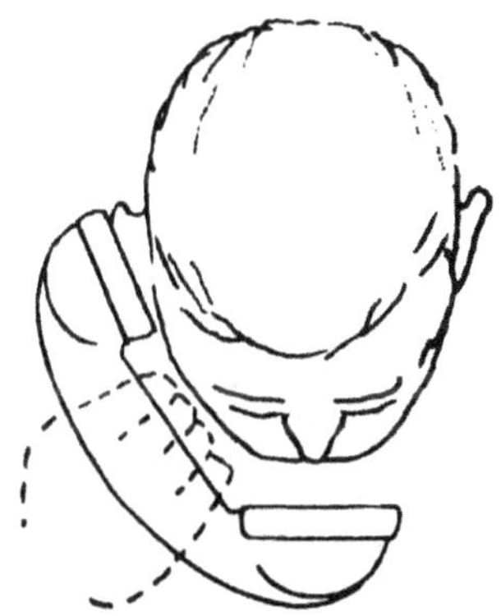

FIGURE 29.1 Handset should accommodate user's head.

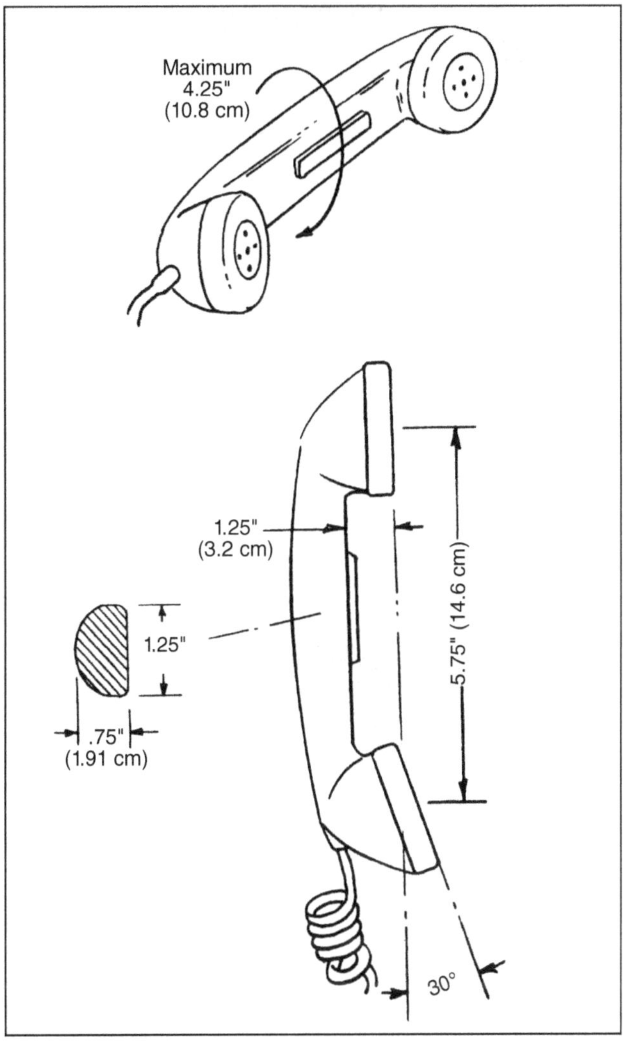

FIGURE 29.2 Recommended handset dimensions.

8. In-handle channel and line selection concepts may be helpful for some applications as long as the following do not occur:

    a. The dial or push buttons cause the handle to be awkward to hold onto.

    b. The dial or push buttons are too small for accurate finger manipulation, or the push buttons are placed too close together.

    c. The handset becomes too heavy.

9. Except in special cases, any color should be acceptable for the handset package. A given color can be used to identify a special channel, i.e., a "hot line."

10. Car phones must be designed to reduce distractions from the primary task of driving.

WALL-MOUNTED PUBLIC AND PRIVATE TELEPHONES

1. The height of the wall-mounted telephone should be determined by the convenience of the critical use elements, i.e., the handset, the dial or push-button pad, a fixed mouthpiece, a coin slot, etc. (See Fig. 29.3.)

   a. For adult users, the phone package typically should be located between waist and eye height. Generally this would place the phone box about 4.0 ft (1.2 m) above the ground or floor.

   b. For public places frequented by wheelchair users, at least one phone should be placed at a lower level so that it can be operated easily by the person sitting in a wheelchair. The highest element to be used by the person in a wheelchair (dial, coin slot, etc.) should not be more than about 4 ft (1.2 m) above ground or floor level.

2. From a human engineering point of view, dials are less preferred than push-button selector types, either for the typical desk-mounted phone or for other types, including wall-mounted phones, handset-mounted phones, etc. (See Fig. 29.4.)

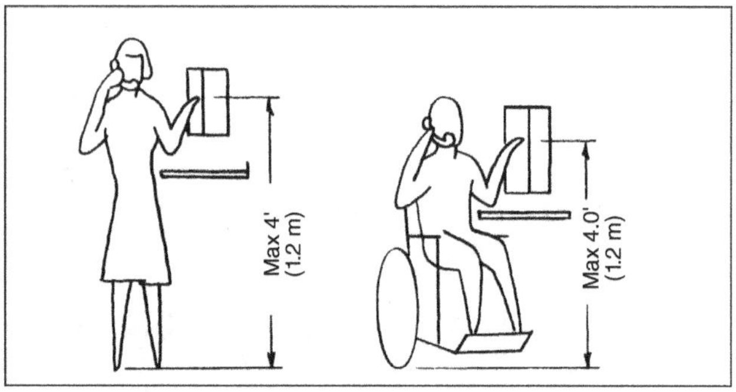

FIGURE 29.3 Recommended heights for wall-mounted phones.

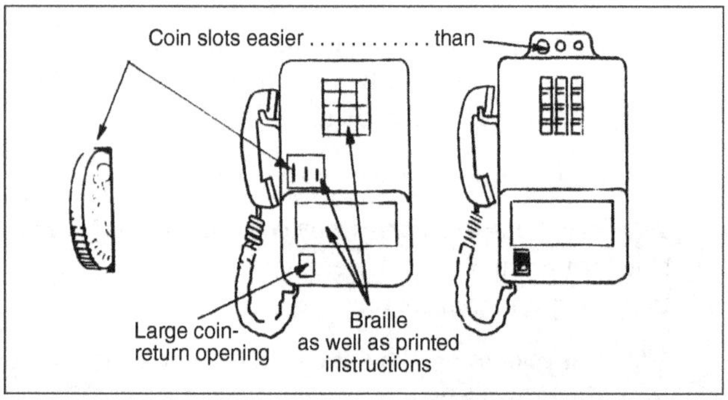

FIGURE 29.4 Wall-mounted phone design recommendations.

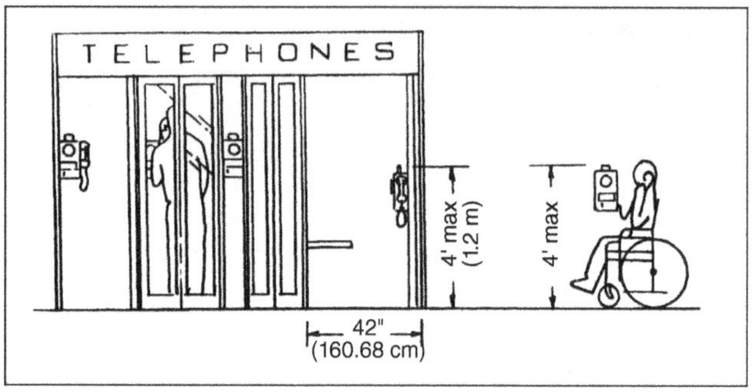

FIGURE 29.5 Minimum phone booth opening widths.

3. Use a push-button phone rather than dial phone to minimize errors and difficulties for the orthopedically handicapped.

4. Provide coin insert slots that permit the coin to be inserted vertically rather than laid into a flat, disk opening. This allows the user to maintain a more secure grasp of the coin.

5. Mount the phone so that there is sufficient space for a wheelchair user to approach parallel to the front of the phone box. (See Fig. 29.5.)

HEAD-MOUNTED MICROPHONES AND EARPHONES

Operating Characteristics

The microphone (mouthpiece) should be (see Fig. 29.6):

1. Highly sensitive to acoustic speech signals.
2. Of sufficiently high quality to provide faithful transduction of the speech signal into an electrical signal.
3. Capable of rejecting other acoustic signals and noise.

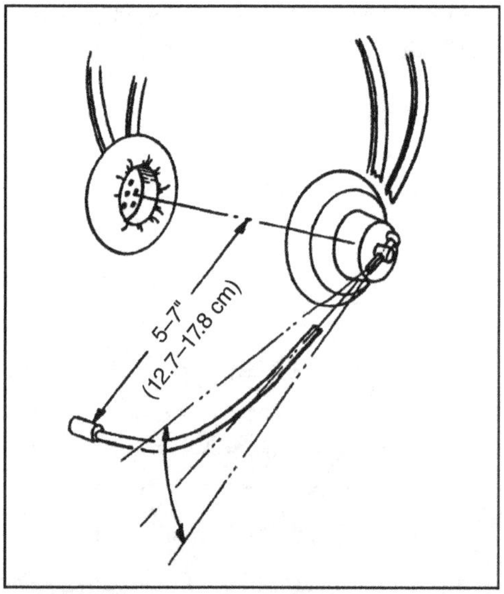

FIGURE 29.6 Recommended microphone placement.

4. Designed for "close talking." Do not use large pressure-gradient velocity type microphones; instead, use noise-canceling types in low-frequency and loud environment (100 dB overall SPL). Avoid "Ribbon Mikes" for close talking.

5. Designed so that the talker has to talk into the axis of the microphone, so that the talker's lips naturally fall close to the microphone, and so that the operator's hand never covers the microphone holes inadvertently.

6. Designed to have a smooth frequency response between 200 and 6100 Hz.

7. Designed to provide a dynamic range (when working with the selected amplifier) great enough to admit a minimum of a 50-dB variation in signal input. A close-talking microphone should not overload with signals as high as 130 dB.

8. Equipped with a noise shield for applications where the ambient noise level is extremely high (above 100 dB). The shield volume should be at least 250 cm^3, but should not be so large as to be unwieldy for the operator to handle and use. The shield should fit tightly against the operator's face with normal hand pressure or head-strap tension; i.e., the operator should not have to press the shield with considerable force in order to get it to fit. Provide exhalation holes in the shield, but place these as far from the actual microphone as practicable. Sound-absorbing material should be enclosed in acoustically transparent, waterproof material. The shield (if worn) should not impede the talker's mouth or jaw motion, thus causing speech distortion.

Packaging (See Fig. 29.7)

1. Provide complete adjustability of headset components.

2. There should be lightweight, minimum tension to hold the earphones snugly against the head.

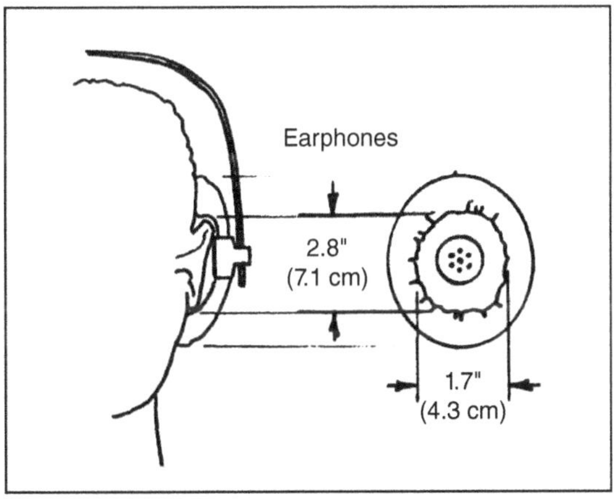

FIGURE 29.7 Recommended earphone dimensions.

3. Provide a soft cushion that is large enough so that it does not press on the user's ear but which seals around the ear to minimize external noise interference.

Design Recommendations

1. Provide smooth frequency response characteristics that are as broad as those of the remainder of the system.

2. Provide a dynamic range, without appreciable distortion (at least 40 dB), capable of power-handling capacity for peak amplifier output.

3. Provide a combination of earphone and socket (or cushion) for which the earphone sensitivity and the earcup attenuation together will provide an adequate signal-to-noise ratio of 5:1.

4. Provide an earphone cushion which is comfortable for long-duration use and which is not prone to developing fungus or collecting moisture.

5. Connect the earphones to operate out of phase.

6. Delay the signal reaching one ear about 500 μs.

7. Use binaural rather than monaural headsets if listeners will be in an intense noise environment.

8. Provide a sidetone (delay of 0.05 s) to make the talker increase his or her voice level of effort.

9. Provide sufficient electrical power to drive the peak sound pressure level to 131 dB when using two earphones.

10. Provide a gain control with a dynamic range sufficient to make the signal at least 15 dB more intense than the ambient noise.

11. Provide a reduction in frequency range below 500 Hz and above 4000 Hz if it results in an increase in the average power of the audio signal.

12. Provide a uniform frequency response of receiver and headset between 300 and 4000 Hz to avoid unpredictable distortions.

13. Provide a pressure-operated gain control switch to compensate for the effects of altitude in aircraft cabins.

LOUDSPEAKERS, AMPLIFIERS, TRANSMITTERS, AND RECEIVERS

Loudspeakers

General. Loudspeakers should be subject to the same frequency response characteristics as other elements of the communication system, except that loudspeakers for use in multispeaker installations (i.e., where several speech channels are to be monitored simultaneously) should respond uniformly over the range of 100–8000 Hz.

Use Considerations

1. Where an individual loudspeaker is provided a single operator, locate the speaker as close as practicable to the operator (within 3 ft, 0.9 m) so that the operator can reduce the signal of his or her particular speaker and thus refrain from adding to the overall noise level of an operating area.

2. Locate speakers and microphones so as to minimize speaker feedback into the microphone.

3. Where several channels are to be monitored simultaneously by means of loudspeakers, separate the speakers by at least 10° with respect to the principal listener so that he or she can localize the separate speaker sounds. If additional channel differentiation becomes necessary (i.e., if the 10° does not provide clear definition), utilize low-pass filtering (fc = 1800 Hz) to signals fed to loudspeakers on one side of the operator, high-pass filtering (fc = 2000 Hz) to signals fed to loudspeakers on the opposite side of the operator, and unfiltered signals to loudspeakers in front of, or behind, the operator.

4. Position speakers approximately at the head (or ear) height of listeners, except in auditoriums, where the speaker should be at podium height or higher, and in multilevel work areas, where it may be important to recognize that a sound emanates from a position below the operator.

Amplifier, Transmitter, and Receiver Characteristics

1. Bandwidth: Provide an essentially "flat" audio frequency response between 200 and 6100 Hz.

2. Gain: Provide sufficient dynamic range and gain to handle the range of instantaneous pressures of the primary input objectives (speech, music, etc.) and to develop the necessary signal level at the headset or loudspeaker terminals.

3. Noise: Amplifier background noise should be less than that introduced by the microphone.

4. Control: Provide automatic (nonlinear) gain control (AGC) if noisy environments are anticipated. Talker sidetone should be taken from the system ahead of the AGC. Use an attack time of 0.1 and a release time of about 10 s in "limiter" amplifiers.

5. Frequency Pre-Emphasis: Speech system input devices should employ frequency pre-emphasis wherever practicable. This should consist of a positive-slope frequency characteristic no greater than 9 dB per octave, over the range 150–4800 Hz. Where transmission equipment utilizes pre-emphasis and where peak clipping is not used, reception equipment should employ frequency de-emphasis of characteristics that are complementary to those of the pre-emphasis feature. De-emphasis should be a negative-slope frequency response not greater than 9 dB per octave over the frequency range 150–4800 Hz.

6. System Limitations: Where system engineering necessities require speech-transmission bandwidths narrower than 150–4800 Hz, the frequency range should be limited to the band 200–3000 Hz.

7. Clipping: Where speech signals are to be transmitted over channels showing less than 15 dB peak-speech-to-RMS-noise ratios, peak clipping of 12–20 dB should be employed at system input and should be preceded by frequency pre-emphasis as specified above, except that peak clipping should not be employed in single sideband voice-transmission systems.

CHAPTER 30

CONTROLS

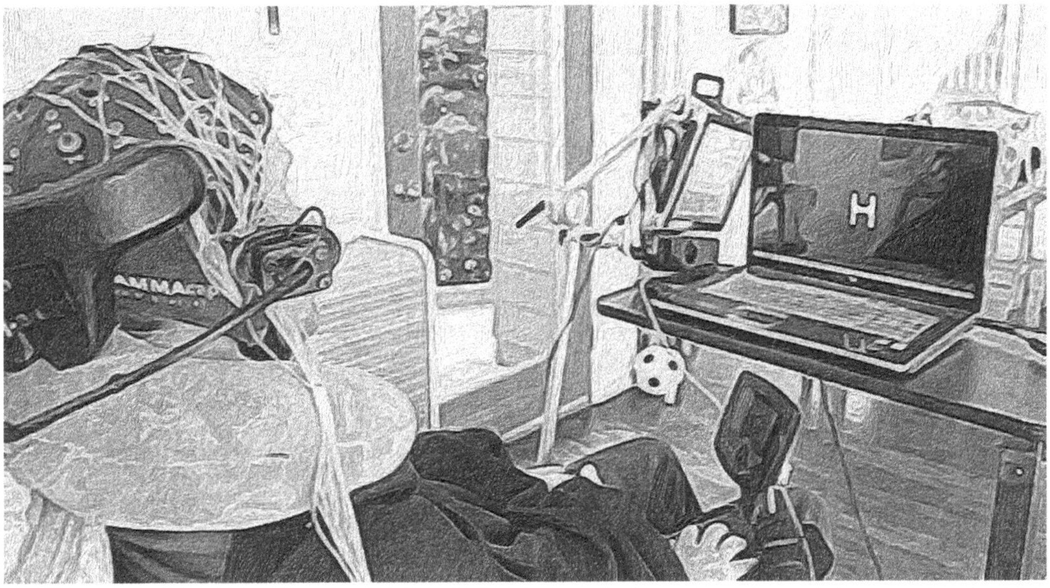

The technology for controlling systems has evolved dramatically. Controls have moved from direct physical contact to levers, nobs, dials, and now to computer control of remote devices. Control levers are replaced by keyboard and monitors. Computer control is now trending to touchscreens, vocal controls, ocular controls, and even direct brain–computer interfaces.

This chapter will first address the design of computer control systems and then the classic mechanical controls (levers, knobs, etc.). Information about direct physical manipulation of a system can be found in the next section that describes human strength and physical capabilities.

COMPUTER CONTROLS

Keyboard

Use. Arrangements of push buttons in the form of keyboards should be used when alphabetic, numeric, or special function information is to be entered into the system.

Layout of Keys. Keys should be grouped according to their function, based on convention. Groupings can include numeric keys, alphabetical keys, and function keys.

The specific arrangement of keys should match the standards used in keyboards with similar functions. For example:

1. Alphanumeric Keys—The QWERTY layout is the standard for alphabetical keys. Other keyboard formats exist (such as the Dvorak layout) and have ergonomic advantages, but are n ot as widely used as the QWERTY layout. The Dvorak layout is defined in the *American National Standards Institute. (1991) Alternate Keyboard Arrangement.*

2. Numeric Keys—Two standard layouts are used for numeric keys: the telephone layout and the calculator layout. (See Table 30. 1.)

TABLE 30.1 Standardized Numeric Keyboard

Layouts					
Telephone Keypad Layout			Calculator Keypad Layout		
1	2	3	7	8	9
4	5	6	4	5	6
7	8	9	1	2	3
	0			0	

Key Displacement. Recommendations for keyboard key dimensions, separation, resistance, and displacement are in Fig. 30.1.

Keystroke Feedback. Keystroke feedback can be auditory, tactile, or both. Tactile feedback is preferred in noisy environments because the background noise may prevent the user from hearing auditory feedback. If the feedback is auditory, it should occur at a consistent point in the displacement of all the keys. Changes in the key resistance can provide tactile feedback. Figure 30.2 shows the recommended force profile for user feedback.

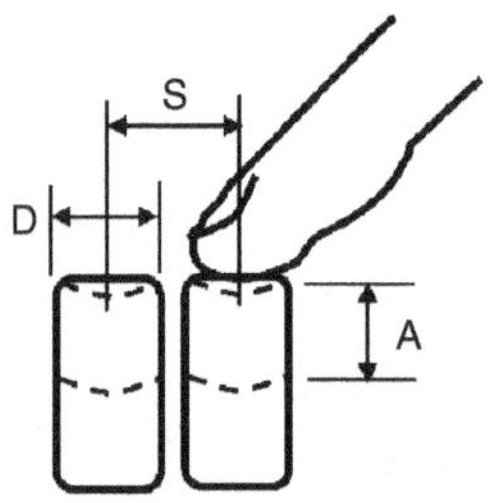

	Dimensions (D)	Resistance (R)		
		Permitted at snap point	Preferred at snap point	Initial resistance
Min	12 mm (0.47 in.)	0.25 N* (0.9 oz)	0.5 N (1.8 oz)	25% of R at snap point
Max	–	1.5 N (5.4 oz)	0.6 N (2.2 oz)	75% of R at snap point

Displacement (A)

	Permitted (continuous entry use)	Preferred	Acceptable for non-continuous entry
Min	1.5 mm (0.04 in.)	2.0 mm (0.08 in.)	0.7 mm (0.03 in.)
Max	6.0 mm (0.24 in.)	4.0 mm (0.16 in.)	1.6 mm (0.06 in.)

Separation (S)

Measured between centerline of any two adjacent keys (without gloves)

Min	18 mm (0.70 in.)
Max	20 mm (0.78 in.)

FIGURE 30.1 Requirements for keys.

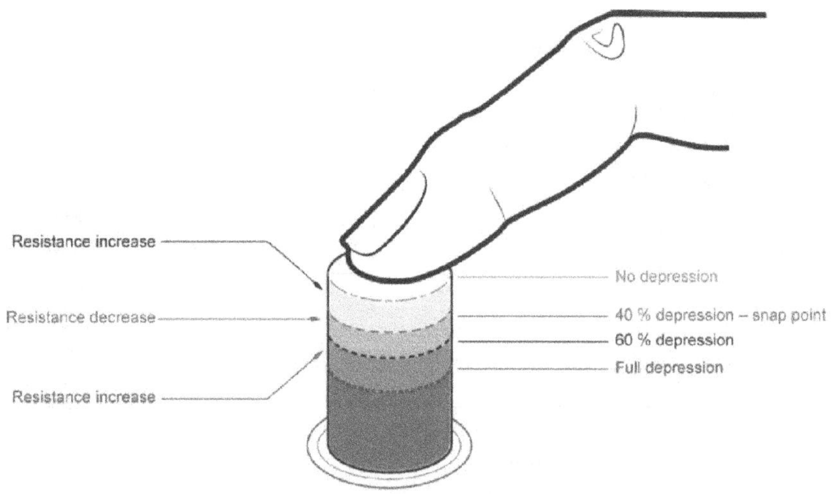

FIGURE 30.2 Tactile feedback recommendations.

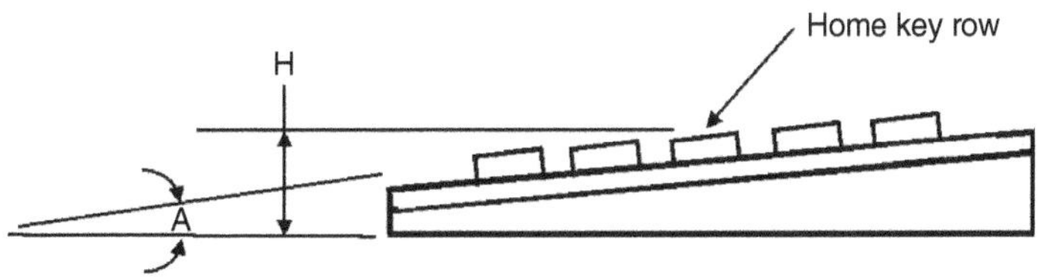

	Keyboard slope (A)		Keyboard height (H)	
	Permitted	Preferred	Permitted	Preferred
Min	0°	5°	–	–
Max	15°	12°	35 mm (1.4 in)	30 mm (1.2 in)

FIGURE 30.3 Requirements for keyboard slope and height. (From International Organization for Standardization. (1994). Anesthesia and respiratory care alarm signals (ISO 9703-2). Geneva, Switzerland: ISO.)

Keyboard Height and Slope. Figure 30.3 shows recommendations for keyboard slope and height.

Key Repeat Rate. If a key is held for longer than 500 ms, the character should repeat at a rate of approximately 10 characters per second. This character repeat rate should be adjustable.

Home Row Locator. The keyboard should have at least one tactile feature that helps initial finger positioning. For the QWERTY keyboard, this could be a dimple or raised area on the "F" and "J" keys. Similarly, this could be a tactually distinguishable feature at the center "5" on a numeric keyboard.

Pencil Joystick

The pencil joystick gets its name from the fact that it is generally held like a pencil or pen. It is useful for console situations in which an operator manipulates cursors or other electronic graphics on an electronic display.

The more "pencil-like" the device is, the easier it is to operate. Figure 30.4 shows that the ideal approach would be to provide the operator with a "wand" with which the basic controller arm can be moved through a cone about 2 in. (5.0 cm) in diameter.

A cup positioned approximately at desk surface height receives the wand, making it possible to "write" with the controller.

Perhaps the next best alternative is to place the control pivot point below the desk surface, as shown in Fig. 30.4.

Various manufacturers offer other alternatives, as shown in the figure. Although these do not provide optimum "writing" characteristics previously discussed, they are generally satisfactory. Avoid selecting devices that have handle diameters considerably different from those shown here.

Although the switch-handle configuration is acceptable, the smaller joystick with a separate, left-hand-operated switch on the console is preferred.

Joy-Ball Controller

Although slightly less desirable than the pencil joystick from the point of view of operator accuracy, the joy-ball controller is sometimes preferred in order to eliminate the interference problem occasionally associated with the protruding joystick.

A separate switch should be provided to disengage the joy ball when it is not being used.

Joy balls are not self-centering and therefore have the disadvantage that one can easily leave the controlled in a position in which the display hook or cursor is completely out of view. When joy balls are used, it is desirable to design the control system so that the displayed elements will always remain in view on the monitor.

Figure 30.5 illustrates available joy-ball hardware.

Mouse Controller

The mouse is a computer input device. When experimentally compared with other devices, the mouse was found to be easier to use, faster and more accurate.[8] The mouse is moved across a flat surface to separate X and Y coordinate values. The mouse senses location by operating as an upside down joy ball or using a grid and optical sensors. Figure 30.6 shows that buttons on the mouse should have the following characteristics:

Stylus and Light Pen

Both the stylus and light pen are hand-held devices used to enter points or pick off data from a computer display screen. The light pen is used directly on the display screen and a stylus is used on a grid on the screen or remote from the screen. Figure 30.7 shows recommended dimensions for the stylus and light pen. Provide a hook or receptacle near the display to mount the stylus or light pen when not in use.

Touchscreen Computer Controls

Touchscreen computers are controlled by using the fingers or a stylus to make gestures on specific graphic controls, and by making gestures to the screen as a whole. The following discusses the touchscreen graphic controls and then the general types of gestures used.

[8] S. K. Card et al. (1978). Evaluation of mouse, rate-controlled isometric joystick, step keys, and text keys for text selection on a CRT, *Ergonomics*, 21. 8.

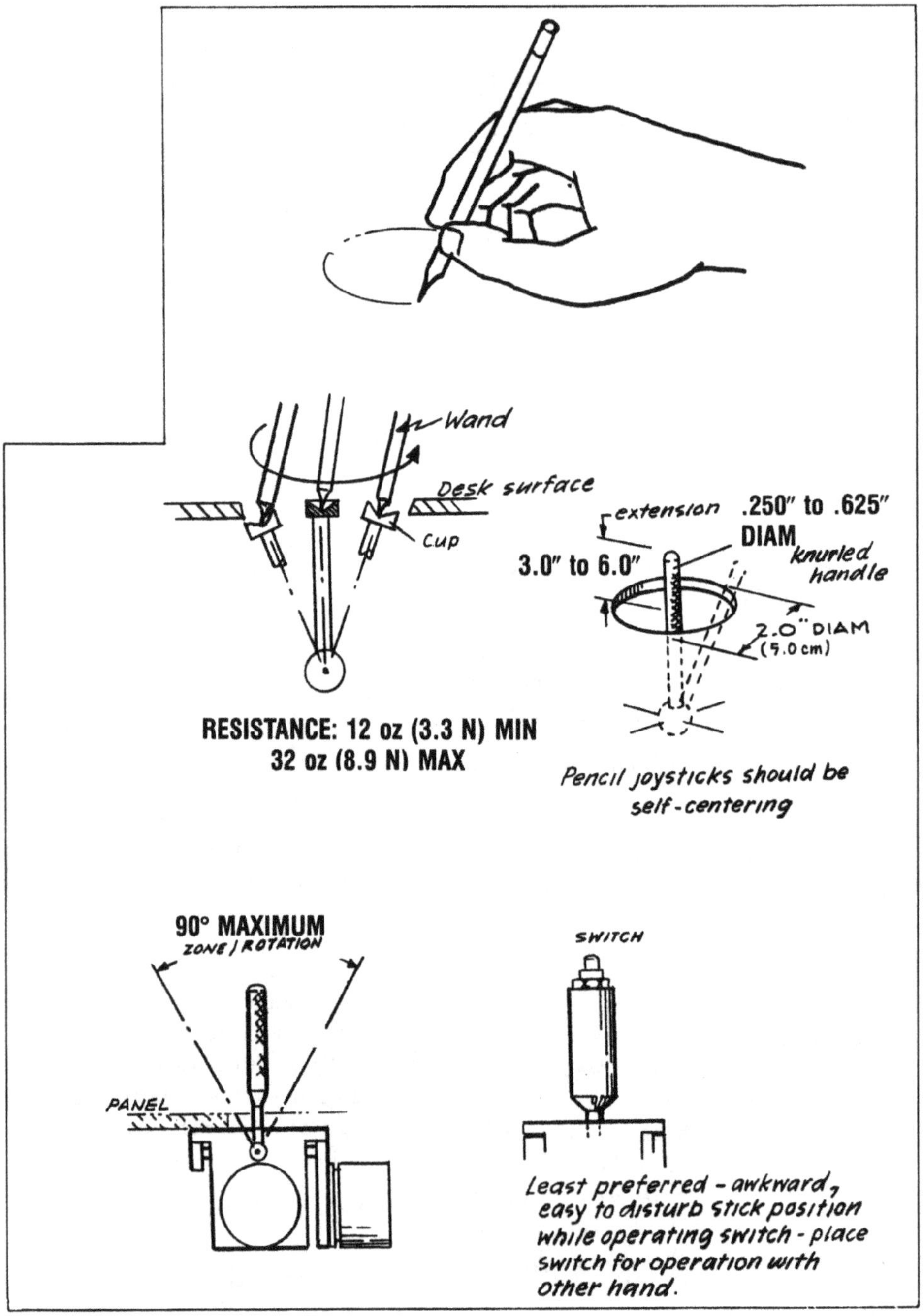

FIGURE 30.4 Pencil joystick design recommendations.

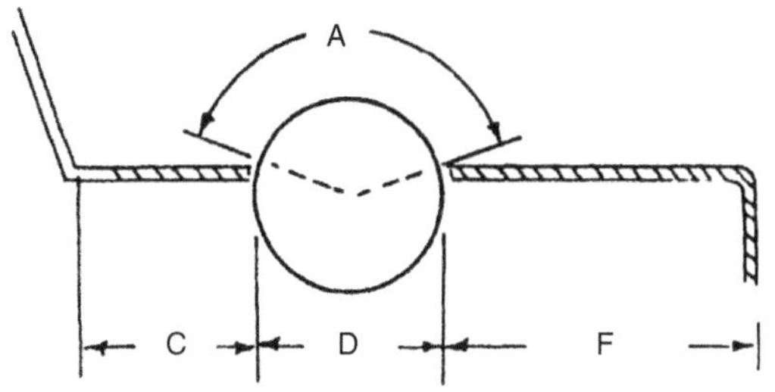

	Dimensions		Resistance		Clearance		
	D Diam	A Surface exposure	Precision required	Vibration or accel conditions	S Display CL to ball CL	C Around ball	F Ball to shelf front
Minimum	50 mm (2")	1745 mrad (100°)			0	50 mm (2")	120 mm (4–3/4")
Maximum	150 mm (6")	2445 mrad (140°)	1.0 N (3.6 oz.)	1.7 N (6 oz.)	320 mm (12–5/8")		250 mm (9–3/4")
Preferred	100 mm (4")	2095 mard (120°)	0.3 N (1.1 oz.)				

Figure 30.5 Joy-ball design recommendations (MIL-STD-1472D).

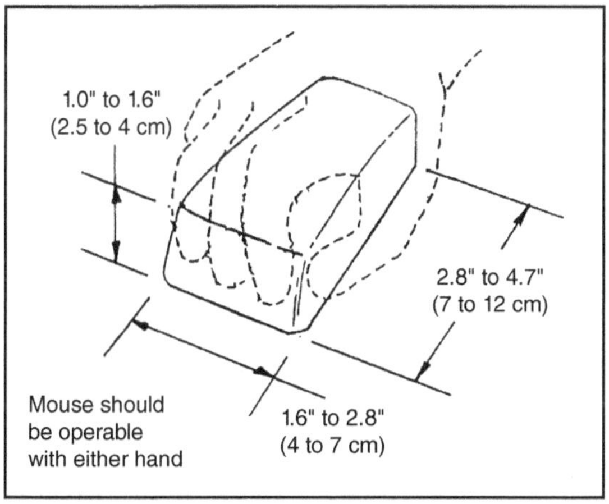

Figure 30.6 Mouse design recommendations (MIL-STD-1472G).

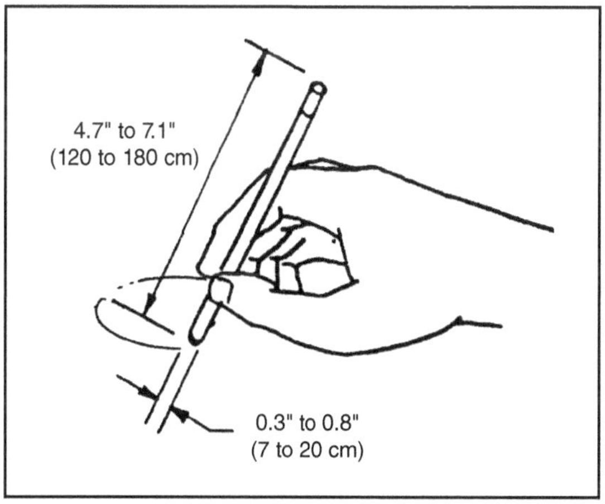

Figure 30.7 Stylus and light pen design recommendations (MIL-STD-1472G).

	Button Characteristics	
	Resistance	**Displacement**
Min	0.5 N (1.8 oz.)	5.0 mm (0.02 in.)
Max	1.5 N (5.4 oz.)	6.0 mm (0.24 in.)

Detail Disclosure or Info Button. A Detail Disclosure or Info Button reveals a separate view that contains additional information or functionality related to a specific item. (See Fig. 30.8.)

Figure 30.8 Detail disclosure or info button.

Contact Add Button. A Contact Add button lets the user add any existing contact to a text field or other text-based view. (See Fig. 30.9.) The Contact Add button gives users an easy way to access a contact without using the keyboard because the button functions as an alternative to typing contact information, it is not appropriate to use the button in a view that also accepts keyboard input.

FIGURE 30.9 Contact Add button.

Picker. A picker displays a set of values on a "wheel" from which the user picks one. (See Fig. 30.10.) Use a picker when users are familiar with the entire set of values. Because many of the values are hidden when the wheel is stationary, it is best when users can predict what the values are. If you need to provide a large set of choices that aren't well known to users, consider a table instead.

FIGURE 30.10 Picker.

Page Control. A page control indicates the number of open views and which one is currently visible. (See Fig. 30.11.) A page control displays an indicator dot for each open view in an app (from left to right, the dots represent the order in which the views were opened). The page control also uses an opaque dot to represent the currently visible view and translucent dots to represent all other open views.

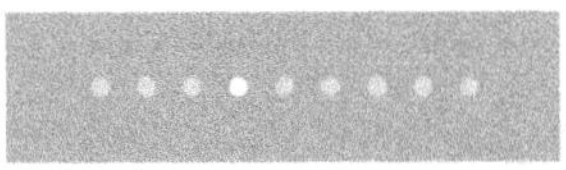

FIGURE 30.11 Page control.

Use a page control when it is more important to show users how many views are open than it is to help them choose a specific view. Avoid displaying too many dots. More than about 10 dots are hard for users to count at a glance and more than about 20 open views are time consuming to visit in sequence. If users can open more than about 20 page views in your app, consider displaying the views in a different arrangement that provides more information about the views and enables nonsequential navigation. Position a page control so that it is always visible without getting in the users' way.

Segmented Control. A segmented control is a linear set of segments, each of which functions as a button that can display different views. (See Fig. 30.12.) A segmented control consists of two or more segments that display text or images. Avoid mixing text and images in a single control. Use a segmented control to offer choices that are closely related but mutually exclusive.

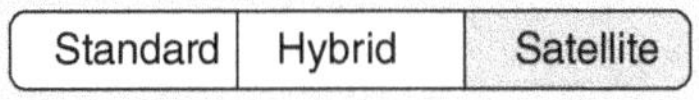

FIGURE 30.12 Segmented control.

Stepper. A stepper increases or decreases a value by a constant amount. (See Fig. 30.13.) A stepper is a two-segmented control in which one segment displays a plus symbol and the other segment displays a minus symbol by default. Use a stepper when users might need to make small adjustments to a value. Avoid using a stepper when users are likely to make large changes to a value. Make it obvious which value the stepper affects. A stepper doesn't display any values, so you need to make sure that users know which value they are changing when they use a stepper.

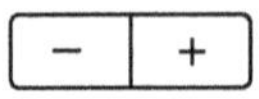

FIGURE 30.13 Stepper.

Switch. A switch presents two mutually exclusive choices or states (like "on" or "off"). (See Fig. 30.14.) A switch indicates the binary state of an item and is used in table views only. You can use a switch control to change the state of other user interface elements in the view. Depending on the choice users make, new list items might appear or disappear, or list items might become active or inactive.

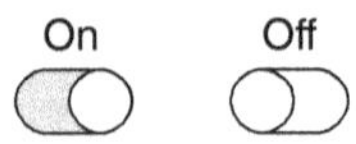

FIGURE 30.14 Switch.

System Button. Use a system button to initiate an action. (See Fig. 30.15.) When you supply a title for a system button, use a verb or verb phrase to describe the action the button performs. An action-specific title shows users that the button is interactive and tells them what will happen when they tap it. Commonly used button icons are as follows:

Figure 30.15 Button icons.

Sliders. A slider allows users to make adjustments to a value or process throughout a range of allowed values (shown as follows with custom images on the left and the right). (See Fig. 30.16.) A slider consists of a horizontal track and a thumb (a circular control that users can slide). Sliders can include optional images that convey the meaning of the right and left values and what the slider does. Use a slider to give users fine-grained control over values they can choose or over the operation of the current process.

Figure 30.16 Slider.

Touchscreen Graphic Control Design

The specifications of touchscreen graphic control design are shown in Fig. 30.17 and Table 30.2.

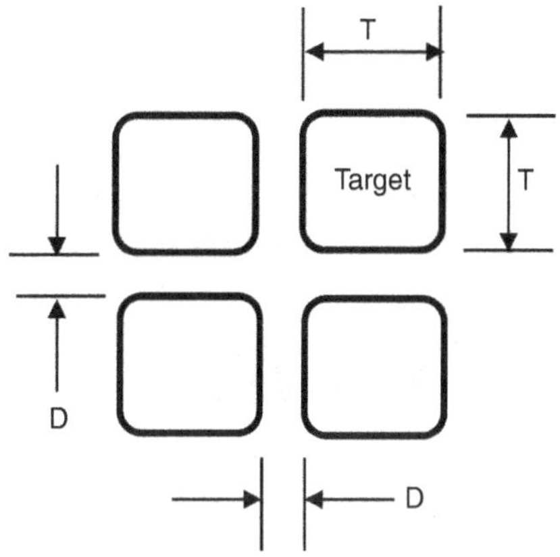

Figure 30.17 Touchscreen.

TABLE 30.2 Touchscreen Graphic Control Design

		T (Target Area)	S (Separation)
Alphanumeric/ numeric keyboards	Minimum	–	0
	Preferred	13 mm by 13 mm (0.5 in. by 0.5 in.)	–
	Maximum	–	6.0 mm (0.25 in.)
Touchscreen graphic controls	Minimum	15 mm by 15 mm (0.6 in. by 0.6 in.)	3.0 mm (0.12 in.)
	Maximum	38 in. by 38 (1.5 in. by 1.5 in.)	6.0 mm (0.25 in.)

Touchscreen Standard Gestures

Gestures allow direct manipulation of onscreen objects. People should expect gestures to work the same in all programs they use. On some devices, multiple fingers can be added to expand the menu of gesture functions (e.g., swiping) with multiple fingers instead of just one.

Tap to press or select a control or item. (See Fig. 30.18.)

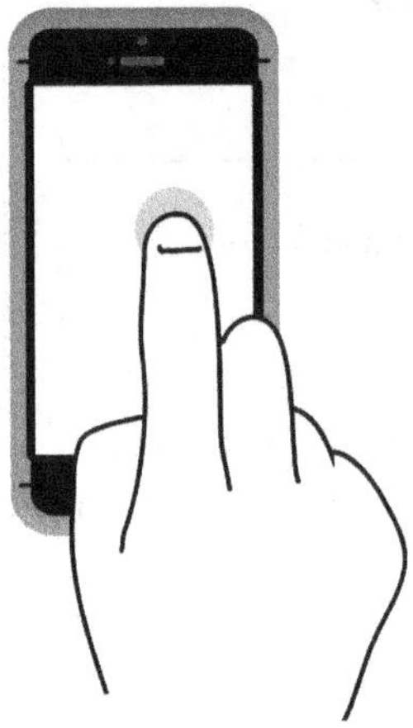

FIGURE 30.18 Tap.

Drag to scroll or pan up and down or side to side. Drag to move an element. (See Fig. 30.19.)

FIGURE 30.19 Drag.

Flick to scroll or pan quickly. (See Fig. 30.20.)

FIGURE 30.20 Flick.

Swipe with finger to return to a previous screen. (See Fig. 30.21.)

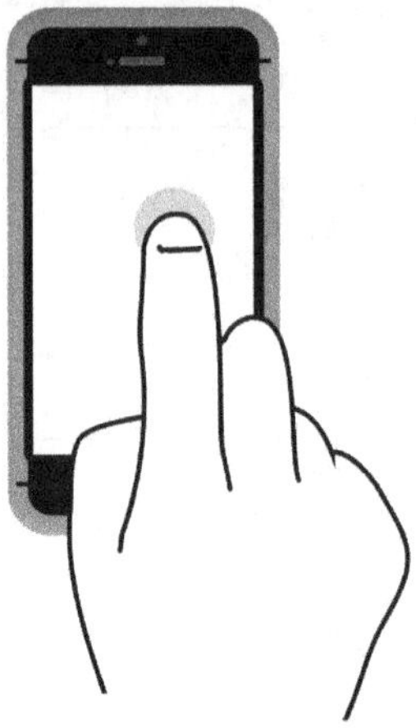

FIGURE 30.21 Swipe.

Double tap to zoom in and center a block of content or an image, or to zoom out (if already zoomed in). (See Fig. 30.22.)

FIGURE 30.22 Double tap.

Pinch open to zoom in; pinch close to zoom out. (See Fig. 30.23.)

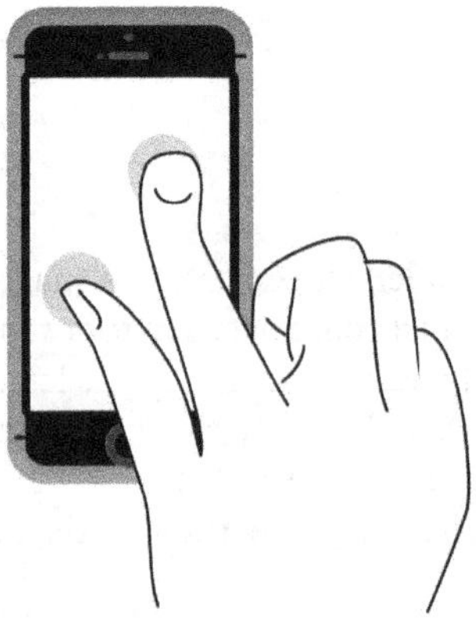

FIGURE 30.23 Pinch.

Touch and hold in editable or selectable text, to display a magnified view for cursor positioning. (See Fig. 30.24.)

Figure 30.24 Touch and hold.

Portable Computers

Computing devices are carried and worn as glasses, wrist bands, and patches. Figures 30.25–30.29 provide guidance on the design of portable computing devices. Figures 30.25–30.28 show the dimensions for the design of portable computers.

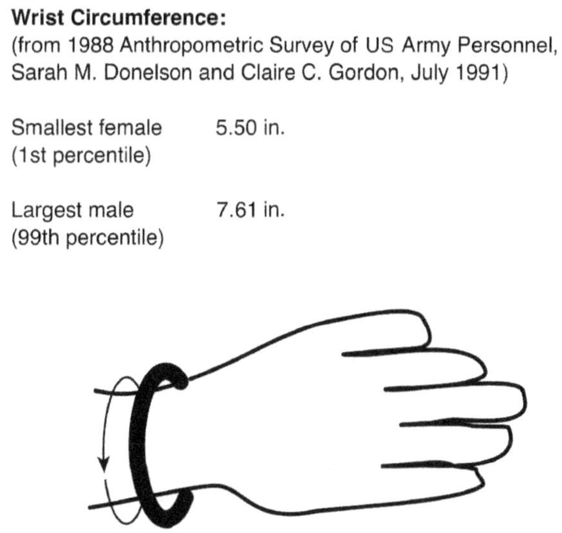

Figure 30.26 Wrist.

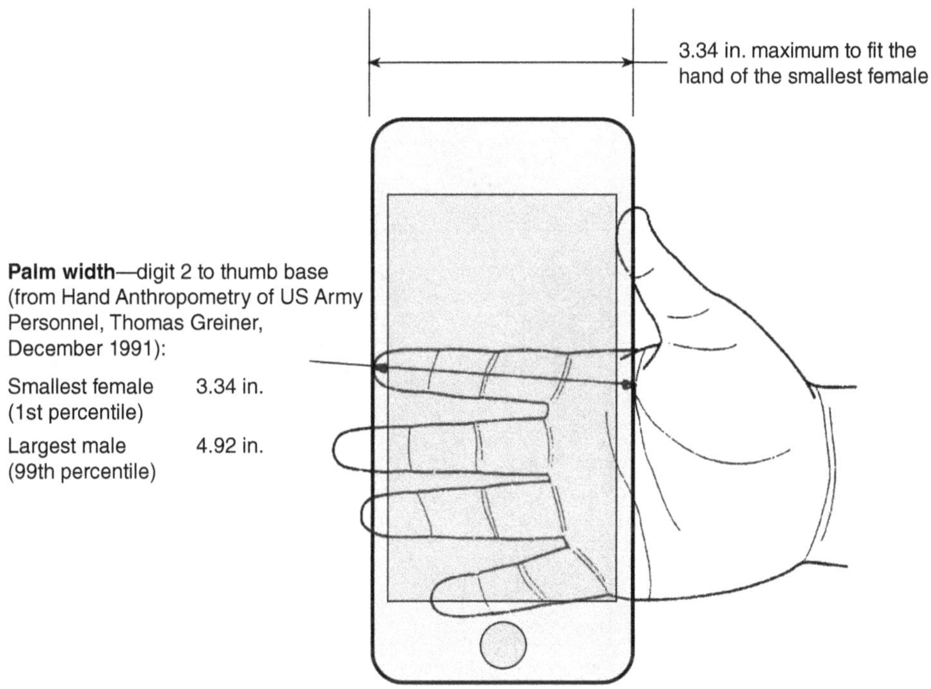

Figure 30.25 Hand-held width.

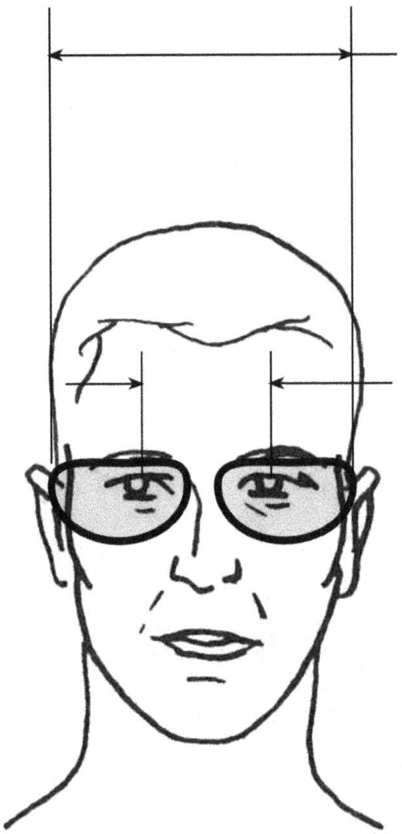

FIGURE 30.27 Goggles.

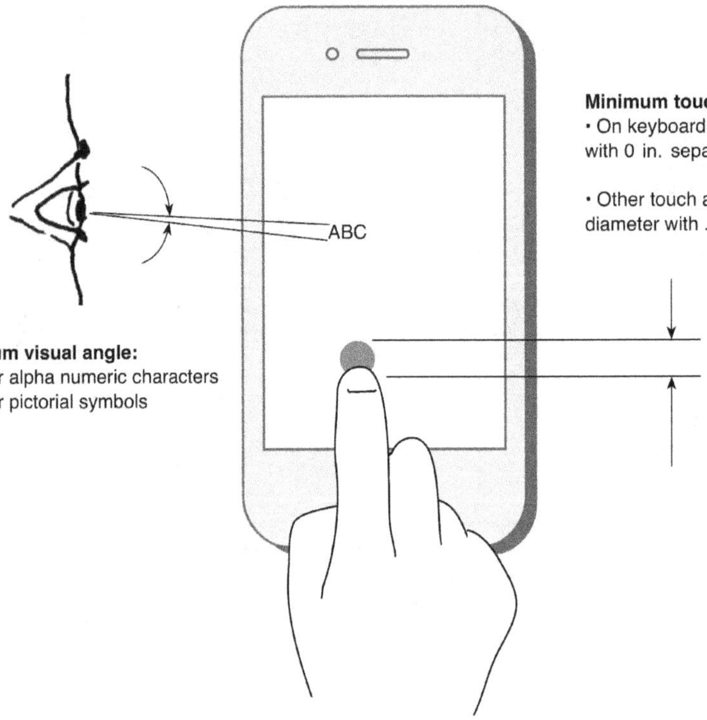

FIGURE 30.28 Hand-held design.

Figure 30.29 shows the best areas for placing touchscreen contact areas that can be operated with the thumb.

Figure 30.29 Thumb performance. Key tap performance using the thumb. "Performance" includes speed and accuracy. Performance scores are sequentially shown starting with the best (1) at the thumb resting position. (*Source: Thumb Motor Performance Varies by Movement, Orientation, and Device Size During Single-handed Mobile Phone Use, Matthieu Trudeau and Tawan Udtamadilok, Harvard School of Public Health.*)

MECHANICAL CONTROLS—SELECTION, DESIGN, AND USE

Type of Control

The control should be chosen as though it were an extension of the operator's limb; i.e., it should be operable in terms of the natural motions of the arm, wrist, finger, leg, ankle, or foot, and it should not require awkward and unnatural positioning, extension, or motion on the part of the operator.

Feedback

The control interface and basic controller system should provide feedback so that the operator knows at all times what his or her input is accomplishing.

Resistance

There should be sufficient resistance to operator inputs to dampen spurious inputs, but not so much that the operator has to put great force into the control, that his or her muscles are quickly fatigued, or that the operator has difficulty maintaining the nominal operating position.

Position of the Control

Controls should be placed where they do not require the operator to assume awkward body positions or make frequent long reaching movements. The position should reflect consideration of the excursion requirements of the control system so that there is no chance that the operator will be unable to reach a critical point in the control movement path.

Size and Shape

The size and shape of control interfaces (handles, knobs, buttons, etc.) should be compatible with the size of the operator's hands, fingers, or feet. Consideration should also be given to the additional implications of shoes, gloves, or mittens. Critical considerations are ability to maintain proper grip, sufficient space to preclude inadvertent contact, and adequate clearance to prevent inadvertent contact with adjacent controls or structures.

The shape of a control should also be compatible with the kind of grip or motion required to operate the control interface; e.g., knobs that are rotated should be round, oblong handles must not accidentally turn in the hand or fingers, and a control that is supposed to point at a certain scale mark or numeral should be in the shape of a pointer.

Interface Surface

The surface of a control handle should depend on the type of operation required; i.e., in some cases the operator may need a smooth surface in order to change positions on the handle or knob without having his or her fingers stick to the surface, while in other cases the operator may need serrations or knurling in order to apply a firm grip for maximum force.

One-Hand versus Two-Hand Operation

Two hands often provide more precision or force. However, two-handed controls should not be used if one hand is required to operate another control simultaneously with the other.

Control Selection Guidelines

Three control selection aids follow. Table 30.3 will help if you must decide between using a hand or foot control.

TABLE 30.3 Hand-Operated vs. Foot-Operated Controls

Hand-Operated Controls	Foot-Operated Controls
Very precise inputs possible	Limited precision capability
Limited force capability	Great force capability
Because hand controls are usually located closer to the eyes, controls are easier to find and observe during the operation	Foot controls usually are located far from the yes and therefore are difficult to find and generally cannot be observed easily during operation
Hand controls usually take up less space	Foot controls usually require more space
Wide range of motions available, e.g., push, pull, rotate, press, and squeeze	Only a few motions are available, e.g., leg push and angle-toe press
Two hands can perform entirely difference operations with relative ease	Two feet generally can perform only similar operations (although one can press while the other is being withdrawn)
Can usually be operated as easily from a standing as from a sitting position	Operator can use only one foot at a time while standing, and precision while standing is poor
Operator can perform rapid, repetitive motions	Repetitive inputs are difficult and very slow
Operator can utilize different fingers to make extremely rapid and complex, successive inputs; also, complex combinations of fingers can be used at the same time	Use of toes is impractical; thus, the operation is limited to successive, alternate, or coincident two-foot operations

Tables 30.4 and 30.5 are organized by a function. They will help you select a control based on the function you want to accomplish. The data in the tables are from the Department of Army publication, MIL-STD-1472G. Table 30.6 is organized by types of controls. The table lists use criteria and human engineering considerations for each control type.

TABLE 30.4 Control Selection Guidelines Based on Functional Needs

	Control									
Function	Selector Switch	Round Knob	Discrete Thumb Wheel	Cont. Thumb Wheel	Crank	Push Button	Toggle Switch	Rocker Switch	Lever	Joystick, Lever, Ball, Mouse
Select power state ON-OFF	3					1	2	2	1a	
3-state (OFF-STBY-ON)	1	2					2	3		
Select between OFF/Prime Mode/Secondary Modes(s)	1					2	3	3	1a	
Select one or more of N-related functions						1	2	2		
Select one of N mutually exclusive functions of any order						1				
Select one of 3 to 24 discrete alternatives—sequential order	1									
Select digit—discrete	2b		2b			1c				
Set value on—continuous scale		1		2	3				3	
Select value in—discrete steps	1		1			1				
Select operating condition	2					1	1	1	2	
Enter alphanumeric data						1c				
Initiate test subfunction (momentary hold)	3					1	1	2		
Initiate directional function	3			3		2d	1	1d	1	2
Generate stepping impulse (momentary hold)						1	1	2		
Slew counters or other numeric readout		1e			1f	1	1			
Reset mechanical counter, manual		1	3	1						
Interrupt sequence, "hold"						1	2	2		
Engage-disengage mechanical function									1	
Adjust light level, continuous		1	3	1				2	3	
Adjust sound level, continuous		1	3	1				2	3	
Coarse adjustment		1g		2h	2i				2j	2
Fine adjustment		1k		2l	2m				3n	2
Adjust to null position		1		2	3				3	2
Single-coordinate tracking		3			2				1	
Two-coordinate tracking					3					1

NOTES:

1: most preferred; 2: secondary preference; 3: least preferred

a: Lever for heavy duty power circuits
b: Only if sequential selection is acceptable
c: Keyboard
d: Multiple controls
e: Rate control
f: Manual only
g: Small diameter
h: Small motion
i: Few turns
j: Short throw
k: Large diameter
l: Large motion
m: Many turns
n: Long throw

From MIL-STD-1472G.

TABLE 30.5 Control Selection Guidelines Based on Functional Needs and Force Levels

Control Function	Control Type
Small actuation force required:	
2 discrete positions	Keylock Legend switch Push button Slide switch Toggle switch
3 discrete positions	Push button Rotary selector switch Toggle switch
4 to 24 discrete positions	Rotary selector switch
Continuous setting (linear and less than 360°)	Continuous rotary knob Joystick or lever
Continuous slewing and fine adjustment	Continuous rotary knob Crank Joystick or lever
Large actuation force required:	
2 discrete positions	Detent lever Foot push button Hand push button
3 to 24 discrete positions	Detent lever Rotary selector switch
Continuous setting (linear and less than 360°)	Crank Handwheel Joystick or lever Two-axis grip handle
Continuous setting (more than 360°)	Crank Handwheel Two-axis grip handle Valve

TABLE 30.6 Control Selection Guide

CONTROL CONFIGURATION	USE CRITERIA	HUMAN ENGINEERING CONSIDERATIONS
Hand-operated push buttons: Square push button 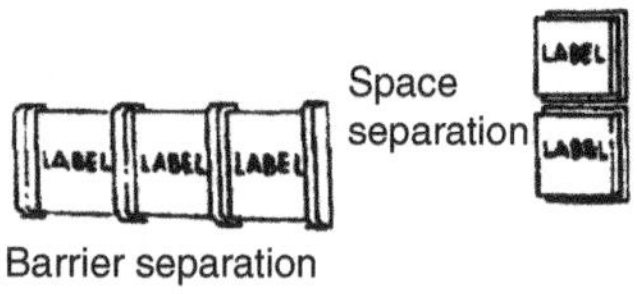	Use for noncritical operations which start an action, select a channel, or turn on a piece of equipment. Use for "event stack" combination lighted push buttons.	Locate within easy reach of the operator, considering the constraints of pressurized garments, acceleration restraint systems, etc.
Round (guarded) push button 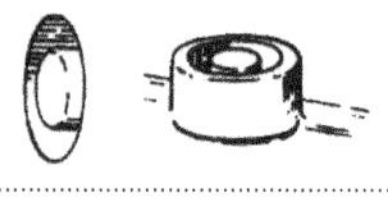	Use for critical operations (such as for a "panic" button) to stop a "runaway" equipment operation, fire and emergency escape sequence, etc.	Locate so that inadvertent operation is impossible (i.e., so that the switch cannot be actuated by bumping it accidentally).
Round (extended) push button 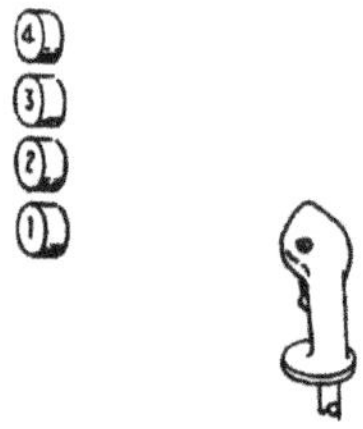	Use in conjunction with a primary hand controller for changing control mode, initiating communication, etc. Use for multibutton keyboard arrays and circuit-breaker panels.	Locate within easy reach of the operator's thumb, considering the constraints of a pressurized glove. Locate so that the switch will not be activated accidentally during normal manipulation of the parent controller.
Bar-type push button 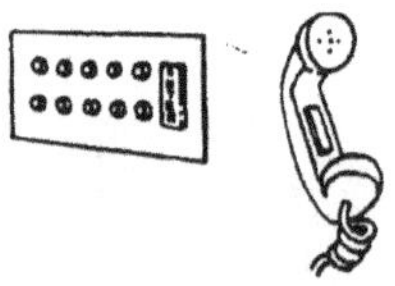	Use for either finger or thumb operation of a communication handset. Use as an "entry" tab for inserting preset numbers into a computer from a keyboard.	Consider location within keyset configuration for optimum operational sequence. Consider position on the handset for convenient use by fingers or thumb for both left- and right-hand operation.
Toggle switch 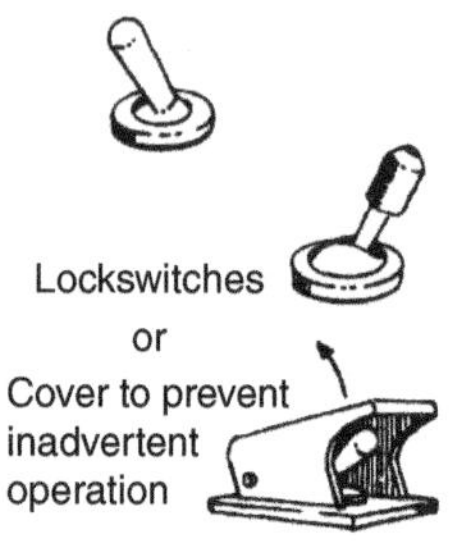	Use a standard two-position toggle switch for START/STOP, ON/OFF, etc. Use a "momentary" (spring-return) toggle switch for check-reading an instrument or circuit or use a spring-return toggle switch for a "slewing" operation. Use a three-position toggle switch for combining alternatives of a single function such as AUTOMATIC, MANUAL, or OFF. *Note:* Do not use four-position switches unless absolutely necessary because of space limitations.	Consider particularly the hazards associated with accidental contact with the toggle switch, and also the difficulties of operating in a pressurized glove.
Rocker Switch 	Use in lieu of a toggle switch when the protruding toggle handle would be a considerable hazard, i.e., when a crew member could be injured by bumping or striking the handle or handle cover accidentally. Use an illuminated rocker switch in lieu of a toggle switch where illumination conditions make it impossible to see the toggle position, assuming this requirement to be critical. Use in lieu of a toggle switch when panel space is inadequate.	Special precautions should be taken in placement of rocker switches to obviate accidental operation. Minimize intermixing of toggle switches and rocker switches in the same panel area; i.e., use one or the other.
Rotary selector control (mechanical detents) 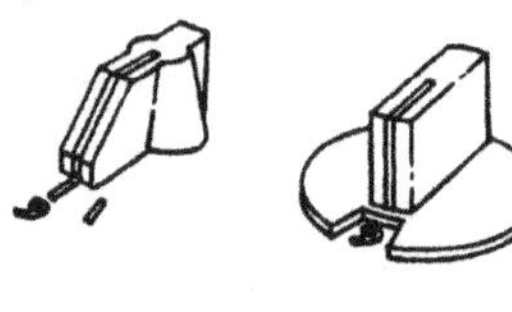	Use (where it is desirable to select two or more—but no more than 24—channels, circuits, etc., with a single control) to save operator time and panel space which would be required by separate control switches. *Note:* Do not use for continuous frequency selection, audio intensity control, panel illumination control, etc.	Make the "pointing" end of the switch knob obviously different from the nonpointing end. Provide mechanical detents so that the operator can "feel" and "hear" when the control is properly positioned. The knob should be shaped like a bar or pointer to differentiate this type of control from continuous-function controls, which have round knobs. Provide an index mark on both the top and the end of the pointer or provide a cutout in the knob skirt that exposes the numerical setting.

TABLE 30.6 Control Selection Guide (*Continued*)

Control Configuration	Use Criteria	Human Engineering Considerations
Rotary adjustment control (continuous motion) 	Use where smooth, continuous adjustment of audio intensity, light intensity, random radio frequency selection, etc., is required. *Note:* Do not use on detent-type switches.	Use a round knob, which allows ease of finger positioning. Use suitable knob surfaces which provide necessary finger security but which will not interfere with manipulation dexterity. Use a larger-diameter knob when "fine" adjustment is required.
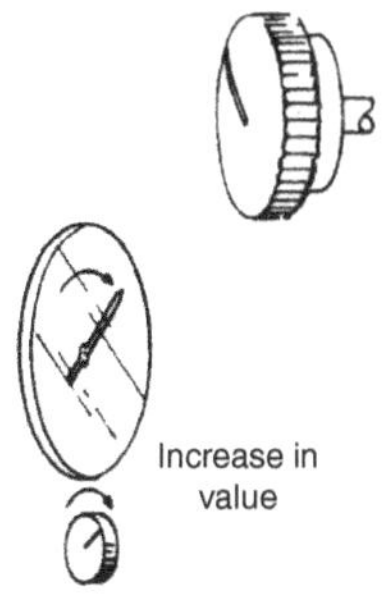	Use to provide continuous control of a related display pointer.	Maintain proper direction-of-motion relationships. Place the control where the operator's hand will not cover up the display when he or she is adjusting the display. Use the proper CD-ratio to ensure ease and accuracy of display adjustment.
Rotary valve control 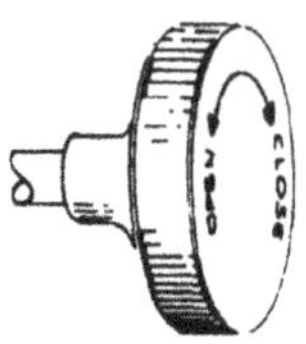	Use to control flow of materials, fuel, air, liquid oxygen, etc.	Use a round handle configuration suitable for gripping by the hand as a whole, but with sufficient finger clearance for pressurized glove operation. Whenever possible, locate valve handles far enough from other control knobs (which are similar in shape) to avoid direction-of-motion confusion (i.e., accepted usage for valves is for a clockwise turn to provide a decrease in flow, whereas the same motion in other types of controls provides an increase in functional value). In addition to placement separation, emphasize clear, unambiguous marking and other appropriate code techniques to minimize the chance of direction-of-motion of operator error.
Thumb-wheel controls (continuous smooth motion) 	Use for limited-range, continuous adjustment functions (panel light dimmer, audio volume, etc.). *Note:* Use only as a second choice to a normally mounted finger knob in the event some advantages may accrue in terms of saving panel space.	Coarse serrations should be provided on the edge of the knob. Sufficient exposure of the knob rim must be provided for convenient manipulation with pressurized gloves, but guarded against accidental activation.
Thumb-wheel controls (discrete stepping motion) 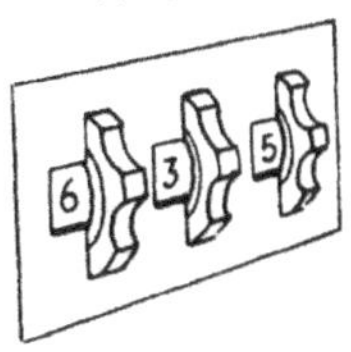	Use in conjunction with detented counter-type selector control and display combinations.	The edge of the thumb wheel should provide protrusions, or "teats," which coincide with the position of the numbers on the appropriate counter drum. These devices should be placed and/or guarded so that they cannot be accidentally displaced, or bumped, causing injury to personnel or damage to garments.
Push-pull controls: Stirrup handle 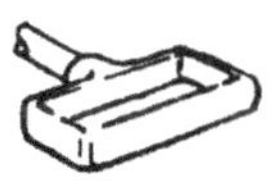	Use for an ON/OFF control which (because of mechanical reasons) requires more force than can be applied by a finger-operated control.	Locate controls so that they cannot be accidentally struck by personnel and cause injury or garment damage or interfere with normal or emergency crew movement. Make the stirrup opening large enough to accommodate four fingers (including gloves when worn).
T-shaped handle 	Use in lieu of a stirrup handle when space is not available for whole-hand operation.	Make handle extensions on each side of the central shaft long enough to accommodate the index and middle fingers on one side and the third and fourth fingers on the other.
Ball handle 	Use in place of the above when the orientation of a stirrup or T-shaped handle would cause serious inconvenience to the operator and possibly delay his or her total reaction time in the event of an emergency.	Select a ball diameter that will comfortably accept a "fist grip" if the force requirement is high and the thumb and at least two fingers if the force requirement is low.

(*Continued*)

TABLE 30.6 Control Selection Guide (*Continued*)

Control Configuration	Use Criteria	Human Engineering Considerations
Levers (console type)   	Use for main power setting to provide forward vehicle thrust or machine acceleration. Use for vehicle attitude control. Use for discrete (bang-bang type) power application (e.g., translation).	The handle length should be long enough to accommodate a four-finger grip. All lever-type controls must be designed and located so that the full control motion (excursion) is within the limits of reach of the restrained operator wearing a pressurized garment.
Levers (extended arm operation) 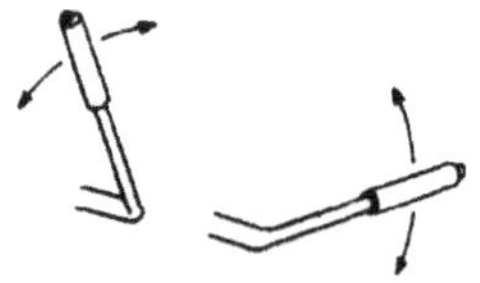	Use where medium to heavy manual force application is required.	
Hand wheel  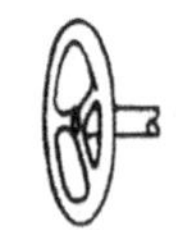  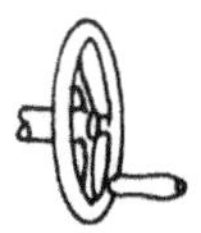	Use for opening and closing pressure-tight hatches where extremely heavy force application may be required, for high-torque valves, etc.* Use for low-torque, multirevolution cranking operations where many turns are made in rapid fashion.*	The diameter of the wheel and the diameter of the wheel rim must be compatible with operator dimensions, considering the constraints of pressurized garments. The operator must be restrained in a weightless environment. Select an optimum diameter for the expected force, rate, and operator position variables involved.
Steering column levers 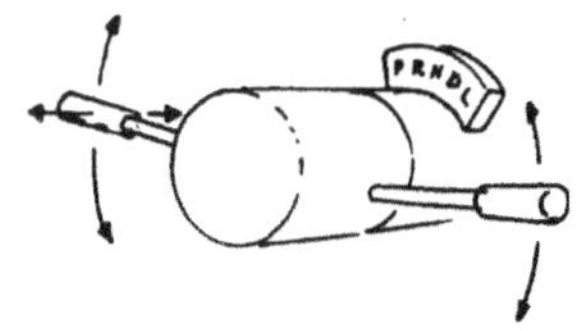 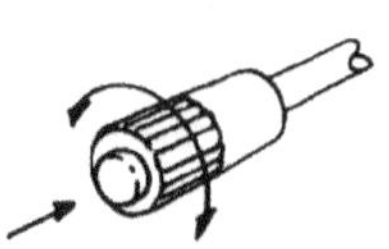	Use for automobiles, trucks, buses, etc. Use when it is desirable to bring critical control functions within "fingertip" reach of the driver (i.e., for easy reach from the normal hands-on-steering-wheel position). Typical functions: Automatic gear selection Turn signal operation	Consider typical driver control placement and motion expectancy, i.e., certain functions for right-hand operation and others for left-hand operation. Avoid too many different functions or motions on a single lever—especially if inadvertent "cross-coupling" might occur. Use only one lever per hand. Use rotation action sparingly. If used, provide knurling to indicate to the driver that this motion is possible.

TABLE 30.6 Control Selection Guide (*Continued*)

Control Configuration	Use Criteria	Human Engineering Considerations
Slide-type panel control† 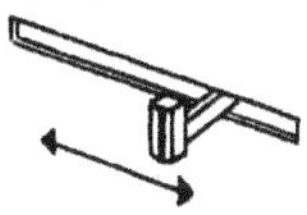	Typical functions: Environmental control	
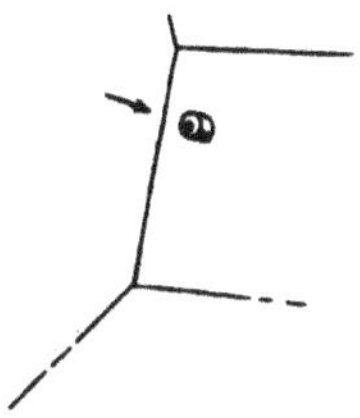 	Use foot-operated controls to free the driver's hands and allow simultaneous functional operations. Typical functions: Foot-operated dimmer switch Accelerator pedal Service brake pedal Clutch pedal	Optimize pedal and switch locations to simplify the driver's problem of "finding" a particular control without looking at it and to help ensure that the driver will not inadvertently operate a control that is adjacent to the one he or she wants to operate. Place controls within the driver's leg reach capabilities, considering the required control excursion. Mount the control so that its major motion axis is compatible with the natural motion axis of the operator's limbs.
Aircraft rudder pedals† 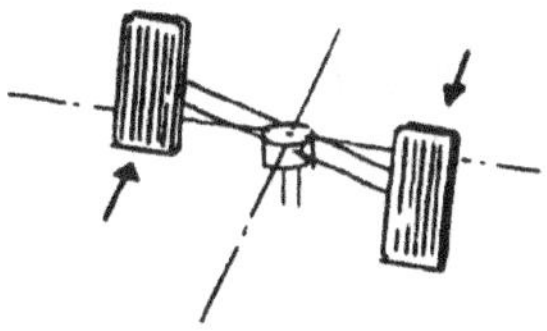	Use for directional control of aircraft.	Provide an optimum force regime, i.e., enough resistance to minimize inadvertent input, but not so much as to cause strain or fatigue.
Hip-operated switch bar† 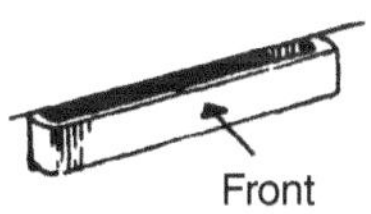	Use for machine operation where both hands are normally busy at the same time (commercial laundry press, industrial stamping press, etc.) either to start or to stop an operation.	Locate the control appropriately for the user's hip height. Minimize resistance and/or make the contact surface large in order to avoid bruising the user's hip.
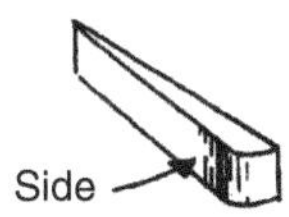	Use for transportation system turnstile so that passengers can pass through while holding bundles.	
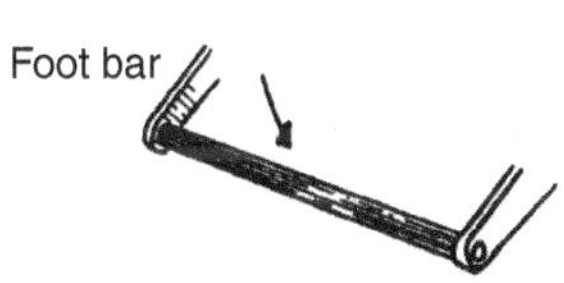	Use as above for machine operation, but only when it is expected that the operator can afford to stand safely on one foot while the other foot operates the foot bar. (As an alternative, a switch mat can also be used to open doors.)	Determine convenient location on the basis of the nominal working position of the operator. Optimize switch position in relation to door opening speed to accommodate typical user walking speed.

Note: This table was initially created for application to manned space vehicle system design. It has been expanded here to include additional systems and product applications.

* These suggestions originally were made with reference to space systems, in which operators normally would be "weightless."

† These are additions to the original space systems guidelines to provide suggestions for typical terrestrial systems, e.g., highway vehicles, aircraft, and industrial machines.

CONTROL MOTION EXPECTANCY

Motion Expectancy for Common Controls

Some control motions seem more natural to people as a result of either innate characteristics (musculoskeletal configuration, handedness, etc.) or habit patterns that have been learned because most products are designed in a certain way. (See Fig. 30.30.)

1. Perhaps the most reliable motion stereotype is that associated with a steering wheel; practically everyone (both experienced and novice drivers) expect to turn the wheel clockwise to cause the vehicle to turn right.

2. Lever motion stereotypes vary depending on the location and position of the lever. When the lever is positioned as shown in the figure, a forward position usually implies to the operator that whatever he or she is controlling will increase in speed. Pulling the lever back is expected to reduce speed and stop the machine or vehicle.

3. When the lever is placed alongside the operator, he or she normally expects to pull up on the lever to "brake" the machine or vehicle. An exception to the "forward-to-increase" and "back-to-decrease" expectancy is the gear-shifting lever.

4. Other lever-type controls generally relate to the movement of some subcomponent (as opposed to control of vehicle motions, noted above). The direction of motion of the lever is expected to reflect the direction in which the operator wants the subcomponent to move (as illustrated in the figure).

5. Special vehicle controls, such as those in an airplane, have been standardized because of the way in which the controlling elements (e.g., the rudder, elevators, and ailerons) are placed on the vehicle. A typical lever control (joystick) is expected to operate the vehicle or change vehicular direction when it is moved as shown in the figure; i.e., the pilot moves the stick forward to cause the nose of the aircraft to pitch down and pulls back on the stick to cause the nose to pitch up. Right-left roll is accomplished by very "natural" right-left stick movements.

6. Another typical aircraft control raises and lowers the landing gear. The pilot expects the control motion to coincide with the desired position of the landing gear. In order to ensure this expected relationship, the control should be positioned as shown in the figure.

7. Aircraft throttle levers should be positioned as shown in Fig. 30.30 in order to provide the expected changes in aircraft speed, i.e., forward for faster and aft for slower. However, when engines are used for breaking, a secondary "aft" motion of the controller is required. This motion must be thought of as braking, although there is an actual increase in engine power as the control is pulled farther aft. To preclude inadvertent use of the reverse thrust, the control is usually designed so that the pilot has too "pull up" before pulling the control aft.

8. Certain high-performance aircraft have adjustable wing positions; i.e., the wings are normally placed in the forward position for landing (slower speeds) and in an aft position for cruising (high speeds). This creates a direction-of-motion conflict in control operation. Although the novice operator might expect the direction of motion of both the control and the wings to be alike, experienced pilots generally expect most of the lever-type controls in the cockpit to move forward when they want to increase the speed of the aircraft. Thus, in this special case, the controller should move forward to reflect increasing aircraft speed, even though the actual control action sweeps the wings aft. For all aircraft control conventions, refer to MIL-STD-203 and MIL-STD-250.

9. Tiller bars are in actuality "levers." Typically associated with directional control of marinecraft, they cause the craft's rudder to move right or left and thus change the direction of boat movement. As illustrated in Fig. 30.30, the tiller control could be positioned in at least two different ways. The first configuration provides the more natural control motions for novice boat handlers because they use the tiller as a "pointing device." On the other hand, experienced boat handlers are more used to the second tiller configuration, and use of the first configuration could very well confuse the experienced sailor.

Whenever practical, a steering-wheel type of controller is recommended because it provides a natural controller vehicle movement relationship for both the novice and the experienced boat handler.

10. The well-established rudder bar stereotype shown in Fig. 30.30 applies to foot-operated directional control of aircraft and other, similar vehicular control configurations (e.g., sleds). Although there is an obvious discrepancy between controller motions and controlling element movement (e.g., the rudder), the natural expectancy of moving the right foot to effect a turn to the right and the left foot to effect a turn to the left is more significant (primarily because the operator seldom has knowledge of the actual control system characteristics). If pedals replace the bar, toe pressure should control vehicle direction, and heel pressure should control braking.

11. A common type of panel-mounted control is the pull-type knob illustrated in the figure. Although the typical control movement direction seems to be opposite to that described earlier for other controls (i.e., one might expect to push the control forward for ON, GO, or OPERATE conditions), longstanding convention has created an opposite expectancy; i.e., the control is pulled

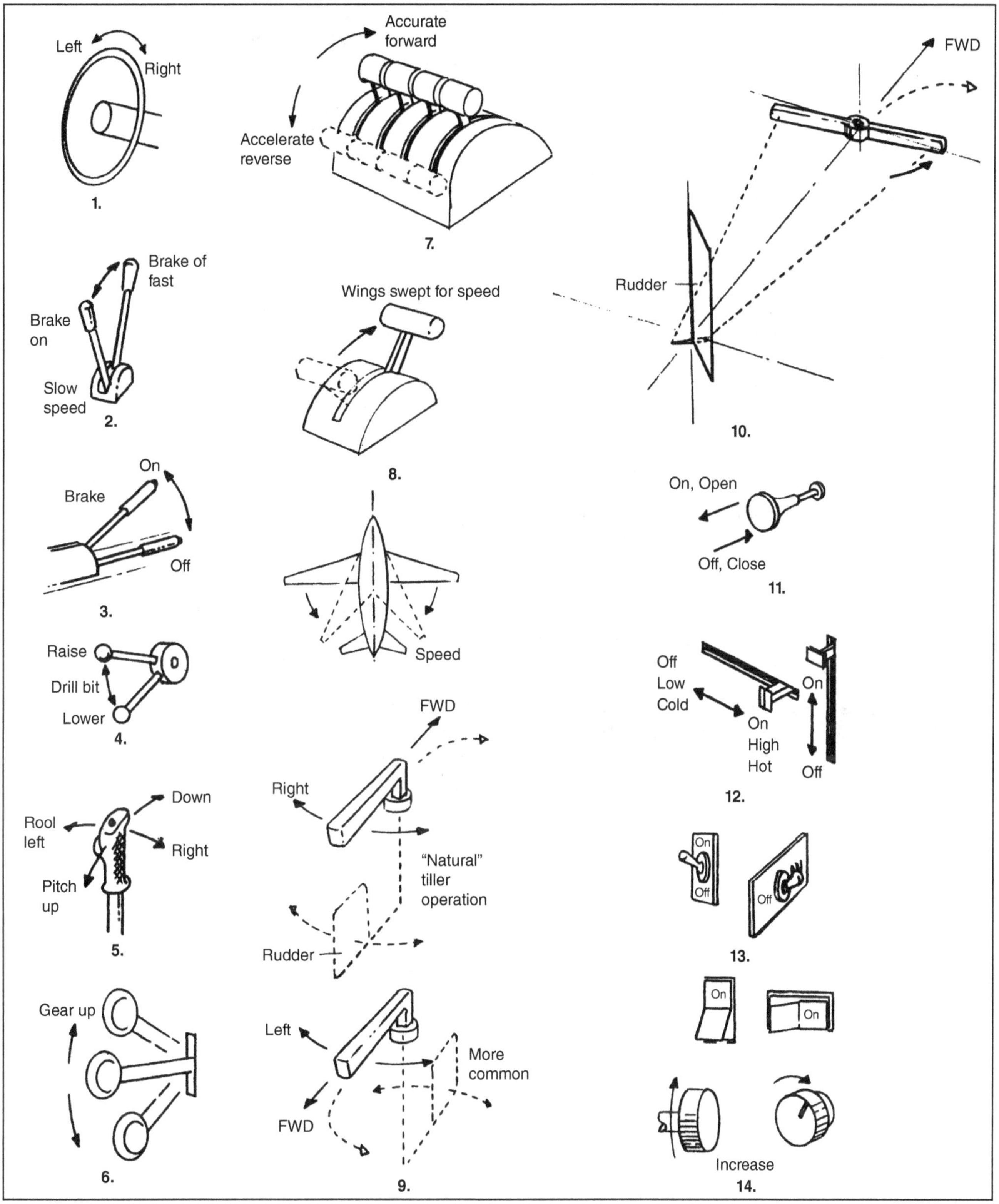

Figure 30.30 Expected control motions.

out for an operational mode and is pushed in to cause the operation to cease.

12. Recently, slide-type controls have been introduced for a variety of control operations (including control of vehicular functions, such as the environmental system, and control of radios, e.g., station selection and volume control). When slide controllers are used (not necessarily recommended as the preferred type of controller), the direction of control movement should be as shown in the figure.

13. ON/OFF function switches, regardless of their particular configuration (toggles, rockers, etc.), should operate as shown in the figure; i.e., movement of the control upward or to the right or depression of the top or right portion of the rocker switch should cause an ON condition.

14. Rotary-motion switches (including both the detented, discrete-positioning and the continuously variable potentiometer types) should move clockwise for an increase in "value." However, this is true only when the control knob is operated only by the right hand and/or is located on a panel facing the operator.

Control Motions

Recommended Control Motions. Table 30.7 shows general control motion recommendations.

TABLE 30.7 Recommended Control Motions

Function	Control Action
On	Up, right, forward, press
Off	Down, left, rearward, pull
Right	Clockwise, right
Left	Counterclockwise, left
Up	Up, rearward
Down	Down, forward
Retract	Rearward, pull, counterclockwise, up
Extend	Forward, push, clockwise, down
Increase	Right, up, forward, clockwise
Decrease	Left, down rearward, counterclockwise

From Human Engineering Guide for Equipment Designers.

Control position-direction-of-motion relations may vary depending on the location of the control with respect to the operator.

Although, as a general rule, one can translate control motions from one operating plane to another merely by assuming that the operator "faces" the control panel, this does not carry through to all conditions (as shown in Fig. 30.31). Typically, controls alongside an operator are not "faced" until the control position moves aft of the operator's chest. Once aft of this point, the operator typically turns to face the control.

Rotary controls for discrete positioning should use the well-established principle of "a moving pointer versus a fixed scale," as shown in the first illustration of Fig. 30.32. The second illustration shows that when values are imprinted on the knob skirt, they progress in the opposite direction to that in which one normally expects to read them (i.e., the principle of "a fixed index versus a moving scale"). The latter configuration leads to frequent positioning errors.

Some control relationships have been confused by the fact that various hardware manufacturers created devices without considering the operator expectancies. Although some people have sufficient background to "see" the mechanical relationships between the controller and the device being controlled, others rely entirely on instinct. Thus, we must depend on extrapolation of laboratory research dealing with direction-of-motion expectations using controls other than those shown in Fig. 30.33. The following recommendations are therefore made in terms of "expert opinion" derived from this research.

Assume that the typical user will have at least a reasonable amount of mechanical background for judging control operation relationships; i.e., he or she will expect to move the rotary latching pawl as shown in the first sketch, and to move the doorknob and key as shown in the second sketch. Note that the principle of "moving the securing element away from its secured position" is carried out for each type of controller and that the direction of movement varies depending on which hand is used to operate the control.

Valves. Valve operation is a special case of predetermined convention creating expectancies opposite to those established by human behavioral research. (See Fig. 30.34.) The single valve direction of motion stereotype convention is mechanically related; i.e., the valve is turned to "unscrew" a valve seat and thus allow liquid, gas, or other substances to "flow." (See Fig. 30.35.) In other words, to increase the functional factor, the valve handle is turned counterclockwise—as opposed to the behaviorally oriented principle of a clockwise motion to effect an increase in functional value or factor.

This stereotype is so well established now that any attempt to reverse the standard should be avoided. This applies to single valves only, as will be explained below.

When dual valve handles are involved, i.e., when one is controlled with the right hand and the other is controlled with the left hand, the above convention ceases to remain constant. The average user becomes somewhat confused and tends to revert to another direction-of-motion referencing philosophy. Although this stereotype is not as well established as others, it involves the following interpretive factors:

1. Motion toward the user to obtain flow increase

2. Mirror-image operation, i.e., both hands performing similar motions

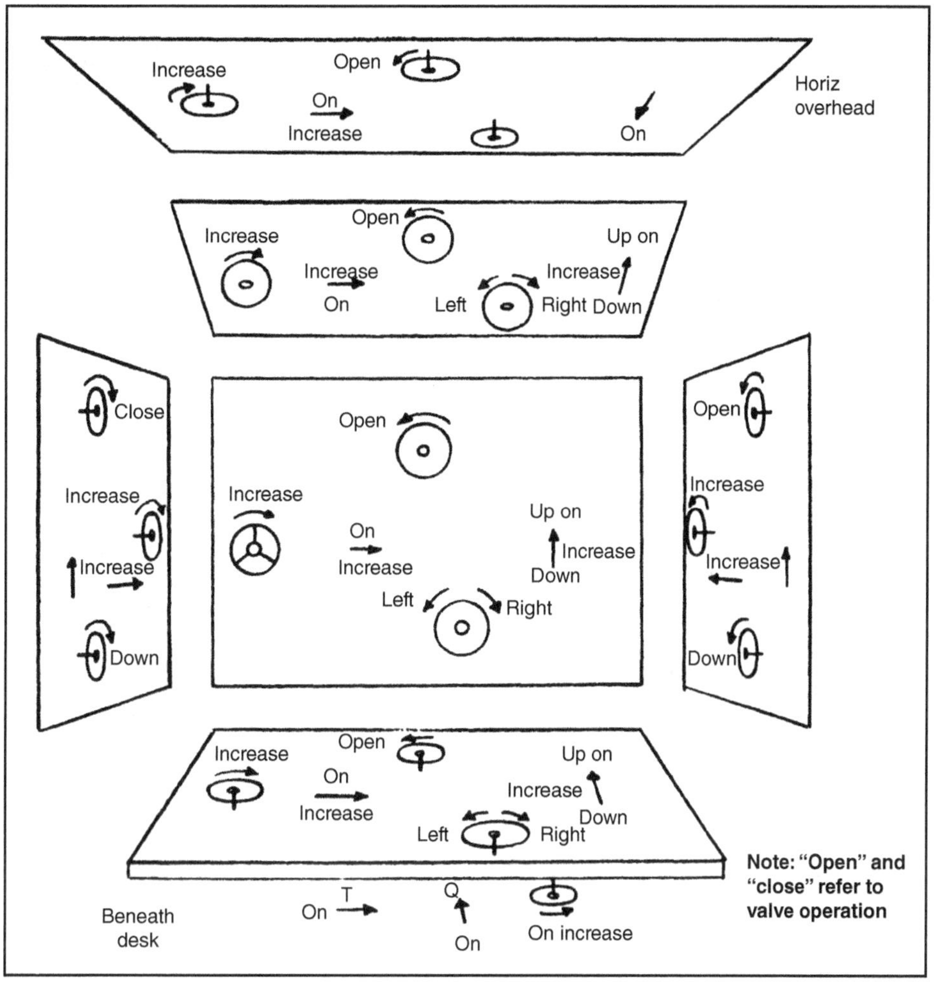

Figure 30.31 Recommended control motions for various panel orientations.

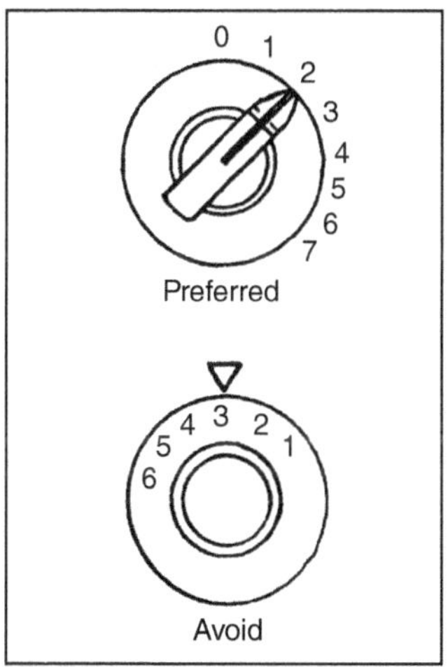

Figure 30.32 Avoid moving scale—causes "reversed" control motion.

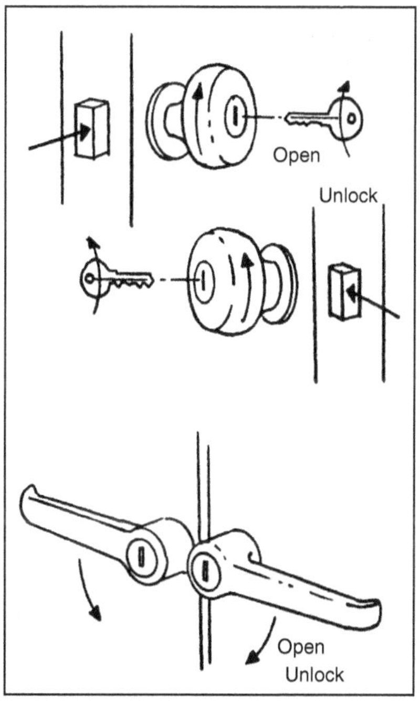

Figure 30.33 Expected control motions.

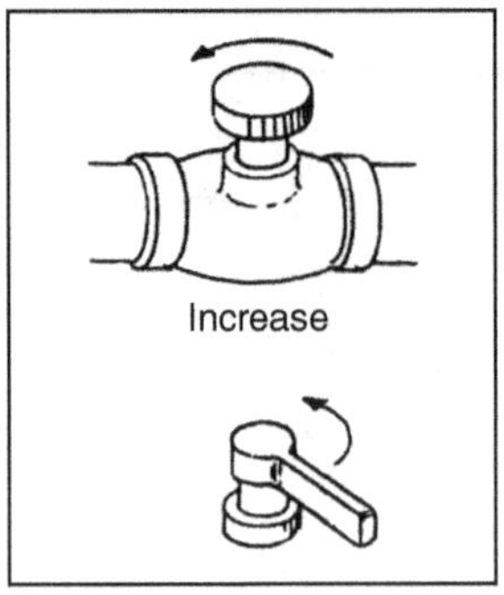

FIGURE 30.34 Valve operation expectancy.

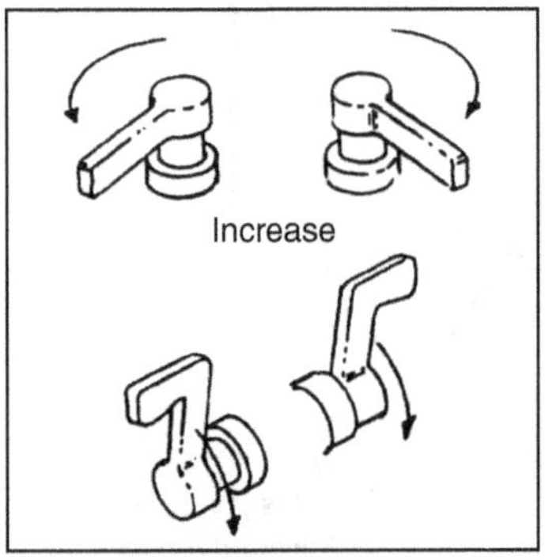

FIGURE 30.35 Conflicting conventions with two valve controls.

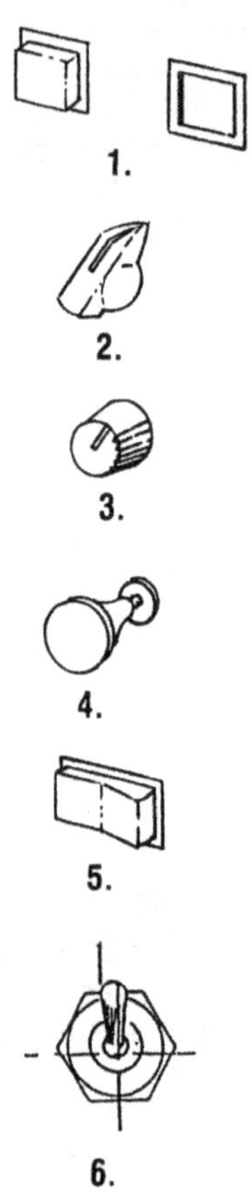

FIGURE 30.36 Controller appearance.

CONTROL OPERATION EXPECTATIONS VERSUS CONTROLLER APPEARANCE

General

One should not use a controller configuration that is confusing to the potential operator. The following typical expectations should be observed (see Fig. 30.36):

1. A push button should extend out from the panel and thus imply that it is to be pressed or pushed in. Some push button and indicator displays are made to look alike; i.e., they both are flush with the panel, making it difficult to tell which one is a control and which is merely an indicator.

2. A pointer-shaped knob implies that the device is a positioning control that can be positioned in discrete steps.

3. A circular-shaped knob implies that the knob turns continuously. It also makes it easier for the operator to reposition his or her fingers for multiturn manipulation.

4. The wafer-like knob indicates that it is to be pulled out. If it also is to be turned, it should have serrations around the edges, implying rotational possibilities. (Do not use serrations if the knob is not intended to be turned.)

5. The rocker-shaped control implies that either end of the switch cap can be pushed. However, the angular cues may not be apparent if the control is positioned at a poor visual angle. Make sure that the operational cues are apparent in the use position before selecting this type of control.

6. The toggle-switch handle provides a readily apparent cue that the handle can be alternately positioned from one side to the other—especially when several of the switches are arranged in a horizontal or vertical pattern.

Confusing Designs

Sometimes deliberate confusion is created by the appearance of a control device.

1. Lack of an obvious "pointing" feature makes it possible for the operator to turn a selector knob to the wrong position. Symmetrical bar handles on rotary knobs should be avoided.

2. Confusion with other panel-mounted features often leads the operator to believe that the control is something other than a controller; i.e., it may look like a lighting fixture or a panel fastener.

3. An otherwise effective pointer-selector knob may be so distorted that when the operator manipulates the control in the dark or without actually looking at it, he or she tends to point the handle at the desired setting because the handle feels like a pointer.

4. A sleeve handle that is designed to slide inward is seldom recognized as performing that type of movement.

5. An extremely thin, wafer-like rotating knob is often mistaken for a push button.

6. Push buttons deeply inset in the end of a control handle are usually not observed when the operator views the handle from the side. Thus, the operator fails to find the button switch.

CONTROL DYNAMICS

Feedback

All control systems should be designed to provide maximum information to the operator concerning the effects of the operator's input. This may consist of intrinsic feedback, extrinsic feedback, or both. Intrinsic feedback is that which is generated by the controller device, e.g., by "feel." Extrinsic feedback is that which emanates either from observation of an external visual condition or from a visual display especially provided to indicate the effect of controller inputs.

Control Input Mode

Continuous controlling or tracking is accomplished by either a compensatory or a pursuit mode. Compensatory tracking consists of adjusting a controller to maintain a vehicle, reticle, or pointer coincident with some visual reference. Pursuit tracking consists of "following" some randomly deviating reference. Compensatory tracking modes are more effective for acceleration control or when feedback inputs are infrequent. Otherwise, pursuit tracking modes are more efficient.

Controller Characteristics

1. Elastic resistance is desirable in a controller because it provides useful feedback from the controller movement or displacement; i.e., displacement can be quickly related to the changes in resistance. Design the controller system so that it provides distinct resistance gradients at critical control positions.

2. Viscous damping should be provided to aid in executing smooth control input at a prescribed rate.

3. Inertial resistance in the control system is desirable to enhance smooth control input and minimize the possibility of accidental or inadvertent activation. It should not be excessive, however, or it will become difficult to change adjustment direction. A certain amount of inertial resistance can, however, compensate for frictional drag that may be difficult to eliminate from the system.

4. Static friction should be minimized (e.g., resistance to control movement initiation). Not only does it provide no feedback, but also it often causes an excessive input force to overcome, which the operator may have difficulty correcting once he or she has observed the "overshoot."

5. In order for an operator to judge resistance, the input force required should be at least in the 5- to 10-lb (2.3- to 4.5-kg) range or above. Obviously, the resistance for manual controllers should be less than that for pedal controllers. Resistances above 30 to 40 lb (13.6 to 18 kg) should be avoided, however, in order to prevent fatigue.

6. Continuous tracking controllers should be designed to automatically return to a neutral reference position.

7. Sliding (coulomb) friction should be minimized in controller systems since it provides no meaningful input to the operator and may contribute to fatigue.

Control/Display Ratio

A low control/display (CD) ratio occurs when one unit of movement of the controller equals one unit of movement of the controlled element (e.g., the instrument pointer). (See Fig. 30.37.) Low CD ratios are desirable when it is important to obtain a gross system change with minimum controller movements. A high CD ratio is one in which the controller adjustment unit is greater than the unit motion of the system. High CD ratios are desirable for "fine" adjustment. Although specific control systems may require either a low or a high CD ratio because the nature of the control task is fairly constant, other systems may require a varying CD ratio program, such as for steering a vehicle or in using a joystick controller to manipulate an electronic "hook" to capture a target on a display.

Although it should be optimized for each situation whenever possible, some general guidelines are shown in Fig. 30.37.

1. A complete rotation of a control knob should cause about 1–2 in. (2.5–5.1 cm) of movement of the displayed element, e.g., the pointer on the scalar instrument.

2. In the case of an auditory signal used to find a station, about 30° to 60° of rotation of the knob should be used to "null out" the station position with minimum necessity for unnecessary searching.

3. A dual-mode sensitivity capability may be required for some kinds of control-display relationships. For example, if a pencil joystick is being used to control an electronic hook on a display, a single CD ratio would be unsatisfactory since the operator typically needs to "get into the vicinity of a target" as rapidly as possible, after which he or she requires a more sensitive CD ratio in order to finally pinpoint the target location. As a general rule, the gross CD ratio should be approximately 1:1; i.e., a full sweep of the joystick should translate the hook all the way across the screen. However, for the fine

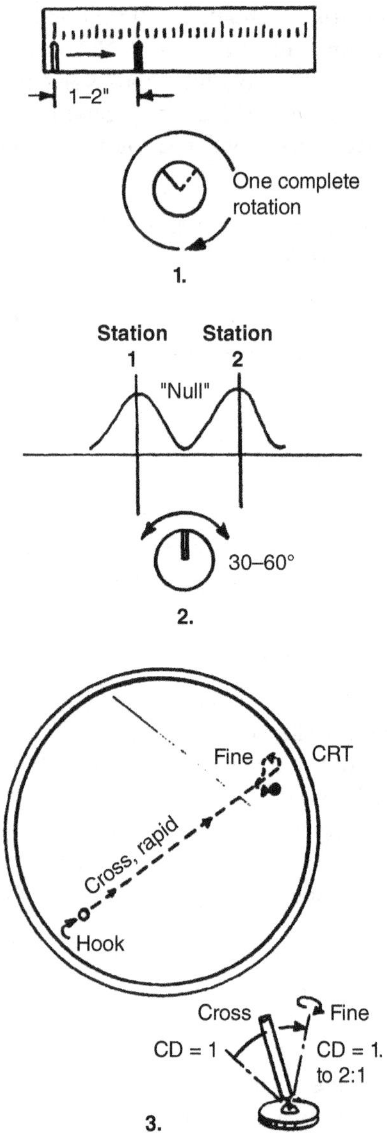

FIGURE 30.37 Control/display ratio.

adjustment around the target site, the control level should move approximately 1½ to 2 times as far (relatively speaking) as the hook moves on the scope.

Detrimental Characteristics

Dead space, lag, backlash, and excessive resistance should be avoided in controller systems. Dead space occurs when movement of a controller does not immediately result in movement of the vehicle or visual display element. Dead space should be eliminated as much as possible, especially for sensitive or high CD ratio systems. Lag is a constant delay that may be inherent in the system design. With high CD ratios (1:6 to 1:3), lag is more serious for compensatory tracking. However, when CD ratios are low, lag may improve performance. Backlash is the tendency of a system to reverse when the control input ceases. If backlash cannot be minimized, the system display gain should be as low as possible.

System Aiding

Aided acceleration or rate tracking (i.e., allowing a single control movement to affect two variables, such as position and rate or acceleration) should be considered in order to shift the burden of differentiation, integration, and algebraic addition from the operator. Quickening (or display augmentation) should be considered for vehicle control systems where the "apparent" response of the system to control adjustments is delayed (for example, in ships or submarines). The quickened display shows the operator what control action should be taken. Predictor display concepts should be considered when the future state of the system is not directly predictable, i.e., when the control system computes and displays the predicted position of the vehicle at some selected future time. This permits operators to assess the predicted results of any current input they may desire to make, thus allowing them to readjust the input in the event the first input would have an undesired result. The predictor display provides a surrogate for operator prediction (e.g., when a driver looks ahead and predicts that his or her steering input is too much or too little to properly negotiate a curve at the vehicle's current speed).

CONTROL CODING

Common methods for distinguishing controls are as follows:

1. Location
2. Shaping coding
3. Size coding
4. Mode of operation coding
5. Labeling
6. Color coding

Table 30.8 from MIL-STD-1472D shows the advantages and disadvantages of each coding method.

Shape Coding

Shape-coded knobs can be differentiated by sight and by feel. Figures 30.38 and 30.39 show knob and handle shapes that can be easily distinguished by touch. If operators will use gloves, test the shapes with gloves.

Size Coding

The examples in Fig. 30.40 illustrate size coding rules.

TABLE 30.8 Advantages and Disadvantages of Various Types of Coding

	Types of Coding					
	Location	Shape	Size	Mode of Operation	Labeling	Color
Advantages						
Improves visual identification	X	X	X		X	X
Improves nonvisual identification (tactual and kinesthetic)	X	X	X	X		
Helps standardization	X	X	X	X	X	X
Aids identification under low levels of illumination and colored lighting	X	X	X	X	(When transilluminated)	(When transilluminated)
May aid in identifying control position (settings)		X		X	X	
Requires little (if any) training: is not subject to forgetting					X	
Disadvantages						
May require extra space	X	X	X	X	X	
Affects manipulation of the control (ease of use)	X	X	X	X		
Limited in number of available coding categories	X	X	X	X		X
May be less effective if operator wears gloves		X	X	X		
Controls must be viewed (i.e., must be within visual areas and with adequate illumination present)					X	X

From MIL-STD-1472D.

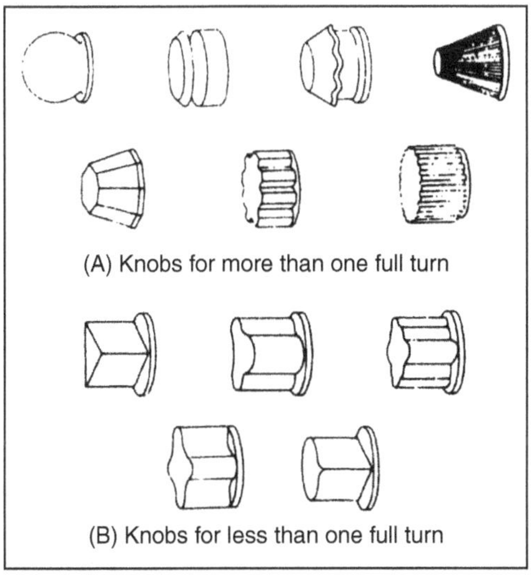

Figure 30.38 Recommended knob shapes for coding (from MIL-HDBK 759a).

Figure 30.39 Recommended handle shapes for coding (from MIL-HDBK-759a).

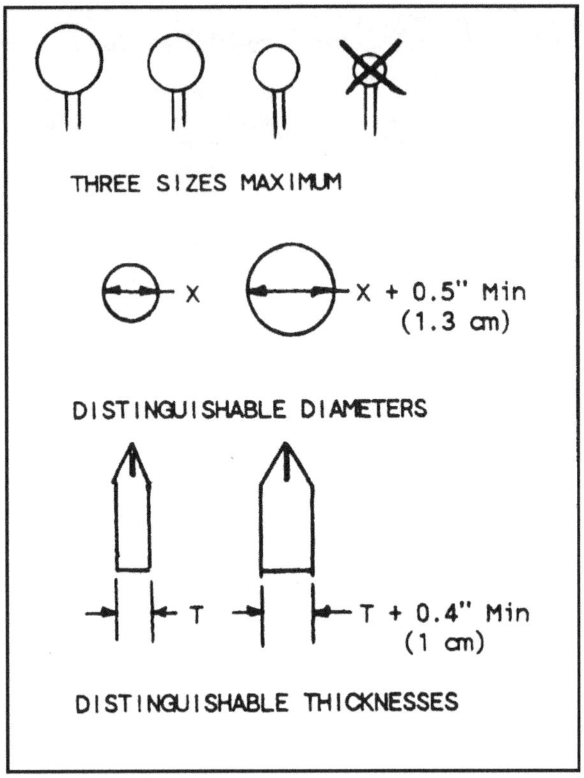

FIGURE 30.40 Size coding rules.

Color Coding

Table 30.9 shows distinguishable colors for controls (adapted from MIL-STD-1472D).

TABLE 30.9 Distinguishable Colors to Code Controls

COLOR	COLOR NUMBERS (FED-STD-595)	NOTES
Black	17038, 27038, 37038	1
Gray	26231, 36231	1
Red	11105, 21105, 31105	2
Green	14187	
Orange-yellow	13538, 23538, 33538	
White	17875, 27875, 37875	2
Blue	15123	3

Notes:

Use colors for controls without color coding

Under red light, use orange-yellow with black stripes

Use blue only if absolutely necessary

CONTROL LABELING

Label contents:

1. Primary: control functions
2. Secondary: control identification
3. Do not label controls when function and identification are obvious to the operator.

Examples of controls that do not need labels:

1. Automobile steering wheels
2. Aircraft steering yoke or joystick
3. Automobile foot pedals
4. Aircraft rudder pedals
5. Door and equipment cover
6. Latches

Locate labels where they can be read from right to left. Avoid locating control labels where they will be blocked by equipment or the operator's hand. The examples in Fig. 30.41 illustrate rules for locating labels.

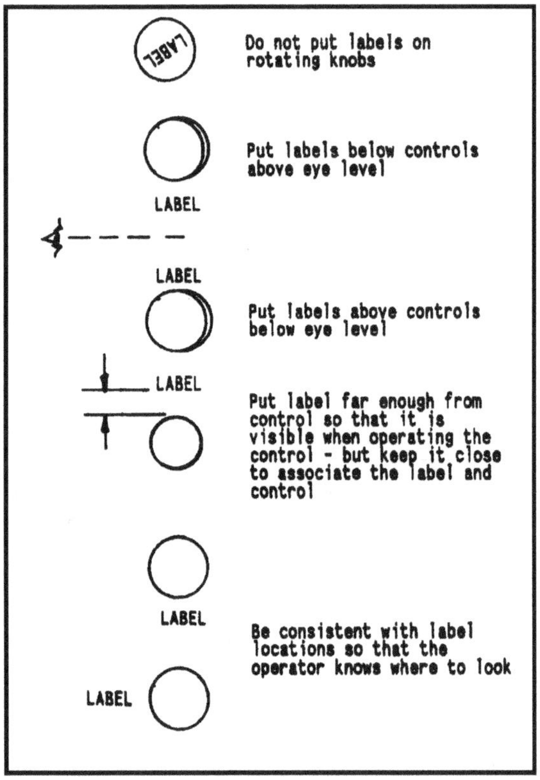

FIGURE 30.41 Rules for control labeling.

When a control can be operated in the dark, provide illumination for all labels. A nonilluminated control may be inadvertently operated and confuse an operator because he or she assumes that the illumination is malfunctioning.

CONTROL HANDLE AND KNOB SIZE

Both the size and shape of a control handle or knob can affect the ease with which an operator can manipulate the control. (See Fig. 30.42.) The size and shape of the handle must be

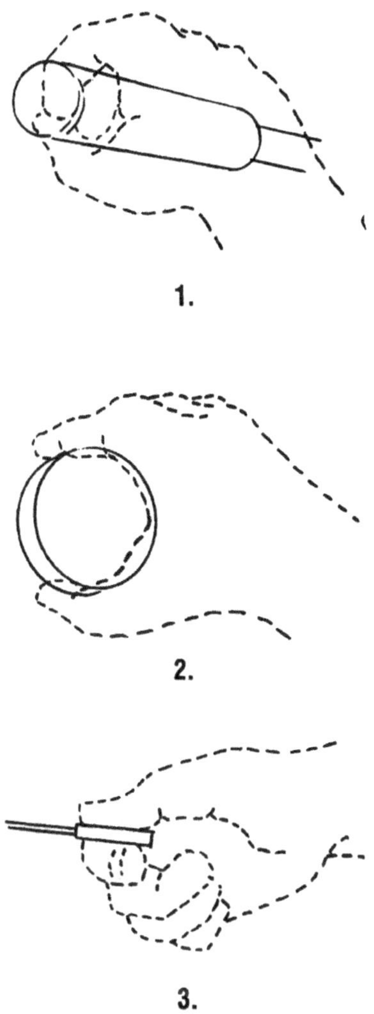

Figure 30.42 Control handle and knob size.

compatible with the size of the hand. The following general considerations are important when selecting and/or designing control handles and knobs:

1. When considered force has to be applied to a lever or steering wheel, the diameter of the handle must be large enough so that the user's hand and finger surface contact is maximized—but not so large that a firm grip cannot be maintained.

2. If a precise knob rotation is required, the knob should be large enough so that all the user's fingers and the thumb can be placed on the knob rim and/or surround the knob edges if considerable torque is also required.

3. If limited panel space requires the use of very small knobs, makes sure that there is sufficient knob depth so that the operator can have as much surface contact as possible; for knobs with very small diameters, add serrations and knurling to improve the operator's grip.

Always analyze the relative position that the operator's hand will be in when he or she manipulates the knob or handle: this may suggest a different type of knob or handle.

CONTROL HANDLE SHAPE

The shape of a control handle should be determined on the basis of how the operator is expected to grip the handle (depending on the relative position of the handle control placement, position, and expected control motions). Some typical considerations are illustrated in Fig. 30.43:

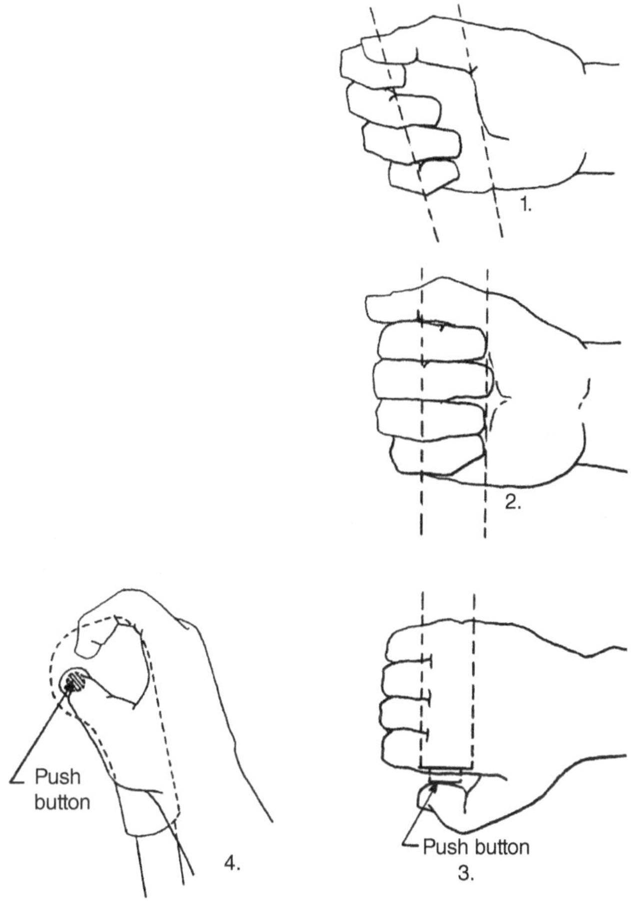

Figure 30.43 Control handle shape.

1. A pistol-grip handle should taper toward the bottom, as illustrated, because of the fore-shortening of the operator's fingers.

2. For maximum grip strength and security, use a round or oval handle of equal dimensions (cross-section). The typical practice of providing finger convolutions on the handle is not desirable because few people need to position their fingers in exactly the same place. To improve grip, add knurling or serrations to the handle's sides.

3. A round grip provides greater flexibility for the operator if his or her orientation with respect to the handle must be changed.

4. When finger- or thumb-operated push buttons are required on the handle, position them where the intended finger or thumb "naturally" falls.

FUNCTIONAL PRINCIPLES FOR DECIDING ON THE LOCATION AND ARRANGEMENT OF CONTROLS

Functionally Related Controls

Consider those controls that relate to a common system to be functionally related and arrange them so that they are in the same general area on a panel. When a panel display is affected by a particular control, locate the display and control near each other.

Sequence of Use

When several controls are normally operated in a certain sequence, arrange them in some systematic order, i.e., from left to right, from top to bottom, etc. This is particularly important when "missing" a step might be critical to an operating sequence.

Most Frequently Used Controls

To reduce operator fatigue and/or to improve control operation performance, locate the most frequently used or continuous-operation controls where they are most convenient for the operator.

Emergency Controls

Any control that requires rapid response time in an emergency should be located where the operator does not have to search for or reach the control. For example, if the operator's hands are typically on a steering control most of the time, emergency controls should be within a few inches reach from the steering control, except when the emergency control could be foot operated, in which case the principle applies to any normally continuous foot control operation. The process for arrangement and layout of controls and displays is summarized in Fig. 30.44.

Precision Controls

Although limited precision is potentially possible for foot-operated control, as a general principle precise controlling functions should be performed by the operator's hands.

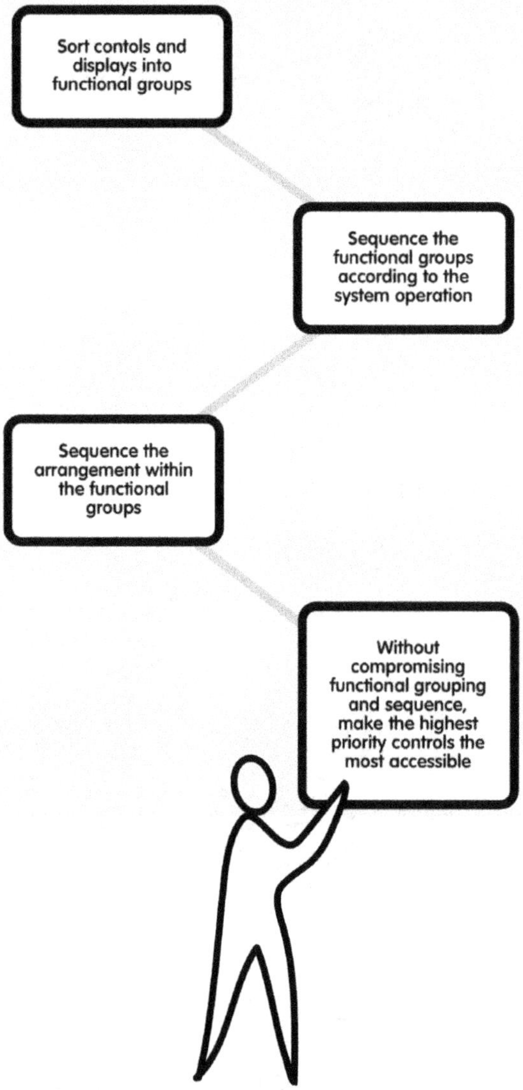

FIGURE 30.44 Control and display layout process.

Alternate Hand Operation

Consider the potential emergency in which an operator may not be able to use one hand or the other, e.g., when a battle wound has incapacitated a pilot's right or left hand. In this situation, location of critical controls such as the joystick for operation by either hand would be vital to the pilot's safe return.

Control Interference

Sufficient clearance should be provided between adjacent controls, between controls and the adjacent structure, and between the operator's own body and the equipment so that critical controls can be easily grasped and manipulated in the normal manner. Special attention must be given to separating critical controls whose inadvertent or accidental operation could lead to loss of control or damage to a system. Controls that the operator normally reaches for without looking at them (blind reach) must be separated more than others.

Hidden Controls

All controls normally operated by a pilot, driver, or other operator should be visible from the nominal operating position; i.e., avoid hiding operational controls behind the seat, underneath work surfaces, or behind unlabeled cover panels.

Work-Load Distribution

As a general principle, equalize the workload of the control operator; i.e., distribute the control activity between his or her two hands and two feet (recognizing the limitations of limbs and preferred hand, etc.). The primary objective should be to avoid overloading the operator's right hand just because he or she is right handed and is more agile and dexterous when using this hand.

Also, avoid assigning control operations to two limbs when a simultaneous motion geometry or motion pattern may cause motor response disruption, e.g., "patting one's head while rubbing one's stomach." Typical interference patterns occur when one hand is required to move a lever forward, while the other hand has to move a similar lever aft; when one hand operates a crank in the clockwise direction, while the other hand has to rotate a potentiometer also in the clockwise direction; or when an operator has to perform intricate hand manipulations and foot manipulations at the same time (a feat requiring considerable skill and practice).

Caution: The reader will no doubt recognize that there are possible conflicts among the various principles enumerated above. The temptation is to ignore the principles because of these apparent conflicts. Rather than ignoring them, think of them as having varying priorities; i.e., at times one principle has a higher priority than others, depending upon the particular design problem and operating situation. Thus, the designer should examine all the principles and place them in proper perspective according to his or her system's operating situation. Above all, do not ignore the principles, for they are individually valid objectives.

CONTROL LOCATION AND REACHING REQUIREMENTS

Avoid the temptation to place controls where operators have to reach for them, just because they are "within reach." Several typical errors are illustrated in Fig. 30.45:

Do not make operators shift from their normal operating position; it often requires that they "look for the control" and take their eyes and attention away from the roadway or displays they should be looking at. Do not place one control behind another so that operators have to take the time to reach around the intervening control.

STEERING CONTROLS

Automotive Steering Wheels

Wheels are traditionally provided for steering automobiles, trucks, and buses. See Fig. 30.46 for design criteria.

1. For maximum comfort the wheel should be tilted about 60°, and the wheel diameter should be 14–15 in. (35.6–38 cm).

2. When power steering is not provided for large, heavy vehicles, more force may be required to turn the wheel. Thus, the steering wheel should be oriented horizontally, and the wheel diameter should be increased to 18–20 in. (45.7–50.8 cm).

3. When power steering is provided for larger vehicles, the steering wheel should be tilted to a more comfortable position for long-duration driving, and the wheel diameter should be reduced so that the driver can reach more easily across the wheel.

4. For typical passenger cars, vans, and smaller trucks, the steering-wheel angle and size should approximate the optimum configuration noted above; i.e., the wheel tilt should be between 45° and 60°, and wheel diameter should be between about 14 and 16 in. (35.6 and 40.6 cm).

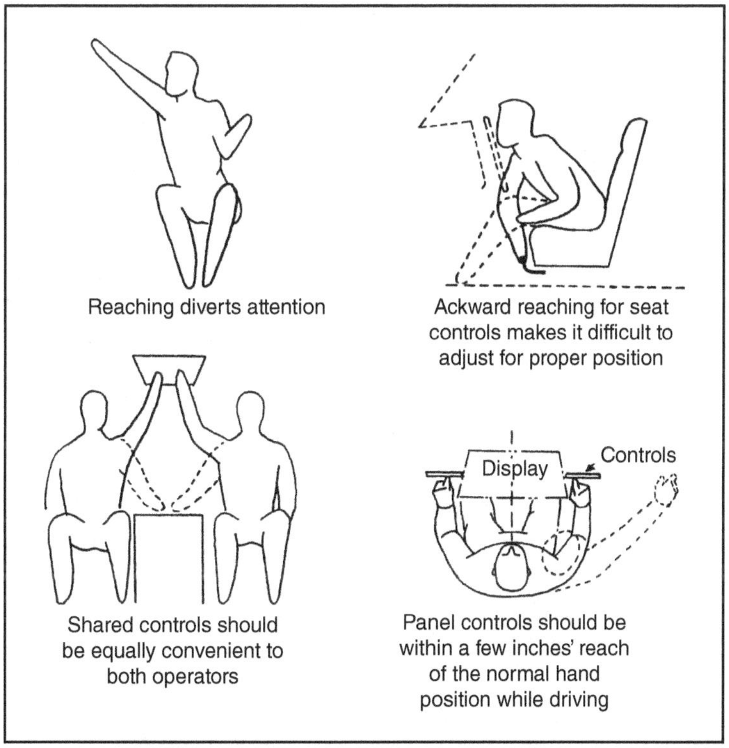

FIGURE 30.45 Control location and reaching requirements.

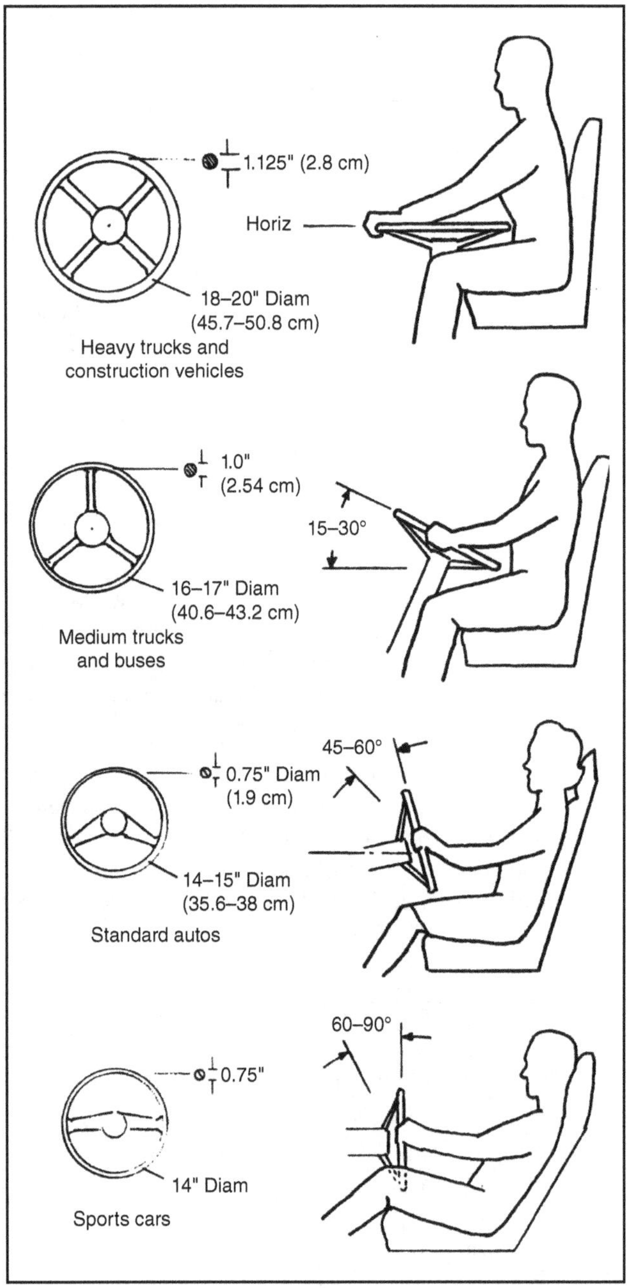

Figure 30.46 Steering wheel design criteria.

5. For special vehicles that are low to the ground (e.g., sports cars), the wheel tilt should be between 60° and 90°, and the diameter should be 14–15 in. (35.6–38 cm). *Note: The spoke arrangements illustrated in Fig. 30.46 for each of the vehicle classes should be appropriate for the type of steering-wheel manipulation; i.e., the number, positioning, and size of spokes and wheel rims should vary to provide gripping ease and force.*

6. Steering-wheel rims should be round or oval. Avoid special decorative convolutions or ridges that interfere with gripping. (See Fig. 30.47.)

7. However, provide gripping aids on the underneath side of the rim to reduce the possibility that the driver's grip will be disturbed by a sudden wheel motion or slipping when he or she attempts to apply force during a turning maneuver.

8. Wheel spokes should be oval and of the recommended dimensions so that the driver can be assured of a firm, comfortable grip if he or she elects to use the spokes for steering.

9. Note that the oval dimensions for high-force wheels for large vehicles are larger than those recommended for smaller vehicles and vehicles that are provided with power steering assist.

10. When practicable, limit steering-wheel displacement to about 120° for the primary turning range so that the driver is able to keep both hands on the wheel during the maneuver (without repetitive hand replacement).

11. Lock-to-lock limits for maximum turn should be between 1.5 to 1 (steering-wheel versus vehicle-wheel turns) and 3.0 to 1.

12. Although very strong drivers can exert more force than indicated above, they may not apply it in time to avoid an accident. If power steering is provided to reduce turning force requirements, be sure to provide sufficient emergency, mechanical advantage to keep force requirements below 50 lb (23 kg).

The National Highway Traffic Safety Administration maintains safety information on crash tests of steering wheels.

When steering wheels are tilted, they should be designed so that the driver cannot be impaled on the steering column in the event of a crash. (See Fig. 30.48.)

1. The steering-wheel hub should be recessed with respect to the wheel rim (see the figure).

2. The hub should be large, flat, and padded; i.e., avoid decorative hubs with protruding or pointed profiles. If an air-bag (passive restraint) system is installed in the center of the steering wheel, make sure that there is sufficient clearance between the wheel rim and the rim of the bag housing so that the driver can easily grasp and grip the wheel.

3. The steering-wheel system should include a collapsible column to help absorb the energy of a driver who is thrown into the wheel during a crash.

4. Tiltable wheels are not generally recommended; they make it extremely difficult to ensure that a driver will not position the wheel where the rim may obscure an instrument, and/or the driver may try to adjust the wheel tilt while the car is in motion and thus interfere with steering efficiency. When the wheel tilt option is provided, however, limit the tilt to no more than 15°.

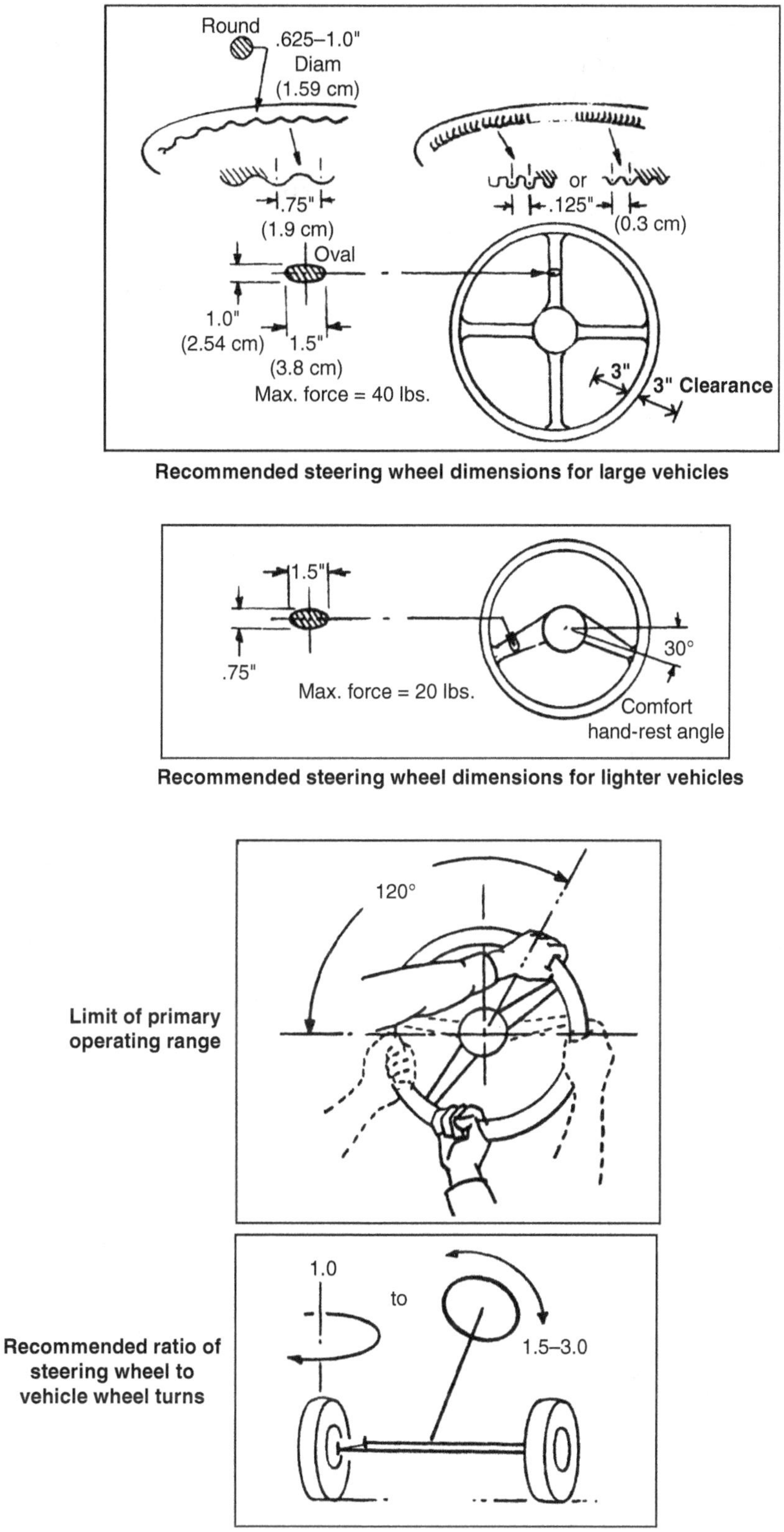

Figure 30.47 Steering wheel dimensions, operating range, and ratio to wheels.

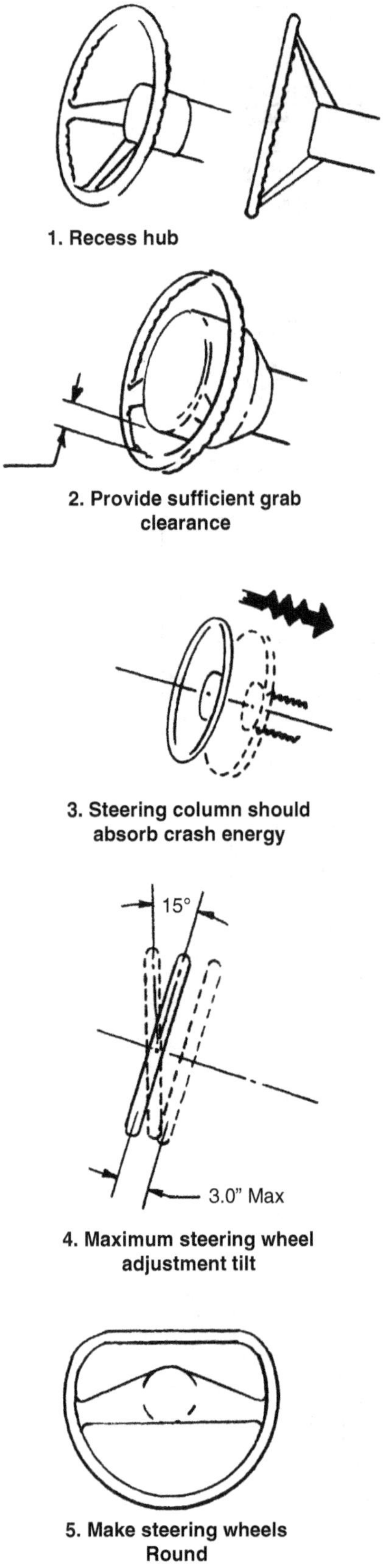

Figure 30.48 Tilted steering wheels.

5. Make oddly shaped wheels only if the maximum wheel deflection is less than 120°; drivers cannot perform multiturn manipulations as efficiently with such wheels as they can with round wheels.

Wheel-Type Joystick

The wheel-type joystick is generally used for most aircraft other than military fighters where lack of lateral space within the cockpit dictates a single lever-type joystick. (See Fig. 30.49.) The so-called wheel generally is configured as a half wheel in order that the pilot can see panel instruments beyond the wheel. Key factors in designing and positioning the wheel control are the following:

1. Fore-aft excursion should be within comfortable reach of the pilot; the pilot should not have to lean forward for pitch-down movements or rear back for pitch-up movements, and his or her normal position for flight control should be comfortable (i.e., the upper arm should hang approximately vertically). An arc pattern is acceptable, but a straight-line pull is easier for the pilot.

2. Clearance must be provided between the wheel and the pilot's knees and between the vertical control shaft and the pilot's seat, and there must be clearance aft of the pilot's elbow when pulling the wheel aft.

3. When possible, design the system so that full rotation is no more than 40°.

4. For small aircraft, where a wheel of minimum size is desirable, be sure that the handles are at least 12 in. (30.5 cm) apart. Handgrips should be round or oval. A 1.125-in. (2.858-cm) diameter is optimum. (Oval dimensions are illustrated.) The handgrip should be at least 4 in. (10 cm) long and should be tilted inward, as shown in the figure.

5. A very effective wheel configuration is shown in the figure. Not only is it very comfortable, but it also provides useful feedback in terms of control position. Note, however, that the lateral handle spacing must be at least 15 in. (38.1 cm) to prevent the handles from striking the pilot's knees during a turn.

LEVERS AND JOYSTICKS

There are two types of levers and joysticks:

1. Displacement—Lever or joystick moves away from some neutral position.

2. Isometric—Lever or joystick has no perceptible movement but output is a function of applied force.

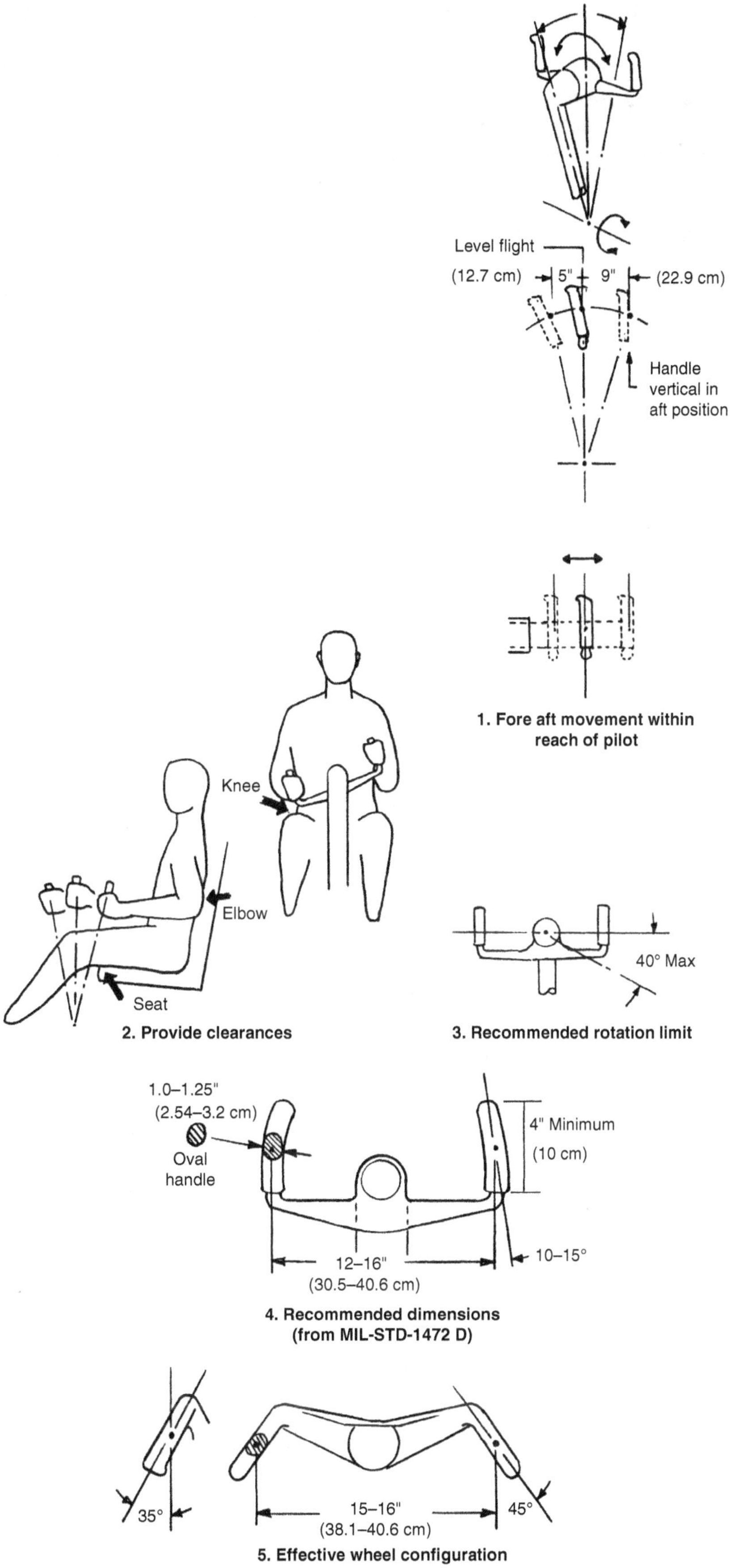

Figure 30.49 Wheel-type joystick.

The relative advantages and disadvantages of each are included in Table 30.10.

TABLE 30.10 Comparison of Displacement Isometric Controls

TYPE OF LEVER OR JOYSTICK	ADVANTAGES	DISADVANTAGES
Isometric	Positioning Accuracy Less force required Less fatiguing Best for secondary switches	Slower operation Requires more operating room
Displacement	Faster Center or null position is simple to locate Requires less operating room	Provides less tactile feedback of control position Does not provide tactile feedback

Center-Post-Type Joystick

The simple post-type joystick is generally used for very small aircraft where the fuselage width does not permit use of a wheel. Although many of the clearance problems are similar to those described for the wheel-type controller, one special problem must be addressed, i.e., lateral clearance between the pilot's legs for moving the joystick from side to side. The guidelines shown in Fig. 30.50 may help the designer develop a satisfactory joystick arrangement in the cockpit. Note especially the height of the joystick handle with respect to the seat. Although the handle can be located higher without affecting controllability, it can easily interfere with the pilot's view of the panel instruments. The control system should not require the operator to employ more than about 20 lb (9 kg) of force in order to move the joystick to maximum excursions in any direction. (See Fig. 30.51.)

The joystick handle shown in Fig. 30.52 is typical of pistol-grip types often designed for use in military aircraft, i.e., where additional switches and/or controls are added to the handle so that they are convenient during critical flight modes. Note especially the dimensional suggestions for the gun or bomb-release trigger. Also note the suggested knurling on each side of the grip to aid the pilot in maintaining a secure grip on the handle.

Switches should be placed far enough apart to minimize inadvertent switch actuation.

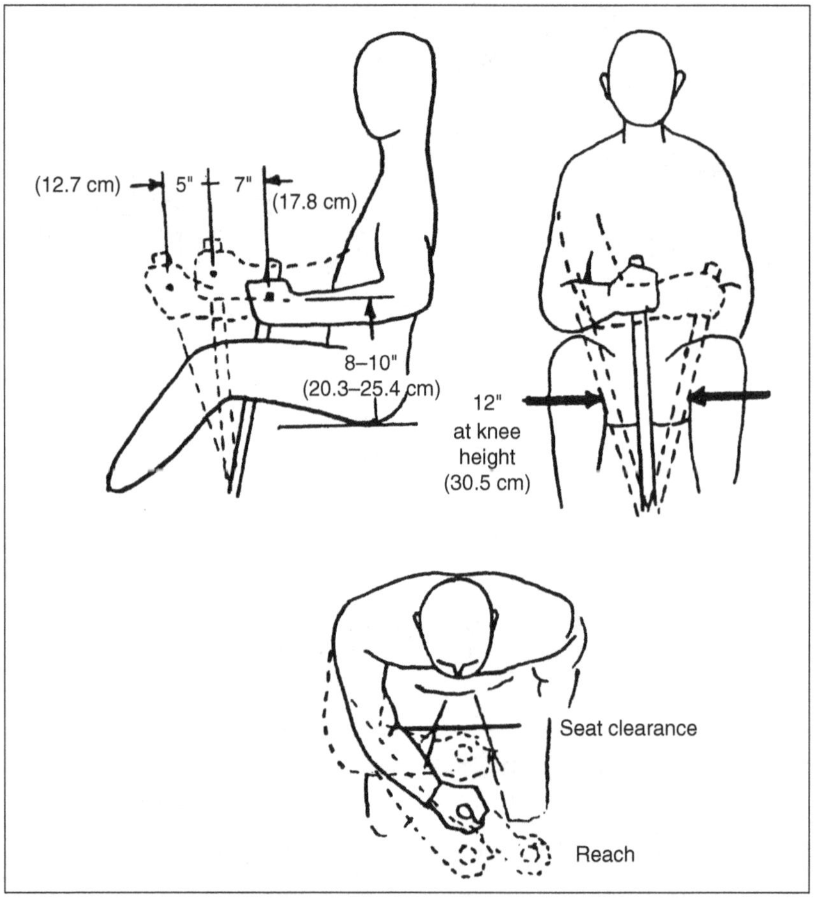

FIGURE 30.50 Control movement must clear obstructions and remain within reach.

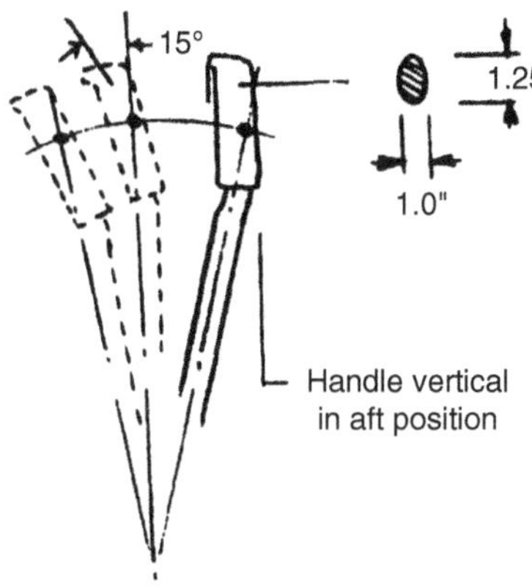

Figure 30.51 Recommended control handle design.

A "heel rest" is recommended at the bottom of the handle so that the pilot can rest his or her hand and reduce the necessity to grip the handle tightly during relaxed flight conditions.

Six-Function Joystick

Six functions are probably the maximum number that should be considered for a single controller; more functions make it difficult for pilots to keep from actuating the wrong one, either because they forget which switch is which or because of inadvertent motion inputs (e.g., cross-coupling). Figure 30.53 shows one possible six-function configuration.

Note: The specific geometry and dimensions are suggested guidelines and should not be construed as absolute.

Side-Stick Controller

A side-stick type of joystick may be required when sophisticated display and control packages must be placed between the pilot's knees (thus eliminating the possibility of the normal joystick). Obviously, this greatly reduces the range of controller movement. Instead of full arm motions, pilots must rely almost entirely on wrist motion, which also reduces the amount of force they can apply. A typical electric stick is shown in Fig. 30.54. Fore-aft movement should be limited to plus or minus 30° and side-to-side motion to plus or minus 45°. Control movement resistance should range from 10 to 20 oz. (283 to 567 g), and the control should be spring centered so that it will return to the neutral position.

Theoretically, the most efficient side controller would be one that does not require the pilot to make any arm motion; i.e., the controller should be designed so that the pilot's arm can remain firmly on an armrest, with only natural wrist motions required to manipulate the controller. (See Fig. 30.55.)

The shape and actual size of handles for side-stick controllers should be approximately the same as those suggested for the normal joystick (center type), as shown in Fig. 30.56.

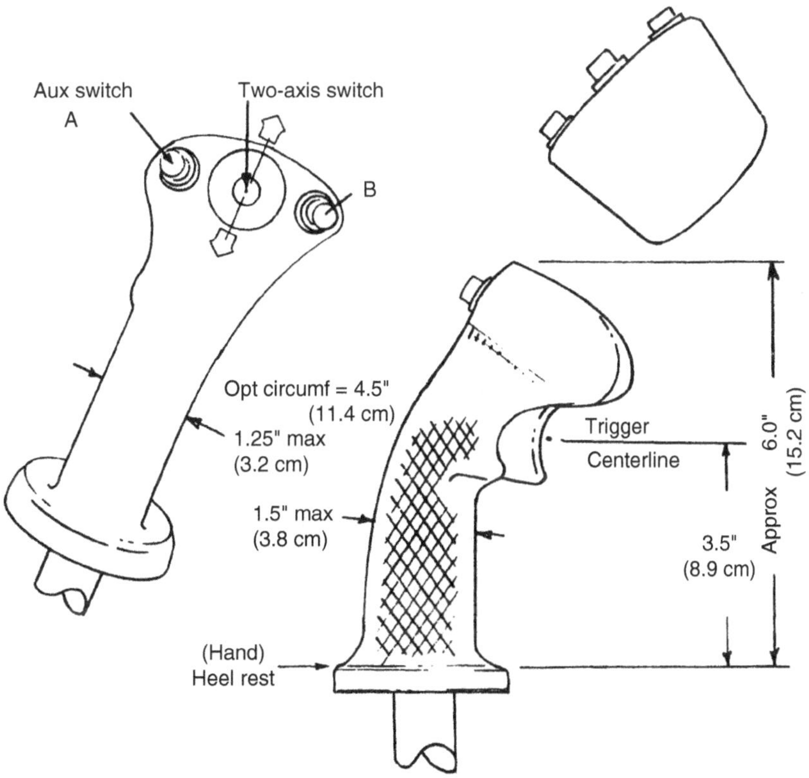

Figure 30.52 Pistol grip joystick handles.

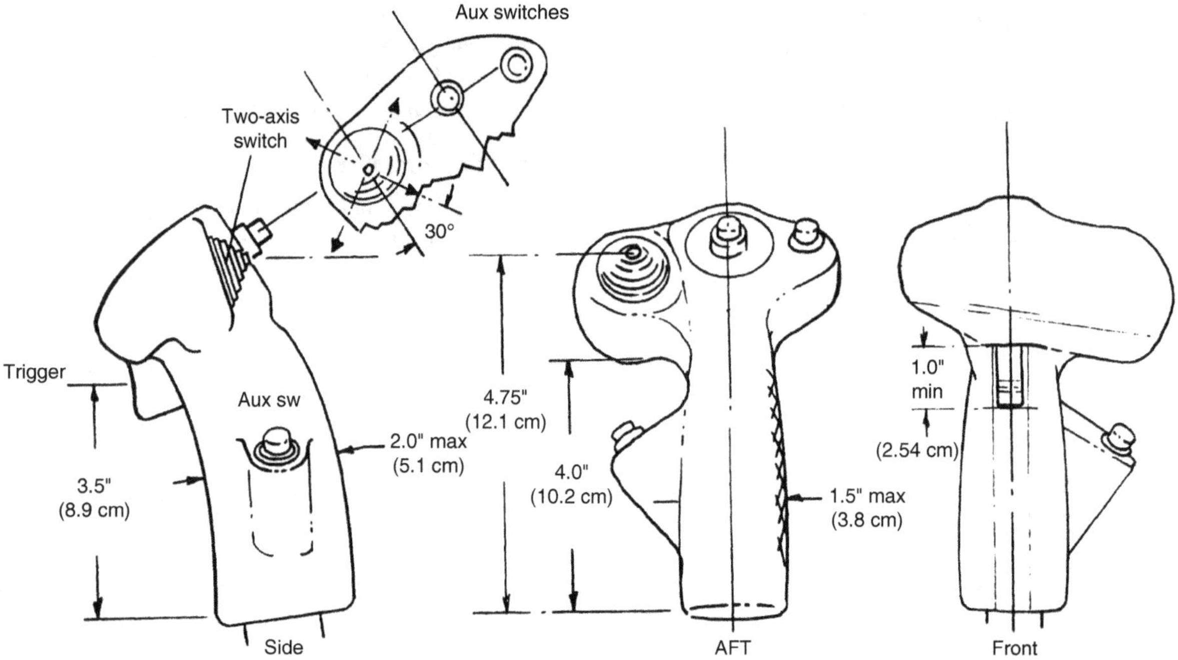

Figure 30.53 Examples of six-function joysticks.

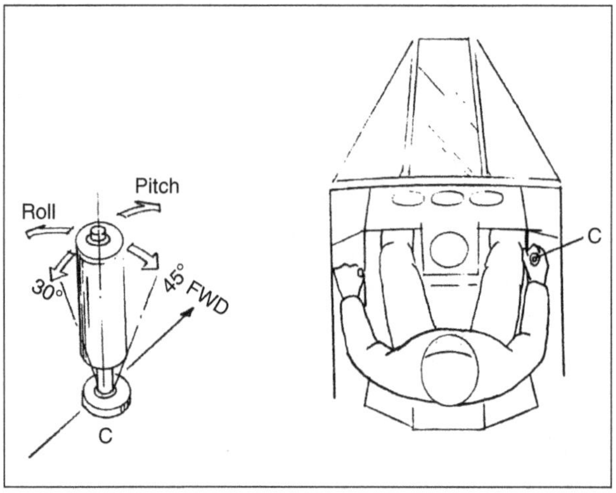

Figure 30.54 Recommended limits of wrist control movement.

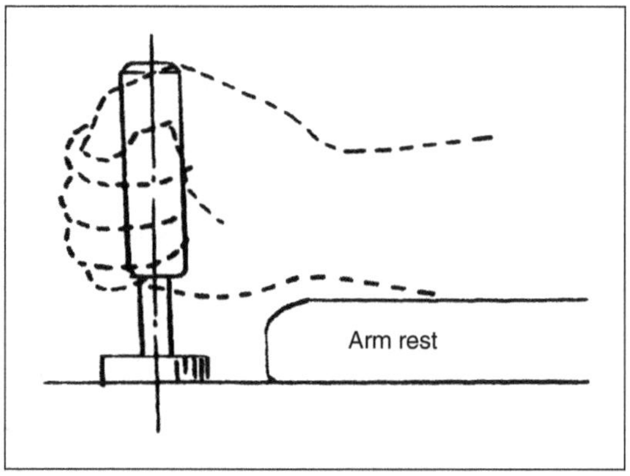

Figure 30.55 Use rest to support arm.

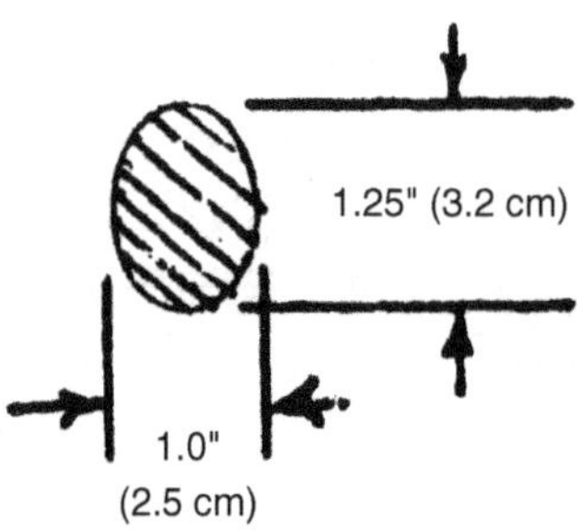

Figure 30.56 Recommended control handle size.

GEAR-SHIFTING CONTROLS (AUTOMOTIVE)

General

Although there are inherent conflicts with recognized human engineering principles regarding the direction of movement of lever controls, automotive gear-shifting patterns have been in use sufficiently long that people have come to expect them to articulate in relatively standard patterns. Mechanical and automatic gear systems are illustrated in Fig. 30.57.

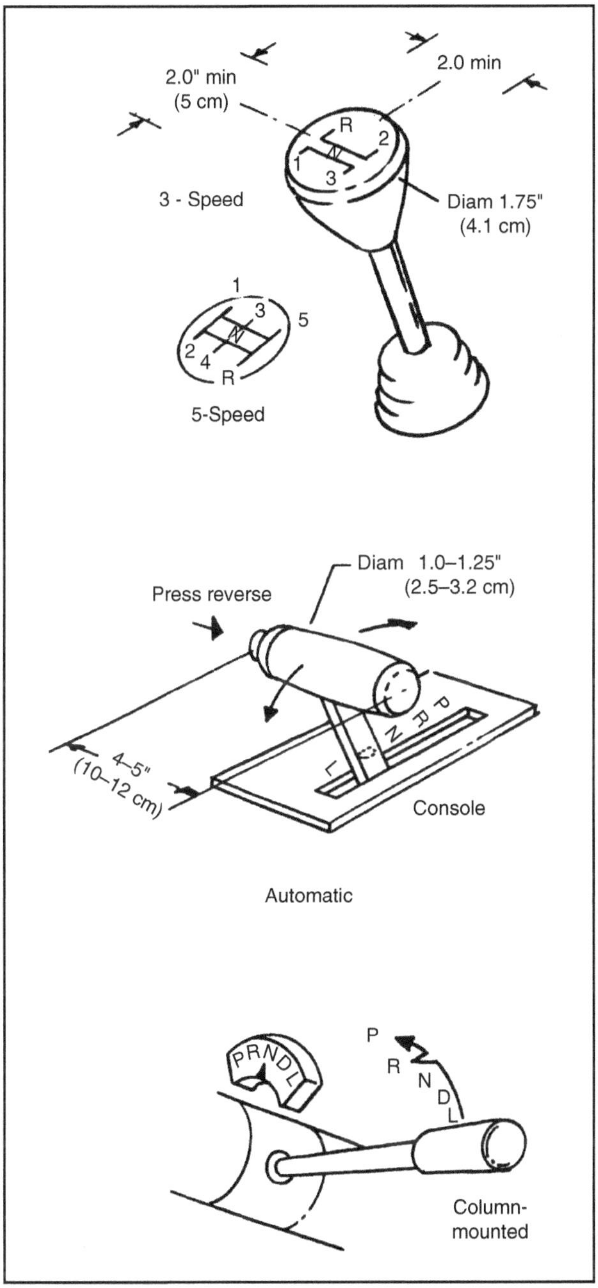

FIGURE 30.57 Auto shift control design guidelines.

The several dimensional guidelines shown in the figure are not to be taken as hard-and-fast requirements; rather, they are guidelines for designing control handles that will be acceptable to most drivers, regardless of their hand size.

1. There should be enough spacing between gear positions so that the operator can "feel" the change in position, i.e., so that he or she can locate the new position from an adjacent one in the same direction. In addition, the lever should also be designed so that the lever pawl is spring loaded, which gives additional "feel" to the positioning action.

2. A separate action should be required of the operator in order to select the reverse position (to preclude inadvertently placing the lever into reverse gear when the vehicle is moving forward). A push button is recommended for the console-mounted system, and a "pull toward the driver" for the column-mounted system.

3. The column-mounted automatic lever position is preferred over the console configuration because in this position the position displays are easy to see from the driver's normal "eyes-on-the-road" sight line.

4. The force required to move a gear-shifting lever should not exceed about 10 lb (4.5 kg), nor should it be less than 2 lb (0.9 kg) to preclude inadvertent actuation.

Other Gear-Shifting Lever Handle Configurations

The key factors to consider in the design of any gear-shifting lever handle interface are the following:

1. Make the motion compatible with the user's natural expectations; i.e., the user should push forward or lift up on the lever to cause an increase in function, value, or forward motion and should pull back for the opposite effects.

2. To prevent inadvertent movement into a particular gear, provide some mechanical deterrent that requires the operator to perform a special, secondary action such as moving the lever in and out of a slotted channel or pressing an auxiliary lever. (See Figs. 30.58 and 30.59.)

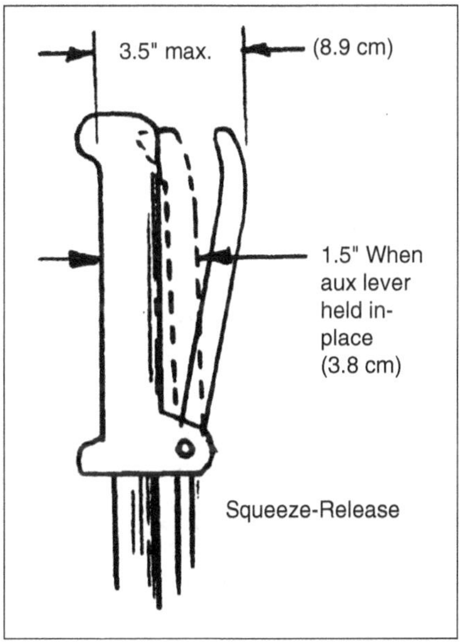

FIGURE 30.58 Prevent inadvertent activation.

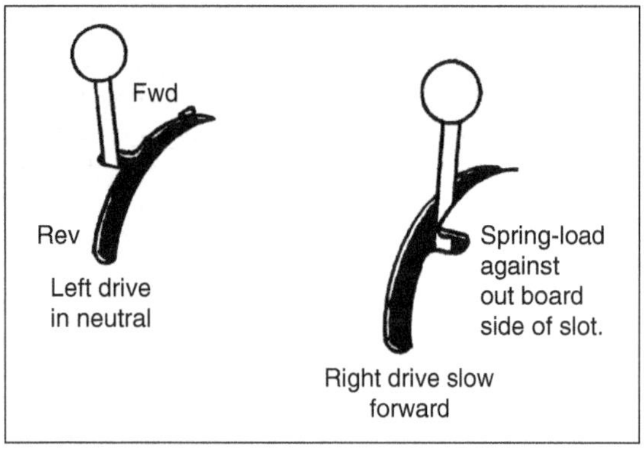

FIGURE 30.59 Slotted neutral position for dual drive system.

3. Provide a handle design that is compatible with the user's hand shape and size and with the strength of his or her fingers, such as for squeezing.

4. Avoid certain types of release concepts, such as those illustrated in Fig. 30.60.

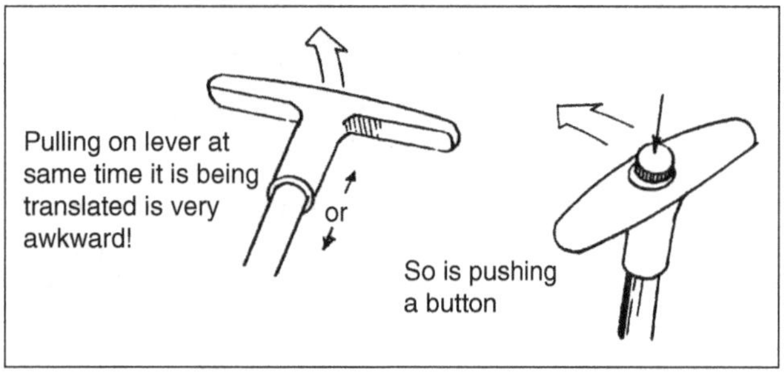

FIGURE 30.60 Avoid awkward release mechanisms.

HIGH-FORCE MANUAL CONTROLS[9]

When there is no alternative and an operator is required to apply considerable force to a controller, consider the following guidelines (see Fig. 30.61):

1. Position the levers between the operator's waist and shoulder levels.

2. Provide something the operator can pull or pull against, or secure the operator so that his or her body is not displaced.

3. Increase the lever length or the wheel diameter, but not to the point where the driver cannot reach the lever or wheel comfortably.

4. Provide the proper size of handle to maximize gripping capability.

5. Allow the operator to use two hands.

6. Consider the use of a foot-operated control.

[9] See Chap. 35 for human strength capabilities.

FOOT-OPERATED CONTROLS

Maximum Force Applications

When operators are required to apply the absolute maximum force of which they are capable, use a foot-operated controller. In order for such operators to obtain their maximum capability, they should be seated so that they can push against a solid backrest, as illustrated in Fig. 30.62. The point at which the ball of the operator's foot contacts the pedal should be approximately at the same height as the operator's seat reference point, with the seat backrest at the angle shown.

It is necessary to provide a seat that is adjustable fore and aft, however, so that operators with different leg lengths can be in a position that will allow them to obtain the optimum upper and lower leg angles shown. Remember, however, that people cannot apply maximum force for sustained periods, nor can they apply maximum force repeatedly over a long period of time.

Vehicular Pedal Relationships

A typical array of clutch, service brake, and accelerator pedals should provide an approximately common pedal level when the pedals are in the nonactuated position. (See Fig. 30.63.) Since the brake and clutch pedals generally involve some leg motion, they should be suspended from above; the accelerator pedal involving only ankle movement must pivot from below. The centerlines of the brake and clutch pedals should be at least 8–10 in. (20–25 cm) apart and should be placed symmetrically in front of the driver. If the vehicle has an automatic transmission (no clutch), place the brake pedal on the driver's centerline; the pedal should be approximately 6–8 in. (15–20 cm) wide so that it can be operated by either foot.

Brake Pedals

Since many brake pedal controls require some amount of force, the geometric relationship between the operator's leg and foot and the position and angle of the pedal is an important consideration; i.e., depending on the height of the seat, not only must foot pedals be placed within reach, but they also must operate in a direction that is compatible with the force application vector.

A—A high seat position wherein the operator presses down more than he or she presses forward. (See Fig. 30.64.) Brake forces should not exceed about 20 lb (9.0 kg). The braking stroke should articulate downward and forward in a straight line, as shown.

B—A mid-position seat height wherein the operator presses about equally forward and down. (See Fig. 30.65.) Brake forces should not exceed about 40 lb (18.0 kg). The stroke should follow a reversed curvilinear pattern, as shown.

C—A low seat height wherein the operator can obtain maximum force because he or she can take maximum advantage of the seat back. (See Fig. 30.66.) However, the

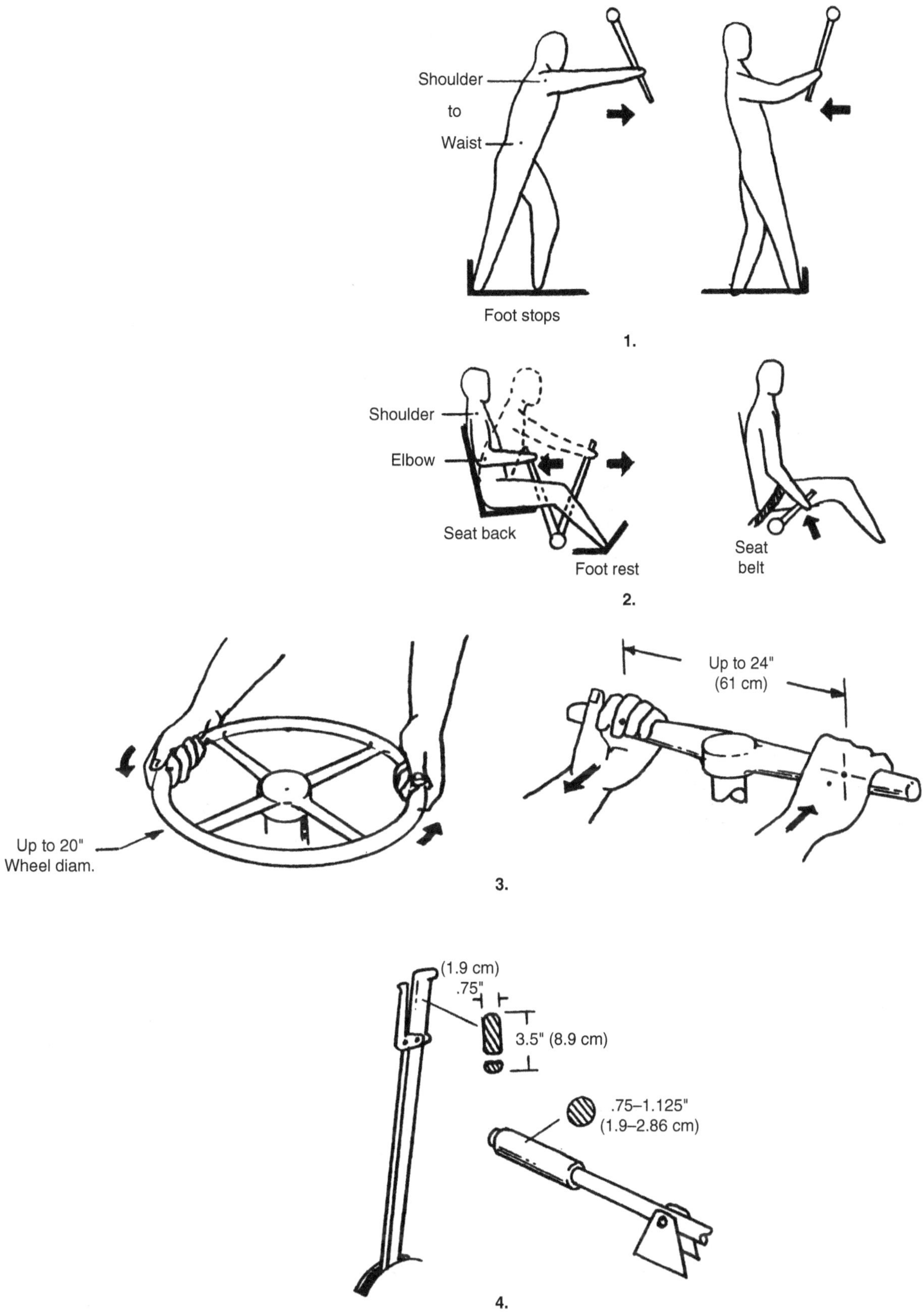

Figure 30.61 High-force manual controls design considerations.

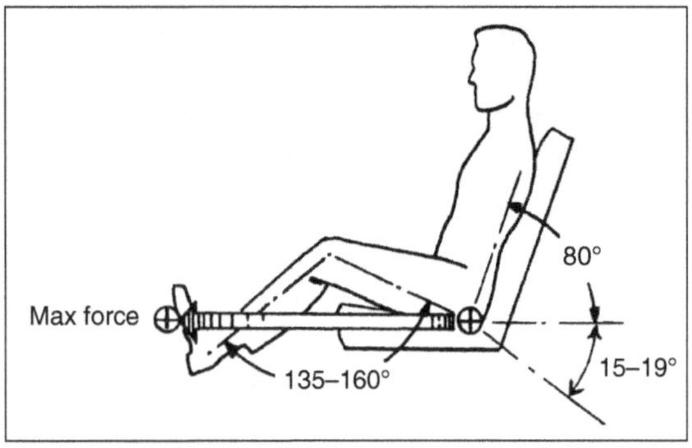

Figure 30.62 Leg angles for maximum force.

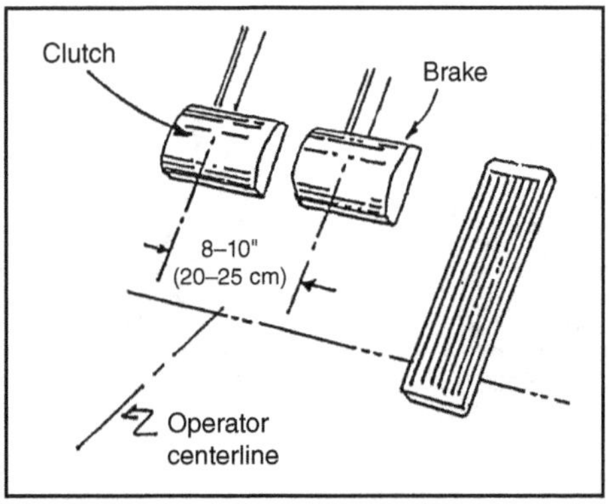

Figure 30.63 Location of brake and clutch pedals.

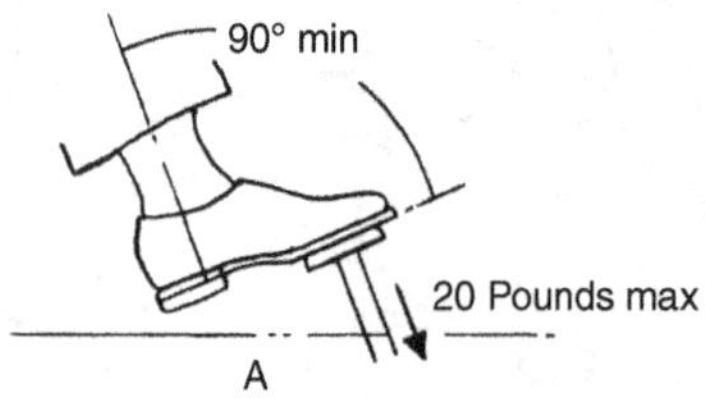

Figure 30.64 High seat position.

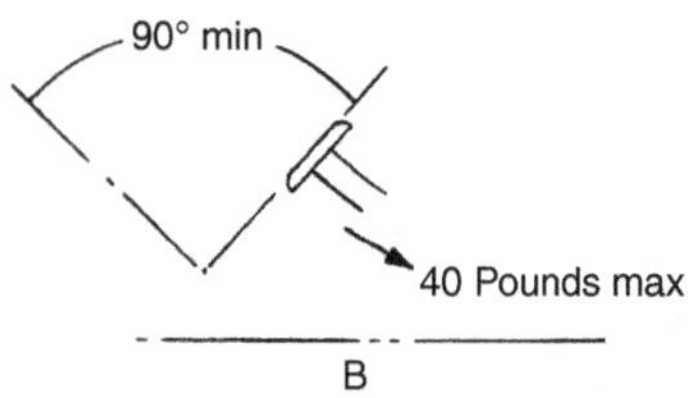

Figure 30.65 Midposition seat height.

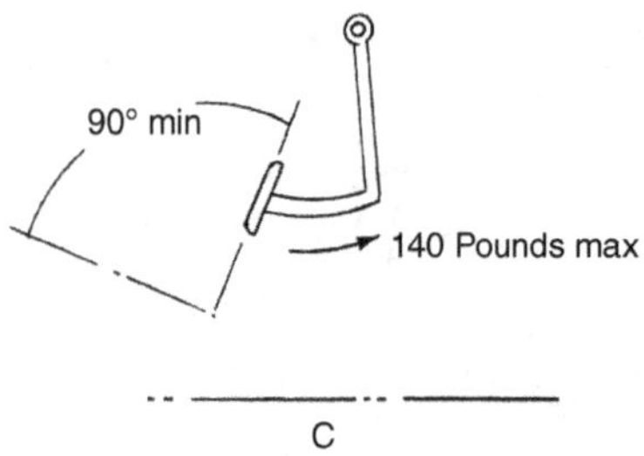

Figure 30.66 Low seat height.

maximum braking requirement should not exceed about 140 lb (63 kg) for vehicles operated by males and females and 180 lb (82 kg) for vehicles operated only by males. The stroke should follow a reversed curvilinear pattern, as shown.

Accelerator

The experienced driver has an uncanny ability to find and operate a variety of accelerator pedal configurations, as long as they are generally located slightly to the right and forward of the seat. However, if the placement and angle of the pedal depart appreciably from the guidelines shown in the Fig. 30.67, the driver may soon become fatigued because of the necessity to twist the body or extend the leg and ankle to fit a poorly configured accelerator pedal arrangement. Key considerations are the following (see Fig. 30.68):

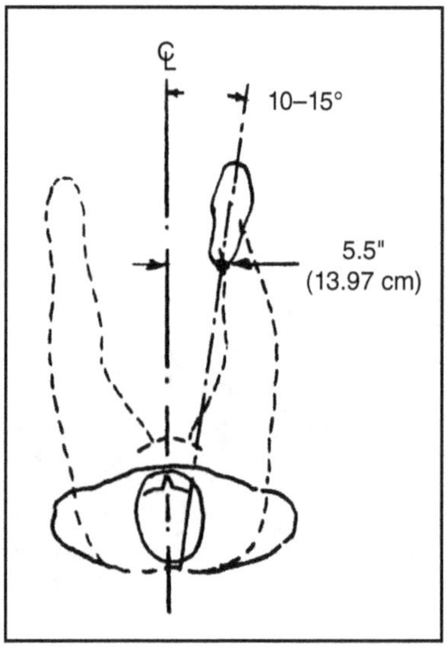

Figure 30.67 Accelerator pedal location.

1. The driver should always be able to rest his or her heel on the floor or on a heel stirrup while holding or depressing the pedal.
2. The pedal should be big enough so that both the small- and the large-footed driver can press the pedal with the ball of the foot.
3. The "release angle" (nonactive) of the pedal must not be too steep; otherwise, the driver's ankle tires easily during the time he or she releases pressure on the pedal.
4. The force required to depress the pedal should not exceed about 10 lb (4.5 kg).
5. Displace the accelerator pedal from the body centerline and angle it slightly.

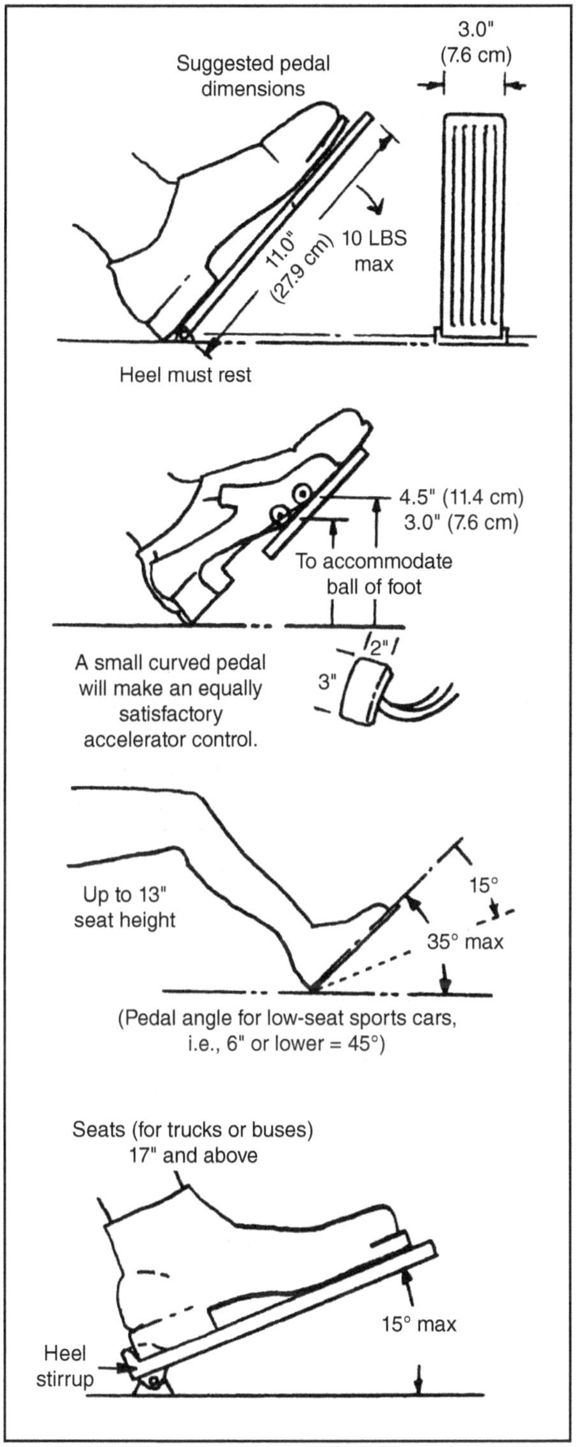

Figure 30.68 Accelerator pedal design recommendations.

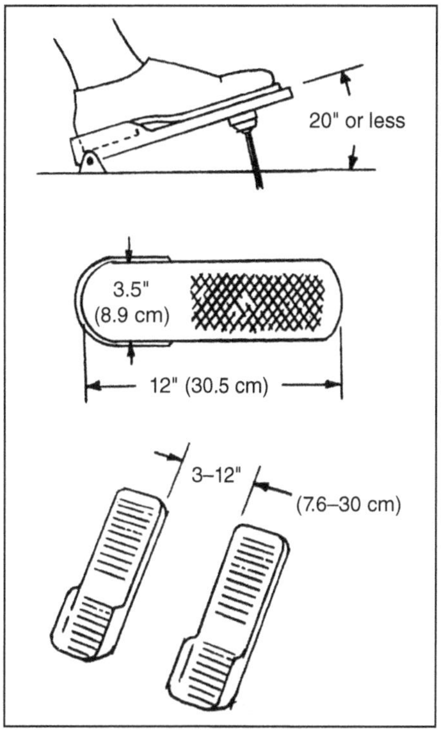

Figure 30.69 Pedals with stirrups.

Pedal Considerations for Vehicles with High Driver Positions

For some applications, such as the driver station of buses or heavy cargo tractors, the driver's position is typically quite high, and the plane of foot pedals is at a low angle (20° or less). In such cases, it is desirable to select a pedal which includes a heel stirrup, as shown in Fig. 30.69, because operators tend to "miss" placing the heel at the pivot point of the pedal and therefore have to apply too much force because the foot is too far back on the pedal.

Heel stirrup designs are also appropriate when, for some reason, one cannot place the fulcrum or heel reference on the floor of the vehicle, i.e., when the foot is suspended in midair. If two stirrup-type pedals are used together (as for aircraft rudder pedals), the spacing between the pedals should be at least 3.0 in. (7.6 cm) but not more than 12.0 in. (30.5 cm). The preferred and most comfortable spacing is about 6.0 in. (15 cm). Lateral separation between the pedals can vary acceptably between 4 and 8 in. (10 and 20 cm). The pedals should rotate freely on spindles and should be made of rubber or provided with some other nonskid feature. (See Fig. 30.70.)

Foot Switches (Standing Operation)

Generally speaking, foot-operated controls are not recommended because it is difficult for standing operators to operate a pedal without losing their balance. Proper location requires that the foot pedal be placed so that a minimum of movement is required to reach it. By the same token, the choice of location should take into account possible inadvertent actuation.

Portable foot pedals are suggested for many operations since they allow individual operators to place the pedal where it is most convenient for the particular task. There are no current standards regarding switch pedal size, either the mechanically operated type or the pressure-sensitive type

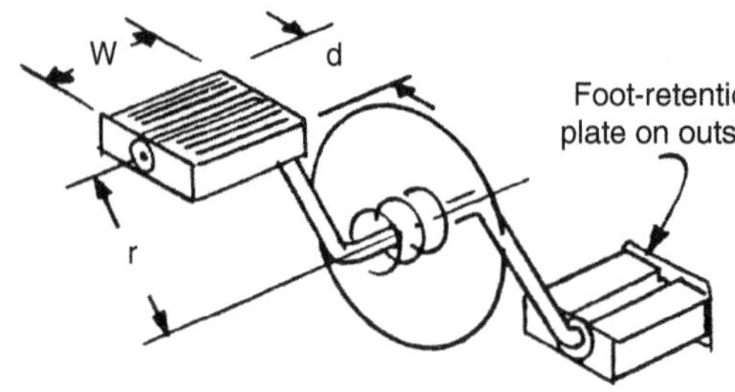

Bicycle pedals

Adult bicycle	Youth bicycle	Toddler tricycle
r = 7.0 in. (17.6 cm)	h = 6.0 in. (15.2 cm)	r = 4.0 in. (10.2 cm)
w = 4.0 in. (10.2 cm)	w = 3.5 in. (8.9 cm)	w = 2.5 in. (6.4 cm)
d = 3.0 in. (7.6 cm)	d = 2.5 in. (6.4 cm)	d = 2.0 in. (5.0 cm)

FIGURE 30.70 Bicycle pedal design recommendations.

(see Fig. 30.71). However, avoid pedals that are too small, since they are difficult to locate without looking for them. The best angle for the mechanically operated type is about 10° or less.

If it appears advantageous to operate a foot switch alternately with both feet because of some other requirement to change operator position, either a bar-type foot or a knee switch configuration should be considered. This type of alternate foot or leg access is particularly important for applications where a safety cutoff feature is necessary. The significant advantages are that an operator does not have to look for the switch and does not have to be very accurate. The figure notes the important dimensional considerations. The pedal-bar height should not exceed about 6.0 in. (15 cm), and it should not extend into the normal footprint zone.

Knee-Operated Switch (Seated Operator)

A knee-operated control is useful to tree an operator's hands for other tasks. The primary objective is to position the knee bar so that operators of various sizes can use it easily and also operate it regardless of variations in their seating positions. The knee bar should be made large enough to account for these variations. Figure 30.72 shows recommended distances forward of the nominal seat reference (seat-pan–backrest juncture) and above the floor for the center of the knee bar. Usually the knee bar should be about 6.0 in. (15.2 cm) tall by 8.0 in. (20.3 cm) long in order to accommodate both small female and large male operators.

Note: All corners facing the operator should be radiused, and the leading edge should curve outward slightly to minimize the possibility that operators will injure their knees as they slide into place (see the figure).

PUSH BUTTONS

A variety of push-button sizes and shapes are acceptable as long as certain conditions are maintained. The smallest diameter should not be less than about ¼ in. (0.64 cm) for bare-hand operation or ½ in. (1.3 cm) when gloves are worn. (See Fig. 30.73.) Normally, the button should extend far enough above a panel so that when the button is in the depressed position, a reasonable amount of it is still exposed (otherwise, the operator may not be able to press the switch far enough to make contact).

When buttons are approximately ¾ to 1 in. (1.9–2.5 cm) in diameter, it sometimes helps to provide a concave top to the button (finger centering). An additional "location aid" may be provided in the form of a tapered, recessing bezel (such as is found in the typical doorbell). Illuminating the push button is useful when the button needs to be located in the dark.

The push button is an effective emergency control device for shutting off machines in a hurry. In such applications, the button should be larger so that the operator does not have to be so accurate, i.e., so that the button can be operated either with the end of a finger or with the heel of the hand. As the button is made larger, consider a concave surface.

When it becomes important to preclude inadvertent push-button actuation, recess the button so that the button is actually below the adjacent panel surface plane. (See Fig. 30.74.) The "well" for such configurations needs to be at least 1 in. (2.5 cm) in diameter for bare-hand operation and at least 2 in. (5.1 cm) for gloved-hand use. Tapering the sides of the well helps the operator find the button without having to be so accurate.

Grouping Push Buttons

As indicated by these guidelines, adequate button separation is required to avoid inadvertent operation of adjacent switches. (See Fig. 30.75.)

Push-Button Shape

From a biomechanical point of view, it makes no difference whether one selects a round, square, or rectangular push button. However, the square or rectangular button provides

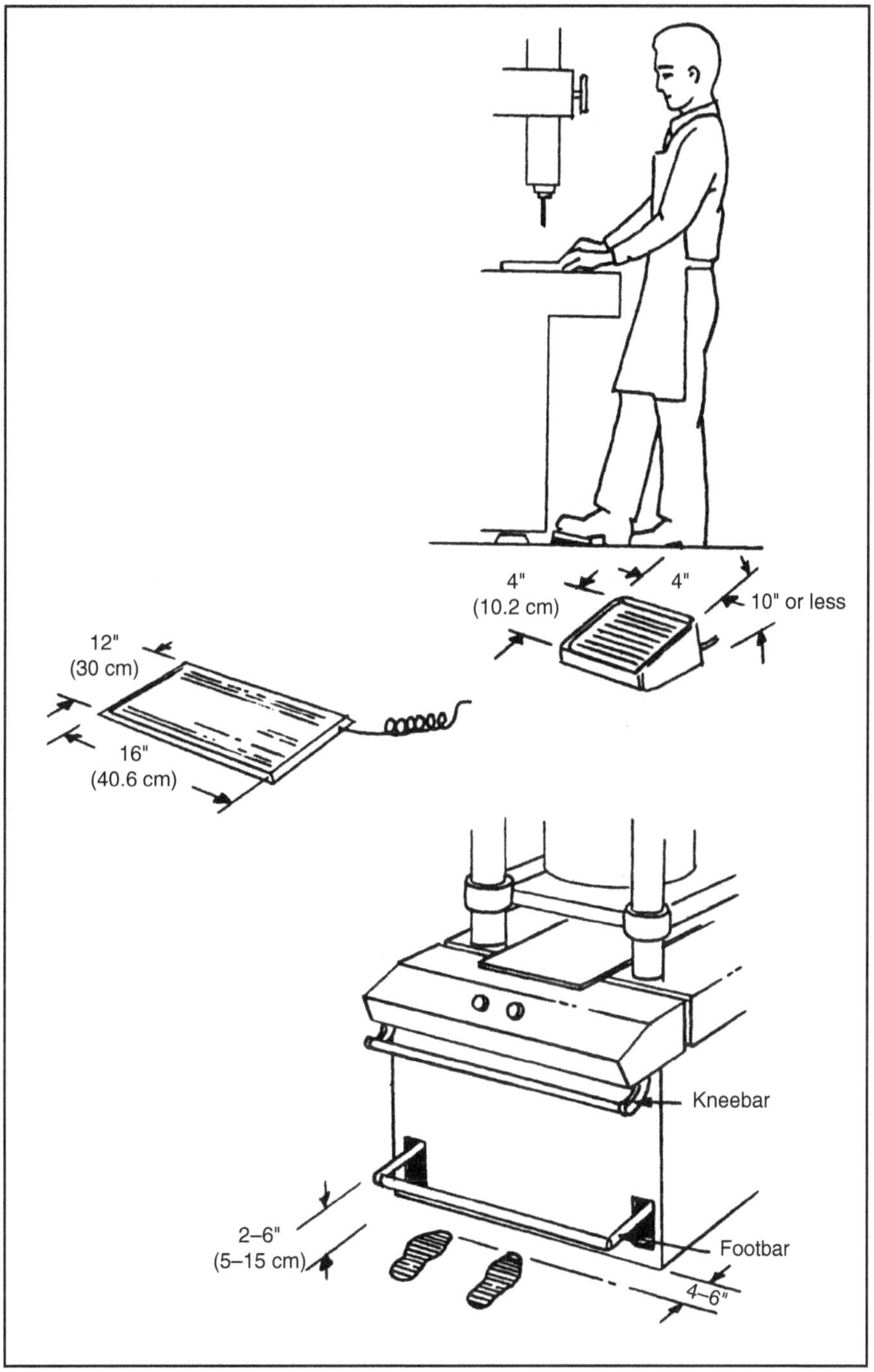

Figure 30.71 Foot and leg switch design recommendations.

more useful area for labeling, and obviously a rectangular button laid on its side provides more label space than one that is standing on end.

Shape can also be used as a cue for determining the functional significance of a particular push button. (See Fig. 30.76.) This is especially important when advisory indicators are used in combination with push buttons. A round push button is seldom mistaken for an advisory indicator.

Push buttons are useful as safety controls for hand tools, but they must be located within reach of the operator's thumb or finger. Push buttons should be at least 0.25 in. (0.64 cm) in diameter and should extend about 0.125 in. (0.318 cm). (See Fig. 30.77.)

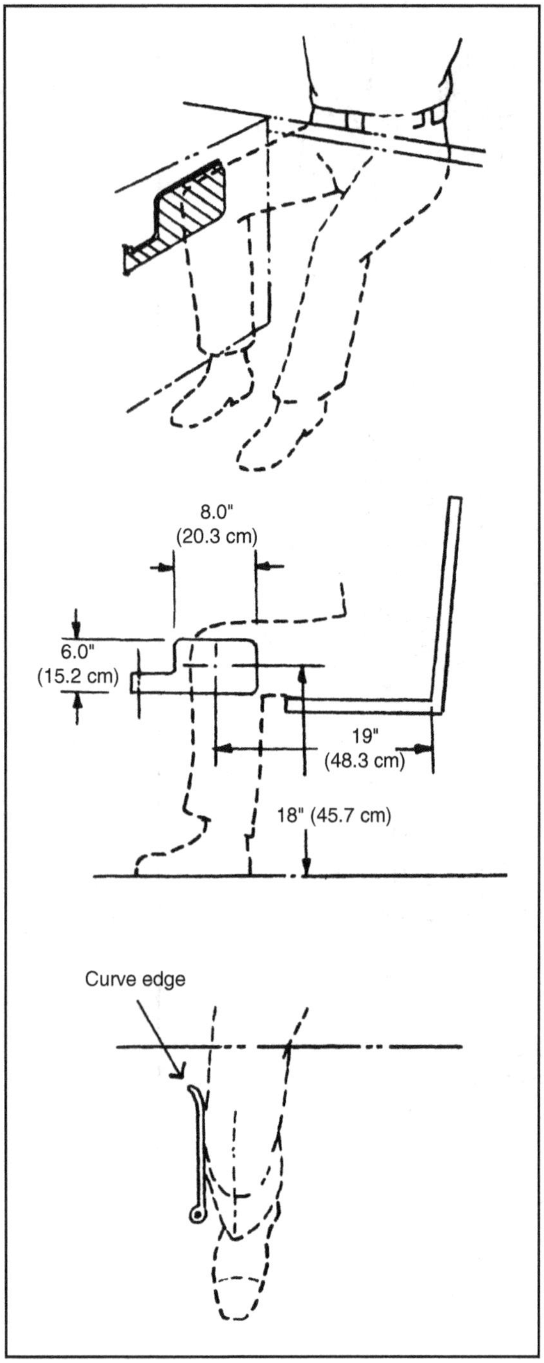

FIGURE 30.72 Knee switch design recommendations.

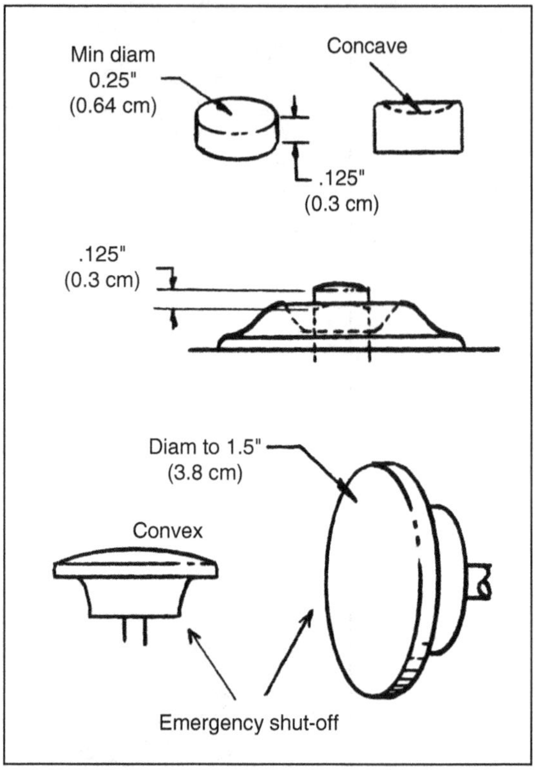

FIGURE 30.73 Recommended push button dimensions.

A push button located in the end of a handle should be visible from the operator's normal eye reference position. (See Fig. 30.78.)

Foot-operated push buttons are often useful to alleviate the workload placed on an operator's hands. (See Fig. 30.79.) More than one button per foot is not recommended, however. Make sure that there is space between any adjacent structure and the button so that there is plenty of room for the foot. However, too much space may not be desirable since an operator may use the adjacent structure as a guide in finding the button.

Push-Button Forces

Push-button forces should be within the range of 10 to 20 oz. (283 to 567 g) (although they can be as high as 80 oz., 1134 g)

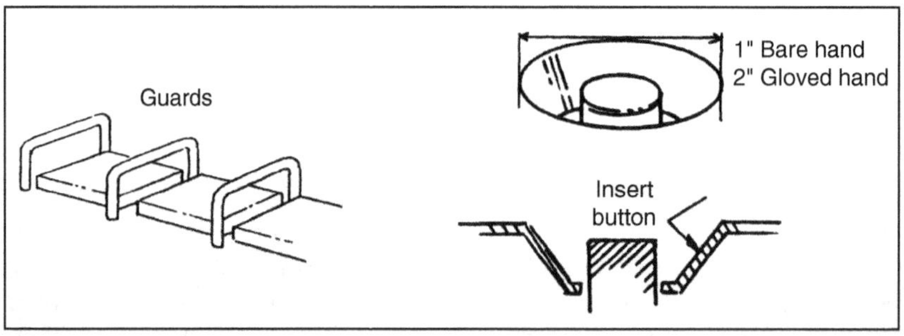

FIGURE 30.74 Ways to prevent inadvertent activation.

Arrangement	A	B	C	D	E	F
Vertical plane:						
No gloves	0.75 in (1.9 cm)	0.625 in (1.6 cm)	0.1875 in (0.476 cm)	0.25 in (0.64 cm)	1.25 in (3.17 cm)	0.75 in (1.9 cm)
With gloves	1.75 in (9.45 cm)	1.625 in (4.12 cm)	0.375 in (0.95 cm)	1.25 in (3.17 cm)	2.25 in (5.72 cm)	1.75 in (4.45 cm)
Horizontal plane:						
No gloves	1.0 in (2.54 cm)	0.875 in (2.22 cm)	0.4375 in (1.11 cm)	0.50 in (1.27 cm)	1.50 in (3.81 cm)	1.0 in (2.54 cm)
With gloves	2.0 in (5.1 cm)	1.875 in (4.76 cm)	0.375 in (0.95 cm)	1.25 in (3.17 cm)	2.25 in (5.72 cm)	2.0 in (5.1 cm)
Under severe vibration or oscillation	3.0 in (7.6 cm)			3.0 in (7.6 cm)		3.0 in (7.6 cm)
For blind selection	6.0 in (15 cm) apart in front of operator; 12 in (31 cm) apart when buttons are located in the peripheral areas.					

Note: The above guidelines apply to any shape of button.

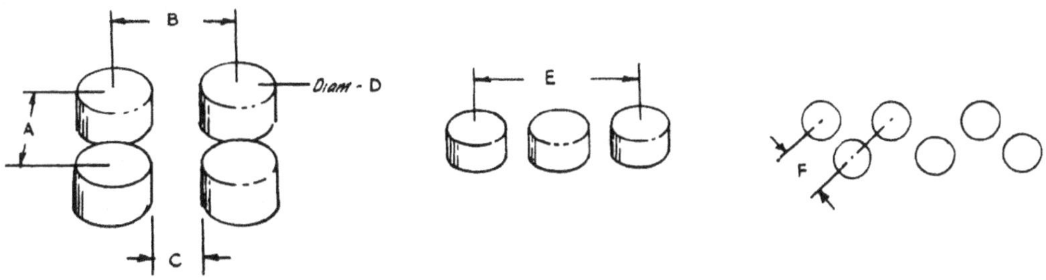

Figure 30.75 Minimum button separation.

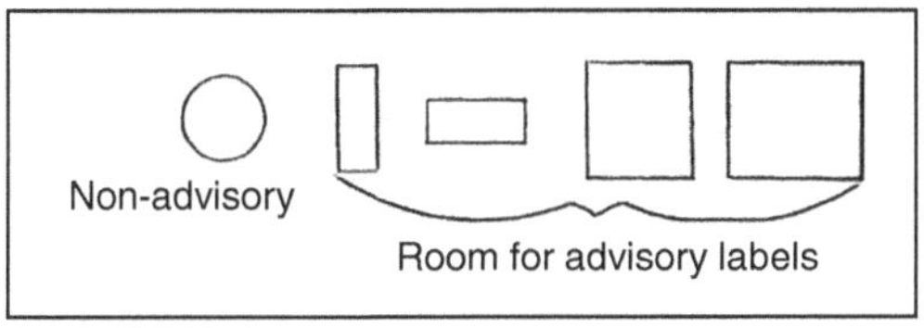

Figure 30.76 Button shape indicates function.

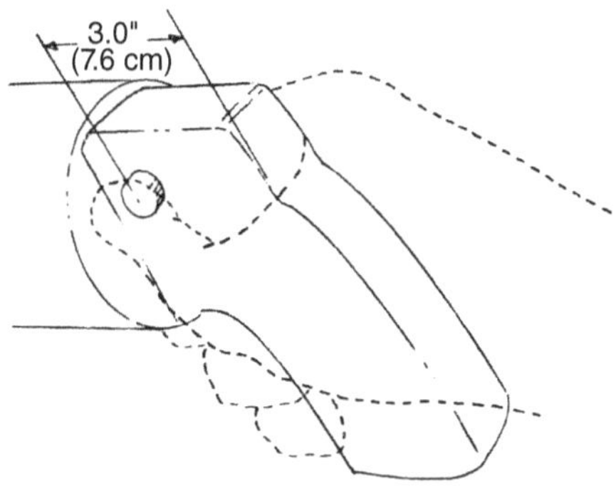

Figure 30.77 Push button for hand tool safety control.

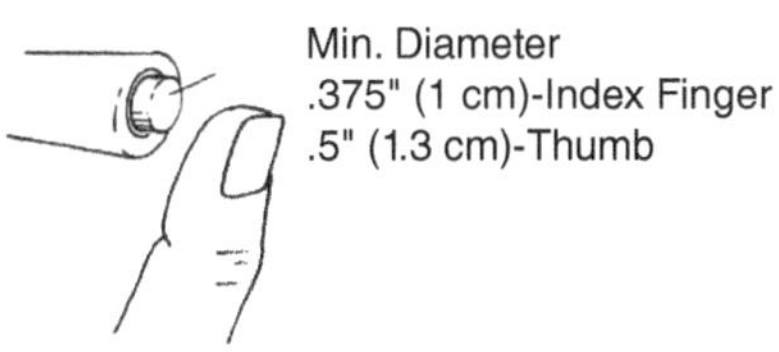

Figure 30.78 Push button at end of handle.

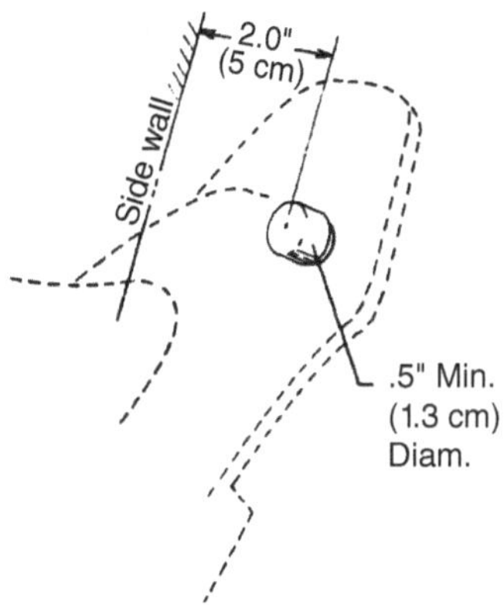

Figure 30.79 Foot-operated push button.

in order to reduce the possibility of inadvertent actuation. (See Table 30.11.)

TABLE 30.11 Recommended Push Button Forces

Resistance		
Single Finger	**Different Fingers**	**Thumb or Palm**
2.8 N (10 oz.)	1.4 N (5 oz.)	2.8 N (10 oz.)
11 N (40 oz.)	5.6 N (20 oz.)	23 N (80 oz.)

From MIL-STD-1472D.

Push-button action should be positive, i.e., there should be elastic resistance aided by slight sliding friction, starting slowly, building rapidly, and with a final sudden drop to indicate activation. An audible snap action is helpful under high-noise-level conditions, although the snap action must not be too heavy if the button is to be operated frequently over extended periods of time.

TOGGLE SWITCHES

Figure 30.80 illustrates typical handle shapes that are generally available.

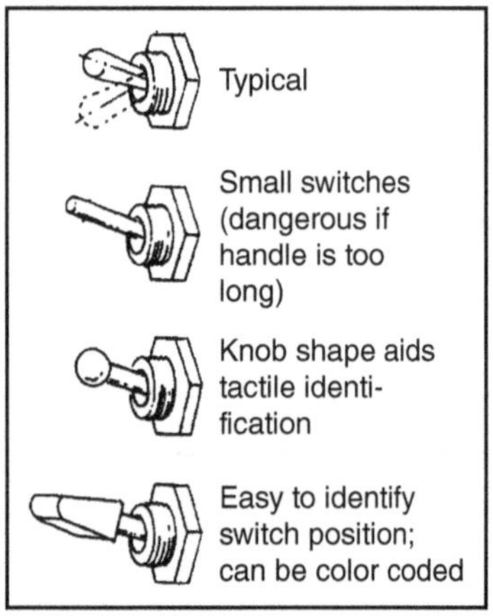

Figure 30.80 Toggle switch handle shapes.

Although three-position toggle-switch hardware is available, it is not recommended. It is not possible to maintain the 30° "throw" when more than two positions are provided. It is suggested that, when three or more switch positions are required, some other type of switch be selected. Figure 30.81 shows recommended dimensions for toggle switches.

ROCKER SWITCHES

Rocker switches are a good substitute for the more common toggle switch to provide good physical indication of switch position. Like toggle switches, rocker switches, to be effective with regard to visual identification of position, should be designed so that the slope of the handle from the nominal plane is about 30°, and the nominal plane should be at least 0.125 in. (0.32 cm) above the plane of the panel on which the switch is mounted.

Rocker switches may be broad and thus provide space for labeling, or they may be made narrower, as illustrated in Fig. 30.82. In either case, however, the switches should be mounted so that the switch centerlines are no closer than 0.75 in. (1.9 cm) when the switches are arranged side by side.

Switch resistance should be between about 8 and 12 oz. (227 and 340 g) so that the switch is easy to operate and yet is not easily disturbed inadvertently.

Rocker switches typically can be obtained in different colors and thus can be color coded for ease in differentiation. (See Fig. 30.83.) In fact, the two faces of a single switch usually can be obtained in two different colors, which makes it even easier to tell which switches are activated and which are not.

Rocker switches (like toggle switches) should be designed to snap into position with an audible click.

The rocker switch is preferred over the toggle switch when switches are laid out on a horizontal panel because the toggle switch tends to protrude in this arrangement and is easy to become snagged on if an operator has to reach across it.

ROUND CONTROL KNOBS

Round knobs should be used only for applications in which the rotation is continuous and smooth and/or when multirevolutions are required.

The smallest useful knob should have minimum dimensions of 0.25 in. (0.6 cm) in diameter × 0.75 in. (1.9 cm) in depth, and the knob should be either serrated or knurled. (See Fig. 30.84.) A maximum resistance of 2 to 4 oz. (57 to 113 g) should be the most force expected by an operator. Such configurations should be used only sparingly, such as for a clock setting or for gross adjustment of certain noncritical displays. The preferred minimum size of the knob for convenient operation of volume or illumination potentiometer controls is 0.50 in. × 0.50 in. (1.3 cm × 1.3 cm), with appropriate serrations to ensure comfortable and effective gripping. (See Fig. 30.85.) The upper limit to ensure ability to make fine adjustments is approximately 4 oz. (113 g) of resistance. The edge of the knob can be tapered slightly, although this adds nothing to the effectiveness of the knob. Avoid severe tapering, however, for the operator's fingers may tend to slip from the knob.

For most normal panel layouts, select a knob whose dimensions are approximately as shown in Fig. 30.86. These are easily accommodated even by an operator who is wearing gloves. Resistance can vary from 4 to 10 oz. (113 to 283 g) as the diameter of the knob is increased.

Be particularly careful in accepting some of the decorative knobs produced by some manufacturers, for they sometimes possess features that make the knob difficult to handle. An acceptable alternative to the previous shape is shown in Fig. 30.87.

Note: Choose serration patterns that provide good gripping characteristics, i.e., sharper peaks or points rather than "humps."

Acceptable Alternate Shape

Control knob or switch or potentiometer operating characteristics should include viscous damping as opposed to static

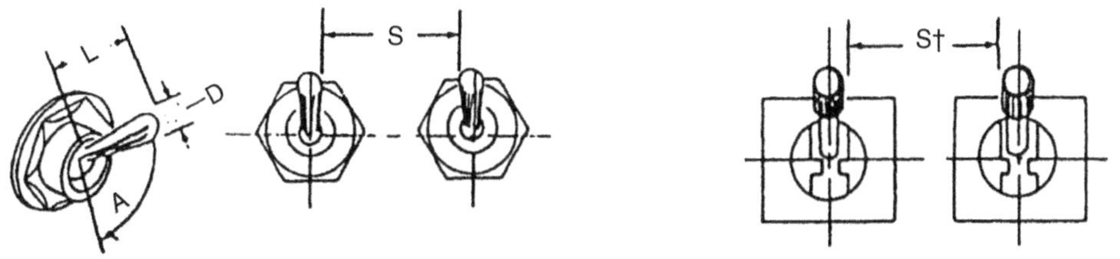

	Dimensions: L Arm length *	Dimensions: L Arm length **	Dimensions: D Control tip	Resistance: Small switch	Resistance: Large switch
Minimum	13 mm (1/2 in.)	38 mm (1-1/2 in.)	3 mm (1/8 in.)	2.8 N (10 oz.)	2.8 N (10 oz.)
Maximum	50 mm (2 in.)	50 mm (2 in.)	25 mm (1 in.)	4.5 N (16 oz.)	11 N (40 oz.)

	Displacement between positions: A, 2 Position	Displacement between positions: A, 3 Position
Minimum	525 mrad (30°)	295 mrad (17°)
Maximum	1400 mrad (80°)	700 mrad (40°)
Desired	....	435 mrad (25°)

	Separation S: Single finger operation	Separation S: Single finger operation †	Separation S: Single finger sequential operation	Separation S: Simultaneous operation by different fingers
Minimum	19 mm (3/4 in.)	25 mm (1 in.)	13 mm (1/2 in.)	16 mm (5/8 in.)
Optimum	50 mm (2 in.)	50 mm (2 in.)	25 mm (1 in.)	19 mm (3/4 in.)

* Use by bare finger ** Use with heavy handwear † Using a lever lock toggle switch

Figure 30.81 Toggle switches. (From MIL-STD-1472D.)

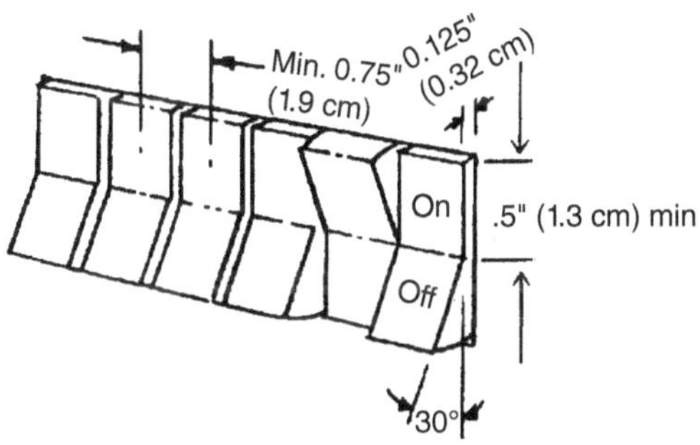

Figure 30.82 Recommended rocker switch dimensions.

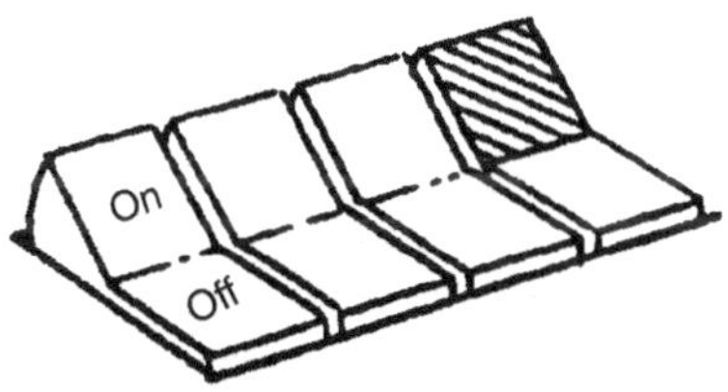

Figure 30.83 Use contrasting colors for "on" and "off" illuminate "on."

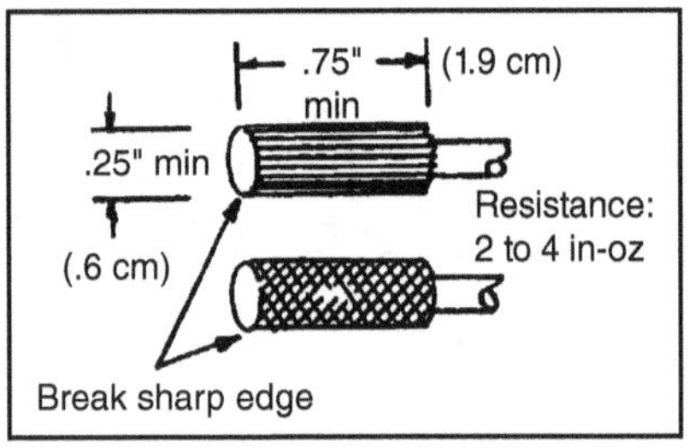

FIGURE 30.84 Smallest useful knob.

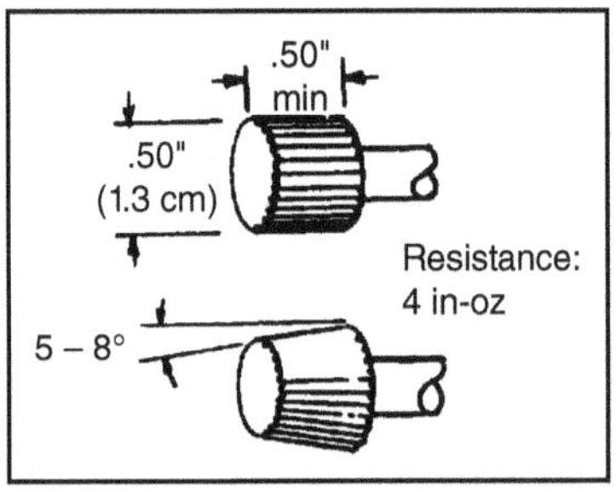

FIGURE 30.85 Preferred minimum size.

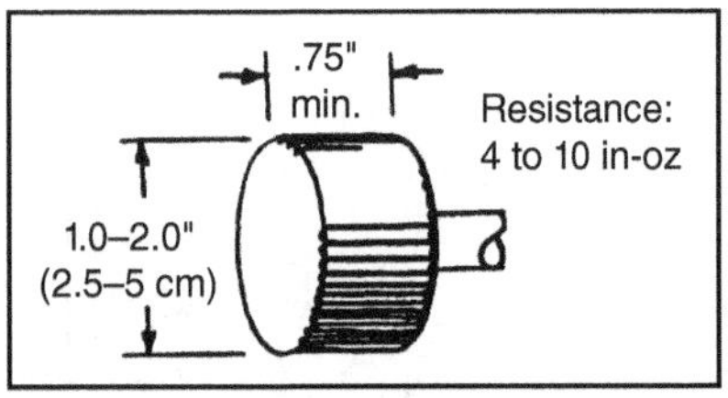

FIGURE 30.86 Recommended for common application.

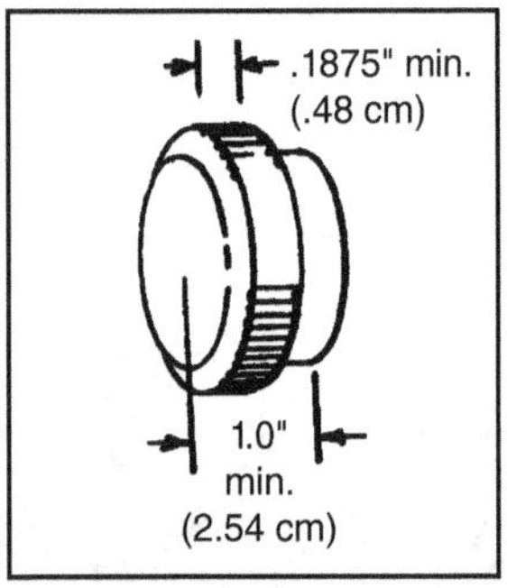

FIGURE 30.87 Acceptable alternate shape.

resistance. A certain amount of resistance is desirable (approximately 2.0 in./oz. in order to minimize the possibility of inadvertent knob movement).

1. Generally, more precise knob movement is possible with larger knob diameters (1.5–3.0 in., 3.8–7.6 cm) because more fingers can be brought in contact with the knob surface. The knob torque should not exceed about 4 in./oz. (See Fig. 30.88.)

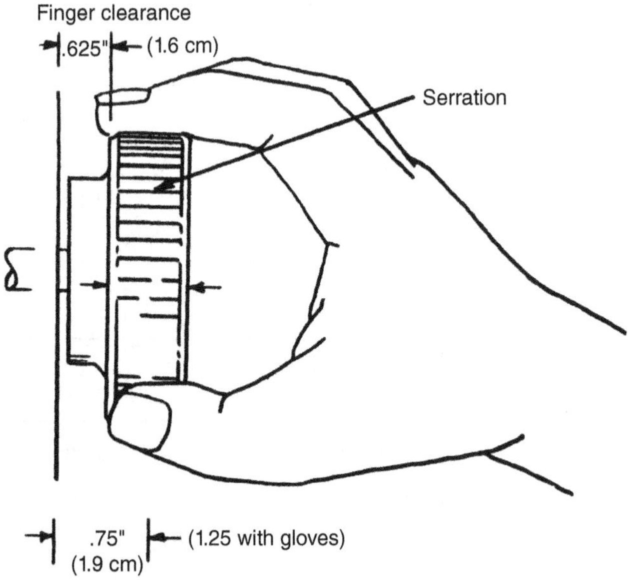

FIGURE 30.88 Precise movement control.

2. For high-torque applications, select knob diameters between 2 and 3 in. (5.1 and 7.6 cm) or 3.25 in. (8.2 cm) maximum. (See Fig. 30.89.) Maximum torque should not exceed 15 ft/lb. Instead of serrations, use knurling on the edges of high-torque knobs for maximum gripping. The edges should be slightly beveled so as not to injure the operator's hand.

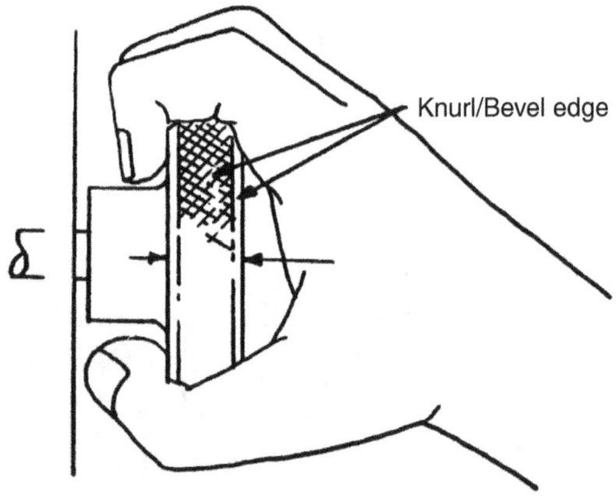

FIGURE 30.89 High-torque control.

Stacked (or Ganged) Knobs

Up to three knobs may be ganged on the same shaft center to save space. This should not be done, however, unless space is really at a premium. When this concept seems necessary, consider the following in order to effect the best operation and the least operator error (see Fig. 30.90):

1. The largest knob should be used for operation 1, as shown in the figure, so that the operator works outward and thus is less apt to disturb a previous setting.

2. Gross tuning should be assigned to the smaller knob, and fine tuning to the larger one.

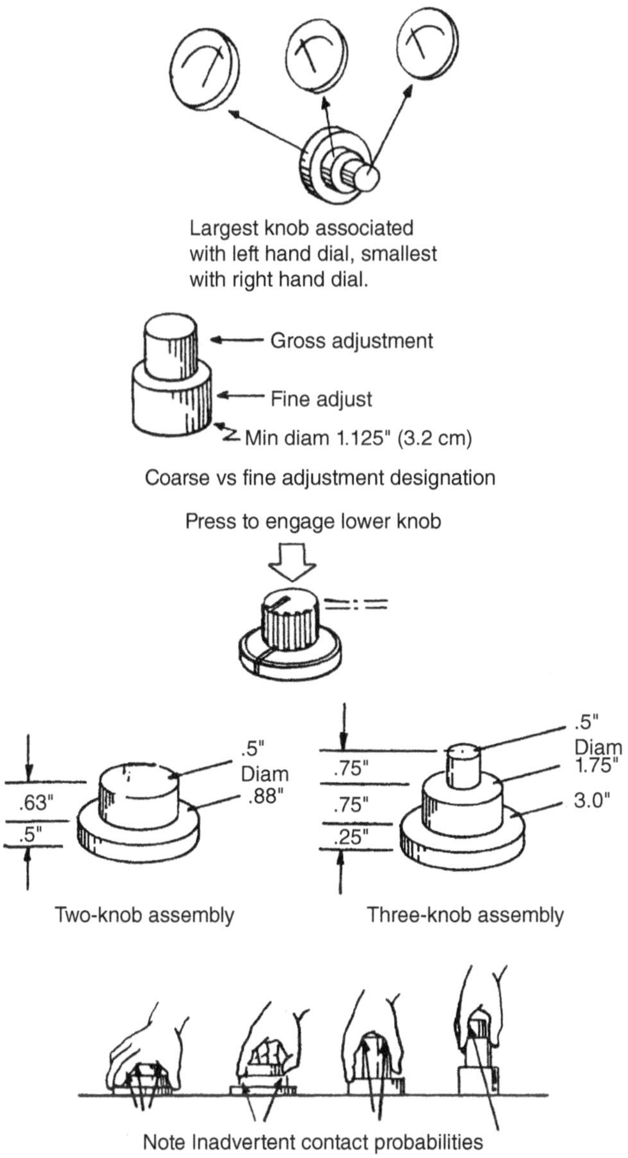

FIGURE 30.90 Examples 1 to 5.

3. If the two knobs work sometimes together and sometimes independently, design them so that the operator depresses the top knob to engage the bottom one, with spring loading to return the upper knob to its independent status.

4. The "ideal dimensioning" for ganged controls has been determined experimentally, as illustrated.

5. The figure shows unintentional contact conditions to look out for.

CRANKS

Design recommendations for cranks are in Fig. 30.91.

THUMB WHEELS

Although operable with either the thumb or the forefinger, the control configuration illustrated in Figs. 30.92–30.94 is commonly referred to as a "thumb wheel." From a human engineering point of view, thumb wheels are seldom considered the most desirable type of control. In cases where they are used, they are applicable only to smooth, continuous functions (audio volume, light intensity, etc.) where nonprecise adjustment is required.

The principal guidelines for selecting or designing these devices are illustrated in the figures:

1. Resistance should not be more than about 6 oz. or less than 2 oz. (170 to 579 g).

2. Acceptable multiple arrangements and spacing minimums are shown in Fig. 30.93. Note that horizontally arrayed wheels should not be arranged one above the other or one in front of another.

3. Thumb wheels should not be used if the operator has to wear gloves.

4. Wheel motions upward or to the right should provide "increasing function."

POINTER KNOBS

In contrast to the smooth, round knob recommended for continuous, multiturn controls, a "pointer-shaped" knob should be used for controls when the operator may wish to switch from one discrete position to another. The switch should have mechanical detents so that the operator can "feel" and perhaps hear when the knob is properly seated or positioned.

A prime objective of the pointer knob is to provide (1) a sufficiently discriminable pointing shape so that the operator knows which end of the knob is the pointing end and

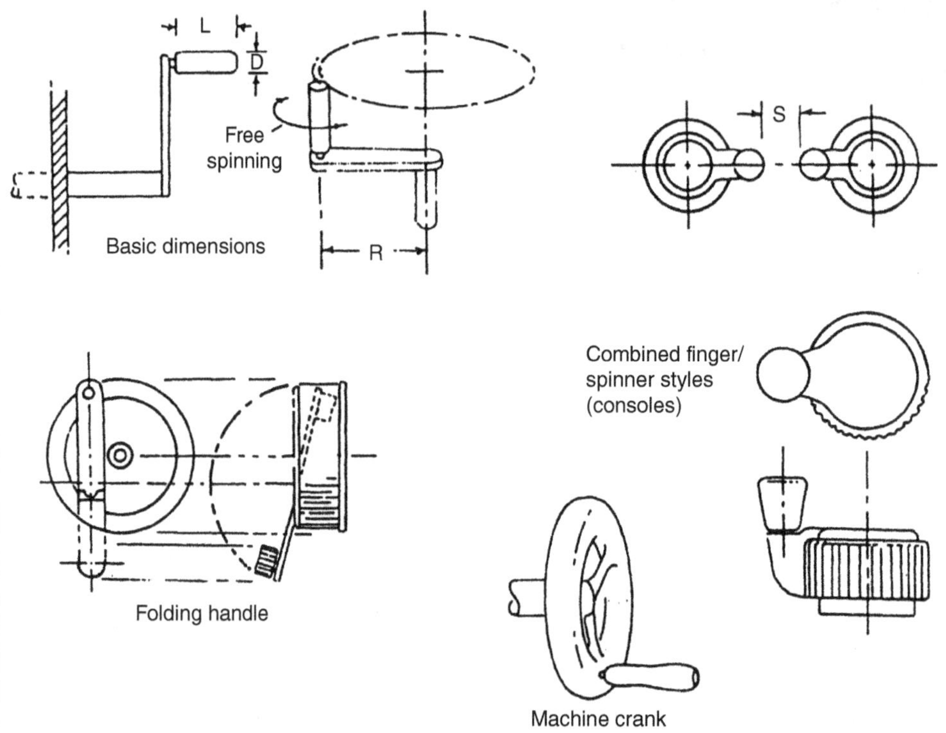

Load	Specification	Handle: L, Length (mm)	Handle: L, Length (in.)	Handle: D, Diameter (mm)	Handle: D, Diameter (in.)	R, Turning Radius: Rate below 100 RPM (mm)	R, Turning Radius: Rate below 100 RPM (in.)	R, Turning Radius: Rate above 100 RPM (mm)	R, Turning Radius: Rate above 100 RPM (in.)
Light loads: Less than 22 N (5 1b). (Wrist and finger movement)	Minimum	25	1	10	3/8	38	1½	13	1/2
	Preferred	38	1½	13	1/2	75	3	65	2½
	Maximum	75	3	16	5/8	125	5	115	4½
Heavy loads: More than 22 N (5 1b). (Arm movement)	Minimum	75	3	25	1	190	7½	125	5
	Preferred	95	3-3/4	25	1	–	–	–	–
	Maximum	–	–	38	1½	510	20	230	9

S, Separation between adjacent controls: 75 mm (3") minimum.

Figure 30.91 Crank design recommendations. (From MIL-STD-1472D.)

(2) adequate depth so that the operator can get a firm enough grip to turn the knob.

Various knob shapes are generally acceptable as long as certain details are not compromised. For example, Fig. 30.95 shows typical "bar-shaped" controls that provide adequate pointing characteristics. The d dimension should always be at least 0.50 in. (1.3 cm); the 1 dimension should be at least 0.50 in. (1.3 cm); and the w dimension should not be less than 0.25 in. (0.6 cm). The third knob shape is generally a better configuration than the first two because of the sloping pointer end, which makes it easier for the operator's eye to follow the pointer directly to the point at which it becomes adjacent to an index mark on the panel. The fourth pointer shape provides additional length to assist the operator when more force is required to position the knob.

The minimum torque requirement for pointer knobs should be about 10 in./oz. (283 g); the maximum should be around 45 in./oz. (1275 g). The switches should have elastic resistance which builds up and then decreases as the switch "seats." When panel space dictates use of a very small switch knob, slight variations are possible, as shown in the figure. The left-hand knob is not especially desirable, but it can be used under extreme conditions. The others are preferred because they are more comfortable to use.

When it is known that a pointer knob will be used at night or under darkened ambient lighting conditions, the knob

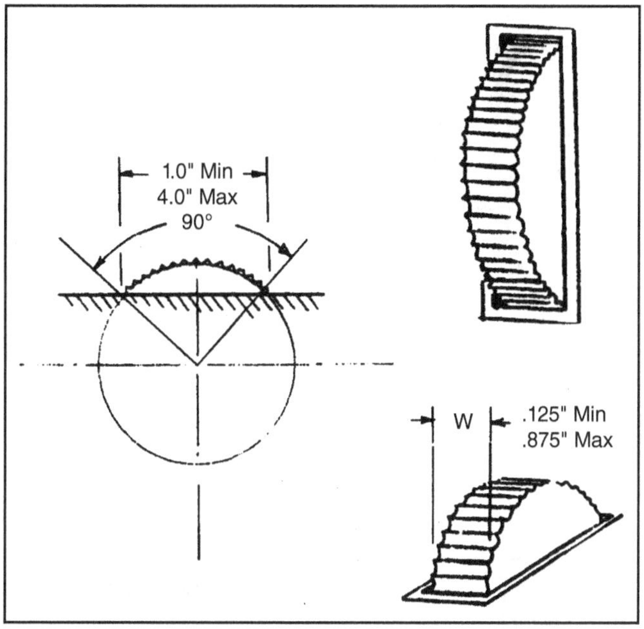

Figure 30.92 Design recommendations for continuous adjust thumb wheel.

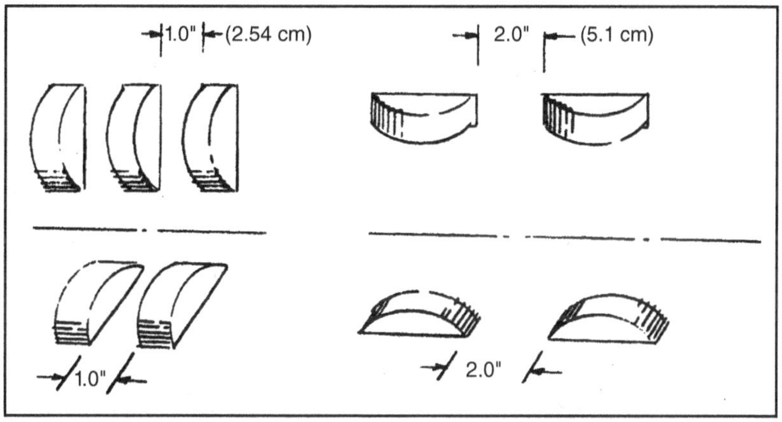

Figure 30.93 Acceptable multiple thumb wheel arrangements.

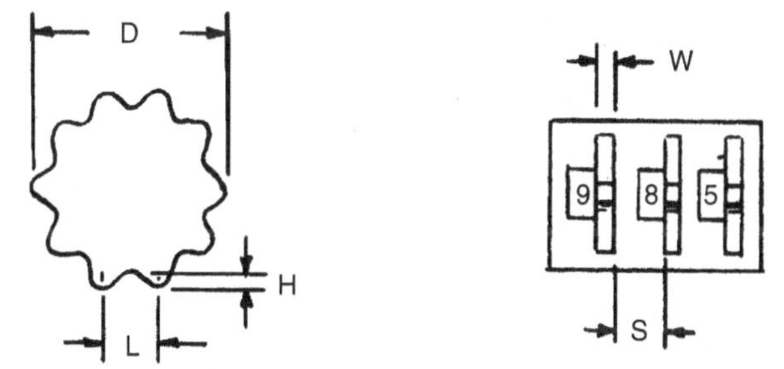

	D (Diameter)	L (Trough Distance)	W (Width)	D (Depth)	S (Separation)	Resistance
Minimum	1.5 in (3.8 cm)	0.45 in (1.14 cm)	0.1 in (2.54 cm)	1/8 in (0.3 cm)	0.4 in (1.0 cm)	1 in-lb (0.12 m · N)
Maximum	0.5 in (6.4 cm)			1/2 in (1.3 cm)		3 in-lb (0.34 m · N)

Figure 30.94 Design recommendations for discrete positioning thumb wheels.

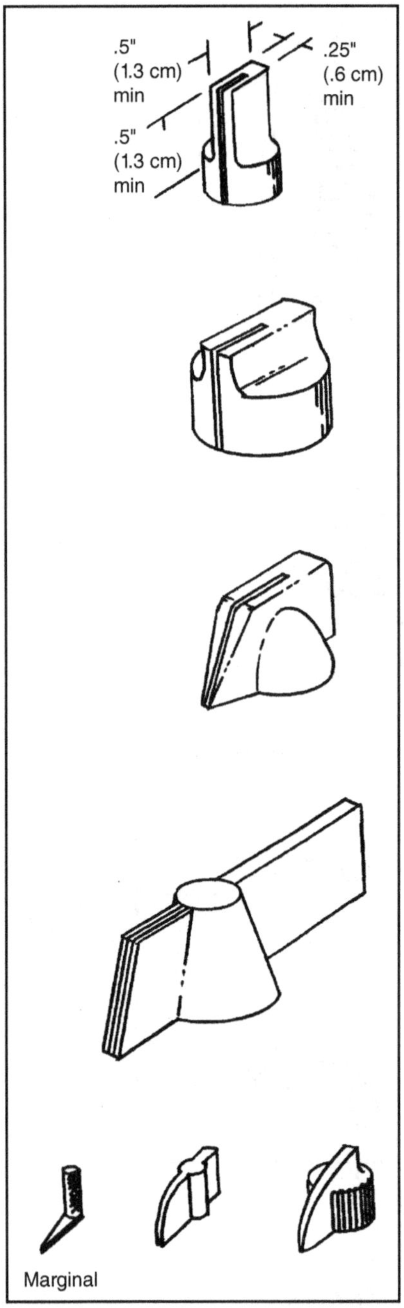

FIGURE 30.95 Acceptable pointer knob shapes.

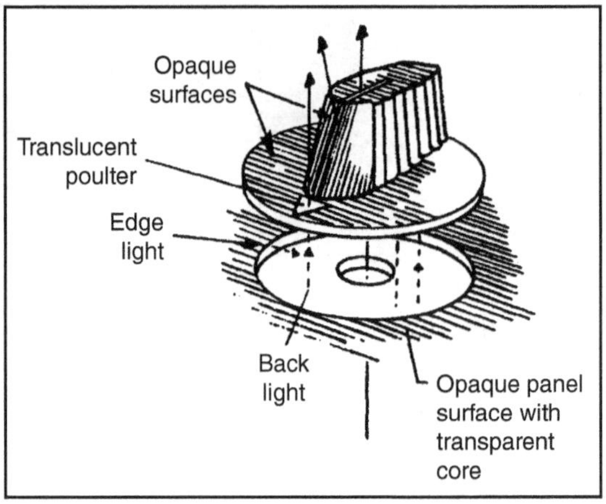

FIGURE 30.96 Lighted pointer knob.

pointer should be illuminated. An effective method is to use a transparent plastic knob, cover it with a coat of translumi-nescent white paint or material, and then place an opaque coating over that. The pointer should then be inscribed through the outer opaque coating so that the white inner coating shows through and so that either edge lighting or back lighting from the control panel can shine through the pointer areas (see Fig. 30.96).

Idealized Pointer Knob

The idealized pointer-knob shape shown in Fig. 30.97 illustrates several desirable features to look for when selecting a knob:

1. A 45° pointer slope ensures a better view of the pointer tip from most viewing positions.

2. The long, tapering handle provides good leverage for rotating the knob, without having to cover the pointer tip with the fingers.

3. All sharp corners are eliminated.

4. The skirt provides a finger rest so that the fingers do not scrub the panel.

5. Not only is the configuration attractive, but also it will never be confused with round knobs, nor will the operator be in doubt as to which is the pointing end of the knob, whether he or she is looking at the knob or not.

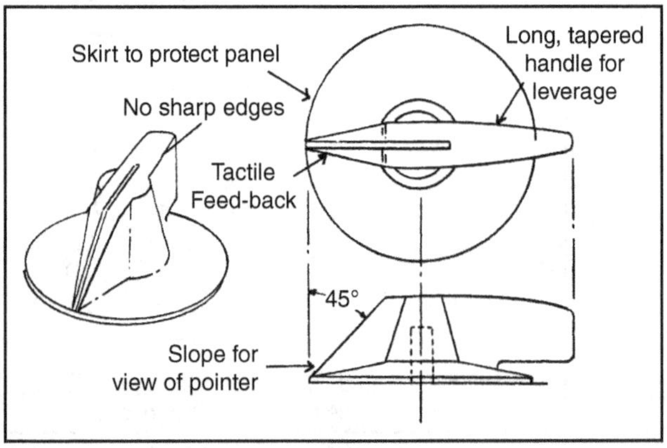

FIGURE 30.97 Ideal pointer knob shape.

Recommended Pointer-Knob Dimensions (See Fig. 30.98)

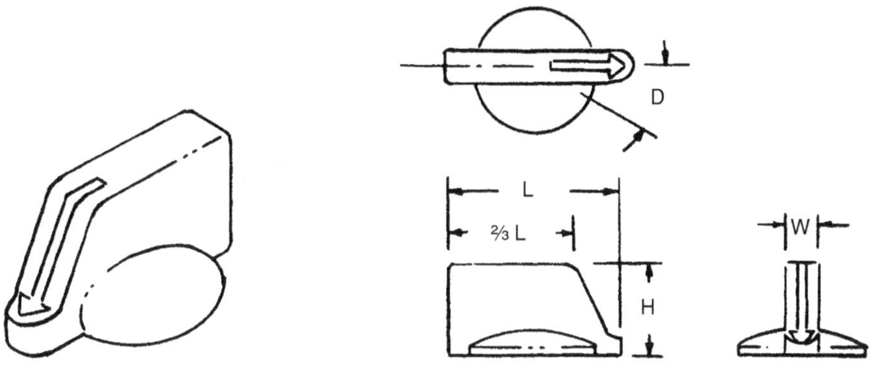

	Pointer, in			D (Displacement in degrees)		
Operating method	L (Length)	W (Width)	H (Depth)	Visual positioning	Nonvisual positioning	Resistance, in-lb*
Minimum	1.00	0.25	.625	15	30	1
Maximum	4.00	1.00	3.00	40	40	6
				(90 when large separations are necessary)		

* Use elastic resistance which builds up and then decrease as each detent is approached so that the control will snap into position and cannot easily stop between adjacent positions.

Figure 30.98 Recommended pointer knob dimensions. (From MIL-STD-1472D.)

OTHER FINGER-OPERATED CONTROLS

Although a variety of control shapes are suitable for operation by the finger, it is important to select controls that are large enough so that the operator can manipulate them conveniently and with a minimum of discomfort. The guidelines shown in Figs. 30.99 and 30.100 provide useful information for selecting appropriate control devices.

1. Finger-operated levers should have resistances of 30 oz. (850 g) or less. The action should include definite positional detents so that the operator knows the control is properly positioned.

2. The handle-length minimum for resistances of 30 oz. (850 g) or less should be about 1.0 in. (2.5 cm). When the resistance exceeds 30 oz. (850 g), the handle length must be increased. The maximum resistance should not exceed about 3 lb (1.4 kg), in which case the handle length needs to be increased to at least 1.5 in. (3.8 cm).

3. The slide switch resistance should not exceed about 10 oz. (283 g).

4. A single "push-push" type of wall switch is preferred over the more typical two-button configuration because it removes the necessity for the user to decide which position the switch is in. The larger switch plate makes it possible to operate the switch with an arm or elbow when the hands are occupied.

Lever length	Maximum resistance (ounces, grams)
under 1"	30 oz (850 g)
1.5"	48 oz (140 g)

Figure 30.99 Resistance for finger-operated controls.

Fingertip-Operated Slide Controls

Although these controls can be used either for continuous-motion or for discrete, detented switching functions, they are not generally recommended for either one. However, they can be used for fairly coarse operating functions when operating error may be of little consequence and/or when the time taken by an operator to get the control positioned does not introduce other response delays. The basic problem with using these controls is that, in order to operate them precisely, the operator must steady his or her hand against the panel,

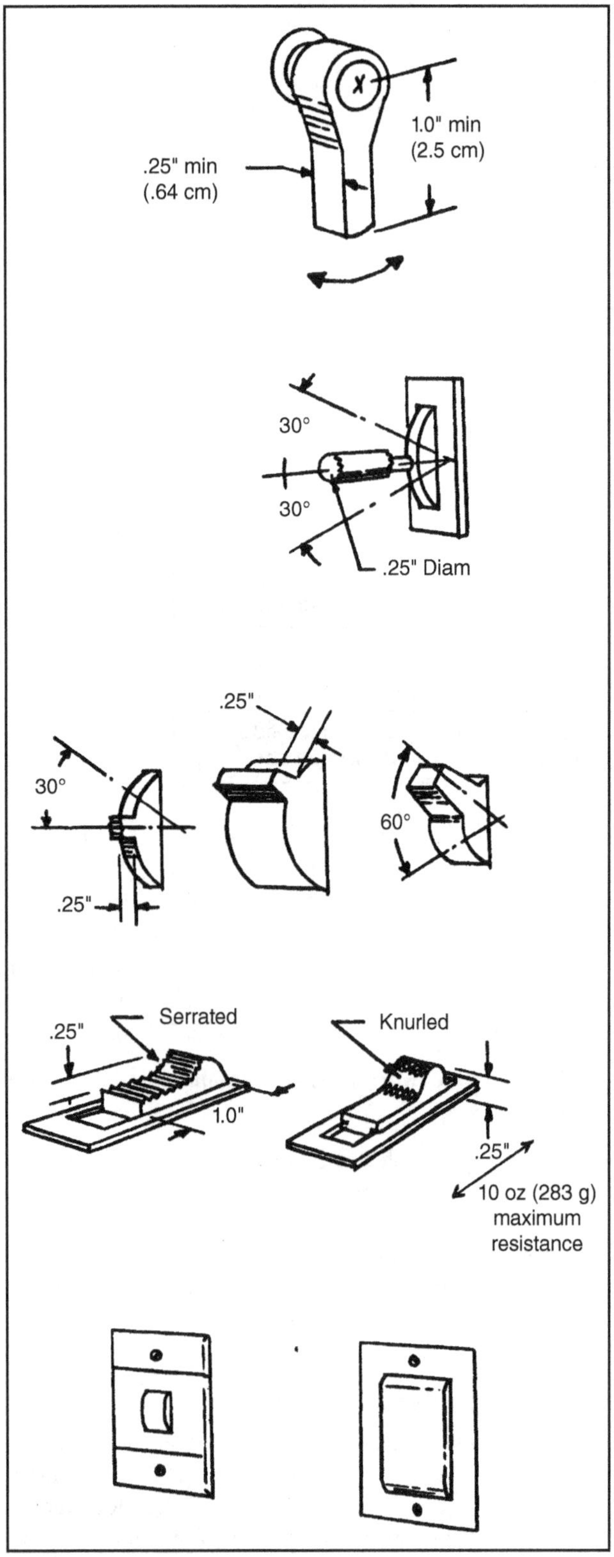

FIGURE 30.100 Other finger-operated control shapes.

using a long, total-arm movement from the shoulder, which is exceedingly difficult and very inaccurate. Nevertheless, if such controls are selected, consider the following guidelines (see Fig. 30.101):

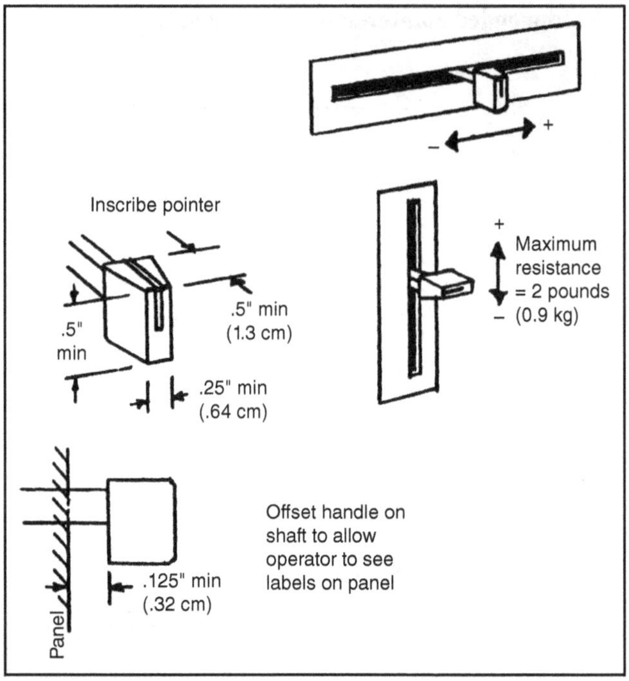

FIGURE 30.101 Slide control design recommendations.

1. The control knob or tab should be large enough to get hold of, e.g., minimums for a = 0.50 in. (1.3 cm); b = 0.25 in. (0.64 cm); and c = 0.50 in. (1.3 cm).

2. The handle should be offset on its shaft in the direction that will allow the operator to read whatever labeling is provided on the panel; i.e., the labeling will normally be placed above the horizontal slide or to the left or right of the vertical slide, depending on which hand is used to operate the slide.

3. The slide resistance should not exceed about 2 lb (0.9 kg).

4. The handle should be oriented as shown in the figure, and there should normally be a pointer inscribed in the handle edges.

5. When multiple slide arrangements are made, sufficient separation must be maintained so that labels can be seen and easily associated with the slide they apply to. (See Fig. 30.102.) (This is one of the most common errors observed in the design of slide assemblies. It occurs because of the desire to save space.)

Vehicular Column-Mounted Controls

Variously referred to as "stalks," "wands," and "fingertip controls," column-mounted controls are used primarily to bring certain critical control functions within reach of the driver's fingertips as his or her hands hold the steering wheel. In recent years, more and more control functions have been located here, and the potential for operator

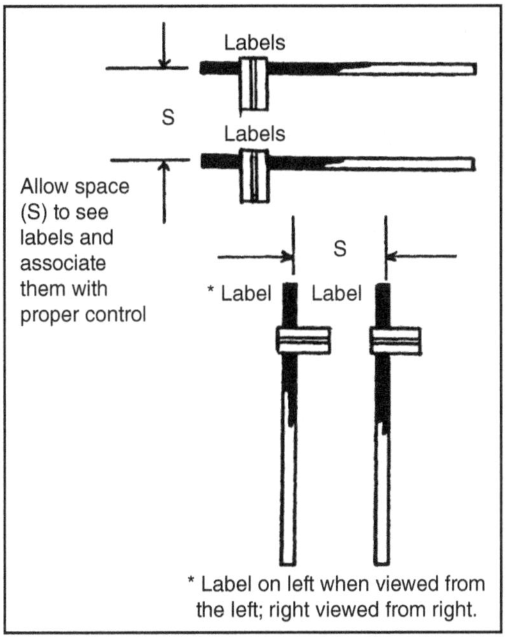

FIGURE 30.102 Recommendations for placement of slide control labels.

error has increased. Consider the following guidelines (see Fig. 30.103):

1. Do not employ more than one lever on each side of the column.

2. Do not employ more than two lever motions per lever, plus use of an "end-mounted" push button, except when it can be demonstrated that another motion (i.e., rotation of the lever handle) can be accomplished without inadvertent input of one of the other lever motions.

3. Mount the levers and handles so that they are approximately within fingertip reach (e.g., 3.0 in., 7.6 cm, from the steering-wheel rim), but not closer, so that the lever handle cannot be bumped as the driver manipulates the steering wheel. When labels are required on the lever handle, these must be visible from the driver's nominal eye reference position; i.e., the driver should not have to shift his or her body to see around the steering wheel.

4. The minimum force resistance should be about 6 oz. (170 g); the maximum should be 20 oz. (567 g).

DRIVING CONTROLS FOR THE ORTHOPEDICALLY HANDICAPPED

Steering aids generally require a custom fit for each particular handicap; i.e., although since individuals may be able to use the simple spinner knob, others cannot grip such a device and therefore require a "stirrup" device, as shown in Fig. 30.104. Example of hand controls for handicapped drivers is shown in Fig. 30.105.

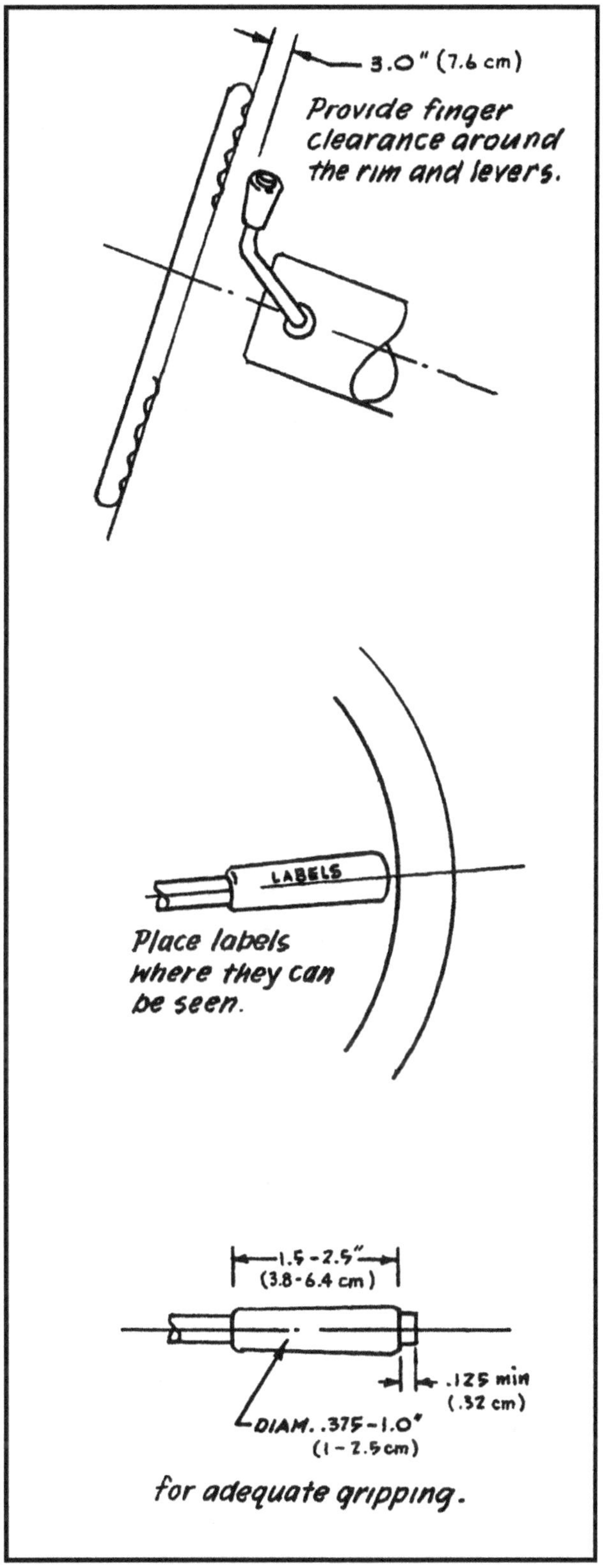

FIGURE 30.103 Design guidelines for column mounted controls.

Note: All spinner-type aids should have some means for adjusting the position of the spinner on the steering-wheel rim (although for 95 percent of the user population, a four-o'clock position is usually best).

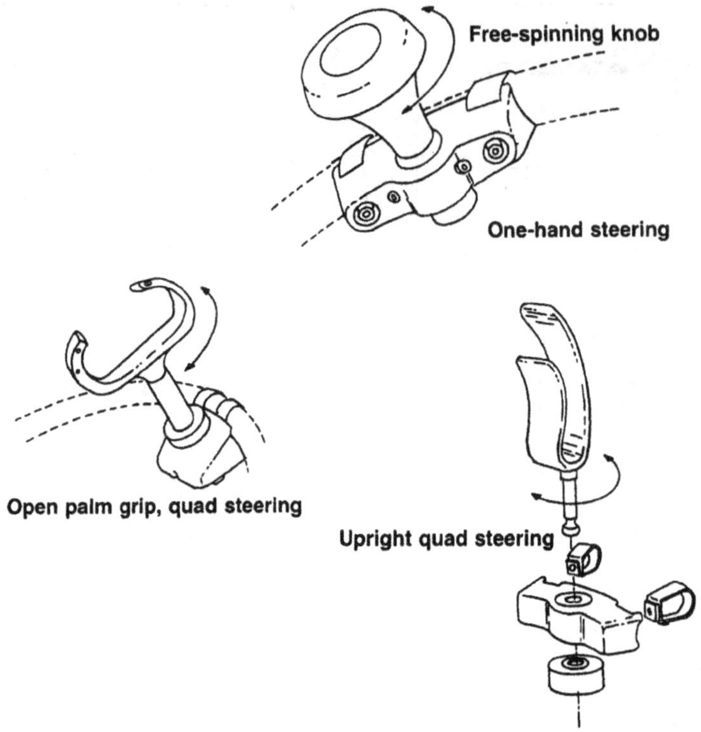

FIGURE 30.104

Guide for Determining Special Control Requirements for Handicapped Drivers

Handicap	Complete Hand Control	Hand Control Extension	Complete Foot Control	Spinner Knob	Steering Cuff	Right-Side Directional	Left-Side Gear Shift	Left-Foot Gas Pedal	Parking Brake Extension	Hand Clutch	Hand Brake	Hand Dimmer Switch	Automatic Transmission	Power Steering	Power Breaks	Power Windows	Power Six-Way Seat Adjustment	Full-View Mirror
Left leg									X	X		X	X					
Right leg								X			X	X	X					
Both legs	X			X					X			X	X	X	X	X	X	
Right arm				X			X						X	X				
Left arm				X									X	X				
Both arms			X										X	X	X	X		
Left leg and left arm				X					X			X	X	X				
Right leg and right arm				X			X	X				X	X	X				
Left arm and right leg				X				X				X	X	X				
Right arm and left leg				X			X		X			X	X	X				
Quadriplegic (all four limbs)	X			X	X				X			X	X	X	X	X	X	
Balance high paraplegic	X			X					X			X	X	X	X			
Neck rotation restriction																	X	X
(lack of height)		X										X	X	X	X			

Source: M. Less et al., *Hand controls and Assistive Devices for the Physically Disabled Driver,* Human Resources Center, Albertson, N.Y., 1977

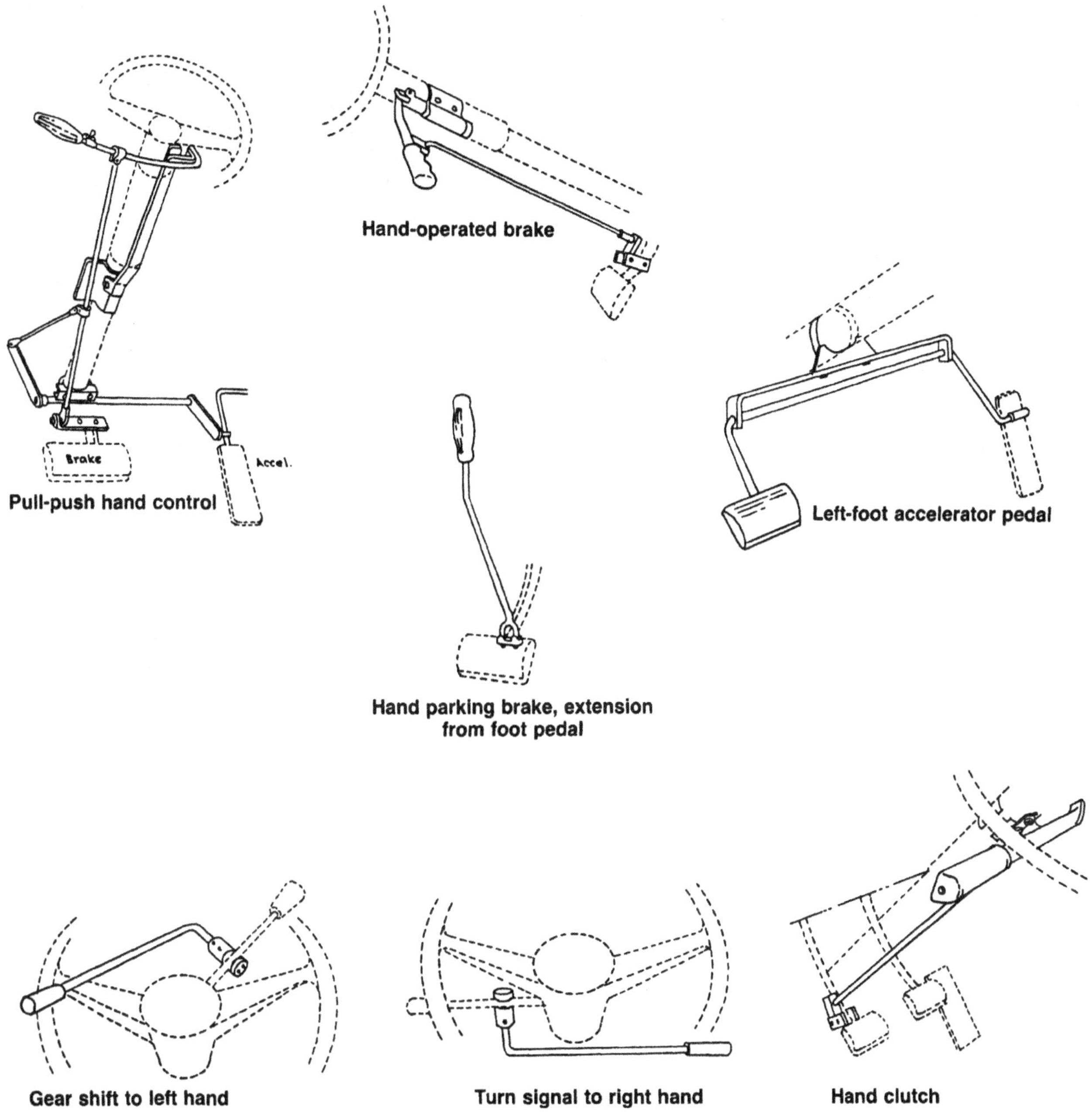

Figure 30.105 Example of Hand Controls for Handicapped Drivers.

VALVE HANDLES

Since valve handles generally are associated with the possibility of considerable force application on the part of the user, the handle should be designed so that the operator can place his or her hand in a comfortable position and grip the handle with a minimum possibility of pain or injury.

For very high force valve requirements, consider a wheel that is large enough so that the operator can use both hands. To assist the operator, select a rim that has ribbing or serrations to provide additional gripping power. (See Fig. 30.106.)

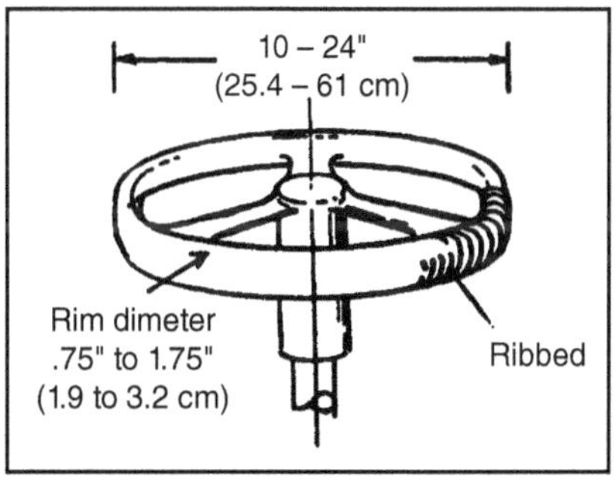

Figure 30.106 Valve handle design for very high forces.

For higher force applications, the round handle with knurled edges is preferred. The edges should be radiused, as shown in Fig. 30.107, and the dimensions shown should be followed in selecting an appropriate handle.

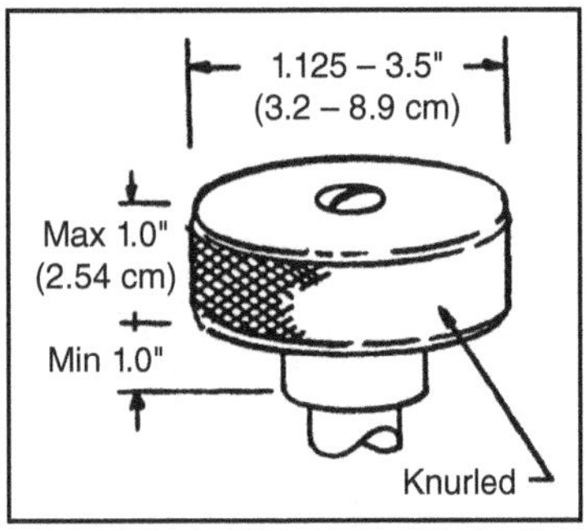

Figure 30.107 Valve handle design for high forces.

For lower force applications, consider the handle shapes shown in Fig. 30.108.

PULL-TYPE CONTROLS

The size and shape of a pull-type control handle should be compatible with the manner in which the hand can most easily grasp and manipulate the handle, i.e., with the fingers or the whole hand. Even for very light forces, the handle should be large enough so that the user is not forced to struggle to get hold of, and maintain proper grip on, the device. A simple "fluted" handle, as shown in Fig. 30.109, provides the best gripping characteristics. Such handles are suitable for forces between 1.0 and 4.0 lb (0.45 and 1.6 kg).

The T-shaped handle is effective for light to medium forces. Although the T-shaped handle can be operated with two fingers (i.e., one on each side of the shaft), it should not be used if the force requirement exceeds about

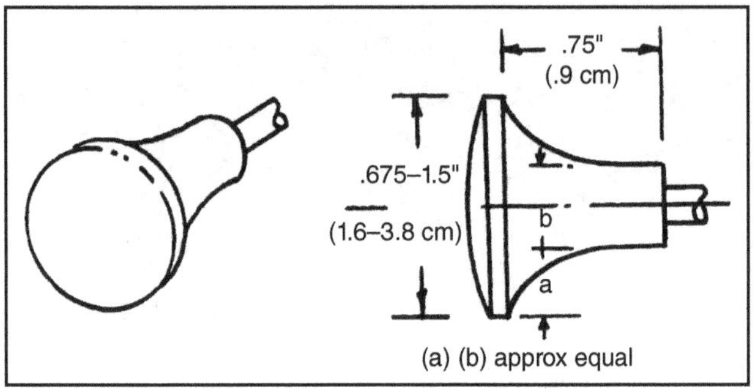

Figure 30.109 Pull control for light forces (1.0 to 4.0 lb).

15 lb (6.8 kg). For higher forces, use the dimensional guidelines shown in Fig. 30.110; i.e., the handle should be large enough so that the user can get two fingers on each side of the shaft.

A stirrup-shaped handle is sometimes useful to help keep objects such as belts or straps or the operator's clothing from becoming caught behind the handle (which may occur with the T-shaped handle). In the case of this handle, since the ends are closed, it is critical to provide enough space inside the stirrup so that the operator's hand and fingers are not crushed when force is applied.

Note: Both the T-shaped handle and the stirrup-shaped handle should be either round or oval; i.e., avoid rectangular shapes with sharp edges, which may injure the user's hand.

DOOR HANDLES

General

Although round knobs are used extensively for architectural purposes in the United States, L-shaped handles are used more frequently in other countries. (See Fig. 30.111.) The

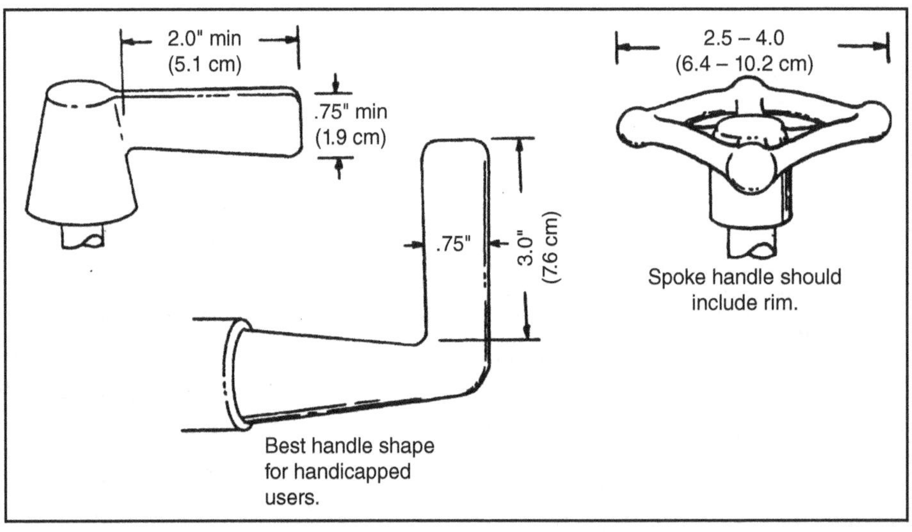

Figure 30.108 Valve handle design for lower forces.

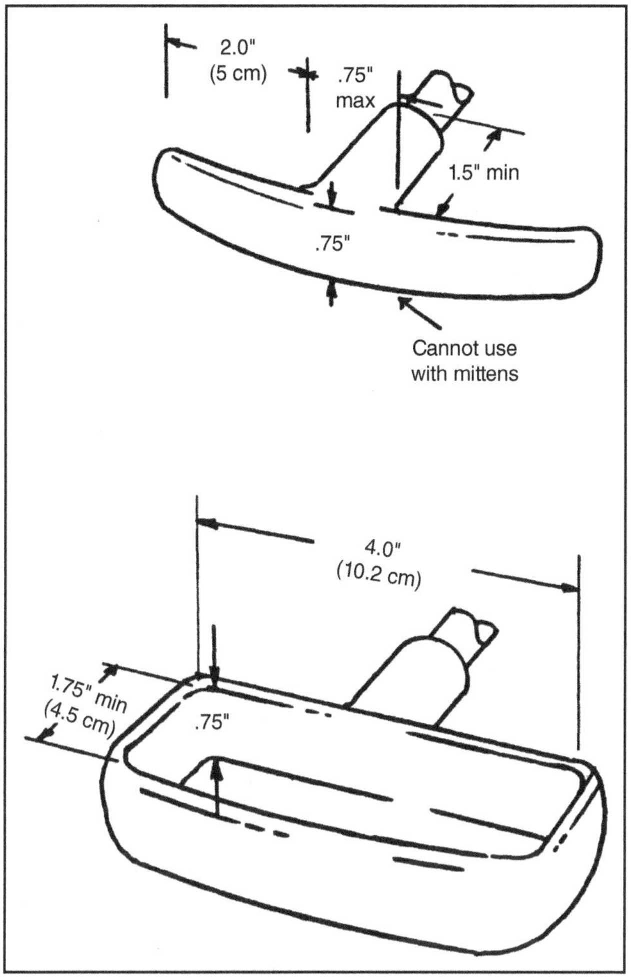

Figure 30.110 Pull controls for high forces.

L-shaped handle is preferred because it is more easily used by handicapped persons and children, who have difficulty gripping and applying force to the round knob.

Emergency bar controls requiring two-hand operation, such as those illustrated in Fig. 30.112, are difficult for children and many handicapped persons to use.

Door latching controls that utilize a thumb latch lever, as shown in Fig. 30.113, are also difficult for children and handicapped persons to operate. If used, the critical factor is thumb lever force, which should not exceed 1.0 lb (0.45 kg).

Handles and latching hardware used for sliding doors are often poorly designed because the manufacturer was not concerned about the user's manipulatory capabilities. The major problem invariably relates to a latching or locking device that is too small for the operator to obtain sufficient gripping area. As a result, many people, especially children and the elderly, cannot apply enough force to operate the latch. The latch handle should provide a surface sufficiently large for adequate gripping. Although not particularly critical, the door handle should be oval for maximum comfort. (See Fig. 30.114.)

Special Note: With regard to the selection of door handles for architectural purposes, it should be noted that opening and closing a door may often be an emergency matter; thus designers should be particularly careful not to allow a desire to select aesthetically pleasing but difficult-la-operate designs to overshadow the basic functional objective. Many acceptable handle designs can be found that are also pleasing. Above all, avoid knobs or handles that have sharp edges or decorative patterns. (See Fig. 30.115.)

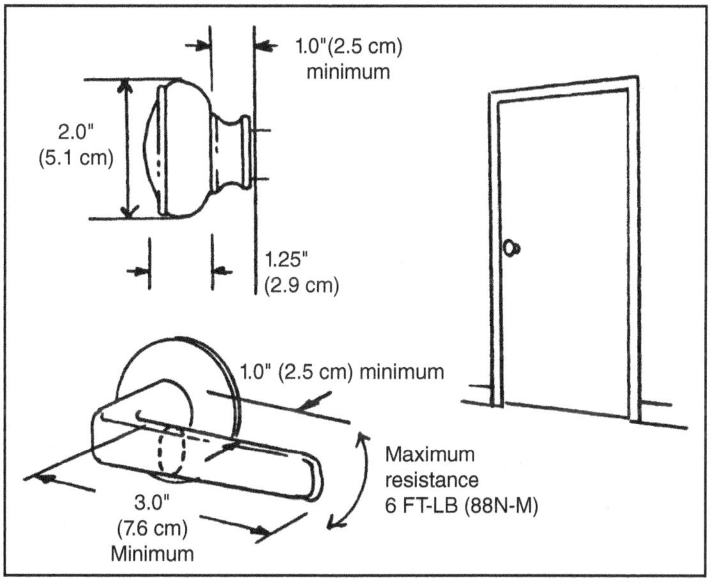

Figure 30.111 Door knobs and handles.

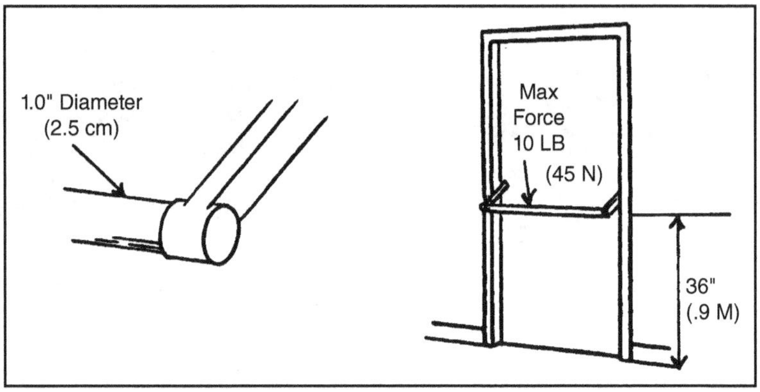

Figure 30.112 Emergency bar controls.

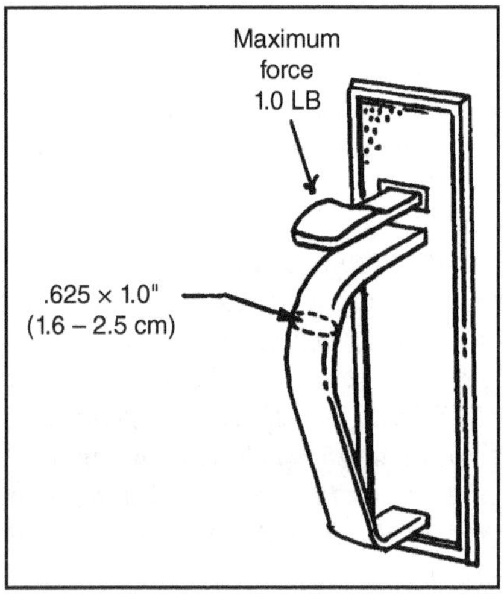

Figure 30.113 Thumb latch lever.

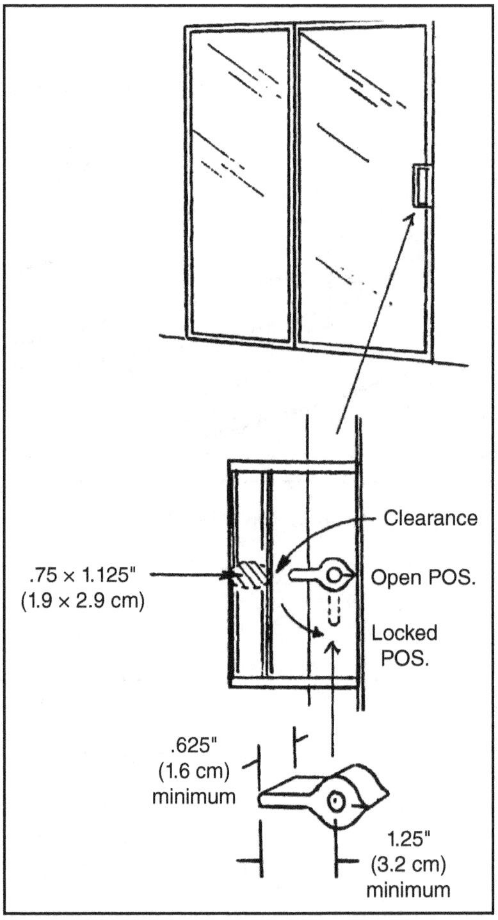

Figure 30.114 Sliding door hardware design.

Figure 30.115 Avoid sharp edges.

Vehicle Door Handles

The shape of the exterior vehicular door handle shown in Fig. 30.116 is preferred, although other shapes may be acceptable in certain circumstances.

1. The handle shown in Fig. 30.116 is a closed loop in which the latch is operated by pushing the button with the thumb. The desirable feature of this type of handle is that the operator is not confronted with a movement of the handle when the latch releases the door. Note that the handle for the right-hand door is oriented differently from that for the left-hand door. This forces users to position themselves properly to keep the door from knocking them over when it is pulled open.

2. Although more awkward to use, the lift-type handle shown in the figure is reasonably acceptable.

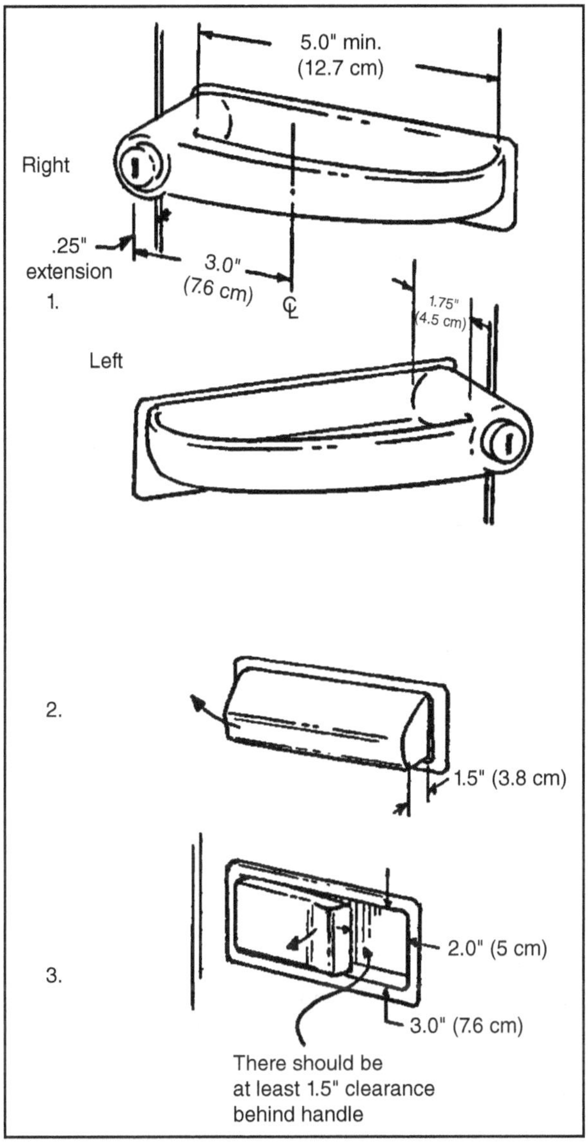

Figure 30.116 Recommended dimensions for exterior vehicle door handles.

3. Although this handle is often found on some vehicles, it is extremely hard to use because it has to be grasped from an awkward angle, and the pulling motion is even more awkward. When it is used, the handles on each side must be mounted so that the operator is not forced to stand in front of the door. The handle shown is for the right side of the car and is intended to be operated by the right hand, as the operator stands aft of the door.

Interior vehicular door handles should be placed where they are easily located by the user (sometimes in an emergency), but they should not become a hazard in the event the occupant is inadvertently thrown against the handle. (See Fig. 30.117.)

1. A preferred handle is illustrated in the figure. Note that it is inset and that the open end is toward the front.
   The important hand and finger clearance dimensions are shown for an optimum handle. In addition, however, there must be sufficient space behind the handle (i.e., inside the housing) so that the occupant can slip his or her fingers behind it. This requires a minimum of 1.0 in. (2.5 cm) of clearance.

2. The "loop handle," which is a commonly used interior vehicular handle, is shown in the figure. Although this is an acceptable configuration, many manufacturers tend to make the handle and space around it too small. As a minimum, the user should be able to get at least two fingers either inside the loop or past the end of the handle. The dimensions shown are recommended.

SAFETY-BELT BUCKLES

Safety-belt buckles have become more or less standardized for all vehicles; i.e., they usually are operated by means of a push button to release the catch. It has been shown, however, that buckles that approximate the dimensions illustrated are more acceptable. (See Fig. 30.118.) First, the overall buckle dimension is easy to grasp. Second, the button release is large enough to be actuated by the thumb or finger with minimum difficulty. Note that the button is inset (so that the thumb or finger naturally slides into the depression and onto the button). Third, the button cannot be inadvertently depressed and thus unlock the buckle at the wrong time.

Button pressure should not exceed about 4 lb (1.8 kg).

COMBINATION CABINET HANDLE AND LATCHING MECHANISMS

Equipment racks and consoles generally have a removable chassis and panels that require a combination handle and latching mechanism. Since the handle is also used to carry the chassis around after it is removed, it should be fixed. To provide the best combination of handle and latch release, the handle concept shown in Fig. 30.119 is suggested; i.e., the handle is fixed, but it contains a push-button release in a position that is convenient for thumb operation. For ease of carrying, the handle cross-section should be oval, and the dimensions shown should be used to make it easy to reach the release button while the hand is placed firmly on the handle.

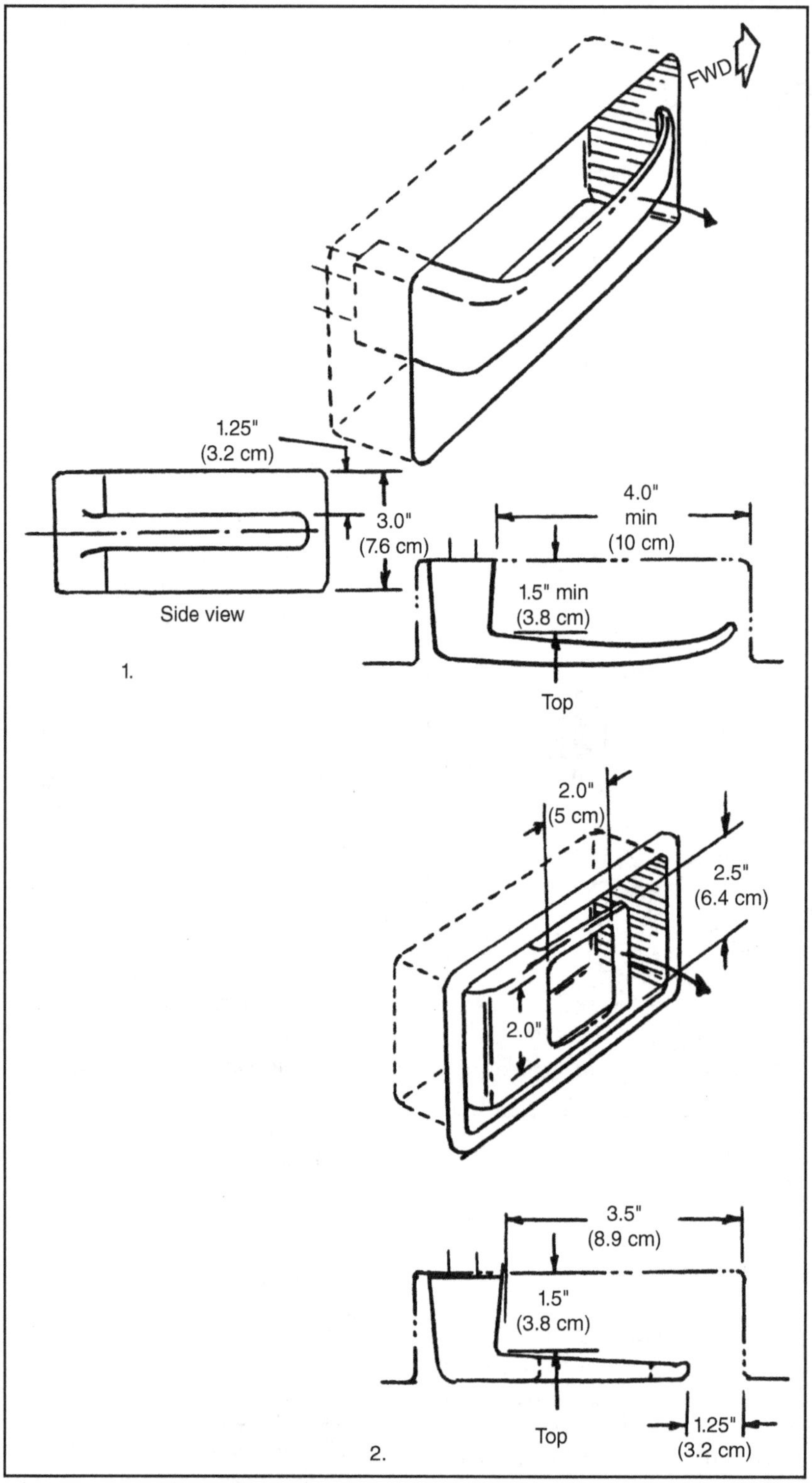

FIGURE 30.117 Recommended dimensions for interior vehicle door handles.

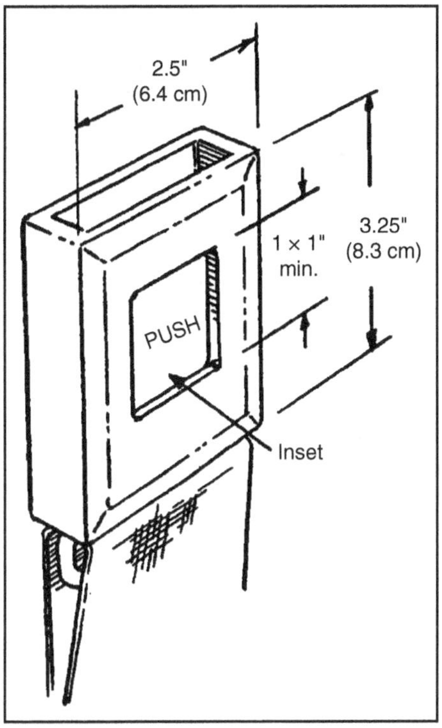

FIGURE 30.118 Recommended safety-belt buckle dimensions.

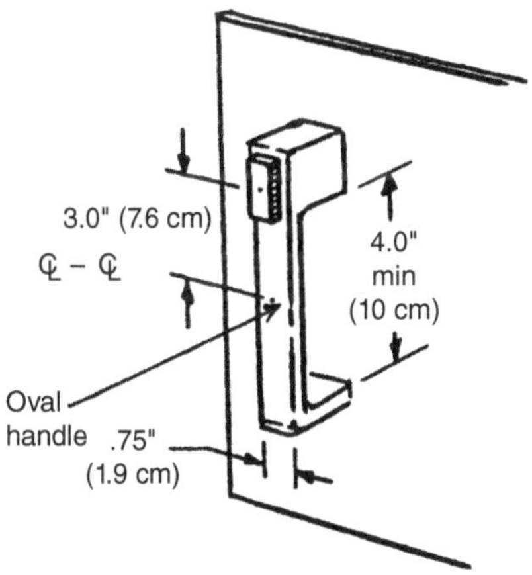

FIGURE 30.119 Handle for removable chassis and panels.

Filing cabinets also require a combination of handle and latch release. Since the cabinet drawer is not generally carried around, a single handle is sufficient. A convenient combination of handle and latch release tab is illustrated in Fig. 30.120. Note the dimensions suggested for the tab release handle. A common fault of most similar hardware is that the tab is too small and shallow for easy operation. The handle should be oval in shape, and/or the edges of the rectangular handle should be radiused so that the user does not injure his or her fingers while pulling out the drawer.

SUITCASE AND BRIEFCASE LATCHES

The primary problem with most of these devices is that the manufacturer has not provided adequate finger clearance to "lift" the primary locking tab. The dimensional guidelines illustrated in Fig. 30.121 will provide a more acceptable latching device.

OTHER LATCHING CONTROLS

Folding handle configurations are often desirable in order to provide unobstructed stacking of equipment or cases. The design shown in Fig. 30.122 is typical. The critical dimensions required to make such handles accessible are provided. A folding handle should, however, be designed with a strong spring detent to prevent it from inadvertently folding while the user is trying to use the handle to move the equipment. Although such handles are not ordinarily intended to be used for this purpose, they often are.

Although the ½-in. (1.3-cm) clearance behind the folded handle is generally adequate, more clearance is desirable if it will not compromise the basic package interior space.

Small fasteners are sometimes required to eliminate the necessity to find a special tool such as a screwdriver. The wing-tip fastener shown in Fig. 30.122 is commonly used for this purpose. Avoid selecting fasteners that are so small that it becomes difficult to get hold of them or apply the necessary force to rotate them.

Note: Folding types are also available, so that the fasteners do not stick out and become a hazard to passersby.

SPECIAL INTEGRATED HANDLES AND CONTROLS

Slide Switches

Key considerations in the design and/or selection of slide switches for applications such as a flashlight are sufficient contact surface area, vertical extension above the operating surface, and serration and surface shaping to ensure that the operator can apply proper force to move the switch.

1. The concave switch shape shown in Fig. 30.123 should be at least 0.75 in. (1.9 cm) long by 0.25 in. (0.63 cm) wide. The switch travel should be at least 0.25 in. (0.63 cm), but no more than 0.625 in. (1.588 cm), to provide tactile feedback without requiring too long a throw to make electrical contact.

2. The alternate tab and slide shown in the figure should extend about 0.375 in. (0.953 cm) above the working surface, but should be at least 0.375 in.

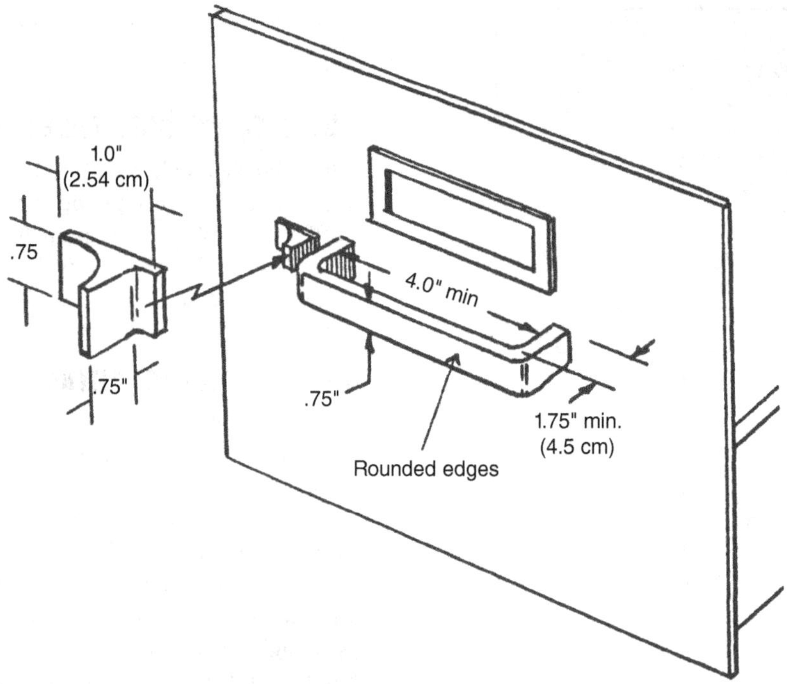

Figure 30.120 Filing cabinet drawer handle.

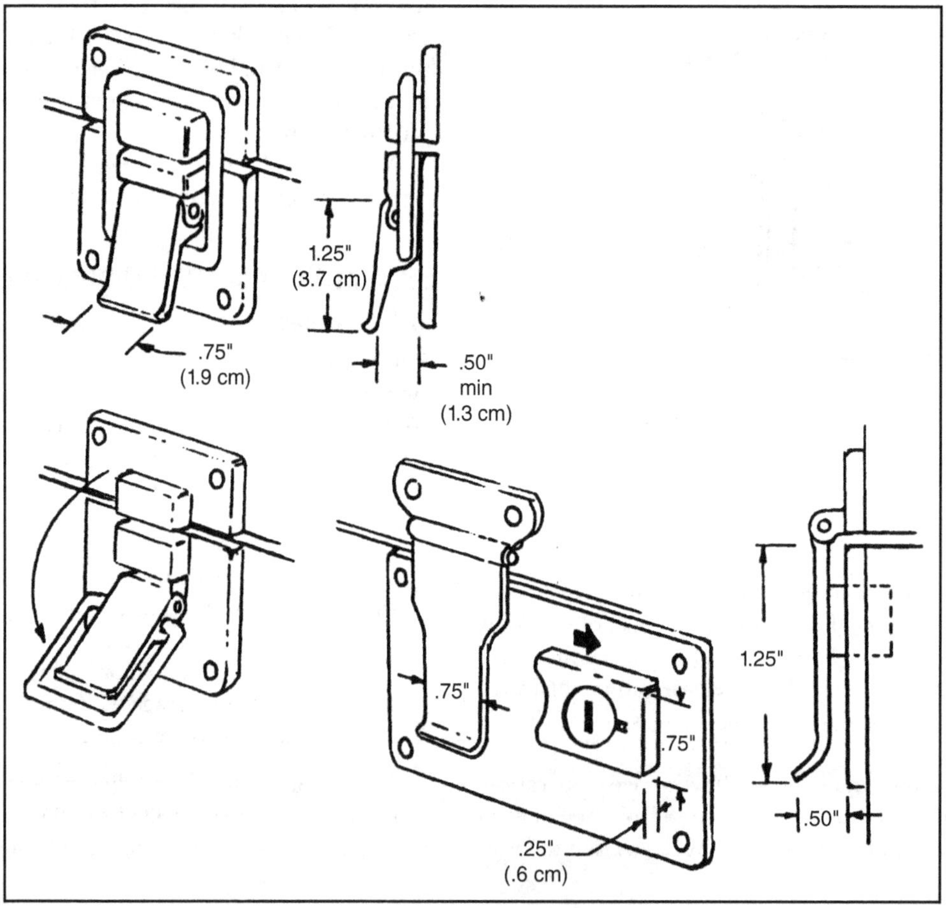

Figure 30.121 Recommended dimensions for suitcase and briefcase latches.

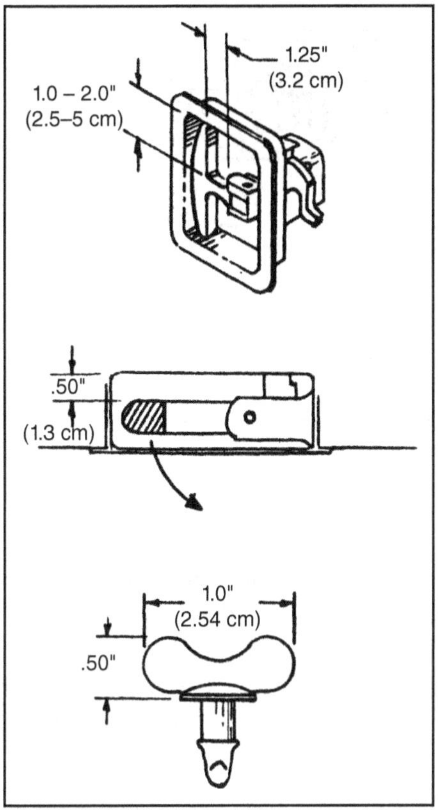

Figure 30.122 Recommended dimensions for folding handles and wing fasteners.

(0.953 cm) wide to provide adequate thumb-contact surface. Neither switch configuration should require more than about 10 in./oz. (283.5 g) force to operate.

Trigger Controls

Triggers used on power tools and hand-held weapons should be shaped and should provide clearance dimensions approximately as shown in Figs. 30.124 and 30.125. Other handle and handle-trigger dimensional relationships are also important to effective operation. In selecting or designing trigger installations, take special precaution against potential finger entrapment and snagging of gloves.

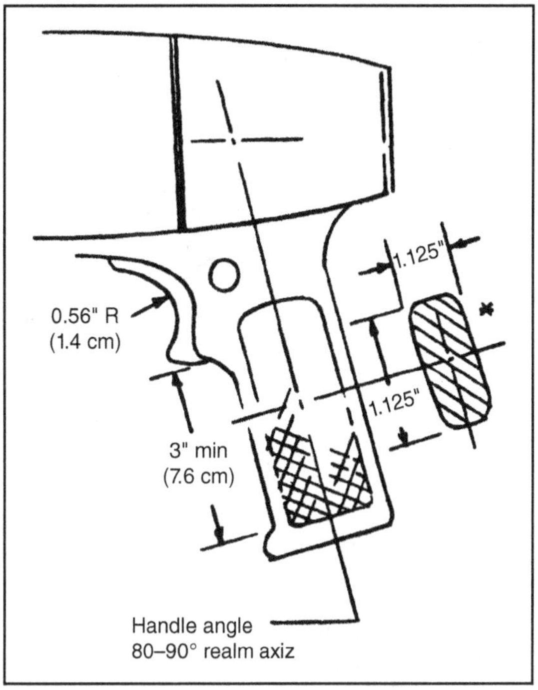

Figure 30.124 Power tool trigger dimensions.

AUTOMATED AND ROBOTIC SYSTEMS

Portions of many systems are automated. This may occur because certain operations are simply too dangerous for humans or the operators are overloaded by tasks, and machines can carry some of the workload. The following

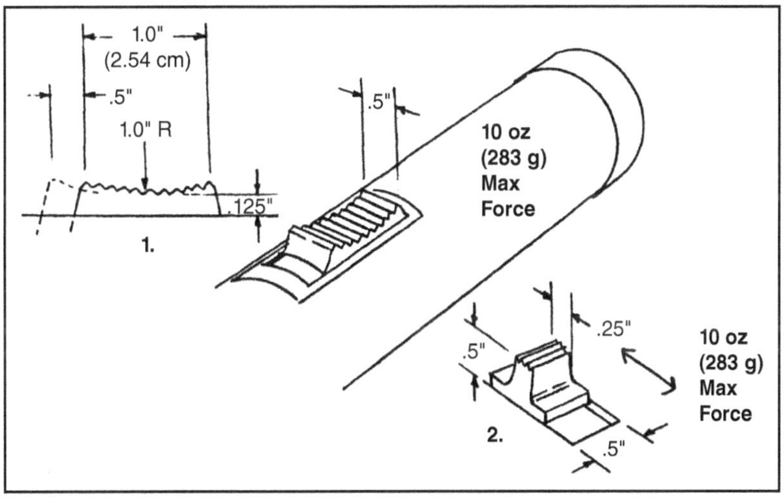

Figure 30.123 Slide switch design recommendations.

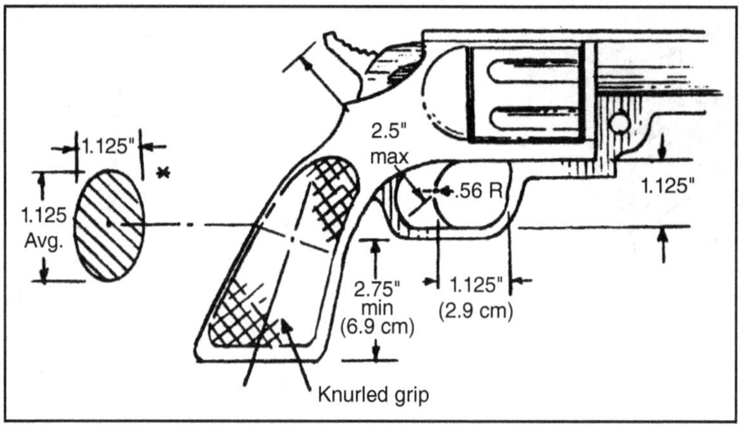

FIGURE 30.125 Weapon trigger dimensions.

are guidelines for the design of automated systems and robots.

1. Design an automatic and robotic system to be self-monitoring and regulating to avoid mission degradation, equipment damage, or injuries to people.

2. Operators of automatic robotic systems must be provided with information on the status of the system including when the system changes between levels of automation.

3. For a given robotic system with multiple control locations, all controls and displays will be the same regardless of their locations.

4. Robotic systems with multiple operators should be able to accept the input from the operators and arbitrate between the multiple operators so as to perform safely and without degradation.

5. Operators must be provided with the ability to override and shut down automated and robotic systems.

6. Operator interfaces to robotic systems must account for the different frames of reference between the operator and the robot to minimize spatial disorientation.

CHAPTER 31

FASTENERS

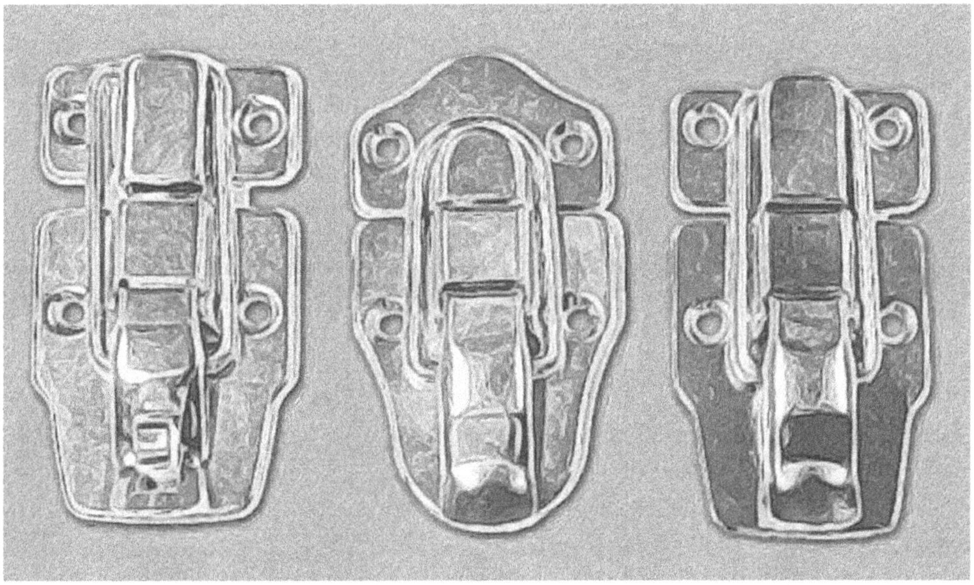

GUIDELINES FOR DESIGN AND SELECTION

1. Use the minimum number of fasteners compatible with requirements for securing components. In designing equipment packages and covers, use tongue and slot or similar techniques to help reduce the number of fasteners required.

2. Use "captive" fasteners to avoid loss of all or part of the fastening device.

3. It should be readily apparent when a fastener has been "released."

4. Work space should be provided around a fastener for fingers, tools, or "wrenching space."

5. Standardized fasteners should be used wherever possible to minimize stocking requirements and the number of different tools required.

6. Use manually operated fasteners where suitable to avoid obtaining or carrying tools.

7. Use fasteners that are operable by commonly used rather than special tools.

8. Use the same type and size of fastener consistently for a given application.

9. Screws, bolts, or nuts with different threads also should have clearly different physical sizes to minimize their being interchanged.

10. Consider how a worn or stripped fastener can be removed before selecting it; avoid stud fasteners that are an integral part of the machine or housing. Use screws rather than rivets.

11. Select safety fasteners for applications where vibration could result in loosening and perhaps loss of the fastener during operation.

12. Fasteners that are operable by hand should also be rugged enough to withstand operation by some type of tool.

13. Spring-load catches so that they lock on contact, rather than requiring a separate locking step.

14. Use "long latch" catches to minimize the possibility of inadvertent latch release.

15. When a latch has a handle, locate the latch release near the handle so that it can be operated with one hand.

16. Screw heads should have deep slots to reduce tool slippage and damage to the screw head.

17. Place the screw head so that offset screwdrivers are not required.

18. When bolts and nuts are used, make sure there is clearance to both ends.

19. Keep bolts as short as possible so that they will not snag personnel or equipment. Do not locate bolts where inadvertent replacement by a longer bolt might interfere with the movement of an adjacent mechanism or puncture adjacent cover or "skin."

20. Coarse threads are preferable to fine threads for low torques because they reduce the possibility of cross-threading.

21. Avoid the use of left-hand threads unless the system requirements demand them; then identify both bolts and nuts clearly by a suitable marking, shape, or color.

22. Cotter keys or pins should fit snugly, but they should not have to be "driven out."

23. Use safety wire only when self-locking fasteners cannot withstand the expected vibration or stress. Attach safety wire so that it is easy to remove and replace.

24. Use retainer rings that hold with a positive snap action. Avoid rings that become difficult to remove when they are worn.

25. Provide retainer chains to capture fasteners that have to be completely removed but which could easily become lost.

26. Avoid the use of fastener systems in which a captive spring could suddenly drive the fastener into the face of the technician once the fastener is released or in which the spring itself pops out and becomes lost. Similarly, avoid systems in which a washer easily becomes detached and lost without the technician's knowing about it. Avoid springs that are so strong that a technician has to use a special tool to depress the spring before engaging the basic fastener.

TYPES AND DESIGN CONSIDERATIONS

Tables 31.1–31.5 describe various types of fasteners and latches and their recommended application and design characteristics.

CHASSIS AND DRAWER SLIDE FASTENERS

General

Slide-mounted equipment chassis systems (and file-drawer slide systems) using ball or roller bearing slides are often improperly designed in terms of the locking and fastening features. (See Fig. 31.1.) Such slide systems should, of course, provide an automatic locking system to prevent a drawer or chassis from being accidentally pulled all the way out of a cabinet. Care should be taken, however, to design the slide-locking system so that it is possible to override the lock and fully remove the chassis or drawer. More often than not, this is an extremely complex and confusing operation because the basic procedure is not evident in the hardware itself. Obvious features (i.e., latches or buttons) should be provided on the slide-locking system, and/or clear instructions should appear directly on the hardware. Remember that people seldom remember where they have put an instruction booklet.

More flexible slide-rotational systems may appear even more complex to the user, and thus careful consideration must be given to the release latching concepts and to the visibility and recognition factors associated with locking and unlocking both the horizontal and rotational elements of the slide system mechanisms.

Consider the following guidelines:

1. The chassis should be easily and quickly detachable for removal from a cabinet without the need for hand tools.

2. Detaching the chassis from the slide should not require placing the hands in a position that might lead to injury if the chassis suddenly moves.

3. The slides should automatically lock in the open position and/or in a rotated position securely enough to prevent dislodging the chassis.

4. The slides should be easily removable from the cabinet for replacement.

5. Special tools should not be required to remove or repair slides.

See also Fig. 31.2 and Table 31.6, Fig. 31.3 and Table 31.7, Fig. 31.4 and Table 31.8, Fig. 31.5 and Table 31.9, and Fig. 31.6 and Table 31.10.

Separable Hinges

When frequent removal of a chassis is necessary, the time required for dismounting a hinged cover may be significant. For example, demounting a chassis using two standard hinges may require the removal of four to six screws, taking

TABLE 31.1 Fastener and Latch Types and Design Considerations

Type	Description
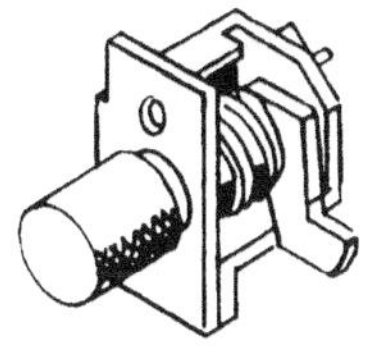	Adjustable pawl fastener. As knob is tightened, the pawl moves along its shaft to pull back against the frame. 90" rotation locks, unlocks fastener.
	"Dzus"-type fastener with screwdriver slot. Three-piece one-quarter–turn fastener. Spring protects against vibration. 90" rotation locks, unlocks fastener.
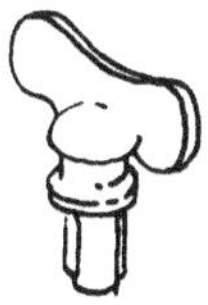	Wing head. "Dzus" type. 90" rotation locks, unlocks fastener.
	Captive fastener with knurled, slotted head. Retaining washer holds the threaded screw captive.
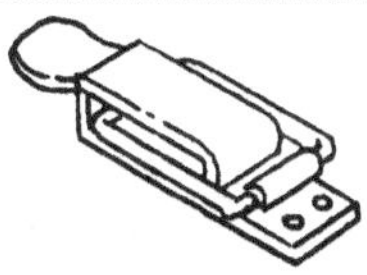	Draw-hook latch. Two-piece, spring latch, base unit, and striker. When engagement loop is hooked over striker, depressing lever closes unit against force of springs. Lever is raised to unhook.
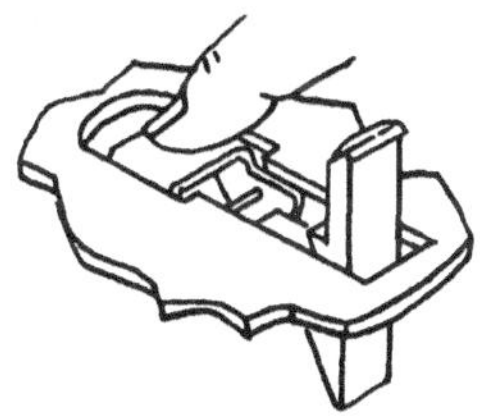	Trigger-action latch. One-piece, bolt latch. Depressing trigger releases bolt, which swings 90" under spring action and opens latch. To close, move bolt back into position.
	Snapslide latch. One-piece snapslide. Latch is opened by pulling lever back with finger to engage release lever.
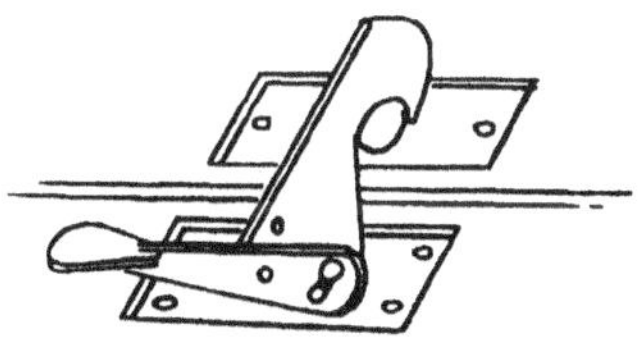	Hook latch. Hook engages knob on striker plate. Handle is pulled up locking in place. To release, reverse procedure.

TABLE 31.2 Types of Fasteners

Type	Description	Maintainability Considerations	Approximate Operating Time
* 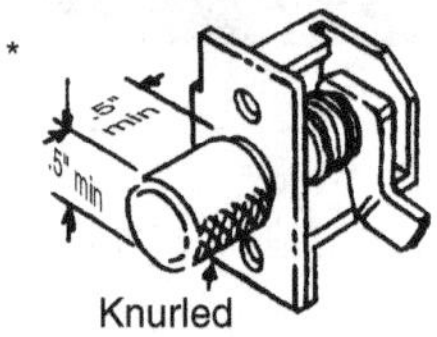	Adjustable pawl fastener. As knob is tightened the pawl moves along its shaft to pull back against the frame. 90" rotation locks, unlocks fastener.	1. No tools required.	01 min
	"Dzus" type fastener with screwdriver slot. Three-piece, one-quarter–turn fastener. Spring protects against vibration 90" rotation locks, unlocks fastener.	1. Tools may be required. 2. Should not be used for front panel fasteners or in structural applications. Preferred type for lightweight panels other than front panels.	05 min
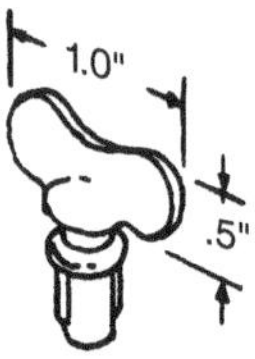	Wing head. "Dzus" type. 90" rotation locks, unlocks fastener.	1. No tools required. 2. Should not be used for front panel fasteners or in structural applications. Preferred type for lightweight panels other than front panels.	04 min
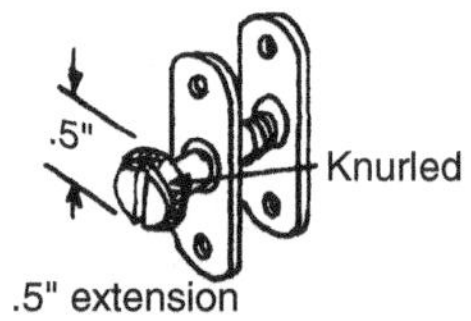	Captive fastener with knurled, slotted head. The threaded screw is made captive by a retaining washer.	1. Tools may be required. 2. Operating time depends on number of turns required.	04 min

*When gloves may be worn, increase all dimensions by 0.125 in. (0.318 cm).

Each of the fasteners shown in the accompanying table is evaluated from the point of view of ease of use in the context of equipment accessibility for maintenance.

TABLE 31.3 Types of Latches

Type	Description	Maintainability Considerations	Approximate Operating Time
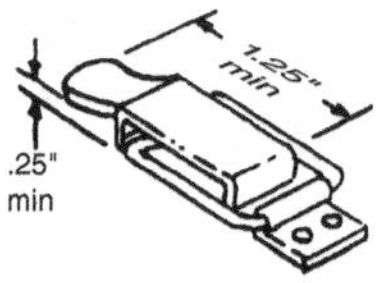	Draw hook latch. Two-piece, spring latch, base unit, and striker. Engagement loop is hooked over striker and lever is depressed, closing unit against force of springs. Lever is raised to unhook.	1. Relatively slow action. 2. Requires considerable force to disengage loops.	03 min
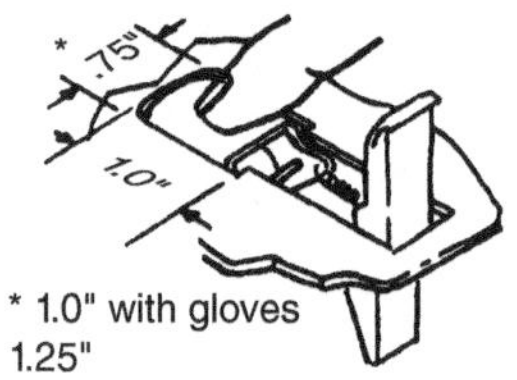	Trigger action latch. One-piece, bolt latch. Latch is opened by depressing a trigger to release bolt which swings 90" under spring action. To close move bolt back into position.	1. Extremely fast action type latch. 2. Strong spring action might cause personal injury.	01 min
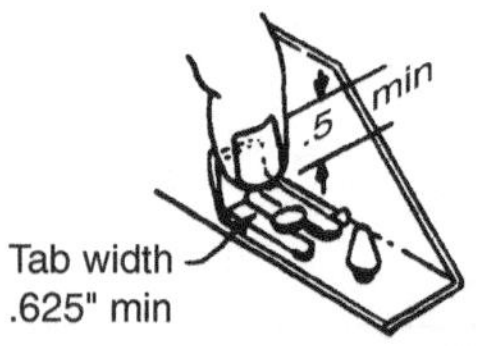	Snapslide latch. One piece snapslide. Latch is opened by pulling lever back with finger to engage release lever.	1. Fast action.	02 min
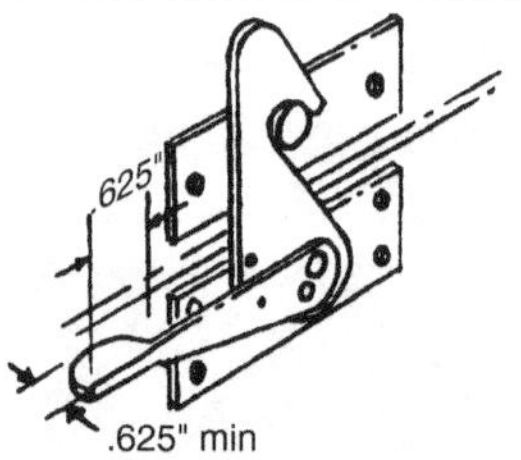	Hook latch. Hook engages knob on striker plate. Handle is pulled up locking in place. To release reverse procedure.	1. Relatively slow action. 2. Takes up room on equipment.	03 min

TABLE 31.4 Types of Fastener Head Styles

Type	Description	Maintainability Considerations
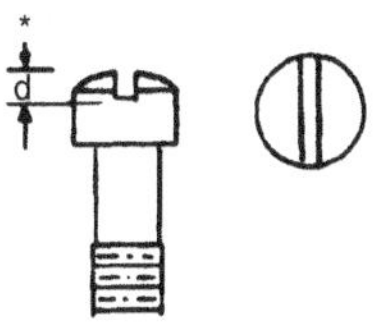	Fillister head, slotted. Smaller in diameter than round head, but has deeper slot. (FF · S · 92 Type 1, Style 4S)*	1. Requires common screwdriver. 2. Deep slot not easily stripped.
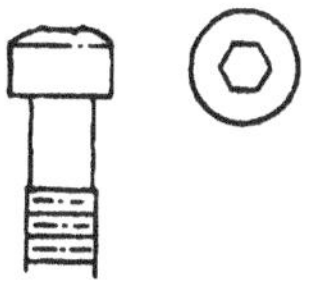	Fillister head, hexagon socket. Same as above except with "Allen" type socket.	1. Required "Allen" wrench not always available. 2. "Allen" wrench usually takes longer to use than common screwdriver.
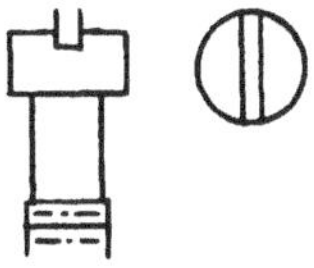	Knurled head, slotted. This is an exaggerated fillister with a knurled head and a screwdriver slot.	1. Can be used with fingers or common screwdriver. 2. Preferred type for retaining front panels or removable cover plates.
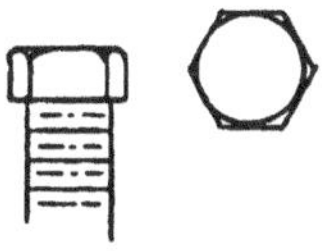	Hexagon head. Standard head for machine bolts and screws. (FF-S-92. Type II. Style 10P)*	1. Requires use of wrench or "spintite."
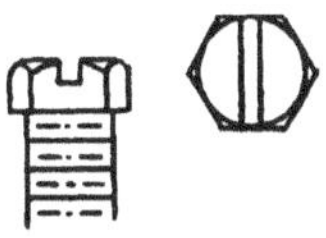	Hexagon head, slotted. Same as above except with added screwdriver slot. (FF-S-92. Type I. Style 10S)*	1. Wrench, "spintite," or common screwdriver can be used to remove or install.
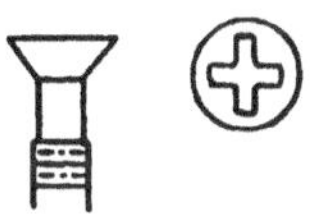	82° flathead, cross-recessed. (FF-S-92. Type III. Style 2C)*	1. Cross-recessed screwdriver not always available.
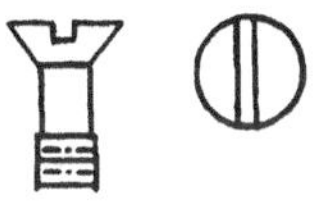	82° flathead, slotted. (FF-S-92. Type I. Style 2S)*	1. Not suitable on thin panel.
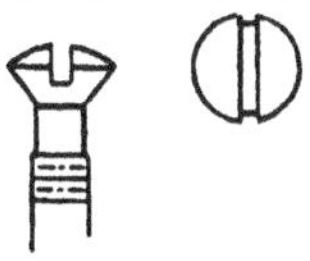	92° oval head, slotted. Similar to flathead but with rounded head. (FF-S-91. Type I. Style 6S)0*	1. Allows deeper slot than flathead. 2. MIL-E-16400 specifies this type with cup washer for rack-mounted panels.
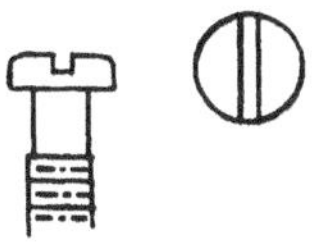	Pan head, slotted. Large diameter with high outer edges for maximum driving power. (FF-S-91, Type I, Style 9S)0*	1. Standard type for panels other than rack-mounted panels and front panels.

*Styles recommended by Federal Specification FF-S-92 for use wherever possible, in order to keep inventories to a minimum.

d = minimum slot depth = 0.0625 in. (0.1588 cm).

TABLE 31.5 Types of Captive Screws

Type	Description	Design Considerations
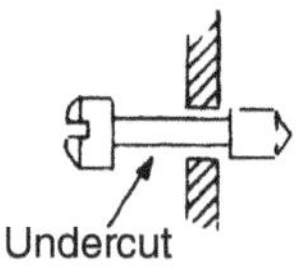	Captive screw with undercut. Panel is tapped to screw size.	1. Easily installed and removed.
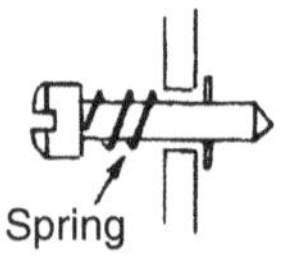	Captive screw, spring loaded.	1. Easily installed. 2. Fast action.
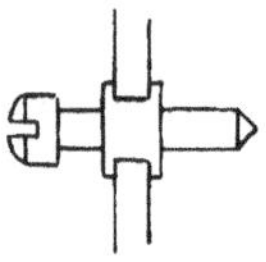	Captive screw which is forced through rubber or plastic grommet.	1. Frequent use can cause excessive wear on rubber or plastic grommet resulting in a loose fit.
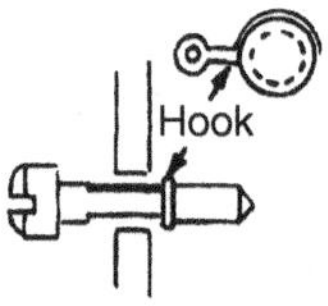	Captive screw with hook which falls into undercut of screw.	1. Hook can be easily jarred loose.
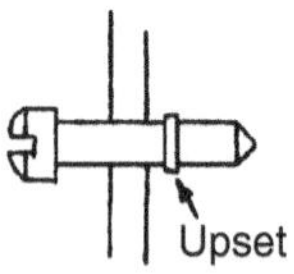	Captive screw with upset thread.	1. No removal features.
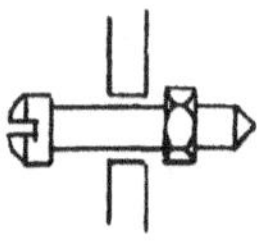	Captive screw with nut staked in place after assembly.	1. Time consuming to put in place and remove.
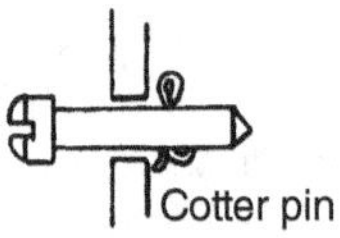	Captive screw with cotter pin forced in place after assembly.	1. Cotter pin can work loose if subjected to excessive vibration.

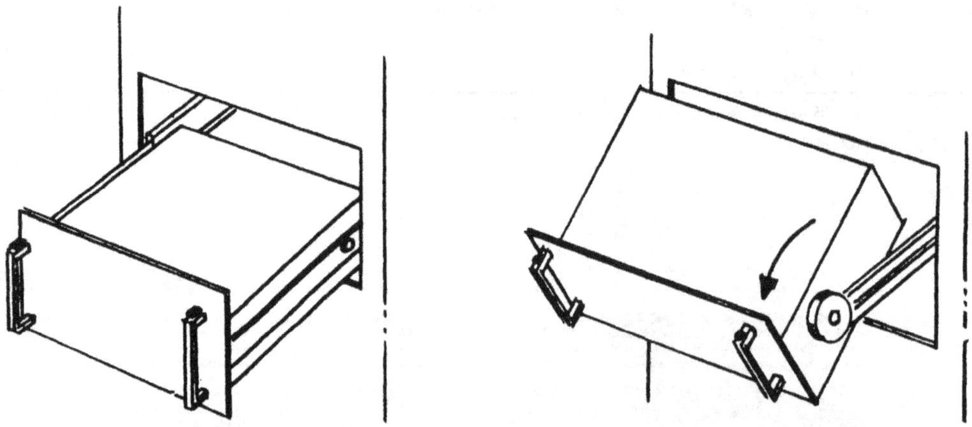

FIGURE 31.1 Improved access using slides and pivots.

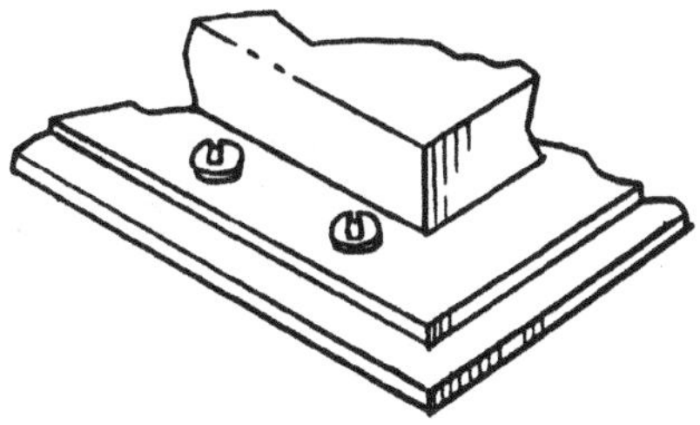

FIGURE 31.2

TABLE 31.6 Chassis Mounted on Horizontal Shelf (Secured by Screw Fasteners through Flange)

Securing Method	Maintainability Considerations
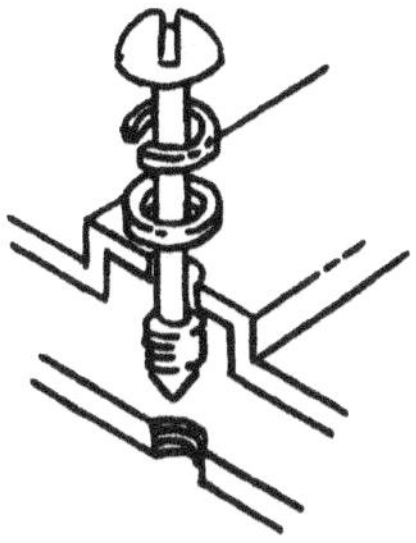	Description: Captive screw. Advantages: No loss of screws and washers. All work performed from one side. Disadvantage: Space required for captive device. Tools Required: Screwdriver. Operating Time: Approximately 6 min per fastener.
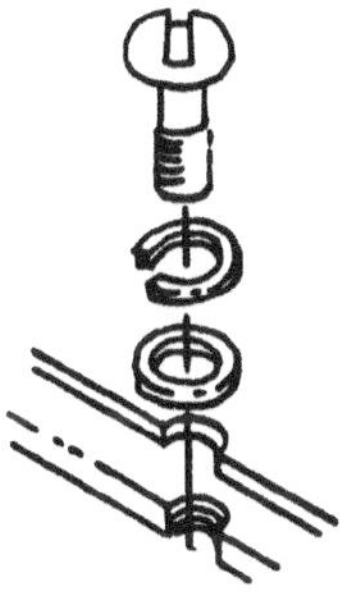	Description: Screw into tapped note with flat washer and lock washer. Advantage: All work performed from one side. Disadvantages: Time required to position washers. Possible loss of screws and washer. Tools Required: Screwdriver. Operating Time: Approximately 8 min per fastener.
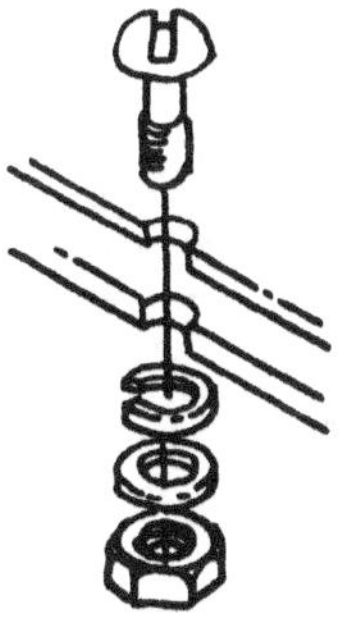	Description: Screw through clearance holes with flat washer, lock washer, and nut. Advantages: No significant advantages. Disadvantages: Possible loss of screws, washers, and nuts. Requires access to both sides of shelf. Two-handed operation. Time required to position washers and nuts. Tools Required: Screwdriver and wrench or "spintite." Operating Time: Approximately 8 min per fastener.
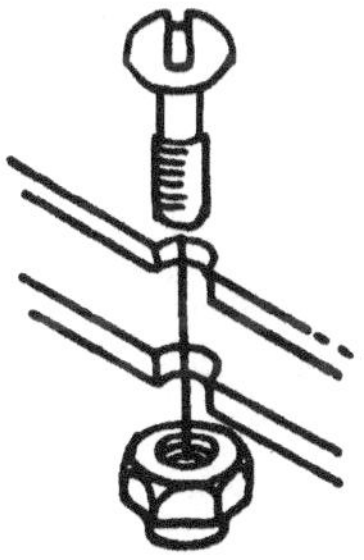	Description: Screw through clearance holes with lock nut. Advantage: No washers required. Disadvantages: Lock nut difficult to turn. Requires access to both sides of shelf. Two-handed operation. Possible loss of screws and nuts. Time required to position nut Tools Required: Screwdriver and wrench or "spintite." Operating Time: Approximately 14 min per fastener.

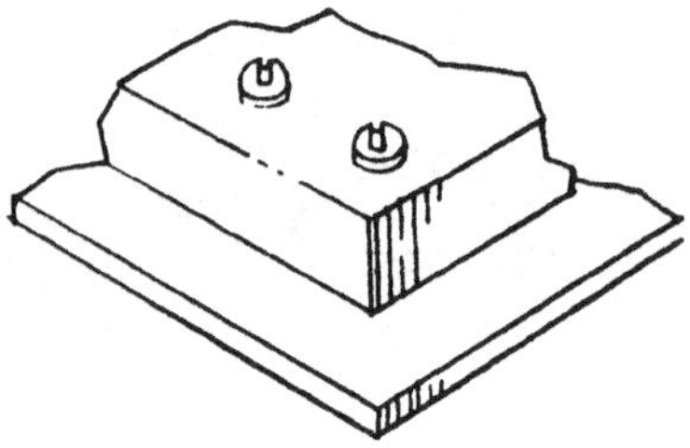

Figure 31.3

TABLE 31.7 Chassis Mounted on Horizontal Shelf (Secured by Screw Fasteners through Chassis)

Securing Method	Maintainability Considerations	
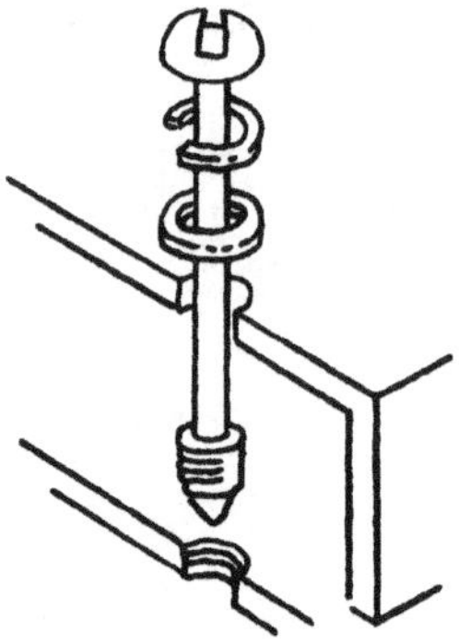	Description: Advantage: Disadvantage: Tools Required: Operating Time:	Captive screw. No loss of screws or washers. All work performed from one side. No separate handling washers. One-handed operations. Alignment of screw can be difficult. Space required for captive device. Screwdriver. Approximately 6 min per fastener.
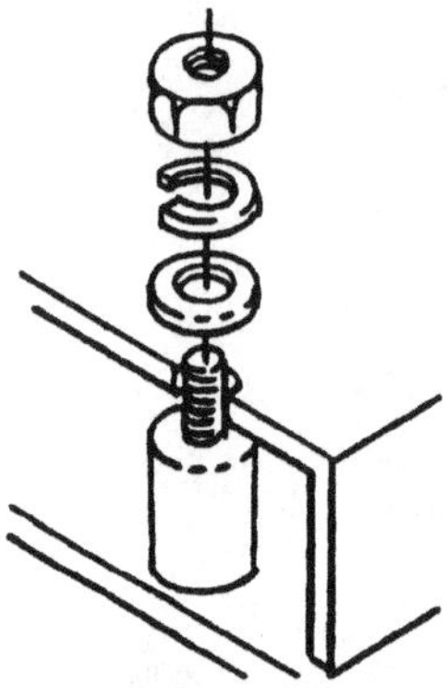	Description: Advantage: Disadvantages: Tools Required: Operating Time:	Stud through chassis with flat washer, lock washer, and nut. No screw alignment problem. Studs act as locating pins. Possible loss of nuts and washers. Chassis must be lifted over studs. Wrench or "spintite." Approximately 7 min per fastener.
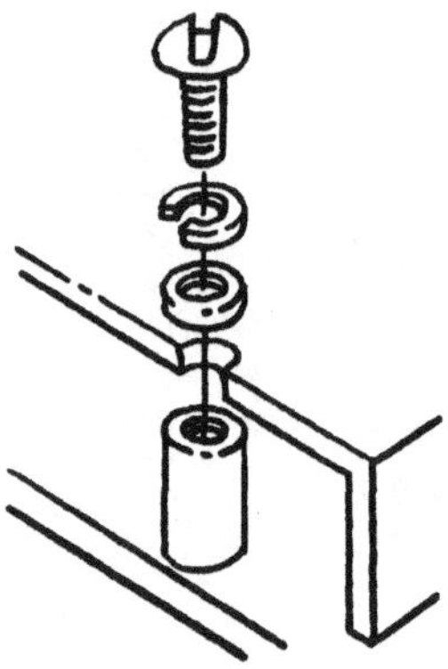	Description: Advantage: Disadvantages: Tools Required: Operating Time:	Screw into stand-off with flat washer and lock washer. Very little screw alignment problem. Possible loss of screws and washers. Chassis must be lifted over stand-off. Screwdriver. Approximately 8 min per fastener.

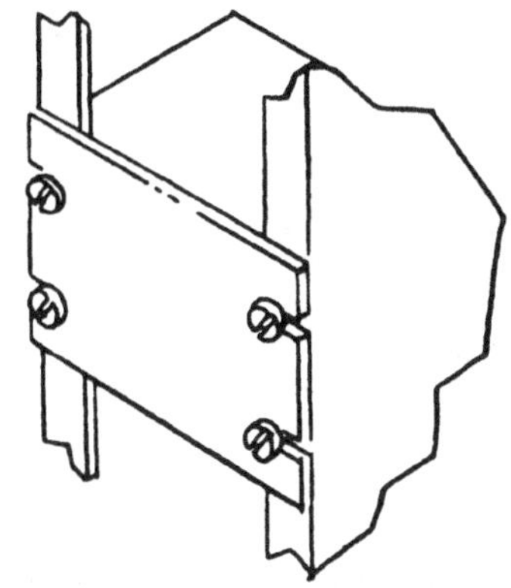

Figure 31.4

TABLE 31.8 Chassis Mounted on Vertical Rack (Secured by Screw Fasteners into Frame)

Securing Method	Maintainability Considerations	
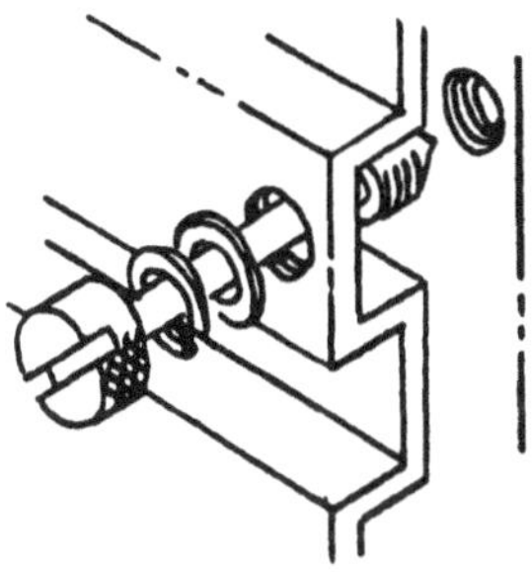	Description: Advantages: Disadvantage: Tools Required: Operating Time:	Captive screw. Slotted head-type preferred for front panels. No loss of screws and washers. All work performed from one side. No separate handling of washers. One-handed operation. Space required for captive device. Screwdriver. None if thumb screw is used. Approximately 6 min per fastener.
 1:0 × .75" min	Description: Advantage: Disadvantage: Tools Required: Operating Time:	Thumb screw with lock washer and flat washer. No tools required. One-sided operation. Possible loss of screws and washers. None. Approximately 8 min per fastener.
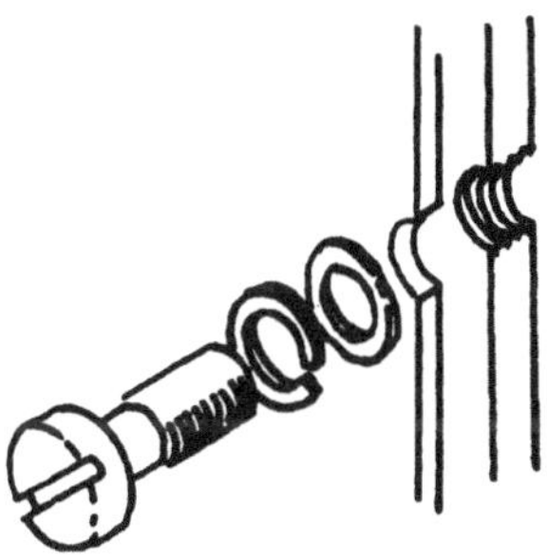	Description: Advantage: Disadvantage: Tools Required: Operating Time:	Screw into tapped hole with flat washer and lock washer. Preferred for panels other than front panels. All work performed from one side. Time required to position washers. Possible loss of screws and washers. Screwdriver. Approximately 8 min per fastener.

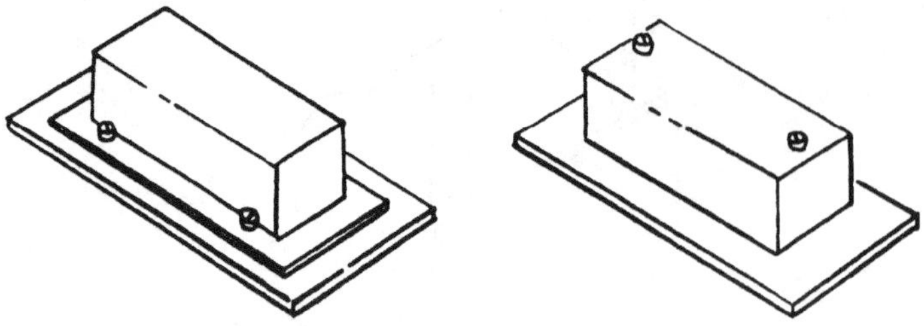

FIGURE 31.5

TABLE 31.9 Chassis Mounted on Horizontal Shelf (Secured by Quick-Acting Fasteners)

SECURING METHOD	MAINTAINABILITY CONSIDERATIONS	
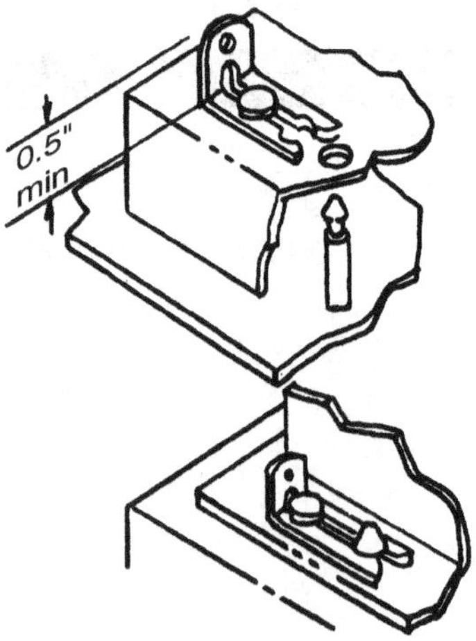	Description:	Snap-slide latch.
	Advantages:	Fast action. No tools required.
	Disadvantages:	Flange-mounted method may be difficult to operate when side clearance is limited.
	Tools Required:	None.
	Operating Time:	Approximately 0.02 min per fastener.
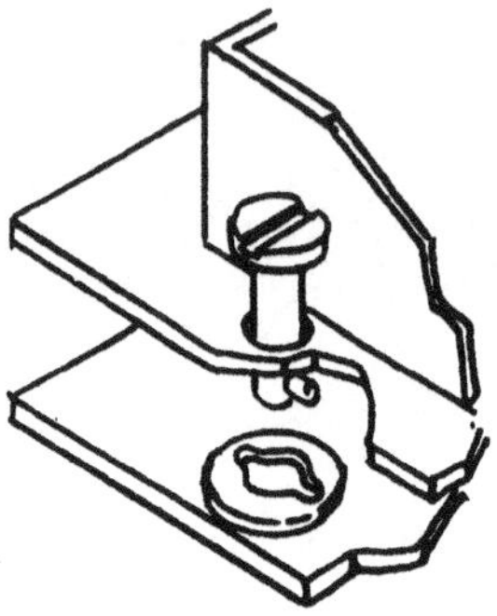	Description:	"Dzus"-type fastener.
	Advantages:	Operating time approximately one-tenth that of captive screw-type fastener.
	Disadvantage:	Requires use of screwdriver.
	Tools Required:	Screwdriver.
	Operating Time:	Approximately 0.05 min per fastener.
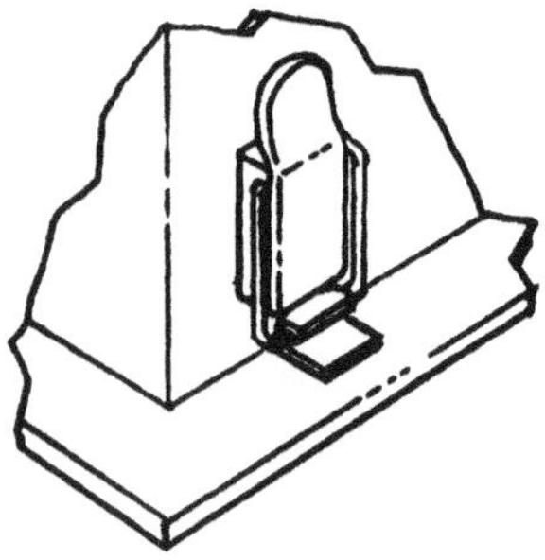	Description:	Spring or drawhook latch.
	Advantage:	Requires no tools.
	Disadvantages:	Requires space to operate. Closing may require more time due to difficulty in engaging drawhook. Strong spring action might cause personnel injury.
	Tools Required:	None.
	Operating Time:	Approximately 0.03 min per fastener.

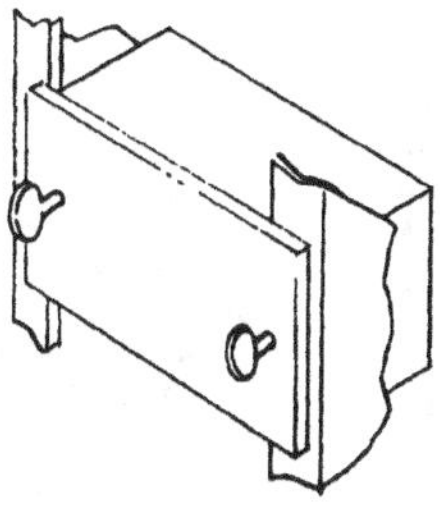

Figure 31.6

TABLE 31.10 Chassis Mounted on Vertical Rack (Front Panel Secured by Quick-Acting Fasteners)

Securing Method	Maintainability Considerations	
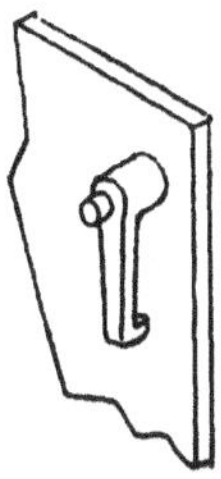	Description:	Push-button latch.
	Advantages:	Integral handle provides easy removal of chassis. Fast release. Closes with snap action.
		Requires outside space for handle. Does not pull panel against gasket as well as other types.
	Disadvantages:	None.
		Approximately 0.03 min per fastener.
	Tools Required:	
	Operating Time:	
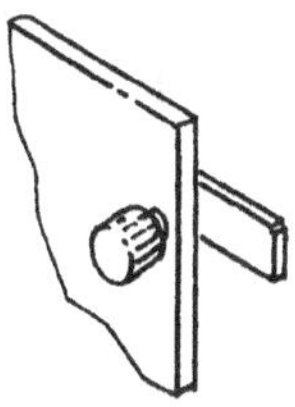	Description:	Pawl latch. Operates with 90° turn.
	Advantage:	Requires minimum inside space.
	Disadvantage:	Difficult to operate when much pull-up is required. None.
	Tools Required:	Approximately 0.05 min per fastener.
	Operating Time:	
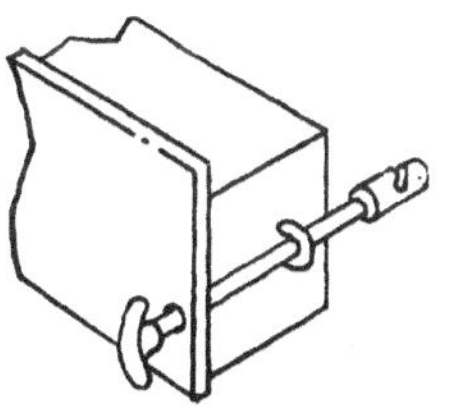	Description:	Cam-action. Operates with 90–180° turn of handle.
	Advantage:	Fairly fast release operation.
	Disadvantage:	Initial engagement of cam may be difficult.
	Tools Required:	None.
	Operating Time:	Approximately 0.05 min per fastener.
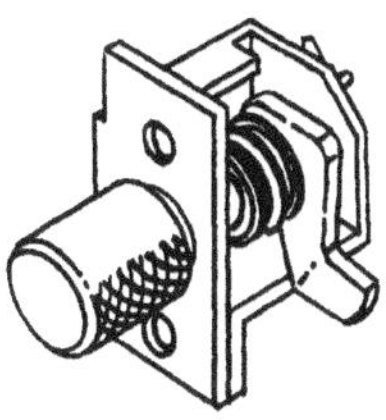	Description:	Adjustable pawl latch. Operating handle pulls pawl toward panel.
	Advantage:	No significant advantage over the other types.
	Disadvantages:	Requires more time than other types. Excessive inside space.
	Tools Required:	None.
	Operating Time:	Approximately 0.1 min per fastener.

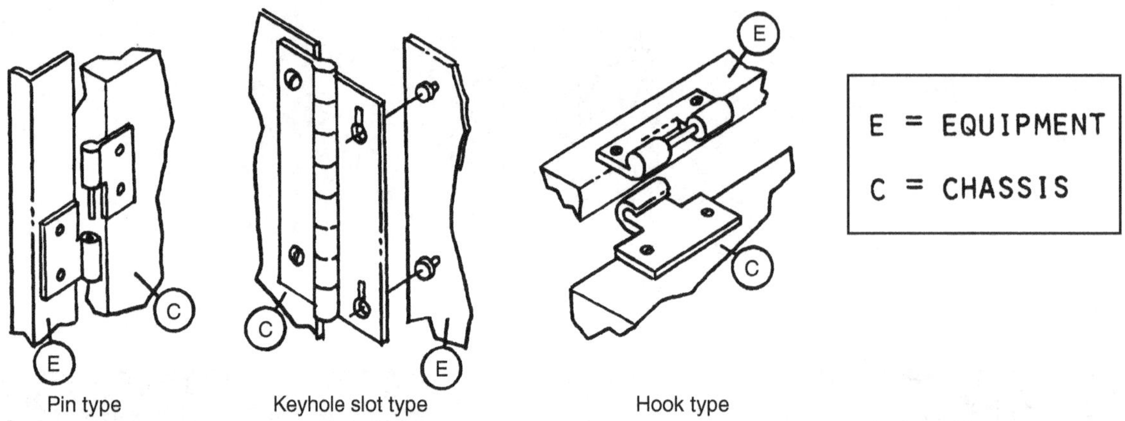

FIGURE 31.7 Examples of separable hinges that will reduce maintenance times.

up to 5 minutes more. Use of separable hinges can reduce this time to a fraction of a minute.

Any separable hinge must, however, include features to prevent accidental separation under shock or vibration conditions. The hinges shown in Fig. 31.7 should not be used to bear the weight of a heavy chassis when the equipment is closed; i.e., guide pins or screw fasteners should be provided to support a heavy chassis.

BLIND FASTENERS

Blind fasteners are, of course, commonly used where the designer assumes that the fastening device may spoil the aesthetic appearance of a product. (See Fig. 31.8.) From a user's point of view, a readily apparent and accessible screw-type fastener will save time and frustration and will prevent damage to the panels. (See Fig. 31.9.)

Another desirable type of blind fastener is used on cabinet doors where there is no dynamic environment involved, in which case a magnetic catch is preferred over the spring-clip type of fastener.

When panels or components are located overhead, fastener concepts should be chosen that will prevent the panel or fixture from falling on someone's head when the last device holding the component is loosened. (See Fig. 31.10.)

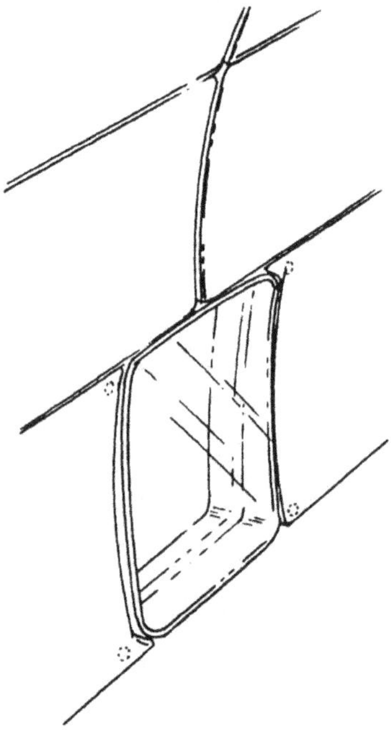

FIGURE 31.8 Blind fasteners are used in automobile and aircraft interiors.

FIGURE 31.9 Mechanics may lose time and damage equipment when finding clips.

FIGURE 31.10 Fluorescent fixtures.

CHAPTER 32
TOOLS

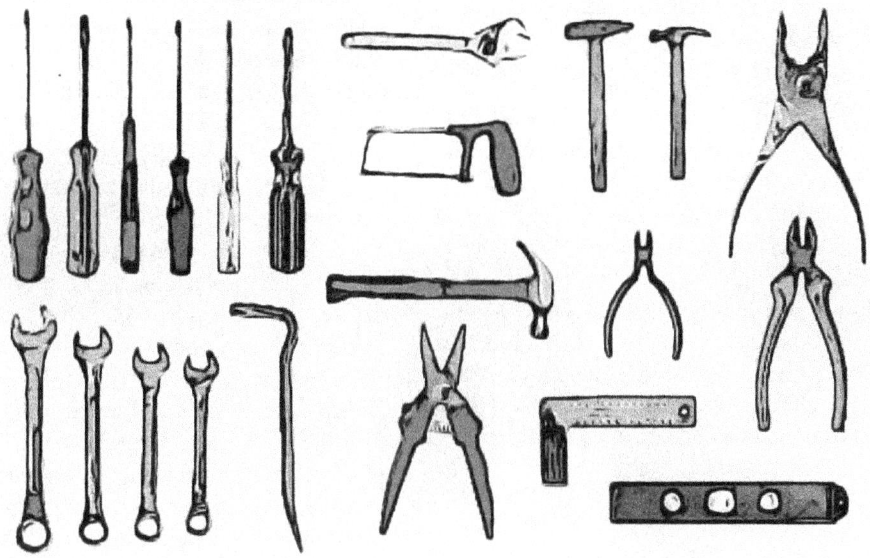

GUIDELINES FOR DESIGN AND SELECTION

The following are key factors to consider:

1. Mechanical Factors: A hand tool is used to extend and reinforce the strength and effectiveness of the operator's limb or limbs while he or she is performing a specific task. Forces generated within the human musculoskeletal system are transmitted to the workpiece through the tool. Reactive forces are generated from the workpiece to the human structure via the tool. Stresses on the human result from excessive forces and poor posture, which frequently are due to poor tool design. Worse yet, when operators are applying a large force on a tool and it slips or breaks, they often find themselves in an unstable position, or they may be injured from the recoil of force.

2. Anthropometric Factors: The dimensional characteristics of human operators are critical to the effectiveness with which they can grasp, hold, manipulate, or guide a hand tool. Particularly important to effective tool use is the design of tool handles, so that not only they can be gripped properly, but also the tool can be properly guided during a given application of force.

3. Anatomical Considerations: The tool design must reflect adequate consideration of the fact that the operator's wrist and arm have specific rotational characteristics and that whenever tool action requires thrust, rotation, or both, limb rotational axes are therefore important to proper tool manipulation. Basic to the tool handling is the posture that must be taken in order to permit the best mechanical advantage. Also, consider both right- and left-handed persons. The tool should be safe and effective for both.

4. Physical Characteristics of the Tools: The basic tool characteristics to be considered include the following:

   a. The tool weight should balance close to the point of support.

   b. The tool material should not conduct electricity.

   c. Tool handles should have a fairly high coefficient of friction to minimize slipping.

   d. Nonporous, hard materials should be used to preclude the retention of toxic substances or the collection of sharp materials, which may become embedded in grip areas.

e. The overall weight should be compatible both with the need to reduce the potential for fatigue when lifting or holding the tool and with the need to minimize the amount of pressure the operator is required to exert in order to make the tool perform as intended.

5. Ergonomic Considerations: The matter of human energy expenditure and how this energy is best exerted should underlie the conceptualization of each new tool design. Such factors as human mobility, dexterity, equilibrium, and posture, as well as consideration of potential ballistic movement advantage, all should receive the full attention of the tool designer.

6. The Work Environment: Not all work environments may lend themselves to a single tool-operator interface concept; i.e., the tool may have to be used under less than optimum conditions, such as overhead, near the floor or ground, with inadequate lighting, or on an unstable platform. Consideration should be given to the potential difficulties imposed by radical temperature, ventilation, and humidity conditions and/or by the requirement to wear heavy clothing and gloves. Power hand tools subject an operator to additional stresses that must be considered.

7. Rotating Components: Proper guards against inadvertent contact with moving blades or gears must be provided.

8. Electrical Shock: Appropriate grounding and/or contact isolation between the tool and the operator must be provided.

9. Heat: Heat generated either by motors or by the interaction between the tool blade and the workpiece poses a dual hazard for operators; i.e., they may receive actual burns, or their escape reaction when they touch a hot surface may cause them to drop the tool or enter into a series of dangerous motions (falling from a ladder, backing into an obstacle, dropping the tool against the body, etc.).

10. Flying Particles or Sparks: Power tools frequently generate flying particles or sparks, thus creating hazards not only for the operator but also for persons nearby. Consideration should be given early in the tool conceptualization to providing appropriate protection from such hazards.

11. Power Control: Unintentional power activation and the lack of an expedient power cutoff are both critical to the safety of power tool operators.

12. Hazards to the Nonoperating Hand: Many injuries caused by power tools result from the fact that the nonoperating hand is accidentally placed where it should not be; e.g., the operator may be using his or her nonoperating hand to hold material, an electric cord, or some other object out of the way. In conceptualizing a hand-held power tool, one should carefully examine the various procedural sequences wherein an operator may place the nonoperating hand in an exposed position. Such an examination may suggest that the tool be specifically designed so that the operator has to place the nonoperating hand on the tool in order to ensure that the hand is kept out of the way.

Other types of special tools create unique problems. Typical of these are soldering tools, which can cause burns if the operator touches a tip that is still hot or lays the tool on flammable materials, for example. Welding operations and the tools used to perform them create another hazard in terms of the "blinding" effects of the extremely bright torch and also the attendant possibility of burns.

It should be noted that the standard approach has been to accept many of the above hazards as part of the job and to place the burden of protection on users, i.e., to recommend wearing eye protectors, using special electrical grounding devices, wearing gloves, etc. Obviously, in many cases these are the only methods available to reduce the hazard potential. However, the designer should, in each new tool design, review such hazards and attempt to remove them whenever possible in the design itself. When this cannot be accomplished, the designer should assume the responsibility for providing appropriate warning labels on the tool and/or properly worded warning instructional materials to accompany the tool. After all, the designer should know better than anyone else what hazards a new tool presents.

GENERAL PRINCIPLES FOR TOOL DESIGN[10]

1. Make the tool easy to carry close to the body with one hand.

2. Make the tool easy to set down or hang up.

3. When the tool includes a "weight-holding function," separate this from force, guidance, and control functions.

4. Provide specific hand holds (otherwise, the operator may take hold of the tool in a hazardous manner).

5. Make the tool compact and light so that it is easy to carry, handle, and store.

6. Align the tool's center of gravity with the grasping hand so that the operator will not have to overcome rotational movement or torque of the tool.

7. Make tool grasping surfaces resilient and slip-resistant.

[10]L. Greenberg and Don B. Chaffin, *Workers and Their Tools: A Guide to the Ergonomic Design of Hand Tools and Small Presses*, Pendell Publishing Co., via an educational gift from AMP incorporated of Harrisburg, PA.

8. Round off all edges and corners of the tool package and avoid adding sharp, protruding elements.

9. Make flywheels solid and without spokes; round the edges.

10. Provide nonaccess guards around all moving parts.

11. Use low-voltage electrical power and double insulation.

12. Use safety-secured couplings with high-pressure, compressed air or fluid power tools to prevent inadvertent disconnection and line whipping.

13. Provide flanges on the tool handle to prevent the hand from slipping forward as force is applied and/or on the aft part of the handle to keep a heavy tool from slipping out of the hand when it is being carried.

14. Orient the tool handle so that the operator's wrist can remain in the most natural position while he or she is applying force or guidance inputs.

15. When an auxiliary control must be manipulated while the operator is holding the tool, place the control where the thumb or fingers of the holding hand can reach and operate it without disturbing the holding position or place the control where it can be operated by the other hand.

16. If sleeve-type handle covers are provided, make sure they do not inadvertently slip, rotate, or come off.

17. Color code certain tool components, e.g., safety switches.

18. Insulate contact surfaces to preclude electrical shock or high-temperature interface with the operator's hand or body.

19. Use a battery-powered design where practical to avoid entanglements with an electric cord.

20. Although it may not be possible to control noise produced by the tool-work interface, the designer should attempt to minimize internal noise that may be produced by bearings and other internal components.

TOOL DESIGN CHECKLIST (SEE FIG 32.1)

1. Work position
2. Work material holding
3. Tool grasp interface
4. Visibility
5. Tool operation and guidance
6. Prevention of debris and sparks

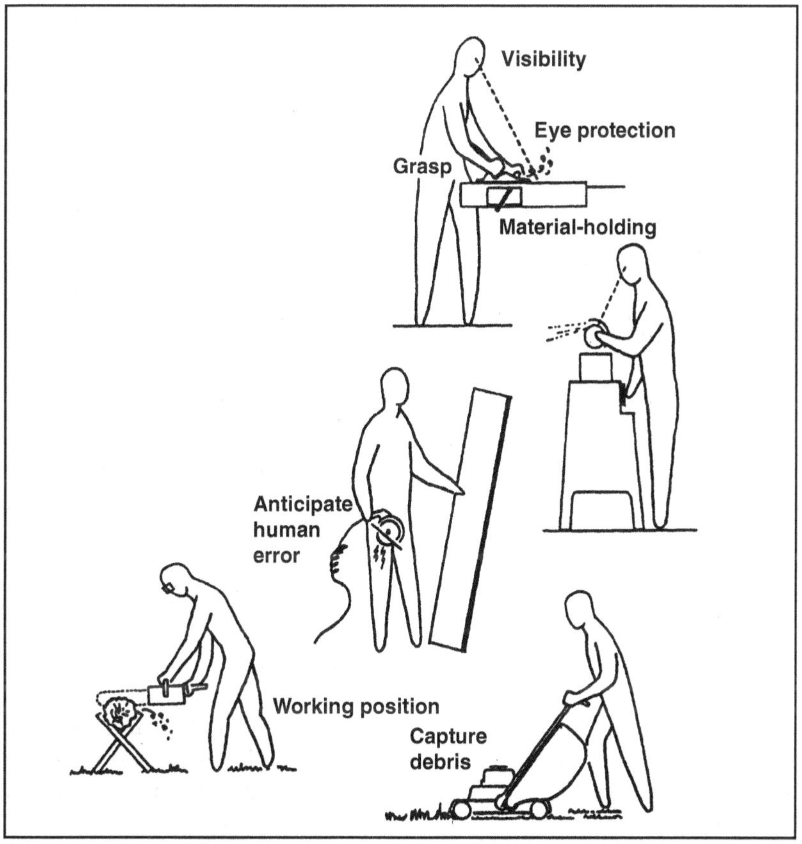

FIGURE 32.1 Factors to consider in tool design.

7. Tool balance and weight
8. Tool holder
9. Electric cord position and control
10. Guards for moving parts
11. Tool storage
12. Tool adjustment
13. Vibration and heat attenuation
14. Emergency safety switch
15. Prevention of parts loosening
16. Prevention of electrical shorts
17. Tool maintenance and repair

In considering various tools, it is almost always important to examine the relationship between the handle and the method or manner in which the tool will be used; i.e., the designer should consider whether it will be used merely to hold the tool or whether it must be used to rotate, push, pull, or guide the tool, for example. Equally important are the problems associated with gripping and applying pressure. Some important handle considerations are illustrated in Figs. 32.2–32.4:

1. Not only must screwdrivers and chisels be gripped tightly, but also they usually require that pressure be exerted toward the work. Generally, it is desirable to place the handle-tool longitudinal axis in line with the operators arm in order to minimize fatigue and reduce the chance of "skewing" the tool and causing it to slip off the work. (See Fig. 32.2.)

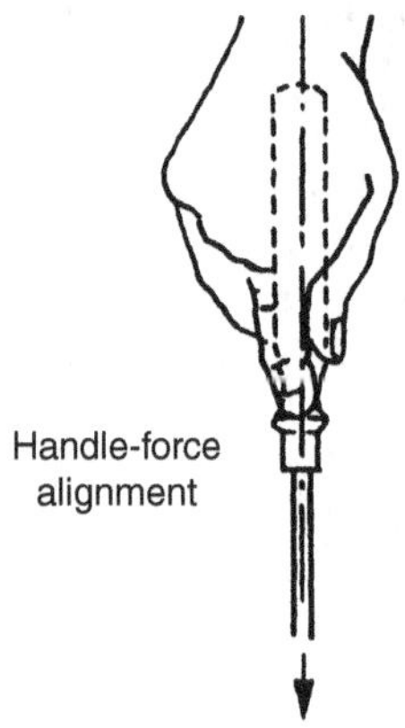

FIGURE 32.2 Align the handle axis with the operator's arm to allow tool control and application of force.

2. Screwdrivers have to be rotated, and thus they should have round handles. Slight grooves or knurling helps keep the hand from slipping. Screwdrivers should have handle diameters that are compatible with the force expected to be applied. Except for very small, finger-operated screwdrivers, handles should be about 1.0–1.5 in. (2.5–3.8 cm) in diameter. (See Fig. 32.3.)

FIGURE 32.3 Groove or knurl screwdriver handles.

3. Other tools should have semiflat sides. In the case of a hammer, the flat side aids in guiding the stroke. In the case of a chisel, the flat side helps control the lateral position of the blade. (See Fig. 32.4.)

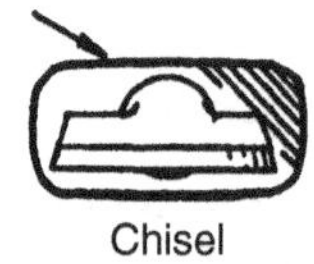

FIGURE 32.4 Flat sides on handles help control lateral position and rotation.

DESIGN RECOMMENDATIONS

Tools which are held in the hand for a considerable time or which require continuous gripping and possible pressure applications should be designed not only so that they fit the hand and fingers but also so that stresses are not placed on the wrist and forearm.

A typical surgical instrument shown in Fig. 32.5 tended to roll in the surgeon's fingers and produced considerable

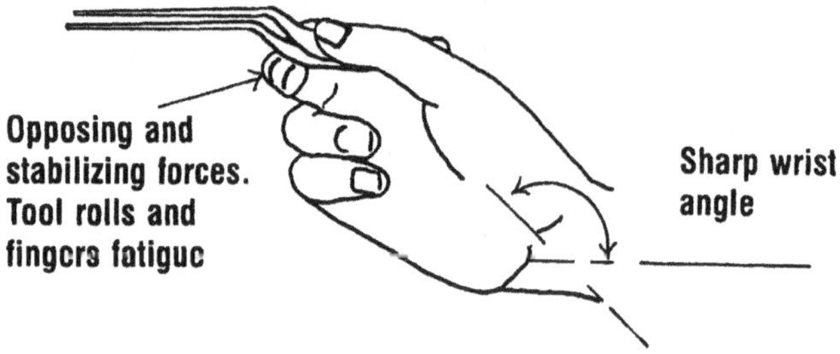

FIGURE 32.5 Surgical tool.

fatigue in the fingers and thumb. A redesign (shown in Fig. 32.6) by Professor Tichauer of New York University eliminated most of the problems by changing the tool geometry. The main action of the thumb opposing the second, third, and fourth fingers and the stabilizing action of the second, fourth, and fifth fingers were made easier. In addition, the redesigned instrument lessened sharp wrist angle. (See Fig. 32.6.)

Consider the readability of the tool markings. A typical caliper tool can be anodized so that the background surface

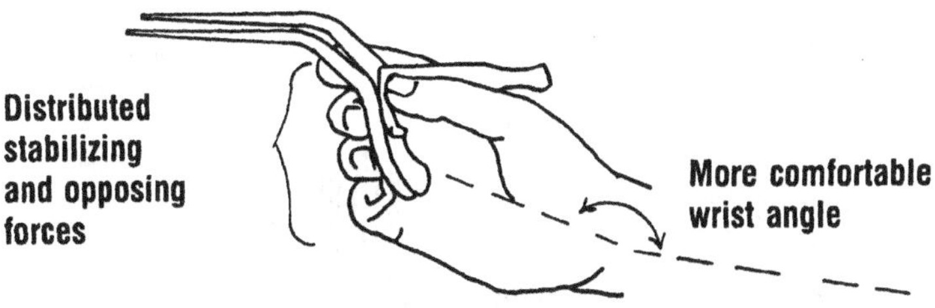

Figure 32.6 Surgical tool after redesign.

provides good contrast for black scale markings and numbers (see Fig. 32.7.)

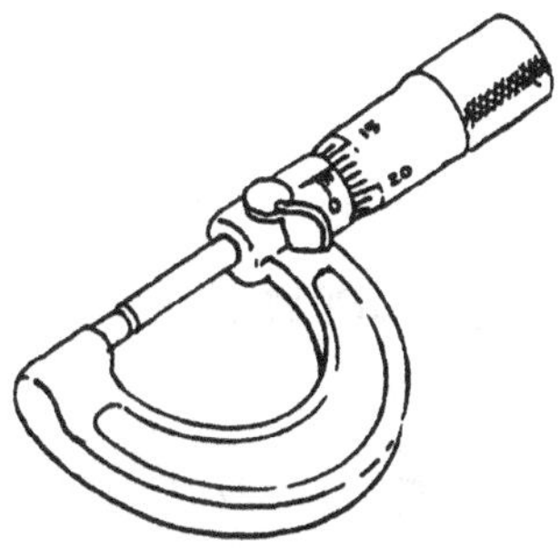

Figure 32.7 Avoid glossy surfaces on scales—provide good contrast.

TOOL DESIGN TO MINIMIZE OPERATOR INJURY

Typical hazards include the following (see Fig. 32.8):

1. Laceration due to exposed blades, points, edges, rotating wheels or gears, rough edges, burrs, etc.

2. Bruises and contusions due to coming in contact with accessible pinch points and bumping against sharp corners or edges or protruding structures or fasteners

3. Abrasions due to exposure to rapidly moving machine components, brushing against a rough surface, or slipping of the hand from a rough handgrip

4. Skeletal fractures or breaks due to inadvertent entrapment between two articulating machine components, gears, or belts or due to unforeseen component reaction against the operator's limb

5. Eye penetration due to a lack of eye protection from flying particles

6. Burns due to access to hot surfaces

7. Electrocution due to lack of insulation from inherent electrical current and/or possible failures

8. "White fingers" (Raynaud's syndrome) due to excessive, continued exposure to tool and hand vibration

9. Injuries due to movements and forces that are not within the normal range of human capabilities.

10. Tool designs that require repeated aggravating actions and motions causing injuries such as carpal tunnel syndrome.

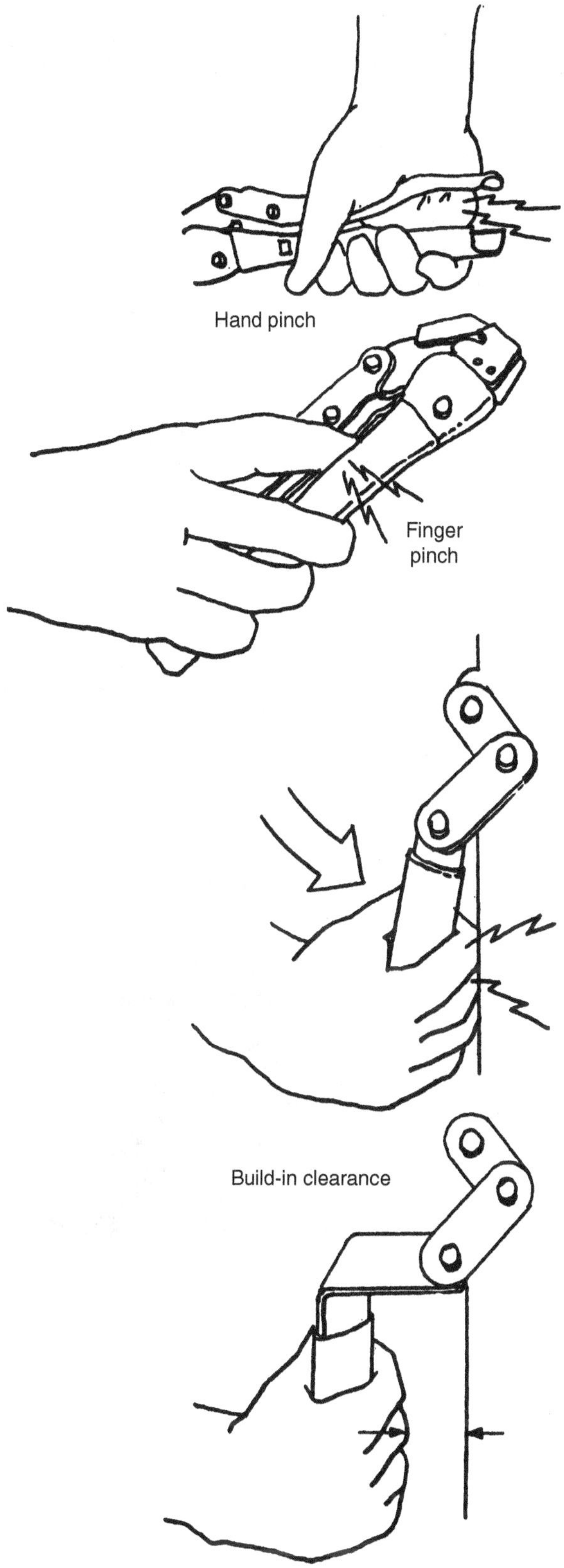

Figure 32.8 Hazards in tool design.

Figures 32.9 and 32.10 illustrate a number of typical problem solutions that, it is hoped, will provide the reader with ideas regarding an even wider array of tools.

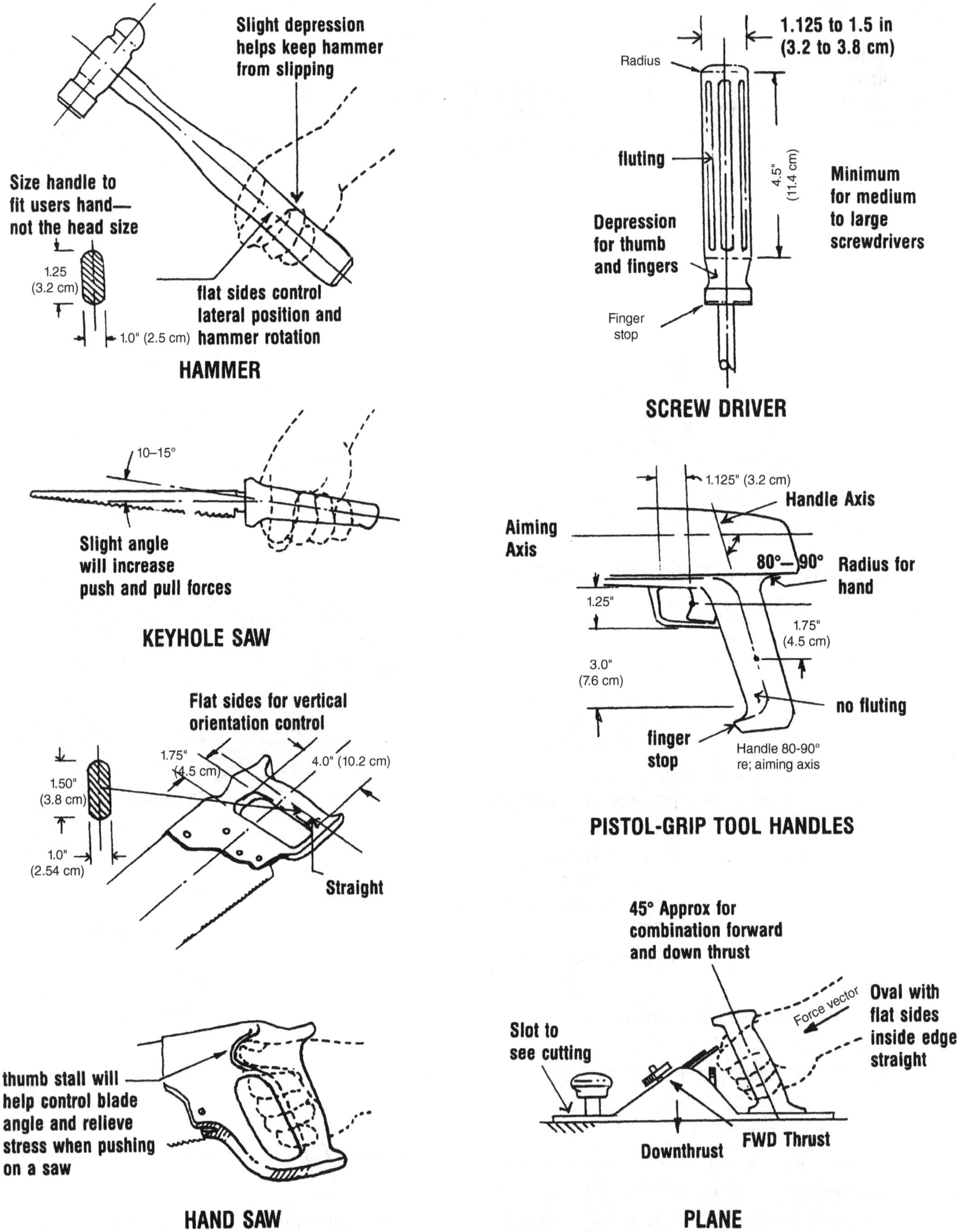

FIGURES 32.9 Tool design solutions.

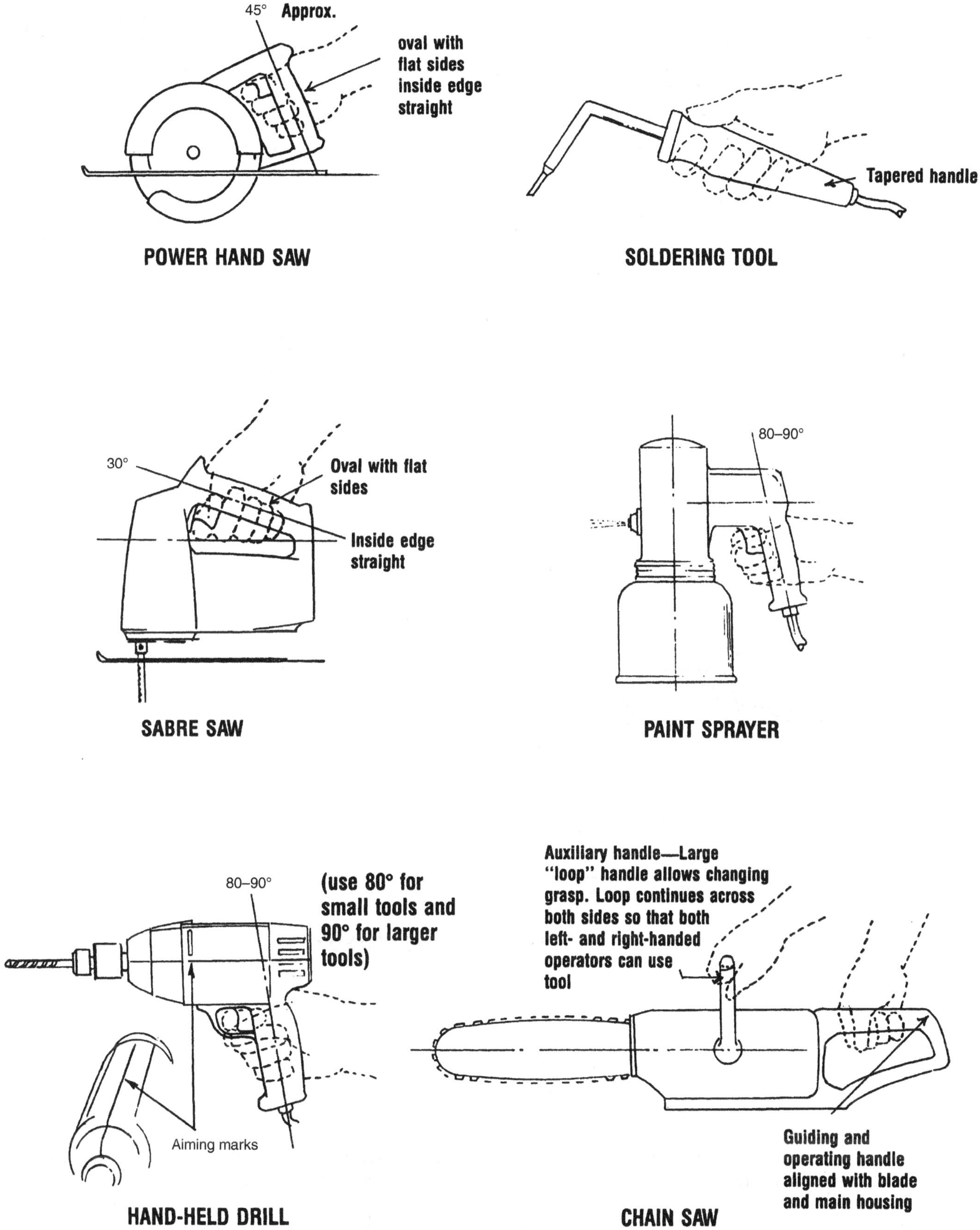

Figures 32.9 *(Continued)*

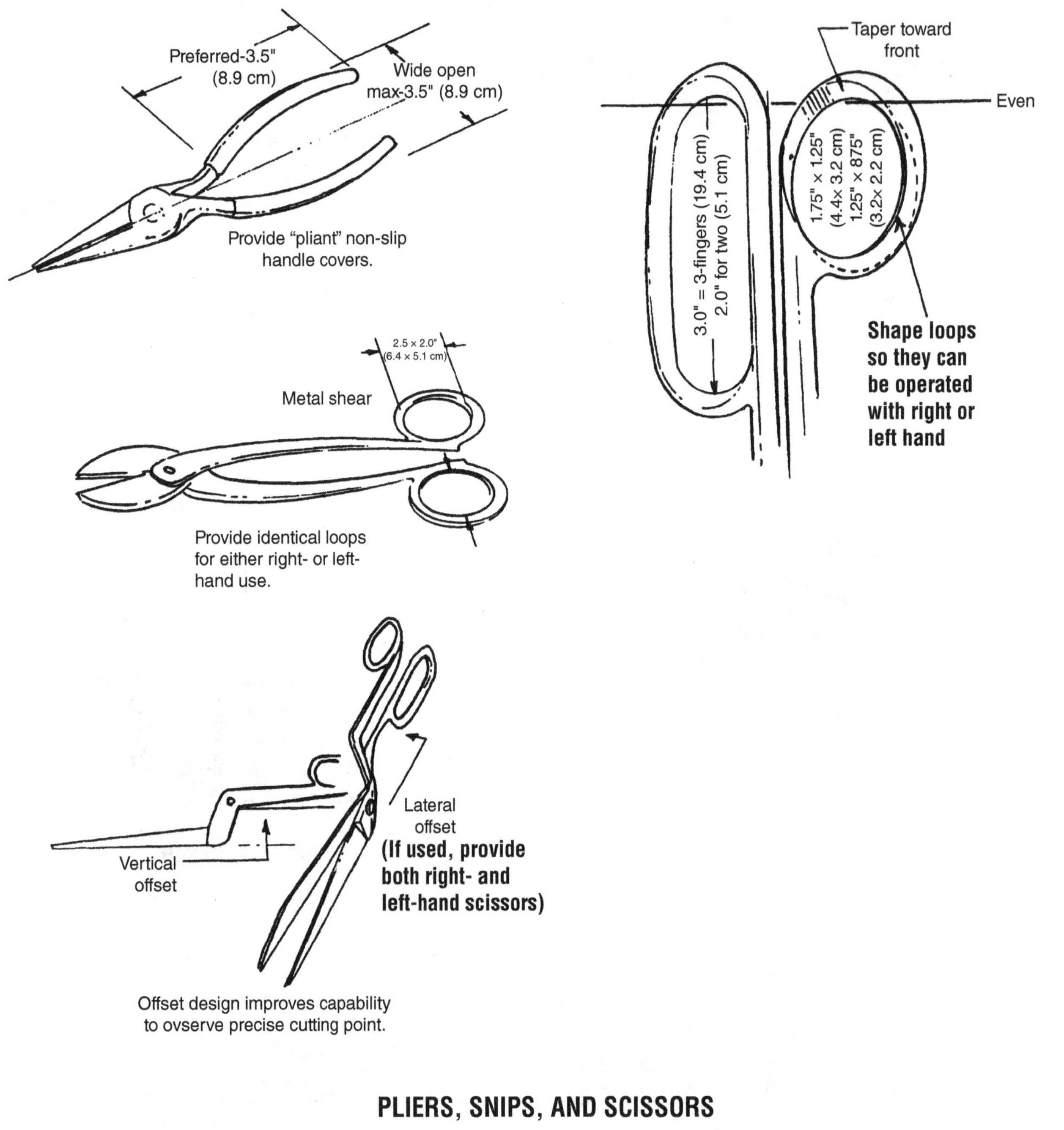

FIGURES 32.9 *(Continued)*

Details for tool design and worker protection are published in the Occupational Safety and Health Administration (OSHA). The standards are in the Code of Federal Regulations, Title 29, Part 1910.

Whether the tool being planned is a very simple hand-actuated device or a complex powered device, a complete analysis should be made of the potential personal modes. Although it is not always possible to eliminate all the injury potential (i.e., one must hope that users will follow instructions for their own safety), many fairly simple design modifications can change an otherwise unsafe tool into a relatively safe one. Be particularly conscious of the hazards that the average user will not see or expect. Handles that close completely catch the operator unaware. A simple expedient is to create a "stop" so that the handles do not fully close and thus pinch the operator.

Equipment Guarding

A larger tool, such as a press or ram (as illustrated in Figs. 32.11, 32.12, and 32.13), can be especially dangerous,

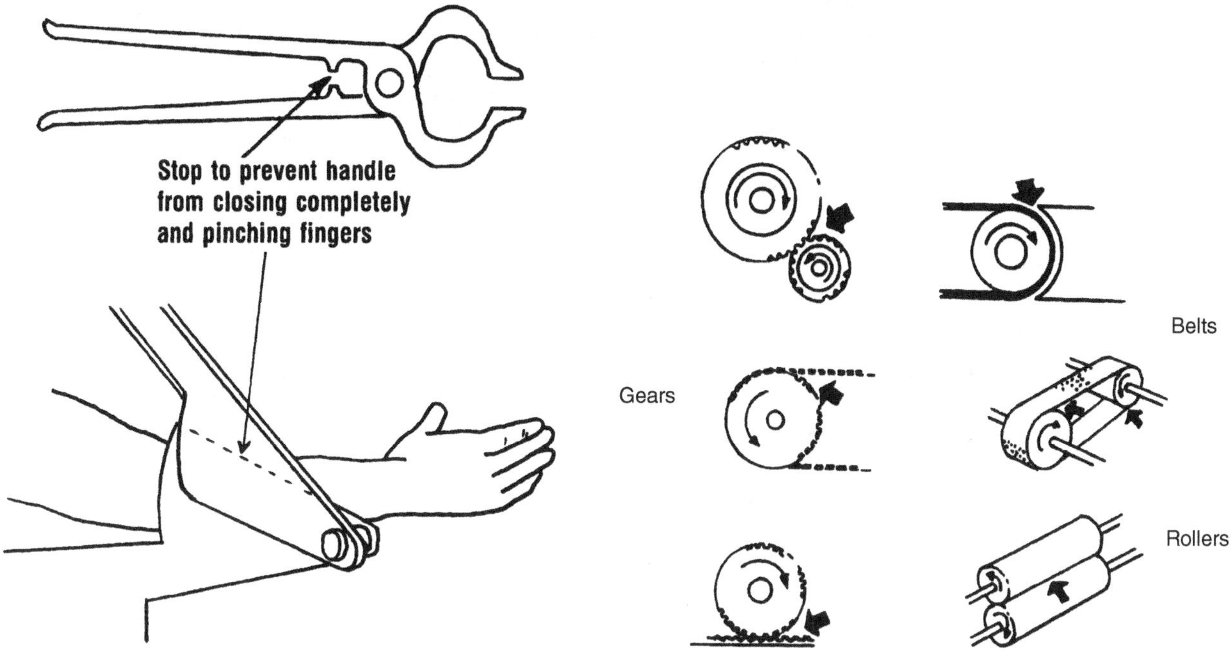

FIGURE 32.10 Guard pinch points.

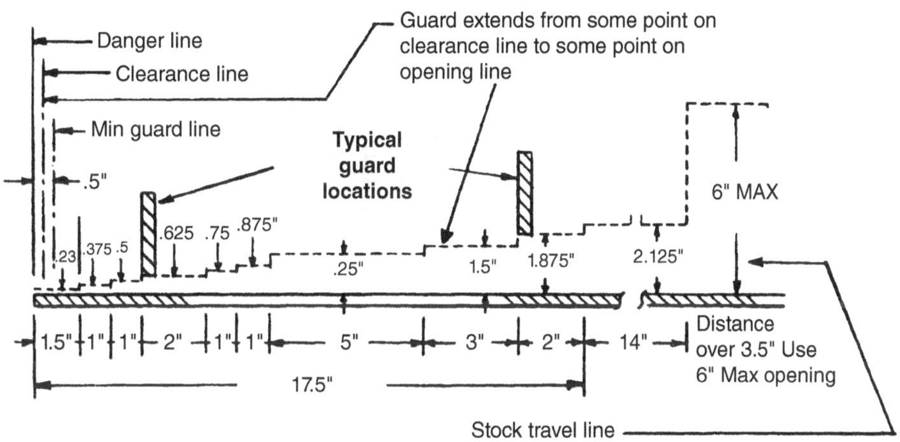

GUARDING LIMITATIONS FOR RAM OPENING

DISTANCE OF OPENING FROM POINT OF OPERATION HAZARD, INCHES	MAXIMUM WIDTH OF OPENING, INCHES
½ to 1½	¼
1½ to 2½	3/8
2½ to 3½	½
3½ to 5½	5/8
5½ to 6½	¾
6½ to 7½	7/8
7½ to 12½	1¼
12½ to 15½	1½
15½ to 17½	1⅞
17½ to 31½	2⅛

FIGURE 32.11 Relationship between allowable opening in barrier guard and its distance from the point of operation. (From OSHA Standards, Sec. 1910.217.)

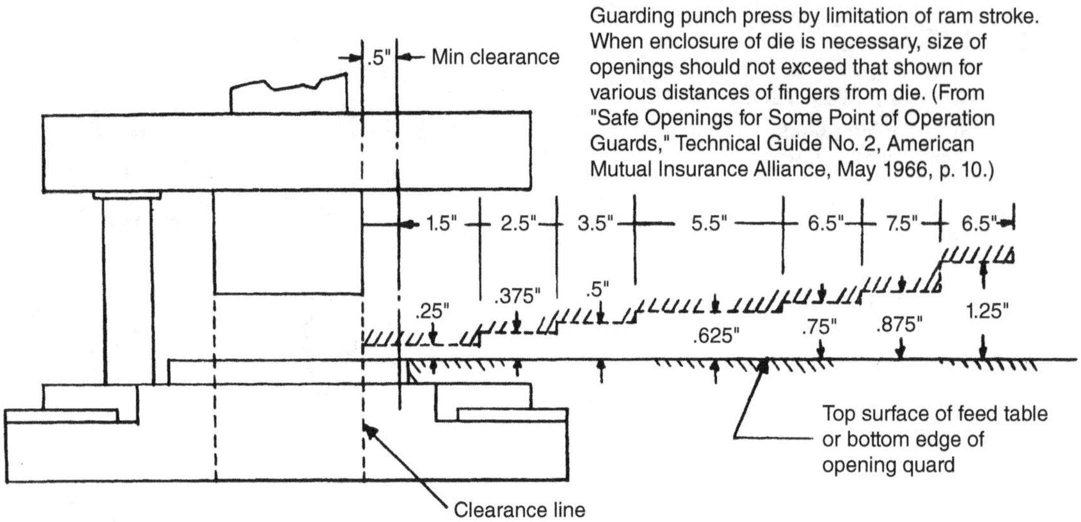

FIGURE 32.12 Guarding limitations for punch press ram stroke.

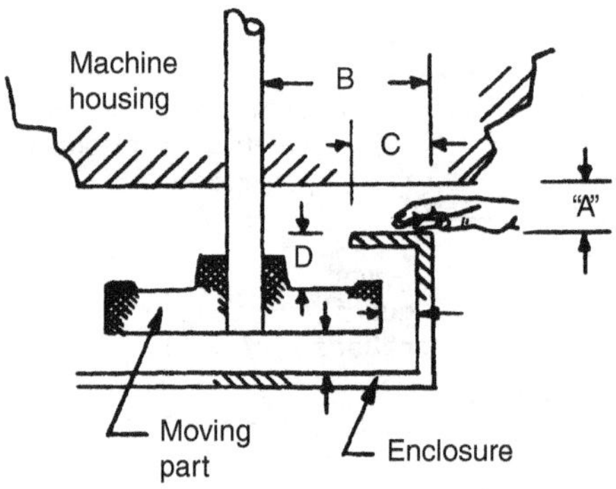

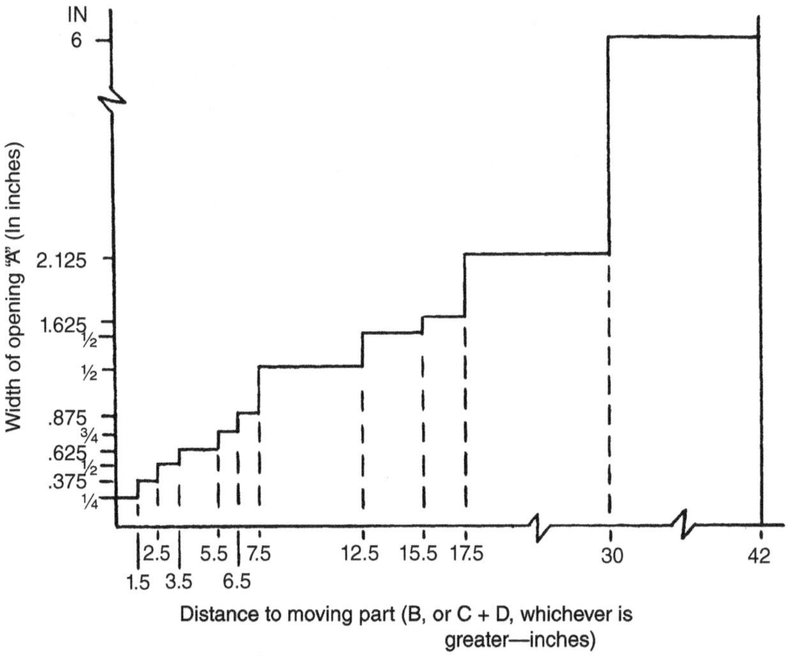

FIGURE 32.13 Guarding access to moving parts.

not so much because the operator cannot see the hazards, but because people tend to hurry and focus not on the hazards but on the work activity. Guards must be designed to minimize operator injury. Figures 32.11–32.17 show design solutions for large tool guarding.

A good solution to the problems of guarding a typical rotating component, providing an opportunity to inspect the component during operation, and making it accessible for part changes and/or other servicing. (See Fig. 32.15.)

A properly designed guard for an alligator shear permits observation for inspection and ease of access for maintenance. The hinged section of the guard must be electrically interlocked to prevent operation of the shear if the guard is not in place. (See Fig. 32.16.)

A photoelectric guard is completely static; thus there is no need for signal devices or adjustments. (See Fig. 32.17.)

Vibrating Tools

When selecting or designing reciprocating or vibrating tools, the frequency and amplitude spectrum of the vibrations transmitted onto the operator should be evaluated. The risk of somatic resonance response is highest when the amplitude

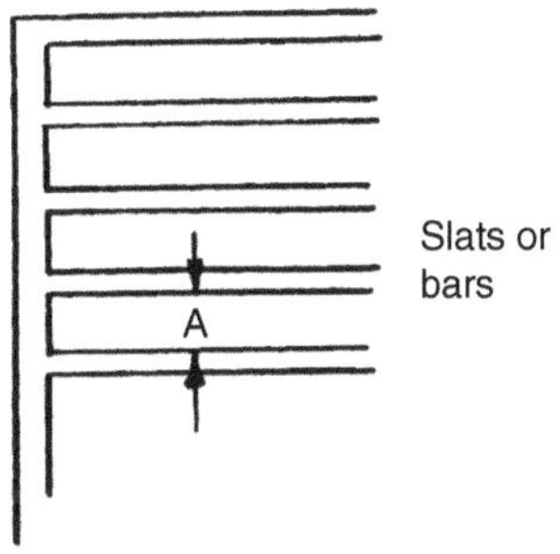

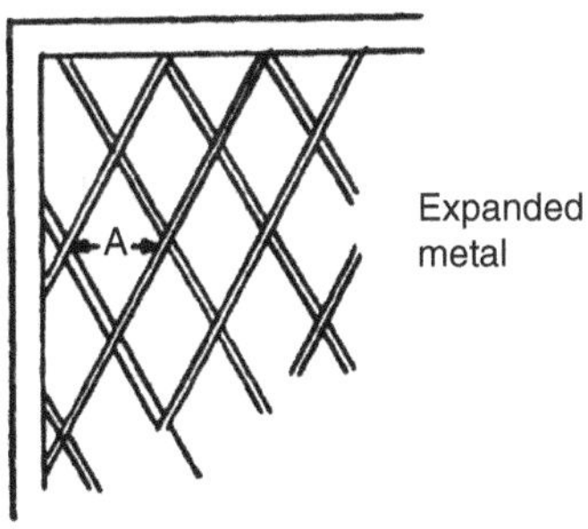

Material	Clearance from moving part at all points, in.	Largest mesh or opening allowable *B*, in.	Minimum gauge (U.S. standard) or thickness
Woven wire	Under 2	⅜	No. 16-⅜ in.
	2–4	½	No. 16-½
	4–15	2	No. 16-2
Expanded metal	Under 4	½	No. 18-⅓ in.
	4–15	2	No. 13-2
Perforated metal	Under 4	⅓	No. 20-⅓ in.
	4–15	2	No. 14-2
Sheet metal	Under 4		No. 22
	4–15		No. 22
Wood or metal strips crossed	Under 4	⅜	¾-in. wood or No. 16 metal
	4–15	2	
Wood or metal strips not crossed	Under 4	½ the width	
	4–15	One width	
Plywood, plastic, or equivalent	Under 4		¼ in.
	4–15		¼ in.
Standard railing	Minimum of 15		
	Maximum of 20		

Figures 32.14 Standard materials and dimensions for machinery guards.

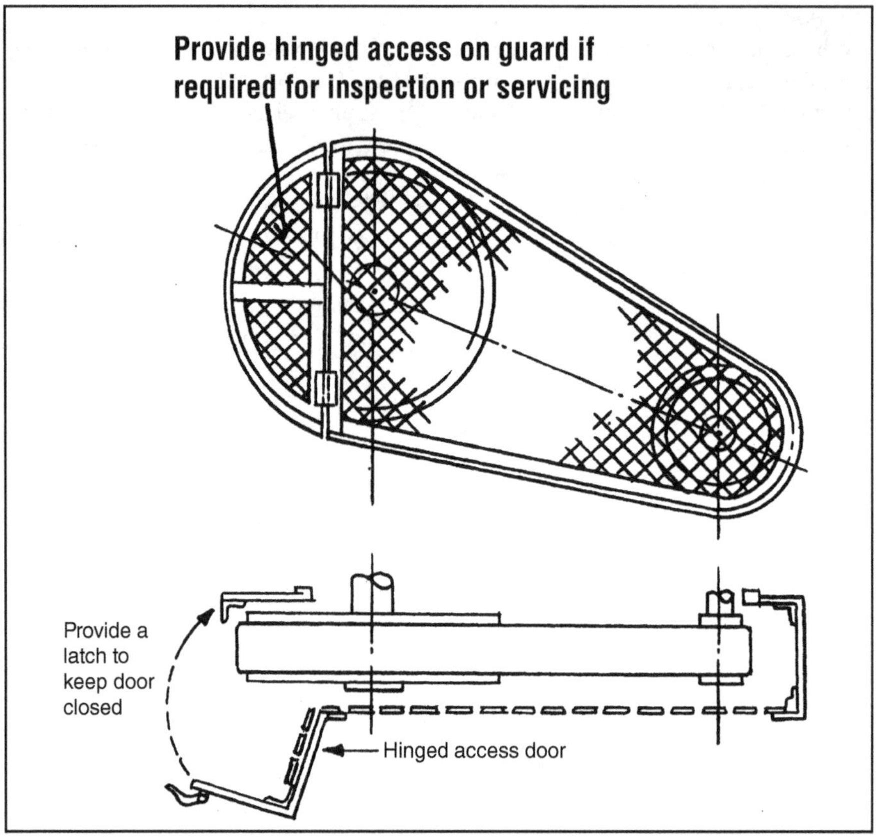

FIGURE 32.15 Example of design of machine guard with hinged access.

exceeds 100 μm within the frequency band 3 to 125 Hz. (See Fig. 32.18.)

When tools are activated by a rotating mechanism, the maximum torque transmitted on the axis of rotation of the forearm should be below 12 in. lb (1.4 N m). A two-handed power tool should be designed with an angle of 120° between the gripping axes of both hands, and the handles should be isolated to minimize as much vibratory transmission as possible.

Electrical Hazards

Table 32.1 gives human response to electrical shock. Protective measures are outlined in the Federal Occupational Safety and Health Administration Standards (Title 29, Part 1910) and the military standard MIL-STD-454, "Standard General Requirements for Electronic Equipment."

FIGURE 32.16 Guard allows visibility has hinged access and electrical safety interlock.

FIGURE 32.17 Photo electric interlocks do not require adjustments.

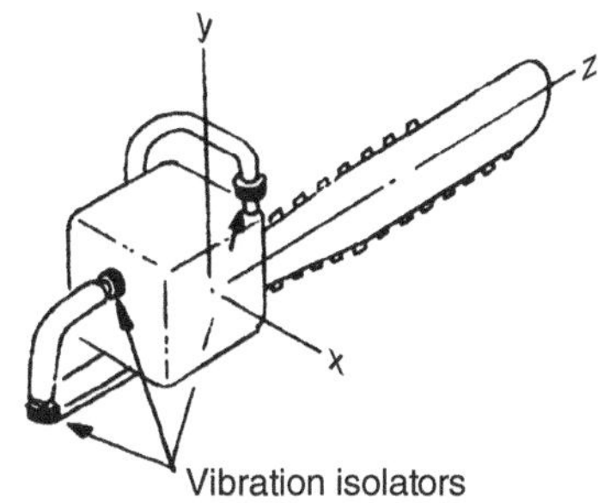

Figure 32.18 Vibrating tools should have handles that isolate vibrations.

TABLE 32.1 Shock Current Intensities and Their Probable Effects

Current (milliamperes)		
AC (60 Hz)	**DC**	**Effects**
0–1	0–4	Perception
1–4	4–15	Surprise
4–21	15–80	Reflex Action
21–40	80–160	Muscular Inhibition
41–100	160–300	Respiratory Block
Over 100	Over 300	Usually Fatal

Chapter 33
Furniture and Appliances

GENERAL GUIDELINES

In designing or selecting furnishings, the following general considerations should be kept in mind:

1. User fit: Careful consideration should be given to the dimensions and anatomic characteristics of users to make sure that the furniture fits them, supports them properly, and adjusts to their activities, i.e., rest, work, and recreation.

2. User efficiency: Consideration should be given to what users have to do to and with the furniture in terms of arrangement of elements, articulation of components, operation of special controls, accessibility within components, and visual impact of finishes.

3. Interactive characteristics: Consideration should be given to how one piece of furniture interacts with another positionally and in terms of orientation and interference.

4. Safety: Consideration must always be given to potential safety hazards, including sharp contact points, fragility of the structure, balance and stability, and flammability.

5. Housekeeping: Someone has to clean around, move, store, and rearrange furniture at some time. The design should aid, not hinder, such operations.

Good furniture design can be created for any styling objective, but good human engineering must come first.

DESIGN CHARACTERISTICS OF A GOOD SEAT

Whether designing a chair, a sofa, or a pilot or vehicle driver seat, one should start with the following critical features as a base line:

1. Good posture: To provide good sitting posture, i.e., one in which the back and neck muscles are least strained, the seat pan should tilt back approximately 5° to shift the upper torso weight and thus cause the torso to rest fully against the seat back. The angle between the seat pan and the seat backrest must be approximately 105° to keep the torso against the backrest and yet not force the occupant to lean his or her head forward in order to balance it property.

2. Surface support: The seat should be approximately 19 in. (48.3 cm) wide to keep large occupants from

"lapping over." The seat-pan length should be about 17 in. (43.2 cm). A longer seat strikes the back of the short person's legs; a shorter one provides too little support for the longer-legged person's thighs. The seat back should be about 20 in. (50.8 cm) high so that even the tallest person's shoulder blades are supported.

3. Seat-pan height: The best compromise seat-pan height for a full range of male and female adults for normal sitting (i.e., chairs) is about 17 in. (43.2 cm) at the leading edge of the seat pan. If the seat pan is higher, the short female occupant's legs will dangle. Even the 17 in. (43.2 cm) assumes that she will be wearing shoes with a 1.5- to 2.0-in. (3.8- to 5.1-cm) heel. Although a slightly lower seat pan would be desirable for very short-legged people, the lower height would make it very difficult for tall people to get up out of the seat.

4. For other types of seats, all the above guidelines, except for the seat height, should be used.

5. Stools can be higher, but a footrest should always be provided to maintain the basic foot-to-seat-height relationship.

6. Vehicle seats should generally be lower, at least for the driver, because the driver needs to extend his or her legs to operate the pedals. The optimum height for an automobile driver's seat should vary from about 12 to 13 in. (30 to 33 cm) above the nominal floor (heel resting position for accelerator pedal operation). In order to accommodate the range of leg reaches from that of the 5th-percentile female driver to that of the 95th-percentile male driver, the seat should have a minimum of 8 in. (20 cm) of fore-aft adjustment; i.e., when the seat is all the way back, the seat height should be 12 in. (30 cm) above the heel point, and when it is all the way forward, the seat height should be 13 in. (33 cm). This seat-height variation, coupled with the fore-aft adjustment, allows the short driver to see over the hood and still reach the foot pedals, and the lower aft position allows the taller driver to sit low enough to clear his or her head.

Seat Design: Cautions and Suggestions

No two people need contours in the same place, and a contour in the wrong place is much worse than none at all. Fortunately, most people are appropriately padded so that they can accommodate well to a flat seat pan or seat backrest. (See Fig. 33.1.)

The so-called bucket seat should be avoided, especially for the seat pan. The concave seat pan causes the occupant's thighs to roll inward; as a result, although the occupant may think the seat is fine at first, he or she will soon discover that it causes discomfort and, in the case of a vehicle operator's seat, it may even make it difficult to move the legs laterally.

On the other hand, a concave backrest may help reduce the effects of sidesway in automobile driving. The concave feature should be applied only to the lateral axis of the seat back; the backrest should be flat or straight up and down.

Deep cushions and/or softly sprung cushions should be avoided for both the seat pan and the backrest. As Fig. 33.1 illustrates, the upper part of the backrest pushes the occupant's torso forward, his or her own weight compresses the lower back and seat cushions to reduce the desired included angle, and the forward cushion cuts off circulation in the legs.

Designed-in Hazards

Designers often unknowingly create hazards for the consumer, especially those associated with chairs and seats. A few of the typical designed-in hazards are illustrated in Fig. 33.2.

HOUSEHOLD SEATING

The following suggestions will aid in the design and/or selection of appropriate furnishings. (See Figs. 33.3–33.5.)

1. All seating should have approximately the basic dimensional characteristics noted previously.

2. Chairs and/or stools used in conjunction with eating should be selected at the same time that tables are chosen so that the seat-height–table-height relationship is proper. A 30-in. (76.2-cm) table requires an 18-in. (45.7-cm) seat; a 29-in. (73.7-cm) table requires a 17-in. (43.2-cm) seat. Snack bars and counters should be designed with a particular seat height in mind. Since production chairs or counter stools come in several standardized heights, select the desired height of chair first and design the table to fit.

3. Sofas, and other overstuffed seats almost invariably suffer from a lack of concern for good seating principles. Either look for the following or design to these guidelines:

   a. Observe the previously defined seat-length criteria—most sofa seats are too long.

   b. Select firm cushions—most production furniture is too soft.

   c. Provide kick room at the front of the seat or divan; people have to place their feet as near their center of gravity as they can in order to get out of the typically low overstuffed chair or divan.

d. If the chair is a reclining type, make sure that the base of the backrest coincides with the surface of the seat; i.e., make sure that as the backrest folds back, the hinge point does not cause the lower edge of the backrest to raise higher than the seat cushion.

OFFICE SEATING

Although considerably more attention has been paid to the improvement of certain kinds of office seating (especially workstation chairs) in recent years, there is still a tendency to glamorize office furniture to the detriment of its basic

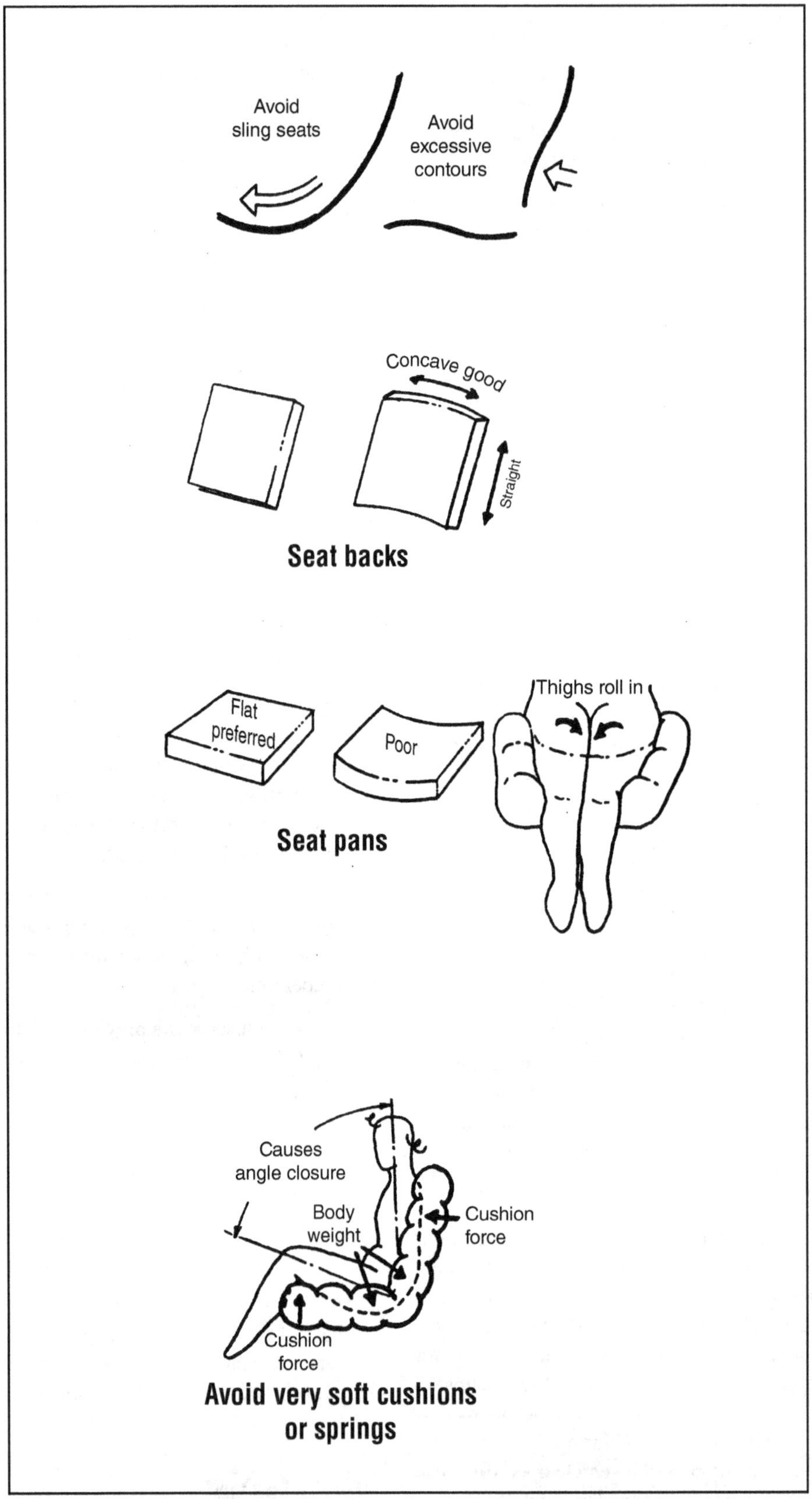

FIGURE 33.1 Seat designs to avoid.

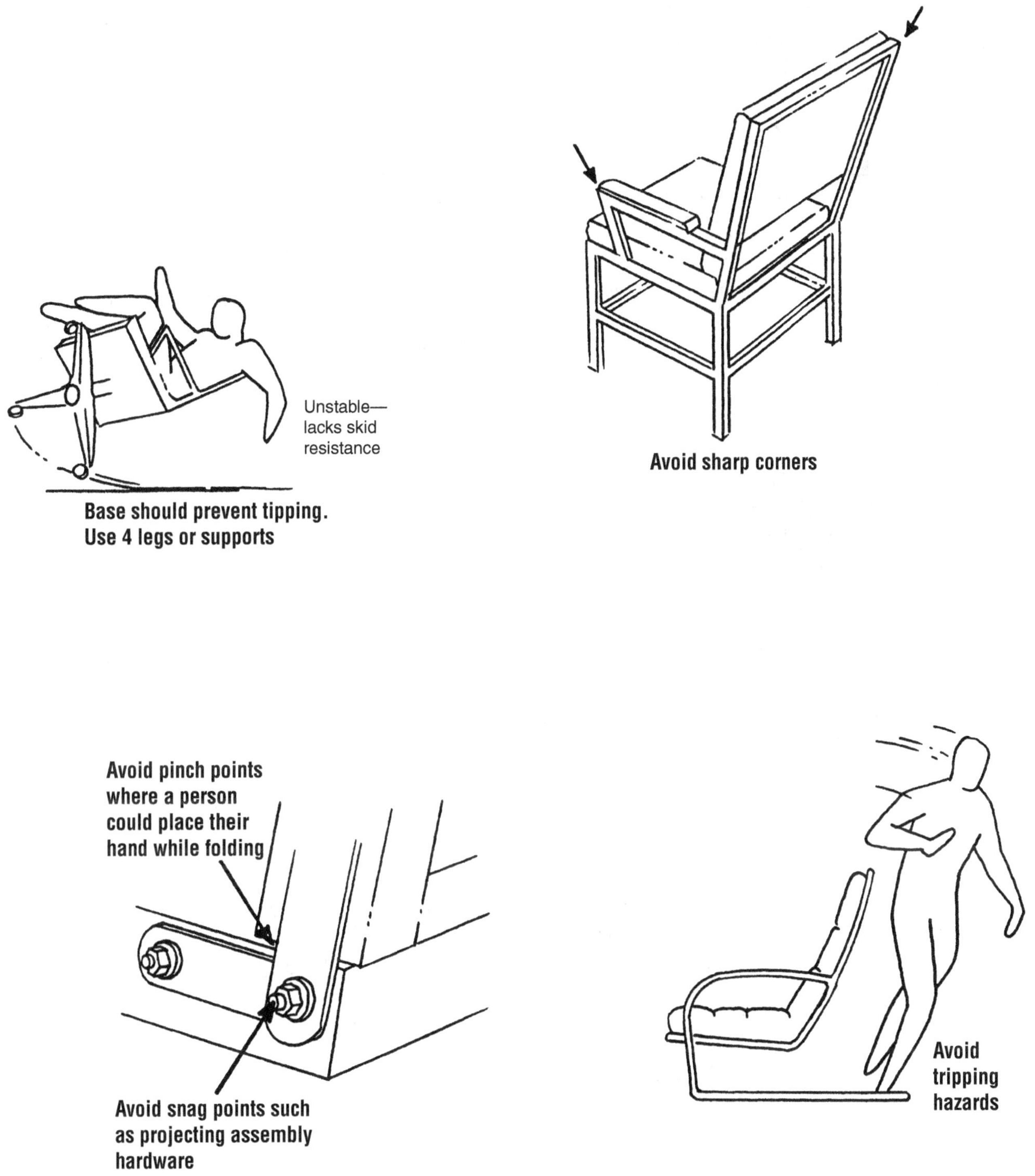

FIGURE 33.2 Designed-in hazards.

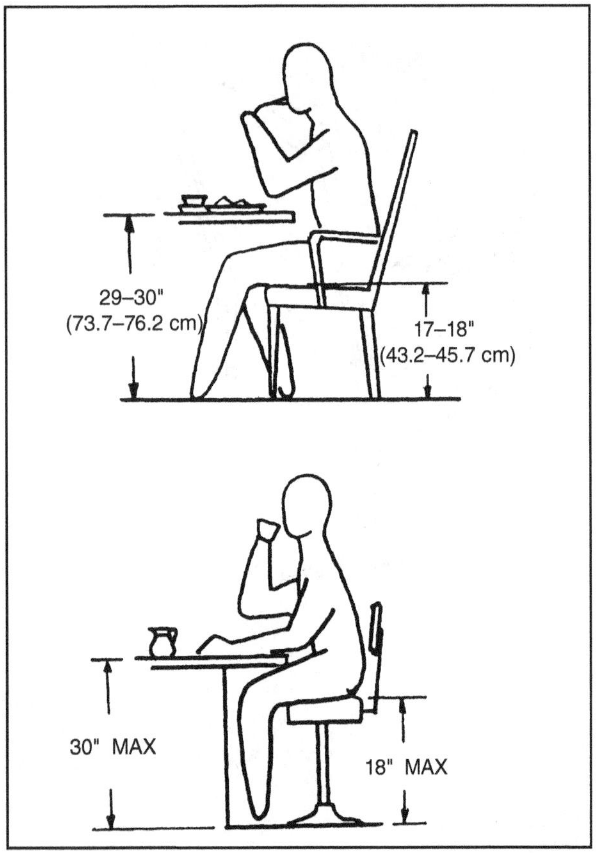

FIGURE 33.3 Dining chair height should be compatible with table height.

purpose, i.e., to improve the comfort and productivity of the office worker.

Desk Chairs

Most desk chairs have seat pans that are too long—the previously noted dimensional and angular guide lines apply here as well as to any other seat. (See Fig. 33.6.) Since the desk chair generally becomes the personal chair of the person to whom it is assigned, provide height adjustment. Also, provide casters for added mobility and a spring-adjustable backrest so that the user can either maintain an alert working position or lean back for more relaxed conversation.

Workstation Chairs

Provide personal height adjustment for both the seat and the backrest. The backrest should also tilt so that it can conform to the individual worker's lumbar support needs. The workstation chair should have dimensions and seat and backrest angles similar to those shown in the Fig. 33.7. The height adjustment should be the secure plunger type that is spring-loaded so that the adjustment is easy to make.

Workbench Stools

Provide essentially the same adjustments for personalization that were suggested for the workstation chair, except do not provide castors; i.e., the seat must "sit still." (See Fig. 33.8.) Make sure that the vertical adjustment systems remain secure if the chair is lifted by the backrest. Note: Some

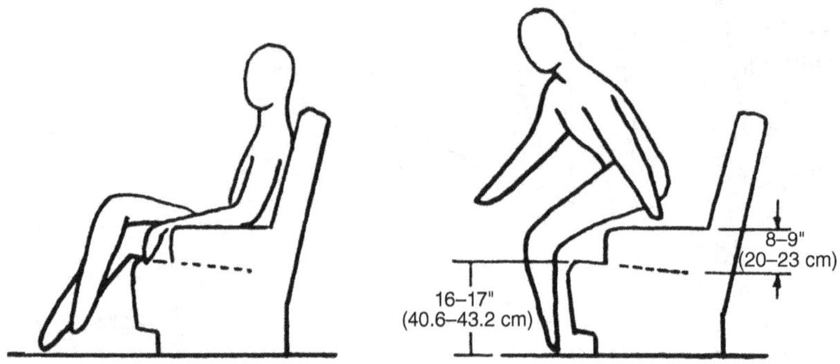

FIGURE 33.4 Recommended seat dimensions (note kick room in front).

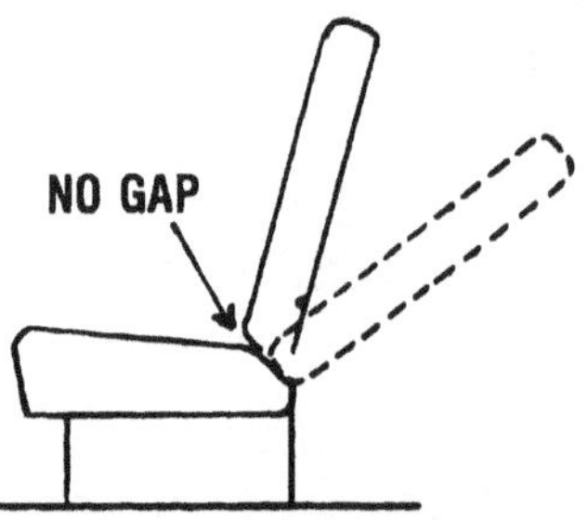

FIGURE 33.5 Backrest should not separate from seat cushion.

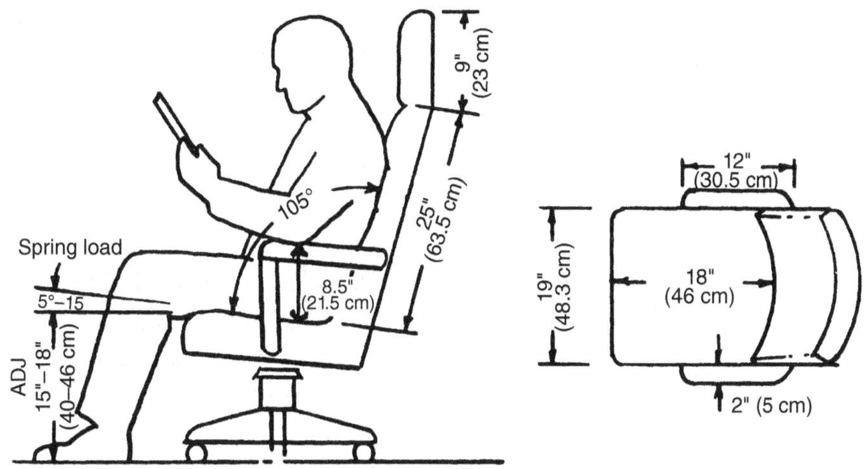

Figure 33.6 Desk chairs.

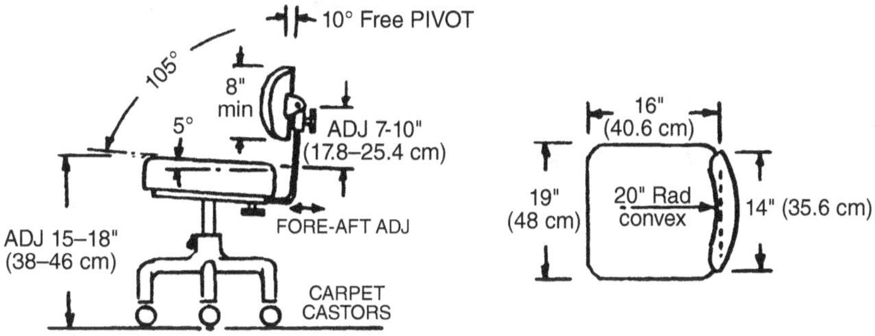

Figure 33.7 Workstation chairs.

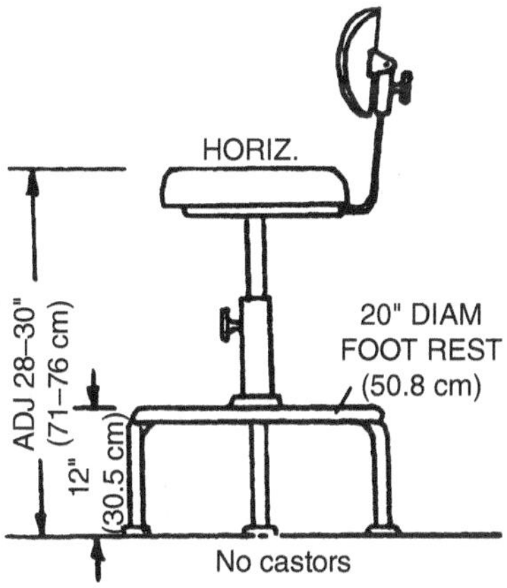

Figure 33.8 Workbench stools.

experimental stools have been suggested that have tilting seats (i.e., for a proposed "half-sit, half-lean" position). There is no evidence that such a device improves a worker's productivity.

SPECIAL WORK SEATS AND CHAIRS

When special working situations suggest the need for special seating, consider the possibility of designing the work seat so that it can be adjusted more specifically to each individual worker's size and to the workplace (task situation).

Regardless of the several possibilities for making a seat adjustable, certain basic dimensions should be retained. These include the ones shown in Fig. 33.9, i.e., a proper seat-pan size and a minimum back support area.

In addition, consider the following possibilities in terms of adjustable seating:

1. Either a mobile, caster-borne chair or a swivel seat if the chair has to be secured to a deck or other structure. Swivel seats should be designed so that the seat can be secured in specific positions for situations such as aboard ship. An adjustment lock (spring-loaded) every 45° or so is usually suitable.

2. Fore-aft adjustment in the seat pan may also be helpful for a seat that is secured to a deck (so that the worker can move closer to the work).

3. Vertical adjustment should be provided so that each worker can seek the best working height for his or her own height.

4. Although the nominal seat-pan–backrest angle should be about 105°, provide a swivel backrest that will allow the cushion to conform to the occupant's own back contour.

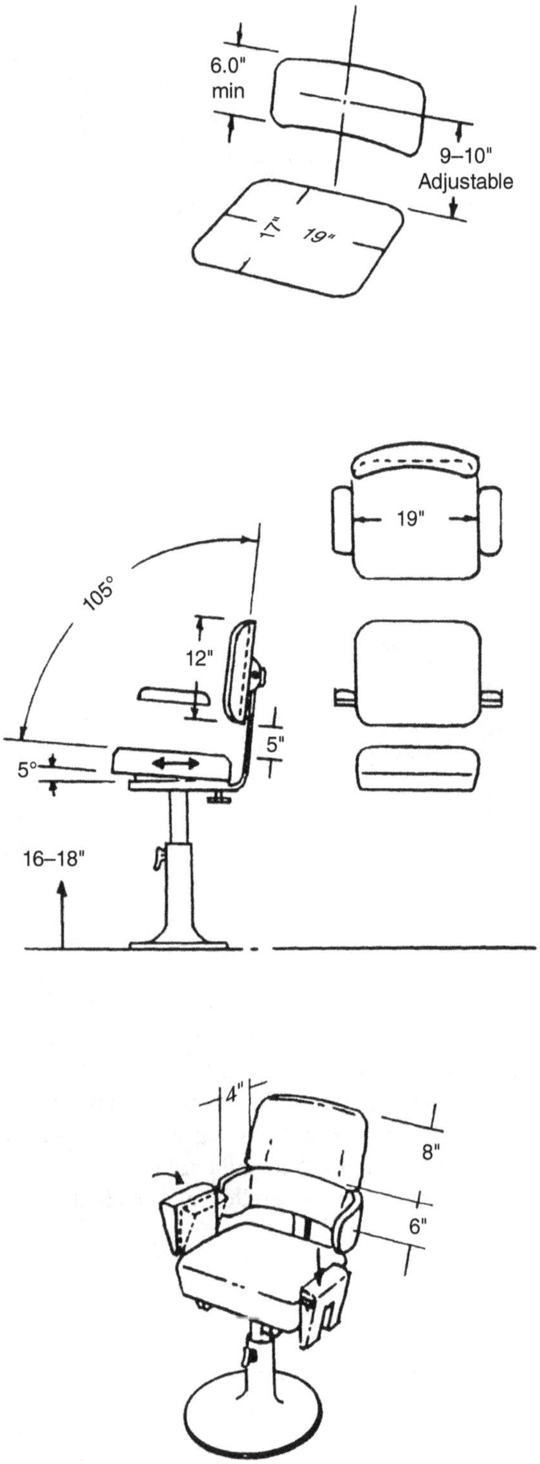

FIGURE 33.9 Special work seats and chairs.

5. Armrests should also be considered if the seat is used for long durations. Armrests should adjust up and down so that the occupant can easily slip in and out of the seat. Armrests (see Fig. 33.9) can also tilt inward to serve as a semirestraint system if the seat is used on a moving platform such as a ship or tractor. The addition of a "wraparound" section to the backrest also helps secure the occupant against lateral forces. Also, arm rest can be removed or folded out of the way.

6. Adjustment handles should always be designed and located so that the operator can use them easily, but the controls should not get in the way and possibly cause personal injury.

DESK/CHAIRS FOR CLASSROOMS

Using Table 33.1 and the reference numbers in Fig. 33.10, one can determine the general recommended dimensions for various aspects of a typical classroom desk chair.

TABLE 33.1 Recommended Dimensions for Classroom Desk Chairs

	Dimensions						
	1	2	3	4	5	6	7
Kindergarten	11 in.	18 in.	14 in.	9 in.	10 in.	11 in.	7 in.
Grades 1 through 3	13 in.	19 in.	14 in.	10 in.	11 in.	11 in.	10 in.
Grades 4 through 6	13 in.	22 in.	14 in.	15 in.	13 in.	12 in.	10 in.
Grades 7 through 9	15 in.	25 in.	15 in.	17 in.	15 in.	12 in.	12 in.
Grades 10 through 12	17 in.	29 in.	17 in.	19 in.	20 in.	16 in.	12 in.
Adults	17 in.	29 in.	17 in.	19 in.	20 in.	16 in.	12 in.

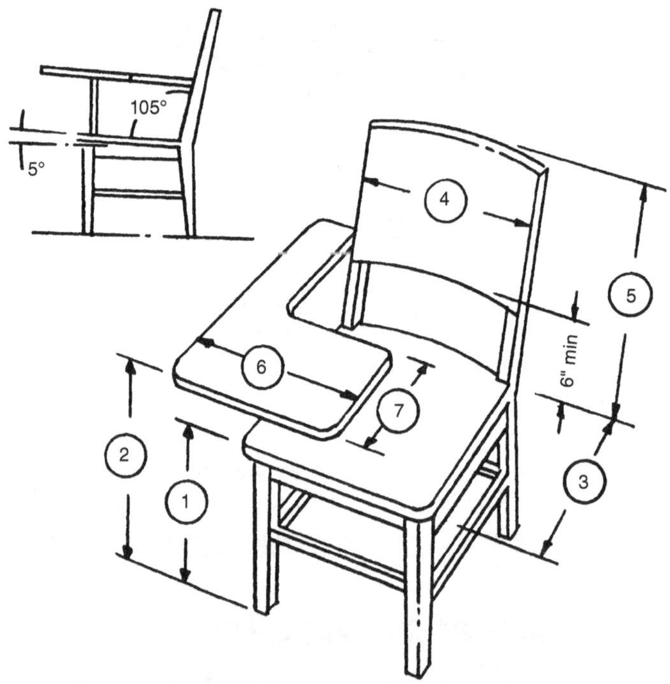

FIGURE 33.10 Desk chair dimensions.

The desk chair shown is for right-handed students only. Over 10 percent are left handed and should have a desk with elbow support for the left arm.

Although these dimensions provide a fairly good compromise solution, it is highly recommended that adjustable chairs be used for at least the lower grades, since children vary greatly in size at these ages.

AUDITORIUM SEATING

General

Although more austere seating dimensions and interseat spacing are typically specified for auditorium seating, the following suggestions are considered to be more adequate in terms of providing patrons with the proper convenience and comfort. (See Fig. 33.11.)

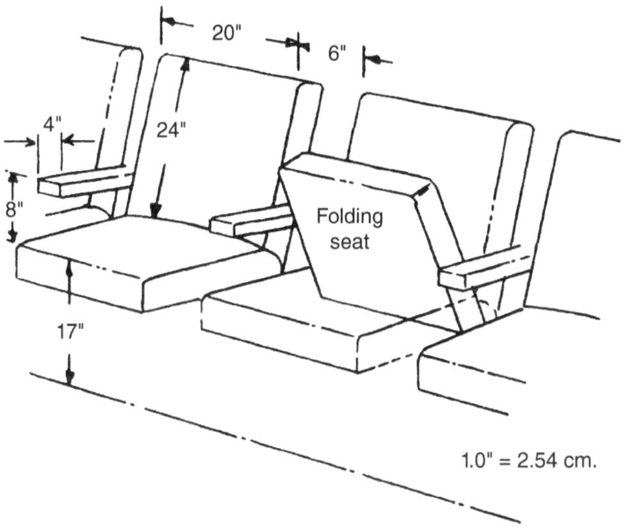

FIGURE 33.11 Recommended auditorium seat design.

1. Seats should be reasonably wide [20 in. (51 cm)], backrests should be high [24 in. (61 cm)], and the spacing between seats should be about 6 in. (15 cm) to minimize interelbow interference.

2. Armrests should be about 8 in. (20 cm) above the seat-pan surface level to prevent the occupant's shoulders from becoming hunched. Armrests should be wide [a minimum of 4 in. (10 cm)].

3. Maximum comfort is realized when the seat-pan surface is angled upward about 5°, and flat (no contours). A flat backrest is best, although a concave shape [radius of 20 in. (50 cm)] is acceptable.

4. Seats should be padded, but not to the extent that occupants sink into soft, billowy cushions, which become uncomfortable after a lengthy occupancy.

Recommended Auditorium Seat Design

Seats should be staggered as shown in Fig. 33.12 so that patrons have a good view between the heads of persons sitting in front of them. To calculate a given row spacing (x) and seat spacing (y) for unrestricted view of the stage (a) from any seat, proportional to the distance (d), apply the following formula: $a = kd$ [where k is a constant and equals $(y - t)/x$, t being the thickness of one head]; i.e., if x = 36", y = 19", and t = 6", then k = 0.33. Thus, 9 m (30 ft) from the front of the stage, $a = 0.33 \times 9 = 3$ m (9.8 ft); i.e., 3 m (9.8 ft) of the width of the stage can be seen without interruption, which is one-third of the average 9-m (30-ft) proscenium opening.[11]

[11] R. Ham, *Theatre Planning,* University of Toronto Press, Toronto.

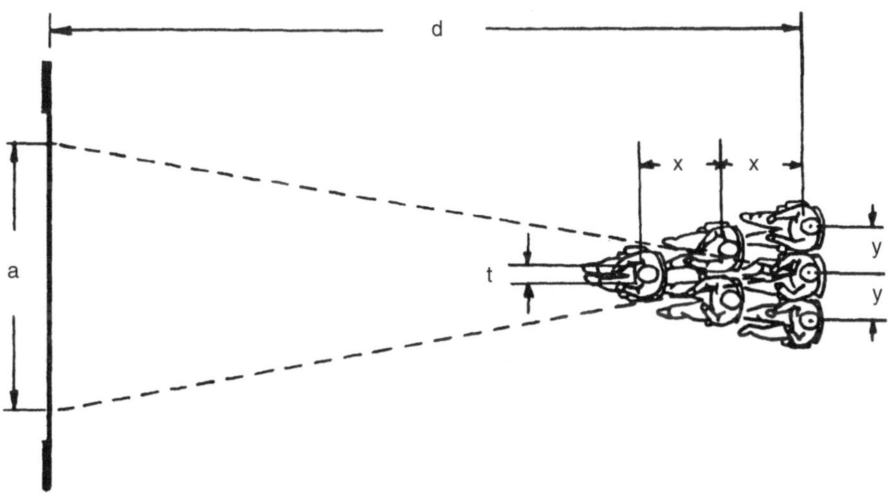

FIGURE 33.12 Calculation of seat spacing to permit unrestricted viewing.

AUTOMOTIVE SEATING

General

The basic criteria noted earlier apply to the automobile seat. In addition, several other features should be considered (see Fig. 33.13):

1. The frequently provided fully tiltable seat back may be desirable to allow a front-seat passenger to lie back to sleep or rest, but it obviously places that passenger in an unsafe position in the event of a crash. Seat belts and/or passive air bags will not be effective if such a passenger is not in the right position.

2. The major portion of the seat should be covered with fabric so that the seat "breathes." Leather or plastic may be used on the edges to make it easier to slide in and out of the seat and to minimize wear.

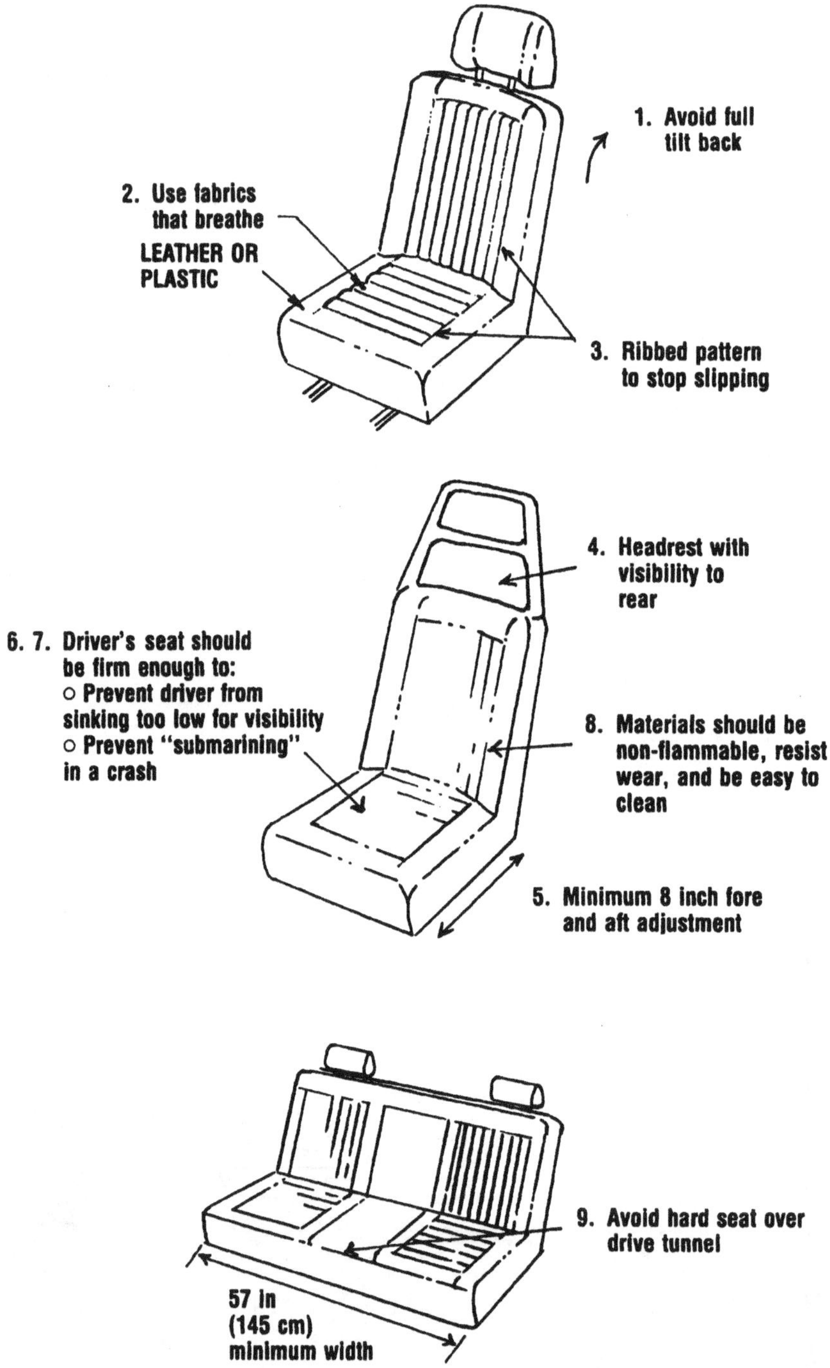

FIGURE 33.13 Automotive seating.

3. A ribbed pattern should be sewn into the seat pan and backrest covers as illustrated in the figure. Note that the seat-back pattern runs vertically. This helps control sidesway resistance. The seat-pan cover pattern runs in the opposite direction. This helps keep the occupant from sliding forward during a sudden deceleration.

4. A headrest should be provided for all passengers to reduce the possibility of whiplash injury during a rear-end collision. However, the driver needs to see past the headrest in order to look for traffic in the rear. The figure shows one method for improving rear vision. Headrests should be provided on back seats too.

5. The driver's seat should be adjustable fore and aft at least 8 in. (20 cm). Although other adjustments are appreciated by many who can afford them, they may place the occupant in the wrong position for a safety belt or air bag to be effective. The back of the driver's seat should never be designed to lay back more than a few degrees.

6. The driver's seat should be constructed so that the occupant will not sink into the cushion more than about 1 in. (2.5 cm) (unless the seat is equipped with a height adjustment control). Otherwise, it is very possible that the driver's eye position will not always be appropriate for the viewing requirements (i.e., seeing the instruments or seeing past the pillars and mirror).

7. All seats should be firm enough so that, in the event of a crash, the occupant will not sink deeply into the seat cushion and be subject to "submarining." (Submarining is the rotation of the buttocks beneath the seat belt, which allows slack in the belt and the high probability that the belt will be across the occupant's abdomen at the moment of a crash impact and cause serious or fatal internal injuries.)

8. Materials used in the fabrication of vehicular seats (especially upholstery) should be nonflammable, resistant to wear, and easy to clean, since people are typically careless about smoking and eating in a vehicle. Although the latter problem may seem minor, the flammability factor is important.

9. When bench seats are used, follow the same general guidelines suggested for any other type of single seat. Although most manufacturers design their so-called three-occupant bench seats for two adults and a child in the center, it is recommended either that bench seats be made wide enough for three adults (using the 19-in., 48-cm, width criteria per person) or that only bucket or individual seats be provided. Avoid a bench seat in which the center position sits atop a drive tunnel or other hard structure. Sudden bottoming of this portion of the seat can easily injure the occupant's spine.

Example of the Application of Human Engineering Principles to the Design of an Automobile Seat

Figure 33.14 illustrates the following points:

1. A tubular steel frame provides the basic dimensional and seat-pan–backrest geometry.

2. A thin, concave steel plate is welded to the frame to preserve the desired backrest form, i.e., one that reduces the effects of sidesway.

3. Rubberized upholsterer's strips are laced tightly across the seat-pan frame, and a 1-in. (2.5-cm) sheet of high-density fire-resistant padding is laid over both the seat pan and the backrest. This is covered with upholsterer's muslin to hold the pads in place and prevent scrubbing and sloughing of the padding.

4. A transparent plastic sacklike cover is drawn tightly over the headrest frame and secured to the frame. This provides a somewhat resilient, fixed headrest that fits all occupants and still provides "see-through" capability so that the driver can look to the rear.

5. A combination fabric and Naugahyde cover is used over the seat pan and seat back. Fabric is used in the center sections and is "ribbed" as shown in order to allow the seat to breathe and to minimize lateral sway and forward slippage of the occupant. The borders of the cover are smooth to allow the occupant to slip in and out of the seat easily and to provide good wear capabilities.

6. The seat adjustment control is placed in the mid-front position to be accessible to either hand. A simple upward pull deactivates a spring-return latch so that the seat can be moved fore and aft. Note that this construction is unlike typical contemporary construction. A 220-lb (100-kg) occupant would not sink into the seat pan more than 0.5 in. (1.3 cm), except if the car hits a bump—in which case the lacing would absorb the shock without causing the occupant to "bottom out." By not allowing heavy occupants to sink into a soft cushion, as opposed to allowing light-weight occupants to sink less, the designer was able to hold eye references and head clearance tolerances to a fine degree. Many current soft-cushioned seats allow the eye reference to vary as much as 2.0 in. (5.1 cm) as a result of the varying weights of different occupants.

COMMERCIAL TRANSPORT PASSENGER SEATING

General

In spite of the fact that people probably will continue to travel even though seats are uncomfortable, public reaction would improve if aircraft, train, and bus passenger seats were designed to "fit." Instead of designing the seats after the

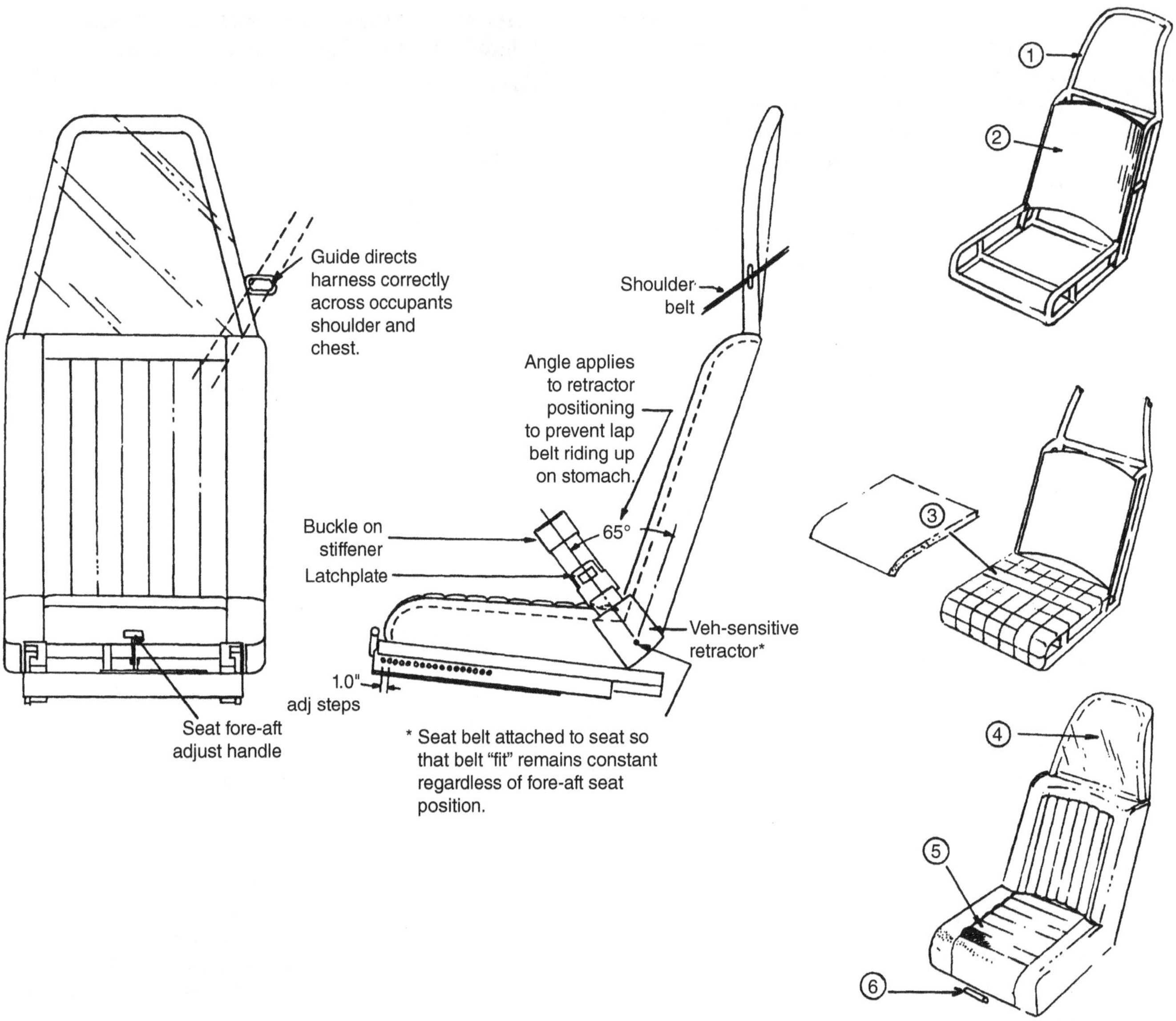

FIGURE 33.14 Automotive seat design.

fuselage size has made it necessary to crowd people, basic seating criteria (along with other human factors) should dictate how big the fuselage should be.

The basic dimensional and seat angle guidelines noted earlier apply to passenger seats as well as to other types of seating. Several other factors should be considered, some of which are illustrated in Fig. 33.15. Key considerations should include the following:

1. A 5° seat-pan pitch with a 105° seat-to-backrest angle for normal upright seat position.

2. Continuous seat recline adjustment to 45°.

3. A backrest height of 24 in. (61 cm) before the bottom of the headrest begins.

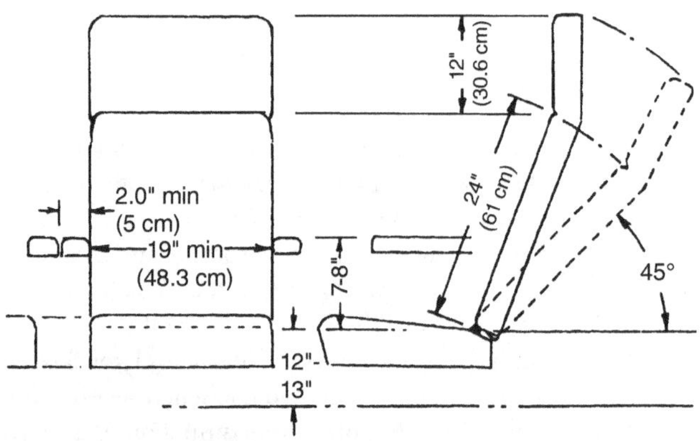

FIGURE 33.15 Recommended seat design for commercial transport passenger.

4. A minimum space between armrests of 19 in. (48 cm) (more is desirable for long-haul applications).

5. The seat-pan angle should automatically adjust to a horizontal position as the seat backrest reclines so that the occupant can stretch out his or her legs.

6. Side-by-side passengers (when provided with intermediate armrests) should be given either separate armrests 2 in. (5 cm) wide or a large 4- to 5-in. (10- to 13-cm) common armrest between them. When a minimum seating width has to be observed, design the intermediate armrests so that they can be folded back out of the way.

7. If tray tables are integrated into the back of the seats, design them so that the tray table will remain level even though the seat to which it is attached is reclined.

8. Passenger seat controls, flight attendant call buttons, and light and communications controls should be located where passengers can see and find them easily without getting out of their seats.

9. Seat cushions should not be overly soft.

10. They should be flat and should be made of, and covered with, nonflammable materials that "breathe."

11. No sharp projections, corners, or edges should be exposed that would create a hazard either during the normal movement of people within the vehicle or during an emergency escape.

12. As a minimum, provide lap belts to secure occupants against rough air or crash conditions. This applies also to intercontinental buses. Although lap belts are also recommended for rail vehicles and local buses, it would perhaps be unrealistic to expect passengers to use them.

Bus Seating

Levis and others studied several aspects of bus design and developed the accompanying recommendations regarding seating. (See Fig. 33.16 and Table 33.2.) The principal issue leading to these recommendations was ease of access, with special emphasis on the problems of the older passenger.

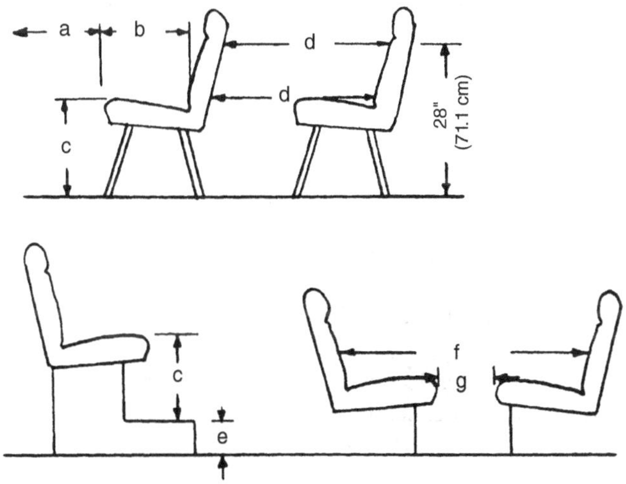

Figure 33.16 Recommended bus seat dimensions.

TABLE 33.2 Recommended Bus Seating Dimensions

Dimension	Preferred, mm	Acceptable, mm
a	310	280
b	400	380
c	432	400–460
d	720	680
e	200	100–250
f	1460	1360
g	700	600

Source: J. A. Levis, *The Seated Bus Passenger—A Review,* Applied Ergonomics, vol. 9, no. 3. pp. 143–150, 1978.

TABLES, DESKS, COUNTERS, AND WORKBENCHES

Guidelines for Design and Selection

The key issues that should be addressed to ensure compatibility between the user and the table, desk, counter, or workbench are the following:

1. Proper working height
2. Adequate leg and foot clearance
3. Adequate work area

Figures 33.17–33.19 provide nominal dimensional guidelines for tables that are used for writing, reading, or eating; for desks that are used for writing, storage, and conferences.

The work surface height shown represents the maximum, and the knee clearances represent minimums. The dimensions given for the tops of tables and desks obviously are some-what arbitrary. However, they are suggested for the following good reasons:

1. The width must be sufficient so that the worker is not continually knocking things off the surface.

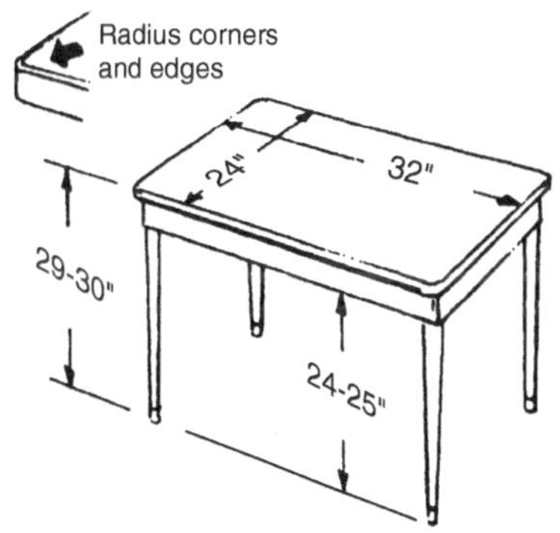

Figure 33.17 Basic table.

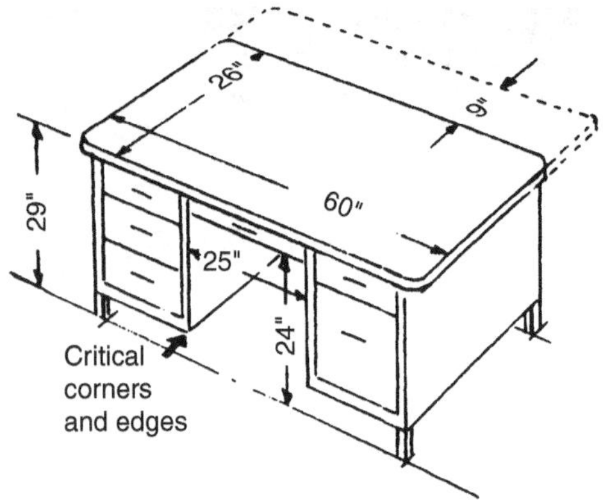

FIGURE 33.18 Office desk.

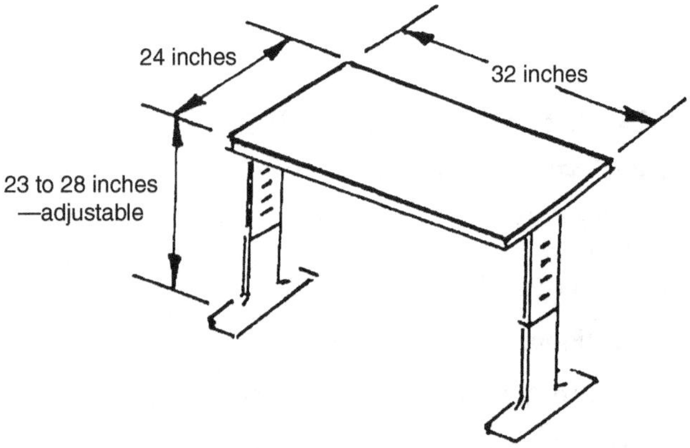

FIGURE 33.19 Computer table.

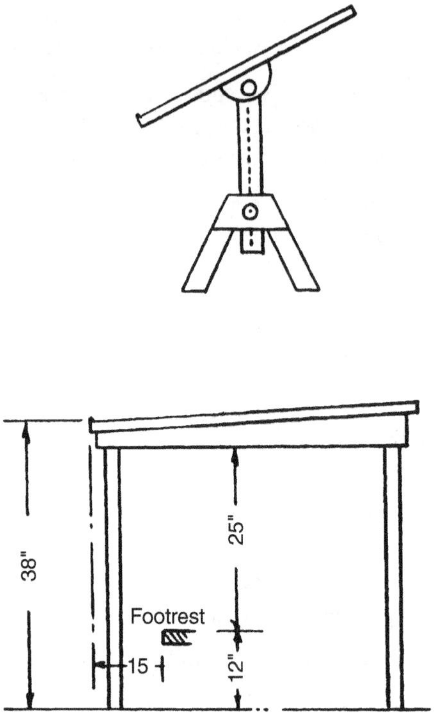

FIGURE 33.20 Drafting table.

Note the footrest, which implies that the table serves for both a sitting (with stool) and a standing work position. A 28- to 29-in. (71- to 74-cm) stool height works well with the specified table dimensions. (See Fig. 33.21.)

2. Depths should be sufficient so that the user's legs and feet do not extend beyond the opposite side and inadvertently trip someone who might be passing by. All potential contact points (corners, edges, handles, etc.) should be rounded so that if the user or a passerby happens to bump into the furniture, he or she will not receive a serious injury.

Drafting tables generally should be as adjustable as possible in order to satisfy a variety of work requirements. (See Fig. 33.20.) Consider the possibility of the user's choosing between a sitting and a standing position, working on a flat versus a vertical surface, or being able to work from several sides of the table. When adjustment hardware is designed, make sure that it provides the necessary ease and the necessary security so that the user's valuable time is not wasted and the table position cannot shift at a critical moment (e.g., because the individual leaned a little too heavily on the edge of the tabletop).

If a single fixed (all-around) drafting table is desired, consider the general dimensional guidelines shown in Fig. 33.21. This set of specifications will provide a good compromise to suit a wide range of drawing activities and sizes of workers.

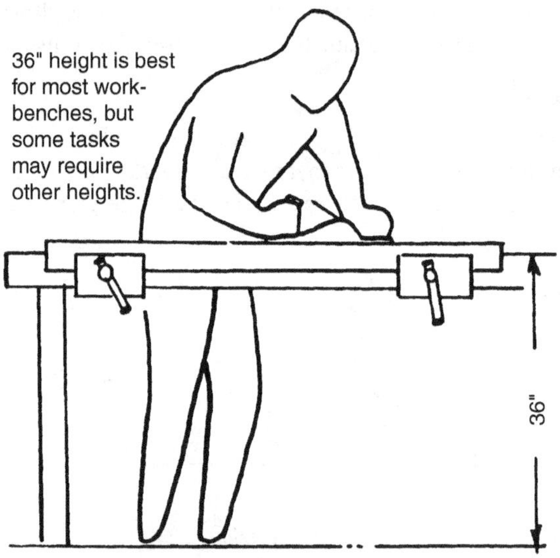

FIGURE 33.21 Workbenches.

A typical service counter at which people stand across from one another should be between 40 and 41 in. (102 and 104 cm). (See Fig. 33.22.)

The typical household counter top should be approximately 36 in. (91 cm) high in order for it to be reasonably

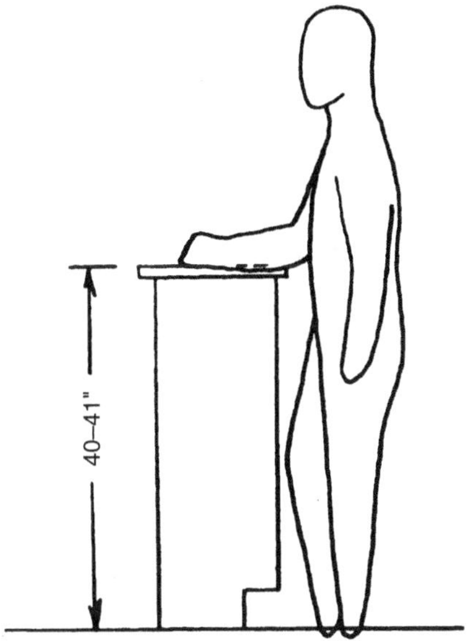

Figure 33.22 Counters.

comfortable for both the male and female members of the family and to accommodate appliances. (See Fig. 33.23.) Studies have shown, however, that people prefer a working height that specifically fits them, which suggests that designers should consider ways to create adjustable counter heights or have varied height counter tops. This would be especially helpful to alleviate the problems that children and handicapped obviously have in using the bathroom sink.

Counter tops should be about 24 in. (61 cm) deep to provide a good compromise between maximum useful work space and the ability to reach easily to the back of the

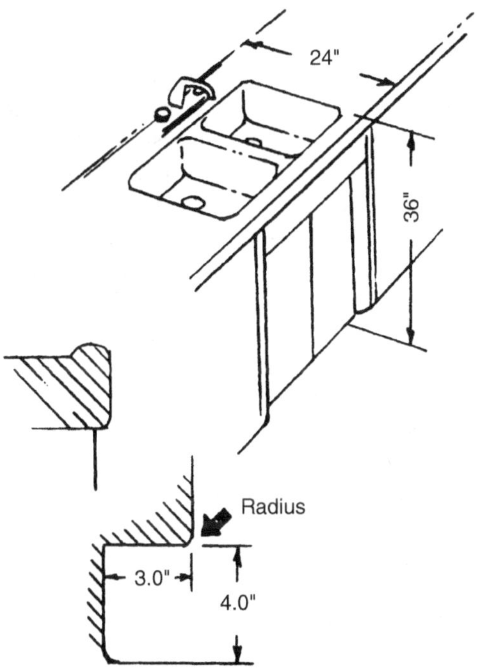

Figure 33.23 Counter dimensions.

counter. Counter edges around sinks should contain a lip to catch liquids. The edges should be radiused, however, so that anyone bumping against the edge is not injured. Provide kick space below the front of the supporting cabinet.

There are fewer restrictions with regard to many other types of tables that are used in the typical household or lobby, such as end tables and coffee tables. (See Fig. 33.24.) Some considerations that might be observed in designing or selecting such tables are the following:

1. End tables should not be higher than the armrests of an adjacent chair.

2. Coffee tables should not be lower than about 12 in. (30 cm) or higher than about 18 in. (46 cm).

3. Very low tables must definitely not have sharp corners or edges. This is especially true of glass-topped tables.

4. The legs of single-pedestal occasional tables should not extend beyond the perimeter of the top and thus create tripping hazards.

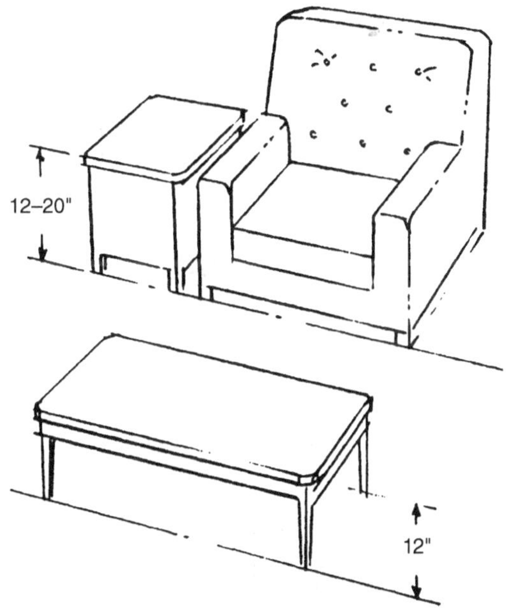

Figure 33.24 End table and coffee table dimensions.

The Critical Table Leg

Table leg concept, design, and positioning often cause the consumer as many problems as any other aspect of the table, or more. A few of the most common errors are discussed below.

The legs on dining tables are obviously not in view when diners attempt to take their places at the table. Placing the legs where diners are least likely to sit is an important design objective. A second objective is to avoid creating fancy styling features on the leg that protrude precisely at knee height. (See Fig. 33.25.)

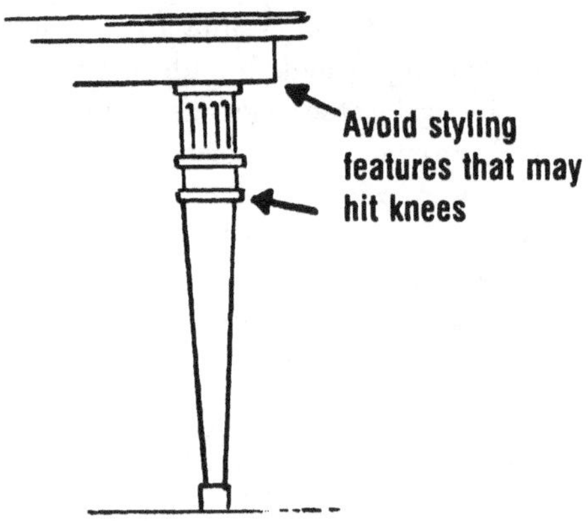

FIGURE 33.25 Dining table leg.

Although single-pedestal tables are probably the least annoying to the user (see Fig. 33.26), special styling such as that illustrated creates an extremely annoying problem for the diner. However, it may be more annoying for the table's owner when he or she discovers that the legs have been damaged (because they provide an ideal place for people to rest their feet).

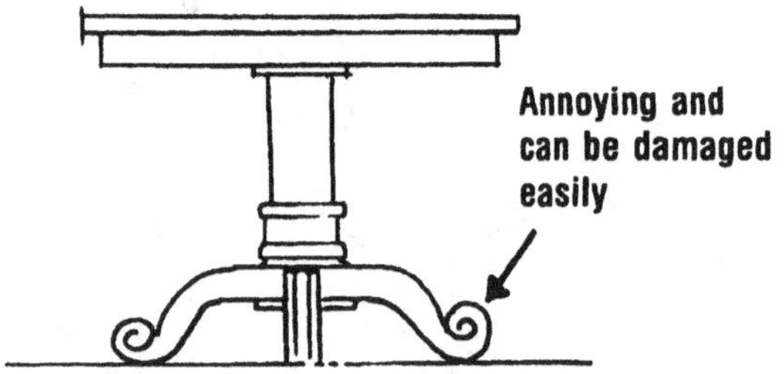

FIGURE 33.26 Special styling features on dining tables.

Small occasional tables such as the one shown in Fig. 33.27 are easily tipped because the legs often have been extended beyond the perimeter of the top.

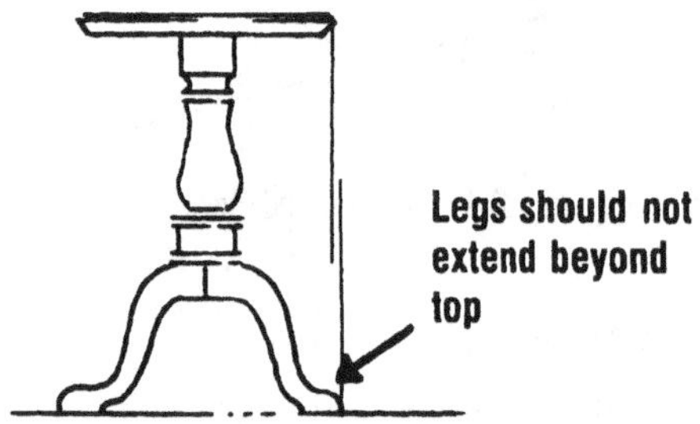

FIGURE 33.27 Occasional table.

Horizontal Dimensions of Tables

People must have room for their feet and elbows as they sit around a table. Figure 33.28 provides important guidelines for designing or selecting the proper size of table for the particular application.

The guidelines shown are for adults. The dimensions can be reduced for special tables to be used by children. For children between the ages of 10 and 13, the dimensions can be reduced by about 15 percent; for younger children, they can be reduced by about 20 percent.

CABINETS AND SHELVING

In designing or selecting cabinetry or shelf-type storage units, a primary consideration to keep in mind is the nature of the potential storage and retrieval tasks involved. This requires attention to the following (see Fig. 33.29):

1. The size and weight of the articles to be stored.
2. Whether the user needs to see the article.
3. How high and how far in the particular user can reach.
4. Strength and mobility limitations of the user at various positions, e.g., stretching, bending, stooping, and kneeling.
5. As indicated in the figure, one must keep in mind the size differences among the possible users.
6. Avoid the temptation to assume that shelf height is the only consideration. The figure illustrates the importance of shelf depth as well as shelf height. Although it may appear wasteful not to take advantage of the total available depth of a cabinet, this is not necessarily an advantage to the user.

Storage Cabinets

General philosophies regarding storage cabinet selection and design are gradually changing to provide more efficiency for the user as well as more effective use of space. (See Fig. 33.30.) The need to rearrange one's immediate work space from time to time suggests mobile cabinetry. In addition to allowing the cabinet to be brought to a better position, this type of filing arrangement positions the stored materials so that the user does not have to reach or strain to inspect them.

Typical filing cabinets generally are relatively deep and narrow. Not only do they take up valuable floor area, but they also are easily upset when the user inadvertently opens the upper drawers without thinking about the imbalance they are creating.

It is generally more desirable to take advantage of available wall space without losing floor area, as is illustrated in Fig. 33.29. Special cabinets are available in this format and should be considered whenever floor space is at a premium.

Key operating issues to address are the following:

1. Reach
2. Materials visibility
3. Flexibility for rearrangement

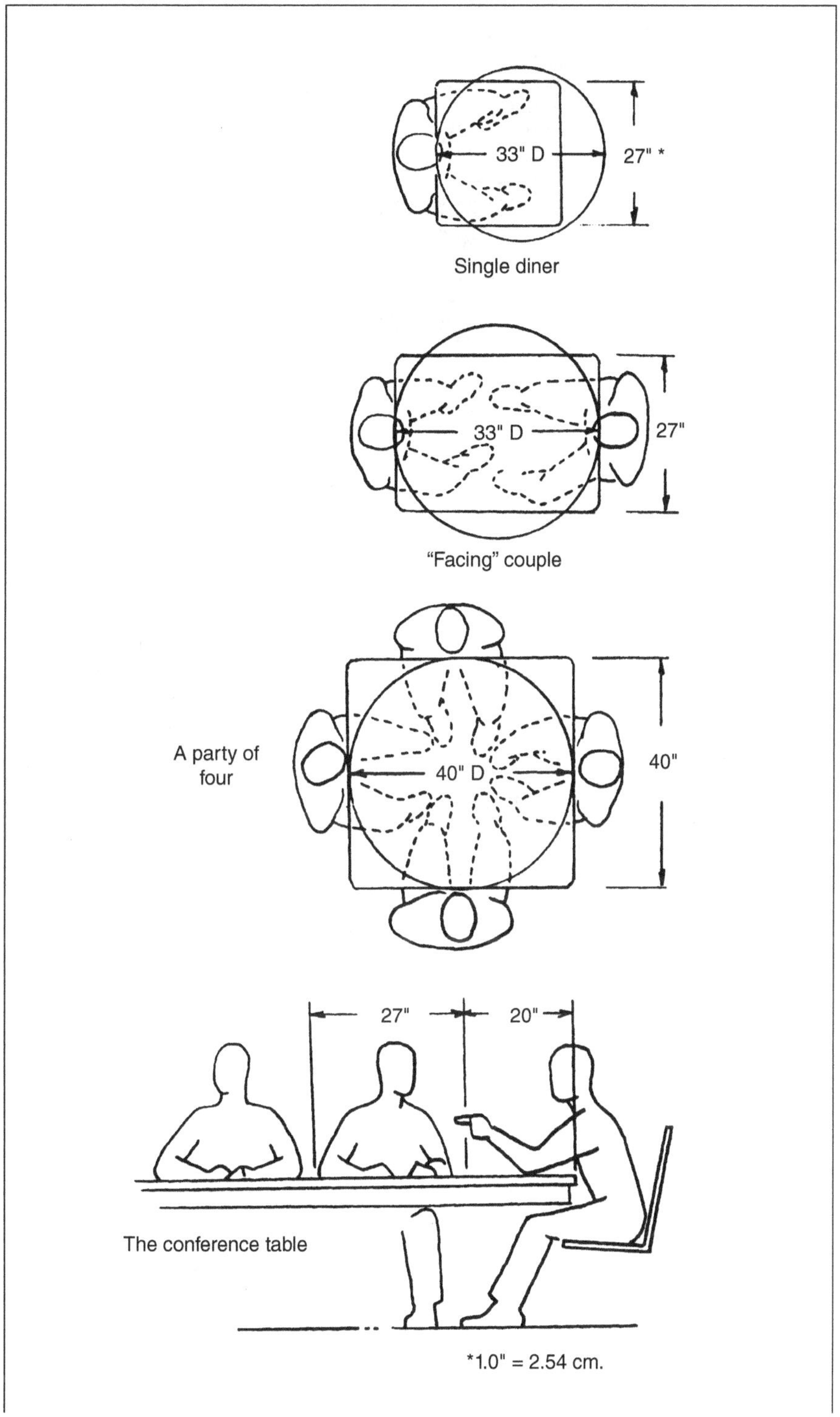

Figure 33.28 Recommended table surface dimensions.

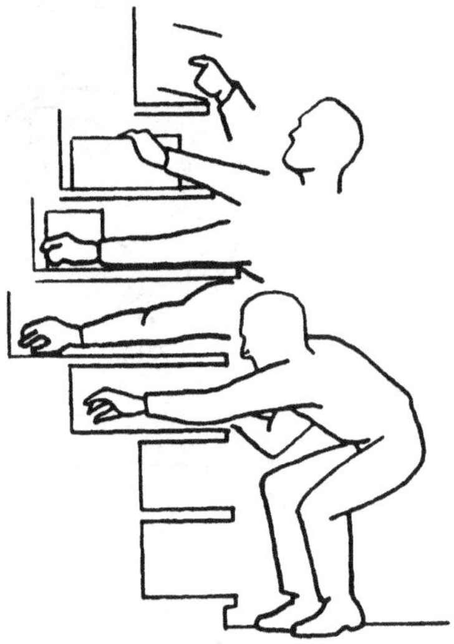

FIGURE 33.29 Consider shelf depth as well as height.

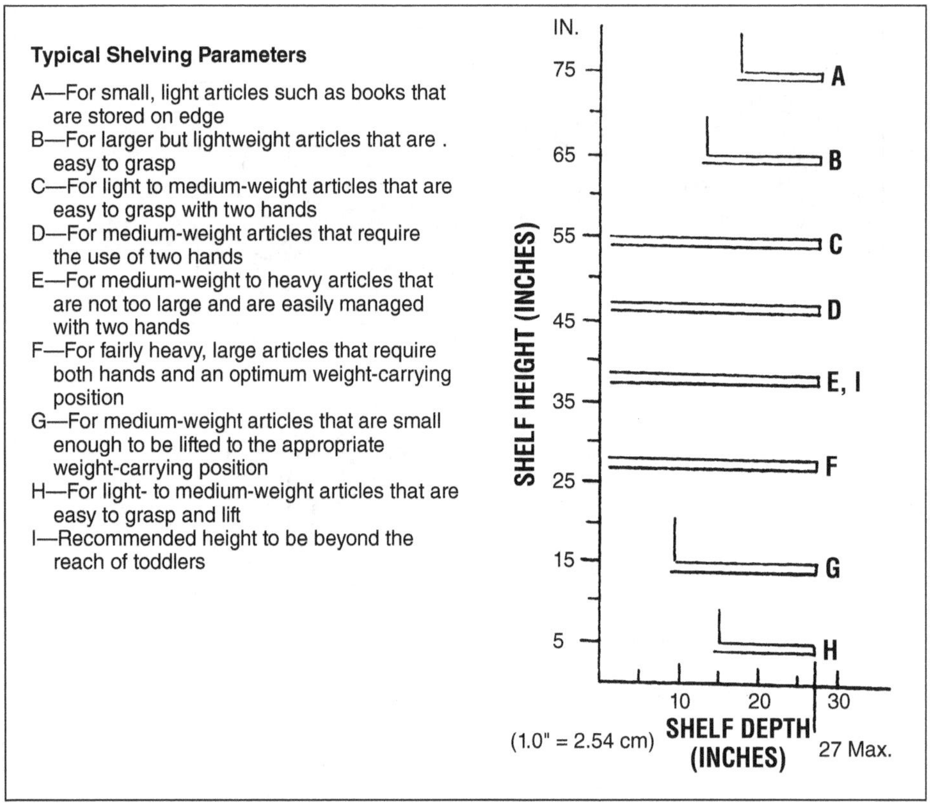

FIGURE 33.30 Storage shelf design recommendations.

4. Cabinet stability
5. Drawer control
6. Handle and latch operability
7. Compartment identification
8. In-cabinet materials separation and control
9. Cabinet mobility

Specialized Furniture[12]

Figures 33.31 and 33.32 illustrate general dimensional guidelines used by architects for the development and/or procurement of display shelving for libraries and supermarkets.

[12] J. De Chiara and J. H. Callender, *Time-Saver Standards for Building Types*, McGraw-Hill Book Company. New York, 1973, pp. 276, 616.

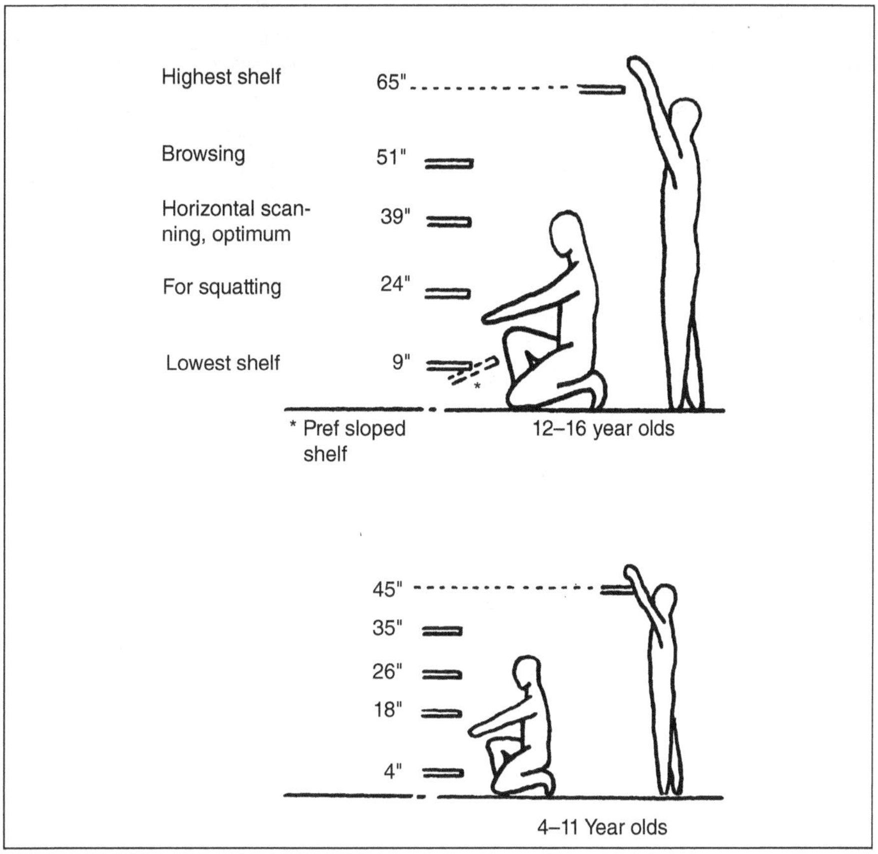

FIGURE 33.31 Library shelves.

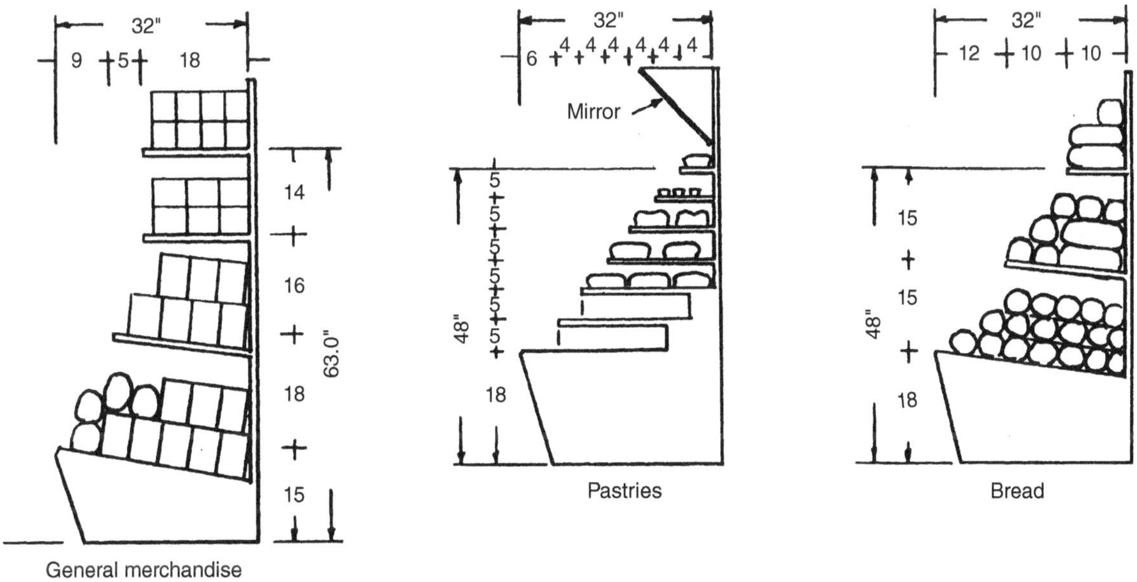

FIGURE 33.32 Market shelves.

BED DESIGN

The following are standard U.S. bed sizes:

Twin, 6 ft 3 in. × 3 ft 3 in. (1.9 × 0.99 m)

Double (full), 6 ft 3 in. × 4 ft 6 in. (1.9 × 1.37 m)

Queen, 6 ft 8 in. × 5 ft (2.0 × 1.5 m)

King, 6 ft 8 in. × 6 ft 6 in. (2.0 × 2.0 m)

Recent sales records, however, show that most people now purchase queen- or king-size beds. There is good reason for wanting a larger bed. The standard length of 6 ft 3 in. (1.9 m) does not provide the space needed at the ends of a bed; e.g., many people sleep with their arms above their heads, and most people who are of average height or taller do not like to have covers cramp their feet. Beds (or, more appropriately, mattresses) should be at least 80 in. (203.2 cm) long. The width of a double bed is more a matter of choice. Two people who do not like to touch each other need a bed that is at least 72 in. (182.9 cm) wide. (See Fig. 33.33.)

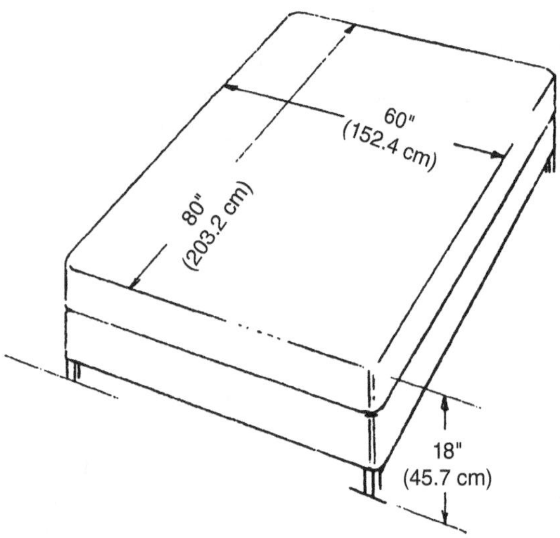

FIGURE 33.33 Queen-size bed dimensions.

The top of the mattress of an adult's bed should be approximately 18 in. high (45.7 cm) above the floor. If the bed is lower, it is difficult to get out of and to make up. If it is higher, it is difficult to get into.

There should be at least 30 in. (76.2 cm) of vertical clearance between bunk beds. A ladder should be provided to aid in getting in and out of the upper bed.

Children's beds should be designed so that the side rails (when up) are too high for the youngster to crawl over [at least 29 in. (73.7 cm) above the mattress; see Fig. 33.34]. One side rail should be adjustable so that the child can be easily retrieved from the bed. Vertical open rail sides are suggested so that an adult can see into the bed to make sure that everything is all right, but the rails must be spaced close together so that the youngster cannot get his or her head through the openings [a maximum of 3 in. (7.6 cm)]. All structure and hardware must be "childproof" and nontoxic.

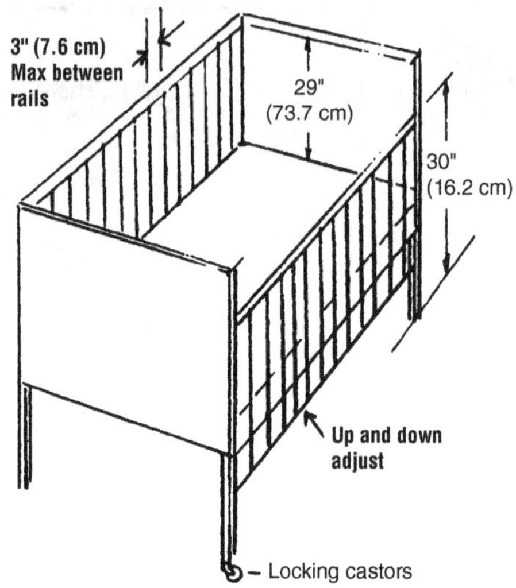

FIGURE 33.34 Recommended child's bed design.

Headroom is a significant consideration in the design of bunk beds. Head clearance (C) should actually be sufficient so that a person can sit down on the bed to don or doff shoes and sit up in the bed (at least partially) to read. (See Fig. 33.35.) For youth beds, this head clearance should be at least 26 in. (66 cm); for adult bunk beds, it should be at least 30 in. (76 cm).

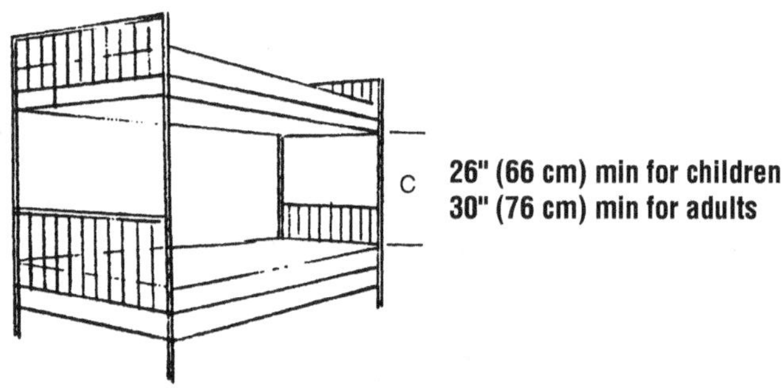

FIGURE 33.35 Recommended head clearances in bunk beds.

In a recent study of shipboard berthing conditions, it was found that head clearance was essentially absent; vertical clearances as little as 18 in. (46 cm) were found.

In the design of bunk beds, it is also critical that the support for the upper bed be sufficiently rigid to prevent it from sagging under the weight of a heavy upper occupant and encroaching on the head clearance of the lower occupant.

FURNITURE MOBILITY

The ability to move and adjust furniture is an important human factor since most furniture owners do not have the special aids available to moving specialists.

In considering the mobility problem of the user and the eventual design required to improve mobility, keep in mind the typical problems itemized below (see Fig. 33.36):

1. To move furniture or appliances, one must first be able to get hold of the article. Although it is perhaps undesirable to put handholds where they will detract from the appearance of the furniture, appropriately concealed handles or handholds should be provided.

2. A heavy piece of furniture should be provided with casters so that it can be moved away from a wall without damage to the wall or adjacent furniture. Select caster hardware that is especially designed to roll easily on carpet. On furniture items with an attached mirror (see the figure), design the mirror so that it is easily removed and replaced—preferably without the need for tools. Mirrors should be provided with a height adjustment.

3. Heavy appliances that ordinarily will be positioned in close quarters (i.e., beside or in between built-in cabinetry) should be provided with a built-in slide-roller system so that the item can be rolled straight in or out (see the sketch of a refrigerator in the figure). Although hardware add-ons can be purchased by the homeowner,

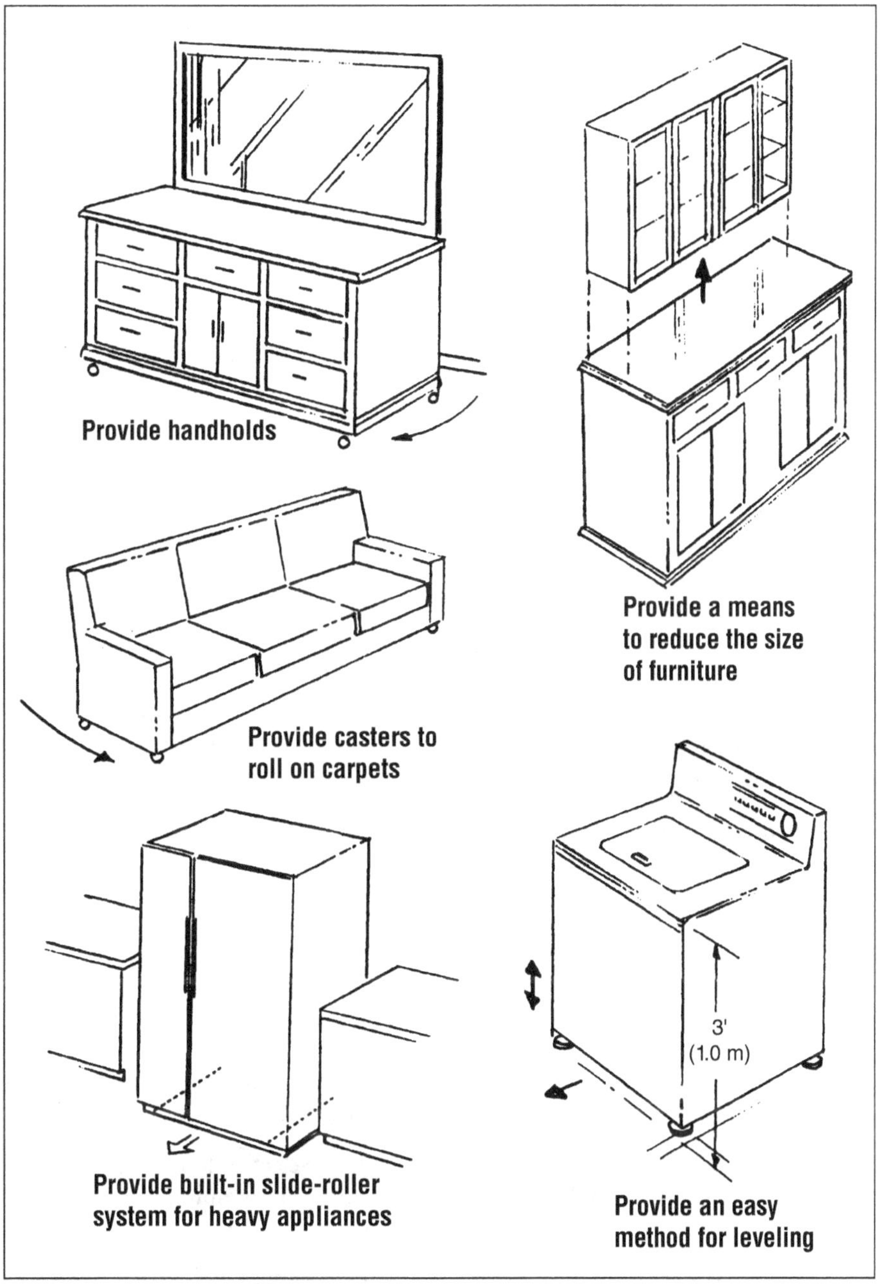

Figure 33.36 Furniture mobility aids.

such devices seldom work well. Besides, it is the manufacturer's responsibility to design and build a complete appliance, not one the buyer has to add to.

4. All furniture should be designed with standard access dimensions in mind since architectural doorways are well standardized and are not likely to change. The clearance on each side of a piece of furniture when it is being moved through a doorway is critical to those handling the furniture. The owner also worries about movers scratching furniture when it has to be passed through a doorway.

5. If the overall dimensions of a particular piece of furniture are going to be too large for it to pass in one piece through the typical door, up a staircase, or down a hallway, design the furniture to be easily disassembled so that it can be taken in and out in several pieces.

6. Certain items of furniture and/or appliances often need to be leveled in order for them to operate properly or even to keep them from rocking on an uneven floor. The standard practice of providing a leveling screw is not considered good human engineering. In the first place, it creates an awkward working position. Second, most people do not understand the system and do not know which way to turn the screw or bolt. Third, to make the setting secure, a wrench is required, and it is impossible to arrive at a proper adjustment without much trial and error. A leveling system should be designed that can be operated by the foot.

SAFETY CONSIDERATIONS IN THE DESIGN AND SELECTION OF FURNITURE AND APPLIANCES

1. Avoid sharp corners and edges.

2. Avoid cabinet hardware (i.e., handles) with sharp corners or edges.

3. Provide drawer stops so that the user cannot inadvertently pull the drawer completely out of the cabinet.

4. Provide latches to preclude drawers from drifting open.

5. Place the hinge point of adjacent cabinet doors (or structures) so that, when opening a door, the user will not pinch his or her fingers between the handle and the adjacent structure or door handle.

6. Provide hand and finger clearance at the base of cabinets or appliances that may be picked up and transported from one place to another.

7. Provide locks on cabinets that are used to store drugs, medicine, or other products that should be kept from children.

8. Avoid the use of glass shelves. If they are used, make sure that the glass is thick enough to support the expected loads and/or provide enough support to prevent the glass from breaking.

9. Eliminate all sharp edges and corners on glass shelves.

10. If slide-out shelves are provided, make sure that there are mechanical stops to prevent the shelf from being inadvertently pulled all the way out. Provide an edge around a slide-out shelf so that stored objects will not slide off the shelf when it is being manipulated.

11. Provide framing around all glass doors. Provide friction hinges or other devices to prevent glass doors from swinging shut and breaking.

12. Avoid glass-top tables without suitable edge frames. If edge frames are not used, make sure that all corners and edges are rounded. A slight color tint is recommended so that everyone can see the glass.

13. Avoid placing glass doors in the lower portions of any cabinet unless the glass is shatterproof.

14. Provide adequate insulation within the surfaces or structure of all appliances that produce heat.

15. Place controls for heating appliances where children cannot get to them or design the controls so that children cannot operate them.

16. Make sure that all handles used on heating equipment are insulated.

17. Provide proper insulation to preclude shock from any electrical appliance.

18. Design all electrical appliances so that electrical shorts are not probable under expected use conditions. Provide fuses within the equipment to prevent external electrical hazards (i.e., lightning) from entering the equipment.

19. Design electrical appliances so that potentially dangerous electrical exposures are inaccessible for the service or repair person.

20. Provide a visual indication that an appliance is on or off. This includes flush stove-top burners that do not appear to be on.

21. Provide heavy-duty electrical lead-ins and plugs with sufficient surface area to encourage the use of the plug rather than the wire to disconnect the lead from a wall outlet.

22. Provide implosion-proof covers over TV display tubes.

23. Select and/or design freestanding tables, cabinets, stools, etc., to minimize the probability of their tipping over.

24. Design folding furnishings to preclude pinched fingers as the elements are being retracted or deployed.

Make sure that there is an easily recognized catching "feel" or sound to let the user know when lids, legs, etc., are securely fastened.

25. Use only nonflammable upholstery materials on furniture.

26. Locate the controls for such appliances as refrigerators, washing machines, and ironers where children are not able to reach them and/or place such controls under well-latched covers.

27. Design portable packages so that they are not heavy enough to cause strain from carrying, the package can be carried easily without injury to the person's fingers, and the package content is properly balanced throughout the typical retrieval–transport–deposit cycle. Provide suitably sized and positioned carrying handles for such internal components as equipment chassis.

28. Provide automatic electrical disconnects for servicing so that when a cabinet drawer is unlatched, all electrical circuitry is shut down (until the technician is ready to perform further testing, in which case provide a separate bypass switch within the cabinet or chassis).

PART 4

HUMAN FACTORS DATA

INTRODUCTION

It would obviously take a small library to provide all the human factors data required for a full understanding of how the human functions. Previously, an attempt has been made to include design-related interpretations of the way the human interacts with various design situations and to recommend solutions to problems where research and/or experience has provided guidelines.

It would be impossible to anticipate all the possible design problems with which the designer may be confronted. The purpose of Part 4 is to acquaint the reader with the basic human factors data that designers commonly need to assist them in seeking their own solutions. This part does not provide a complete exposure to basic human factors data; rather, it presents carefully selected information that experience has shown to be related to the majority of design questions. It is hoped that this brief compilation will aid the reader when he or she is unable to find a particular design guideline earlier in this book.

Organization

Part 4 first defines the basic characteristics of the human body, including its various subsystem responses in a physical sense. This is followed by a series of chapters dealing with the more significant human–system interactions, e.g., motor response, work output response, response to the thermal environment, visual response, and auditory response.

Additional References

Since Part 4 provides only a very limited coverage of human factors data, the following list of additional references has been included. Readers are encouraged to refer to these sources to expand their personal understanding of the various human factors that may be extremely important to their design work.

Damon, A., et al. *The Human Body in Equipment Design,* Harvard University Press, Cambridge, MA, 1966.

DeGreene, K. B. *Systems Psychology,* McGraw-Hill Book Company, New York, 1970.

McCormick, E. J. *Human Factors in Engineering Design,* 4th ed., McGraw-Hill Book Company, New York, 1975.

Murrell, K. F. H. *Human Performance in Industry,* Reinhold Publishing Corporation, New York, 1965.

Olishifski, J. B., and F. E. McElroy (eds.). *Fundamentals of Industrial Hygiene,* National Safety Council, Chicago, 1971.

Parker, J. F., and V. R. West (eds.) *Bioastronautics Data Book,* 2nd ed., National Aeronautics and Space Administration, 1973.

Roebuck, J. A., et al. *Engineering Anthropometry Methods,* John Wiley & Sons, Inc., New York, 1975.

Steen, E. B., and A. Montagu. *Anatomy and Physiology,* Vols. I and II, Barnes & Noble, Inc., New York, 1961.

Van Cott, H. P., and R. G. Kinkade (eds.). *Human Engineering Guide to Equipment Design,* Tarn's Books, Inc., Los Angeles, 1972.

Woodson, W. E., and D. W. Conover. *Human Engineering Guide for Equipment Designers,* 2nd ed., University of California Press, Berkeley, 1964.

Anthropometric Source Book, NASA Reference Publication 1024, Vols. I, II, and III, 1978 (prepared for the National Aeronautics and Space Administration, Scientific and Technical Information Office, by Webb Associates, Yellow Springs, OH).

NASA-STD-3000, Man-Systems Integration Standards, Chapter 3.

Martin, A. D., D. T. Drinkwater, and J. P. Clarys. *Human Body Surface Area: Validation of Formulae Based on a Cadaver Study,* Wayne State University Press, 1984.

Churchill, E. *Sampling and Data Gathering Strategies for Future USAF Anthropometry,* Webb Associates, Inc., A/F Aerospace Medical Res., 1976.

McConville, J. *Anthropometric Relationships of Body and Body Segments Moments of Inertia,* Anthropology Research Projects, Inc., 1980.

POINTS TO CONSIDER

The information in Part 4 is unique—it focuses on the human side of the human–system interface. The human factors engineer must be aware of the range of human capabilities and limitations and ensure that the design accommodates this range. The following points will be helpful in understanding how to successfully apply the information in this part.

1. Human Variability: Humans vary in dimensions and capabilities. And the range of variability changes across different population groups. Scandinavian are generally larger than the other populations, for example. And a system designed to be used by younger and physically fit soldiers may not be usable by a large portion of the general public. Factors that commonly separate user groups include age, sex, ethnic origin, physical fitness, and education. It is important that system developers understand who the user will be. Anthropometric and performance data, for example, are available for specific user groups. In testing designs the test subjects should be representative of the user group. Also, the user population may change over the life-changing of the system. For example, over the past decades there has been a growth trend for each new generation. A Mars mission vehicle designed today might not actually be used for another 20 years. It is feasible that the user will be larger than today. Ways to accommodate this secular growth are detailed in *NASA Handbook: Human Integration Design Handbook (HIDH)*, NASA/SP-2010-3407. Consider different systems—cars, clothing, toys, airplanes, etc. Are they used by everyone, or just specific groups of people?

2. Application of Anthropometric Data: Accommodation of human size usually involves two factors: clearance and reach. The large person will have clearance problems and the smaller will have reach problems. The following discusses techniques for addressing both these factors. The following steps will be helpful to ensure that a design meets clearance requirements:

a. Identify the critical physical clearance dimension (a door opening, for example).

b. Define possible human body movements and positions relative to the critical dimension.

c. Select the worst-case body position(s) that would cause clearance to be an issue.

d. Determine the body dimensions that are associated with the worst cases.

e. Use the appropriate database to determine the value of the worst-case dimension(s) for the largest expected person.

f. Define the worst (bulkiest and largest) possible clothing and equipment worn or carried by the human and add this dimension to the worst-case body dimension. The results will define the design dimensions required to meet the clearance requirements.

The following steps will help ensure that a design meets reach or access requirements:

a. Identify the critical physical access dimension.

b. Determine possible human body movements and positions relative to the critical dimension (consider the task being performed).

c. Determine the worst-case body posture(s) or dimension(s) that will cause a reach or access problem.

d. Determine relevant factors such as clothing, pressurization, and any other environmental factors before using the data charts for the relevant data.

e. Use the appropriate database to determine the value of the body dimensions for the smallest expected person.

Look at all the items around you and think about each task step in using the items. Do the steps involve possible reach or clearance problems?

3. Once the user characteristics are determined, the system or hardware must be designed to those dimensions. Three general methods for doing this are described below:

a. Single Solution for All: In the case of anthropometry, a single size may accommodate all members of the population. For example, usually if a workstation has a switch located within the reach limit of the smallest person, everyone will be able to reach the switch. In the case of lighting, sufficient illumination will be provided to accommodate the oldest workers with dimming vision and yet not be glaring.

b. Adjustment: The design can incorporate an adjustment capability. A common anthropometric example of this is the automobile seat.

c. Several Solutions: Several sizes of equipment may be required in order to accommodate the full population size range. This is usually necessary for equipment or personal gear that needs to closely conform to the body, such as clothing.

Can you find more examples of each type of design accommodation?

CHAPTER 34

THE BODY

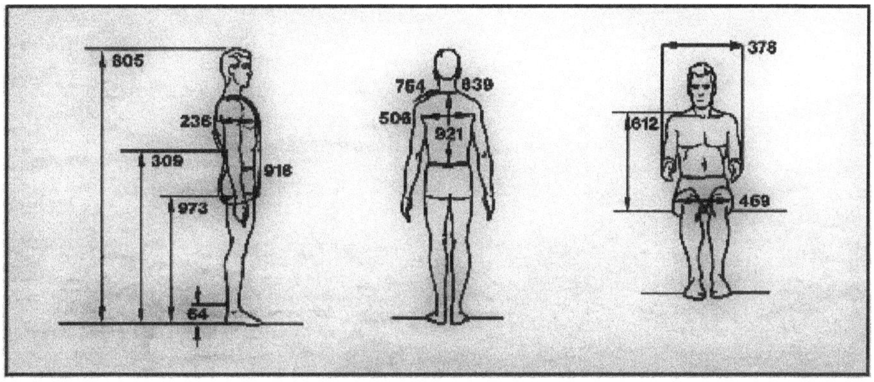

GENERAL BODY CHARACTERISTICS

Temperature

As indicated in Fig. 34.1, normal body temperatures range from a low skin temperature of about 80°F (26.6°C) to a high internal temperature of about 99.6°F (37.5°C). At about 79.5°F (26.4°C), the heart fails.

Weight

Adult males normally weigh from about 110 to 215 lb (50–97 kg); adult females weigh from about 92 to 175 lb (42–79 kg). The adult brain weight varies from about 38.8 to 60 oz (1100–1700 g) for males and from about 37 to 55 oz (198–1560 g) for females. The adult male heart weighs about 10 oz (280 g); the adult female heart weighs about 8 oz (226 g). Water makes up about 54–60 percent of total body weight; muscles, 36–42 percent; fat, 18–20 percent; and bone, about 18 percent.

Other elementary composition percentages are as follows:

Carbon	50 percent
Oxygen	20 percent
Hydrogen	10 percent
Nitrogen	8.5 percent
Calcium	4 percent
Phosphorus	2.5 percent
Potassium	1 percent
Sulfur	0.8 percent
Sodium	0.4 percent
Chlorine	0.4 percent
Magnesium	0.1 percent

Additional trace elements including:

Iron	0.01 percent
Manganese	0.001 percent
Iodine	0.00005 percent

Quantity of Blood

Males—1.5 gal (567 cm^3); females—0.875 gal (330 cm^3).

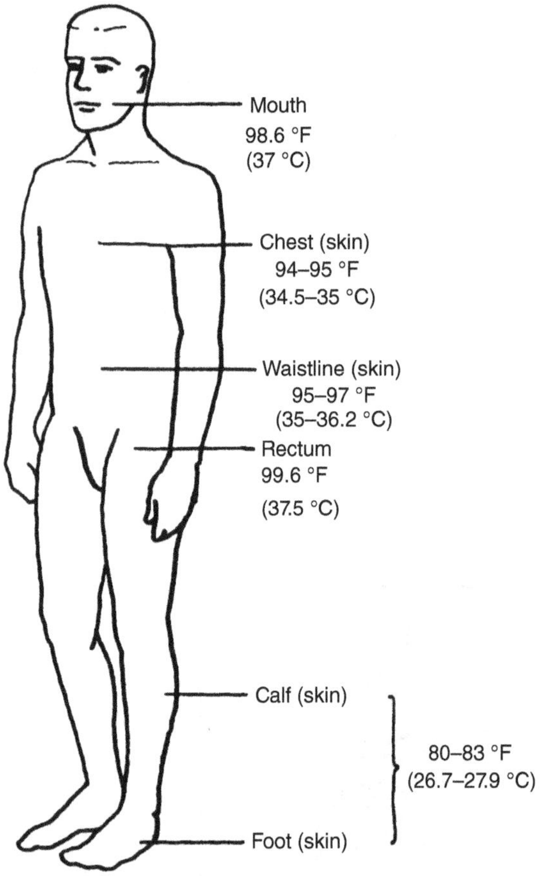

FIGURE 34.1 Normal temperature.

Surface Area of the Skin

Males—2.2 m^2; females—1.9 m^2

Lung Capacity

Males—4.5 to 9.5 qt; females—3.3 to 5.7 qt

Air Intake per Breath

Resting: males—0.79 qt; females—0.36 qt

Light work: males—1.77 qt; females—0.91 qt

Heavy work: males—2.15 qt; females—0.93 qt

Number of Breaths per Minute at Rest

Males—14 to 18; females—20 to 22.

The deepest intake (vital capacity) for males is about 5.18 qt; for females it is 3.17 qt.

Heartbeats per Minute

Sleeping—55 to 60 beats per minute

Awake—70 beats per minute

Blood Pressure

120/80 mm at age 20.

BODY DEVELOPMENT

Certain general body proportion relationships are observable between children and adults. Similar relationships are observable between males and females during the maturing or growth cycle. Figures 34.2 and 34.3 illustrate some of these relationships.

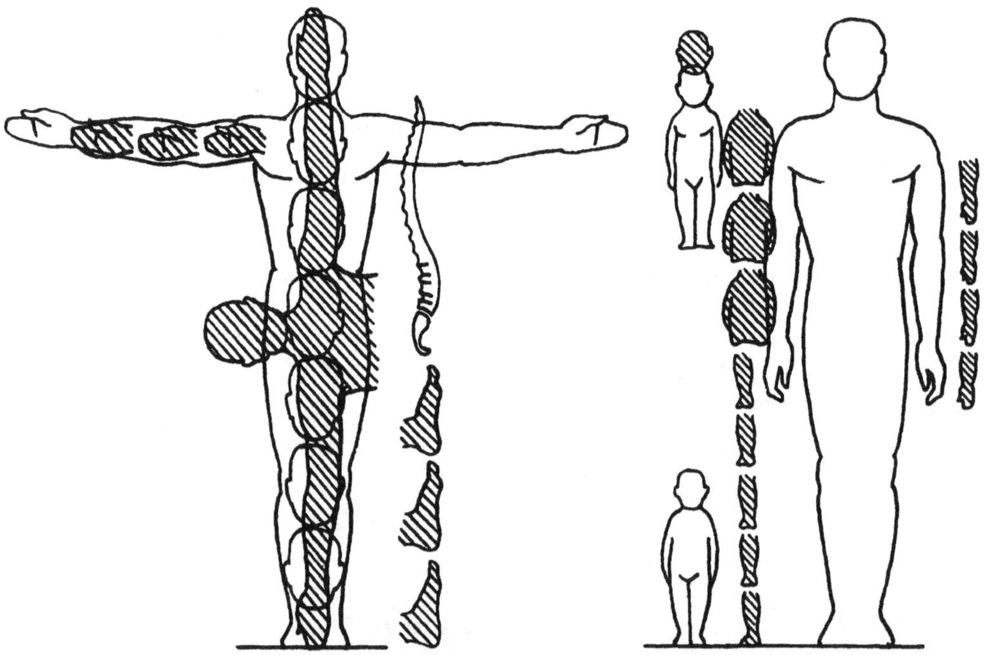

FIGURE 34.2 Proportional relationships between body components and between child and adult size.

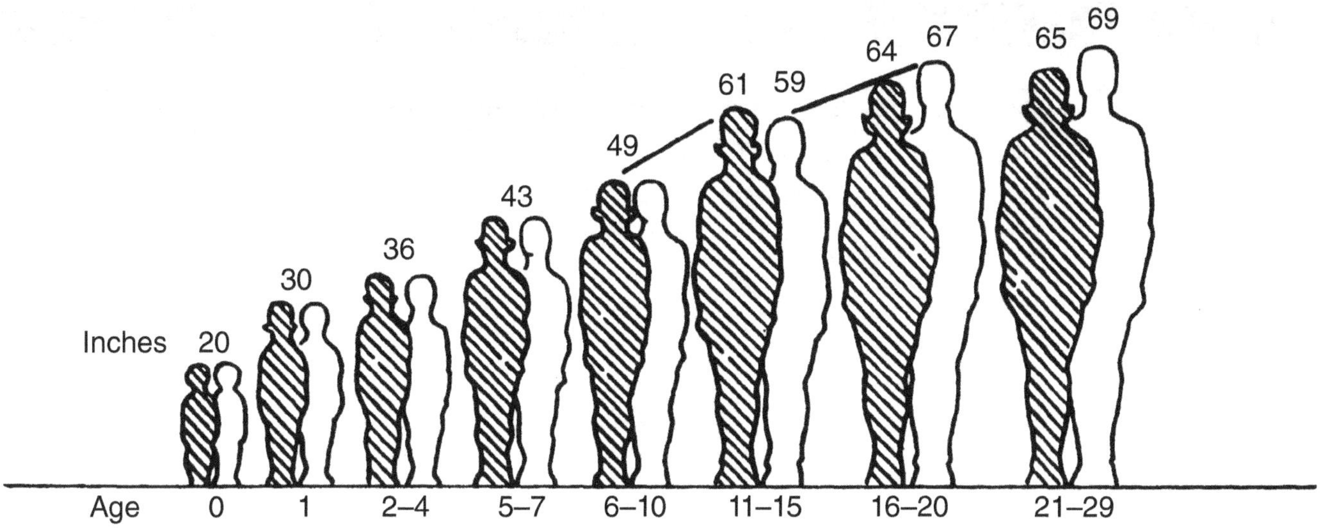

Figure 34.3 Male–female growth pattern.

BODY SURFACE AREA

The body surface area in relation to stature and weight is shown in Table 34.1.

SKELETAL LINKAGE SYSTEM

Although not completely accurate, it is often useful to think of the human body as mass segments linked by pin hinges as shown in Fig. 34.4. Analysts can make estimates using this simplified model, including:

- Reach range of the arms and legs (using estimates of the length of the segments between the hinges and the range of hinge movement).
- Whole body center of gravity (using estimates of the center of mass location and value for each segment).

Two major potential problems with using this simplified model are:

- Segment hinges are not simple pins. The shoulder joint, for example, is very complex and far from a simple pin hinge.
- The lengths of segments listed in tables cannot be added. For example, a 95th percentile length forearm and a 95th percentile arm would likely be much smaller than the total of the two segments. This is because each segment is measured individually and the chance that both arm segments are 95th percentile is very small.

The only reliable estimate of total arm length would be from direct measurement of the total arm.

SKELETAL NOMENCLATURE AND CHARACTERISTICS

The skeleton is the basic support structure for the body, and thus it is an important consideration when designing equipment that interfaces with the body. The dry, fat-free adult male skeleton weighs approximately 8 lb (3.6 kg); the adult female skeleton weighs approximately 6 lb (2.7 kg). Selected skeletal nomenclature is shown in Figs. 34.5 and 34.6. These are the terms that have been found to be most useful when working with special scientific references dealing with typical body functions and equipment design. In addition, certain terms are commonly used in connection with particular design problems; e.g., the lumbar vertebra is a key reference when working with seating design. Of the total weight of the skeleton, the skull makes up about 20 percent; the ribs, 18 percent; the arms, 18 percent; and the legs, about 45 percent.

BODY MOBILITY: HINGE POINTS AND CENTERS OF GRAVITY

Mobility and posture control depend on the skeletal hinge-point locations and the weight and centers of gravity (c/g) of various body components. Table 34.2 and Fig. 34.7 present key hinge-point and c/g locations and the approximate weight of key body segments.

Body mobility characteristics are shown in Table 34.3.

TABLE 34.1 Body Surface Area in Relation to Stature and Weight[1]

Body	Height, cm																								
Weight, kg	20	30	40	50	60	70	80	90	100	110	120	130	140	150	160	170	180	190	200	210	220	230	240	250	260
5	.18	.20	.23	.26	.29	.33	.37	.42	.48	.55	.62														
10		.35	.36	.38	.41	.44	.48	.52	.57	.64	.69	.76													
15					.54	.57	.60	.63	.67	.72	.77	.83	.89												
20							.68	.72	.76	.80	.85	.91	.97	1.03											
25								.80	.84	.88	.93	.98	1.03	1.09	1.15										
30									.92	.96	1.01	1.05	1.10	1.16	1.22	1.28									
35										1.04	1.08	1.12	1.17	1.23	1.29	1.35	1.42								
40										1.11	1.15	1.20	1.25	1.30	1.36	1.42	1.48	1.55							
45											1.23	1.27	1.32	1.37	1.43	1.48	1.54	1.61							
50											1.30	1.34	1.39	1.44	1.49	1.54	1.60	1.67	1.74						
55											1.37	1.42	1.46	1.50	1.55	1.61	1.67	1.73	1.80						
60											1.44	1.48	1.52	1.57	1.62	1.67	1.73	1.79	1.85	1.92					
65												1.54	1.58	1.63	1.68	1.73	1.79	1.85	1.91	1.97					
70												1.61	1.65	1.70	1.75	1.80	1.85	1.91	1.96	2.02	2.08				
75												1.68	1.72	1.76	1.81	!.86	1.91	1.96	2.02	2.07	2.13				
80												1.74	1.78	1.82	1.86	1.91	1.96	2.02	2.07	2.13	2.18	2.25			
85												1.81	1.84	1.88	1.92	1.97	2.02	2.07	2.13	2.18	2.24	2.31			
90												1.87	1.90	1.94	1.98	2.03	2.08	2.13	2.18	2.24	2.30	2.36			
95													1.97	2.01	2.05	2.09	2.14	2.18	2.24	2.30	2.36	2.42	2.48		
100													2.03	2.07	2.12	2.16	2.20	2.24	2.30	2.35	2.41	2.47	2.54		
105													2.10	2.14	2.18	2.22	2.26	2.31	2.35	2.41	2.47	2.53	2.60		
110													2.17	2.21	2.24	2.28	2.32	2.36	2.41	2.47	2.53	2.58	2.65	2. 73	
115													2.23	2.27	2.30	2. 33	2.38	2.42	2.47	2.53	2.58	2.64	2.71	2.78	
120														2.33	2.36	2. 39	2.43	2.48	2.53	2.58	2.63	2.70	2.77	2.84	2.93
125														2.39	2.42	2.45	2.49	2.53	2.58	2.63	2.69	2.76	2.83	2.90	2.97
130														2.44	2.47	2.51	2.54	2.59	2.63	2.68	2.75	2.82	2.88	2.95	3.02
135														2.50	2.53	2.56	2.60	2.64	2.69	2.74	2.81	2.87	2.93	3.00	3.08
140														2.55	2.58	2.62	2.66	2.70	2.74	2.80	2.87	2.93	2.98	3.06	
145														2.61	2.63	2.67	2. 71	2.75	2.80	2.86	2.92	2.98	3.04		
!50														2.66	2.69	2.73	2.77	2.81	2.86	2.92	2.97	3.03	3.09		
155														2.72	2.74	2.78	2.83	2.87	2.92	2.97	3.03	3.08			
160														2.77	2.80	2.83	2.88	2.92	2.97	3.02	3.08				
165															2.86	2.89	2.93	2.97	3.02	3.07					
170															2.91	2.94	2.98	3.03	3.07						
175															2.96	2.99	3.03	3.08							
180															3.0l	3.04	3.08								
185															3.06	3.09									

[1]Albert Damon el al., *The Human Body in Equipment Design*, Harvard University Press, Cambridge, Mass, 1966, p. 155. Area in square meters.

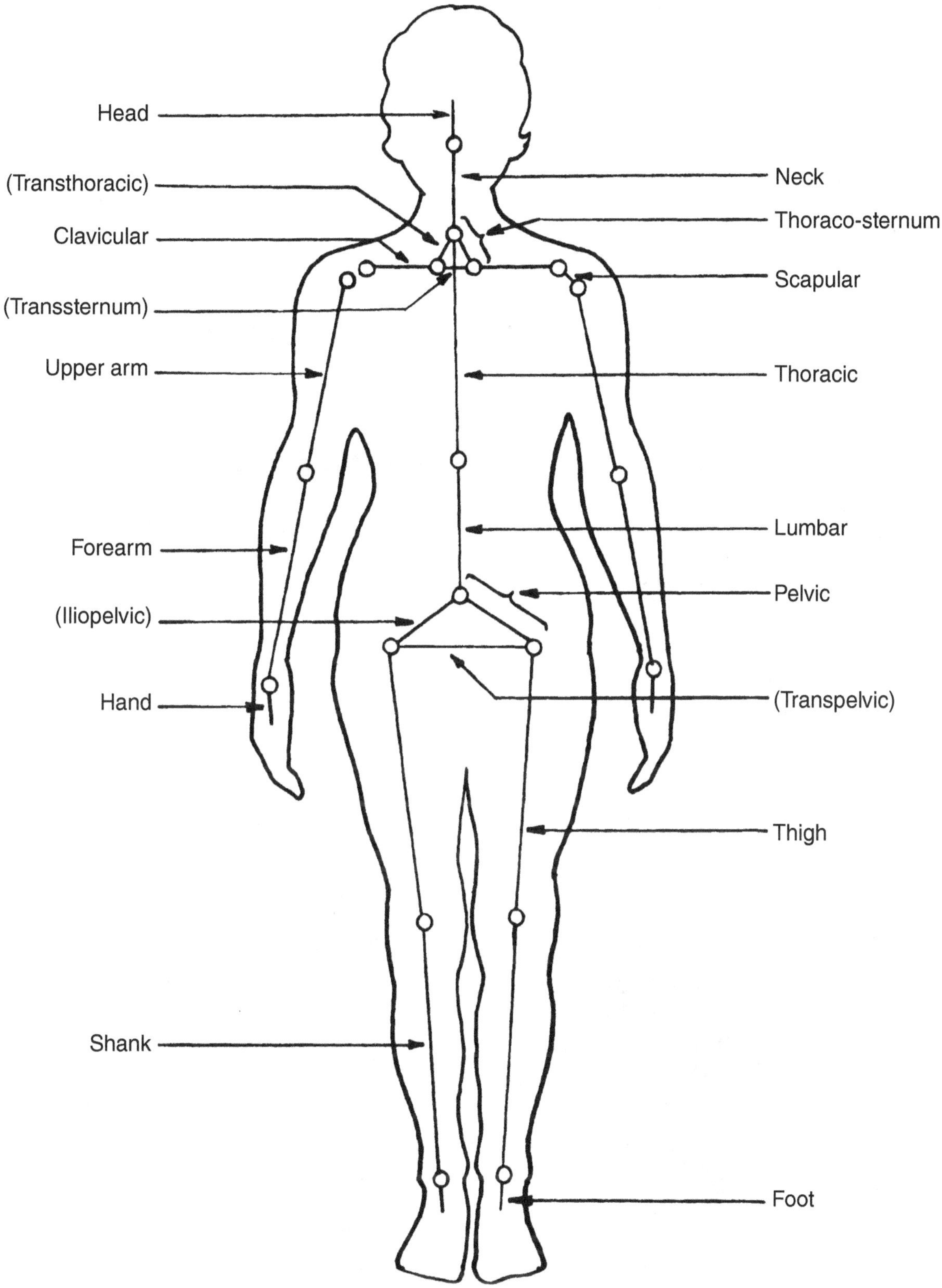

Figure 34.4 Skeletal linkage system.

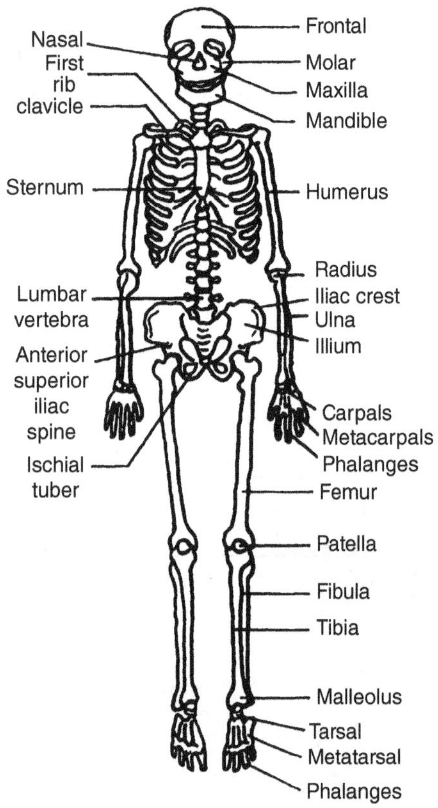

Figure 34.5 Anatomical and anthropometric landmarks.

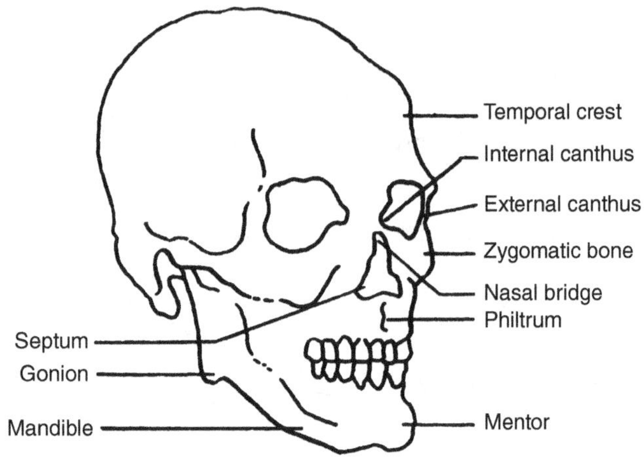

Figure 34.6 Anthropometric landmarks of the head.

TABLE 34.2

	Weight, lb (kg)	Percent of Total Weight
Head	10.7 (4.86)	6.9
Trunk and neck	70.7 (32.1)	46.1
Upper arms	10.1 (4.59)	6.6
Lower arms	6.4 (2.9)	4.2
Hands	2.6 (1.18)	1.7
Upper legs	33.0 (14.98)	21.5
Lower legs	14.7 (6.67)	9.6
Feet	5.2 (2.36)	3.4
Note: Example Total	*153.4*	*100.0*

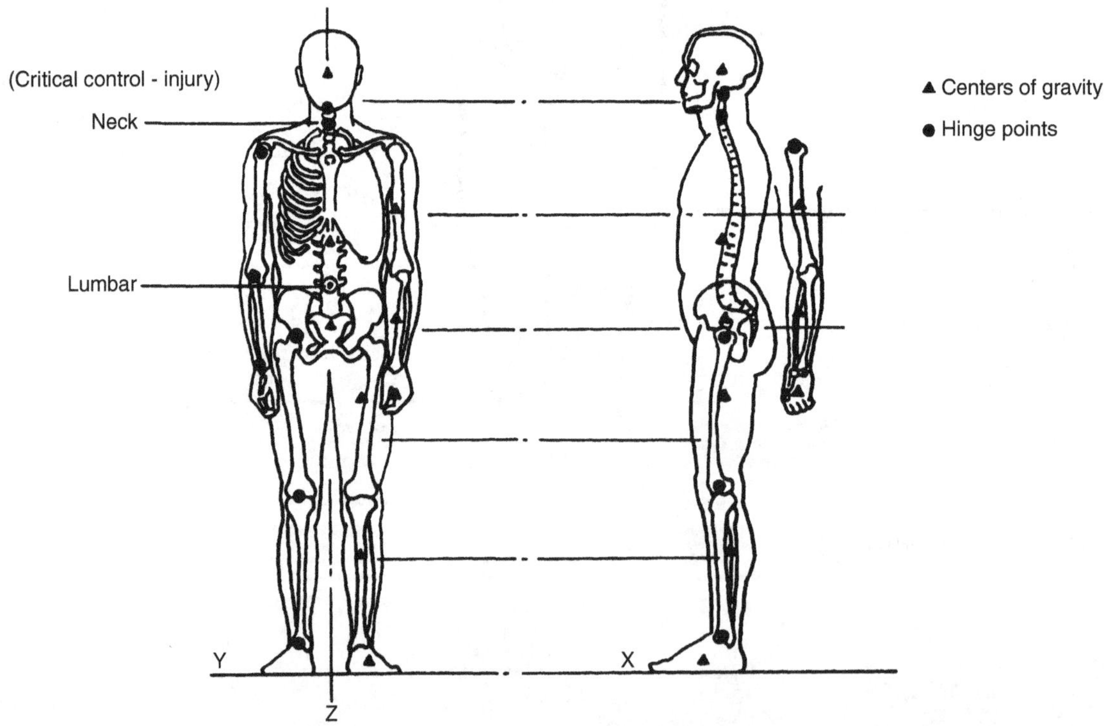

Figure 34.7 Body hinge points and centers of gravity.

TABLE 34.3 Body Mobility: Child Development

Age	Mobility Characteristic
2 months	Can turn from side to back.
4 months	Can turn from back to side.
6 months	Can turn over completely.
16 weeks	Can pull self to sitting position.
20 weeks	Can sit with erect back, when supported.
28 weeks	Can sit momentarily without support, after being placed in a sitting position.
6 months	Rolling and hitching occur.
7 to 9 months	Creeping occurs.
10 to 11 months	Crawls; can stand alone.
1 year	Can stand alone; typically can shift from merely standing to walking.
14 months	Walks well without support.
18 months	Walks like an adult (i.e., is not stiff-legged).

Note: The above are, of course, general approximations that can be used to establish operational conditions for products that may relate to child use and/or safety.

TYPICAL LIMITS OF BODY MOVEMENTS[1]

Other body movement characteristics are illustrated in Figs. 34.8 and 34.9 and Tables 34.4 and 34.5. Although joint mobility decreases only slightly with age for healthy individuals between the ages of 20 and 60, the incidence of arthritis beyond age 45 suggests that designers give serious consideration to this limiting factor.

Note: Females generally exceed males with respect to the range of joint movements, except in the case of the knee joint (see Table 34.6). Ordinarily, fat men and women have less movement capability than slender ones. People with muscular body builds have less than those with average builds; however, because of the effect of special exercise or training, one should not assume that these generalizations always apply.

[1]C. T. Morgan et al. (eds.), *Human Engineering Guide to Equipment Design*, McGraw-Hill Book Company, New York, 1963, pp. 553–555.

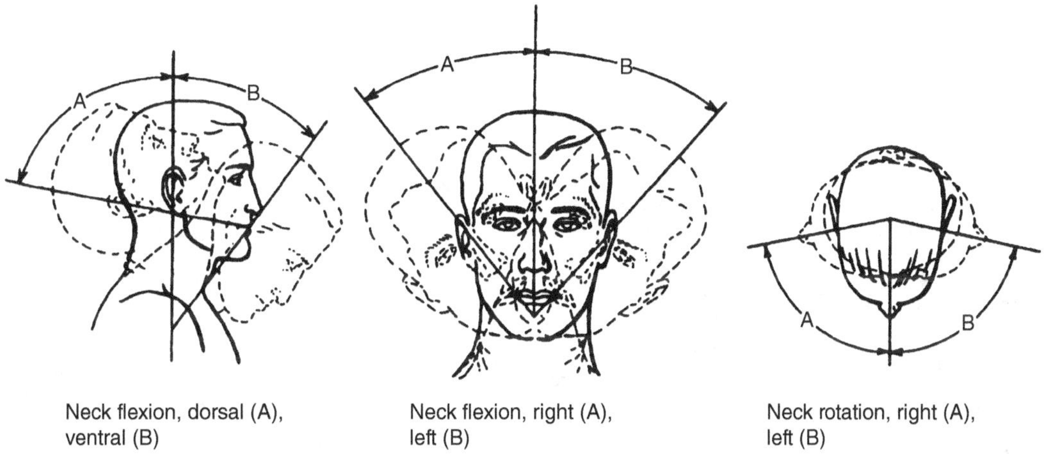

Figure 34.8 Neck movement.

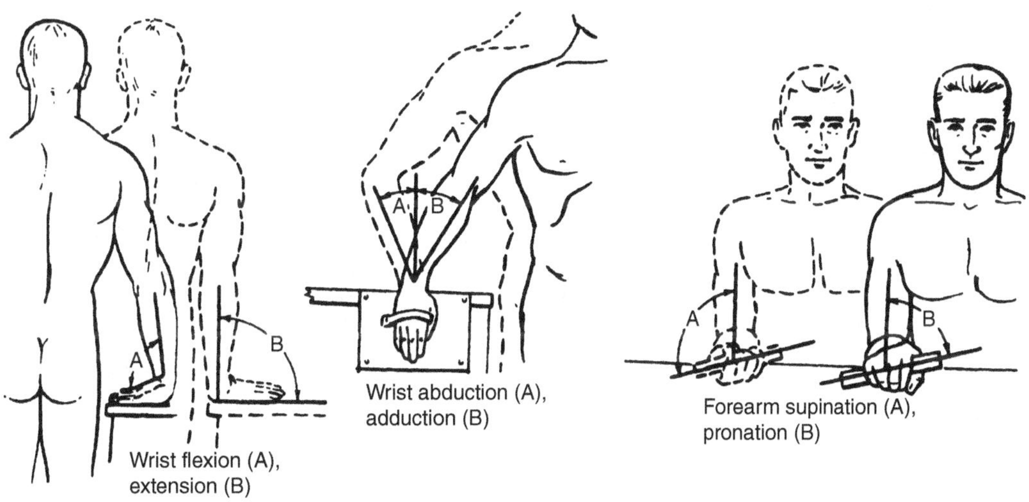

Figure 34.9 Body joint movements.

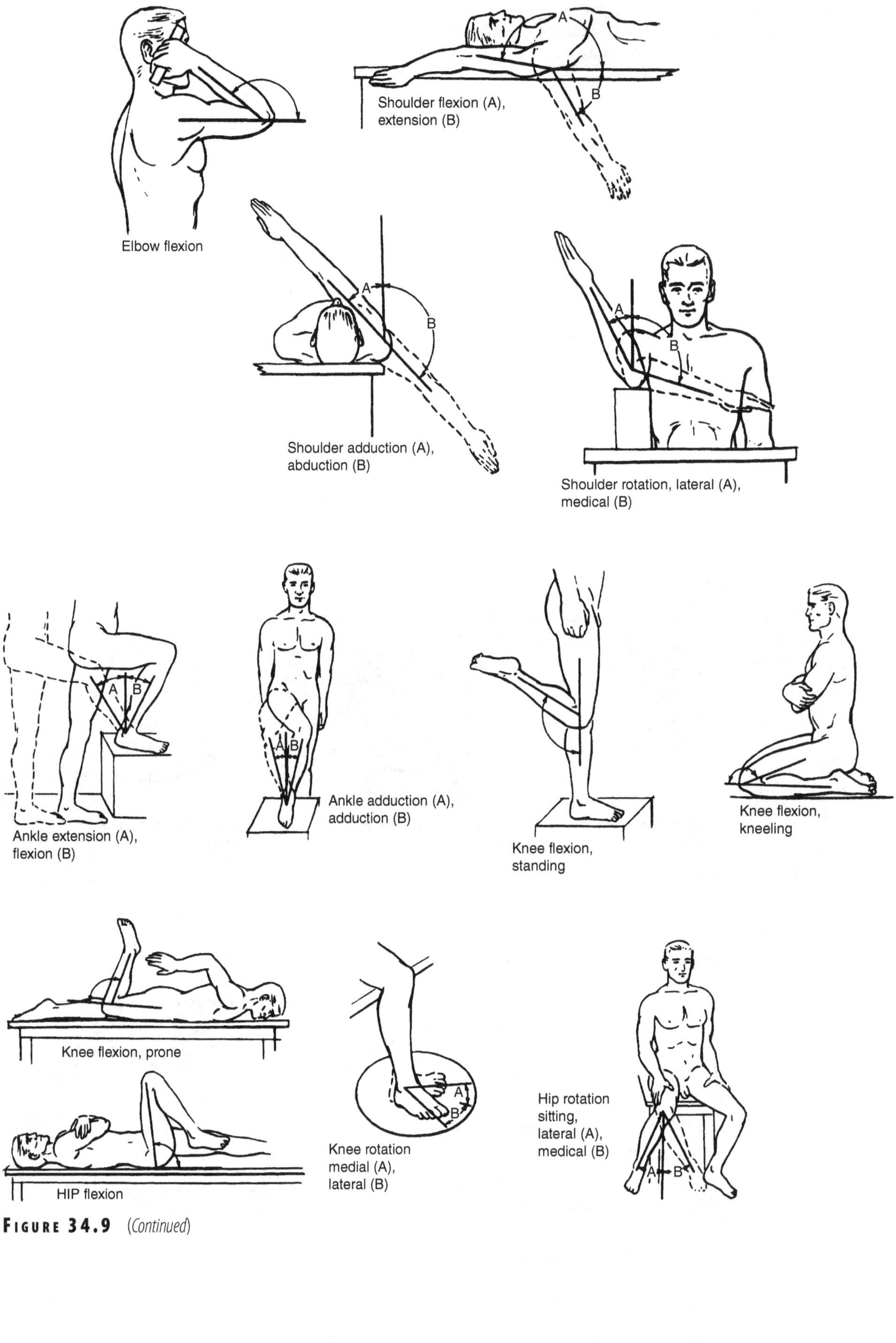

FIGURE 34.9 *(Continued)*

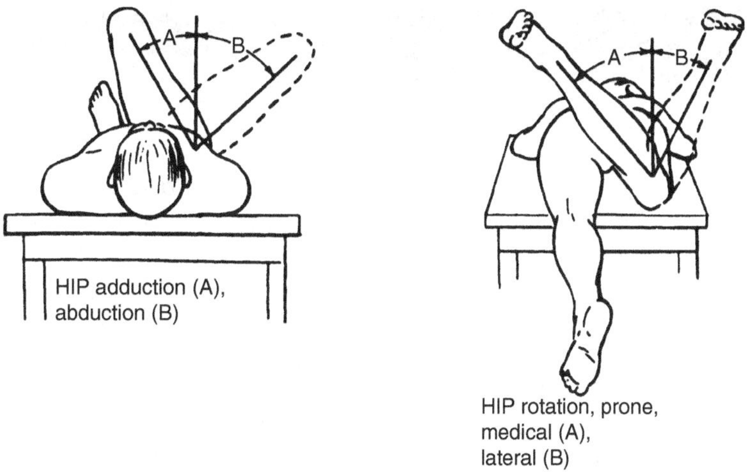

Figure 34.9 *(Continued)*

TABLE 34.4 Range of Movement at the Neck Joint*

	Average	SD
Ventral flexion	60°	12
Dorsal flexion	61°	27
Right-left flexion	41°	7
Right-left rotation	79°	14

*Male civilians.

TABLE 34.5

Range of Movement at the Joints of the Hand and Arm of Male Air Force Personnel

	Range (deg.)	
Movement[2]	**Avg.**	**S.D.**
Wrist flexion	90	12
Wrist extension	99	13
Wrist adduction	27	9
Wrist abduction	47	7
Forearm supination	113	22
Forearm pronation	77	24
Elbow flexion	142	10
Shoulder flexion	188	12
Shoulder extension	61	14
Shoulder adduction	48	9
Shoulder abduction	134	17

TABLE 34.5 *(Continued)*

Range of Movement at the Joints of the Foot and Leg of Male Air Force Personnel

	Range (deg.)	
Movement	**Avg.**	**S.D.**
Ankle flexion	35	7
Ankle extension	38	12
Ankle adduction	24	9
Ankle abduction	23	7
Knee flexion		
Standing	113	13
Kneeling	159	9
Prone	125	10
Knee rotation		
Medial	35	12
Lateral	43	12
Hip flexion	113	13
Hip adduction	31	12
Hip abduction	53	12
Hip rotation (sitting)		
Medial	31	9
Lateral	30	9
Hip rotation (prone)		
Medial	39	10
Lateral	34	10

TABLE 34.6 Average Increase in Range of Joint Movement of Women over Men

MOVEMENT	DIFFERENCE (DEG.)
Wrist flexion and extension	14
Wrist adduction and abduction	11
Elbow flexion and extension	8
Shoulder abduction (rearward)	2
Ankle flexion and extension	4
Knee flexion and extension	0
Hip flexion	3

ORTHOPEDIC HANDICAPS RELATIVE TO MOBILITY

In spite of either congenital or permanent or temporary physical handicaps, many people are still capable of moving about and utilizing various products and facilities. However, designers must recognize and understand the limitations of handicapped consumers and take special pains to make things accessible to, and operable by, these individuals.

The definitions in Table 34.7 may help cope with the problem of designing for the handicapped.

TABLE 34.7

TERM	DEFINITION
Abduction	The movement of an extremity (arm or leg) away from the middle of the body.
Adduction	The movement of an extremity (arm or leg) toward the middle of the body.
Amputee	A person who has suffered the loss of a limb.
Arthritis	A chronic disease of the joints that may limit limb positioning direction or extent and/or the use of the hand and fingers for gripping.
Ataxia	Loss of coordinate movement-position sense as a result of malfunction of the lower part of the brain.
Athetosis	Repetitive, involuntary movements due to brain damage (impossible to control even with treatment).
Bilateral	On both sides.
Cerebral palsy	A complex handicap with variations resulting from brain damage and manifested in poor muscle control and/or paralysis.
Congenital	Existing at birth.
Cord lesion	A disease of, or injury to, the spinal cord.
Diplegia	Paralysis affecting similar parts on both sides of the body.
Dysautonomia	A genetic disease affecting the autonomic nervous system.
Dystonia musculorum deformans	A rare chronic disease marked by involuntary, irregular contortions of the muscles of the trunk and extremities. Symptoms appear chiefly during walking, at which time the contortions twist the body forward and sideways in a grotesque fashion.
Friedrich's ataxia	Progressive paralysis of the lower limbs. Symptoms depend upon the area or areas of the nervous system involved.
Hemiparesis	Partial paralysis of one side of the body.
Hemiplegia	Complete paralysis of one side of the body.
Little's disease	The general term for all forms of cerebral spastic diplegia.
Multiple sclerosis	Deterioration of the brain or spinal cord; involves weakness in coordination, strong jerking movements of the arms and legs, euphoria, and scanning speech.
Muscular dystrophy	A progressive disease of the muscular tissue, resulting in diminished strength.
Paraparesis	Partial paralysis of the legs, such as might be caused by polio.
Paraplegia	Total paralysis of the legs due to spinal cord or brain injury.
Prosthesis	A biomechanical device fitted to the human body to replace a part that is missing, e.g., an artificial limb.
Quadriparesis	Complete paralysis of the body below the arms and some dysfunction in the arms and hands short of paralysis.
Rheumatoid arthritis	A systemic disorder resulting in inflammatory changes, atrophy, and rarefaction of the bones. Leads to crippling deformities.
Spastic paraplegia (infantile)	Spastic paralysis occurring in early childhood as a result of injuries during birth.
Vertebrae cervical	The vertebrae in the neck area of the backbone. (C-6 is the sixth vertebra from the head.)
Vertebrae, thoracic	The 12 vertebrae in the chest area of the backbone. (T-11 is the eighteenth vertebra from the head.)
Vertebra, lumbar	The fifth vertebra in the abdominal area of the backbone. (L- 5 is the twenty-fourth vertebra from the head.)
Vertebra, sacral	The "tailbone," which is a combination of the sacrum and coccyx fused together.
Erb's palsy	Paralysis of a group of muscles of the shoulder and upper arm. The arm hangs limp, the hand rotates inward, and normal movements are lost.
Epilepsy	An irregular electrical discharge of the brain, causing a temporary lapse of consciousness (known as a "seizure").
Encephalitis	A group of infectious diseases of the central nervous system, characterized by inflammation of brain tissues.

Note: The commonly used terms "hemiplegia," "paraplegia," and "quadriplegia" are often misused. "Hemiparesis," "paraparesis," or "quadriparesis" is the correct term if some function remains; i.e., "plegia" denotes that no function remains.

WHOLE-BODY WEIGHT (NUDE)[2]

Whole-body weights for various populations are shown in Table 34.8. (See also Table 34.9.) Values for only the 5th, 50th, and 95th percentiles are provided, since these are the most significant for the majority of design requirements. For additional population percentiles, refer to the references provided in the section on body dimensions.

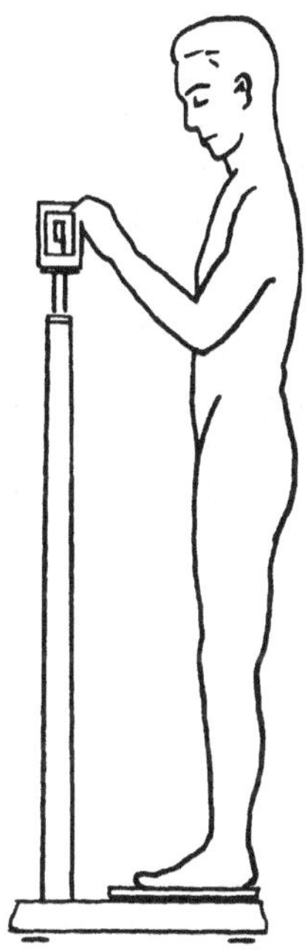

TABLE 34.8 Whole Body Weight

	Percentile		
	5th	**50th**	**95th**
Adults:			
Males	124 lb	168 lb	224 lb
Females	104 lb	139 lb	208 lb
Boys:			
Age 17	120 lb	149 lb	195 lb
Age 14	84 lb	112 lb	153 lb
Age 12	67 lb	84 lb	121 lb
Age 6	37 lb	44 lb	59 lb
Age 2	23 lb	27 lb	31 lb
Girls:			
Age 17	92 lb	120 lb	162 lb
Age 14	81 lb	108 lb	146 lb
Age 12	62 lb	88 lb	120 lb
Age 6	35 lb	44 lb	54 lb
Age 2	22 lb	24 lb	31 lb
Adults age 70 and over:			
Males	105 lb	144 lb	196 lb
Females	93 lb	135 lb	191 lb
Truck and bus drivers:			
Males	138 lb	175 lb	227 lb
Females	100 lb	128 lb	181 lb
Airline pilots (Male)	134 lb	168 lb	201 lb
Flight attendants (Female)	102 lb	117 lb	133 lb
Law enforcement officers:			
Males	146 lb	181 lb	230 lb
Females	—	—	—
U.S. Army			
Males	136	171	216
Females	109	135	170

[2]These data have been compiled from the sources listed at the end of this section.

TABLE 34.9 Average Male Adult Weights for Other Populations

NATIONALITY	WEIGHT, LB
Chinese	120
English	143
Filipino	118
French	149
Greek	143
Hawaiian	172
Indian	123
Japanese	115
Korean	124
American Indians:	
Hopi	135
Navajo	139
Zuni	125

Note: 1.0 lb = 0.45 kg.

PHYSICAL DIMENSIONS

The science of measuring the human body is *anthropometry*.

Anthropometrists have been measuring a wide variety of people in many ways for many years. More than 350 different measurements have been taken at one time or another. Although measuring techniques have become more or less standardized, there is no assurance that each anthropometrist measured his or her particular sample of subjects, or each particular dimension, in the same manner. Thus, one has to assume a certain caution in comparing data from various populations.

A great share of anthropometric data has been generated for the purpose of scientific comparison of populations, and not necessarily for the purpose of providing design-related information. This is evident from the fact that subjects are measured in stiff and unnatural positions, and in the nude. Also, since anthropometric surveys are extremely expensive, they are not repeated periodically to update information as people and populations change. Because of these and other unique conditions, it is important that the user of anthropometric data recognize the limitations of such data and treat the information merely as a point of departure. That is, anthropometric data are most useful as an aid in selecting a sample of "live" test subjects, who then should be used to test any proposed design.

Unfortunately, not all surveys include all the dimensions one might need for a given design problem. The most complete surveys have almost always been conducted on military personnel, and most of these have dealt with males.

In spite of these shortcomings, designers should make a point of referring to anthropometric data whenever a design involves a "fit" problem. Also, care should be taken to refer to data that are representative of the pertinent user population, since a particular survey may not have examined the dimension in question and thus may provide no data on it. To assist the designer, the following information on body dimensions has been organized according to the population groups in which product designers are most commonly interested. It will be noted that there are some obvious gaps in the tables where data for a given dimension and population are unavailable. Also, it should be pointed out that, in some cases, a substitute value has been entered when (in the author's opinion) a similar population seemed to justify using a particular dimensional value for the population in question. Although such liberty is often frowned upon by conservative anthropometrists, it is excused here on the basis that *none of these data should be used directly for design purposes, but rather as starting points to establish requirements for selecting a dynamic (live) test subject sample.* There is no substitute for a good, full-scale mockup evaluation to validate "fit" of the human body.

U.S. ANTHROPOMETRY SURVEY DATA

Every five years since 1960 the U.S. Department of Health and Human Services measures thousands of Americans. Some of the results for the 2010 survey are summarized in Table 34.10. Researchers made the following measurements.

1. Stature.
2. Weight.
3. Upper leg length.
4. Upper arm length.
5. Waist, arm, and head circumference.
6. Skinfold measurements.
7. Body mass index.
8. The data are categorized by sex, age, and racial and ethnic groups.

TABLE 34.10 Anthropometric Reference Data for Children and Adults

	Percentile		
	5th	50th	95th
Male 10 years old			
Weight (pounds)	56.8	83.8	131.7
Stature (inches)	51.8	56.1	60.2
Female 10 years old			
Weight (pounds)	60.2	86.1	135.2
Stature (inches)	52.7	56.9	61.2
Male 15 years old			
Weight (pounds)	106.8	145.5	254.2
Stature (inches)	63.1	68.4	73.9
Female 15 years old			
Weight (pounds)	97.5	131.1	203.4
Stature (inches)	52.7	56.9	61.2
Male 20 years old and greater			
Weight (pounds)	135.5	189.8	273.6
Stature (inches)	64.4	69.3	74.1
Female 20 years old and greater			
Weight (pounds)	110.7	157.2	250.9
Stature (inches)	59.3	63.8	68.4

Anthropometric Reference Data for Children and Adults: United States 2007–2010, October 2012, U.S. Department of Health and Human Services.

STANDING HEIGHT (STATURE)

Standing height is the primary indicator by which one selects general test subjects to evaluate various designs when there is no specific aspect of the design pertaining to an individual body component. The dimension is pertinent for adjusting head clearances. (See Table 34.11.)

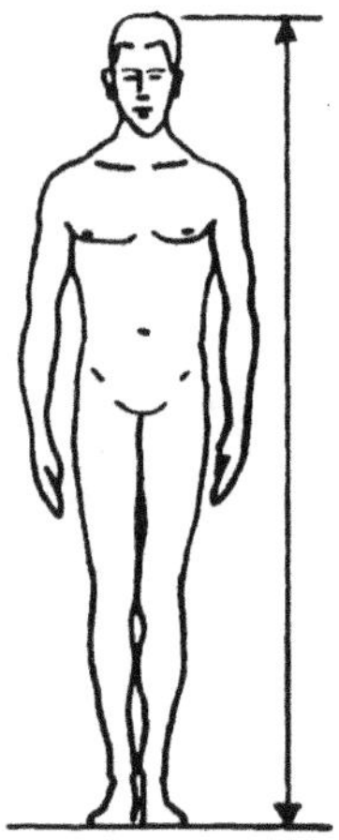

TABLE 34.11

	PERCENTILE		
	5TH	50TH	95TH
Adults:			
Males	63.6 in.	68.3 in.	72.8 in.
Females	59.0 in.	62.9 in.	67.1 in.
Boys:			
Age 17	65.1 in.	69.4 in.	72.6 in.
Age 14	58.0 in.	63.2 in.	69.1 in.
Age 12	54.4 in.	58.0 in.	63.5 in.
Age 6	41.8 in.	45.1 in.	49.1 in.
Age 2	31.7 in.	33.2 in.	35.8 in.
Girls:			
Age 17	60.0 in.	64.1 in.	67.6 in.
Age 14	57.8 in.	62.3 in.	66.5 in.
Age 12	54.1 in.	58.8 in.	63.0 in.
Age 6	41.3 in.	45.0 in.	47.9 in.
Age 2	30.4 in.	33.0 in.	35.4 in.
Adults age 70 and over:			
Males	61.3 in.	66.2 in.	69.5 in.
Females	55.3 in.	61.8 in.	64.9 in.
Truck and bus drivers:			
Males	65.1 in.	69.8 in.	74.3 in.
Females	58.9 in.	63.5 in.	68.0 in.
Airline pilots (Male)	66.0 in.	70.0 in.	73.9 in.
Flight attendants (Female)	62.5 in.	65.4 in.	68.8 in.
Law enforcement officers:			
Males	66.6 in.	70.0 in.	74.0 in.
Females	—	—	—
U.S. Army			
Males	64.8	69.1	78.5
Females	60.2	64.1	68.4

EYE HEIGHT (STANDING)

This dimension is pertinent to the location of visual displays and/or the sizing of visual obstructions, where a small person may have to see over the obstruction or over someone who is taller. As a general "rule of thumb," the eye height of males is about 5.2 in. (13 cm) less than their standing height; for females it is about 4.8 in. (12.2 cm) less. Similarly, when people stand normally (i.e., with some slump), their eye height lowers by about 1.6 in. (4.1 cm) for males and by about 1.2 in. (3.0 cm) for females.[3] (See Table 34.12.)

[3] Rules of thumb apply only to adults.

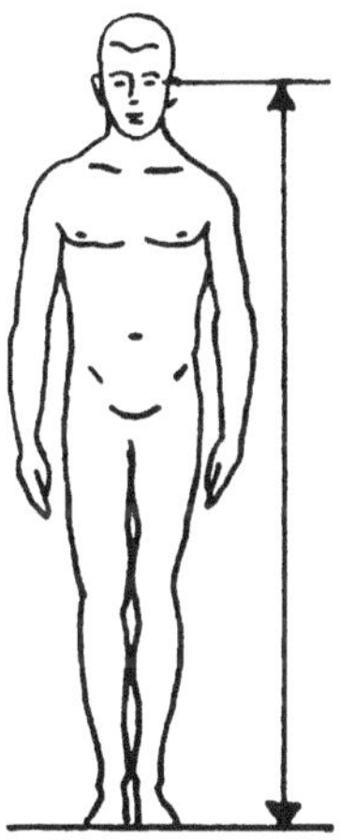

TABLE 34.12

	Percentile		
	5th	50th	95th
Adults:			
Males	60.8 in.	64.7 in.	68.6 in.
Females	57.3 in.	60.3 in.	65.3 in.
Boys:			
Age 17	—	—	—
Age 14	—	—	—
Age 12	—	—	—
Age 6	—	—	—
Age 2	—	—	—
Girls:			
Age 17	—	—	—
Age 14	—	—	—
Age 12	—	—	—
Age 6	—	—	—
Age 2	—	—	—
Adults age 70 and over			
Males	58.3 in.	61.1 in.	64.9 in.
Females	50.8 in.	55.8 in.	60.8 in.
Truck and bus drivers:			
Males	59.6 in.	63.4 in.	67.5 in.
Females	—	—	—
Airline pilots (Male)	61.0 in.	65.0 in.	68.9 in.
Flight attendants (Female)	58.0 in.	60.9 in.	64.3 in.
Law enforcement officers:			
Males	61.6 in.	65.0 in.	69.0 in.
Females	—	—	—
U.S. Army			
Males	60.2	64.3	68.6
Females	55.7	59.6	63.8

OVERHEAD REACH (STANDING)

This dimension is pertinent to locating controls that are overhead. It should be used in conjunction with stature because, although a short person must be able to reach a control, it should not be so low that it becomes an obstruction for the taller person. Use this dimension to select subjects for evaluating the accessibility of objects on high shelves. (See Table 34.13.)

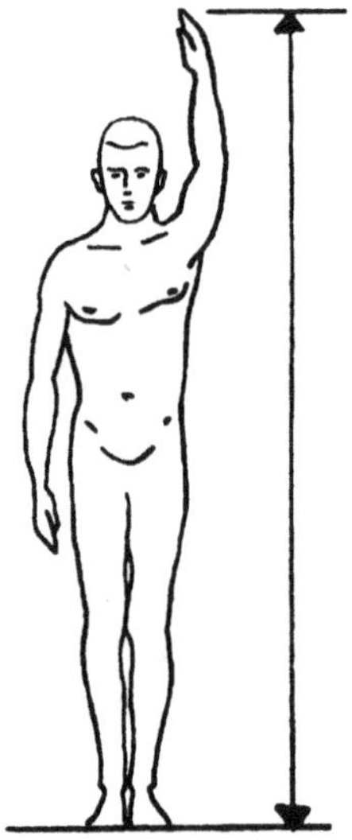

TABLE 34.13

	Percentile		
	5th	**50th**	**95th**
Adults:			
Males*	82.0 in.	88.0 in.	94.0 in.
Females*	73.0 in.	79.0 in.	86.0 in.
Boys:			
Age 17	77.7 in.	84.1 in.	88.6 in.
Age 14	70.5 in.	77.3 in.	83.1 in.
Age 12	64.7 in.	70.7 in.	78.9 in.
Age 6	50.3 in.	53.3 in.	58.2 in.
Age 2	38.2 in.	41.7 in.	48.1 in.
Girls:			
Age 17	72.6 in.	77.7 in.	81.9 in.
Age 14	66.1 in.	75.2 in.	83.1 in.
Age 12	65.0 in.	72.1 in.	77.4 in.
Age 6	48.1 in.	53.0 in.	56.8 in.
Age 2	38.7 in.	41.6 in.	45.6 in.
Adults age 70 and over:			
Males	—	—	—
Females	69.5 in.	75.5 in.	82.5 in.
Truck and bus drivers:			
Males*	81.0 in.	88.1 in.	93.7 in.
Females	—	—	—
Airline pilots* (Male)	85.0 in.	91.5 in.	95.1 in.
Flight attendants* (Female)	65.8 in.	68.9 in.	72.3 in.
Law enforcement officers:			
Males*	85.0 in.	89.7 in.	95.2 in.
Females	—	—	—
U.S. Army*			
Males	85.3	91.6	98.2
Females	78.8	84.7	91.1

*Estimated.

FORWARD REACH (STANDING)

This dimension is pertinent to the selection of test subjects who will be evaluating reach conditions within the design situation. This dimension should be used in conjunction with leg-reach dimensions when the design situation calls for a seated operator to operate both hand and foot controls. (See Table 34.14.)

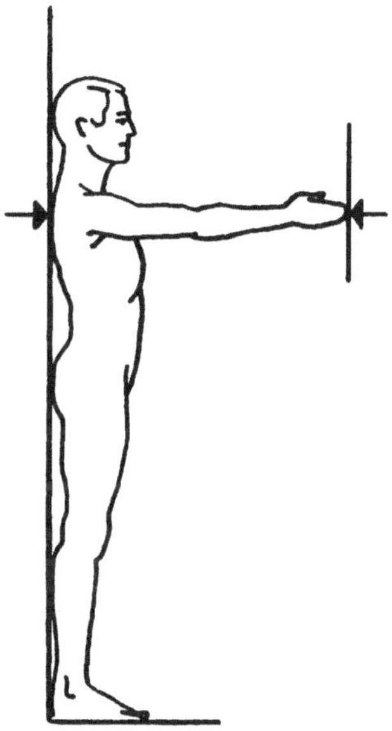

TABLE 34.14

	Percentile		
	5th	**50th**	**95th**
Adults:			
Males	31.9 in.	34.6 in.	37.3 in.
Females	29.7 in.	31.8 in.	34.1 in.
Boys:			
Age 17	25.7 in.	28.7 in.	32.0 in.
Age 14	23.8 in.	26.9 in.	29.3 in.
Age 12	22.2 in.	24.2 in.	26.8 in.
Age 6	17.6 in.	19.1 in.	20.9 in.
Age 2	14.0 in.	15.8 in.	18.1 in.
Girls:			
Age 17	23.9 in.	26.7 in.	28.8 in.
Age 14	22.8 in.	25.5 in.	28.4 in.
Age 12	21.8 in.	24.3 in.	26.4 in.
Age 6	16.6 in.	18.7 in.	20.6 in.
Age 2	13.7 in.	15.1 in.	16.8 in.
Adults age 70 and over:			
Males	—	33.5 in.	—
Females	—	—	—
Truck and bus drivers:			
Males	33.0 in.	35.8 in.	38.4 in.
Females	—	—	—
Airline pilots (Male)	32.9 in.	34.3 in.	37.0 in.
Flight attendants (Female)	29.0 in.	31.0 in.	33.3 in.
Law enforcement officers:			
Males	32.3 in.	34.8 in.	37.7 in.
Females	—	—	—
U.S. Army*			
Males	29.1	31.5	34.1
Females	26.6	28.9	31.4

*To thumbtip.

MAXIMUM BODY WIDTH

As indicated by the lack of data in Table 34.15, this measurement is seldom taken during typical anthropometric surveys. However, it is pertinent to lateral clearance requirements in design work. It would be useful, for instance, in selecting test subjects to evaluate lateral corridor clearance.

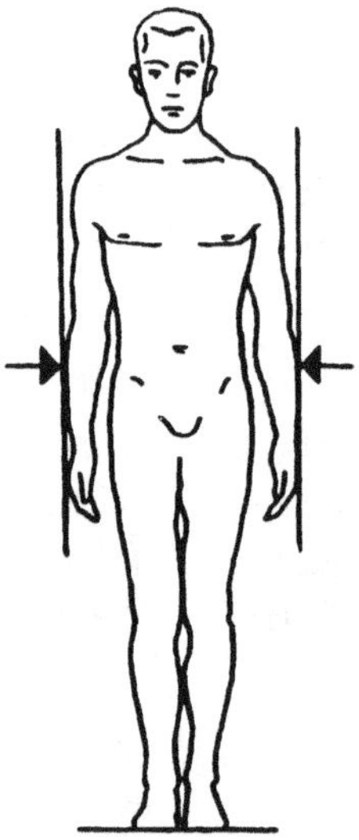

TABLE 34.15

	Percentile		
	5th	50th	95th
Adults:			
Males*			
Females	18.8 in.	20.9 in.	22.8 in.
Boys:			
Age 17			
Age 14			
Age 12			
Age 6			
Age 2			
Girls:			
Age 17			
Age 14			
Age 12			
Age 6			
Age 2			
Adults age 70 and over:			
Males			
Females			
Truck and bus drivers:			
Males			
Females			
Airline pilots			
Flight attendants			
Law enforcement officers:			
Males			
Females			

*Combined Air Force personnel and college students.

CROTCH HEIGHT

This dimension is a secondary indicator of leg length and is pertinent to design situations in which a person is required to step over an obstacle. (See Table 34.16.)

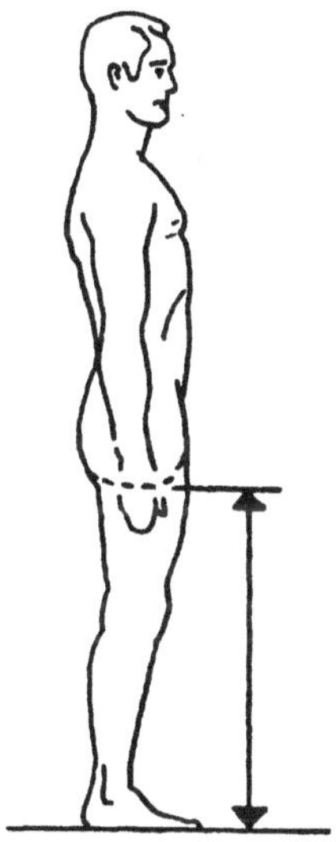

TABLE 34.16

	Percentile		
	5th	50th	95th
Adults:			
Males	30.0 in.	33.0 in.	36.1 in.
Females	26.8 in.	29.3 in.	32.0 in.
Boys:			
Age 17	—	—	—
Age 14	—	—	—
Age 12	25.1 in.	27.8 in.	30.7 in.
Age 6	17.9 in.	19.9 in.	21.7 in.
Age 2	11.5 in.	13.4 in.	17.3 in.
Girls:			
Age 17	—	—	—
Age 14	—	—	—
Age 12	24.9 in.	28.0 in.	30.6 in.
Age 6	17.8 in.	20.2 in.	21.8 in.
Age 2	11.5 in.	13.4 in.	17.3 in.
Adults age 70 and over:			
Males			
Females			
Truck and bus drivers:			
Males			
Females			
Airline pilots			
Flight attendants			
Law enforcement officers:			
Males			
Females			
U.S. Army			
Males	30.1	32.9	36.1
Females	27.6	30.3	33.3

SITTING HEIGHT

This dimension is pertinent to the establishment of proper overhead clearances for seated persons. It is particularly important in the design of driver workstations. (See Table 34.17.)

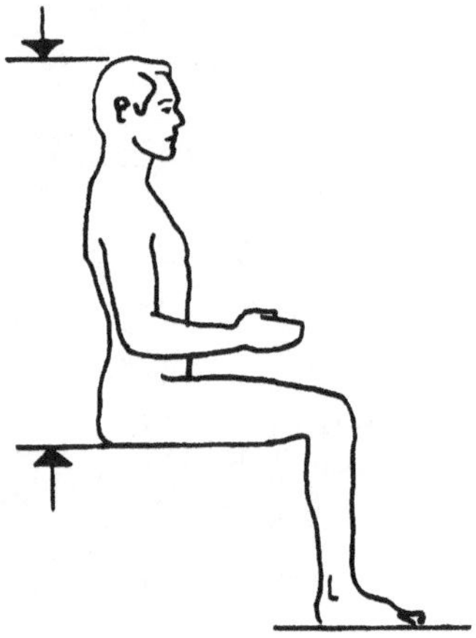

TABLE 34.17

	Percentile		
	5th	**50th**	**95th**
Adults:			
Males	33.2 in.	35.7 in.	38.0 in.
Females	30.9 in.	33.4 in.	35.7 in.
Boys:			
Age 17	33.7 in.	35.9 in.	37.8 in.
Age 14	29.4 m	32.3 in.	35.6 in.
Age 12	28.2 in.	29.9 in.	32.5 in.
Age 6	23.0 in.	25.0 in.	26.8 in.
Age 2	19.0 in.	20.6 in.	21.9 in.
Girls:			
Age 17	31.8 in.	33.7 in.	35.9 in.
Age 14	29.9 in.	32.4 in.	34.8 in.
Age 12	28.0 in.	30.4 in.	32.9 in.
Age 6	22.8 in.	24.9 in.	26.5 in.
Age 2	18.9 in.	19.8 in.	21.5 in.
Adults age 70 and over:			
Males	31.8 in.	34.3 in.	36.7 in.
Females	28.1 in.	32.1 in.	34.8 in.
Truck and bus drivers:			
Males	34.3 in.	36.3 in.	38.2 in.
Females	—	—	—
Airline pilots	—	—	—
Flight attendants	32.4 in.	34.3 in.	36.1 in.
Law enforcement officers:			
Males	34.1 in.	36.3 in.	38.5 in.
Females	—	—	—
U.S. Army			
Males	33.6	36.0	38.3
Females	31.3	33.5	35.8

EYE HEIGHT (SITTING)

This dimension is pertinent to the design of workstations in which visual displays and/or outside viewing requires accommodation of a range of operator sizes. As a rule of thumb, the eye height of males is about 5.2 in. (13 cm) less than their sitting height; that of females is about 4.8 in. (12.2 cm) less. Similarly, when people sit normally (with some slump), their eye height lowers by about 1.2 in. (3.0 cm) for males and by about 1.2 in. (3.0 cm) for females. (See Table 34.18.)

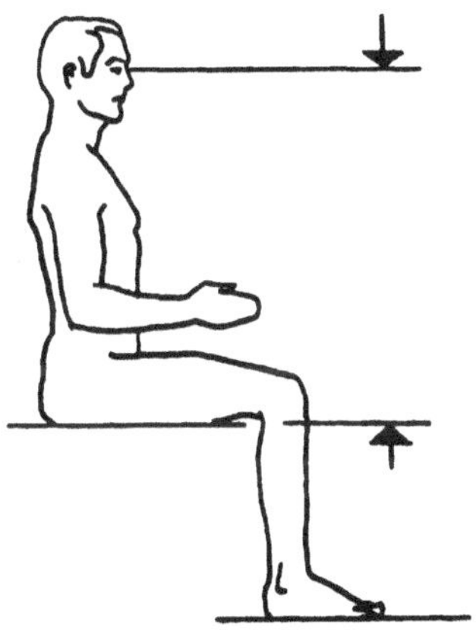

TABLE 34.18

	Percentile		
	5th	50th	95th
Adults:			
Males	28.7 in.	31.3 in.	33.5 in.
Females	27.4 in.	29.3 in.	31.0 in.
Boys:			
Age 17	29.7 in.	31.6 in.	33.2 in.
Age 14	25.2 in.	28.2 in.	30.7 in.
Age 12	23.6 in.	25.4 in.	27.0 in.
Age 6	18.3 in.	20.5 in.	22.2 in.
Age 2	15.9 in.	17.6 in.	18.8 in.
Girls:			
Age 17	26.9 in.	29.0 in.	31.1 in.
Age 14	25.1 in.	28.0 in.	30.4 in.
Age 12	23.6 in.	25.6 in.	27.6 in.
Age 6	17.8 in.	20.6 in.	22.4 in.
Age 2	15.2 in.	16.7 in.	18.1 in.
Adults age 70 and over:			
Males	—	26.8 in.	—
Females	—	—	—
Truck and bus drivers:			
Males	27.7 m	29.6 in.	31.6 in.
Females	—	—	—
Airline pilots (Male)	29.4 in.	31.5 in.	33.5 in.
Flight attendants (Female)	28.1 in.	29.9 in.	31.7 in.
Law enforcement officers:			
Males	29.1 in.	31.3 in.	33.5 in.
Females	—	—	—
U.S. Army			
Males	28.9	31.2	33.4
Females	27.0	29.1	31.3

BUTTOCK-TO-POPLITEAL LENGTH

This dimension is pertinent to seat length. Although it is desirable to provide adequate support for the larger person, it is the shorter person who will have the most problems if this dimension is ignored in the test subject sample. (See Table 34.19.)

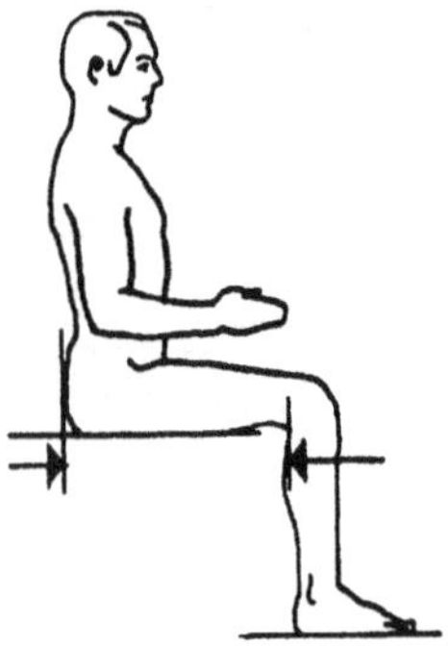

TABLE 34.19

	Percentile		
	5th	50th	95th
Adults:			
Males	17.3 in.	19.5 in.	21.6 in.
Females	17.0 in.	18.9 in.	21.0 in.
Boys:			
Age 17			
Age 14			
Age 12			
Age 6			
Age 2			
Girls:			
Age 17			
Age 14			
Age 12			
Age 6			
Age 2			
Adults age 70 and over:			
Males	17.0 in.	18.9 in.	21.2 in.
Females	17.0 in.	18.7 in.	20.0 in.
Truck and bus drivers:			
Males	20.2 in.	22.2 in.	24.0 in.
Females	18.4 in.	20.5 in.	22.3 in.
Airline pilots			
Flight attendants (Female)	17.4 in.	19.0 in.	20.6 in.
Law enforcement officers:			
Males	—	—	—
Females	—	—	—
U.S. Army			
Males	18.0	19.7	21.5
Females	17.3	18.9	20.8

MIDSHOULDER HEIGHT (SITTING)

This dimension is pertinent to the location of controls (i.e., most controls should not be located above the shoulder) and to the location of the anchor point for seat belts (i.e., the belt should not depart aft of the shoulder horizontally or at a negative angle; otherwise, it will create discomfort and tend to push the occupant downward during a frontal-impact crash). (See Table 34.20.)

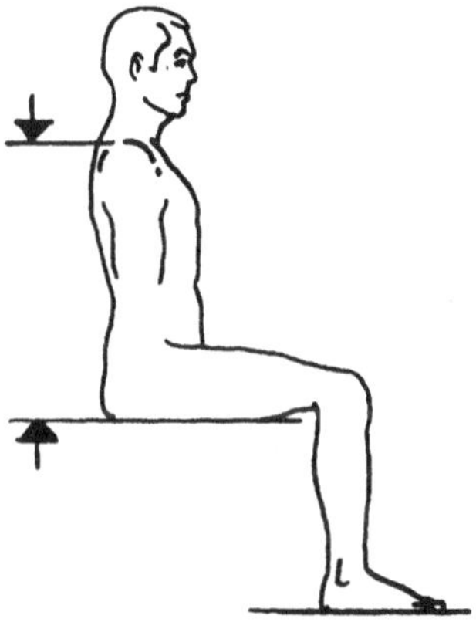

TABLE 34.20

	Percentile		
	5th	50th	95th
Adults:			
Males	21.0 in.	—	25.0 in.
Females	18.0 in.	—	25.0 in.
Boys:			
Age 17	—	—	—
Age 14	18.1 in.	20.1 in.	21.2 in.
Age 12	17.8 in.	19.4 in.	21.5 in.
Age 6	13.7 in.	15.4 in.	17.0 in.
Age 2	11.1 in.	12.3 in.	13.3 in.
Girls:			
Age 17	—	—	—
Age 14	19.1 in.	212 in.	22.6 in.
Age 12	17.6 in.	19.5 in.	21.0 in.
Age 6	13.6 in.	15.1 in.	16.5 in.
Age 2	11.1 in.	12.3 in.	14.0 in.
Adults age 70 and over:			
Males	20.9 in.	24.0 in.	27.2 in.
Females	—	—	—
Truck and bus drivers:			
Males	21.2 in.	24.3 in.	27.5 in.
Females	—	—	—
Airline pilots	—	—	—
Flight attendants (Female)	20.1 in.	22.9 in.	25.7 in.
Law enforcement officers:			
Males	—	—	—
Females	—	—	—
U.S. Army			
Males	23.0	24.8	26.6
Females	21.2	23.0	24.8

BUTTOCK-TO-KNEE LENGTH

This dimension is pertinent to establishing knee clearance for the seated operator. It should be used in conjunction with knee-height and thigh-clearance dimensions. (See Table 34.21.)

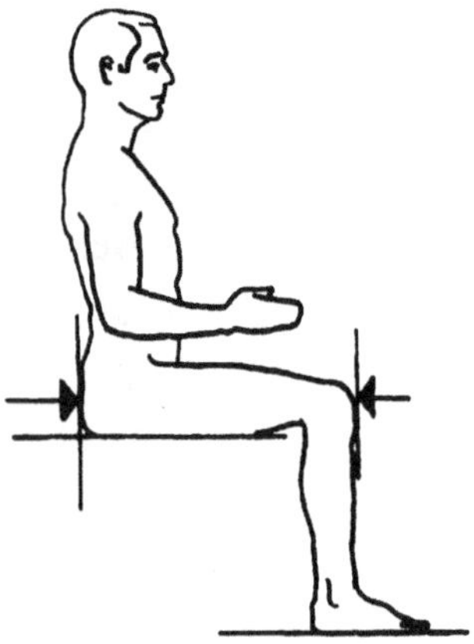

TABLE 34.21

	Percentile		
	5th	50th	95th
Adults:			
Males	21.3 in.	23.3 in.	25.2 in.
Females	20.4 in.	22.4 in.	24.6 in.
Boys:			
Age 17	21.2 in.	23.1 in.	25.0 in.
Age 14	19.2 in.	21.4 in.	23.5 in.
Age 12	17.7 in.	19.5 in.	21.5 in.
Age 6	12.9 in.	14.1 in.	15.6 in.
Age 2	8.5 in.	9.3 in.	10.5 in.
Girls:			
Age 17	20.0 in.	21.7 in.	23.5 in.
Age 14	19.4 in.	21.2 in.	23.2 in.
Age 12	17.9 in.	20.1 in.	22.0 in.
Age 6	12.5 in.	14.2 in.	15.4 in.
Age 2	7.4 in.	9.4 in.	10.6 in.
Adults age 70 and over:			
Males	21.0 in.	22.6 in.	24.4 in.
Females	19.9 in.	22.2 in.	23.9 in.
Truck and bus drivers:			
Males	22.7 in.	24.6 in.	26.8 in.
Females	20.6 in.	22.9 in.	24.9 in.
Airline pilots (Male)	22.0 in.	23.6 in.	25.6 in.
Flight attendants (Female)	21.2 in.	22.6 in.	24.2 in.
Law enforcement officers:			
Males	22.6 in.	24.2 in.	26.1 in.
Females			
U.S. Army			
Males	22.4	24.2	26.3
Females	21.3	23.1	25.2

POPLITEAL HEIGHT (SITTING)

This dimension is pertinent to the establishment of appropriate seat heights. It is also pertinent to the selection of test subjects who will be used to evaluate the relationships between a vehicle seat and foot controls. (See Table 34.22.)

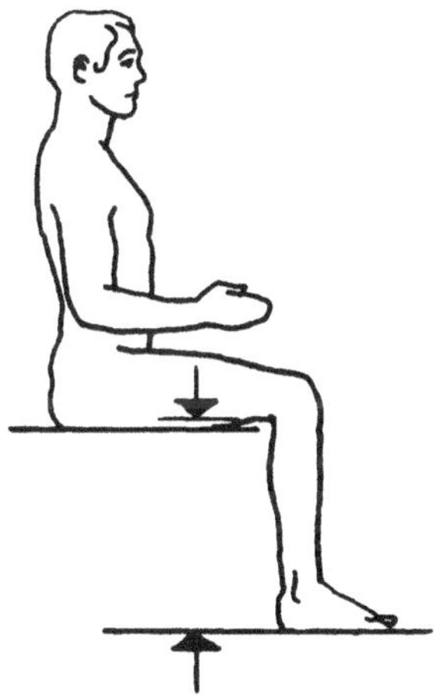

TABLE 34.22

	Percentile		
	5th	50th	95th
Adults:			
Males	15.5 in.	17.3 in.	19.3 in.
Females	14.0 in.	15.7 in.	17.5 in.
Boys:			
Age 17	—	—	—
Age 14	—	—	—
Age 12	13.2 in.	14.6 in.	16.1 in.
Age 6	10.4 in.	11.5 in.	12.6 in.
Age 2	—	—	—
Girls:			
Age 17	—	—	—
Age 14	—	—	—
Age 12	13.0 in.	14.7 in.	16.3 in.
Age 6	10.2 in.	11.3 in.	12.5 in.
Age 2	—	—	—
Adults age 70 and over:			
Males	15.2 in.	16.6 in.	17.9 in.
Females	13.5 in.	15.6 in.	17.2 in.
Truck and bus drivers:			
Males	15.7 in.	17.5 in.	19.7 in.
Females	—	—	—
Airline pilots (Male)	15.7 in.	17.0 in.	18.2 in.
Flight attendants (Female)	15.9 in.	17.1 in.	18.5 in.
Law enforcement officers:			
Males	—	—	—
Females	—	—	—
U.S. Army			
Males	15.5	17.1	18.8
Females	13.8	15.3	16.9

KNEE HEIGHT (SITTING)

This dimension is pertinent to the establishment of knee clearance. It should be used in conjunction with buttock-to-knee length and thigh-clearance dimensions. (See Table 34.23.)

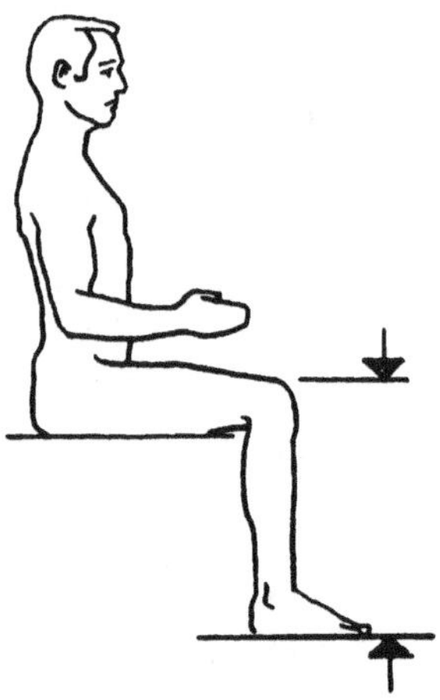

TABLE 34.23

	Percentile		
	5th	**50th**	**95th**
Adults:			
Males	19.3 in.	21.4 in.	23.4 in.
Females	17.9 in.	19.6 in.	21.5 in.
Boys:			
Age 17	19.4 in.	21.8 in.	23.4 in.
Age 14	18.5 in.	20.3 in.	22.6 in.
Age 12	17.2 in.	18.6 in.	20.7 in.
Age 6	12.4 in.	13.9 in.	15.3 in.
Age 2	8.1 in.	9.4 in.	10.2 in.
Girls:			
Age 17	17.8 in.	19.5 in.	21.2 in.
Age 14	17.7 in.	19.5 in.	21.1 in.
Age 12	17.0 in.	18.6 in.	20.2 in.
Age 6	12.4 in.	13.5 in.	14.8 in.
Age 2	8.0 in.	9.4 in.	10.0 in.
Adults age 70 and over:			
Males	19.0 in.	20.7 in.	22.7 in.
Females	17.3 in.	19.4 in.	20.9 in.
Truck and bus drivers:			
Males	20.1 in.	21.7 in.	23.5 in.
Females	—	—	—
Airline pilots (Male)	20.1 in.	21.7 in.	23.3 in.
Flight attendants (Female)	19.1 in.	20.4 in.	21.9 in.
Law enforcement officers:			
Males	20.4 in.	22.0 in.	23.7 in.
Females	—	—	—
U.S. Army			
Males	20.2	22.0	23.9
Females	18.7	20.3	22.1

THIGH CLEARANCE (SITTING)

This dimension is pertinent to the establishment of the proper clearance between a seat and the lower edge of a desk and to the establishment of the space between the driver's seat and a vehicle steering wheel. (See Table 34.24.)

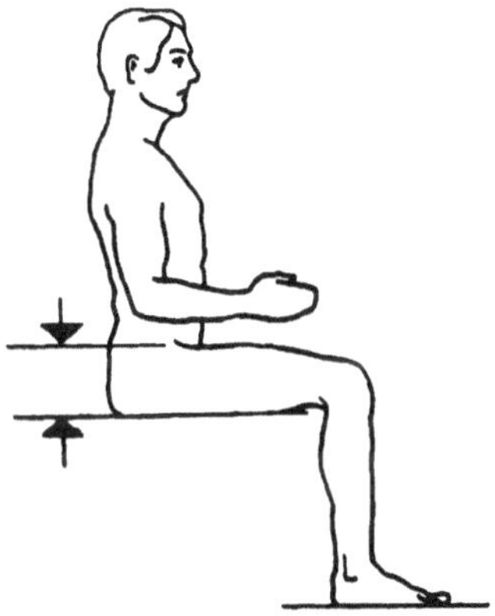

TABLE 34.24

	Percentile		
	5th	50th	95th
Adults:			
Males	4.3 in.	5.7 in.	6.9 in.
Females	4.1 in.	5.4 in.	6.9 in.
Boys:			
Age 17	4.8 in.	6.1 in.	6.9 in.
Age 14	4.3 in.	5.5 in.	6.4 in.
Age 12	4.2 in.	5.0 in.	5.8 in.
Age 6	3.1 in.	3.6 in.	4.3 in.
Age 2	2.5 in.	3.3 in.	3.9 in.
Girls:			
Age 17	4.6 in.	5.4 in.	6.8 in.
Age 14	4.6 in.	5.4 in.	6.1 in.
Age 12	4.2 in.	4.9 in.	6.1 in.
Age 6	3.0 in.	3.7 in.	4.5 in.
Age 2	2.1 in.	3.1 in.	3.7 in.
Adults age 70 and over:			
Males	4.1 in.	5.2 in.	6.6 in.
Females	4.0 in.	5.2 in.	6.5 in.
Truck and bus drivers:			
Males	5.0 in.	5.9 in.	6.9 in.
Females	—	—	—
Airline pilots (Male)	4.8 in.	5.6 in.	6.5 in.
Flight attendants (Female)	—	—	—
Law enforcement officers:			
Males	—	—	—
Females	—	—	—
U.S. Army			
Males	5.9	6.6	7.5
Females	5.3	6.2	7.1

ELBOW-TO-FINGERTIP LENGTH

This dimension is pertinent to special problems such as evaluating whether a control may have been placed too close to the operator, with the result that (because of the seat) the operator cannot get his or her arm back tar enough to use the control properly. (See Table 34.25.)

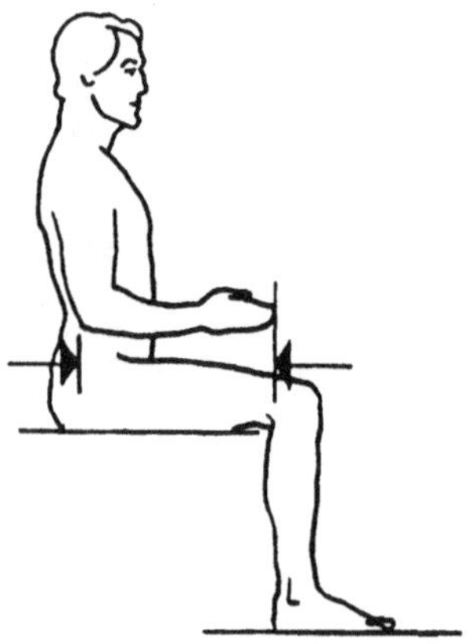

TABLE 34.25

	Percentile		
	5th	50th	95th
Adults:			
Males	—	—	—
Females	—	—	—
Boys:			
Age 17	17.5 in.	18.7 in.	20.0 in.
Age 14	15.6 in.	17.3 in.	19.1 in.
Age 12	14.4 in.	15.6 in.	17.3 in.
Age 6	10.9 in.	12.0 in.	13.2 in.
Age 2	7.6 in.	8.9 in.	9.9 in.
Girls:			
Age 17	15.5 in.	16.8 in.	18.0 in.
Age 14	15.2 in.	16.5 in.	17.9 in.
Age 12	14.0 in.	15.7 in.	17.3 in.
Age 6	10.7 in.	11.7 in.	12.8 in.
Age 2	7.7 in.	8.8 in.	9.4 in.
Adults age 70 and over:			
Males	17.2 in.	18.3 in.	19.5 in.
Females	—	—	—
Truck and bus drivers:			
Males	17.4 in.	18.8 in.	20.4 in.
Females	15.1 in.	16.7 in.	18.2 in.
Airline pilots (Male)	17.6 in.	18.9 in.	20.2 in.
Flight attendants (Female)	16.0 in.	17.2 in.	18.3 in.
Law enforcement officers:			
Males	—	—	—
Females	—	—	—
U.S. Army			
Males	17.6	19.0	20.6
Females	16.0	17.4	19.0

WAIST DEPTH

This dimension is pertinent to the establishment of clearance requirements between the backrest of a chair or seat and the leading edge of a worktable, desk, or steering wheel. (In certain working situations, one should consider the additional requirement for clearance when the pregnant woman must be accommodated. Unfortunately, to our knowledge, there are no data on this additional waist-clearance factor.) (See Table 34.26.)

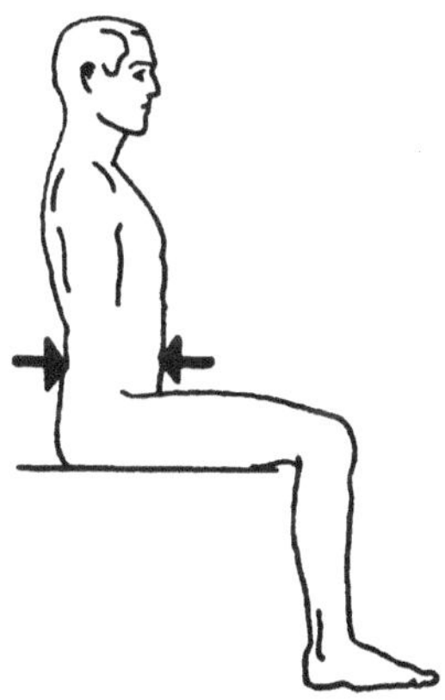

TABLE 34.26

	Percentile		
	5th	50th	95th
Adults:			
Males	7.1 in.	9.7 in.	12.3 in.
Females	7.1 in.	6.6 in.	7.9 in.
Boys:			
Age 17			
Age 14			
Age 12			
Age 6			
Age 2			
Girls:			
Age 17			
Age 14			
Age 12			
Age 6			
Age 2			
Adults age 70 and over:			
Males			
Females			
Truck and bus drivers:			
Males			
Females			
Airline pilots (Male)	7.6 in.	9.9 in.	10.4 in.
Flight attendants (Female)	5.2 in.	5.8 in.	6.5 in.
Law enforcement officers:			
Males			
Females			
U.S. Army			
Males	7.8	9.3	11.4
Females	7.3	8.6	10.7

ELBOW REST HEIGHT

Not only is this dimension pertinent to the establishment of armrest heights, but it also provides a basis for establishing the level of a writing surface and/or the approximate position of the middle row of a keyboard, the location of a joystick handle or control wheel, etc. (See Table 34.27.)

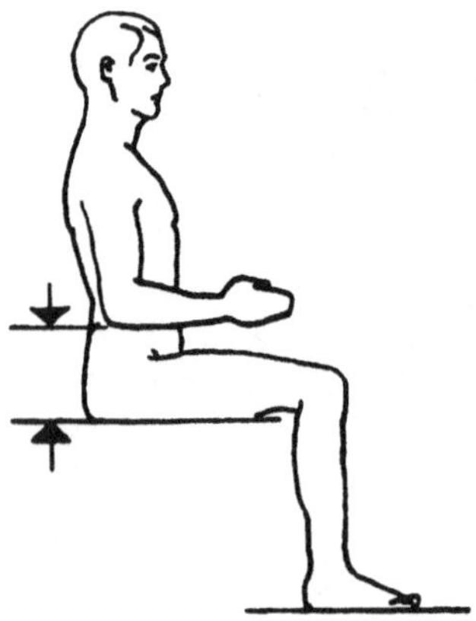

TABLE 34.27

	PERCENTILE		
	5TH	50TH	95TH
Adults:			
Males	7.4 in.	9.5 in.	11.6 in.
Females	7.1 in.	9.2 in.	11.0 in.
Boys:			
Age 17			
Age 14			
Age 12			
Age 6			
Age 2			
Girls:			
Age 17			
Age 14			
Age 12			
Age 6			
Age 2			
Adults age 70 and over:			
Males	6.5 in.	8.6 in.	10.6 in.
Females	6.4 in.	8.4 in.	10.0 in.
Truck and bus drivers:			
Males	—	—	—
Females	—	—	—
Airline pilots (Male)	7.4 in.	9.1 in.	10.8 in.
Flight attendants (Female)	7.7 in.	9.6 in.	11.0 in.
Law enforcement officers:			
Males	—	—	—
Females	—	—	—
U.S. Army			
Males	7.3	9.1	10.8
Females	6.9	9.7	10.4

BUTTOCK-TO-HEEL LENGTH

This dimension is pertinent to the establishment of the distance of foot-operated controls from the seat reference point. (See Table 34.28.)

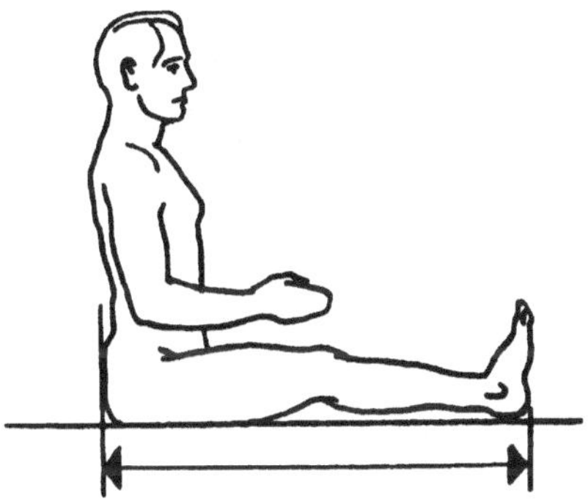

TABLE 34.28

	Percentile		
	5th	50th	95th
Adults:			
Males	39.0 in.	42.0 in.	46.0 in.
Females	34.0 in.	37.8 in.	41.2 in.
Boys:			
Age 17	—	—	—
Age 14	31.7 in.	34.1 in.	39.1 in.
Age 12	30.4 in.	33.7 in.	37.3 in.
Age 6	22.5 in.	24.7 in.	26.4 in.
Age 2	15.1 in.	16.5 in.	18.3 in.
Girls:			
Age 17	—	—	—
Age 14		36.8 in.	39.0 in.
Age 12		33.7 in.	37.1 in.
Age 6		24.6 in.	27.2 in.
Age 2		17.0 in.	18.0 in.
Adults age 70 and over:			
Males	—	—	—
Females	—	—	—
Truck and bus drivers:			
Males	—	—	—
Females	—	—	—
Airline pilots (Male)	38.3 in.	42.3 in.	46.3 in.
Flight attendants (Female)	—	—	—
Law enforcement officers:			
Males	—	—	—
Females	—	—	—
U.S. Army			
Males	39.4	42.4	46.0
Females	36.7	39.8	43.1

SHOULDER BREADTH

This dimension is pertinent to the establishment of lateral clearance between persons who may be required to sit side by side and to the establishment of the lateral clearance requirement for a worker who may have to squeeze into a tight space to work on an item of equipment. (See Table 34.29.)

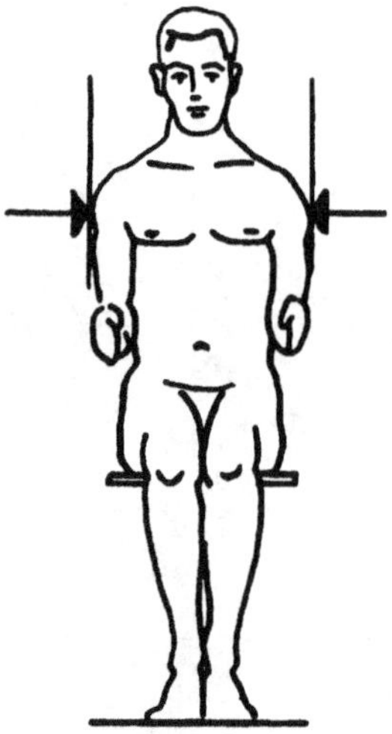

TABLE 34.29

	Percentile		
	5th	50th	95th
Adults:			
Males	16.4 in.	17.9 in.	19.6 in.
Females	14.4 in.	15.7 in.	17.6 in.
Boys:			
Age 17	15.5 in.	17.4 in.	19.2 in.
Age 14	13.7 in.	15.3 in.	17.4 in.
Age 12	12.3 in.	13.8 in.	15.6 in.
Age 6	9.3 in.	11.0 in.	12.5 in.
Age 2	8.0 in.	8.9 in.	9.8 in.
Girls:			
Age 17	14.2 in.	15.6 in.	17.3 in.
Age 14	13.4 in.	14.8 in.	16.7 in.
Age 12	12.2 in.	13.8 in.	15.8 in.
Age 6	10.0 in.	10.9 in.	11.9 in.
Age 2	7.6 in.	9.0 in.	9.7 in.
Adults age 70 and over:			
Males	15.6 in.	17.0 in.	18.5 in.
Females	—	—	—
Truck and bus drivers:			
Males	16.9 in.	18.3 in.	19.9 in.
Females	—	—	—
Airline pilots (Male)	16.5 in.	17.9 in.	19.4 in.
Flight attendants (Female)	14.9 in.	16.0 in.	17.0 in.
Law enforcement officers:			
Males	17.6 in.	19.4 in.	21.4 in.
Females			
U.S. Army			
Males	17.7	19.4	21.1
Females	15.6	17.0	18.6

HIP BREADTH (SITTING)

This dimension is pertinent to the establishment of seat widths, particularly the lateral clearance between armrests. (See Table 34.30.)

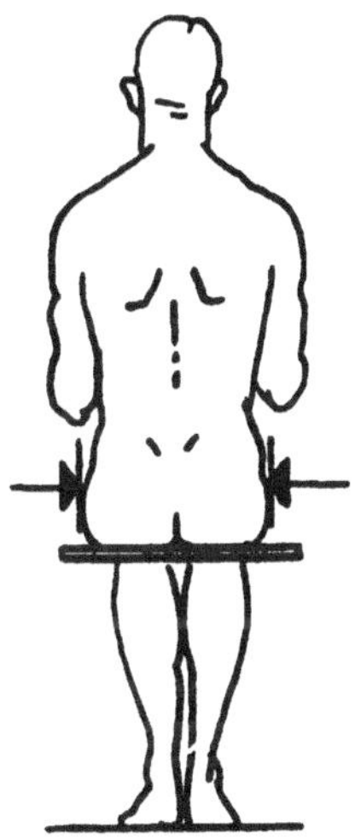

TABLE 34.30

	Percentile		
	5th	**50th**	**95th**
Adults:			
Males	12.2 in.	14.0 in.	15.9 in.
Females	12.3 in.	14.3 in.	17.1 in.
Boys:			
Age 17	11.9 in.	13.1 in.	14.8 in.
Age 14	10.4 in.	11.8 in.	13.6 in.
Age 12	9.2 in.	10.7 in.	12.8 in.
Age 6	7.5 in.	8.3 in.	9.4 in.
Age 2	6.9 in.	7.5 in.	8.4 in.
Girls:			
Age 17	12.0 in.	13.5 in.	15.7 in.
Age 14	11.2 in.	12.7 in.	14.4 in.
Age 12	9.6 in.	11.3 in.	13.5 in.
Age 6	7.4 in.	8.4 in.	9.3 in.
Age 2	6.6 in.	7.4 in.	8.4 in.
Adults age 70 and over:			
Males	12.1 in.	13.6 in.	15.5 in.
Females	11.7 in.	14.0 in.	16.8 in.
Truck and bus drivers:			
Males	13.2 in.	14.5 in.	16.3 in.
Females	—	—	—
Airline pilots (Male)	12.1 in.	13.2 in.	14.4 in.
Flight attendants (Female)	13.3 in.	14.5 in.	15.6 in.
Law enforcement officers:			
Males	—	—	—
Females	—	—	—
U.S. Army			
Males	12.9	14.4	16.2
Females	13.5	15.1	17.0

FOREARM-TO-FOREARM BREADTH

This dimension is pertinent to the establishment of seat and armrest separation. It also should be considered along with shoulder breadth to establish lateral body clearance for side-by-side seating. (See Table 34.31.)

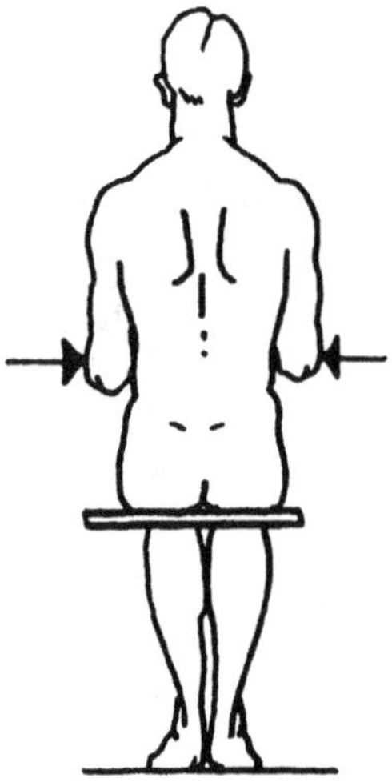

TABLE 34.31

	Percentile		
	5th	50th	95th
Adults:			
Males	13.7 in.	16.5 in.	19.9 in.
Females	12.3 in.	15.1 in.	19.3 in.
Boys:			
Age 17			
Age 14			
Age 12			
Age 6			
Age 2			
Girls:			
Age 17			
Age 14			
Age 12			
Age 6			
Age 2			
Adults age 70 and over:			
Males	14.0 in.	16.4 in.	18.7 in.
Females	13.1 in.	15.7 in.	19.1 in.
Truck and bus drivers:			
Males	16.6 in.	19.4 in.	23.1 in.
Females	14.1 in.	16.5 in.	20.2 in.
Airline pilots (Male)	15.2 in.	17.2 in.	19.8 in.
Flight attendants (Female)	11.6 in.	13.0 in.	14.6 in.
Law enforcement officers:			
Males	—	—	—
Females	—	—	—
U.S. Army			
Males	18.8	21.4	24.4
Females	16.3	18.4	20.8

INTERPUPILLARY BREADTH

This dimension is pertinent to the design of eyeglasses, binoculars, and other optical aids. (See Table 34.32.)

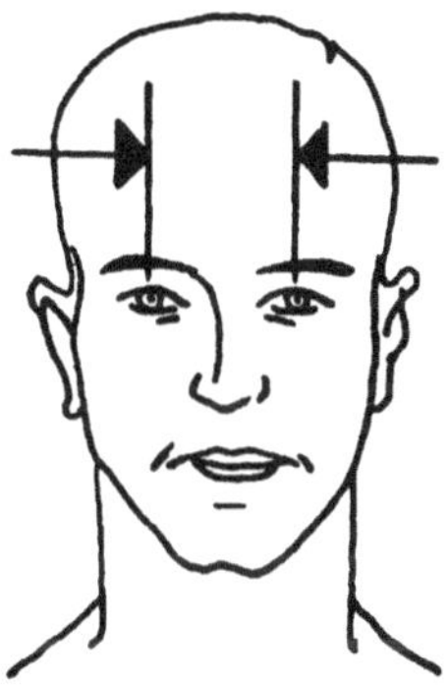

TABLE 34.32

	Percentile		
	5th	50th	95th
Adults:			
Males	2.27 in.	—	2.74 in.
Females	1.41 in.	—	2.94 in.
Boys:			
Age 17			
Age 14			
Age 12			
Age 6			
Age 2			
Girls:			
Age 17			
Age 14			
Age 12			
Age 6			
Age 2			
Adults age 70 and over:			
Males			
Females			
Truck and bus drivers:			
Males	2.28 in.	2.44 in.	2.80 in.
Females	—	—	—
Airline pilots (Male)	2.27 in.	2.49 in.	2.74 in.
Flight attendants (Female)	—	—	—
Law enforcement officers:			
Males	—	—	—
Females	—	—	—
U.S. Army			
Males	2.31	2.54	2.80
Females	2.23	2.45	2.70

HEAD BREADTH

This dimension is pertinent to the design of helmets and other head-mounted gear. (See Table 34.33.)

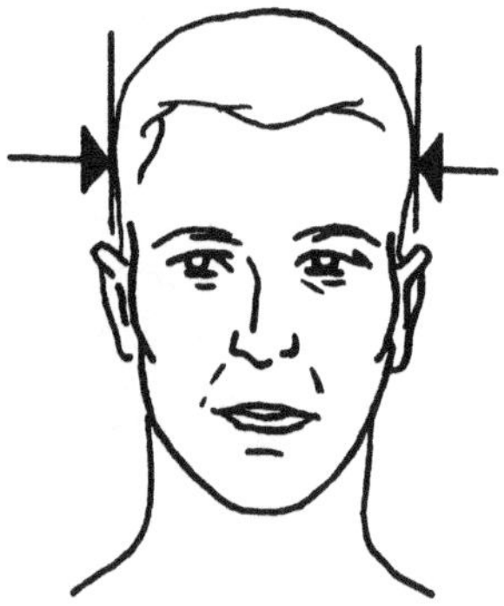

TABLE 34.33

	PERCENTILE		
	5TH	50TH	95TH
Adults:			
Males	5.6 in.	6.0 in.	6.4 in.
Females	5.4 in.	5.7 in.	6.1 in.
Boys:			
Age 17	5.7 in.	6.0 in.	6.4 in.
Age 14	5.5 in.	5.9 in.	6.2 in.
Age 12	5.4 in.	5.7 in.	6.1 in.
Age 6	5.2 in.	5.5 in.	5.9 in.
Age 2	4.9 in.	5.2 in.	5.5 in.
Girls:			
Age 17	5.5 in.	5.8 in.	6.1 in.
Age 14	5.5 in.	5.7 in.	6.1 in.
Age 12	5.4 in.	5.7 in.	5.9 m
Age 6	5.1 in.	5.4 in.	5.7 in.
Age 2	4.7 in.	5.0 in.	5.4 in.
Adults age 70 and over:			
Males	5.8 in.	6.1 in.	6.4 in.
Females	—	—	—
Truck and bus drivers:			
Males	5.7 in.	6.0 in.	6.3 in.
Females	—	—	—
Airline pilots (Male)	5.7 in.	6.1 m	6.4 in.
Flight attendants (Female)	5.4 in.	5.7 in.	6.1 in.
Law enforcement officers:			
Males	5.7 in.	6.1 in.	6.5 in.
Females	—	—	—
U.S. Army			
Males	5.6	6.0	6.3
Females	5.4	5.7	6.0

HEAD LENGTH

This dimension is pertinent to the design of helmets and other head-mounted gear. It should be used in conjunction with head breadth. (See Table 34.34.)

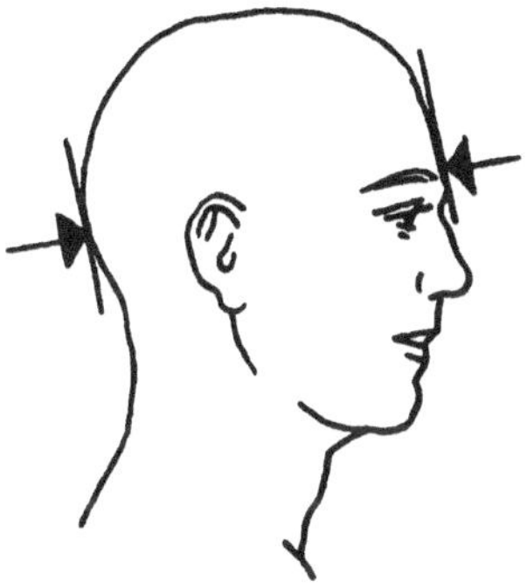

TABLE 34.34

	Percentile		
	5th	50th	95th
Adults:			
Males	7.3 in.	7.7 in.	8.2 in.
Females	7.0 in.	7.4 in.	7.7 in.
Boys:			
Age 17	7.2 in.	7.7 in.	8.0 in.
Age 14	7.2 in.	7.6 in.	8.1 in.
Age 12	6.9 in.	7.5 in.	7.7 in.
Age 6	6.6 in.	7.2 in.	7.6 in.
Age 2	6.5 in.	6.9 in.	7.3 in.
Girls:			
Age 17	6.9 in.	7.4 in.	7.8 in.
Age 14	7.0 in.	7.4 in.	7.9 in.
Age 12	6.8 in.	7.3 in.	7.8 in.
Age 6	6.6 in.	7.1 in.	7.4 in.
Age 2	6.1 in.	6.7 in.	6.9 in.
Adults age 70 and over:			
Males	7.3 in.	7.7 in	8.1 in.
Females	—	—	—
Truck and bus drivers:			
Males	7.2 in.	7.6 in.	8.1 in.
Females	—	—	—
Airline pilots (Male)	7.2 in.	7.8 in.	8.6 in.
Flight attendants (Female)	6.9 in.	7.3 in.	7.7 in.
Law enforcement officers:			
Males	7.3 in.	7.8 in.	8.2 in.
Females	—	—	—
U.S. Army			
Males	7.3	7.8	8.2
Females	6.9	7.4	7.8

HEAD LENGTH (MAXIMUM)

This dimension is useful in the design of special headgear that requires a full face mask. These values include an arbitrary 1-in. (2.5-cm) additional clearance. (See Table 34.35.)

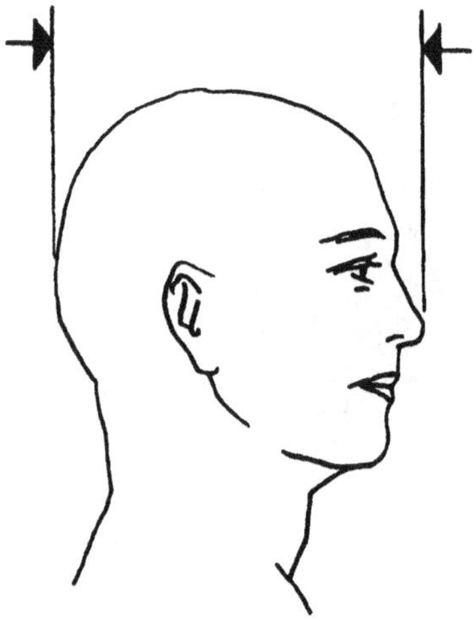

TABLE 34.35

	Percentile		
	5th	50th	95th
Adults:			
Males			
Females			
Boys:			
Age 17			
Age 14			
Age 12			
Age 6			
Age 2			
Girls:			
Age 17			
Age 14			
Age 12			
Age 6			
Age 2			
Adults age 70 and over:			
Males			
Females			
Truck and bus drivers:			
Males	8.1 in.	8.7 in.	9.2 in.
Females			
Airline pilots			
Flight attendants			
Law enforcement officers:			
Males			
Females			
U.S. Army			
Males	8.1	8.6	9.1
Females	7.8	8.3	8.7

HEAD HEIGHT (MAXIMUM)

This dimension is pertinent to the design of fully enclosed headgear or helmets. As suggested by the lack of data for various populations, the values indicated in Table 34.36 were derived from a very small sample of male and female subjects selected at random and provide some idea of the larger head height characteristics.

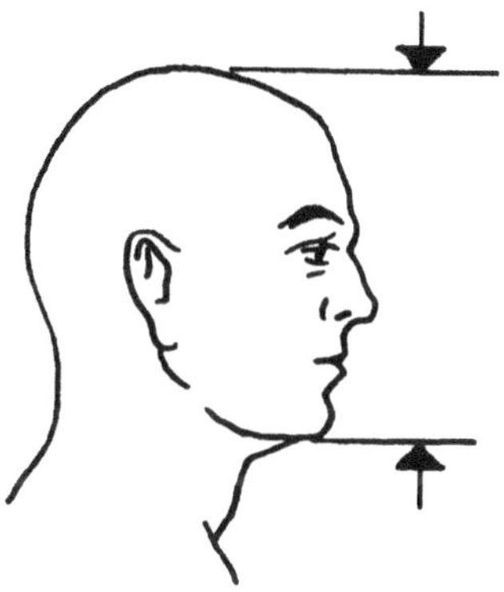

TABLE 34.36

	Percentile		
	5th	50th	95th
Adults:			
Males	—	—	10.2 in.
Females	—	—	9.0 in.
Boys:			
Age 17	7.8 in.	8.6 in.	9.1 in.
Age 14	7.5 in.	8.1 in.	8.9 in.
Age 12	7.5 in.	7.8 in.	8.5 in.
Age 6	6.6 in.	7.3 in.	8.0 in.
Age 2	6.3 in.	6.9 in.	7.5 in.
Girls:			
Age 17	7.5 in.	8.1 in.	8.7 in.
Age 14	7.3 in.	8.0 in.	8.5 in.
Age 12	7.0 in.	7.8 in.	8.5 in.
Age 6	6.3 in.	7.2 in.	7.6 in.
Age 2	6.1 in.	6.6 in.	7.2 in.
Adults age 70 and over:			
Males			
Females			
Truck and bus drivers:			
Males			
Females			
Airline pilots			
Flight attendants			
Law enforcement officers:			
Males			
Females			
U.S. Army			
Males	8.6	9.1	9.7
Females	8.0	8.6	9.1

CHIN-TO-EYE HEIGHT

This dimension is pertinent to the design of headgear in which eye protective elements are integrated into the helmet. Although these values represent a very small subject sample, they provide some idea of the larger chin-to-eye heights. They should be considered in conjunction with head height. (See Table 34.37.)

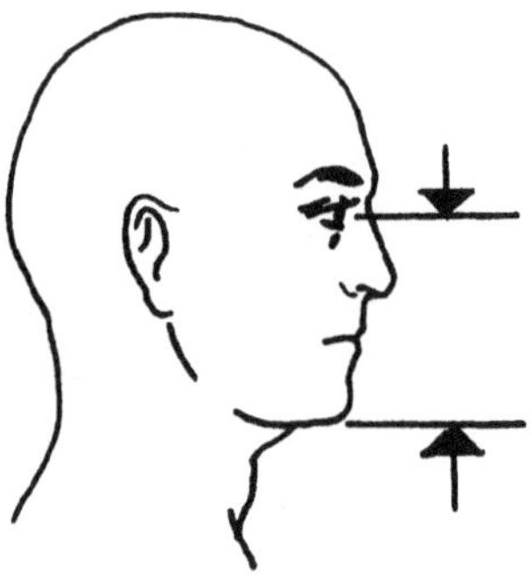

TABLE 34.37

	Percentile		
	5th	50th	95th
Adults:			
Males	4.3 in.	—	5.0 in.
Females	3.8 in.	—	4.25 in.
Boys:			
Age 17	4.1 in.	4.4 in.	4.8 in.
Age 14	3.8 in.	4.3 in.	4.7 in.
Age 12	3.7 in.	4.0 in.	4.3 in.
Age 6	3.2 in.	3.5 in.	3.8 in.
Age 2	2.9 in.	3.2 in.	3.6 in.
Girls:			
Age 17	3.7 in.	4.2 in.	4.5 in.
Age 14	3.9 in.	4.1 in.	4.5 in.
Age 12	3.6 in.	4.0 in.	4.4 in.
Age 6	3.1 in.	3.5 in.	3.8 in.
Age 2	2.8 in.	3.1 in.	3.4 in.
Adults age 70 and over:			
Males			
Females			
Truck and bus drivers:			
Males			
Females			
Airline pilots (Male)	4.3 in.	—	5.1 in.
Flight attendants (Female)	3.8 in.	—	4.6 in.
Law enforcement officers:			
Males			
Females			

HAND LENGTH

This dimension is pertinent to the design of gloves, mittens, and other devices for protecting the hands. This dimension is also useful in selecting test subjects for evaluating hand-held or hand-manipulated devices such as handrails and gun and joystick grips. (See Table 34.38.)

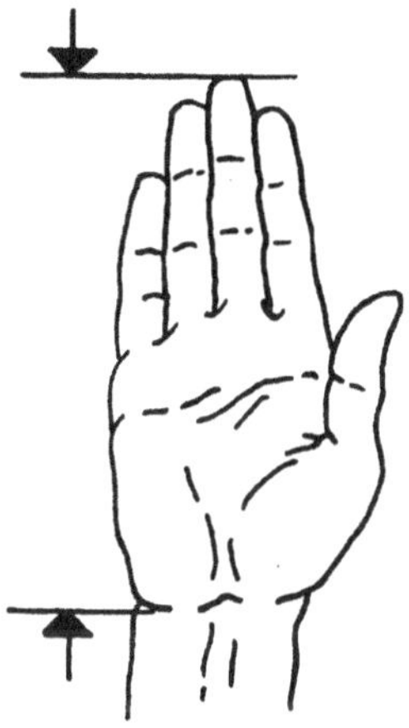

TABLE 34.38

	Percentile		
	5th	50th	95th
Adults:			
Males	7.0 in.	7.6 in.	8.2 in.
Females	6.4 in.	6.9 in.	7.4 in.
Boys:			
Age 17	6.8 in.	7.4 in.	7.9 in.
Age 14	6.3 in.	7.0 in.	7.6 in.
Age 12	5.7 in.	6.3 in.	7.0 in.
Age 6	4.6 in.	5.0 in.	5.7 in.
Age 2	3.6 in.	3.9 in.	4.2 in.
Girls:			
Age 17	6.1 in.	6.7 in.	7.2 in.
Age 14	6.1 in.	6.7 in.	7.2 in.
Age 12	5.6 in.	6.4 in.	7.0 in.
Age 6	4.5 in.	4.9 in.	5.4 in.
Age 2	3.3 in.	3.8 in.	4.3 in.
Adults age 70 and over:			
Males	7.0 in.	7.4 in.	8.0 in.
Females	—	—	—
Truck and bus drivers:			
Males	7.1 in.	7.4 in.	8.0 in.
Females			
Airline pilots (Male)	6.9 in.	7.5 in.	8.0 in.
Flight attendants (Female)	6.3 in.	6.8 in.	7.3 in.
Law enforcement officers:			
Males	7.0 in.	7.6 in.	8.2 in.
Females	—	—	—

HAND BREADTH AT THUMB

This dimension is pertinent to the design of hand protective gear and is also useful in defining lateral hand clearance. (See Table 34.39.)

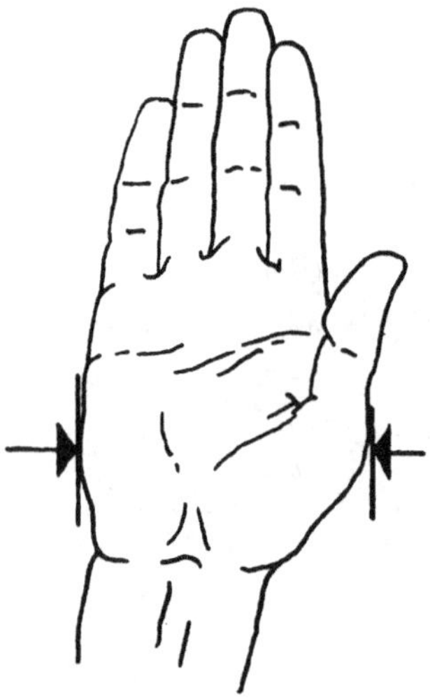

TABLE 34.39

	Percentile		
	5th	50th	95th
Adults:			
Males			
Females			
Boys:			
Age 17			
Age 14			
Age 12			
Age 6			
Age 2			
Girls:			
Age 17			
Age 14			
Age 12			
Age 6			
Age 2			
Adults age 70 and over:			
Males			
Females			
Truck and bus drivers:			
Males			
Females			
Airline pilots (Male)	3.7 in.	4.1 in.	4.4 in.
Flight attendants (Female)	3.2 in.	3.6 in.	4.0 in.
Law enforcement officers:			
Males			
Females			

HAND BREADTH AT METACARPAL

This dimension is pertinent to the design of hand protective gear. (See Table 34.40.)

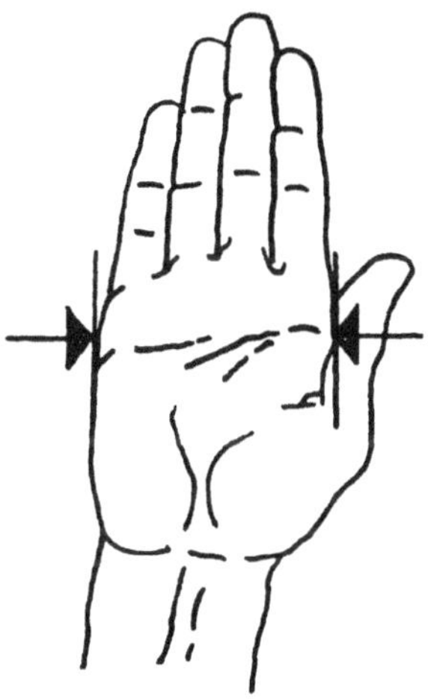

TABLE 34.40

	Percentile		
	5th	50th	95th
Adults:			
Males	3.1 in.	3.4 in.	3.8 in.
Females	2.7 in.	3.0 in.	3.4 in.
Boys:			
Age 17	3.2 in.	3.5 in.	3.7 in.
Age 14	2.8 in.	3.2 in.	3.6 in.
Age 12	2.6 in.	2.9 in.	3.3 in.
Age 6	2.1 in.	2.4 in.	2.6 in.
Age 2	1.7 in.	1.9 in.	2.2 in.
Girls:			
Age 17	2.7 in.	2.9 in.	3.3 in.
Age 14	2.7 in.	2.9 in.	3.2 in.
Age 12	2.5 in.	2.8 in.	3.1 in.
Age 6	2.0 in.	2.3 in.	2.5 in.
Age 2	1.6 in.	1.8 in.	2.1 in.
Adults age 70 and over:			
Males	3.1 in.	3.3 in.	3.6 in.
Females	—	—	—
Truck and bus drivers:			
Males	3.2 in.	3.5 in.	3.8 in.
Females	—	—	—
Airline pilots (Male)		3.5 in.	3.8 in.
Flight attendants (Female)	2.7 in.	2.9 in.	3.1 in.
Law enforcement officers:			
Males	3.3 in.	3.5 in.	3.8 in.
Females	—	—	—
U.S. Army			
Males	3.3	3.6	3.8
Females	2.9	3.1	3.4

HAND THICKNESS AT METACARPAL

This dimension is pertinent to the design of hand protective gear and also to the establishment of hand clearance. (See Table 34.41.)

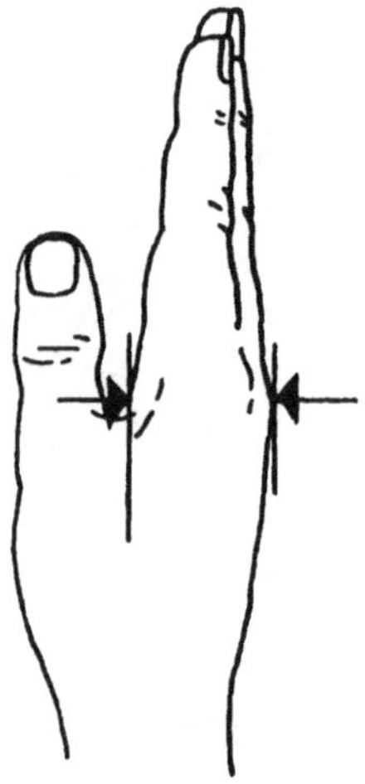

TABLE 34.41

	Percentile		
	5th	50th	95th
Adults:			
Males	1.1 in.	1.2 in.	1.3 in.
Females	0.8 in.	1.0 in.	1.1 in.
Boys:			
Age 17			
Age 14			
Age 12			
Age 6			
Age 2			
Girls:			
Age 17			
Age 14			
Age 12			
Age 6			
Age 2			
Adults age 70 and over:			
Males			
Females			
Truck and bus drivers:			
Males			
Females			
Airline pilots			
Flight attendants			
Law enforcement officers:			
Males			
Females			

FOOT LENGTH[4]

This dimension is pertinent to the design of footgear and also to the establishment of clearances for the foot (see the discussion of clothing effects in the section "Effects of Clothing on Human Body Dimensions"). (See Table 34.42.)

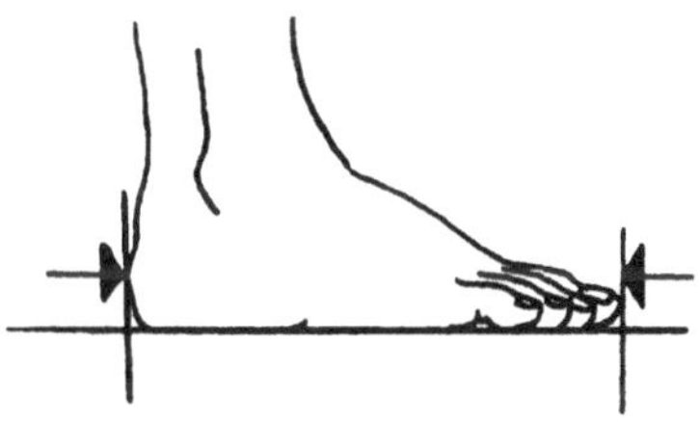

TABLE 34.42

	Percentile		
	5th	**50th**	**95th**
Adults:			
Males	9.7 in.	10.4 in.	11.2 in.
Females	8.7 in.	9.4 in.	10.2 in.
Boys:			
Age 17	9.5 in.	10.6 in.	11.1 in.
Age 14	9.0 in.	10.0 in.	11.0 in.
Age 12	8.4 in.	9.2 in.	10.1 in.
Age 6	6.5 in.	7.0 in.	8.0 in.
Age 2	4.5 in.	5.1 in.	5.7 in.
Girls:			
Age 17	8.6 in.	9.2 in.	10.1 in.
Age 14	8.5 in.	9.2 in.	10.2 in.
Age 12	8.3 in.	9.1 in.	9.8 in.
Age 6	6.3 in.	7.0 in.	7.6 in.
Age 2	4.2 in.	5.0 in.	5.5 in.
Adults age 70 and over:			
Males	9.7 in.	10.2 in.	10.9 in.
Females	—	—	—
Truck and bus drivers:			
Males	9.6 in.	10.4 in.	11.3 in.
Females	—	—	—
Airline pilots (Male)	9.8 in.	10.5 in.	11.3 in.
Flight attendants (Female)	8.7 in.	9.4 in.	10.1 in.
Law enforcement officers:			
Males	—	—	—
Females	—	—	—
U.S. Army			
Males	9.8	10.6	11.5
Females	8.8	9.6	10.4

[4]Measured with subjects standing.

BALL OF FOOT WIDTH[5]

This dimension is pertinent to the design of footgear and also to the establishment of foot clearance. (See Table 34.43.)

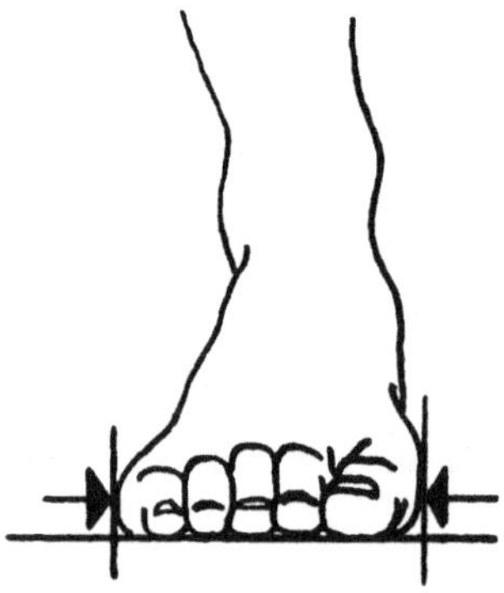

TABLE 34.43

	Percentile		
	5th	50th	95th
Adults:			
Males	3.5 in.	3.9 in.	4.3 in.
Females	3.2 in.	3.6 in.	4.0 in.
Boys:			
Age 17	3.7 in.	4.1 in.	4.4 in.
Age 14	3.4 in.	3.9 in.	4.3 in.
Age 12	3.1 in.	3.5 in.	4.0 in.
Age 6	2.5 in.	2.8 in.	3.1 in.
Age 2	1.8 in.	2.1 in.	2.4 in.
Girls:			
Age 17	3.2 in.	3.5 in.	3.9 in.
Age 14	3.1 in.	3.5 in.	3.9 in.
Age 12	3.0 in.	3.4 in.	3.8 in.
Age 6	2.4 in.	2.7 in.	3.0 in.
Age 2	1.6 in.	1.9 in.	2.2 in.
Adults age 70 and over:			
Males	3.6 in.	3.9 in.	4.3 in.
Females	—	—	—
Truck and bus drivers:			
Males	3.7 in.	4.0 in.	4.3 in.
Females	—	—	—
Airline pilots (Male)	3.5 in.	3.8 in.	4.1 in.
Flight attendants (Female)	3.2 in.	3.4 in.	3.8 in.
Law enforcement officers:			
Males			
Females			
U.S. Army			
Males	3.6	4.0	4.3
Females	3.2	3.5	3.9

[5]Measured with subjects standing.

SHOULDER CIRCUMFERENCE[6]

This dimension is pertinent to the design of garments. It also is useful in establishing the general description of a particular test subject who may be used for dynamic testing of various physical mockups. (See Table 34.44.)

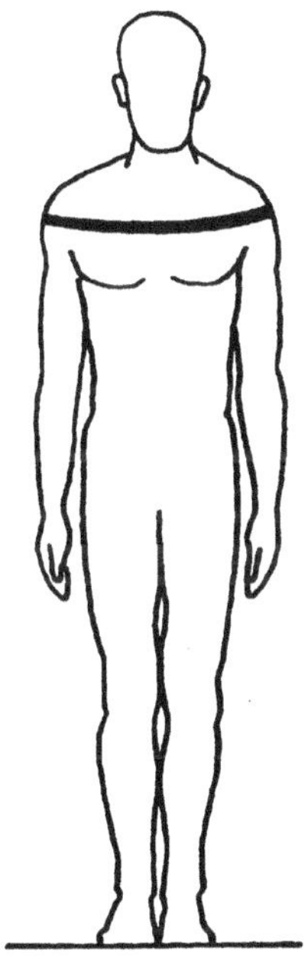

TABLE 34.44

	Percentile		
	5th	50th	95th
Adults:			
Males			
Females			
Boys:			
Age 17			
Age 14			
Age 12			
Age 6			
Age 2			
Girls:			
Age 17			
Age 14			
Age 12			
Age 6			
Age 2			
Adults age 70 and over:			
Males			
Females			
Truck and bus drivers:			
Males			
Females			
Airline pilots (Male)	41.9 in.	45.3 in.	47.7 in.
Flight attendants (Female)	35.6 in.	37.6 in.	39.7 in.
Law enforcement officers:			
Males			
Females			
U.S. Army			
Males	42.4	46.2	50.2
Females	37.3	40.3	44.3

[6]Measured with subjects standing.

CHEST CIRCUMFERENCE

This dimension can be used in conjunction with shoulder circumference for garment design and the definition of the body characteristics of test subjects. (See Table 34.45.)

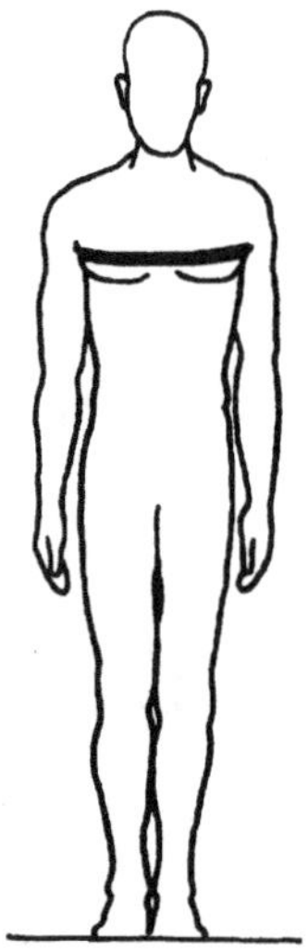

TABLE 34.45

	Percentile		
	5th	50th	95th
Adults:			
Males	34.3 in.	39.0 in.	44.9 in.
Females	30.4 in.	34.3 in.	40.5 in.
Boys:			
Age 17	32.5 in.	36.2 in.	40.8 in.
Age 14	27.5 in.	31.3 in.	36.1 in.
Age 12	25.6 in.	28.1 in.	32.1 in.
Age 6	20.7 in.	22.6 in.	25.3 in.
Age 2	18.0 in.	19.0 in.	20.6 in.
Girls:			
Age 17	29.7 in.	32.3 in.	36.3 in.
Age 14	27.6 in.	30.8 in.	34.6 in.
Age 12	25.2 in.	28.7 in.	33.4 in.
Age 6	20.3 in.	22.3 in.	24.3 in.
Age 2	17.3 in.	18.4 in.	20.5 in.
Adults age 70 and over:			
Males	33.0 in.	38.1 in.	44.1 in.
Females	30.1 in.	35.0 in.	39.9 in.
Truck and bus drivers:			
Males	—	—	—
Females	—	—	—
Airline pilots (Male)	36.3 in.	38.8 in.	41.3 in.
Flight attendants (Female)	31.2 in.	33.7 in.	36.4 in.
Law enforcement officers:			
Males	35.4 in.	40.1 in.	45.6 in.
Females	—	—	—
U.S. Army			
Males	34.9	38.9	43.8
Females	32.1	35.5	40.3

WAIST CIRCUMFERENCE

This dimension can be used in conjunction with shoulder and chest circumference for garment design and the definition of the body characteristics of test subjects. (See Table 34.46.)

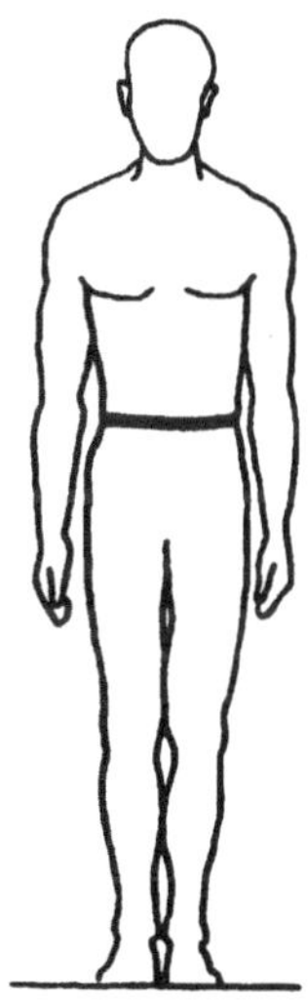

TABLE 34.46

	Percentile		
	5th	50th	95th
Adults:			
Males	28.4 in.	34.8 in.	42.9 in.
Females	24.1 in.	29.2 in.	39.1 in.
Boys:			
Age 17	26.5 in.	29.6 in.	35.9 in.
Age 14	23.7 in.	26.8 in.	32.7 in.
Age 12	22.1 in.	24.8 in.	29.7 in.
Age 6	18.0 in.	20.3 in.	23.0 in.
Age 2	15.5 in.	17.3 in.	19.3 in.
Girls:			
Age 17	24.9 in.	27.8 in.	34.4 in.
Age 14	23.1 in.	26.8 in.	31.9 in.
Age 12	21.8 in.	25.1 in.	32.0 in.
Age 6	18.1 in.	20.2 in.	23.1 in.
Age 2	14.8 in.	17.4 in.	19.4 in.
Adults age 70 and over:			
Males	28.3 in.	35.8 in.	41.3 in.
Females	25.9 in.	33.2 in.	39.9 in.
Truck and bus drivers:			
Males	—	—	—
Females	—	—	—
Airline pilots (Male)	29.0 in.	32.0 in.	35.0 in.
Flight attendants (Female)	22.8 in.	24.5 in.	26.4 in.
Law enforcement officers:			
Males	30.1 in.	35.3 in.	42.3 in.
Females	—	—	—
U.S. Army			
Males	28.7	32.8	38.2
Females	25.1	28.2	33.2

HIP (BUTTOCK) CIRCUMFERENCE (STANDING)

This dimension can be used in conjunction with hip circumference considerations for garment design and the definition of the body characteristics of test subjects. (See Table 34.47.)

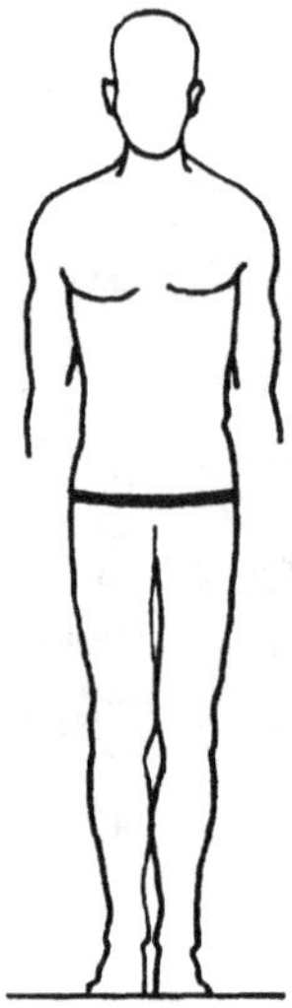

TABLE 34.47

	PERCENTILE		
	5TH	50TH	95TH
Adults:			
Males	—	—	—
Females	—	—	—
Boys:			
Age 17	32.8 in.	35.7 in.	40.7 in.
Age 14	28.3 in.	32.4 in.	37.4 in.
Age 12	26.0 in.	29.2 in.	34.5 in.
Age 6	20.5 in.	22.6 in.	26.0 in.
Age 2	18.6 in.	20.5 in.	22.7 in.
Girls:			
Age 17	32.1 in.	36.1 in.	41.6 in.
Age 14	29.9 in.	34.1 in.	38.8 in.
Age 12	26.1 in.	30.7 in.	35.9 in.
Age 6	20.5 in.	23.1 in.	25.5 in.
Age 2	18.1 in.	20.2 in.	22.1 in.
Adults age 70 and over:			
Males	—	—	—
Females	—	—	—
Truck and bus drivers:			
Males	—	—	—
Females	—	—	—
Airline pilots (Male)	35.5 in.	37.8 in.	40.1 in.
Flight attendants (Female)	33.4 in.	35.4 in.	37.8 in.
Law enforcement officers:			
Males	—	—	—
Females	—	—	—
U.S. Army			
Males	35.0	38.6	42.9
Females	34.3	38.0	42.1

HIP (BUTTOCK) CIRCUMFERENCE (SITTING)

This dimension is pertinent to the design of seats and seatbelts. When it is used in conjunction with hip breadth, subjects can be selected to represent a larger user population for evaluating seat widths and lap belt lengths. (See Table 34.48.)

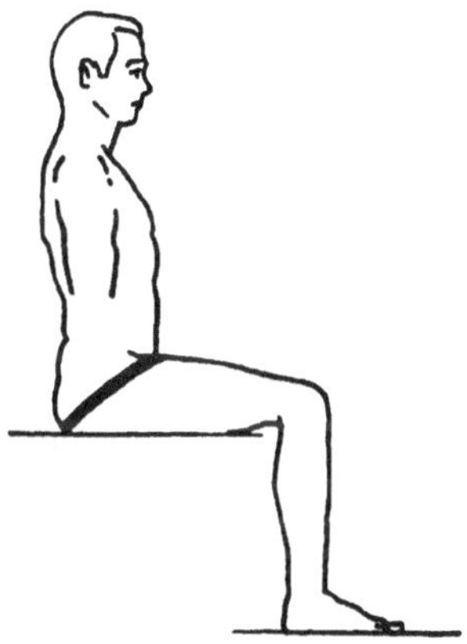

TABLE 34.48

	Percentile		
	5th	50th	95th
Adults:			
Males			
Females			
Boys:			
Age 17			
Age 14			
Age 12			
Age 6			
Age 2			
Girls:			
Age 17			
Age 14			
Age 12			
Age 6			
Age 2			
Adults age 70 and over:			
Males			
Females			
Truck and bus drivers:			
Males			
Females			
Airline pilots (Male)	38.9 in.	41.7 in.	44.5 in.
Flight attendants (Female)	37.0 in.	39.2 in.	41.4 in.
Law enforcement officers:			
Males			
Females			

HEAD CIRCUMFERENCE

This dimension is pertinent to the design of headgear. (See Table 34.49.)

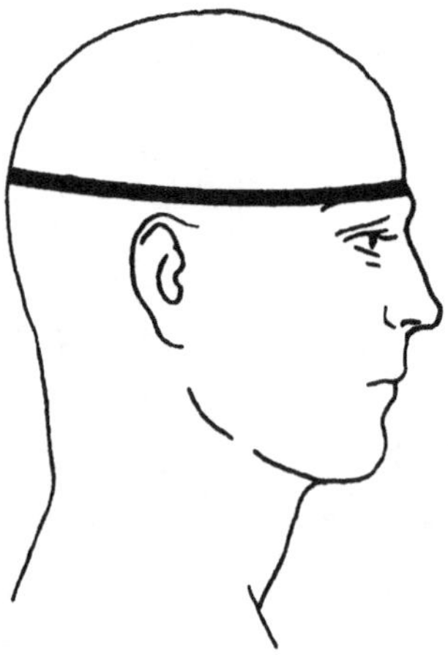

TABLE 34.49

	Percentile		
	5th	50th	95th
Adults:			
Males	—	—	—
Females	—	—	—
Boys:			
Age 17	21.1 in.	22.3 in.	23.3 in.
Age 14	20.4 in.	21.5 in.	22.8 in.
Age 12	20.1 in.	21.1 in.	22.3 in.
Age 6	19.2 in.	20.4 in.	21.2 in.
Age 2	18.5 in.	19.3 in.	20.4 in.
Girls:			
Age 17	20.4 in.	21.5 in.	22.6 in.
Age 14	20.3 in.	21.2 in.	22.3 in.
Age 12	19.8 in.	20.9 in.	21.9 in.
Age 6	19.1 in.	19.8 in.	20.8 in.
Age 2	18.1 in.	18.8 in.	19.7 in.
Adults age 70 and over:			
Males	—	—	—
Females	—	—	—
Truck and bus drivers:			
Males	—	—	—
Females	—	—	—
Airline pilots (Male)	21.9 in.	22.5 in.	23.1 in.
Flight attendants (Female)	20.8 in.	21.5 in.	22.2 in.
Law enforcement officers:			
Males	21.6 in.	22.6 in.	23.6 in.
Females	—	—	—
U.S. Army			
Males	21.4	22.3	23.4
Females	20.6	21.5	22.5

NECK CIRCUMFERENCE

This dimension may be pertinent to the design of special garments and/or devices worn about the user's neck. (See Table 34.50.)

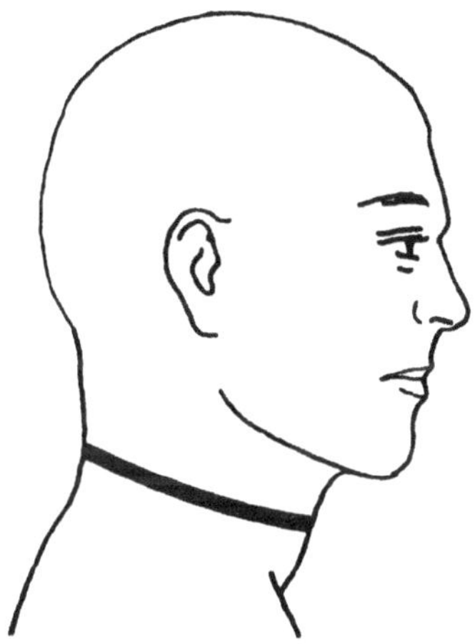

TABLE 34.50			
	Percentile		
	5th	50th	95th
Adults:			
Males	—	—	—
Females	—	—	—
Boys:			
Age 17	12.4 in.	14.2 in.	15.0 in.
Age 14	11.5 in.	12.7 in.	13.8 in.
Age 12	10.7 in.	11.6 in.	12.6 in.
Age 6	9.2 in.	10.2 in.	11.0 in.
Age 2	8.6 in.	9.6 in.	10.1 in.
Girls:			
Age 17	11.3 in.	12.2 in.	13.3 in.
Age 14	10.9 in.	11.7 in.	13.1 in.
Age 12	10.4 in.	11.4 in.	12.8 in.
Age 6	9.1 in.	10.0 in.	10.9 in.
Age 2	8.2 in.	9.2 in.	9.7 in.
Adults age 70 and over:			
Males	—	—	—
Females	—	—	—
Truck and bus drivers:			
Males	—	—	—
Females	—	—	—
Airline pilots (Male)	14.3 in.	15.0 in.	15.7 in.
Flight attendants (Female)	11.1 in.	11.8 in.	12.6 in.
Law enforcement officers:			
Males	—	—	—
Females	—	—	—
U.S. Army			
Males	13.7	14.9	16.3
Females	11.5	12.4	13.5

EYE TO TOP OF HEAD HEIGHT

This dimension is valuable for providing head clearance. Allow additional clearance for head wear or vehicle movement (roughly 2 in.). The measurement was made from the top of the head to the level at the outer corner of the eye. (See Table 34.51.)

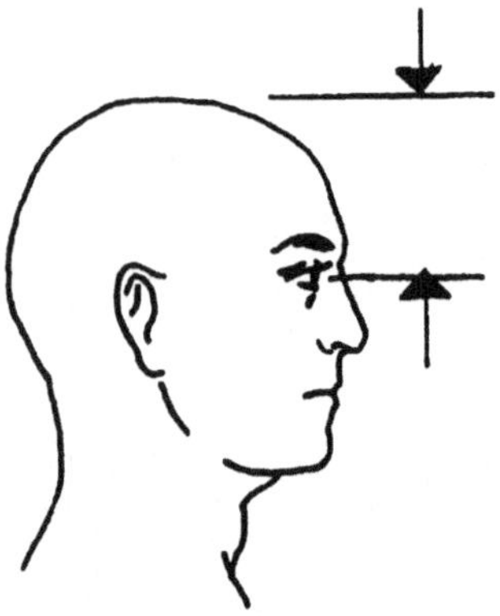

TABLE 34.51

	Percentile		
	5th	50th	95th
Adults:			
Males			
Females			
Boys:			
Age 17			
Age 14			
Age 12			
Age 6			
Age 2			
Girls:			
Age 17			
Age 14			
Age 12			
Age 6			
Age 2			
Adults age 70 and over:			
Males			
Females			
Truck and bus drivers:			
Males			
Females			
Airline pilots (Male)			
Flight attendants (Female)			
Law enforcement officers:			
Males			
Females			
U.S. Army			
Males	4.29 in.	4.65 in.	5.03 in.
Females	4.02 in.	4.39 in.	4.76 in.

STATURE STATISTICS FOR OTHER POPULATIONS

With the exception of anthropometric surveys conducted by military organizations, most population surveys include only a few dimensional statistics. The most common (and one which provides a point of departure for comparing populations) is stature. The compilation of statures for different populations presented in Table 34.52 is useful when a particular product design is intended for one or more of the specific populations noted in Table 34.53.

TABLE 34.52 Stature Statistics for Various Populations

Group or Nationality	Percentile, cm 5th	50th	95th
Australian Army males	164.0	173.0	184.0
Bulgarian civilian males	163.6	170.2	176.8
Bulgarian women	151.6	157.8	164.0
Canadian civilian males	153.9	159.8	173.5
Chinese civilian males	—	166.4	—
English, RAF male pilots	166.1	177.0	185.9
English, male Navy recruits	164.7	174.5	184.2
Filipino civilian males	—	163.1	—
French civilian males	—	172.5	—
French civil pilots	168.1	173.1	178.0
German civilian males	167.1	176.8	187.5
Greek Army males	160.9	170.5	180.3
Hawaiian civilian males	—	171.2	—
Indian civilian males	—	167.9	—
Iranian Army males	157.5	166.7	176.6
Israeli civilian males	—	168.7	
Italian Army males	160.2	170.7	180.7
Japanese Air Force, male pilots	159.4	166.7	175.0
Korean Army males	154.5	164.0	172.7
Korean Air Force, male pilots	159.6	165.6	173.7
Latin American Army males	157.0	165.9	177.0
Norwegian (unknown males)	171.5	177.5	183.5
Swedish civilian males	—	174.0	—
Swedish civilian females	164.1	—	—
Thai Army males	155.0	163.5	172.0
Turkish Army males	160.6	169.0	179.2
Vietnamese Army males	151.6	160.4	169.6
U.S. military:			
Army ground forces, male	163.8	174.4	185.6
Army ground forces, female	152.6	162.8	174.1
Army aviators, male	164.2	174.6	185.0
Air Force, male pilots	167.2	177.3	187.7
Air Force, ground, female	156.8	162.7	168.6
Marine Corps, male	168.2	174.5	180.8
Navy, enlisted, male	171.7	177.6	183.5
Navy, pilots, male	168.2	177.5	187.7
Navy, divers, male	170.0	176.2	182.2
U.S. FAA civil pilots, male	170.4	176.7	183.2

TABLE 34.53

Race	Number	Percentage of Total Population	Percentage of Alone or in Combination
Total population	**308,745,538**	**100.0**	**(X)**
White			
Alone or in combination	231,040,398	74.8	(X)
Alone	223,553,265	72.4	96.8
In combination	7,487,133	2.4	3.2
Black or African American			
Alone or in combination	42,020,743	13.6	(X)
Alone	38,929,319	12.6	92.6
In combination	3,091,424	1.0	7.4
American Indian and Alaska Native			
Alone or in combination	5,220,579	1.7	(X)
Alone	2,932,248	0.9	56.2
In combination	2,288,331	0.7	43.8
Asian			
Alone or in combination	17,320,856	5.6	(X)
Alone	14,674,252	4.8	84.7
In combination	2,646,604	0.9	15.3
Native Hawaiian and Other Pacific Islander			
Alone or in combination	1,225,195	0.4	(X)
Alone	540,013	0.2	44.1
In combination	685,182	0.2	55.9
Some Other Race			
Alone or in combination	21,748,084	7.0	(X)
Alone	19,107,368	6.2	87.9
In combination	2,640,716	0.9	12.1

(X) Not applicable.

Note: The total population is equal to the number of respondents. In the 2010 Census, there were 308,745,538 respondents. The total of all race categories alone or in combination with one or more other races is equal to the number of responses; therefore, it adds to more than the total population.

Source: U.S. Census Bureau, *2010 Census Redistricting Data (Public Law 94-171) Summary File*, Table P1.

ANTHROPOMETRIC REFERENCES

The following references were used to compile the previous data on weight and body dimensions. Note that some liberties have been taken; i.e., when a certain dimension was not available for a population, a substitution was made from the most likely (similar) population. This was done, however, only when it was considered likely that the borrowed data would be quite similar to that for the population designated.

Anthropometric Source Book, NASA Reference Publication 1024, Vols. I, II, and III, 1978 (prepared for the National Aeronautics and Space Administration, Scientific and Technical Information Office, by Webb Associates, Yellow Springs, OH).

Damon, A., et al. *The Human Body in Equipment Design,* Harvard University Press, Cambridge, MA, 1966.

Garrett, J. W., and K. W. Kennedy. *A Collation of Anthropometry,* AMRL-TR-68-1, Vols. I and II, Aerospace Medical Research Laboratory, Aerospace Medical Division, Air Force Systems Command, Wright-Patterson Air Force Base, OH, 1971.

Hamill, P. V. V., et al. *Body Weight, Stature, and Sitting Height: White and Negro Youths 12-17 Years,* U.S. Department of Health, Education, and Welfare, Public Health Service Ser. 11, no. 126, 1973.

Malina, R. M., et al. *Body Dimensions and Proportions: White and Negro Children 6-11 Years,* U.S. Department of Health, Education, and Welfare, Public Health Service Ser. 11. no. 143, 1973.

Malina, R. M., et al. *Selected Body Measurements of Children 6-11 Years,* U.S. Department of Health, Education, and Welfare, Public Health Service Ser. 11, no. 123, 1973.

Martin, J. I., et al. *Anthropometry of Law Enforcement Officers,* Technical Document 442, U.S. Naval Electronics Laboratory Center, San Diego, 1975.

Snyder, Richard G., et al. *Physical Characteristics of Children as Related to Death and Injury for Consumer Product Safety Design,* UM-HSRI-BI-75-5, Highway Safety Research Institute, Ann Arbor, MI, 1975.

White, R. M. Anthropometric Measurements on Selected Populations of the World, in Alphonse Chapanis (ed.). *Ethnic Variables in Human Factors Engineering,* The Johns Hopkins Press, Baltimore, 1975.

The reader is referred to the following for selected military population anthropometrics:

1988 *Anthropometric Survey of U.S. Army Personnel:* Summary Statistics Interium Report, U.S. Army Natick Technical Reports, NATICK/TR-89/027 March 1989.

Churchill, E., et al. *Anthropometry of U.S. Army Aviators—1970,* U.S. Army Natick Laboratories Technical Report no. 72-52-CE, Natick, MA, 1971.

Churchill, E., et al. *Anthropometry of Women of the U.S. Army—1977,* Report no. 2. Natick-TR-77-024, U.S. Army Natick Research and Development Command, Natick, MA.

Clauser, C. E., et al. *Anthropometry of Air Force Women,* AMRL-TR-70-5, Aerospace Medical Research Laboratory, Aerospace Medical Division, Air Force Systems Command, Wright-Patterson Air Force Base, OH, 1972.

Gifford, E. C. *Compilation of Anthropometric Measures on U.S. Navy Pilots,* NAMC-AGEL-437, Naval Air Material Command, Air Crew Equipment Laboratory, Philadelphia, 1960.

White, R. M., and E. Churchill. *The Body Size of Soldiers: U.S. Army Anthropometry—1966,* U.S. Army Natick Laboratories, Natick, MA, 1971.

Special acknowledgment is made of the invaluable assistance of Mr. Robert M. White in reviewing and contributing to the foregoing special anthropometric tables. Mr. White, formerly with the U.S. Army Laboratory in Natick, Massachusetts, has specialized for many years in the collection of anthropometric data for various populations and is recognized as an authority in this area.

CORRELATION AMONG BODY DIMENSIONS[7]

Some body dimensions correlate rather well with others, making it possible to select test subjects for special design evaluation activities.

However, although ordinarily one would expect tall persons to have long legs, arms, and torsos, and short persons to have short legs, arms, and torsos, to ensure that a particular design does not cause problems for special members of the expected user population, it is important not to make too many assumptions about dimensional correlations. Table 34.54 presents those correlations that indicate safe assumptions and those that do not.

TABLE 34.54

	Correlation with	
Dimension	**Stature**	**Chest Circumference**
Weight		0.820
Torso		
Sitting height	0.722	
Trunk height	0.587	
Cervical height	0.948	
Waist circumference		0.745
Hip circumference		0.744
Bideltoid		0.706
Cross-back width		0.486
Hip breadth		0.649
Shoulder circumference		0.806
Neck circumference		0.627
Arm		
Inside arm length	0.718	
Sleeve length	0.624	

(Continued)

[7]W. E. Woodson and D. W. Conover, *Human Engineering Guide for Equipment Designers,* University of California Press, Berkeley, 1964.

TABLE 34.54 (*Continued*)

Dimension	Correlation with	
	Stature	Chest Circumference
Inseam	0.819	
Forearm-hand length	0.640	
Shoulder-elbow length	0.660	
Elbow breadth		0.677
Wrist circumference		0.556
Leg		
Lower leg	0.688	
Patella height	0.795	
Buttock-knee height	0.751	
Outseam	0.886	
Total crotch length		0.467
Crotch-thigh circumference		0.731

In addition, consider the following suggestions:

For reach problems select a variety of subjects who variously have short stature, short legs, and short arms.

For clearance problems select a variety of subjects who variously are tall, have long legs and/or sitting heights, and have wide shoulders and hip breadths.

If the weight of a subject is pertinent (in addition to the other dimensional characteristics noted above), consider, for example, one or two heavy but small subjects, for they may sink into a seat and spoil your plans for stabilizing a range of eye references or head clearances.

EFFECTS OF CLOTHING ON HUMAN BODY DIMENSIONS[8]

As noted previously, most human body dimensions were obtained using nude subjects. For practical applications, therefore, one must consider the effect that clothing, shoes, gloves, hats, or other items of apparel may have on specific base-line dimensions. Table 34.55 provides rough

[8] W. E. Woodson and D. W. Conover, *Human Engineering Guide for Equipment Designers*, University of California Press, Berkeley, 1964.

TABLE 34.55 Effect of Various Types of Clothing on Human Body Dimensions

Dimension	Street Clothes		Winter Clothes		Heavy Flight Clothing	Pressure Suits*	
	Men	Women	Men	Women		Unpressurized	Pressurized
Weight	5 lb	3 ½ lb	10 lb	7 lb	12–15 lb	21 lb	21 lb
Stature	1 in.	½–3¾ in.	1 in.	½–3¾ in.	3 in.	3½ in.	2½ in.
Vertical reach	1	½–3¾	1	½–3¾	1	(−2½)	(−16½)
Eye height, standing	1	½–3¾	1	½–3¾	1	(−3½)	2½
Crotch to floor	1	½–3¾	1	½–3¾	1	(−1)	(−1)
Foot length	1¼	½	1½	½–¾	1	1	1
Foot width	½–¾	¼–(½)†	½-1	¼–½	¾	¾	¾
Head length	—	—	—‡		4½	4½	4½
Head width	—	—	—‡		4½	4½	4½
Hand length	—	—	¾	½	½	½	¼
Hand width	—	—	½	¼	½	½	1
Hand thickness	—	—	½	¼	½	¾	1¼
Fist circumference	—	—	1	¾	1	1¼	3
Shoulder width	½	¼	2–3	1	1½	1	½
Hip width	½	¼	2–3	1	1½	1	2¾
Elbow-to-elbow width	¾	¼	2–3½	1–1½	1	6	9
Thigh clearance	½	¼	1	¾	2	1¾	2
Forearm-to-fist length	½	¼	¾	½	1	1½	5½

*Certain pressurized garments are designed for seated position and therefore shorten certain dimensions. The helmet, however, tends to rise under pressure.

†Women's dress shoes confine and shrink foot width.

‡An Army steel helmet is approximately 12 by 10¼ in.

approximations of how much clothing adds to, or subtracts from, given body dimensions. However, when a particular design problem indicates the possibility that some particular clothing item may influence clearance, reach, etc., one should select an appropriate subject sample and have subjects put on the specific clothing items pertinent to the expected user environment. Remember that clothing not only increases clearance dimensions but also often creates considerable restrictions to mobility and reach capability.

Key questions with regard to what type of clothing should be considered include whether an operation will be performed in shirt sleeves, at low temperatures, under reduced pressure, or in other special environments.

FUNCTIONAL DIMENSIONS[9]

Grasping Reach for Air Force Men to a Horizontal Plane at the Seat Reference Level (SRL) (Seat-Pan Angle—6°; Backrest—13°) (See Fig. 34.10 and Table 34.56)

[9]*Anthropometric Source Book,* NASA Reference Publication 1024, 1978.

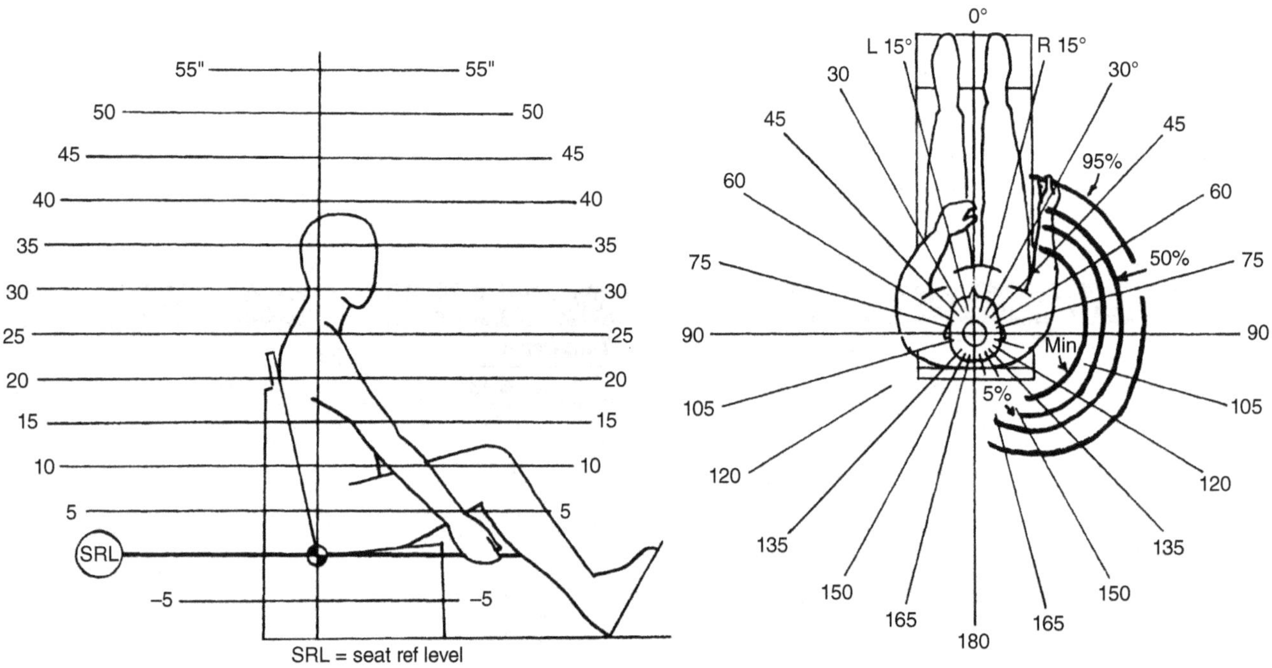

FIGURE 34.10 Functional dimensions.

TABLE 34.56

Angle to Left or Right	Minimum	Percentile 5th	50th	95th
L 165				
L 150				
L 135				
L 120				
L 105				
L 90				
L 75				
L 60				
L 45				
L 30				
L 15				
0				
R 15				
R 30		17.50 in.	20.75 in.	25.00 in.
R 45	16.25 in.	19.50 in.	21.75 in.	26.00 in.
R 60	17.50 in.	20.50 in.	22.25 in.	26.25 in.
R 75	17.25 in.	20.00 in.	22.25 in.	26.00 in.
R 90	17.00 in.	19.50 in.	22.25 in.	25.50 in.
R 105	16.25 in.	18.75 in.	22.00 in.	25.25 in.
R 120	15.00 in.	18.25 in.	20.75 in.	24.50 in.
R 135	13.00 in.	16.50 in.	19.00 in.	23.50 in.
R 150		14.00 in.	16.50 in.	20.25 in.
R 165			13.00 in.	17.00 in.
180				

Grasping Reach for Air Force Women to a Horizontal Plane at the SRL (See Table 34.57)

TABLE 34.57

Angle to Left or Right	Percentile 5th	50th	95th
L 165			
L 150			
L 135			
L 120			
L 105			
L 90			
L 75			
L 60			
L 45			
L 30			
L 15			
0			
R 15			22.0 in.
R 30		16.2 in.	21.7 in.
R 45	14.0 in.	17.5 in.	22.2 in.
R 60	15.2 in.	18.7 in.	23.0 in.
R 75	16.2 in.	19.0 in.	23.7 in.
R 90	16.7 in.	19.5 in.	23.7 in.
R 105	16.0 in.	19.0 in.	23.0 in.
R 120	15.2 in.	18.2 in.	22.0 in.
R 135	13.0 in.	16.5 in.	20.5 in.
R 150		13.0 in.	18.7 in.
R 165			15.7 in.
R 180			

Grasping Reach for Air Force Men to a Horizontal Plane 5 in. (13 cm) above the SRL (See Fig. 34.11 and Table 34.58)

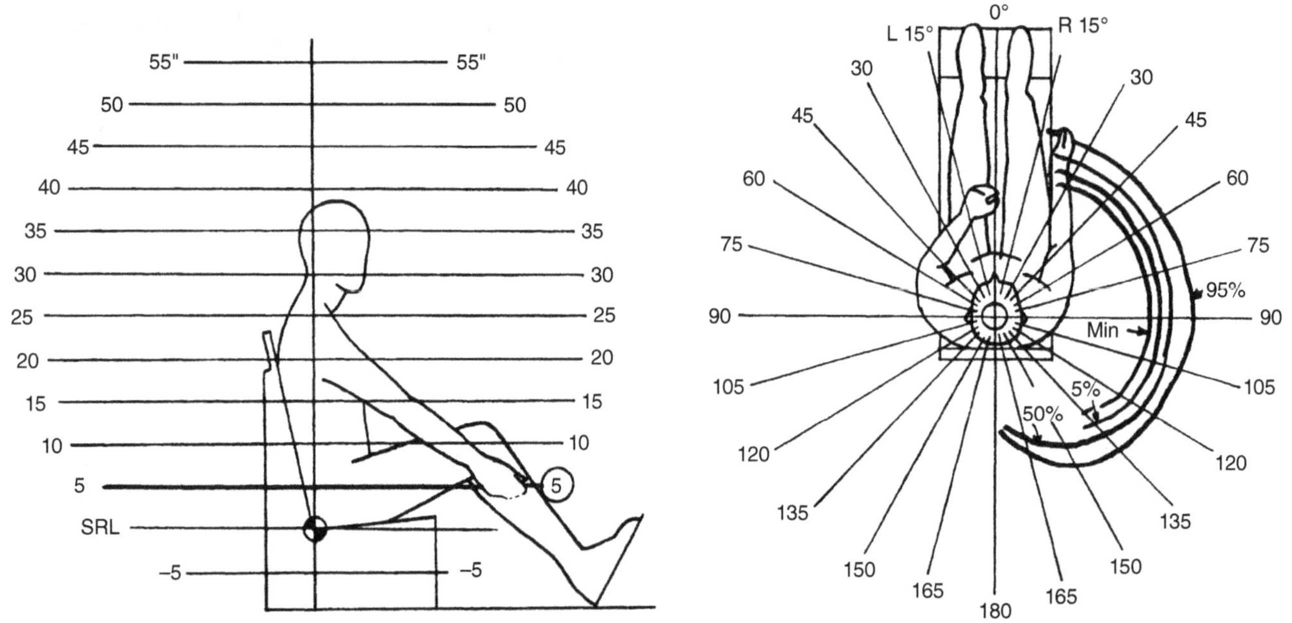

FIGURE 34.11

TABLE 34.58

ANGLE TO LEFT OR RIGHT	MINIMUM	PERCENTILE 5TH	50TH	95TH
L 165				
L 150				
L 135				
L 120				
L 105				
L 90				
L 75				
L 60				
L 45				
L 30				
L 15				
0				
R 15				
R 30	22.00 in.	23.75 in.	26.00 in.	29.50 in.
R 45	23.50 in.	25.25 in.	27.25 in.	30.00 in.
R 60	23.75 in.	25.75 in.	27.75 in.	30.00 in.
R 75	24.00 in.	25.75 in.	27.50 in.	30.25 in.
R 90	24.00 in.	25.75 in.	27.50 in.	30.75 in.
R 105	23.75 in.	25.25 in.	27.00 in.	30.00 in.
R 120	23.00 in.	24.50 in.	26.50 in.	29.00 in.
R 135	21.50 in.	22.75 in.	25.00 in.	28.00 in.
R 150			22.25 in.	25.75 in.
R 165			19.25 in.	21.25 in.
180				

Grasping Reach for Air Force Women to a Horizontal Plane 6 in. (15 cm) above the SRL (See Table 34.59)

Only a single illustration of the male has been provided for both male and female data. However, it is important to note that the reference levels above the SRL for the data on females are not quite the same as those for the data on males. For example, the data on males in Table 34.58 were taken at 5 in. (13 cm) above the SRL, while the data on females in Table 34.59 were taken at 6 in. (15 cm) above SRL. This variation in reference level occurs throughout the remaining pairs of data.

TABLE 34.59

Angle to Left or Right	Percentile		
	5th	50th	95th
L 165			
L 150			
L 135			
L 120			
L 105			10.5 in.
L 90			11.5 in.
L 75			14.5 in.
L 60			16.0 in.
L 45			18.0 in.
L 30			20.0 in.
L 15			—
0			
R 15	20.0 in.	22.5 in.	26.5 in.
R 30	21.0 in.	23.0 in.	27.5 in.
R 45	21.5 in.	23.7 in.	28.0 in.
R 60	23.2 in.	25.0 in.	28.0 in.
R 75	23.7 in.	25.0 in.	28.5 in.
R 90	23.7 in.	25.2 in.	28.5 in.
R 105	23.2 in.	25.0 in.	27.7 in.
R 120	22.0 in.	24.0 in.	26.2 in.
R 135	20.7 in.	23.0 in.	25.5 in.
R 150	—	20.0 in.	24.0 in.
R 165	—	16.2 in.	21.0 in.
R 180			

Grasping Reach for Air Force Men to a Horizontal Plane 10 in. (25 cm) above the SRL (See Fig. 34.12 and Table 34.60)

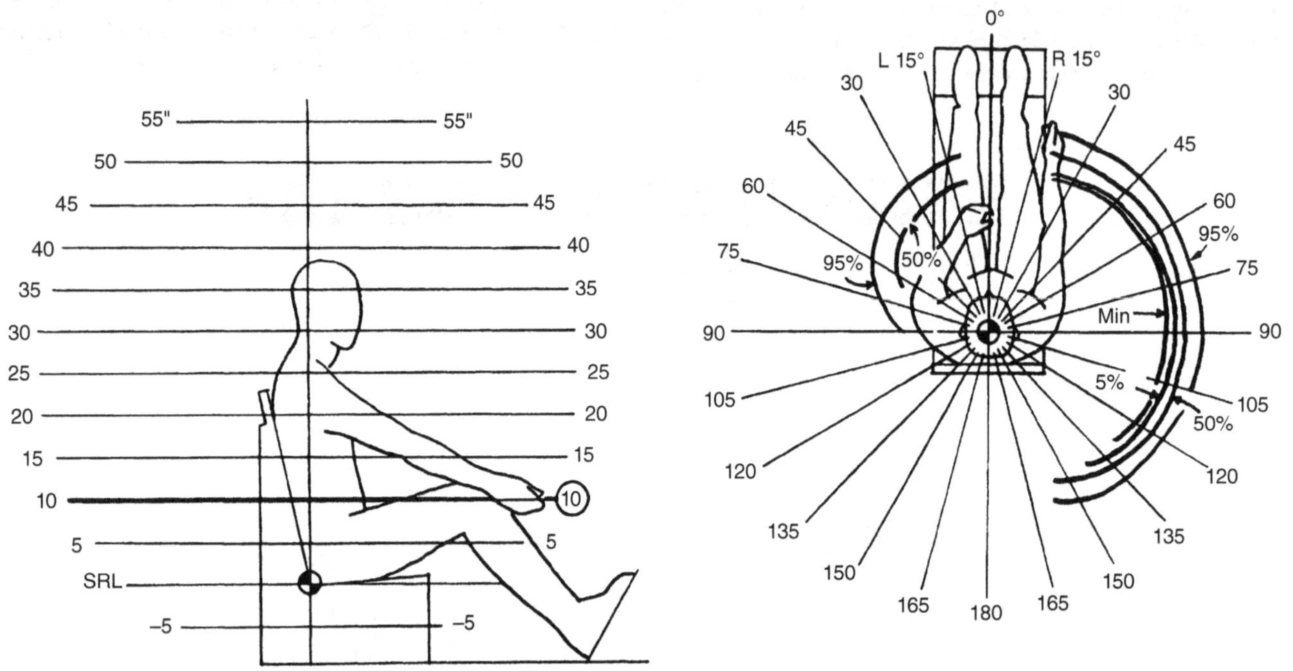

FIGURE 34.12

TABLE 34.60				
		PERCENTILE		
ANGLE TO LEFT OR RIGHT	MINIMUM	5TH	50TH	95TH
L 165				
L 150				
L 135				
L 120				
L 105				
L 90				13.50 in.
L 75				17.25 in.
L 60			16.50 in.	21.00 in.
L 45			19.50 in.	23.25 in.
L 30			21.00 in.	24.75 in.
L 15			22.00 in.	26.25 in.
0				
R 15				
R 30	26.25 in.	27.00 in.	29.25 in.	33.00 in.
R 45	27.25 in.	28.25 in.	30.50 in.	33.75 in.
R 60	28.00 in.	29.00 in.	30.75 in.	33.50 in.
R 75	28.25 in.	29.25 in.	30.75 in.	33.50 in.
R 90	28.25 in.	29.25 in.	31.00 in.	33.50 in.
R 105	27.75 in.	28.75 in.	30.50 in.	32.75 in.
R 120	26.75 in.	27.75 in.	29.75 in.	31.50 in.
R 135	—	26.25 in.	28.25 in.	30.75 in.
R 150	—	—	25.25 in.	28.75 in.
R 165				
180				

Grasping Reach for Air Force Women to a Horizontal Plane 12 in. (31 cm) above the SRL (See Table 34.61)

TABLE 34.61

Angle to Left or Right	Percentile		
	5th	50th	95th
L 165			
L 150			
L 135			
L 120			12.7 in.
L 105			14.0 in.
L 90		11.0 in.	15.5 in.
L 75		13.0 in.	17.5 in.
L 60	12.2 in.	15.0 in.	20.0 in.
L 45	14.5 in.	17.7 in.	21.5 in.
L 30	16.5 in.	10.0 in.	22.7 in.
L 15	19.0 in.	21.7 in.	24.5 in.
0	21.5 in.	23.5 in.	26.0 in.
R 15	23.0 in.	25.0 in.	28.0 in.
R 30	24.0 in.	26.0 in.	29.2 in.
R 45	25.5 in.	27.2 in.	30.0 in.
R 60	26.5 in.	28.2 in.	30.7 in.
R 75	26.7 in.	28.2 in.	31.0 in.
R 90	27.2 in.	28.5 in.	31.0 in.
R 105	26.5 in.	28.5 in.	31.0 in.
R 120	—	27.5 in.	29.5 in.
R 135	—	25.5 in.	28.2 in.
R 150	—	19.0 in.	25.0 in.
R 165			22.5 in.
R 180			

Grasping Reach for Air Force Men to a Horizontal Plane 15 in. (38 cm) above the SRL (See Fig. 34.13 and Table 34.62)

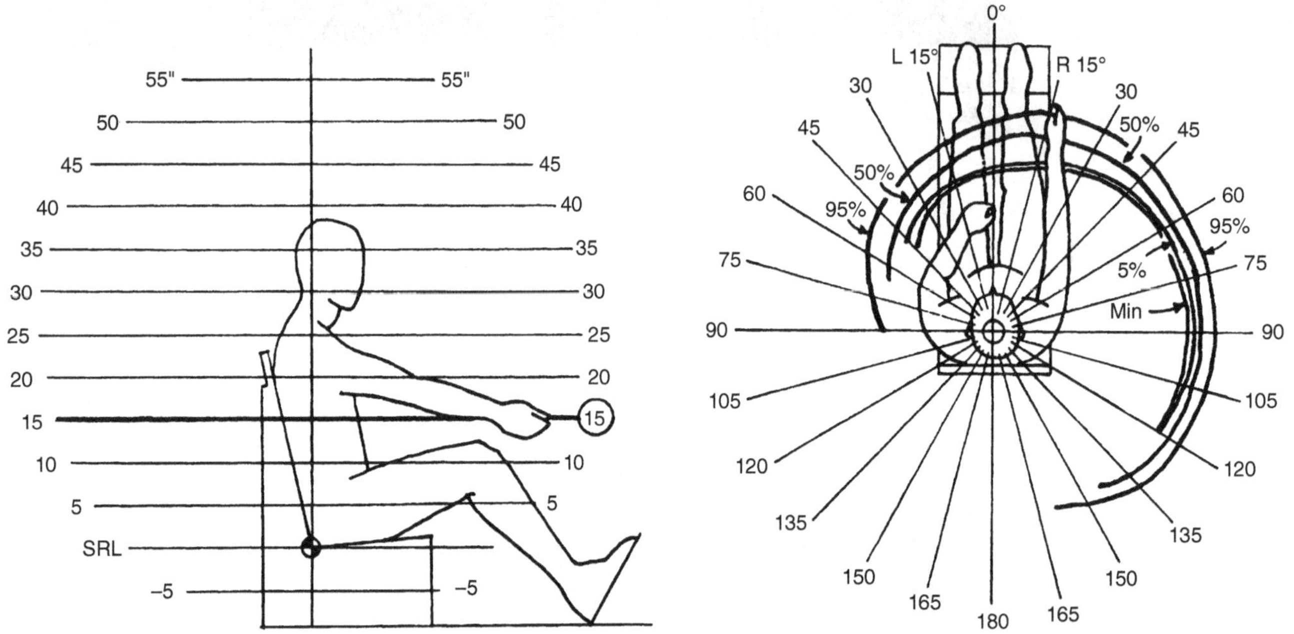

FIGURE 34.13

TABLE 34.62				
ANGLE TO LEFT OR RIGHT	MINIMUM	PERCENTILE		
		5TH	50TH	95TH
L 165				
L 150				
L 135				
L 120				
L 105				
L 90				17.50 in.
L 75				20.00 in.
L 60			19.25 in.	23.00 in.
L 45		19.00 in.	21.50 in.	25.75 in.
L 30	21.00 in.	21.75 in.	24.00 in.	27.25 in.
L 15	22.50 in.	23.25 in.	26.00 in.	28.75 in.
0	24.25 in.	24.75 in.	28.75 in.	31.00 in.
R 15	26.00 in.	26.50 in.	30.50 in.	34.00 in.
R 30	28.25 in.	28.50 in.	31.50 in.	35.00 in.
R 45	29.50 in.	30.00 in.	32.75 in.	35.50 in.
R 60	30.00 in.	31.00 in.	32.50 in.	34.75 in.
R 75	30.00 in.	31.50 in.	32.50 in.	34.75 in.
R 90	30.25 in.	31.00 in.	32.50 in.	34.75 in.
R 105	30.00 in.	30.75 in.	32.25 in.	34.50 in.
R 120	29.00 in.	29.50 in.	32.00 in.	33.75 in.
R 135	—	—	30.00 in.	32.50 in.
R 150	—	—	—	29.50 in.
R 165				
180				

Grasping Reach for Air Force Women to a Horizontal Plane 18 in. (46 cm) above the SRL (See Table 34.63)

TABLE 34.63

Angle to Left or Right	Minimum	Percentile		
		5th	50th	95th
L 165				
L 150				
L 135				
L 120				
L 105				
L 90			14.00 in.	18.75 in.
L 75			18.00 in.	21.50 in.
L 60	17.00 in.	17.50 in.	20.50 in.	24.50 in.
L 45	18.25 in.	19.50 in.	22.75 in.	26.75 in.
L 30	20.25 in.	21.50 in.	24.75 in.	28.25 in.
L 15	22.50 in.	23.50 in.	26.75 in.	29.75 in.
0	25.00 in.	25.50 in.	28.75 in.	31.75 in.
R 15	27.25 in.	28.00 in.	30.50 in.	34.00 in.
R 30	29.00 in.	30.00 in.	32.00 in.	35.75 in.
R 45	30.50 in.	31.00 in.	33.50 in.	36.25 in.
R 60	31.50 in.	32.00 in.	33.75 in.	36.25 in.
R 75	31.50 in.	32.25 in.	34.00 in.	36.50 in.
R 90	31.75 in.	32.25 in.	34.00 in.	36.00 in.
R 105	31.50 in.	31.75 in.	33.50 in.	35.75 in.
R 120	—	30.50 in.	33.00 in.	35.50 in.
R 135	—	—	—	34.50 in.
R 150				
R 165				
180				

Grasping Reach for Air Force Men to a Horizontal Plane 20 in. (51 cm) above the SRL (See Fig. 34.14 and Table 34.64)

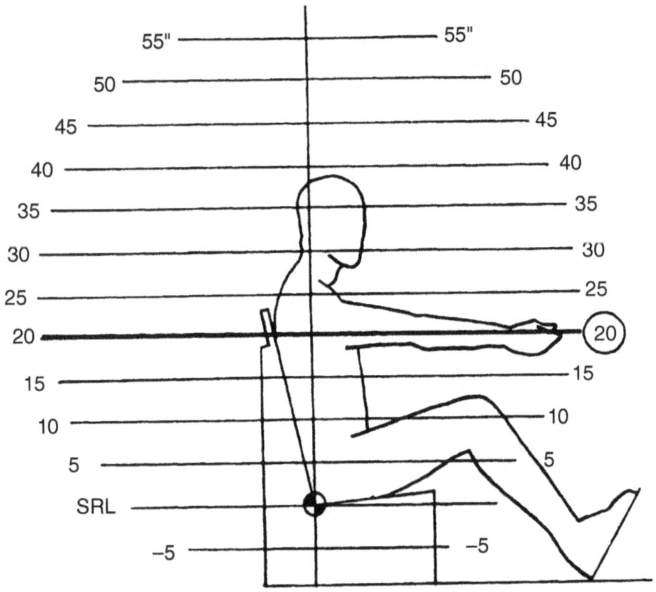

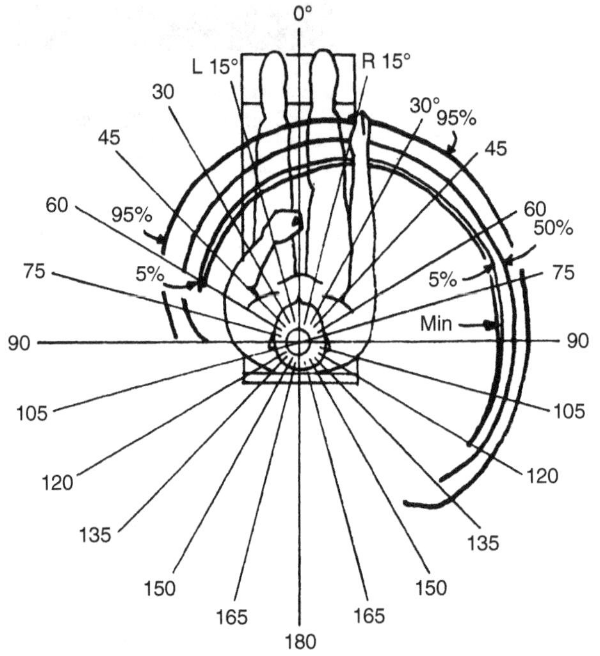

FIGURE 34.14

TABLE 34.64

Angle to Left or Right	Minimum	Percentile 5th	50th	95th
L 165				
L 150				
L 135				
L 120				
L 105				
L 90			14.00 in.	18.75 in.
L 75			18.00 in.	21.50 in.
L 60	17.00 in.	17.50 in.	20.50 in.	24.50 in.
L 45	18.25 in.	19.50 in.	22.75 in.	26.75 in.
L 30	20.25 in.	21.50 in.	24.75 in.	28.25 in.
L 15	22.50 in.	23.50 in.	26.75 in.	29.75 in.
0	25.00 in.	25.50 in.	28.75 in.	31.75 in.
R 15	27.25 in.	28.00 in.	30.50 in.	34.00 in.
R 30	29.00 in.	30.00 in.	32.00 in.	35.75 in.
R 45	30.50 in.	31.00 in.	33.50 in.	36.25 in.
R 60	31.50 in.	32.00 in.	33.75 in.	36.25 in.
R 75	31.50 in.	32.25 in.	34.00 in.	36.50 in.
R 90	31.75 in.	32.25 in.	34.00 in.	36.00 in.
R 105	31.50 in.	31.75 in.	33.50 in.	35.75 in.
R 120	—	30.50 in.	33.00 in.	35.50 in.
R 135	—	—	—	34.50 in.
R 150				
R 165				
180				

Grasping Reach for Air Force Women to a Horizontal Plane 24 in. (61 cm) above the SRL (See Table 34.65)

TABLE 34.65

Angle to Left or Right	Percentile		
	5th	50th	95th
L 165		9.0 in.	15.0 in.
L 150		9.0 in.	16.0 in.
L 135		10.7 in.	14.0 in.
L 120		10.0 in.	16.7 in.
L 105	8.0 in.	12.2 in.	19.0 in.
L 90	10.0 in.	14.7 in.	17.7 in.
L 75	11.5 in.	16.0 in.	21.0 in.
L 60	14.2 in.	18.5 in.	21.5 in.
L 45	17.0 in.	20.0 in.	23.5 in.
L 30	19.0 in.	21.7 in.	24.7 in.
L 15	20.5 in.	23.0 in.	26.0 in.
0	22.0 in.	25.0 in.	28.0 in.
R 15	23.5 in.	26.2 in.	29.5 in.
R 30	25.0 in.	27.5 in.	30.2 in.
R 45	26.2 in.	28.5 in.	31.0 in.
R 60	26.7 in.	29.2 in.	32.0 in.
R 75	27.0 in.	30.0 in.	32.0 in.
R 90	27.5 in.	30.5 in.	32.0 in.
R 105	27.2 in.	30.2 in.	32.2 in.
R 120	13.0 in.	28.5 in.	31.0 in.
R 135	11.0 in.	14.0 in.	27.0 in.
R 150	9.0 in.	12.0 in.	22.0 in.
R 165	8.2 in.	11.1 in.	18.0 in.
180	—	11.0 in.	16.0 in.

Grasping Reach for Air Force Men to a Horizontal Plane 25 in. (64 cm) above the SRL (See Fig. 34.15 and Table 34.66)

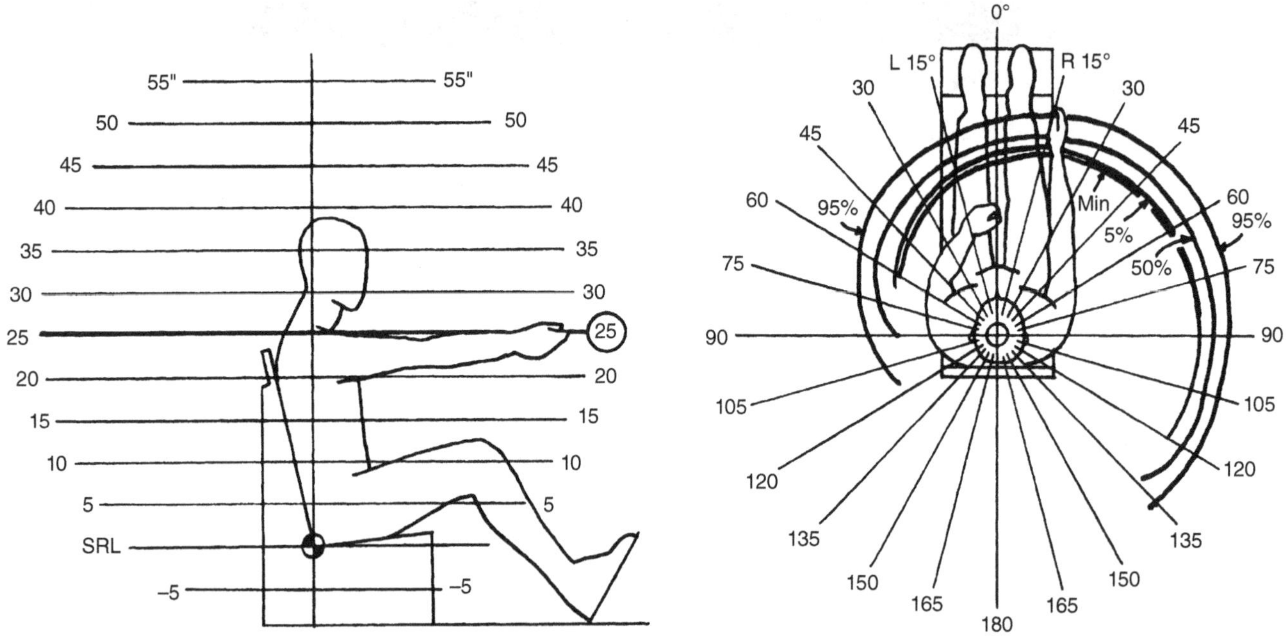

FIGURE 34.15

TABLE 34.66				
		PERCENTILE		
ANGLE TO LEFT OR RIGHT	MINIMUM	5TH	50TH	95TH
L 165				
L 150				
L 135				
L 120				
L 105				17.75 in.
L 90			15.75 in.	20.25 in.
L 75			19.25 in.	22.25 in.
L 60	17.75 in.	18.25 in.	21.50 in.	24.75 in.
L 45	19.25 in.	20.00 in.	23.25 in.	27.25 in.
L 30	21.50 in.	22.50 in.	25.00 in.	28.50 in.
L 15	23.25 in.	24.00 in.	27.00 in.	29.75 in.
0	25.00 in.	26.25 in.	28.50 in.	31.50 in.
R 15	27.25 in.	28.25 in.	30.25 in.	33.50 in.
R 30	29.25 in.	30.25 in.	32.50 in.	35.25 in.
R 45	30.50 in.	31.00 in.	33.50 in.	35.75 in.
R 60	31.00 in.	31.50 in.	33.75 in.	37.00 in.
R 75	31.50 in.	32.00 in.	33.50 in.	36.50 in.
R 90	31.75 in.	32.25 in.	33.75 in.	36.25 in.
R 105	31.25 in.	31.50 in.	33.50 in.	36.00 in.
R 120		30.50 in.	33.25 in.	35.50 in.
R 135				35.00 in.
R 150				
R 165				
180				

Grasping Reach for Air Force Men to a Horizontal Plane 30 in. (76 cm) above the SRL (See Fig. 34.16 and Table 34.67)

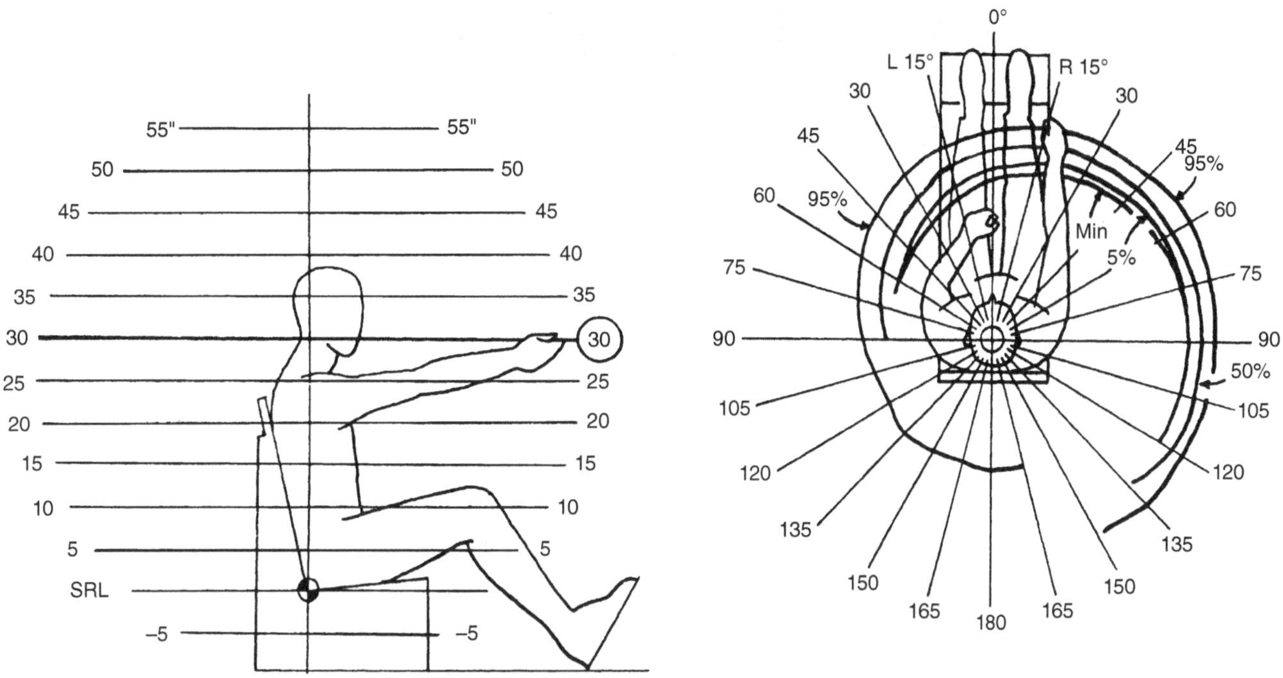

Figure 34.16

TABLE 34.67

Angle to Left or Right	Minimum	Percentile 5th	50th	95th
L 165				18.75 in.
L 150				19.25 in.
L 135				20.00 in.
L 120				18.75 in.
L 105				19.00 in.
L 90			16.75 in.	20.75 in.
L 75			18.75 in.	22.50 in.
L 60	17.00 in.	17.25 in.	20.75 in.	24.50 in.
L 45	18.25 in.	19.00 in.	22.50 in.	26.50 in.
L 30	19.75 in.	21.50 in.	24.50 in.	28.25 in.
L 15	22.00 in.	23.75 in.	26.75 in.	29.50 in.
0	23.75 in.	25.50 in.	28.50 in.	31.00 in.
R 15	26.00 in.	27.25 in.	29.75 in.	33.00 in.
R 30	27.75 in.	29.00 in.	31.50 in.	34.25 in.
R 45	28.75 in.	30.25 in.	32.25 in.	34.75 in.
R 60	30.00 in.	31.00 in.	32.75 in.	35.75 in.
R 75	30.75 in.	31.25 in.	33.00 in.	35.50 in.
R 90	31.00 in.	31.25 in.	33.25 in.	35.75 in.
R 105	30.75 in.	31.00 in.	33.00 in.	35.25 in.
R 120	—	30.25 in.	32.50 in.	34.75 in.
R 135				34.50 in.
R 150				—
R 165				19.50 in.
180				20.25 in.

Grasping Reach for Air Force Women to a Horizontal Plane 30 in. (76 cm) above the SRL (See Table 34.68)

TABLE 34.68

Angle to Left or Right	Percentile		
	5th	50th	95th
L 165	7.2 in.	12.5 in.	19.2 in.
L 150	6.2 in.	12.0 in.	16.5 in.
L 135	6.7 in.	8.7 in.	15.2 in.
L 120	7.0 in.	10.7 in.	17.0 in.
L 105	6.5 in.	12.0 in.	18.0 in.
L 90	8.7 in.	13.0 in.	17.2 in.
L 75	10.0 in.	15.5 in.	20.0 in.
L 60	13.0 in.	17.5 in.	21.0 in.
L 45	15.0 in.	19.0 in.	22.0 in.
L 30	17.0 in.	20.5 in.	24.2 in.
L 15	18.2 in.	22.0 in.	25.2 in.
0	20.0 in.	23.0 in.	27.0 in.
R 15	21.5 in.	24.5 in.	28.2 in.
R 30	22.5 in.	25.7 in.	29.0 in.
R 45	23.2 in.	27.5 in.	29.7 in.
R 60	24.5 in.	27.7 in.	30.5 in.
R 75	25.2 in.	28.5 in.	30.2 in.
R 90	25.7 in.	28.7 in.	31.0 in.
R 105	26.0 in.	29.0 in.	31.0 in.
R 120	16.2 in.	26.2 in.	29.5 in.
R 135	12.7 in.	19.5 in.	27.5 in.
R 150	11.0 in.	16.2 in.	23.5 in.
R 165	10.5 in.	15.5 in.	22.0 in.
180	9.5 in.	15.0 in.	20.0 in.

Grasping Reach for Air Force Men to a Horizontal Plane 35 in. (89 cm) above the SRL (See Fig. 34.17 and Table 34.69)

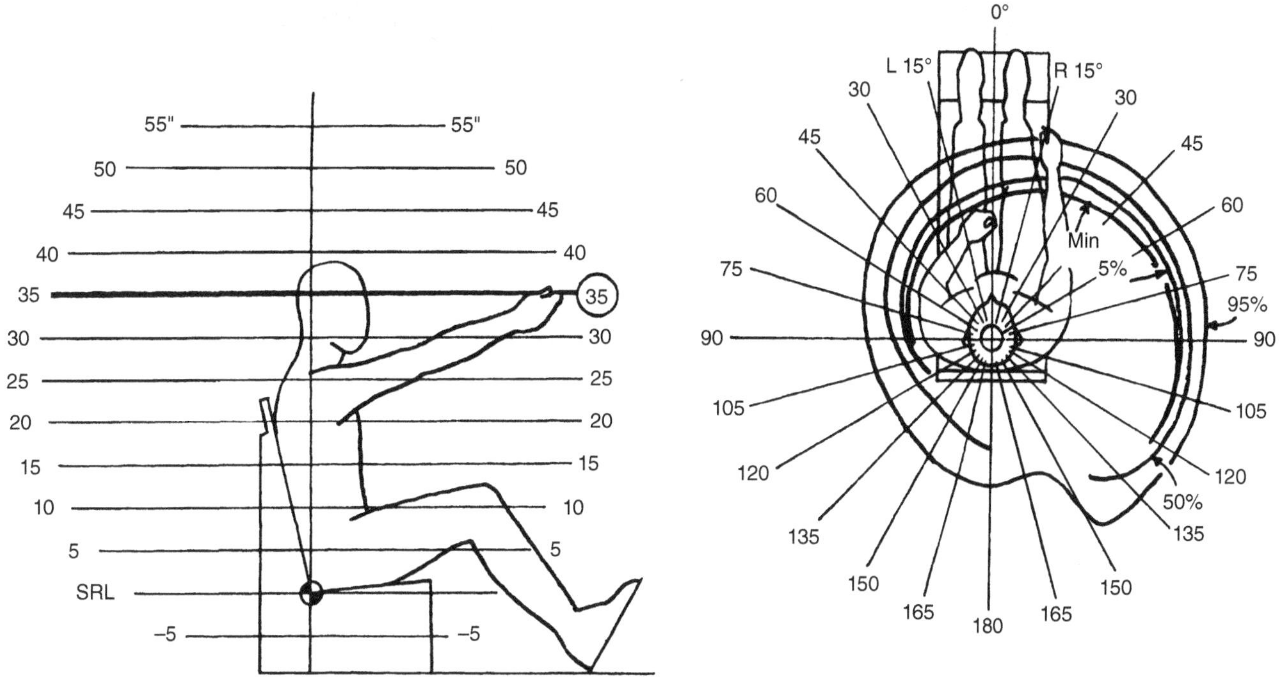

Figure 34.17

TABLE 34.69

Angle to Left or Right	Minimum	Percentile 5th	50th	95th
L 165			14.75 in.	21.00 in.
L 150			13.75 in.	20.00 in.
L 135			13.25 in.	19.00 in.
L 120		10.75 in.	13.25 in.	18.75 in.
L 105		12.25 in.	14.00 in.	18.75 in.
L 90	12.75 in.	13.75 in.	15.50 in.	20.00 in.
L 75	14.25 in.	15.00 in.	17.25 in.	21.00 in.
L 60	15.25 in.	16.00 in.	18.75 in.	21.50 in.
L 45	16.25 in.	17.25 in.	20.50 in.	24.75 in.
L 30	18.00 in.	19.25 in.	22.50 in.	26.25 in.
L 15	19.25 in.	21.00 in.	24.75 in.	27.00 in.
0	20.75 in.	22.25 in.	26.50 in.	28.50 in.
R 15	22.75 in.	24.75 in.	27.75 in.	31.00 in.
R 30	24.50 in.	26.75 in.	29.25 in.	32.75 in.
R 45	26.75 in.	28.25 in.	30.50 in.	33.75 in.
R 60	28.00 in.	29.00 in.	31.00 in.	33.75 in.
R 75	28.75 in.	29.50 in.	31.25 in.	34.00 in.
R 90	29.00 in.	29.75 in.	31.25 in.	33.50 in.
R 105	29.00 in.	29.75 in.	31.50 in.	33.50 in.
R 120	28.50 in.	29.00 in.	31.00 in.	33.50 in.
R 135			28.50 in.	33.50 in.
R 150				31.50 in.
R 165				21.75 in.
180			16.50 in.	22.25 in.

Grasping Reach for Air Force Women to a Horizontal Plane 36 in. (91 cm) above the SRL (See Table 34.70)

TABLE 34.70

Angle to Left or Right	Percentile		
	5th	50th	95th
L 165	9.0 in.	13.0 in.	19.5 in.
L 150	8.0 in.	11.5 in.	17.7 in.
L 135	7.2 in.	10.2 in.	16.0 in.
L 120	7.2 in.	10.0 in.	15.5 in.
L 105	7.2 in.	10.5 in.	15.2 in.
L 90	7.7 in.	11.5 in.	16.0 in.
L 75	8.2 in.	13.0 in.	17.2 in.
L 60	10.0 in.	14.2 in.	18.0 in.
L 45	11.5 in.	15.5 in.	19.5 in.
L 30	13.2 in.	17.2 in.	21.5 in.
L 15	14.2 in.	19.0 in.	22.7 in.
0	16.2 in.	20.5 in.	24.0 in.
R 15	17.5 in.	21.5 in.	14.7 in.
R 30	18.5 in.	22.5 in.	26.0 in.
R 45	19.2 in.	24.0 in.	27.0 in.
R 60	20.7 in.	25.0 in.	27.7 in.
R 75	21.0 in.	25.5 in.	28.0 in.
R 90	22.2 in.	26.2 in.	28.7 in.
R 105	21.2 in.	26.2 in.	28.7 in.
R 120	18.2 in.	25.0 in.	27.7 in.
R 135	12.5 in.	19.0 in.	25.7 in.
R 150	10.0 in.	17.2 in.	23.5 in.
R 165	10.2 in.	16.0 in.	22.0 in.
180	9.5 in.	15.2 in.	21.2 in.

Grasping Reach for Air Force Men to a Horizontal Plane 40 in. (102 cm) above the SRL (See Fig. 34.18 and Table 34.71)

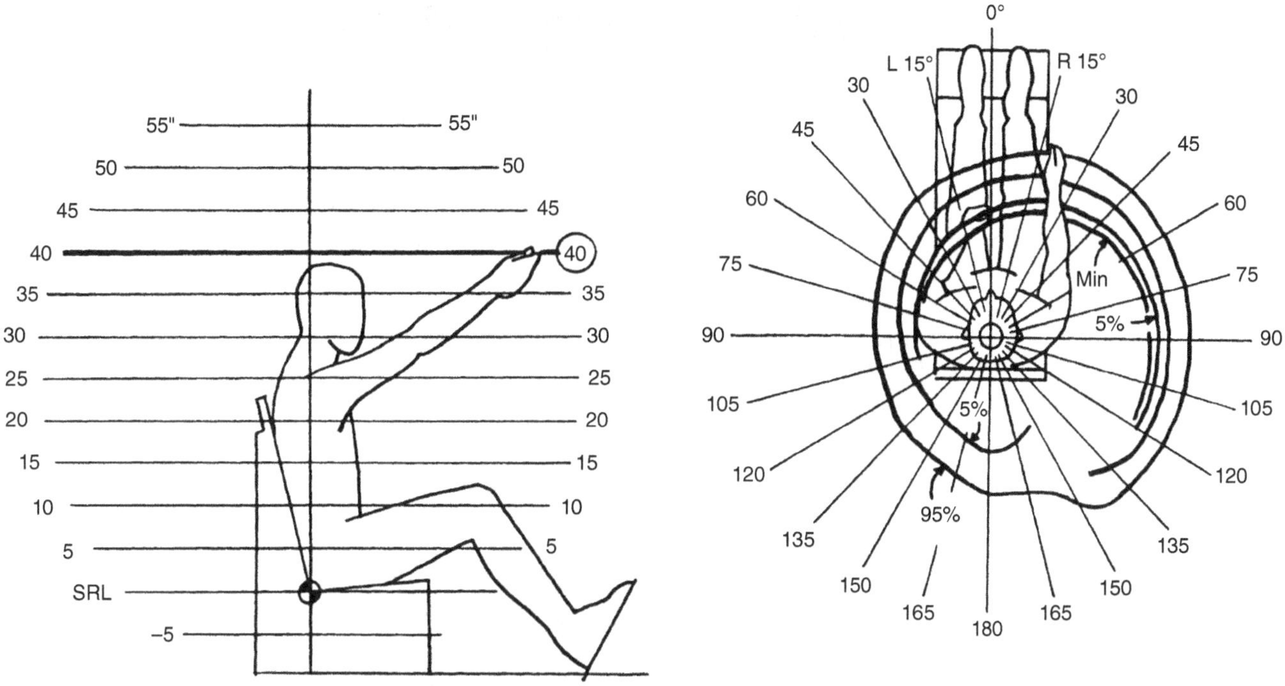

Figure 34.18

TABLE 34.71

Angle to Left or Right	Minimum	Percentile 5th	50th	95th
L 165			15.50 in.	21.50 in.
L 150			14.75 in.	20.00 in.
L 135			14.00 in.	19.25 in.
L 120		11.25 in.	13.25 in.	18.50 in.
L 105		11.75 in.	13.25 in.	18.25 in.
L 90	12.00 in.	12.25 in.	13.75 in.	18.25 in.
L 75	12.25 in.	12.50 in.	15.00 in.	18.75 in.
L 60	12.50 in.	13.25 in.	16.25 in.	20.00 in.
L 45	13.00 in.	14.00 in.	17.75 in.	21.50 in.
L 30	13.75 in.	15.50 in.	19.50 in.	23.50 in.
L 15	15.25 in.	17.00 in.	21.25 in.	24.50 in.
0	17.00 in.	19.00 in.	23.00 in.	25.75 in.
R 15	18.75 in.	21.00 in.	24.50 in.	28.50 in.
R 30	21.00 in.	22.75 in.	26.25 in.	30.50 in.
R 45	23.25 in.	24.75 in.	27.75 in.	31.50 in.
R 60	24.25 in.	25.50 in.	28.00 in.	31.25 in.
R 75	25.00 in.	26.00 in.	28.00 in.	31.50 in.
R 90	25.00 in.	26.25 in.	28.25 in.	31.50 in.
R 105	25.75 in.	26.75 in.	28.50 in.	31.75 in.
R 120		26.25 in.	28.75 in.	31.50 in.
R 135			27.00 in.	31.00 in.
R 150				29.25 in.
R 165			16.75 in.	23.75 in.
180			17.75 in.	23.50 in.

Grasping Reach for Air Force Women to a Horizontal Plane 42 in. (107 cm) above the SRL (See Table 34.72)

TABLE 34.72

Angle to Left or Right	Percentile		
	5th	50th	95th
L 165	5.0 in.	10.2 in.	17.0 in.
L 150	4.2 in.	9.0 in.	15.0 in.
L 135	3.7 in.	8.5 in.	13.7 in.
L 120	3.5 in.	8.0 in.	13.0 in.
L 105	3.2 in.	8.0 in.	12.5 in.
L 90	3.5 in.	8.0 in.	13.0 in.
L 75	3.7 in.	8.7 in.	14.5 in.
L 60	4.0 in.	9.5 in.	16.2 in.
L 45	4.7 in.	10.5 in.	16.0 in.
L 30	5.5 in.	11.5 in.	17.0 in.
L 15	6.5 in.	12.5 in.	17.7 in.
0	7.5 in.	14.0 in.	18.5 in.
R 15	9.0 in.	16.0 in.	19.0 in.
R 30	10.0 in.	17.0 in.	20.5 in.
R 45	11.2 in.	17.5 in.	22.0 in.
R 60	12.0 in.	19.0 in.	22.5 in.
R 75	13.0 in.	20.0 in.	23.5 in.
R 90	14.0 in.	20.0 in.	24.0 in.
R 105	14.0 in.	20.5 in.	24.0 in.
R 120	12.0 in.	18.5 in.	23.5 in.
R 135	9.2 in.	15.5 in.	21.2 in.
R 150	7.5 in.	14.0 in.	19.7 in.
R 165	6.5 in.	12.2 in.	19.0 in.
180	5.5 in.	11.0 in.	18.7 in.

Grasping Reach for Air Force Men to a Horizontal Plane 45 in. (114 cm) above the SRL (See Fig. 34.19 and Table 34.73)

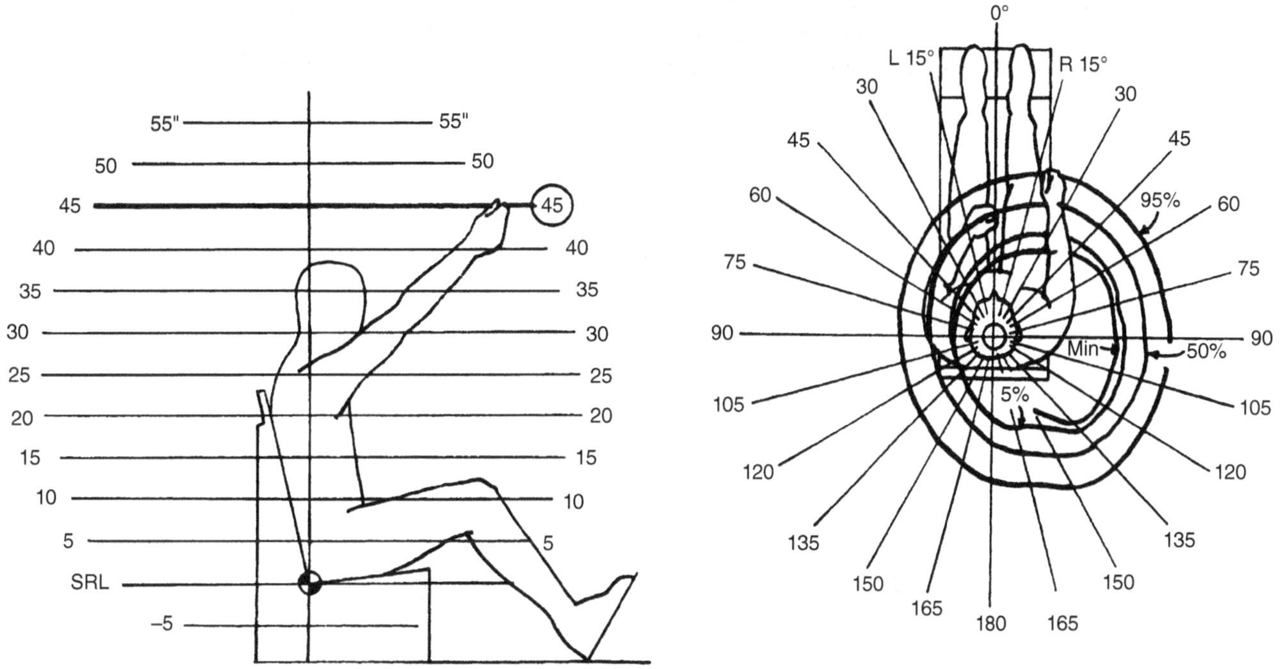

Figure 34.19

TABLE 34.73				
		Percentile		
Angle to Left or Right	Minimum	5th	50th	95th
L 165		10.50 in.	14.00 in.	20.00 in.
L 150	8.50 in.	8.75 in.	12.25 in.	18.25 in.
L 135	7.50 in.	7.75 in.	11.00 in.	16.75 in.
L 120	7.00 in.	7.50 in.	10.50 in.	15.50 in.
L 105	6.75 in.	7.25 in.	10.25 in.	15.00 in.
L 90	6.75 in.	7.25 in.	10.50 in.	15.00 in.
L 75	6.75 in.	7.50 in.	11.00 in.	15.25 in.
L 60	7.00 in.	7.75 in.	12..00 in	16.25 in.
L 45	7.50 in.	8.50 in.	13.50 in.	18.25 in.
L 30	8.50 in.	9.50 in.	15.00 in.	19.75 in.
L 15	10.00 in.	11.00 in.	16.50 in.	21.25 in.
0	11.25 in.	12.75 in.	18.25 in.	22.75 in.
R 15	13.00 in.	15.50 in.	20.00 in.	24.75 in.
R 30	14.75 in.	17.50 in.	22.00 in.	26.25 in.
R 45	17.25 in.	19.00 in.	23.50 in.	27.00 in.
R 60	19.25 in.	20.50 in.	24.00 in.	27.25 in.
R 75	19.50 in.	20.50 in.	24.00 in.	27.50 in.
R 90	19.75 in.	21.00 in.	24.25 in.	27.75 in.
R 105	20.25 in.	21.50 in.	24.50 in.	28.00 in.
R 120	19.75 in.	21.25 in.	24.50 in.	27.75 in.
R 135	18.75 in.	20.00 in.	23.25 in.	27.75 in.
R 150		15.50 in.	20.75 in.	26.00 in.
R 165		14.75 in.	18.00 in.	22.75 in.
180		12.75 in.	16.50 in.	21.50 in.

Functional Reach Limits for the Typical Automobile Driver (5th-Percentile Female)[10]

On the basis of automobile seating geometry (see Fig. 34.20, which shows a typical family vehicle), maximum reach limits of a representative (U.S. driver population) 5th-percentile female driver were defined (see Table 34.74). Reach limits are described with reference to the standard H point used by the automobile industry. Limits are shown for both the left and right hands. Because of the constraints of the door panel on the left, values are included only for three distances

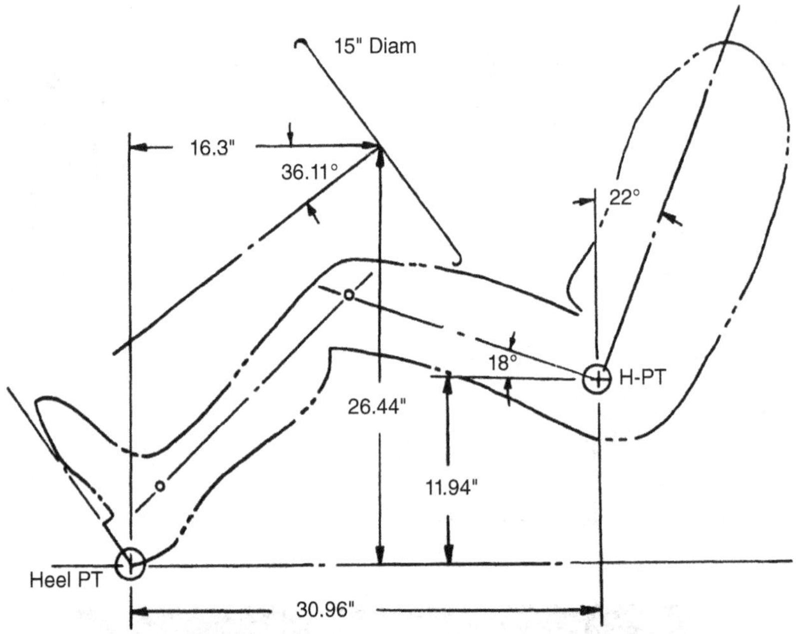

FIGURE 34.20 Typical family vehicle geometry.

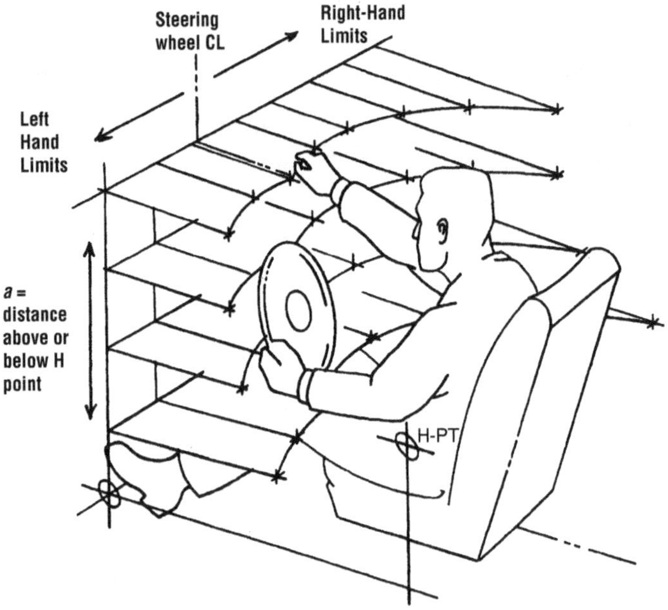

TABLE 34.74 Maximum Reach Limits (5th Percentile Female)

	LEFT HAND				RIGHT HAND				
*a**	16 IN.	8 IN.	4 IN.	C/L	C/L	4 IN.	8 IN.	16 IN.	24 IN.
35 in.	21.8	25.1	26.0	25.5	24.4	25.4	25.1	22.4	16.1
	15.6	19.2	20.0	19.0	21.2	21.3	20.7	16.0	0
20 in.	28.6	30.1	31.2	29.6	29.3	30.2	30.5	29.2	25.2
	23.8	25.9	25.7	24.9	26.2	26.6	26.4	23.1	20.0
15 in.	29.8	31.7	31.1	29.8	27.9	30.1	30.6	30.2	26.7
	24.4	26.2	25.0	23.5	23.0	25.9	26.7	25.0	20.1
10 in.	29.5	31.8	31.3			29.8	30.4	30.5	26.0
	24.0	26.1	25.2			26.0	26.3	25.1	19.9
5 in.	29.7	30.5					29.6	28.6	25.2
	22.8	24.7					25.0	23.5	18.9
−3 in.	26.4	28.1					27.8	27.5	23.6
	16.7	20.0					21.8	0	0

*a** = reach level above or below the H point.

[10] W. E. Woodson el al., Driver Eye Position and Control Reach Anthropometrics, MFI Report 71-117, 1971 (DOT Contract FH-11-7619).

from the steering-wheel centerline. The two central value sets represent reach around a typical steering wheel with a 15-in. (38-cm) diameter. The first value represents reach limits when the subject is not constrained by a shoulder belt; i.e., she can lean forward and/or to the side. The second value represents limits when a tight shoulder belt is worn (i.e., the belt does not allow the driver to lean forward or to the side).

Reach limits for other conditions, e.g., different seating configurations, obviously would be different.[11]

Figure 34.21 combines the reach limits of a 5th-percentile woman for the inside arm versus the combined outside arm reaches.

[11]For a complete description of reach limits for other seating configurations, see "Driver Hand Control Reach," SAE-J287, in *SAE Handbook*, Vol. 4.

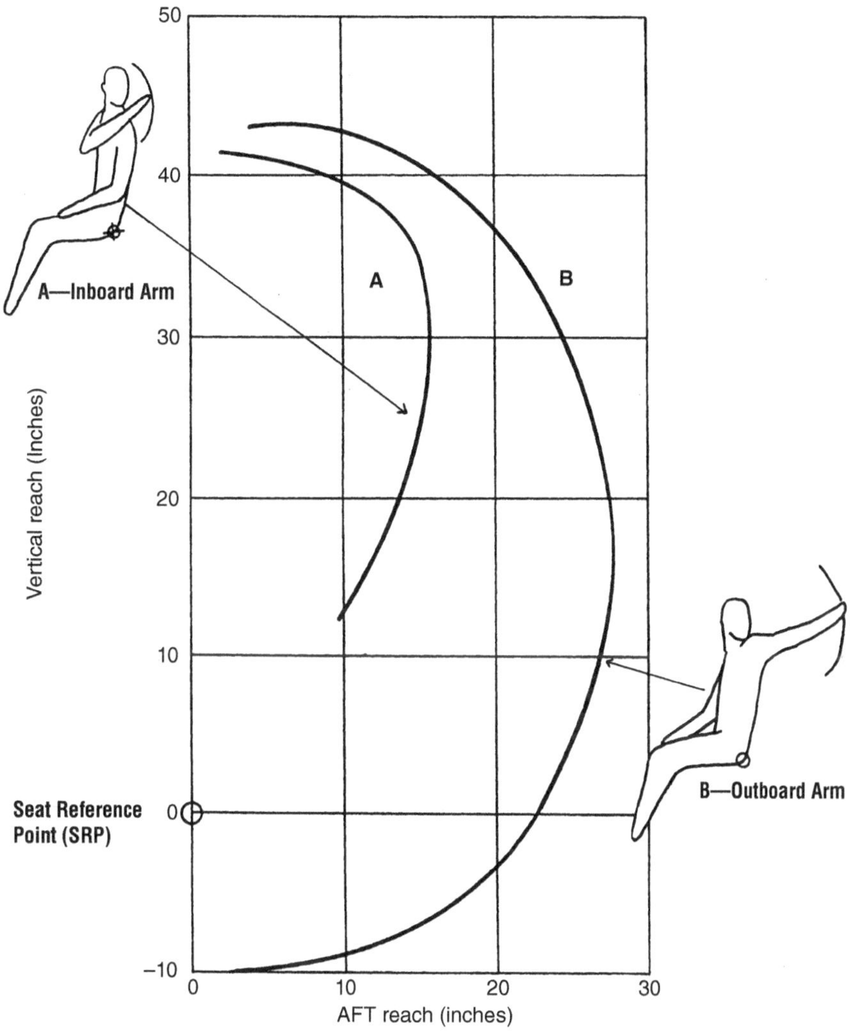

Figure 34.21 Reach Limits.

Functional Reach from a Wheelchair[12]

The approximate reach-limit values shown in Figs. 34.22–34.24 were derived on the basis of a sample of 91 male and 36 female subjects confined to a wheelchair. (See Table 34.75.) Note the differences between the maximum and the comfortable reach limits, a subjective but important consideration in design.

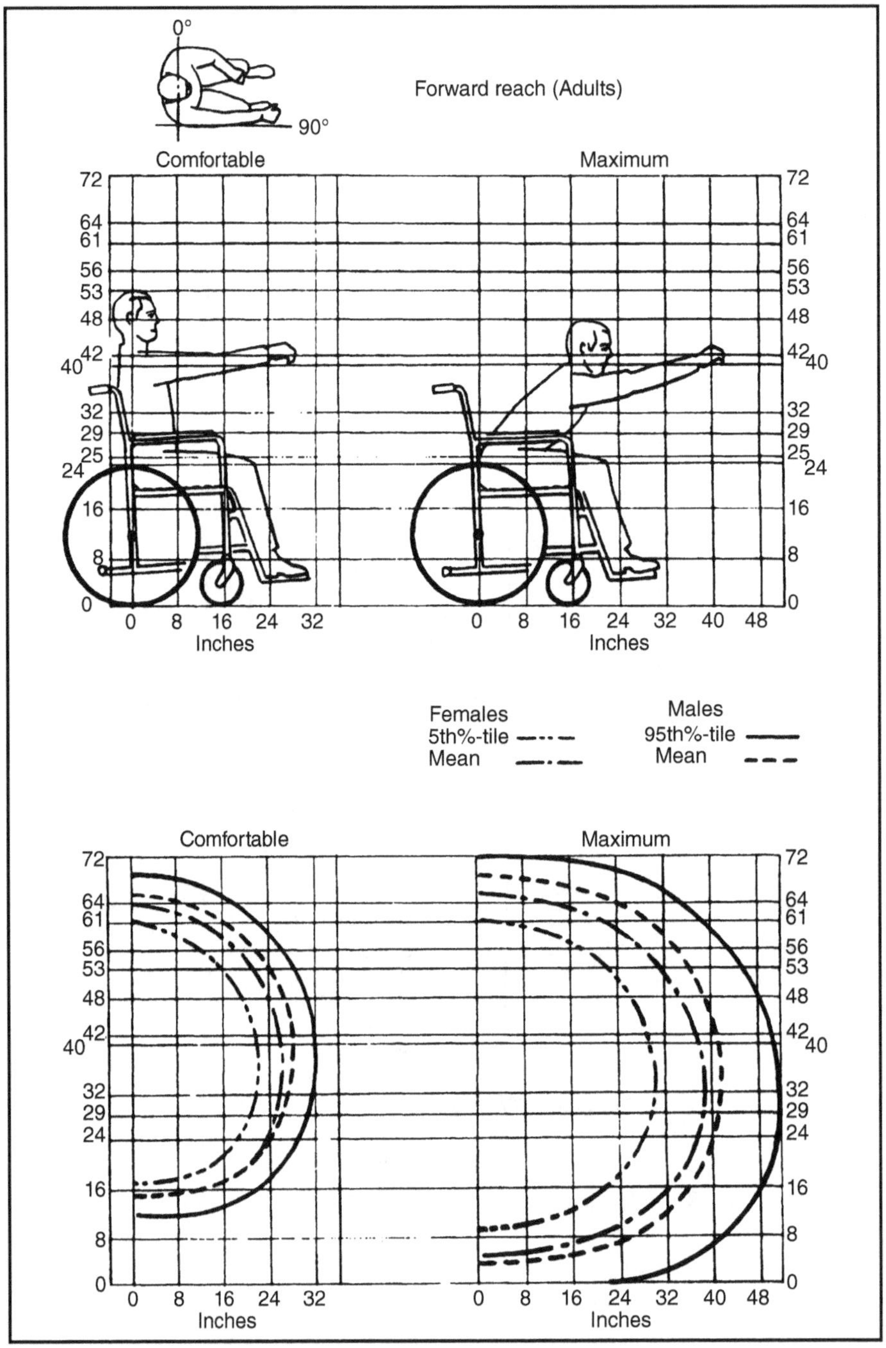

FIGURE 34.22 Forward reach limits (adults).

[12]W. F. Floyd et al., "A Study of the Space Requirements of Wheelchair Users," *Paraplegia*, Vol. 4, No. 1, pp. 24–37, May 1966.

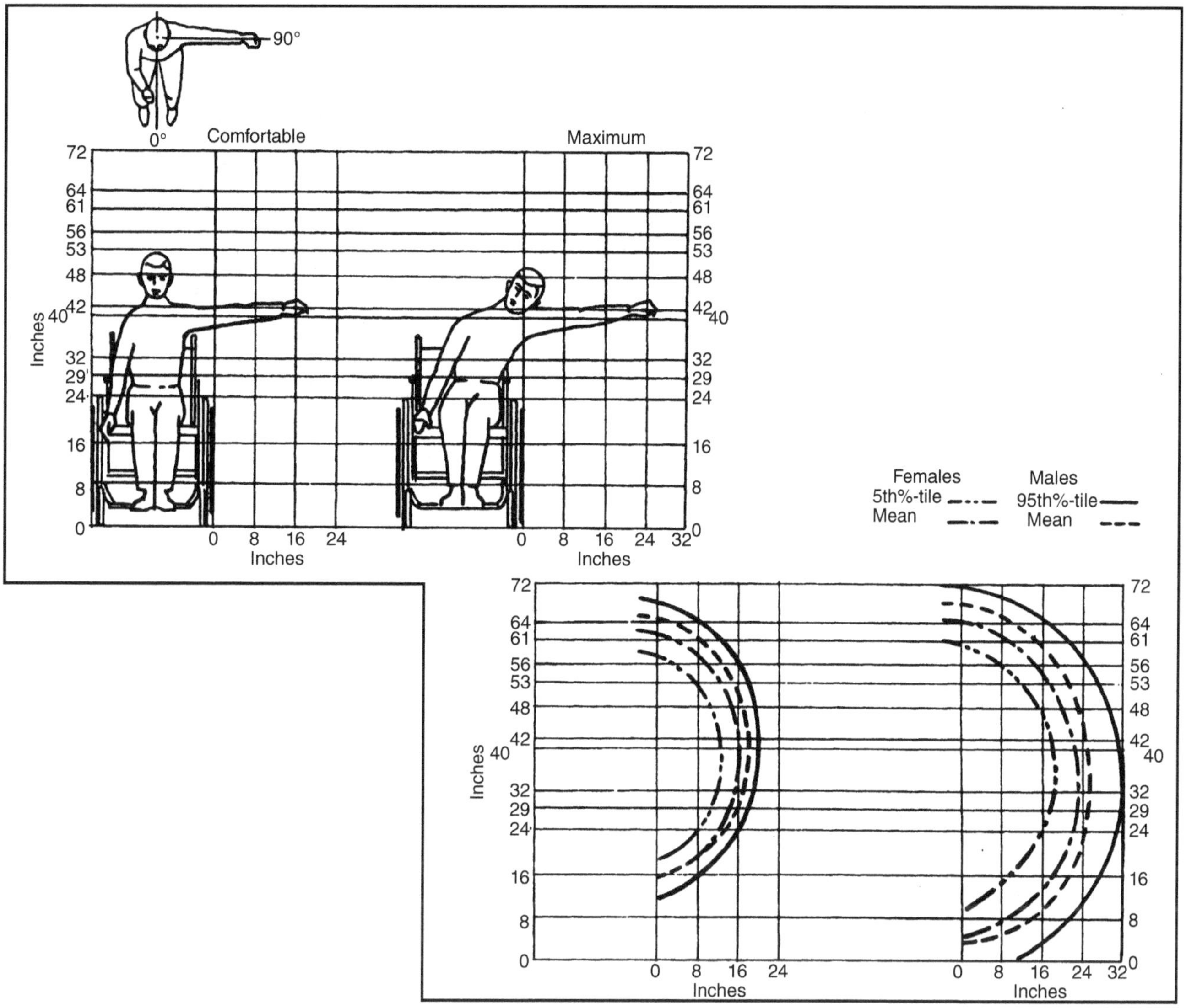

Figure 34.23 Lateral reach (adults).

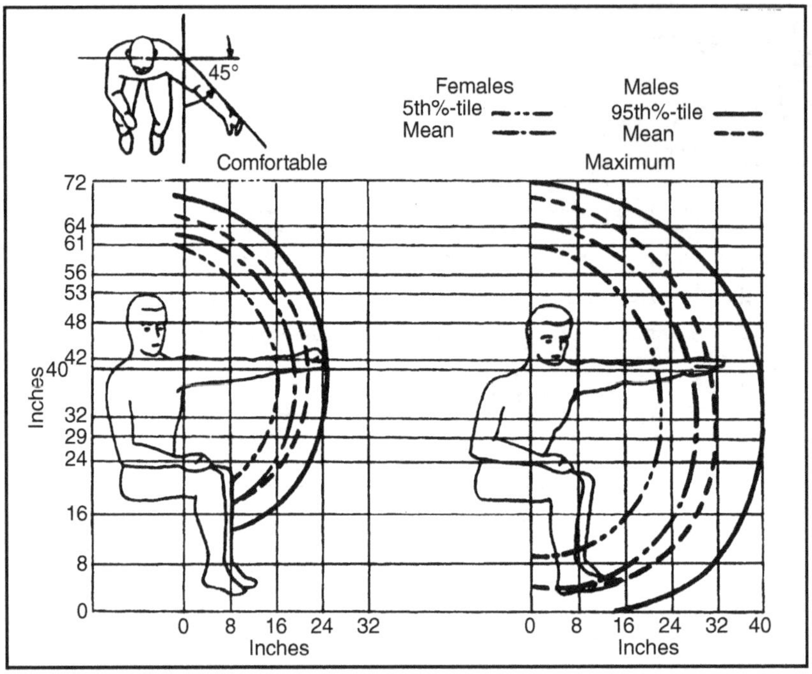

FIGURE 34.24 Diagonal reach limits (adults).

TABLE 34.75 Test Population for Wheelchair Reach Data

Male	5th Percentile	Mean	95th Percentile	SD	Female	5th Percentile	Mean	95th Percentile	SD
Floor to vertex	49.1	52.4	55.7	2.0	Floor to vertex	46.9	50.4	53.8	2.1
Floor to eye	44.7	48.1	51.5	2.1	Floor to eye	42.7	46.4	50.1	2.2
Floor to shoulder	38.4	40.8	43.2	1.5	Floor to shoulder	35.4	39.3	42.2	1.8
Floor to elbow	24.9	27.3	29.7	1.5	Floor to elbow	23.2	26.7	30.2	2.1
Floor to thigh	23.3	25.4	27.5	1.3	Floor to thigh	22.8	24.7	26.6	1.2
Shoulder width	14.3	16.8	19.3	1.5	Shoulder width	13.5	15.1	16.7	1.0
Hip width	12.2	14.1	16.0	1.2	Hip width	12.6	14.3	16.6	1.8
Toe projection	3.7	5.1	6.5	0.9	Toe projection	2.6	4.1	5.6	0.9
Lower leg length	15.7	17.3	18.9	1.0	Lower leg length	14.4	16.2	18.0	1.1
Age in years		34.7			Age		32.1		

WORKING POSITIONS

Anthropometric Data Relative to Typical Working Positions[13]

Figures 34.25–34.30 and Tables 34.76–34.81 provide useful information about work-space dimensioning. However, keep in mind that these data were derived from measurements of male Air Force personnel, and although they represent a somewhat functional dimensional guide, each work space must be given specific attention in terms of its own unique requirements.

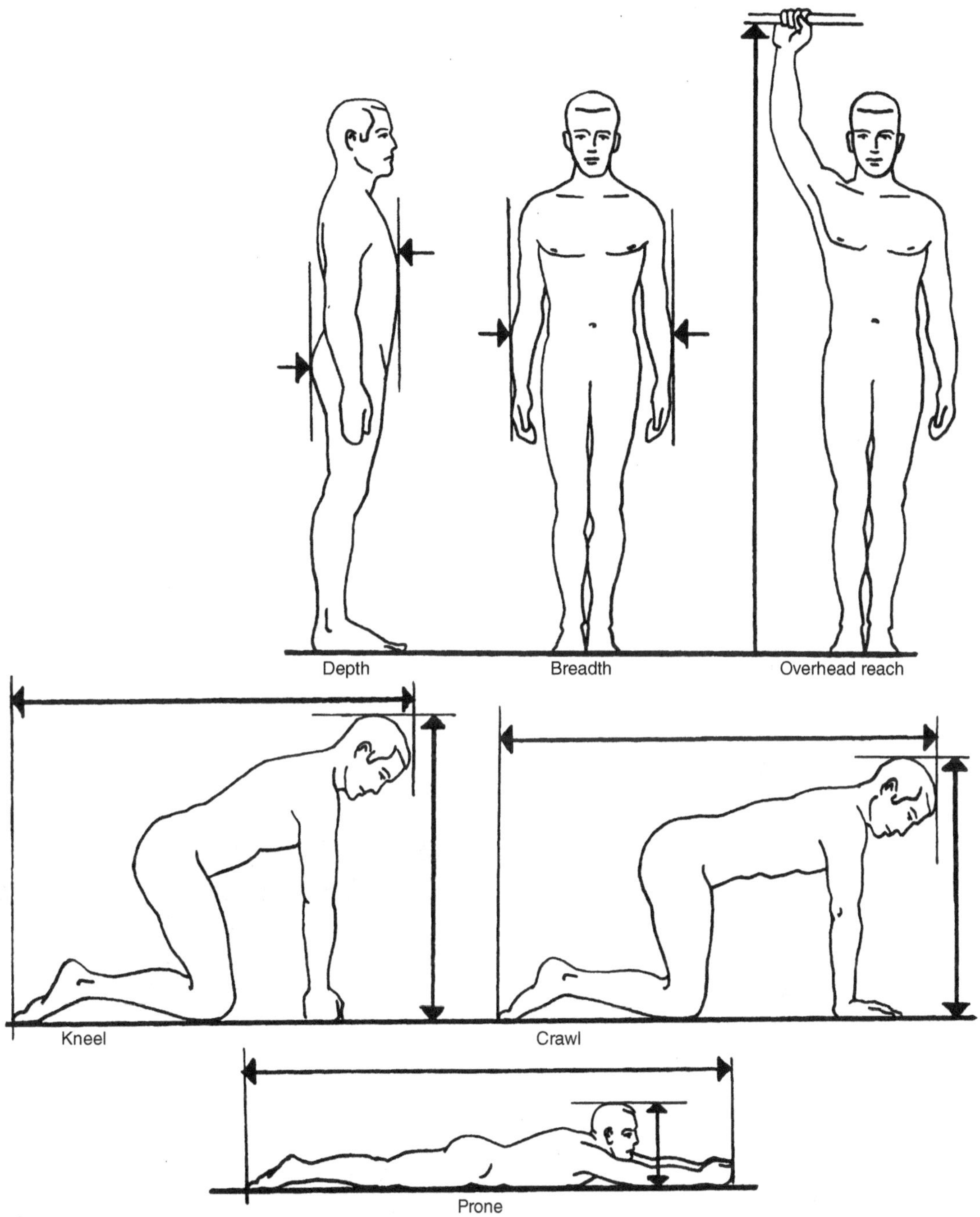

FIGURE 34.25 Typical working positions.

[13]H. T. E. Hertzberg et al., *The Anthropometry of Working Positions,* WADC Technical Report 54-520, Wright-Patterson Air Force Base, OH; and K. W. Kennedy and B. E. Filler, *Aperture Sizes and Depths of Reach for One- and Two-Handed Tasks*, AMRL-TR-S6-27, Aerospace Medical Research Laboratory, Wright-Patterson Air Force Base, OH, 1966.

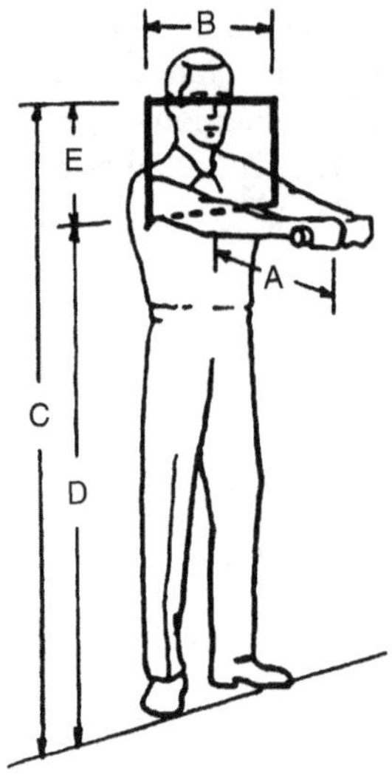

FIGURE 34.26 Standing, forward reach (both arms).

FIGURE 34.29 Seated forward reach (both arms).

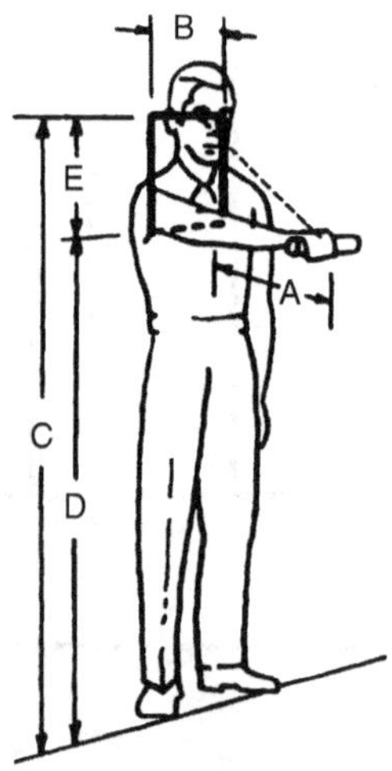

FIGURE 34.27 Standing forward reach (preferred arms).

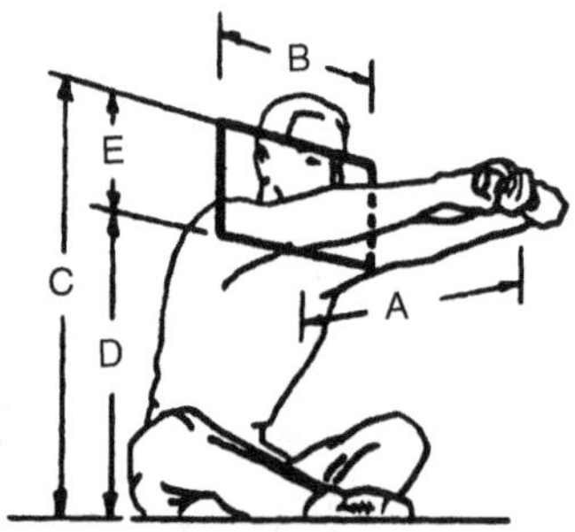

FIGURE 34.30 Cross-legged seated, forward reach (both arms).

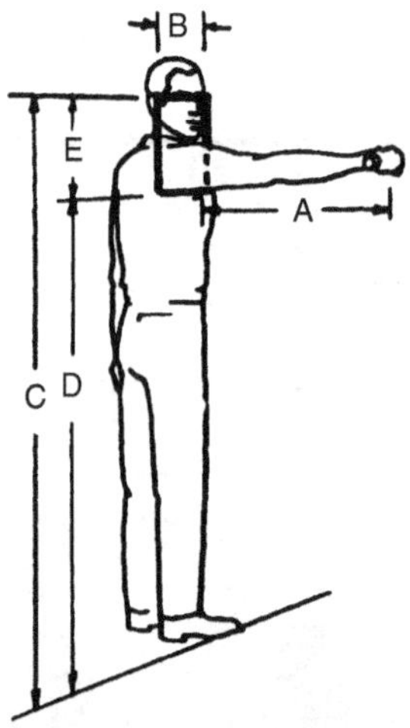

FIGURE 34.28 Standing, lateral reach (preferred arms).

TABLE 34.76 Typical Working Position Dimensions (Male)

Measurement	Number of Subjects	5th Percentile	Mean	95th Percentile	SD
Maximum body depth	118	10.1	11.5	13.0	0.88
Maximum body breadth	40	18.8	20.9	22.8	1.19
Overhead reach	40	76.8	82.5	88.5	3.33
Kneeling height	40	29.7	32.0	34.5	1.57
Kneeling length	40	37.6	43.0	48.1	3.26
Crawling height	40	26.2	28.4	30.5	1.30
Crawling length	40	49.3	53.2	58.2	2.61
Prone height	40	12.3	14.5	16.4	1.28
Prone length	40	84.7	90.1	95.8	3.41

TABLE 34.77 Standing, Forward Reach (Both Arms)

	Percentiles				
	5th	25th	50th	75th	95th
A. Depth of reach Range: 17.50 to 25.25 SD: 1.50	19.25 in.	21.00 in.	22.25 in.	22.75 in.	24.50 in.
B. Breadth of aperture Range: 15.00 to 20.25 Mean: 17.69 SD: 1.19	15.50 in.	17.00 in.	17.75 in.	18.50 in.	19.50 in.
C. Floor to top of aperture Range: 58.75 to 70.50 SD: 2.34 D.	61.00 in.	63.50 in.	65.25 in.	66.50 in.	69.00 in.
D. Floor to bottom of aperture Range: 51.25 to 61.75 Mean: 56.09 SD: 2.05	52.25 in.	54.75 in.	56.00 in.	57.25 in.	59.00 in.
E. Vertical dimension of aperture	(1)	(1)	(1)	(1)	(1)

(1) 16.75 in.

TABLE 34.78 Standing, Forward Reach (Preferred Arm)

	Percentiles				
	5th	25th	50th	75th	95th
A. Depth of reach Range: 19.50 to 27.50 Mean: 23.61 SD: 1.82	20.25 in.	22.25 in.	23.75 in.	25.00 in.	26.75 in.
B. Breadth of aperture: 12.00					
C. Floor to top of aperture Range: 58.25 to 70.50 Mean: 64.88 SD: 2.36	61.00 in.	63.25 in.	65.00 in.	66.25 in.	69.00 in.
D. Floor to bottom of aperture Range: 51.25 Mean: 56.09 SD: 2.05	52.25 in.	54.75 in.	56.00 in.	57.25 in.	59.00 in.
E. Vertical dimension of aperture	(1)	(1)	(1)	(1)	(1)

(1) 16.75 in.

TABLE 34.79 Standing, Lateral Reach (Preferred Arm)

	Percentiles				
	5th	25th	50th	75th	95th
A. Depth of reach Range: 21.75 to 28.63 Mean: 24.65 SD: 1.51	22.00 in.	23.50 in.	24.75 in.	25.75 in.	26.75 in.
B. Breadth of aperture 10.00					
C. Floor to top of aperture Range: 58.25 to 70.00 Mean: 64.70 SD: 2.32	60.75 in.	63.25 in.	64.25 in.	66.00 in.	68.75 in.
D. Floor to bottom of aperture Range: 51.25 to 61.75 Mean: 56.09 SD: 2.05	52.25 in.	54.75 in.	56.00 in.	57.25 in.	59.00 in.
E. Vertical dimension of aperture	(1)	(1)	(1)	(1)	(1)

(1) 16.5 in.

TABLE 34.80 Seated, Forward Reach (Both Arms)

	Percentiles				
	5th	25th	50th	75th	95th
A. Depth of reach Range: 14.00 to 23.50 Mean: 18.26 SD: 2.15	15.00 in.	16.50 in.	17.75 in.	19.50 in.	22.25 in.
B. Breadth of aperture Range: 13.50 to 18.75 Mean: 16.12 SD: 1.25	13.75 in.	15.25 in.	16.00 in.	17.00 in.	18.25 in.
C. Floor to top of aperture Range: 39.25 to 51.00 Mean: 43.25 SD: 2.05	19.75 in.	41.75 in.	43.00 in.	44.25 in.	46.50 in.
D. Floor to bottom of aperture[†] Range: 32.50 to 41.75 Mean: 36.59 SD: 1.59	34.25 in.	35.50 in.	36.50 in.	37.50 in.	39.00 in.
E. Vertical dimension of aperture	(1)	(1)	(1)	(1)	(1)

(1) 12.25 in.

TABLE 34.81 Cross-Legged Seated, Forward Reach (Both Arms)

	PERCENTILES				
	5TH	25TH	50TH	75TH	95TH
A. Depth of reach Range: 13.50 to 22.25 Mean: 17.08 SD: 1.91	13.75 in.	15.75 in.	16.75 in.	18.25 in.	20.00 in.
B. Breadth of aperture Range: 13.50 to 18.50 Mean: 15.89 SD: 1.54	13.75 in.	14.75 in.	16.00 in.	16.75 in.	17.75 in.
C. Floor to top of aperture Range: 22.25 to 30.50 Mean: 25.30 SD: 1.54	22.75 in.	24.25 in.	25.25 in.	26.25 in.	28.00 in.
D. Floor to bottom of aperture Range: 17.00 to 23.25 Mean: 19.23 SD: 1.19	17.00 in.	18.50 in.	19.25 in.	20.00 in.	21.25 in.
E. Vertical dimension of aperture	(1)	(1)	(1)	(1)	(1)

(1) 11.0 in.

Selected Working-Position Data for U.S. Army Men and Women (Wearing Typical Army Clothing)[14]

Standing with heels 23 cm apart and toes 15 cm from wall; arms extended overhead with fists touching and against wall; first phalanges horizontal. Measured from floor to highest point on first phalanges. (See Fig. 34.31.)

FIGURE 34.31 Overhead reach height.

[14]Figures 34.31–34.41 refer to position specifications and values in Table 34.82.

Standing with heels 23 cm apart and toes 15 cm from wall; arms extended overhead with fists touching and against wall; first phalanges horizontal. Measured horizontally across arms or shoulders, whichever is wider. (See Fig. 34.32.)

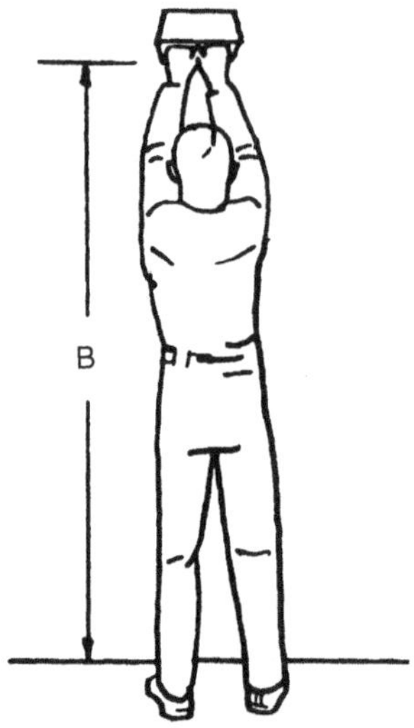

FIGURE 34.32 Overhead reach breadth.

Standing with feet 30 cm apart; bending over and placing palms of the hands on kneecaps; elbows and knees locked; looking forward; head tilted as far back as possible. Measured from floor to top of head. (See Fig. 34.33.)

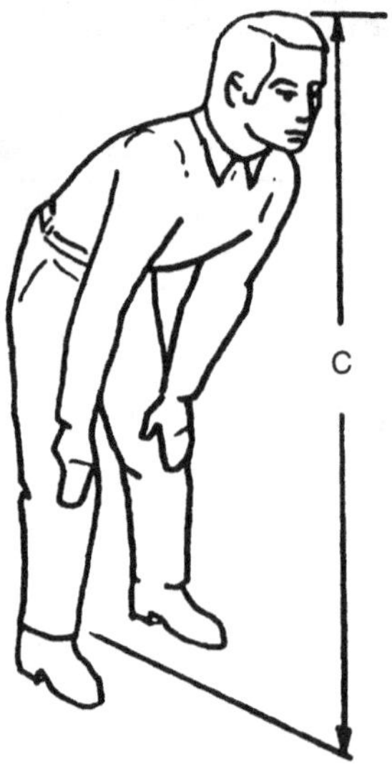

FIGURE 34.33 Bent torso height.

Standing with feet 30 cm apart; bending over and placing the palms of the hands on kneecaps; elbows and knees locked; looking forward; head tilted as far back as possible. Measured as maximum horizontal distance across shoulders. (See Fig. 34.34.)

Figure 34.34 Bent torso breadth.

Kneeling with toes extended and lightly touching rear wall; torso erect with arms hanging loosely at sides. Measured from floor to top of head. (See Fig. 34.35.)

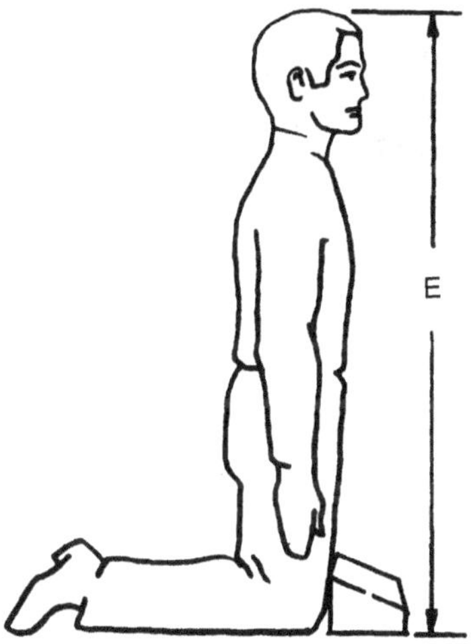

Figure 34.35 Kneeling height.

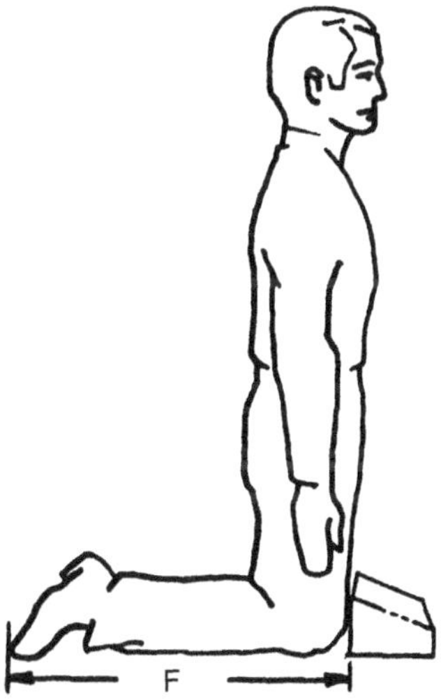

Figure 34.36 Kneeling leg length.

Kneeling with toes extended and lightly touching rear wall; torso erect with arms hanging loosely at sides. Measured from wall to anterior portion of both knees. (See Fig. 34.36.)

Figure 34.37 Overhead reach sitting.

Sitting erect; right side against wall; right arm extended upward with palm flat against wall and fingers extended. Measured from sitting surface to tip of middle finger. (See Fig. 34.37.)

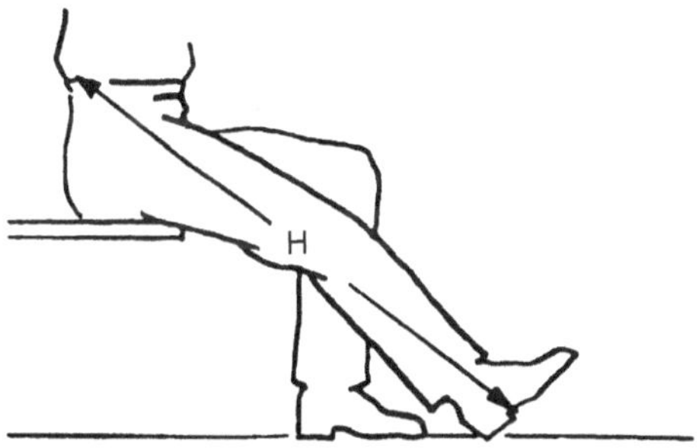

Figure 34.38 Functional leg length.

Sitting erect on edge of chair; right leg extended forward with knee straightened. Measured from heel along axis of leg to posterior waist. (See Fig. 34.38.)

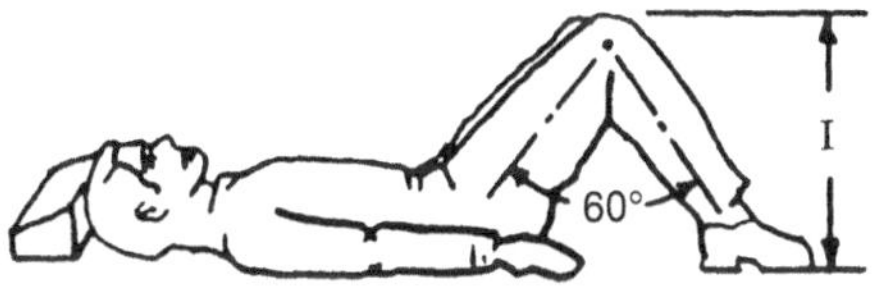

Figure 34.39 Bent knee height, supine.

Lying supine; knees raised until angle between upper and lower legs approximates 60°; toes lightly touching wall. Measured from to highest point on knees. (See Fig. 34.39.)

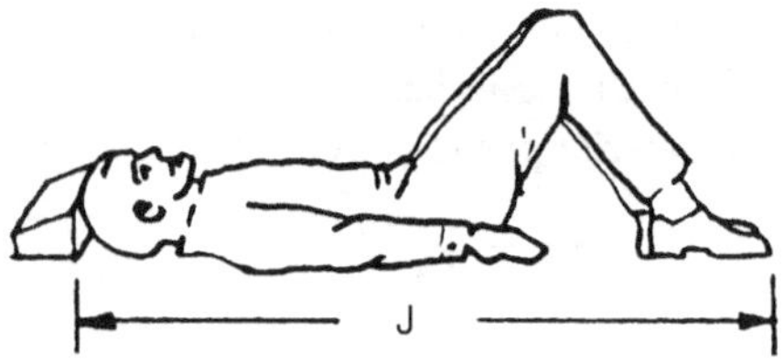

FIGURE 34.40 Horizontal length, knees bent.

Lying supine; knees raised until angle between upper and lower legs approximates 60°; toes lightly touching wall. Measured from wall to top of head. (See Fig. 34.40.)

Standing erect; looking straight ahead; right shoulder extended as far forward as possible while back of left shoulder firmly against wall; arm horizontal. Measured from wall to tip of index finger. (See Fig. 34.41.)

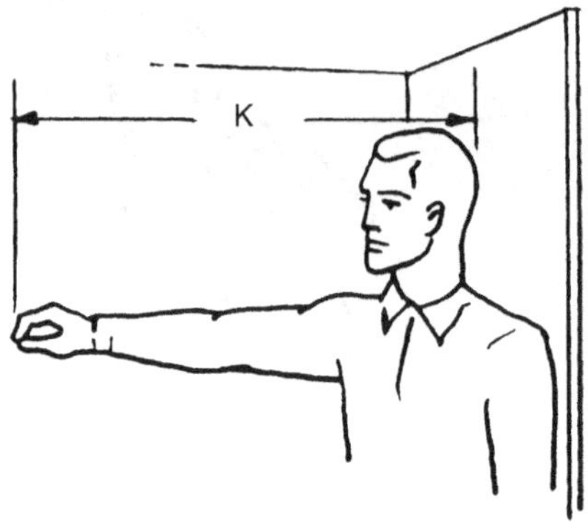

FIGURE 34.41 Functional reach, extended.

Some anthropometric data for U.S. male and female personnel are given in Table 34.82.

TABLE 34.82 Anthropometric Data for U.S. Male and Female Personnel: Common Working Positions, Percentile Values

	5th Percentile		95th Percentile	
	Men	Women	Men	Women
A. Overhead reach	78.9 in.	73.0 in.	90.8 in.	84.7 in.
B. Overhead reach, breadth	13.9 in.	12.4 in.	16.5 in.	14.9 in.
C. Bent torso height	49.4 in.	44.4 in.	59.0 in.	54.6 in.
D. Bent torso breadth	16.1 in.	14.5 in.	19.0 in.	17.1 in.
E. Kneeling height	48.0 in.	45.1 in.	53.9 in.	51.3 in.
F. Kneeling leg length	25.2 in.	23.3 in.	29.7 in.	27.8 in.
G. Overhead reach, sitting	50.3 in.	46.2 in.	57.9 in.	54.9 in.
H. Functional leg length	43.5 in.	39.2 in.	50.3 in.	46.7 in.
I. Bent knee height, supine	17.6 in.	16.3 in.	21.1 in.	19.5 in.
J. Horizontal length, knee bent	59.4 in.	55.2 in.	68.1 in.	64.5 in.
K. Functional reach	33.2 in.	28.9 in.	39.8 in.	36.5 in.

Source: MIL-STD-1472B, Notice—2, 10 May 1978.

SELECTED FUNCTIONAL DIMENSIONS[15]

Few definitive studies have been conducted for the purpose of establishing the functional dimensions of the human as he or she pursues day-to-day living and working activities. One study of shipboard habitability requirements (using 5th-percentile males) produced the following general dimensional criteria shown in Fig. 34.42.

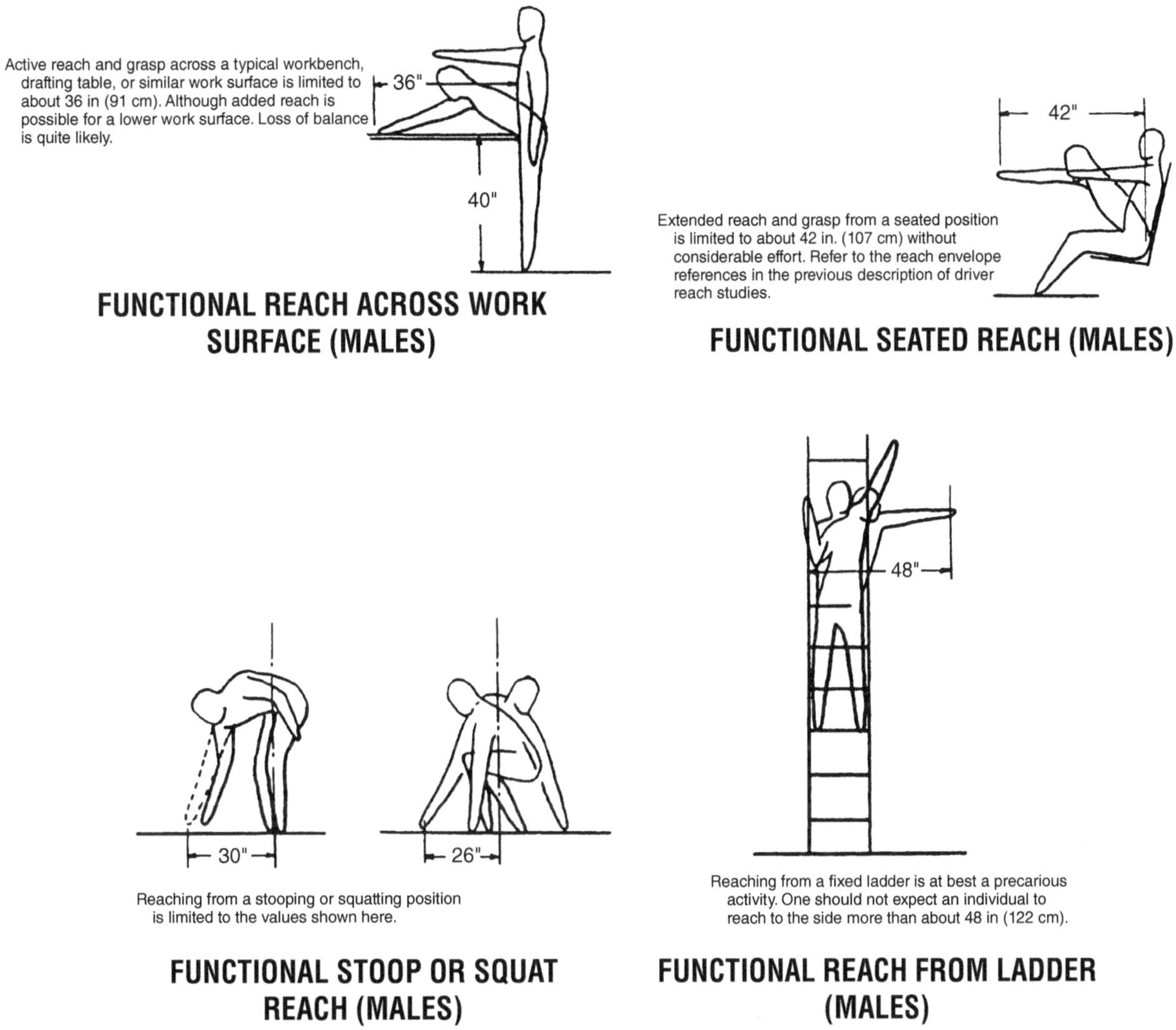

FIGURE 34.42 Functional reach achievable by 95 percent of an all-male population.

[15] W. E. Woodson et al., *Habitability in Shipboard Living Spaces*, Vol. II, *Supporting Data: General*, prepared for the U.S. Naval Ships Engineering Center, Hyattsville, MD, 1972.

Functional Clearance Requirements (See Fig. 34.43)

Berthing space should be large enough to permit stretching the legs, raising the arms overhead, and turning easily without rolling out of bed. The dimensions shown are suggested.

The clearance between bunks should be sufficient to allow the individual to sit up. There should be enough clearance in front of bunks so that the individual can make up the bed without bumping into the opposite wall.

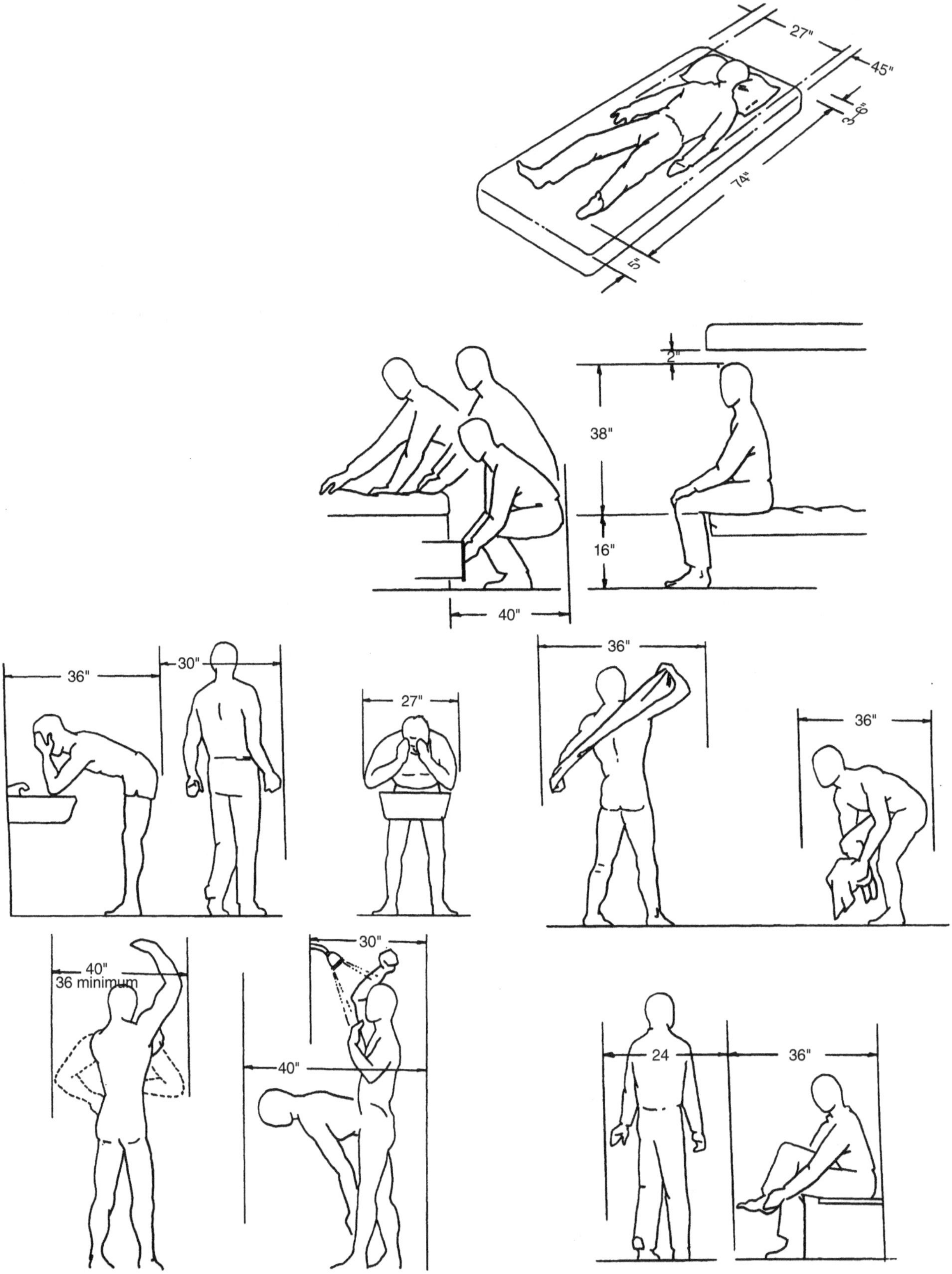

FIGURE 34.43 Functional clearance requirements.

SPACE ANTHROPOMETRICS

The human body floats in zero gravity. The relaxed body assumes a slightly stooped position called the "neutral body posture." People do not sit, stand, or lie down in zero gravity. It requires muscular effort to assume these positions. Figure 34.44 shows the body joint angles in the neutral body posture. NASA measured the data in space. The angles are mean angles and standard deviations about the mean. Space vehicle designs should accommodate this position. This includes workstations, sleeping quarters, and bathroom facilities. Restraints, such as foot straps, are usually used to hold the astronaut in one place.

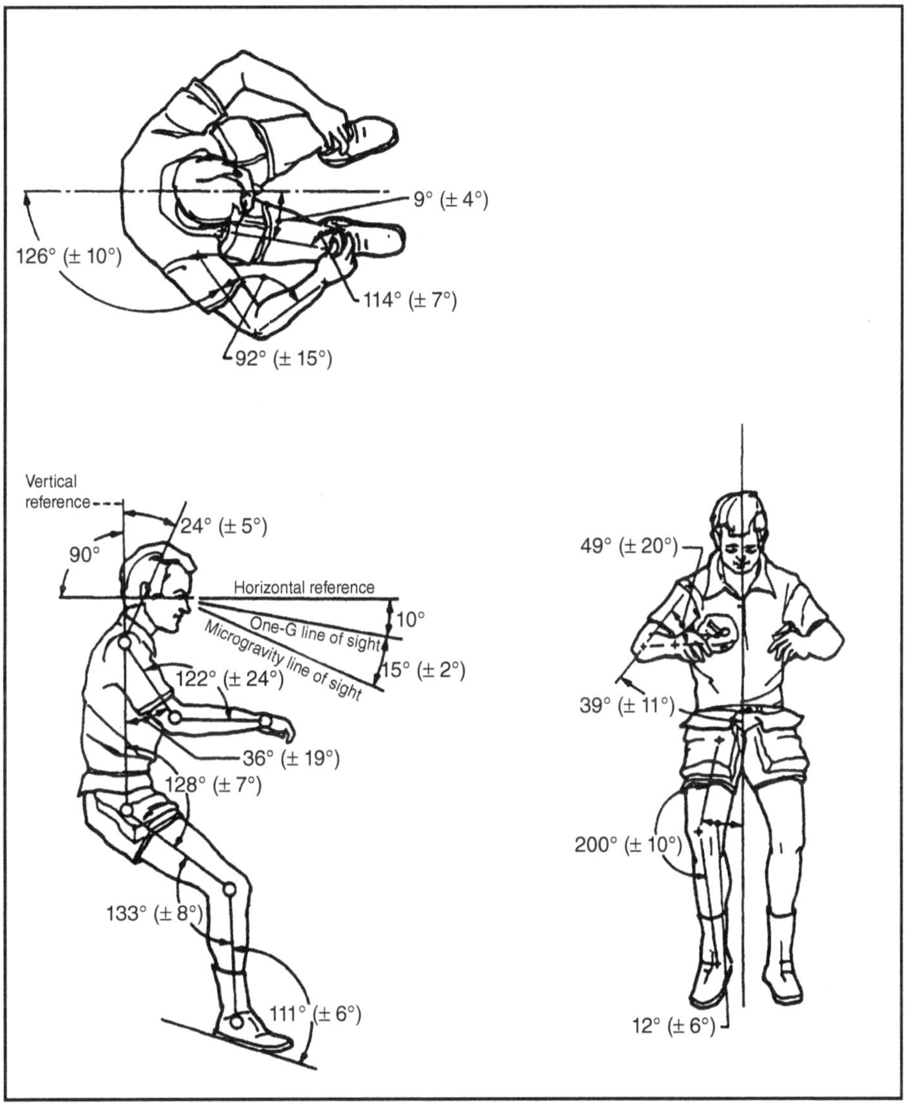

Figure 34.44 Neutral body posture in zero gravity space. (From NASA-STD-3000.)

Chapter 35
Human Strength

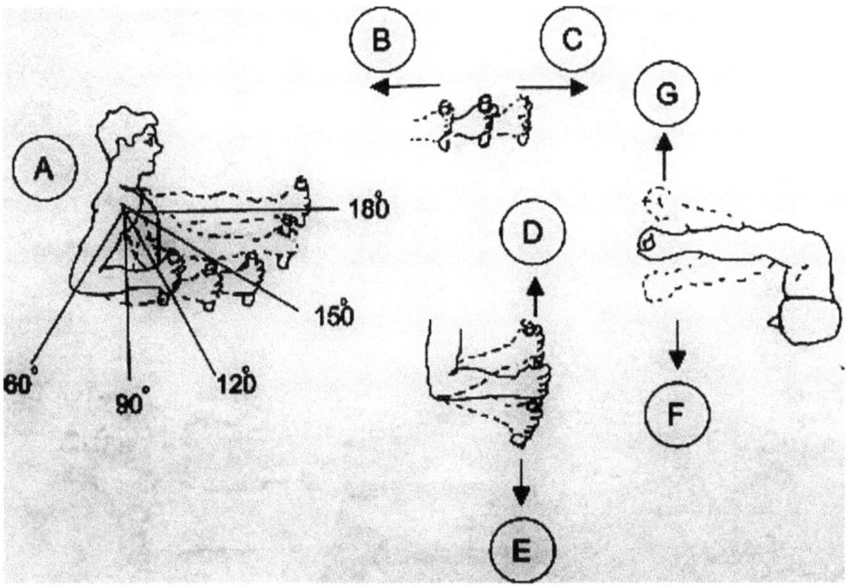

BASIC DEFINITIONS

Isometric Strength (static): The maximum force that muscles can exert when muscle length remains constant during contraction.

Isotonic Strength: The maximum force that muscles can exert when muscle length changes.

Concentric Force: The force exerted when the muscle is shortened against an external resistance.

Eccentric Force: The force exerted when the muscle lengthens passively against an external force.

Effort: Physiological strain, both static and dynamic.

Work: Dynamic effort (i.e., force times displacement).

Endurance and Static Effort: The ability to continue work or exert force. (See Fig. 35.1.)

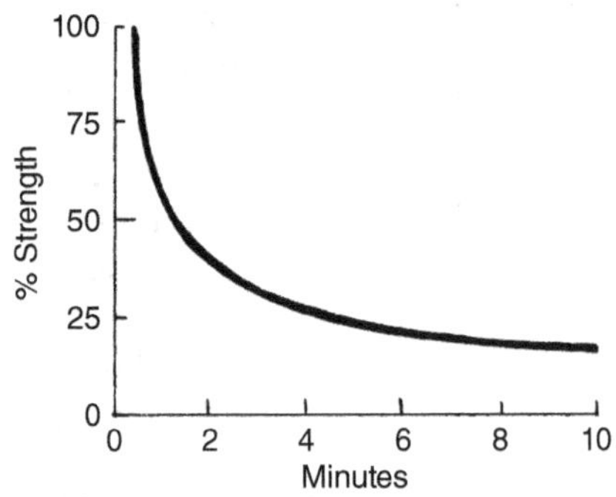

Figure 35.1 Typical endurance time in relation to force requirements.

MUSCLE FORCE

Muscle force is a function of the following.

Muscle Tension

Muscle tension is maximum when the length of the muscle is greatest and momentarily there is no change in length. Muscle force decreases as the muscle shortens and as its rate of shortening increase.

Mechanical Advantage

Power is applied at the point of muscle attachment (i.e., the long bones are the lever arms, and the joints are the fulcrums). Thus, for example, when muscle force of extension is applied at the elbow, power is greatest when the elbow is flexed. However, optimum mechanical advantage occurs at the midpoint of full elbow travel.

Optimum mechanical advantage more than compensates for the shortened muscle, therefore providing maximum strength at the midpoint. Human muscles in maximum contraction can exert considerable force [as much as 1000 lb (453 kg)], but such forces cannot be fully utilized directly because all muscles work at some mechanical disadvantage, thus reducing output but increasing rate of movement.

Age and Sex in Relation to Strength

In general, adult females are only about two-thirds as strong as adult males. In terms of age, individuals have maximum strength between the ages of 30 and 40. Usually there is a rapid development in strength between the ages of 13 and 19, and this development slows somewhat between the ages of 20 and 25. This is followed by a slower increase in strength to the maximum between 25 and 30 years of age. People begin to lose about 10 percent of their strength by age 40, 15 percent by age 50, 20 percent by age 60, and at least 25 percent by age 65.

Effect of Body Build

As a rule, people with larger body builds have more strength, although less powerfully built individuals may require less oxygen for a given task requiring strength. Slender persons often are best at performing rapidly fatiguing tasks involving strenuous exercise. Physique does not necessarily correlate with the ability to perform moderate exercise. Normal persons usually show a 30–50 percent increase in strength after about 12 weeks of training.

Strength is affected by health, diet, and the use of drugs, and strength often varies with diurnal conditions (e.g., people usually have maximum strength at about midmorning).

Effect of the Thermal Environment

Heat affects strength adversely [e.g., when temperatures exceed 85°F (29°C), especially under conditions of high humidity]. In general, however, low temperature has little effect except in terms of body mobility and finger dexterity. When individuals become acclimated to a hot environment, they generally gain back a great deal of their normal strength.

Effect of Altitude

Although there usually is no loss in handgrip strength as one goes from sea level to altitudes of about 12,000 ft (3660 m), endurance for muscular activity declines progressively above about 8000 ft (2440 m). Strength endurance declines to about one-half by 20,000 ft (6100 m). Partial handgrip paralysis begins at around 24,000 ft (7320 m), and an individual reaches the point of unconsciousness at about 26,000 ft (7930 m). Although some acclimatization occurs, this requires several weeks or months and therefore is of little practical consequence. Lack of oxygen is a major factor in the decrease of muscle efficiency.

Effect of Acceleration

Although accelerations up to 5*g* do not affect strength, endurance is affected. Arm movements are effective up to about 6*g*, and wrist and finger movements are effective up to about 12*g*. Practical considerations include the following:

Forces acting against the direction of acceleration are decreased.

Forces acting with the direction of acceleration are increased.

Forces acting perpendicular to the direction of acceleration are least affected.

Strength in Relation to Emotional Condition

Strength may increase under stress (i.e., fear, panic, rage, or even excitement). However, skill and accuracy generally are degraded. It has been demonstrated that, with hypnosis, pull with forearm flexion may increase as much as 26 percent. Increases may also occur when the maximum effort is preceded by a pistol shot or when the subject shouts during the effort.

Generally speaking, psychological rather than physiological factors determine maximum strength in the "real world." It has been noted that white-collar workers generally are about 10–20 percent weaker than manual or blue-collar workers. The implication is that the latter are used to a rougher and more strength-demanding environment.

Body Position

When individuals are not restricted in terms of body position and are provided with appropriate supporting and/or anchoring facilities, they generally will assume a position from which they can apply their maximum force capability. However, this does not necessarily mean that this is always the best position for maintaining lesser force applications for extended periods of time.

Since there is usually a reciprocal response during force applications (e.g., lifting, pushing, and pulling), it is important to provide appropriate support and anchoring conditions, such as a flat, level floor or deck; a solid, stable seat and seat backrest; or a footrest.

Limb Position

Both limb position and direction of force application are important variables in determining the amount of force an individual can apply. Handgrip forces generally are greater if the gripping task is close to the individual's body than if it is at arm's length. Arm strength is greater if the individual can push against a backrest or footrest. Maximum leg force occurs when the individual's knee is slightly bent (in a seated position with the leg just "short" of maximum extension and with the ball of the foot at approximately the same level as the individual's buttocks). Maximum arm force occurs when the force can be applied approximately at shoulder level. For the seated individual, pull force is greatest when the object is positioned at nearly maximum arm length; push force is greatest when the object is positioned at about half the full arm extension.

Lifting capabilities depend on the size, shape, and gripping characteristics of the package being lifted and on the distribution of weight within the package. For example, a package that is too large to allow the individual to wrap his or her arms around it, to grip it securely, or to offset poorly distributed weight (the package's c/g is too far from the individual's own c/g) cannot be lifted or carried without the probability that the individual will drop it, will lose his or her balance, and/or will suffer strain and possibly some semipermanent or permanent injury.

Selected strength capabilities for various lifting and force-application situations are provided on the following pages. Since some of the data pertain only to adult males, a good rule of thumb for applying the guidelines to females is that females generally are about one-third weaker than males.

GENERAL HUMAN STRENGTH-LIMIT EXPECTATIONS

Since people do not always use the best method for applying force, one should not plan on using maximum capability values for design. The values shown in Fig. 35.2 are suggested by the military for various common force-application activities (reference MIL-STD-1472G).

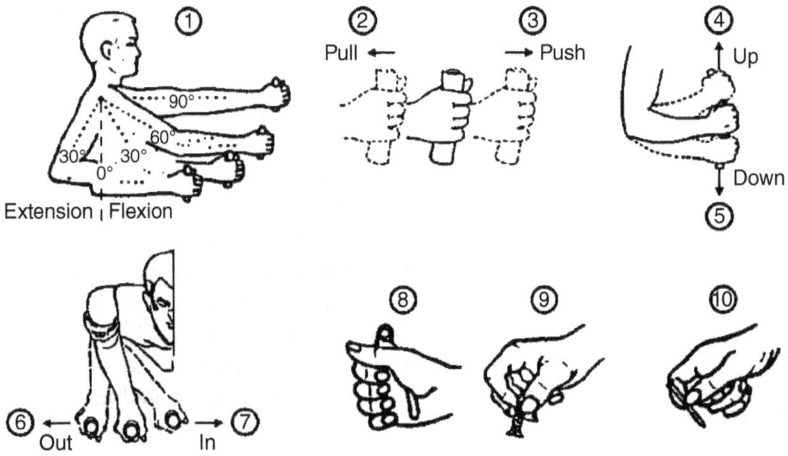

Arm strength in Newtons (pounds of force)														
(1)	(2)		(3)		(4)		(5)		(6)		(7)	
Degree of shoulder flexion/extension	Pull		Push		Up		Down		Out		In	
	L	R	L	R	L	R	L	R	L	R	L	R
90 deg (flexion)	222 (50)	231 (52)	187 (42)	222 (50)	40 (9.0)	62 (14)	58 (13)	76 (17)	36 (8.0)	62 (14)	58 (13)	89 (20)
60 deg (flexion)	187 (42)	249 (56)	133 (30)	187 (42)	67 (15)	80 (18)	80 (18)	89 (20)	36 (8.0)	67 (15)	67 (15)	89 (20)
30 deg (flexion)	151 (34)	187 (42)	116 (26)	160 (36)	76 (17)	107 (24)	93 (21)	116 (26)	45 (10)	67 (15)	89 (20)	98 (22)
0 deg (neutral)	142 (32)	165 (37)	98 (22)	160 (36)	76 (17)	89 (20)	93 (21)	116 (26)	45 (10)	71 (16)	71 (16)	80 (18)
30 deg (extension)	116 (26)	107 (24)	98 (22)	151 (34)	67 (15)	89 (20)	80 (18)	89 (20)	53 (12)	76 (17)	76 (17)	89 (20)

Hand and pinch strength Newtons (pounds of force)				
	(8)		(9)	(10)
	Hand grip		Palmer pinch grip	Tip pinch grip
	L	R	(Thumb pad to index & middle finger pads)	(Thumb tip to index finger)
Momentary hold	250 (56)	260 (59)	60 (13)	60 (13)
Sustained hold	145 (33)	155 (35)	35 (8.0)	35 (8.0)

FIGURE 35.2 Strength.

Although people can lift more than the values shown in the figure, one should not expect more from an operator (i.e., these are 5th-percentile values when the individual tries to lift using only back-muscle power).

The lower illustrations in Fig. 35.2 represent maximum force expectations for adult males. For females, multiply the values by 2/3.

FORCE LIMITS

The sketches and force values in Figs. 35.3–35.8 represent approximate upper limits for situations involving typical civilian user populations (male and female adults).

Two-Hand Scissor-Action Force

The suggested force limit is based on the assumption that the two hands are about 6 to 8 in. (15 to 20 cm) apart. (See Fig. 35.3.) If the action takes place above or below normal elbow level, force capabilities diminish rapidly.

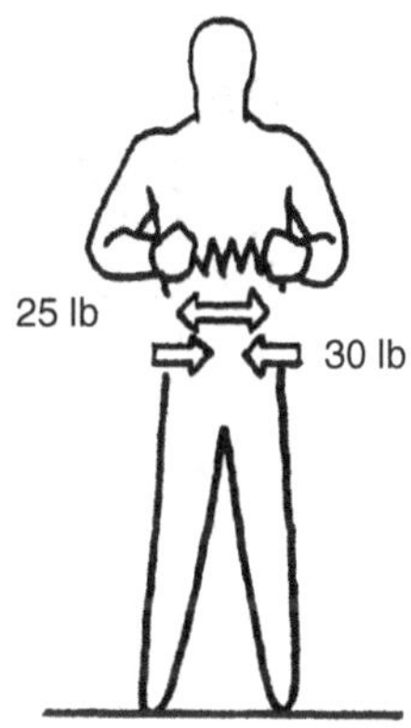

Figure 35.3

Pedal Force from a Seated Position

In the typical driving scenario, a driver often is surprised and must apply brake force quickly. (See Fig. 35.4.) Thus the suggested upper force limit accounts not only for the weaker members of the user population but also for the psychological disadvantages of being faced with an emergency. When power assist can be provided, this force can and should be considerably reduced.

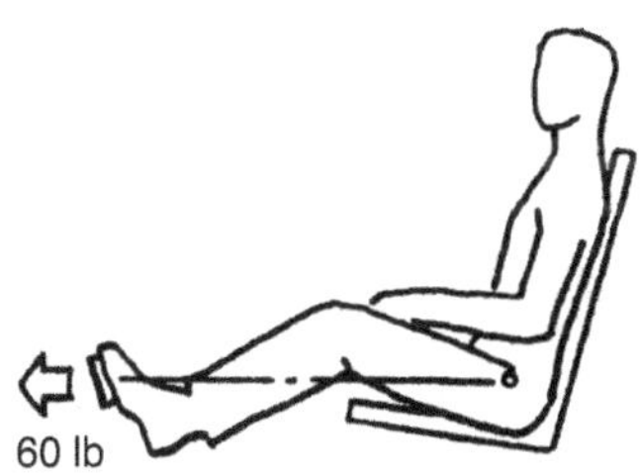

Figure 35.4

Lever or Joystick Forces from a Seated Position

Note that fore-aft force capacities are greater than lateral capacities because of the support advantages provided by a seat and toe board or rudder pedals. (See Fig. 35.5.)

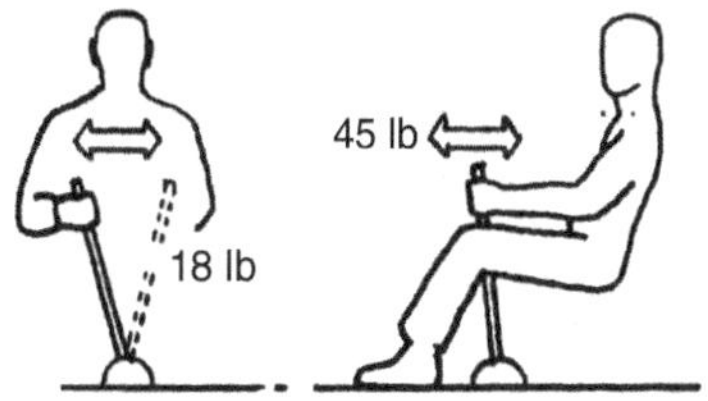

Figure 35.5

Vehicle Steering Forces

For vehicles operated by both males and females, the upper force-limit objective should be 25 lb (11.3 kg) or less in order to account for an emergency response (e.g., failure of the power steering). For heavier vehicles, a higher force may be acceptable, but the wheel angle and diameter should be increased as indicated in Fig. 35.6.

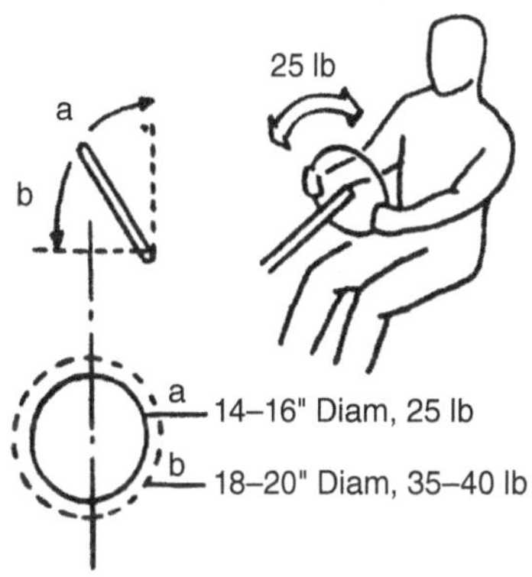

Figure 35.6

Standing Force Limits for Lever-Type Controls

Although slightly higher values could be expected if both hands were used, the values are representative of the upper design limit, assuming that the operator has no special aids to help in retaining his or her primary body position (i.e., something to push against). (See Fig. 35.7.)

The lever is assumed between the waist and shoulder height above the floor.

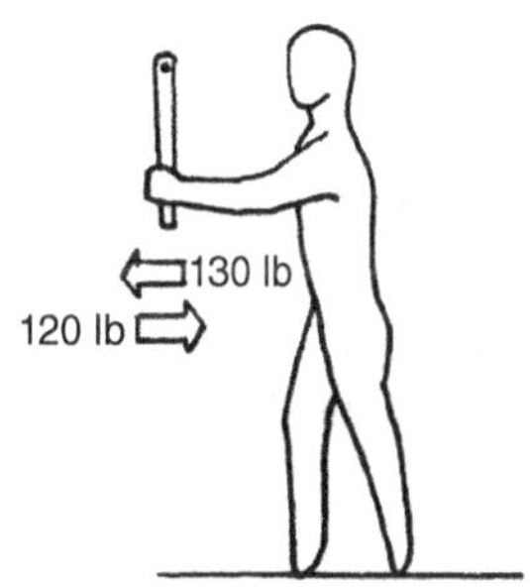

Figure 35.7

Standing Force Limits for Hand Cranking

The best hand-cranking positions are illustrated in Fig. 35.8. The value shown assumes a wheel diameter between about 4.5 and 8.0 in. (11.4 and 20 cm). Cranking efficiency and capability to apply maximum forces diminish rapidly when wheel diameters are either smaller or greater than noted. *Note that force limits should be lower when the cranking plane is parallel to the operator's own frontal plane.* If hand cranks are located overhead, the upper force-limit values should be reduced by 40–50 percent.

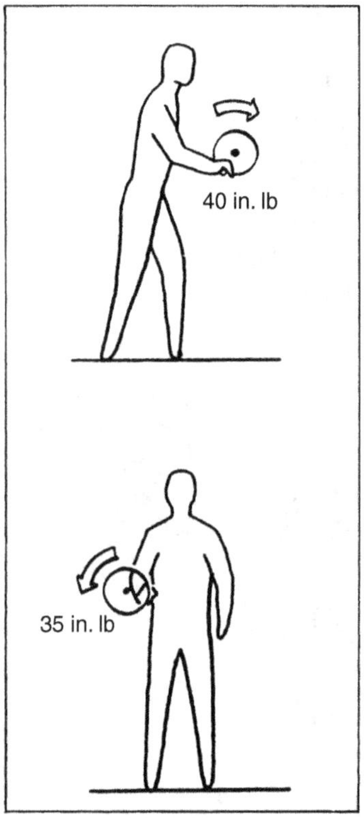

FIGURE 35.8

Note: Physical activity incurs some physical debt as a consequence of the physiological work load and/or the work environment. If the work rate is low enough, a muscle group can function almost indefinitely. However, at high rates muscles soon fatigue and, in some cases, may cease to function effectively altogether. The values illustrated in the figure apply to reasonably slow and infrequent manipulation of the particular control.

STRENGTH IN RELATION TO CONTROL OPERATION

Hand Controls[16] (See Fig. 35.9 and Table 35.1, Fig. 35.10 and Tables 35.2 and 35.3, Fig. 35.11 and Tables 35.4 and 35.5, Fig. 35.12 and Tables 35.6 and 35.7)

[16] C. T. Morgan et al. (eds.), *Human Engineering Guide to Equipment Design,* McGraw-Hill Book Company, New York, 1963, pp. 560, 562–567.

Hand Strength

Control knob or switch or potentiometer operating characteristics should include viscous damping as opposed to static resistance. A certain amount of resistance is desirable (approximately 2.0 in./oz in order to minimize the possibility of inadvertent knob movement).

1. Generally, more precise knob movement is possible with larger knob diameters [1.5–3.0 in. (3.8–7.6 cm)] because more fingers can be brought in contact with the knob surface. The knob torque should not exceed about 4 in./oz.

2. For high-torque applications, select knob diameters between 2 and 3 in. (5.1 and 7.6 cm) or 3.25 in. (8.2 cm) maximum. Maximum torque should not exceed 15 ft/lb. Table 35.8 provides approximate values for knobs of different diameters and depths. Instead of serrations, use knurling on the edges of high-torque knobs for maximum gripping. The edges should be slightly beveled so as not to injure the operator's hand.

Hand Dynamometer Strength

Although there is some disagreement on this point, hand dynamometer strength is generally considered a fairly good indicator of an individual's overall ability to apply force and thus is useful for selecting specific subject groups to test designs involving human force applications. Table 35.9 provides general dynamometer strength capabilities for various populations. Values are for the right hand only.

Handwheels (Partial Turn)

Considerably higher forces are possible for two-handed operation of handwheels with large diameters. For short turns using a vertical or horizontally mounted wheel with a diameter of, say, 20 or more in. (51 cm), a strong male adult can exert as much as 120 in./lb. The wheel column and/or rim should be mounted approximately at the operator's elbow level. Table 35.12 provides guidelines for establishing handwheel force requirements.

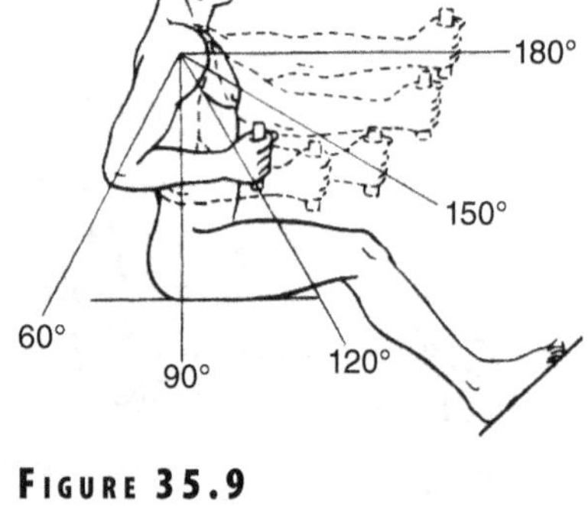

FIGURE 35.9

TABLE 35.1 Maximum Force Exerted in the Sitting Position on a Vertical Handgrip at Various Elbow Angles by the (a) Right Arm of Male College Students and (b) Left Arm of Male College Students

(a)						(b)					
Arm Strength in Sitting Position											
		Percentiles (lb)						Percentiles (lb)			
Direction of Force	Elbow Angle (deg)	5th	50th	95th	S.D.	Direction of Force	Elbow Angle (deg)	5th	50th	95th	S.D.
Push	60	34	92	150	38	Push	60	22	79	164	31
	90	36	86	154	33		90	22	83	172	35
	120	36	103	172	43		120	26	99	180	42
	150	42	123	194	45		150	30	111	192	48
	180	50	138	210	49		180	42	126	196	47
Pull	60	24	63	74	23	Pull	60	26	64	110	23
	90	37	88	13	30		90	32	80	122	28
	120	42	104	154	31		120	34	94	152	34
	150	56	122	189	36		150	42	112	168	37
	180	52	120	171	37		180	50	116	172	37
Left	60	20	52	87	19	Left	60	12	32	62	17
	90	18	50	97	23		90	10	33	72	19
	120	22	53	100	26		120	10	30	68	18
	150	20	54	104	25		150	8	29	66	20
	180	20	50	104	26		180	8	30	64	20
Right	60	17	42	82	20	Right	60	17	50	83	21
	90	16	37	68	18		90	16	48	87	22
	120	15	34	62	17		120	20	45	89	21
	150	15	33	64	18		150	15	47	113	27
	180	14	34	62	24		180	13	43	92	22
Up	60	20	49	82	18	Up	60	15	44	82	18
	90	20	56	106	22		90	17	52	100	22
	120	24	60	124	24		120	17	54	102	25
	150	18	56	118	28		150	15	52	110	27
	180	14	43	88	22		180	9	41	83	23
Down	60	20	51	89	21	Down	60	18	46	76	18
	90	26	53	88	20		90	21	49	92	20
	120	26	58	98	23		120	21	51	102	23
	150	20	47	80	18		150	18	41	74	16
	180	17	41	82	18		180	13	35	72	15

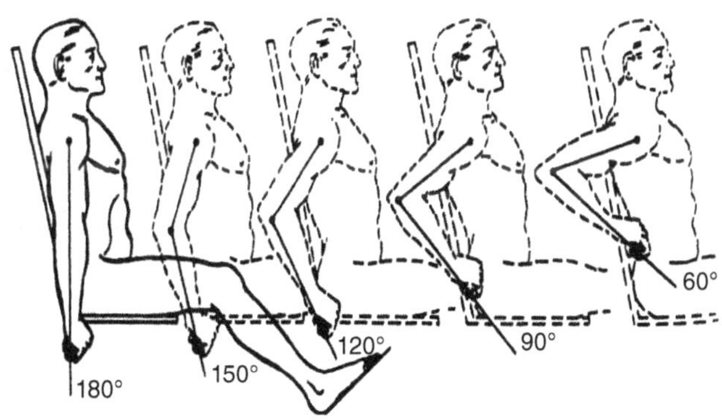

Figure 35.10

TABLE 35.2 Maximum Force Exerted in the Sitting Position with the Wrist Pronated on a Horizontal Handgrip at Various Elbow Angles by the (a) Right Arm of Male College Students and (b) Left Arm of Male College Students

ARM STRENGTH IN SITTING POSITION

(a)						(b)					
		Percentiles (lb)						Percentiles (lb)			
Direction of Force	Elbow Angle (deg)	5th	50th	95th	S.D.	Direction of Force	Elbow Angle (deg)	5th	50th	95th	S.D.
Push	60	40	94	156	36	Push	60	33	86	138	35
	90	25	65	100	24		90	27	60	93	28
	120	23	46	70	15		120	17	43	71	17
	150	18	40	66	18		150	15	37	69	18
	180	17	32	59	12		180	12	32	59	13
Pull	60	13	37	50	16	Pull	60	20	39	64	18
	90	14	32	54	13		90	17	37	65	18
	120	13	26	43	10		120	12	30	56	14
	150	12	29	48	10		150	15	32	52	13
	180	11	28	48	12		180	16	34	61	15
Right	60	19	41	72	19	Right	60	20	42	66	15
	90	12	31	64	15		90	17	38	60	12
	120	9	26	53	13		120	17	34	53	8
	150	9	21	39	11		150	17	31	54	11
	180	10	19	34	7		180	15	28	41	8
Left	60	16	48	73	18	Left	60	18	36	51	15
	90	16	39	59	15		90	11	27	54	11
	120	15	34	47	11		120	10	22	39	10
	150	18	32	45	7		150	9	23	53	16
	180	16	31	57	13		180	10	20	49	13
Up	60	23	49	79	20	Up	60	22	57	100	22
	90	28	69	112	29		90	37	77	123	24
	120	41	91	138	30		120	45	91	145	30
	150	43	99	165	38		150	58	100	159	32
	180	35	95	156	35		180	47	101	171	11
Down	60	23	81	158	35	Down	60	18	74	139	35
	90	22	83	142	35		90	23	75	136	34
	120	37	92	161	35		120	29	75	148	40
	150	40	90	154	34		150	39	79	136	29
	180	41	87	143	31		180	34	76	138	31

TABLE 35.3 Maximum Force Exerted in the Sitting Position with the Wrist Supinated on a Horizontal Handgrip at Various Elbow Angles by the (a) Right Arm of Male College Students and (b) Left Arm of Male College Students

Arm Strength in Sitting Position

(a)

Direction of Force	Elbow Angle (deg)	Percentiles (lb) 5th	50th	95th	S.D.
Push	60	34	96	172	39
	90	25	65	117	24
	120	20	43	71	17
	150	17	36	59	14
	180	12	32	58	15
Pull	60	16	51	93	25
	90	13	43	74	19
	120	11	40	63	17
	150	11	37	66	17
	180	15	39	73	19
Right	60	18	44	73	19
	90	18	39	72	24
	120	17	34	64	15
	150	15	32	60	14
	180	14	29	48	12
Left	60	13	36	70	17
	90	13	31	48	12
	120	12	30	46	11
	150	12	31	52	14
	180	10	28	44	10
Up	60	17	45	78	22
	90	21	63	107	27
	120	41	88	143	33
	150	37	103	161	40
	180	51	113	165	34
Down	60	20	59	132	35
	90	17	80	143	37
	120	29	92	148	13
	150	37	93	150	35
	180	44	87	135	32

(b)

Direction of Force	Elbow Angle (deg)	Percentiles (lb) 5th	50th	95th	S.D.
Push	60	35	89	176	42
	90	25	59	104	27
	120	15	40	80	18
	150	13	38	69	30
	180	14	30	47	10
Pull	60	23	54	87	23
	90	13	42	68	21
	120	14	40	66	18
	150	16	40	62	15
	180	17	40	70	18
Right	60	16	38	64	12
	90	12	32	46	12
	120	14	31	55	13
	150	12	32	62	15
	180	12	29	43	9
Left	60	17	42	81	20
	90	16	33	52	12
	120	14	28	45	8
	150	12	26	43	10
	180	8	27	44	10
Up	60	20	49	89	22
	90	24	75	131	29
	120	38	94	152	33
	150	44	104	164	36
	180	45	111	173	40
Down	60	20	58	138	41
	90	23	80	160	43
	120	35	84	136	33
	150	43	84	136	29
	180	36	78	124	28

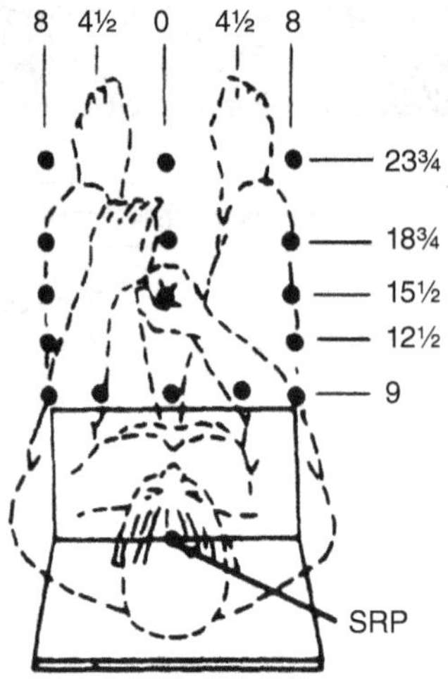

Figure 35.11

TABLE 35.4 Maximum Force Exerted in (a) Pushing an Aircraft Control Stick by the Right Arm of Male Air Force Personnel in the Sitting Position and (b) Pulling an Aircraft Control Stick by the Right Arm of Male Air Force Personnel in the Sitting Position

(a)

Control Distance from SRP (in.)	Control Distance from Mid Plane of Body (in.)	Percentiles (lb) 5th	50th	95th
9	0	26	46	67
	4½ (left)	18	33	54
	8 (left)	12	29	44
	4½ (right)	34	58	82
	8 (right)	37	65	95
12½	8 (left)	18	36	68
	8 (right)	43	74	102
15½	0	43	86	160
	8 (left)	23	60	118
	8 (right)	53	100	164
18¾	0	64	124	177
	8 (left)	36	72	114
	8 (right)	70	125	198
23¾	0	54	106	141
	8 (left)	29	64	104
	8 (right)	56	100	147

(b)

Control Distance from SRP (in.)	Control Distance from Mid Plane of Body (in.)	Percentiles (lb) 5th	50th	95th
9	0	34	57	86
	4½ (left)	28	45	66
	8 (left)	26	40	67
	4½ (right)	39	62	88
	8 (right)	39	58	86
12½	8 (left)	33	53	77
	8 (right)	49	80	108
15½	0	54	83	113
	8 (left)	39	64	98
	8 (right)	55	89	119
18¾	0	56	86	127
	8 (left)	45	74	108
	8 (right)	58	99	126
23¾	0	62	102	138
	8 (left)	51	90	129
	8 (right)	58	103	133

[1]Stick is grasped 13½ in. above the SRP.

TABLE 35.5 Maximum Force Exerted to (a) the Left on an Aircraft Control Stick by the Right Arm of Male Air Force Personnel in the Sitting Position and (b) the Right on an Aircraft Control Stick by the Right Arm of Male Air Force Personnel in the Sitting Position

Arm Strength in Sitting Position Using Aircraft Control Stick

(a)

Control Distance from SRP (in.)	Control Distance from Mid Plane of Body (in.)	Percentiles (lb)		
		5th	50th	95th
9	0	30	47	66
	4½ (left)	31	49	67
	8 (left)	24	44	65
	4½ (right)	26	46	78
	8 (right)	26	44	72
12½	8 (left)	23	44	70
	8 (right)	22	39	59
15½	0	24	38	52
	8 (left)	20	35	58
	8 (right)	24	40	70
18¾	0	8	32	53
	8 (left)	16	30	56
	8 (right)	22	39	70
23¾	0	14	29	46
	8 (left)	11	21	49
	8 (right)	20	37	66

(b)

Control Distance from SRP (in.)	Control Distance from Mid Plane of Body (in.)	Percentiles (lb)		
		5th	50th	95th
9	0	23	38	49
	4½ (left)	31	48	64
	8 (left)	34	55	74
	4½ (right)	15	27	51
	8 (right)	12	22	43
12½	8 (left)	31	48	70
	8 (right)	16	24	46
15½	0	20	28	39
	8 (left)	25	43	63
	8 (right)	13	22	49
18¾	0	15	25	35
	8 (left)	22	36	61
	8 (right)	14	24	50
23¾	0	13	20	30
	8 (left)	19	31	48
	8 (right)	12	22	51

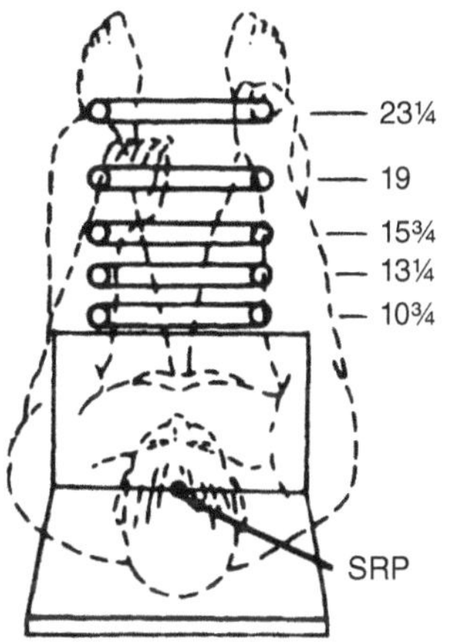

FIGURE 35.12

TABLE 35.6 Maximum Force Exerted in (a) Pushing on an Aircraft Control Wheel by the Right Arm of Male Air Force Personnel in the Sitting Position and (b) Pulling on an Aircraft Control Wheel by the Right Arm of Male Air Force Personnel in the Sitting Position

(a)

Control Distance from SRP (in.)	Control Position (deg)	Percentiles (lb) 5th	50th	95th
10¾	0	52	86	135
	45 (left)	48	84	149
	90 (left)	32	67	125
	45 (right)	40	67	128
	90 (right)	19	52	112
13¼	90 (left)	32	54	93
	90 (right)	25	51	83
15¾	0	61	90	155
	90 (left)	32	59	139
	90 (right)	32	53	102
19	0	64	121	235
	90 (left)	37	88	171
	90 (right)	33	67	140
23¼	0	105	171	242
	90 (left)	82	131	211
	90 (right)	49	117	197

(b)

Control Distance from SRP (in.)	Control Position (deg)	Percentiles (lb) 5th	50th	95th
10¾	0	44	66	102
	45 (left)	40	67	111
	90 (left)	23	55	109
	45 (right)	39	67	97
	90 (right)	18	43	87
13¼	90 (left)	33	67	120
	90 (right)	31	60	102
15¾	0	66	94	145
	90 (left)	42	71	144
	90 (right)	49	80	130
19	0	73	106	169
	90 (left)	60	88	127
	90 (right)	61	94	149
23¼	0	77	125	182
	90 (left)	73	117	162
	90 (right)	74	110	186

[1]Wheel grips are 18 in. above the SRP and 15 in. apart.

TABLE 35.7 Maximum Force Exerted to (a) the Left on an Aircraft Control Wheel by the Right Arm of Male Air Force Personnel in the Sitting Position and (b) the Right on an Aircraft Control Wheel by the Right Arm of Male Air Force Personnel in the Sitting Position

Arm Strength in Sitting Position Using Aircraft Control Wheel

(a)

Control Distance from SRP (in.)	Control Position (deg)	Percentiles (lb) 5th	50th	95th
10¾	0	26	46	88
	45 (left)	21	54	123
	90 (left)	23	47	91
	45 (right)	31	54	120
	90 (right)	21	42	104
13¼	90 (left)	26	44	86
	90 (right)	25	45	99
15¾	0	27	46	112
	90 (left)	27	43	82
	90 (right)	29	50	86
19	0	25	44	95
	90 (left)	22	43	76
	90 (right)	33	52	104
23¼	0	20	39	86
	90 (left)	21	38	73
	90 (right)	26	55	109

(b)

Control Distance from SRP (in.)	Control Position (deg)	Percentiles (lb) 5th	50th	95th
10¾	0	20	48	96
	45 (left)	24	69	121
	90 (left)	27	59	101
	45 (right)	24	51	118
	90 (right)	15	54	112
13¼	90 (left)	21	52	98
	90 (right)	19	51	111
15¾	0	27	59	97
	90 (left)	19	53	96
	90 (right)	20	46	91
19	0	30	63	104
	90 (left)	27	46	94
	90 (right)	22	41	87
23¼	0	35	60	98
	90 (left)	26	42	82
	90 (right)	22	40	68

TABLE 35.8 Turning Hand Strength

	KNOB DEPTH (INCHES)			
	FINGER GRIPPED 		FULL HAND GRIPPED 	
KNOB DIAMETER (INCHES)	0.50	1.0	0.50	1.0
0.50	5 in. lb	6 in. lb	11 in. lb	16 in. lb
0.75	6 in. lb	8 in. lb	20 in. lb	29 in. lb
1.00	8 in. lb	10 in. lb	5 ft lb	6 ft lb
1.50	13 in. lb	15 in. lb	7 ft lb	10 ft lb
2.00	20 in. lb	24 in. lb	11 ft lb	13 ft lb
3.00	6 ft lb	6 ft lb	14 ft lb	16 ft lb

TABLE 35.9 Dynamometer Strength Capabilities

		MAXIMUM SQUEEZE				
SUBJECT POPULATION	AGE RANGE OR MEAN	RANGE	5TH PERCENTILE	MEAN 95TH	PERCENTILE	SD
U.S. Air Force, male personnel	18–25	90–203 lb	105 lb	134 lb	164 lb	18
U.S. Navy, female personnel	22	46–119 lb	58 lb	73 lb	87 lb	8.8
U.S. civilian males	20–30		74 lb	108 lb	142 lb	21
U.S. industrial workers (female)	32	46–119 lb	57 lb	74 lb	91 lb	10.3
U.S. truck and bus drivers (male)	37	80–190 lb	91 lb	121 lb	151 lb	18.1
U.S. elderly males	81		35 lb	64 lb	92 lb	17.3

Source: Albert Damon et al., *The Human Body in Equipment Design*, Harvard University Press, Cambridge, Mass., 1966, pp. 219–220.

GRIP STRENGTH FOR CHILDREN AGES 3 TO 10 (KP)

AGE	10TH PERCENTILE	MEDIAN	90TH PERCENTILE	MAXIMUM
3	3.2	4.3	6.8	8
5	5.1	7.2	10.1	12.1
7	6.2	11.0	14.8	17.9
10	12.9	16.7	22	26.2

Source: Clyde L. Owings, *Strength Characteristics of U.S. Children for Product Safety Design*, 011903-F, University of Michigan, Ann Arbor, for the Product Safety Commission, 1975.

Steering-Wheel Forces [16-in.-Diameter (41-cm) Wheel][17]

Although numerous studies of steering-wheel force have been performed in static laboratory settings, the true, expected force response of drivers while driving is considerably different because emergency conditions requiring a sudden application of force—say, to overcome a failure of the power steering—introduce a psychological factor. The steering-wheel force data shown in Table 35.13a represent typical responses of randomly selected male and female drivers in the U.S. civilian population. Steering-wheel turning rates attained by the same subjects are shown in Table 35.13b.

[17] B. F. Pierce et al., *Human Force Considerations in the Failure of Power Assisted Devices*, MFI Report 73-105, 1972 (DOT-HS-230-2-396).

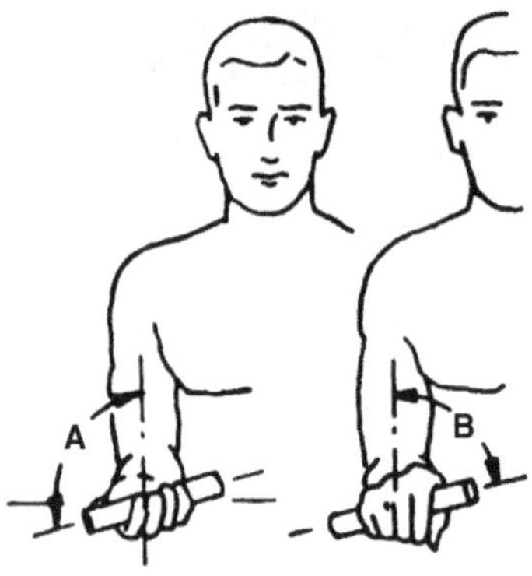

FIGURE 35.13 Wrist turning strength.

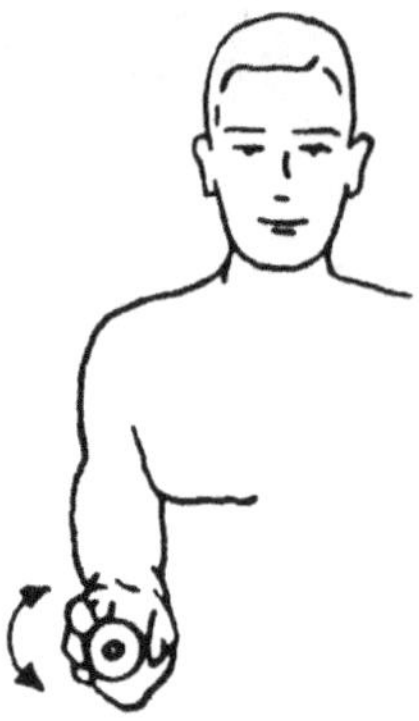

FIGURE 35.14 Hand turning strength.

TABLE 35.10 Maximum Force on a Vertical Handgrip by Wrist-Forearm Supination (A) and Pronation (B)*

WRIST TURNING STRENGTH

		PERCENTILE			
HAND	MOVEMENT	5TH	50TH	95TH	SD
Right	A	35 lb	64 lb	93 lb	18
	B	29 lb	71 lb	119 lb	28
Left	A	30 lb	62 lb	88 lb	16
	B	31 lb	71 lb	132 lb	31

*Male college students.

TABLE 35.11 Maximum Force on Various Sizes of Round Knobs with Knurled Edges*

HAND TURNING STRENGTH

KNOB DIAMETER, IN.	MEAN TORQUE, IN.-OZ	SD
0.25	19.6	5.4
0.50	45.9	13.5
0.75	93.1	33.1
1.00	116.0	35.5
1.25	132.9	37.7
1.50	146.8	37.5
2.00	210.2	48.9
3.00	477.7	136.6
4.00	698.0	173.9
5.00	973.4	262.8

*Male Air Force personnel.

TABLE 35.12 Optimum Size of Handwheels for Partial Turns (Handwheel Facing Standing Operator)

EFFECT OF CRANK DIAMETER AND LOCATION OF TORQUE

TORQUE, IN.-LB	WHEEL DIAMETER, IN. (38 TO 48 IN. ABOVE THE FLOOR)	WHEEL DIAMETER, IN. (BELOW 38 IN. OR ABOVE 48 IN.)
20-40	6.0	10.0
40-60	10.0	16.0
60-90	10.0	16.0
Above 90	16.0	16.0

Source: K. F. H. Murrell, *Human Performance in Industry*, Reinhold Publishing Corporation, New York, 1965, p. 242.

TABLE 35.13 Forces on Steering Wheel (Male and Female)

(a)				(b)			
TRIAL	N	MEAN	5TH %-TILE	TRIAL	N	MEAN	5TH %-TILE
1	182	41.6 lb	14.4 lb	1	182	4.3 lb	0.5 lb
2	182	36.2 lb	17.1 lb	2	182	6.6 lb	1.8 lb
3	182	38.3 lb	19.5 lb	3	182	7.3 lb	2.2 lb

Pedal Force-Limit Expectations

The maximum amount of force that an individual can apply to foot pedals depends on several factors, including the angular relationships between the seat back and the seat pan, the distances between the seat and the pedal, and the angle and size of the pedal. Perhaps the most important consideration, however, is whether the operator is expected to operate the pedal using only ankle flexion. Figure 35.15 and Table 35.14 provide a number of force-limit values for various seating geometries and force-application conditions. Data are for young male populations.

See also Fig. 35.16.

In actual practice, however, consider the following guidelines:

For typical automobile braking, limit force requirements to approximately 60 lb (27 kg).

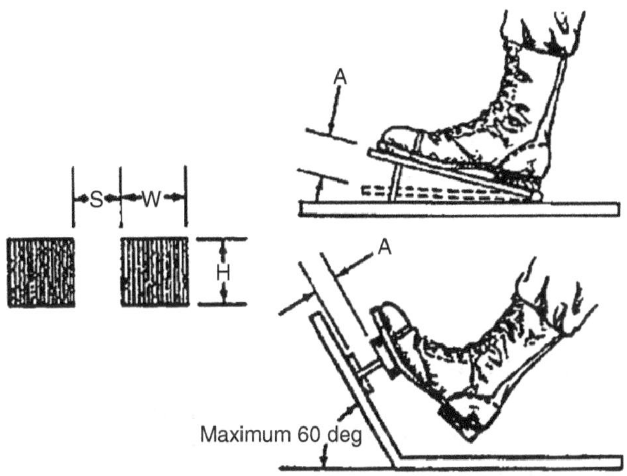

Figure 35.15

For typical accelerator pedal operation, limit forces to 20 lb (9.1 kg) [minimum 10 lb (4.5 kg)].

For aircraft rudder pedals, limit forces to 20 lb (9.1 kg) for ankle operation and to 150 lb (68 kg) for full-leg operation.

LIFTING AND CARRYING

Variables to be considered are the following:

1. Task Variables
   a. Location of the object to be lifted (accessibility)
   b. Size and shape of the object
   c. Height from and/or to which the object is lifted
   d. Weight and weight distribution of the object
   e. Relation between the object's c/g and points of grasp
   f. Whether the object is lifted by handles
   g. Working-position characteristics (e.g., awkward rather than normal)
   h. Manipulatory accuracy requirements
   i. Frequency and duration of lifting and carrying movements
2. Human Variables
   a. Age and sex of the person or persons doing the lifting
   b. Body dimensions (stature, arm length, etc.)
   c. Physical fitness
   d. Experience and training

TABLE 35.14

	Dimensions		Displacement			
	H, Height	W, Width	Normal Operation	Heavy Boots	Ankle Flexion	Total Leg Movement
Minimum	25 mm (1.0 in.)	75 mm (3.0 in.)	13 mm (0.5 in.)	25 mm (1.0 in.)	25 mm (1.0 in.)	25 mm (1.0 in.)
Maximum	—	—	65 mm (2.5 in.)	65 mm (2.5 in.)	65 mm (2.5 in.)	180 mm (7.0 in.)

	Resistance			
	Foot Not Resting on Pedal	Foot Resting on Pedal	Ankle Flexion Only	Total Leg Movement
Minimum	18 N (4 lb)	45 N (10 lb)	—	45 N (10 lb)
Maximum	90 N (20 lb)	90 N (20 lb)	45 N (10 lb)	800 N (180 lb)

	Separation (S)	
	One Foot Random	One Foot Sequential
Minimum	100 mm (4.0 in.)	50 mm (2.0 in.)
Maximum	150 mm (6.0 in.)	100 mm (4.0 in.)

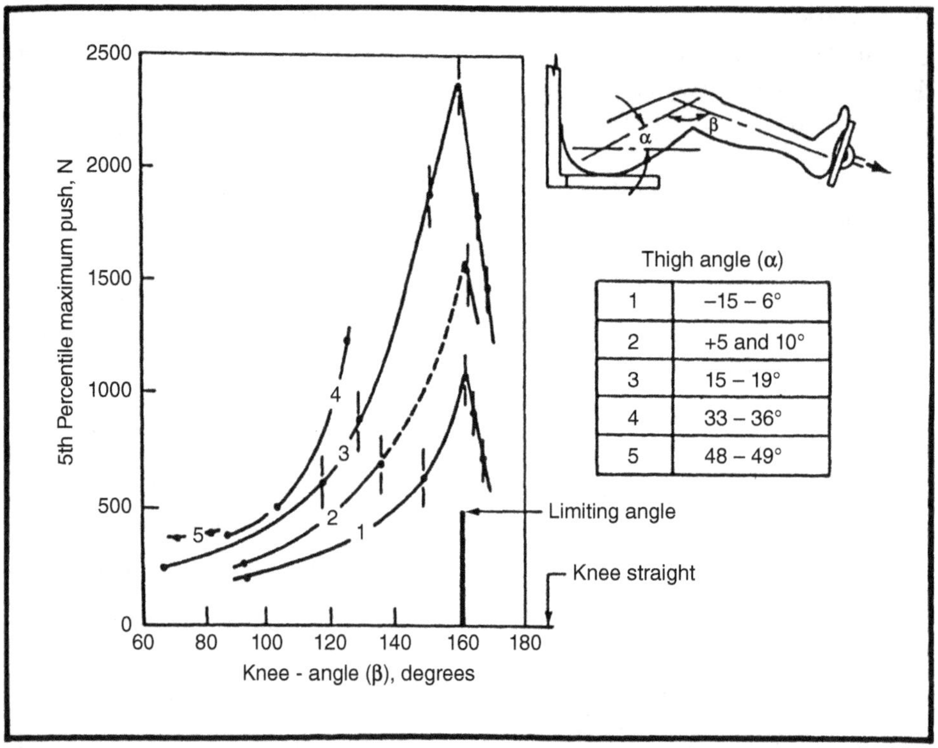

Figure 35.16 Leg strength at various knee and thigh angles.

3. Environmental Variables
   a. Thermal (temperature, humidity, and ventilation)
   b. Dynamic (platform motion and vibration)

The most important determinant of lifting force is the distance of the feet from the grasping axis (lifting force is greatest if the lift weight is in the same vertical plane as the body, and it decreases rapidly as the weight moves away from the body). The best height for lifting is approximately fingertip height, with the hands hanging freely at the sides. Above this height, lift capacity decreases very rapidly; below this height, it decreases more slowly. With the load near the floor, capacity is only about 75 percent of best height capacity.

The maximum individual loads carried in either hand by means of handgrips should be about 60 lb (27 kg) for short distances and 35 lb (14 kg) for longer distances. Bulky articles should not weigh more than about 30 lb (13 kg). In general, a package is considered heavy when it reaches about 35 percent of a person's own body weight. Less strain occurs if equal weights are carried in a balanced manner, i.e., half in one hand and half in the other. Carrying one heavy weight a given distance incurs less physiological cost than carrying two weights, each weighing half as much, the same distance in two trips.

Handle size and shape, as well as position, make a considerable difference in the individual's acceptance of maximum weight.

Load Lift Limits

The following guidelines are in the military standard MIL-STD-1472G. (See Fig. 35.17.) Additional guidelines are published by the National Institute of Safety and Health (NIOSH) in the Applications Manual for the Revised NIOSH Lifting Equation and Ergonomic Guidelines for Manual Material Handling.

Weight in Relation to Carrying Mode

The ease with which various packages can be carried depends not only on the weight but also on the type and placement of the handles. Figures 35.18–35.20 illustrate common situations and suggested upper weight limits for typical package-carrying configurations:

1. When the package (e.g., a suitcase) has smooth sides, it can be carried against the person's hip or leg without too much stress. The weight should not exceed more than about 45 lb (20 kg) for males and 35 to 40 lb (14 to 18 kg) for females.

2. When the package surfaces are not smooth (as is the case with electronic equipment chassis, etc.), limit weight to about 35 lb (14 kg).

3. Although the typical male can carry an equipment package weighing up to about 60 lb (27 kg) for a short distance, this much weight should involve two persons

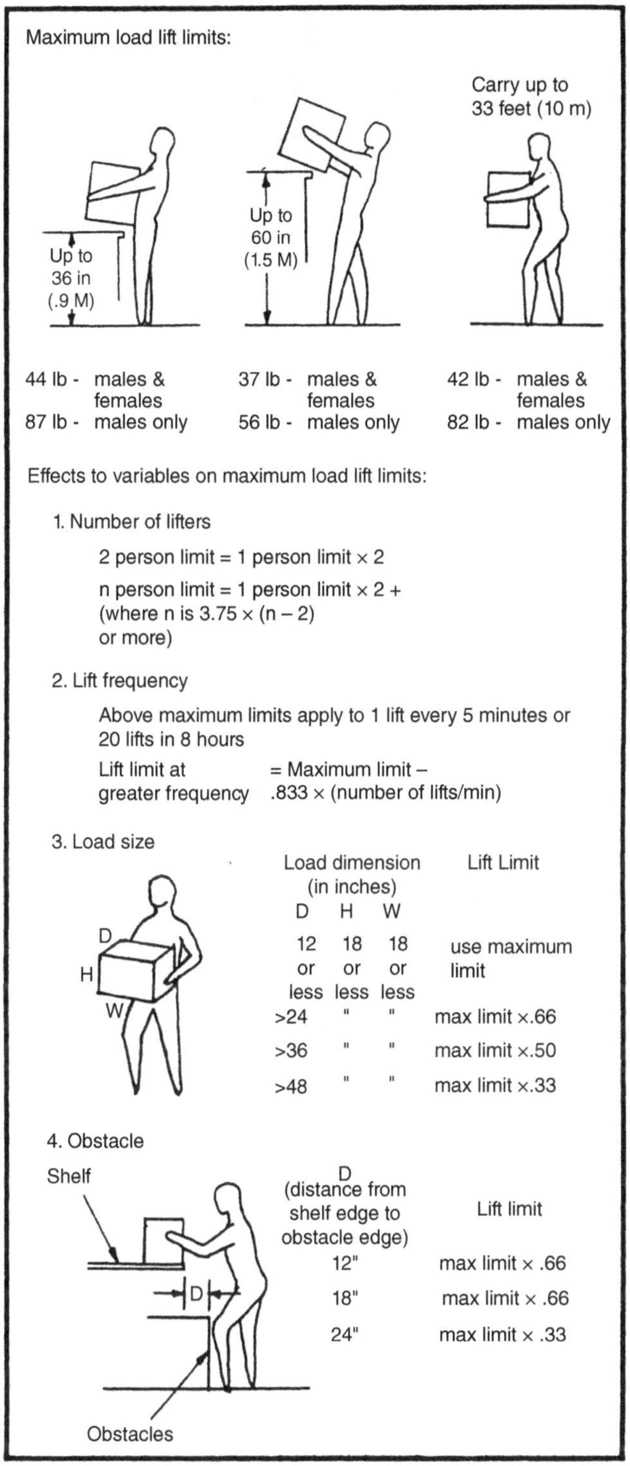

Figure 35.17 Load lift limits.

carrying the package by means of appropriately located dual handles. Two men should not be required to carry 100 lb (45 kg) very far or more than 200 lb (90 kg) more than a short distance.

4. The removal and replacement of equipment units from a rack at different heights require consideration of

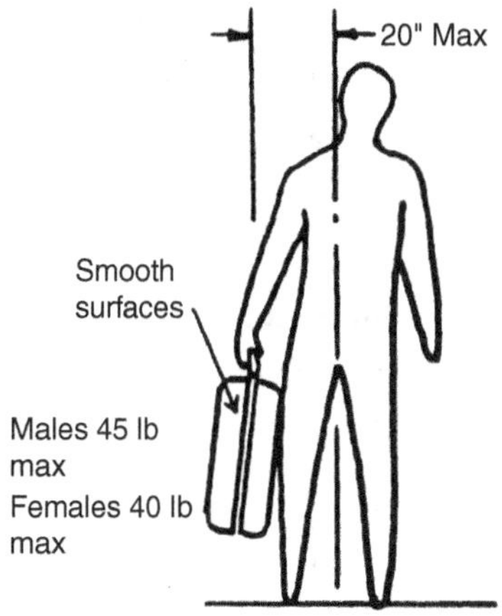

Figure 35.18

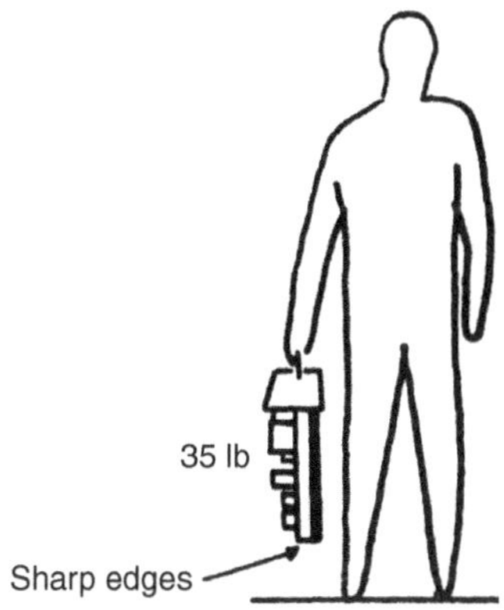

Figure 35.19

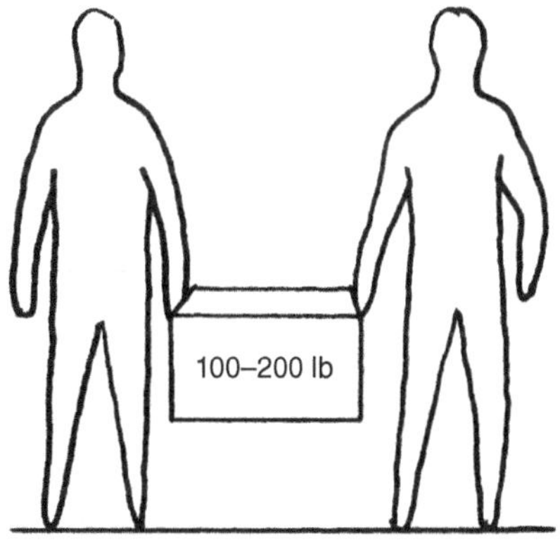

Figure 35.20

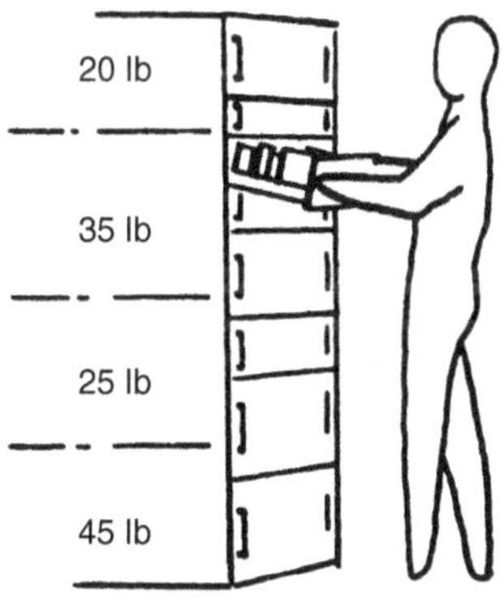

Figure 35.21

the suggested maximum package weights in relation to working height, as indicated in Fig. 35.21.

5. Backpack weight-carrying guidelines are indicated in Fig. 35.22.

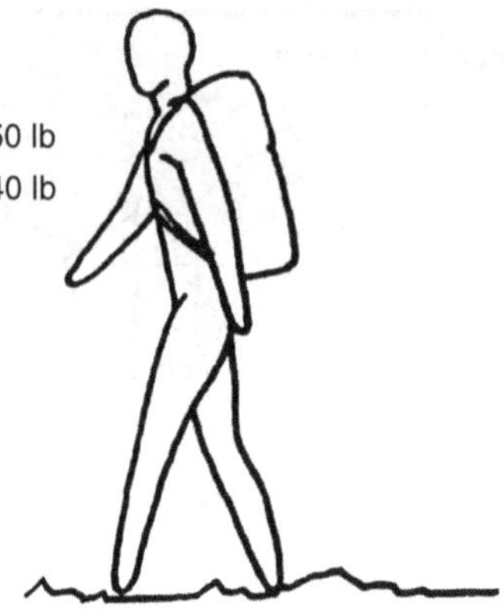

FIGURE 35.22

Horizontal Push and Pull Forces

Maximum push and pull forces for males are in the table. Table 35.15 lists approximate maximum horizontal push and pull forces for males under various pushing conditions. The data assume the surface pushed against is vertical, at least 16 in. wide, and is between 20 and 50 in. above the floor.

TABLE 35.15 Horizontal Push and Pull Forces Exertable Intermittently or for Short Periods of Time (Male Personnel) (From MIL-STD-1472D)

Horizontal Force*; At Least	Applied With	Condition (μ: Coefficient of Friction)
100 N (25 lb) push or pull	both hands or one shoulder or the back	with low traction $0.2 < \mu < 0.3$
200 N (45 lb) push or pull	both hands or one shoulder or the back	with medium traction $\mu \sim 0.6$
250 N (55 lb) push	one hand	if braced against a vertical wall 510–1525 mm (20–60 in.) from and parallel to the push panel
300 N (70 lb) push or pull	both hands or one shoulder or the back	with high traction $\mu > 0.9$
500 N (110 lb) push or pull	both hands or one shoulder or the back	if braced against a vertical wall 510–1780 mm (20–70 in.) from and parallel to the panel or if anchoring the feet on a perfectly nonslip ground (like a footrest)
750 N (165 lb) push	the back	if braced against a vertical wall (600–1100 mm) (23–43 in.) from and parallel to the push panel or if anchoring the feet on a perfectly nonslip ground (like a footrest)

*May be doubled for two and tripled for three operators pushing simultaneously. For the fourth and each additional operator, not more than 75% of their push capability should be added.

CHAPTER 36

EQUILIBRIUM

BODY EQUILIBRIUM (BALANCE AND BODILY ORIENTATION)

Maintenance of body equilibrium results from sensory inputs from visual, auditory, semicircular canal, skin, and kinesthetic and visceral sources.

Body verticality can be fairly accurately judged when the inputs from the various sources represent an accurate picture of the relationships between the body and the earth's gravity. However, if for some reason information from one or more of the sources becomes distorted (e.g., abnormal dynamic conditions affect the semicircular canals, and high g forces from a rapid vehicular movement cause high skin pressures, movement of the internal organs, or even confusing visual patterns or auditory sound delays), one's assessment of his or her body position may reflect considerable variation from the true condition of the body.

Perception of Body Verticality

Both visual and postural verticality are determined largely by the joint action of visual and gravitational forces. Visually, it is important that external visual references be available and that they be compatible with the earth reference. Gravitation (g) force exerts a considerable influence on our perception of verticality (which is in the direction of action of the resultant g force). The average threshold for sensing body tilt in any direction is 2° to 3°, although under certain conditions tilts of as much as 15° may not be detected (e.g., an aircraft banking or pitching at night or under foggy or cloudy conditions).

Blindfolded subjects generally can reorient themselves to a vertical position within about 4° to 5° (the arc of uncertainty). (See Fig. 36.1.) When subjects sit on soft seats, this increases to about 7.5°. The error is usually in the direction of the initial tilt away from verticality. One tends to adapt to a slight tilt, which explains in part why some pilot trainees continue to "fly with one wing down."

BODY SWAY

The body is ordinarily moving to some degree; i.e., it is not rigid, but is swaying in one direction or another. Fore-aft sway is about 50–75 percent greater than side-to-side sway when the body is in a standing position. Sway is independent of body height or weight. It increases if the body is held tense. Sway obviously increases when there is no visual input. There is no relationship between static equilibrium and the

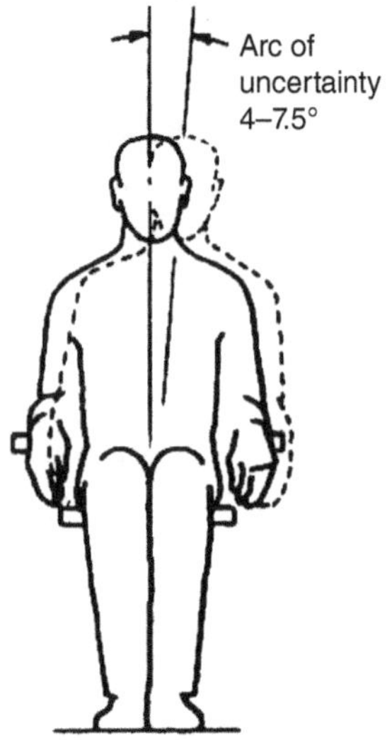

FIGURE 36.1 Ability to sense if body is vertical.

ability to maintain dynamic equilibrium, i.e., to remain upright under unstable platform conditions.

Sway can be induced by certain noncompatible visual cues, e.g., illusions of motion (nystagmus). An interrelationship between vision and labyrinthine (inner ear) cues occurs with body movements (rotation). Nystagmus is a reflex eye movement interrupted by saccadic eye movements in the opposite direction to rotational movement. Upon rotation of the body at a constant, maintained rate, the following occurs:

1. Compensatory eye movement after about a 50- to 80-ms delay from the start of rotation
2. Opposite eye movement after the preceding body motion has decelerated to zero (called "postrotational nystagmus")
3. Eye movements in the original direction once postrotational nystagmus or eye movement has decelerated (called "inverse nystagmus")

This sequence may take as long as 10 s after the start of rotation.

Some illusions related to body movement are identified in Table 36.1, including several that are not necessarily associated with nystagmus. It should be noted, however, that these are not universally observed by all people or described in the same manner by those who do observe them.

VISUAL AND GRAVITATIONAL CONFLICT

Both visual and postural orientation are influenced by the visual framework. Both are likewise influenced by resultant *g* forces acting on the body. There is some controversy over which cues, visual or gravitational, are predominant as determinants of our orientation, but it may safely be said that our judgments, in cases of conflict between the two cues, represent a compromise. Such a conflict is present, for example, in an airplane flying in conditions of zero visibility, having an angular acceleration giving a resultant *g* force on the body, say 45° from the vertical. The cockpit, say 90° from the vertical when this force is acting, represents the whole visual field. Under this conflict of visual and gravitational cues (90° versus 45°), one feels the vertical to be somewhere between them, with the bulk of the evidence indicating that the judgment is closer to the gravitational than to the visual forces. The condition illustrated in Fig. 36.2, in which a pilot's orientation with respect to the earth is disturbed, is called "aviator's vertigo." Visual indicators which show the plane's true attitude to the pilot, and proper training in the use of these instruments, help suppress any actions on his part that might be taken on the basis of erroneous orientation. (See Figs. 36.3 and 36.4.)

TABLE 36.1 Illusions Associated with Body Movement**

Condition of Observer	Condition of Visual Stimulus Object(s)	Nature of Illusion
Starts motion	Stationary	Observer sees visual stimulus moving in opposite direction, feels stationary.
Stationary	Starts motion	Observer sees object stationary, feels self moving in direction opposite to object's motion.
Accelerating rotation to 15 rpm, dark room with only target illuminated	Slow revolution around observer's head, in same direction as his acceleration	Object loses motion, is displaced in direction opposite to its direction of movement.*
Above rotation continued at 15 rpm	As above	Observer loses own sensation of motion, object appears to move in opposite direction with greater velocity than before.*
Strong rotation suddenly stops	Fixed	At first, object and subject both seem to rotate. Then observers may feel fixed while visual stimuli rotate about them in a direction opposite to that of their previous motion. This illusion is called the *oculo-gyral illusion*. After all apparent motion stops, subject still feels vaguely unpleasant sensations.*

*When experiments are conducted in a lighted room these illusions do not occur.
**W. E. Woodson and D. W. Conover, *Human Engineering Guide for Equipment Designers*, University of California Press, Berkeley, 1964.

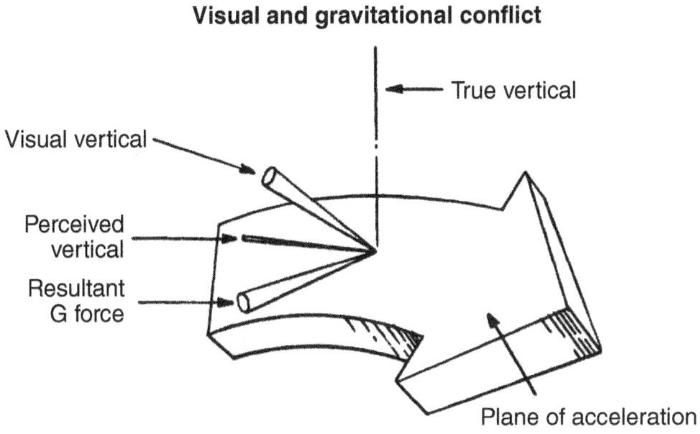

Figure 36.2 Effect of angular acceleration on perceived vertical.

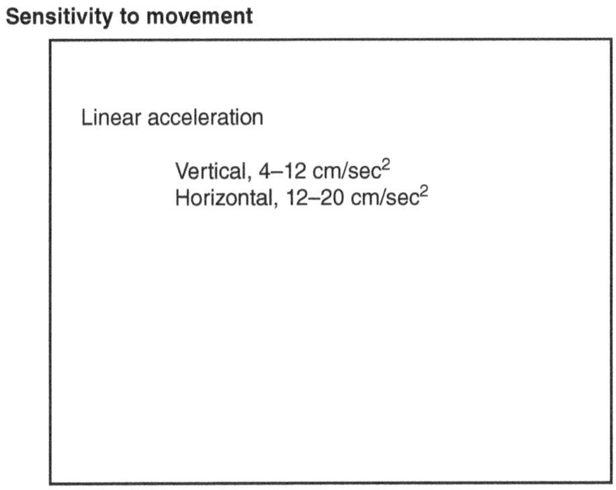

Figure 36.3 Threshold sensitivity to linear accelerations.

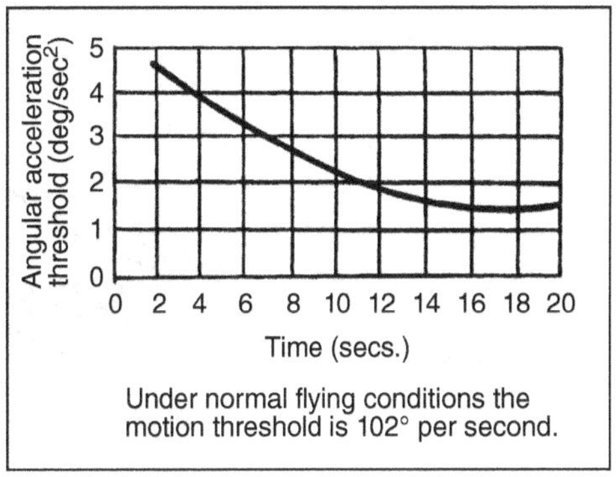

Figure 36.4 Threshold sensitivity to angular accelerations.

CORIOLIS FORCES

Although most people are aware of the discomfort sometimes associated with riding on a rotating platform such as a merry-go-round, they are not aware of what would happen to them if they could not see the stationary earth reference outside the rotating platform. This particular phenomenon poses an important question for the space station designer who may be considering "spinning" the station to create artificial gravity conditions for astronauts on board. Although the idea of reproducing normal earth gravitational effects will aid astronauts in several ways (both physiologically and operationally), the Coriolis forces reacting on the station's occupants will introduce some strange conditions which must be overcome by careful

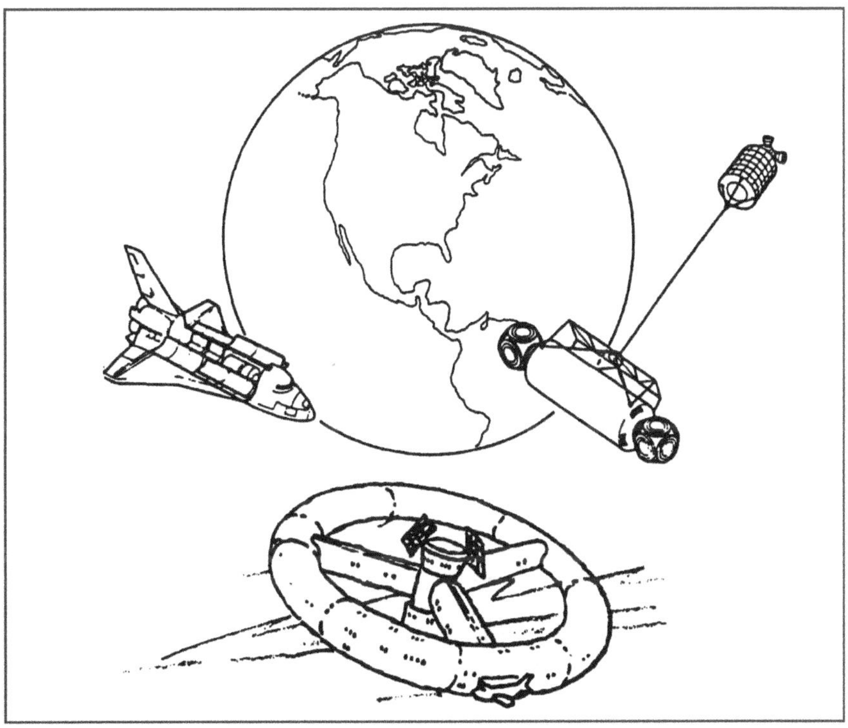

Figure 36.5 Design concepts for artificial gravity space stations.

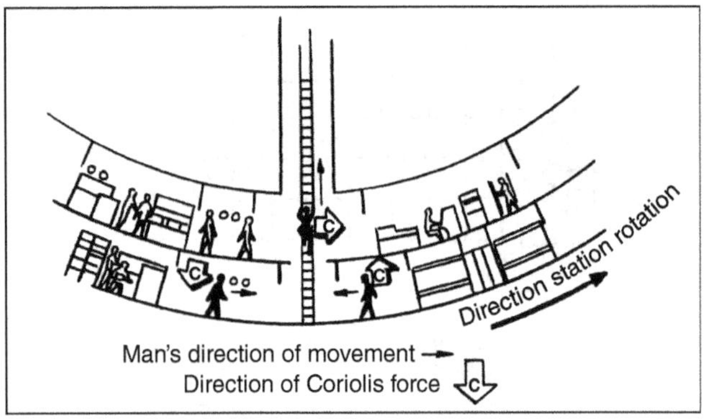

FIGURE 36.6 Coriolis forces in artificial gravity.

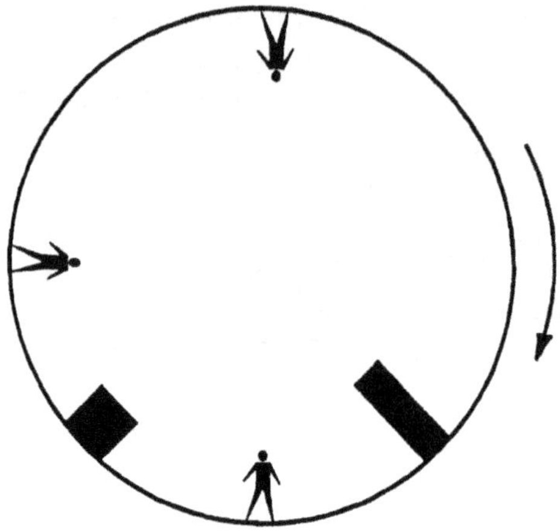

FIGURE 36.7 Disorienting visual environment in spinning space station.

arrangement of on-board facilities. For example, the Coriolis force will affect astronauts differently, depending on the direction in which they are moving and also on how far from the hub of the station they are located. Figure 36.5 shows at least three different conditions for a station rotating counterclockwise. Experiments have indicated that although some people can adapt to this unique condition, others cannot. Even assuming that only astronauts who can adapt will be aboard, the on-board facilities and equipment should be designed and arranged to minimize rather than add to the astronauts' difficulties in coping with Coriolis forces. (See Fig. 36.6.)

Other potential human factors problems in the design of an artificial gravity environment include (see Fig. 36.7):

1. Living on a curved plane will be disorienting. The astronauts in the illustration will see their coworkers upside down across a spinning station. Cabinets and furniture will "lean" toward the astronaut.

2. Head movements perpendicular to the axis of rotation can cause severe dizziness.

3. The astronaut working outside will be in danger of being thrown from the spinning space station.

4. The astronaut may be disoriented by looking out windows and watching the universe spin past.

CHAPTER 37

REACTION TIME

SIMPLE VERSUS COMPLEX REACTION TIME

In practical terms, one should think more in terms of total response time than in terms of simple reaction time. The term "reaction time" generally is considered at two levels. The first is called "simple reaction time," which is the shortest time between the moment a sensory receptor is stimulated and the time a body element reacts, as when a speck of dust strikes an eyelash, and the eyelid immediately closes to protect the eye. Practically speaking, however, scientists measuring simple reaction time typically have a test subject respond to a stimulus by pressing, say, a key—thus allowing the scientist to measure and compare the reaction time of various sensory channels (vision, audition, etc.).

The second level is called "complex reaction time," in which the scientist attempts to include human information-processing time; e.g., the subject may be asked to recognize one stimulus from among several and to respond by selecting one of several response modes.

Since complex reactions are more akin to real-world situations, the designer is more concerned with complex reaction time or, more realistically, total response time. In a sense, response time is a function of several factors, including the following:

1. The sensory channel through which the stimulus is initiated
2. The signal or stimulus characteristics
3. The complexity of the signal
4. The signal rate
5. Whether anticipatory provisions are present
6. The response mode, e.g., the body member used

EFFECTS OF SIGNAL CHARACTERISTICS ON REACTION TIME

1. Signal Intensity: The greater the intensity of the signal, the faster will be the reaction time.
2. Signal Anticipation: When the signal is anticipated, reaction time is typically shorter.

3. Practice: Reaction time tends to be reduced with practice.

4. Pacing: If operators can set their own pace, they can often react faster to known signals.

5. Signal Quality: Operators generally can react faster to a high-pitched sound, a brighter light, a larger visual target, a longer-duration signal, and a signal emanating from a particular location (i.e., a visual signal that strikes the center rather than the periphery of the eye). There is no difference in simple reaction time between a steady and a flashing light signal. However, when one intermittent light signal must be distinguished on the basis of flash rate, reaction time is related to the length of the flash.

6. Likelihood of Signal Appearance: The least likely signals will have the longest reaction times.

7. Signal Format: When signals are arranged sequentially or are meaningfully grouped, reaction time is typically shortened; i.e., reaction time is a function of the number of signals that occur in the next sequential step or is proportional to the number of signal groups. When two successive signals occur within 0.1 s, they are generally treated as a single signal. When the interval between signals is 0.5 s or greater, the operator is capable of responding to each signal separately.

8. Overload: Although an operator can adjust to excessive signal rates by relying on memory for short bursts, total response failure may occur when rates are too high for too long.

Simple Reaction-Time Comparisons of Sensory Input Channels

Using the typical key response technique, scientists have determined the reaction times for various sensory input channels shown in Fig. 37.1. Signals should not occur at rates faster than about two per second unless some means are provided for anticipating the signal. Avoid alerting periods shorter than 0.1 s.

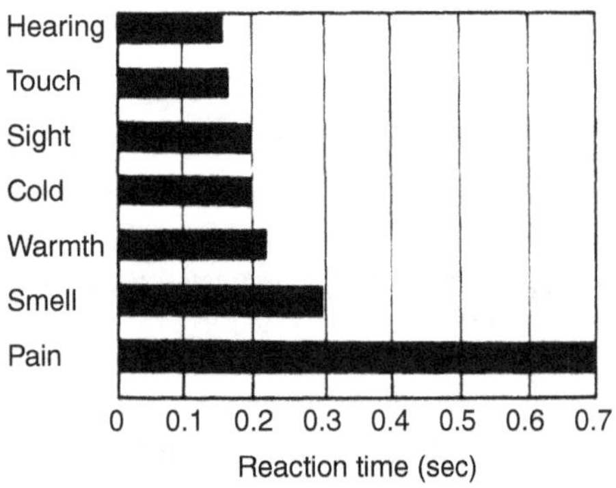

FIGURE 37.1 Simple reaction-time comparisons of sensory input channels. (From T. Morgan et al. (eds.), Human Engineering Guide to Equipment Design, McGraw-Hill Book Company, New York, 1963, p. 229.)

COMPLEX REACTION TIME

The following estimates have been made regarding the several components of a basic, complex reaction time, i.e., speed of perception to action:

1. Stimulus detection and neural transit time, 0.1 s
2. Brain recognition time, 0.4 s
3. Decision-making time, up to 4.0 s
4. Motor response time (minimum), 6.0 s

Age and Sex Differences

The reaction time of adult males is generally faster than that of adult females, and, as one might expect, older people have a longer reaction time. Typical comparisons are shown in Table 37.1.

TABLE 37.1

	Males		Female	
Age	Light, s	Sound, s	Light, s	Sound, s
20	0.24	0.23	0.32	0.31
30	0.22	0.19	0.26	0.20
40	0.26	0.24	0.34	0.30
50	0.27	0.25	0.36	0.30
60	0.38	0.37	0.44	0.42

Some Common Response Times

Other generalizations of interest include the following (see Table 37.2):

1. It generally takes about 20 percent longer to respond with the feet than with the hands.
2. The preferred hand is usually about 3 percent faster than the nonpreferred one.

TABLE 37.2 Some Common Response Times

Simple Movement Response	Time, s
Moving eyes (focusing on new spot)	0.48
Walking (one pace)	0.60
Sitting down in a chair	1.32
Standing up from a sitting position	1.62

3. Everyone has a refractory period of about 2 to 3 s, regardless of a stimulus demand rate, which means that a second stimulus arriving within 0.5 s will be treated together with the first (e.g., two successive signals arriving 0.1 s apart will be treated as a single signal).

4. Simple reaction times usually can be reduced by as much as 40 percent by providing an alerting signal (approximately 2 to 8 s in advance for isolated response tasks and 2 to 3 s in advance for sequential tasks).

Effect of Number of Response Choices

As one might expect, when the number of response choices increases, the reaction time is lengthened. Table 37.3 illustrates this point.

TABLE 37.3

Number of Choices	Mean Reaction Time, s
1	0.20
2	0.35
3	0.40
4	0.45
5	0.50
6	0.55
7	0.60
8	0.60
9	0.65
10	0.65

Selected Manual Response-Time Examples

Although many studies have been carried out for the purpose of establishing response times for a variety of manual tasks, space permits citing only a few examples to demonstrate the variability that must naturally accompany the specifics of a given task. (See Table 37.4.)

TABLE 37.4 Reach Distance and Terminal Movement Condition

	Terminal Conditions, s	
Reach Distance, in.	Always the Same	Varied
10	0.03	0.041
20	0.045	0.066
30	0.062	0.093

Expected versus Unexpected Stimulus Response Times

1. Simple reaction-time reference, 0.20 s
2. Forewarned, 0.51 s
3. Surprised, 0.61 s

Movement Response End-Point Control

When the operator has to control the final movement end point, his or her total response time is greater than when the end point is automatically controlled (e.g., the operator moves a stylus or control to a mechanical "stop"). (See Table 37.5.)

TABLE 37.5

Distance, in.	Mechanical Stop, s	Operator-Controlled
6	0.28	0.47
12	0.36	0.57
20	0.46	0.67
30	0.58	0.79

End-Point Precision

The more precise or accurate the operator has to be in making a control movement, the longer his or her response time will be. (See Table 37.6.)

TABLE 37.6

Precision, in.	Time, s
0.50	0.11
0.25	0.27
0.125	0.48
0.0625	0.67

Effect of Added Manual Force on Response Time

TABLE 37.7 Effect of Added Manual Force on Response Time

Force Required, lb	Time, s
2	0.02
4	0.04
6	0.06
8	0.08
10	0.10
50	0.33

Reaction Time to Warning Light

TABLE 37.8 Reaction Time to a Warning Light Relative to Its Angular Variation from the Operator's Nominal Line of Sight (While Tracking)	
ANGULAR VARIATION	TIME, S
20	0.40
40	0.40
80	0.80
100	1.30

Effects of Distraction

TABLE 37.9 Effects of Distraction		
TYPE OF DISTRACTION	WITHOUT DISTRACTION	WITH DISTRACTION
Continuous	0.218	0.220
Continuous with periodic interruptions	0.202	0.206
Random/intermittent	0.204	0.220

Effect of Reach Distance on Toggle-Switch Operation

The response times shown in Table 37.10 are those of an operator who, in addition to making the simple movement to the switch location in response to a warning light, had to decide which hand to take off a joystick and then proceed to reach for and manipulate a toggle switch.

TABLE 37.10	
DISTANCE, IN	TIME, S
6	0.76
9	0.66
15	0.84
18	0.77

Note the randomness of the responses, which is due, of course, to the decision-making step in the total response scenario.

Effect of Toggle-Switch Separation or Spacing

TABLE 37.11 Effect of Toggle-Switch Separation or Spacing	
DISTANCE APART, IN	TIME, S
1.0	0.28
2.0	0.24
4.0	0.35
8.0	0.40

Note that when switches are too close, response time is slightly longer; when they are too far apart, response time is significantly longer.

Comparison of Response Times for Various Digit-Input Devices

The task was to enter 10 different digits.

1. 10-key pad (calculator type), 12.0 s
2. 10 × 10 matrix keyboard, 13.0 s
3. 10 separate levers, 17.0 s
4. 10 separate rotary knobs, 18.0 s

Visual Scanning Response Time

Two examples of visual scanning response time are shown below. The first involves the common task of pilots when they are required to transfer their line of sight from looking at instruments inside the cockpit to scanning outside the cockpit to look for other aircraft:

1. Vision shifts from outside to inside, 1.5 s
2. Vision shifts from inside to outside, 2.39 s

The difference is explained by the greater problem of accommodating from the near-vision instrument-watching task to the far-vision blank-sky, target search task.

A further example of flying response deals with the difference between making a recovery flight maneuver using instruments and making the same recovery using a VFR (visual flight reference):

1. Instrument recovery, 11.0 s
2. VFR recovery, 9.2 s

The difference here is explained by the naturalness of a VFR, making it unnecessary for the pilot to perform extra data processing from abstract instrument graphics.

A second scanning example relates to the positioning of panel displays in the typical automobile. When instruments are clustered high, near the normal roadway line of sight, response to their indications is faster than when the instruments are farther down and are spread out on the instrument panel:

1. Close to line of sight, 1.9 s
2. Normal panel position, 2.2 s
3. Bottom of panel, spread out, 2.9 s

Other Factors

Various other factors, of course, can obviously affect operator response time. These include physical and psychological stress, constraints within the workplace, and constrictions caused by clothing. Often, too, response times are shorter when the task is paced; i.e., operators can usually operate faster when they are paced than when they pace themselves. Response times vary less under paced conditions.

Finally, one must always consider the response of the mechanical system, since in many cases the operator's response is conditioned by feedback received from the machine.

SYSTEM RESPONSE TIME CONSIDERATIONS

Whereas the discussion so far has dealt with the general question of how human reaction time affects a system, the following comments deal with how system response time affects the human—or how machine delays are perceived by the human operator. Gallaway[18] notes that "silences longer than about 4 s are perceived as unacceptable breaks" in both human (conversation) and man–machine operations (especially in interactive human–computer systems). He defines the following general response-time considerations:

1. As a delay (response time) is extended beyond the user's concept of the amount of work being done by a system, frustration increases and affects the quantity and quality of the user's performance.

2. Response-time delays are more than a mere inconvenience to the user; i.e., rather than a straight-line decrease in efficiency as delay increases, there are sudden drops in mental efficiency.

3. Performing most tasks requires an operator to "hold" a body of information in mind, and the longer this information has to be held, the greater is the chance of forgetting (after 2.0 s, most individuals become aware of waiting, and short-term memory failure begins to occur).

4. Response times that exceed 2.0 s should follow only a task that an operator perceives as being completed or one that is structured to appear as a completion point.

5. Longer response times are more acceptable to an operator following completion of an important or long task than following routine, short tasks.

6. Although a response signal can communicate more than one message concurrently, the response time that should be sought is that demanded by the task in the group which demands the fastest response time.

7. When a response to an operator input exceeds about 5.0 s, it is desirable to indicate to the operator that some process is being performed by the machine, e.g., "transaction being processed."

8. Systems with a high degree of variation in response times for a given task are more frustrating to operators. In principle, the range of acceptable variation in delay is that range within which the human cannot detect differences, i.e.,

   a. Response time of 0–2 s: variance should be about +5 percent.

   b. Response time of 2–4 s: variance should be about +10 percent.

   c. Response time of 4–30 s: variance should be about +15 percent.

9. In general, delays greater than 15 s rule out conversational interactions. When delays exceed this amount, the operator should be freed from physical or mental captivity (waiting on the machine) and allowed to turn to another activity until some display requests that he or she return to the operation.

10. Response times greater than 4–15 s generally are too long for conversation requiring short-term memory and probably inhibit problem-solving activity and/or create frustration for a data-entry task.

11. Response times greater than 2–4 s may be inhibiting for tasks requiring a high level of concentration. A 2-s wait at a terminal when the operator is absorbed and emotionally committed to task completion will seem an interminable wait for most operators.

12. Response times less than 2 s are desirable when the operator has to remember information through several responses.

13. Response to a signal (e.g., feedback to a pressing key) should be set at less than 100 ms.

[18]G. R. Gallaway, Response *Times to User Activities in Interactive Man-Machine Computer Systems,* Report no. 277-0014601. NCR Corp., Dayton, OH, 1978.

REFERENCES FOR HUMAN REACTION TIME

Bailey, G. B., and R. Presgrave. *Basic Motion Timestudy,* McGraw-Hill Book Company, New York, 1958.

Broadbent, D. E. *Successive Responses to Simultaneous Stimuli,* Report FPRC-934, Flying Personnel Research Committee, RAF Institute of Aviation Medicine, Farn-borough, England, 1955.

Brown, J. S., et al. *Discrete Movements toward and away from the Body in the Horizontal Plane,* Report no. 6, Project no. 2, U.S. Naval Training Devices Center, Port Washington, N.Y., 1948.

Brown, J. W., and A. T. Slater-Hammel. Discrete Movements in the Horizontal Plane as a Function of Their Length and Direction, *Journal of Experimental Psychology,* Vol. 39, no. 84, 1949.

Damon, A., et al. *The Human Body in Equipment Design,* Harvard University Press, Cambridge, MA, 1966.

Deupree, R. K, and J. R. Simon. Reaction Time and Movement Time as a Function of Age, Stimulus Duration and Task Difficulty, *Ergonomics,* Vol. 6. no. 403, 1963.

Forbes, G. The Effect of Certain Variables on Visual and Auditory Reaction Times, *Journal of Experimental Psychology,* Vol. 35, no. 153, 1945.

Gibbs, C. B., et al. *Reaction Times to the Flashing Light Signals of Cars: The Effects of Varying the Frequency and Duration of the Flash,* Report APU-245, Applied Psychology Research Unit, Medical Research Council, Cambridge, UK, 1955.

Hyman, R. Stimulus Information as a Determinant of Reaction Time, *Journal of Experimental Psychology,* Vol. 45, no. 188, 1953.

Mackworth, J. F., and N. H. Mackworth. The Overlapping of Signals for Decisions, *American Journal of Psychology,* Vol. 69, no. 26, 1956.

Milton, J. L., et al. *Pilot Reaction Time: The Time Required to Comprehend and React to Contact and Instrument Recovery Problems,* Report TSEAA-694-13a, Aerospace Medical Research Laboratory, Wright Air Development Center, Wright-Patterson Air Force Base, OH, 1947.

Morgan, C. T., et al. (eds.). *Human Engineering Guide to Equipment Design,* McGraw-Hill Book Company, New York, 1963.

Provins, K. A. *A Study of Some Factors Affecting Speed of Cranking,* Report no. 53/755, Royal Naval Personnel Research Committee, Medical Research Council, London, 1953.

Reid, L. S., and J. G. Holland. *The Influence of Stimulus Similarity and Stimulus Rate: Third of a Series of Reports on Experimental Analysis of Complex Task Performance,* WADC-TR-54-146, Aerospace Medical Research Laboratory, Wright-Patterson Air Force Base, OH, 1954.

CHAPTER 38

CONTROL RESPONSE

VOLUNTARY CONTROLS

Voluntary control of human body components, as distinguished from involuntary movement, is achieved by means of the human brain (cortex). Voluntary movements are generally classified as "tension movement" or "ballistic movement." In the first case, there is a slow, intense movement created by contraction of antagonistic muscles operating against one another with unequal tension. In the second case, movements are free and generally more rapid, since the simultaneous operation of antagonistic muscles is at a minimum.

Voluntary movements are part of a perceptual-motor process involving coordination between one or more sensors (visual, auditory, tactile, and/or proprioceptive), the cortex, and the musculoskeletal (motor response) system. The perceptual-motor system is a complex feedback system involving at least three major loops (see Fig. 38.1).[19] As this figure illustrates, the human control system involves such concepts as oscillation, inertia, damping, reverberation, and lag, and its characteristics at any given time depend on many factors, including mental set, practice, instructions, motivation, the physical characteristics of the workplace, and the environment (temperature, light, sound, vibration, etc.).

In general, the human functions as a nonlinear system but becomes more linear as the task becomes more predictable.

The human system often introduces its own "noise," which may affect the ultimate response. Motor coordination cannot be generalized from task to task, although steadiness in any given individual remains constant.

GENERALIZATIONS

1. Eye–hand coordination is generally more accurate and reliable than ear–hand coordination.

2. Sensorimotor control between any of the sensory channels and the hands is generally more accurate and reliable than that between the sensory channel and the feet.

3. Hand and arm movement coordination is better when these movements are close to the body and symmetrical.

[19] W. E. Woodson and D. W. Conover, *Human Engineering Guide for Equipment Designers*, University of California Press, Berkeley, 1964.

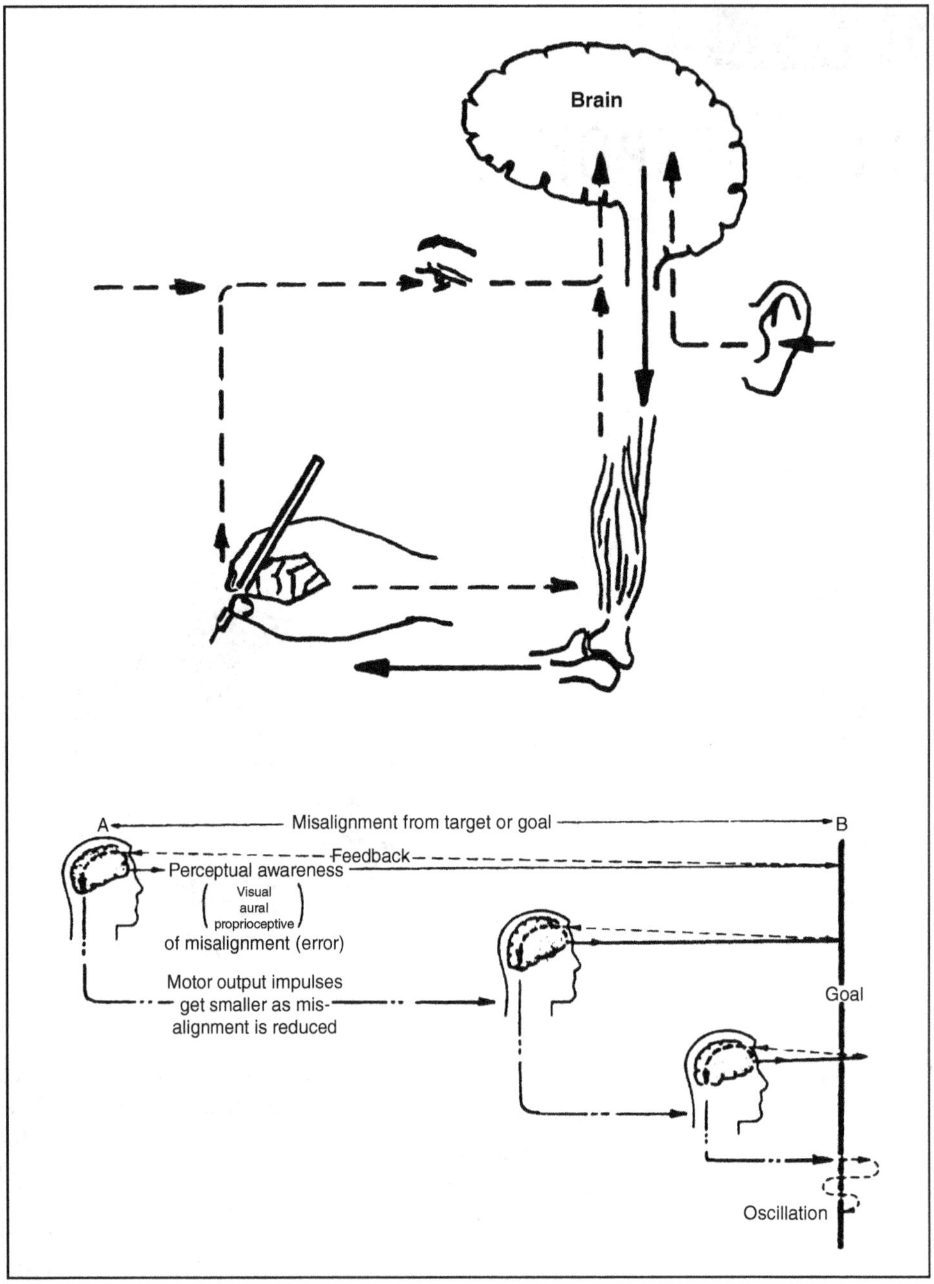

FIGURE 38.1 Perceptual-motor system models.

4. Arm movements that progress forward and/or away from the body are more accurate than arm movements that are directed toward the body. The same is true of leg movements.

5. Right-handed individuals are more proficient at making clockwise movements than at making counter-clockwise movements; left-handed individuals are just the opposite. However, all people make clockwise movements better with the right hand, and counterclockwise movements better with the left hand.

6. Generally, a person can rotate his or her hand and wrist more precisely in one direction than another, i.e., the right hand clockwise, and the left hand counterclockwise.

7. Multiple arm and/or leg translatory movements are more efficient when they are similar, i.e., moving the left hand to the left or forward while at the same time moving the right hand to the right and forward.

8. A person can apply force more accurately to two simultaneously operated controls (i.e., one for each hand) when the controls are located symmetrically with respect to the body and when the directions of movement are similar.

9. When separate and different kinds of controls are operated simultaneously, there is a high probability that the operation of one or another control will suffer in terms of operator input efficiency.

10. Combined movements often result in "cross talk," e.g., trying to push and precisely rotate a control at the same time almost invariably introduces inaccuracy in one of the motions.

11. Excessive disparity between two manual operations (e.g., patting one's head while at the same time rubbing the stomach) often results in a complete breakdown of the sensorimotor response of one of the activities.

12. Continuous feedback is desirable in order for most control movements to remain optimized, i.e., to maintain an accurate direction, force application, and/or rate of movement.

13. On the basis of proprioceptive feedback, individuals judge extent of movement more accurately than movement force applied, and force more accurately than duration of movement.

14. Although visual control is more important while an individual is learning a new perceptual-motor task, as performance becomes habitual, proprioceptive feedback (or "feel") may become the more important feedback resource.

15. When less than 0.5 s per movement is required, blind movements are as accurate as when using visual positioning.

16. When the rate of movement is constant, accuracy diminishes as the interval between movements increases.

17. It takes about 0.04 s to develop maximum tension in the muscles of the human arm.

18. People tend to compensate for changes in the rate of sensory inputs versus movement rates; i.e., they move a control at a higher rate when the control display ratio is less sensitive.

19. Friction that is proportional to the rate of movement (viscous friction), in contrast to coulomb friction, is often advantageous in many control systems. Similarly, inertia in the form of a heavy control moved directly by the operator may also improve performance.

20. The control ratio (i.e., the ratio of control movement to indicator movement) generally has more effect on the time required to make settings within a given degree of tolerance than inertia, friction, or size of the control. Primary adjustment times decrease with increasing distance moved by a controlled marker (pointer or "hook") for each revolution of a control. Secondary adjustment times, however, increase with increases of marker movement per control movement or revolution. Overall time is minimum for gear ratios (e.g., pointer movement in inches per control revolution is between about 1 and 2). Although this ratio is optimum when fine tolerances are required, it may be advantageous to use higher ratios (6:1) in order to arrive at the general, desired setting. For joystick controls, the optimum condition for tolerances of 0.10 in. (0.25 cm) is about 2½ units of stick movement for one unit of indicator movement.

HAND-POSITIONING ACCURACY OF THE BLIND[20]

Although people become extremely accurate at positioning their hands with practice (as in the case of a concert pianist), without benefit of sight the accuracy with which one can point or position the arm or hand is reduced. Figure 38.2 illustrates this point. Note that an individual's blind positioning response is most accurate when he or she points straight ahead. Accuracy diminishes in all other directions. Several values are shown to give an approximate idea of how accuracy in peripheral points may vary from the most accurate point directly in front of the operator. The dark circles within each larger circle denote accuracy within general areas; i.e., the smaller circles denote greater accuracy.

The study from which these data were derived required subjects to sit in a level seat. When an individual is sitting or standing on a tilting platform, inaccuracies increase rapidly and nonuniformly.

FINGER DEXTERITY AND CONTROL OPERATION ACCURACY

Although certain people develop considerable dexterity with practice, the average person is able to perform certain types of control manipulations more accurately than others, as Figs. 38.3 and 38.4 illustrate.

1. Rotational manipulation is more accurate than sliding manipulation or movement of thumb or finger wheels, although the latter is more accurate than the sliding manipulation.

2. Rotation in the horizontal plane is generally more accurate than rotation in the vertical plane, although the horizontal accuracy depends on the ability of the operator to rest his or her hand on the adjacent surface.

3. A push button is located and pressed more accurately when it is positioned as shown in Fig. 38.3 than when it is positioned as shown in Fig. 38.4.

4. A pencil-sized joystick is manipulated more precisely than one requiring a full fist grip, and the accuracy is increased significantly if the operator can rest his or her arm on a nearby horizontal surface.

[20] P. M. Fitts and C. Crannell, *Location Discrimination: Accuracy of Reaching Movements to Twenty-Four Different Areas*, U.S. Air Force Air Materiel Command Technical Report no. 5833, 1950.

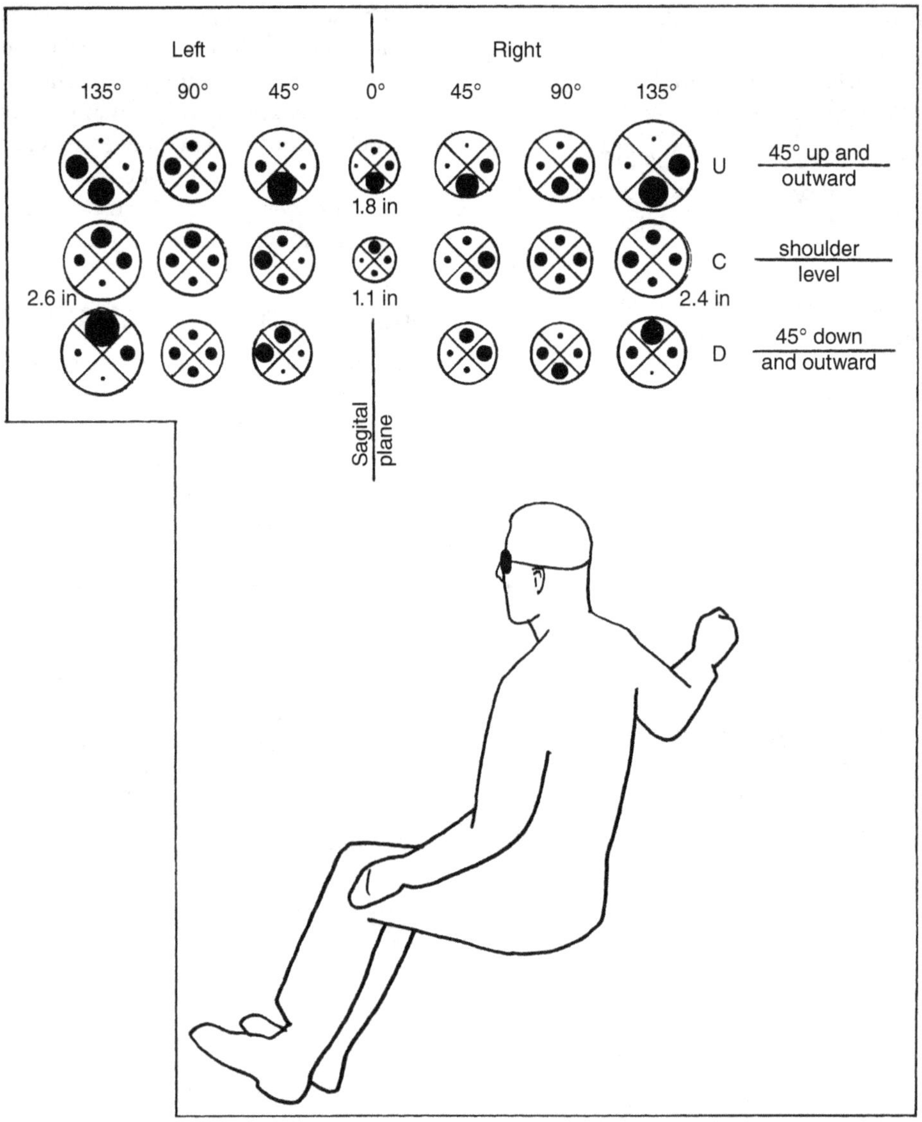

FIGURE 38.2 Blind reaching accuracy (smaller circles denote greater accuracy).

5. An L-shaped handle is more accurately positioned than a round knob, such as a doorknob.

VISUAL POSITIONING ACCURACY

Visual positioning accuracy is a function of motor patterns such as direction and distance of movement. Figure 38.5 illustrates the three important components: (A) primary movement time, (B) average reaction time (250 ms), and (C) secondary movement time. Primary movement time increases with distance, but less than in direct proportion to distance. Secondary movement time (e.g., small corrective adjustments) is relatively constant for distance in excess of about 10 cm.

In general and within reason, the relative time required for visual positioning movements is about the same for short and long distances because a higher velocity is typically used as distance increases.

CRANKING MOVEMENTS

Maximum turn or cranking rate is obtained when the cranking radius is between 3 and 5 cm. However, maximum turn rate decreases with increasing load. For example, the curve shown in Fig. 38.6 is for zero load where the maximum turn rate is about 275 r/min. At 5000 g/cm, the turn rate becomes about 240 r/min.

In general, friction aids the discrimination of small load changes. However, inertia usually provides smoother cranking performance.

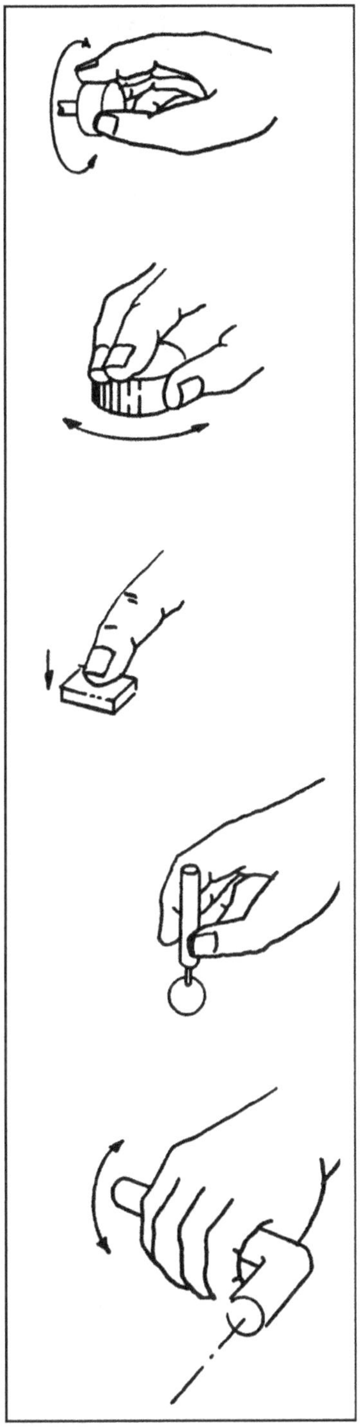

Figure 38.3 More accuracy.

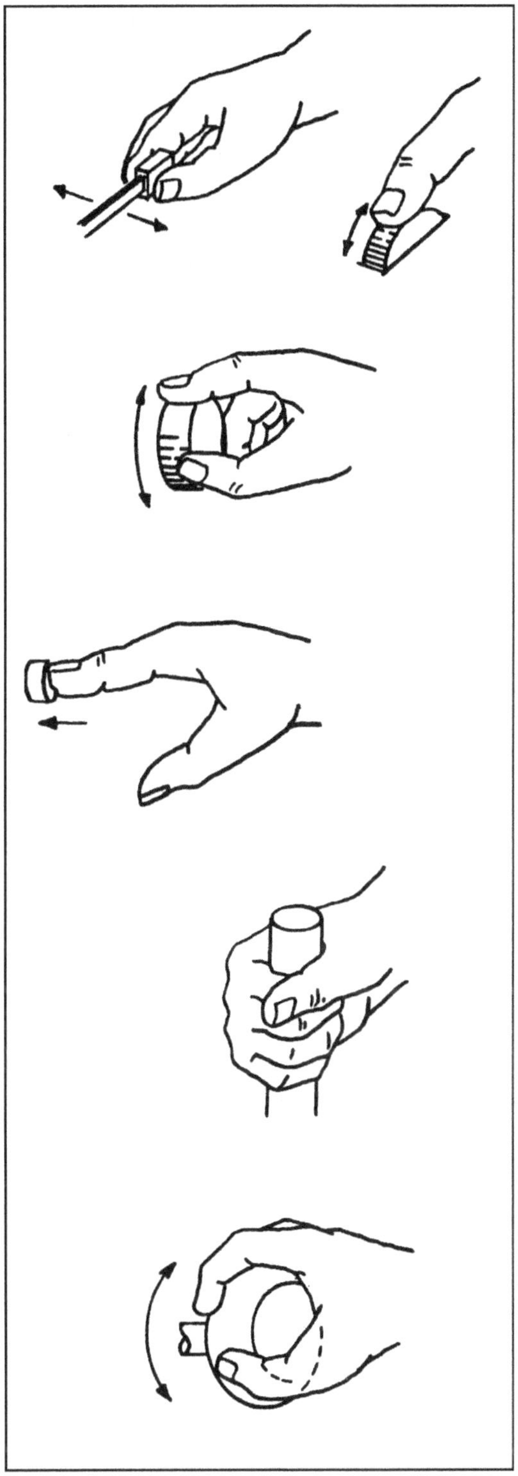

Figure 38.4 Less accuracy.

CONTINUOUS-ADJUSTMENT CONTROL RESPONSE

Common continuous-adjustment controlling tasks include steering and controlling speed and altitude. The controlled system may be a highway vehicle, an aircraft, a spacecraft, a ship, or a submarine, or the movement and positioning of an external object may be involved, as when a crane operator acquires, transports, or deposits a package that is attached to the crane cable system.

Control effectiveness depends on at least the following factors:

1. The ability of the operator to *anticipate* what is going to happen when he or she provides input to the system

2. The ability of the operator to predict what will happen when he or she makes specific system inputs

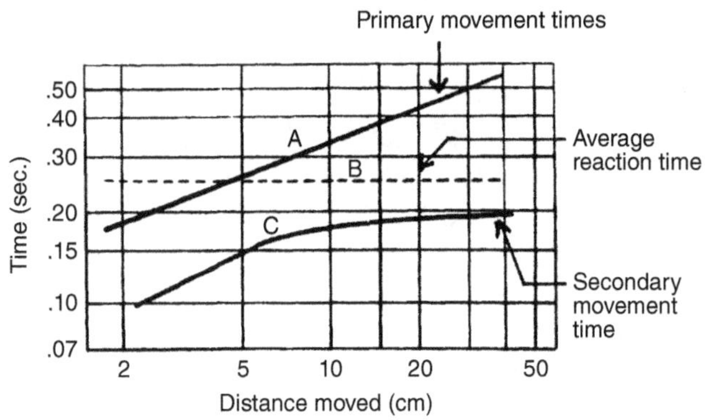

FIGURE 38.5 Visual positioning accuracy.

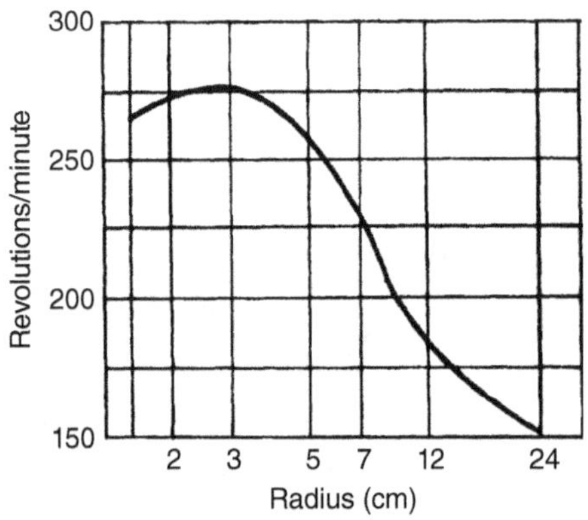

FIGURE 38.6 Cranking speed.

3. Feedback on a timely basis about what is happening as the operator makes control inputs

4. How much differentiation, integration, and/or algebraic addition the control and display task requires of the operator

5. How well the specific control and display devices provide compatible relationships between the operator's sensory, perceptual, and motor and physical characteristics

Control effectiveness is degraded by the following factors:

1. Long delays between inputs, e.g., perceived changes in incoming information, results of operator inputs on system condition, or direct feedback from controller manipulation

2. Too much noise in the system, e.g., extraneous signals, dynamic disturbances, or mechanical artifacts such as "dead space," "stiction," and force irregularities

3. Incompatibilities between control and display direction and rate of motion

4. Controller force requirements that are either too high or too low

5. Incompatibility of the position, direction, and range of movement of the controller with the operator's musculoskeletal system and operating position

6. A requirement that the wrong body element be used, e.g., the hand versus the foot, the left hand versus the right hand, or the whole limb versus the hand and fingers

The following general statements about human efficiency with respect to continuous-adjustment control activities should be considered. Humans are more efficient:

1. When they can make larger motions, since their own proprioceptive feedback mechanisms impart information about what they are doing.

2. When the movements they make are in the *same* direction in which the object, system, or displayed element moves.

3. When the rate of change of their control movement is similar to that of the controlled object or displayed element.

4. When sufficient information is supplied (naturally or artificially) to allow them to "predict" what is going to happen if they maintain their present control input and/or modify it to some extent. To predict, they must also know the general limits of their control system's response range.

5. When the control forces are not too high and are approximately equal throughout the controller movement range; i.e., high initiating forces (stiction) require the operator to suddenly compensate, once the controller is put into motion.

6. When there is some friction in the control system to minimize the effect of external dynamic disturbances and/or inequalities in their own musculoskeletal systems (asymmetry, tremor, fatigue, etc.). However, variable force within certain systems may be desirable to provide the operator with cues relative to the position or condition of the system, e.g., increasingly higher force as a brake pedal approaches "lock-up," toward the limits of a steering-wheel excursion to minimize overcontrolling, or force cues to indicate position of a transmission shift lever.

7. When they are properly positioned and secured; i.e., a seated operator with appropriate handrests or armrests is less influenced by dynamic disturbance and by problems of maintaining body equilibrium.

8. In terms of small, accurate inputs, when they primarily rely on hand and finger movement to manipulate a controller.

9. When they are not required to manipulate too many separate controls in an integrated manner, i.e., when they do not have to perform sequential operation of several hand and/or foot controls while at the same time carrying

out a primary continuous-adjustment task (several switching inputs on a joystick or steering wheel, etc.).

10. When they do not have to hold a control device (and their arms or legs in suspension) for long periods of time or hold the control in a fixed position for extended periods.

11. When they are provided appropriate system aids to relieve them of complex mental information processing to differentiate, integrate, extrapolate, or perform algebraic additions during the control tasks.

12. When the controller-display setup allows them to see what they are doing, e.g., when their hands do not cover up displayed information, and when vehicle elements do not obscure the external view.

PURSUIT VERSUS COMPENSATORY TRACKING

When performing a tracking task that involves matching a continuously variable input signal with an output of the system, the operator requires information indicating the input to be matched and the output with which it is to be matched. When an operator's control and display system provides separate indications for input and for output, this is referred to as a "pursuit task"; i.e., the operator makes inputs into the control and then to a display element to "follow" a target (system input). (See Fig. 38.7.) On the other hand, when input and output signals are presented to the operator in terms of a difference between the system and the operator's control input, the task is called "compensatory" or "error-reducing."

Pursuit tracking is generally a more "natural" task. In addition, the operator is better able to keep track of input changes and to anticipate them, and the operator can also keep better track of the position of the controlled element.

TABEL 38.1 When to Use Compensatory and Pursuit Tracking

A pursuit display should be used when:
1. The course contains high frequencies.
2. It is a zero-order control system.
3. The operator must know actual output, not just error.

A compensatory display should be used when:
1. The system is aided or quickened.
2. The display must be kept small, but output range is large, and/or the precision requirements are high.

Either display can be used when:
1. The course is simple.
2. The display needs to be kept small, but the output range is large, and/or the precisions requirements are high.*

*For a more complete discussion of human tracking capabilities and limitations and for detailed recommendations regarding the design of closed-loop systems, see E. J. McCormick. *Human Factors in Engineering and Design*, 4th ed., McGraw-Hill Book Company, New York, 1976, p. 220.

Pursuit tracking is the preferred mode for simple systems, particularly where input frequency is relatively high. (See Table 38.1.)

Compensatory tracking is generally less natural (and thus requires training). However, when the dynamic characteristics

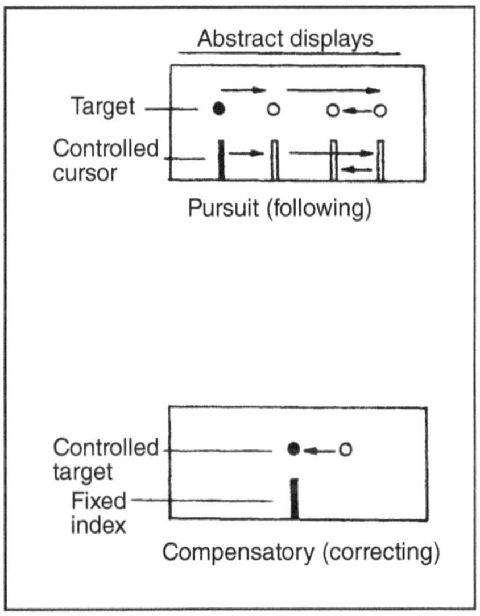

FIGURE 38.7 Illustration of pursuit versus compensatory tracking.

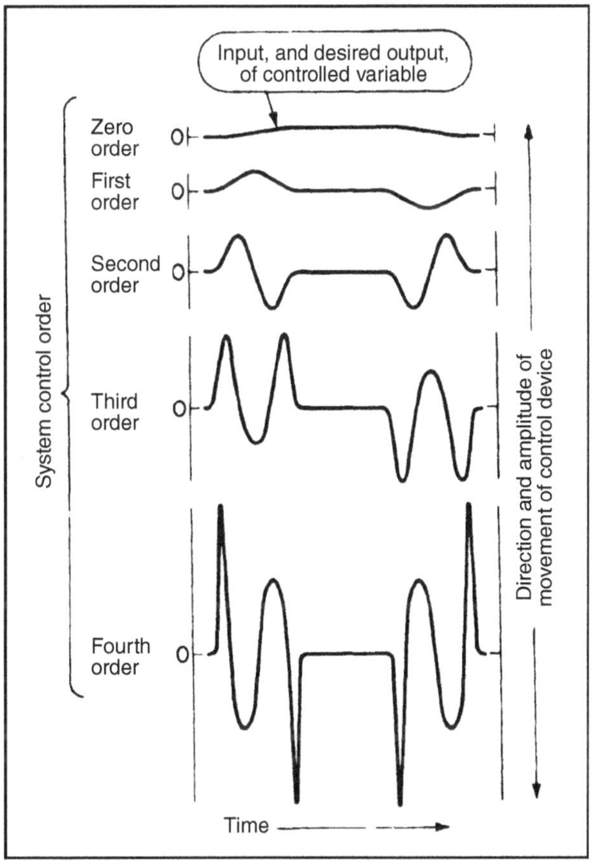

FIGURE 38.8 Changes over time that would have to be made with various control orders to make the controlled variable correspond to the input.

of the system are complex, compensatory tracking modes minimize the probability that operators will confuse input with the results of their own control actions. Another important advantage is that compensatory tracking information can be displayed in less space because in error-reducing formats the total output range does not have to be displayed.

In general, a pursuit display is best when high-frequency inputs are present, and a compensatory display is best when low-frequency inputs are present. (See Fig. 38.8.) A pursuit display allows the operator to anticipate. It is best for higher order control systems; the higher the frequency of desired output, the more superior a pursuit mode is to a compensatory mode. With rate controls, a pursuit display is superior as long as the desired output and input are time-varying. With rate controls, a pursuit display is superior except when the cutoff frequency is about 0.1 Hz or less. When the display must be small (because of lack of space) and when the change in desired output is very slow, a compensatory display mode is superior. A compensatory display is best when an "aided system" supplies derivative terms where positional errors are corrected (i.e., the operator performs better when making corrective responses proportional to only the error).

CHAPTER 39

WORK OUTPUT AND COST

WORK EFFICIENCY

The efficiency with which work is performed is a ratio of output to input. The following equation describes the process:

$$\text{Work efficiency} = \frac{\text{Work done (e.g., foot-pounds)}}{\text{Work done} + \text{heat produced}}$$

It is estimated that human efficiency generally does not exceed about 20 to 25 percent. According to Shilling,[21] a typical 154-lb (70-kg) adult male has the fuel and energy requirements for typical activities shown in Table 39.1. The approximate amount of energy required for one day of typical activities is estimated to be as shown in Table 39.2. Energy expenditures for various occupations, activities, and tasks are shown in Tables 39.3, 39.4, and 39.5.

[21]C. W. Shilling, *The Human Machine,* U.S. Naval Institute, Annapolis, MD, 1955.

COMPARISON OF ENERGY EXPENDITURES

TABLE 39.1 Energy Consumption for Various Activities

Activity	Calories per Hour
Sleeping	65
Resting (awake but lying still)	77
Standing at attention	115
Typewriting rapidly	140
Sweeping the floor (38 strokes per minute)	169
Walking slowly (2.6 mi/h)	200
Performing carpentry, sheetmetal work, or painting	240
Walking moderately fast	300
Sawing wood	480
Swimming	500
Running (5.3 mi/h)	570
Walking up steps	1100

TABLE 39.2 Energy Consumption in a Typical 24-Hour Day

	Calories
8 h of sleep	520
3 h of light work	510
8 h of medium to heavy work	1920
5 h of sitting at rest	500
Day's total	3450

TABLE 39.3 Energy Expenditures by Occupations

		Kilocalories per Day	
Occupation	Type of Work	Men	Women
Bookkeeper	Sedentary, light manual	2400	2000
Secretary	Sedentary, light manual	2700	2250
Bus driver	Sedentary, medium manual	3000	2500
Letter carrier	Walking, climbing stairs, and carrying mailbag	3000	2500
Machine operator	Sedentary, heavy manual	3300	2750
Stonemason	Heavy manual	3600	3000
Laborer	Heavy manual, lifting and carrying	3900	3250
Carpenter	Manual, continuous (hammering and sawing)	3900	3250
Ballet dancer	Vigorous, continuous whole-body movements	3900	3250
Coal miner	Standing, extremely heavy body and limb movements	4200	—
Lumberjack	Standing, extremely heavy body and limb movements	4500	—
Athlete	Marathon running, swimming, and football	4800	4250

WORK, HEAT, AND OXYGEN COSTS

The human machine oxidizes its fuel in order to obtain energy. The body oxidizes fuel at temperatures of about 37°C and converts chemical energy directly into work.

Human energy exchange is expressed primarily in terms of heat. When work is performed, 10–30 percent of the energy expenditure is applied to the external load, and the rest comes out as heat.

Oxygen costs (e.g., heat production) for typical activities are listed in Tables 39.6 and 39.7.[22]

[22]P. Webb, "Work, Heat, and Oxygen Cost," in J. F. Parker and V. R. West (eds.), *Bioastronautics Data Book* (2nd ed., National Aeronautics and Space Administration), 1973.

TABLE 39.4 Energy Expenditures in Relation to Activity

Posture and Mobility Involvement	Kilocalories per Minute	Kilocalories per Hour
Sitting	0.3	20.0
Kneeling	0.5	30.0
Crouching	0.5	30.0
Standing	0.6	35.0
Bending over while standing	0.8	50.0
Walking (medium rate)	1.7–3.5	100–200
Climbing (10° grade minus load)	0.75-m climbing height	0–400
Work involving one arm	0.7–2.2	40–120
Work involving two arms	1.5–3.0	80–160

Note: The above values represent general estimates and should not be used except to establish approximations for evaluating the consequence of proposed design or task configurations.
Source: G. Lehmann, *Praktische Arbeitsphysiologie*, Georg Thieme Verlag, Stuttgart, 1953.

TABLE 39.5 Specific Task Output Costs

Activity	Working Conditions	Kilocalories per Minute
Walking	Without load, level surface (3 km/h)	1.7
	(4 km/h)	2.1
Walking	With load carried on back (10 kg/4 km/h)	3.6
	(30 kg/4 km/h)	5.3
	(50 kg/4 km/h)	8.1
Climbing	16°/11.5 m/min (no load)	8.3
	(20-kg load)	10.5
	(50-kg load)	16.0
Cycling	16 km/h	5.2
Housekeeping	Cooking	1.0–2.0
	Light cleaning, ironing	2.0–3.0
	Scrubbing floor	4.0–5.0

WORK AND REST

Approximately 4 kcal/min has been established as a desirable upper limit of the average energy cost of work. This requires that there be some rest in order for the body to compensate for the excess above this limit. A formula for estimating the amount of rest required for a given amount of work is

$$R = \text{rest} = \frac{T(K-S)}{K-1.5}$$

TABLE 39.6 Everyday Activities

Activity	Oxygen Consumption, L/min	Equivalent Heat Production, kcal/min
Asleep		
Sleeping, men over 40	0.22	1.1
Sleeping, men aged 30–40	0.24	1.2
Sleeping, men aged 20–30	0.24	1.2
Sleeping, men aged 15–20	0.25	1.3
Resting		
Lying fully relaxed	0.24	1.2
Lying moderately relaxed	0.26	1.3
Lying awake, after meals	0.28	1.4
Sitting at rest	0.34	1.7
Very light activity—seated		
Writing	0.36	1.8
Riding in automobile	0.40	2.0
Typing	0.46	2.3
Polishing	0.48	2.4
Very light activity—standing		
Relaxed	0.36	1.8
Drafting	0.38	1.9
Taking lecture notes	0.40	2.0
Peeling potatoes	0.42	2.1
Light activity—seated		
Playing musical instruments	0.58	2.9
Repairing boots and shoes	0.60	3.0
At lecture	0.60	3.0
Assembling weapons	0.72	3.6
Light activity—standing		
Entering ledgers	0.52	2.6
Washing clothes	0.74	3.7
Ironing	0.88	4.4
Scrubbing	0.94	4.7
Light activity—moving		
Slow movement about room	0.50	2.5
Vehicle repairs	0.68	3.4
Slow walking	0.76	3.8
Washing	0.84	4.2
Moderate activity—lying		
Creeping, crawling, prone resting maneuvers	1.14	5.7
Crawling	1.22	6.1
Swimming breaststroke at 1 mi/h	1.36	6.8
Swimming crawl at 1 mi/h	1.40	7.0
Moderate activity—sitting		
Rowing for pleasure	1.00	5.0
Cycling at 8–11 mi/h	1.14	5.7
Cycling rapidly	1.38	6.9
Trotting on horseback	1.42	7.1
Moderate activity—standing		
Gardening	1.16	5.8
Chopping wood	1.24	6.2
Baseball pitching	1.30	6.5
Shoveling sand	1.36	6.8
Moderate activity—moving		
Golf	1.08	5.4
Table tennis	1.16	5.8
Tennis	1.26	6.3
Army drill	1.42	7.1
Heavy activity—lying		
Leg exercises, average	1.50	7.5
Swimming breaststroke at 1.6 mi/h	1.64	8.2
Swimming backstroke at 1.0 mi/h	1.66	8.3
Lying on back, head raising	1.76	8.8
Heavy activity—sitting		
Cycling rapidly, own pace	1.66	8.3
Cycling at 10 mi/h, heavy bicycle	1.78	8.9
Cycling in race (100 mi in 4 h 22 min)	1.96	9.8
Trotting on horseback	1.96	9.8

(Continued)

TABLE 39.6 Everyday Activities (*Continued*)

Activity	Oxygen Consumption, L/min	Equivalent Heat Production, kcal/min
Heavy activity—standing		
Chopping wood	1.50	7.5
Shoveling sand	1.54	7.7
Sawing wood by hand	1.60	8.0
Digging	1.78	8.9
Heavy activity—moving		
Skating at 9 mi/h	1.56	7.8
Playing soccer	1.66	8.3
Skiing at 3 mi/h on level	1.80	9.0
Climbing stairs at 116 steps/min	1.96	9.8
Very heavy activity—sitting		
Cycling at 13.2 mi/h	2.00	10.0
Rowing with two oars at 3.5 mi/h	2.20	11.0
Galloping on horseback	2.28	11.4
Sculling (97 strokes/min)	2.52	12.6
Very heavy activity—moving		
Fencing	2.10	10.5
Playing squash	2.10	10.5
Playing basketball	2.28	11.4
Climbing stairs	2.40	12.0
Extreme activity		
Wrestling	2.60	13.0
Marching at double	2.66	13.3
Endurance marching	2.96	14.8
Harvard Step Test	3.22	16.1

TABLE 39.7 Specialized Activities

Activities	Oxygen Consumption, L/min	Equivalent Heat Production, kcal/min
Engineering tasks		
Medium assembly work	0.58	2.9
Welding	0.60	3.0
Sheet metal work	0.62	3.1
Machining	0.66	3.3
Punching	0.70	3.5
Machine fitting	0.90	4.5
Heavy assembly work–noncontinuous	1.02	5.1
Driving vehicles and piloting aircraft		
Driving a car in light traffic	0.26	1.3
Night flying—DC-3	0.32	1.6
Piloting DC-3 in level flight	0.34	1.7
Piloting helicopters	0.36	1.8
Instrument landing—DC-4	0.50	2.5
Piloting light aircraft in rough air	0.54	2.7
Taxiing DC-3	0.58	2.9
Piloting bomber aircraft in combat	0.58	2.9
Driving car in heavy traffic	0.64	3.2
Driving truck	0.66	3.3
Driving motorcycle	0.70	3.5
Moving over rough terrain on foot		
Flat, firm road 2.5 mi/h	0.56–0.98	2.8–4.9
Grass path 2.5 mi/h	0.64–1.02	3.2–5.1
Stubble field 2.5 mi/h	0.80–1.22	11.0–6.1
Deeply plowed field 2.0 mi/h	0.98–1.38	4.9–6.9
Steep 45° slope 1.5 mi/h	0.98–1.38	4.9–6.9
Plowed field 3.3 mi/h	1.56	7.8
Soft snow, with 44-lb load 2.5 mi/h	4.2	21.0
Load carrying		
Walking on level with 58-lb load, trained men		
2.1 mi/h	0.38	1.9
2.7 mi/h	0.58	2.9
3.4 mi/h	0.92	4.6
4.1 mi/h	1.66	8.3

TABLE 39.7 Specialized Activities (*Continued*)

Activities	Oxygen Consumption, L/min	Equivalent Heat Production, kcal/min
Walking on level with 67-lb load, trained men		
2.1 mi/h	0.46	2.3
2.7 mi/h	0.58	2.9
3.4 mi/h	1.02	5.1
4.1 mi/h	1.68	8.4
Walking on level with 75-lb load, trained men		
2.1 mi/h	0.50	2.5
2.7 mi/h	0.68	3.4
3.4 mi/h	1.04	5.2
4.1 mi/h	1.72	8.6
Walking up 36% grade with 43-lb load, sedentary men		
0.5 mi/h	1.34	6.7
1.0 mi/h	2.46	12.3
1.5 mi/h	3.20	16.0
Swimming on surface Breaststroke		
1 mi/h	1.40	7.0
2 mi/h	5.80	29.0
3 mi/h	19.40	97.0
Crawl		
1 mi/h	1.80	9.0
2 mi/h	3.60	18.0
3 mi/h	9.60	48.0
Butterfly		
1 mi/h	2.40	12.0
2 mi/h	5.80	29.0
3 mi/h	15.00	75.0
Walking under water		
Walking in tank minimal rate	0.58	2.9
Walking on muddy bottom minimal rate	1.10	5.5
Walking in tank maximal rate	1.44	7.2
Walking on muddy bottom maximal rate	1.68	8.4
Movement in snow		
Skiing in loose snow 2.6 mi/h	1.62	8.1
Sled pulling—hard snow 2.2 mi/h	1.72	8.6
Snowshoeing (bearpaw) 2.5 mi/h	1.74	8.7
Skiing on level 3.0 mi/h	1.8	9.0
Sled pulling—low drag, medium snow 2.0 mi/h	1.94	9.7
Snowshoeing—trail type 2.5 mi/h	2.06	10.3
Walking, 12–18 in snow,		
breakable crust 2.5 mi/h	2.54	12.7
Skiing on loose snow 5.2 mi/h	2.92	14.6
Snowshoeing—trail type 3.5 mi/h	2.96	14.8
Skiing on loose snow 8.1 mi/h	4.12	20.6
Measured work at different altitudes		
Bicycle ergometer		
Workload Altitude		
430 kg • m/min 720 mm Hg	1.02	5.1
430 kg • m/min 620 mm Hg	0.98	4.9
430 kg • m/min 520 mm Hg	1.08	5.4
Mountain climbing		
880–1037 kg • m/min 610 mm Hg	1.84–2.20	9.2–11.0
566–786 kg • m/min 425 mm Hg	1.54–1.90	7.7–9.5
393–580 kg • m/min 370 mm Hg	1.28–2.10	6.4–10.5

where R = rest (in minutes), T = total working time, K = the average kilocalories per minute of work, and S = kilocalories per minute as the adopted standard. The value 1.5 approximates the resting level in kilocalories per minute.

If an S value of 4 kcal/min is used and we want to calculate R for a 1-h period (e.g., T = 60 min), the formula becomes

$$R = \frac{60(K-4)}{K-1.5}$$

Then one can project a curve such as that shown in Figs. 39.1–39.4. Of course, individual differences in energy considered as an acceptable standard make this formula purely an approximation.

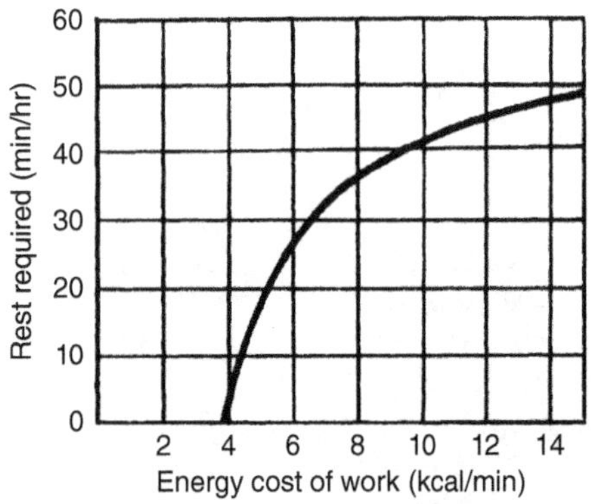

FIGURE 39.1 Rest requirements after work.

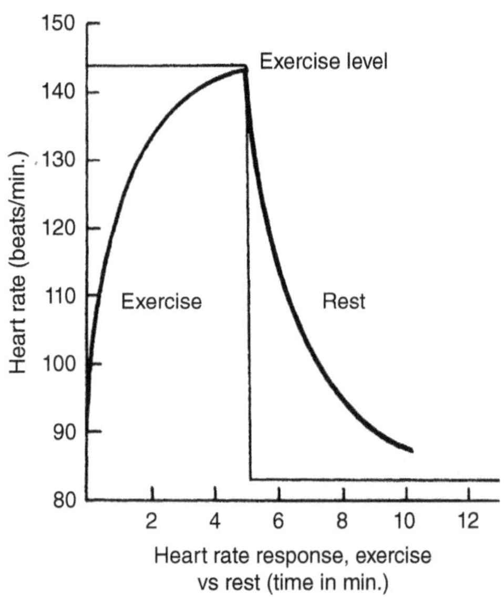

FIGURE 39.2 Heart rate: exercise versus rest.

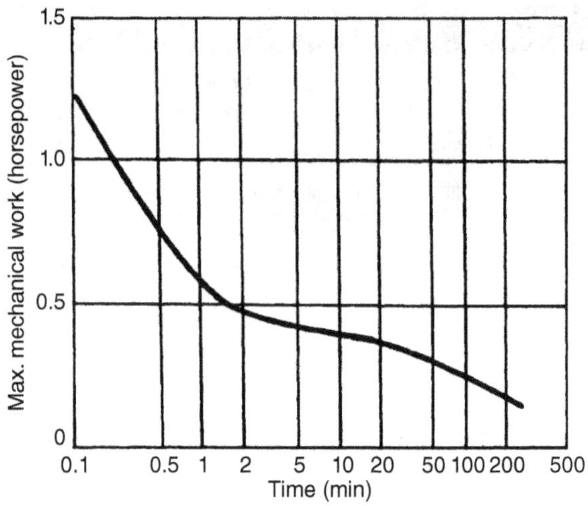

FIGURE 39.3 Work output versus work duration.

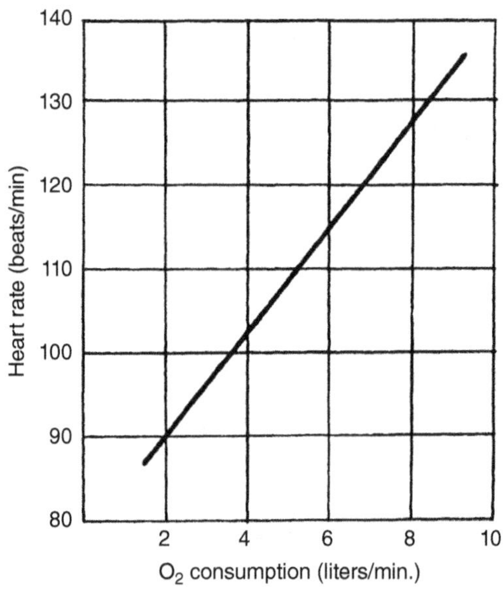

FIGURE 39.4 Heart rate versus oxygen consumption.

CALCULATING OXYGEN COSTS

To determine heat output from respiratory data:

1. Calculate oxygen consumption from respiratory ventilation volume and the difference in O_2 concentration between inspired and expired air.
2. Correct volume to 0°C, 760 mmHg, dry (STPD).
3. Select heat output corresponding to each unit volume of O_2 by approximating the subject's diet or on the basis of his or her measured respiratory quotient.

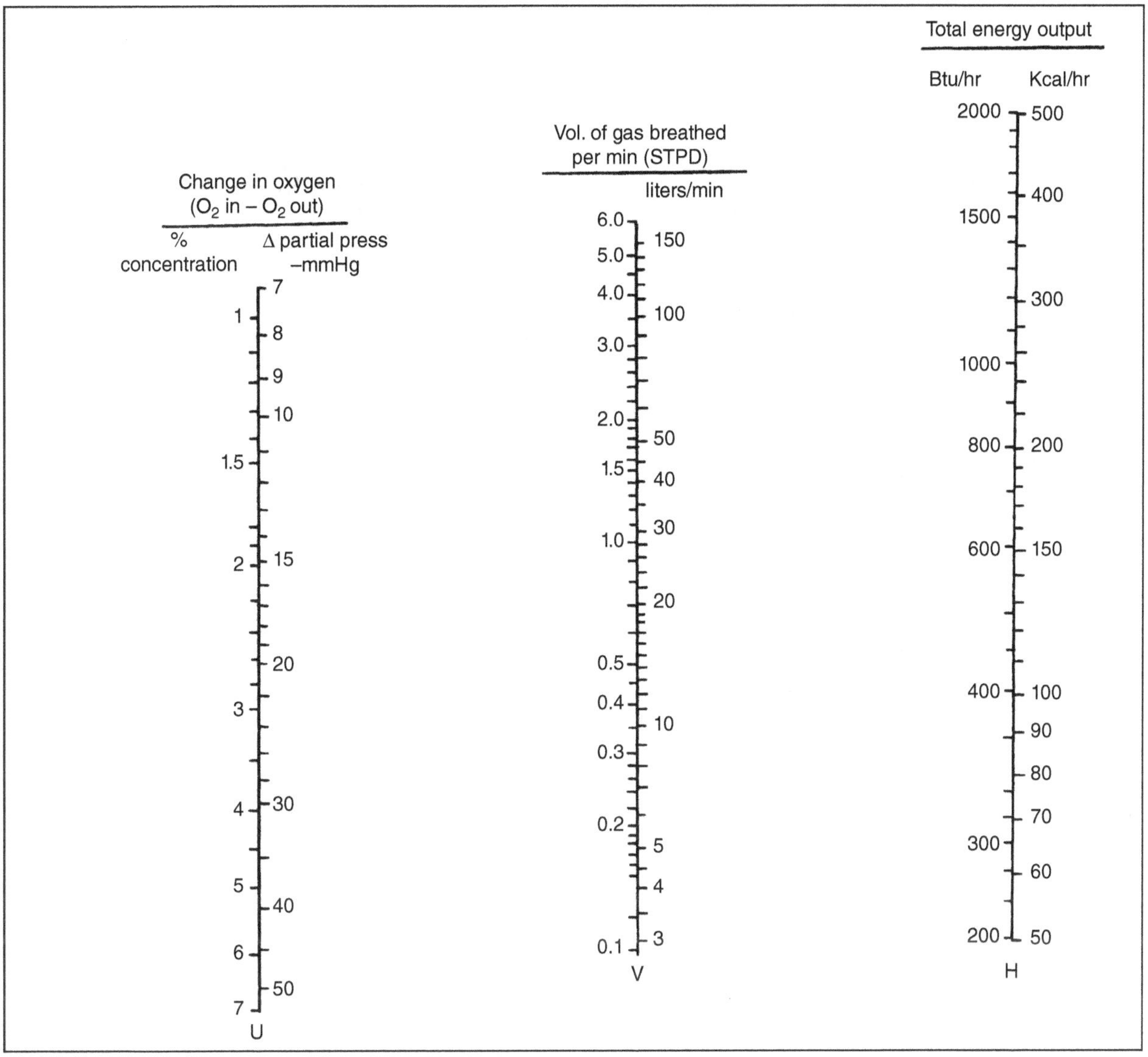

Figure 39.5 Nomogram relating respiratory data and total energy output.

Figures 39.5 and 39.6 help simplify the calculations.

Nomogram A (Fig. 39.5) uses the following values: RQ = 1.00, and 1 L of O_2 is equivalent to 5.0 kcal, permitting direct calculation of heat output (H) in Btu per hour from oxygen uptake (U) and ventilation rate (V). Alternatively, U can be calculated from H and V, or V can be calculated from U and H.

Nomogram B (Fig. 39.6) uses the following values: RZ = 0.82, and 1 L of O_2 is equivalent to 4.825 kcal. This nomogram allows one to interrelate, by drawing straight vertical lines, the values for oxygen consumption (O), heat output (H), external work output (W), and carbon dioxide production (C) at typical conversion rates. H may be as much as 3 percent lower or 5 percent higher than the quoted value at any specific oxygen consumption, depending on the RQ, which equals 0.7 for a pure fat diet and 1.00 for a pure carbohydrate diet. The values given in the third and fourth lines should be modified if the efficiency changes. Typical ranges are 5–25 percent, with an average of 20 percent; thus, the listed work output may increase by three-fourths if the task is one that can be performed at high efficiency (e.g., bicycling). Conversely, the true value may be reduced by three-fourths if the function is inefficiently performed (e.g., high-speed walking).

See also Fig. 39.7.

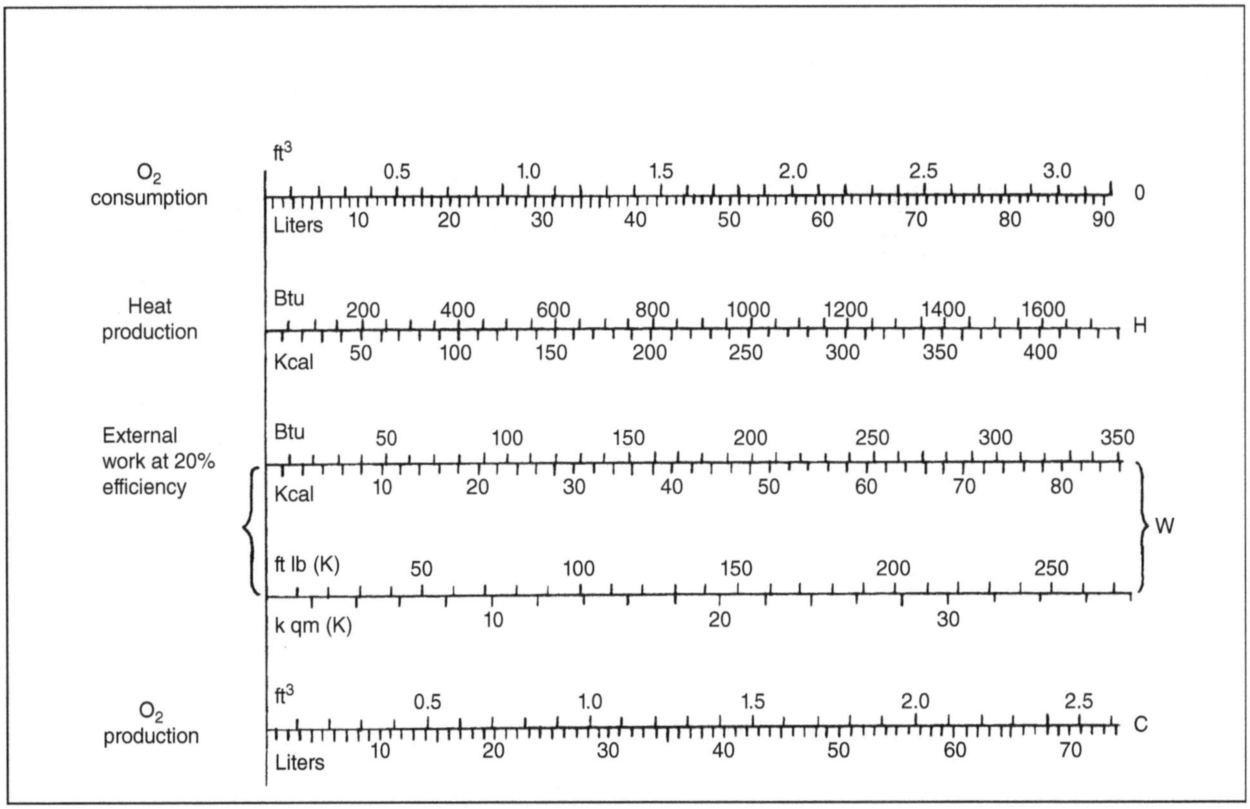

FIGURE 39.6 Nomogram relating respiratory data, external work, and heat production.

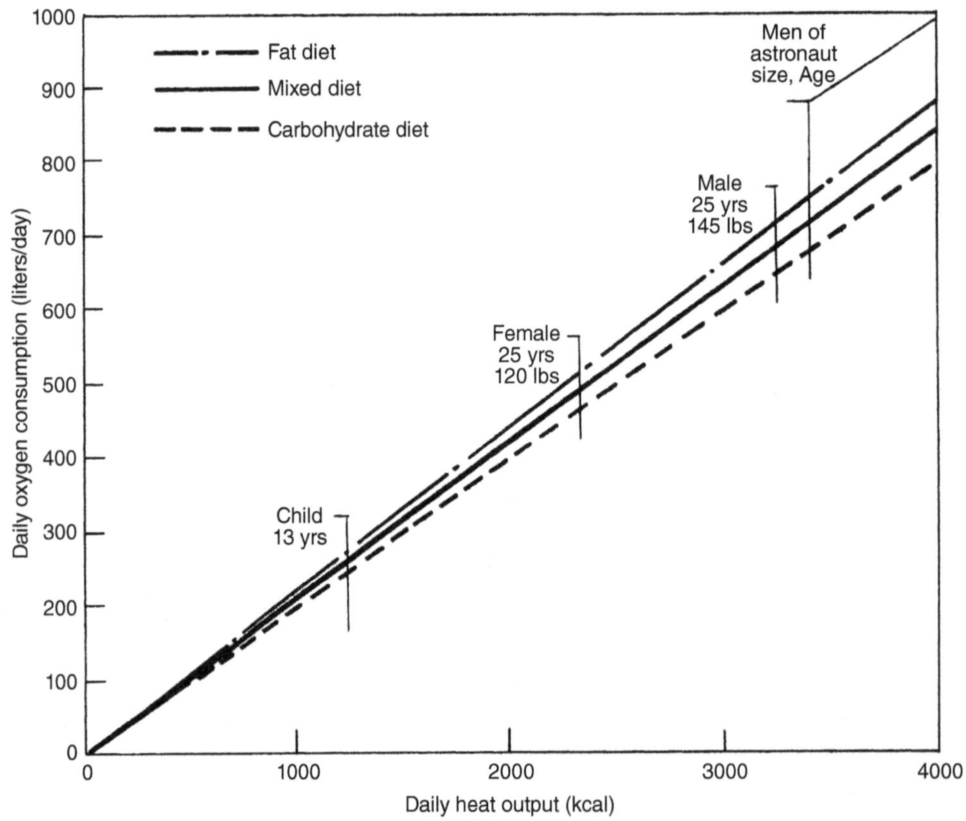

FIGURE 39.7 Estimate of total daily O_2, cost and heat produced for normally active people, as affected by body size. (J. F. Parker, and V. R. West [Eds.], *Bioastronautics Data Book, 2nd ed.*, National Aeronautics and Space Administration, 1973, p. 852.)

CHAPTER 40

THERMAL RESPONSE

THERMAL RESPONSE OF THE BODY

Normal body temperatures and cooling requirements for various regions of the body are given in Table 40.1. At 62°F (16.7°C) body cooling begins. Above 78°F (25.5°C) the body gains heat. The body adapts to changes as great as 20 percent within the comfort range through evaporative cooling when it is hot and through exercise when it is cold.

RESPONSE IN RELATION TO SKIN TEMPERATURES, AMBIENT AIR, AND DIRECT SURFACE TEMPERATURES

Tolerance to High Temperatures[23]

Elevated temperatures not only affect comfort (which can lead to high absenteeism and turnover), but can degrade the performance of even motivated personnel. Comfort and impaired performance limits are shown in Tables 40.2 through 40.5 and Figs. 40.1 through 40.4.

[23]Criteria for Recommended Standard-Occupational Exposure to Hot Environments, National Institute for Occupational Safety and Health, USGPO-HSM72-10269, 1972.

Tolerance to Low Temperatures

In a cold-stress situation, the body will rapidly reduce peripheral circulation in an attempt to conserve core heat. As the core and skin temperatures continue to drop, shivering begins and discomfort is continually present. Eventually shivering may become violent and uncontrollable. Comfort and impaired performance limits are shown in Tables 40.2 through 40.5 and Figs. 40.1, 40.5, and 40.6.

If the surface temperature of the hand falls below 12°C–14°C (54°F–57°F), manipulative ability begins to fail. As the hand temperature drops lower, more serious loss of manipulative ability occurs, partly from stiffness and partly from the loss of tactile sensitivity. The final decrement is caused by the loss of brain functioning due to a drop in core temperature. The brain loses the capacity for cognitive functions if its temperature drops much below 34°C–35°C (93°F–95°F) even though the body is still capable of responding to instruction.

As the core continues to lose heat, shivering eventually lessens and then stops altogether. At this point, complete loss of thermoregulation is imminent. Death, however, may not

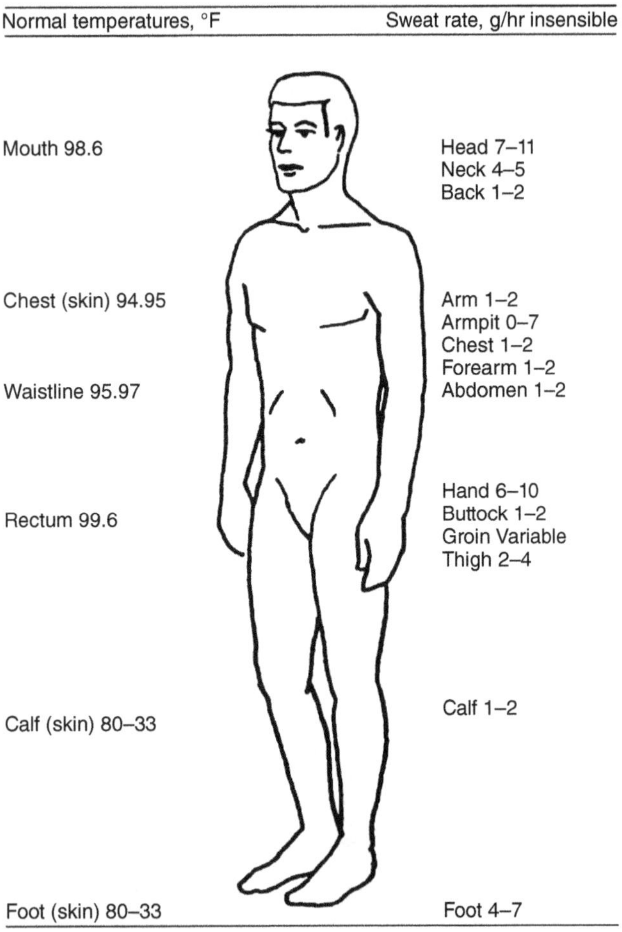

TABLE 40.1 Normal Body Temperatures and Heat Loss

Region	Preferred Temperature, °F	Heat Loss, Btu/h	Area, ft²	Skin Conductance, Btu/ft²·°F
Head	94.4	15.9	2.15	1.61
Chest	94.4	32.6	1.83	3.87
Abdomen	94.4	17.9	1.29	3.02
Back	94.4	49.3	2.43	4.31
Buttocks	94.4	33.0	1.94	3.70
Thighs	91.4	47.7	3.55	1.76
Calves	87.5	58.0	2.15	2.06
Feet	83.6	39.7	1.29	1.96
Arms	91.4	33.4	1.07	4.10
Hands	83.5	63.5	0.75	6.45

TABLE 40.2 General Responses

Temperature of Skin, °F	Response
98	Very hot
96	Unpleasantly warm
94	Slightly warm
93	Comfortable
91	Comfortably cool
88	Slightly too cool
86	Unpleasantly cool
84	Very cold

TABLE 40.3 Temperature and Its Effect on Comfort of the Extremities

	Skin Temperature, °F	
	Hands	Feet
Minimum acceptable	68	73
Tolerable	68-59	73-64
Intolerable	59-50	64-55
Numbness sets in	<50	<55

TABLE 40.4 Effects on Skin of Contact with Surface of Different Temperatures

Temperature, °F	Sensation or Effect
212	Second-degree burn on 15-s contact
180	Second-degree burn on 30-s contact
160	Second-degree burn on 60-s contact
140	Pain; tissue damage
120	Pain; burning heat
91	Warm; neutral (physiological zero)
54	Cool
37	"Cool heat" sensation
32	Pain
Below 32	Pain; tissue damage (freezing)

come quickly. The core temperature can be drastically reduced, to 26°C (78.6°F) or lower, and the body can still survive. If the core experiences such extreme cooling, however, death can occur when attempts are made to re-warm the body, cardiac fibrillation being the common cause of death.

Figure 40.5 shows tolerance times for endurance to cold temperatures. Subjects were wearing various types of clothing, were

TABLE 40.5 Effect of Temperature on Human Performance

Effective Temperature °F*	Performance Effects
90	Upper limit for continued occupancy over any reasonable period of time.
80–90	Expect universal complaints, serious mental and psychomotor performance decrement, and physical fatigue.
80	Maximum for acceptable performance even of limited work; work output reduced as much as 40–50 percent; most people experience nasal dryness.
78	Regular decrement in psychomotor performance expected; individuals experience difficulty falling asleep and remaining asleep; optimum for bathing or showering.
75	Clothed subjects experience physical fatigue, become lethargic and sleepy, and feel warm; unclothed subjects consider this temperature optimum without some type of protective cover.
72	Preferred for year-round sedentary activity while wearing light clothing.
70	Midpoint for summer comfort; optimum for demanding visual-motor task.
68	Midpoint for winter comfort (heavier clothing) and moderate activity, but slight deterioration in kinesthetic response; people begin to feel cool indoors while performing sedentary activities.
66	Midpoint for winter comfort (very heavy clothing), while performing heavy work or vigorous physical exercise.
64	Lower limit for acceptable motor coordination; shivering occurs if individual is not extremely active.
60	Hand and finger dexterity deteriorates, limb stiffness begins to occur, and shivering is positive.
55	Hand dexterity is reduced by 50 percent, strength is materially less, and there is considerable (probably uncontrolled) shivering.
50	Extreme stiffness; strength applications accompanied by some pain; lower limit for unprotected exposure for more than a few minutes.

*These temperature effects are based on relatively still air and normal humidity (40 to 60 percent). Higher temperatures are acceptable if airflow is increased and humidity is lowered (a shift of from 1 to 4°); lower temperatures are less acceptable if airflow increases (a shift upward of 1 to 2°).

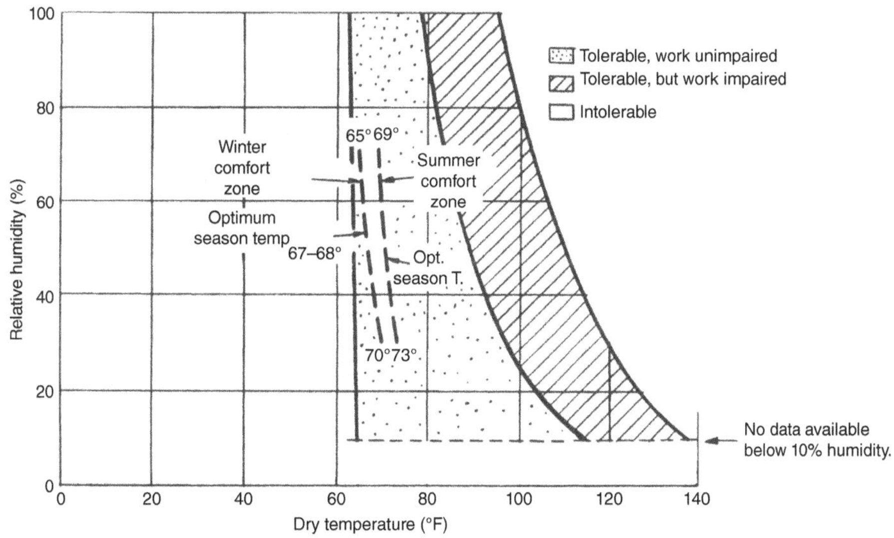

Figure 40.1 Tolerable temperature–humidity requirements (with conventional clothing).

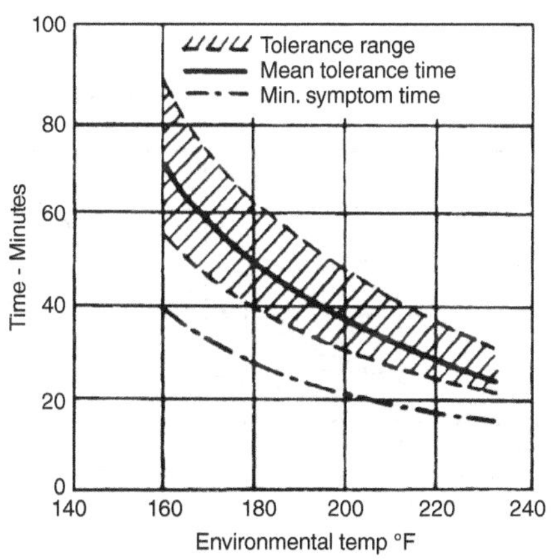

Figure 40.2 Mean and range of tolerance time and minimum symptom time for high temperature.

seated, and were performing light work [air velocity, approximately 200 ft/min (61 m/min); barometric pressure, 1 atm].

A—1-clo (light coveralls)

B—2-clo (woolen underwear, coveralls, and jacket)

C—3-clo (intermediate-weight flight clothing)

D—4-clo (heavy flight clothing)

Effects of Cold on Selected Task Performances

A—tactile sensitivity, bare hand (see Fig. 40.6)

B—simple visual reaction time

C—manual skill

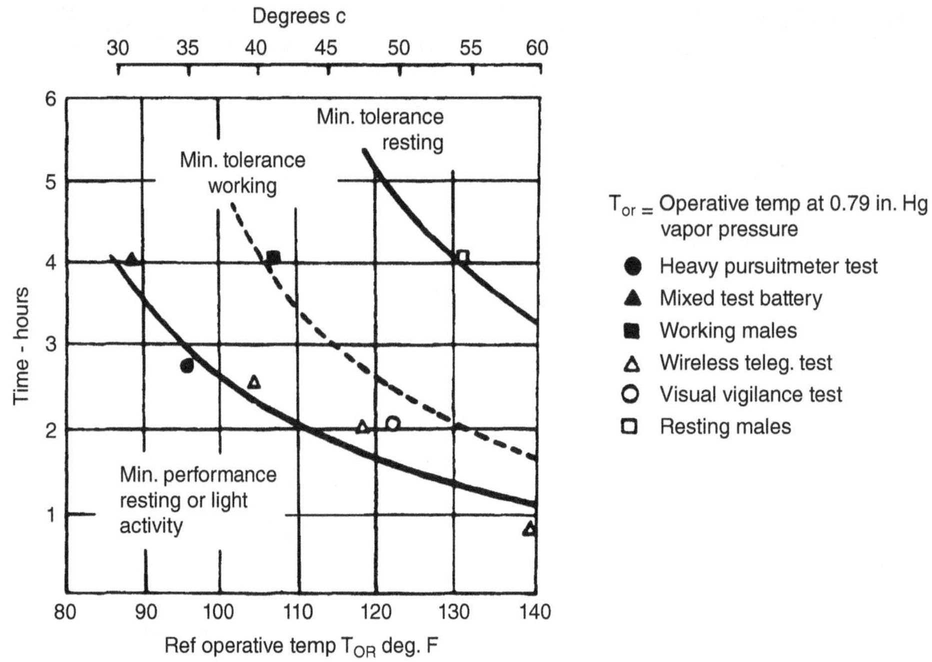

FIGURE 40.3 Performance and tolerance limits in the quasi-compensable zone for lightly dressed men.

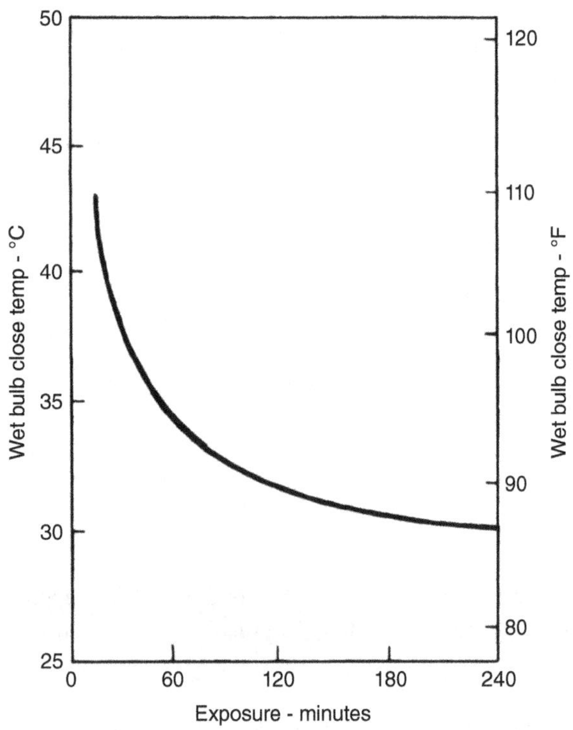

FIGURE 40.4 Upper limits of exposure for unimpaired mental performance.

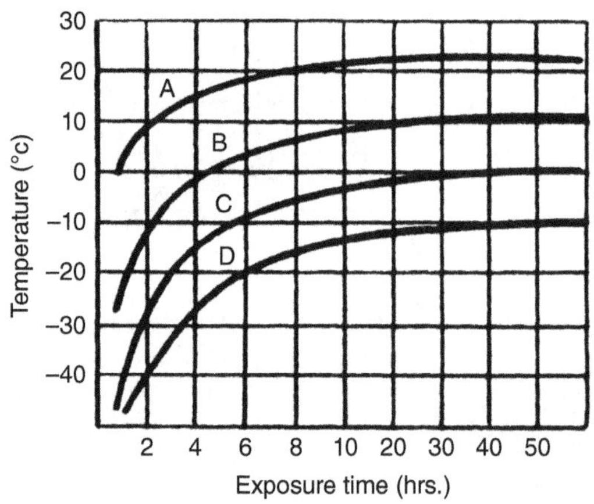

FIGURE 40.5 Tolerance times to cold.

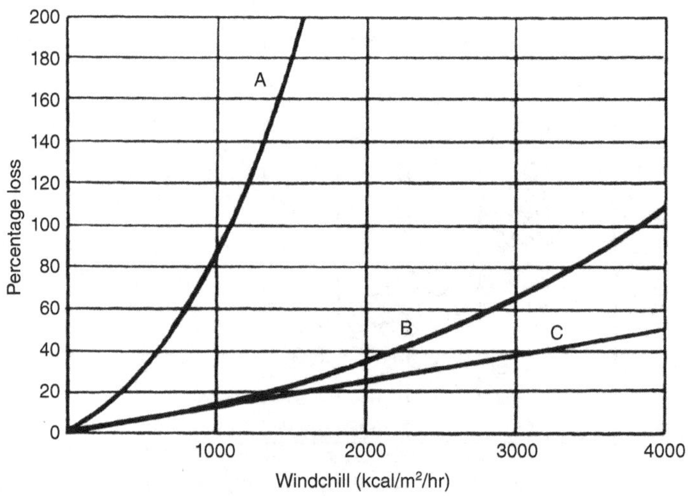

FIGURE 40.6 Effects of cold on tasks.

Wind-Chill Nomograph[24] (See Fig. 40.7)

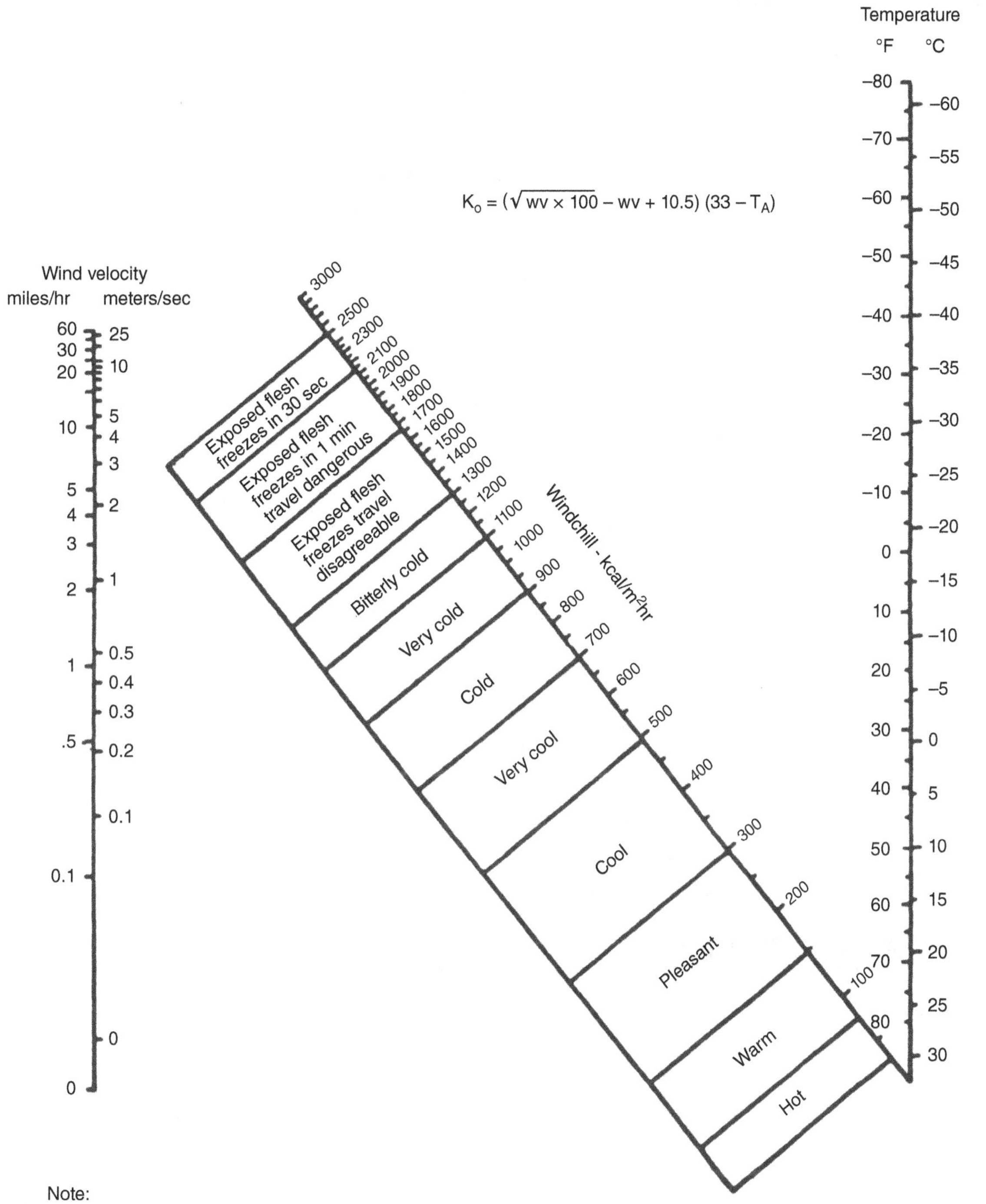

Note:

The windchill index does not account for physiological adaptations or adjustments and should not be used in a rigorous manner. It is based on field measurements during world war II of the rate of cooling of a container of water.

FIGURE 40.7 Nomogram for calculating windchill.

[24]J. F. Parker and V. R. West (eds.), *Bioastronautics Data Book,* 2nd ed., National Aeronautics and Space Administration, 1973.

COMFORT ZONE

Figure 40.8 shows the overall comfort zone for humans, considering broad ranges of temperature, humidity, pressure, activity, and clothing.

BODY CORE TEMPERATURES

Table 40.6 identifies body core temperature range limits and associated performance decrements (from NASA Human Interface Design Handbook).

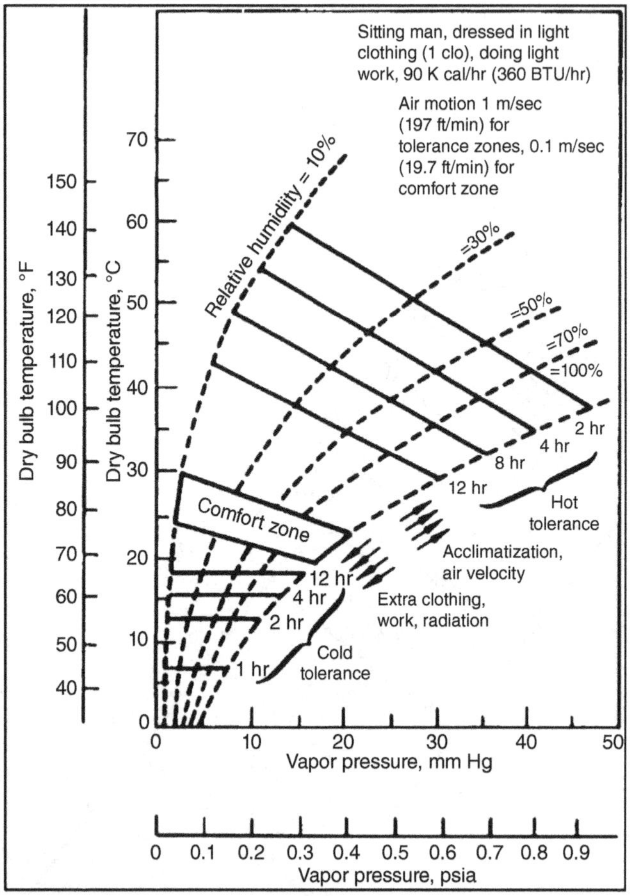

FIGURE 40.8 Comfort zone.

TABLE 40.6 Body Core Temperatures

Core Temperature, °C (°F)	Equivalent Heat Storage, kilojoules (BTU)	Medical Condition
33.7–38.2 (99.9–100.8)	317–422 (300–400)	Onset of decrement in performance of cognitive tasks Decreasing manual dexterity Discomfort Hyperthermia (or heat stress)
38.2–39.2 (100.8–102.6)	422–633 (400–600)	Slowed cognitive function Increased errors in judgment Loss of tracking skills 25% risk of heat casualties Possible heat exhaustion
39.2–39.6 (102.6–103.3)	633–844 (600–800)	Functional limit of physical tasks 50% risk of heat casualties Possible heat exhaustion Possible heat stroke
>40 (>104)	>844 (>800)	100% risk of heat casualties Probably heat stroke

TEMPERATURE AND VENTILATION REQUIREMENTS

Body metabolism, airspace per person, and ventilation rate effect the intensity of human body odor. Ventilation requirements in relation to net airspace and the removal of body odor are shown in Fig. 40.9, which illustrates how the intensity of body odors in a given area depends on the rate of flow of odor-free air. Table 40.7 suggests minimum ventilation rates for various levels of crowding (airspace per person) and human metabolic levels (type of occupancy.)

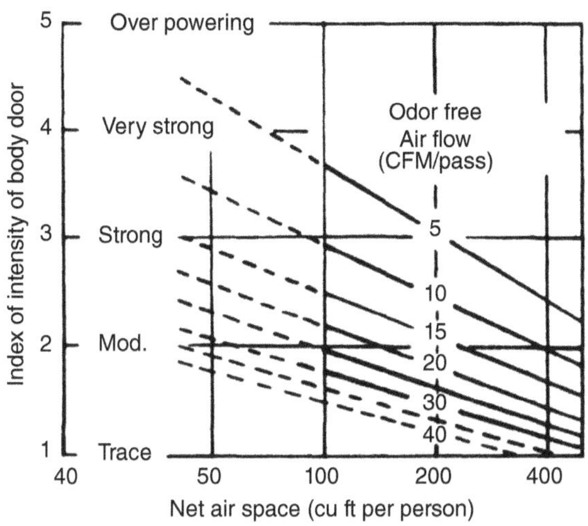

Figure 40.9 Ventilation requirements.

TABLE 40.7 General Minimum Fresh-Air Requirements for Typical Ventilation Needs

Type of Occupancy	Airspace per Person, ft³	Ventilation per Person, ft³/min
Sedentary adults	100	25
	200	16
	300	12
	500	7
Adult laborers	200	23
Grade school children	100	29
	200	21
	300	17
	500	11

Note: These are minimum requirements during the normal heating season, with or without recirculation and without air conditioning. Norms for rooms at sea level range from 5 to 50 ft³/min per person, e.g., 7.5 for theatres; 15 for restaurants, hospital rooms, and apartments; and 30 for private offices, boardrooms, etc.

Source: R. A. McFarland, *Human Factors in Air Transport Design*, McGraw-Hill Book Company, New York, 1946, pp. 119, 121.

CHAPTER 41
VISUAL RESPONSE

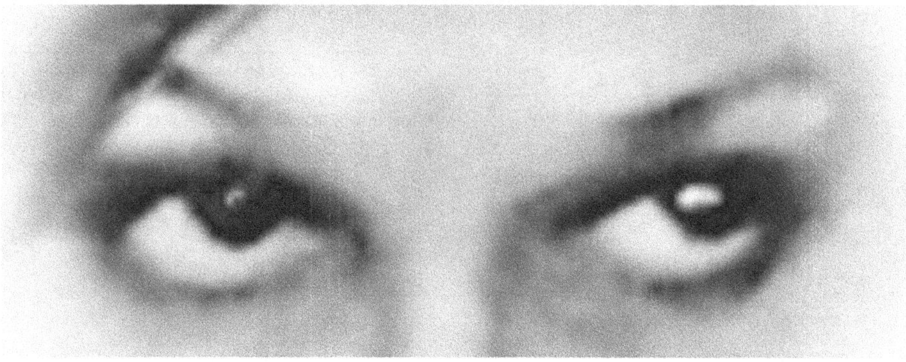

VISUAL SYSTEM CHARACTERISTICS

The primary organ through which visual responses are made is, of course, the eye. Figures 41.1 and 41.2 show the key elements of the visual receptor and liken them to the camera and the photographic process.

Image and Retinal System

The image of a moving object moves across the retina when the eyes are held still, providing information of the movement through sequential receptor stimulation along the path.

Eye and Head Movement System

When the eye follows a moving object, the image remains more or less stationary upon the retina, although the person still sees movement. (See Fig. 41.3.)

The image and retinal system and the eye and head movement system sometimes disagree, leading to visual illusions that are often confusing and misleading.

Size Constancy

The image of an object on the retina halves in size with each doubling of viewing distance of the object. However, it does not appear to shrink so much because the brain compensates for the shrinkage of the image with distance. (See Fig. 41.4.)

ACCOMMODATION AND CONVERGENCE

The size of a retinal image is calculated by using the equation $AB/ab = An/an$, when the size of the object and its viewing distance are known. (See Fig. 41.5.)

Amplitude of Accommodation

Although a child can focus on an object that is as close as 2.4 in. (6.1 cm), an adult can focus no closer than 6 in. (15.2 cm). Hardening of the lens with age reduces the focusing capacity of older persons' eyes. (See Fig. 41.6.)

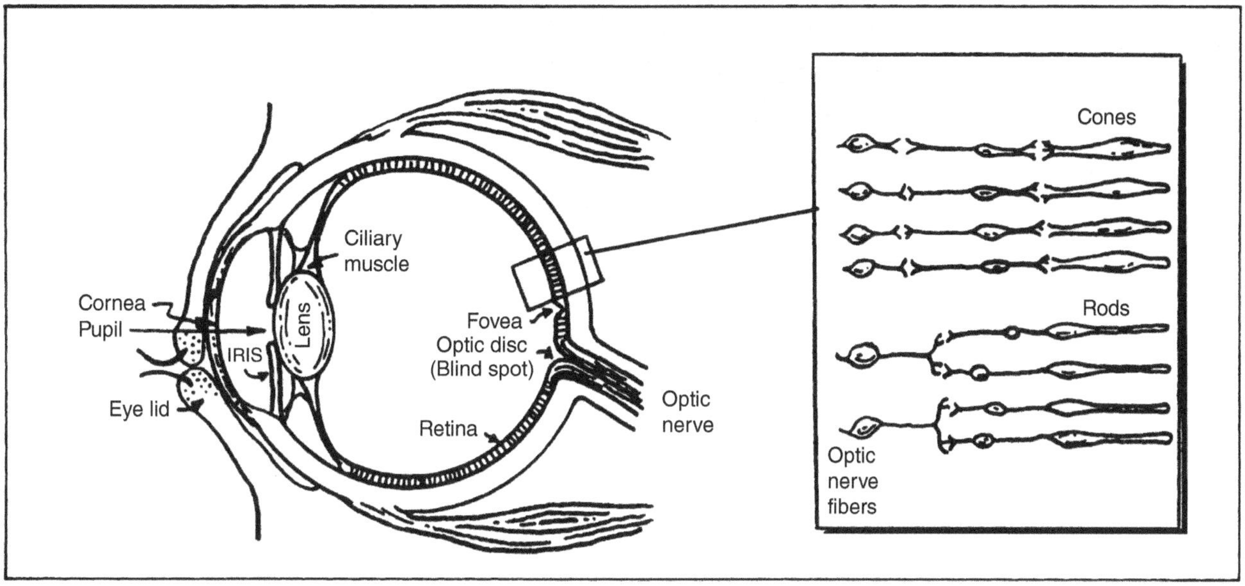

Figure 41.1 Eye anatomy.

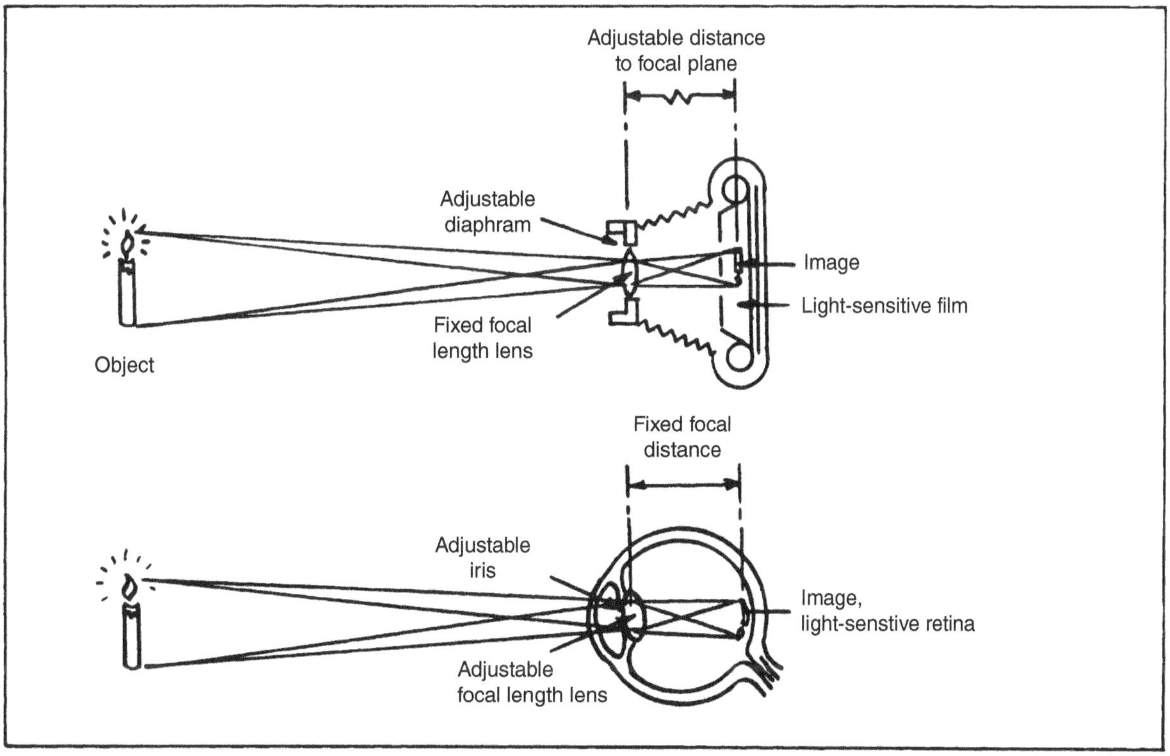

Figure 41.2 Comparable characteristics of eye and camera operation.

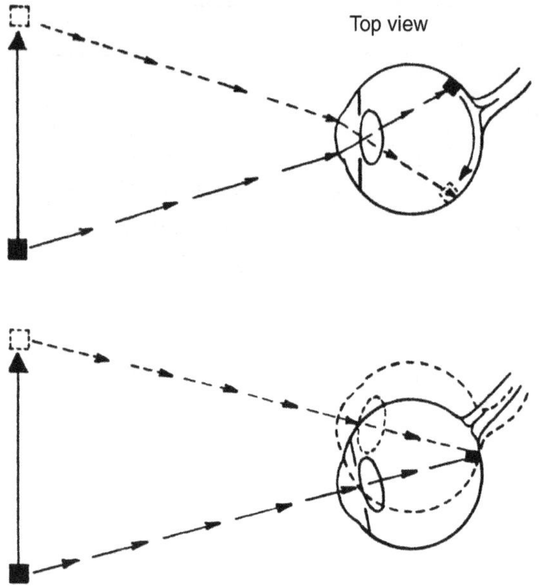

FIGURE 41.3 When the eye follows a moving object the image on the retina remains stationary. The brain interprets motion.

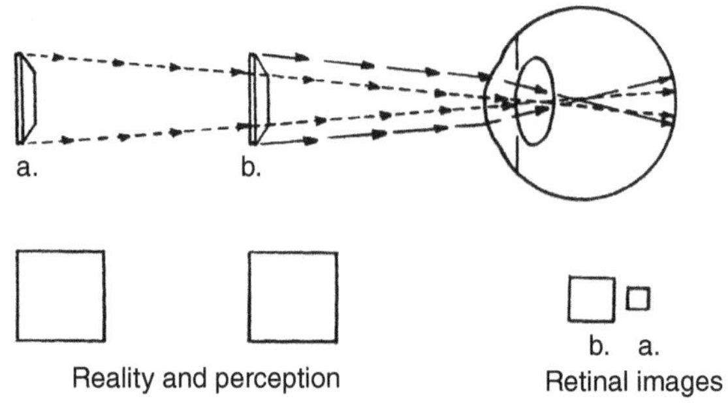

FIGURE 41.4 The brain interprets changes in size as a change in distance.

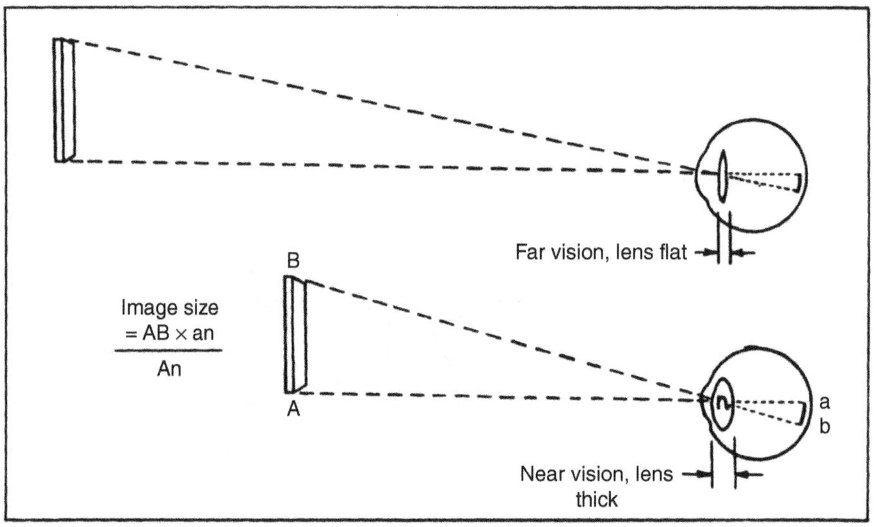

FIGURE 41.5 Calculation of retinal image size.

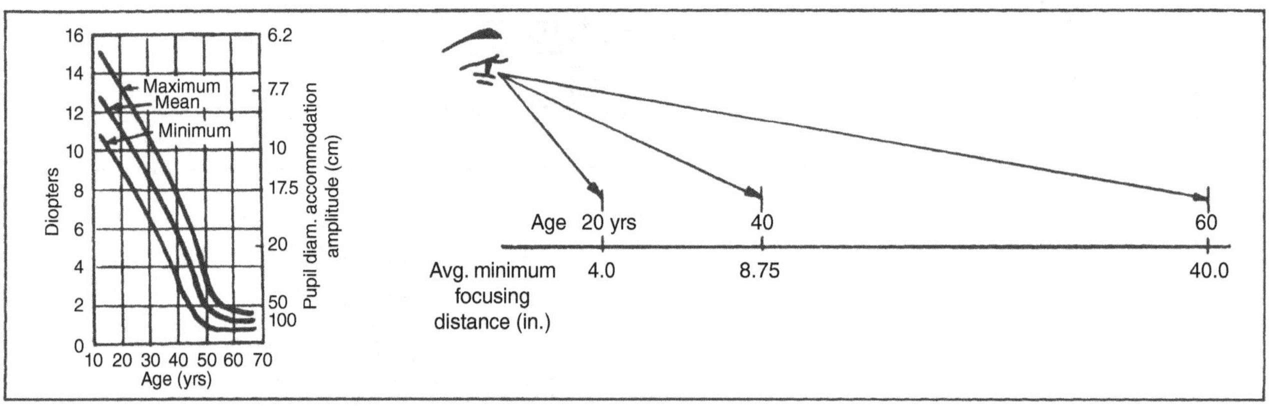

FIGURE 41.6 Accommodation versus age.

See also Table 41.1.

TABLE 41.1 Diameter of the Pupil in Millimeters

AGE	IN DAYLIGHT	AT NIGHT	DIFFERENCE
20	4.7	8.0	3.3
30	4.3	7.0	2.7
40	3.9	6.0	2.1
50	3.5	5.0	1.5
60	3.1	4.1	1.0
70	2.7	3.2	0.5
80	2.3	2.5	0.2

DEPTH PERCEPTION (SEE TABLE 41.2)

TABLE 41.2 External and Internal Cues to Depth and Distance

EXTERNAL CUES	
Linear perspective	Apparent convergence of parallel lines and related effects
Apparent size	A strong cue to distance of objects of known size and texture
Motion parallax	Relative angular motion as either head or objects move
Interposition	Nearer objects eclipse more distant ones
Aerial perspective	Contrast and color loss due to aerosols; useless in free space
Shading	A cue to three-dimensional form of objects (not to distance)
Apparent intensity	A cue only to distance of effective "point sources"
INTERNAL CUES	
Accommodation	Relatively unimportant
Convergence	Useful limit is about 20 m
Binocular disparity	Most important intrinsic cue to depth and distance

Note: All cues except for the last two can be utilized by a single eye, and by extension, in uniocular optical devices.

RADIANT ENERGY AND THE EYE

Light is visually evaluated radiant energy. As shown in Fig. 41.7, the visible portion of the electromagnetic radiant energy spectrum is extremely narrow—about 380 to 770 nm (1 nm = 0.001 μm = 10^{-9} m). (See Table 41.3.)

Intensity Relationships

It takes only one-billionth of a lambert to excite the eye (absolute threshold). The minimum ratio at which light can be seen is called the "contrast threshold." A light of very low intensity can be seen against a dark background. However, the light must be much brighter to be seen against a bright background, as shown in Table 41.4.

Visibility also depends on light target area (visual angle), as shown in Table 41.5.

The threshold visibility of a flashing light depends on the total energy in the flash, which is the intensity of the flash multiplied by the duration. (For flashes longer than 0.1 s, visibility depends on intensity alone.) (See Fig. 41.8.)

PHOTOPIC AND SCOTOPIC RESPONSE FUNCTIONS

Photopic (foveal) vision occurs under high illumination conditions, where the luminosity function is mediated by the cones. Scotopic (peripheral) vision occurs under low illumination conditions, where the luminosity function is mediated by the rods. A third type of vision (mesopic) occurs when both cones and rods are activated, i.e., under low or intermediate illumination conditions, such as at dusk or dawn.

In the fovea, where visual acuity is at a maximum, vision is mediated entirely by cones. Although cones are distributed throughout the peripheral retina, cone density is considerably less than in the fovea region, and therefore

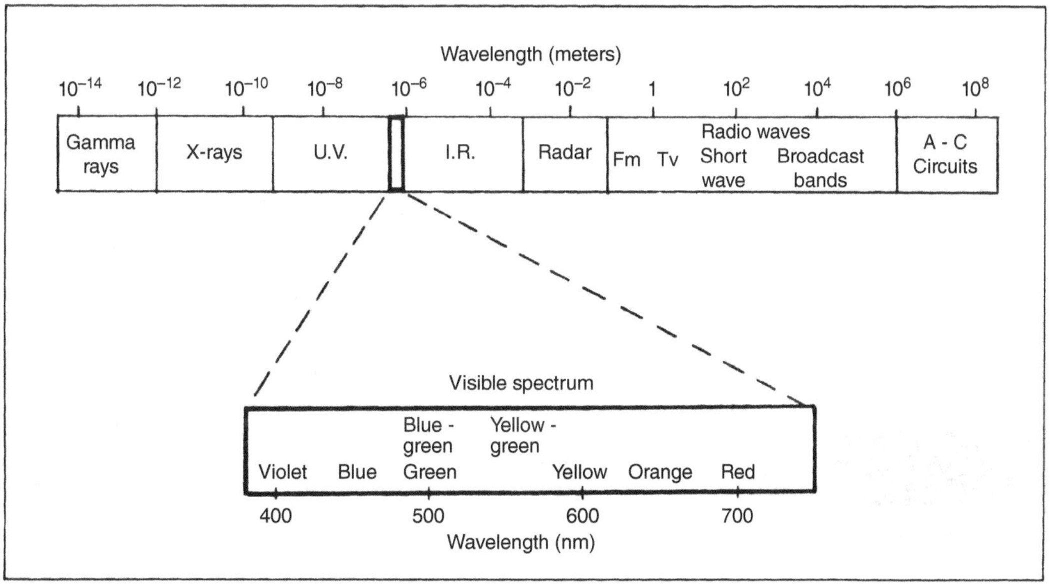

FIGURE 41.7 Radiant energy visible to the eye.

TABLE 41.3 Characteristics of Vision

PARAMETER	VISION
Sufficient stimulus	Light-radiated electromagnetic energy in the visible spectrum
Spectral range	Wavelengths from 400 to 700/μ. (violet to red)
Spectral resolution	120 to 160 steps in wavelength (hue) varying from 1 to 20/μ
Dynamic range	~90 dB (useful range) for rods = 0.00001 to 0.004 mL; cones = 0.004 mL to 10.000 mL
Amplitude resolution ΔI/I	Contrast = ΔI/I = 0.015
Response rate for successive stimuli	~0.1 s
Reaction time for simple muscular movement	~0.22 s
Best operating range	500 to 600/μ (green-yellow) 10 to 200 fc
Indications for use	1. Spatial orientation required. 2. Spatial scanning or search required. 3. Simultaneous comparisons. 4. Multidimensional material presented. 5. High ambient noise levels.

TABLE 41.4 Visibility Thresholds for Different Contrasts

LOG OF ILLUMINANCE JUST VISIBLE AT THE EYE, FC	LOG OF BACKGROUND BRIGHTNESS, ML
Photopic vision:	
−5	4
−6.25	2
−7.60	0
Mesopic vision:	
−8.3	−2
−8.8	−4
Scotopic vision:	
−9.6	−6
−9.8	Total darkness

TABLE 41.5 Visibility Thresholds for Different Target Sizes

Just visible brightness, log fL	−2.2	−2.7	−3.25	−4.5	−4.75	−5.6	−5.8
Diameter of target, min/arc	0.3	0.6	1.0	6.0	10.0	60.0	100.0

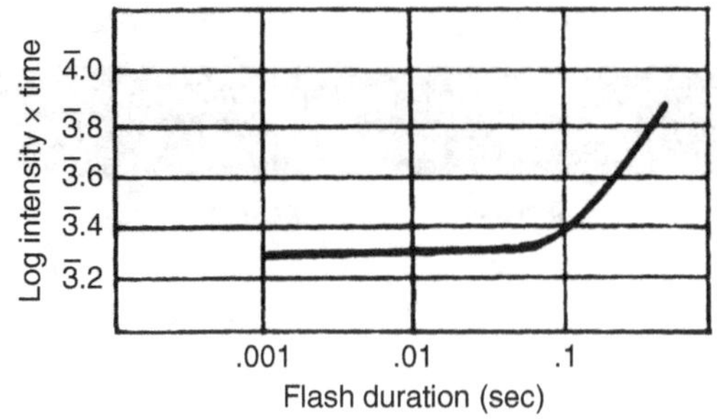

FIGURE 41.8 Visibility thresholds for different flash durations.

acuity is reduced in the periphery. At the limits of peripheral vision, cone vision may be completely absent, and only the rods mediate. As a result, the eye detects only light. Therefore, a faint light will disappear if one looks directly at it.

It should be noted that while each of the curves is shown at the same luminosity scale in Fig. 41.9, this does not mean that the eye has equal peak luminous responses in the photopic and scotopic modes.

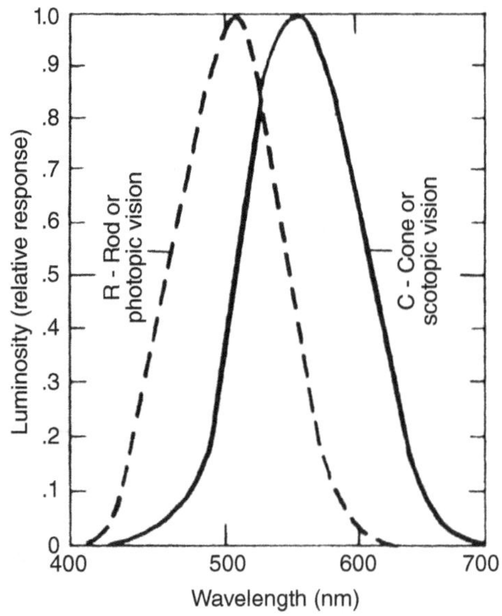

FIGURE 41.9 Standard luminosity functions.

See also Figs. 41.10–41.12.

VISUAL ACUITY[25]

The term "visual acuity" refers to the resolving power of the eye or its ability to detect fine detail.

There is a marked relationship between visual acuity and the distribution of retinal rods and cones, as illustrated in

[25] J. W. Wulfeck et al., *Vision in Military Aviation*, WADC Technical Report 58-399, 1958.

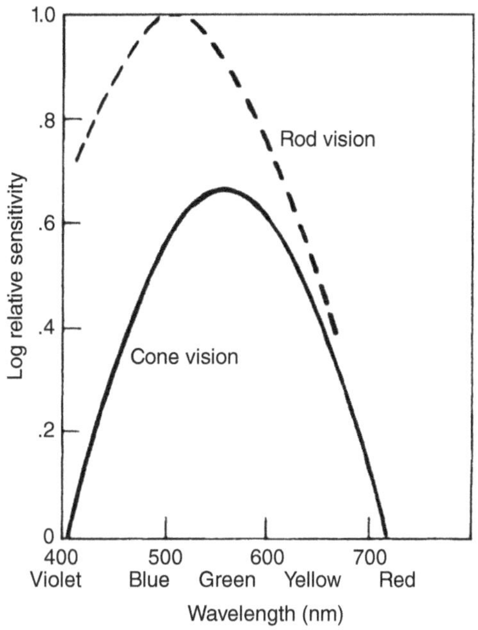

Figure 41.10 Spectral response of retinal rods and cones, with relative amount of radiant energy for absolute threshold vision as a function of the wavelength of light.

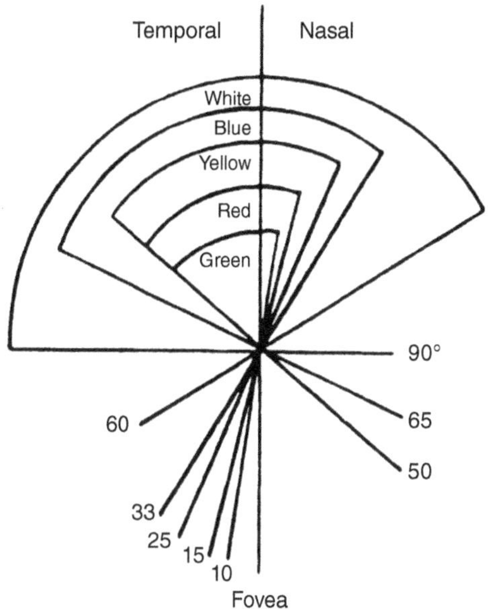

Figure 41.12 Sensitivity of different parts of the retina in daylight.

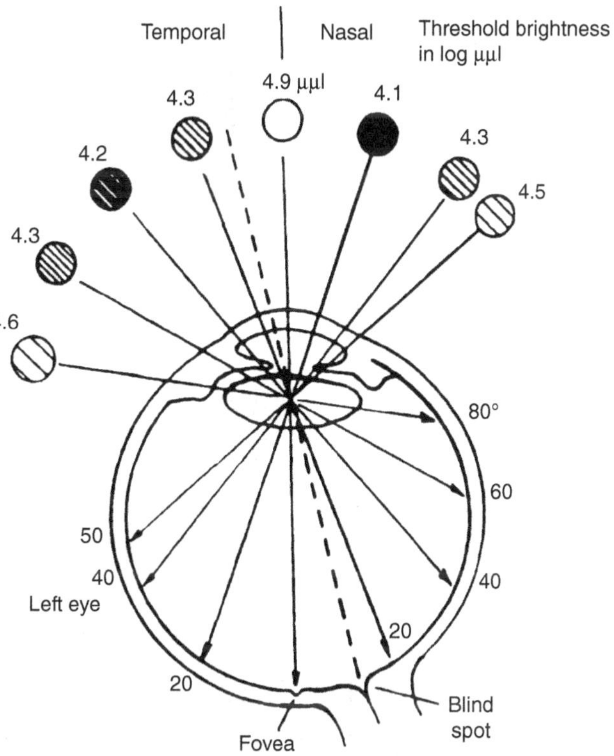

Figure 41.11 Sensitivity of different parts of the retina at night.

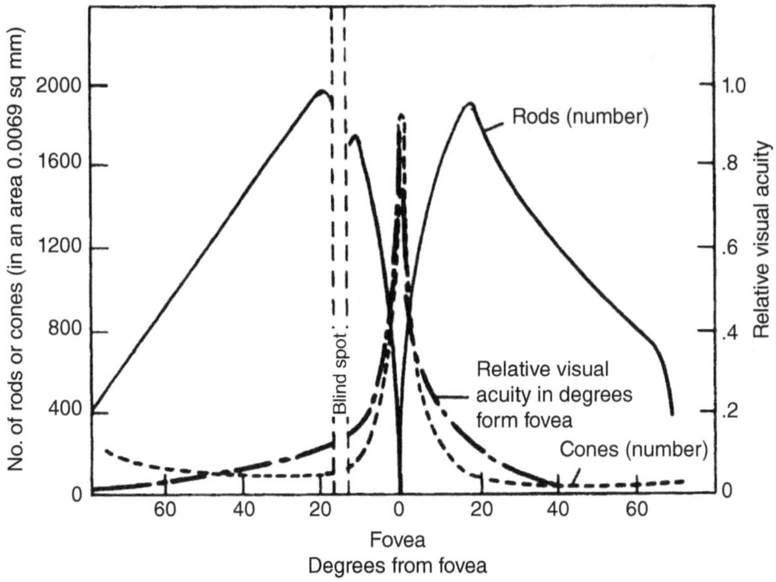

FIGURE 41.13 The relationship of visual acuity to the distribution of rods and cones.

Fig. 41.13. Since there are more cones in the central area, acuity (photopic vision) is better in this area. Photopic vision, in which rods are most used, is better farther from the foveal area of the retina.

The relationship between field brightness and minimum perceptible brightness difference (another measure of visual acuity) is shown in Fig. 41.14. Although there is no appreciable change in the sensitivity of the cones (B) used for daylight conditions, there is a considerable change in the sensitivity of the rods (A), which are used for nighttime conditions. (See also Fig. 41.15.)

At the lowest intensity of light, the eye can see a line whose thickness subtends a visual angle of 10 min. At very high intensities, the eye can see a line subtending less than 1 s of

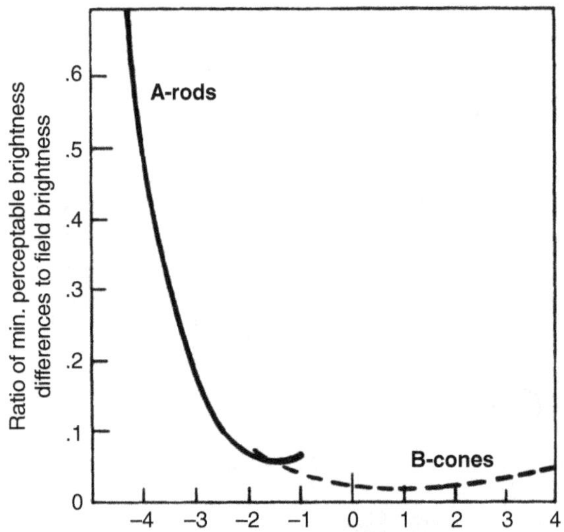

FIGURE 41.14 Relationship between field brightness and minimum perceptible brightness differences.

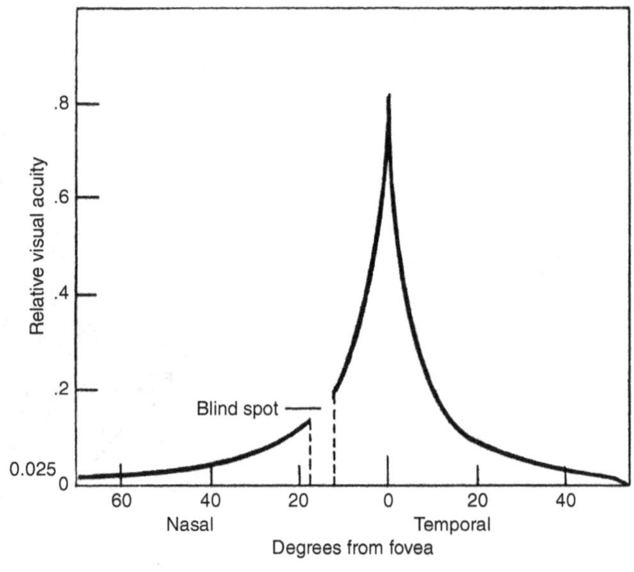

FIGURE 41.15 Daylight visual acuity for different parts of the eye.

the visual angle. These values assume maximum contrast between the lines and the background.

The relationship between contrast and image size is illustrated by a standard "parallel bar" target under a brightness level of 30 mL. As shown in Table 41.6, as contrast is increased, minimum size and spacing between parallel bars can be decreased without rendering the separation invisible. However, with decreasing contrast, target size and separation must be increased in order to maintain threshold acuity.

TABLE 41.6

Contrast, %	45	8	5	3	2.8
Visual angle min	1	2	4	10	16

Visual acuity also varies under different spectral illuminants. (See Fig. 41.16.) When the background brightness level is 0.075 fc, the relationships shown in Table 41.7 hold true.

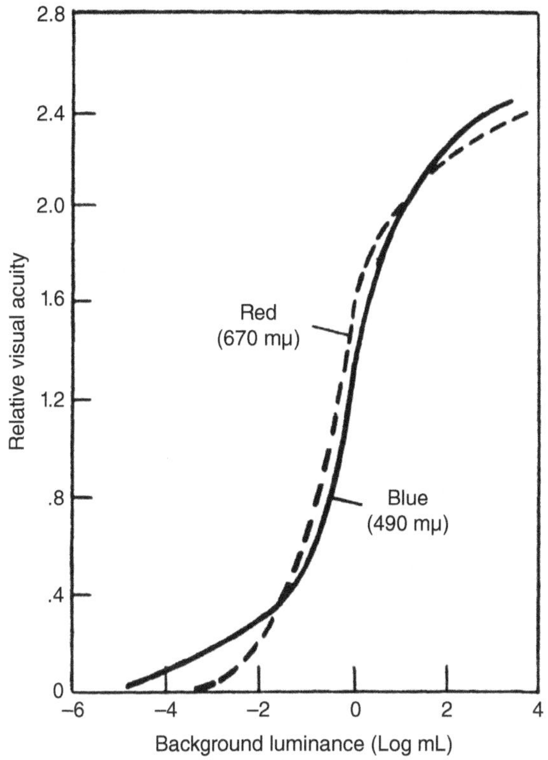

Figure 41.16 Acuity versus color of illuminant.

TABLE 41.7

Visual acuity, %	52	70	75	68	63
Wavelength (Mμ)	485	520	590	625	665

When higher illumination levels are available, however, the relationship between illuminant color and acuity for black-and-white targets is negligible. Luminance contrast, color contrast, illumination level, and exposure time are more important to acuity than color is.

Moving targets need to be larger as the rate of movement increases in order to provide the necessary acuity to distinguish target detail (see Table 41.8).

TABLE 41.8

Visual acuity, %	100	56	30	19	10
Velocity, %	0	50	100	150	200

Acuity deteriorates with age, as shown in Fig. 41.17. The designer must remember that older persons' eyes require more light and larger detail.

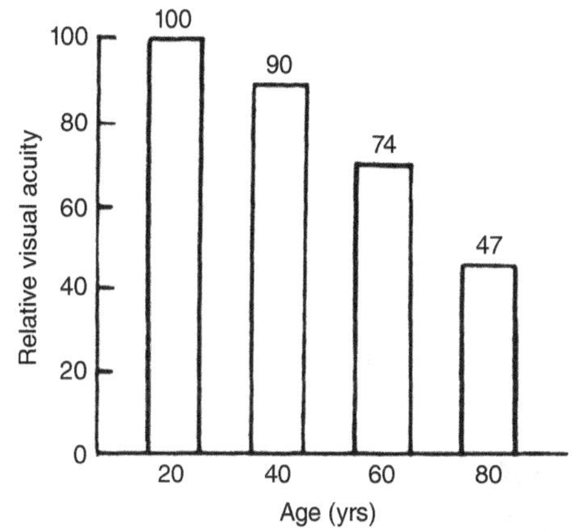

Figure 41.17 Deterioration of acuity with age.

CONTRAST, VISUAL ANGLE, AND BRIGHTNESS

Three-dimensional Fig. 41.18 illustrates the relationship between three critical variables that determine whether a

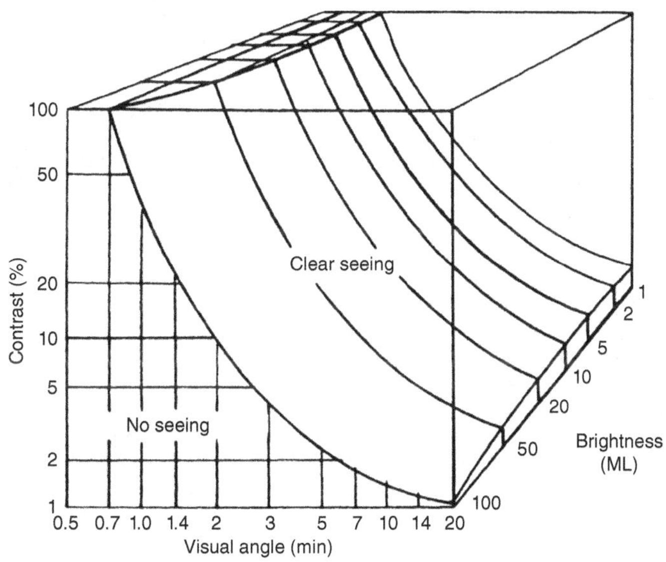

Figure 41.18 Relationship of contrast, visual angle, and brightness to seeing.

person sees or does not see an object. When the combination of variables intersects behind the curve, the eye cannot discriminate, whereas any intersecting point in front of the curve provides good seeing conditions.

Good seeing is also affected by the brightness relationships between all elements within the visual field. As Table 41.9 shows, it is desirable to maintain different brightness relationships between the primary visual task and immediate and distant visual phenomena.

TABLE 41.9 Recommended Brightness Ratios within the Visual Field

	Recommended Maximum Luminance Ratio	
Areas	**Office**	**Industrial**
Task and adjacent surroundings	3:1	
Task and adjacent darker surroundings		3:1
Task and adjacent lighter surroundings		1:3
Task and more remote darker surfaces	5:1	10:1
Task and more remote lighter surfaces	1:5	1:10
Luminaries (or windows, etc.) and surfaces adjacent to them		20:1
Anywhere within normal field of view		40:1

VISUAL FIELD

Monocular Field

The general field that can be seen by the single eye is illustrated in Fig. 41.19, which shows how the brow and the nose limit the field above and toward the nose. This figure also shows the extent to which some colors are more perceptible than others as a result of the distribution of retinal sensors; i.e., one can perceive blue farther out on the periphery than red. The outer limit denotes that only white light is recognizable.

Binocular Field

Overlapping of the two binocular fields is illustrated in Figs. 41.20 and 41.21. Note that the 15° cone about the major visual axis is always recommended as the limit for locating warning lights that must be seen while watching a specific display. Figure 41.22 illustrates the general limiting variable for peripheral vision.

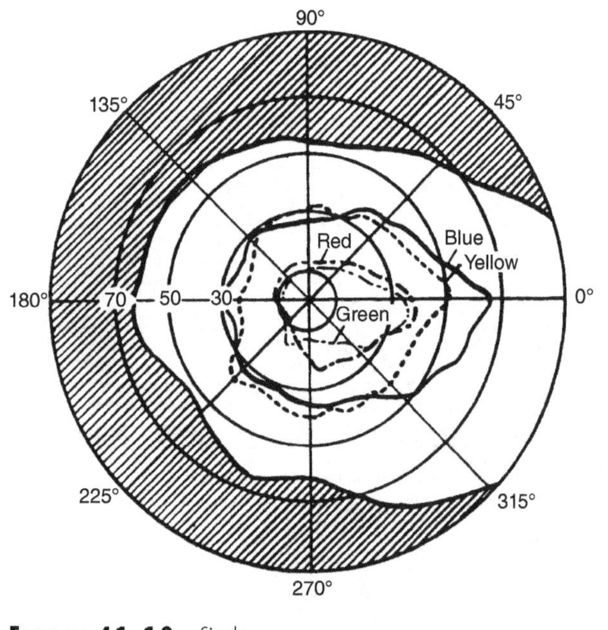

FIGURE 41.19 Single.

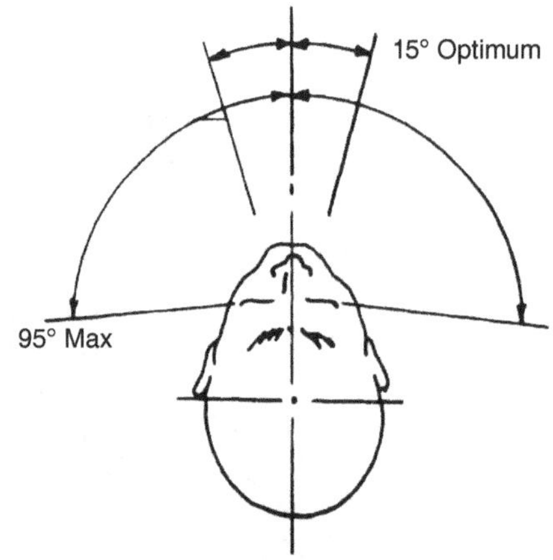

FIGURE 41.20 Visual field.

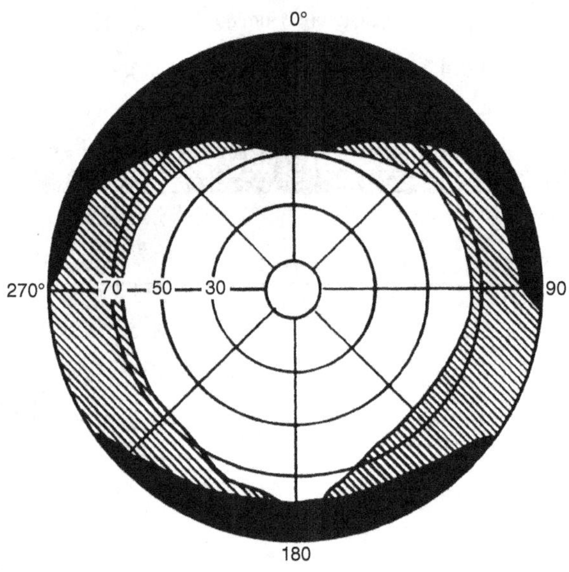

FIGURE 41.21 Binocular field.

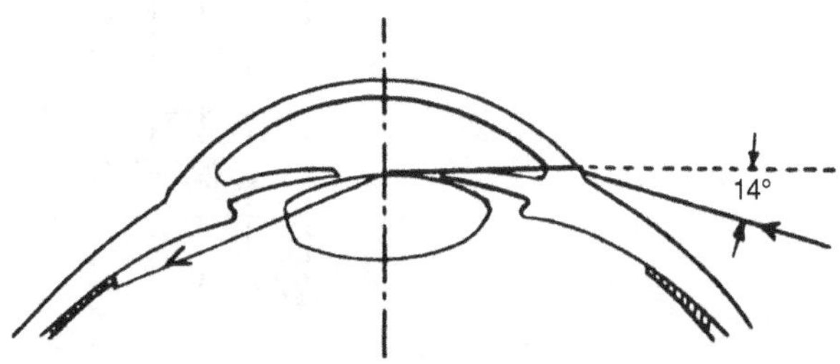

FIGURE 41.22 Eye anatomy limits peripheral vision.

TABLE 41.10 The Limits of the Visual Field under Various Kinds of Restraint

		Horizontal Limits		Vertical Limits	
Movement Permitted	**Type of Field and Factors Limiting Field**	**Temporal Ambinocular Field (Each Side)**	**Nasal Binocular Field (Each Side)**	**Field Angle Up**	**Field Angle Down**
Moderate movements of head and eyes assumed as:	Range of fixation	60°		45°	
Eyes: 15° right or left	Eye deviation (assumed)	15°	15°	15°	15°
15° up or down	Peripheral field from point of fixation	95°	(45°)	46°	67°
Head: 45° right or left	Net peripheral field from central fixation	110°	60°	61°	82°
30° up or down	Head rotation (assumed)	45°	45°	30°	30°
	Total peripheral field (from central body line)	155°	105°	91°	112°
Head fixed					
Eyes fixed (central position with respect to head)	Field of peripheral vision (central fixation)	95°	60°	46°	67°
Head fixed					
Eye maximum deviation	Limits of eye deviation (= range of fixation)	74°	55°	48°	66°
	Peripheral field (from point of fixation)	91°	Approx (5°)	18°	16°
	Total peripheral field (from central head line)	165°	60°	66°	82°
Head maximum movement	Limits of head motion	60°		45°	
Eyes fixed (central with respect to head)	(= range of fixation)	72°	72°	80°	90°
	Peripheral field (from point of fixation)	95°	60°	46°	67°
	Total peripheral field (from central body line)	167°	132°	126°	157°
Maximum movement of head and eyes	Limits of head motion	72°	72°	80°	90°
	Maximum eye deviation	74°	55°	48°	66°
	Range of fixation (from central body line)	146°	127°	128°	156°
	Peripheral field (from point of fixation)	91°	Approx (5°)	18°	16°
	Total peripheral field (from central body line)	237°	132°	146°	172°

J. W. Wulfeck et al., *Vision in Military Aviation*, WADC Technical Report 58-399, 1958.

Table 41.10 shows the limits of the visual field under various kinds of restraints.

MONOCULAR DAYLIGHT ACUITY RELATIVE TO CENTRAL ACUITY[26]

The relative acuity for one-eye daytime vision in various parts of the visual field is shown in Fig. 41.23. Note that the best vision is within the small ring at the center of the field. Under the most favorable conditions, ½ min of arc may be discriminated. The figures on the larger rings illustrate how many times larger an object must be to be seen clearly.

DARK ADAPTATION

The transition from high- to low-level light requires adaptation time. Adaptation from low to higher light levels is approximately instantaneous, although it can be annoying or even slightly painful.

Although the time required to become dark-adapted varies both among subjects and within an individual eye, the general form of the dark adaptation curve remains somewhat characteristic. (See Fig. 41.24.) Note the transition point at which adaptation is fairly rapid, after which it levels off and then proceeds farther, but at a slower pace.

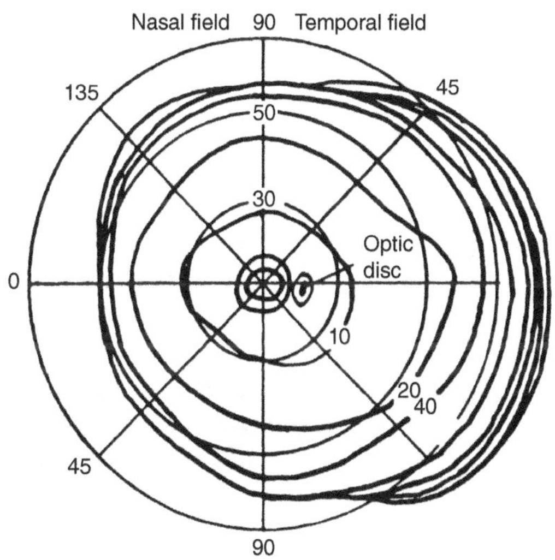

FIGURE 41.23 One-eye daytime acuity.

[26] After W. S. Duke-Elder, *Test Book of Ophthalmology*, The C. V. Mosby Company, St. Louis, 1934.

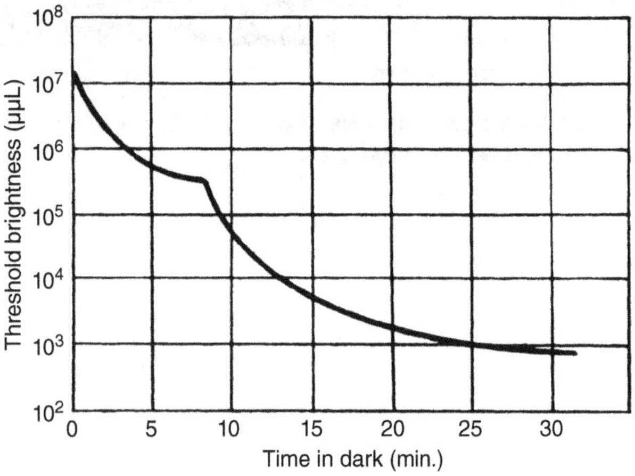

FIGURE 41.24 Course of dark adaptation showing intensity just barely discernible as a function of time spent in dark.

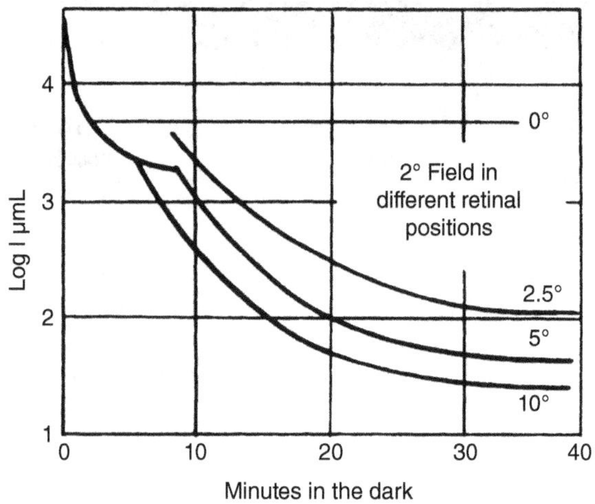

FIGURE 41.26 Dark adaptation.

Dark adaptation patterns vary depending on the color of the light, as illustrated in Fig. 41.25.

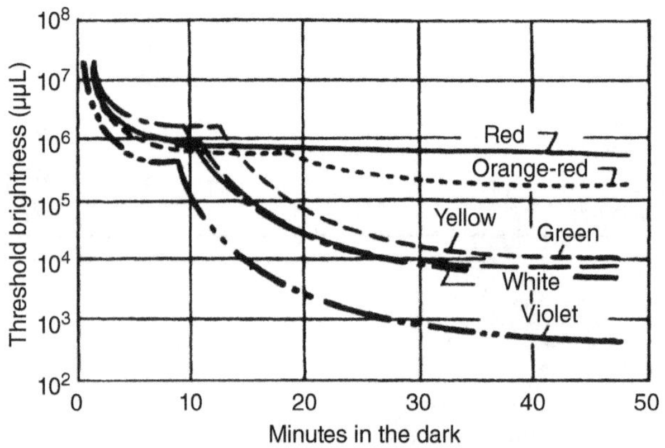

FIGURE 41.25 Dark adaptation measured with light of different wavelengths.

Although lights were all equated in brightness initially, they are no longer equally bright even at cone threshold. The differences are further exaggerated during rod dark adaptation. Figure 41.26 shows dark adaptation as a function of the region of the retina being stimulated.

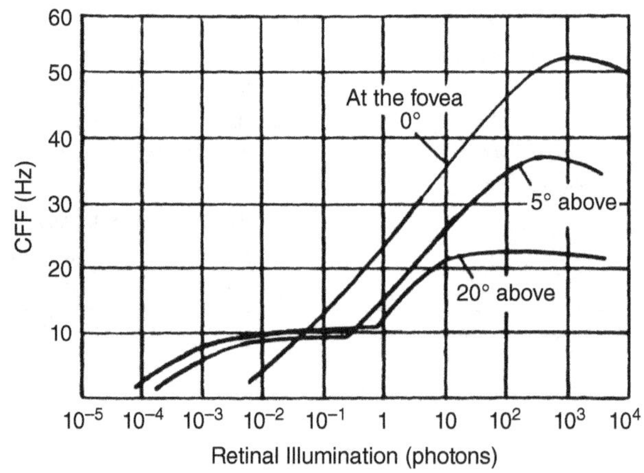

FIGURE 41.27 CFF vs. retinal illumination for different target locations.

CRITICAL FUSION FREQUENCY

Critical fusion frequency (CFF) is the point at which successive light flashes blend into a continuous light. A flickering light that is on 50 percent of the time and off 50 percent of the time flashes at the rate of 10 Hz; it appears brighter than a steady light.

The relation between CFF and retinal illumination *(I)* with white light for three retinal locations is shown in Fig. 41.27. Crossing of the curves is due to the shift from rod to cone vision.

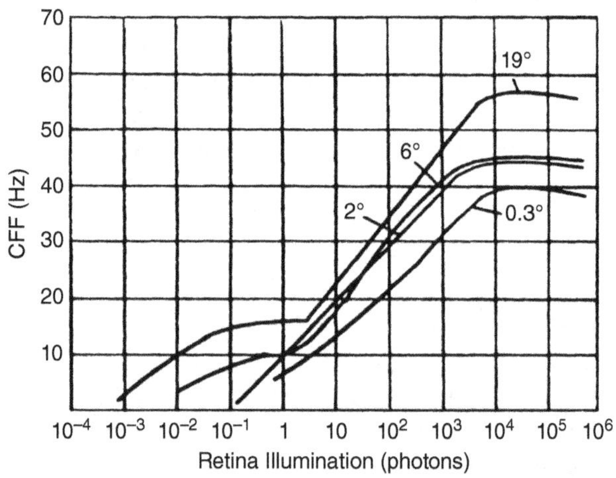

FIGURE 41.28 CFF vs. retinal illumination for different target sizes.

The relation between CFF and various target-area sizes is shown in Fig. 41.28. The small areas (2° to 3°) stimulate only the cones, whereas for larger areas (6° and 19°) a double function yields a higher CFF.

Figure 41.29 shows the relation between CFF and retinal illumination for different spectral regions.

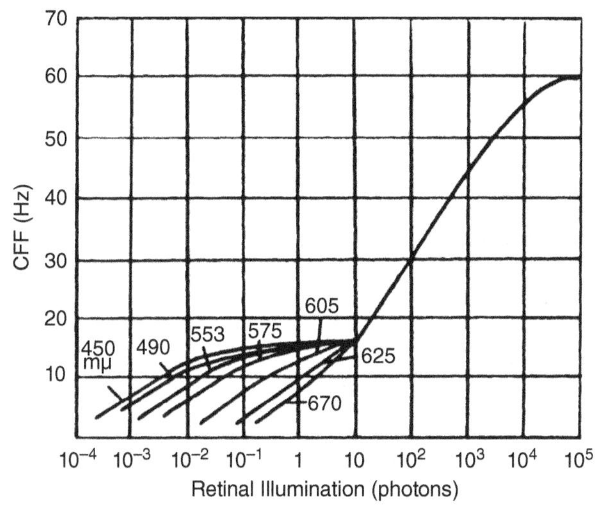

Figure 41.29 CFF vs. retinal illumination for different light colors.

GLARE (SEE TABLE 41.11)

COLOR PERCEPTION

We can only generalize with respect to what each individual perceives in the case of a particular color. The chromaticity diagram shown in Fig. 41.30 has been created to identify particular wavelength characteristics that normally lead to an individual's identification of colors. The points A, B, C, and E represent standardized light sources under which colors are designated. The dominant wavelength and purity of a color are determined by reference to a particular illuminant. This diagram is based on the concept that the eye actually has three separate mechanisms of color response and that any wavelength stimulates all these mechanisms, not just one, in varying degrees.

Color Identification Capability

People use the term "color" to identify only one aspect of color, i.e., hue. As the color model in Fig. 41.31 illustrates, color consists of three attributes—hue, brightness, and saturation. A fully saturated *red hue* can be modified considerably and still be perceived as red by color-normal observers; i.e., when one moves toward the center of the model, the so-called red color becomes less saturated (less intense), and when one moves toward the upper end of the model, the color becomes pale. In addition, as one moves around the circumference of the model, a given color tends to blend with its neighboring hue; e.g., red moving toward yellow first appears orange-red, then orange, then reddish yellow, and finally yellow.

TABLE 41.11 Effect of Glare on Performance

Function	Effects
Visibility	Glare of 5 fc located 5° above the line of vision reduces visibility about as much as replacing a 100-fc lighting system with a 1-fc system. An object that can be seen at 20 ft with 5-fc glare at 5° could be seen at 45 ft with no glare. Glare becomes less significant as the brightness of an object is raised, and is greater for a large object seen in low contrast to its background than for a small, high contrast object, such as a printed letter.
Size threshold	Glare greatly increases size threshold, particularly for objects of low contrast. The effect of glare diminishes as contrast and illumination of surrounds are increased, and thus is more easily controlled at high levels of illumination. Threshold size for an object of 5% contrast under 1 fL with a 5-fc glare 5° above line of vision = 30.0 min visual angle; for an object at 100 fL with no glare = 0.63 min.
Contrast threshold	Glare greatly increases the contrast threshold. The degree of increase depends in complex fashion upon the brightness of task background, the glare illumination produced by each glare source at the eye on a plane perpendicular to the line of sight to the task, and the angle of separation between the glare source and the line of sight.
Muscular tension	Glare significantly increases muscular tension—in some cases up to 30%. Although visual tasks may be carried out under severe handicaps with no loss in immediate accuracy, a decrease in efficiency is apparent with a considerable increase in nervous muscular tension. Subjects not compensating for glare effects may show a considerable drop in performance but no increase in muscular tension.
Vision through eyeglasses	The number of persons wearing glasses is sufficiently large and the glare effect sufficiently severe so that reflections from eyeglasses should be considered in placing light sources. When the light source is behind the head, no glare from eyeglasses will exist if the light source is 30° or more above the line of vision, 40° or more below the line of vision, or at an angle of 15° or more with the axis of symmetry of the head.

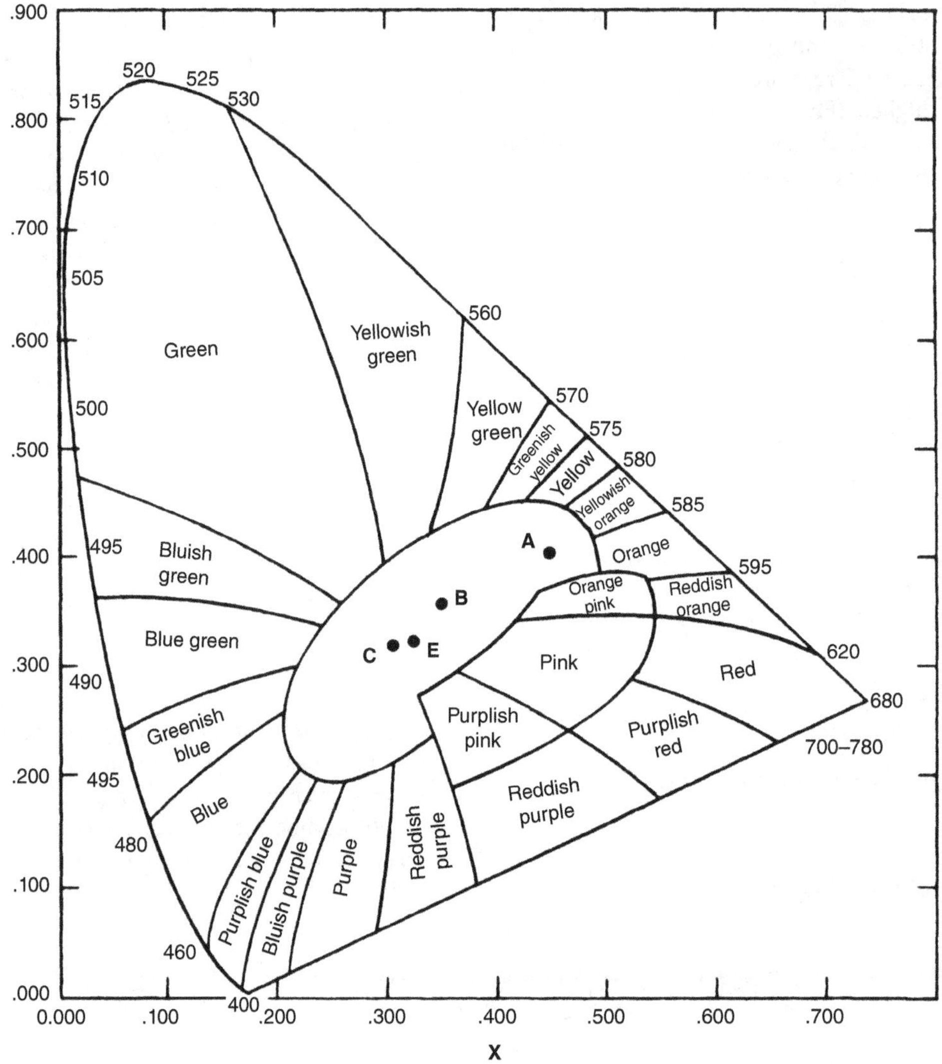

FIGURE 41.30 International Commission on Illumination (C.I.E.) coordinate system for color identification.

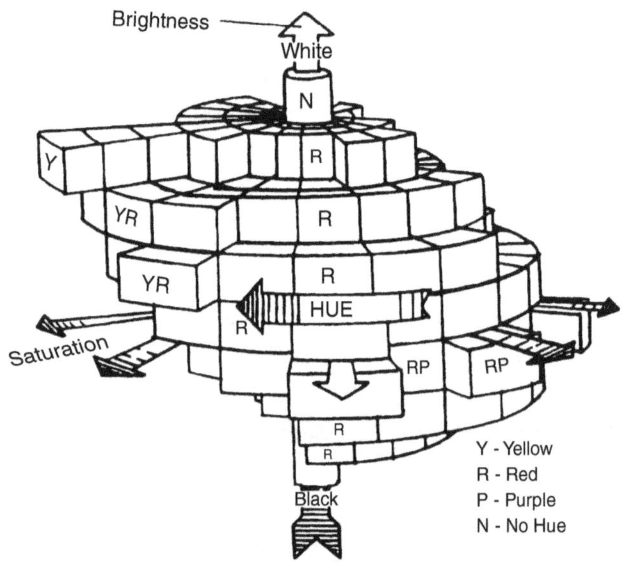

FIGURE 41.31 Color component model.

Although some observers can discriminate 150 or more hues under proper viewing conditions, the average person can accurately and reliably verbalize only about eight or nine hues. Table 41.12 and Fig. 41.32[27] indicate probable color discrimination limits.

Since many people have some level of color deficiency, it is desirable to select colors that both color-normal and color-deficient individuals can safely identify. Table 41.13 defines such colors.

What People Do Perceive versus What They Can Perceive

Surface color recognition depends on three distinct factors: the color of the light source, the color of the reflecting surface or surfaces, and the state of the observer's visual system. (See Fig. 41.33.) Even though the object—say, a red

[27] C. T. Morgan et al. (eds.), *Human Engineering Guide to Equipment Design*, McGraw-Hill Book Company. New York, 1963, p. 83.

TABLE 41.12 Nine Equally Discriminable Surface Colors

Hues	Code Number	Munsell Book Number	Excitation Purity	Dominant Wavelength
1	1.5	3R	37.2	629
2	3	9R	65.8	596
3	5.5	9YR	81.8	582
4	8.5	1GY	76.0	571
5	11.5	3G	27.5	538
6	15	7BG	35.0	491
7	18	9B	56.5	481
8	20.5	9PB	52.7	460
9	24	3RP	36.5	510

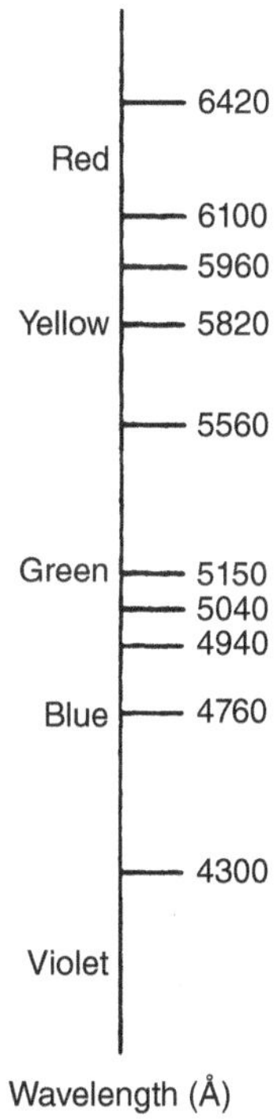

Figure 41.32 Color names.

box—may be a single basic color, this color can be modified both by the color of the light falling directly on it and/or by the color of the light reflected by nearby surfaces. And, of course, the color of the top surface (which receives the full impact of the light source) will not appear the same as the color of the side of the box that is in shadow or the side that is receiving bounce light from another surface (which may be of a different color). Fortunately, there is

TABLE 41.13 Surface Colors for Color-Normal and Color-Blind People

Color	Spec. No.*	Color	Spec. No.*
Red	1110	Gray	1625
Orange	1210	Buff	1745
Yellow	1310	White	1755
Blue	10B 7/6	Black	1770
Purple	2715		

*From Fed. Spec. II-C-595 except for blue, which is from Munsell (1942).

usually enough color constancy so that these variations are still accounted for, and thus the observer compensates and calls the box "red."

The paler a color is, the more easily it is influenced by the color of the light source and by the color of the light being reflected by nearby surfaces. In addition, colors are greatly influenced by the level of illumination and by the inherent reflectivity characteristics of the surface being viewed. For example, in a dim light, colors tend to take on a grayish and less vital appearance. A highly polished surface may reflect high-intensity illumination so effectively that the basic hue is indistinguishable.

Effect of Monochromatic Light on Colored Objects

If the wavelength distribution of the light source is modified so that certain wavelengths are missing or considerably reduced (red light, yellow light, green light, etc.), the inherent color characteristics of a surface reflect a different color impression. Thus, as indicated in Table 41.14, a yellow object will appear to be red-orange when viewed under a red light.

Effect of Standard Luminaries[28]

Tables 41.15–41.17 give a general description of the effects that common fluorescent and mercury or filament Luminaries have on colored surfaces. Both the lighting level and the color of the light affect the appearance of colored surfaces. Filament lamps and warm fluorescent lamps, which are deficient in blue, emphasize the redness of a surface color and thus accent warm hues.

PSYCHOLOGICAL RESPONSE TO COLOR

Although color researchers have not always been able to quantify the precise effects of various colors or light levels on humans, their research and experience seem to indicate that certain colors and light conditions often elicit typical repeatable reactions. Illumination level and quality do contribute to impressions of compartment size, to an atmosphere of

[28] A. O. Halse, *The Use of Color in Interiors,* McGraw-Hill Book Company, New York, 1968, pp. 41, 42.

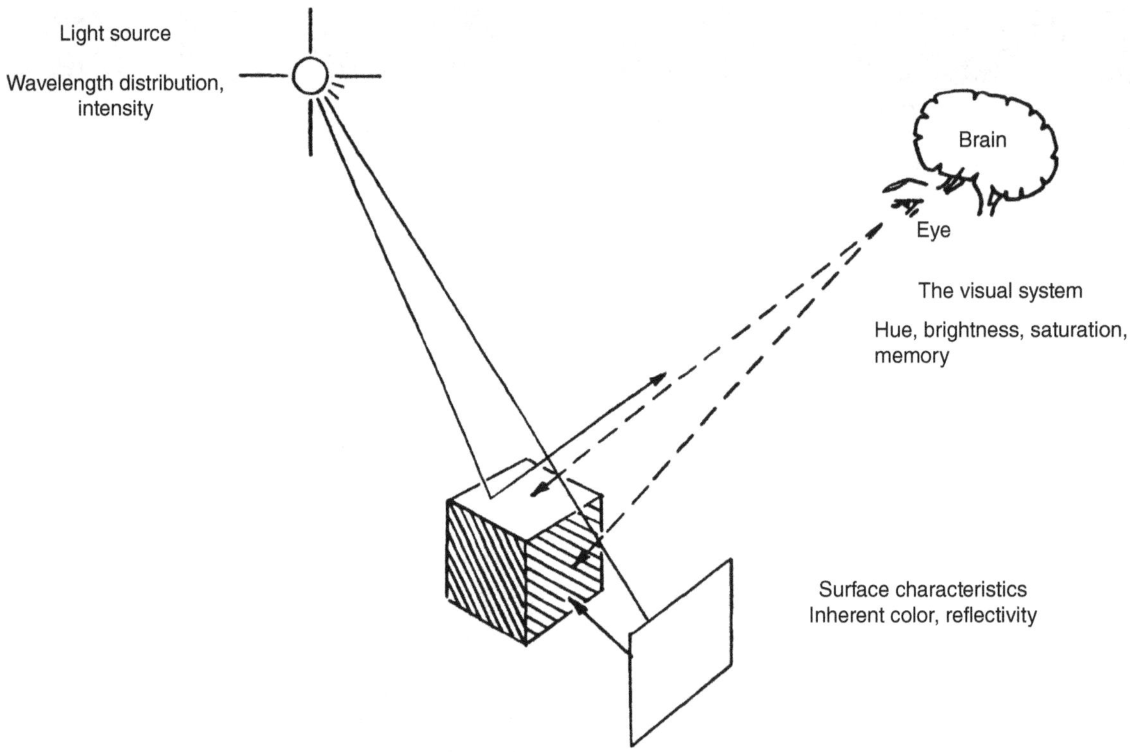

Figure 41.33 Components of the visual system.

TABLE 41.14 Effect of Colored Light on Colored Objects

Object Color	Red Light	Blue Light	Green Light	Yellow Light
White	Light pink	Very light blue	Very light green	Very light yellow
Black	Reddish black	Blue black	Greenish black	Orange black
Red	Brilliant red	Dark bluish red	Yellowish red	Bright red
Light blue	Reddish blue	Bright blue	Greenish blue	Light reddish blue
Dark blue	Dark reddish purple	Brilliant blue	Dark greenish blue	Light reddish purple
Green	Olive green	Green blue	Brilliant green	Yellow green
Yellow	Red orange	Light reddish brown	Light greenish yellow	Brilliant light orange
Brown	Brown red	Bluish brown	Dark olive brown	Brownish orange

warmth or coolness, or generate moods or feelings of feeling repose, pleasantness, excitement, boredom, or depression. Color can have a considerable bearing on whether food looks appetizing and on the appearance of one's skin.

The following five basic psychological response characteristics are perhaps most important to the development of appropriate color and lighting schemes:

1. Certain colors make a space appear larger than it actually is, while others cause a space to "close in" on the observer.

2. Certain colors cause a space to seem warm, while others make it seem cool.

3. Some colors appear to have a definite effect on the mood of the observer, i.e., some colors may be stimulating, while others are quieting.

4. Some colors seem to clash with each other and therefore produce a feeling of irritation to observers who are especially sensitive to color incompatibilities.

5. Everyone does not respond to colors in the same way. At the extreme, some (mainly males) have physiological abnormalities that prevent full color vision. Tables 41.18 and 41.19 describe these color vision deficiencies.

Color in Relation to Size of Space

Average observers perceive dark and/or saturated colors as advancing toward them, and light and/or desaturated colors as receding away from them. If we wish to make a compartment appear less crowded, we should use light and unsaturated colors. Any structure or piece of furniture that is dark in color will

TABLE 41.15 Color Effects of White Fluorescent Lamps

	Cool White	Deluxe Cool White	Warm White	Deluxe Warm White	Daylight	White	Soft White—Natural
Lamp appearance; effect on neutral surfaces	White	White	Yellowish white	Yellowish white	Bluish white	Pale yellowish white	Pinkish white
Effect on "atmosphere"	Neutral to moderately cool	Neutral to moderately cool	Warm	Warm	Very cool	Moderately warm	Warm, pinkish
Colors strengthened	Orange, yellow, blue	All nearly equal	Orange, yellow	Red, orange, yellow, green	Green, blue	Orange, yellow	Red, orange
Colors grayed	Red	None appreciably	Red, green, blue	Blue	Red, orange	Red, green, blue	Green, blue
Remarks	Blends with natural daylight	Best over all color rendition; simulates natural daylight	Blends with incan descent light	Excellent color rendition; simulates incandescent light	Usually replaceable with CW	Usually replaceable with CW or WW	Usually replaceable with CWX or WWX

TABLE 41.16 Color Effects of Mercury and Filament Lamps

	Mercury	White Mercury	Color-Improved Mercury	Deluxe While Mercury	Filament
Lamp appearance; effect on neutral surfaces	Greenish blue white	Greenish white	Yellowish white	White	Yellowish white
Effect on "atmosphere"	Very cool, greenish	Moderately cool, greenish	Warm, yellowish	Moderately cool	Warm
Colors strengthened	Yellow, green, blue	Yellow, green, blue	Yellow, green	Orange, yellow, blue	Red, orange, yellow
Colors grayed	Red, orange	Red, orange	Blue	Green	Blue
Remarks	Poor overall color rendering		Color rendering often acceptable, but not equal to any white fluorescent	Color rendering good; compares favorably with CWX fluorescent	Excellent color rendering

TABLE 41.17 Room Color Recommendations (from NASA-STD-3000 Volume I, Rev. A. October 1989)

Space Module Areas	Only Use These Colors in Small Amounts or as Trim in Areas Indicated with Dot																										Use in Any Amounts in Areas Indicated With Dot															
	Pale Pink	Reddish Pink	Red	Red Orang	Maroon	Orange	Apricot	Yellow Orange	Chamois	Yellow	Yellow Green	Green	Blue Green	Blue	Violet	Royal Blue	Purple	Magenta	Reddish Brown	Buff	Burnt Sienna	Brown	Olive Drab	Gray	Dark Gray	Black	Champagne	Cinnamon	Beige	Salmon	Peach	Straw	Ivory	Cream	Maise	Pale Yellow	Pale Green	Pale Blue	Lavender	White	Off White	Light Gray
Living areas																																										
■ Rest area/crew quarters	•				•	•	•	•	•	•	•	•	•	•	•	•			•	•	•	•					•	•	•	•	•	•	•	•	•	•	•	•	•	•	•	
■ Recreation	•	•	•	•	•	•	•	•	•	•	•	•	•	•	•	•	•	•	•	•	•	•		•	•		•	•	•	•	•	•	•	•	•	•	•	•	•	•	•	•
■ Lounge	•	•	•	•	•	•	•	•	•	•	•	•	•	•	•	•	•	•	•	•	•	•	•	•	•		•	•	•	•	•	•	•	•	•	•	•	•	•	•	•	•
■ Personal hygiene, shower	•	•		•		•	•	•	•	•	•	•	•	•		•				•		•					•	•	•	•	•	•	•	•	•	•	•	•		•	•	•
■ Passageways					•	•	•	•	•	•	•	•	•	•		•			•	•	•	•		•	•		•	•	•	•	•	•	•	•	•	•	•	•		•	•	
Work areas																																										
■ General workstations						•	•	•	•	•	•	•	•	•		•			•	•	•	•		•	•		•	•	•	•	•	•	•	•	•	•	•	•		•	•	•
■ Data processing					•	•	•	•	•	•	•	•	•	•		•			•	•	•	•		•	•		•	•	•	•	•	•	•	•	•	•	•	•		•	•	•
■ Communications					•	•	•	•		•	•	•	•	•		•			•	•	•	•		•	•		•	•	•	•	•	•	•	•	•	•	•	•		•	•	•
■ Maintenance													•	•		•			•	•	•	•		•	•		•	•	•			•	•	•	•	•		•		•	•	•
■ Mechanical equipment/power generation													•	•		•			•	•	•	•		•	•			•	•			•	•	•	•	•		•		•	•	•
■ Security									•	•	•	•	•	•		•			•	•	•	•		•	•		•	•	•	•	•	•	•	•	•	•	•	•		•	•	•
■ Logistics									•	•	•	•	•	•		•			•	•	•	•		•	•		•	•	•	•	•	•	•	•	•	•	•	•		•	•	•
■ Administration					•	•	•	•	•	•	•	•	•	•		•		•	•	•	•	•		•	•		•	•	•	•	•	•	•	•	•	•	•	•		•	•	•
Food Preparation and serving																																										
■ Kitchen			•			•	•	•	•	•	•					•			•	•	•	•	•				•	•	•	•	•	•	•	•	•	•		•		•	•	
■ Dining					•	•	•	•	•	•	•					•			•	•	•	•	•				•	•	•	•	•	•	•	•	•	•		•		•	•	•
Service areas																																										
■ Laundry					•	•	•	•	•	•	•	•	•	•		•				•					•		•	•	•	•	•	•	•	•	•	•	•	•		•	•	•
■ Health maintenance										•	•	•	•	•														•	•	•	•	•	•	•	•	•		•		•	•	•
Assembly areas																																										
■ Conference/briefing			•		•	•	•	•	•	•	•	•	•	•		•			•	•	•	•		•	•		•	•	•	•	•	•	•	•	•	•	•	•		•	•	•
■ Training						•	•	•	•	•	•	•	•	•		•			•	•	•	•		•			•	•	•	•	•	•	•	•	•	•	•	•		•	•	•
Storage areas																																										
■ Food storage						•	•	•	•																				•				•	•	•					•	•	•
■ General Storage/supplies											•	•																	•				•	•	•					•	•	•

TABLE 41.18 Classification of Color Vision

Type of Color Vision	Incidence (Men),* %	Number of Color Receptors	Spectral Colors Confused	Colors Probably Seen
Trichromatic:				
Normal		3	Some reduction in wavelength discrimination	Reds look darker than normal
Deuteranomalous	4.69	3	Some reduction in wavelength discrimination	Reds look darker than normal
Protanomalous	0.89	3	Some reduction in wavelength discrimination	Reds look darker than normal
Tritanomalous	Not known			
Dichromatic:				
Deuteranope	1.42	2	Red, yellow, and green	Yellow, white, and blue
Protanope	1.21	2	Red, yellow, and green	Green, white, and blue
Tritanope	0.008	2	Blue and green	Reds look darker than normal
Monochromatic	0.001	1	All	White and black

*Average of several surveys.

†Assuming trichromatic color vision theory.

TABLE 41.19 Average Reaction Times for Each Class of Color Vision

Subjects	Reaction Time, s
Normals	0.497
Deuteranomals	0.788
Protanomals	0.708
Deuteranopes	0.855
Protanopes	0.984
All abnormals	0.945

make a space appear crowded. The level of illumination contributes to this appearance of being crowded in that a dimly lit space generally appears smaller than one that has a higher, more evenly distributed level of light. Therefore, with few exceptions, good illumination and lighter colors are desirable.

Color in Relation to Temperature

So-called cool colors (e.g., blues and greens) make a space seem cooler, while so-called warm colors (such as reds, browns, and oranges) make a space seem warmer. Although little quantitative evidence of any direct correlation between cool or warm colors and actual physiological temperature variation exists, cool colors usually are preferred for facilities located in hot climates, and warmer colors are preferred for facilities located in cold climates. However, since many colors considered cool or warm also have other mood-generating properties, it is important not to overemphasize these temperature-related characteristics. Thus a warm color may be more important in terms of its ability to generate a feeling of comfort or relaxation, especially if a compartment is well air-conditioned and therefore does not need "cooling off" by means of color.

Color in Relation to Mood

Most individuals tend to associate various "artificial" colors with their personal experiences with nature's own color schemes. Table 41.20 provides a fairly complete list of known color–mood associations as well as other psychological responses in terms of food, odors, etc. Perhaps the most important consideration in selecting a color scheme is avoidance of traditional institutional gray and white. A variety of colors helps enliven the environment and provide relief from the boredom of neutral colors.

Psychological Response to Amount and Color of Light

A certain amount of confusion has been generated as a result of conflicting specifications for illumination for good seeing versus desired psychological response. The designer of interiors, in particular, must accommodate consumers in terms of both aspects; i.e., good seeing is essential to ensure

TABLE 41.20 Typical Colors and Their Effect on Humans

Description	Response
Pale pink	Slightly warm feeling, pleasing when associated with odors.
Pink	Warm, soft, pleasing when associated with odors and taste.
Reddish-pink	Warm feeling, slightly stimulating, pleasing association with taste.
Scarlet	Very warm feeling, stimulating and exciting.
RED	Warm feeling, stimulating, generally exciting, but calm to the extent of
	indicating a protective quality, pleasing to the appetite, associated
	with danger.
Reddish-orange	Warm, stimulating, exciting, cheerful, pleasing association with taste.
ORANGE	Warm, stimulating, exciting, cheerful, pleasing association with taste.
Any of the above reflect light that is flattering to skin and complexion, thus making people appear healthy and normal (as opposed to cool colors which create an appearance of being pale). Scarlet through ORANGE colors advance slightly toward the observer.	
Yellowish-orange	Warm, somewhat exciting, cheerful, express a feeling of comfort, associated with pleasing taste.
YELLOW	Warm, somewhat exciting, cheerful, comfortable, associated with pleasing taste.
Pale yellow	Warm, cheerful, associated with pleasing odor and taste, softness and comfort.
Greenish-yellow	May be associated with feeling of slight warmth or coolness depending on other colors used; not associated with good taste.
Yellowish-green	May be associated with feeling of slight warmth or coolness depending on other colors used; calming, somewhat neutral, may or may not be associated with good taste.
GREEN	Generally cool, slightly cheerful, comfortable, calming, associated with pleasing refreshing odor.
Bluish-green	Cool, calming, associated with good taste.
BLUE	Cool, comfort, protective, calming, although may be slightly depressing if other colors are dark, associated with bad taste.
GREEN through BLUE will advance toward observer if dark shades are used.	
Pale blue	Cool, soft, calming, tends to neutralize if other colors are pale, reflected light makes skin appear pale.
Lavender	Slightly cool, calming, soft, associated with pleasing odor, but with bad taste.
Violet	Slightly warm, calming, associated with bad taste.
Royal blue	Rich, substantial, may be slightly depressing if used with other dark colors, associated with bad taste.
Purple	Rich, protective, calming, may be depressing, associated with pleasing odor, but bad taste.
Violet through Purple will advance toward observer, creates a feeling of heaviness.	
Hot pink (a yellowish cast)	Very warm, stimulating, exciting, cheerful.
Rose (a bluish cast in pink	Neutral relative to warmth, comfortable, calm, associated with pleasing
range)	odor.
Fluorescent orange	Warm, stimulating, exciting, cheerful, extremely conspicuous.
Fluorescent red	Warm, stimulating, exciting, cheerful, extremely conspicuous.
Fluorescent yellow	Warm, stimulating, slightly irritating but cheerful.
Chartreuse	Slightly warm, cheerful, associated with bad taste.
Olive	Warm, comfortable, slightly depressing, associated with bad taste.
Cream	Slightly warm, comfortable, calming, clean, reflected light enhances skin tone.
Buff	Warm, comfortable, calm, soft; good blend with other colors.
Tan	Warm, very comfortable; good blend with other colors.

TABLE 41.20 Typical Colors and Their Effect on Humans (*Continued*)

Description	Response
Reddish-brown	Warm, comfortable, cheerful, slightly stimulating.
Brown	Warm, comfortable, rich, substantial, protective, may be slightly depressing.
Fluorescent Orange and Red, Olive, Reddish-brown and Browns advance toward observer.	
WHITE	Neutral, sterile, clean, fresh, stark, crisp; may appear hard or soft depending on lighting color; may appear harsh, glaring.
Off-white	Neutral, clean, fresh.
Light gray	Neutral, clean, fresh, calming, soft, comfortable; critical that the tint is compatible with other colors (e.g., bluish, pinkish, yellowish, etc.)
GRAY	Neutral, comfortable, calming, slightly hard; critical that the tint is compatible with other colors used (e.g., bluish, pinkish, yellowish, etc.)
Dark gray	Neutral, comfortable, may be depressing; critical that the tint is compatible with other colors used (e.g., bluish, pinkish, yellowish, etc.); substantial, heavy, advances toward the observer.
Flat black	Solid, heavy, comfortable, generally neutral, advances toward the observer, gives the impression of being dirty, vagueness, recedes.
Deep black	Protective, depressing, heavy, substantial, advances toward observer in small amounts, but may recede in large amounts (e.g., painting a ceiling black).
Gold/brass	Rich, comfortable, warm (tint important for compatibility with other colors); slightly advancing.
Silver/aluminum	Neutral, cold, hard, clean, stiff and uncomfortable, sterile, lifeless, recedes.
Light wood grain	Warm, comfortable, quiet.
Dark wood grain	Warm, comfortable, quiet, protective, slightly depressing.
Light leather grain	Warm, comfortable, cheerful, soft.
Dark leather grain	Warm, comfortable, protective, soft, slightly depressing, advancing.

Source: U.S. Navy Shipboard Color Coordination Guidance Manual, NAVSEA 0929-002-7010 (prepared by Man Factors, Inc., 1975).

adequate visual task performance, but it is also necessary to create a lighting environment that produces a desired psychological response. For example, although a high level of illumination may not be necessary in order for the occupants of a space to perform a visual task, the illumination level may have to be high so that they will "feel" that there is adequate light.

An important aspect of the lighting of interior spaces is the use of interior surfaces that provide adequate reflectance, as indicated in Table 41.21.

TABLE 41.21 Suggested Reflectance Ranges for Interiors

Interior Surface or Element	Percentage Reflectance Desired
Ceilings	60–90%. Surface should be white, or a very pale tint that is compatible with the hue of the interior wall. A minimum of 80% is required for effective visual performance of indirect lighting methods.
Walls	50–85%. Higher reflectance is required above standing waist height when the wall consists of two different reflectance values (i.e., a lower "dado" is used to reduce the expected soiling of lower wall surfaces).
Windows or glass walls	Fabric or other drapery provisions should maintain an average reflectance of 15–45% (or more).
Furniture	30–40%. It is important to use higher reflectance on the tops of desks or tables to avoid extreme brightness contrast between the top and reading materials. Matte finishes are required to prevent glare.
Floors	15–35%. Middle to high values are preferred in order to ensure visual detection of possible obstructions on the floor that might cause tripping.

TABLE 41.22 Effects of Light on Alertness

Alertness-Level Requirement	Alertness Light Level, fL
Maximum mental alertness required for highly complex mental task performance	50 and above
Medium mental alertness required for routine manual tasks, leisure reading, and/or stimulating social activity	40
Minimum mental alertness required for nondemanding social intercourse and/or perceptual-motor performance (dining, dressing, personal hygiene, etc.)	30
Rest, mental alertness for minimal interaction with other people (e.g., as in a bar or private dining environment)	15
Sleep	Below 3

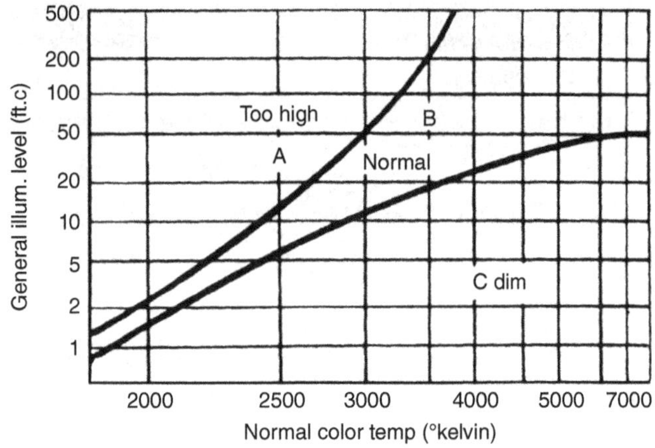

FIGURE 41.34 Effect of color on interpretation of brightness.

Although little quantitative evidence exists regarding the psychological effect on general alertness of light levels, Table 41.22 indicates suggested requirements.

The color of the light source also affects an individual's interpretation of brightness or intensity (see Fig. 41.34). If the level is too high (A), surface colors seem to be faded and unnatural; if the level is too low, the colors appear dim and cold (C). Illuminants falling within the area (B) appear normal. The graph suggests that warmer light is more acceptable to most people when the brightness level is low.

CHAPTER 42
AUDITORY RESPONSE

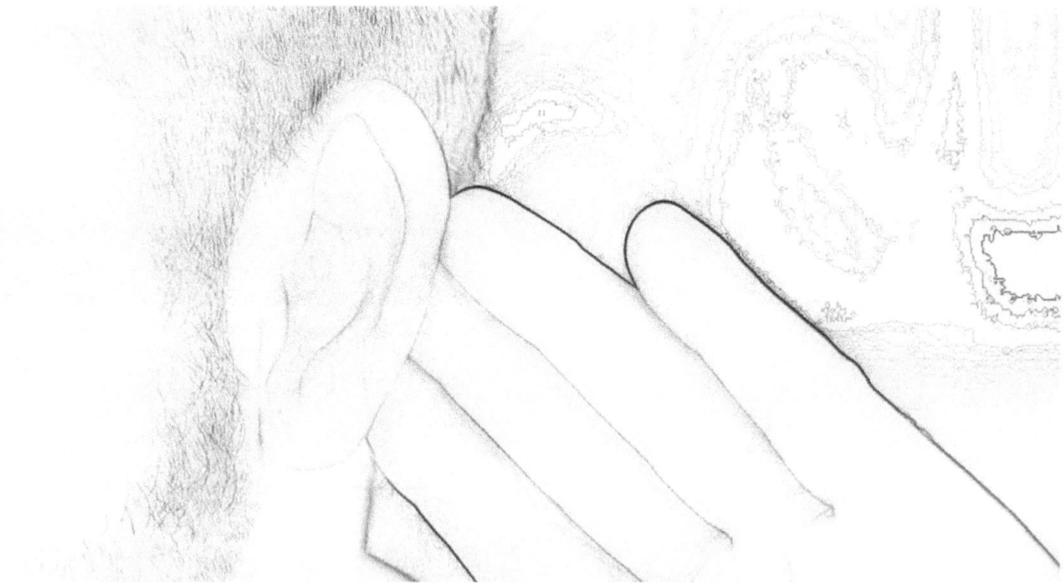

AUDITORY SYSTEM CHARACTERISTICS[29]

Figures 42.1 through 42.4 illustrate the properties of sound: rapid changes in air pressure levels. The pressure changes can be thought of as waves. The frequency of the waves defines the pitch—higher frequency makes a higher pitch. Uninterrupted frequencies make a pure tone. Consistent interruptions to the frequency will make a periodic sound and randomly timed interruptions will produce a non-periodic sound. Completely random frequencies produce a hissing sound, called white noise. When these air pressure changes strike the ear, a series of mechanical structures within the ear converts the pressure changes to auditory nerve stimulations which are perceived as sounds (see Figs. 42.5 and 42.6).

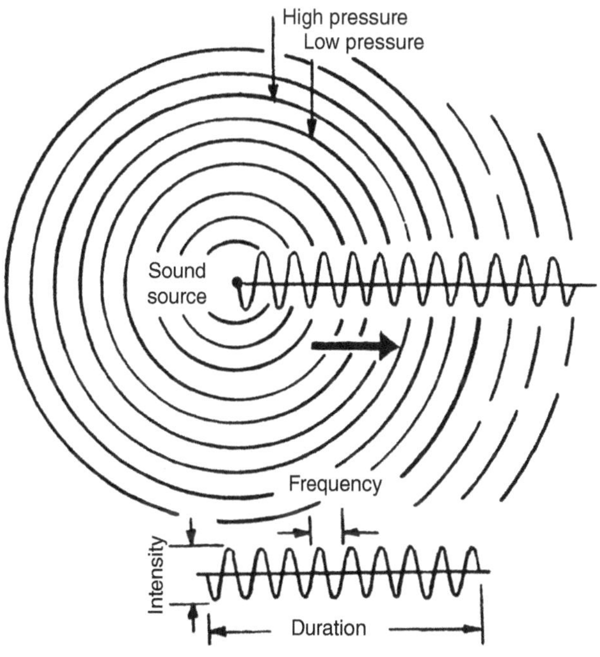

FIGURE 42.1 Pure tone.

HEARING PARAMETERS

Figure 42.7 shows thresholds for hearing (B) and feeling (A). Generally, hearing is considered a subjective phenomenon, while sound is considered an objective phenomenon. The

[29]After W. E. Woodson and D. W. Conover, *Human Engineering Guide for Equipment Designers*, University of California Press, Berkeley, 1964.

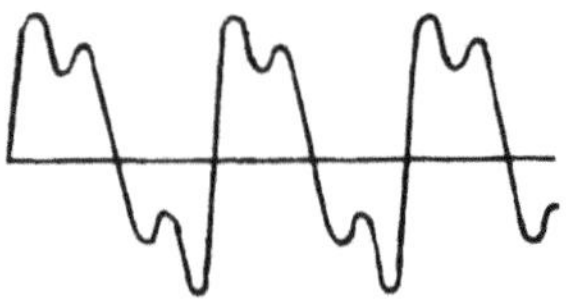

FIGURE 42.2 Periodic sound.

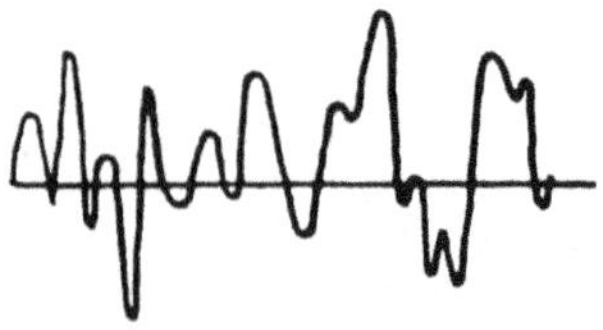

FIGURE 42.3 Non-periodic sound.

FIGURE 42.4 White noise.

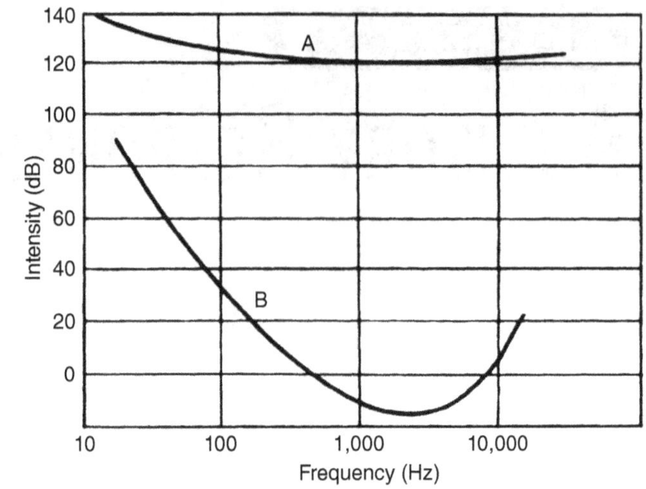

FIGURE 42.7 Human audibility curve.

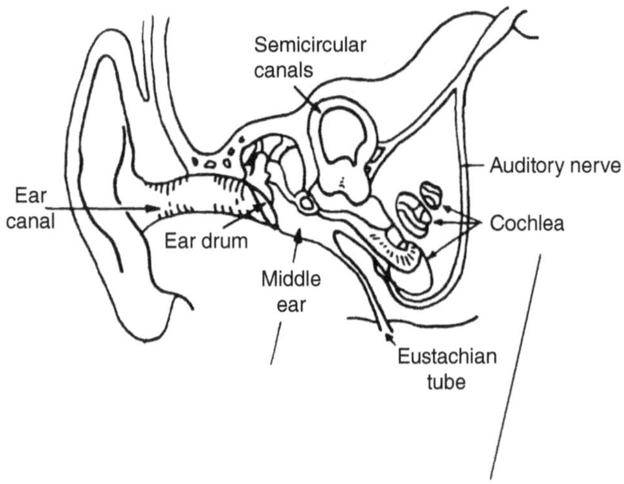

FIGURE 42.5 Cross-section of the ear, with major structures.

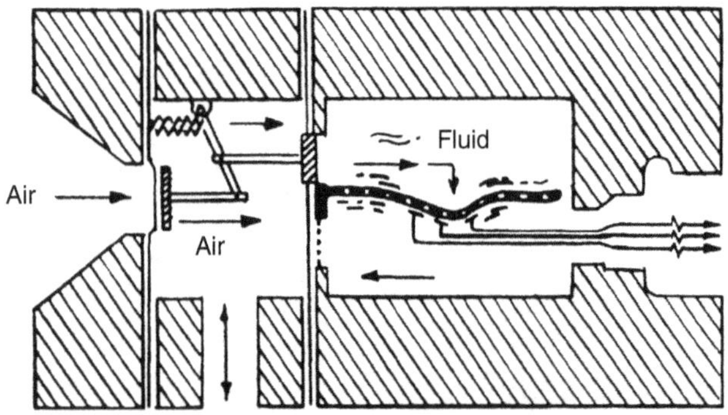

FIGURE 42.6 Functional diagram of the ear. Sound waves impinge on outer ear and cause ear drum to vibrate. Vibrations are conducted via lever action of middle-ear bones to oval window, which actuates fluid-filled inner ear. Portions of nerve endings in cochlea are selectively excited, and their outputs terminate in brain.

TABLE 42.1 Units for Sound Properties and Their Response

PHYSICAL SOUND PROPERTIES (UNITS)	SUBJECTIVE RESPONSE (UNITS)
Properties (units)	Response (units)
Intensity (decibels)	Loudness (phons, sones)
Frequency (cycles/second, Hertz)	Pitch (mels, barks)

ear typically responds in a relatively predictable manner to physical sounds. Table 42.1 shows the units commonly used for sound properties and their subjective response.

Loudness is the physical response to sound pressure and intensity, and it varies directly with the sound pressure and intensity, as indicated in Fig. 42.8.

Contours of constant loudness level are shown in Fig. 42.9. The curves illustrate the amount by which the pitch of a pure tone is shifted when the loudness level is changed from 40 dB to the level of contour. For example, a 100-Hz tone will seem 10 percent lower in pitch when the loudness level shifts from 40 to 100 dB, but one of 500 Hz will shift only 2 percent lower.

Loudness-level contours showing all pure tones on each curve sound equally loud. Observations with tones that differ considerably from the reference tone, however, are more difficult to make, and the error for the combined result is on the order of 1–2 dB. Table 42.2 shows that a 10-dB change might make a sound seem twice as loud, no matter what the initial level was.

Loudness as a function of intensity and frequency is illustrated in Fig. 42.10. Subjective loudness in "sones" is represented vertically above the intensity–frequency plane. The heavier curves progressing from front to rear are equal-loudness contours for pure tones.

Pitch as a function of frequency is illustrated in Fig. 42.11. The upper curve shows that subjective pitch (in "mels")

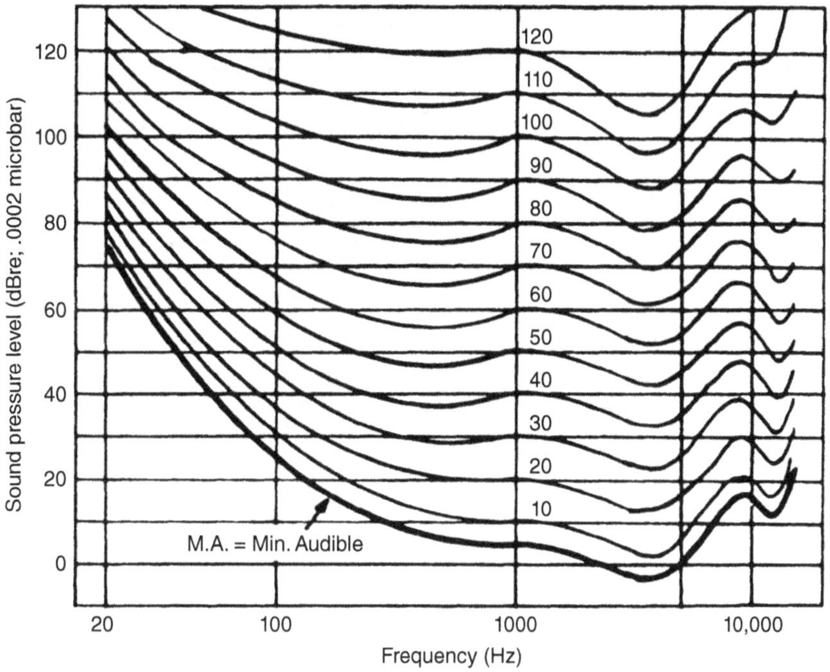

Figure 42.8 Equal loudness contours.

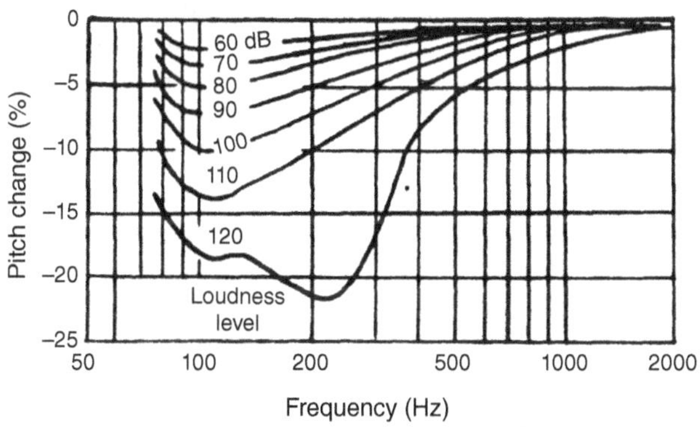

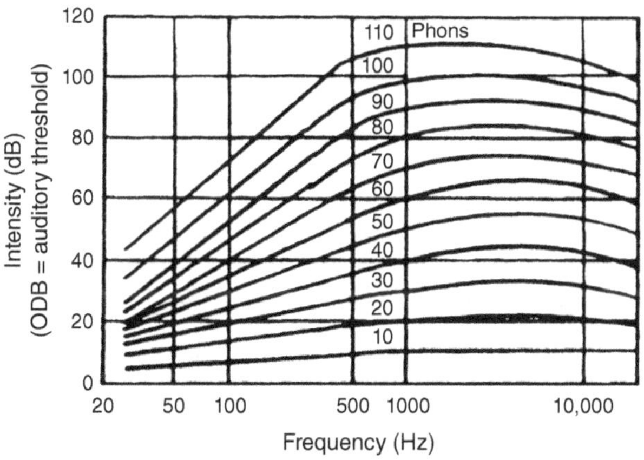

Figure 42.9 Equal loudness contours.

TABLE 42.2 Sensitivity to Changes in Intensity	
Sound-Level Change, dB	**Change in Apparent Loudness**
3	Barely perceptible
5	Clearly noticeable
10	Either twice as loud or twice as soft
20	Much louder or softer

increases less and less rapidly as the stimulus frequency increases in a linear fashion. The lower curve shows that subjective pitch increases as stimulus frequency is increased logarithmically. Our tempered musical scale is a logarithmic scale. The pitch of a 1000-Hz tone, 40 dB above threshold, is defined as 1000 mels.

Effect of Frequency on Differential Threshold

$\Delta f/f$ as a function of frequency for five different sensation levels is shown in Fig. 42.12a. Between 500 and 8000 Hz, $\Delta f/f$ remains relatively constant. Between 62 and 500 Hz, Δf is approximately constant. At 45 Hz, $\Delta f/f$ versus f shows an inflection point and the tendency to bend downward, with the tendency being more marked at higher sensation levels (because of subjective harmonics).

Effect of Intensity of Test Tones

$\Delta f/f$ as a function of sensation level for seven different frequencies is shown in Fig. 42.12b. Sensitivity increases as sensation increases, up to 40 dB (0 = threshold of hearing at

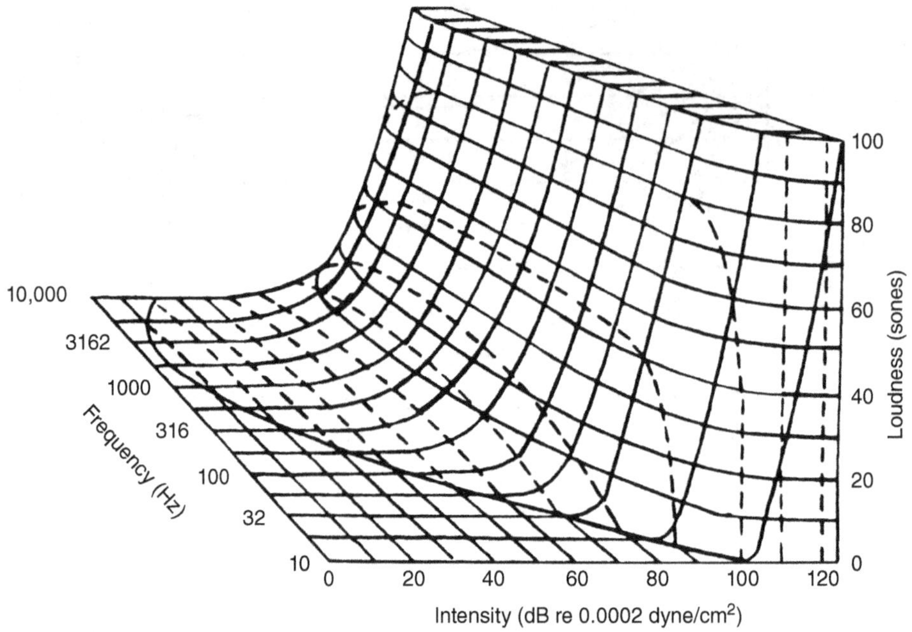

FIGURE 42.10 Loudness as a function of intensity and frequency.

that frequency). Note that the effect is more pronounced for lower tones. When the loudness level of a 1024-Hz tone (as well as its duration) is varied, accuracy of discrimination is reduced by the decrease in loudness; the effect is most pronounced for short durations.

Sensitivity to Differences in Loudness: Effect of Frequency

Sensitivity to differences in loudness of pure tones as a function of frequency for seven different intensity levels is shown in Fig. 42.12c (energy of the wave is the intensity measure). On the basis of these data, it is apparent that the greatest number of loudnesses that are distinguishable at any frequency is 370 at 1300 Hz.

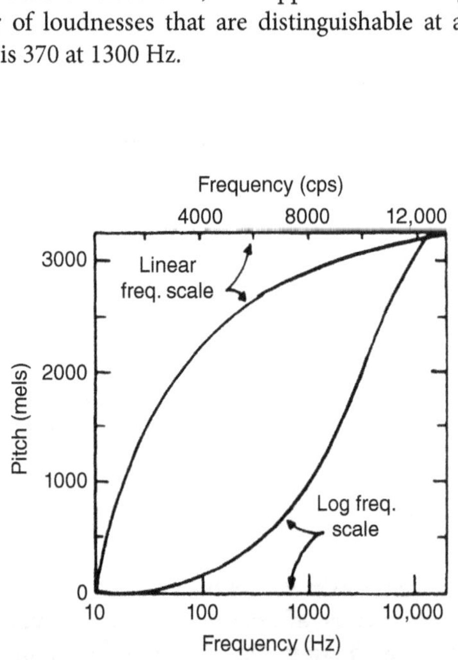

FIGURE 42.11 Pitch as a function of frequency.

RESPONSE TO COMBINED TONES

When two tones are combined, even though they are of the same frequency, their loudness is the summation of both tones. However, if the phase of one tone is altered, the same pitch is heard, but with less loudness. If two tones are identical in terms of phase, they cancel each other out, and neither is heard.

Beats will occur (i.e., loudness builds and fades) when two tones that are slightly different in frequency are presented at the same time. Beats occur at a frequency equal to the difference between the two tones. The pitches of the two tones merge and will be heard as a single tone, the pitch of which is halfway between the pitches of the separate tones.

When the separation of tones is varied, beats will occur in three distinct stages:

1. At close frequencies (6 beats per second) beats will be very distinct and will appear as smooth changes in loudness.

2. As the frequency interval between tones increases (8 beats per second), beats become faster, and thus the resultant tone appears to throb or pulsate.

3. The two tones become more prominent as the frequency differential becomes greater (20 beats per second), resulting in a "buzz" rather than a throb.

Figure 42.13 illustrates the several phenomena that occur when tones are combined.

The effect of different intensity interactions upon the experiencing of beats is shown in Fig. 42.14. At the upper and lower extremes, only one tone is heard (a strong tone masking out a weaker one). Between 20 and 60 dB, the absolute

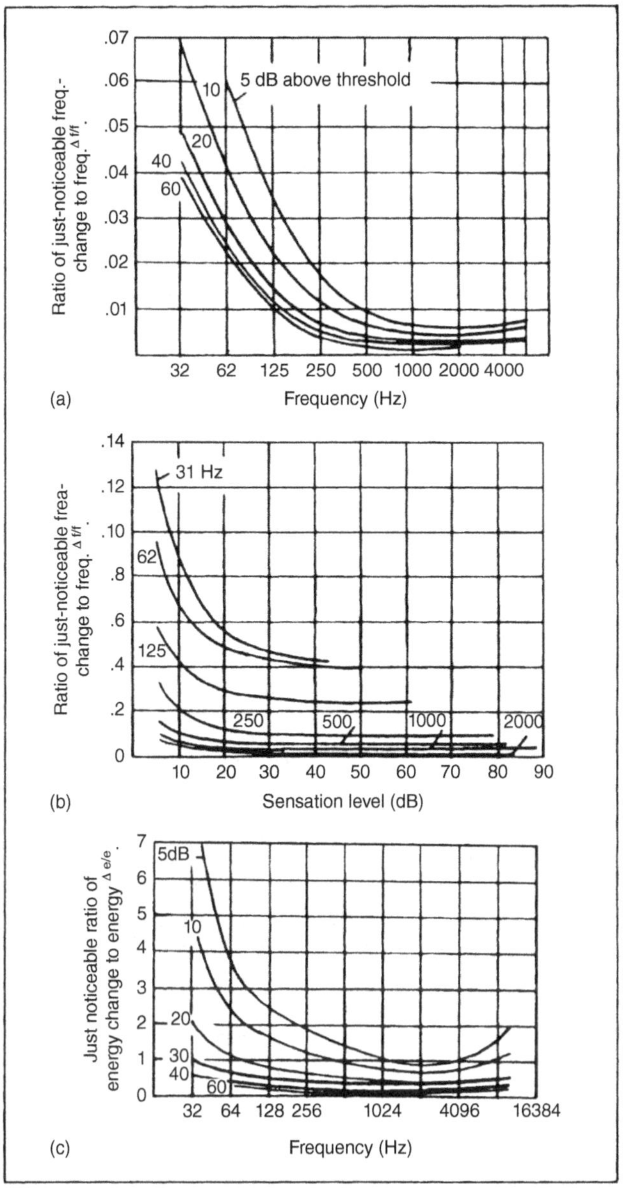

Figure 42.12 Sensitivity to changes in frequency and intensity.

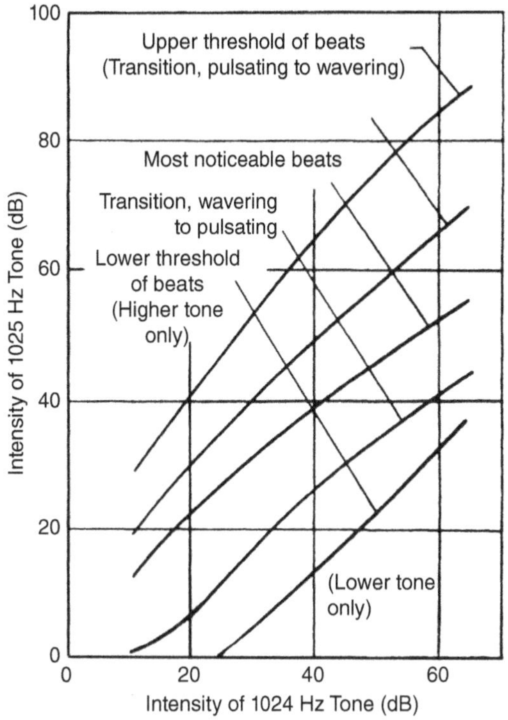

Figure 42.14 Effect of tone intensity on perception of beat.

intensity is unimportant. Beats are most apparent when the two generators are approximately equal in intensity.

THRESHOLDS FOR SHORT TONES WITH DIFFERENT RATES OF REPETITION

Threshold signal-to-noise ratio for masked thresholds as a function of repetition rate of the sound is shown in Fig. 42.15. Differences between curves for different durations are almost exactly equivalent to the differences in total energy produced (difference is independent of repetition rate). Note that the shape of curves is similar regardless of frequency and that threshold decreases continuously as repetition rate increases. Discontinuity between the 2- and 5-per-second rates is more marked, indicating that the ear does little integrating of energy beyond 200-ms duration.

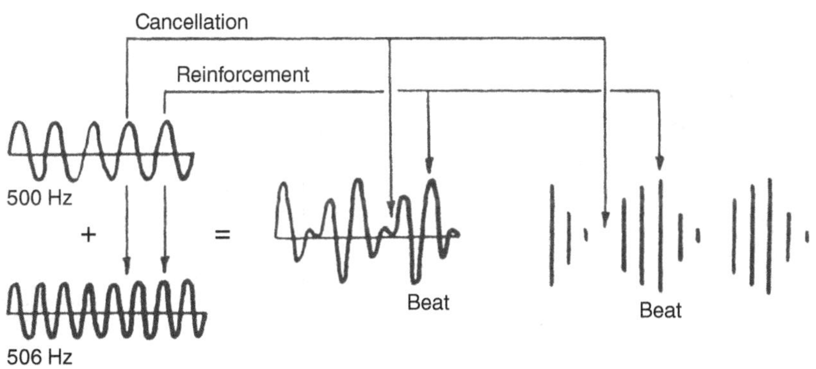

Figure 42.13 Tones at close frequencies combine to form a beat.

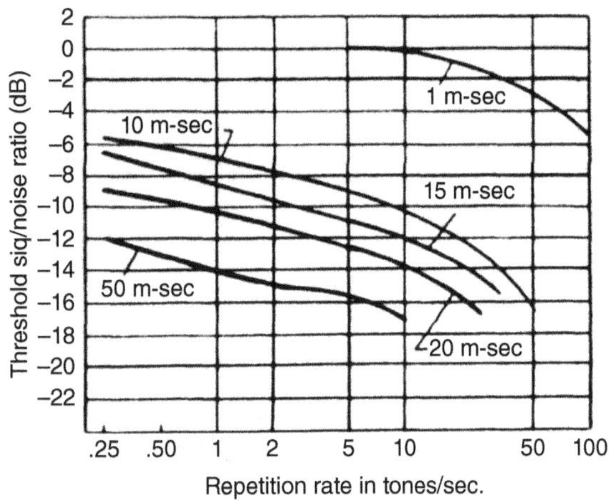

FIGURE 42.15 Repeated short tones.

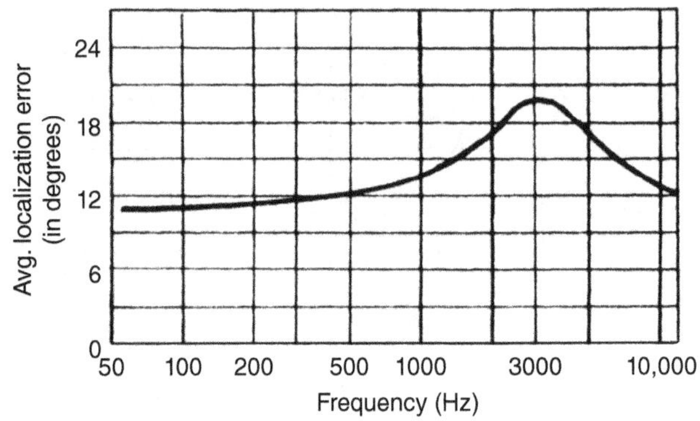

FIGURE 42.17 Localization accuracy vs. sound frequency.

LOCALIZATION OF SOUND

Listeners can determine the direction from which a sound emanates accurately when the sound source is to their right or left or directly above or below them but less accurately when the sound source is in front of them, to the rear, or at intermediate levels above or below their heads. It is often disconcerting to listeners if they can see the lips of an orator or the motions of an orchestra but the sound emanates (via loudspeakers) from another direction. (See Fig. 42.16.)

Dependence of Localization upon Frequency

The average error of localization in degrees at various frequencies is illustrated in Fig. 42.17. Accuracy is worst between 2000 and 4000 Hz; otherwise, localization is relatively constant. Note that the average error is smallest nearest the median plane and increases toward the side of the listener's head. For stimuli of 200 and 2000 Hz, the least amount of intensity difference at the two ears required to produce a shift in localization is 0.9 dB at 200 Hz and 0.2 dB at 2000 Hz.

MASKING

When one sound masks another, intensity and frequency relationships determine the extent of the masking. Figure 42.18 indicates that masking the function is essentially independent of the frequency of the masked noise.

A pure tone more effectively masks a tone higher than itself than one that is lower. Masking is greatest when the masking frequency is close to the "wanted" signal frequency.

Masking increases with increases in the intensity of the masking tone. Figure 42.19 illustrates masking at frequencies from 200 to 1200 Hz produced by a 1200-Hz tone.

Loudness Threshold of a Masked Tone

Contours for masked threshold (i.e., the sound pressure necessary to make a pure tone just audible in the presence of an

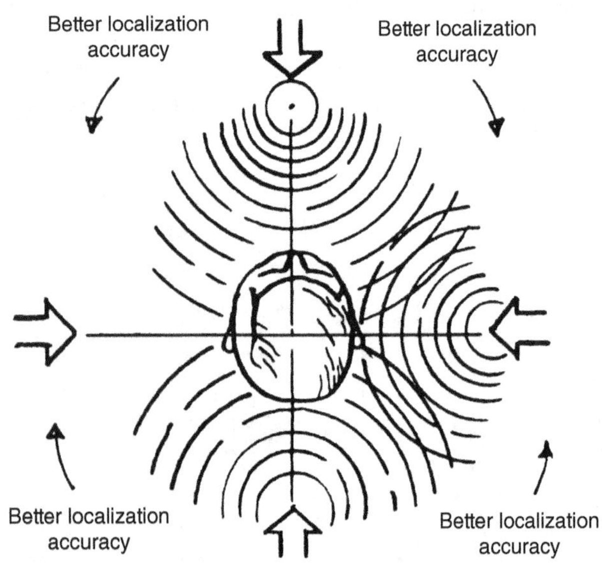

FIGURE 42.16 Localization accuracy vs. sound source.

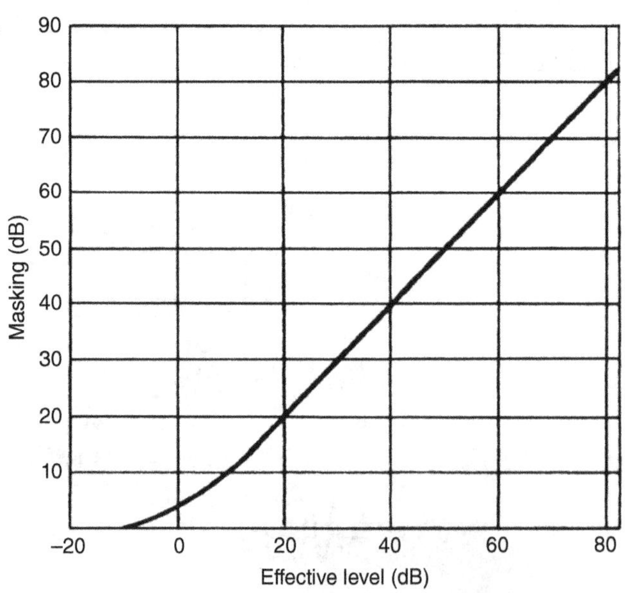

FIGURE 42.18 Masking effectiveness for any frequency.

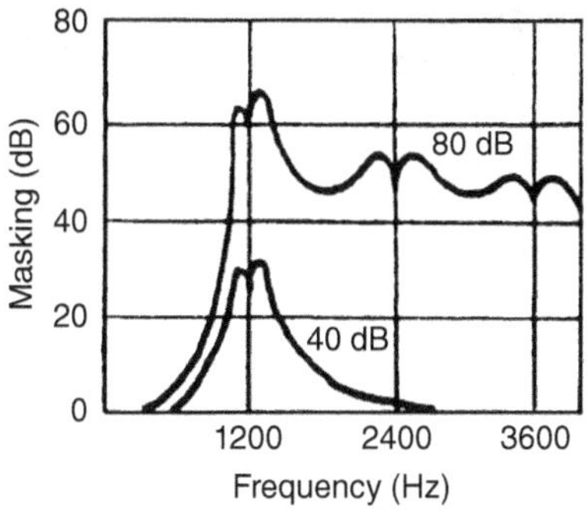

Figure 42.19 Masking produced by a 1200-Hz tone.

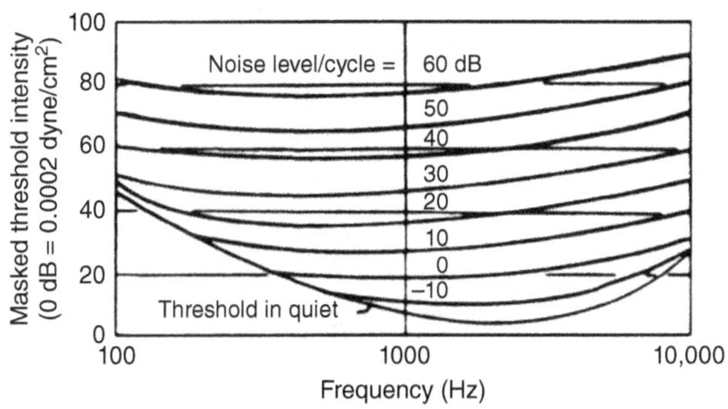

Figure 42.20 Threshold of masked tones.

interfering sound) at various noise levels are shown in Fig. 42.20. Note that thresholds generally parallel one another, with approximately 20-dB intervals between masking noise levels.

Pitch discrimination becomes worse as the signal-to-noise ratio changes from infinity (no noise) to zero. Pitch discrimination is, however, better at some frequencies than others; e.g., 500 Hz is superior to 800 Hz.

EFFECTS OF NOISE ON HUMAN PERFORMANCE

Noise can have both positive and negative effects on the performance of a tasks; noise stimulates the listener, but arousal results in the narrowing of attention. As the noise intensity increases, increased arousal causes an improvement in task performance up to a point; beyond that level of intensity, over-arousal sufficient to degrade task performance occurs. Accompanying the degraded task performance, psychological effects of noise can include anxiety and learned helplessness.

It is important to take into consideration factors such as the person's level of familiarity with the work, the degree to which verbal communication is necessary, the extent to which several parts of a task make conflicting demands on attention, and duration of the noise or the task.

Continuous noise as high as 100 dBC does not normally affect simple tasks, visual acuity, tool use, or short-term intellectual function. This is particularly true when discriminations are easy, time pressure is absent, or the response manner or time is predictable. However, below are specific task performance areas that noise does affect:

a. Short-Term Memory: If several operations are connected that require short-term memory, for instance to solve a mathematical problem that appears only briefly, noise can have an impact.

b. Comparison and Discrimination: At levels as low as 85 dBC, there may be an effect on confidence in the nature of unexpected visual signals that are difficult to discriminate, or on tasks that require remembering and comparing signals.

c. Multitasking: Continuous noise also has an adverse effect on cognitive performance that may involve "multitasking." If the same material can be analyzed in more than one way, the most dominant and obvious means will be adopted during continuous noise. Given two memory tasks under relatively high (75–100 dBC) continuous background noise levels, noise can even improve efficiency at the dominant task, but with a corresponding degradation in a secondary task.

d. Spatial Memory: Tasks involving the learning of lists, e.g., from a repair manual, may involve verbal or spatial memory tasks; noise hinders those who use a spatial strategy.

e. Visual Tracking: Continuous regular periodic and aperiodic noise reduces performance on a complex visual tracking task. At levels of 50, 70, and 90 dB of white noise, the greatest decrement occurred with the highest noise level.

Short-term changes in level that are infrequent can cause temporary disturbance of performance of 2 to 30 seconds, with a loss of efficiency proportional to the change in level. This can affect changes in reaction time, ability to focus on visual tasks or take information from visual displays, and other cognitive activities. Irregular and unpredictable bursts of noise outside of the control of the subject are more disturbing to performance than those that are perceived to be within the subject's control, even when the subject does not control them.

Elevated noise levels above background noise or exposure to constant levels of moderate noise can delay the onset of sleep, awaken one from sleep, and interfere with rest. This sleep loss can lead to degraded performance.

Table 42.3 presents a general description of how various noise levels affect humans and their ability to perform typical tasks.

When noise levels exceed 100 dB, potentially serious consequences occur as shown in Table 42.4.

EFFECTS OF NOISE ON COMMUNICATIONS

Typical speech intensities are shown in Table 42.5. The effect of distance and noise levels on speech intensities are shown in Fig. 42.21.

TABLE 42.3 Noise Effects on Performance

Noise Level, dB	Effects
100	Serious reduction in alertness. Attention lapses occur, although attention duration is usually not affected. Temporary hearing loss occurs if no protection is provided in the region 600–1200 Hz. Most people will consider this level unacceptable, and 8 hr is the maximum duration they will accept.
95	Considered to be the upper acceptance level for occupied areas where people expect the environment to be noisy. Temporary hearing loss often occurs in the range of 300–1200 Hz. Speech will be extremely difficult, and people will be required to shout, even though they may be talking directly into a listener's ear.
90	At least half of the people in any given group will judge the environment as being too noisy, even though they expected a noisy environment. Some temporary hearing loss in the range of 300–1200 Hz occurs. Skill errors and mental decrements will be frequent. The annoyance factor is high, and certain physiological changes often occur (e.g., the pupils dilate, the blood pressure increases, and the stroke volume of the heart may decrease). Listening to a radio is impossible without good earphones. The maximum duration that most people will accept is 8 hr.
85	The upper acceptance level (noise expected) in the range of 150–1200 Hz. Some hearing loss occurs in the range of 300–1200 Hz. This is considered the upper comfort level, although some cognitive performance decrement can be expected, especially where decision making is necessary.
80	Conversation is difficult (i.e., people have to converse in a loud voice less than 1 ft apart). It is difficult to think clearly after about 1 hr. There may be some stomach contraction and an increase in metabolic rate. Strong complaints can be expected from those exposed to this level in confined spaces, and 8 hr is the maximum duration acceptable within the frequency range 1200–4800 Hz.
75	Too noisy for adequate telephone conversation. A raised voice is required for conversants 2 ft apart. Most people will still judge the environment as being too noisy.
70	The upper level for normal conversation, even when conversants are close together (at a distance of 6 ft people will have to shout). Although persons such as industrial workers and shipboard personnel who are used to working in a noisy environment will accept this noise level, unprotected telephone conversation will be difficult (upper phone level is 68 dB).
65	The acceptance level when people expect a generally noisy environment. Intermittent personal conversation is acceptable. About half of the people in a given population will experience difficulty sleeping.
60	The upper limit for spaces used for dining, social conversation, and sedentary recreational activities. Most people will rate the environment as "good" for general daytime living conditions.
55	The upper acceptance level for spaces where quiet is expected (150–2400 Hz). People will have to raise their voices slightly to converse over distances greater than 8 ft. This level of noise will awaken about half of a given population about half of the time. It is still annoying to people who are especially sensitive to noise.
50	Acceptable to most people where quiet is expected. About 25 percent will be awakened or delayed in falling asleep. Normal conversation is possible at distances up to 8 ft.
40	Very acceptable to all. The recommended upper level for quiet living spaces, although a few people may still have sleep problems.
30	Necessary for specialized listening tasks (e.g., threshold signal detection).
Below 30	Introduces additional problems; i.e., low-level intermittent sounds become disturbing. Some people have difficulty getting used to the extreme quiet, and a few may become psychologically disturbed.

Note: The above represents an amalgamation of many studies and contains a general interpretation of a wide variety of subject samples and testing conditions.

TABLE 42.4 Physical Effects of Very High Noise Levels

Noise Level, dB	Spectrum	Duration	Effects
105	Jet engine	2 min	Reduced visual acuity, stereoscopic acuity, and near-point accommodation and permanent hearing loss when exposure continues over a long period (months)
110	Machinery noise	8 hr	Chronic fatigue and digestive disorders
120	Broadband	1 hr	Loss of equilibrium
150	1-100 Hz	2 min	Reduced visual acuity, chest-wall vibration, changes in respiratory rhythm, and a "gagging" sensation*

*Subjects were wearing so-called protective aids to prevent hearing loss.

TABLE 42.5 Typical Speech Intensities

Description	Typical Intensity of Source, dB	Distance	Message Characteristic
Soft whisper		3–6 in	Secret communication
Audible whisper	44–69	8–20 in	Confidential or personal
Normal voice	50–75	20–60 in	Personal or small group
Loud voice	56–81	5–8 ft	Nonpersonal, intermediate group
Very loud voice	62–87	To 20 ft	Group address
Shouting	68–93	Upper limit	Hailing or emergency communication

The concept of signal-to-noise ratio (expressed in decibels) is fundamental to understanding general statements regarding speech intelligibility. Normal speech has a level of 65 dBA at a distance of 1 meter from a talker to listener. If the background noise is 60 dBA at the head position if a listener, for example, the a positive signal-to-noise ratio of 5 dB is said to exist. Generally speaking, at least a 10-dB signal-to-noise ratio in the frequency region of speech (200 Hz–7 kHz) is optimal for making speech intelligible (approximately 90% word recognition).

NOISE EXPOSURE LIMITS

Noise limits have been established based on the anticipated tasks and working environments.

Occupational Safety and Health Administration Noise Limits

The goal of the Occupational Safety and Health Administration (OSHA) limits are to prevent permanent hearing loss due to exposure throughout a working career. Table 42.6 shows the limits established by OSHA. Figure 42.22 allows

TABLE 42.6 Occupational Safety and Health Act Permissible Daily Noise Exposure

OSHA Noise Standards for Steady-State Noise

Duration, hr	Sound Level, dBA
8	90
6	92
4	95
3	97
2	100
1.5	102
1	105
0.5	110
0.25	115

Note: When exposure is intermittent at different levels, the fraction $C_1/T_1 + C_2/T_2 \ldots C_n/T_n$ should not exceed unity to meet the exposure limit. C_n = total exposure time at the specified level; T_n = total exposure time permitted at the specified level.

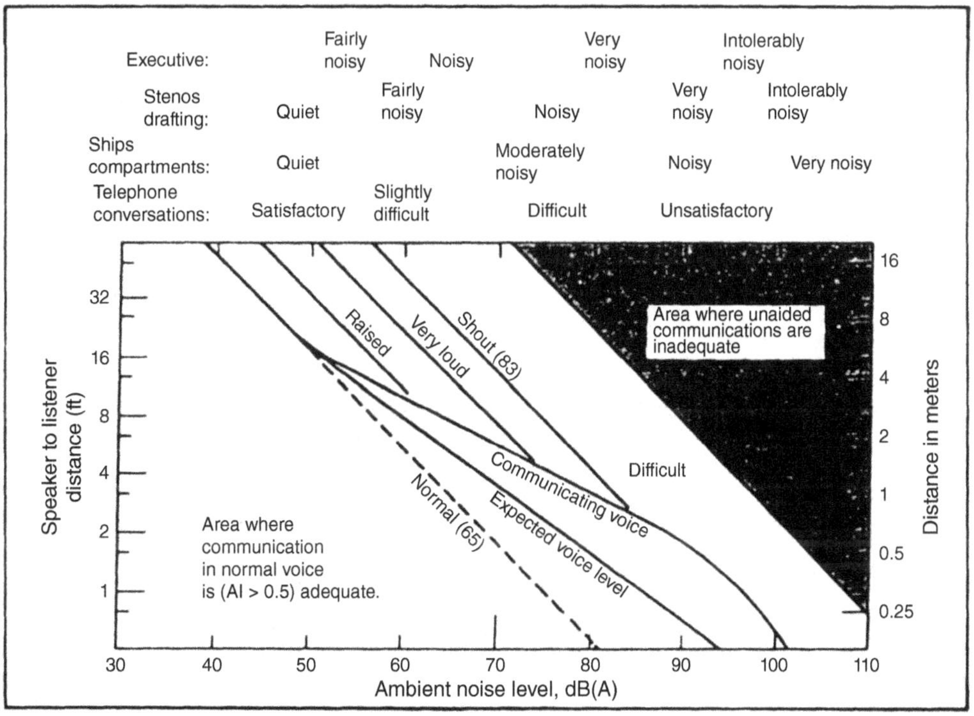

Figure 42.21 Permissible distance between a speaker and listeners.

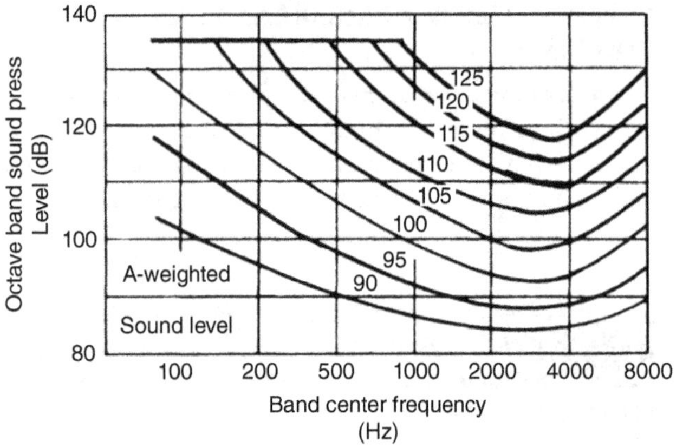

FIGURE 42.22 Contours for determining equivalent A-weighted sound level. This graph is used interpreting octave-band sound levels according to OSHA provisions.

the user to convert noise levels measures at octave frequencies to the overall dBA weighted values used in the limits.

Military Systems Noise Limits

The noise limits for military working environments are defined in MIL-STD-1474. The limits depend on the working environments and the task requirements.

Various working environments are lumped into steady-state noise categories according to Table 42.7. Table 42.8 then defines the maximum noise levels for each working category.

Table 42.9 shows limits for noise levels in various military facilities.

NASA Systems Noise Limits

The continuous noise limits for NASA systems are defined in NASA-STD-3001. The limits use the noise criterion (NC) curve as a base. Figure 42.23 shows the NC curve and Table 42.10 has the NASA limits for both general working and sleeping areas.

Commercial Work-Space Noise Criteria[30]

NC curves are widely used as work-space design criteria. Table 42.11 identifies typical work spaces with the appropriate NC curves that should be applied.

NC curves are referenced to preferred octave bands (lower abscissa) and to commercial bands (upper abscissa). When using the commercial frequency, the NC number is also the speech interference level for that particular spectrum.

[30] D. C. Hodge and G. R. Garinther, in J. F. Parker and V. R. West (eds.), *Bioastronautics Data Book*, 2nd ed., National Aeronautics and Space Administration, 1973, pp. 693–750.

TABLE 42.7 Steady-State Noise Categories

CATEGORY[1]	SYSTEM REQUIREMENTS
A	No direct person-to-person voice communication required. Maximum design limit. Hearing protection required.
B	Electrically-aided communication via attenuating helmet or headset required. Noise levels are hazardous to unprotected ears.
C	No frequent direct person-to-person voice communication required. Occasional shouted communication may be possible at a distance of 30 cm. Hearing protection required.
D	No frequent person-to-person voice communication required. Occasional shouted communication may be possible at a distance of 60 cm. Levels in excess of Category D require hearing protection.
E	Occasional telephone or radio use or occasional communication at distances up to 1.50 m required. (Equivalent to NC-70)
F	Frequent telephone or radio use or frequent communication at distances up to 1.50 m required. (Equivalent to NC-60)

[1]Category A, B, C, and D are based primarily on hearing conservation priorities; the remaining categories are based primarily on communication requirements. For fixed-plant facilities, see MIL-STD-1472.

TABLE 42.8 Steady-State Noise Limits (dBA) for Personnel-Occupied Areas

Limit Category:	A[1]	B[1]	C[1]	D[1]	E[1]	F[2]
A-Weighted Limit (dBA):	108	100	90	<85	75	65
SIL-4 Limit:					67	57

[1]These limits are based on a maximum daily exposure time of 8 hours. In those cases where the mission profile for the equipment being developed or acquired exceeds 8 hours of operation in each 24-hour day, the limits specified in Categories A, B, C, and D shall be reduced to allow for an exposure of longer than 8 hours, as approved by the procuring activity in conjunction with the appropriate military Surgeon General's Office.
[2]Categories E and F are defined by either the sound level (dBA) or the speech interference level (SIL-4). If meeting the dBA requirement has been documented to be infeasible (see 5. 1.3.2), then this requirement shall be replaced with the corresponding SIL-4 limit.

TABLE 42.9 Upper Noise-Level Limits Recommended for Military Facilities

Sound Level, dBA	Type of Activity	Communication Equivalent	Office Application
108	Maximum design limit for AMC equipment (hearing protection required)	No direct communication	Not recommended
100	Armored vehicles (hearing protection required)	Electrically aided communication satisfactory with attenuating helmet or headset; limited shouted communication possible with difficulty	Not recommended
90	Materiel which is beyond the state of the art of meeting 85 dBA (hearing protection required)	Shouted communication possible at short distances (1–2 ft)	Not recommended
85	Acceptable level for unprotected hearing for 8-hr exposures	Shouted communication possible at several feet (3–4 ft); telephone use difficult	Not recommended
75	Maintenance shops, garages, and keypunch areas	Occasional telephone use and occasional direct communication at up to 5 ft is acceptable	Not recommended
65	Operation centers, mobile command and communication centers, computer rooms, word processing centers, kitchens, and laundries	Frequent telephone use and frequent direct machine communication at up to 5 ft is acceptable	Business machine offices
55	Drafting rooms, laboratories, and conferences with two or three people	No difficulty with telephone use and occasional direct communication at up to 15 ft	Shop offices and general secretarial areas
45	Libraries, conference rooms, command and control centers, theatres, and sleeping areas	No difficulty with direct communication	General offices
35	Recording studios and large conference rooms	Areas requiring unusually extreme quiet	Executive offices

These recommended levels are in consonance with the requirements of MIL-STD-1472 and MIL-STD-1474.

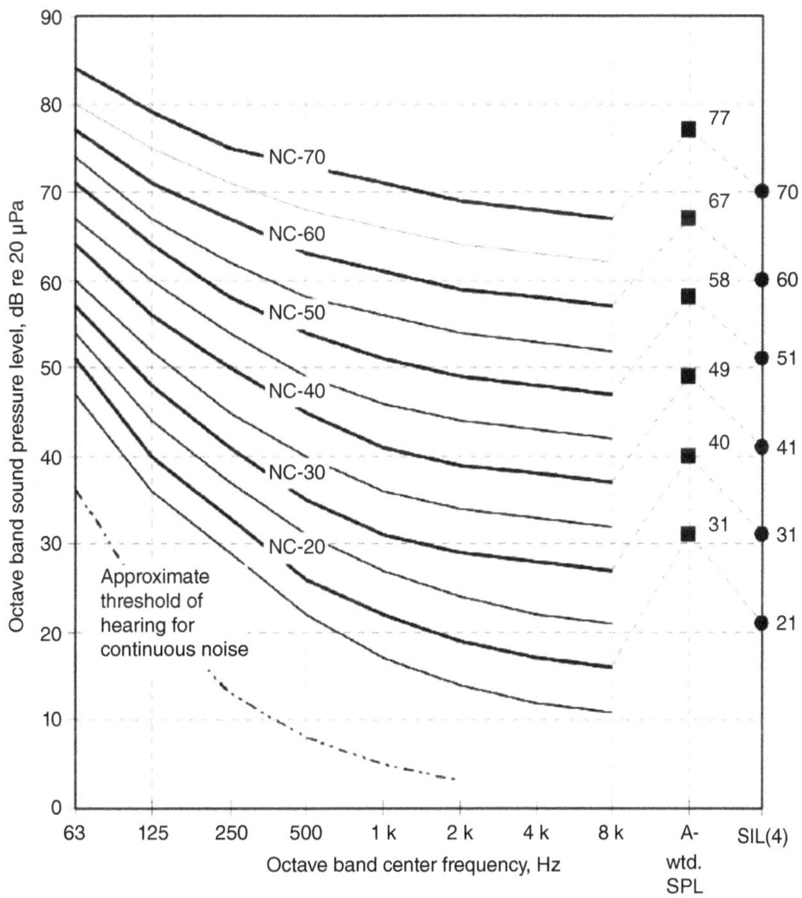

FIGURE 42.23 Noise criterion curves. Corresponding A-weighted sound pressure levels and speech interference levels are given for reference only (Beranek, 1992). SIL(4) is speech interference level, 4-band method.

TABLE 42.10 Octave Band SPL Limits for Continuous Noise, dB re 20 μPa

Octave Band Center Frequency (Hz)	63	125	250	500	1 K	2 K	4 K	8 K	16 K	NC
Work Areas Maximum (NC-50)	71	64	58	54	51	49	48	47	46	50
Sleep Areas Maximum (NC-40)	64	56	50	45	41	39	38	37	36	40

TABLE 42.11 Recommended NC Curves for Various Work Spaces

		Noise Criteria Exposure Limits	
NC Curve	**Type of Work Space**	**Communication Equivalent**	**Office Application**
90		Noise-attenuating headset required	Not recommended
80		Communication very difficult; telephone use unsatisfactory	Not recommended
70-80		Raised voice range 1-2 ft; shouting range 3-6 ft; telephone use very difficult	Not recommended
60-70		Raised voice range 1-2 ft; telephone use difficult	Not recommended
55-60		Very noisy, not suited for office; telephone use difficult	Not recommended
55	Spacecraft during nonpowered flight		
50-55		Unsatisfactory for conferences of over 3 people; telephone use slightly difficult; normal voice at 2 ft; raised voice at 3 ft	Areas with typists and accounting machines
40-50	Restaurants, sports coliseums	Conferences at 4- to 5-ft table; telephone use slightly difficult; normal voice at 3-6 ft; raised voice at 6-12 ft	Large drafting rooms
35-40		Conferences at 6- to 8-ft table; telephone use satisfactory; normal voice at 6-12 ft	Medium-sized offices
30-35	Libraries, hospitals, motion picture theatres, home sleeping areas, assembly halls	Quiet office; conferences at 15-ft table; normal voice at 10-30 ft	Private or semiprivate offices; reception rooms; conference rooms for up to 20 people
25-30	Courtrooms, churches, home sleeping areas, assembly halls, hotels and apartments, TV studios, music rooms, schoolrooms	Very quiet offices; large conferences	Executive offices; conference rooms for 50 people
20-25	Legitimate theatre, concert halls, broadcasting studios		

Impulse Noise Exposure Limits

An impulse noise has a duration less than 1 second. Impulse noise can be a problem in the military environment. The exposure limits shown in Fig. 42.24 are from MIL-STD-1474, *Noise Limits for Army Material.*

NASA limits for impulse noise simply state that "the system shall limit impulse noise measured at the crewmember's head to 140 dB peak SPL..." (NASA-STD-3001, Volume 2).

Infrasonic (Low-Frequency) Noise Limits

Infrasonic (from 1 to 20 cycles per second) noise levels must be limited to less than 150 dB (NASA-STD-3001, Volume 2).

Ultrasonic (High-Frequency) Noise Limits

Ultrasonic (over 20 cycles per second) noise levels must be limited to the values in Table 42.12 (NASA-STD-3001, Volume 2).

REVERBERATION

Reverberating sounds tend to mix with direct sounds, thus affecting the clarity with which the sounds are perceived by the listener.

Reverberation time in relation to speech intelligibility is illustrated in Fig. 42.25.

Reverberation time in relation to room size for both speech and music sounds is illustrated in Fig. 42.26.

Suggested reverberation objectives for specific listening requirements are noted in Table 42.13.

Impulse noise limit selection criteria

Maximum expected number of exposures in a single day	Impulse noise limit		
	No protection	Either plugs or muffs	Both plugs and muffs
1000	W	X	Y
100	W	Y	Z
5	W	Z	Z

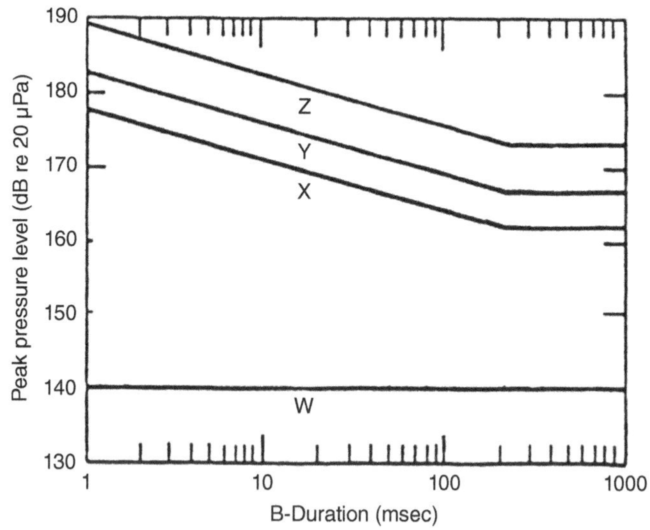

Peak pressure level, and B-duration limits for impulse noise

Note: B-duration is the time that the noise remains within 20 dB of the peak pressure level.

Figure 42.24 Impulse noise exposure limits. (From MIL-STD-1474B.)

TABLE 42.12 Ultrasonic Noise Limits Given in One-Third Octave Band SPLs

One-Third Octave Band Center Frequency (kHz)	Limits [One-Third Octave Band SPL (dB)]	
	Not to Exceed	8-hr TWA*
10	105	89
12.5	105	89
16	105	92
20	105	94
25	110	—
31.5	115	—
40	115	—
50	115	—
63	115	—
80	115	—
100	115	—

*TWA = time-weighted average

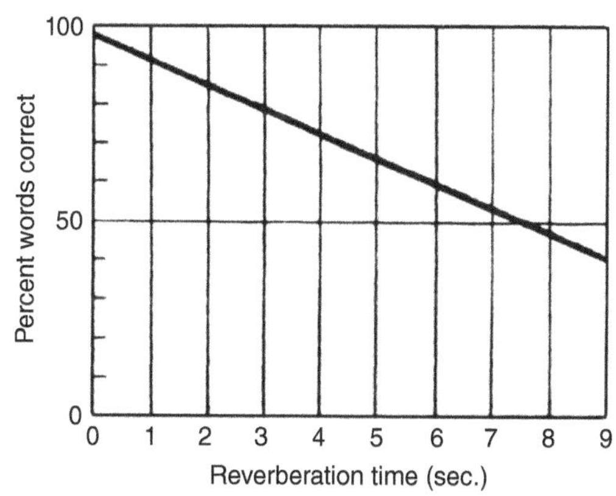

Figure 42.25 Effect of reverberation on intelligibility.

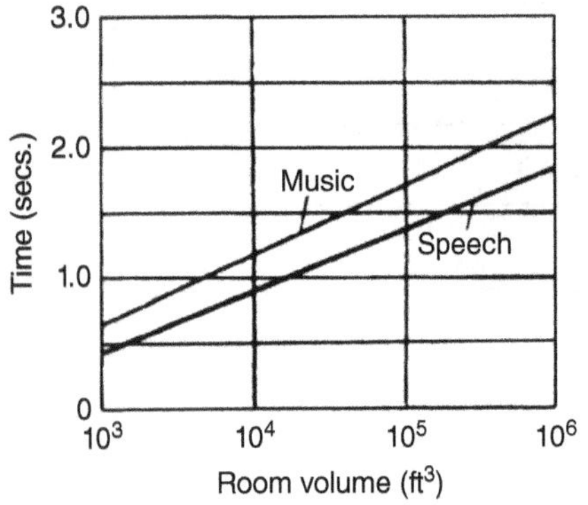

Figure 42.26 Estimating reverberation by room size.

TABLE 42.13 Recommended Reverberation Environments

Listening Requirement	Reverberation Times, s
Optimum for speech (too dead for music)	Below 1.0
Good for speech, fair for music	1.0–1.5
Fair for speech, good for music	1.5–2.0
Poor for speech, good for liturgical music	2.0 +

CHAPTER 43

TACTILE RESPONSE

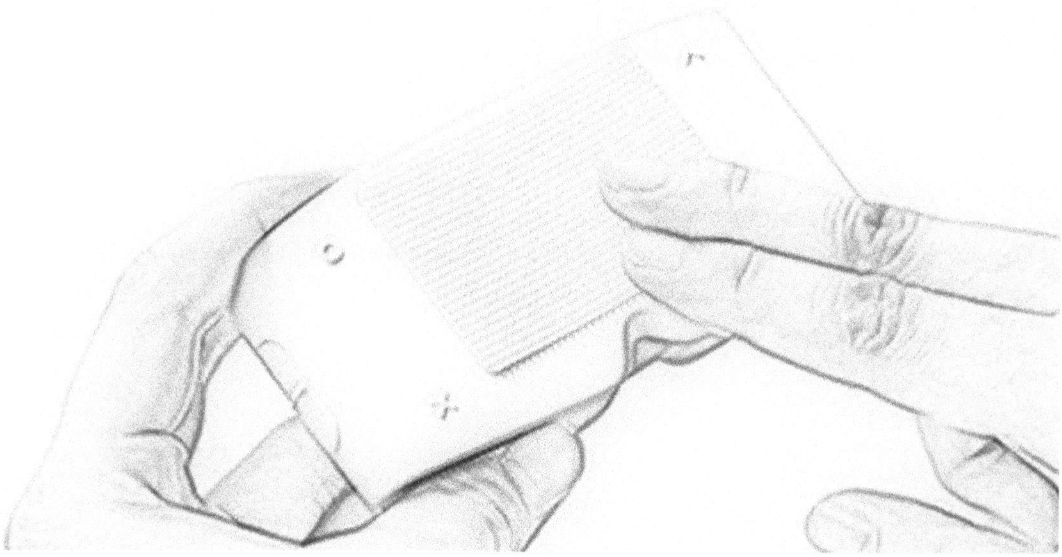

RESPONSE TO TACTILE PRESSURE

Stimulus thresholds for pressure on significant regions of the body are as shown in Table 43.1.

Relative accuracy of judging location and spatial separation and number of pressure stimuli on the volar surface of the forearm (2-s interval between successive stimulations) are as shown in Tables 43.2 and 43.3.

TABLE 43.1 Threshold Sensitivity to Tactile Pressure

REGION	PRESSURE, g/mm²
Tip of the finger	3
Back of the finger	5
Front of the forearm	8
Back of the hand	12
Abdomen	26
Back of the forearm	33
Thick parts of the sole of the foot	250

TABLE 43.2 Ability to Judge Separation of Pressure Points

	POINT SEPARATION, mm	
AXIS	SUCCESSIVE POINTS	SIMULTANEOUS POINTS
Longitudinal	9.16–9.64	12.67–15.73
Transverse	9.40–9.55	14.60–15.02

Note: Weight of probe = 20 g.

COMFORT CONSIDERATIONS

A different kind of tactile response, and one that is often ignored, involves the pressure of seat-belt webbing on the user's body. The most critical complaints concern too much pressure on the occupant's shoulder or chest. As is illustrated in Fig. 43.1, a typical torso belt crosses the chest at a diagonal, and although it should approximately bisect the chest near the sternum, it often is poorly laid out; as a result, the belt rides across the individual's breast (which is especially annoying to females). A recent study indicates that the

TABLE 43.3 Amount of Pressure Relative to Accuracy of Judging Location

Region	Weight 12g	27 g	40 g	57.5 g	67.5 g
Back of hand:					
Mean	7.21	6.93	6.86	6.69	6.76
SD	4.7	4.6	4.4	4.3	4.4
Volar surface of forearm:					
Mean	7.19	7.13	6.14	6.75	6.59
SD	4.5	4.5	3.7	4.1	3.8

Note: Tactile adaptation time tends to occur and is longer for heavier pressures. People generally improve with practice.

FIGURE 43.1 Shoulder belt comfort limits.

pressure of a torso belt on the occupant's chest or shoulder should not exceed about 0.9 lb (4 N); otherwise, at least 50 percent of typical automobile passengers will complain. (Note that this is not the force coming from a retractor, but the pressure, or force, measured at the point where the webbing presses against the occupant's chest or shoulder.)

Chapter 44
Acceleration Response

LINEAR ACCELERATION

Figures 44.1 and 44.2 show possible whole-body sustained duration accelerations and their physiological effects: acceleration upward, downward, forward, and backward, and side to side. Depending on their direction, magnitude, and duration, accelerations can injure the body. Figures 44.2 through 44.6 show severe, moderate, and voluntary exposure limits for upward, downward, forward, and backward short duration accelerations. The plots show the acceleration magnitude on the vertical axis (in *g* units, where 1*g* equals 32.2 feet per second squared) and the acceleration duration on the horizontal axis (in seconds).

ROTARY ACCELERATION

Most people can tolerate rotation rates up to 6 r/min (revolutions per minute) in any axis or combination of axes, but they cannot tolerate rates in the region of 12–30 r/min and rapidly become disoriented and nauseated above 6 r/min unless they have been prepared by gradual rate increases.

However, rotation rates of 60 r/min for up to 3–4 s around the *y* axis (pitch) and around the *z* axis (spin) not only are tolerated but also elicit responses of pleasantness. The rate becomes intolerable again at around 80 r/min in the pitch mode and at about 90 r/min in the spin mode. There are, of course, exceptional people, such as professional skaters, who may spin for as long as 12 s at about 420 r/min.

Long-duration exposures to the pitch mode have been endured for as long as 60 min at 6 r/min by selected subjects.

The center of rotation within the body determines the nature of the resulting effects. It has been estimated that unconsciousness from circulatory effects probably occurs in humans after 3–10 s at 160 r/min when the center of rotation is about the heart and at 180 r/min when the center of rotation is about the iliac crest (hip). (See Figs. 44.7 and 44.8.)[31]

Figure 44.9 estimates tumbling tolerances for the specific case of tumbling combined with deceleration from 35 to 15 *g*.[32] Figure 44.10 summarizes the effects of rotation on the body.

[31] T. M. Fraser, in J. F. Parker and V. R. West (eds.), *Bioastronautics Data Book,* 2nd ed., National Aeronautics and Space Administration, 1973, pp. 191–219.

[32] *Human Factors Engineering,* AFSC DH 1-3, ser. 1-0, General Air Force System Command, Wright-Patterson Air Force Base, Ohio, 1977.

	**Effects of sustained +G, acceleration (eyeballs down)**
+1g	Equivalent to the erect or seated terrestrial posture
+2 to 2.5g	Increased weight; increased pressure on buttocks; drooping of face and body tissues; hypotension; difficult to raise oneself at 2.5g
+3 to +4g	Impossible to raise oneself; difficult to raise arms and legs; movement at right angles extremely difficult; progressive dimming of vision (grayout) after 3–4 seconds; progressive tunneling of vision
+4 to +6g	Total loss of vision (blackout) after about 5 seconds; hearing and then consciousness lost if exposure continued; mild to sereve convulsions in about 50% of the subjects after unconsciousness, frequently with dreams; occasionally paresthesias (abnormal nerve sensations, such as tingling or burning); confused states; pain not common, but tension and congestion of lower limbs with cramps and tingling; inspiration difficult; loss of orientation of time and space for up to 15 seconds post acceleration; after unconsciousness return to purposeful action takes an average of 24 seconds
>+6g	Protection is needed to preserve health
	**Effects of sustained –Gz acceleration (eyeballs up)**
–1Gz	Tolerable; sense of pressure and fullness in the head; congestion of eyes
–2 to 3 Gz	Severe facial congestion; bradycardia; dysrhythmia; throbbing headache; blurring, graying, or occasional reddening of vision after 5 seconds; congestion disappears slowly; may leave petechial hemorrhages, swollen eyelids.
>–3Gz	The 5-second tolerance limit is rarely reached; causes mental confusion and unconsciousness

FIGURE 44.1 Physiological effects of up and down sustained whole body accelerations.

	**Effects of sustained $+G_x$ acceleration (eyeballs in)**
$1G_x$	Slight increase in abdominal pressure; respiratory rate increase
+2 to $+3G_x$	Difficulty in spatial orientation; +2g tolerable for at least 24 hours
+3 to $+6G_x$	Progressive tightness in chest and abdomen; cardiac rhythm disturbances; loss of peripheral vision; difficulty in breathing and speaking; blurring of vision, effort required to maintain focus; 4g tolerable up to at least 60 minutes
+6 to $+9G_x$	Chest pain and pressure; shallow respiration from position of nearly full inspiration; decreased oxygen uptake during acceleration; pulmonary vascular pressures increase toward dorsal part of chest and fall in alveolar pressure on the ventral part; arterial oxygen saturation falls below 85% which can lead to cognitive impairment; further reductions in visual acuity and depth perception, increased blurring, occasional tunneling, great concentration required to maintain focus; occasional lacrimation (tears); body, legs, and arms cannot be lifted at +8g; head cannot be lifted at +9g; precise manual control compromised
+9 to $+12G_x$	Increase severity of symptoms; severe breathing difficulty, increased chest pain, marked fatigue, loss of peripheral vision, diminution of central acuity, lacrimation
$> +12G_x$	Extreme difficulty in breathing and speaking, severe viselike chest pain, loss of tactile sensation; total loss of vision possible
	**Effects of sustained $-G_x$ acceleration (eyeballs out)**
All levels	Similar to those of forward acceleration with modifications produced by reversal of the force vector; pulmonary vascular pressures reversed from those in $+G_x$, total body restraint system is critical and has a direct relationship to the ability to tolerate rearward acceleration exposures
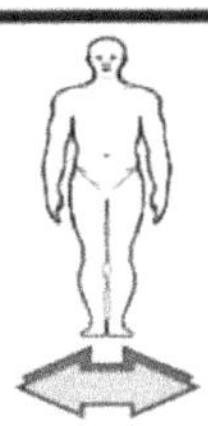	**Effects of sustained $+/-G_y$ acceleration (eyeballs left/right)**
+/–1 to $2G_y$	Difficulty maintaining head and shoulders upright without restraints; difficulty of precise manual control
$+/-3G_y$	Discomfort after 10 seconds; pressure on restraint system; feeling of supporting entire weight on clavicle; inertial movement of hips and legs; yawing and rotation of head toward shoulder; petechiae and bruising; engorgement of dependent elbow with pain; total body restraint system is critical
$+/-5G_y$	Conjuctival hemorrhage has been reported; severe headache after exposure.

Figure 44.2 Physiological effects of horizontal sustained whole body acceleration.

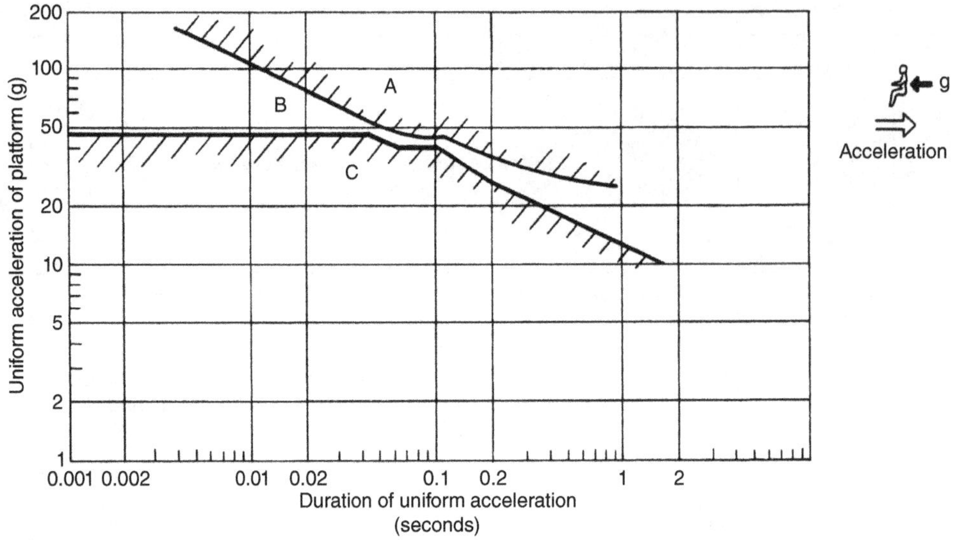

FIGURE 44.3 Maximum tolerance limits to short duration backward acceleration.

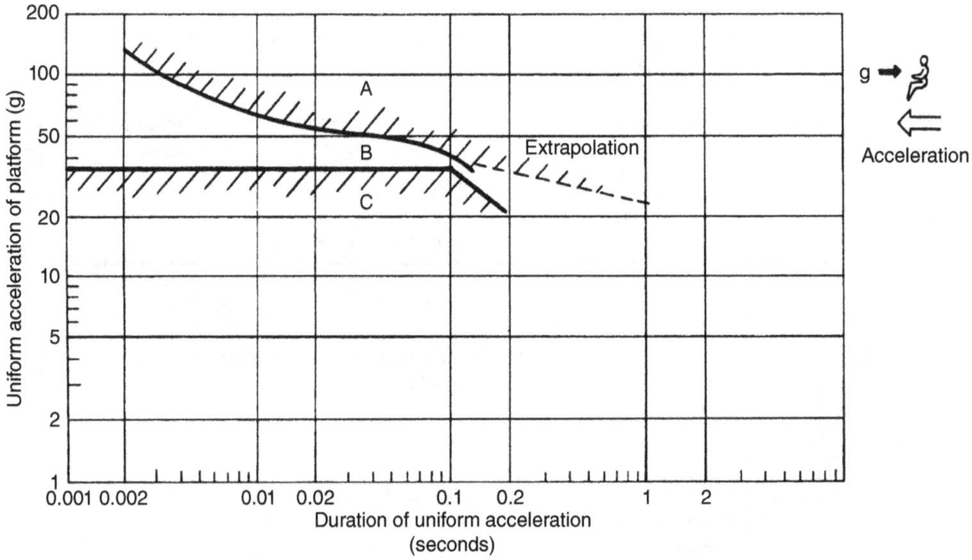

FIGURE 44.4 Maximum tolerance limits to short duration forward acceleration.

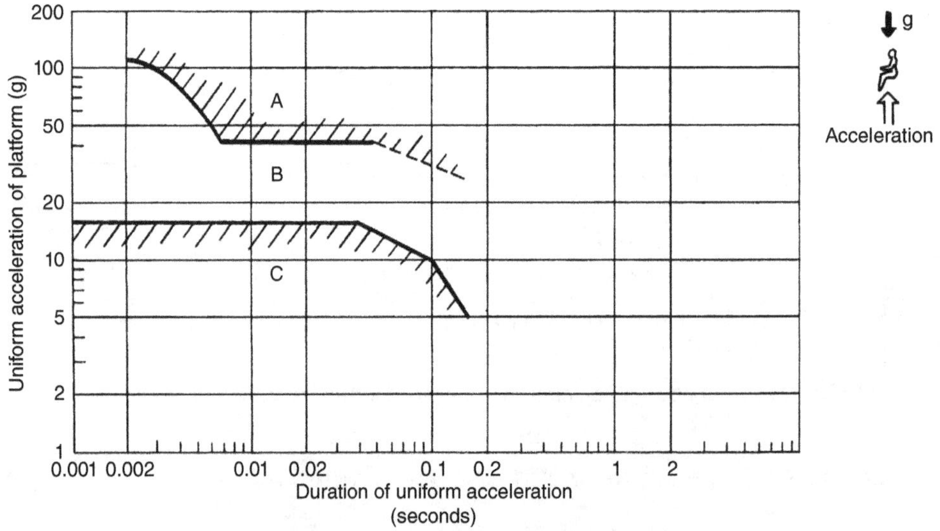

FIGURE 44.5 Maximum tolerance limits to short duration upward acceleration.

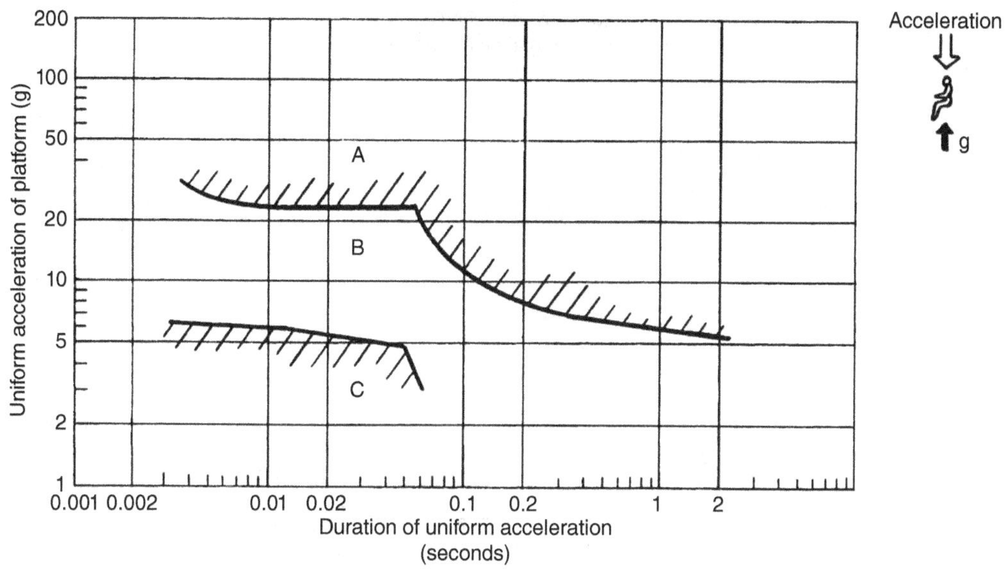

Figure 44.6 Maximum tolerance limits to short duration downward acceleration.

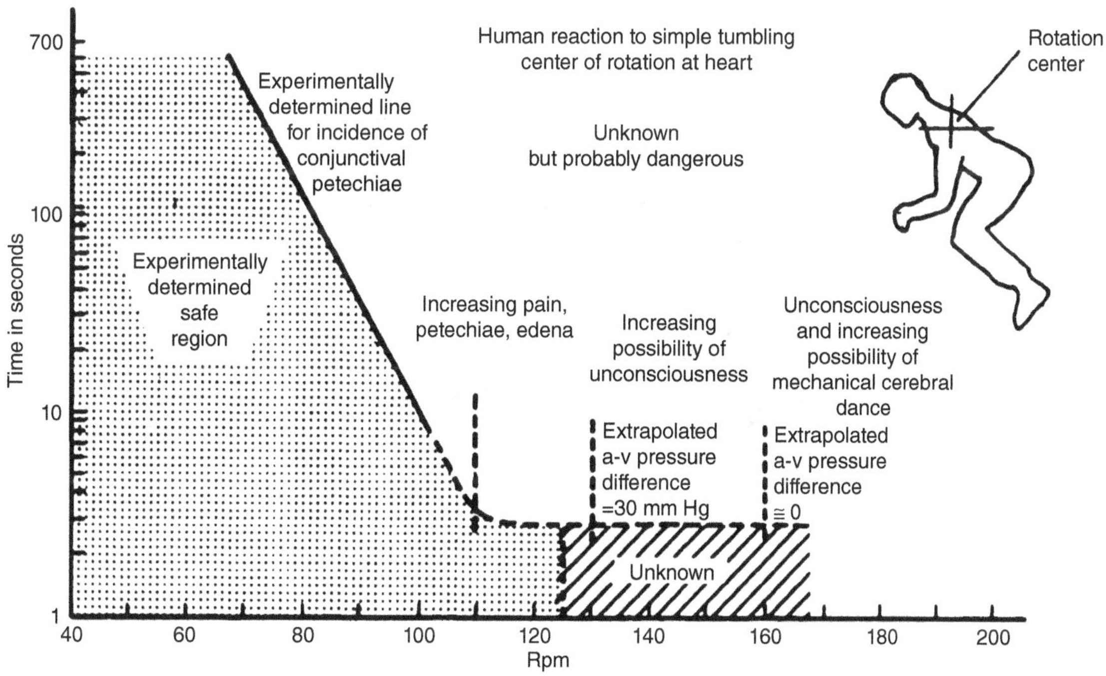

Figure 44.7 Human reaction to simple tumbling–center of rotation at heart.

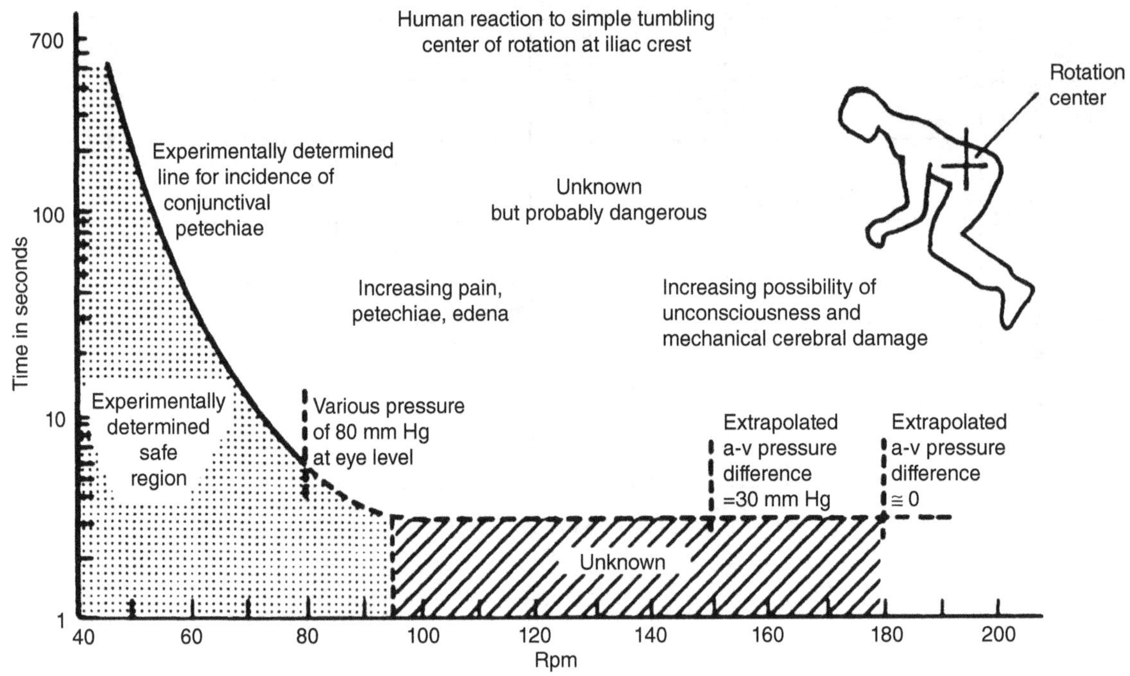

FIGURE 44.8 Human reaction to simple tumbling—center of rotation at iliac crest.

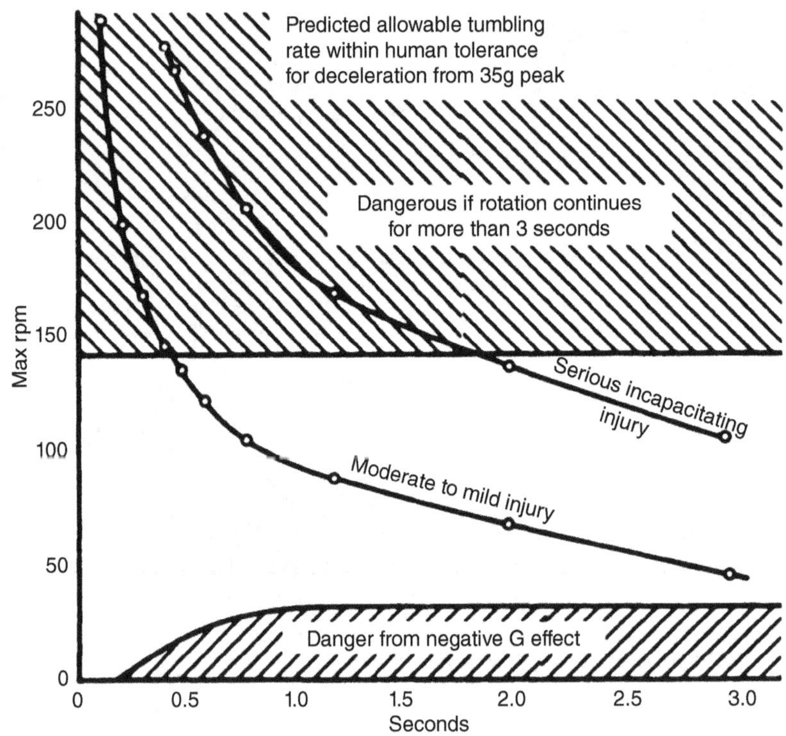

FIGURE 44.9 Tolerance to tumbling and deceleration.

ROTATION AXIS	EXPOSURE LEVELS	EFFECTS ON MOST INDIVIDUALS
Any axis	6 rpm	Most individuals without previous experience can tolerate this rotation in any axis or combination of axes
	>6 rpm	Most individuals rapidly become sick and disoriented unless carefully prepared by a graduated program of exposure
	12 to 30 rpm	Most individuals cannot initially tolerate rotation
	Random tumbles	Severe disorientation; reach and manipulative performance degradation ultimately interfere with the ability to make corrective actions
Pitch 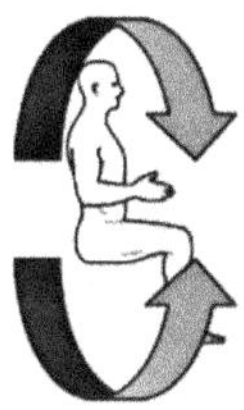	6 rpm	Some individuals have endured 60-minute runs
	80 rpm	Generally intolerable; with the center of rotation at hear level, symptoms of backward acceleration ($-G_x$) are demonstrated and are tolerable for only a few seconds
	90 rpm	Some effects of forward acceleration ($+G_x$), namely numbness and pressure in the legs, are also observed but develop slowly, with pain. No confusion or loss of consciousness is found, but in some individuals, disorientation, headache, nausea, or mental depression are noted for several minutes after a few minutes of exposure
	160 rpm	Unconsciousness from circulatory effects alone occurs after 3-10 seconds with the center of rotation at the heart
	180 rpm	Unconsciousness from circulatory effects alone occurs after 3-10 seconds with the center of rotation at the iliac crest (hip bone)
Yaw 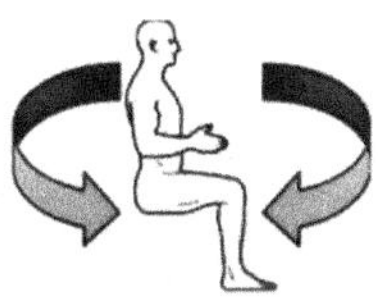	60 to 90 rpm	When the head and trunk are inclined forward out of the z-axis, rotation becomes close to limiting for 4 minutes, although some motivated individuals have endured 90 rpm in the same mode. Except for unduly susceptible individuals, tolerance tends to improve with exposure
	90 rpm to 100 rpm	Intolerable

Figure 44.10 Physiological effects of rotation on the body (NASA Human Interface Design Handbook.).

CHAPTER 45

VIBRATION RESPONSE

EFFECTS OF VIBRATION ON HUMAN PERFORMANCE

Depending on their direction, magnitude, frequency, and duration, vibrations can degrade body movement, motor performance, speech, hearing, and cognition. This section describes those effects and the vibration environment that generates them. Tables 45.1 through 45.3 describe research studies on the effects of vibration on human performance, tracking ability, and operation of aviation controls.

RESPONSE TO VEHICULAR VIBRATION

Oborne[33] has suggested that subjects' responses to vibration provide a reasonable prediction of the acceptability of various vibration levels for transportation systems. Table 45.4 categorizes the responses of subjects in Oborne's latest studies.

VIBRATION EFFECTS ON HEALTH

Low-frequency vibration (0.1–0.5 Hz) can cause motion sickness. Long-term exposure to higher frequencies (0.5–80 Hz) can damage the lumbar spine and the connected nervous system. Metabolic and other factors originating from within may have an additional effect on the degeneration of discs and vertebrae. Vibrations above 0.01 g (0.1 m/s 2) RMS (such as on a passenger ship or train) can also disturb sleep.

Different frequencies and directions impact the effect of the vibration. Measurement and weighting of vibrations are specified in ISO 2631, "Mechanical vibration and shock—Evaluation of human exposure to whole-body vibration."

Motion Sickness Limits

Vibration should be limited such that the frequency-weighted acceleration between 0.1 and 0.5 Hz in each of the X, Y, and Z axes is less than 0.05 g (0.5 m/s^2) RMS, when calculated in accordance with ISO 2631.

Health Limits

To maintain health, vibration between 0.5 and 80 Hz (when measured and calculated in accordance with ISO 2631) must be within the limits shown in Fig. 45.1 (from MIL-STD-1472G).

Limits for Sleep

Vibration acceleration between 1.0 and 80 Hz in each of the X, Y, and Z axes, weighted in accordance with ISO 6954:2000(E) ("Mechanical vibration—Guidelines for the measurement,

[33]D. J. Oborne, "Vibration and Passenger Comfort: Can Data from Subjects Be Used to Predict Passenger Comfort?" *Applied Ergonomics*, vol. 9, no. 3. pp. 155–161, 1978.

TABLE 45.1 Summary of Effects of Vibration on Biodynamics, Psychomotor Performance, Speech, Hearing, and Higher Mental Processes

Vibration Conditions	Measures	Effect	Source
		Biodynamic Mechanisms	
± 0.15–$0.35\,g_z$ at 0.9–6.5 Hz, low amplitude	Whole body vertical vibration, hand tremor, body equilibrium, foot pressure	Foot pressure constancy impaired at 3.5 to 6.5 Hz, error increase with intensity; no residual effects	Schmitz, Simons, and Boettcher, 1960
$\pm ng_x$, $\pm ng_y$ for ½ hr	Body sway equilibrium	No effects	Hornick, Boettcher, and Simons 1961
$\pm g_z$, 2–20 Hz (intensities = 1/3 short-term tolerance limits)	Control of pitch and roll of a chair	Wide individual differences; decrement between 3 and 12 Hz, worst at 6 Hz	Coermann, Magid, and Lange, 1962
$\pm g_z$, at 0, 2, 5, and 8 Hz	Orientation (orienting body position to face targets at 15, 30, and 60° from reference plane)	Only small decrement in accuracy; mean error < 0.5°	Ayoub, c. 1969
$\pm 0.03 - \pm 0.41\,g_z$ at 0, 3, 5, and 8 Hz	Leg muscular power (on bicycle ergometer)	No effects	Harrison, 1969
Various peak-to-peak accelerations at 1 Hz with 3 Hz, and 2 Hz with 6 Hz	Arm-hand steadiness	Positional errors significantly related to rms and frequency of vibration; 90% of error was periodic; 1 Hz with 3-Hz combination produced larger error; small (0.5–1 g) differences in acceleration had no effect	Clarke et al., 1965
		Psychomotor Performance	
$+0.25\,g$, at 2.4–9.5 Hz	Time to pick up markers and place in small circular areas	Completion time worst at 3.4 and 4.8 Hz	Guignard and Irving, 1960
$+0.5$ rms g_z at 2–30 Hz (13-Hz peak power)	Digital decimal input with push button, toggle switch, rotary switch, and thumbwheel controls	Accuracy unaffected; insert times increased by 4%; push buttons and toggle switches were most rapidly used, with the former preferred; thumbwheels were most accurate	Dean et al., 1967
0, 0.2, 0.4, 0.6, and 0.8 rms g_z, for 5 min	Same	No effects for 0.2 and 0.4 rms g_z; significant increase in insert time for 0.6 and 0.8 rms g_z; speed: push buttons > *rotary switches > thumbwheels; error rate: push buttons highest and thumbwheels lowest for high intensity vibrations	Dean, Farrell, and Hitt, 1967
$\pm g_x$ and $\pm g_y$ at 0.33 and 0.80 Hz at amplitude of ± 6.3 and ± 7.0 in.	Nut and bolt assembly and disassembly; placement of probe through various sized holes	No effects at 0.33 Hz; time required increased by 30% at 0.80 Hz with no increase in accuracy	Seeman and Williams, 1966
		Speech Intelligibility	
$+g_z$ at 10, 20, 30, 40, and 50 Hz	Intelligibility	Most effect at 10 and 20 Hz	Nixon, 1962
$0.5\,g_z$ sinusoidal at 6 Hz; $0.75\,g_z$ at 4 and 8 Hz; $1.0\,g_z$ at 2–20 Hz	Intelligibility and quality	No effect on intelligibility at 65 dB; "quality" poorer than control condition	Nixon and Sommer, 1963
		Audition	
5-Hz sinusoidal, 5-Hz random amplitude, 4- to 12-Hz random frequency	Frequency (pitch) change (1200 for 1600 Hz) at 86-dB tones of 0.25-s duration every second—detection	No effect	Weisz, Goddard, and Allen, 1965
$\pm g_z$	1200 Hz at 86 dB presented every 0.25 s for 1 s against a 74-dB, 30- to 3000-Hz white noise; pitch change at 86 dB (1600 for 1200 Hz)—detection	No effect	Holland, 1966

(Continued)

TABLE 45.1 Summary of Effects of Vibration on Biodynamics, Psychomotor Performance, Speech, Hearing, and Higher Mental Processes (*Continued*)

Vibration Conditions	Measures	Effect	Source
	Audition		
$+1\,g_z \pm 0.7\,g_z$ at 15 Hz (amplitude 0.036 in) for 30 min	TTS determined as function of variation and noise versus noise alone (acoustical frequencies from 250–6000 Hz)	Extremely small vibration effect at low tone frequencies only	Guignard and Coles, 1965
	Higher Mental Processes		
$\pm 0.15–0.35\,g_z$ at 2.5 and 3.5 Hz	Mental addition	No effect	Schmitz, Simons, and Boettcher, 1960
$+g_z$ at 5, 7, and 11 Hz	Pattern matching and discrimination	No effect	Buckhout, 1964
0.40 rms g_z random vibration	Navigational tasks in simulated low-altitude, high-speed flight	No effect	Schohan, Rawson, and Soliday, 1965; Soliday and Schohan, 1965
No vibration; no noise, no vibration; noise only; vibration plus noise; postvibration $+4.0\,g_z$ at 70 Hz	Continuous counting at a given rate	Decrement, especially during 5–7 min of exposure; residual effects noted; 70% of decrement attributed to vibration (30% to noise) Subjects over 36 showed greater decrement	Ioseliani, 1967

*Symbol > here indicates faster than.

[42]R. J. Hornick, in J. F. Parker and V. R. West (eds.), *Bioastronautics Data Book*, 2d ed., National Aeronautics and Space Administration, 1973, pp. 297–348.

TABLE 45.2 +1 *Gz* + *ngz* Vibration Effects on Tracking Ability, Seated Subjects

Vibration Conditions	Tracking Task	Effect	Source
2.5 Hz, ±0.18 and 0.35 *g*; 3.5 Hz, ±0.15 and 0.30 *g*, 1.5-hr duration	Subjects and wheel on shaker; static display; single axis compensatory tracking task in horizontal axis	Error is a function of frequency and intensity; error greater at 3.5 Hz; incomplete recovery after exposure	Schmitz, Simons, and Boettcher, 1960
0.9–5.5 Hz at ±0.15. 0.25, and 0.35 *g*, 10-min duration	Same	Error elicited, but no effect of frequency or intensity (possibly due to short duration); incomplete recovery after exposure	Schmitz, Simons, and Boettcher, 1960
±0.20 *g* at 5 Hz, 1-hr duration	Side-arm controller, single axis (horizontal) compensatory tracking	40% decrement in task; incomplete recovery in 15-min postvibration period	Ayoub, c. 1969
1-27 Hz at four subjective reaction levels—from "perceptible" to "alarming"	Aircraft wheel and column; two axis compensatory; horizontal tracking with wheel, vertical with column	Error related to subjective intensity level; greatest error at 10–18 Hz; greater error in horizontal axis in vibration and nonvibration conditions, but vibration has greater relative effect on vertical control	Chaney and Parks, 1964
2, 4, 6, 8, 11, and 15 Hz with single amplitudes of 0.06 and 0.13 in.	Two axis compensatory; static base display; control stick	Slight improvement for 0.06 in; over 40% error increase at 0.13 in.; greatest error at 8 Hz	Catterson, Hoover, and Ashe, 1962
5 Hz, ±0.26–0.36 *g*; 7 Hz, ±0.29–0.41 *g*; 11 Hz, +0.55–0.77 *g*; 5-min duration	Two axis compensatory; side-mounted stick controller, second-order dynamics	Greatest error produced in vertical tracking axis; 5 and 11 Hz have greatest error in vertical axis; 5 Hz produces greatest error in horizontal; vertical error about 40% greater in vibration compared to static condition; amplitude levels equivalent to 25–35% of the 1-min human tolerance level degrades tracking	Buckhout, 1964
5 Hz, ±0.10–0.26 *g*, 20-min duration	Two axis compensatory; side-mounted controller	Horizontal tracking not affected; vertical tracking significantly impaired at and above ±0.21 g, about 40% greater than in a nonvibration condition	Harris, Chiles, and Touchstone, 1964
5 Hz, ±0.10–0.26 *g*; 7 Hz, ±0.15–0.30 *g*; 11 Hz, ±0.25–0.62 *g*; 20-min duration	Two axis compensatory; rate control, side-stick	Horizontal tracking not affected; vertical axis tracking, first impaired at ±0.20 *g* for 5 Hz, ±0.25 *g* for 7 Hz, and ±0.37 *g* for 11 Hz, these levels being about 20% of the 1-min human tolerance limits; decrements in tracking were up to 40% greater than nonvibration error	Harris and Shoenberger, 1966

TABLE 45.3 Effects of Long Duration Random gz Vibration

Vibration Conditions	Tracking Task	Effect	Source
Aircraft F-111 dynamics; input 0.017-0.294 rms *g* but closed-loop control inputs resulted in 0.028–0.291 rms *g*; 1.5-hr duration	Center-stick and side-stick controllers, two axis terrain following and heading changes	Horizontal control not affected; higher rms *g* inputs result in greater pitch control error; no trend for pitch control error to increase as a function of time; pitch error and altitude error for side-stick is about 50% of that with the center-stick, with resulting rms *g* reduced by 20% with the sidestick; control error increases as function of task difficulty (displayed terrain)	Soliday and Schohan, 1964
1–5 Hz, shaped PSD	Two axis, with several controller configurations and dynamics	Best performance with side-stick controller with arm support	Torle, 1965
Aircraft, fighter dynamics, input 0.33 rms *g*, but closed-loop control inputs resulted in 0.40 rms g; 1-hr duration	Center-stick simulated terrain following; three task difficulty levels—simulated flat, rolling, and mountainous terrain	Primary error relationship with tracking task difficulty, greatest error with roughest terrain; altitude error does not increase as function of time	Soliday and Schohan, 1965
Closed-loop rms *g* levels varied between 0.05 and 0.40 rms *g* as function of simulated gust, speed, and pilot inputs; 3-hr duration	Similar to preceding	No altitude control error trend as function of time; increase in simulated "crashes" at end of runs suggesting fatigue	Schohan, Rawson, and Soliday, 1965
1–12 Hz, peak energy at 1 and 7 Hz, with 0.10, 0.15, and 0.20 rms *g*; 4-hr duration	Side-stick controller for two axis terrain following and avoidance over three simulated terrain types—flat, rolling, mountainous; open-loop platform dynamics	No trend for error to increase as function of time for two easiest terrains; error increase as a function of time for only the mountainous terrain after 2.5 hr of exposure; error not related to rms *g* intensity; error residual in postvibration period; greatest error for most difficult displayed terrain	Hornick and Lefritz, 1966
1–6 Hz, 0.12 and 0.16 rms *g*; two PSD shapes, one with power peak at 2 Hz, one at 5 Hz; 6-hr duration	Two axis, open-loop platform dynamics; primarily side-stick controller, with occasional use of center-stick	No trend for error to increase as function of time; error not a function of rms *g* intensity; absolute error greater in vertical axis; greater error in spectrum with 5-Hz power peak	Holland, 1967
CH-46A helicopter spectrum, variable rms *g* and PSD shapes; seven 40-min tests during 6-hr period	Side-stick controller, two axis	No error with this very simple tracking task	Dean, McGlothlen, and Monroe, 1964
UH-1B helicopter spectrum, and frequency sweeps, 4 Hz to over 100 Hz	Variety of controls and servo systems	Greatest tracking error near 5 Hz; position control more accurate in 5- to 25-Hz range; force control more accurate above 25 Hz	Rosenberg and Segal, 1966

TABLE 45.4 Comfort Response to Hovercraft Vibration

		RMS Acceleration, m/s²	
Label	Definition	0–4 Hz	8–16 Hz
Very comfortable	Willing to put up with this level for a long journey	Below 0.48	Below 0.15
Comfortable	Willing to put up with the present level for about 1½ hr	0.48–0.95	0.15–0.85
Just comfortable	Not be willing to put up with this level for more than about ½ hr	0.95–1.30	0.85–1.30
Uncomfortable	Would only tolerate this level again for a short journey	1.30–1.80	1.30–1.95
Very uncomfortable	Would not consider traveling in any form of transport with so high a level	Above 1.80	Above 1.80

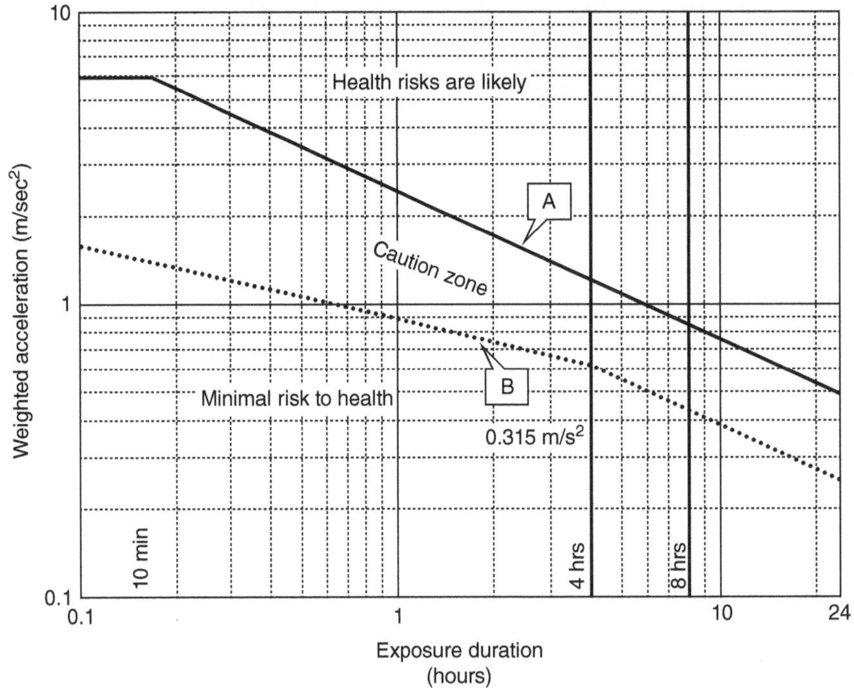

FIGURE 45.1 Vibration.

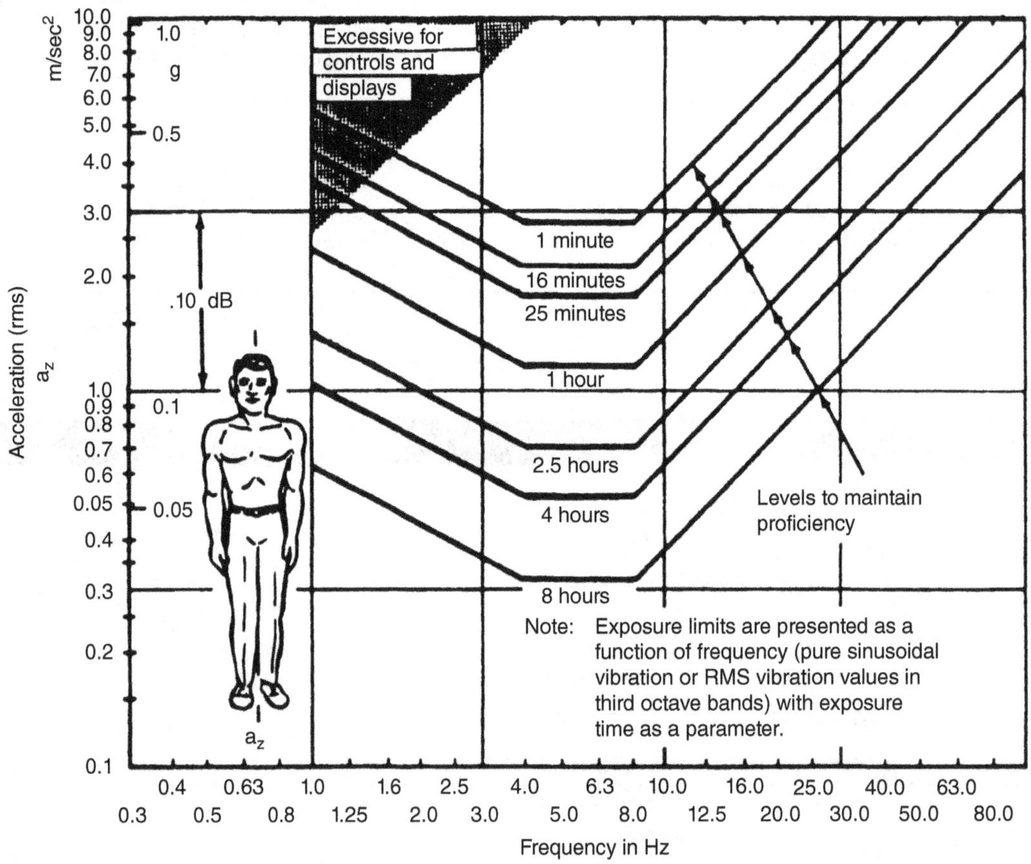

FIGURE 45.2 Limit for down vibration.

reporting and evaluation of vibration with regard to habitability on passenger and merchant ships, Annex A") should be less than 0.01 g (0.1 m/s²) RMS for each 2-minute interval during sleep periods.

VIBRATION TOLERANCE LIMITS[34]

Vibrations in land, air, and sea vehicles should not exceed the vibration limits in Figs. 45.2, 45.3, and 45.4.

Factor to consider	Vibration limit
Safety	Use the values in the charts multiplied by 2
Performance	Use the values in the charts directly
Comfort	Use the values in the charts multiplied by 0.317

Figure 45.3 Exposure limit selection guide.

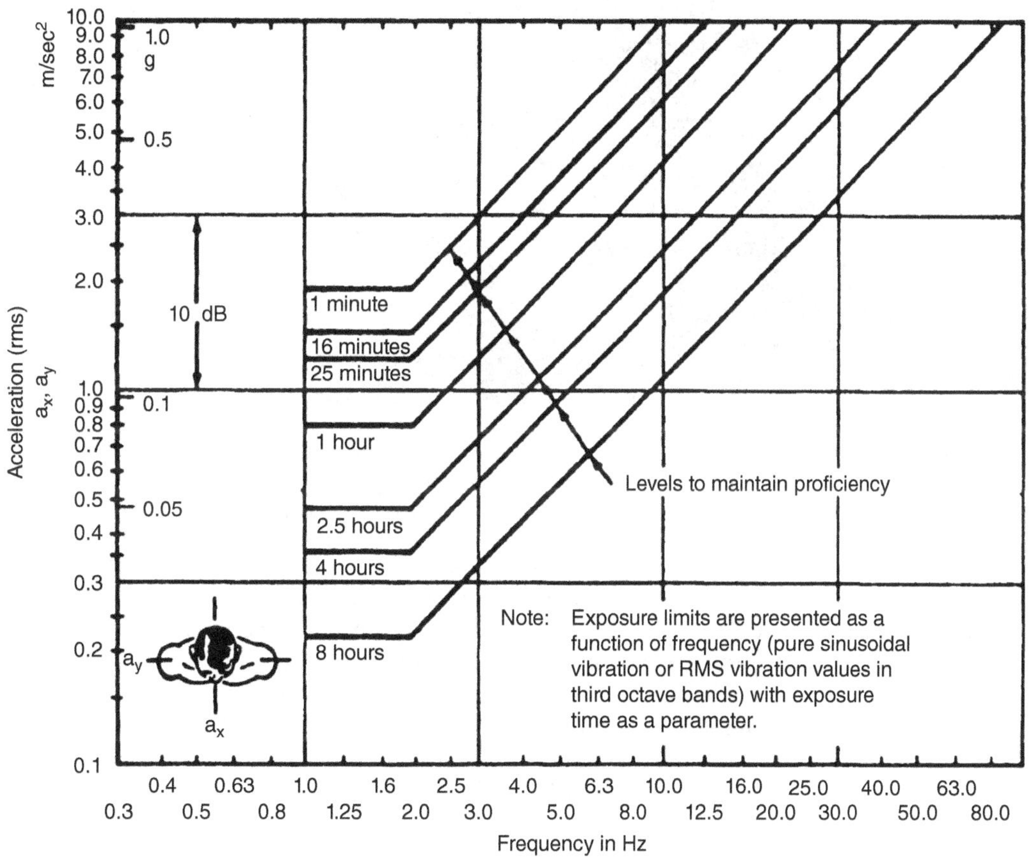

Figure 45.4 Limit for side to side and forward and back vibration.

[34]From ISO DIS 2631, *Guide to the Evaluation of Human Exposure to Whole Body Vibration*, International Standardization Organization.

CHAPTER 46

ATMOSPHERIC FACTORS

OXYGEN[35]

Figure 46.1 and Tables 46.1 and 46.2 show that a lack of oxygen results from reduced atmospheric pressure. At altitudes above 2000 m, the fall in atmospheric pressure requires compensation in the inspired air, particularly when physical activity is required. Oxygen can also manifest toxic effects at high partial pressures. The symptoms will generally be respiratory. At pressures of oxygen around 251 mm Hg, changes in red blood cell fragility and cell wall permeability have been reported after long period of exposure.

Chronic oxygen poisoning can occur at or above 460 mm of mercury. Short-term memory deteriorates after about 10 min of breathing pure oxygen. (See Fig. 46.2.)[36]

[35] *Human Factors for Designers of Naval Equipment,* Medical Research Council, Royal Naval Personnel Research Committee, 1970, chap. 1.

[36] *Human Factors for Designers of Naval Equipment,* Medical Research Council, Royal Naval Personnel Research Committee, 1970, chap. 1.

CARBON DIOXIDE

The air we breathe normally contains about 0.03 percent of carbon dioxide. The concentration of carbon dioxide should be maintained below 0.5 percent. Tables 46.4 and 46.5 show the carbon dioxide limits and its effects for both short- and long-term exposures. The necessary ventilation rate to maintain a safe level is illustrated in Table 46.3. (See also Fig. 46.3.)

CARBON MONOXIDE

The effects of carbon monoxide poisoning are outlined in Table 46.6. Carbon monoxide is particularly dangerous because it is odorless and colorless.

Carboxyhemoglobin levels of over 5 percent have been known to produce intellectual deterioration, and excessive exposure over long periods often leads to arterial and heart disease.

$$\frac{\text{CO in vol/million}}{6} = \text{carboxyhemoglobin \%}$$

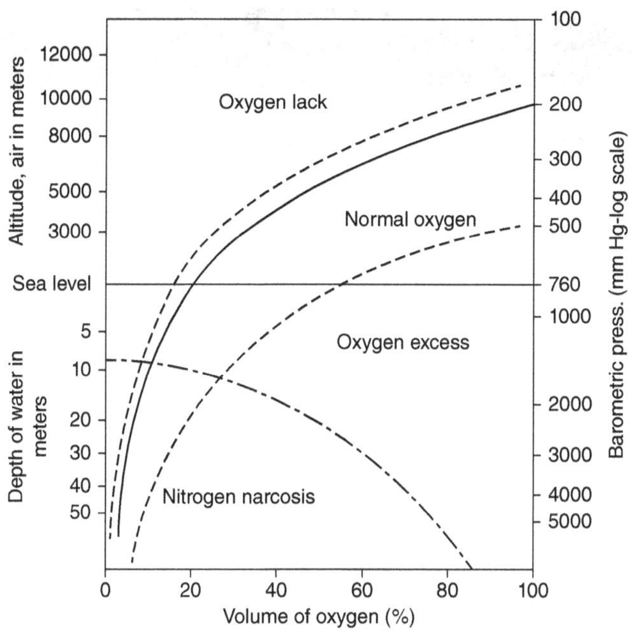

Figure 46.1 The volume of oxygen in the atmosphere necessary to maintain proper oxygen levels for habitation.

OZONE

Under certain conditions three atoms of oxygen combine to form one atom of ozone, which is poisonous in both short- and long-term exposure. An individual who is exposed for about 10 min to 1 volume per million of ozone is likely to feel irritation in the nose and throat, to cough, and to experience visual degradation; 100 volumes per million may injure body tissues, and 1000 volumes per million produces fatigue, depression, and pain in the head and body. Eventual unconsciousness and death may result.

OTHER TOXIC GASES

Table 46.7[37] shows permissible limits for other toxic gases, excess exposure to which may endanger human health, both temporarily and permanently. The American Conference of Government Industrial Hygienists has established threshold limit values on the basis of an 8-hour working-day exposure. Maximum permissible concentrations for 90-day exposures have been established by the British Royal Navy for use in nuclear submarine environments.

TABLE 46.1 The Effects of Excess Oxygen and Lack of Oxygen

Partial Pressure of Oxygen, mm of Hg	Percent of Oxygen in Dry Air at Sea Level Pressure	Effect
		Oxygen Excess
456	60	Onset of oxygen poisoning after some hours.
167	22	Limit set in RN to control fire hazard in charcoal filters in nuclear submarines.
		Normal
160	21	Normal atmospheric level
		Oxygen Lack
137	18	Accepted limit of alertness. Loss of night vision. Earliest sign—dilation of the pupils.
114	15	Performance seriously impaired. Hallucinations, excitation, apathy.
100	13	Coordination impaired. Emotional upset.
84	11	Paralysis, loss of memory. Irreversible unconsciousness.
46	6	Death before symptoms apparent.

Note: The effect of falling oxygen is insidious, because it dulls the brain and prevents realization of danger.

[37]After *Human Factors for Designers of Naval Equipment,* Medical Research Council, Royal Naval Personnel Research Committee, 1970, chap. 1.

TABLE 46.2 Major Effects of Altitude Without Supplemental Oxygen or Pressurization, Approximate Time of Consciousness, and Recommended Altitude Limits for Various Situations

Altitude, ft	Major Effects	Approximate Time of Consciousness, s	Recommended Limits
5,000	Loss of central night vision initiated	—	
8,000		—	Begin supplemental oxygen for routine flights
10,000	Loss of peripheral night vision, complex coordination, orientation, hearing, memory, and intellectual skills initiated	—	Maximum without supplemental oxygen in routine situation
15,000	Loss of tracking skills and reaction (decision) time initiated	—	
16,000	Loss of emotional control initiated	—	
17,000	Loss of simple coordination and vision initiated	—	
18,000		—	Maximum without supplemental oxygen in emergency
20,000		—	Begin cabin pressurization for routine flights
23,000	Decompression sickness initiated	—	
25,000		116	
28,000		69	Maximum without cabin pressurization
30,000		54	Begin positive-pressure breathing for routine flights
35,000		32	Maximum for demand oxygen in routine situation
40,000		23	
42,000		—	Maximum for positive-pressure breathing in routine situation
			Maximum for cabin pressurization
43,000		—	Maximum for demand oxygen in emergency
45,000		—	Maximum for positive-pressure breathing in emergency Begin use of pressure suit

Source: Adapted from U.S.A.F. Manual 80-1.

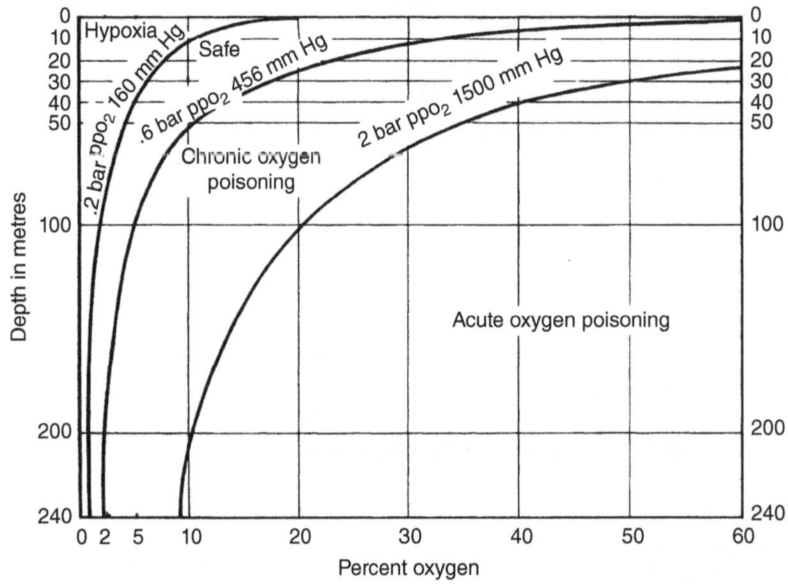

FIGURE 46.2 Oxygen requirements for undersea habitation.

TABLE 46.3 Fresh Air Ventilation Requirements to Maintain Safe CO_2 Levels

	Oxygen Consumption per Person at Sea Level	Ventilation Rate per Person			
		Sea Level	5000 ft	10,000 ft	15,000 ft
At rest	0.008 ft^3/min	1.2 ft^3/min	1.4 ft^3/min	1.7 ft^3/min	2.1 ft^3/min
Moderate activity	0.028 ft^3/min	3.9 ft^3/min	4.7 ft^3/min	6.7 ft^3/min	6.9 ft^3/min
Vigorous activity	0.056 ft^3/min	8.7 ft^3/min	9.7 ft^3/min	11.7 ft^3/min	14.5 ft^3/min

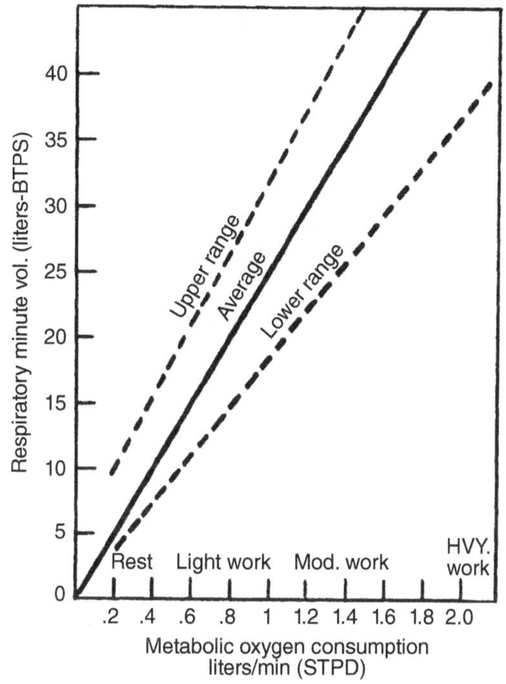

Figure 46.3 Estimating oxygen consumption from respiratory volume.

TABLE 46.4 Maximum Limits for Carbon Dioxide Exposure

Maximum Limit, %	Situation	Remarks
0.5	General limit for 8-hr industrial exposure	Threshold limit value (TLV) laid down by the American Conference of Government Industrial Hygienists (ACGIH).
1	90-day limit for continuous exposure in nuclear submarines (to be reduced to 0.5%)	Future design limit for removal machinery is 0.5%.
3	Limit for short-term exposure in conventional submarines	Brain functions deteriorate above this level. Heavy breathing.

TABLE 46.5 Effects of Continuous Exposure to Carbon Dioxide Over A Number of Weeks

Level, %	Effect
0.5–1	Probably no significant physiological, psychological, or adaptive changes occur.
1–3	Gradual changes in the composition of the blood and body fluids, which may lead to irreversible changes in the tissues on prolonged exposure.
3 and above	Deterioration in performance, alteration of basic body functions, and irreversible changes in the tissues.

TABLE 46.6 Effects of Carbon Monoxide Exposure

Atmosphere, Volumes per Million	Carbon Monoxide in Blood, % Carboxyhaemoglobin	Effects
0–60	0–10	None subjectively noticeable, but initial visual and psychomotor impairment is revealed in objective tests.
60–120	10–20	Tightness across forehead, slight headache, flushed complexion.
120–180	20–30	Headache with throbbing in temples, breathlessness from any exertion.
180–240	30–40	Severe headache, weakness, dizziness, dimness of vision, nausea, and vomiting with possibility of collapse.
240–300	40–50	All preceding symptoms with increased pulse rate and respiration and greater possibility of collapse.
300–360	50–60	Loss of consciousness, with increased or irregular respiration, rapid pulse, and possibility of coma with convulsions.
360–480	60–80	Coma, convulsions, depressed heart action, respiratory failure, and possibility of death.

TABLE 46.7 Maximum Permissible Concentrations of Toxic Gases

Substance	Formula	Threshold Limit Value for an 8-hr Working Day,* TLV	Maximum Permissible Concentration for a 90-day Exposure,† MPC90
Ammonia	NH_3	25 vpm (to be reduced)	25 vpm
Arsine	AsH_3	0.05 vpm	0.01 vpm
Carbon dioxide	CO_2	0.5%	1% (to be reduced to 0.5%)
Carbon monoxide	CO	50 vpm	25 vpm
Hydrocarbons (not methane)		Not listed as a group	40 mgm/m^3 (8 mgm/m^3 aromatic)
Hydrogen chloride	HCl	5 vpm	0.1 vpm
Hydrogen fluoride	HF	3 vpm	0.1 vpm
Mercury	Hg	Alkyl (skin effect) 0.01 mgm/m^3 Non-alkyl 0.05 mgm/m^3	0.05 mgm/m^3
Chlorine	Cl	1 vpm	0.1 vpm
Methane	CH_4	Explosive	1.2% (risk of explosion)
Methyl alcohol (including formaldehyde and formic acid)	CH_3OH	200 vpm	10 vpm
Nitrogen dioxide	NO_2	5 vpm	0.5 vpm
Ozone	O_3	0.1 vpm	0.02 vpm
Phosgene	$COCl_2$	0.1 vpm	0.05 vpm
Stibine	SbH_3	0.1 vpm	0.05 vpm
Sulphur dioxide	SO_2	5 vpm	1 vpm
Benzene	C_6H_6	25 vpm	1 vpm
Toluene	C_7H_8	100 vpm	4 vpm
Xylene	C_9H_{12}	100 vpm	4 vpm

Note: Vpm—volumes of gas or vapor per million of air. Mg/m^3—milligrams of solid material, particles, or dust per cubic meter of air.
*Standard of the American Conference of Government Industrial Hygienists.
†MODN Book of Reference 1326, 1973.

CHAPTER 47
RADIATION

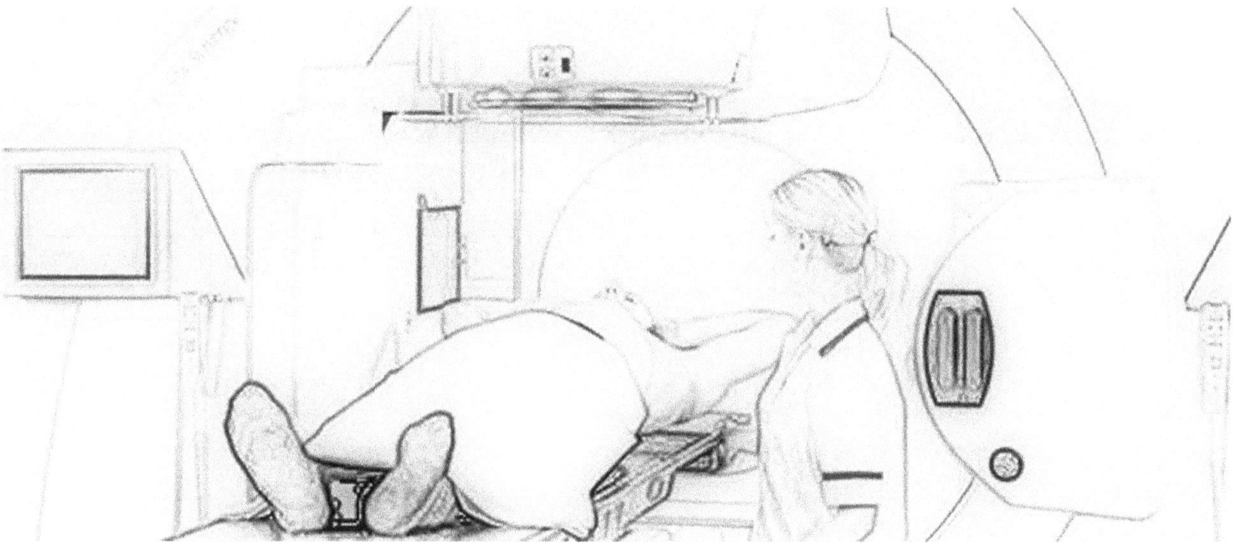

IONIZING RADIATION[38]

Figure 47.1 shows wavelengths of electromagnetic radiation and how they are used, indicating the region in which damage to living tissue can occur.

Nuclear radiation may damage the tissues of the body. An acute dose of 500 rad of penetrating ionizing radiation is likely to kill about half of an exposed population within 30 days. However, if the dose is distributed slowly over 10 years, there is only an increased risk that an exposed person will develop cancer. An individual who receives an acute dose of 200 rad of penetrating ionizing radiation equally distributed over his or her body is likely to feel sick for a day or two before recovering. With doses of over 5000 rad, the brain is directly affected; the person may collapse within minutes and die within hours.

Ultraviolet light, after repeated exposure, may produce localized cancer of the skin. Laser beams, microwaves, and very intense radar waves may produce temporary tissue damage.

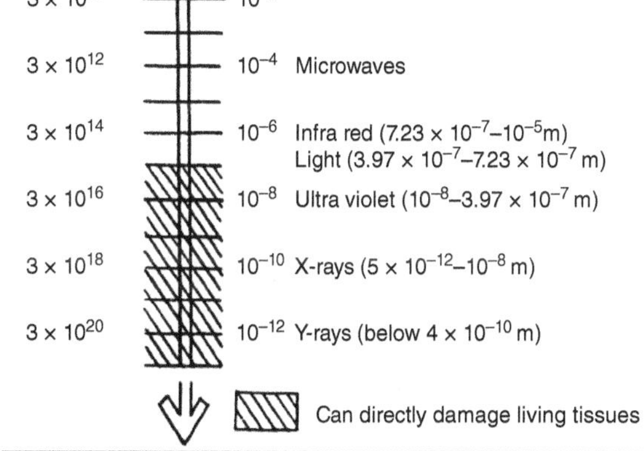

FIGURE 47.1 Wavelengths of electromagnetic radiation.

[38]*Human Factors for Designers of Naval Equipment,* Medical Research Council, Royal Naval Personnel Research Committee, 1970, chap. 1.

Exposure Limits (See Table 47.1)

TABLE 47.1 U.S. Ionizing Radiation Exposure Limits

	REMS PER CALENDAR QUARTER
Whole body: Head and trunk; active blood-forming organs; lens of eyes; or gonads	1¼
Hands and forearms; feet and ankles	18¾
Skin of whole body	7½

From OSHA, 29 CFR 1910.1096, 2005.

Effects on Body

Table 47.2 shows the effects of acute, whole-body, external radiation exposure, and the probable therapy required.[49] (A rem is the dosage of an ionizing radiation that will have the same biological effect as 1 part R of x-ray or gamma-ray dosage.)

NON-IONIZING (RADIO FREQUENCY) RADIATION

Exposure Limits

The following limits are from NASA-STD-3001, Volume 2, Human Factors, Habitability, and Environmental Health. (See Figs. 47.2 and 47.3.)

TABLE 47.2 Effects of Ionizing Radiation on the Body

0.25 REMS	25–100 REMS	100–200 REMS	200–300 REMS	300–600 REMS	600 REMS OR MORE
IMMEDIATE EFFECTS					
No detectable clinical effects	Slight transient reductions in lymphocytes and neutrophils. Disabling sickness not common; exposed individuals should be able to proceed with usual duties.	Nausea and fatigue with possible vomiting above 125 rems.* Reduction in lymphocytes and neutrophils with delayed recovery.	Nausea and vomiting on first day. Latent period up to 2 weeks or perhaps longer.	Nausea, vomiting, and diarrhea in first few hours. Latent period with no definite symptoms, perhaps as long as 1 week.	Nausea, vomiting, and diarrhea in first few hours. Short latent period with no definite symptoms in some cases during first week.
DELAYED EFFECTS					
Delayed effects may occur	Delayed effects possible, but serious effects on average individual very improbable.	Delayed effects may shorten life expectancy in the order of 1 %.	Following latent period, the following symptoms appear but are not severe: loss of appetite, and general malaise, sore throat, pallor, petechiae, diarrhea, moderate emaciation. Recovery likely in about 3 months unless complicated by poor previous health or superimposed injuries or infections	Epilation, loss of appetite, general malaise, and fever during second week, followed by hemorrhage, purpura, petechiae, inflammation of mouth and throat, diarrhea, and emaciation in the third week. Some deaths in 2 to 6 weeks. Possible eventual death to 50% of the exposed individuals for about 500 rems.	Diarrhea, hemorrhage, purpura, inflammation of mouth and throat, fever toward end of first week. Rapid emaciation and death as early as the second week, with eventual death of up to 100% of exposed individuals.
THERAPY					
	Reassurance is probably the only therapy needed.	Patients should be kept under hematological surveillance.	Antibiotics should be administered as indicated.	Antibiotics and blood transfusions should be administered as indicated.	Bone marrow transplantations may be tried. Electrolyte balance should be maintained.

*Occurs on the first day or so following irradiation, followed by a "latent period" up to 2 weeks or more, during which the patient has no disabling illness and can proceed with his regular occupation. The usual symptoms, such as loss of appetite and malaise, may reappear, but if they do, they are mild.

***Design Handbook. Human Factors Engineering*, AFSC DH 1-3. Ser. 1-0. Department of the Air Force, Headquarters Aeronautical Systems Division, Wright-Patterson Air Force Base. Ohio

Frequency range (MHz)	RMS electric field strength (E)[a] (V/m)	RMS magnetic field strength (H)[a] (A/m)	RMS power density (S) E-field, H-field (W/m²)	Averaging time[b] $\|E\|^2$, $\|H\|^2$, or S (min)	
0.1–1.34	614	$16.3/f_M$	$(1{,}000,\ 100{,}000/f_M^2)^C$	6	6
1.34–3	$823.8/f_M$	$16.3/f_M$	$(1{,}800/f_M^2,\ 100{,}000/f_M^2)$	$f_M^2/0.3$	6
3–30	$823.8/f_M$	$16.3/f_M$	$(1{,}800/f_M^2,\ 100{,}000/f_M^2)$	30	6
30–100	27.5	$158.3/f_M^{1.668}$	$(2,\ 9{,}400{,}000/f_M^{3.336})$	30	$0.0636 f_M^{1.337}$
100–300	27.5	0.0729	2	30	30
300–5000	–	–	$f/150$	30	
5000–15,000	–	–	$f/150$	$150/f_G$	
15,000–30,000	–	–	100	$150/f_G$	
30,000–100,000	–	–	100	$25.24/f_G^{0.476}$	
100,000–300,000	–	–	100	$5048/(9f_G-700)f_G^{0.476}$	

Figure 47.2 Maximum permissible exposure (MPE) to RF electronic fields. (Modified from IEEE C95.1, lower tier.)

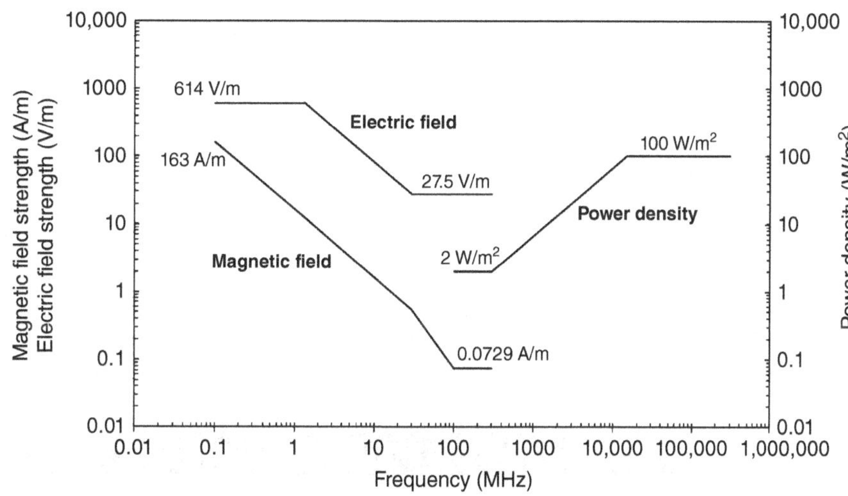

Figure 47.3 RF electromagnetic field exposure limits. (Modified from IEEE C95.1.)

Effects on Body (See Table 47.3)

TABLE 47.3 Effects of Non-Ionizing Radiation on the Body (From NASA-STD-3000, Rev A)

Response	Human Response and Description
Perception of heat	13 to 59 mW/cm^2 for 4 sec at 3000 and 10,000 MHz (observations in humans).
Pain threshold	1800 mW/cm^2 for 60 sec at 3000 MHz (observations in humans).
Cataracts	Lens clouding when temperature of lens increases by 4°K (4°C). Accumulation of subclinical damage at low power densities for short durations may yield cataracts. Evidence still equivocal.
Reproductive detriment	
Testes	Intrascrotal temperatures rise of > 1°K (> 1°C) by rf radiation or any other means reduces viable sperm count; this effect is usually reversible. Exposure to 2880 MHz at 5 mW/cm^2 for an indefinite period is the "threshold" for evidence of testicular damage in the most sensitive dog out of 35 dogs tested. Exposure to 3000 MHz at 8 mW/cm^2 did not effect mating of mice or rats.
Ovaries	No evidence that exposure to 10 mW/cm^2 or even somewhat greater Interfere with reproduction in female mice.
Visceral effects	
Gastric ulcers	> 100 mW/cm^2 for ≥ 10 min.
Delay of gastric secretion and emptying	0.05 to 1 mW/cm^2 for 30 min. Reversible
Hematopoietic effects—Leukocytosis, lymphocytopenia, eosinopenia, red blood cell lifespan alteration, impaired bone marrow function, hemoglobin decreases, platelet decrease, reticulocytosis , etc.	Generally, long exposures to > 10 mW/cm^2 are required to yield an effect. Effects are generally reversible.
Cardiovascular effects—Blood flow changes, blood pressure decrease, heart rate increase, etc.	Effects generally attributable to peripheral vasodilation and hemodilution in response to heat stress.
Central nervous system—Agitation, drowsiness, muscular weakness, electroencephalogram changes, avoidance behavior, altered conditioned response, decreased endurance, headache, etc.	Large number of studies; some with occupationally exposed humans; some conflicting results. Eastern Europeans claim effects < 10 mW/cm^2; Investigators in Western countries have not always observed these effects even at higher exposure levels. This is the area of greatest controversy.

LASERS (SEE TABLE 47.4)

TABLE 47.4 Laser Exposure Limits (From NASA-STD-3000, Rev A)

Laser Type	Primary Wavelength (nm)	Exposure Limit: Eye	Exposure Limit: Skin
Helium-Cadmium Argon	441.6 488/514.5	a) 2.5 mW • cm^{-2} for 0.25 s b) 10 mJ • cm^{-2} for 10 to 10^4 s c) 1 μW • cm^{-2} for t > 10^4 s	0.2 W • cm^{-2} for t < 10s
Helium-Neon	632.8	a) 2.5 mW • cm^{-2} for 0.25 s b) 10 mJ • cm^{-2} for 10 s c) 170 mJ • cm^{-2} for t > 453 s d) 17 μW • cm^{-2} for t > 10^4 s	0.2 W • cm^{-2} for t > 10 s
Krypton	647	a) 2.5 mW • cm^{-2} for 0.25 s b) 10 mJ • cm^{-2} for 10 s c) 280 mJ • cm^{-2} for t > 871 s d) 28 μW • cm^{-2} for t > 10^4 s	0.2 W • cm^{-2} for t > 10 s
Neodymium: YAG Gallium-Arsenide at room temp	1.064 905	1.6 mW • cm^{-2} for t > 1000 s 0.8 mW • cm^{-2} for t > 1000 s	1.0 W • cm^{-2} 0.5 W • cm^{-2} for t > 10 s
Helium-Cadmium Nitrogen	325 337.1	1 J • cm^{-2} for 10 to 3 × 10^4 s	a) 1 J • cm^{-2} for 10 to 1000 s b) 1 mW • cm^{-2} for t > 1000 s
Carbon-dioxide (and other lasers 1.4 μm to 1000 μm)	10,600	0.1 W • cm^{-2} for t > 10 s	0.1 W • cm^{-2} for t > 10 s

CHAPTER 48

EFFECT OF ELECTRICAL CURRENT ON THE HUMAN BODY

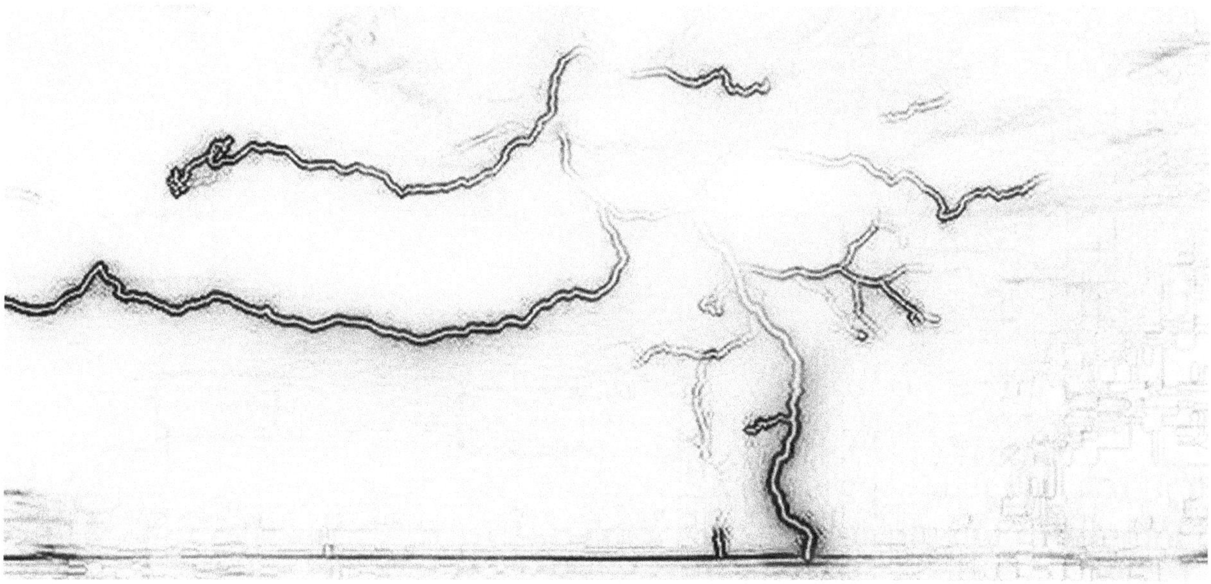

Body resistance to electrical current varies over a wide range, making it difficult to define dangerous potentials exactly. Hand-to-hand resistance varies from 1000 to 4000 Ω even with good contact. Hand-to-foot resistance is slightly less. With dry skin, resistances as high as 250,000 Ω are possible. Under ordinary working conditions when the hands are sweaty, a resistance of 5000 Ω is commonly expected. Safety recommendations suggest 1000 Ω as the highest resistance value. Table 48.1 identifies some of the potential effects that can be expected for various current levels.

Table 48.2 shows a rough guide for electrical exposure limits in terms of voltage. However, precise limit values depend on details of the exposure environment.

Precise voltage exposure limits are defined in IEC/TR 60479-5 Technical Report, Effects of current on human beings and livestock, 2007. This document specifies current limits depending on the exposure environment. In applying this document, the user must go through a series of exposure descriptions in order to find the applicable limit values. Figure 48.1 outlines the exposure environment categories.

TABLE 48.1 Potential Effects for Various Current Levels

Current, mA	Effects
Less than 1	Usually not felt, i.e., no sensation.
1–2	There is a sensation of shock, but not pain. The individual can let go because muscular control is not lost.
3–15	Painful shock occurs, but the individual can still let go.
15–20	Painful shock occurs, and the individual may not be able to let go because control of the immediately adjacent muscles is affected.
20–50	Very painful shock occurs, plus severe muscular contractions. Breathing typically becomes difficult.
50–100	Ventricular fibrillation (a heart condition that results in instant death) is very likely.
100–200	Same as above, except that the results are certain.
200 and over	Severe burns occur as well as muscular contractions so severe that the chest muscles stop the heart during the duration of the shock.

TABLE 48.2 Voltage Limits—Typical

	DC CURRENT	AC CURRENT
Seated	45	20
Standing	67	40
Free Floating in Space (current flows hand to hand)	89	40

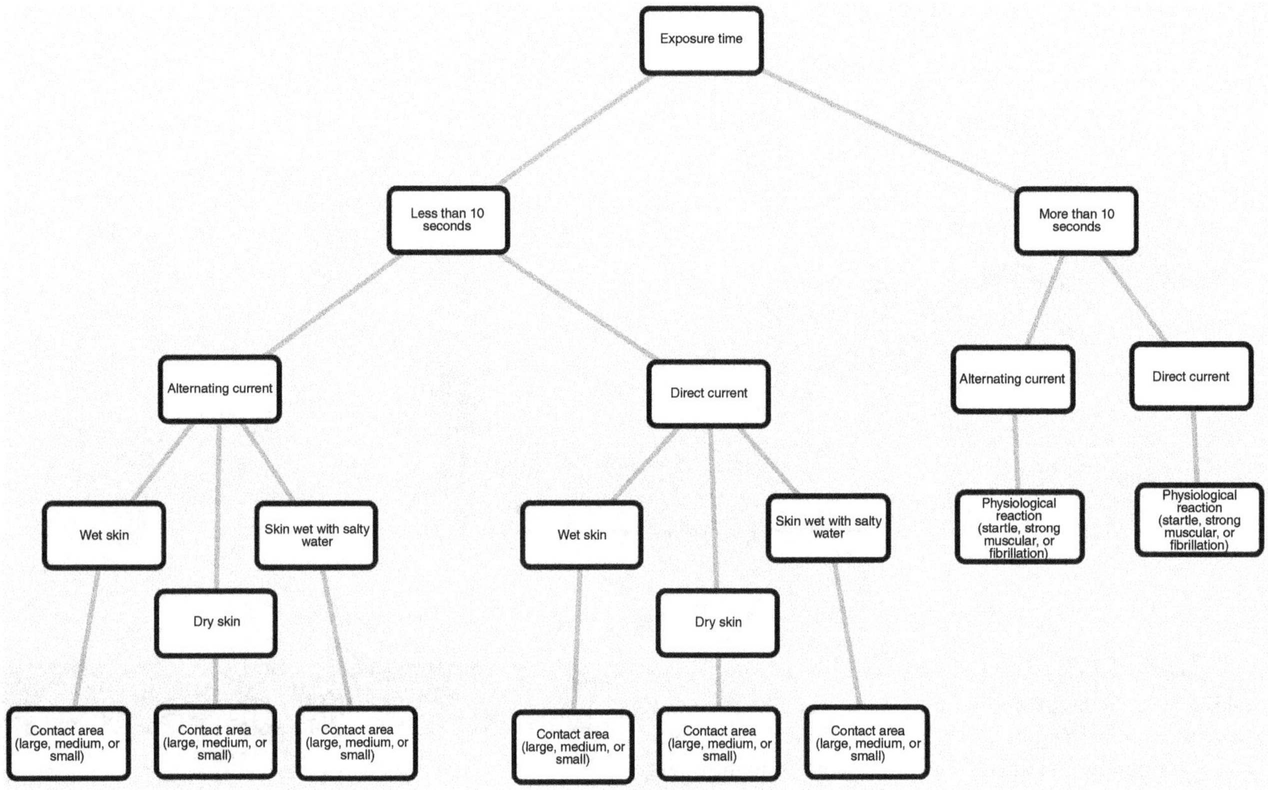

FIGURE 48.1 Electrical.

Chapter 49

Skin Reaction to Chemical Substances

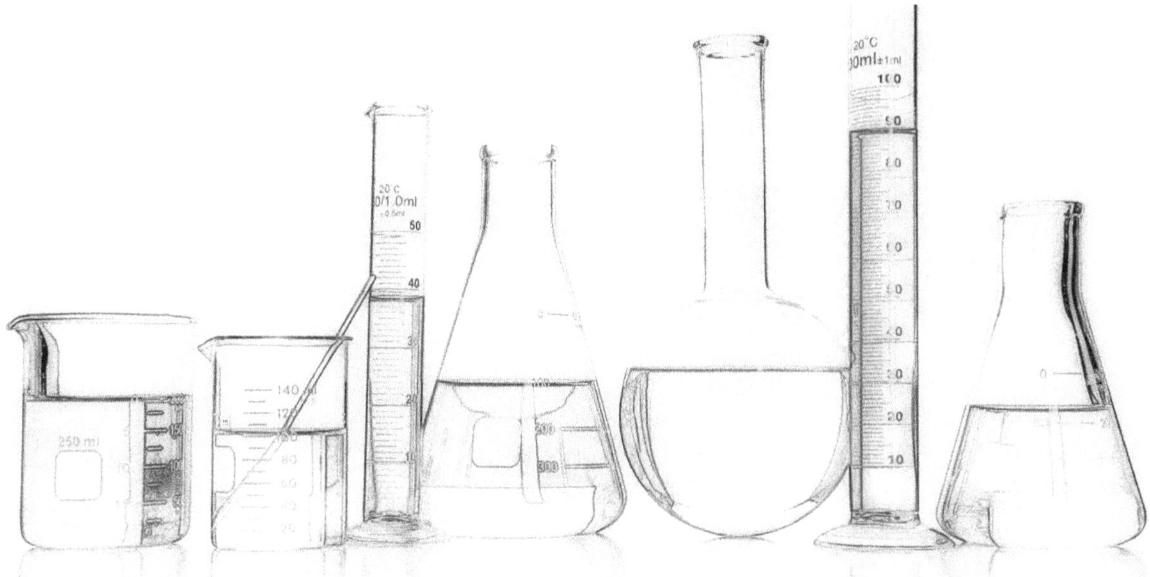

Table 49.1 presents a partial list of chemical substances and their reaction on the skin.

TABLE 49.1 Skin Reactions to Chemicals

Agent	Reaction
Acids:	
Acetic	Dermatitis and ulcers
Carbolic	Irritation and erosion, eczema, and anesthesia
Chromic	Ulcers (chrome holes on the skin), inflammation, and perforation of the nasal septum
Hydrochloric	Irritation and ulceration
Hydrofluoric	Severe burning, erosion ulcers, and blisters
Lactic	Ulcers (if strong solution)
Nitric	Severe burns and ulcers
Oxalic	Local caustic action on the skin
Sulfuric	Corrosive action on the skin and severe inflammation of the mucous membranes

(Continued)

TABLE 49.1 Skin Reactions to Chemicals (*Continued*)

Agent	Reaction
Alkalis:	
Calcium cyanamide	Irritation and ulceration
Calcium oxide	Dermatitis, burns, and ulcers
Potassium hydroxide	Severe burning, persistent ulcers, and loss of fingernails
Sodium hydroxide	
Sodium silicate	Thickening of the skin and ulcers on the fingers
Sodium or potassium cyanide	Blisters and ulcers
Salts:	
Antimony and its compounds	Irritation and eczematous eruptions
Arsenic	Skin darkening, perforation of the nasal septum, eczema around the mouth and nose, and possible loss of nails or hair
Barium	Eczema and cyanosis of skin
Bromine	Brownish stains and skin eruptions
Chromium (hexavalent compounds)	Chrome holes on the skin, perforation of the nasal septum, and eczematous eruptions
Mercury compounds	Corrosion and irritation and mercurial eczema
Sodium	Burns and ulcers
Zinc chloride	Ulcers of the skin and nasal septum
Solvents:	
Acetone	Dry (defatted) skin
Benzene	Dry (defatted) skin
Carbon disulfide	Dry (defatted) skin
Chlorinated phenols	Severe eruptions
Petroleum distillates	Acne and epithelioma
Trichlorethylene	Dry, cracked skin
Turpentine	Red, blistered skin and eczema
Dyes:	
Chlorinated compounds	Blisterlike eruptions
Dinitrochlorobenzene	Blisterlike eruptions
Nitro and nitroso compounds	Red skin and eczematous eruptions
Phenyl hydrazine	Blisterlike skin eruptions
Insecticides:	
Chlorophenols	Red skin, and blisters
Creosote	Pustular eczema, warts, and epithelioma
Fluorides	Severe burns and dermatitis
Pyrethrum	Red skin, blisters, and pimples
Rotenone	Red skin and blisters
Resins:	
Coal tar, pitch, and asphalt	Acute dermatitis, acne, inflammation, epitheliomatous cancer, eczema, and ulcers
Synthetics, e.g., phenol-formaldehyde	Extremely red and itchy skin
Synthetic waxes, e.g., chloronaphthalenes and chlorodiphenyls	Dermatitis and acne

Source: Julian B. Olishifski and Frank E. McElroy (eds.). Fundamentals of Industrial Hygiene, National Safety Council. Chicago. 1971.

CHAPTER 50

INFORMATION PROCESSING

HUMAN INFORMATION-PROCESSING CAPABILITIES

Humans can transmit only about 5 to 10 bits of information per second (b/s). They can transmit about 2 b/s when the stimuli they receive are fairly well structured, although this can often be doubled by adding appropriate coding or anchoring of the input. The relationship between bits per item and the bits-per-second limitation depends on what is referred to as "attention switch time" (e.g., 0.1–0.2 s). Thus an operator can accept no more than two or three items of data per second.

As the speed of information-processing demand increases, the number of errors typically increases. Thus the overall information transfer rate tends to remain constant at about 10^{-2} b/s. Through careful input display design, however, this rate can be increased to 10^{-3} or even 10^{-6} under good conditions. The following error classification scheme, which was suggested by Kidd,[39] helps characterize types of information-processing errors and why they occur:

1. Failure to detect a signal: Input overload or underload and/or actual interference
2. Misidentification: Insufficient cues
3. Improper weighting of informational factors and/or selection of input factors: Poor or inadequate conceptualizations or evaluation of action choices
4. Action failure: A wrong action at the right time or a right action at the wrong time

[39] J. S. Kidd, "Some Sources of Load and Constraints on Operator Performance in a Simulated Radar Air Traffic Control Task," USAF WADD TR-60-612, 1961.

Information Storage Capacity

Information storage is of two distinct types: long-term and short-term. Short-term memory storage capacity is generally limited to about three to eight individual items. The human generally organizes stored information in terms of sensory modality (visual, auditory, etc.). The most significant storage problem occurs because of the potential inference between old ("held") information and new items that present themselves during the holding period. This accounts for the frequent "reversal errors" in information processing. As a general principle, human memory is more effectively utilized as a means of orienting and sequencing information than as a depository for isolated data or symbolic items.

Input Capacity

The total sensory input capacity of the human system is about 10^9 b/s (as compared with output capacity of 10 b/s). The five basic input categories for the visual channel are:

1. Relative position
2. Shapes
3. Brightness
4. Color
5. Movement

For the auditory channel they are:

1. Pitch
2. Loudness
3. Rhythm
4. Timbre (the quality that allows one to distinguish different voices, instruments, or special auditory displays)

Decision-Making

Three basic kinds of information must be available for an operator to make good decisions:

- Information regarding policies and objectives
- Information regarding possible alternatives and consequences
- Information about the current state of the system

It is important to recognize that, in making decisions, people pursue two different courses of action:

1. Evaluating likely outcomes
2. Keeping options open as long as possible

Stress Susceptibility

Human Information processing is subject to a variety of stresses that may affect the efficiency with which information is received, processed, and acted upon. Two basic factors are important to consider:

1. The state of arousal of the human system (alertness)
2. Potential skill deterioration due to disorganization, cumulative disruption, and/or fatigue

The effects of, and reactions to, skill breakdown have been categorized as shown in Table 50.1.

TABLE 50.1 Skill Breakdown

Effect	Reaction
Failure of selective attention fails	Narrowing of attention
Perceptual disorganization	Reduction in the size of the data sample and actions upon it and reduction in filtering efficiency
Cumulative disruption	Temporary halts (complete stop pages) with start-overs

INFORMATION DISPLAY IN RELATION TO DECISION-MAKING[40]

Although research has shown that, on the basis of information acquisition only, added display complexity generally degrades operator performance, decision-making performance is not always similarly affected.

The relationship between "fact density" and decision adequacy may be curvilinear; i.e., where there is a decreasing slope under low-density conditions and/or an increasing slope under high-density conditions, actual decision-making performance may improve. For example, although complexity of information acquisition degrades overall performance when fact density is low, decision-making performance may actually increase with increases in fact density.

When information is compressed, an inverted "U-shaped" relationship occurs, with moderate levels of compression producing the best decision-making performance; i.e., the critical variable is not merely symbol count, etc., but also the subjective weighting of the particular facts being compressed.

At levels of low fact density, high compression, and high display clutter, simple coding (e.g., color) may be superior to more elaborate encoding combinations. However, at other levels of fact compression and clutter, a double coding is usually more effective (e.g., color plus size).

Subject motivation (incentive) works only at low informational levels. At high levels of motivation, degraded performance is often mediated by scattered attention.

Increases in perceptual clutter sometimes increase performance rather than degrading it, as one might expect, because we are used to working in a particularly cluttered type of information-transfer environment. The more random an

[40]After W. T. Singleton, "The Ergonomics of Information Presentation," *Applied Ergonomics*, vol. 2, no. 4, pp. 213–220, 1971.

irrelevancy, the more performance is facilitated because of our need to have irrelevancies for a "figure-ground" decision-making background.

It is important, however, to recognize the importance of input validity to decision-making, since humans quickly recognize and react to the futility of trying to decide on information that destroys or degrades the quality of their decisions.

INFORMATION CAPACITY

"Although we cannot neatly separate different sequential phases of information processing (i.e., sensation, perception, decision, response, etc.), it should be noted that the bottleneck probably is the brain rather than the sensory mechanisms."[41] (See Table 50.2.) It has been estimated that information reduction from reception through intermediate processes to permanent storage is as shown in Table 50.3.

[41]E. J. McCormick, *Human Factors Engineering*, 3rd ed., McGraw-Hill Book Company, New York, 1970, p. 92.

TABLE 50.2 Information Capacity of Human Senses

Stimulus Dimension	Number of Levels Which Can Be Discriminated	Bits of Information Transmitted*
Audition (single dimension)		
Pure tones	5	2.3
Loudness	5	2.3
Audition (combination of six variables including frequency and intensity, on-time fraction, duration, spatial location, and rate/interruption)	150	7.2
Vision (single dimension), e.g., pointer and linear scale	9	3.1
Visual size	7	2.8
Hue	9	3.1
Brightness	5	2.3
Vision (combinations):		
Size, brightness, and hue	17	4.1
Hue and saturation	15	3.9
Position of dot in square	24	4.6
Odor (single dimension)	4	2.0
Odor (combination, e.g., kind, intensity, and number)	16	4.0
Taste:		
Saltiness	4	1.9
Sweetness	3	1.7

*Lack of correspondence of bit values is due to rounding to the nearest whole number.

TABLE 50.3 Information Flow Rate in Humans from Reception to Storage

Process	Maximum Flow of Information, b/s
Sensory reception	1,000,000,000
Nerve connections	3,000,000
Consciousness	16
Permanent storage	0.7

VISUAL AND AUDITORY CODING METHODS[42] (SEE TABLE 50.4)

TABLE 50.4 Summary of Certain Visual and Auditory Coding Methods

Alphanumeric	Single numerals, 10; single letters, 26; combinations, unlimited. Good; especially useful for identification; uses little space if there is good contrast. Certain items easily confused with each other.
Color	Hues, 9; hue, saturation, and brightness combinations, 15–24. Particularly good for searching and counting tasks; poorer for identification tasks; trained observers can use many codes (up to 24). Affected by some lights; problem with color-defective individuals.
Geometric shapes	15 or more. Generally useful coding system, particularly in symbolic representation; good for CRTs. Shapes used together need to be discriminable; some sets of shapes more difficult to discriminate than others.
Visual angle	24. Generally satisfactory for special purposes such as indicating direction, angle, or position on round instruments like clocks, CRTs, etc.
Size of forms (such as squares)	5. Takes considerable space. Use only when specifically appropriate; preferably use less than 5.
Visual number	6. Use only when specifically appropriate, such as to represent numbers of items. Takes considerable space; may be confused with other symbols.
Brightness of lights	3–4. Use only when specifically appropriate. Preferably limit to two levels; weaker signals may be masked.
Flash rate of lights	4. Limited applicability. Preferably limit to two levels; combination of individual flashes and controlled time intervals may have special application, such as lighthouse signals and naval communications.
Sound frequency	5. For untrained listeners, use less than 5 levels; space widely apart, but avoid multiples and low and high frequencies; intensity should be 30 dB above threshold. Frequency changes easier to detect than single frequencies; combinations usually require training except for clearly distinguishable sounds such as bells, buzzers, and sirens.
Sound intensity	4. Preferably use less than 4. Intensity changes easier to detect than single intensities; for pure tones restrict to 1000–4000 Hz, but preferably use wide band.
Sound duration	Use clear-cut differences, preferably 2 or 3.
Sound direction	Difference in intensity to two ears should be distinct; particularly useful for directional information (i.e., right versus left).

(Numbers refer to number of levels which can be discriminated on an absolute basis under optimum conditions.)

KEEPING TRACK OF SEQUENTIAL EVENTS[43]

At slow event-presentation rates (one per second or less), individual events falling into two categories can be discriminated one at a time and mentally tallied with considerable accuracy.

As either the number of categories or the trial length increases, the number of errors per trial increases, generally very rapidly. For a given trial length, one should present the operator with fewer categories with multiple occurrences rather than large numbers of categories with fewer occurrences.

At fast presentation rates, the on/off ratio has little effect on performance as long as the on time is sufficient for accurate stimulus perception. Intermediate-rate-short-stimulus on time tends to optimize performance. Slow-rate-short-stimulus on time degrades performance.

It is generally desirable to "pace" operator performance by means of cues as to the expected occurrence of a next event.

In general, the slower the rate of presentation, the better the performance, and a truism that nevertheless is important to remember. Irrelevant information along with relevant information does not necessarily degrade performance, provided that it is introduced in a manner that allows operators to pace themselves. Irregular rates, on the other hand, disrupt performance because the operator cannot anticipate the occurrence of the next stimulus. Stimuli with a "natural" or built-in order result in better performance. Whenever possible, sort out information categories prior to displaying information to the operator and align visual information elements spatially so that each category appears in its own unique location each time it is presented.

Subtractions are more difficult than additions in setting up informational content, and they should be kept to a minimum in any given sequence.

Information of greater value generally is "remembered" best, although typically at the expense of the remaining information.

The number of multiple categories displayed simultaneously can be increased to three with no significant degradation when the information rate is held constant.

[42] F. J. McCormick, *Human Factors Engineering, 3rd ed.*, McGraw-Hill Book Company, New York, 1970, p. 113.

[43] After R. A. Monty, "Keeping Track of Sequential Events: Implications for the Design of Displays." *Ergonomics*, vol. 16, no. 4, 1973.

SUMMARY GUIDELINES

1. Multidimensional coding typically results in more information transmission than single-dimension coding.

2. Information-processing capacity is adversely affected by both load and speed; i.e., the number of sources should be limited.

3. When time sharing of sensory inputs occurs, signals should be separated temporally (preferably by 0.5 s or more).

4. When operators can control the input rate (are self-pacing) and have some method of identifying the more important input (if they have a choice), auditory signals are generally more durable than visual signals.

5. Simultaneous presentation of the same information via two sensory modalities usually increases the probability of reception.

6. Signals presented on one channel (such as audition) can serve as cues to facilitate the use of another channel.

7. Time sharing of visual tasks adversely affects primarily those tasks with greater uncertainty and those which depend on short-term memory; thus speed stress does not affect all tasks equally.

INFORMATION TRANSFER

Certain information transfer facilitators are at the disposal of the designer and should be utilized whenever possible to maximize the human's processing response. Table 50.5 reviews some of the key facilitators that can be used.

TABLE 50.5 Design for Information Transfer

FACILITATOR	REMARKS
Familiar patterns	Present information to operators in a form that is already familiar to them.
Visualization	Whenever practical, utilize the natural tendency of most individuals to try to visualize even if the presentation mode is via channels other than the visual one.
Context	Provide as complete a picture as possible in a logically organized manner.
Normal relationships	Present informational components and component-observer relationships that are natural (alphanumeric characters right side up, high-low values oriented according to specific use format, etc.).
Minimal extrapolation	Avoid requiring the observer to extrapolate; i.e., present quantities, values, and patterns in their intended form.
Reference bases	Provide a continuous reference (a scale value, a map, a comparison background, etc.).
Timing	Time inputs to avoid overload and delay; present them in a logical sequence and provide immediate feedback to the operator's responses or queries.
Noise	Control input competition, interference, and distractions.
Conspicuousness	Emphasize informational cues via adequate intensity, contrast, and special encoding to ensure maximum arousal and attention.
Expectation	Anticipate the operator's mental set and prepare the operator to receive the information.
Meaningfulness	Clarify the value, necessity, and urgency of the information transfer.

Note: For additional enlightenment with regard to human information processing, see D. E. Rumelhart, *Introduction to Human Information Processing*, John Wiley & Sons, Inc., New York, 1977.

CHAPTER 51

BEHAVIORAL EXPECTANCY

IMPORTANCE OF BEHAVIORAL EXPECTANCY

Designers should recognize the importance of anticipating what people will do with the equipment or facilities they are about to design. Otherwise, it is very likely that the consumer will misuse the product or facility, possibly suffer injuries that lead to litigation, and/or select a competitive product when next making a purchase.

This chapter presents a number of different types of expectancy data, some of which are based on research studies, and others on the experience of designers and human factors specialists.

BEHAVIORAL CHANGE FROM CHILDHOOD TO ADULTHOOD

Table 51.1 lists the general changes that the human goes through from infancy to adulthood. Although there are obvious individual variations, the generalizations may help the designer understand a particular age group's behavior with respect to the design under consideration.

ATTITUDES OF VARIOUS AGE GROUPS TOWARD SAFETY

Since safety should be of the utmost concern to the designer, it is important to recognize the general nature of human behavior and attitudes as they relate to the general question of safety. (See Table 51.2.)

TABLE 51.1 Behavior Change with Growth

Age	Behavioral and Motor Development Description
1 month	Lifts head when held to another's shoulder. Makes crawling motions when prone. Lifts head unsteadily when prone. Turns head laterally when prone.
2 months	Holds head erect for a short time when held to another's shoulder. Lifts head when held horizontally on back. Lifts chest a short distance when prone. Makes vertical arm thrusts in random play when on back.
3 months	Holds head erect and steady when held to another's shoulder. Turns from back to side. Pushes or elevates self by using arms when in a prone position.
4 months	Holds head steady when carried or when swayed. Tries to sit up when on back. Sits with resistant body pressure when supported by pillows. Hands frequently open. Thumb opposition appears in grasping.
5 months	Rolls from back to stomach. Sits with slight prop. Picks up cube from table on contact.
6 months	Sits momentarily without support if placed in a favorable leaning position. Grasps with simultaneous flexion of fingers. Retains transient hold of two cubes, one in each hand. Can turn from stomach to back, and back again to stomach. Rolling and hitching (in leg region) occur.
7 months	Tends to unilateral reaching and manipulation. Rotates wrist freely in manipulation. Scoops or rakes hand to secure a pellet. Picks cube deftly and directly from table. Begins crawling. Leans toward object of curiosity, reaches for it, and then handles, pulls, sucks, shakes, and rattles it (which often results in damage to the object and harm to the baby but which is valuable as a means of learning).
8 months	Sits momentarily without support. Raises self to sitting position. Picks up pellet with partial finger prehension. Thumb opposition appears in grasping objects.
9 months	Sits alone. Use thumb opposition in seizing cube. Makes a locomotive reaction in prone position.
10 months	Pulls self to standing position. Crawls for toys.
12 months	Stands with support. Creeps or hitches along. Walks with help. Shows a preference for one hand on reaching. Scribbles imitatively with crayon. Has acquired a fairly mature pattern of reaching. Can use a cup with both hands; later, with practice, can use a cup with one hand. Can pull off socks, shoes, caps, and mittens. Can hold pencil or crayon. Interested in self-feeding.
15 months	Stands alone. Walks alone. Can grasp spoon and insert in dish. Uses toys more. Throws and picks up objects and throws them again; puts them in and takes them out of receptacles. Often destructive in play with toys because of poor muscle coordination.
18 months	Climbs stairs or chair. Throws ball into box. Scribbles spontaneously and vigorously. Tries to put on clothes; wants to bathe self, brush hair, and use a toothbrush. Pulls toys, carries and hugs a doll or stuffed animal, imitates adult activities (such as reading), and actively gets into everything.
21 months	Walks attended on street. Walks backward. Differentiates between stroking scribbles and circular scribbles. Play is solitary. Feels, pats, and pounds toys. Strings wooden beads and pulls them in and takes them out of holes in tops of boxes. Transports blocks in wagons rather than building with them. Imitates activities of persons in the environment.
2 years	Runs. Piles a tower of six blocks with good coordination. Imitates vertical and horizontal strokes. Plays simple catch and toss with ball. Can operate a kiddie car around a chair. Can remove all garments unless buttoned in back. Can open boxes, unscrew lids from bottles and jars, turn leaves of books, insert pegs in a peg-board, string beads, and use scissors. Tends to use one hand more than the other. Makes deliberate swimming movements and tends to remain in prone position in water.
2-1/2 years	Goes up and down stairs alone. Piles seven or eight blocks with coordination. Tries to stand on one foot. Copies vertical line.
3 years	Draws circle from copy. Creases a piece of paper neatly. Aligns a card to an edge. Can feed self with a fork and spread butter with a knife. Cannot yet use a knife to cut. Much given to dawdling. Most vulnerable to accidents (boys more so than girls). Most accidents occur in home. Thursday through Saturday.
4 years	Draws from copy. Traces diamond path. Hand preference established between the ages of 4 and 6. Can ascend and descend ladders. Can learn tricycling, jumping rope, balancing on a rail or top of wall, roller skating, ice skating (double runners), and dancing. Perception of short distances is similar to that of an adult. Ability to perceive differences in form increases gradually from 2 to 6 years.
5 years	Draws triangle from copy. Draws prism from copy. Between 5 and 6 becomes proficient in throwing and catching balls. Between 5 and 6 can make recognizable letters, though writing is slow, laborious, and poor at this age. Can select medium-sized objects from a group of small, medium-sized, and large objects. Concept of roundedness is well established. Is able to estimate weights of different objects with fair accuracy; can tell the difference between 3- and 15-g weights when they are the same size.

(Continued)

TABLE 51.1 Behavior Change with Growth (*Continued*)

AGE	BEHAVIORAL AND MOTOR DEVELOPMENT DESCRIPTION
6–12 years	Can handle all garment fasteners in any way. Should be able to bathe self, dress, tie shoes, and comb hair. Can use tools to make simple objects like boatb dim wagons. Nursery school and kindergarten children can use scissors in following the outlines of pictures, model with clay, make cookies, sew, paint, crayon, paste, copy simple geometric figures, and help with simple household tasks such as carrying glasses and pitchers of milk without spilling. Can roller-skate, ice-skate (single runners), swim, dive, ride a bicycle, and keep time to music by skipping or dancing. Girls surpass boys in skills involving finer muscles (painting, sewing, weaving, and hammering); boys excel in skills involving grosser muscles (throwing a basketball, kicking a soccer ball, and doing a standing and running broad jump). Handedness is established; with guidance (and willingness on the child's part), handedness can be changed before age 6, but not easily after that age. Concepts of space become defined. From using weights and rulers, comes to learn the meaning of ounces, pounds, inches, feet, yards, and even miles. School arithmetic helps formulate more definite ideas of space and distance.
12–14 years	Time of rapid physical change, leading to confusion, feelings of motor development comparable to that of peers is extremely important; otherwise, is unable to take part in games and sports, which in turn has an adverse effect on social adjustments. Tests to measure agility, control, strength, and static balance show the greatest increase for boys after 14 years of age. For girls, improvement comes up to age 14 and then lags, mainly because of change of interests. Speed of voluntary movement increases continually from beginning of early adolescence to the end of that period. A 13-year-old has six-sevenths as much speed as a 17-year-old. Among boys, physical strength doubles between the ages of 12 and 16. Among girls, the greatest increase in strength comes near the time of the menarche. At all ages after puberty boys are stronger than girls, and this superiority increases with age. Girls usually attain their maximum strength at about age 17, and boys at about age 21 or 22.
17–21 years	A transitional period during which the adjustments to a mature status and to mature levels of behavior (begun during early adolescence) normally are gradually completed. The increase in physical strength that accompanies growth of the muscular system motivates the older adolescent to make use of his or her newly acquired strength. Complicated skills can be acquired during this period.

Source: Design Reference Guide (prepared by Man Factors, Inc., for the Consumer Product Safety Commission, Contract no. CPSC-C-75-010s. 1976).

TABLE 51.2

AGE GROUP	ATTITUDINAL DESCRIPTION
Infants and young children	Have a great curiosity about everything; anything new or unusual will motivate them to explore, unless the newness is so pronounced that it frightens them.
School-age children	Pressures to adjust to peer groups, to others at home and at school, to extracurricular activities, and to community activities tend to make them more cautious. However, the importance of conforming and of being accepted by the peer group tempts them to take chances that result in more serious accidents than those suffered by younger children.
Teen-agers	Because of the physiological changes that are occurring, tend toward restlessness, social antagonism, resistance to authority, erratic behavior, instability, and often daring behavior.
Young adults	Because of a gain in control over their larger bodies, plus the increase in physical strength that accompanies growth of the muscular system, are motivated to make use of their newly acquired strength, take more chances, and attempt more daring and demanding feats.
Persons in the early and middle years of adulthood	Adjusting to new life patterns and playing new roles such as spouse, parent, and breadwinner make them more cautious, more oriented toward efficiency and cost effectiveness, and willing to take only well-calculated risks.
Older adults, retirement age	Because of decreasing strength, lowered efficiency of sensory organs, reduced reaction speed, and increased physical handicaps, are less willing to take risks and generally exercise greater caution.
The elderly	For a variety of reasons (including physical infirmities and general perceptual narrowing), move more slowly and cautiously when engaging in all activities, their caution sometimes creating dangers when they drive, travel on foot, and use products and facilities. Most significantly, older people tend to lack patience, to be unaware of obvious hazards, and to expect others to "look out" for them.

Source: Design Reference Guide (prepared by Man Factors, Inc., for the Consumer Product Safety Commission, Contract no. CPSC-C-75-0102. 1976).

GENERAL BEHAVIORAL EXPECTANCY DATA FOR DESIGNERS (SEE TABLE 51.3)

TABLE 51.3 General Behavioral Expectancy Data for Designers

Behavior Pattern	Safety Implications	Design Considerations
Built-in habits and/or natural behaviors and associations:	Some appearances, operations, and procedures seem more natural than others to people; hence their first impulse is to interpret, operate, or use a device according to these expectations.	If the product is designed so that it either is or appears to be contrary to natural expectations, the user frequently makes mistakes, some of which could lead to hazardous situations.
1. People tend to assume that an object, device, or package is small enough to get hold of and also light enough to pick up.	If an object actually is very heavy, users may suffer severe strain or lose their balance and fall.	Either the user should be warned or the design should include some type of fastening to prevent the object from being lifted.
2. People assume that a vehicle is designed to "fit" their physical characteristics and thus try to adapt to it whether they can do so effectively or not.	Lack of proper fit often causes users to assume a precarious position or attempt a reach effort that not only makes their situation insecure but also forces them to assume a fatiguing position or apply uncontrolled forces and manipulatory patterns.	The vehicle should be adjustable when it cannot be made to fit all users by a single configuration.
3. People have learned (and therefore expect) that water faucets and liquid valve handles rotate counterclockwise to increase the flow of a liquid, gas, or steam.	If such valves are designed contrary to expectations, when such valves have to be turned off rapidly (in an emergency), individuals frequently will turn them in the wrong direction first and then make a correction—which could be too late.	Valve handles should be marked and mounted according to standard practice. No liquid valve should be incorporated into a design that does not operate according to this standard.
4. Electrically powered control switches are expected to move upward, to the right, or clockwise to effect electrical continuity, i.e., to turn power on.*	If such switches are designed contrary to expectations, people may turn an electrical device on when they think they have turned it off, thus exposing themselves or others to a functioning system which they assume is off.	Design or select electrical switches and control handles that are compatible with the expected directions of motion.
5. People expect that when they move a vehicle control to the right, or clockwise, the vehicle will move in that direction.	Incompatibilities with user expectations result in confusion and often in loss of control.	In the design of vehicles, vehicle control motions should be similar. If the operator must turn around to face a control, the vehiclecontrol relationships become confusing. Avoid such arrangements if at all possible.
6. People expect that when they actuate a knob or switch that causes an instrument pointer and/or some part of a machine to move, the directions of motion of both the control and the pointer (or part) will be the same.	Incompatibilities between the motion of the instrument pointer (or part) and that of the control element lead to erroneous dial settings and movements of subelements, often causing delays to correct the error and sometimes considerable danger to the user.	Knob and switch motion directions should be consistent with motion compatibility rules, and no control relationship should be used that does not produce a consistent natural response.†
7. People have become conditioned to certain color meanings (e.g., red for "danger," "fire," and "hot"; green for "OK," "go," and "acceptable"; yellow or amber for "caution," "look out for," and "yield"; and blue (in the case of electrical devices) for "cold" or "cool."	Misuse of colors may cause observers to ignore a caution or danger signal or to assume the presence of one that does not actually exist. Either condition creates confusion and delay.	Observe current color-coding standards (which may differ for various product categories).
8. People's attention is drawn to bright and vivid colors, bright lights, loud noises, flashing lights, and repeated and undulating sounds.	People may be distracted by an inadequately selected display and fail to attend to an important task detail, or an important task may not be given attention when needed because the device, signal, or sound was not designed to be conspicuous. When their attention is drawn to a bright light, individuals may disturb their own visual adaptation level to the point where they cannot see necessary visual detail (as when they face oncoming automobile headlights while driving).	Use stimuli of adequate intensity when attention needs stimulation; do not use high-intensity stimulation when distractions should be minimized.
9. Dark, dull colors appear to advance toward the observer; lighter, brighter colors appear to recede.	People may react instinctively (i.e., "duck") to miss a dark object above their heads, even though there is ample clearance.	Observe normal brightness and color impression conventions in selecting colors for interior surfaces.

(Continued)

TABLE 51.3 General Behavioral Expectancy Data for Designers (*Continued*)

Behavior Pattern	Safety Implications	Design Considerations
10. People expect to face auditory signals in order to corroborate and/or reinforce reception.	Placing an auditory warning signal behind an individual could cause the person to displace his or her eye reference as a result of turning the head or body. If this occurs at the wrong moment, it could cause a vehicle operator to miss seeing something in front of the vehicle.	Locate auditory warning signals so that they appear to come from in front of the operator and also locate them close to head height.
11. People assume a relationship between objects on the basis of their spatial proximity (i.e., they assume that things are related somehow when they are located or grouped together).	If a label or control appears to be associated with a display or another control, the individual's inclination is to use that control to effect a modification of the nearby display, whereas it may be the wrong control.	Careful positioning and/or arrangement of controls, displays, labels, or instructions is required to ensure the proper associations.
12. People generally regard products as being safe; that is, they: a. Assume that a system is *ready* to go and hence do not check. b. Grab a handrail assuming that it is strong enough to support them. c. Assume that a system is not on and hence proceed to touch or manipulate without caution. d. Assume that no one else has turned the system on and therefore do not check to make sure.	Although the assumption of "safe" should not be made, the fact that it is tends to preclude users from thinking safety, thus reducing the likelihood that they will anticipate possible hazards and make the necessary check.	Try to design the product so that it cannot be used improperly. If that is impractical, provide guards or other means to cause the user to think about the hazards. Finally, if neither of the above is practical, take steps to warn the user by means of labels or instructions, appropriately coded to imply caution or warning.
13. When about to lose their balance or fall, people instinctively reach for and grab the nearest thing.	Sometimes the things most convenient for grasping are not strong enough to support the individual, or they may be hot or slippery or cause an electrical shock.	Configure the design to provide appropriate emergency supports, or else make sure that hazardous components that might be used for support are not within reach.
14. People instinctively use their hands first to test or explore.	Many injuries occur because users touch an object with their hands or fingers, even when it was intended that the task be performed by means of an intermediate device of some kind.	Either make the product suitable for handling or else make it quite clear that its use requires a device supplied to obviate the necessity for using one's hands.
15. People expect a seat height to be at a level they are used to; a stair angle or tread pattern similar to one with which they are familiar and equally spaced in terms of riser height and tread depth; and a handhold or railing of a size and shape that can be gripped securely.	Individuals tend to adjust their posture, weight, mass, and gait to what they expect. When their expectations are not fulfilled, they are apt to lose their balance, take a wrong step, or trip.	Adhere to recognized standards and guidelines so that furnishings and architectural features are consistent with learned use patterns and are compatible with human physical dimensions.
Lack of knowledge and/or experience: 1. Across the broad consumer population there is a lack of knowledge about, and experience with, the technological features of many products, e.g.: a. Structural integrity b. Mechanical relationships c. Electricity d. Thermal characteristics e. Gaseous and combustible materials and conditions leading to explosion or fire f. Implications regarding center of gravity and balance g. Acceleration, velocity, and inertial factors	Since consumers generally are not prepared either to anticipate or to analyze conditions or the possible results of incompatibilities among physical elements and phenomena, they inadvertently initiate events that lead to improper and hazardous interactions. And because many consumers do not know what makes structures collapse or tip over, what causes materials to ignite or explode, or what inertial forces will do either to others or to themselves, they do not approach the product prepared to deal with these hazards.	Analyze the potentially hazardous interactions between the naive user and the various product elements to make sure that every means possible of precluding misuse has been designed into the product, that misuse has been warned against, and/or that the potential severity of the hazard has been minimized.

TABLE 51.3 General Behavioral Expectancy Data for Designers (*Continued*)

Behavior Pattern	Safety Implications	Design Considerations
2. Most people know very little about their own physical, sensorimotor, or cognitive characteristics or limitations. Thus they do not recognize: a. How they might cause muscle strain, sprains, or skeletal fracture b. How easily they can be thrown off balance c. How vertigo is induced d. How momentum and centrifugal forces affect their balance or ability to reach or apply force e. What is required for them to see effectively and reliably, hear effectively, and feel (touch) accurately f. The importance of adequate sensory feedback in order to control their actions properly or how various internal and external stresses can modify their interpretation of the feedback phenomena	Product users often do not recognize that they are not getting adequate inputs through their sensory channels or that they are reacting to illusory phenomena, upon which they are basing their decisions, rather than to the true informational input.	Learn more about how the human body works, and about its basic characteristics and limitations and then design in such a way as to minimize overstressing, creating illusory stimuli, or creating sensory contradictions and uncertainties.
3. The average consumer is not familiar with much of the technical terminology used for identification	Typical users do not understand the terminology, or they may misinterpret it and thus proceed on an erroneous basis to misuse the product.	Avoid highly technical terms (i.e., use standard nomenclature and abbreviations) and/or test new terms to make sure that they are clear to nontechnically oriented individuals.
Inattention: 1. Inattention is a phenomenon exhibited by everyone at one time or another, either as a natural trait or because of external and internal stresses, fatigue, boredom, or lack of motivation.	Lapse of attention often results in failure to observe a hazardous condition, read a label or instruction correctly, understand an instruction fully, see a warning light, or watch where one's feet, hands, head, or other body and limb components are going and whether they are clearing obstructions.	Anticipate the possibility of inattention in conceptualizing product design; i.e., think of the design in use terms. Where the possibility of inattention may be critical to the proper operation and safety of the product-user system, steps must be taken to preclude misuse and/or to provide means for attracting the user's attention.
Distraction: 1. Many people are easily distracted, either by certain aspects of the product's features or by the procedures for its use.	Distraction from the primary product use task may cause users to miss seeing an important element or event, to fail to observe the path of a part of their bodies or the bodily position they have assumed, or to break their train of thought, causing them to omit a step or fail to remember what they have completed.	Analyze the product design and potential use procedures to eliminate distractions and/or to provide a stimulus to maintain user procedural continuity and reliability, or else design to minimize the possible effects of procedural errors. The environment within which the product will be used must be known, and affective influences must be guarded against.
Hurrying: 1. Most people tend to hurry at one time or another.	When individuals hurry, they tend not to observe or to react carefully; they may not have time to think out or arrive at decisions based on a full analysis of a given situation. Thus, oversights may lead to inadvertent omissions, inappropriate responses, or even inaccuracies in sensorimotor performance.	Recognize the probability that a user will be in a hurry and design the product so that the use procedures help pace the user's response and/or so that the possibility of a hazardous consequence of an error or omission is minimized.
Deliberate risks and shortcuts: 1. Although risk taking and frequent shortcutting of proper procedures are more typical of certain users (e.g., young people), almost everyone takes some risk or shortcut at one time or another.	When a calculated risk cannot be based on a full evaluation of the potential failure modes and consequences, a product use failure probably will occur. Usually the user has neither the knowledge nor the time to make such an evaluation; hence the probability of failure is high.	Anticipate the possible risktaking and shortcutting behaviors and design the product in such a way as to preclude shortcutting or make the consequences of omissions less serious.
2. Many people refuse to look at and read signs or other visual warnings.	Failure to see or read labels, instructions, or signs leaves the individual uninformed and therefore more prone to error.	Make critical signs and signals conspicuous, legible, visible, and understandable and place them where the user is expected to be looking.

(*Continued*)

TABLE 51.3 General Behavioral Expectancy Data for Designers (*Continued*)

Behavior Pattern	Safety Implications	Design Considerations
Complacency, neglect, and overconfidence: 1. Because of overconfidence, individuals often proceed without thinking, observing, checking, or reading instructions or labels. 2. Complacency tends to dull people s sensitivity and alertness. 3. Some people fail to exercise caution because of a neglectful and disorganized approach to life.	When people assume that they know how something should be operated, they often make faulty judgments that lead to errors. In some product use situations, alertness is vital to safe performance.	Anticipate possible lack of interest or overconfidence on the part of the user and design in such a way as to alert and stimulate the user to possible critical aspects of product use and operation. The design should be made as simple as possible, and potential hazards should be conspicuously identified and brought to the user's attention. The design should preclude the operational steps' being performed out of sequence.
Forgetting: 1. Although everyone occasionally forgets things, some people by their very nature seem to forget more regularly than others or to forget because of stress or fatigue	Forgetting a critical step, for whatever reason, can have disastrous consequences. People also can forget about something they ordinarily know is dangerous or potentially injurious.	Anticipate the possibility of forgetfulness on the part of the user and design out critical features where such procedural omissions could lead to injury. Where necessary and possible, the design should provide some technique to monitor whether the user has forgotten a critical step.
Confusion: 1. People are easily confused by things that are unfamiliar.	Lack of familiar features not only lengthens the time required for a user to respond (perhaps making the response too late for safety) but also may freeze the individual into inaction, resulting in an unsafe condition.	Avoid designing products which are so entirely new that nothing about them (shapes, colors, nomenclature, etc.) is familiar to the user.
2. People are easily confused by complexity (in numbers, sizes, shapes, arrangements, concomitant operations, etc.).	Complexity places considerable demand on human perceptual-motor and cognitive processes, and under these circumstances any outside stress could lead to a collapse of the human system response.	Simplify all operator involvements; i.e., complex decisions should be made by the designer of the product.
3. Most people are confused by lack of identification of operating interface elements and by the inability to see or find a display or control or to locate the spot where the hand should be placed in order to pick up the device.	Not only will proper identification improve understanding, but it will also reduce the time necessary to figure out what to do and how to do it. Too much time and/or faulty deductions can lead to misuse or other human errors affecting operator and equipment safety.	Do not assume that people can recognize product features and understand them without labels or instructions (often labels or instructions are not included because of aesthetic design objectives or cost economies).
4. Many people are unable to perceive mechanical relationships, geometric relationships, or pictorial representations and to interpret these correctly.	If the designer assumes that all people have his or her skill in interpreting relationships and graphics and if the designer uses these to communicate with the typical consumer, the result will be confusion and misinterpretation by the user, resulting in possible misuse.	Avoid using special graphics or depending on relationships that are familiar only to technical people (i.e., design for the nontechnical user).
Lack of manual dexterity, skill, or practice: 1. Many people either lack the innate capacity for dextrous manipulation or control of their bodies and limbs or else have not developed (through practice) the manual skills needed for many tasks involving the fingers, hands, arms, legs, or feet.	When the use of a product (and the ultimate safety of the operation) is highly dependent on manual dexterity or skill and practice, there is a great opportunity for error at one time or another—error that could lead to an unsafe manipulation or operation.	Try to design products in such a way that they do not require great dexterity, precision, or speed or highly sensitive response to various feedbacks. Designers should understand the innate characteristics of the human sensorimotor servosystem and design so that lags in the human system are taken into account.
2. Some manipulatory and other manual skills are especially degraded when the specific human-product relationship is not optimum (e.g., the position of the operator in relation to the task, the extent and direction of movement, and the rate and rate of change of movement).	Increased physical and mental strain reduces the ability to coordinate body, limb, hand, and finger movements and to make precise direction, rate, and force inputs, and it reduces attention and perceptual awareness of errors.	Human-product design relationships should be "natural," convenient, minimally demanding, and not subject to potential misuse.
3. Some manipulatory skills are especially degraded by gloves and clothing.	Clothing may snag and preclude completion of movement; bulk may cause inadvertent contact.	Provide *ample*, not just minimum, clearance between components.

TABLE 51.3 General Behavioral Expectancy Data for Designers (*Continued*)

Behavior Pattern	Safety Implications	Design Considerations
Spurious autonomic responses: 1. Fear, anxiety, and uncertainty often lead to irregular, uncontrolled motor responses. 2. Panic often leads to unexpected responses (e.g., withdrawal, inaction, and violent and uncontrolled input to a device or control). 3. Pain leads to automatic recoil in most cases, although the opposite may be true of small children.	When responses cannot be predicted under these unusual conditions, many undesirable reactions may occur that have adverse consequences in terms of safety; e.g., a driver may freeze to the wheel or accelerator or suddenly exert forces or movements opposite to those which should be exerted, or a person may jump back after receiving an electrical shuck, a burn, a sharp cut, or a blow to the body, twisting his or her body suddenly in an effort to recover from a shift in body mass, etc.	Be careful about deciding how users may react to very sudden disturbances in normal situations; these usually do not allow time for thinking, since they are the result of completely involuntary and unpredictable reactions. The design should preclude the possibility or probability of such situations and provide "wiggle room" for the possible reactions so that they do not set off a chain of hazardous events.
Special behaviors: 1. Children: a. Put things into their mouths.	Choking and strangulation.	Make products too large to reach the child's throat.
b. Put their hands and fingers in holes and other openings (including electric outlets).	Entrapment, burns, cuts, shock, electrocution.	Make openings too small and provide covers that children cannot remove.
c. Try to squeeze into or through openings.	Entrapment.	Make opening too small for children to fit through.
d. Climb up on anything they can (e.g., boxes, furniture, and ladders).	Falling.	Eliminate climbing aids, i.e., handles or openings that can be used as a foothold or handhold.
e. Crawl into boxes, sacks, closets, pipes, etc.	Exposure to hazards within.	Provide secure closures and locks that cannot be operated by children.
f. Pull on cords, clothes, and tablecloths.	Pulling objects on top of child.	Use cordless systems and place electrical outlets where cords will not be within reach.
g. Pick things up and shake them or pound them on other people or other things.	Poking an object into the eyes and breaking windows.	Use soft material that is too heavy to pick up.
h. Pound on all vertical surfaces with their hands (including mirrors and glass doors).	Cuts.	Use shatterproof glass or plastic.
i. Try to open containers, doors, and drawers and to push all such things closed.	Exposure to hazardous substances or medicine; mashing, pinching, or cutting fingers: and dumping contents of a drawer on self.	Use lids that children cannot open, drawer stops, and large, obvious handles.
j. Try to push chairs, stools, and low tables.	Pushing into glass and tipping furniture over on self.	Add sufficient weight and provide proper support and balance.
k. Try to reach for things on the tops of counters, stoves, and pieces of furniture.	Burns, cuts from falling glass objects, and bruises from falling objects.	Increase the height beyond the reach of children and place controls and burners out of their reach.
l. Try to take everything apart, including toys, books, and all kinds of devices.	Swallowing small parts.	Make as indestructible as possible and difficult to disassemble.
m. Try to turn handles on appliances, doors, and other equipment.	Being exposed to hazards outdoors, turning on electrical power or gas, and starting machines, the operation of which is hazardous to the child.	Place handles out of the reach of children and design them so that children cannot operate them (i.e., design handles that require great force, complex coordination, etc.).
n. Try to ride, operate, or drive any kind of vehicle they can climb upon or get into.	A variety of injuries if machine moves.	Provide locks to keep children out.
o. Push all buttons and pull all levers.	Starting an operation that could lead to hazards if uncontrolled.	Provide locks to keep children out. Cover push-button panel, and require special secondary movement before lever will move.
p. Put things into any other things that have holes in them.	Equipment malfunction that leads to a variety of hazards.	Cover openings.
q. Like to push things over (to see them fall).	Cuts and bruises.	Design so that weight and/or balance precludes tipping.
r. Like to poke their fingers and other objects into other people's (especially other children's) mouths, ears, and eyes.	Various physical injuries.	No known precaution except to make all objects less pointed.

(*Continued*)

TABLE 51.3 General Behavioral Expectancy Data for Designers (*Continued*)

Behavior Pattern	Safety Implications	Design Considerations
2. The aged and handicapped:		
a. Shuffle when they walk and therefore trip over things.	Falling and striking objects.	Provide smooth walkway surfaces and mark steps with a highly contrasting color.
b. Hold objects with a weak grip and often unsteadily and therefore tend to drop them.	Cuts while trying to retrieve a broken object.	Use large handles, shatterproof materials.
c. Do not see well or observe carefully and hence bump into, trip and fall over, or fall from irregular surfaces or stairs.	Falling, striking objects, and tumbling on stairs.	Provide good illumination and high contrast at changes of grade.
d. Do not exert a normal effort to control their posture and hence tend to lose their balance easily.	Falling and bumping into objects, receiving injuries on contact.	Make all walking surfaces level, keep all related objects free from sharp corners or edges, and provide strategically located handholds.
e. Do not hear well or listen carefully and hence often are unaware of impending danger indicated by sounds.	Entering a hazardous zone unaware of impending danger, e.g., an approaching vehicle.	Provide visual warning and use an audio warning signal that uniquely penetrates consciousness.
f. Cannot apply as much force as younger or unimpaired persons and hence approach and try to manipulate things in an unconventional manner.	Misapplication of force, leading to operational hazard and potential injury.	Reduce operating force, optimize control to preclude misuse, and encourage proper use by making it the only way in which a product can be used.
g. Have a limited perceptual awareness and are less apt to think problems through; hence they observe less and attend less to what they are doing.	Failure to notice a potential hazard, leading to a variety of consequences.	Maximize perceptual cues (i.e., more contrast and more conspicuous features) and minimize complexity.

*These expectancies are peculiar to Americans. Europeans expect the reverse, i.e., to turn a switch down or to the left in order to turn power on.

†See W. E. Woodson and D. W. Conover, *Human Engineering Guide for Equipment Designers*, University of California Press, Berkeley, 1964.

TYPES OF SPACE

Besides needing enough space in order to move about and perform various tasks, people react to space in a variety of ways. Several researchers have defined the space surrounding the individual in terms of the limits within which people categorically respond (see Fig. 51.1). *Intimate space* is that area in which a person tends not to allow anyone to intrude unless intimate relationships are expected. *Personal space* is that area within which a person allows only selected friends or fellow workers with whom personal discussion is mandatory. *Social space* is that area within which the individual expects to make purely social contacts on a temporary basis. And, finally, *public space* is that area within which the individual does not expect to have direct contact with others. Obviously, the more intimate the spatial relationship becomes, the more people resist intrusion by others. Personal space factors are important in establishing the privacy requirements for architectural design.

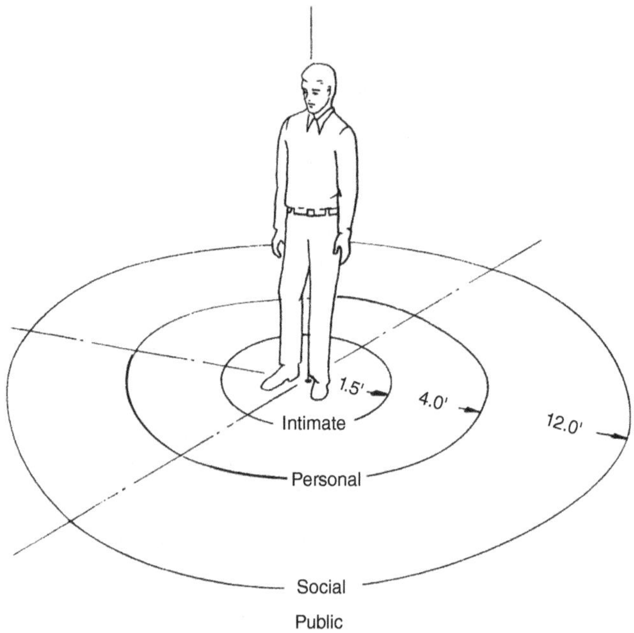

FIGURE 51.1 Spatial responses to people (U.S. culture).

RESPONSES TO SPATIAL FEATURES

Typical Subjective Responses to Selected Spatial Features

Although few research data have been generated with regard to how people respond to specific spatial factors (at least in terms of being able to prescribe precise, quantitative guidelines), it is important for the designer to reflect on potentially negative reactions that often result when a given space is not made compatible with what the user expects in terms of the size, shape, organization, color, and illumination of a particular space. The considerations listed in Table 51.4 are suggested as a checklist for the designer.

TABLE 51.4 Environmental Effects on People

Space Characteristic	Probable Response
Size (generally volume)	If the space is too small for the number of people, furnishings, equipment, or other objects that occupy it, people will consider it to be crowded. Although they may accept a crowded condition on a temporary basis, they will object to living or working in such a space for extended periods of time. If the space is too large for the people, furnishings, equipment, or other objects that occupy it, people will consider it "unfriendly," inconvenient, and/or overly demanding in terms of communicating, travel distance, maintenance, etc. Although they may accept the "barnlike" atmosphere for temporary periods, they will object to living or working in such a space for extended periods of time.
Shape (generally proportion)	If the space is out of proportion (too narrow, wide, long, high, etc.) for the intended use, people will consider it awkward and often distracting or oppressive. Although they may accept proportional distortion on a shortterm basis (i.e., as they pass through briefly), they will object to living or working in such a space for extended periods of time. If the space contains such distortions as all curved surfaces, acute wall junctures, and too many projections or surface changes, people will consider it confusing and difficult to maneuver in and/or furnish. Although they may accept such distortions (or even consider them interesting) on a temporary or one-time basis, they will object to living or working in such a space for extended periods of time. It should also be noted that blind people depend on the constant proportions of right-angle corners to aid them in negotiating a space; such individuals are easily confused by curved surfaces, walls that are not at right angles, and periodic projections that imply they may have reached a turning point. When a ceiling is extremely high relative to the lateral dimension of a space, people feel as though they are working in a pit and that the walls are closing in on them. When a ceiling is extremely low and the space in front of the observer is very long, people feel as though the room is "endless" or as if they will hit their heads unless they duck.
Color and illumination	If a space is dark (unless this is required for a particular operation, such as a motion picture presentation), people tend to become lethargic and less active, or they may feel anxious. As a rule, the less bright a room is, the less cheerful it seems. A small space will seem even smaller. If a space is too bright, people will feel overly exposed, or they will complain of glare or thermal discomfort (even though actual glare in terms of accepted light levels or inappropriate thermal conditions for comfort are not present). If there are too many different colors, too large expanses of very saturated color, or too many and too "busy" patterns of color within a space, most people become irritated after more than a brief exposure to the space. If there is too little color, no visual pattern, or no other decorative "break" in the visual environment, people will find the space monotonous, boring, and eventually irritating to the point of wanting to escape. Although isolated points of highly reflective surface provide interest, all-metallic and highly reflective surface treatments create both subjective and directly objective interference for most people who have to work in the space.
Windows	Generally, most people do not like to live and work in a space that is devoid of windows. First and foremost, people seem to need visual contact with the outside world. Too many windows, on the other hand, can cause the following possible negative reactions: too much glare, too much exposure (fishbowl effect), lack of protection from outside elements, true anxiety (caused by floor-to-ceiling glass at high elevations).
Space organization	The internal components within a space and the traffic corridors and entrance and exit locations will seem either well organized or badly organized. The furnishings, partitions, decorative objects, etc., will appear as being either organized or disorganized, depending on the observer's ability to comprehend what things are and where they are with respect to his or her vantage point. Key behavioral response issues are: apparent capability to find one's way to specific locations, apparent ease for interacting and communicating with others with whom the individual must associate, apparent privacy provisions necessary to perform individual tasks. Although these are sometimes conflicting needs, the people who use a space will perform on the basis of how well each of these factors has been executed for them, not for the designer or the boss. The organization of internal space components obviously interacts with all the other space characteristics; i.e., the individual perceives and reacts to the combined effects of size, shape, color and illumination, windows, and organization simultaneously. A significant behavioral response will be an individual's interpretation of whether sufficient options are available for local modification of his or her own portion of the space. Even though people may never require a modification, they react to their own space in terms of permanently established restrictions that eventually elicit the feeling that the space is too small, the wrong shape, too dark, or isolated from the rest of the world, for example.
Furnishings	As a general rule, people are sensitive to improperly proportioned furniture, i.e., furniture that is too large, too small, or the wrong shape for the space in which it is placed. Although the designer normally tries to select furnishings that are properly proportioned for the space he or she has created, this may ultimately restrict the efficiency of the individual (e.g., a desk or storage cabinet may be too small). Thus, although the general visual proportions of furniture in relation to space must be taken into account to avoid negative observational responses, shortchanging the individual in terms of specific furniture and use requirements soon stimulates an even stronger negative response.

Unique Responses to the Individual's Location within a Space

Observation and study of what people do when they occupy spaces provide several important thoughts with respect to where a designer expects to place people. For example:

1. People in auditoriums will not sit next to a side wall, especially if the wall is very high.

2. People generally dislike sitting facing a wall that is too close, unless they can look out a window.

3. People generally prefer to sit in a position from which they can observe the entrance to the room and also observe other people, but they do not like to sit so that they are directly observed by others (or think they are).

4. People dislike having to face other people (in a waiting room airplane, train, etc.) as they come out of a restroom.

5. People coming into an auditorium from the rear will tend to take the back seats, not those farther toward the front. Similarly, they will take the seat nearest an aisle, rather than one toward the center of a pew or row of seats.

6. In a conference situation, most people will take a seat at the corner of the conference table, not an end position or one halfway along the side of the table.

7. People generally will sit facing a light-colored rather than a dark or highly (saturated) colored or patterned wall.

8. When there is a row of chairs or stools, most people will seek out one that is not next to one that is occupied.

PSYCHOLOGICAL MOOD

Many attempts have been made to define basic human needs in relationship to designing an environment that is satisfying to the eventual customer, user, or occupant of living and working spaces. Figure 51.2 illustrates one such attempt to formalize human needs.

Complete fulfillment probably is beyond the expectation of most people. However, certain negative factors may seriously affect moods, attitudes, and eventual response to the environment. Some of these factors are the following:

1. Boredom
2. Apathy
3. Anger
4. Fatigue
5. Depression
6. Distress
7. Frustration
8. Anxiety
9. Pain

Although any or all of these may not be controllable through design, they can be influenced by it, i.e., by the nature of the task, by the environment within which the task is performed, and by how much stress or lack of stimulation the task involves.

EXPECTANCY WITH RESPECT TO PRODUCT APPEARANCE

Although there are apparent relationships between various aspects of a product's appearance and the attitudes individuals may express toward it (especially in terms of what people

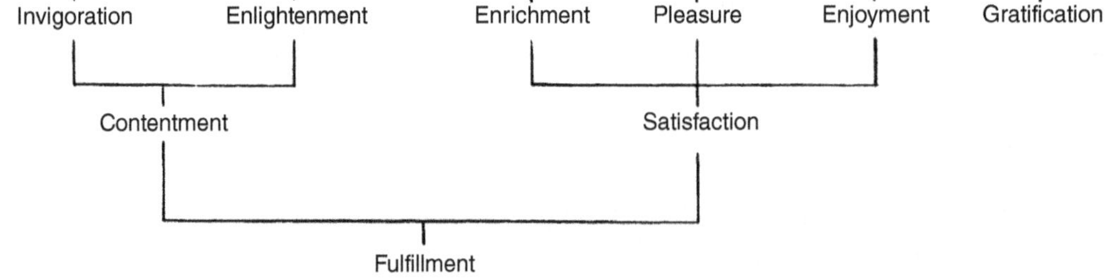

FIGURE 51.2 Model of human needs.

decide to buy), there is little quantitative information to assist the designer in knowing what to do to make a product more acceptable. Key issues regarding product appearance include the following:

1. Size, shape, and proportion
2. Color
3. Texture
4. Apparent functionality
5. Compatibility with contemporary concepts (and/or with the style of a particular period, e.g., early American)

It becomes obvious that each of these characteristics hinges upon the particular visual environment within which the product will operate. For example, the size of furnishings needs to be compatible with the size of the space within which the furniture will be placed; the color of the product should be compatible not only with its general surroundings but also with the colors of the other products within those surroundings; the texture of the product should fit with the other textures used in the operating environment; the product must appear to have a purpose within the given environment; and, finally, the product should not appear outdated, unless it is obvious that it is meant to look like an antique, for example.

There is some validity to the theory that an attractive product will be treated with more respect than one that looks "cheap" or poorly designed. However, such concepts are difficult to quantify and write specifications for. On the other hand, it is important to recognize that most people soon neglect products that are hard to maintain in terms of good appearance.

A balance is required between making a product conspicuous or gaudy (to attract the user) and making it so dull and functional that the user tends to neglect it or tire of it after a short period of time.

Almost any product can be made to look attractive *after* its basic functionality has been established. That is, although attractiveness is important at the point of initial purchase, the consumer soon becomes disenchanted if the product is not functional and easy to use.

A product should be attractive from all expected viewing angles; i.e., there should be apparent continuity and completeness from each point of view.

A product should have the appearance of being substantial and sturdy; i.e., it should not appear too fragile and easily damaged.

Chapter 52

Human Fatigue

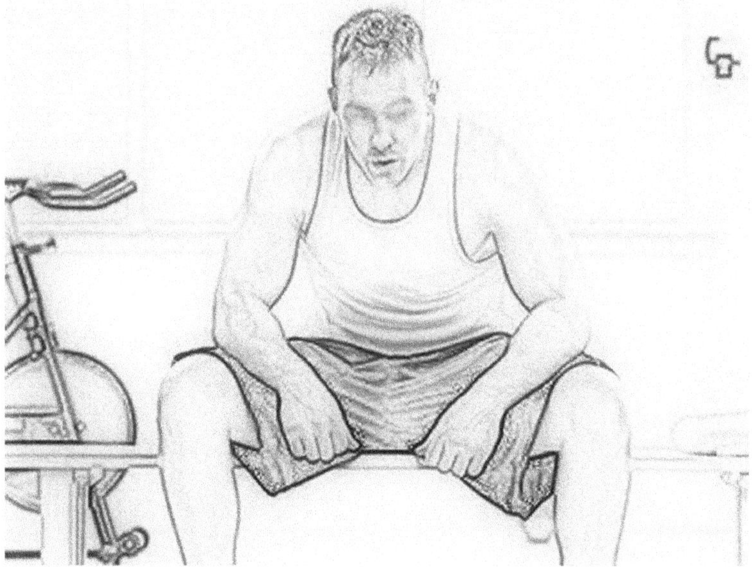

The problem of fatigue is very important to the consumer, and therefore it should be of concern to the designer; i.e., at least a design should be such that it does not result in activities that are unnecessarily fatiguing in terms of operation or maintenance.

A design may create two different types of fatigue: physiological fatigue, in which the operator's muscles are overstressed, and psychological fatigue (or mental fatigue), which may be caused by design-induced stress, i.e., complexity, high accuracy demands, or environmental implications, such as noise.

MUSCLE FATIGUE

Muscle fatigue occurs because of certain biochemical reactions within the muscle. Nerve impulses acting on a muscle fiber initiate a series of chemical reactions which result from, and contribute to, muscle contractions. Adenosine triphosphate breaks down under the influence of enzymes to create adenosine diphosphate, which in turn releases energy to enable the muscle to perform work. The triphosphate has to be regenerated before another contraction can occur. Energy required for this to happen is supplied by a breakdown of glycogen to lactic acid. Lactic acid is a poisonous byproduct that must be removed by oxidation to carbon dioxide and water. The oxygen thus removes the byproducts of the energy-producing reaction, which may continue some time after the muscular activity has ceased. The energy for muscular activity thus comes from a reaction which does not depend primarily on the presence of oxygen and which allows work to be done even when the immediate supply of oxygen is insufficient. This often permits the body to make a sudden extreme effort that might otherwise be impossible if the energy always had to be obtained directly from the oxidation of some substance within the muscle fiber.

Oxygen may come either from that stored in red muscle fiber or from the blood supply. As long as the supply is adequate to preclude lactic acid buildup, work is classified as "aerobic." If, however, the rate of work exhausts the reserve of oxygen, the work is said to be "partly anaerobic," and the muscle builds up an oxygen debt; i.e., there is an accumulation of lactic acid in the muscle and bloodstream, causing pain or muscular fatigue.

PSYCHOLOGICAL FATIGUE

The exact nature of psychological or mental fatigue is less clearly understood since it is extremely difficult to isolate and quantify causes and effects. Mental fatigue also confounds the researcher because opposite conditions may produce essentially the same mental or psychological aberration; i.e., one can become just as fatigued by boredom as by overwork. Unfortunately, we know of many instances in which individuals have been able to summon some reserve during a crisis, and thus we tend to believe that, with the proper stimulation, most people can continue mental activity indefinitely. Considerable evidence to the contrary, however, shows us that when an individual works too near mental capacity for long periods, almost any emergency that suddenly occurs may push the individual beyond his or her capacity to cope, the result often being a complete collapse or disorientation. Finally, the question of the threshold of psychological fatigue is further confounded by the fact that an individual's threshold may be stressed to within a few degrees of tolerance by preoperating conditions (prior activities), with the result that he or she has no tolerance to cope with an overly demanding mental task or situation.

DESIGN IMPLICATIONS RELATIVE TO MINIMIZING POTENTIAL MUSCLE FATIGUE

Avoid the following (see Fig. 52.1):

1. Designs that require operators to apply near-maximum force capacities over many cycles and for long periods of time
2. Designs that require continuous, rapid, repetitive muscle contractions for long periods, e.g., pounding, tapping, cranking, or push-pull cycling
3. Designs that force operators to "hold" some device in a fixed position for long periods without intermittent rest periods
4. Designs that require operators to maintain an upright posture for long periods without adequate body support (as in the case of a seat)
5. Designs that require operators to make very long reaches, frequently and for extended periods of time
6. Designs that require operators to stand or sit in an awkward position and to hold their arms above their heads for a long period of time
7. Designs that require operators to work in a "bent-over" or squatting position or in a position on their stomachs or backs, with the accompanying stress of holding the head and arms in a strained position
8. Designs that require operators to bend over and straighten up frequently and over a long period of time
9. Workplace layouts that require many steps, repeated again and again over a long period of time
10. Workplace layouts that require operators to sit "askew" (in a twisted position) in order to watch a display and at the same time operate some control (especially a foot control)
11. Workplace layouts that require operators to hold one foot above a foot control (between pedal depressions) for long periods of time
12. Workplace layouts that require operators to continuously move their heads from side to side or up and down

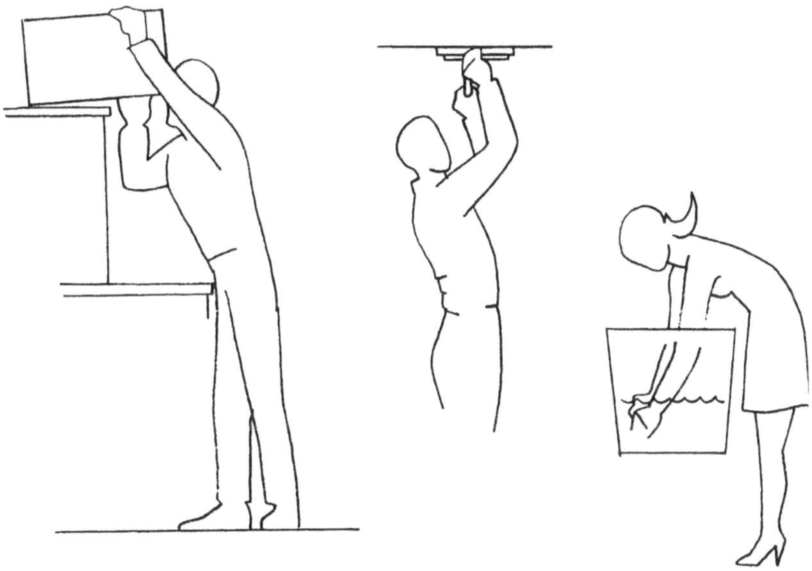

Figure 52.1 Repetitive lifting to an awkward position. An extended work task located above the worker's head. Bending over for extended periods.

13. Workplace layouts that require operators to step up and down frequently for long periods of time

14. Tool designs that require operators to hold and push a tool against a work surface or component to maintain contact pressure

15. Tool designs that require operators to hold a very heavy tool in a precise position for a long period of time

16. Tool or other equipment designs that require operators to maintain a very tight grip to keep the tool in place (especially if the grip must be maintained for a long period of time)

Lack of attention during workstation design planning often results in the worker's being subjected to severe strain, especially if his or her task requires assumption of these poor postures for a considerable length of lime. Neural and skeletal disorders often can be traced to the workplaces illustrated. (See Fig. 52.2.)

EQUIPMENT DESIGN AND OPERATING FEATURES KNOWN TO CONTRIBUTE TO MENTAL FATIGUE

1. Too many separate visual displays have to be monitored simultaneously.

2. Visual display formats require extrapolation rather than providing directly usable information.

3. Visual display detail is considerably greater than required; e.g., there are more scale marks than are warranted by the inherent accuracy of the instrumentation.

4. The legibility of visual display details is borderline, requiring unnecessarily close scrutiny in order to detect, recognize, and interpret what is being displayed.

5. Visual displays vibrate because they are not property shock-mounted.

6. Visual displays are not adequately illuminated, or there are uncontrolled glare sources within the critical viewing envelope.

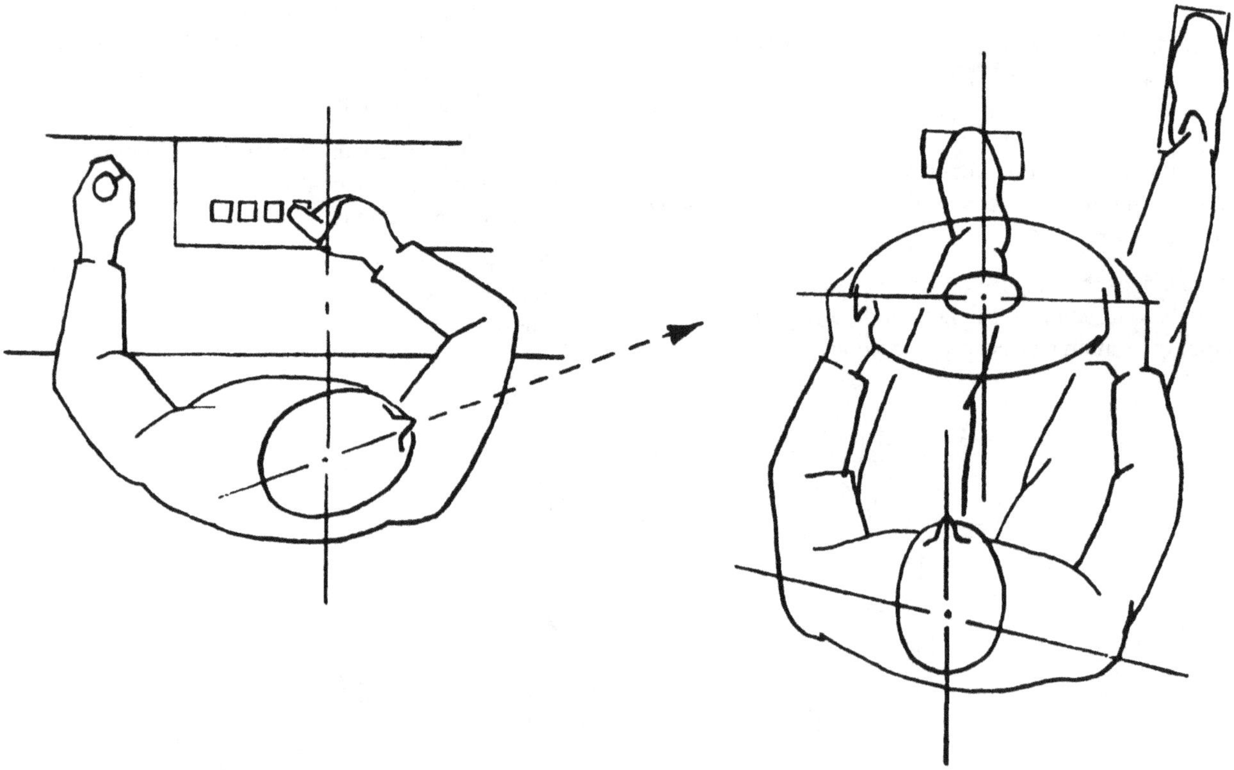

FIGURE 52.2 Sitting for long periods in a skewed position.

7. Simultaneous audio communications and/or excessive background noise is present.

8. Overly precise control adjustments are required.

9. Poof control dynamics, in terms of force-displacement, control-display, direction-of-motion, and/or movement ratio incompatibilities, are present.

10. There are long delays in informational feedback, i.e., long periods between signals or changes in equipment status.

11. There is a lack of timely indication of whether the equipment is functioning properly.

12. Continuous manual monitoring or control tasks are required that could just as well be automatic, with periodic operator alerting.

13. There is a lack of standardization among various similarly operated pieces of equipment, thus requiring operators to shift their point of reference.

14. The control panel layout is poorly organized, making it necessary for the operator to search for appropriate panel elements.

15. The workplace environment is inadequately controlled in terms of:
   a. Lighting, temperature, humidity, ventilation, noise, vibration, acceleration, pressure, etc.
   b. Support furnishings (standing platform, seating, writing surface, reference storage, restraint system, etc.).
   c. Space, e.g., clearance.

WORK OVERLOAD VERSUS BOREDOM

Mental stress and/or fatigue may be due either to an overload condition or to sheer boredom. Generally, it is rather obvious when a design creates an overload condition, i.e., when the operator is required to do too many things at the same time. However, boredom is more difficult to cope with in terms of what can be done to a particular design. It has been suggested that one of the primary methods for dealing with the problem of potential boredom is to make sure that the human is given tasks that are more suitable for his or her unique capabilities, namely, planning and decision-making. Alternatively, the machine or equipment should be given those tasks which are more difficult for the human and/or which are monotonous and unchallenging. The following general guidelines are suggested:

To prevent work overload:

- Sequence tasks rather than creating overlaps.
- Make individual tasks short.
- Minimize task precision requirements.

To prevent boredom:

- Provide task variety.
- Distribute tasks equally throughout the work period.
- Assign the operator only significant tasks and make it clearly evident that the operator rather than the machine is in control.

PART 5

HUMAN FACTORS ENGINEERING METHODS

POINTS TO CONSIDER

1. There is a broad range of human factors engineering applications and specialists. Some are focused primarily on the office workplace and the comfort, health, and performance of the workers in that environment. Others specialize in software development and the human–computer interface. Still others are responsible for the human interface with hardware designs—such as trucks, trains, and military vehicles. The common thread among these human factors specialists is their understanding of human capabilities and limitations and the core steps that they can use to apply this understanding to system development:

- Function definition
- Function allocation
- Task analysis
- Development and validation testing

Think of the specific types of questions and problems that the various human factors specialists have when they go through the core steps to apply human factors engineering to a design.

2. The human factors engineering methods described in this section can essentially be applied to systems of any size and complexity. The methods can be applied to something as small as the shape and size of a hammer. And on the other extreme, scores of human factors engineers worked on the development and operation of the International Space Station—probably one of the most complex systems humans have developed. It could even be argued that human factors engineering analytical methods can be applied to the design of systems for governing people.

Imagine the issues that arise in the application of human factors engineering analytical methods to various sized systems. Are there any steps more or less significant due to system size?

3. In the real world, it is rare to create an entirely new system. Designs are usually upgrades or repairs to existing systems. It is up to the analyst to define the scope of the work and then proceed with the analysis within that scope. Questions to be addressed include:

- What were the functions of the original configuration?
- What were the tasks and task problems?
- Exactly what parts can be modified?
- How will the new and old parts of a system impact each other? Will workers need retraining, will the overall working environment change (lighting, noise, temperature, etc.), or will the overall task flow be modified?

Once a boundary is defined, analysts can apply methods in this section just as they would for any "new" system.

What would be the effects of changing all computers in an office to touchscreens? Would it simply mean that software would have to be redesigned and evaluated to a touchscreen interface? Or should analyses consider redesign of desks and chairs and training for people unfamiliar with touchscreens? What should be the scope of analysis if an auto manufacturer wanted to replace the steering wheel with a joystick? Should analysis stop at just looking at the size and shape of the joystick? What are roles human factors should play in adding totally automated driving for some roadways or in evaluating the effects of unmanned aircraft on flight control operators? What methods can be brought to bear in all of these scenarios for applying human factors engineering?

Chapter 53

Human Factors Engineering Methods Overview

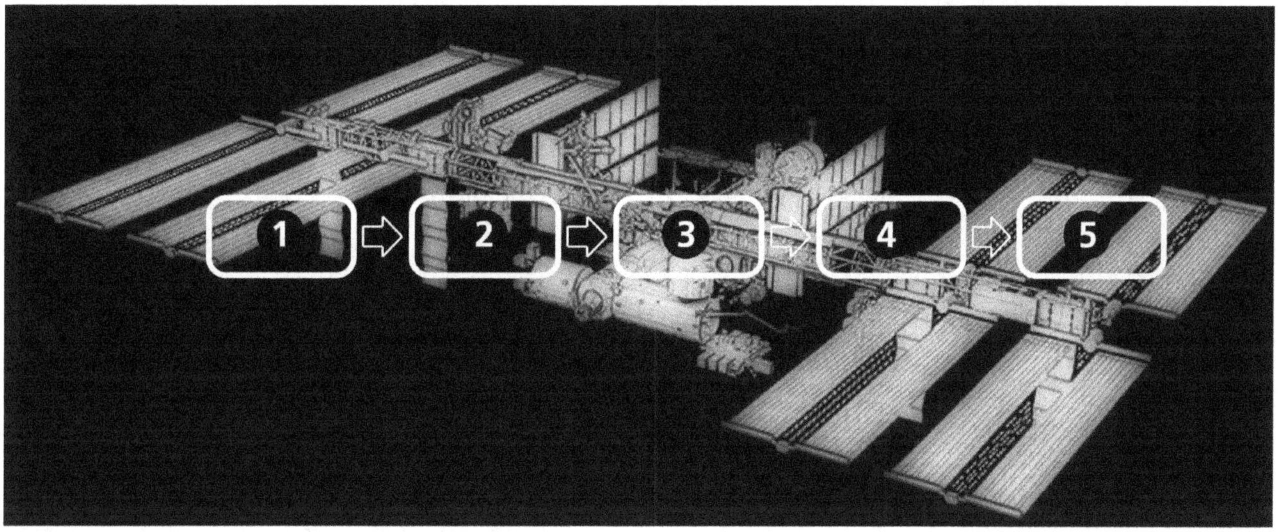

INTRODUCTION TO HUMAN FACTORS ENGINEERING METHODS

Human factors engineering (HFE) specialists should be used to help identify and solve human engineering problems during product or system development. However, this is not always possible, particularly on small projects, and thus one purpose of Part 5 is to provide a number of guidelines to assist managers, developers, and designers in doing their own human engineering. Also, Part 5 is intended to help any member of a design, development, or production team understand the importance and methods of HFE.

The materials presented in Part 5 derive from multiple sources, particularly the human engineering program practices of the U.S. military services and other U.S. government agencies since these have been well documented in HFE specifications and standards. They deal, for instance, with requirements to prepare plans for HFE involvement and to perform certain analyses during the conceptualization of systems; with defining HFE's responsibility for monitoring the human engineering aspects of hardware design; and with the performance of essential HFE testing and evaluation of the design at various stages during the design and development processes through to product delivery and operational validation.

Although some of these guidelines were originally developed for use in military systems development, they have been found to be equally applicable to any hardware development program, large or small. Because the methods are based on a logical and systematic process of (1) establishing the proper role of the human in the system, (2) designing the human–machine interfaces to fit the human's capabilities and limitations, (3) evaluating and testing to see that the design does fit, and (4) properly training the human to finally close the loop and thus assure reliable, total human–machine performance effectiveness, they should apply to all products that are used or operated by humans. Note that these four steps of the HFE process will be repeatedly mentioned throughout Part 5.

A General Philosophy

There is a tendency to believe that HFE occurs naturally as part of the typical engineering process. However, the truth is that engineering specialists often focus on the hardware and software aspects of system design and development to the detriment of keeping in mind their human operators, maintainers, and assemblers. Human interactions with hardware and software are essential elements of any operational system. To produce operationally sound and efficient systems, the project manager, engineer, and/or designer must be alert to the value of good human engineering principles, and the human factors (HF) practitioner must be sufficiently cognizant of the designer's technical problems to perform HFE in a manner that is applicable to a particular project's design challenges. The so-called human engineers are those individuals who have gained the necessary experience and knowledge to perceive design needs and interpret scientific information on human capabilities and limitations in a manner conducive to adding value to a project's design and development process. In other words, the human engineering specialist is part engineer-designer and part HF scientist.

The methods and techniques discussed in Part 5 have been developed over a period of many years and have been designed specifically to solve human engineering problems. The HFE methods and techniques to be covered include:

- Project Management
    - HFE and Project Management
    - Systems Engineering
    - Human Systems Integration (HSI)
- HFE Planning
    - HFE Methodology
    - HFE and the Project Life Cycle
    - Estimating the HFE Effort
    - Preparing an HFE Plan
- Concept Development
    - Concept of Operations
    - Function Allocation
    - Requirements Definition
    - Performance Criteria
- System Design
    - Task Analysis
    - Design Trade Studies
    - Design and Evaluation
- System Evaluation
    - Testing
    - Mockups
    - Modeling
- System Operation Documentation

Suggested Reading

For more help with HFE methods, consult:

- ISO/AWI TR 9241 Series, *Ergonomics of Human-System Interaction*
- ISO 9241-210:2010, *Ergonomics of Human-System Interaction—Part 210: Human-Centred Design for Interactive Systems*
- ISO/CD 9241-220, *Ergonomics of Human-Computer Interaction—Part 220: Processes for Enabling, Executing and Assessing Human-Centred Design within Organizations*
- MIL-STD-46855A, *DoD Standard Practice: Human Engineering Requirements for Military Systems, Equipment, and Facilities*
- MIL-STD-1472G, *DoD Design Criteria Standard: Human Engineering*
- NUREG-0711, *Human Factors Engineering Program Review Model*
- ISO 13407, *Human Centered Design Processes for Interactive Systems*

HUMAN FACTORS AND PROJECT MANAGEMENT

Design and production projects generally occur within process and organizational frameworks. Particularly for large projects, this framework is implemented in program or project management (PM)—the planned processes and structures for project leadership, participant responsibilities, and accountability. The project manager is ultimately responsible for the execution of the PM plan but is accountable to his or her programmatic and institutional sponsors to deliver results within agreed upon budgets, schedules, and performance parameters.

It is critical to the success of HF participation in a project that the HF practitioner understands the larger PM framework and his/her role in it—i.e., HF's position in the organizational structure, HF's agreed upon roles and responsibilities, HF's reporting process and structure through the PM hierarchy, and, ultimately, the activities that HF must perform, deliverables HF must produce, and their role in the success of the resulting final product or design. HF participation is much more likely to be valued and successful if everyone in the PM framework understands the role of HF and the actions of the HF practitioner are being performed as part of a larger, interactive choreography.

To promote clarity and appreciation for HF's involvement in a program or project, it is essential for the HF participants—whether an individual or a team—to work with the PM at the earliest outset of a project's formation to integrate the plans for HF into the greater PM plan.

Program and Project Management

Programs are generally the largest organizational framework of a planned design and product delivery. Projects are manageable subsets of the larger program. For example, the program for a large automobile manufacturer might be the current year delivery of new makes and models. The production of individual models may then be designed as individual projects and meted out to individual managers. Subset systems of an individual model may be managed as individual subsystem projects—e.g., body, engine, power train, etc.—and so on. The establishment of each hierarchical management structure requires conscious decision-making to determine the degree to which every participating discipline—including HF—is involved and is required for success.

It is wise to baseline the program or project in a management plan to promote compliance and ensure consistency. (See Fig. 53.1.) Some significant considerations for any PM plan include:

1. Project life cycle. What are the phases that the project must mature through? Does the life cycle end at product delivery or is the project responsible for operations and operational costs as well? How much time and financial budget is planned for each project maturation phase?

    a. Define the phases of the project and the metrics that determine a phase has been completed such that the project may move into the next phase.

    b. Identify key reviews as the project moves through the life cycle. Reviews should include performance, budget, and risk evaluations.

    c. Clearly identify the purpose and value of reviews and metrics and identify how they help ensure the project's success.

2. Roles and responsibilities. Who are the required participants in a project and what are the expectations of their involvement?

    a. Define the team structure and the roles of the key participants.

    b. Define the PM framework, whether informal (small projects) or formal (large projects). Clearly identify responsibility.

    c. Who has authority? How are decisions made? Is there an appeal process for key decision-making?

3. Risk management. How will threats and opportunities be recognized, documented, and addressed?

4. Cost management. How will product design, development, and manufacturing costs be tracked and held within a planned budget?

5. Performance management. What metrics and key parameters must be continuously tracked and managed to ensure that the project will deliver the product(s) it set out to deliver?

[*PROJECT NAME*] PROJECT PLAN

1. Project Overview
    - 1.1. Introduction
    - 1.2. Objectives
    - 1.3. Mission Description and Technical Approach
    - 1.4. Project Authority, Governance Structure, Management Structure and Implementation Approach
    - 1.5. Stakeholder Definition
2. Project Base lines
    - 2.1. Requirements Baseline
    - 2.2. Work Breakdown Structure Baseline
    - 2.3. Schedule Baseline
    - 2.4. Resources
    - 2.5. Joint Cost and Schedule Confidence Level
3. Project Control Plans
    - 3.1. Technical, Schedule, and Cost Control Plan
    - 3.2. Safety and Mission Assurance Plan
    - 3.3. Risk Management Plan
    - 3.4. Acquisition Plan
    - 3.5. Technology Development Plan
    - 3.6. Systems Engineering Management Plan
    - 3.7. Information Technology Plan
    - 3.8. Software Management Plan
    - 3.9. Verification and Validation Plan
    - 3.10. Review Plan
    - 3.11. Mission Operations Plan
    - 3.12. Environmental Management Plan
    - 3.13. Integrated Logistics Plan
    - 3.14. Science Data Management Plan
    - 3.15. Integration Plan
    - 3.16. Configuration Management
    - 3.17. Security Plan
    - 3.18. Project Protection Plan
    - 3.19. Export Control Plan
    - 3.20. Lessons Learned Plan
    - 3.21. Human Rating Certification Package
    - 3.22. Planetary Protection Plan
    - 3.23. Nuclear Safety Launch Approval Plan
    - 3.24. Range Safety Risk Management Plan
4. Waivers or Deviations Log
5. Change Log
6. Appendices
    - 6.1. Acronyms
    - 6.2. Definitions

Figure 53.1 Example outline of a project management plan. (From NASA Procedural Requirements 7120.5D, "*NASA Space Flight Program and Project Management Requirements*", Appendix H, "Project Plan Template.")

Tailoring Human Factors Involvement

The HF practitioner should be involved with the program and/or project manager(s) in the conscious decision-making required to define the planned inclusion of HF's disciplines in programs and projects. This is essential to ensure that there aren't undue expectations of what HF contributes, the HF participation is appropriately sized and scoped, and HF participation is designed to add value to each project's outcome.

Perhaps the most common perceived failure of HF involvement on projects and on successfully influencing project outcomes occurs when HF practitioners are brought in only at the end of a project to evaluate the usability and consumer desirability. To cost-effectively affect the outcome of a product, the HF practitioner must be actively involved throughout the product's design and development. Design must be a process of incremental decision-making and evaluation. When brought in only at the end of a product's design—to, for example, give the product's "human interface" an appropriate look and feel—the overall product is not as likely to succeed as when the product's desired inter-relationship and interaction with its planned human user communities is considered and evaluated at every step of the design process.

HUMAN FACTORS AND SYSTEMS ENGINEERING

Systems Engineering Overview

Systems engineering is a formal, interdisciplinary methodology for managing the design, development, and operations of large, complex systems over their life cycles. Whereas program or PM addresses team structure and responsibilities, cost management and program/project authority, systems engineering focuses on managing the technical, engineering, and design aspects of a project. In this sense, a project's systems engineering plan may be considered a subset or component of the overall PM plan, though in actual implementation there is often overlap between the two. The project manager is ultimately responsible for the development and execution of a project's systems engineering plan although this responsibility is often delegated to a chief systems engineer and—on large projects—to his/her team.

The systems engineering management plan for a project outlines the technical expectations for each phase of the project and establishes design and performance metrics that must be met before the project can proceed from one phase to the next. (See Fig. 53.2.) Each organization and each project may define their own set of project phases, but the following sequence is common:

- Project Formulation
- Conceptual Design
- Preliminary System Design
- Detailed Design and Development
- Production and Construction
- Utilization and Support (i.e., "Operations")
- Phase-Out and Disposal

[*PROJECT NAME*] SYSTEMS ENGINEERING MANAGEMENT PLAN

1. Title Page
2. Purpose and Scope
3. Applicable Documents
4. Technical Summary
    - 4.1. System Description
    - 4.2. System Structure
    - 4.3. Product Integration
    - 4.4. Planning Context
    - 4.5. Boundary of Technical Effort
    - 4.6. Cross-References
5. Technical Effort Integration
    - 5.1. Responsibility and Authority
    - 5.2. Contractor Integration
    - 5.3. Analytical Tools that Support Integration
6. Common Technical Processes Integration
7. Technical Insertion
8. Additional Systems Engineering Functions and Activities
    - 8.1. System Safety
    - 8.2. Engineering Methods and Tools
    - 8.3. Specialty Engineering (*including Human Systems Integration/HFE*)
    - 8.4. Technical Performance Measures
    - 8.5. Heritage
    - 8.6. Other Systems Engineering Functions and Activities
9. Integration with the Project Plan and Technical Resource Allocation
10. Compliance Matrix
11. Appendices
    - 11.1. Acronyms
    - 11.2. Definitions

FIGURE 53.2 Example template for a project's systems engineering management plan. (From NASA Procedural Requirements 7123.1B, "*NASA Systems Engineering Processes and Requirements*", Appendix D, "Systems Engineering Management Plan.")

The International Council on Systems Engineering (INCOSE) is an excellent resource for learning more about systems engineering methodologies. The *INCOSE Systems Engineering Handbook* is a particularly useful guidelines document. Additionally, individual organizations and agencies typically document their own approaches to systems engineering, e.g., the National Aeronautics and Space Administration's (NASA's) Procedural Requirements 7123.1B, *NASA Systems Engineering Processes and Requirements* and the accompanying NASA/SP-2007-6105, *NASA Systems Engineering Handbook.*

It is relevant to note the importance of systems engineering, the responsible systems engineer, and a particular project's systems engineering plan because increasingly, HFE is viewed as a subset of the overall engineering approach to large systems developments and the role and scope of HFE is increasingly documented in a program's or project's systems engineering management plan. This has been the case not only for the U.S. Department of Defense (DoD), but for many other U.S. government agencies—e.g., the NASA, the Federal Aviation Agency, the Nuclear Regulatory Commission, and the Department of Homeland Security. In these agencies, it is critical for the HFE practitioner to understand the agency's systems engineering and program management approaches in order to most effectively embed and engage HFE into the core of a program's or project's organizational structure.

It is generally a key concept in any definition of systems engineering or outline of systems engineering implementation that a key purpose of the discipline is to integrate the three most critical components of a total system—i.e., hardware, software, and humans. This broad view of the definition of "system" makes systems engineering an excellent context and conduit through which to integrate HFE, as the responsible discipline for understanding human/system interaction, into a program or project.

To thoroughly capture and agree upon HFE's involvement and the HFE approach to be taken in a given project, it is prudent to document an HFE—or human systems integration, HSI—plan as either a subsection of the project's systems engineering management plan or as a standalone document that is specifically invoked by the systems engineering management plan. The HSI plan exists to document the strategy for and planned implementation of HFE/HSI through a particular program's/project's life cycle. In particular, an HSI plan should define the responsibility of HFE/HSI in the program or project

- To ensure that the human elements of the total system are effectively integrated with hardware and software elements
- To ensure that all human capital required to develop and operate the system is accounted for in life cycle costing
- To ensure that the system is built to accommodate the characteristics of the user population that will operate, maintain, and support the system

Developing a Human Systems Integration Plan

Each project's HSI plan is project-specific and should be tailored to fit the project's size, scope, and purpose. (See Fig. 53.3.) The HSI plan should address:

- Roles and responsibilities for integration across HSI domains
- Roles and responsibilities for coordinating integrated HSI domain inputs with the project team and stakeholders

[*PROJECT NAME*] HUMAN SYSTEMS INTEGRATION PLAN

1. Introduction
    1.1. Purpose
    1.2. Scope
    1.3. Definitions
2. Applicable Documents
3. Human Systems Integration (HSI) Objectives
    3.1. System Description
    3.2. HSI Relevance
4. HSI Strategy
    4.1. Summary of Strategy
    4.2. Relevant HSI Domains
5. HSI Requirements, Organization and Risk Management
    5.1. HSI Requirements
    5.2. HSI Organization, Roles and Responsibilities
    5.3. HSI Issues and Risk Processing
6. HSI Implementation
    6.1. Summary of HSI Implementation
    6.2. HSI Activities and Products by Project Phase
    6.3. HSI Plan Update Process

Figure 53.3 Example template for a project's human systems integration plan. (From the 2015 update to NASA/SP-6105, "*NASA Systems Engineering Handbook.*")

- HSI goals and deliverables for each phase of the life cycle
- Entry and exit criteria with defined metrics for each phase, review, and milestone
- Planned methods, tools, requirements, processes, and standards for conducting HSI
- Strategies for identifying and resolving HSI risks
- Alignment strategy with the systems engineering management plan

The party or parties responsible for program/project HSI implementation—e.g., an HSI integrator and/or team—should be identified. The HSI integrator and/or team develops and maintains the HSI plan with support from and coordination with the project manager and system engineer.

Implementation of HSI on a project utilizes many of the tools and products already required by systems engineering—e.g., development of a concept of operations, clear functional allocation across the elements of a system (hardware, software, and human), and the use of key performance measurements through the life cycle to validate and verify HSI's effectiveness. Establishing and implementing the HSI Plan is not intended to duplicate other systems engineering plans or processes. The HSI Plan works in harmony with systems engineering processes to define uniquely the HSI effort being taken to ensure that the human element is given consideration equal to the hardware/software elements of a project.

HUMAN SYSTEMS INTEGRATION

In recent years, an emphasis on HFE has been expanded in the U.S. DoD and other government agencies as an emphasis on HSI. HSI has specific meaning to the agencies that have embraced it. Generally, then tenets of HSI include:

- Grouping human-centered disciplines together and coordinating their integration into programs and projections under the umbrella of HSI
- A comprehensive systems engineering approach to the human elements required to make a system operational including, essentially, every human who touches the system—i.e., operators, maintainers, logistics suppliers, ground controllers, assemblers, trainers, etc.
- A mandate to incorporate HSI into future programs and projects with the specific goal of reducing the life cycle costs of future systems' acquisition and operations costs

The number and types of human-centered disciplines covered by HSI varies from agency to agency but generally includes:

- HFE
- Environmental factors

- Habitability
- Human-centered safety and occupational health
- Force protection and survivability
- Manpower—i.e., the numbers of people required to make a system operational
- Personnel—the types of skills needed to make a system operational
- Training—the specialized training required for personnel to operate a system

HSI was first mandated by the DoD in 2003 with Enclosure 7, "Human Systems Integration (HSI)" of DoD Instruction 5000.02, "Operation of the Defense Acquisition System," which has subsequently been renewed, most recently in 2015. Enclosure 7 states that:

> The Program Manager will plan for and implement HSI beginning early in the acquisition process and throughout the product life cycle. The goal will be to optimize total system performance and total ownership costs, while ensuring that the system is designed, operated, and maintained to effectively provide the user with the ability to complete their mission. The Program Manager will ensure that HSI is considered at each program milestone during the program life cycle.

Enclosure 7 goes on to address some of the essential program/project planning that must be performed by HSI for the domains listed above.

The Economic Rationale for HSI

The rationale for mandating HSI stemmed from recognition in the DoD that roughly 80% of the life cycle cost of their systems were occurring in the operations phase of deployment and that most of these costs were human-related—e.g., (1) additional highly trained personnel were needed to operate, repair, field assemble, train, or otherwise support complex systems' operations; (2) the cost of rework of systems to compensate for poor HSI in systems' original delivery; and (3) system down-time due to inability to support complex logistics or maintenance. By 2003, HSI was seen as an influence that could help reduce unanticipated operations expenditures through better design for operability, usability, and efficient use of the human elements critical to systems operations. The arguments are classic HF—i.e., that usability is an important and ultimately cost-effective design consideration—and the emphasis on HSI by major government agencies has been one of the most significant efforts to institutionalize HFE and human-centered design practices.

Figure 53.4 illustrates the intended economic impact of HSI. The figure was compiled from the experience of multiple DoD systems development projects. It shows that though actual design and development expenses—"% Completed (Cost Expended)"—seem to ramp up gradually, design decisions quickly lock in the true life cycle costs of a project—"Committed Life Cycle Costs." For example, by the time only 8% of the design and development has been spent on conceptual design, 50% or more of the life cycle costs may be locked in or determined due to the decisions that have been

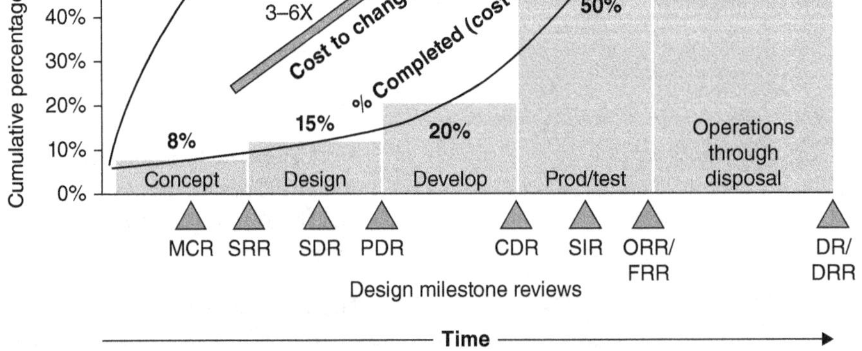

Figure 53.4 Life cycle cost impacts from early phase decision-making. (Adapted from the 2015 update to NASA/SP-6105, "NASA Systems Engineering Handbook" and from Defense Acquisition University data, 1993.)

made. The notion of "locking in" design is reinforced because the cost of "going back" and making design modifications as the design and development processes are maturing is constantly escalating—"Cost to Change Design Direction." Late design changes mean changing design drawings and documentation, investigating flow-down impacts, changing production tooling, etc.

Examples of design decisions that may radically impact operations might include consciously putting off design challenges for the operation implementers to resolve—e.g., how large or delicate subsystems will be moved to the assembly site, how a system will be maintained in the operational arena, or how operators will acquire needed training. Making invalid assumptions about operations is quite common—e.g., assuming that operators will intuitively understand the system's user interfaces or assuming that design of maintenance is another organization's responsibility. A serious and ongoing challenge to realizing the cost-savings intended through HSI implementation is that the managers of the design and development phases of a system's life cycle are seldom the same managers responsible for the operations of the system. Systems are likely to spend much more time in the operations phase of the life cycle than they spent in development, thus even small cost impacts deferred to operations will aggregate over time.

When participating in design and development of systems, the HSI/HFE practitioner will be under the same pressures as the rest of the design team to deliver products on time and within budget. But the HSI/HFE practitioner has a responsibility to always keep in mind the long view for operability of systems and for impacts to total life cycle costs. The HSI/HFE practitioner may need to argue that design funding may need to be expended in the short term to investigate a design concern, with the rationale that the result could be considerable savings down the road. There is a challenge extant in the systems development world to ensure that the program/project manager or the parties to which he/she answers put a value on total life cycle costs.

Suggested Reading

For more information on HSI and various agencies' implementations of HSI, see the following:

- AFD-090121-054, Air Force Human Systems Integration Handbook: Planning and Execution of HSI
- DoD Instruction 5000.02, Operation of the Defense Acquisition System, January 7, 2015
- FY011 Department of Defense Human Systems Integration Management Plan
- DoD HSI and Environmental, Safety, and Occupational Health Handbook for Pre-Milestone A Joint Capabilities Integration and Development System and Analysis of Alternatives Activities
- Army Regulation 602-2, Manpower and Personnel Integration in the Systems Acquisition Process
- MIL-STD-46855A, DoD Standard Practice: Human Engineering Requirements for Military Systems, Equipment, and Facilities
- MIL-HDBK-46855A, DoD Handbook: Human Engineering Program Process and Procedures

CHAPTER 54

PLANNING FOR HUMAN FACTORS INPUT

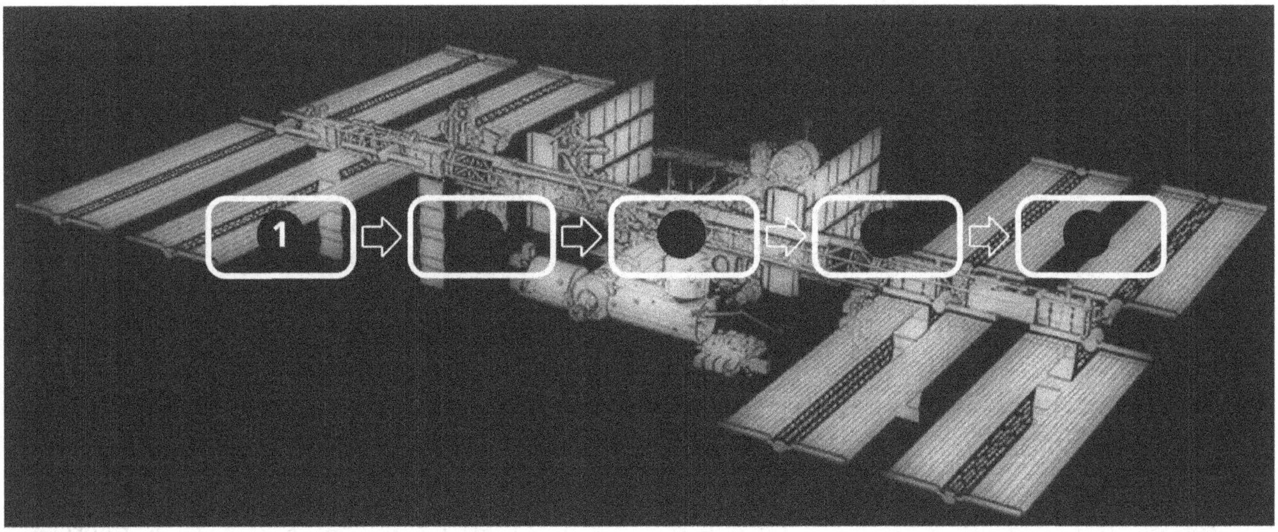

HUMAN FACTORS ENGINEERING METHODOLOGY

As noted in the previous chapter, it's highly desirable for the human factors engineering (HFE) practitioner to participate with project managers and chief engineers in designing the involvement of HFE on a project—i.e., the scope, role, responsibilities, and products of the human factors (HF) engineer's participation. The HFE practitioner is most knowledgeable of how to scope the required level of HFE participation to produce the desired project outcome.

Generally practiced HFE participation is based on a logical and systematic process of:

1. Establishing the proper role of the human in the system

2. Designing the human–machine interfaces to fit the human's capabilities and limitations

3. Evaluating and testing to see that the design does fit

4. Training the human elements required to operate, deploy, and maintain the system

This sequence can be tracked to the sequential life cycle phases of most systems development projects. Following are some of the inputs, methods, and tools used by HFE practitioners when participating in a systems development project.

1. Establish the role of the human in the system and of HFE in the project

- Human Factors Management Plan: As noted in Chap. 53, it is highly desirable to document the agreements and expectations of the HFE discipline in a project to ensure that there is team buy-in and understanding of HFE's contributions to the overall design and development process. An example outline for a human systems integration (HSI) plan was presented in Chap. 53. Also, it has become a way of life in modern design to document how and why one has arrived at a specific design concept and configuration. This is necessary because of the number of approval steps required before the design can be released to production. It may also be an important resource later if questions concerning product liability arise. In any case, it makes good sense to systematically examine and document each step of the design process. The *Human Factors Management Plan* should be a living document through the life of the project as well as a record of HFE's contributions.

■ Concept of Operations: The *Concept of Operations* (or "ConOps") is a dynamic, operating scenario that should be developed as early as possible utilizing the full membership of the project management (PM) and design teams. The ConOps can be an extremely useful tool for reaching a collective understanding and illustration of how the final system design is expected to accomplish its planned mission objectives when put into operation. The ConOps is particularly useful for communicating and reaching agreements on project intent with non-engineering-oriented team members, clients, and/or managers. The ConOps will be a useful product throughout the entire project life cycle and it should be continually revisited as validation that the team is indeed building the right product for the mission. Significantly more detail may be added to the ConOps as the design and mission understanding mature during the project's development phases.

■ Function Allocation

Function Definition: All of the functions and subfunctions required to accomplish the proposed mission should be identified and described. These should include both developmental and operational (i.e., mission use) functions.

Function Allocation: Trade-off analyses should be performed to determine the most cost-effective assignment of basic mission functions—i.e., to hardware, to software, and/or to human operators or maintainers. Significant HF expertise is required for quality function allocation to identify if the functions being allocated to the human elements of the system are in line with human capabilities and limitations and in line with expected levels of resourcing for operations. Like the other HFE products that are first developed during initial project formulations, the functional allocation should continually be revisited and updated as the design matures.

■ Performance Criteria (Operational Resources Definition): Target metrics should be proposed for the system's operations or use objectives. In systems engineering terms, target "performance metrics" should be established that define expected or desired operations resourcing. For the HFE disciplines, these should include expected levels of human resourcing for operations—i.e., manpower, personnel, and training. These will eventually be the measure of how successful the design is.

Manning and Personnel Analysis: Gross manning requirements are examined during the early planning and conceptual stage of system development. However, it is not until operator and maintainer tasks have been defined in more detail that one has sufficient information to clearly define what kinds of people and how many people will be required to man the system once it is put into operation. In this manner, setting target requirements for levels of manpower and personnel is a task that is performed in close conjunction with ConOps development, functional allocation, and task analysis studies.

■ Requirements Definition: The full set of requirements for a system defines the desired total performance of the interacting system hardware, software, and human elements. Individual requirements define singular physical or performance characteristics.

Customer Requirements: The general purpose of the proposed product or system should be clearly defined at the very outset of a project, before design work commences. Even though one may have some preconceived design ideas, it is important to make sure that the basic purposes for which the product is to be used are clearly understood by all stakeholders involved in the project.

Environmental Constraints: On the basis of the specific mission, all constraints within which the system must operate must be identified. Note that these may include resource constraints such as cost, manpower, etc.

Operational Requirements: Detailed performance specifications must be developed that establish quantitative requirements for the sensitivity, accuracy, speed, tolerance, reliability, delivery schedule, and service and maintenance of the system.

Detailed Functional Requirements Definition: For the planned mission and within all environmental constraints, hardware, software, and human functions are examined both independently and collectively to define the specific requirements that end items must accomplish. HFE principles should be engaged to make sure that proposed design concepts are compatible with human capabilities and limitations.

The full set of system requirements are continually visited during design and are used in the verification of the final system development.

2. Design human/system interactions to fit human capabilities and limitations

- Task Analysis: As design concepts are formulated, operator, maintainer, and user tasks should be described in sufficient detail to illustrate how, where, and with what operators and maintainers will be required to perform salient operating and maintaining tasks in order to exercise control of the hardware/software system. Task analyses should be continually revisited and matured during design as concepts mature into solid design proposals. There are many task analysis techniques and their applicability depends on the nature of the system. Two very useful techniques are time-line analysis and link analysis.

  Time-Line Analysis: Although the early mission ConOps should provide some general description of time relationships, it is not until design details begin to become available that one can estimate the operating and maintaining timing characteristics of system operation. At this point, it becomes evident whether the planned use of the operator or maintainer is feasible, i.e., whether the tasks imposed on the human are within his or her capability to respond or whether the operator or maintainer is apt to be overloaded or underutilized.

  Link Analysis: Link analysis allows one to examine the organization of human tasks in terms of logical grouping and sequencing of activities. It should be done at both the general workplace and the individual control interface levels.

- System, Equipment, and Facilities Design Concept Definition: As preliminary design tradeoffs are conducted and initial hardware designs are formulated, a "first look" at the total human–machine system operations concept develops and conceptual assumptions can be validated. It is important that HFE be given high priority during trade-off studies, particularly as human/system functional allocation assumptions can be studied. The initial system baseline provides the first significant hardware review point, at which major decisions will be made concerning whether to proceed with the concept, rework it, or drop it entirely because of technical or economic difficulties. Good design practice at this point should be open to iterative investigation of multiple concepts and trades for achieving the desired total system performance.

  Human Test and Evaluation: Best HFE practice entails an ongoing program of design concept mockup and evaluation using human-in-the-loop testing during system development. Human engineering evaluations at each stage of analysis and design are not only tests of the design but also confirmations of the adequacy of the human engineering effort. A robust human/system evaluation program will identify system issues while they may be still cost-effectively modified. Problems should be identified and resolved long before final system testing and final requirements verification.

3. Evaluate and test to verify that the design fits the human/system intent

- Detailed Design: Close interaction between hardware/software designers and HF engineers remains vital to successful system design since the hardware/software subsystems and the user community must eventually join to demonstrate a successful match between human and machine and to ultimately produce effective system performance. Human engineering principles should be applied, tested, and validated at each stage of design. HF engineers should be included in all critical reviews of design concepts, approval of drawings before release, evaluation of mockups, and validation of human–machine interactions. Some basic HF research may be required when previously established human engineering practices and criteria are insufficient to make appropriate design decisions as unique design challenges are encountered. All previously initiated HFE methods should be continually matured, validated, and eventually verified by the final design. At some point, the design should be sufficiently mature that HF specialists can begin to identify needed personnel development activities—e.g., procedure development and training requirements.

  Hazards Analysis: Although gross analyses should have been made during concept formulation, it is not until design details begin to emerge that one can identify the critical probabilities of certain hazards due to equipment failures, operator failures, or unique interactions between the hardware system and the environment within which it will operate. These should at a minimum be identified and proposed to be mitigated through design modification or operational procedure development.

4. Train the human elements of the system as needed
Plans and requirements for selection and training of operators, maintainers, and general users derive

from task descriptions and personnel expectations. Training objectives and a training plan should be developed as quickly as possible following detailed design because early development of key training equipment or facilities may be involved.

Training Aids, Equipment, and Facilities Design: As soon as training requirements are defined, training support items can be defined, including classroom aids, part task trainers, training simulators, special training facilities, and supporting documentation and other materials. As noted above, because it is necessary to provide certain long-lead-time items at the same time that prime hardware is delivered, one may have to perform an initial training requirements analysis prior to having all the design information that normally is generated by the task analyses and detailed hardware human engineering. These items include such things as complex simulators and special training facilities.

Technical Publications: Special publications must be prepared in time to be delivered along with the product or before the beginning of training. These include operations and maintenance service manuals which may become teaching aids. Human engineering attention should be given to the conceptualization and final preparation of these documents so that they will be made maximally helpful to the user. In military system development, it is required that these manuals be field-tested to ensure their effectiveness.

The final verification of the system brings together all elements of the total system—including hardware, software, documentation, and trained personnel—in integrated testing to confirm the ultimate measure of total system performance and effectiveness.

HUMAN FACTORS ENGINEERING THROUGH THE PROJECT LIFE CYCLE

The phases of a program's or project's life cycle were noted in Chap. 53's sections on program/PM and systems engineering. This section draws parallels and maps between the HFE methods described in the previous section and typical program/project life cycle phases.

1. Project Formulation: This project phase is sometimes called by other names—e.g., "Pre-Concept." The ideal point of entry for HFE on any program is before the customer has made any specific hardware designations and every aspect of the project is still in the formative stage, including the project's management structure. At this point, the design team has an opportunity to work with the customer to define the basic operational requirements and the goals of creating the product—i.e., the mission. The customer, the PM team, and the design team should consider starting the project without preconceptions and using a "total systems analysis program" to begin systematically conceiving and defining:

    a. The stakeholder, mission, and operational objectives of the project

    b. The functions that are required to accomplish each mission event, sometimes called "functional requirements decomposition"

    c. The performance requirements for each function

    d. The allocation of functions to hardware, software, or human elements

    e. The manner in which the project will be structured and managed should gel during this period

    f. The roles and responsibilities of the HFE participants should be clearly defined and agreed upon

2. Concept Development Phase: The next most opportune point of entry for HFE is in this phase in which conceptual design solutions begin to be proposed and the design team, working with the general directions defined by the customer, management, and other stakeholders, is allowed to perform trade studies on how to integrate various hardware, software, and human elements into a (preliminary) system. In this phase the design team—including HFE—will mature early preconceptions into baselines of:

    a. A ConOps defining the projected manner in which the project design solution is to operate throughout its life cycle

    b. The stakeholder, mission, operational, functional, performance, environmental, and quality requirements of the system

    c. Definition of the functions that are required to accomplish each mission event and allocation of those functions to hardware, software, and human elements of the system

    d. Performance criteria for all elements of the system including projected criteria for the operational efficiency including the role of human manpower, personnel skills, and required expertise/training

e. Preliminary operator, maintainer, and user task descriptions

f. Design concepts sufficiently mature to develop into hardware and software design solutions in the next project phase

3. Preliminary Design Phase: The next phase for the design team matures the design into hardware, software, and operation procedures sufficiently to test and evaluate key functions, performance criteria, and interfaces. The design team, though constrained by preconditions and by stakeholder expectations, has the opportunity and responsibility to perform more detailed design trade studies and explore any need for possible modification and improvement of initial concepts and designs. During this stage, the design team not only should review and update the previously noted analyses but also should begin the following:

a. Human–machine mockup or modeling studies

b. Fundamental (but applied) HF research studies where needed to clarify particular project design concerns

c. Baseline the operator, maintainer, and other system user task descriptions and analyses

d. Total human–system operations performance studies relative to baselined system performance criteria

4. Detailed Design Phase: As the design is being fleshed out to full maturation, the design team should ramp up test and evaluation as needed to confirm there will be no surprises for final product delivery—i.e., that the design will be fully compliant with all system requirements—and to ensure that all projected project goals are being met. During this phase the design team—including HFE—should consider the following:

a. Update all design process tools and methods noted above to track the design's maturation and document any design or requirements compliance concerns that arose during development

b. Use the ongoing updates of system design documentation to begin drafting operations documentation products—e.g., preliminary operating procedures that will be required for critical operations, deployment, assembly, maintenance, etc.

c. Begin to identify critical skill requirement specifications that imply the need for operator, maintainer, and user indoctrination and/or training

d. Simulate and evaluate human–machine interactions using mockups and models for critical key operator, maintainer, and user interfaces—e.g., control stations, equipment handling, facilities layouts, etc.

e. Participate in detailed hazards and safety analyses conducted by safety specialists working within a project's systems engineering and management framework.

5. Product Delivery and Documentation: For final product delivery, a demonstration, test, or analysis must be performed to show that the result of the development phase of the project meets every project requirement. Also, the program or project must complete some form of final report on the history of decisions made during design, any challenges or perceived compromises, and how hazards encountered were resolved. For HFE, this is a crucial time to transition from development to operations and to participate in the development of operational era documentation—e.g., operating and maintenance procedures, training requirements, etc. Note that this is also a good time for every discipline involved in the design and development process to capture "lessons learned" to help build their knowledge and experience base for involvement in future programs and projects.

6. Operations Phase: Development teams are not always the same management and support teams that deploy and operate the system, and for the building and maintenance of "corporate knowledge," that is often unfortunate. Ideally, the members of the design team—particularly HFE—will have an opportunity to directly see in a system's operations how design and development decisions panned out, whether assumptions or preconceptions about operations were accurate, etc. HFE discipline(s) may be more likely to participate in the transition from development to operations and should use operational experience to build their knowledge and experience. Another aspect of a design team's maintaining a role in operations is that many complex systems are often updated—e.g., to improve some aspect of operations, to increase safety, to cut operating costs, or to incorporate new technology. A design team that's maintained familiarity with a deployed system is likely to be an asset to any upgrade or renovation project.

A graphic mapping of the functions that might be performed by the responsible parties for HFE on a project to the various project phases is illustrated in Fig. 54.1.

Product \ Phase	Project formulation	Concept development	Preliminary design	Detailed design	Product delivery	Operations
Human engineering plan	Baseline	Update	Update	Update	Final & lessons learned	Use to baseline Ops documents
Concept of operations	Preliminary	Baseline	Update	Update	Final	Use to baseline Ops documents
Function allocation	Preliminary	Baseline	Update	Update	Final	Use to baseline Ops documents
Requirements definition	Preliminary	Baseline	Update	Update	Verification report	
Performance criteria	Preliminary	Baseline	Update	Update	Verification report	
Task analysis	Initiate	Preliminary	Baseline	Update	Final	Use to baseline Ops documents
Design concept reviews		Ongoing				
Preliminary design test & evaluation			Ongoing			
Detailed design test & evaluation				Ongoing		
Product verification documentation					Final	Use to baseline Ops documents

Figure 54.1 Mapping HFE inputs to project life cycle phases.

PREPARING A HUMAN FACTORS ENGINEERING PLAN

As noted in Chap. 53 in the section on HFE and systems engineering, a plan for integrating HFE or HSI into a program or project should define the responsibility of the HFE/HSI involvement. That is,

- To ensure that the human elements of the total system are effectively integrated with hardware and software elements
- To ensure that all human capital required to develop and operate the system is accounted for in life cycle costing
- To ensure that the system is built to accommodate the characteristics of the user population that will operate, maintain, and support the system

An example HFE/HSI plan template is also included in that Chap. 53 section.

Establishing an HFE plan during the earliest project formulation discussions cements the role of HFE into the formal structure of the program or project. Though the HFE plan should be a living document throughout the life of the project, the initial baseline and agreements on the scope of HFE's responsibility and engagement should be documented very early in program/project formation. An HFE plan not only captures the scope, role, responsibilities, expectations, products and deliverables for the program manager and—if the project is large enough to have one—the systems engineer, but also serves to inform the entire design and management team membership what HFE is bringing to the project.

In large organizations, the HFE team on a program or project may consist of in-house HFE practitioners working with contracted HFE specialists. Also, large organizations—particularly governmental agencies—may typically contract out entire programs or projects with in-house managers, engineers, designers, and HFE practitioners serving oversight roles. When such large programs or projects are contracted, military and large government agency system development typically calls for submission of a formal human engineering plan as part of a contractor's proposal. Although the content and level of detail may vary, any HFE/HSI plan should consider inclusion of the following types of information:

1. Introduction: This is a general statement of the purpose and scope of the proposed effort and includes references to specific regulatory documents to which the plan

relates. Key among these documents are HFE and HSI process and requirements documents of the governing agency.

2. Agreements from initial guidance meetings: Usually it is required that, upon award of a contract, an initial proposal review will be conducted during which the agency's and the contractor's human engineering personnel will resolve any issues with the proposed human engineering plan.

3. Personnel and organization: The agency may want to know who in the contractor's organization will work on the program and where these people are in the contractor's organization. It is expected that the human engineering function will be in a position to have an appropriate impact on the entire development program and that it will not be relegated to an ineffectual level.

4. Coordinate with subcontractors: Many systems are built with established "off-the-shelf" subsystems and component. Human engineering analysts must have a plan to ensure that these subsystems individually and collectively meet HFE design criteria.

5. System and task analysis: Human engineering is expected to play an important role in the entire basic system, subsystem, and design analysis effort to ensure that the human elements of the system's design and operations are given appropriate emphasis at each stage of development.

6. Human engineering design: It is important to indicate where and when human engineering will be applied, by whom, and how this effort is to be documented.

7. Mockups: The military, in particular, expects human engineering to play a significant role in deciding not only what mockups will be constructed but how they will be used. The military expects mockups to be used as design tools, not merely sales gimmicks.

8. Human engineering tests: The agency wants to know what kinds of human engineering tests will be conducted and how and when they will be conducted.

9. Human engineering design verification plan: A design verification plan is often required as a standalone document to identify the points at which specific HF requirements are to be evaluated. The relationship and interaction of this verification plan with the overall project's verification plan should be a strong consideration. Often a program or project has one set of verification managers who will consider verification of HFE requirements to be one subset of their responsibilities.

10. Documentation and reporting: Any government agency will expect to have complete documentation of key phases of the human engineering effort, including, for example, monthly reports, special study reports, and human engineering contributions to all key design reviews. Appropriate engineering drawings are to be attached to all design reviews. If human engineering standards have been compromised for any reason, this is to be reported, including reasons for the compromise. Risks and hazards should be documented in accordance with the overall program's risk and hazard control process. Additionally, near the end of the program or project lessons learned from the experience may be required as part of the agency's overall lessons learned and process improvement systems.

11. Schedule: A detailed schedule for accomplishing the human engineering effort is to be provided. It should be keyed to the contractor's overall milestone schedule, and the schedule should indicate the level of effort throughout the program.

Finally, it should be emphasized again that though it is critical to baseline the HFE/HSI plan as early as possible during project formulation, the HFE/HSI plan should be considered a living document to be updated as the program or project matures. Any updates must be coordinated with larger formal management contexts—e.g., with the project manager and the systems engineer—since the HFE/HSI plan is part of the overall program/project's formal set of agreements and impacts every team members' responsibilities.

ESTIMATING THE HUMAN FACTORS ENGINEERING EFFORT REQUIRED FOR SYSTEM DEVELOPMENT

Obviously, the same amount of HFE effort is not required for every product design, either because the inherent need for HFE in an individual product's design and development varies or because of the amount of time and economic resources that can be made available by a particular project team varies. The following suggestions may be helpful as a guide in determining the extent to which HFE should be proposed and performed.

1. Need: Need should always be the first consideration. In the case of some products, lack of proper preparatory analysis may lead to costly consequences in terms of redesign, loss of sales potential, and the even more serious possibility of litigation in the event of personal injury suits. Some guidelines are as follows:

   - If a proved product design is merely being updated or refined and if its in-use track record indicates reliable operation, good human–machine performance, and no apparent safety problems, extensive human engineering analysis is not usually cost-effective, unless the proposed modifications could possibly introduce hardware- or user-induced failures. In the latter case, human engineering analysis generally is required only on those aspects of the product modification which are new and suspect in terms of potential user misunderstanding, performance failures due to previously learned habit patterns, and/or unique hazards that the modification may introduce.

- If the product is brand new but fairly simple in terms of concept and operation, the human engineering analysis requirement is probably minimal. Here it may be necessary to develop only a simple descriptive scenario of how the product will be used. This must be done in order to force the designer to review all the basic operations and maintenance interfaces in order to determine what human engineering design principles should be applied during the design process.
- If the product is generally new and relatively complex, a more complete human engineering analysis may be required in order (a) to ensure that the basic concept is compatible with user capabilities and limitations and (b) to derive a more complete set of human engineering design requirements for each subelement that has to be designed. As the system becomes more complex (particularly a new one), there is generally a greater need to perform these human engineering analyses for products that must define manpower, personnel, and training requirements.
- Either modified or new products or systems that pose special hazards for users (operators and maintainers) usually require more human engineering analysis—especially analyses of hazards and of failure modes and effects. Typical of these are systems involving excessive energy-producing features; extremes of vibration, noise, or temperature; radiation; toxic contamination; and/or unusual environmental conditions such as space or deep sea systems. Such systems require the most thorough and complete human engineering analysis because their basic abnormality requires that special precautions be taken concerning the manner in which humans participate in the system and the way they must be protected, not only from the system and the environment but also from their own performance failures.

2. Time Allowance: Time often is a constraining factor in terms of the kind and amount of human engineering analysis effort that can be accomplished before actual product design is commenced. On the other hand, time constraints are too often used as an excuse for not performing the necessary analysis—e.g., arbitrary decisions to forego human engineering analysis in order to meet arbitrary design schedules. The following are suggested guidelines with respect to evaluating human engineering analysis time:

- A design modification or even a new design that has sufficient operating performance precedence to indicate that there are no significant reasons to doubt that users can operate and maintain the new product as efficiently and safely as the previous or similar one generally does not require extensive preliminary human engineering analyses and therefore much time for analysis. The human engineering concern here is chiefly one of monitoring detailed hardware selection and design on the basis of recognized human engineering design practices, principles, and criteria.
- In the case of a design that, although somewhat new, is required as quickly as practical to meet an urgent operating need, the time allotted for human engineering analysis may be limited—i.e., although one would prefer to take more time, conditions exist that make it mandatory to produce the new or modified product as quickly as possible. Typical examples might include a special weapon needed for use in urgent battle conditions, a device to control traffic at a particularly dangerous intersection, or a safety device to prevent unintended operation of an agricultural machine. In such cases, the urgency is established by a higher authority on the basis of high probability that delaying production will add to already mounting problems within the operational arena. Not only is there insufficient time to perform the desired human engineering analyses, but also the chances are great that the product will already be designed and in use before any reasonably efficient analysis could be completed. Typically this kind of urgency should dictate eliminating human engineering analysis only when the item being designed is for specific, limited application and very limited production runs.
- When there is even a shred of doubt about a new design with respect to whether the human's role is compatible with his or her inherent capabilities or about possible hazards that could exist for the user, time should not be used as an excuse to eliminate desirable human engineering analyses. Rather, the approach should be to examine the product design challenges realistically and recommend that sufficient time be provided to perform the analyses. This is particularly important for new designs which push hardware or software state of the art, which are particularly complex in nature (involving many hardware/software elements and many human participants), or which appear to require personnel skills that are borderline relative to the capabilities of the expected user population. Generally, if enough time has been provided to do typical systems engineering analyses, there is also enough time to include appropriate human engineering analyses.

3. Level of Effort: The level of human engineering analysis effort—e.g., the number of actual work hours—is or

should be a function of need versus time, not just cost. At one extreme one should avoid "make-work" analyses costing many work hours; at the other extreme one should not expect to derive much benefit from an analysis which is limited to the efforts of too few analysts or which was performed too quickly. Some of the factors to consider in making decisions about the level of effort for human engineering analyses are as follows:

- If the design involves a military or large governmental agency customer, the level of human engineering analysis effort will likely be specified by the customer. If this is not clearly spelled out in a request for proposal, it is best to provide a complete estimate based on the requirements of MIL-HDBK-46855, *Human Engineering Requirements for Military Systems, Equipment and Facilities.* Ordinarily, in reviewing the proposal, the military customer will define how much analysis is required. If the customer does not do so, steps must be taken to clarify this matter before signing the contract.

- For designs that do not involve a military or other governmental customer, one should estimate the requirements for a human engineering analysis effort on the basis of the following:

- If the operating concept is not radically different from that of previous designs demonstrated to be effectively and safely operated by the expected user, the human engineering analysis can probably be limited to a brief identification of operating sequences and/or communications- or information-flow diagrams to help identify significant user–hardware interface requirements.

- If only one or two new elements are being added to an otherwise older and proven design, analyses should be performed only on the new additions. That is, it should not be necessary to perform full-system, in-depth function allocation, information-flow, or operational sequence analyses.

- On the other hand, if the new design includes untried techniques or procedural concepts, advanced hardware, or applications in unusual environments, the human engineering analysis should be as complete as necessary to ensure that the concept is compatible with the human user's characteristics, capabilities, and limitations. It is wise in this case to make sure that there are enough *trained* human engineering analysts to complete a thorough analysis of all design elements before these are committed in the project's plan.

- A good rule of thumb for estimating the amount of effort that should be devoted to human engineering analysis is that such analysis should be roughly 50 percent of the total projected engineering analysis time. This is required because the human engineering analyst generally has to be involved not only in his or her own analysis efforts but also in those of the engineering analysts—i.e., the human engineering practitioner should be an equal participant in every aspect of the design team's efforts.

4. Cost: Actual cost predictions for any kind of analysis activity cannot, of course, be made without addressing the specific system or product or without knowing specific manpower, personnel, and salary rates. However, certain cost factors can be noted that are perhaps useful in approaching the task of costing an analytic program. Some of these are as follows:

- Costs will vary depending on the types of analysts required. Human engineering analysis may require a variety of analysts, depending on the type of system under consideration. To start with, the required analyses may require not only a special kind of generalist who has had experience and is familiar with the objectives, parameters, and subject matter of human–machine system or product interface design, but may also require unique experts to be involved who provide special backgrounds for a particular project's needs. For example, in a manned space system, many of the analyses will involve the expertise of an aerospace surgeon or physiologist. Complex information systems may require the expertise of a senior-level psychologist. As a minimum, a qualified HFE analyst is usually a senior-level scientist. The salary levels of such personnel will not be low. This is not to say, however, that lower-salaried individuals cannot learn to be qualified human engineering analysts. In many cases, junior analysts can be used when only minimal analysis is required.

- Cost is generally determined to be proportional to the size and complexity of the system—i.e., the number of subsystems within the system that involve interactions between equipment elements and human elements. The more elements there are, the greater the number of analyses and thus the greater the costs.

The checklist in Fig. 54.2 may be used as an example reference for preparing the human engineering effort for most system development projects. A minimum of 2 percent of the total program cost is required for effective human engineering participation.

Service category	Deliverable end product	Labor, in days, by subsystem								Total labor, in days
		A	B	C	D	E	F	G	H	
A. Program planning	HF prog plan									
B. Requirements determination										
1. Documentation review	Report									
2. Field studies	Report									
C. Concept development										
1. Systems	Report									
2. Equipment	Report									
3. Facilities	Report									
D. Analysis										
1. Information flow	Info. fl. chart									
2. Time line	T. L. analysis									
3. Performance prediction	Report									
4. Gross task	Gross T. A.									
5. Fine task	Fine T. A.									
6. Operation sequence and interface	OSD 1-4									
7. Human-machine	HM fl. chart									
11. Hazards	Report									
12. Habitability	Report									
13. Trade studies	Report									
14. Allocation studies	Alloc. table									
15. Information requirements	Info. Req. table									
16. Cont./disp./com.	C/D/C table									
E. Design and development										
1. Performance specs	Spec.									
2. Hardware specs	Spec.									
3. Software specs	Spec.									
5. Mock-up preparation	Mock-up (1–4)									
6. Dynamic simulation	Report									
7. Consultation										
8. Design guidance	Design guide									
9. Design review										
10. Cont. disp. hardware select										
11. Panel layout	Drawings									
12. Work space arrangement	Drawings									
13. Equipment procedures	Procedures man'l									
14. Detail design	Dwgs. & doc.									
15. Training programs	Training pkg.									
16. Operator manuals	Manual									
17. Maintenance manuals	Manual									
Design verification										
1. Checklist preparation	Checklist									
2. Appraisal by checklist	Report									
3. Test planning	Test plans									
4. Test and evaluation	Report									
5. Test support										
6. Follow-up studies	Report									
Applied research	Report									
Totals										

Figure 54.2 Typical, expected human engineering services in system development.

Chapter 55

Concept Development

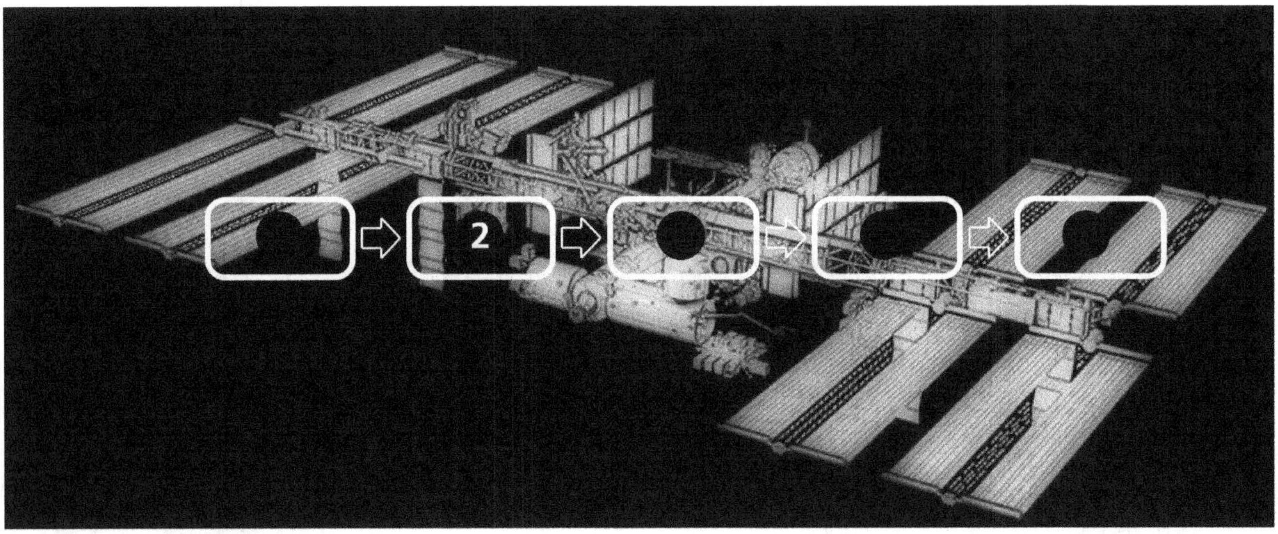

CONCEPT OF OPERATIONS

The Concept of Operations (or "ConOps") captures the collective design and development team understanding of the operations scenarios in which the system hardware, software, and human elements—operating within a realistic operational context and environment—will interact to accomplish mission goals and objectives. It is, essentially, the collective team agreement on how the system will work. The ConOps is important for capturing the expectations of all stakeholders, for development system requirements, and for early definition of system architecture. It is important to baseline the ConOps as early during project concept development as possible since the agreements in the ConOps give all team members their initial understanding of what to plan. The act of thinking through the ConOps may stimulate realizations about requirements and design functions that might not have been previously considered. The baseline of the ConOps should be a key deliverable of the concept development phase of the program or project. The ConOps should continually be revisited and updated as the program/project matures, however, and as design decisions lock in the reality of operational scenarios.

At its baseline the ConOps may include many assumptions, but the vision it provides serves as the basis for developing initial system requirements and conceptual designs. Late in the project the ConOps will serve as the basis for operations plans, handbooks, procedures, facility requirements, and operational constraints.

Because of the criticality of the human operators, maintainers, assemblers, ground controllers, and logistics suppliers in making the system operational, human engineering plays a key role in the development of the ConOps. Any assumption in the ConOps for human engagement that exceeds human capabilities and limitations or exceeds planned levels of manpower, personnel, and training puts the entire program or project at risk of succeeding.

ConOps should address not only nominal operations, but to be fully effective should address potential risks, hazards, and failure scenarios that may occur, particularly if they are life-threatening. Risk mitigations must be designed into the system to prevent loss of life or loss of mission.

No matter the size or scope of a project, the ConOps should be considered an essential tool. Essentially, how can any designer or design team achieve success without a clear understanding of what the system is supposed to do and how the system must perform in a real world scenario?

The ConOps is a foundational document for any program or project. The allocation of functionality across the hardware, software, and human elements of the system can't occur without a collective understanding of how the system will operate and perform. System requirements and performance targets are developed based on the collective understanding of operational performance. Detailed task analyses evolve from the ConOps. The ConOps may be the primary tool used by the program or project to ensure ongoing validation of the design effort—i.e., is the development team building the right product to accomplish the planned mission?

The ConOps provides an early and ongoing perspective on the complexity of operations and on how the hardware, software, and human elements must integrate and function together to fulfill the mission. (See Fig. 55.1.) The ConOps description of operations should be dynamic and interactive. It should describe activities and tasks and their sequencing. It should help identify operations timelines, modalities, and transitions across modalities. The understanding of operational timelines should mature in parallel with the system design and development.

Together the ConOps and resulting operations timelines provide a basis for system and subsystem component configurations and for their interactions. By walking through operations scenarios, the components that support operations and the roles the human elements contribute to mission success—all considered in realistic operational sequence—a vision of the total system will begin to unfold. (See Fig. 55.2.)

For additional information on the development of ConOps documents, the ANSI/AIAA publication "Guide to the Preparation of Operational Concept Documents" (ANSI/AIAA G-043A-2012e) is recommended.

Performing System Mission Planning

Several of the human factors engineering (HFE) techniques discussed in this chapter actually occur concurrently and interactively, not one after the other. For example, in conceiving the system ConOps the design team will concurrently be considering customer and other stakeholder expectations, environmental constraints, human/system function allocation, and system and mission requirements development. The example in Fig. 55.3 of a mission analysis provides insight into the types of information one

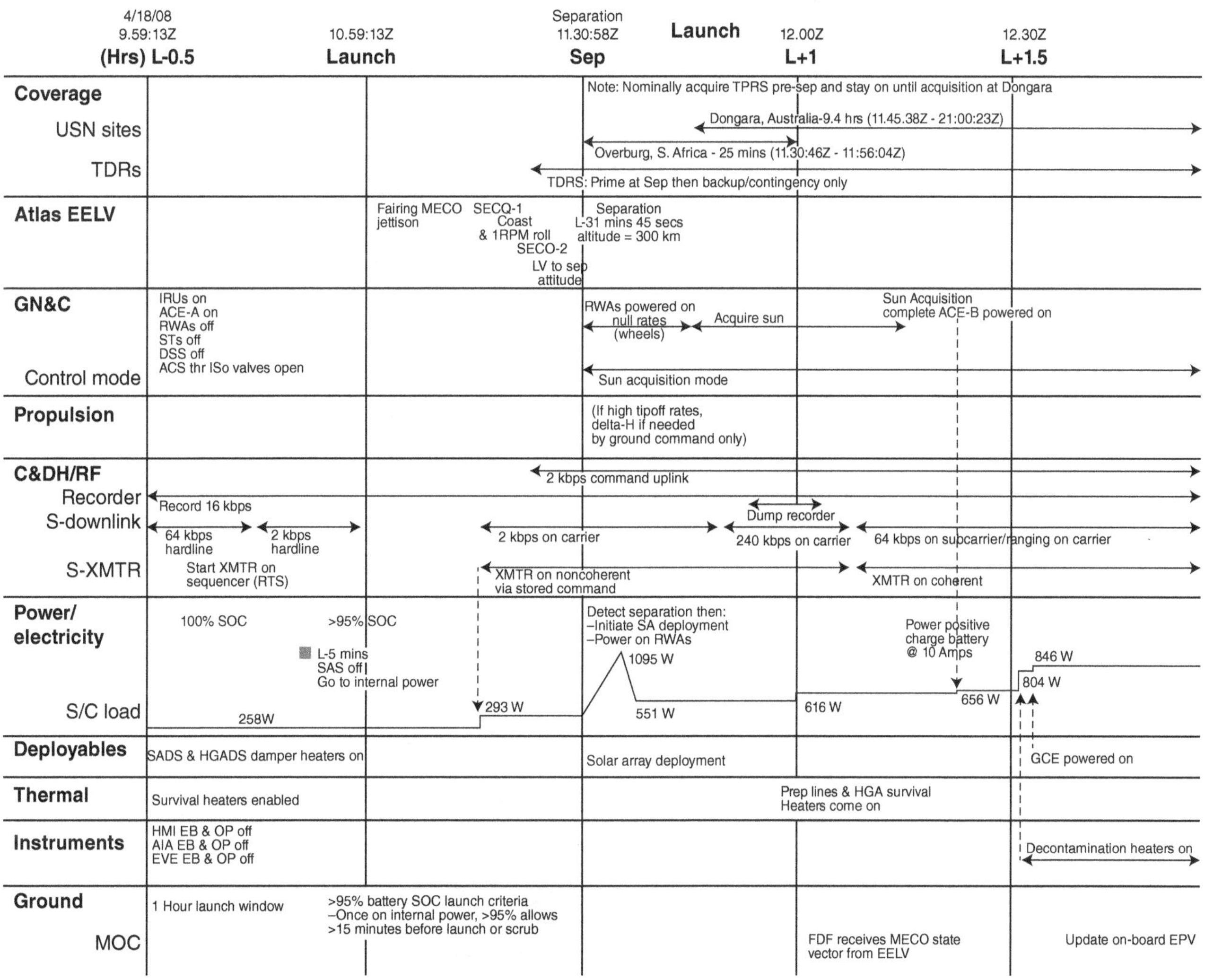

FIGURE 55.1 Example of a mature integrated timeline for a NASA science mission. (From NASA/SP-6105, *NASA Systems Engineering Handbook*.)

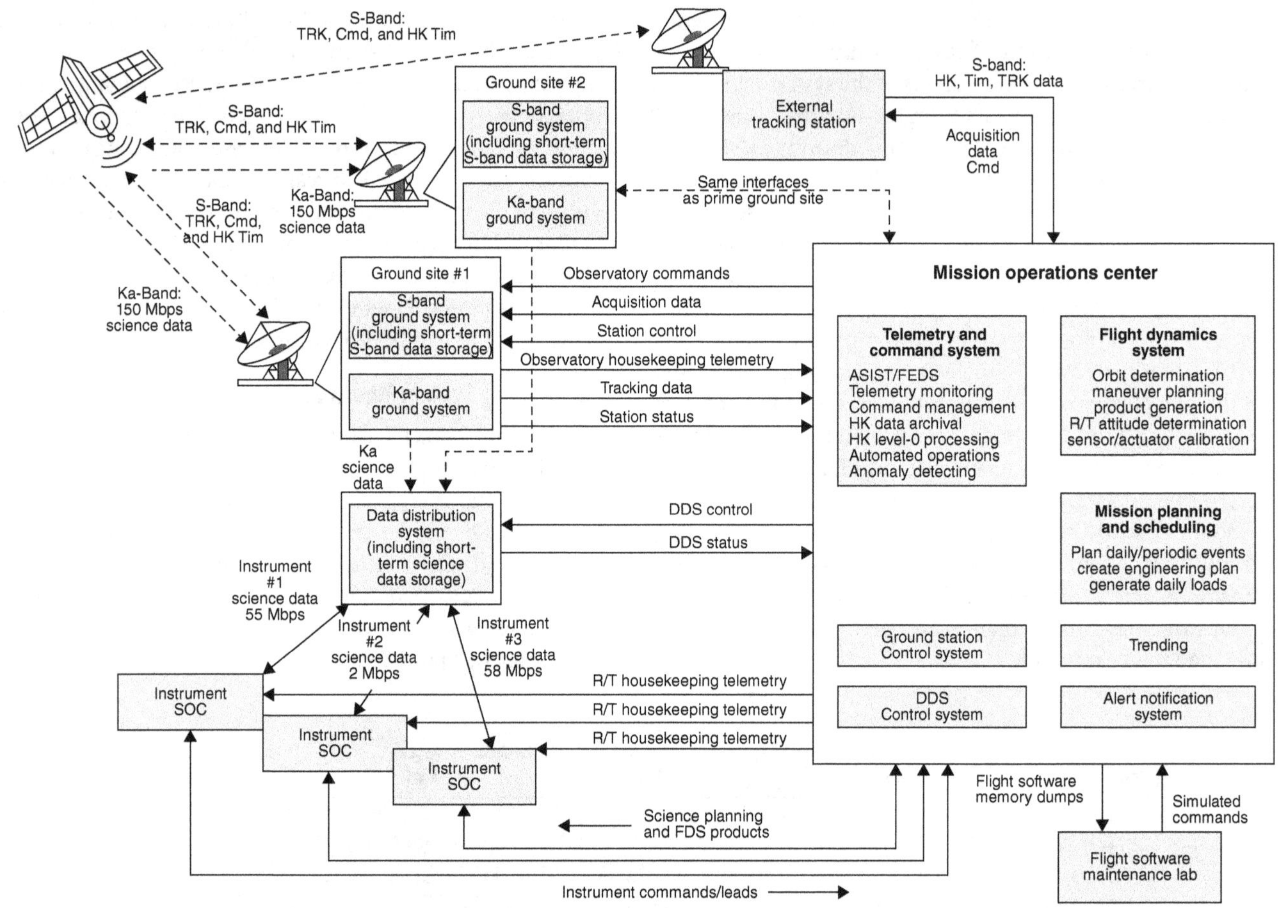

FIGURE 55.2 Example early block diagram for a NASA operational architecture. (From NASA/SP-6105, *NASA Systems Engineering Handbook*.)

	System description
General:	Midrange ballistic missile
Range:	Maximum 1200 mi, minimum 400 mi
Payload:	10,000 lbs
Propulsion:	Solid
Guidance:	Inertial; not susceptible to ECM jamming
Launch capability:	All missiles off ground within 5 min following strike order
Areas of deployment:	All climatological and geographic conditions north of 45° N. latitude
Mobility:	Temporary site, 15-min lead time to initiation of redeployment
Missile:	Roadable, transportable on trailer-erector vehicle
Nature of site:	Soft with provision of mobile living quarters for squadron personnel; surveyed bench mark required at each site location
Logistics:	Supply from closet air base, maximum separation 200 mi; helicopter requirement for personnel and components

Mission	Mission requirements
1. Strategic bombardment of military, industrial, and urban targets with high-yield weapons	*a.* System performance such as to minimize enemy capability for: (1) Detection of missile (2) Adequate early warning and dispersal (3) Interception of missile or ECM (4) Retaliation (5) Strategic support of military *b.* Satisfactory guidance accuracy *c.* Effective control of warhead burst altitude *d.* Short reaction time-strike order to launch *e.* Mobility of weapon system *f.* Low initial dollar cost per operational missile (and associated equipment) *g.* Low operations and maintenance costs *h.* Moderate manpower requirements

FIGURE 55.3 General mission—strategic bombing of targets within 1200-mi range / mission segments / system functions.

Mission requirements	Performance requirements
Missile flight performance	*a*. Engine ignition upon firing signal *b*. Acceleration and velocity per program *c*. Attitude control within tolerance limits *d*. Nose cone separation per program (following engine burnout)
Guidance accuracy	*a*. Circular error probability (.50) radius of 3 mi
Warhead burst altitude control	*a*. Detonation at preselected altitude ± 5000 ft
Reaction time	*a*. All operational missiles launched within 15 min following strike order (1) Possible requirement for automated launch subsystem (2) Continuous monitoring of missile readiness required, implying need for fairly extensive display subsystem (3) Possible requirement for highly skilled maintenance personnel *b*. Missiles continuously on alert (combat standby) *c*. Downtime for maintenance not in excess of 10%
Mobility of system	*a*. Site capable of initiating redeployment within 15 min *b*. Total time required for site setup (to launch time) not in excess of 30 min

System constraints	Description
Dollars	Limited; total available for R&D—$100 million
Schedule time	R&D complete by December 1960; production by 1964
Physical resources	Unlimited for purposes of this system
Availability of manpower	Extremely limited in re AF 5 to 7 level personnel with missile and electronic AFSCs

Environmental programs	Description
Climate and weather	Arctic, continental, and marine climates as found in Europe north of 45° N latitude. Because of alert requirements, system must be capable of all-weather 24-h operation.
Wind	System must be capable of launch in winds up to 45 mi/h.
Temperature and humidity	See "climate and weather."
Geography	Installation may be located on any terrain accessible by roads of less than 7% grades. Level area of finite dimensions required for site. Tundra may provide a problem. Trees may provide some cover with respect to aerial reconnaissance.
Atmospheric composition and contaminants	Not relevant to mobile open site and solid propellant operations.
Lighting and audition	Lighting and auditory noise anticipated as design problems in trailer interiors. Communication system requires lines from living quarters to operations trailer. Also need alternative means of communications between wings, squadrons, and sites.
Safety hazards	Accidental detonation of warhead, premature ignition of missile power plant, toppling of erect missile, falls from work platform, electric shock. Requirement for "buddy" system may increase manpower requirements.

Mission segments

Segment Start time, min	Segment	Segment demarcation
Variable by distance, time starts at site	Transport system equipment to launching site	Deployment order
	Assemble missile and prepare pad	Reach site
	Erect missile	Missile moved to pad
	Activate and check out missile guidance and control subsystems	Missile in launch attitude
	Insert mission and target data	Arm and fuse test OK
	Maintain alert status*	Warhead burst altitude set
	Arm warhead	Arm order—redeploy order
	Launch missile	Fusing check complete strike order
	Prepare for redeployment†	Missile launched Trailers ready to move

*This segment is not required if missile is to be launched immediately.
†This segment is not required if missile is launched.

FIGURE 55.3 *(Continued)*

Mission segment	System function
Transport system equipment to launching site	1. Supply automotive energy to transport: *a*. Missile *b*. Warhead and nose cone *c*. Guidance modules and spares *d*. Transporter erector *e*. Operations trailer *f*. Electrical power generator trailer *g*. Crew quarters, mess, administrative, and security trailers 2. Control above prime movers as necessary for trip from air base to site. Communications equipment and procedures required for communicating among vehicles during move. Will status of missile and warhead have to be monitored during move? If so, how will movement of vehicles affect monitoring displays? What must and can be monitored during move?
Assemble missile and prepare pad	1. Transport W/H, nose cone, and guidance modules to missile. 2. Mate components. Mating equipment required. Implications for weights to be lifted by personnel. 3. Attach components and physical links. Stress importance of identifying components for fast assembly. 4. Lay portable pad support for missile.
Erect missile	1. Control angular movement of missile from horizontal to vertical position. 2. Secure missile on pad.
Activate and check out missile guidance and control subsystems	1. Activate gyros and other components. 2. Activate test instruments. 3. Check yaw accuracy. 4. Check roll accuracy. 5. Check pitch accuracy. 6. Identify, remove, and replace malfunctioning units. 7. Communicate possible no-go situation to launch trailer. 8. Test warhead arming and fusing components.
Insert mission and target data	1. Orient missile in azimuth. 2. Insert trajectory tape. 3. Set warhead burst altitude (bombardment mission).
Maintain alert status*	1. Periodically monitor all loops for in-tolerance functioning. 2. Provide warning when malfunction exists. 3. Isolate out-of-tolerance condition to specific module. Because of requirement for fast reaction, malfunction diagnosis will have to be highly automated. 4. Remove and replace module. 5. Report existing and anticipated missile out-of-commission time.
Arm warhead upon receipt of strike order	1. Arm warhead safety precautions. 2. Recheck fusing system.
Launch missile upon receipt of strike order	1. Make final subsystem checks (probably all automatic). 2. Activate firing circuit. 3. Clear launch area. 4. Ignite engine.
Prepare for site redeployment† upon receipt of order	1. Remove missile from pad to TV. 2. Disassemble missile, nose cone, and guidance. 3. Load nose cone, guidance, and spares in helicopters. 4. Disassemble and stow launch pad. 5. Ready all trailers for movement.

*This function is not required if missile is to be launched immediately.
†This function is not required if missile is launched.

Figure 55.3 *(Continued)*

might consider in defining the mission of the proposed system, while concurrently beginning to assess performance expectations, system and environmental constraints, mission segment sequencing, and basic insights regarding probable functions.[1]

Example: Semipictorial Operating-Flow Diagramming Technique

Formats for ConOps may take many forms. In Fig. 55.4, the analyst attempts to graphically illustrate the flow of transactions within a banking system. Figure 55.4 uses pictorial techniques to illustrate a current banking system in order to provide a baseline for comparison with a proposed new system (not shown). The analyst divides operating sequences into functional cycles and proceeds to depict the flow of information and activity functions commencing with the initial cue to begin daily operations within a branch bank facility. In this analysis, it should be noted that proposed new equipment elements have already been established. However, the purpose of the analysis was to clarify where, when, and how these would function within the desired flow of information, decision, and action protocol of a current banking institution.

FUNCTION ALLOCATION

Function allocation is the process of assigning performance of individual system tasks or functions across the elements of the total system—i.e., hardware, software, and human. For example, regular monitoring of potable water quality on a long-duration human spaceflight mission might be performed automatically by hardware/software systems or it might be performed manually by human crews using a set of hardware/software tools designed entirely differently. Either option has its own set of opportunities, constraints, development costs, operational impacts/costs, risks, etc. Extensive design trade studies must be performed during the design phases of development to weigh and resolve many such function allocation issues for the total system.

For the human factors (HF) engineer the most important consideration should be to ensure that functions allocated to the human elements of a system are realistic—i.e., within human capabilities and limitations and within planned human resourcing targets for the system's operations and maintenance. Technical designers may be prone to make unrealistic assumptions about what human operators and maintainers can accomplish, how they adapt to new systems, and the ease of operability/maintainability that their system affords. The HF engineer must be quick to validate and verify assumptions made about the human element. Numerous seemingly small decisions allocating system functionality to the assumed capabilities of the human elements may, when aggregated, pose huge operational issues for system success and/or for growth in numbers or skills of people required to operate the system.

Functional Block Diagramming

Before allocating functionality, one must first understand the array of required system functions. Block diagramming is a commonly used method of indicating basic system organization and function. Often, however, what purports to be functional block diagramming is really equipment block diagramming, as is readily evident, for example, in the appearance of blocks labeled "display console," "data entry panel," "tape recorder," "drum storage," etc. Functional blocks should be concerned with what is *done* rather than with the specific realization of the means to do it. Functional block diagrams should not be allowed to evolve into equipment block diagrams prematurely; i.e., the decision concerning which functions should be performed by a piece of equipment rather than a human operator should not be made until the full scope of the functional layout has been clearly defined.

For example, "detection" is a functional term, whereas "detector's console" is not and already assumes an allocation of a function to human and machine. Since allocation of functions should always follow development of the initial system concept, it is essential to avoid equipment representation and its implication that the function has already been allocated, to either a human *or* a machine. A premature human–machine allocation may overlook the possibility that a human can perform a given task with greater cost-effectiveness than a machine. By the same token, to assign a human a function that is really beyond his or her reliable limits just to save the cost of another machine is equally foolish. However, as the various trade-offs are considered during design maturation, the original block diagrams may be refined for each of the alternatives under consideration.

The example in Fig. 55.5 of a functional block diagram shows elements of a hierarchical block structure. Some of the essential features to note are as follows:

1. An expanding series of levels gives successively more detailed information on each functional block. This detail may be carried to as many levels as are required to drive out some end item (operator, machine, component, etc.).

2. Functions are numbered from level-to-level in a hierarchical manner that preserves the continuity of function and logical breakout from function origin.

3. The top-level diagram should show the operational functioning of the system to be developed—i.e., the layout of the top-level diagram indicates the sequential relationship of all functions that must be accomplished by the system.

4. Branching can and should be shown as indicated in the top-level diagram—i.e., nominal operations versus "Contingency Operations." Once the particular system in

[1] D. Meister et al. *The Impact of Manpower Requirements and Personnel Resources Data on System Design*, AMRL-TR-68-44, Air Force Systems Command, Wright-Patterson Air Force Base, Ohio.

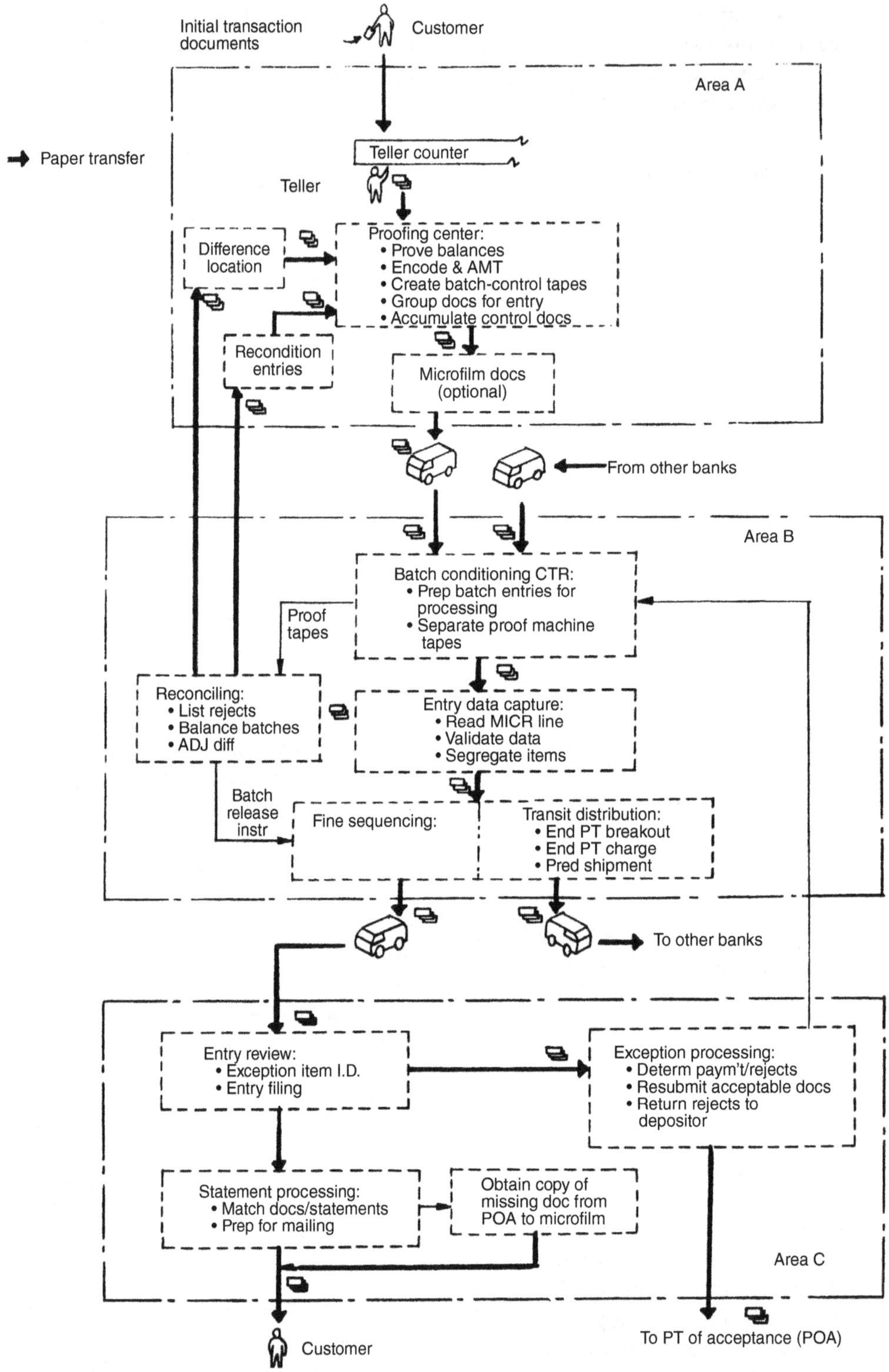

FIGURE 55.4 Original banking operation.

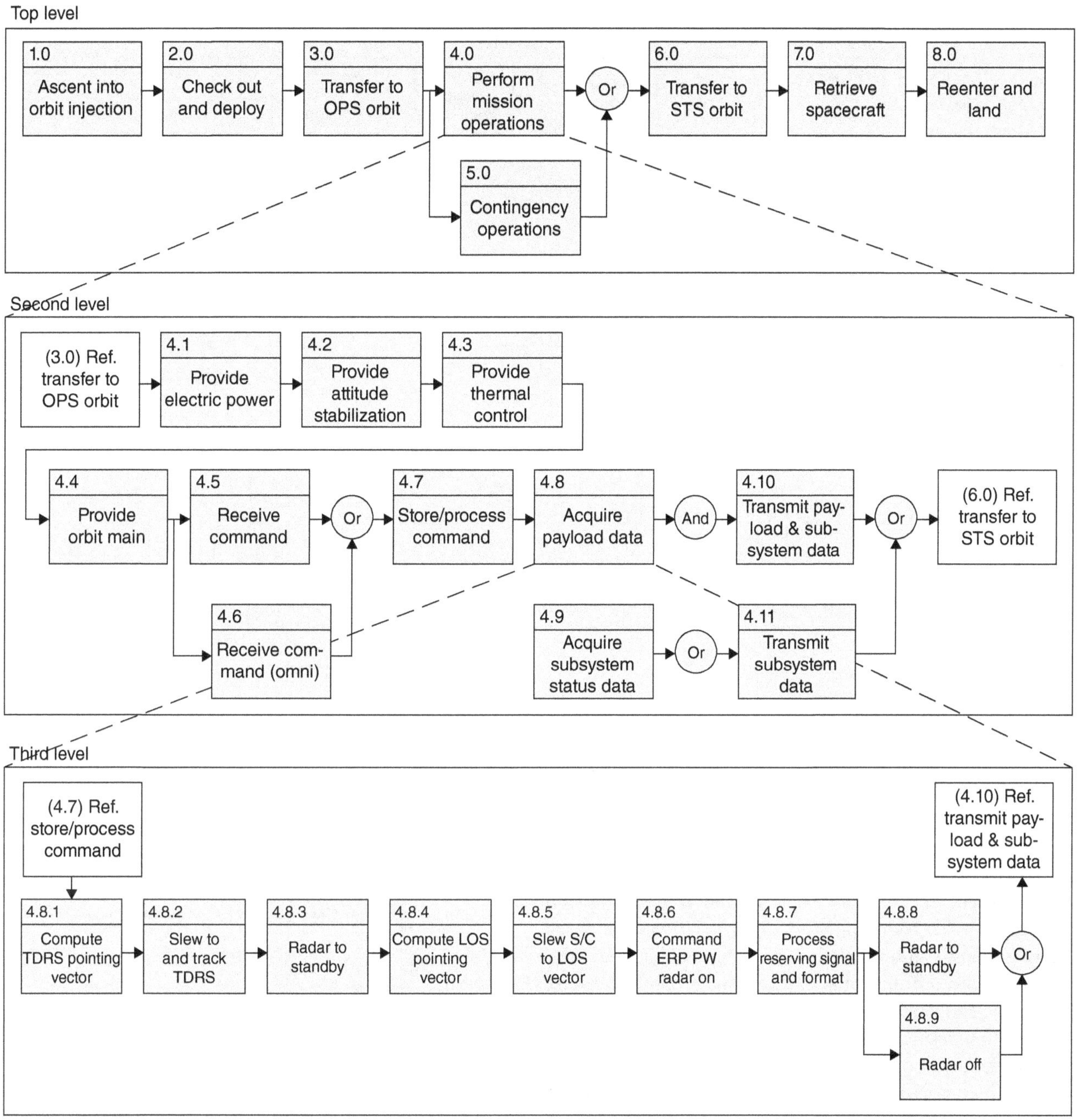

Figure 55.5 Example of a functional flow block diagram. (From NASA/SP-6105, *NASA Systems Engineering Handbook*.)

the example becomes operational, it is either in service or in an impaired/maintenance state, both of which require additional breakout with finer-level diagrams. The introduction of branching in functional block diagramming provides great flexibility and allows one to address multiple modalities of system operation.

5. The diagram should be organized so that one can easily find the input and follow through the function blocks to the resulting output. This should be evident at each level. In fact, if one finds that at a particular level it is impossible to follow from left to right, from input to output, an error in functional designation has been made (a useful method of cross-checking).

Note in the illustrated example that there is not yet any implication of human–machine function allocation.

Figure 55.6 illustrates a second example of functional block diagram for the common experience of residential planning, design, and living.

Operations Information Flow

Some of the most egregious mistakes in system design—mistakes that impact final system efficiency, effectiveness,

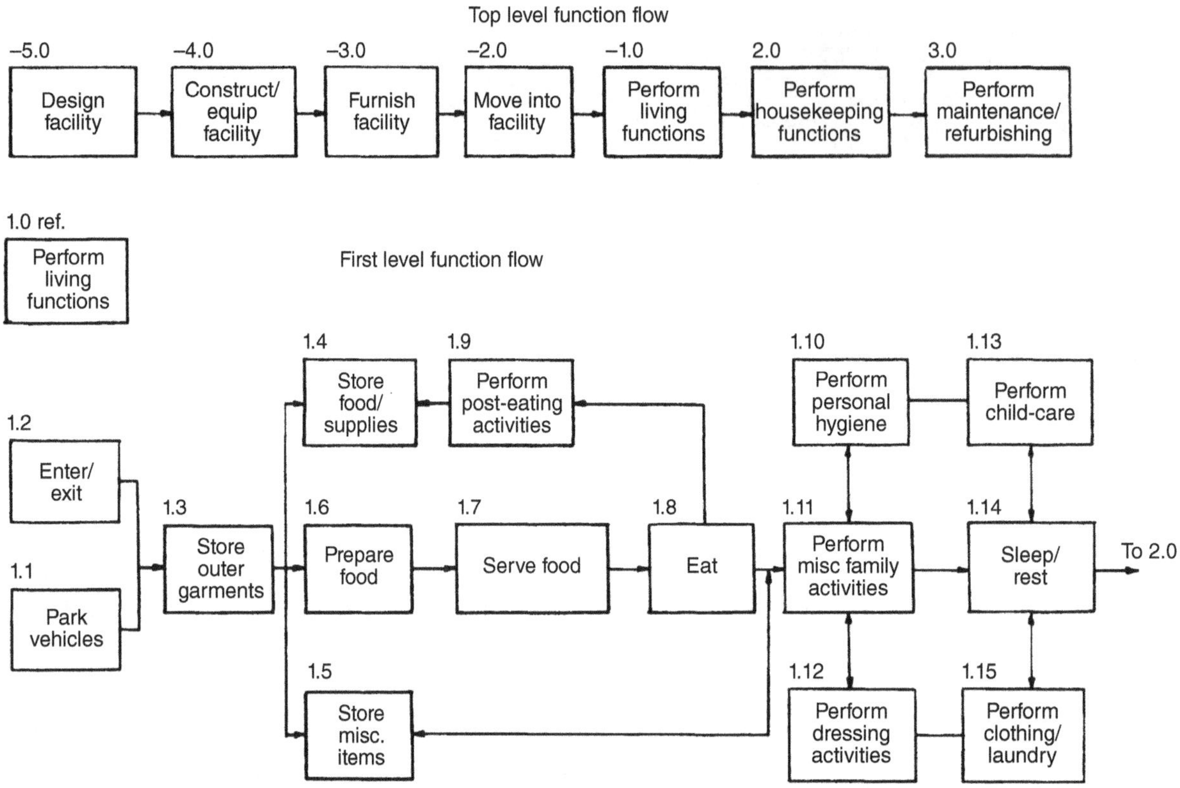

FIGURE 55.6 Example: functional flow diagramming applied to residential planning.

performance, and life cycle cost—occur when designers and developers make invalid assumptions about the operations of a system—i.e., the history of prior design solutions, the environmental context, the user community, operations lessons learned, etc. Operations information flow may be employed as a technique to consciously counter poor assumptions about operations. The technique illustrates the "reverse flow" of information on operations that must inform design decisions about what is to be accomplished by each function identified in a system function block diagram. Like block diagramming, information flow charting may be used at various levels of detail. The initial information flow charts should be concerned with gross functions without regard to whether functions are performed by humans or by machines. (See Fig. 55.7.)

In that it records the sequence of operations and decisions that must be performed in order to satisfy a definite system function, the information flow chart may be similar to the types of logical flow charts used by computer programmers. Both charts are based on binary choice decisions and intervening operations. That most decisions can be reduced to a binary situation is evidenced by the vast array of problems that can be computerized using simple binary logic. There are three important reasons for using binary decision logic in all information flow charting:

1. It expedites communication through the use of simple yet universally applicable conventions.

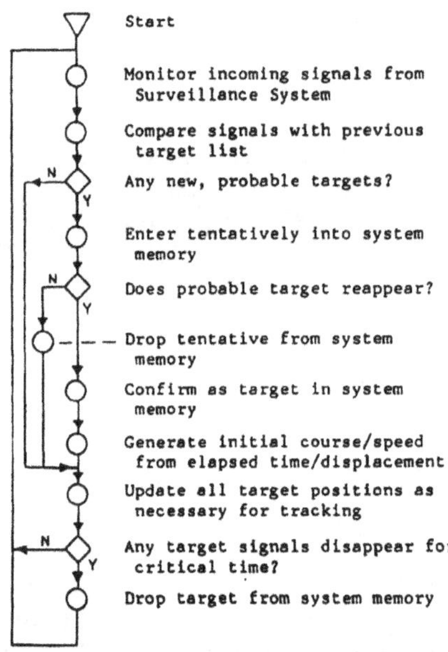

FIGURE 55.7 Example: Gross-level flow chart for detection and tracking.

2. It provides for easy translation of information flow charts into logic flow charts for computerized sections of the system.

3. The binary format forces the analyst to focus on the decision details and the information required by operators to make the decision.

Like block diagramming, information flow charting can be used at various levels of specificity. A decision at a general level may split into several decisions at a more detailed level. For example:

General Level

- Any targets need identification processing?

Specific Level

- Any newly entered targets need ID processing?
- Any target tracks need confirmation of tentative ID?
- Any confirmed IDs need rechecking?

Each of these more detailed decisions may have associated with it one or more detailed operations. Similarly, an *operation* at a general level may break down into more detailed decisions and operations.

It is not necessary that the flow chart be prepared to a uniform level of specificity. For many situations it may be entirely appropriate to treat certain parts of a process in only general terms and focus in on other, more critical aspects by going into greater detail. The analyst must keep his or her purpose focused on informing system functionality and perform the level of detail accordingly. Often parts of a flow chart presented in general terms are broken into greater detail only as system development progresses. The example above illustrates gross-level detection and tracking functions. Note that at this level the chart is applicable to virtually any detection and tracking system—i.e., the listed decisions and operations are common to all such systems. However, the utility of the flow chart may become apparent as it helps one think of implementation alternatives, such as:

1. By what means can any given signal set be compared with known targets in the system?
2. How can probable targets be marked so that their appearance can be recognized readily?

Note that in the illustrated example the format for the information flow utilizes a narrow column on the left-hand side consisting only of symbols and connecting lines, and a right-hand column for accompanying text statements. Note also that flow paths are always complete—every path either reiterates or eventually terminates in a valid exit with no ends left dangling. Logical operations information flow charting imposes a discipline upon analysts, requiring them to consider alternatives that might otherwise be overlooked.

Guidelines for Data-Flow Diagramming Analysis[2]

1. Show components in the same relative position as in equipment to the extent that this is practical.
2. Include only information that is needed to perform specified operational control, service, and/or maintenance.
3. Present only enough information to identify components; they are neither pictured nor schematized.
4. Emphasize chiefly the nature of the signal flow between components, particularly of points where power or energy checks are made.
5. Be concerned only with the information- and energy-flow characteristics of the interface actions, not with the electromechanical characteristics of components.
6. Strive for similarity of appearance (e.g., graphics), whether data flow among major components, assemblies, or subassemblies is being depicted.
7. Base the analysis on a particular configuration of control settings for the prime equipment and/or test equipment or on induced special conditions such as broken feedback loops or servos set to a particular position (these control settings or special conditions are indicated on the diagram). Note that all loops should close on each type or level of diagram.
8. Show electrical characteristics of the signal to be monitored or checked at each operating or maintenance test point; complex time and shape characteristics of the signal are keyed to tables at the edge of the diagram or to separate, referenced pages.
9. Omit information about the nature of the signal flow within units which are replaceable or can be thrown away, at the level of maintenance for which the diagram was prepared.
10. Every functional system, subsystem, piece of equipment, and component should be interrelating numbering systems so that each end item is traceable to its original functional requirement.

Relating the Functional Block Diagram and Operations Information. Breaking down the system life cycle into program management and/or systems engineering phases is itself a form of functional block diagramming and can be used to relate the flow of data that informs a design. At its top level, the life cycle process moves in sequence from concept development through design, production, delivery, and into operations. There are particular functions that must occur in each step of the process and these can be broken out into more detailed levels of process flow (in the manner they have been elaborated upon in numerous texts on systems engineering process). Such a sequential model of function can also be used to show how a "reverse flow" of operational data must exist to inform the functions, processes, and products of the design phases of the life cycle. (See Fig. 55.8.)

[2] Based on original material from *Data Flow: The General Problem and a Cognitive Model.* MRL-TDR-62-42, Behavioral Sciences Laboratory. 6570th Aerospace Medical Research Laboratories. Air Force Systems Command. Wright-Patterson Air Force Base, Ohio, 1962.

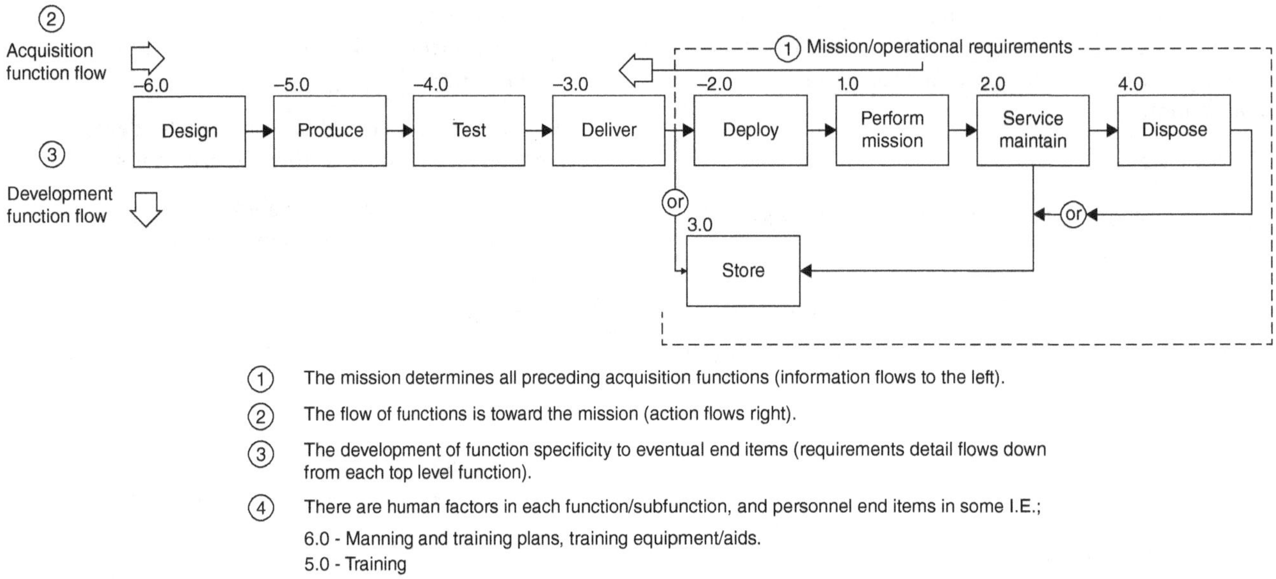

FIGURE 55.8 General model for system life cycle.

This is in essence what the U.S. Department of Defense (DoD) has intended to emphasize in making human systems integration (HSI) a mandatory part of all acquisition programs and projects—i.e., in order to produce designs and products that are cost-effective to operate after they are delivered, the system designers must work closely with the experienced operations community to understand the opportunities and challenges that have worked well (and not worked well) in the operations, logistics, deployment, and maintenance of legacy systems.

As much as creativity should be encouraged and promoted during design, most systems build on prior systems, technologies, design solutions, and operational needs. Every design team—including the membership responsible for HFE—should, during conceptual design, investigate the following as part of good design practice:

- Fundamental analysis of projected operations. Engage existing system operators and maintainers to obtain information on their manner of performing their jobs and contingencies that might arise.

- Similar systems analysis. Examine previous systems and/or systems currently in use. Obtain lessons learned and best practices. HFE should obtain data on the operability, maintainability, manpower, personnel, and training required to staff legacy systems and relate the information to projected systems' stakeholder expectations.

- Environmental factors of the operational arena. What environmental traits affect personnel? What environmental protections are required to mitigate hazards and how do they impact requirements for safety, facilities, etc.?

- Workload analysis. Are current system operator and maintainer workloads acceptable or are they stressful? Are times and numbers of people allotted for operational tasks in line with stakeholder projections for the new system? What specialized training has been required for all operators and maintainers?

- Critical incident investigations. Have legacy systems had major failures or major failure modes that made them susceptible to off-nominal performance or excessive down-time? Obtain investigation reports from prior incident investigations and obtain metrics on the numbers of incidents, resulting injuries, system down-time, design and operations workarounds, etc.

- System failure mode and fault analysis. What shortcomings do legacy systems have that might conceivably be designed out of the system through technology advances or alternate human/system function allocation?

Human–Machine Function Allocation Analysis

Theoretically, decisions regarding how to accomplish each function established by the functional block diagramming and operations information charting should not be made until one is sure that all the necessary functions have been specified and agreed upon. The goal should be to prevent some preconceived idea of how to accomplish a function without serious consideration of alternate possibilities, particularly whether the function is more appropriate to human or to machine implementation. There are, of course, exceptions, i.e., when past experience has clearly demonstrated that people cannot do certain things—e.g., in the time allotted or with the strength required—and therefore, some hardware/software device *must be used.* The reality of program/

project schedule and cost constraints typically drives the need for early concept development answers to some key human/system function allocation concerns long before design concepts have fully matured. Nonetheless, the program's/project's function allocation decisions should be clearly documented—including rationale and any trade or legacy system study information available—and should be iteratively revisited during the maturation of the design to determine if early concept assumptions are proving reliable, valid, and accurate.

The human–machine function allocation step is extremely important because it allows knowledgeable individuals to contemplate the pros and cons of using a human rather than a machine to accomplish each of the proposed functions. Additional contextual information must also be considered such as risks to human welfare and overall project cost-effectiveness.

HF specialists are essential to function allocation analysis since they are generally more knowledgeable about what humans can or cannot do well, risks that may be too great to expose the human to, severe problems with regard to selecting the right individuals to perform the function, problems associated with training the individual to do an adequate job, and so on. The following guidelines and analytic examples are provided to aid the reader in addressing the function allocation analysis task.

General Considerations for Human versus Machine Function Allocation

- Environmental Constraints: The human's physiological tolerance to certain operating environments is limited; therefore, one has to make an early decision whether to incur the necessary costs and complexity to protect and support the human under severe environmental conditions (extreme atmospheric pressure, acceleration, temperature, noise, vibration, or radiation and/or potential emergency situations produced by explosive blasts, fire, atmospheric or chemical contamination, etc.). Also, the impact of human protection equipment on the human operator's/maintainer's ability to perceive and react must be considered—e.g., does a deep sea diving suit impair the diver's field of vision and range of motion?

- Sensory Isolation: To perform useful tasks within a control environment, humans must be able to receive sensory inputs (information) at levels commensurate with their inherent sensory channel threshold capabilities; e.g., humans can see only so far and hear a signal that is only so soft, and these direct perceptions are easily degraded by various interfering environments. In fact, sensory inputs may be distorted, causing humans to make perceptual errors, i.e., to misinterpret what they see, hear, or feel.

- Speed and Accuracy: Human response cannot compete with the capacity of a machine in terms of speed and accuracy; thus, functional allocations to humans must be made on the basis of their capacity and the required speed at which a decision or action must be taken.

- Overload: Humans are fairly limited compared with machines in terms of how much information they can absorb and handle at one time, how many things they can monitor or control at one time, and how effectively they can maintain cognizance of a situation for extended periods of time or under severe physiological and psychological stress conditions.

- Physical Strength: Humans are extremely limited compared with machines in terms of how much force they can apply, and for how long.

- Storage Capacity: Humans' capacity to store large amounts of information over the long term is extremely great, but their ability to retrieve information quickly is sometimes extremely limited and unreliable; a machine, however, can store almost any amount of data and recall it almost immediately. On the other hand, the machine's capacity to store and retrieve is entirely limited to what is designed into it.

- Human–Machine Performance Surveillance: Humans (compared with machines) are relatively poor "self-monitors" and are easily influenced by emotional factors and by environmental and operational distortions.

- Interpretation of, and Response to, Unexpected Events: Humans possess the unique capacity to constantly reevaluate a situation, change their approach, and invent new ideas on the basis of unexpected events and operating conditions. They often can continue with an alternative or less-than-perfect procedure, whereas a machine may quit completely. As noted earlier, a machine does only what it was designed to do; i.e., its capability is limited by the designer's capacity to anticipate all events and conditions of operation.

- Fatigue: Humans' capacity and functional capabilities are subject to short- and long-term fatigue effects, whereas machines *can* be designed to be almost fatigue-resistant.

- Learning: Humans generally require some finite learning period to perform a new function. A machine begins its operation immediately and theoretically requires neither initial training nor proficiency refreshment.

- Cost: As long as humans are used properly (within their basic physiological and psychological limits), they are often perceived to be the least expensive component of a system. System designers must be careful not to try to duplicate human capabilities by machines since such duplication may end up costing much more. A thorough cost comparison analysis—particularly one that looks at total life cycle cost rather than development cost along—is the only sound method for deciding whether to use the human or the machine for given functions. In the case of military systems, the customer must provide complete support for the human. Military services must account for the costs of supporting personnel (housing, pay, medical expenses, etc.). On the other hand, domestic systems and products typically do not require such complete support for the human.

Checklist for Making Decisions Relative to Human–Machine Coupling. In order for humans to complement the capability of a system, they must be coupled with machines in a way that will allow them to utilize their capabilities to the maximum. Thus the following should be considered.

1. Select the sensorimotor link which makes the best use of human capacity, sensitivity, and reliability. Avoid coupling via a particular link merely on the basis of tradition or because it may appear that a particular hardware implementation is less expensive, easier to design, or already available.

2. Choose a coupling approach that maximizes total system effectiveness; do not choose an approach on the basis of whether it is easy or hard to automate a function.

3. Couple humans with machines in such a way that they are not compelled to work at peak limits all or most of the time.

4. Couple humans with machines in such a way that they can recognize or feel that their contribution is meaningful and important. Avoid giving humans machine-serving responsibilities.

5. Couple humans with machines in such a way that information flow and information processing are natural; this minimizes learning time and the probability of confusion or errors.

6. Select coupling methods that do not require extremely precise manipulations; continuous, repetitive movements; frequent, laborious, and lengthy calculations where accuracy is critical; or physical contributions that demand reaching one's upper strength limits.

7. Couple humans with machines as though they might at some time have to assume control (even though the nominal mode may be automatic).

8. Use hardware to aid the human; do not use the human to complement a predetermined hardware concept.

General Physiological Considerations in Human–Machine Allocation Analysis[3]. Actual physiological hazards to operator health obviously must be considered while determining function allocation to human or machine. (See Figs. 55.9 and 55.10.) However, one must also examine the other boundaries in terms of how (although they may not lead directly to personal injury) conditions might stress operators sufficiently to induce them to make errors and in terms of how these errors might in turn lead to potential system loss and/or eventual injury to operators and others.

It is suggested that the designer prepare and have at hand a list of potential injury modes (cuts, bruises, fractures, amputations, burns, internal ruptures, eye penetration, asphyxiation, etc.). This should be used as a checklist to evaluate the proposed human-machine allocation decisions. (See Fig. 55.11.)

REQUIREMENTS DEFINITION

Requirements are descriptive statements of the needs that a system must fulfill. They are often the most definitive tool for defining the performance and functionality of the system during design. Requirements are statements that may identify a system attribute characteristic, capability, interface, and/or other quality that must be met for the system to meet all stakeholder expectations. Good requirements are written in a clear, concise style that avoids nuance and misinterpretation sufficient that they may become legally binding when written into a system specification and/or legally binding acquisition contract. Requirements differ from ConOps and function allocation in that they may be seen as an end product themselves, not just a methodology to identify an end. They serve as the basis for final verification at the completion of product design, though they should be iteratively reviewed during design and the in-process design products continually tested for compliance at every project phase and milestone to ensure that the project is building the right product for system success and to avoid waiting to find any deviations from full compliance at project close-out.

All of the work that fed the ConOps and function allocation efforts also supports requirements development. System requirements definition begins with some of the same investigations—i.e., understanding and cataloging the forces that drive and constrain the system design. Capturing stakeholder expectations is an essential foundation, as well as documenting mission goals and environmental, political, safety, and budget constraints. From that starting point, the requirements definition process transforms these drivers and constraints into a salient problem definition expressed as a collection of "shall" statements taking the general form, "The system shall…." For more information on what constitutes a good requirement, see the excerpt in Fig. 55.12.

Work on requirements begins as soon as system definition begins. Requirements should be complete before entering into the design phases of a project and there should be agreement among all stakeholders that the requirements are indeed complete. A systems requirements review should be a project major milestone review. Note, however, that requirements are used throughout design and development

[3]R. A. McFarland, *Human Factors in Air Transportation Occupational Health and Safety,* McGraw-Hill Book Company, New York, 1952, p. 705.

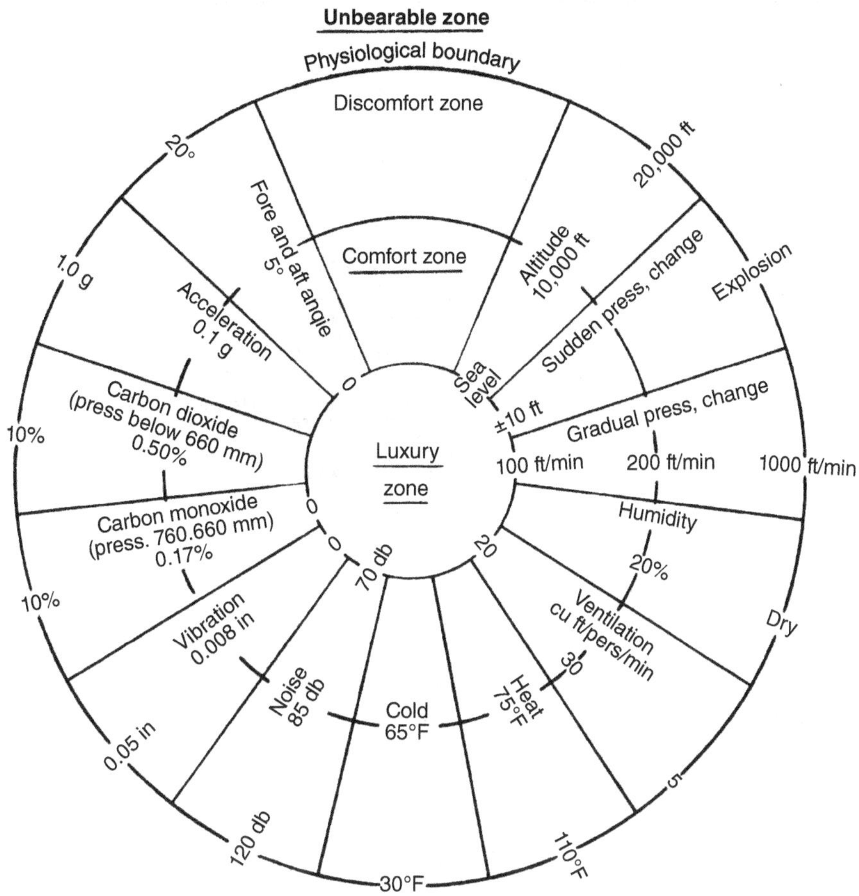

FIGURE 55.9 Physiological considerations in human–machine allocations.

Hypothetical tracking functions	Inherent operator capabilities: Detecting signals in the presence of high-noise environment (X5)	Recognizing objects under varying conditions of perception (X4)	Handling unexpected occurrences or low-probability events (x4)	Reasoning inductively (X1)	Profiting from experience (X2)	Inherent equipment capabilities: Responding quickly to signals (X3)	Performing precise routine repetitive operations (X2)	Computing and handling large amounts of stored information quickly and accurately (x4)	Total Score: Operator	Machine	Proposed allocation: Operator	Both	Machine: Equipment	Software
1. Determine if target tracks in system	25	8	12	3	6	9	8	4	81	41	X			
2. Actuate sequence	5	4	4	1	2	3	2	4	20	24		X		
3. Put next target in track list under close control	5	4	4	1	2	9	10	4	21	43			X	X
4. Advance hook on display to track coordinates	5	4	4	1	4	9	10	4	21	43			X	X
5. Determine if target video present	20	8	8	3	4	9	8	4	70	39	X			
6. Determine if hook lines up with present target position, etc.	20	8	8	3	6		8	4	73	40	X			

FIGURE 55.10 Example: function allocation analysis and screening worksheet.

Function number	Function title	Function requirements	Environmental considerations	Operational considerations	Potential human role	Desirable equipment role	Human function description	Equipment function description
1.0	Prepare system for deployment.	Verify system operational readiness, and prepare system for shipment.	Temperature, humidity. ventilation. illumination. noise.	Time urgency	Inspect equipment visually. Use test equipment. Test operating condition of all system components. Remove and replace components in shipping on vehicles.	Measure subsystem component functioning Provide lift and load of system components onto transport vehicles.		
1.3	Verify system operational readiness.	Verify weapon. Test set and trainers' operation readiness.	(Same as above)	(Same as above)	Visual inspection and test of weapon using test set visual inspection and test of test set and trainers using portable test equipment: Replacement of lost accessories such as covers, seals, etc.	Measure subsystem/ component functioning.		
1.3.1	Verify weapon operational readiness.	Remove weapon from container. Prepare weapon for testing. Prepare test set. Perform Post test. Shut down and Remove weapon. Return weapon to container.	(Same as 1.0)	(Same as 1.0)	Remove weapon from container. Prepare weapon for test. Prepare test set. Perform test. Perform post-test shutdown and remove weapon. Return weapon to container.	Test set = provide go-no/go indications of weapon functional readiness.		
1.3.1.4	Test impulse generator.	Connect weapon impulse-generator circuit to test set impulse generator circuit. Actuate weapon impulse-generator circuit. Measure by weapon impulse generator.	(Same as 1.0)	(Same as 1.0)	Operate a switch to connect weapon/test set impulse circuits. Operate weapon warm-up lever. Read weapon impulse generator output on test set indicator.	Switch to connect weapon/test. Set impulse circuits. (See 4.2.1.2) Visual indication of impulse generator output.	Actuate impulse generator circuits switch interface device on test set. Actuate weapon warm-up lever. Observe impulse-generator indicator on test set.	Provide switch knob. (See 4.2.1.2.1) Provide impulse generator generator indicator on test set.

FIGURE 55.11 Example: alternative man–machine function allocation analysis.

should be continually revisited during design development, and may—with formal consensus from project management and the design team—be modified as new findings from design and development are revealed.

Types of Requirements

Requirements for a project should comprise a complete set that fully defines the system. In drafting requirements, the stakeholder expectations, system ConOps and functional allocation are helpful tools for considering the desired end-to-end functionality and performance of the system. A complete set of requirements should include the following types:

- Functional: Technical requirements that define what functions need to be performed to accomplish the system objectives. Functional requirements need to be specified for all intended uses of the product over its entire lifetime. In structuring the entire set of project requirements, similar functions should be grouped.

- Performance: Technical requirements that quantitatively define how well the system needs to perform the functions.

The information in Fig. 55.13 illustrates differences between functional and performance requirements.

Additional Requirement Types

- Interface: Technical requirements that define all interfaces for the system—e.g., field of operations, existing command and control centers, existing test equipment, transportation systems, and other mechanical, electrical, data, etc., interfaces. All product life cycle phases should be considered. Three types of system interfaces of particular importance to HFE are the following:

  1. Environment. Every system operates within an environment which the system must accommodate. The environmental envelope(s) within which the system will be expected to be transported, deployed, and to operate during all life cycle phases must be defined. Example environments include thermal, acoustic, vibration, shock, acceleration, static loads, radiation, atmospheric qualities, etc.

Use of Correct Terms

- Shall = requirement
- Will = facts or declaration of purpose
- Should = goal

Editorial Checklist

Personnel Requirement

1. The requirement is in the form "responsible party shall perform such and such." In other words, use the active, rather than the passive voice. A requirement must state who shall (do, perform, provide, weigh, or other verb) followed by a description of what must be performed.

Product Requirement

1. The requirement is in the form "product ABC shall XYZ." A requirement must state "The product shall" (do, perform, provide, weigh, or other verb) followed by a description of what must be done.
2. The requirement uses consistent terminology to refer to the product and its lower level entities.
3. Complete with tolerances for qualitative/performance values (e.g., less than, greater than or equal to, plus or minus, 3 sigma root sum squares).
4. Is the requirement free of implementation? (Requirements should state WHAT is needed, NOT HOW to provide it; i.e., state the problem not the solution. Ask, "Why do you need the requirement?" The answer may point to the real requirement.)
5. Free of descriptions of operations? (Is this a need the product must satisfy or an activity involving the product? Sentences like "The operator shall..." are almost always operational statements not requirements.)

Example Product Requirements

- The system shall operate at a power level of...
- The software shall acquire data from the...
- The structure shall withstand loads of...
- The hardware shall have a mass of...

General Goodness Checklist

1. The requirement is grammatically correct.
2. The requirement is free of typos, misspellings, and punctuation errors.
3. The requirement complies with the project's template and style rules.
4. The requirement is stated positively (as opposed to negatively, i.e., "shall not").
5. The use of "To Be Determined" (TBD) values should be minimized. It is better to use a best estimate for a value and mark it "To Be Resolved" (TBR) with the rationale along with what must be done to eliminate the TBR, who is responsible for its elimination,and by when it must be eliminated.
6. The requirement is accompanied by an intelligible rationale, including any assumptions. Can you validate (concur with) the assumptions? Assumptions must be confirmed before baselining.
7. The requirement is located in the proper section of the document (e.g., not in an appendix).

Requirements Validation Checklist

Clarity

1. Are the requirements clear and unambiguous? (Are all aspects of the requirement understandable and not subject to misinterpretation? Is the requirement free from indefinite pronouns (this, these) and ambiguous terms (e.g., "as appropriate," "etc.," "and/or," "but not limited to")?)
2. Are the requirements concise and simple?
3. Do the requirements express only one thought per requirement statement, a standalone statement as opposed to multiple requirements in a single statement, or a paragraph that contains both requirements and rationale?
4. Does the requirement statement have one subject and one predicate?

FIGURE 55.12 How to write a good requirement, perhaps with a border to distinguish it. The source document can be found here: http://ntrs.nasa.gov/archive/nasa/casi.ntrs.nasa.gov/20080008301.pdf. (From NASA/SP-2007-6105, *NASA Systems Engineering Handbook*, Appendix C: How to Write a Good Requirement.)

Completeness

1. Are requirements stated as completely as possible? Have all incomplete requirements been captured as TBDs or TBRs and a complete listing of them maintained with the requirements?
2. Are any requirements missing? For example have any of the following requirements areas been overlooked: functional, performance, interface, environment (development, manufacturing, test, transport, storage, operations), facility (manufacturing, test, storage, operations), transportation (among areas for manufacturing, assembling, delivery points, within storage facilities, loading), training, personnel, operability, safety, security, appearance and physical characteristics, and design.
3. Have all assumptions been explicitly stated?

Compliance

1. Are all requirements at the correct level (e.g., system, segment, element, subsystem)?
2. Are requirements free of implementation specifics? (Requirements should state what is needed,not how to provide it.)
3. Are requirements free of descriptions of operations? (Don't mix operation with requirements: update the ConOps instead.)

Consistency

1. Are the requirements stated consistently without contradicting themselves or the requirements of related systems?
2. Is the terminology consistent with the user and sponsor's terminology? With the project glossary?
3. Is the terminology consistently used throughout the document?
4. Are the key terms included in the project's glossary?

Traceability

1. Are all requirements needed? Is each requirement necessary to meet the parent requirement? Is each requirement a needed function or characteristic? Distinguish between needs and wants. If it is not necessary, it is not a requirement. Ask, "What is the worst that could happen if the requirement was not included?"
2. Are all requirements (functions, structures, and constraints) bi-directionally traceable to higher level requirements or mission or system-of-interest scope (i.e., need(s), goals, objectives, constraints, or concept of operations)?
3. Is each requirement stated in such a manner that it can be uniquely referenced (e.g., each requirement is uniquely numbered) in subordinate documents?

Correctness

1. Is each requirement correct?
2. Is each stated assumption correct? Assumptions must be confirmed before the document can be baselined.
3. Are the requirements technically feasible?

Functionality

1. Are all described functions necessary and together sufficient to meet mission and system goals and objectives?

Performance

1. Are all required performance specifications and margins listed (e.g., consider timing, throughput, storage size, latency, accuracy and precision)?
2. Is each performance requirement realistic?
3. Are the tolerances overly tight? Are the tolerances defendable and cost-effective? Ask, "What is the worst thing that could happen if the tolerance was doubled or tripled?"

Interfaces

1. Are all external interfaces clearly defined?
2. Are all internal interfaces clearly defined?
3. Are all interfaces necessary, sufficient, and consistent with each other?

Maintainability

1. Have the requirements for system maintainability been specified in a measurable, verifiable manner?
2. Are requirements written so that ripple effects from changes are minimized (i.e., requirements are as weakly coupled as possible)?

FIGURE 55.12 *(Continued)*

Reliability

1. Are clearly defined, measurable, and verifiable reliability requirements specified?
2. Are there error detection, reporting, handling, and recovery requirements?
3. Are undesired events (e.g., single event upset, data loss or scrambling, operator error) considered and their required responses specified?
4. Have assumptions about the intended sequence of functions been stated? Are these sequences required?
5. Do these requirements adequately address the survivability after a software or hardware fault of the system from the point of view of hardware, software, operations, personnel and procedures?

Verifiability/Testability

1. Can the system be tested, demonstrated, inspected, or analyzed to show that it satisfies requirements? Can this be done at the level of the system at which the requirement is stated? Does a means exist to measure the accomplishment of the requirement and verify compliance? Can the criteria for verification be stated?
2. Are the requirements stated precisely to facilitate specification of system test success criteria and requirements?
3. Are the requirements free of unverifiable terms (e.g., flexible, easy, sufficient, safe, ad hoc, adequate, accommodate, user-friendly, usable, when required, if required, appropriate, fast, portable, light-weight, small, large, maximize, minimize, sufficient, robust, quickly, easily, clearly, other "ly" words, other "ize" words)?

Data Usage

1. Where applicable, are "don't care" conditions truly "don't care"? ("Don't care" values identify cases when the value of a condition or flag is irrelevant, even though the value may be important for other cases.) Are "don't care" conditions values explicitly stated? (Correct identification of "don't care" values may improve a design's portability.)

Figure 55.12 *(Continued)*

Example of functional and performance requirements

Initial function statement
The thrust vector controller (TVC) shall provide vehicle control about the pitch and yaw axes.

This statement describes a high-level function that the TVC must perform. The technical team needs to transform this statement into a set of design-to-functional and performance requirements.

Functional requirements with associated performance requirements

- The TVC shall gimbal the engine a maximum of 9 degrees, ±0.1 degree.
- The TVC shall gimbal the engine at a maximum rate of 5 degrees/second ±0.3 degrees/second.
- The TVC shall provide a force of 40,000 pounds, ±500 pounds.
- The TVC shall have frequency response of 20 Hz, ±0.1 Hz.

Be careful not to make performance requirements too restrictive. For example, for a system that must be able to run on rechargeable batteries if the performance requirements specify that the time to recharge should be less than 3 hours when a 12-hour recharge time would be sufficient, potential design solutions are eliminated.
In the same sense, if the performance requirements specify that a weight must be within ±0.5 kg, when ±2.5 kg is sufficient, metrology cost will increase without adding value to the product.

Figure 55.13 Example of functional and performance requirements. (From NASA/SP-2007-6105, *NASA Systems Engineering Handbook*.)

2. Human. The primary interface of concern to HFE is the human. All human operators, maintainers, assemblers, ground controllers, users, etc., comprising "the human element" with which the system hardware and software must interact should be of concern and critical expectations and needs identified in requirements.

3. Operations. The system will need to operate within an operational framework that, where critical to describe, must be included in requirements. For example, the system might work with an existing infrastructure of maintenance depots, might need to be shipped in accordance with definable existing constraints (e.g., "shall fit on a standard size rail car"), might plan to use existing logistics (e.g., "shall operate on regular gasoline"), or might be expected to meet defined operational resources (e.g., "shall be deployable and fully operable by no more than two people").

■ Quality: A complete set of requirements must include characteristics that are typically difficult to define within strict requirements structure but nevertheless must be addressed. These are sometimes called the "non-functional requirements"—i.e., they can't be strictly categorized as defining the basic expected system functionality or performance—or as "the -ilities"—e.g., maintainability, reliability, operability, producibility, trainability, survivability, habitability, etc. These requirements apply such domain discipline expertise to the system as safety engineering, reliability engineering, and design for maintenance. HFE might be considered by some as one of these specialty disciplines although HFE requirements—like some of the other specialty disciplines—can often be written as strict function and performance requirements. Note that these types of requirements typically apply to the entire, integrated system whereas function and performance requirements are often allocated to particular elements or subsystems of the design.

Additional considerations for a complete set of project requirements include:

- Rationale Statements: Note also that in addition to the four types of requirements listed above, a complete and thorough package of project requirements will include rationale statements for each requirement. The rationale statement captures in language understandable to every stakeholder and every member of the design team what the reason and origin is for each requirement and any associated data—e.g., performance numbers.

- Standards: Many large organization—particularly government agencies—have established HFE or HSI standards that they require to be applied to programs and projects. These may either be invoked into a program or project with a single "shall" statement in the requirements document or they may serve as the starting point for all HFE/HSI requirements to be derived and tailored to suit the needs and mission of a particular project.

Suggested Reading

A good example of a completed human/systems requirements document is the following, used for NASA's Constellation human spaceflight program:

- CxP 70024, NASA Constellation Program Human-Systems Integration Requirements (December 15, 2006)

Human Factors Engineering Requirements

The capabilities and limitations of the human elements of any system must be considered equally and similarly to the manner in which the capabilities and limits of hardware and software must be considered by the project's technology designers. The responsible parties for HFE and/or HSI on a program or project should bring domain knowledge, experience, and expertise to the design team and should be directly engaged in creating definitions, requirements, and ultimately design concepts for the human–system interfaces of the system. The analyses conducted in early concept formulation of the project—i.e., documented stakeholder expectations, mission analysis, ConOps, and function allocation—provide the foundation for initial HFE/HSI requirements development. As noted earlier, large organizations may already have established HFE/HSI standards that are required to be directly applied to or tailored to support new programs and projects. This route, should it pre-exist for a project, provides strong support and facilitates the efforts of HFE/HSI practitioners.

Initial HFE/HSI requirements are part of the total suite of requirements that define the intent and direction of a project as it matures from the concept development phase and into the initial design phases. The role of the HFE/HSI specialist at this point is far from over as the application, interpretation, validation, and ongoing verification of a project's requirements to design is likely to be a much more involved and time-intensive process than the original baselining of the HFE/HSI requirements. Just as a hardware or software designer's task might be said to begin with requirements, so it is with the HFE/HSI practitioner. As an example, the initial HFE/HSI requirements may be the first formal codification of the allocation of specific mission functionality to humans versus machines but as the design matures, such function allocation and other system decisions and assumptions may be revisited based on better understanding of system complexity, technology development concerns or costs, new data on the workload required of the operator/maintainer, etc. Also, because not everyone on a design team will be familiar with the domain knowledge and concerns of HFE/HSI, the practitioner needs to be a core member of the design team, engaged in all key decisions and design assessments, and may need to help educate the design team at times why a particular HFE/HSI concern is critical to the ultimate success of the system.

In Chap. 56, suggested roles and responsibilities of HFE/HSI practitioners during system design are discussed in the section on design trade studies.

Special Safety Considerations[4]

System safety deserves special attention and should be a strong consideration of the HFE/HSI specialist since safety is usually referenced to the safety and well-being of the system's operator, user, and/or maintainer. A thorough analysis of the procedures required to accomplish each operator and maintainer task is desirable in order to isolate critical interfaces in which (1) the design should be reconsidered, (2) special safety procedures and training requirements should be defined, and/or (3) the steps the manufacturer has taken to provide maximum safety for the customer should be documented (in the event that a legal action is taken by an injured customer in the future). (See Fig. 55.14.)

For some complex systems, it is desirable to extend this analysis by estimating operator and maintainer reliability for each of the more critical tasks wherein human failures have a significant impact on system performance or possible human injury.

Because of the special nature of some products, it is sometimes necessary to consider safety from more than one point of view. These problems are best explained through examples:

[4]Willie Hammer, *Handbook of System and Product Safety*, Prentice-Hall, Inc., Englewood Cliffs, NJ, 1972, p. 253.

Accident prevention

1. Hazard elimination
2. Hazard level limitation
   - *a.* Intrinsic safety
   - *b.* Limit-level sensing control
   - *c.* Continuous monitor and automatic control
3. Lockouts, lockins, and interlocks
   - *a.* Isolation
   - *b.* Lockouts and lockins
   - *c.* Interlocks
4. Fail safe designs
   - *a.* Fail passive
   - *b.* Fail active
   - *c.* Fail operational
5. Failure minimization
   - *a.* Monitoring
   - *b.* Warning
   - *c.* Safety factors and margins
   - *d.* Failure rate reduction
     - Derating
     - Timed replacements
     - Screening
     - Redundancy
6. Backout and recovery
   - *a.* Normal sequence restoration
   - *b.* Aborting entire operation
   - *c.* Inactivating only malfunctioning equipment
     - Automatic
     - Manual

Damage minimization and control

1. Isolation
   - *a.* Distance
   - *b.* Energy absorption
   - *c.* Deflection
   - *d.* Containment
     - Hazard
     - Operation
     - Personnel
     - Material
     - Critical equipment
2. Personal protective equipment
   - *a.* Programmed dangerous operation
   - *b.* Investigations and correction
   - *c.* Emergencies
3. Minor loss acceptance
4. Escape and survival
   - *a.* Point of no-return warning
   - *b.* Crashworthiness designs
   - *c.* Escape and survival equipment
   - *d.* Escape and survival procedures
5. Rescue
   - *a.* Procedures
   - *b.* Equipment

Figure 55.14 Safety measures.

■ A hand-held weapon will always have certain inherent hazards because of the fact that the device is designed purposely to fire a projectile. Experience has shown that such a weapon should always have some type of safety mechanism to help prevent unplanned firings. The position and specific articulation of the safety feature should be such that the device is convenient to use after the weapon is placed in the firing position; this is necessary in order to minimize the possibility of discharging the weapon while the user is carrying it or is mounting it to the firing position. However, the position *and articulation of the safety device should* not be such that it can be activated by inadvertent brushing of the device as the weapon is being carried or lifted to firing position or at any time the weapon is being serviced. In addition, since weapons often may be found by children, the safety device should be designed so that a child *cannot* operate and thus remove the safety feature.

■ A power tool or implement should always have a safety feature to prevent inadvertent operation, and the safety device must be convenient to operate once the tool is in the operating position. Designing and locating the safety device require careful consideration of *all* possible points in the handling cycle in order to make sure that careless handling will not also deactivate the device. In addition, there should be an automatic deactivation of power when the tool is set down temporarily so that no one else could inadvertently pick up the tool and accidentally start the motor, thinking that it had been turned off.

■ Child-proof bottle caps are generally required by law on all medicine bottles. Remember, however, that some medicines must be quickly accessible to certain persons (e.g., an elderly heart patient), and it is therefore vital that the individual needing to open the bottle can do so easily and rapidly. Confusing safety caps or ones that require considerable finger strength may make the medicine inaccessible to the person who needs it as quickly as possible.

■ Safety belts obviously serve a basic safety need. However, in addition to providing security for the passenger during a crash, they must also be easy to remove in an emergency—either by the individual wearer or by a rescuer, in the event the wearer may be unconscious. One must consider both the location and the method of operation of the release device in terms of how difficult it might be to find and operate under abnormal conditions, such as in the dark, when the vehicle is not upright, and when it is not possible to open a door.

Standardization plays an important part in safety in a variety of situations and should be kept in mind during any hazards analysis.

The following considerations serve as examples:

1. Critical safety features (door handles, safety latches, light switches, etc.) should always be illuminated so that they can be located and operated at night.

2. Standard safety devices (light switches, door handles, restraint system release handles, fire extinguishers, first-aid boxes, etc.) should be located where people expect to find them.

3. Safety signs and signals should be located where people normally will be looking, and they must not have the potential of being hidden by an object placed in front of them or by individuals standing in front of them at a critical moment.

4. Safety devices should operate in a standardized manner; i.e., one device should not have a switch or handle that moves in one direction, while another, similar device has a handle that moves in the opposite direction. The split second required to correct an action may result in injury.

5. A safety device should never be used that may malfunction because of loss of electrical power. A safety system should provide maximum security throughout any anticipated operating cycle; i.e., avoid a system which may provide an initial safety condition but which, because of the nature of the device, expends its safety capability in "one shot" and thus is unavailable for continuing hazardous conditions.

Toys present a particularly difficult safety problem in that they are subject to breakage by the child and thus present hazards after they are broken as well as before.

In order to resolve some of the special problems exemplified by the above, one should systematically analyze both the normal use and potential abnormal or misuse modes of the particular product. In doing this, keep in mind the following critical human characteristics:

1. The users' level of intelligence (both intended and unintended users)

2. The users' experience with this particular device

3. The various environmental conditions under which the device may be used

4. The physical characteristics of the individual user population (both intended and unintended users)—e.g., size, strength, mobility, dexterity, visual and auditory capacity, reaction time, and possible handicaps

To correct safety deficiencies, consider the following:

1. Remove the hazard altogether if possible.

2. If that is not possible, try to minimize the potential effects of the hazard.

3. Provide a barrier between the hazard and the user.

4. Warn the user of the potential hazard.

Finally, do not simply assume that your solution to a safety problem works. Test it.

HUMAN FACTORS ENGINEERING PERFORMANCE CRITERIA

Near the end of the concept development phase of a project's formulation it is likely that the human factor engineer (HFE) has enough information on the system to form a vision of what's needed for full mission success. At this point, he or she may establish system performance criteria—i.e., technical metrics that define successful performance for the overall system.

Performance criteria differ from requirements in respect to that (1) the measure of their success is not as black and white, and (2) they apply to the overall system, not to a particular aspect of the system's functionality or performance. In this latter respect, they are more like quality requirements—and in fact some quality requirements may be derived from performance criteria—though they are meant to be measures of a range of successful outcomes. Examples of types of performance criteria may include:

- Measures of Effectiveness (MOEs): MOEs define successful outcomes against which the "goodness" of a system design solution may be evaluated. Perhaps most important for use by the HFE membership of the design team, MOEs may measure system success as operations metrics. Examples might include:

  a. The system shall be fully operable in the field with a crew of no more than 40 personnel.

  b. The system shall be able to be repaired and refurbished to perform its next mission within seven days.

  c. The system shall place nominal maintenance and habitability demands on the crew such that a full 50% of the crew's waking time is available to perform mission science.

  HFE disciplines may generally focus on expected levels of human resourcing for operations—i.e., manpower, personnel, and training. Also, these may be key customer measures of the design's success.

- Measures of Performance (MOPs): MOPs are measures that characterize attributes of the system, often at high integration levels—e.g., combinations of subsystems or (for HFE concerns) the performance of human interaction with systems-within-systems. In the above examples, MOPs may be used to more specifically define and quantify "operable" or to what extend "repair and refurbishment" must be conducted for mission readiness.

MOPs might be allocated to the overall system, to subsystems, or even to the human elements of a total system—e.g., to define operator skill levels or training required.

■ Technical Performance Measures (TPMs): TPMs derive from MOEs and MOPs as items that can be measured and tracked during development. They may be generic—e.g., system mass, system power, or onboard crew time for nominal maintenance—or be focused on key systems elements or parameters—e.g., main engine reliability or astronaut selection criteria. Since TPM tracking requires design team resources, the set of key TPMs for a project should be succinct and deemed key factors for project, mission, or safety/reliability success.

The use of performance criteria is nothing new to the concepts and practices of systems engineering, particularly for complex system development. With HFE being increasingly integrated into the practices of systems engineering the use of performance criteria to gauge overall system success from the HFE's perspective is evolving in application and maturity. Performance criteria should be determined as metrics that can be measured, evaluated, and used to impact design options as early in development as the early conceptual phase. Setting early system target criteria for levels of manpower, personnel, training, workload, etc., is a task that may be performed in close conjunction with ConOps development, function allocation, and task analysis studies.

The value of performance criteria may be particularly important for meeting the goals, for example, of the U.S.DoD's HSI program in which they have mandated the use of HSI principles on every acquisition program and project with the express intent of reducing operational costs, particular manpower, personnel, and training. When written as a requirement, for example, "The system shall be able to be repaired and refurbished to perform its next mission within seven days" may traditionally have been deemed by the design and development team to be an unverifiable requirement—i.e., as an unknown until the system is completed or as a responsibility of a follow-on operations manager and his or her team. Yet this seemingly simple to understand need of the customer stakeholder of the system may be one of the stakeholder's most critical needs and one that should be nearly first and foremost in the vision of the development team's designers and evaluators. When written as a performance measure, however, it should become critical to the system engineer—and his or her HFE support team—that "time to refurbish" is a critical measure that must be assessed and evaluated for success at every project development milestone prior to proceeding with the design. Early studies of this metric might be "paper" analyses that break down the system into subsystems and assign allowable refurbishment times to each. This accounting might then be evaluated in increasing levels of detail as the design matures. Also, additional means of evaluating success for this performance criteria should become apparent to the skilled HFE practitioner as the design matures sufficiently to look at the integrated system as a whole, including its interfacing systems (which may be existing and beyond the scope of the project at hand) and the expected environments in which the system must perform.

Similarly, "The system shall place nominal maintenance and habitability demands on the crew such that a full 50% of the crew's waking time is available to perform mission science" should be seen as a metric critical to the success of the overall system—e.g., the design of an Arctic outpost. After all, the primary purpose that drives the project's existence might be the gathering of scientific data and the performance of science analyses. Delivering a system in which the developers have allocated all maintenance functions to the habitat's crew, with little time left for science, might lead to the overall mission being deemed a failure or at least an inefficient, ineffective success. In this example, the waking time available to a known or predetermined crew size should be deemed a "given" from the earliest project formulation. Allocating that time to maintenance—of, perhaps, both the hardware/software systems and to the crew itself (e.g., food preparation, eating, exercise, etc.)—should not be a difficult exercise and should start at the earliest project conceptualization. Then, crew time for systems maintenance should be viewed as a metric that can be allotted, measured through evaluation, and summed in the same manner that other critical physical attributes of the system—e.g., mass, power, thermal load, etc.—are traditionally measured and tracked through the systems engineering process.

Identification and tracking of key HFE performance criteria for use as measures during a project's design and development by the project manager, the systems engineer, and the entire design team are powerful tools that enable HFE concerns to drive the operational, life cycle success of a project.

Chapter 56

System Design

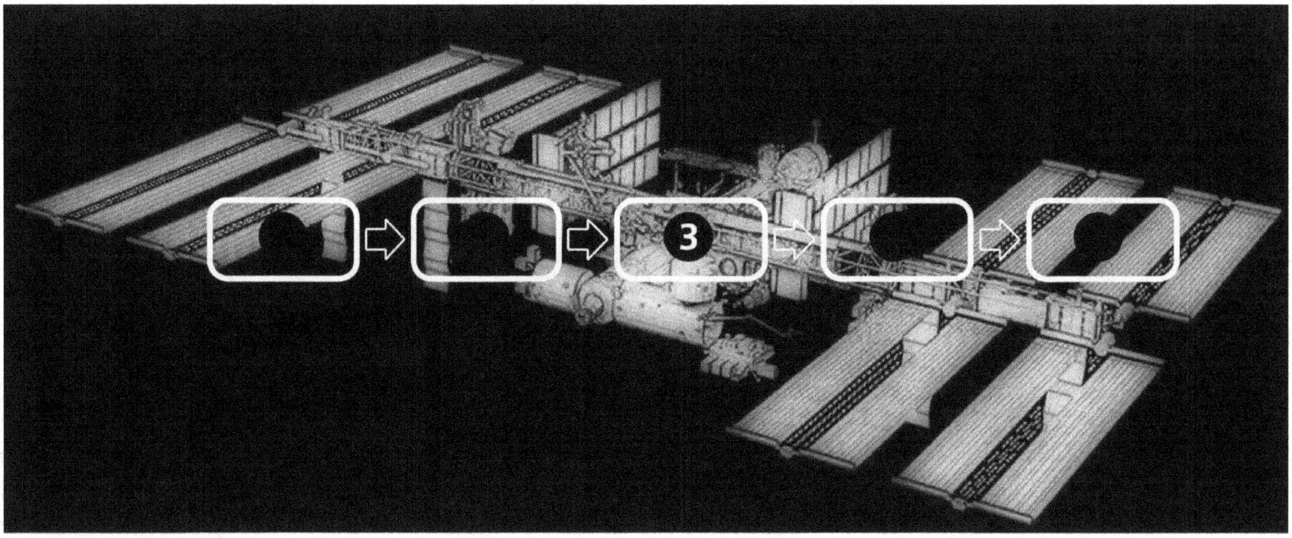

TASK ANALYSIS

As the term implies, task analysis involves the examination of what a particular design requires of an operator, maintainer, or product user who must perform a monitoring, control, or maintenance task. The process also closely examines the operation procedures and operability of all hardware/software systems the operator or maintainer uses to accomplish the task. The sooner one can examine and define the design implications of a proposed task, the sooner he or she will know whether the design is compatible with the user's capabilities and limitations.

Task analyses and task descriptions serve two main purposes, among others. First, as noted above, the task analysis tells us not only whether the design is desirable but also whether it is optimum from the user's standpoint. Second, the task analysis provides important information for the personnel subsystem specialist who must plan and devise personnel acquisition and training programs for the system.

Preliminary task analyses should be performed as soon as possible during or directly following the initial conceptual phases of a development program; thereafter, they should be regularly refined and updated until the task description becomes a procedural guide. Note that task analysis for a complex system will actually comprise a hierarchy of analyses from top-level statements of overall system functionality and performance to subtiered levels of increasingly more detailed single-item or single procedure analyses of individual subsystems and subtasks.

The task analysis begins with detailed descriptions of actions the human operators, maintainers, etc., need to perform. Descriptions should be sufficiently detailed to define the expected system response, error messages or alarms, graphic or aural feedback, names and locations of input controls, and any other supporting information needed to accomplish the task. As applicable, example task description topics to be addressed are listed in Fig. 56.1.

As design concepts are formulated, operator, maintainer, and user tasks should be described in sufficient detail to illustrate how, where, and with what operators and maintainers will be required to perform salient operating and maintaining tasks in order to exercise control of the hardware/software system. Task analyses should be continually revisited and matured during design as concepts mature into solid design proposals.

The information in Figs. 56.2–56.7 is meant to provide guidance in task analysis through several examples of descriptive

Topic	Example
Alerts	• Alarms and warnings
Information	• Parameters (units, precision, and accuracy) • Feedback needed to indicate adequacy of actions taken
Decision-making	• Decision type (relative, absolute, probabilistic) • Evaluations to be performed
Response	• Actions to be taken • Task frequency and required accuracy • Time available and temporal constraints (task ordering) • Physical position (stand, sit, squat, etc.) • Biomechanics –movements (lift, push, turn, pull, crank, etc.) –forces needed
Teamwork and communication	• Coordination needed between the team performing the work • Personnel communication for monitoring information or taking control actions
Workload	• Cognitive • Physical • Overlap of task requirements (serial vs parallel task elements)
Task support	• Special and protective clothing • Job aids, procedures or reference materials needed • Tools and equipment needed
Workplace factors	• Ingress and agress paths to the worksite • Workspace needed to platform the task • Typical environmental conditions (such as lighting, temp, noise)
Situational and performance shaping factors	• Stress • Time pressure • Extreme environmental conditions • Reduced staffing
Hazard identification	• Identification of hazards involved, e.g., potential personal injury

Figure 56.1 Task considerations. (From Nuclear Regulatory Commission, NUREG-0711, *Human Factors Engineering Program Review Model*.)

Function 4.1.2:
Load xx projectile into
x-1 breech block

Preliminary operator functional requirements worksheet

System/OP constraints: loading rate = 2/minute; sea state 4

Task description: Monitor ammunition shaft delivery signal, open door, position projectile carriage, transfer projectile to carriage, move loaded carriage to loading position, open gun breech, transfer projectile to breech, check position, adjust fuse, close breech, notify fire control officer "ready".

Physiological considerations: Maximum lift capability for one man is 60 pounds; higher weights require special sling or method for direct roll-transfer of projectile.

Psychological considerations: Visual codes for identification and special calibration markings must be legible as well as visible under battle (red) lighting environment.

Safety considerations: Automatic door closure on ammunition delivery system and possible effects of ship roll require special indoctrination and drill to prevent accidents (e.g., closure of door on personnel) or loss of control of projectile.

Figure 56.2 Example: preliminary task analysis.

materials and documenting formats. The examples show different approaches to documenting task analysis and use different formats for describing tasks. There is no one way to perform task analysis; therefore, these examples merely illustrate the types of information that various analysts felt were necessary to provide useful data for assisting in design review and/or for developing manpower, personnel, and training requirements. The examples start with very high-level task analysis considerations as shown in Fig. 56.2.

As the design process matures, sufficient system information becomes available to perform task analysis with increasing levels of detail and at lower hierarchical levels. Figure 56.3 illustrates an example of the information fields that might be determined from a more detailed task analysis.

Task Design Considerations

Task Criticality. Criticality is deduced from the nature of the task and is presented to the designer with the task description. There are three major steps in the derivation of task criticality:

1. *Identify the Potential Errors Which Can Be Made in Performance of the Task.* This is largely a matter of considering the elements of the task and the perceptual, motor, and decision-making demands imposed on the operator and/or the speed and accuracy of automated systems. Thus, in the case of a simple task which involves (a) reading a pressure gauge regulating the internal pressure of a rocket and (b) stopping a pump at

Item	Description
Operator and maintainer position	Position-type title or speciality rating that it is believed the task should be assigned.
Job operation	Job operation performance description, i.e., operation performed in support of a mission function, usually performed at a single location as a unit of work, in the sense that it has a well-defined beginning and ending.
Duty	Larger units of work under a job operation assigned to an individual in execution of a position. Duties are made up of operationally similar tasks within a given operator or maintainer position. An example might be "carries out overall subsystem test on XX radar."
Task title	A short title that indicates what an individual does (in functional terms).
Task index number	Task-identifying number, usually corresponding with OSD numbering system.
Data sources	Major sources of data useful for later verification (equipment, location, performance criteria, etc.).
Work-area location	Where task is accomplished.
Task description	Includes: Equipment used Displays used Controls used Actions (detailed, sequential description of each task step) Support equipment and aids used Feedback re individual's actions
Type of task	Qualities or characteristics of task (fixed procedure, variable procedure, motor skill, system analysis, circuit analysis, display interpretation, etc.).
Frequency of performance	Once, twice, etc.; hourly; per shift, daily, weekly, monthly.
Performance time	Estimated hours and minutes job or task will take; cite maximum permissible time, etc.
Criticality	Effect of failure to perform on the success or failure of job or mission; potential for personal injury.
Newness	Extent to which the task is new to the using agency.
Other positions	Other tasks whose performance depends on, or interacts with, the performance of other personnel position types (indicate functional interrelations).
Safety factors	Any known or suspected hazards, sources, or preventive requirements.
Tools and equipment	Tools, test equipment, protective gear, reference manuals, etc.
Skills and knowledge	Types and level of skill or knowledge pertinent to the selection and training of individuals for the job of tasks.
Physical characteristics	Special physical requirements necessary for effective task performance (e.g., body size, strength, dexterity, coordination, physiological tolerance to specific and anticipated stresses).

FIGURE 56.3 Suggested information requirements for expanded task analyses.

Gross task analysis

Task	Display: Description	Display: Critical values	Decisions	Subtasks listed	Characteristic errors or malfunctions
A. Set up before first run	Position of brake	Not in 9 o'clock position	Release brake—move to 9 o'clock position	1	
	Recall of last link	Copy light since last inking (est.)	Measure ink—add if necessary	18, 19, 20, 21, 22	

Analysis of subtasks

Subtask or task	Display-control descriptions	Control action	Indication of response adequacy	Objective criterion of response adequacy	Characteristic errors or malfunctions
A. Release brake	Brake	Turn clockwise (up)	Brake stop in 9 o'clock position		
B. Attach the stencil	Wheel	Turn	Stencil head clamp available		
	Stencil head clamp	Lift left end	Stencil head clamp loosens		

Function	Human functional element	Task	Subtask
Activate gyros and other components	Make power on input	1. Turn on ac and dc missible guidance power	a. Check to assure that all circuit breakers are "in" b. Position ac missile power guidance switch at on c. Position dc missile guidance power switch at on
Activate test instruments	Make power on input	2. Turn on missile preflight analyzer power	a. Check to assure that test patchboard is inserted and locked b. Position analyzer master power switch at on c. Insert test tape in tape reader and lock in place
	Make analyzer self-test initiating input	3. Start analyzer self-test test sequence	a. Position test switch at self-test
	Monitor self-test readout		b. Check test result window for indication of analyzer self-test result
	Provide corrective inputs and monitoring when self-test yields a "no-go"	Alternative procedure: isolate and replace malfunctioning analyzer module	a. Check code displayed in malfunctioning module window against modular code table b. Remove module from analyzer and take to storage c. Obtain replacement module from storage; take to analyzer and insert it d. Repeat task 3

Figure 56.4 Example: task analysis documentation showing one analyst's approach for defining task hierarchy.

1. Job operation:	2. Task title:	3. Location of task:	4. Source reference:	5. Operational site:
Prepare swimmer delivery vehicle (SDV) for launch	Replenish breathing gas supply	On dock or in water	SDV specification	World wide

Equipment data / **Personnel data**

6. Equipment required	CSE required	7. Source ref none	8. Number of personnel	9. Recommended position title	10. Applicable rating
Breathing gas supply pump & containers, press gauge	Breathing gas supply containers		2	Vehicle maintenance man	BM

Task performance data

11. Time required hours 0 / 25 minutes

12. Frequency — Day Week Month — unknown

13. Probable error minimal on land moderate in water

14. Speed
- ☐ 1. Not critical
- ☑ 2. Moderately critical
- ☐ 3. Highly critical

15. Positioning/handling Equip, special care
- ☐ 1. Little
- ☑ 2. Moderate
- ☐ 3. Considerable

16. Manipulating controls
- ☐ None
- ☑ hand valves
- ☐ Toggle switches
- ☐ Selectors switches
- ☐ Pushbuttons
- ☐ Cont. audio feedback
- ☐ Cont. visual feedback

17. Source of special dangers
- ☐ None
- ☐ Mechanical
- ☐ Electrical
- ☑ Explosive
- ☐ Temperature
- ☐ Volatile fuels
- ☐ Toxic substance
- ☐ Other

Training data

18. Nature of procedure
- ☐ Fixed
- ☑ Variable
- ☑ Motor skill
- ☐ System analysis
- ☐ Circuit analysis

19. Technical manual function
- ☐ Primary
- ☑ Secondary

20. Training difficulty
- ☐ Easy
- ☐ Hard
- ☑ Moderate

21. Training equipment
- ☐ Complex
- ☑ Simple
- ☐ None

22. Technical manual title: life support system, maintenance & replenishment of SDV-TR-013

23. Training course title: life support system, repair and replenishment of SDV

24. Task performance date
- ☐ Estimated
- ☑ Tried in mockup
- ☐ Tried, hardware

25. Equipment development status
- ☐ Design plan
- ☐ Preliminary design
- ☐ Mockup
- ☑ Prototype
- ☐ Production design
- ☐ Production equipment

26. Narrative task information; Remove skin cover plate over life support quick-connect fitting. Attach gas supply nozzle to boat, open supply valve, and monitor flow rate. Shut off valve when pressure reaches 2800 psi. Disconnect supply and check for leaks at nozzle. Replace skin cover.

Figure 56.5 Example: task analysis worksheet.

Position: Pilot
Duty: Vehicle checkout and start

Task	Time, min	Control	Elements activity	Indication	Remarks (Precautions, etc.)
1.1.1 Engine start	0.1		Observe	Ground crew signal	Procedure performed per checklist
	0.05	APU battery SW	Press/monitor	On	Decision to use APU
	0.05	APU fuel SW	Press/monitor	On	
	0.05	APU start	Press		
	0.5		Monitor	Start	
		Engine fuel Master SW	Press	On	
	0.05	AUX fuel pumps	Press	On	
	0.05		Observe	Fuel pressure	Variable time estimate
	0.05	Engine start	Press		
	0.05			Engine parameters	

Figure 56.6 Example: format for analyzing and developing task descriptions.

Approach requirements analysis			Trade-off information and data integration		
Approach-land functional requirements	Action requirements	Information requirements	Related information requirements, sources, and problems	Related accident features	Related survey commentary
1.0 Initiate pre-approach procedures	1.0.1 Review approach information	1.0.1.1 Approach orientation 1.0.1.2 Approach constraints Approach Requirements Obstacles Hazards Weather Minima	Approach plate data Obstacle locations Course and path data Terrain characteristics Hazards Minimum decision altitudes position data	Data misinterpreted not used effectively. Hazards misappraised. Navigation positioning errors.	Cannot remember all details. Study time is limited while setting up approach. Improve to emphasize basic data-critical data holder, e.g., go-around hdg. and altitude. Need clearer picture of position situation.
	1.0.2 Coordinate approach with control	1.0.2.1 Communication path designation Unique limitations and constraints Environmental conditions Barometric pressure	Coordination and confirmation of approach clearance Altimeter setting	Clearances and procedures are misunderstood; not followed; in error. Altimeter misset; misread. Confusion of inches mercury versus millibars sea level versus field elevation reference	Need procedures for better coordination between airplane and traffic control to improve understanding of situation and control intent. Improve altimetry presentation method. Standardize setting references —field elevation preferred for landing. Redundant setting checks. Radio chatter and changes.

Figure 56.7 Example: action and information requirements analysis form.

a specified pressure, errors may manifest themselves in two ways:

a. Failing to stop at the prescribed point

b. Stopping the pump before the prescribed point

2. *Identify the Effect of Each Potential Error on System Operation.* In the example above, failing to stop the pump at the prescribed point may result in overpressurization resulting in destruction of the rocket being pressurized. Stopping before the prescribed point will result in underpressurization with resulting loss of mission.

3. *Estimate the Relative Criticality of the Potential Errors.* Criticality may be scaled in terms of categories such as loss of personnel, destruction of the system, mission failure or abort, mission degradation, and mission delay. In these terms, overpressurization may be more critical than underpressurization, since it may result in explosion of the rocket and destruction of the rocket and launch pad as well as loss of life, while under-pressurization is less critical, since the mission may (though not necessarily will) be degraded.

Pointing out a task as being critical to the design engineer "flags" that task as one requiring special consideration in design. Among the solutions which are possible (certainly the list is not exhaustive) are the following:

1. Replacing the human with an automatic means of accomplishing the function if the desired level of correct performance cannot be achieved with needed speed and precision

2. Providing means to reduce the probability of error, e.g., assigning a special feedback device to warn the operator when the task is being performed incorrectly

3. Assigning the task to only highly skilled personnel

Task criticality is highly related to specification of task difficulty and error likelihood. Since the engineer thinks in terms of physical effects on the system, it is preferable to flag the task as being critical without indicating that it also has a high difficulty or error-likelihood index. The provision of quantitative indices of a highly precise nature, such as probability of operator error to four figures (e.g., .0013), is not advised, since the engineer cannot interpret the quantitative values in design-relevant terms. A gross categorization of task difficulty, such as the following three-part scale, is as much precision as the engineer can handle in design terms:

1. Simple, routine

2. Somewhat difficult

3. Very difficult

Moreover, it is necessary to apply the above scale only to those few very critical tasks which merit design attention, not to all tasks.

Task Difficulty (Error Likelihood)[5]**.** The human factors (HF) specialist is especially concerned about task difficulty because this, in turn, may lead to a higher error probability, with its attendant effects on mission accomplishment. Task

[5] D. Meister et al., *The Impact of Manpower Requirements and Personnel Resources Data on System Design*, AMRL-TR-68-44. Air Force Systems Command, Wright-Patterson Air Force Base, Ohio.

difficulty arises because system requirements are incompatible with and overload the capability of the individuals assigned to perform the task.

Task difficulty is not the same as error likelihood. A difficult task need not automatically have a higher error probability if personnel of higher capability may compensate for the increased task difficulty. The significance of task difficulty is intensified when the task is also critical to the accomplishment of the mission. Such difficult and critical tasks automatically demand reconsideration and possible system redesign because their attendant error probability cannot be accepted. As in all cases that indicate excessive operator workload, it may be necessary to automate the performance of the task, relax the accuracy requirement (thus implicitly accepting a higher error probability), or redesign the task to simplify it.

The determination of task difficulty must be made by analyzing the individual task in terms of the inputs which initiate the task (e.g., an audio message) and the outputs which accomplish the task (e.g., a switch action). The HF specialist will look for the following characteristics, which may indicate an excessively difficult task:

1. The input which initiates the task requires excessively precise sensory or cognitive discriminations or fine motor responses.
2. The operator's response to the initiating inputs must be performed so quickly that he or she has problems keeping up with the initiating inputs.
3. The accuracy demanded of the operator in responding to the initiating inputs is excessive (e.g., heading error must be within 0.5°).
4. The task must be coordinated extremely precisely with other tasks performed by other personnel.
5. The environment in which the task must be performed tends to degrade task performance (e.g., high levels of noise or acceleration).
6. Information from multiple sources (e.g., several displays on a control panel) must be integrated by the operator in order to make a decision.
7. There is less than the desirable amount of information available on the basis of which a decision must be made or an action taken.
8. The task is composed of many subtask elements, the correct performance of which is necessary to task performance, but the amount of feedback provided (knowledge of the correctness or incorrectness of subtask accomplishment) is inadequate.
9. Short-term memory requirements for task performance are excessive (e.g., memory for long sequences or target coordinates).

The design solutions available for reducing task difficulty include the following:

1. Providing additional training or selecting more highly skilled personnel
2. Simplifying the task by such means as combining information sources, providing additional feedback, subdividing the task among several operators, or changing the manner in which the task must be performed
3. Reducing system requirements by accepting a higher error probability, longer response time, etc.

Task Duration. Task duration should be considered in two ways:

1. As a system requirement, i.e., the time within which the task must be performed in order to accomplish a given system function
2. As an anticipated human performance capability, i.e., the time within which the operator can actually perform the task

Item 1 is a criterion against which item 2 can be evaluated as satisfying or failing to satisfy system time requirements.

As a system requirement, a task may have to be accomplished in so short a time either that the operator cannot physically perform it or that the probability of the operator's making an error will be substantially increased because of the time loading. In either case, special attention must be drawn to such a task. If the system time requirement is inflexible, it may be necessary to mechanize and automate the function involved (to eliminate the operator) or else to redesign the manner in which the task can be performed or the equipment is to be operated or maintained.

Task duration is, of course, not critical unless the system's required response time is also critical to the successful accomplishment of the mission. Hence, it is necessary to analyze the mission segment in terms of its time demands before examining any individual task duration. Information required in order to perform task duration analysis will include:

1. System performance time requirements
2. A description of the tasks to be performed by personnel in each mission segment
3. The estimated time required to perform the task

Of these informational requirements, the last is the most difficult to secure because it requires data on the performance time capability of personnel, such as the time required to hook up an umbilical connection. That information can be secured from previous comparable systems in which similar or identical tasks have been timed or from the body of general human performance data in the literature or from specialized, tailored human-in-the-loop studies performed by the project as part of a critical design evaluation.

Time-Line Analysis

Though the early mission concept of operations (ConOps) provides a general description of time relationships, it is not until more design detail becomes available that operations and maintenance task design characteristics—criticality, duration, etc.—can begin to be resolved. As system design concepts mature it becomes evident whether the planned use of the operator or maintainer is reasonable. Time-line analysis is a technique used to help derive human performance requirements by showing (diagrammatically) the functional and temporal relationships between tasks as well as the task loadings for any combination of tasks.

Although numerous formats have been developed and used by various analysts, types known to be useful are shown in Figs. 56.8–56.10. In this format, it can be seen that tasks are grouped by operator and displayed as a two-dimensional graph that indicates the estimated amount of operator's time at various intervals along a common time scale.

With this type of graphic, the analyst can easily spot those intervals during a mission in which a given operator may be overloaded. It is obvious that an operator cannot accomplish two tasks simultaneously if both occupy a single perceptual-motor channel or decision-making response 100 percent of the time. The integrated time-line chart (see Fig. 56.11) exposes such conditions if it is properly developed. When such conditions are spotted, it is apparent that one of two things must be done—either a task will have to be given to another operator, or the operator must be provided some type of machine assist (e.g., an operation becomes automatic rather than manual).

The task load estimates come from several sources; for example, the task may be the same as, or similar to, a task required in another system which is in actual operation. Task time information from previous systems is generally the most reliable since it has been verified in practice. When such information is not documented, the next best source of information is from operators who perform or have performed similar tasks. It is desirable to get estimates from several operators since there is frequent variation in their estimates. The human engineer generally has to probe the

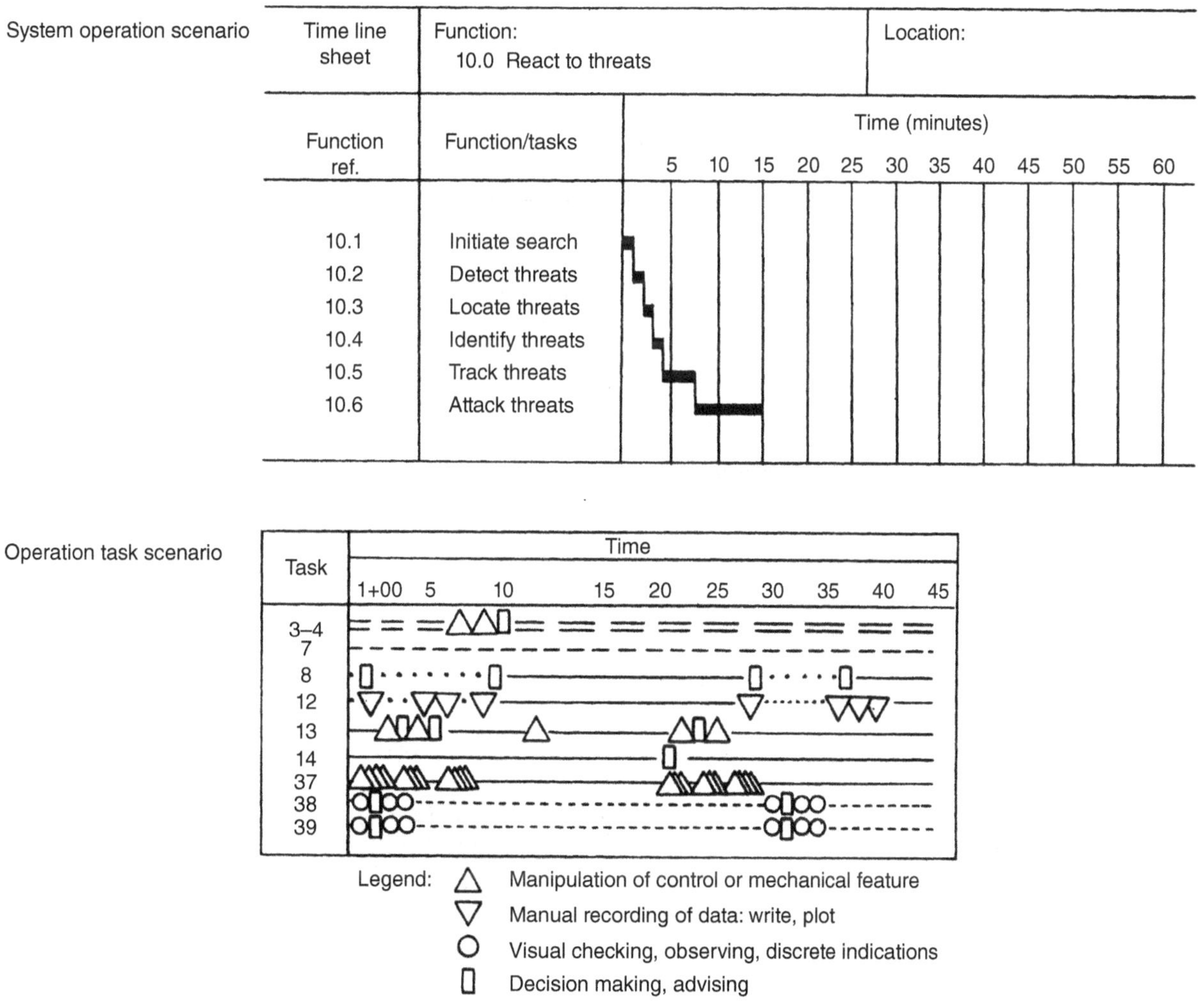

FIGURE 56.8 Example: time-line analyses (gross levels).

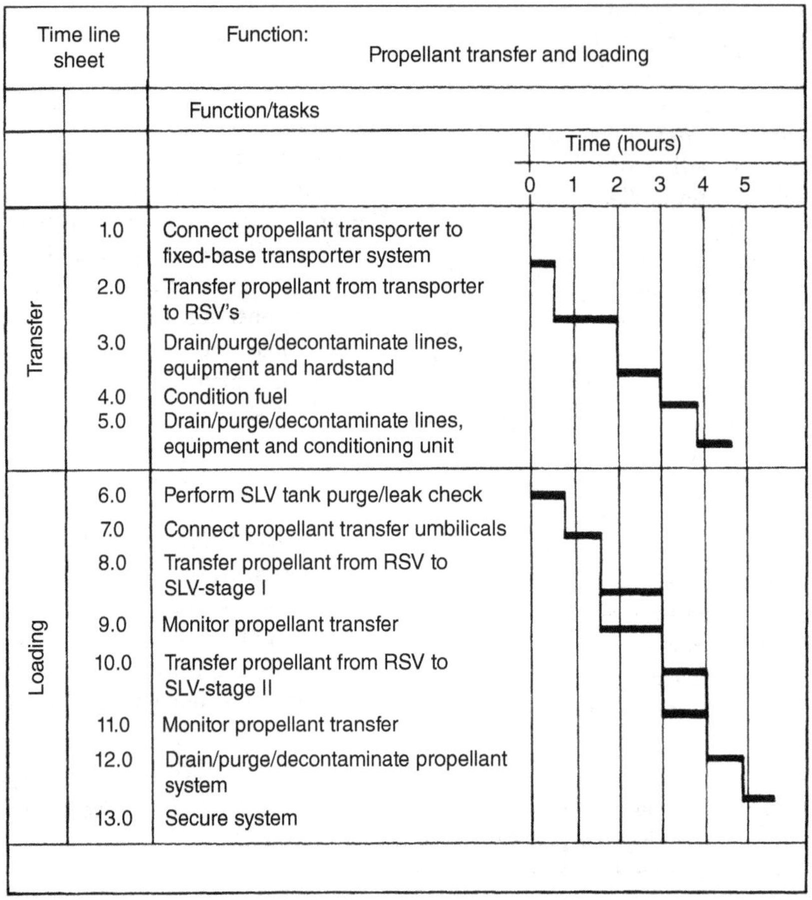

FIGURE 56.9 Example: time line (top level).

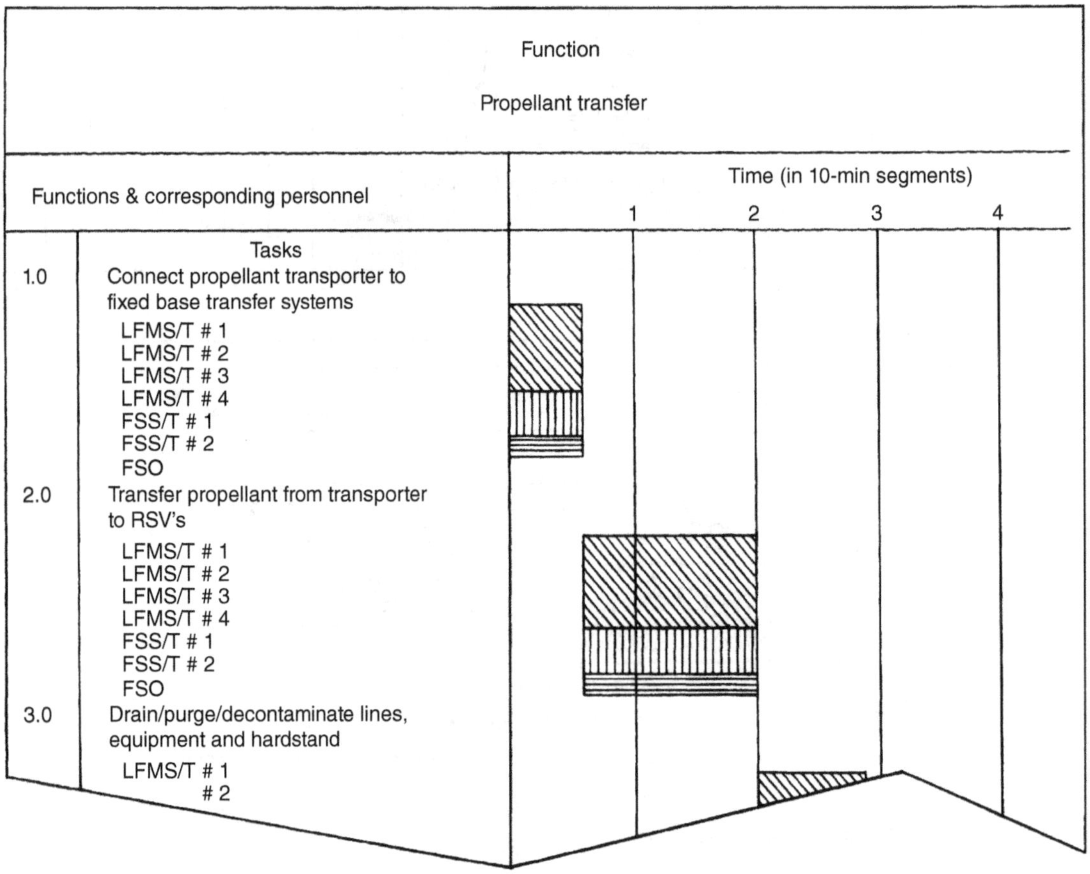

FIGURE 56.10 Example: time line (detailed level).

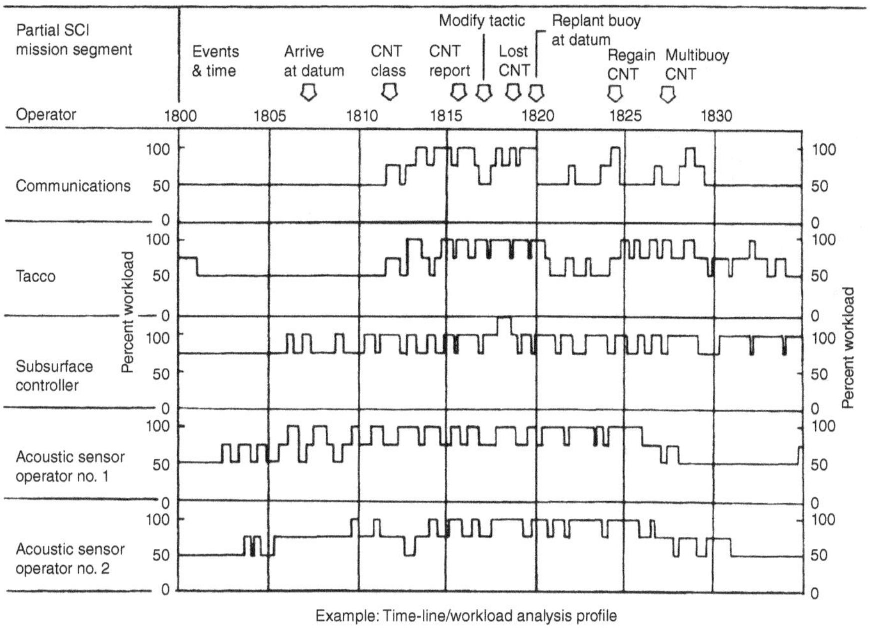

Figure 56.11 Example: integrated time-line–workload analysis profile.

task question with the operator in fairly good detail to provide the operator with a basis upon which to make an estimate. It is important, for instance, to clarify the fact that in some cases two tasks could occupy almost 100 percent of the operator's time if one task involved a different perceptual-motor channel than the other. For example, an operator can usually monitor an aural channel almost full time and still monitor a visual display almost full time. When experienced operators are not available, the human engineer along with knowledgeable equipment engineers may have to make an "expert guess" about the task. The human engineer will have to break the task down into its simplest elements and extrapolate from what he or she knows, based upon human performance studies, about division of attention.

The time line may be made up of a single, continuous chart from the beginning to the end of a mission, or there may be several charts, each of which expands a particularly critical segment of the mission. The time scale should be commensurate with task complexity; i.e., 5-minute intervals may be all that is necessary for simple tasks, while 5-second intervals may be required for more complex tasks. Whatever interval is used, however, should be common for the total group of tasks and operators when they interact.

Additional Task Analysis Considerations

As discussed for the ConOps, task analysis should not only be performed for nominal operation of the system (see Fig. 56.12) but branches of the hierarchical, integrated system task analysis should include all modalities of system performance including degraded system performance, logistics and deployment, maintenance, etc. (See Fig. 56.13.) In short, thorough task analysis should include the expected human/machine interactions for any human using or supporting the system at any stage in its operational life. In particular, this level of understanding of a system's off-nominal operation may become critical where the successful functioning of the system is critical to the survivability of a human crew. In the design of aircraft or spacecraft, for example, it is critical to understand every potential failure mode of a system and its subsystems and to ensure that the actions needed to avoid loss of human life are designed into the system or into the required human and system interaction required for survivability. Every such risk mitigating off-nominal operation should be investigated using task analysis techniques to ensure that the operators—where considered critical to the needed task or action—are available and are capable of performing the task in the time required.

In addition to alternate modes of the system—e.g., off-nominal, degraded, etc.—task analyses should be performed to address every human element. One way to look at this is to ensure that every human who touches the system is considered during the entire system's life cycle. Maintenance, logistics, system deployment and assembly and any other act performed during operations should definitely be addressed. Actions such as producibility during manufacturing might also be included in the most extensive and thorough set of task analyses.

Task analysis is a tool for use during design and development of the system to ensure that the right system is being built for success in the operational phase of its life cycle. The knowledge and documentation created from ongoing maturation of task analysis provides a solid foundation for the HF engineer to assist in building operational procedures, assembly and service manuals, training materials and other documentation to support the transition from development to operations.

Position title: liquid fuel systems maintenance

Specialist/technician

Air force specialty code (AFSC) 56850B/70B

Duties and tasks	Performance reliability	Skill level	Skill type and proficiency availability	Training requirements
1.0 Connect propellant transporter(s) to fixed-base propellant transfer system possible human errors Cross-connection of hoses misalignment of coupling 2.0 Prepare to transfer propellant from transporter(s) to ready-storage vessels (RSV's) Valves in system left open system not completely bled rsv vents not opened 3.0 Transfer propellant from transporter(s) ready storage vessels (RSV's) Leaks in system due to improper connection or maintenance failure to close valves when transporter is empty	.9975 .9960 .9978	5 5 5	1. Position summary The LFMS/T is responsible, under the direction of the fuel Specialist/supervisor, for the conduct of propellant loading and unloading operations at the launch site, including emergency or damage control activities if required. He decontaminates the surface area, the launch duct or the sump, or the rocket as required. He directs emplacement and removal of mobile PTS equipment and seals fixed portions of the PTS. LFMS/T's assigned to maintenance teams perform unscheduled maintenance on PTS fixed equipment at the launch complex. Any LFMS/T's performing PTS procedures at the launch complex should know both the emergency and normal loading and unloading procedures. 2. Environment a. Launch site (1) The LFMS/T controls traffic into and from the launch complex security area. He directs positioning of propellant transporters and the interconnecting of PTS elements.	The training should be designed to train selected air force personnel to: 1. Operate control panels to determine malfunctions in the propellant transfer system. 2. Perform interconnections of the electrical and mechanical mobile equipment. 3. Operate and maintain the propellant transfer system components and the various items of GOE/GSE including the propellant transporters, propellant holding tanks, nitrogen storage tanks, and propellant transfer control unit. 4. Perform emergency unloading procedures for the PTS. 5. Inspect storage tanks, disconnects, valves, piping, pressure regulators, filters, pressure gages, and temperature gages for leakage, corrosion, damage, and wear; also solenoids, relays, switches, chassis of the transfer control panel, sensors, amplifiers, motors, terminals, interconnecting cabling and instrumentation for proper operation. Instruction should include removal and replacement procedures for the above listed items.

FIGURE 56.12 Example: quantitative/qualitative personnel requirements inventory documentation.

Maintenance level	Maintenance task category	Support equipment (AGERD)	Skills and knowledges required to do the job	Category of skills or knowledges to be taught or strengthened	Equipment and/or training recommendations
Organizational Unscheduled Sub-system Troubleshooting (continued)				f. Oral verbalization: Communications procedures between maintenance personnel during task performance.	
Intermediate Unscheduled LRU mechanical and electrical repair	4. LRU (mechanical and electrical) repair: involves disassembly in varying degrees, visual inspections, running of tests and measurements (e.g., clearance tolerances, pressure loss and leakage tests, torque tightening checks). Certain minor servicing functions such as lubrication or dehydrator recharging is accomplished. Repair consists mainly of parts	a. Regulated air supply. b. Power supply. c. Valve and flange caps. d. Pressure and height gauges. e. Depth and thickness micrometers. f. Flowmeter. g. Replacement parts. h. Standard electronic AGE. i. Cleaning materials. j. Anti-G valve test	a. Knowledge of appearance, location, nomenclature and fitting sequences of LRU parts. b. Knowledge of part functioning principles and relationship to LRU and fitting principles. c. Be able to use precision mechanical tools, and other associated AGE. d. Knowledge of visual appearance of acceptable part condition. e. Know how to perform each step of procedures.	a. Nonverbal identification: Identification of LRU parts configuration. b. Nonverbal Detection: Spatial judgements in parts fitting tasks. c. Learning facts: knowledge of part appearance, location nomenclature, fitting sequence, visual cues of acceptable condition, gauge scale readings, test setup. d. Learning principles: Parts fitting, parts and LRU	1. Transparencies (animated and static): For presentation of facts and principles on LRU and precision measuring tools. 2. Self-teaching programmed texts: Using appropriate illustrations to produce identification skills of parts and nomenclature knowledge to aid assembly and disassembly tasks. Also used to teach the

FIGURE 56.13 Example: maintenance training requirements analysis.

Link Analysis

Link analysis allows the responsible human factors engineering (HFE) party to examine the organization of human tasks in terms of logical grouping and sequencing of activities. Associated with the more detailed stages of task analysis, link analysis may be used to derive and test specific aspects of a design's interaction with operators and/or maintainers. For example, link analysis may be used to develop a best arrangement of components on a control panel or of workstations in a maintenance depot.

The term "link" as used here refers to any connection between a human and a machine or between one person and another. If one person must talk or physically contact another (e.g., hand that person a message), this is represented by a link. If a person must see a display or reach a control on a machine, he or she has a link to that machine. Ordinarily, any links between machines and other machines are investigated during other system analyses processes and can be neglected by the HF engineer unless the link possesses some quality or constraint which might cause the system to operate inefficiently (i.e., excessive length of a hardwire connection between two machines might result in high line loss in power). In these cases, links between machines are also included in the link analysis.

Link analysis is used only after decisions have been made regarding what equipment will be used in the system, how many people will be used, etc. Thus, link analysis is used to place the system components—be they items on a control panel or people and machines in a workroom—in the "best arrangement" on the basis of criteria (such as minimum distance between workers or fewest movements between workers and equipment) important for the system under evaluation.

Example Link Analysis. A typical example of a link analysis for a system involving four operators and four pieces of equipment is described through the five steps below.

- **STEP 1:**
  Draw a circle for every person in the system and label it with a code number:

  ①

  Draw a square for every item of equipment and label it with a code letter:

  [A]

- **STEP 2:**
  Determine the type of link between each of the equipments and operators, equipments and equipments, and operators and operators. The different link types should be coded, such as

  __________control links

  --------------visual links

  . . . . talk links

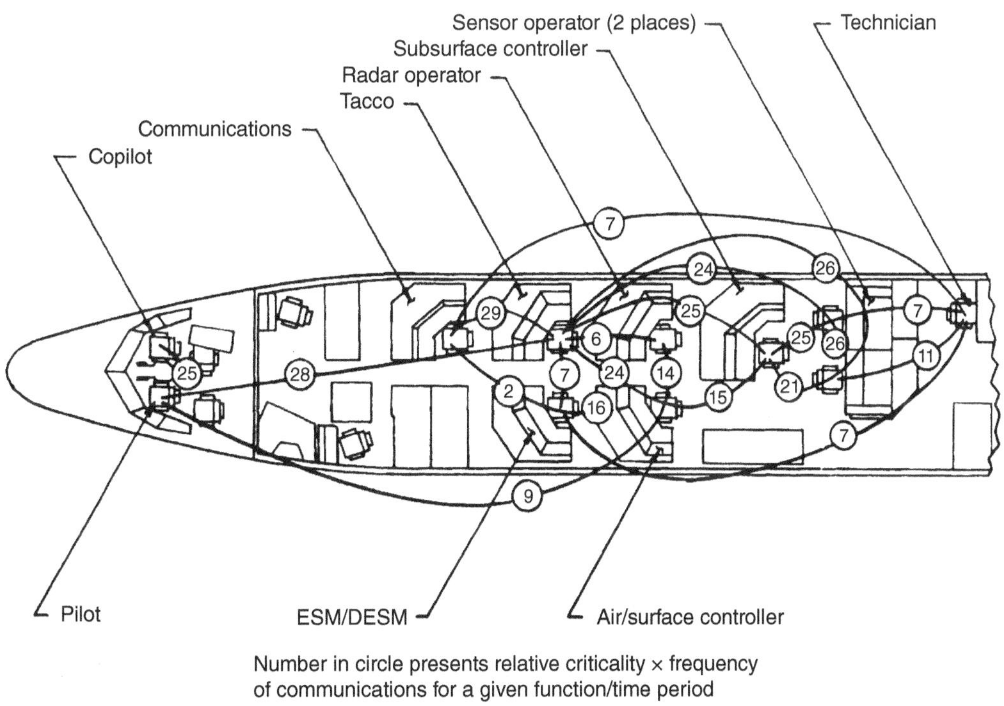

FIGURE 56.14 Example: use of link-analysis technique to evaluate proposed physical layout of components and people.

The three codes are fairly standard for link analysis use. Other codes required can be made up by the individual doing the analysis.

- **STEP 3:**
  Establish the "link value" for each link. The link value is based on two factors: (1) the importance of the link in accomplishing the mission assigned to the system under evaluation and (2) the number of times (frequency) the link is used in completing the mission. A number, between 1 and 3, is assigned for both frequency and importance, with the higher number representing maximum importance and frequency. As an example, if a particular control movement on the panel was very critical in fulfilling the equipment use (3) and it was frequently activated during the equipment use (3), the link value for that particular control link would be 9 (3 × 3). If, on the other hand, a link was infrequently used (1) but was critical when it was needed (3), the link value for that particular link would be 3 (1 × 3). Thus, the link value is found by multiplying the frequency rating by the importance rating.

  Selection of the frequency and importance ratings is usually based on past experience, either of the individual doing the analysis or someone familiar with similar systems already in use. In addition, the engineer responsible for the equipment design should be of assistance.

- **STEP 4:**
  Prepare an analysis chart (as exemplified in Fig. 56.14) of the link values established for the system under evaluation. For each operator, show all the links (with each link value) associated with that operator. Do the same for each piece of equipment. For each operator and equipment item, add up the total of the link values and write this number to the right side. This provides an idea of the priority of equipment use and the operators most active in the system operation.

- **STEP 5:**
  Prepare a schematic diagram (or series of diagrams) of possible arrangements. It is preferable to make this schematic to scale by cutting components out of paper and laying them on a scaled drawing of the space available (panel face, work-space floor plan, etc.). Starting with the operator or machine with the highest total link value, place the remaining components around it, moving them as necessary to minimize link crossings and shorten links, especially those with high link values. If conflicts occur between links, it may be necessary to reassess the original link values. The evaluation and rearrangement continue until the "best fit" solution is obtained. (See the final layout of the sample system in Fig. 56.15.) It should be emphasized that additional changes may be required in system layout once full-scale mockups or early hardware make actual system layout and evaluation possible.

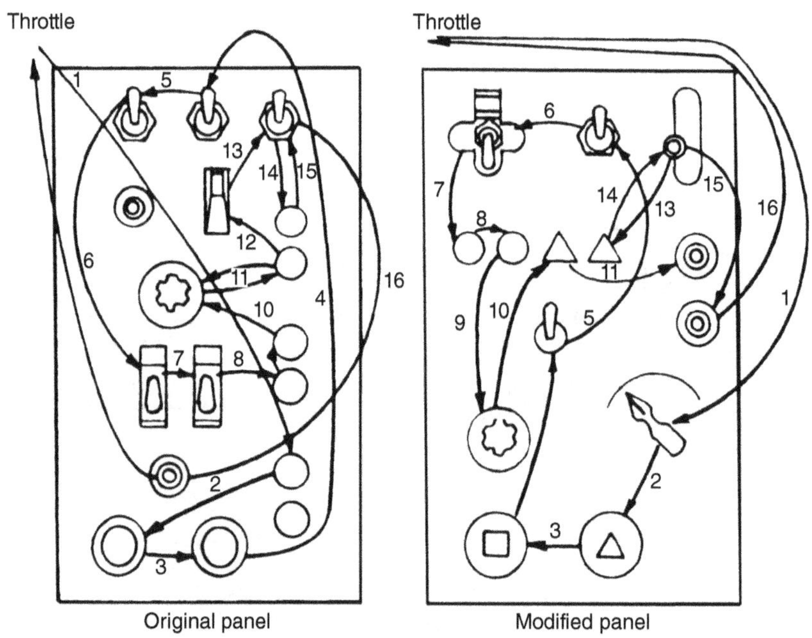

FIGURE 56.15 Example: link analysis of control panel.

***Link Analysis Applied to Control Panels*[6].** Link analysis may be conducted on preliminary drawings of a proposed control panel layout by imagining the operational steps an operator would go through (from the point of turning on power to shutting down the panel or system). An effective approach is to take a pencil and draw lines representing the path the operator's hand would take and/or the points where the operator's eye must observe a visual display. Each step is numbered in order. As the first sketch in Fig. 56.15 illustrates, one can quickly observe some unnecessary steps and repetitious pathways. Although one can make an alternative drawing based on the original link analysis, it is suggested that a more efficient method is to "cut out" the panel components and lay these out on a basic panel drawing (i.e., one that has only the edges shown). In this way, several alternatives can be tried before a second link analysis is performed to see whether the new layout provides an improvement.

During the link analysis one should also remain sensitive to the quality (understandability, legibility, etc.) of proposed panel component labels—e.g., as one proceeds through the link steps, it should become obvious if label nomenclature is confusing or if abbreviations and/or acronyms are inappropriate or misleading.

[6]H. Older, "A Procedure for Representing Results of Human Engineering Studies," paper presented at the Naval Research Human Engineering Conference, October 1953.

Human Error Analysis

Mission success depends on safety and reliability of the systems and the human–system interactions having been a focus of attention during system design and development. Safety and reliability analysis as part of system design is a key concern of general project management and systems engineering. But these should not only focus on the hardware and software elements of the system. The parties responsible for HFE on a project should ensure that human reliability and human error analyses are part of the design process. Human reliability and error analyses are methodologies to understand how human failures might lead to system failure and to estimate the likelihood of such failures. The information from these analyses may affect assumptions and/or decisions on allocation of system functionality to humans versus machines. The goal of using human reliability and error analysis is to promote total system reliability and safety by assuring that the system can tolerate and recover from human errors.

A variety of methods exist for human reliability analysis (HRA) and new methods continue to evolve. It is beyond the scope of this text to delve into the details of HRA, but Figs. 56.16–56.20 provide several examples of products from HRA investigation, some showing their relationship with task analysis.

Procedures analysis

Task	Danger	Effect	Cause	Corrective or preventive measures
Charge nitrogen pressure vessel	1. A loose hose may whip.	Personnel could be injured or equipment damaged.	Hose failure; connection failure; failure to tighten connection adequately.	Tie down, chain, or sandbag hose at close intervals. Personnel wear hard hats and face shields. Establish torque values for tightening connections. Warning and caution notes in procedures.
	2. Vessel bursts.	Fragmentation. Fragments may injure personnel or damage nearby equipment.	Inadequate strength.	Use high safety factor design. provide warning against overpressurizing system. Do not expose pressure vessel to heat. Incorporate relief and safety valves. Test vessel to ensure that it will carry required pressure.
	3. High-velocity gas escapes.	Gas may blow solid particles into eyes or against skin. Loss of gas may cause system to become inoperative due to lack of pressure.	Leak; hose failure; loosening fitting on pressurized system; crack.	Procedures to provide warnings to depressurize system before attempting to disassemble connectors. Personnel to wear face shields.

FIGURE 56.16 Example: task and procedure hazards analysis. (From Willie Hammer, *Handbook of System and Product Safety*, Prentice-Hall, Inc., Englewood Cliffs, NJ, 1972, pp. 197, 199.)

Error Analysis

Control—power toggle

Type of error	Effect	Probability of occurrence	Probability of detection	Recommendations
Inadvertent actuation	Unintentional activation or deactivation	Great—due to location and proximity to other controls	Moderate—feedback from displays	Relocate to periphery of console—add guards
Substitution	Erroneous sequence of operations	Great—switch is located adjacent to other toggles	Moderate—feedback from displays	Relocate to periphery—consider use of pushbutton
Adjustment	Selection of standby instead of ON or OFF	Little—labeling is adequate	Moderate—feedback from displays	None

FIGURE 56.17 Example: format for human error analysis during human engineering design review.

System function	Type of equipment	Task	Potential error	Suggestions for equipment design	Suggestions for procedure
Prepare nose cone for inspection	Crane, sling, hand tools	Remove nose cone afterbody	Removal of bolts in improper sequence	Number bolt holes to indicate proper sequence	Insert caution in tech manual
Check out trajectory control test set	Functional test set	Perform simulated nose cone test	Misinterpretation of indicator lights	Legend lenses should be consistent and unambiguous	Spell out meaning of abbreviations in checklist
Inspect transport trailer for damage	Trailer, checklist	Inspect as indicated by checklist	Personal injury	None	Secure brakes before crawling under trailer
Complete nose cone high-pressure test	Test equipment, hand tools	Disconnect nose cone from test equipment	Exposure to high pressure	None	Ensure that pressure is vented before disconnecting hoses

FIGURE 56.18 Example: procedural analysis for identifying potential human error.

Means and standard deviations of ratings and reliability estimates for the task elements

Task element	Rating Mean	Rating SD	Reliability estimate	Task element	Rating Mean	Rating SD	Reliability estimate
Read technical instructions	8.3	2.2	0.9918	Install nuts, plugs, and bolts	4.6	1.7	0.9979
Read time (brush recorder)	8.2	2.1	0.9921	Install union	4.5	1.8	0.9979
Read electrical or flow meter	7.0	2.8	0.9945	Lubricate "O" ring	4.5	2.5	0.9979
Inspect for loose bolts and clamps	6.4	1.9	0.9955	Remove initiator simulator	4.1	1.9	0.9983
Position multiple position electrical switch	6.3	2.4	0.9957	Install protective cover (friction fit)	4.1	2.2	0.9983
Mark position of component	6.2	2.1	0.9958	Read time (watch)	4.1	2.1	0.9983
Install lockwire	6.0	2.3	0.9961	Verify switch position	4.1	1.9	0.9983
Inspect for bellows distortion	6.0	2.7	0.9961	Inspect for lock wire	4.1	2.1	0.9983
Install Marman clamp	6.0	1.8	0.9961	Close hand valves	4.0	2.6	0.9983
Install gasket	6.0	2.1	0.9962	Install drain tube	4.0	2.1	0.9983
Inspect for rust and corrosion	5.9	2.1	0.9963	Install torque wrench adapter	3.9	1.7	0.9984
Install "O" ring	5.7	2.2	0.9965	Open hand valves	3.8	2.6	0.9985
Record reading	5.7	2.3	0.9966	Position two-position electrical switch	3.8	1.5	0.9985
Inspect for dents, cracks, and scratches	5.6	2.4	0.9967	Spray leak detector	3.7	2.0	0.9986
Read pressure gauge	5.4	2.2	0.9969	Verify component removed or installed	3.5	2.4	0.9988
Inspect for frayed shielding	5.4	2.3	0.9969	Remove nuts, plugs, and bolts	3.5	1.7	0.9988
Inspect for QC seals	5.3	2.6	0.6670	Install pressure cap	3.4	1.6	0.9988
Tighten nuts, bolts, and plugs	5.3	2.6	0.9970	Remove protective closure (friction fit)	3.2	1.6	0.9990
Apply gasket cement	5.3	2.3	0.9971	Remove torque wrench adapter	3.0	1.6	0.9991
Connect electrical cable (threaded)	5.2	2.2	0.9972	Remove reducing adapter	3.0	1.7	0.9991
Inspect for air bubbles (leak check)	5.0	2.2	0.9974	Remove Marman clamp	3.0	1.7	0.9991
Install reducing adapter	4.9	1.6	0.9975	Remove pressure cap	2.8	1.8	0.9992
Install initiator simulator	4.9	2.5	0.9975	Loosen nuts, bolts, and plugs	2.8	1.3	0.9992
Connect flexible hose	4.9	2.4	0.9975	Remove union	2.7	1.4	0.9993
Position "zero in" knob	4.8	1.6	0.9976	Remove lockwire	2.7	1.5	0.9993
Lubricate bolt or plug	4.7	1.6	0.9979	Remove drain tube	2.6	1.4	0.9993
Position hand valves	4.6	1.6	0.9979	Verify light illuminated or extinguished	2.2	1.6	

FIGURE 56.19 Example: predicting human reliability.

Causes of primary errors	Preventive measures (taken by designer or methods engineer)
1. Improvising procedures that are lacking in the field	1. Provide adequate instructions.
2. Following prescribed but incorrect procedures	2. Ensure that procedures are correct.
3. Failure to follow prescribed procedures	3. Ensure that procedures are not too lengthy, too fast, or too slow for good performance, and are not hazardous or awkward.
4. Lack of adequate planning for error or unusual conditions.	4. Provide backout or emergency procedures in instructions.
5. Lack of understanding of procedures.	5. Ensure that instructions are easy to understand.
6. Lack of awareness of hazards.	6. Provide warnings, cautions, or explanations in instructions.
7. Untimely activation of equipment.	7. Provide interlocks or timer lockouts. Provide warning or caution notes against activating equipment unless disconnected or disengaged from load, or other damaging conditions.
8. Error of judgment, especially during periods of stress.	8. Minimize requirements for making hurried judgments, especially at critical times, through programmed contingency measures.
9. Critical components installed incorrectly.	9. Provide designs permitting such components to be installed only in the proper ways. Use asymetric configurations on mechanical equipment or electrical connectors; use female or male threads or different-sized connections on critical valves, filters, or other components in which direction of flow is important.
10. Exceeding prescribed limitations on load, speed, or other parameter.	10. Provide governors and other parameter limiters. Provide warnings on; exceeding limitations, inadequate strength of stressed parts, and use of excessive mechanical leverage.
11. Lack of suitable tools or equipment.	11. Ensure that need for special tools or equipment is minimized; develop and provide those that are necessary; stress their need in instructions.
12. Interference with normal habits.	12. Ensure that recognition and activation patterns are in accordance with usual practices and expectancies.
13. Lack of data on which to make correct or timely decisions.	13. Ensure that response time is adequate for corrective action; if not, provide automatic corrective devices.
14. Hampered activities because of interference between personnel.	14. Ensure that space is adequate to perform required activities simultaneously.
15. Inability to concentrate because of unsafe conditions or equipment.	15. Ensure that personnel must not work close to unguarded moving parts, hot surfaces, sharp edges, or other dangers.
16. Error or delay in use of controls.	16. Avoid proximity, interference, awkward location, or similarity of critical controls. Locate control close to readout. Locate readouts above control so hand or arm making adjustment does not block out readout instrument. Ensure that controls are labeled prominently for easy understanding.
17. Error or delay in reading instruments.	17. Ensure that instruments are labeled and designed for easy understanding; do not require reader to turn head or move body; and that visibility problems due to glare or lack of light, legibility, viewing angle, contrast, or reflections are avoided. Provide direct readings of specific parameters so operator does not have to interpret.
18. Inadvertent activation of controls.	18. For critical functions provide controls that cannot be activated inadvertently; use torque types instead of push buttons. Provide guards over critical switches.
19. Controls activated in wrong order.	19. Place functional controls in sequence in which they are to be used. Provide interlocks where sequences are critical.
20. Control settings by operator not precise enough.	20. Provide controls that permit making settings or adjustments without need for extremely fine movements. Use click type controls.

FIGURE 56.20 Error prevention through good design. (From Willie Hammer, *Handbook of System and Product Safety*, Prentice-Hall, Inc., Englewood Cliffs, NJ, 1972, p. 72.)

Causes of primary errors	Preventive measures (taken by designer or methods engineer)
21. Controls broken by excessive force.	21. Ensure that controls are adequate to withstand maximum stress an operator could apply. Provide warning and caution notes for those devices that could be overstressed.
22. Failure to take action at proper time because of faulty instruments.	22. Provide procedures to calibrate instruments periodically, or provide the means to ensure during operation that they are working correctly.
23. Confusion in reading critical instruments because of instrument clutter.	23. Make critical instruments most prominent or locate in easiest to read area.
24. Failure to note critical indication.	24. Provide suitable auditory or visual warning device that will attract operator's attention to problem.
25. Involuntary reaction or inability to perform properly because of pain due to burns, electrical shock, puncture wound, or impact.	25. Insulate or guard against hot surfaces, "live" electrical conductors, sharp objects, and hard surfaces.
26. Fatigue.	26. Avoid placing on operator severe and tiring physical and mental requirements such as loads, concentration times, vibration, personal stress, and awkward positions.
27. Vibration and noise cause irritation and inability to read meters and settings or to operate controls.	27. Provide vibration isolators or noise elimination devices.
28. Irritation and loss of effectiveness due to high temperature and humidity.	28. Provide environmental control. Prevent entrance or generation of heat or moisture from external sources or from internal equipment or processes.
29. Loss of effectiveness due to lack of oxygen, or to presence of toxic gas, airborne particulate matter, or odors.	29. Prevent generation or entrance of contaminants into the occupied space. Provide suitable life support equipment. Avoid presence near occupied areas of lines or equipment containing hazardous gases or liquids.
30. Degradation of capabilities due to extremely low temperatures.	30. Ensure that design provides for adequate heating or insulation, protective shelter, equipment, or clothing.
31. Fixation or hypnosis.	36. Avoid procedures or designs that require visual concentrations for long periods of time. Avoid humming equipment. Provide alternate reference points. Provide procedures to relieve monotony.
32. Disorientation or vertigo.	32. Provide adequate reference points or means to maintain orientation.
33. Slipping and falling.	33. Incorporate friction surfaces or devices, guard rails, access hole covers on floor openings, or protective harness in designs.
34. Inattention.	34. Avoid long intervals between procedural steps. Provide female voice on audio devices to attract attention. Provide bright, colorful, and pleasant work areas.

FIGURE 56.20 *(Continued)*

DESIGN TRADE STUDIES

Once initial hardware functions are assigned and basic performance requirements are established for each specified hardware end item, the engineer and/or designer commences to explore possible design alternatives. The outline in Fig. 56.21 for a design requirements definition sheet is an example of the considerations that the design team—including HFE—must begin to assemble and integrate into a vision of a system design that meets all functional, performance, interface, and quality requirements. Note that this example indicates that the designer must look not only at the design itself but also at the probable test requirements that will be needed to evaluate the eventual performance of the hardware. It's essential to begin considering evaluation techniques concurrently as design concepts mature. During design, alternative solutions should present themselves and design trade studies will need to be conducted. These trade studies are early design evaluations, using criteria that will be used throughout the project to measure design and project success. As the design solutions move from concepts to a

Design sheet	Nomenclature (a)	CEI no. or critical component code identification ________________ Detail spec no. ________ (b)

<u>Requirements for design and test</u>

- 3.0 Requirements
- 3.1 <u>Performance</u>
  - 3.1.1 Functional characteristics
    - 3.1.1.1 Primary performance characteristics
    - 3.1.1.2 Secondary performance characteristics
  - 3.1.2 Operability
    - 3.1.2.1 Reliability
    - 3.1.2.2 Maintainability
    - 3.1.2.3 Useful life
    - 3.1.2.4 Environment
    - 3.1.2.5 Transportability
    - 3.1.2.6 Human performance
    - 3.1.2.7 Safety
- 3.2 <u>CEI definition</u>
  - 3.2.1 Interface requirements
    - 3.2.1.1 Schematic arrangement
    - 3.2.1.2 Detailed interface definition
  - 3.2.2 Component identification
    - 3.2.2.1 Gov't furnished property list
    - 3.2.2.2 Engineering critical components list
    - 3.2.2.3 Logistics critical components list
- 3.3 <u>Design construction</u>
  - 3.3.1 General design features
  - 3.3.2 Selection of specifications and standards
  - 3.3.3 Materials, parts and processes
  - 3.3.4 Standard and commercial parts
  - 3.3.5 Moisture and fungus resistance
  - 3.3.6 Corrosion of metal parts
  - 3.3.7 Interchangeability and replaceability
  - 3.3.8 Workmanship
  - 3.3.9 Electromagnetic interference
  - 3.3.10 Identification and marking
  - 3.3.11 Storage
- 4.0 Quality assurance provisions
- 4.1 <u>Category I test</u>
  - 4.1.1 Engineering test and evaluation
  - 4.1.2 Preliminary qualification tests
  - 4.1.3 Formal qualification tests
  - etc.

FIGURE 56.21 Example design requirements definition sheet.

preliminary baseline, key project milestone design reviews entail formal points at which major decisions will be made as to whether to proceed with the design, enter into redesign, or drop the ongoing approach entirely due to technical, schedule, or budgetary difficulties. HFE should be given high priority during design trade studies to ensure that the human elements of the system are being given equal consideration to hardware and software design concerns and challenges.

Design is an iterative process of proposing solutions—first conceptual, then maturing into preliminary baselines that can be physically and/or functionally tested—and evaluating these proposed designs for compliance with the needs and constraints defined by the system requirements. ConOps, function allocation, and task analyses. In this manner, design and evaluation are recursive and repeated from concept through preliminary and detailed design. Final verification of the system requirements at product delivery is the last act of the development process and if design and evaluation have been performed iteratively throughout the system's development, there should be no surprises during final verification.

Human Factors Engineering Design Analysis

The HFE responsible party in a project should always be considering human–machine evaluation at every stage of development—e.g., what evaluation technique should be used to assess human–machine interaction assumptions and proposed design solutions, what evaluation techniques can be performed in the time and budget allocated, and which human–system design decisions are the most critical to evaluate to ensure system success. The examples in Figs. 56.22 and 56.23 reflect some of the concerns the HF engineer might consider.

Suggested Steps for Human Factors Engineering during System Design

Following is an example sequence of events and actions the responsible party for HFE on a project will perform or participate in during the design process:

1. Review the mission, purpose, and use scenario for the proposed product or system until the basis for the product or system and the conditions under which it is to be used are fully understood.

2. Where practical and appropriate, engage the services of qualified, experienced human engineering specialists to assist and advise throughout the development of the proposed design and its final production. This may be particularly critical for highly specialized endeavors such as human spaceflight or submarine operations.

3. Acquire and/or develop a "human engineering design checklist" to be used throughout the design development and production cycle. (See the next section in this chapter.) Although general checklist examples can be found—e.g., the one provided in this chapter—take time to create a checklist that is tailored to the nature of the particular system under design. This checklist typically requires modification as the design development progresses since, as one comes to know more about the design features, additional user–hardware interface features generate additional human engineering questions that must be addressed and monitored throughout the design cycle.

4. Acquire appropriate human engineering design guides and references. If a qualified HFE specialist is

Design alternatives	Operator outputs	Evaluation
1. Portable, hand-held illuminator with open sight	Target search activity	Direct vision is the most efficient method. Both pilot and operator will participate. When one detects or identifies the target, he or she will inform the other of the target location. Feasible under daylight conditions.
	Target detection and identification	Visual angle must be at least 5 mils for detection of target. Target detection at maximum range of 20,000 ft will be limited to targets of 200 ft or greater in diameter. Smaller targets will require a reduction in range. This activity is feasible with the stated limitations.
	Aircraft control	Aircraft control is a factor in that the pilot must be aware of the target location in order to properly fly the plane. A comparison of systems must therefore take this activity into account. Given that the pilot is aware of the target location, the activity is feasible.
	Etc.	Etc.

Figure 56.22 Example: design trade-off study, operator implication.

Human engineering design tradeoff

Design criteria	Design approaches 1	2	3
Design for operability			
Number of operations			
Complexity or difficulty of operations			
Duration of operations/sequence of operations			
Relation of operations to crew capability			
Training requirements			
Degree to which operations can be integrated with other mission activities			
Control/display relationships			
Probability of error occurrence			
Probability of error detection			
Information availability			
Information accessibility			
Controllability			
Visibility - Lighting requirements			
Number of operators required			
Design for maintainability			
Ease of malfunction detection			
Ease of malfunction isolation - diagnosis			
Preventive - scheduled maintenance requirements			
Equipment accessibility			
Ease of removal/replacement			
Ease of repair			
Support requirements - tools, test equipment, etc.			
Requirements for job aids or manuals			
Availability of redundant systems and/or spaces			
Number/duration of operations fir checkout, Alignment, Installation, calibration			
Requirements for special handling			
Number of technicians required			
Special tools/equipment requirements			
Design for safety			
Safety hazards			
Safety provisions			

Figure 56.23 Example: human engineering design trade-off analysis.

available, he or she probably will already be familiar with these. Some large organizations—e.g., government agencies—are likely to have developed their own tailored HFE and/or human systems integration (HSI) standards that they require to be applied to the design processes within their organizations. Some examples of standards and guides are:

a. MIL-STD-1472G, *Department of Defense Design Criteria Standard: Human Engineering.*

b. NUREG-0711, Nuclear Regulatory Commission, *Human Factors Engineering Program Review Model*

c. NASA-STD-3001, *Space Flight Human System Standard*, Volume 2: *Human Factors, Habitability, and Environmental Health*, and accompanying NASA/SP-2010-3407, *Human Integration Design Handbook.*

d. Federal Aviation Agency (FAA) HF-STD-001, *Human Factors Design Standard*

5. Review initial design concept ideas to make sure that all potential user interface aspects of the proposed *design concept* have been properly identified and considered.

6. Using the human engineering checklist, monitor each design activity as it progresses from preliminary through detailed design, making sure that the HF requirements are kept constantly in mind and that good human engineering practice, principles, and criteria are

being considered during each step of the design process. Keep a running record of whether human engineering principles and criteria are being applied, how they are being evaluated (see Fig. 56.24), and, if compromises are being made, of how and why these are being introduced in the design. This record may be extremely important, for example, if a military customer requires evidence that HSI has not been ignored or if justification is later required to explain why a human engineering design decision was made in the event of a product liability law suit. A baselined project HFE or HSI plan is likely to serve to document the project's history of HFE/HSI decisions. Such a formal plan should be updated regularly throughout the project's maturation.

Method	Purpose	Typical use
Worker and consumer interrogation	To define general product user needs	When a system or product is to be redesigned or new ones are to be developed, it is desirable to ask people who are using a similar product or system how they feel about it in terms of adequacy to do the job, ease with which the job can be done, and/or special problems that occur with the present hardware, facility, or tools.
Human-machine operational observation	To define general, dynamic, and environmental factors associated with product use	Although interrogation of the user may include demonstration, it is advantageous to observe an operation convertly over an extended period of time in order to have a more abjective assessment of problems the operator may have.
Personal operation experience	To provide the designer with a personal "feel" for the problems a worker may have identified or may have observed from a distance	When practicable, it is recommended that the designer actually operate or use a product similar to the one he or she intends to redesign and/or create in order to experience at firsthand the problems or needs which a worker has pointed out or which the designer may have observed from a distance.
Time and motion study	To measure task performance against a time base and thus identify critical product or task conditions	Especially useful for systems in which worker output is a significant factor, i.e., in which increase in rate of production is a primary objective of the system. The method provides quantitative information that helps establish priorities for design and procedural improvement, either through redesign and/or through new design.
Preliminary design review	To evaluate the general, overall probability that a design will meet original operational objectives	Preliminary design drawings provide the first general picture of what a system or product will look like and how it will probably work. It is especially important at this stage to examine the product-user relationships to make sure that these are compatible with operator, maintainer, and user characteristics and limitations, prior to spending too much time with detailed design efforts.
Mockup evaluation	To evaluate both general and specific three-dimensional relationships between various elements of a proposed system and/or specific relationships between a specific part of the product or system and its user	Miniature scale mockups provide an overall view of a system that could not be viewed effectively in full scale. Full-scale mockups provide one-to-one evaluation of direct product-operator physical relationships and quasi-operating conditions.

Figure 56.24 Example: typical methods for human engineering evaluation of system and component design.

Method	Purpose	Typical use
Simulation	To investigate and verify dynamic aspects of a proposed design prior to final commitment. To provide an opportunity to explore alternatives both during predesign and during detailed design	A proposed design concept may be examined and/or tested at several levels prior to committing it to preparation of preliminary or detailed design. It is important to include the human user during such simulations since the ultimate effectiveness of the product includes the typical user input, both intentional and unintentional. Simulation should be in "real time" and should include representative environmental, dynamic, and personnel elements, depending upon the stage of design at which the simulation effort is being performed. Evaluative methods should include quantitative measures of system or product-operator performance.
Detailed design drawing review	To verify human-machine and product-user compliance with good human engineering practice	Methodical analysis of detailed design drawings prior to their release to production is vital to prevent typical oversights of human-machine interface problems. This is critical because of the probability that changes may have been made since original human engineering inputs were made at the predesign stage.
Prototype testing	To verify the operating and maintenance effectiveness of the system or product under actual operating conditions	Whenever practicable, a prototype of the product or system should be tested prior to committing it to production. Such tests should include use of operator, maintainer, and user personnel who are representative of the final user population. Although initial tests may be made by special test personnel in order to ensure safety of operation, the ultimate question is whether typical users can and will operate the system or use the product as planned. Quantitative measurement of human-machine performance should be accomplished whenever the complexity, integrity, or safety of the system is critical.
Field evaluation	To provide a continuing assessment of system or product effectiveness under typical use and environmental conditions, in order that service requests can be effectively met, design modification requirements can be defined, and/or new concepts can be anticipated	Systematic follow-up using formal data and information retrieval methods provides a means for a manufacturer to maintain adequate customer relations and to develop insight useful for expanding the product's capability. To provide future human engineering guidelines, useful human factors data should be collected at the same time that pure hardware data are being collected.
Human factors experiments	To support current and/or future design requirements	Special human factors experiments may be required at various times, either during a product development and/or independently. Cognizance of applicable human factors research information is required at the beginning of any development program to assist in defining the human role in system operation. When research data do not exist, special studies may be required to define human capabilities with respect to a specific, proposed application. Whether to perform such research depends on the criticality of the question and/or the time available to perform the research. Such research should have as broad an application as possible, as long as it provides the necessary answers to the immediate problem, i.e., when the answer is too specific, the cost is increased over the long run, since the research may have to be repeated for the next program.

FIGURE 56.24 *(Continued)*

7. Use system models or mockups to "test" the efficacy of all user–hardware/–software interface designs, using an appropriate sampling of the end-item user population as subjects. Examine and evaluate the model–/mockup–operator interfaces in terms of human performance efficiency (time, error, inconvenience, comfort, inadvertent hazard potentials, etc.). Record these observations and seek appropriate design modifications. Modify any design modification in the model or mockup and reevaluate.

8. Perform experiments when necessary to establish design criteria where previously cited reference guides or legacy system operations do not provide adequate information for design decisions. This may require development of special, dynamic, real-time simulations of procedural and environmental conditions.

9. Whenever possible, fabricate a hardware prototype and evaluate this under real-life conditions (using typical expected user subjects as well as expert operators). Obtain quantitative performance measures of the total user–hardware operation to prove that the combined human–machine operation is satisfactory. Make sure that all deficiencies are fed back into the design cycle and that appropriate design modifications are made. Retest if necessary.

10. Critically review final production drawings to make sure they are correct before being released for final production fabrication and assembly.

11. Identify critical areas where lack of proper production quality control might result in poor user–hardware interface results because of "sloppy" manufacturing or assembly procedures. Take steps to establish necessary procedural control and inspection to preclude mistakes in the factory.

12. Perform production (field) tests of production hardware before approving it for final delivery to the customer. Ideally, these tests should include "human-in-the-loop" exercising, not merely visual inspection and/or analysis of the hardware and software components alone.

DESIGN AND EVALUATION CHECKLISTS

Suggested Considerations for Human Factors Engineering Product Evaluation

In preparing for and performing design evaluations, the HF engineer is likely to mentally "walk through" a variety of HFE evaluation methods and criteria based on his or her prior knowledge and experience prior to selecting and proposing an HFE test to the design team. This application of HFE domain knowledge is similar to what any experienced designer or engineer would consider before actually proposing a physical or functional test. The following list reflects some concerns that the HF engineer will consider for evaluating whether a product is compatible with user behavioral expectancies:

1. Is it obvious what the product is and what it is to be used for?

2. Is it obvious how the product is to be used?

3. Is the product simple to use, i.e., is it simple to prepare for use, to begin to use or operate, to continue to operate, to stop using or operating, and to place in a stored or nonoperational condition?

4. Are there hidden hazards to the operator or user, i.e., hazards that cannot be readily observed when looking at the product?

5. Will use of the product create hazards for others not involved in its use?

6. Could the product be operated and misused by someone who should not be using it, such as a child?

7. Could the product be used for some purpose for which it was not intended?

8. Should the product be used only under certain conditions, and is this immediately obvious to the potential user?

9. Are potential product failures or conditions under which the product should not be operated or used identified for the user in time to avoid misuse?

10. Would use by someone who is fatigued, under the influence of drugs or alcohol, or otherwise incapacitated be hazardous to the user or others—and if so, does the design provide appropriate use-prevention features?

11. Are there potential hazards associated with the product when it is not being used, when it is stored, or when it is otherwise unattended—and if so, does the design provide appropriate use-prevention features?

12. Is it critical that the product be serviced or maintained in a specific manner—and if so, are the servicing requirements made clear to the user?

13. Is the product prone to damage or failure—and if so, does it contain a failsafe mechanism?

14. Is the product easy and convenient to service and maintain without special training—and if not, is this clear to the user?

15. Can the product be used when it is not functioning properly without creating a hazard to the user?

16. Is the product, in any way, outside the normal user's physical, mental, or normal behavioral capacities (e.g., in terms of size, weight, or operation)?

Example: Preliminary Human Factors Engineering Checklist for Initial Hardware Selection and Design Analysis

The HF engineer is likely to codify his or her concerns and discipline needs for a project into a plan that may take the form of a checklist, tailored to the particular project. The checklist example in Fig. 56.25 provides a starting point for gross evaluation of a project, but as the project matures it will need to be expanded for detailed design evaluation. Evaluation designations are as follows: S—satisfactory; C—compromise but acceptable; and U—unsatisfactory. Various other designations are also possible to include evaluative indications of temporary status.

S | C | U

A. Equipment operation
   1. Console shape/size
      a. Desk height, area
      b. Control reach
      c. Display view
      d. Body, limb clearance
   2. Panel location
      a. Frequency of use
      b. Sequence of use
      c. Emergency response
      d. Multi-operator use
   3. Panel layout
      a. Functional grouping
      b. Sequential organization
      c. Identification
      d. Spacing for clearance
   4. Displays
      a. Functional compatibility for intended purposes
      b. Intelligibility of information content
      c. Control interaction
      d. Legibility; figures, pointers, scales
      e. Visibility; illumination, parallax
      f. Location
      g. Identification
   5. Controls
      a. Functional compatibility for intended purpose
      b. Location, motion, excursion and force
      c. Display interaction
      d. Spacing, clearance, size
      e. Identification

B. Assembly – services – maintenance
   1. Installation, service, & maint accessibility
      a. Location, size of openings
      b. Covers, fastening/removal
      c. Identification
   2. Equipment handling/transport
      a. Size/shape/weight/balance
      b. Handling clearance
      c. Handling aids
      d. Instructions/labels/warnings
   3. Chassis layout, packaging
      a. Ease of handling
      b. Access to components for tests, component replacement
      c. Identification
      d. Hazard/damage protection
   4. Cables/lines/connections
      a. Ease and security of assembly-disassembly
      b. Connection error
      c. Identification
      d. Access, test, trouble-shooting replacement

C. System safety
   1. Personnel hazards
      a. Shock
      b. Burns: direct, chemical
      c. Hearing damage
      d. Tripping/falling
      e. Pinching
      f. Cutting
      g. Bumping
   2. Equipment damage
      a. Electrical overload, short, ground
      b. Mechanical overload, strip, bend, rupture, break
      c. Explosion/fire

D. General
   1. Labels/marking
      a. Intelligibility
      b. Legibility
      c. Location, spacing
      d. Permanence
   2. Equipment finish
      a. Color
      b. Texture
      c. Reflectivity
   3. Storage
      a. Location
      b. Volume
      c. Material accessibility, security
   4. Work area illumination
      a. Light level; range, control
      b. Distribution, contrast
      c. Color

Figure 56.25 Preliminary human engineering checklist.

Human factors checklist
for
design engineers

General dynamics

Quincy division

Example: Detailed Human Factors Engineering Design and Evaluation Checklist

As noted earlier, every project should tailor their human engineering needs and concerns into a project-specific detailed list of design and evaluation intentions. The scope and level of detail will be dependent on many factors with the size and complexity of the project being the most likely driver. The example in Fig. 56.26 may provide a useful point of departure for the individual who is attempting to construct a checklist for the first time. This example is provided courtesy of the Human Factors Group at the Electric Boat Division of General Dynamics.

Pass | Marginal | Reject

i. Plugs and receptacles have painted stripes, arrows, or other indications to show the position of alignment pins for proper insertion.

j. The system is designed so that all "hot" contacts are socket contacts.

5.10 *Fasteners*

a. Maximum use is made of tongue-and-slot catches to minimize the number of fasteners required.

b. The number and diversity of fasteners are the minimum commensurate with requirements for stress, bonding, etc.

c. Where feasible, the same size and type of fastners are used for all covers and cases.

d. Captive fasteners are used where possible.

e. Screws with different threads are of different sizes.

f. If compatible with stress and load considerations, fasteners for mounting assemblies, subassemblies etc., fasten or unfasten with a maximum of one complete turn.

g. Bolts requiring high torque are provided with external grip heads.

h. Captive bolts and nuts are used in situations where the dropping of these small items into the equipment will cause damage or create a difficult removal problem.

5.11 *Test points*

a. Where a unit is not completely self-checking in its operation condition, appropriate test points are provided which provide a measure of unit output.

b. Only such primary test points as are necessary to determine that a unit is malfunctioning are provided.

c. Primary test points are located and coded to be readily distinguishable from secondary test points.

d. Where feasible, primary test points are grouped in a line or matrix reflecting the sequence of tests to be made.

Pass | Marginal | Reject

e. Primary test points used in adjusting the unit are located close to the controls and displays used in the adjustment.

f. Where feasible and not in conflict with other requirements, a secondary tests point us supplied at the input and output of each part or throwaway component.

g. Sufficient test points are provided so that it is not necessary to remove subassemblies from assemblies in troubleshooting.

h. Each test point is marked to be readily identifiable.

5.12 *Lubrication*

a. Units containing mechanical components requiring lubrication is specified by a label at or near the lube port.

b. Where lubrication is required, the type lubricant to be used and the frequency of lubrication is specified by a label at or near the lube port.

5.13 *Unit Labeling*

a. Wherever possible, the outside covering of manufactured parts such as resistors, condensers, and tubes, is stamped or coded with relevant information concerning electrical characteristics of the parts.

b. Where space permits, terminals are labeled with the same code symbol as the wire attached to it.

c. If surface labels must be used, silk-screened or stamped labels are used in preference to stenciled labels or decals.

d. Labels are not hidden by units or parts.

5.14 *Equipment color*

a. The console interior of operational and maintenance equipment is gray. Color 26622, Fed. Std. 595, where maintenance and troubleshooting are required within the console.

b. Where maintenance and troubleshooting within the console will not be required, an economical internal protective finish is used.

FIGURE 56.26 Human factors checklist for design engineers.

Figure 56.27 is a second example of a detailed human engineering design and evaluation project checklist, this one for the design of a vehicle.

General accommodations How appropriate are the general passenger, driver, and package carrying and storage accommodations?	Satisfactory	Unsatisfactory
Ingress and egress clearance for both normal and emergency use and for the handicapped and the elderly	________	________
Interior space dimensions (passenger width, head, and leg clearance)	________	________
Temperature and ventilation	________	________
Ride characteristics (sway, oscillation, and vibration)	________	________
Interior and exterior package storage space (amount, location, and accessibility)	________	________
Crash protection (restraint systems, noninjurious interior, handholds, and shatterproof glazing)	________	________
Noise minimization	________	________
Adequate seating (size, shape, seat-back angle, adjustability, properly secured)	________	________

Driver work station How appropriate are the layout of the driver's work station and the provisions for proper vehicle control?	Satisfactory	Unsatisfactory
Geometric relationship between seat, controls and external viewing (to accommodate 5th-percentile female through 95th-percentile male drivers)	________	________
Control and display location and arrangement and detail interface designs	________	________
Exterior viewing capability (front, side, and rear, i.e., clear angle of view, including normal and adverse weather conditions)	________	________
Control system adequacy (steering force and rate, brake force, clutch force and excursion, accelerator force and excursion, and transmission lever pattern and force characteristics)	________	________

General safety	Satisfactory	Unsatisfactory
Cab integrity (front, rear, and side crash; roll-over)	________	________
Nonflammable interior materials	________	________
Flame barriers between passenger compartment engine and rear fuel compartment or tank	________	________
Fail-safe systems (power windows, seats, steering, and brakes)	________	________
Non-brake-lockup and skid control	________	________
Blowout-proof tires	________	________
Energy-absorption systems (front, rear, and side impact)	________	________
Noninjurious pedestrian-vehicle interfaces (sheet metal, handles, and trim)	________	________
Rupture-proof fuel tank	________	________
Steering-wheel lock	________	________
Rearview mirrors	________	________
Exterior signal lights (stop, turn, running, and backup)	________	________
Audio backup signal (commercial, construction, and agricultural vehicles; buses and trucks)	________	________
Maintenance interfaces (fans, belts, open gears, electrical connections, jacking, fuel service, and sharp corners and edges on sheet metal)	________	________

Sample human engineering checklist (for architectural facilities)

Site planning How appropriate is the layout of structures and related roads and walkways?	Satisfactory	Unsatisfactory
Main visitor reception point obvious from the expected normal approach	________	________
Main visitor and vehicle entry obvious from the expected approach street	________	________
Facility address conspicuous and readable from typical "first-view" point or points	________	________
Visitor parking convenient to visitor entry from the street, pedestrian approach and from visitor parking	________	________
Convenience of employer and employee entrances	________	________
Convenience and separation from visitors of services and delivery entrances	________	________
Interbuilding pedestrian convenience	________	________
Separation of noise-producing or hazard-associated facilities from other occupied buildings	________	________
Access for the handicapped (from parking area to and from buildings)	________	________
Signing adequacy (number, location, and size)	________	________

Building design How appropriate are the general size, shape, interior organization and orientation of the basic building structures?	Satisfactory	Unsatisfactory
Entrances and exits (number, location, size, door manipulability, identification, illumination, and accessibility to the handicapped)	________	________
Exterior and interior entrance and exit thresholds (size, shape, door clearance, protection from adverse weather, accessibility to the handicapped, illumination, steps, ramps, and guardrails)	________	________
Working and living-space size, shape, access, illumination, ventilation, heating and cooling, acoustics and service outlets	________	________
Intraspace organization for ease of traffic flow, control of specifically restricted areas, ease of communication, privacy and flexibility for alternative use	________	________
Convenience of common-use spaces (rest rooms, storage, central utilities, etc.)	________	________
Intralevel traffic convenience (normal, emergency, and for service personnel and the handicapped)	________	________
Interior signing effectiveness (conspicuous, visible from strategic vantage points, braille for the blind, and illumination)	________	________

Stair, ramp, escalator, and elevator design	Satisfactory	Unsatisfactory
Optimized stair configuration (riser and tread dimensions and incline angle, width, number of steps in relation to rest platforms, approach and departure platform dimensions, handrails, illumination, and nonslip tread)	________	________

Figure 56.27 Sample human engineering checklist (for vehicles).

Optimized ramp configuration (ramp angle for the handicapped, width, ramp length in relation to rest platform, handrails, approach and departure platform dimensions, nonslip tread and illumination)	______	______
Escalator characteristics (stair riser and tread dimension, lead-in dimensions, handrails, rate of movement, and visual contrast and illumination)	______	______
Elevator characteristics (interior size, entry dimensions, safety doors, interior handrails, operator control location and layout, including braille identification, ventilation, illumination, and rate of movement)	______	______

Doors, windows, hardware, and fixtures	Satisfactory	Unsatisfactory
Door and doorway characteristics (clearances, door-swing interference and type and location of handles and locks in terms of use by the handicapped)	______	______
Window characteristics (ventilation and drafts; ease of opening, closing and locking; safety relative to inadvertent contact; glare control; and ease of cleaning and repair)	______	______
Plumbing fixture characteristics (accessibility, including to handicapped users; ease of manipulation; absence of injury-producing features in the event of inadvertent contact; accessibility for repair; and appropriate; positioning of fixtures that require perimeter clearance, e.g., sink, toilet, bidet, and urinal)		
External fixture location, i.e., hose bibs, electrical outlets and luminaries	______	______
Interior lighting fixtures (type, number, location, and switch access)	______	______
Environmental system (heating and ventilating distribution, outlet and return locations, automatic temperature sensor control location, and temperature control and display configuration)	______	______

Floors	Satisfactory	Unsatisfactory
Level, nonslip, well-defined grade changes	______	______
Adequate illumination, nonglare surfaces, and nonillusory patterns	______	______
Ramps at level changes for the handicapped	______	______

General safety	Satisfactory	Unsatisfactory
Nonflammable materials	______	______
Adequate number, type, location, and marking of emergency exit routes and exits	______	______
Grounded electrical systems and fixture	______	______
Adequate normal and emergency illumination	______	______
No sharp projections or edges	______	______
View ports in doors that swing in both directions; no revolving doors	______	______
Safety glass or plastic in floor-length windows or doors	______	______

FIGURE 56.27 *(Continued)*

Chapter 57
System Evaluation

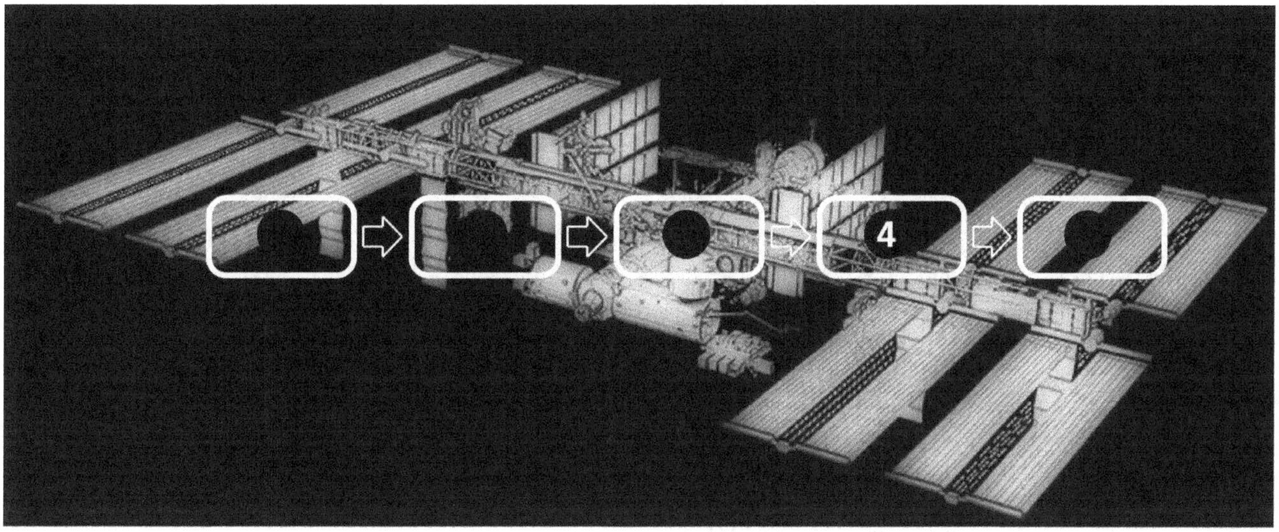

TESTING

General Considerations for Conducting Human Factors Engineering Tests

First, it is highly recommended that an experienced human factors (HF) specialist be employed to assist in the design and conduct of human engineering tests whenever this is practicable. It is particularly important if the proposed test deals with several highly interactive variables that require complex experimental designs in order to isolate and establish the significance of certain variables relative to human performance. It is equally important to acquire help when a particular test may involve possible hazards to the test subjects.

However, designers can conduct certain types of tests on their own as long as they keep in mind certain problems associated with testing that involves human subjects. The following should be considered:

1. A sufficient number of test subjects and trials should be used to ensure that the test results do not represent some sporadic incident that is not representative of the true performance of the subjects.

2. Sufficient procedural control should be provided so that one can safely infer that, "when this adjustment or input occurs, with these conditions, this is the typical response or human output." Generally, one must keep certain features constant while a single variable is adjusted so that, in the end, it is possible to correlate specific conditions with specific human responses.

3. In most cases, test subjects should not be privy to the real purpose or objectives of the test; they should be informed only of the procedures they are to follow. This helps avoid "expectancy biases," wherein subjects may attempt to give the experimenter the response they think the experimenter wants, rather than performing according to their individual or unique response characteristics.

4. The test procedure should cause entry and sequential aspects of the test to be randomly presented; i.e., it should not be possible for the subject to "learn" a sequence or anticipate a step because of start cues. Attempt to remove any possibility that "order effects" show up in the final subject response.

5. When a particular hardware configuration is being evaluated in terms of whether subjects perform better on one version than another, make sure that alternative configurations are equally well represented (in terms of quality of realism, workmanship, and to visual attractiveness).

6. Experimenters should remain as non-conspicuous as possible during evaluations in order to minimize inadvertent cues they might provide the test subjects during a test sequence.

7. Attempt to set up the test so that the trials are long enough to identify the difference between "learning" and informed response. However, trials should not be so long that fatigue produces artifacts in the results that have little to do with the basic evaluation objectives. Note, however, that fatigue effects may be a significant test objective, in which case one must design the test so as to differentiate between design-induced fatigue and fatigue that results merely from the fact that the test subjects are generally tired or bored with the activity.

8. Whenever possible, design all tests in such a way that quantitative (as well as qualitative) results can be documented. Avoid general opinion-oriented evaluations unless these are quantified in terms of forced choice ratings or rankings by test subjects. Incidentally, this is desirable even for general VIP evaluations of a proposed design. Although some visitors may object to being used as test subjects, it is possible to formalize their review in a manner which is not objectionable to them but which is organized sufficiently to allow one to tabulate and compare opinions.

9. Select the right subjects; e.g., avoid using engineers to represent a typical nonengineering user population.

10. Obtain important demographic information about each subject so that, in the event it becomes necessary to evaluate test results in terms of subject differences, you do not have to try to find a particular subject again. In fact, attempt to screen out subjects who for some reason may not be able to complete a test series or who may not be available for a second test.

11. When a test involves visual acuity (visual display, illumination level, etc.), either select subjects on the basis of their having normal acuity (by giving the subjects an eye test), or specifically seek out and include subjects who have the typical visual anomalies you expect among the eventual user population. For example, if the user population will include older people, who normally require higher illumination levels and larger visual detail to see properly, be sure to include subjects who are older and may have to wear corrective lenses.

12. When the design to be tested may be used by persons with various handicaps, select a broad range of handicapped subjects because, although a given subject may have a particular type of handicap, each such individual tends to vary in terms of how he or she copes with various mobility, manipulative, visual, or auditory tasks.

13. Carefully plan any human engineering task in terms of the procedures to be used and practice conducting the test prior to conducting the main test. Remember that, once a test has begun, it should be continued to completion; i.e., avoid having to stop in the middle of a test sequence because you have forgotten something or have lost your place in the test sequence or because the data-recording system has failed. Nothing aggravates test subjects more than feeling that the experimenter is not sure of what he or she is doing.

14. If a particular design evaluation involves special considerations, such as time of day (i.e., both day and night ambient illumination conditions) or special clothing (i.e., light summer clothing and/ or heavy winter clothing), be sure to include these conditions in the test.

15. Make sure that the mockup and or test apparatus is fully checked out and will not fail during the test. Be especially careful to see that there are no unforseen hazards either for the subjects or for the experimenters. If there appear to be certain stress conditions associated with the test, make sure that none of the subjects who are selected have tendencies or conditions that could be compromised during the stressing phases of the test (a potential for epileptic seizure, heart failure, etc.).

Field Testing and Evaluation

Nothing provides a more conclusive demonstration of adequate design than the actual field test of a prototype model. Typically, this is done in the case of airplanes, highway vehicles, sea craft, agricultural machines, etc. Too often, however, such tests rely on the opinions of the so-called expert test personnel, e.g., test pilots or test drivers. Although there is often a good reason to utilize these experts (for reasons of safety and because they are perhaps more skilled and perceptive in their analysis of features that may not be quite right), the real proof of design acceptability comes when the intended user (who lacks professional skill) can also operate the system.

Although the main objective of most field tests is to prove that the system will do what it is supposed to do, too often one may not obtain the type of information from such tests to suggest modifications that will make the system easier to operate by the average user. In addition, the field test should provide other information that is important to the eventual user; e.g., previously described tasks can be verified, training objectives can be confirmed, and training aids can be evaluated.

A field test should be designed with the following objectives in mind:

- It will demonstrate the reliability of the hardware under all mission and environmental conditions.

- It will demonstrate that the training program provided all the necessary skills for the operator or maintainer to cope with the operational conditions under which the system is to be used.
- It will pinpoint any hardware, software, operator interface, or procedural discrepancies in such a way that one knows what to do to correct deficiencies in any of these areas (e.g., hardware, software, human interface, procedures, and training).
- It will allow one to verify whether estimates of manning level are correct.
- It will allow one to evaluate whether estimates of time factors were correct (service turnaround, loading, boarding, emergency escape, etc.).
- It will help determine any safety hazards that may not have been anticipated.
- It will provide an initial impression of consumer acceptability of the product.

Methods and Techniques for Human Factors Engineering Field Testing

- Operator Performance Measurement: A quantitative measure of operator and maintainer performance in terms of time and error should be provided; i.e., in addition to the probable effort to elicit personal evaluations by the test subjects and/or expert test operators, actual measurement of the performance of these participants should always be generated wherever practicable.
- Complete Test Operational Scenario: The field test should be designed and implemented so that one systematically exercises the human–machine system through all phases of the operating scenario. This should include both normal and emergency conditions. The scenario should also be complete in terms of all the human–system interactions: visual, auditory, mobility, dexterity, communications, decision-making, and control.
- Real-World Environment: The test action should occur on site, i.e., on real roadways and streets, in rough water, during the day and at night, and under different environmental conditions (rain, snow, heat, cold, high wind, fog, etc.).
- Performance Recording: In addition to the actual instrumentation of operator interface responses noted above, consider the use of other covert techniques such as voice recorders, motion picture cameras, and human physiological sensors (pulse, heart rate, galvanic skin resistance, etc.).
- Operator Opinion: Provide a means for documenting the opinions of test pilots, drivers, and/or typical test subjects, not only in terms of posttest debriefing, but also in terms of on-line evaluation during the testing (e.g., a voice recorder). Debriefing questionnaires should be designed to elicit specific, design-related comments, as opposed to general verbal descriptions of what the subject felt about the equipment.
- Test Observers: Where appropriate, consider the use of trained independent observers to take notes while the test is under way. The independent observer may be important because the actual test subject is sometimes too busy to observe critical events as they are happening.

  Trained observers are also effective in performing general population activities. For example, strategically located observers can observe and document how many drivers are wearing seat belts, how pedestrians behave with respect to crossing streets or using paths on a campus, or how individuals utilize a particular piece of equipment, such as a ticket vendor, a coin-operated laundry machine, an elevator, or an escalator.

  Equip observers with tabulating devices where appropriate (stopwatch, hand counter, sketch pad, questionnaire and documentation record pad, etc.).
- Test Subject Indoctrination: All field test subjects should be properly indoctrinated before they participate in any field testing. Depending on the complexity of the test, try to develop a brief statement of the purpose and objectives of the test and a general overview of what will be expected of the subject prior to, during, and following the test. This should be written down so that all subjects receive the same instructions. In some cases, it may also be desirable to provide written instructions during the test; i.e., so that each subject receives exactly the same instructions, each step is prepared and presented on a slide or by means of a recording or is read to the subject by the experimenter.
- Test Conduct: Although it is generally best to leave the subjects alone as much as possible in order to reduce the possible influence of an experimenter's method of instruction, voice inflection, or body language, it is sometimes desirable to have the subjects verbalize about what they are doing while the test is in progress. This should be part of the indoctrination, with possible reminders by the experimenter if the subjects seem to forget.

  When several different experimenters are to be used, allow them to practice until they demonstrate that they perform consistently and in a similar fashion.

  In certain kinds of field testing, it will be necessary to assign the responsibility for safety to a given individual; i.e., although each experimenter is responsible for the safety of his or her particular subject during the test, general overall safety monitoring should be provided. This is especially important when several equipment or vehicle components are engaged in an interactive operation.

Safety equipment should be provided to deal with emergencies in certain types of field tests, such as fire trucks, rescue equipment and personnel, and medical personnel.

When extended-duration field tests are being conducted, it may be necessary to provide housing, food, or other support for both the test personnel and the test subjects.

- In-the-Field Test Data Analysis: Depending on the nature, location, and duration of a particular field test, it may be important to provide facilities and equipment for analyzing data on the spot. Such data may be important to the test conductor in terms of deciding whether to make modifications to the test schedule or procedures. Alternatively, it may be possible to transfer data via various telecommunication methods to a base site or to a laboratory where large computer facilities are available. Such remote analysis is often required when the basic system is operating in the air or in space.

- Hazardous-Environment Tests: When a test involves the use of test personnel or subjects in a hazardous environment, it is extremely important to fully instrument both the system and the people involved so that the test conductors can monitor in real time the exact status of the hardware and the individuals involved, such as an astronaut aboard a space vehicle or a pilot performing unusually hazardous maneuvers in an aircraft. Fail-safe communication links must be provided so that information links are not broken as a result of physical conditions or environmental anomalies.

Wherever practicable, rescue personnel must be made available for quick emergency response. In addition, system experts should be at hand to advise the test personnel aboard the system how to correct problems, how to eject or abandon the system, etc. All possible emergency events that can be anticipated should be analyzed, and procedures should be practiced before the test is begun. Above all, provide the test personnel as much onboard capability as possible for taking care of their own emergency; they may not have enough time to communicate with, and receive instructions from, the base test conductor.

The Cooper Rating System. The use of "expert operators" to evaluate the apparent controllability of a complex system can provide an effective method, and possibly the only method, for determining whether a system is acceptable. This requires not only a stable of expert operators but also a method for quantifying the opinions of these experts. The Cooper Rating System was developed to allow experienced pilots to evaluate the quality of a new aircraft in terms of its controllability. It is recommended that some method similar to that outlined in Fig. 57.1 be devised for any system in which the quality characteristics may be critical to the safety and/or general consumer acceptability of the product.

Of special importance, however, is the recognition by the "expert" evaluators that they may have skills that are not particularly representative of those of the general consumer population. Therefore, in evaluating a system, the expert must view the system from the consumer's point of view, i.e., put himself or herself in the position (physically, psychologically, and in terms of experience) of the consumer. The designer should identify the probable differences between the intended user and the expert and make sure that the expert is indoctrinated.

Using a Random Sample of Consumers to Help Identify User/Design Problems. By exposing a large number of typical consumers to currently available systems or equipment, one can generally determine what problems they will have in operating the equipment. The test should be designed so that each subject "uses" the product in systematic steps and then rates the problems he or she has on a scale from "No Problem" to "Serious Problem." The accompanying sample test asks subjects to rate different aspects of automobile seat belts in several different vehicles.

Example Test Procedure. Figure 57.2 illustrates an example subject data, procedures, and data collection questionnaire

Operating conditions	Adjective rating	Numerical rating	Description	Primary mission accomplished	Can be landed
Normal operation	Satisfactory	1	Excellent, includes optimum.	Yes	Yes
		2	Good, pleasant to fly.	Yes	Yes
		3	Satisfactory, but with some mildly unpleasant characteristics.	Yes	Yes
Emergency operation	Unsatisfactory	4	Acceptable, but with unpleasant characteristics.	Yes	Yes
		5	Unacceptable for normal operation.	Doubtful	Yes
		6	Acceptable for emergency condition only.	Doubtful	Yes
No operation	Unacceptable	7	Unacceptable even for emergency condition.	No	Doubtful
		8	Unacceptable—dangerous.	No	No
		9	Unacceptable—uncontrollable.	No	No
	Catastrophic	10	Motion possibly violent enough to prevent pilot escape.	No	No

FIGURE 57.1 Cooper pilot rating system.

PASSIVE RESTRAINT SYSTEM TEST

Subject Demographic Data Record

Subject No. _____

Male:____Female:____Height:______Weight:_______I.D. No:_____

Age Category: Young____, Middle____, Elderly____

Vehicle Driven:

a. Primary; Make______________Model___________Year______

b. Secondary:Make______________Model___________Year______

Primary Car Belt System: Lap Belt Only ______
Separate Lap/Shoulder Belts ______
Combined Lap/Shoulder Belt ______

Primary Seat Belt Usage: Most of the Time ______
Part of the Time ______
Never Use ______

If Not Used, Why? Inconvenient ______
Confusing ______
Uncomfortable ______
Other ______

Detailed Test Procedures for
Passive Belt Systems Evaluation

1.0 INFORMATION HANDOUT

Each subject will be given the following handout which he will read prior to commencing the test sequence:

"Because passive restraint systems will be installed in all cars after the year 1982 models, the U.S. Department of Transportation would like to make sure that such systems will be safe, comfortable, and convenient.

"A passive belt system is one that you, as a vehicle occupant don't have to put on. It is designed so that as you close the car door it goes around you automatically, securing you safely in the seat. And when you open the door the belt is removed from around you automatically so that you can get out of the car easily.

"We have selected several different passive belt systems which we would like you to help us evaluate. We are interested in knowing how comfortable and convenient each of these belt systems seems to you.

"So as you get in and out of each of the cars today be alert to any confusion, inconvenience, or discomfort you experience with each system. We are not evaluating the design of the vehicles! Only the design and operation of the belt system.

"Immediately after each part of the test we will ask you some questions that will allow us to record your observations and reactions while the experience is still fresh in your mind."

2.0 TEST CONDUCTOR'S STATEMENT

The following statement will be read verbatim to each test subject by the Test Conductor:

This is an overview of what you will be asked to do.

- We are going to have you evaluate 5 different passive belt systems.
- You will be asked to get into each car in turn, close the door, adjust the seat to your preferred driving position, then imitate several driving tasks such as: adjust and look into rearview mirrors; look over your shoulder as though you were checking traffic before making a turn; turn the steering wheel; and operate several of the driver controls--after which you will get out of the car.

FIGURE 57.2 Example consumer test forms.

At certain points in this procedure you will be asked specific questions about the effect the belt system had on your ability to perform various tasks and on your general feelings of comfort (or discomfort).

- We will treat the cars and belt systems in pairs testing each one separately, then comparing the second car/system with the previous one.
- Each question should be answered in terms of rating a possible problem. If you feel there was no problem, I will mark a zero for that question. If you experienced a minor problem I will mark the question with a "1", and so on. The evaluation ratings are as follows:

No problem (for you)	= 0
Minor problem	= 1
Moderate problem	= 2
Serious problem	= 3

Your participation in this test will help us decide what is needed to make all passive belt systems more acceptable from the standpoint of comfort and convenience.

Do you have any questions before we begin?

[Note to Examiner: Distinguish between hang-ups on parts of car (door-lock, etc.) vs. hang-ups caused by belt.]

Subject No.______ First or Second Test Car A B C D E F G

Test Conductor_____

Detailed Test Procedures for Passive Belt System Evaluation

Test Conductor Instructions	Questions for Subjects	No Problem			Serious Problem
Read to S: We want your opinion about this passive belt system in terms of how it may (or may not) interfere with your ability to get into this car and get ready to drive. After you are in the car I will ask you a series of questions about how easy it was to open the door, get into the seat, and close the door after you. I also will ask whether the belts got in the way, rubbed your chest or neck or seemed too loose or too tight.	Test Sequence No. 1 - Entering Car				
Now please open the car door, get into the driver's seat and close the door. Rearrange or straighten the seat belt if you feel it is necessary.	1. Did the seat belt make it hard for you to open the car door all the way?	0	1	2	3
	2. Did the belt tend to re-close the door, making it difficult for you to get into the car?	0	1	2	3
Ask this question only when S is entering this car for the first time.	3. When you first saw the belt were you in doubt as to how you were supposed to get into the car?	0	1	2	3
	4. Did you have any difficulty getting past the belt and into the seat?	0	1	2	3

FIGURE 57.2 *(Continued)*

used to conduct a consumer evaluation of various automotive seat belt systems. Although it is uniquely designed for a specific type of test, the general implications are applicable to a variety of tests in which an experimenter conducts a consumer subject test of actual prototype hardware. It is based on complete subject involvement in use of the hardware.

Human Performance Measurement during Design Testing and Evaluation. As a general rule, human performance is measured in terms of how many errors a subject makes and/or how long it takes the subject to perform specific tasks. However, human subjects can also help judge the adequacy of a design either by rating the design against some set of standard criteria or by rating several designs against one another. Figure 57.3 provides a brief description of several techniques for performance and/or design acceptance adequacy.

Method	Technique	Statistic	Application
Adjustment (average error)	The subject adjusts a stimulus until it is subjectively equal to or is in some relationship to a criterion.	Average of settings (average error of settings measures precision).	Absolute threshold; equality; equal intervals; ratios.
Minimal change (limits)	The experimenter varies the stimulus upward and/or downward. The subject signals its apparent relation to a criterion.	Average value of stimulus at transition point of subject's judgment.	All thresholds; equality.
Pair comparison	Stimuli are presented in pairs. Each stimulus is paired with each other stimulus. The subject indicates which of each pair is greater with respect to given attribute.	Proportion of judgment calling one stimulus greater than another. Proportions sometimes are translated into scale values via assumption of normal distribution of judgments.	Order; intervals (under distribution assumption).
Constant stimuli	Several comparison stimuli are paired at random with a fixed standard. The subject indicates whether each comparison is greater or less than the standard. (This is a special case of the pair-comparison technique.)	Size of difference threshold equals stimulus distance between 50 and 75 percentage points on psychometric function.	All thresholds; equality; equal intervals; equal ratios.
Quantal	Various fixed increments are added to a standard, with no time interval between. Each increment is added several times in succession. The subject indicates the apparent presence or absence of the increment.	Size of sensory quantum equals distance between intercepts of rectilinear psychometric function.	Differential thresholds.
Order of merit	A group of stimuli, presented simultaneously, are set up in apparent rank order by subject.	Average or median rank assigned by subjects.	Order.
Rating scale	Each of a set of stimuli is given an absolute rating in terms of some attribute. The rating may be numerical or descriptive.	Average or median rating assigned by subjects.	Order; equal intervals; stimulus rating.

FIGURE 57.3 Techniques to measure human performance and design acceptance.

Organizing and Documenting Field Test Activities. The importance of documenting test results of hardware is well known, and such tests are recognized as absolutely necessary. Documenting how well people perform is equally important. Although this might appear to be an unnecessary admonition, too often it is believed that performance problems should merely be noted in a general way and that the actual performance of the people operating the equipment does not need to be measured. However, if a thorough program of human engineering has been pursued to this point, certain human performance standards will have been established. Therefore, it makes common sense to test the equipment user as well as the equipment.

As noted earlier, a final human–machine system test should examine not only the effectiveness of the hardware design but also the procedures, the supporting documentation (e.g., technical manuals), and the training.

Special test observation and recording forms should be devised to prompt field test observers and to provide them a place to record their observations. The sample format in Fig. 57.4

HUMAN ENGINEERING FIELD EVALUATION FORM

TASK TIME RECORD: TASK: ________ EVALUATOR________

Preparation ____ DATE________
Performance ____
Clean Up ____
Total ____

TASK DIFFICULTY	APPARENT PROBLEM
1. Didn't know how to start. ☐	
2. Didn't know what to do next. ☐	
3. Missed a procedural step. ☐	
4. Got steps out of order. ☐	
5. Repeated step unnecessarily. ☐	
6. Initiated action too late; too soon. ☐	
7. Difficulty coordinating activity of two or more persons. ☐	

RECOMMENDED ACTIONS:

PROCEDURAL PROBLEMS

HUMAN ENGINEERING FIELD EVALUATION FORM

TASK TIME RECORD: TASK:________ EVALUATOR________

Preparation ____ DATE________
Performance ____
Clean Up ____
Total ____

TASK DIFFICULTY	APPARENT PROBLEM
1. Difficulty locating proper display or control. ☐	
2. Confused meaning of label. ☐	
3. Couldn't read display detail. ☐	
4. Mis-read display. ☐	
5. Missed warning light. ☐	
6. Difficulty positioning control. ☐	
7. Set control incorrectly. ☐	
8. Disturbed adjacent control. ☐	
9. Didn't recognize color code. ☐	

RECOMMENDED ACTION:

DISPLAY-CONTROL PROBLEMS

FIGURE 57.4 Sample form for documenting field test findings.

was created for the field testing of an aeronautical system. This test was oriented primarily toward field service and maintenance operations. Although only two of the test sheets are shown, they give an idea of what the field test observers were told to look for. Each type of test should have its own forms, tailored to fit the particular application and the features that will probably be observed.

Commonly Needed Tools for Human Factors Engineering Testing and Design Evaluation. Although there will be times when more elaborate testing facilities and measuring equipment may be needed during particular human engineering design programs, the ones listed in Fig. 57.5 are most commonly required.

MOCKUPS AND USABILITY EVALUATIONS

Use of Mockups for Human Factors Engineering Test and Evaluation

Mockups of products are widely used both as a design tool and for marketing purposes in industry. Unfortunately, management perceives the value of the latter use more readily than the value of the former. This section attempts to champion the use of mockups for purposes of design evaluation, especially for validating design features that involve user interfaces.

Types of Mockups

Two distinct types of mockups, each used for different purposes, are of interest. The first is the miniature-scale mockup,

Type of tool	Typical uses
Anthropometric tools: Flexible tape Anthropometer (designers can fabricate their own less expensively) Adjustable-height chair Adjustable-height footstool Health scale	To select test subjects for evaluating a mockup. The important measurements most often taken include standing height, sitting height, arm length, leg length, hip breadth (sitting), and weight.
Force-measuring tools: Spring scales (0–10, 0–40, and 0–100 lb; 0–4.5, 0–18, and 0–45 kg) Torque meter (0–10 oz, 28 g and 0–40 oz, 1130 g) Strain gauges (0–10 lb, 0–4.5 kg)	To test forces required to operate various manual and pedal controls, forces required to open doors, weights of packages that must be manually moved or transported, or forces that a seat belt imposes on an occupant's body.
Sound-level measurement equipment: Sound-level meter Half-octave band filter	To measure sound levels of noisy environments. Equipment should provide means for measuring both an overall sound level and levels within critical frequency bands.
Light-level measurement equipment: Light meter	To measure light levels falling on surfaces within the visual field and the light output of specific illuminated displays (down to the specific display markings).
Temperature, humidity, and airflow measurement equipment: Wet-and dry-bulb temperature-measuring units volometer	To measure ambient room conditions, surface temperatures, airflow from ventilators, etc.
Oxygen-consumption measurement equipment	To measure the energy demand of the task on the subject.
Heart-rate and pulse-rate measurement equipment	To measure task stress imposed on the subject.
Galvanic-skin-response measurement equipment	To measure task, environment, and psychological stress imposed on the subject.
Time and motion measurement equipment: Stopwatch Automatic timer Video recorder	To obtain and measure task activity information about subject for later analysis.
Audio recorder(s):	To record verbal conversations, make verbal notations, etc., during a test.
Gas chromatographic equipment	To record and identify gaseous conditions in a closed ecological system.
Computer/monitor	To present timed-exposure targets or other information to test legibility and interpretability of labels, nomenclature, type style, and figure-ground contrast ratio.

Figure 57.5 Human engineering testing tools.

which is valuable in examining three-dimensional relationships between several fairly large features that would be difficult to view all at once if they were demonstrated in full scale. The second is the full-scale mockup, which is valuable in examining the immediate three-dimensional relationships between the system being mocked up and an operator or maintainer, in terms of space, reach, visibility, and convenience.

Levels of Mockup Complexity and Detail

Although many people think of a mockup as a fairly substantial structure simulating the three-dimensional attributes of a workstation or space layout, there are several levels of mockup implementation. Each has its place, depending upon the needs of the designer and perhaps the time and money available to create the mockup. The following descriptions and use suggestions provide a guide to help designers decide upon which level of mockup implementation they need to accomplish their objectives.

1. Paper Mockup. The paper mockup is the least expensive and time-consuming to prepare. It is constructed by cutting out scaled, two-dimensional elements and/or a drawing and placing these on a plan view of a work area or on a wall representing the vertical surface of some cabinet that will eventually hold the proposed control panel.

For a miniature-scale workplace or facilities arrangement problem, one can create an excellent first-step evaluation of various alternatives; i.e., one can try out several possible arrangements of consoles, equipment racks, desks, or other furnishings to determine the arrangement that provides for the fewest operator steps, the best clearances for traffic flow, the best viewing angles, and so on. Such mockups can be made out of paper or cardboard (the latter tends to lay flat more readily and is less affected by drafts, which often blow paper cutouts off the table). Typically, one starts out with a substantial baseboard marked off in square feet (to whatever scale is chosen by the designer). Then the equipment items are cut out of either paper or cardboard (to scale) and are labeled. The cutouts are then manipulated into various alternative positions until the best arrangement is obtained, considering all the critical factors of traffic flow, intercommunication between operator stations, access for maintenance, and a general feeling of spaciousness or lack of clutter. This type of mockup is very effective for architectural planning, such as for determining room size or equipment arrangement or even for revamping office space or work space in an existing building or office.

The paper mockup is also suitable for initial examination of full-scale relationships. For example, one can place a full-scale drawing of a proposed control panel on a wall in order to determine at what height the panel should be placed or what the best orientation of the panel would be from a nominal operator position (either standing or sitting). Such a mockup can also be used for initial evaluation of the arrangement of controls and displays; i.e., one can simulate going through the proposed sequence of panel operation and quickly discover that items are arranged in the drawing in such a way that too many hand crossovers occur, that things used most often are too far out of reach, that one tends to cover a display while operating a control, or that the proposed labels and the proposed arrangement of panel elements are confusing, making the panel elements hard to locate quickly. Since no construction is required, it is easy and inexpensive to try several alternatives until the best one is found. The paper wall mockup is also useful in establishing the size of labeling required for viewing from alternate positions, such as from across the room. Finally, this inexpensive approach provides the first opportunity to evaluate space and reach problems using test subjects at the extremes of body size to make sure that both large and small operators are accommodated equally well.

It should be emphasized that the paper mockup is not to be regarded as a plaything that is beneath the dignity of the practicing professional. Experience has demonstrated many times that more costly mockups have not provided any more useful information than a paper mockup would have provided, and considerable time is often wasted on more exotic mockups, not to mention the cost. Finally, it should also be noted that a paper mockup will provide important criteria for developing more sophisticated mockups later on during the design program.

2. Soft Three-Dimensional Mockup. A soft mockup is one that is made of wood and cardboard or of special inexpensive laminates such as Foam Core. The term "soft" relates primarily to the fact that the materials are easy to cut and assemble, as opposed to metal. The general approach consists of creating a simple wood understructure, to which sheets of cardboard or Foam Core are attached to provide the "skin," e.g., a console, panel rack, or workplace enclosure. To these skins are attached panel drawings or simulated controls and displays that may be cut out of stiff cardboard upon which pictorials of instruments or controls and/or function labels are mounted (to give an added three-dimensional appearance not unlike that of the real hardware elements).

The principal value of the full-scale three-dimensional mockup is its ability to represent spatial interrelationships between an operator and the controls and displays he or she must monitor and manipulate, arm reach and viewing angle parameters, and clearances for feet, legs, and body (including seats). For early stages of design analysis, it usually is not necessary to enclose a console completely, for example. Only the front of the console is important. The back of the console need not be enclosed, and thus it is not necessary to spend the time and money required to mock up several alternative console shapes.

Three-dimensional miniature-scale mockups can be made of wood, preferably some lightweight wood that is easy to fabricate (e.g., balsam). Mockup elements (consoles, panel racks, desks, chairs, machine tools, etc.) are more easily manipulated if small magnets are embedded into the base of each piece and if the baseboard representing a floor space or area is made of a suitable board upon which a metal plate is mounted. In this way, the equipment and furnishing models are easily moved about and yet stay in place.

A second-level soft mockup may be desirable once the general pattern of a console has been established. That is, in place of the paper or cardboard display and control simulations, actual instruments can be mounted on the control panels. These displays and controls can in turn be made active to various levels, depending on the purpose of the simulation. That is, instruments can be wired to demonstrate illumination, or they can be instrumented to the extent of tying into a computer program, which allows one to simulate and evaluate alternative display and control system parameter concepts.

A third level of sophistication for the soft mockup consists in mounting the mockup on dynamic platforms that simulate, for example, the motions of an aircraft, space vehicle, or ship. Generally, one would not go to this level of time and cost unless there was some solid indication that the dynamic aspects of the operator-control situation were especially critical to the ultimate functioning of the human–machine system. It becomes obvious that this level of mockup begins to approach the capability of becoming a training simulator. Therefore, one can equate the time and cost in terms of both a design tool and an eventual training tool that may be required by the customer.

Although soft mockups are recognized as a useful tool for examining three-dimensional operator–station problems, it should be noted that they are also extremely useful for examining the problems of maintenance technicians. Once again, both miniature-scale and full-scale mockups have their place, for the reasons noted above. For example, the miniature-scale model provides an excellent method for examining proposed layouts for production and for service and maintenance facilities. Using the same approach discussed above, scale models of prime vehicles, support vehicles, structural features, etc., all can be easily viewed on the small-scale simulation. An added benefit of the scale model is that, when different groups of people are involved (such as the contractor and the customer, who may be many miles away), photographs can be taken of several alternative layouts that have been tried, and these can be mailed back and forth to increase the level of communication between the interested parties. In fact, the photographic technique provides an excellent method for preparing for a design review briefing. For example, the designer can take pictures of each alternative arrangement that he or she has tried and evaluated. By making these into slides, it is possible within a few minutes to run through and explain the rationale for making a final arrangement recommendation. This saves a lot of questions from those who would like to know whether the designer has explored all possibilities.

Finally, in the case of mockups that will not be involved in any kind of dynamic evaluations (i.e., mockups which are static, which will not have to support heavy components, or which will not be sat upon or leaned against by persons involved in the evaluations), it is important to use materials that are easily and quickly modifiable by the designer so that it is not necessary to wait for a mockup specialist to make changes. On the other hand, some mockups will have to have sufficient structural integrity to support heavy components, persons who may climb upon them, or accelerative forces from moving platform dynamics. These generally are all-wood structures and in certain areas are covered with a plywood skin.

3. Hard Mockup. Generally speaking, the so-called hard mockup is used to define the detailed assembly aspects of a final production design. For example, the hard mockup will be made of metal and will include the precise details of the inner structure and outer skin, which the designer, along with the production engineer, will explore and for which they will define types of attachment hardware for internal components, external components, trim, etc. A good example is the fuselage mockup that is used to define the routing and attachment of cables, hydraulic lines, pneumatic lines, control cables, etc. Although the designer generally prescribes where these components go and in general how they will be attached and connected, it is not certain what obstacles there may be within the basic fuselage structure. Thus the hard mockup provides a precisely scaled breadboard, so to speak, on which the designer can determine how best to route and fasten components. Once these decisions are made and final routings and fastenings are actually mounted in the mockup, the mockup becomes a reference for production-line personnel. Typically these hard mockups are retained for the life of a production run and sometimes beyond (to be used for later modification analyses).

The hard mockup also provides one other important contribution. In deciding how to assemble large, complex elements of a hardware system, it becomes important to work out a plan for assembly that minimizes conflicts among assembly operations. For example, in very tight spaces, various assembly operations cannot all work in the space simultaneously, and the mockup allows one to determine the most efficient order in which to assemble components.

From a purely human engineering point of view, the hard mockup provides an excellent opportunity to evaluate specific human–machine interface problems. These include maintenance tasks (accessibility to components and parts that have to be inspected, removed and replaced, calibrated, or adjusted in the field).

4. Other Mockups

Styling Bucks and Models. The vehicular industry has long used and depended on the clay model to “shape” or “style” the exterior of a new-model car, truck, or bus. Over a basic frame, modeling clay is laid and shaped to represent the proposed exterior lines of the new model (and alternatives). Such models are made in both miniature- and full-scale versions. A special clay material allows the model to be painted so that the full scope of appearance elements can be evaluated for eye appeal.

For smaller products, either the structure and clay approach can be used, or package appearance design can be modeled from other more substantial materials such as fiberglass or thermosetting plastics. This type of model provides sufficient structural integrity to allow more specific user testing; i.e., a model of a hand tool, hair dryer, or electric knife can be picked up and manipulated to test manipulability and ease of handling.

Portability Models. Many small products require handling by the user, operator, or maintenance technician; i.e., they must be picked up, positioned, operated, donned or doffed, transported, and set down again. Before final package decisions are made, such models provide an important step in validating concepts before the product is finally produced. Such models should simulate the dimensional and dynamic aspects of the product so that one can determine whether the expected range of users can, in fact, manipulate the package with ease and surety. Often, until the package configuration is tested in full-scale model form, one cannot be sure whether the weight and balance characteristics of the proposed package will stress users or cause them to drop the package or lose their balance.

Accessibility Mockups. Although many military projects require key demonstrations of accessibility for maintenance at certain points during system development, these tend to pertain to major access points in an aircraft fuselage, a tank shell, or other prime hardware unit. These represent only the minimal areas where maintenance access mockups should be considered. For example, any small piece of equipment such as a radio, radar, or sonar unit or a piece of test equipment is a subject for verifying maintenance accessibility by means of a mockup. One should consider early mockups before packaging constraints preclude optimizing access; i.e., by means of an early mockup of various package configuration alternatives, one might be encouraged to change the shape or orientation of the package in order to provide a more convenient opening position when the unit is finally installed within close-fitting quarters. A second level of mockup could examine the positioning order of components so that a technician will not have to remove several interfering components to get at the one that is most often expected to need inspection, removal, and replacement. These mockups are especially useful for evaluating various types of slides, fasteners, latches, and handles, as well as for various approaches to the articulation (for accessibility) and/or the removal and replacement of chassis and printed circuit boards.

How to Determine What Type and Level of Mockup to Use

There are no hard-and-fast rules for deciding what type of mockup to use or to what level of sophistication one should go in developing the mockup and/or operational simulation. The following general guidelines may be helpful.

Available Time. Although one should make every effort to find the time to "be sure" (i.e., by mocking up and verifying a complex, three-dimensional concept before committing it to final design), the time available before a design has to be completed and the product delivered may make it impossible to develop and evaluate the mockup. There have been instances where an extensive mockup and design simulation program was completed months after a system had been delivered to the customer. In order to avoid such errors, it is vital that both the customer and the contractor recognize and prepare reasonable schedules so that, if a mockup or simulator is vital, its capabilities can be used in time to influence hardware design.

Criticality of the Human–Machine Interactions. The following are considered critical interactions:

1. The operator and maintainer working space is limited because of external confining restraints (e.g., a small cockpit, close maintenance quarters, low overhead, or limited crawl space). Such space limitations are particularly critical if the worker or operator may have to wear heavy protective garments. The space should be evaluated with large test subjects wearing the most constraining type of garment.

2. Escape envelope constraints are especially critical for obvious safety reasons. Mockups should be used to evaluate not only the envelope shape and size but also the position of the exit with respect to potentially difficult conditions under which the escape envelope must be traversed and the position of the final exit with respect to opening and departure (e.g., an exit on the top of an aircraft fuselage may be too high for a small person to get to, especially if high *g* forces are working against the escapee). Escape system and procedure mockups are vital for determining whether escape-time criteria can be met by a particular design configuration.

3. Visibility constraints and/or adequacy should be verified using three-dimensional mockups. Although general anthropometric criteria provide ball-park estimates of how to arrange operators and equipment, windows, glare shields, etc., the final verification requires *in situ* examination by selected ranges of subjects of both large and small stature. Sight lines to both internal displays and external viewing requirements need verification before a final operator–station configuration is accepted.

4. Reach constraints should be verified for certain operator and maintainer workplaces, especially if the operator or maintenance technician has to be restrained by a safety belt or limited access space. The question includes the reach limits not only of small individuals (possibly constrained by special garments) but also of large individuals, who may not have sufficient clearance to retract an arm or leg.

5. Lighting parameters are often critical where it is very important to provide low but even illumination of all visual displays, labels, and controls or where it is possible that reflections and glare sources may interfere with proper display or external, out-the-window viewing. Such mockups should be evaluated under all the expected ambient lighting conditions. It is wishful

thinking to believe that one can identify and adjust a design for all possible glare or reflection problems.

6. Ingress and egress mockups are an important consideration, not only for the questions of emergency escape noted above, but also for normal entry and exit. This is especially important in the case of vehicles for which older and disabled individuals are expected to be among the user population.

7. Control and display arrangement for ease of location, identification, and speed of use should be mocked up and evaluated in real time by a number of typical user subjects. This is especially important when the control operation is complex and when mistakes may be costly to the final operation. Although a cursory link analysis of a preliminary layout drawing provides the first indication of whether a panel layout appears to be logical and convenient, it may not tell the full story; i.e., when the panel is associated with other displays and controls and possibly a constraining seat, the otherwise adequate panel arrangement may prove to be less effective than it appeared in the drawing analysis.

8. Component–rack interactions may present difficult mobility problems for the maintenance technician, and therefore, they should be examined in the three-dimensional mockup setup. Typical problems that show up in the mockup evaluation include the following: the technician cannot bend over far enough to lift a chassis out of the lower portion of a rack; there is not enough clearance to open an access door fully; an access opening is so small that the technician's hand or arm blocks the view of the object he or she is trying to reach or manipulate; and there is an intervening hazard when the technician reaches across a moving belt or gear to adjust or remove some component.

Utility of a Particular Mockup Level to Provide Necessary Answers. There are those who swear by mockups, just as there are those who think that mockups are a waste of time and money. Mockups are often made more elaborate than necessary, usually because an individual manager or engineer likes to have the most lifelike representation possible to show a customer. Although this is impressive and may have value from a marketing point of view, it may be a waste in terms of its design verification value. On the other hand, making the mockup too minimal may prevent one from finding answers to the questions outlined above. The designer must decide what kind of mockup will provide the needed answers; any expansion beyond this requirement is probably a waste of effort and money.

This problem is particularly difficult to resolve in the case of mockups that tend to be real-life simulations. That is, how far to go with simulation is always a moot question. In general, one can never completely simulate the real world under any circumstances, short of creating the actual article and testing it in a real-world environment. Basically, that is the role of a prototype article, e.g., a prototype aircraft, a prototype automobile, or a prototype piece of test equipment. Short of this, mockups and simulations of partial task elements are often more cost-effective.

Mockup Adjustability. To be maximally effective as a design evaluation tool, a mockup should be made as adjustable as possible. As noted above, there often is a tendency to feel that a mockup can be made to reflect all the "right" features the first time and that the only purpose is to prove that one's design analysis is correct. Almost without fail, one finds that changes are required after the evaluation. However, unless the mockup is constructed in a fashion that allows for easy modification, it may be necessary to fabricate an entirely new one. Before constructing any mockup, therefore, look ahead and identify all the features that you may suspect could require adjustment during or after the first evaluations. This will allow you to make these changes and reevaluate them with a minimum of time and cost. It will also allow you to demonstrate quickly to customers why you have chosen the final configuration you have—merely by showing them the several versions you tested.

Miniature-Scale Mockups and Models

Miniature-scale mockups and models are especially useful for evaluating the multicomponent layout and arrangement of large elements such as clusters of buildings and large structures, weapons and troops in a battle configuration, multiple aircraft and support equipment on a flight deck, and equipment, machines, and furnishings within a building, office, laboratory, or factory. (See Fig. 57.6.) Depending on the overall number of elements, one should select the scale to fit the evaluation purposes. Models need not be of fine detail, but they should be fairly precisely scaled so that traffic flow and work-space clearance can be realistically assessed. It is

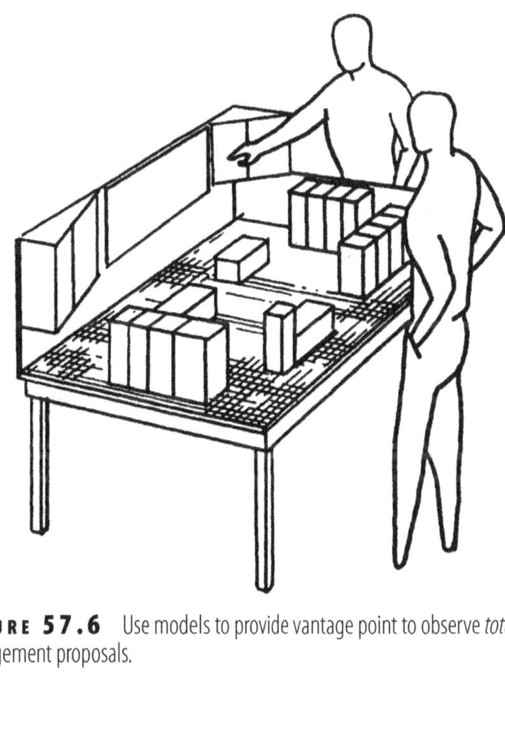

Figure 57.6 Use models to provide vantage point to observe *total* arrangement proposals.

suggested that ⅙-scale models be used for interior room arrangements if it becomes desirable to evaluate any detail on model equipment or furnishings (position of displays, lines of sight, illumination possibilities, etc.).

As long as the mockup representation involves a flat floor, it is suggested that a metal plate inscribed in a 1-ft (0.3-m) grid be used as the primary reference base. This makes it easy to assess clearances quickly.

Model equipment, furniture, and machinery should be magnetized so that, although these elements stay put and are not easily disarranged or knocked over, they are still easily moved from one place to another. (See Fig. 57.7.)

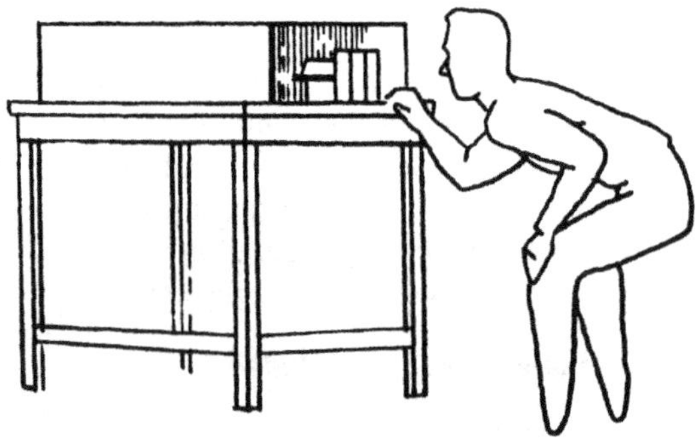

Figure 57.7 Use models for preliminary examination as actual operator would see it.

Use miniature-scale models to:

- Evaluate traffic flow
- Establish visual sight lines
- Route utilities
- Evaluate workplace convenience and maneuvering space
- Evaluate production-line flow
- Position common-access elements
- Make a preliminary lighting and color analysis
- Determine the best escape routes
- Determine access for maintenance and service
- Decide on preliminary décor

Use of Drawings and Paste-Ups

Early evaluation of proposed control panel positioning and layouts can be accomplished inexpensively by pinning a drawing on a wall or, better yet, by pinning or attaching self-adhering panel element patterns on a cork or felt board. If a metal wall is available, similar results can be obtained using two-sided tape or flexible magnetized tape on the back of panel elements. It is recommended that independent panel elements be placed over the panel drawing because this permits moving the elements around on the board.

Figure 57.8 Use drawings to evaluate standing or seated workstations.

As illustrated in Fig. 57.8, both standing- and sitting-operator relationships can be evaluated. The following key issues can be resolved by these techniques:

- Identifying reach problems
- Analyzing operational sequence and functional grouping
- Determining proper spacing of components and positioning of proposed labels
- Establishing the best height for visual displays and determining whether a hand or arm obscures a display

When it appears desirable to change the plane of a particular panel, an inexpensive independent structure can be devised upon which the drawings and/or cutouts can be pasted (as illustrated in Fig. 57.9). A wooden structure can be

Figure 57.9 Use soft three-dimensional mockups to evaluate angled panel.

covered by cardboard or Foam Core to provide sufficient panel rigidity. A light metal panel can be easily attached to the structure and/or Foam Core surfaces, making it possible to utilize the magnet technique for attaching panel components independently.

Reconfigurable Mockups for Operator Control and Display Positioning Definition

As opposed to establishing control and display positioning using general reach criteria, it is highly desirable to establish optimum positions by placing an operator in a three-dimensional framework, unobstructed by preconceived consoles, racks, bulkheads, etc. As Fig. 57.10 illustrates, an open framework provides an ideal method for adapting controls or displays to the operator. Control and display elements should be mounted on fully adjustable supports so that the control or display panel can be moved around until the best position is found in terms of line of sight or reach.

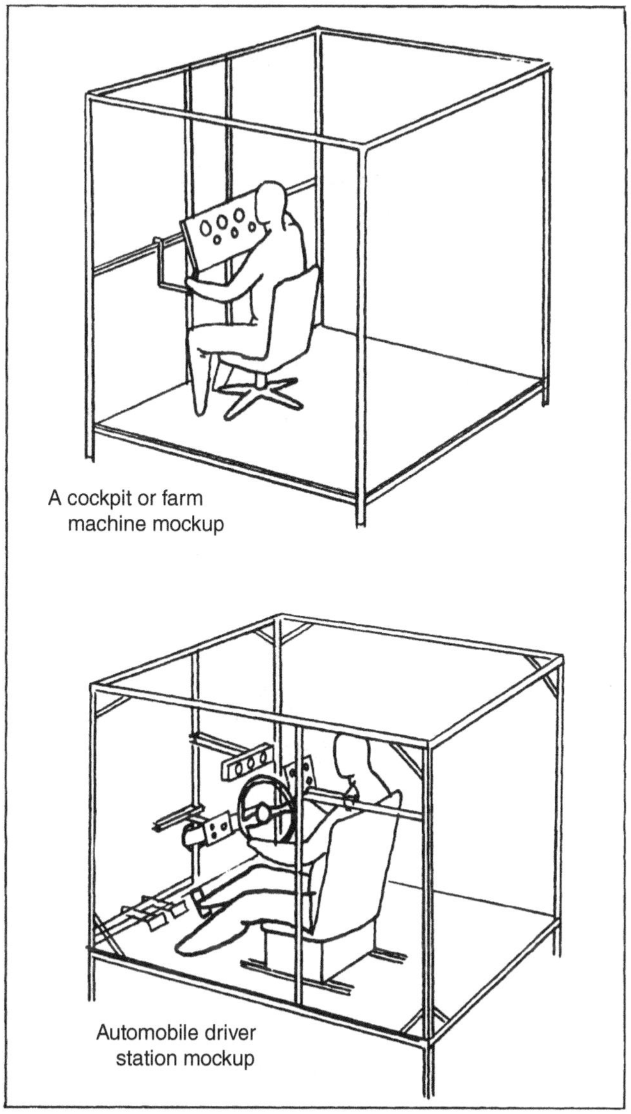

Figure 57.10 Adjustable driver station mockups.

Controls should be mounted in such a way that the control device can be moved through its intended excursion, especially joysticks, steering wheels, foot pedals, etc.

It is also possible to attach key visibility elements (window posts or frames, rearview mirrors, etc.) to evaluate fields of view.

Service and maintenance questions can also be evaluated with a reconfigurable mockup approach. (See Figs. 57.11 and 57.12.)

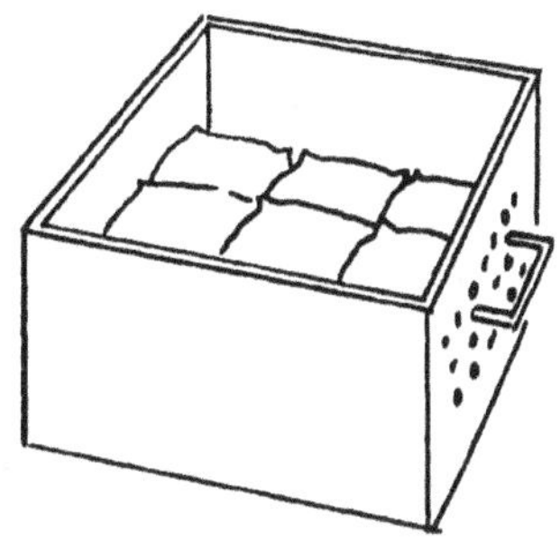

Figure 57.11 Portable package/handle-locating mockup (note sandbag weights).

Figure 57.12 Service equipment rack mockup.

Human–Machine System and Environment Simulation as a Design Tool

Any physical mockup is in a sense a simulation of the human–machine environment. As the system becomes more complex, however, it is often desirable to enlist the aid of more exotic apparatus, coupled with computer-aided scenario generation. As Fig. 57.13 illustrates, one may require simulation (and control) of both external and internal environmental parameters, since all these influence how well the human–machine system functions in various modes. Not shown is the possibility of including dynamic properties such as oscillation and movement of the subject operator.

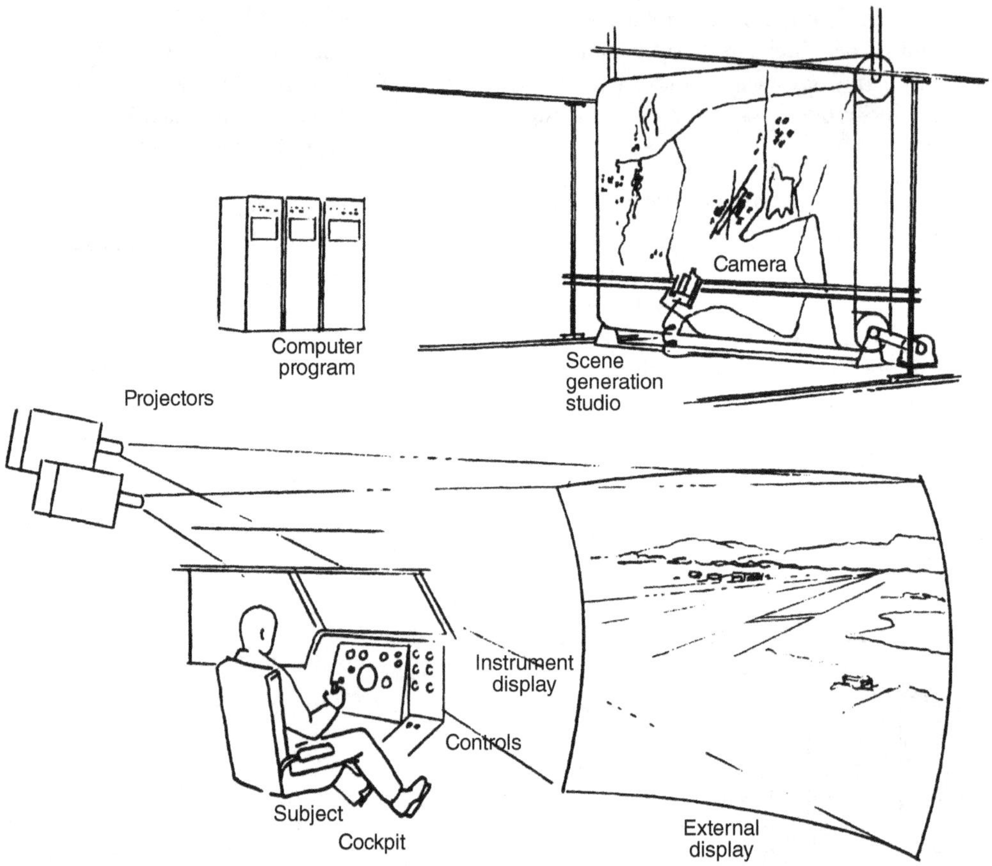

FIGURE 57.13 Adding simulation to a mockup.

The extent of realism depends on several things, including how important each variable may be to the eventual performance of the operator, how well it *can* be simulated (i.e., a poor simulation is worse than none), how much time is available to develop the rather complex simulation system, and, of course, whether the cost is justified. If the cost exceeds the cost of the ultimate hardware system, it usually is not justified. Avoid the temptation to create an exotic simulation just because it is a design challenge.

Mobility Evaluation Mockups

Although there seems to be an abundance of information regarding architectural interfaces wherein the user climbs something, moves about, holds onto something, etc., it is always advisable to evaluate these in each specific application by means of a mockup. This is particularly true whenever the user is encumbered with things that must be carried or by special restrictive clothing that must be worn. In Fig. 57.14, for example, the critical considerations are the exact relationship between the ladder below the opening and between the opening and the hand assist rails above the opening. A difference of just a few inches may be very important in preventing users from missing a step or losing their balance. The positions of these elements may also vary with the direction in which the individual approaches the opening. The best compromise can be determined only by physical trials.

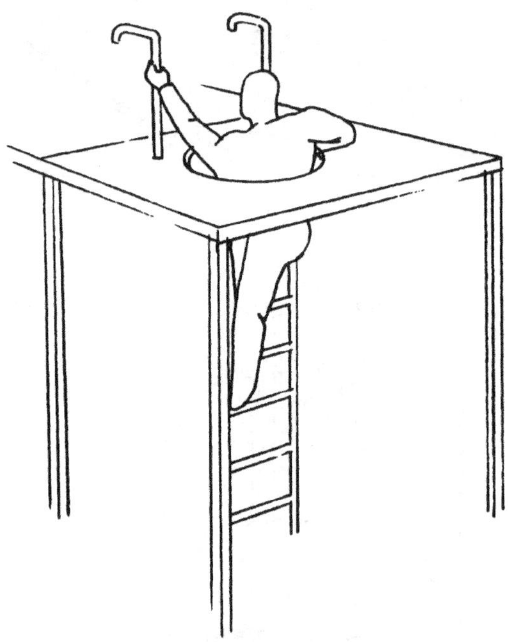

FIGURE 57.14 Mobility evaluation mockup.

Visual Display Element Evaluation

Typically, one should rely on the criteria provided elsewhere in this handbook to determine the type of characters to use for labels. However, display nomenclature may be difficult to establish without evaluating the aspect of understandability. A simple tachistoscopic slide presentation is suggested, wherein one can test several alternative nomenclatures, abbreviations, or pictorial symbols to determine which ones lead to the fewest errors in interpretation. As illustrated in Fig. 57.15, several subjects can be used simultaneously to evaluate the alternative display nomenclature, symbols, or even other visual parameters, such as size and color.

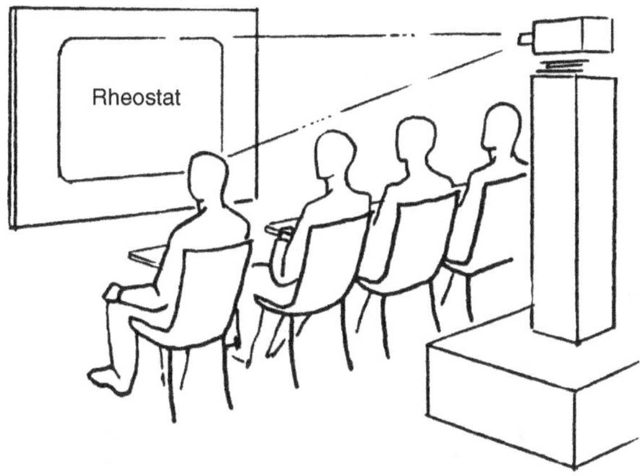

Figure 57.15 Jury evaluation of visual displays.

Dynamic Simulators

When there is good reason to believe that the dynamic characteristics of a system may have a considerable influence on the behavioral response of the eventual system users, consider using a dynamic simulation of the expected environment. In Fig. 57.16, a total integration of both external visual dynamic parameters and the physical movement of the operator provides a realistic simulation of the real world.

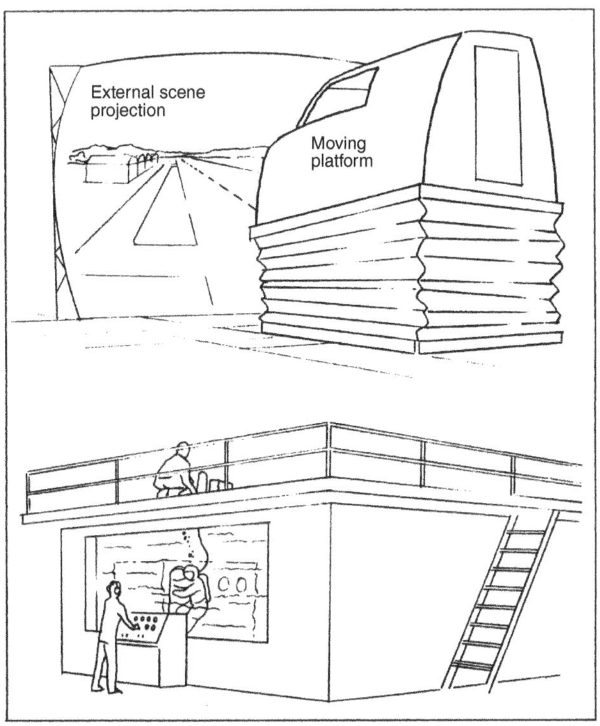

Figure 57.16 Examples of dynamic simulators.

When more critical physical conditions such as positive, negative, or reduced gravity forces may affect the efficiency of an operator, consider the use of special simulators such as the centrifuge, acceleration and deceleration sled, or water tank (for weightless simulation), as illustrated in Fig. 57.16.

Other types of simulations to consider might include the pressure chamber or heat chamber.

Caution: Whenever dynamic simulation may involve conditions that explore stress levels that could be hazardous, enlist the aid of specialists to help design and monitor the test.

Choosing Subjects for Mockup Evaluations

Although there may be good reasons (i.e., time and funding constraints) for limiting mockup evaluation to the designer and his or her immediate associates, the results may be far from satisfactory. There are several reasons for this:

1. Designers may be biased toward a design they have just created; i.e., because it is their own "brainchild," they really do not want to see any problems that the design may create for the user.

2. Designers are not always representative of the user population in terms of either physical or psychological characteristics; i.e., a designer may be extremely large or small, very strong or very weak, or particularly skilled in terms of equipment manipulation and understanding, or the designer may tend to "read into" the operation what the design is intended to demand because he or she already knows what is expected of the user.

3. A male designer may not understand how a woman will react to a design.

4. Designers, because they are adults, may fail to recognize or understand how a child or a younger person will react to a design.

5. Designers typically assume that all consumers will react to a product design in the same way they do (after all, they are also consumers). However, since a design may eventually be used by people with different ethnic, cultural, or language backgrounds, things about the design that appear perfectly clear and/or desirable to the designer may be completely foreign and undesirable to many people who eventually become users of the product.

Guidelines for Selecting Mockup Evaluators

The following guidelines are provided to aid the designer in selecting subjects to use as evaluators.

Population. Select evaluators from the expected user population—e.g., military personnel (as opposed to civilians), adults (as opposed to children), or members of a particular professional or vocational group.

Body Size. Select evaluators from specific or general populations on the basis of recognized anthropometric surveys. For example, several professions (Air Force, Army, and Navy personnel; astronauts; bus drivers; etc.) may be limited by anthropometric selection criteria; i.e., they will be only so tall or so heavy because a particular agency has established selection criteria that preclude taking applicants outside these dimensional ranges. On the other hand, not only do civilian populations include persons with a wider range of body sizes, but also the range is expanded because of inclusions of both males and females and both young and old people.

Strength. When the mockup evaluation involves lifting and carrying the product and/or manipulating controls that may have fairly heavy force requirements, the evaluator population should represent mainly the weaker members of the expected user population.

Intelligence, Experience, and Training. In certain cases, it would be wasteful of human capability to try to make a product acceptable to everyone; i.e., some people would never be able to cope with very complex equipment operations, even with extensive training. When the operation of the proposed product normally would be restricted to specialists already known to have the requisite knowledge and skills, there is no need to compromise the proficiency of the human–machine system (i.e., to make it unduly simple). In such cases, evaluators should be chosen who are already knowledgeable about the proposed design and therefore can evaluate the mockup or simulation from a skilled operator's point of view.

On the other hand, if the product may be operated by unskilled individuals (and therefore may create a critical training requirement), evaluators should be chosen from an unskilled group in order to determine what there may be about the design that is confusing or difficult to manipulate. This provides an opportunity to determine what modifications may be necessary to make the design more "trainable" and less apt to create special personnel selection requirements for the eventual customer.

Product Acceptability. In evaluating a proposed design (via a mockup), one should consider a highly varied sample of evaluators; i.e., the sample should include people of both sexes and of all sizes, ages, and backgrounds in order to allow for the random inclusion of personal attitudes and opinions, physical difficulties, etc. Although such subject samples should be generally random in terms of all these varying characteristics, it is often necessary to make sure that specific human characteristics are represented. This is important whenever the designer suspects that certain characteristics of the design may create special problems for certain members of the user population. For example, in evaluating seat belts, it is known that large and small people have certain kinds of problems. Although one wishes to obtain a general evaluation of how people may expect to use a system or how they may react to certain features of convenience, it is important to make sure that a sufficient number of the large and small evaluators evaluate those features of any belt system that cause belt systems not to fit properly. Thus, although the sample should generally represent the random attitudes of a broad sample of users, one should add a few subjects who are representative of the extreme anthropometric characteristics of the general sample.

Technological Variability. There is a tendency to believe that just because we are a technologically oriented society, all product users understand and can cope with modern machines. Studies have shown that this is not true. In some societies, people have not had the opportunity to grow up with modern technology; therefore, not only are they unable to cope with the demands of modern machines, but they also may completely reject machine-age products. This presents a difficult problem for designers, for seldom do they have an opportunity to study the population in question, or are there typically any study data to help them determine what a particular population will do or expect. The principal recourse under these circumstances is to search for subjects who may be recently from the area in question, using these evaluators as a screening group to determine whether there will be major difficulties as a result of certain characteristics of the design that have been mocked up for evaluation.

Guidelines for Determining Number of Mockup Evaluators

First of all, the designer may be confronted by the problem of convincing management that it is important to evaluate a mockup in more than a cursory sense. For example, many management people believe that *they are the final judge* in determining whether a design is satisfactory. Although it is probably true that some manager will have the final say, it is dangerous to assume that such a person has the knowledge to make judgments regarding effective human–machine performance. Convincing such a manager to rely on a more scientific approach to product evaluation, one that reliably reflects the specific user's response to the design, requires considerable diplomacy on the part of the designer. Consider the following in deciding whom to use as evaluators and how many to use.

Evaluating Reach and Clearance. Although a fairly large, random sample of subjects is required when one wishes to establish some statistical distribution for reach and clearance, in most cases it is satisfactory to evaluate these parameters in a proposed design (via a mockup) using a minimum of about five large subjects to evaluate clearance and about five small subjects to evaluate reach. One should select these groups on the basis of the particular population for which the product is being designed, such as a military or civilian group. For clearance, select five subjects whose stature is at approximately the 95th percentile and whose weight is also at about the 95th percentile. For reach, select five subjects whose stature is at about the 5th percentile (for a mixed user population, use 5th-percentile females). For all practical purposes, selecting five subjects at each end of the stature range will provide sufficient variability in the other characteristics to provide an adequate mix of characteristics that also relate to clearance and reach, e.g., variations in hip breadth, buttock-to-knee length, knee height (sitting), head height (sitting), eye height (sitting), head height, shoulder height, shoulder breadth, abdominal depth, and lengths of upper and lower arm segments. In addition, one usually can depend on a reasonable variation among 5 or 10 subjects in terms of overall body and limb mobility.

Evaluating Weight Characteristics of a Portable Package. The minimum number of subjects required to evaluate the weight characteristics of a package mockup is about 10. One should select relatively small subjects since not only is the size of the subject a general indicator of overall strength, but also the ability to lift is a function of the length of a person's arms, the size of his or her hands, and the general girth of the individual's abdomen, chest, and shoulders. One or two of the subjects should be statistically overweight, since an overweight individual may have difficulty holding the package close to his or her chest and stomach.

Evaluating Control Manipulability. Since a number of significant variables influence how well an individual may be able to manipulate a control device, a minimum of 20 subjects should be used. In general, the subject sample should be selected at random, since one hopes to sample a variety of individual differences, including anthropometric differences and differences in mobility, dexterity, steadiness, strength, sensorimotor proficiencies, and psychological expectations. Depending on the expected user population, both age and sex should be represented about equally throughout the sample; i.e., the subject sample should consist of half males and half females, and there should be a wide distribution of ages.

Team Procedure: Mockup and Simulation Evaluations. When more complex mockups are used to simulate team function and procedural response, a minimum of three complete teams should be used; i.e., as opposed to having an evaluation made by a single team of subjects, repeat the evaluation with at least two additional teams of subjects. This is desirable because one is never sure whether the first team is either unusually good or unusually poor, nor can one be sure, if the performances of the two teams vary significantly, which one's performance is truly representative of the system's quality in terms of design effectiveness. The response of the third team provides a means for breaking a tie, in addition to indicating which response performance is probably more representative of the design quality.

Comfort and Convenience Evaluation. An evaluation of the comfort and convenience aspects of a product or system mockup is perhaps the most difficult type of evaluation to perform, because in spite of all attempts to focus the evaluator's attention on significant design areas, each individual makes a different internal interpretation of what constitutes comfort or convenience. Therefore, it is recommended that a minimum of 50 subjects be used for this type of mockup evaluation. It is further suggested that, in performing this kind of evaluation, a special evaluative approach be taken, i.e., one in which subjects are forced to focus on, and respond to, specific, predetermined aspects of the design, one at a time—or at least short groups of aspects that can be separated in terms of operational steps. An example would be the case of evaluating seat belts in an automobile. Here, one might segregate the several typical use sequences, i.e., entering the vehicle and donning the belt, adjusting the seat and closing the door, simulating operating the vehicle (looking out the windows, reaching for the controls, looking at the displays, etc.), and finally doffing the belt and getting out of the car. Upon completing each of these sequences, the subject would be asked to rate various features of the belt system as they pertained only to that particular operating sequence.

Also important in this area of comfort and convenience are dynamic and environmental factors. Although one can obtain a general first approximation of consumer response in the static mockup, it is probable that a person's responses would be modified after actually experiencing the seat belt under more realistic conditions, i.e., after riding for awhile over various road conditions.

Establishing the Basis for Test Subject Selection

Product Form and Fit Problems. When the problem involves physical relationships between an operator and some hardware element (seat, console, cockpit, furniture, etc.), test subjects should be selected to represent the critical dimensional limits of expected operators. First, it is important to have several subjects who represent the largest people who may have problems of clearance and several subjects who represent the smallest people who may have problems of reach. Second, there should be a scattering of subject sizes in between the largest and the smallest in order to provide assurance that some in-between dimensional incompatibility is not overlooked. In a sense, we have a stratified sample of sizes because we know that certain physical accommodation problems are more critical at the extremes.

It is standard practice in the military services to attempt to fit 90 percent of the population, i.e., from the 5th- to the 95th-percentile population of a particular military sample. This means, however, that, in the event the product is to be used by both male and female operators, the test sample should range from the 5th-percentile female through the 95th-percentile male (as opposed to ranging from the 5th- to the 95th-percentile male when only male operators are expected).

Other Product Factors. When the questions to be answered do not relate to the physical size of the expected user population, a more random sample of test subjects should be used. For example, when we are investigating sensorimotor response relative to some proposed display and control system, body size has less to do with the subject's response, and therefore, we need a random sample of subject skills, experience, motivation, and attitudes in order to assess the average response of an expected user population.

If the problem has something to do with human strength, it may be desirable to select more subjects who are weaker, since these are the users who may have more difficulty with problems of control force, lifting, and so on. It should be noted that one cannot determine how much strength a subject can apply merely on the basis of his or her size, although it can generally be assumed that females are weaker than males.

If the product may be used by older people, it is important to include a number of elderly persons, both male and female, to reflect the compatibilities or incompatibilities of the proposed product design with typical age-related infirmities, e.g., poor vision, reduced mobility and strength, and slower response time.

Similarly, if handicapped users are expected, a variety of handicapped test subjects should be included to reflect the appropriate range of infirmities and difficulties they may have, such as being confined to a wheelchair.

Ethnic differences, such as differences in language or technological background, may be an important aspect of HF testing, and the appropriate selection of test subjects who represent unique design interface barriers must be considered. For example, a test to determine how intelligible certain pictorial symbols are to people of various nationalities might be appropriate.

General Product Acceptance. For most consumer acceptance tests or surveys, it is desirable to take a completely random sampling approach since the normal sales objective is to reach as broad a population of purchasers as possible. In such cases, it is up to the customer to establish the limits to which he or she expects to promote the product, i.e., to the world, to a specific country, or to a specific part of a country. It should be observed that this type of sampling generally relates to consumer opinions and attitudes, which change frequently. Therefore, not only should sampling be approached on an area-by-area basis, but also a successive series of surveys should be carried out to allow for opinion shifts. A good example is provided by the automobile industry. Some people have to get used to a new model of an automobile; i.e., the new body style may be sufficiently radical to "turn them off." However, after 6 months to a year, they are accustomed to the new style, and it becomes the reference by which they measure satisfaction with other new models.

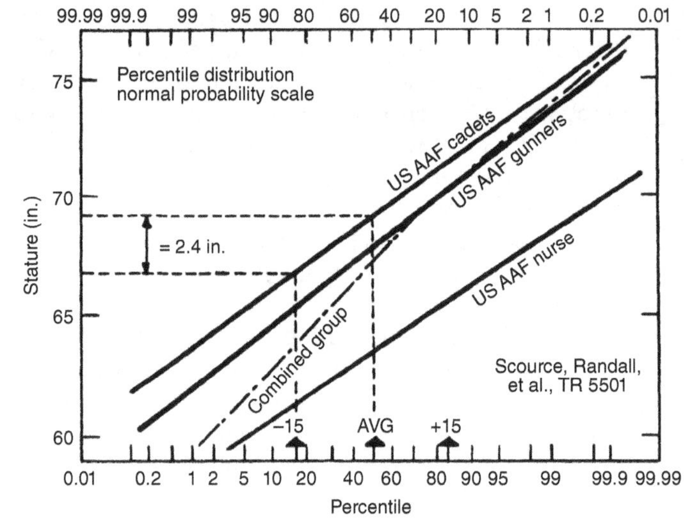

FIGURE 57.17 Illustration of the use of normal probability paper for developing composite populations. (After J. A. Roebuck et al., *Engineering Anthropometry Methods*, John Wiley & Sons, Inc., New York, 1975, pp. 158, 159.)

Developing a Composite Anthropometric Population

When anthropometric survey data on several specific populations are available and when it is desirable to know the distribution of these dimensions as a group rather than as separate populations, a composite population can be synthesized. An example is shown in Fig. 57.17, wherein a certain troop transport is to carry 30 percent cadet pilots, 45 percent gunners, and 25 percent nurses, whose respective heights are given in the graph. The computation is shown in Fig. 57.18, with the final computation plotted as a dashed line in Fig. 57.17.

		Cadets		Gunners		Nurses		Combined
	Stature, in	Percentile	Percent in new population	Percentile	Percent in new population	Percentile	Percent in new population	Population percentile
Column no. operation	(1)	(2)	(3) Col. (2) × .30	(4)	(4) Col. (4) × .45	(6)	(7) Col. (6) × .25	(8) (3) + (6) + (7)
	61	0.03	0.009	0.3	0.135	12.5	3.13	3.27
	64	1.4	0.42	6.0	2.7	59.0	14.75	17.87
	67	18.0	5.4	36.0	16.2	94.6	23.7	45.3
	70	64.0	19.2	80.0	36.0	99.9	24.98	80.18
	73	94.6	28.4	98.0	44.1	100	25	97.5

*Percentile determined from accompanying figure = 67.4 in; 95% = 72.2 in; 5% = 60.4 in.

FIGURE 57.18 A method of forming a composite population dimension distribution from three populations with different means and standard deviations (Roebuck, 1975). (After J. A. Roebuck et al., *Engineering Anthropometry Methods*, John Wiley & Sons, Inc., New York, 1975, pp. 158, 159.)

Measuring and Weighing Test Subjects

In order to verify that one has representative test subjects (i.e., to represent established 5th and 95th percentiles), one should not depend entirely on the subjects' reports of their height or weight. In addition, in many instances the designer needs one or two other measurements that average subjects will not know about themselves. The rules for measuring are quite simple as long as a few rules are followed.

1. Stature: To measure stature, have the subject remove his or her shoes and stand with the head, shoulder blades, buttocks, and heels flush against a wall. Using a draftsman's triangle (see Fig. 57.19), measure the subject's height from the floor.

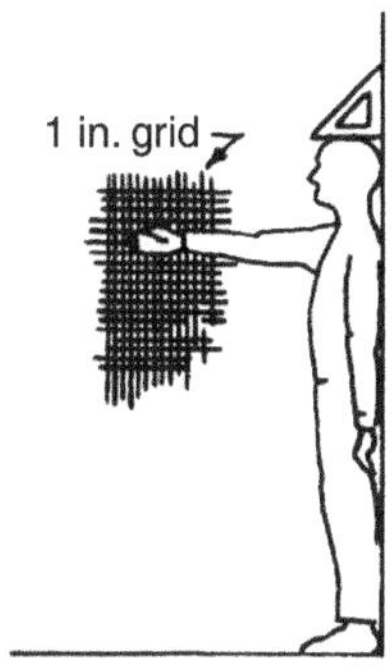

Figure 57.19 Stature and arm reach.

2. Arm Reach: Ask the subject, while he or she is still standing against the wall, to raise and point the arm straight forward with the palm of the hand facing inward and with the fingers extended. Measure the distance from the wall to the tip of the longest finger, usually the middle finger.

3. Leg Reach: Have the subject sit on a flat box with the head, shoulder blades, and buttocks against a wall. (See Fig. 57.20.) An adjustable seat and footstool should be used so that the subject's feet will be flat on the floor. Have the subject raise the right leg and lay it on the footstool. Older people may need help. Measure the distance from the wall to the subject's heel. A small triangle will help define this distance. Do not be disturbed by the fact that some people cannot fully straighten out their legs. However, have the subject straighten the leg as much as possible.

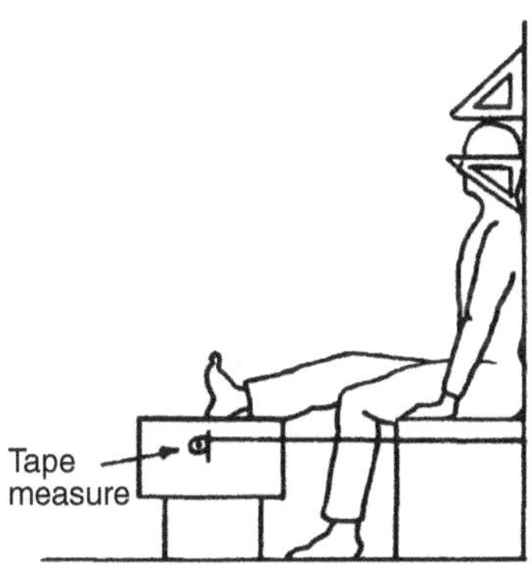

Figure 57.20 Leg reach.

4. Weight: A standard health scale should be used. Subjects should be as lightly clothed as practical. (See Fig. 57.21.)

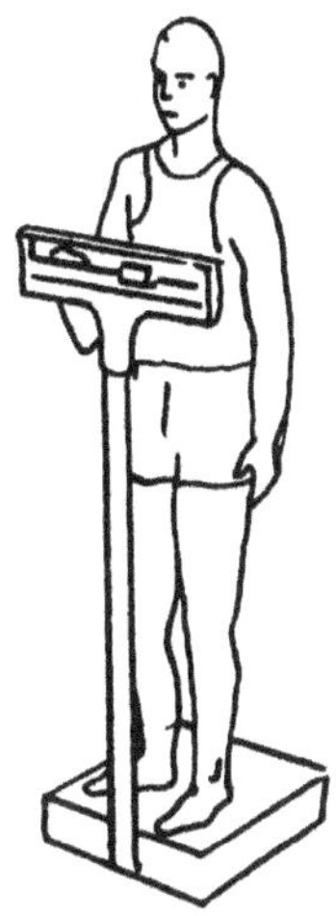

Figure 57.21 Weight.

5. Strength: Although not absolutely valid, hand dynamometer strength is a relatively good indicator of an individual's overall strength capability. Hand dynamometers are not easy to find unless one is purchased for permanent retention. Such instruments often can be borrowed from a college or university that has an anthropology department. Usually one can also be found at a local hospital. Although instructions could be presented here, each dynamometer may be different, and thus one should ask for instructions from those from whom the particular instrument is borrowed.

The above measurements are usually sufficient for most design work where one is trying to make sure that test subjects are representative of a particular population's extremes. If, however, other measurements are required for some special design purpose, it is recommended that the designer seek help from a trained anthropologist.

COMPUTER-BASED MODELING

As with most modern disciplines, human factors engineering (HFE) has been influenced by a rapid growth in computer processing capability, computer graphics, and computer modeling. Computer capabilities have impacted the HF discipline in at least three noteworthy areas:

1. Computer-based HFE tools. Albeit still a field in its early development, many vendors now offer computer-based HFE for data processing, analysis, and modeling. As of this writing, it's impossible to list or review current products because offerings in the field are changing and maturing so rapidly.

The applicability of any particular computer-based tool (or of any HFE tool or methodology) depends highly on the nature of the system and the human/system interface issue that is being developed and evaluated. Computer-based versions of HF tools and methodologies already discussed in this chapter might include: (a) time and motion analysis; (b) function allocation including functional flow block diagramming, data- or information-flow analyses, link analysis, and operational sequence diagrams; (c) task and workload analyses; (d) human error analysis and human probabilistic risk assessment; (e) field evaluation data gathering, user surveying, and data analysis; (f) HF requirements verification and validation; and (g) operations and maintenance procedures development.

As with all of the computer-based tools mentioned in this section, it's recommended that the practitioner survey information resources within the HF discipline to learn about and stay current on recent developments on computer-based HFE tools.

2. Integration of computer-based tools and technologies with system prototyping and mockups. As explained in the previous section "Mockups and Usability Evaluations," the appropriate level of complexity of a mockup for user evaluation is dependent on the specific design questions to be addressed, the time available, and the budget available. While paper, cardboard, and foam core mockups can be cost-effective initial evaluation tools, it's possible to include computer-based components into such mockups to cost-effectively take the level of fidelity and the ability to evaluate more complex interactions to a higher level. Often, the use of computer-based display and control systems can not only enhance the fidelity of the system mockup, they may also provide the HF practitioner the opportunity to integrate real-time data gathering and analysis tools into human-in-the-loop evaluations, tests, and assessments—i.e., not only modeling but also gathering data on human/computer interactions. To be an effective and responsible member of the design team, however, the HFE should always be mindful what the pace and budget of the project will bear. Often the human/system interaction data needed to make a design decision may be elicited by a simple paper representation of a display or control along with an analyst walking a test subject through the presumed interaction. The method and complexity of modeling systems for human-in-the-loop evaluation should be carefully selected to support concise and effective decision-making.

3. Full computer modeling of systems, including their man–machine interactions. As of this writing, it is becoming the norm—particularly for large-scale, complex systems—for the hardware system to be fully and virtually modeled in the computer long before any mockup or prototype is manufactured. Issues of design, form, fit, integration, and component producability can and are typically resolved in virtual space. Models may be static—with full-scale structural analyses able to be carried out in the computer—or dynamically able to interact with other components to perform mechanical, electrical, software, etc., evaluations.

The HFE practitioner must become highly skilled in evolving with this transition toward computer-based design (and manufacturing). The early availability of computer models in the design cycle provides the HF discipline new opportunities to become highly skilled in interactively using computer-based tools to assess the operability and maintainability of every design decision. Tools and methods for such interactive assessment of the role of human interactions in the overall operations of a system are just emerging as of this writing but the field needs developers and innovators.

Model-Based Systems Engineering

As the process of design and development is evolving from pencil and paper drawings to computer-based design and manufacturing, so too is the discipline of systems engineering. The International Council on Systems Engineering (INCOSE; www.incose.org) is an excellent resource for information on the state of the systems engineering discipline. In 2015, INCOSE identified accelerating the transformation of systems engineering to a model-based discipline as one of its core strategies.[7] Model-based systems engineering calls for the formal application of computer-based modeling in all aspects of design and development: requirements development, validation and verification; design and trade study analysis; task analysis and function allocation; system performance analyses; system architecture specification; etc. In its anticipated full implementation, model-based systems engineering methods will not only move design and development programmatics toward computer-based modeling, but will also integrate business concerns such as cost-to-develop and cost-to-operate. Given the comprehensive nature of the transformation of systems engineering to a model-based approach, it is anticipated and expected that HFE will and must similarly transition. If nothing else, HFE must develop a model-based approach in order to stay relevant to and to integrate into the heart of management and systems engineering processes for design and development.

Model-based systems engineering provides the HF discipline an opportunity to develop human models and human/system analysis tools that integrate into a larger model-based system development framework. To fully utilize the promise and potential of model-based systems engineering, the HF discipline must develop tools and analytical methods that integrate measures of human/system effectiveness and technical performance—as discussed in the "Human Factors Engineering Performance Criteria" section of Chap. 55—into a project's core criteria for success. Doing so will drive hardware and software development beyond existing technical measures

[7]INCOSE website: http://www.incose.org/about/strategicobjectives

of performance—e.g., mass, power, speed, etc.—to be inclusive of total human/system performance—e.g., efficient use of manpower resulting in small operating teams, reduced dependence on specialized personnel selection and training, etc.

As of this writing there is no one definitive resource for providing information on the state of the art of model-based HF and its relation to model-based systems engineering. Information is evolving rapidly. The HF practitioner should consult discipline trade, professional and technical groups in HF and systems engineering to become current on recent developments in these fields. Some recommended resources include:

- The International Council of Systems Engineering (INCOSE); www.incose.org
- The Human Factors and Ergonomics Society (HFES); www.hfes.org
- The National Defense Industrial Association (NDIA)/Human Systems Division; http://www.ndia.org/Divisions/Divisions/HumanSystems
- The National Defense Industrial Association (NDIA)/Systems Engineering Division; http://www.ndia.org/Divisions/Divisions/SystemsEngineering

Human Modeling

Human modeling as mentioned here specifically addresses human *performance* modeling. The model-based systems engineer practitioner expects a performance model of each subsystem in order to integrate them into a full system model. Given this mindset, the systems engineer might ideally prefer or even expect to see the system's human elements and interfaces similarly presented as a "plug-in" to his or her comprehensive model to comprise a grand picture of the operational system. The challenge of such an expectation for the HF practitioner is that while hardware and software systems and subsystems typically behave deterministically, with defined inputs leading to predictable outputs, human performance and behavior is not deterministic and is therefore difficult to model. Although human response and behavior might be addressed probabilistically for a given population in a constrained decision space, the behavior of any *single* human under the same conditions can never be predicted. Variations from the desired or expected human response must typically be accounted for in order to ensure that the system is "error tolerant"—i.e., safe under conditions of "human error." Of course, the degree to which any particular system must be error tolerant depends on a number of factors, perhaps most importantly being the safety and health risks to individuals and populations.

Human models that could be "plugged in" to total human/system computer-based models to accurately support design decision-making remain an elusive and challenging prospect as of this writing, certainly to a degree of sophistication that could begin to replace system mockups and human-in-the-loop testing and evaluation. Most often during design the human/system question that needs resolution or analysis is highly tailored to the system, the user population, and to particular operational environmental constraints. At this time the skilled HF practitioner should continue to support human and human/system modeling development but also be mindful of the limitations of the current state-of-the-art. One approach might be to use human and human/system modeling to compare and contrast with data gathered through traditional human population testing and evaluation. In this way, discrepancies between modeling and real-world data may indicate areas that require further testing or analysis. Another methodology when using computer-based modeling tools is to perform modeling on both the system being evaluated and similar modeling of a comparable existing system(s). In this way the modeling of the existing system—which is likely to have a record of prior performance and evaluation—may be used as a control for evaluating the data produced by models of the new or proposed system. In short, independent, standalone modeling of a new system, interface, or design trade may be suspect without supporting data from modeling a comparable concept and/or that correlates to measurable observation.

Human and human/system interaction modeling may well be used for initial trade studies of human/system accommodation. For example, during system design and development, physical size and fit studies might well be assisted by the use of computer-based human models. Rather than perform large human-in-the-loop studies across a wide variety of anthropometric variations within a given user population, physical human modeling tools may well be used to ensure that a given design appropriately accommodates variations in the reach, strength, eye point, etc., of the expected user population. Findings from such a computer-based analysis may well point to a condensed set of investigations into form and fit of the user population that need to be conducted via mockups and human testing. Complete reliance on computer-based models for human/system predictions of performance may be a strategic goal of the HFE discipline but in the interim, as technology evolves, use of computer-based modeling to compare and contrast with real-world analyses is encouraged.

Chapter 58

Human Factors in System Operations

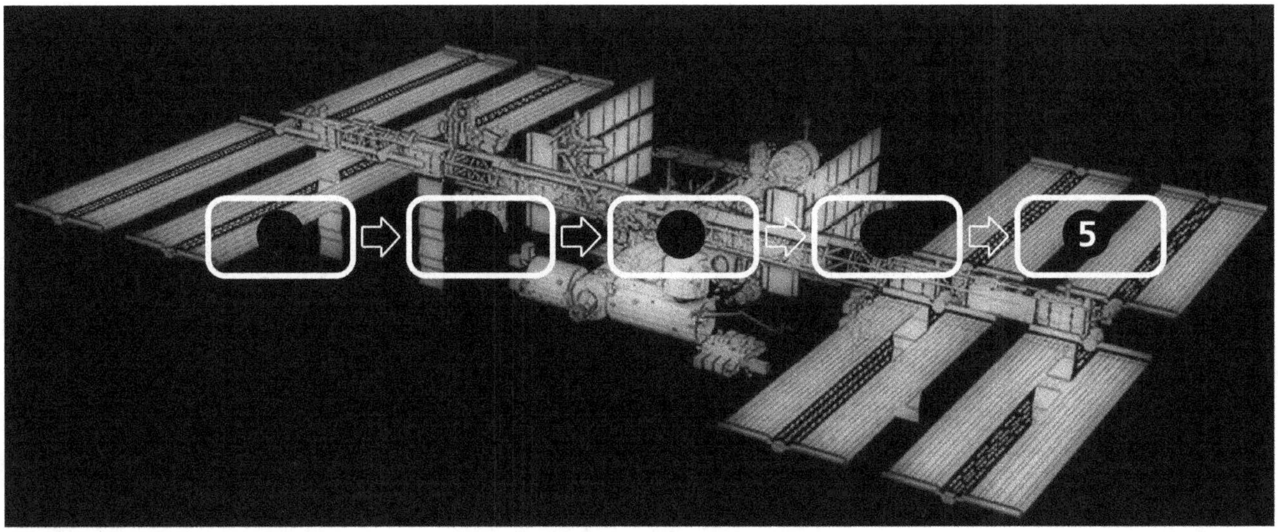

The section on "Human Systems Integration" in Chap. 53 noted that human factors engineering (HFE) is most cost-effective and has the most potential to reduce system life cycle cost when the discipline is fully engaged in the system engineering process as early as possible in the system's design. Then, design recommendations, human engineering tools (such as concepts of operation and task analysis), and feedback from user evaluations can all be brought to bear when design changes and improvements are relatively easy and certainly most cost-efficient. This is not to say, however, that the human factors (HF) discipline doesn't have roles to play late in the system development life cycle.

The HF discipline can fulfill strong needs during the transition to and within a system's operational era. As a system transitions from development to operations it may be too late to make any but safety-critical design modifications. However, there are many system support and management products that need to be developed to ensure complete and effective integration of the system's human elements—i.e., operators, maintainers, etc. Such transition products include final development assessments and reporting, development phase lessons learned, operations procedures development, training material, and personnel selection criteria. In each case, the HF practitioner is uniquely positioned to play a role in understanding and supporting the final integration of human and system.

FINAL HUMAN FACTORS VERIFICATION AND REPORTING

The "Requirements Definition" section in Chap. 55 discussed the importance for every discipline involved in a project to establish requirements that define the outcome's criteria for success. Also noted was that requirements should be continually reviewed and revisited over the course of a project's development to ensure that the end product(s) will be in compliance. As a project's development phases draw to a close, documentation of requirements having been met will need to be documented in a final and formal verification report. It is important for the HF practitioner to document compliance with all requirement related to the human elements of a system's design and to all human/system interface requirements. Almost every project of a substantial size will require such a formal, conclusive report.

Of particular note is that this formal verification report may well be used in the event a system's operational behavior comes into review, especially in the event of a mishap investigation. The types of requirements that a product was held to and the product's compliance with those requirements may be questioned and it is important that any inquiry be able to access a full and accurate account of a project's history. It should be noted that such an accurate history may prove critical in addressing liability questions relating to any and all participating disciplines.

DEVELOPMENT LESSONS LEARNED

Though the composition of design teams may vary, any business or governmental entity involved in system design and development is probably going to be involved in multiple programs and projects. It's important to the "corporate knowledge" of such entities to learn from each design and development exercise in order to improve management processes, become more efficient, and produce increasingly better products. For this reason, it's important to have and maintain a process for recording "lessons learned" from each design and development experience. Also, it's wise to document recommendations for improvement while a lesson is fresh in the team's memory. Lessons learned processes may be called by many names—e.g., continuous product/process improvement. Note that though the value of a lessons learned process certainly applies to a large design and development entity intending to acquire future business, it also applies at the level of each discipline and even each practitioner. At its most basic, a lesson learned process is simply a means of writing down an awareness of and answers to the question, "How can I do better next time?"

OPERATIONS PROCEDURES DEVELOPMENT

Formal requirements verification and lessons learned capture are means of wrapping up the design and development experience. To transition from a system's development to operations, one product that needs to be generated is documentation that informs future system operators and maintainers how best to perform their jobs. These operations procedures documents reflect the observations and assumptions of how the full system is to operate as perceived by the design and development team from the inception of the concept of operations through the completion of formal verification. Although these documents may become living documents, evolving and expanding with the experience of system operators over time, it's critical that some operations manuals and training materials be developed at the outset of operations to inform the operations community.

Having worked on task analyses during design and development, a project's HF practitioners will likely be the most informed personnel on a project to document the operational and maintenance procedures for a system. HF engineers can ensure that human functions and tasks identified during system development through human/system function allocation and human engineering analysis are organized and sequenced for safety, efficiency, and reliability. It's very likely that a draft or baseline set of operational task procedures were developed and used as a foundation for human/system behavior assumptions early in a system's design and development process. These would not only have informed (and been an outcome of) human/system task analysis, they would have been used in making man/machine functional allocation design decisions.

Quality operating procedures should be designed for the defined characteristics of each user population and may ideally minimize system training demands. To the degree practical, the full gamut of HFE techniques should be used to develop and valid operating procedures—i.e., use of system prototypes, mockups and models; human-in-the-loop evaluations; etc.

TRAINING MATERIAL

While operations procedures are step-by-step guides on how particular system functions are to be performed and managed by operators and maintainers, system training material provides more fundamental information on how the system and its subsystems work and, perhaps, interact with other systems of a larger whole. A fully informed operator will be more capable of performing corrective actions if he or she understands the nature of why a system performs the way it does. The HF practitioner is well equipped to help write system training material that is written from and for the user's perspective.

LABELING

All training and operations material written for system users must be consistent in the use of names used for identifying system components, especially those with which the user is expected to frequently interact—i.e., the system displays, controls, and other frequently accessed items. Though this may seem self-evident, it is quite common for a procedure to invoke a control's name that differs from the name actually engraved on the control panel. The HF engineer can and should provide regular reviews of system labels and documentation to ensure that accurate, consistent, legible, and self-evident information is provided to a system's intended user community.

PERSONNEL SELECTION CRITERIA

Some systems are built for a broad general population of retail consumers where others are intended for more closely defined communities—e.g., military personnel, trained system maintainers, or astronauts to name a few. If a system is intended for a defined community, it is highly likely that assumptions were made about that community during design trade studies and decision-making. Any critical assumptions about the user population of a particular system's design needs to be documented as an excursion outside of the intended operations community may well present a hazard, either to the user or to the general population. As the discipline most experienced with documenting human capabilities and limitations, the HF practitioner should be involved in documenting any constraints to be enforced in selecting a system's user community. Note that such restrictions may well apply to physical characteristics—e.g., anthropometric size, strength capabilities, vision capabilities, etc.—and/or to cognitive capabilities—e.g., level of education, prior experience with similar system, etc.

ONGOING SYSTEM RISK ASSESSMENT AND HAZARD ANALYSIS

In large corporations or governmental agencies, it's very likely that a core HF discipline community will be continually engaged and will help monitor the success of systems' operations. This is an excellent means for actively seeking lessons learned on how well assumptions and corporate knowledge about system design actually play out in systems' operational success and efficiency. It also provides an opportunity for a consistent group of like-minded personnel to continually gather data about a system during its operations life cycle. Since the safety and security of the human elements of a system are tantamount in HF discipline practitioners, continually seeking and documenting any unexpected system hazards will likely be a strong consideration of any team monitoring operations. In the event of a mishap, it is essential that HF practitioners be engaged in any investigation. All too often mishaps have been attributed to human operator error when the actual contributing conditions were unexpected system response, incorrect operating procedures, or other total system design error that was built-in to the system or its operating processes long before the operator became involved.

RECOMMENDATIONS FOR SYSTEM UPGRADES AND MODIFICATIONS

Any long-term association with a system will likely lead to observations and lessons learned that may result in recommended system modifications or improvements—either as mandatory safety concerns or as improvements in the next manufacturing iteration. In short, lessons learned shouldn't stop at the end of design and development but should ideally be a part of any system's entire life cycle, including the operational era. Recommendations for improvements may or may not be incorporated into systems "in the field" depending on any number of constraints—e.g., budget, real-time operational demands, logistics, etc. However, if practicable, improvements in a system's design to enhance human/system operational safety, effectiveness, and/or efficiency should be noted and documented for future opportunities.

The HF engineer might be specifically brought into a review of an operating system on whose design he or she did not participate specifically to consult and bring a fresh perspective on seeking potential operational improvements. HF practitioners can bring the capabilities of the profession to bear to objectively monitor and evaluate the efficacy of human/system interactions in real-life operational environments. These data are likely to serve well in the planning and design of any proposed system upgrades.

INDEX

CPSIA information can be obtained
at www.ICGtesting.com
Printed in the USA
JSHW041643250721
17214JS00004B/20

9 780071 702874